BIOLOGICAL REACTIVE INTERMEDIATES III

Mechanisms of Action in Animal Models
and Human Disease

ADVANCES IN EXPERIMENTAL MEDICINE AND BIOLOGY

Recent Volumes in this Series

BIOLOGICAL REACTIVE INTERMEDIATES III

INTERMEDIATES III

Mechanisms of Action in Animal Models
and Human Disease

Edited by

James J. Kocsis

Thomas Jefferson University
Philadelphia, Pennsylvania

David J. Jollow

Medical University of South Carolina
Charleston, South Carolina

Charlotte M. Witmer

Rutgers University
Piscataway, New Jersey

Judd O. Nelson

University of Maryland
College Park, Maryland

and

Robert Snyder

Rutgers University
Piscataway, New Jersey

PLENUM PRESS • NEW YORK AND LONDON

Library of Congress Cataloging in Publication Data

International Symposium on Biological Reactive Intermediates (3rd: 1985: University
of Maryland, College Park, Md.)
Biological reactive intermediates III.

(Advances in experimental medicine and biology; v. 197)
"Proceedings of the Third International Symposium on Biological Reactive Intermediates, held June 6-8, 1985, at the University of Maryland, College Park,
Maryland" — T.p. verso.
Includes bibliographies and index.
1. Poisons — Metabolism — Congresses. 2. Biotransformation (Metabolism) — Con-
gresses. 3. Toxicology, Experimental — Congresses. I. Kocsis, James J., 1920-   . II.
Title. III. Title: Biological reactive intermediates. 3. IV. Series. [DNLM: 1. Biotrans-
formation — congresses. 2. Toxicology — congresses. W1 AD559 v.197 / QV 600 I63
1985b]
RA1220.I58   1985   615.9   86-4973
ISBN 0-306-42264-6

Proceedings of the Third International Symposium on Biological Reactive
Intermediates, held June 6-8, 1985, at the University of Maryland,
College Park, Maryland

© 1986 Plenum Press, New York
A Division of Plenum Publishing Corporation
233 Spring Street, New York, N.Y. 10013

Printed in the United States of America

PREFACE

   This volume contains the proceedings of the third in a series of
conferences entitled, The International Symposium on Biological Reactive
Intermediates.   The first was held at the University of Turku in
Finland, in 1975, the second at the University of Surrey in the United
Kingdom, in 1980 and the most recent at the University of Maryland in
the United States, in 1985.

   The significance of these conferences has been emphasized by the
rapid growth of mechanistic toxicology over the last decade. These
conferences were initially stimulated by the attempt to uncover the
significance behind the observations that the toxicity of carcinogenic
responses produced by many chemicals was associated with the observation
that their metabolism led to the formation of chemcially reactive
electrophiles which covalently bound to nucleophilic sites in cells such
as proteins, nucleic acid or fats. Recently, newer concepts have arisen
which have necessitated the expansion of subjects covered by the
conference. For example, the application of newer knowledge of the role
of active oxygen species in reactive metabolite formation, the concept
of suicide substrates, examination of the function of glutathione in
cells, application of immunological techniques and molecular biological
probes to the solution of toxicological problems all had an impact on
the study of the biological reactive intermediates. Many workers in the
field are now asking questions such as: What is the meaning of covalent
bonding of reactive metabolites to proteins? How does covalent binding
of reactive metabolites to DNA pave the way for a carcinogenic response?
 What role does genetics play in determining toxicological responses?
These and other questions were raised at the symposium.

   The study of biological reactive intermediates began with the
pathfinding observations of Betty and Jim Miller of the McArdle
Institute of the University of Wisconsin, and it is to them that this
symposium was dedicated. They have played an important role in previous
symposia in this series. They are not people who readily sit back on
their laurels and accept acclaim. At the symposium each made a major
presentation, they both attended every session and they were among the
most active questioners. Their overall contribution to the symposium
helped to insure that the scientific quality of the discussions was
excellent and the meeting was an exciting intellectual adventure. We
must thank them for all that they have contributed to our understanding
of these phenomena, as well as, for their vital contribution to this
symposium.

   The meeting was held at the excellent facilities in the The Center
for Adult Education of the University of Maryland at College Park, MD on
June 6-8, 1985. Support for the symposium came from the National
Institute of Environmental Health Sciences, the U.S. Environmental

Protection Agency, Hoffmann-LaRoche, Inc., the International Society for the Study of Xenobiotics, the Drug Metabolism Section of International Union of Pharmacology, Rutgers, the State University of New Jersey, Thomas Jefferson University, the Medical University of South Carolina and the University of Maryland.

The organizing committee for the conference included A.H. Conney (Hoffmann-LaRoche, Inc.), G.G. Gibson (University of Surrey), J.R. Gillettee (National Institutes of Health), D.J. Jollow (Medical University of South Carolina), J.J. Kocsis (Thomas Jefferson University), R.E. Menzer (University of Maryland), J.O. Nelson (University of Maryland), R. Snyder (Rutgers University) and C.M. Witmer (Rutgers University).

The rapid progress that has been enjoyed in the study of biological reactive intermediates in recent years means that complete coverage of the field in a brief symposium is not possible. Nevertheless, active discussion abounded and it fell upon the session chairmen to insure that time was adequate for both presenters and discussers. The highly dedicated and efficient chairmen were: G. G. Gibson (University of Surrey, Guildford), F. DeMatteis (MRC Toxicology Unit, Carshalton), P.N. Magee (Temple University, Philadelphia), D.Y. Cooper (University of Pennsylvania, Philadelphia), H. Greim (Gesellschaft fur Strahlen- und Umweltforschung MBH, Munich), D.J. Reed (Oregon State University, Corvallis), K.J. Netter (University of Marburg, Marburg), H. Kappus (Free University of Berlin, West Berlin), R. W. Estabrook (University of Texas Health Science Center, Dallas), A. Hildebrandt (Max von Pettenkofer Institute, West Berlin), H. Bolt (University of Dortmund, Dortmund).

Modern toxicology is a biological science with chemical overtones, which must ultimately explain how chemicals cause adverse changes in biological systems. As these interactions enlighten us, we will learn more about the underlying principles of life itself. In parallel, we will also apply our new knowledge to attempt to solve the problems of exposure to chemicals in the various environments and media characteristic of life in society. Thus, the small band of investigators that met in Turku in 1975 has not only grown in numbers and accomplishment but is also beginning to have an impact social policy. Of utmost importance, however, is that few if any among these researchers consciously directed their efforts at anything other than pursuing scientific knowledge, thereby enforcing the adage that the intellectual effort involved in the pursuit of excellence in science will bear fruit when properly applied.

It was a pleasure for all attendees to gather together again to continue our discussions. This group has demonstrated the ability to communicate their results to their colleagues and discuss the results in a cordial and cooperative manner which has set an example to their students, many of whom are now working in this area. We plan to continue these discussions in the future and to maintain these close scientific relationships which have potentiated our productivity.

James J. Kocsis
David J. Jollow
Charlotte M. Witmer
Judd O. Nelson
Robert Snyder

CONTENTS

III. SUICIDE SUBSTRATES OF CYTOCHROME P-450

IV. CHARACTERIZATION AND TOXICOLOGICAL SIGNIFICANCE OF PHASE II ENZYMES

SHORT COMMUNICATIONS

YEARS WITH ELIZABETH AND JAMES MILLER: IN APPRECIATION

Van Rensselaer Potter

McArdle Laboratory, University of Wisconsin Medical School

Madison, Wisconsin 53706

When James Watson wrote "The Double Helix" he began by remarking "I have never seen Francis Crick in a modest mood."

When I think about Jim and Betty Miller my feeling is quite the opposite. I have never heard either one of them make an assertive self-serving statement. In thinking about their careers, however, I had another reason for recalling the double helix. Jim and Betty are to each other like the two strands of DNA, each one the complement of the other.

There is a second reason for recalling the DNA story: when Jim and Betty discovered protein binding by a carcinogen in 1947, launching a lifetime of discoveries, the role of DNA as the key substance in gene duplication had not yet been described. It would be six years before the dramatic announcement by Watson and Crick appeared in Nature in 1953.

Yet the 1947 report by the Millers was characteristically not based on speculation. Instead it was a straightforward account of what was actually observed. It led inevitably to the whole sequence of events that followed from the conclusion that a carcinogenic molecule had been modified, and that the modified molecule had combined with protein.

Only the Millers will ever be able to develop a historically accurate account of the step by step process by which they and their collaborators came up with one original discovery after another, how these observations were confirmed by others, and how the work of others led to new findings that set up feedback to the Miller laboratory in an ever expanding network that brings us all together today. Here we celebrate the confluence of the 1953 DNA announcement and the avalanche that began with the Millers in 1947. How did it all happen? I believe it is the outcome of the symbiotic relationship of Jim and Betty Miller both in their personal and professional lives. We are here witnesses to the result of the combined efforts of two people both of whom accepted the high standards of hard work, strong motivation, and commitment to a common goal.

I am indebted to Harold Rusch for the following background information, with interpolations.

"Elizabeth Cavert Miller was born in St. Paul, Minnesota, but while she was still young, her family, including one sister and one brother, moved to Anoka, a suburb of St. Paul, where she attended high school. She grew up in an intellectual environment. Her mother graduated from Vassar and then spent two years at Columbia receiving another degree. Her father had a Ph.D. in Agricultural Economics from Cornell University and was a member of the faculty of the extension division of the University of Minnesota, after which he became director of research for the Seventh District of the Farm Credit Administration in St. Paul.

Elizabeth obtained a bachelor's degree in agricultural biochemistry at the University of Minnesota in 1941 and was influenced by Professor and Chairman Ross Aiken Gortner to apply for a graduate fellowship at the University of Wisconsin. She was awarded a Wisconsin Alumni Research Fellowship and intended to enroll in the Biochemistry Department but was discouraged from doing so when she was informed that, as a woman, it would be difficult for her to get a job in that field. She enrolled in the Department of Home Economics with a joint major in biochemistry. Thus, in her first year she took the required biochemistry courses and thereby destiny played a momentous role in her entire career.

It so happened that James A. Miller, who had enrolled in the Biochemistry Department also on a fellowship from WARF, was serving as a graduate assistant in the biochemistry course in which Elizabeth Cavert had enrolled. Jim had a keen eye for both intellectual ability and for beauty, and it was not long before he convinced his mentor, Prof. Carl Baumann, that Betty had all the highest attributes to more than qualify her for a fellowship in the Biochemistry Department. Thus, after one year in the Department of Home Economics, Betty moved to biochemistry. In 1942 Betty and Jim were married."

"James A. Miller was a youth during the 30's and grew up amidst hard times in Dormont, Pennsylvania, a suburb of Pittsburgh. To make matters worse, his mother died when he was 13. Jim was the fifth of six brothers, and two of his brothers died while Jim was still living at home. Thus, this period was not a happy one. None of his brothers had sufficient funds to go on to college, but Jim's oldest brother read extensively, especially in the field of science, and encouraged Jim to continue his education beyond high school. So when Jim finished high school, he attended night school for six months and worked part-time for two years. He then obtained a job from the National Youth Administra-tion, an organization which was part of the New Deal during the early years of Roosevelt's presidency. This money, plus some help from his brothers, enabled Jim to enroll in the University of Pittsburgh majoring in chemistry. Then a fortunate turn of events occurred which influenced Jim's future career. During his second year he obtained a part-time job with Professor Charles Glen King, who was the first person to obtain pure crystals of Vitamin C. Jim worked for King for three years, graduating in 1939. It so happened that Dr. Max O. Schultze, who obtained his Ph.D. in biochemistry at the University of Wisconsin, was a postdoctoral fellow with Dr. King while Jim was a student, and Schultze encouraged Jim to apply for a fellowship at Wisconsin."

Here I would like to depart from Dr. Rusch's account and insert a comment on the character of Max Schultze as a role model for Jim Miller. I knew Max very well for he was Conrad Elvehjem's teaching assistant when I arrived to work with Connie in 1935. I followed Max Schultze's footsteps when he departed for a postdoctoral appointment with Professor King. Max Schultze was the hardest-working person I had ever met. He never ordered anything he could make or synthesize himself, and in a recent telephone conversation he added with a chuckle, "and never bought anything I couldn't pay for". Max became an Assistant Professor at Pittsburgh, then served for four years in the U.S. Army 1942-46. He then took the position of Professor of Biochemistry in the Department at the University of Minnesota, where Betty Miller had been an undergraduate. Dr. Schultze is now 79, retired, and living in Albuquerque, N.M. I take pleasure in presenting the following excerpt from his letter, which he wrote at my request:

"When I was a post-doc in the Chemistry Department at Pitt, Jim was an eager chemistry student, intent to become a highpowered theoretical physical chemist. He helped me with some of my work in the animal room and in the laboratory. Soon, I recognized not only his ability and curiosity but also his motivation and his ability to do things, be it building cages from chicken wire or making a shaker for extraction of immiscible liquids.

When I suggested that he might find biochemistry an interesting and challenging occupation, he agreed that his curiosity had been aroused. In his Senior year I wrote in his behalf to E. B. Hart and to Connie. That is the way through which he found himself to be a graduate student at Wisconsin. Few have made bettter use of the opportunities offered to them.

As an undergraduate Jim not only excelled in his studies, but he had also an open, pleasant personality, reliable and always cheerful. While well motivated, he never gave me the impression of being a self-centered "pusher". It was a pleasure to have him around. I imagine that these traits stayed with him.

Since then, about 1938, I have followed his work (and recognizing Betty's contribution) with great interest. It is an outstanding example of what chemists can do in the biological areas, showing broad knowledge, purposeful planning, precise execution and careful deductions. Few can equal it over a long professional career."

Returning to Dr. Rusch's account,

"Jim followed Dr. Schultze's advice and was awarded a fellowship made possible with funds from the Wisconsin Alumni Research Foundation. The annual stipend of $700.00 was much sought after in those days. Thus, Jim came to Wisconsin and after the first year, Professor Carl Baumann served as his major professor.

Now a word about how Jim and Betty eventually became involved in cancer research. It started early in the 30's when Dr. Louis Fieser, Professor of Organic Chemistry at Harvard, synthesized methylcholanthrene which proved to be a potent carcinogen and its structure also somewhat resembled cholesterol. Dr. Harold P. Rusch, a Bowman Fellow for Cancer Research in the Department of Physiology in the Medical School of the University of Wisconsin, was working on the carcinogenic activity of both ultraviolet light and of chemical carcinogens. He believed that cholesterol might be converted in the body to form carcinogens and thought that ultraviolet light might accelerate this conversion. Thus, he fed high levels of cholesterol to both mice and rabbits and subjected them to ultraviolet irradiation. This seemed like a good lead and an exciting idea. Rusch realized that he was not an expert in the chemistry of steroids and wanted help, so he asked Prof. Edwin B. Hart, head of the Biochemistry Department, who might be interested in the project. Hart referred Rusch to Prof. Harry Steenbock, also of Biochemistry and the person who demonstrated that the irradiation of milk would produce vitamin D. The patent on this process produced sufficient money to

establish the Wisconsin Alumni Research Foundation, which in turn supported research and fellowships to the University of Wisconsin.

"Dr. Steenbock recommended Dr. Carl Baumann, a former student of
his, who accepted the challenge and began a cooperative venture with
Dr. Rusch in 1936. The arrangement was mutually beneficial to both
partners. After a few years Dr. Baumann, whose chief appointment was in
the Department of Biochemistry, served as mentor for several graduate
students. Among these were Jim and Betty Miller. Dr. Rusch became well
acquainted with Jim and Betty during this period and developed a high
regard for their creative abilities in science. Thus, after the McArdle
Laboratory was constructed and after Jim had obtained the Ph.D. in 1943,
Rusch offered him a position in the McArdle where he served as a
Finney-Howell postdoctoral fellow for one year and then was appointed
instructor in 1944. The same situation held true for Betty. When she
obtained the Ph.D. in 1945, she served as a Finney-Howell Postdoctoral
Fellow at McArdle for two years and then was appointed instructor in
1947. Both Jim and Betty moved up the academic ladder rather rapidly
thereafter based on their outstanding contributions to the field of
cancer research (Table 1). Their work was imaginative, original, and
stimulated many others to follow their lead in a new area. For this
they have received many honors and awards."

Following the background comments by Dr. Rusch and Dr. Schultze I
would like to quote from a letter written to me by Allan H. Conney, who
received the Ph.D. under the Millers' direction in 1956:

"You asked me to give you some of my recollections of Jim and Betty
Miller that might be helpful for your comments honoring them at the Third
International Symposium on Reactive Intermediates. I should start out
by saying that I love Betty and Jim as much as my own mother and father,
and indeed they have been my mother and father in science. The Millers
taught me how to do research, and I first experienced the joys of
discovery with them. Both Jim and Betty tried very hard to instill their
high standards into my research, even though they most certainly found
me to be a very difficult student. In my view, Jim is a little more
theoretical and Betty a little more practical, but both are outstanding
research scientists who work together as a team so that both share
equally in the many outstanding research accomplishments that have flowed
continuously from their laboratory. It is my hope that they will share
a Nobel Prize for their work.

Harold Rusch introduced me to Jim and Betty Miller early in 1952
and suggested them as possible mentors for my graduate studies. The
Millers talked with me about the metabolism of aminoazo dyes and their
efforts to understand how these dyes cause liver cancer. I was intrigued
and delighted by the Millers' carcinogen metabolism approach to the
cancer problem and by Harold Rusch's description of other molecular
approaches to cancer research at McArdle. I was also greatly impressed
by the intense enthusiasm for research that was expressed by the Millers,
and I accepted their offer to be a graduate student. The Millers wanted
me to start working in their laboratory immediately after finishing
pharmacy school at the University of Wisconsin in June of 1952, but I
preferred to start in September, which I did. Although my wishes were
accepted by the Millers, I believe they felt that the summer was wasted
and that it could have been better spent in the laboratory, and to this
day, I feel that I let them down by not starting my graduate work 3
months earlier.

The dedication of their lives to research and the high standards
that were set by Jim and Betty Miller inspired their students to set

Table 1. The Miller Careers I. Vitae

	James A. Miller	Elizabeth C. Miller
Birth	1915	1920
B.S.	1939 Pittsburgh	1941 Minnesota
Entered UW Graduate School	1939	1941
Married	1942	1942
Ph.D.	1943 Wisconsin	1945 Wisconsin
Entered McArdle Group	1943	1945
Finney Howell Post-Doctoral	1943-44	1945-47
Instructor (McArdle)	1944-46	1947-49
Assistant Professor	1946-48	1949-59
Elected, Am. Soc. Biol. Chemists	1949	1951
Associate Professor	1948-52	1959-69
Professor	1952-80	1969-80
Acting Director	--	1972-73
Associate Director	--	1973-present
Elected, National Academy of Sciences	1978	1978
WARF Professor of Oncology	1980-82	1980-82
Van Rensselaer Potter Professor	1982-present	1982-present
WARF Senior Distinguished Professor	1984-present	1984-present

high standards for their own work. It was common for us to work in the
laboratory on Saturdays, Sundays and on holidays. Jim and Betty taught
us by their example to dedicate our lives to research and to strive for
excellence.

My first research project at McArdle was the synthesis of 1-hydroxy-
2-aminonaphthalene, but several attempts failed to result in a pure
product. My failure as a synthetic chemist was extremely fortunate,
because it led me into a new area of research that turned out to have
important implications. In 1952, H. L. Richardson and his colleagues at
the University of Oregon had reported that 3-methylcholanthrene markedly
diminished the carcinogenicity of an aminoazo dye. The Millers had me
work on the mechanism of the antagonistic effect of 3-methylcholanthrene
on the carcinogenicity of aminoazo dyes, and this line of research
resulted in finding that a single injection of 3-methylcholanthrene,
benzo[a]pyrene or several other polycyclic aromatic hydrocarbons caused
a rapid stimulation in the hepatic metabolism of aminoazo dyes and in
the metabolism of the hydrocarbons themselves. These observations, which
were exciting to all of us in the Millers' laboratory and throughout
McArdle, led to additional studies with the Millers that provided strong
evidence that the hydrocarbons induced the synthesis of microsomal monooxygenase and reductase enzymes in the liver. This work with the Millers
provided me with my first taste of the joys of discovery.

My normal laboratory working habits as a student at McArdle resulted
in several experiments per week, and each experiment generated lots of
dirty glassware that I rinsed and placed in sulfuric acid-dichromate
cleaning solution. However, I was not always prompt in cleaning up my
dirty glassware, and my lab bench often had dirty glassware on it, even
though Jim Miller gave me several gentle suggestions that I should keep
my lab bench neat. One Saturday morning, when I came to work, I was
horrified to find Jim cleaning my glassware and lab bench. This approach
to teaching good housekeeping practices made a lasting impression on a
young student!!

Jim and Betty were demanding yet patient teachers. They required
that their students pursue their research programs with great vigor and

dedication, yet they were very patient and encouraging when experiments didn't work. The Millers taught their students and postdocs to be thorough in their research (don't miss something that is important) and to be critical of their own work and the work of others. Jim and Betty provided constructive, helpful advice to their students, to other staff members at McArdle, and to the general scientific community. Jim and Betty have continued to give me advice and encouragement for the almost 30 years that I have been away from their laboratory.

The Millers have given unselfishly of their time to serve on NIH study sections and on large numbers of committees and advisory boards concerned with important national research problems. About 2 years ago Jim and I attended an NCI workshop in Bethesda, and Jim, who recently developed diabetes, had taken too much insulin and needed to be taken to the emergency room at the NIH clinical center. He made a speedy recovery after receiving appropriate amounts of sugar, but the main concern that he expressed to me in the hospital was that he was terribly sorry for having made such a "fuss" and that he was sorry that both of us had missed part of the workshop session. After staying in the hospital for about 2 hours, Jim was completely recovered, and we both returned to the workshop.

On two occasions, I almost left research to pursue a profitable business career in my father's drug store (once after my M.S. degree and again immediately after my Ph.D. degree). On the latter occasion, I did work in my father's drug store for a few months before returning to a full-time career in research. Although the Millers were disappointed about the possibility of my leaving research, their comments were gentle and constructive during these periods. The positive experiences that I had in Jim and Betty's laboratory and seeing these two remarkable individuals and others at McArdle serve humanity in such a noble way were very powerful incentives for me to continue in research."

This ends the letter from Alan Conney.

Turning now to my own association with Jim and Betty Miller, I look back at over 40 years of continuous help and stimulation from these two remarkable people. I began work in the McArdle Laboratory in February 1940 before the building was completely finished. Jim Miller had already begun work in the Biochemistry Department and by 1941 published his first paper with Baumann and Rusch, entitled "Diet and Hepatic Tumor Formation" (Cancer Research, 1: 699-708, 1941). In the same year Betty arrived on the campus and by 1942 they were married, and were regularly participating in the seminars held in the old McArdle building. Jim entered the McArdle group as a post-doctoral fellow with Dr. Rusch in 1943 and Betty came as a post-doctoral fellow in 1945. Their advancement was continuous from that time on, as shown in Table 1.

My association with the Millers has been entirely within the McArdle Laboratory and at annual meetings of the AACR, where we often had dinner together. In Madison we did not share in social activities since the Millers decided early in their marriage that socialization and recreation were not to divert them from their steadfast devotion to research on chemical carcinogenesis, and to each other. Despite their 7-day work week as mentioned by Allan Conney, they parented two fine daughters. Linda is now Mrs. Russell Forbess, and lives in Schofield, Wisconsin. Helen is now Mrs. David Alexander, and lives in Louisville, Kentucky.

My high regard for the Millers is based in the first place on the characteristic of dependability and steadfastness as friends and as colleagues. When I was hospitalized with a broken hip, Betty visited

me. When I had personal sorrow I felt their support. When I have had a
manuscript to referee and needed their opinion, I have always received a
written appraisal within 24 hours. I have the impression that they are
never late for any deadline.

The Millers are devoted friends and colleagues not only to me, but
to Former Director Harold Rusch and to our present Director Henry Pitot.
They are completely dedicated to the McArdle Laboratory, its goals, and
its continued existence. As Associate Director Betty has shepherded
many a grant application for an individual, program project, or core
grant through to completion, with important changes in organization or
wording. Jim has maintained expertise on lab safety, instrumentation,
and newer developments in organic chemistry. I cannot imagine the
McArdle Laboratory without Jim and Betty. They are both so open, so
honest, so modest and unassuming, yet so competent and available that
they provide role models for all the rest of the staff. Every one of us
at McArdle, and every one of you in this meeting, is a unique individual
with a unique genome and a unique history. Very few of us has had the
privilege of "double-strandedness" and it is doubtful whether we could
have maintained the relationship even if we had had the opportunity.

Rarely can two people collaborate professionally for a complete
lifetime, and it is even more rare when a husband and wife can collaborate professionally for a lifetime together, as Jim and Betty have done.
Part of the reason for this is the reluctance of institutions to permit
either of these types of collaboration, with promotion to tenure contin-
gent upon the demonstrated ability of individuals as individuals. It is
very much to the credit of Harold Rusch and even more, I should emphasize, to the patience and determination of Betty and Jim Miller (cf Table
1) that we here today are able to celebrate a collaboration of the first
rank, between two people who have demonstrated their excellence not only
as collaborators but as individuals, serving in various capacities on
the national scene.

Perhaps the academic hierarchy might do well to encourage husband
and wife teams, but we must admit that the Miller phenomenon can occur
but rarely, and we must recognize that its acknowledgement may require
long years of demonstrated performance and recognition by outside peer
groups.

To indicate how the Millers have met these requirements I present
in the briefest possible way a synoptic view of the published record up
to 1969 (Table 2). What has happened in the ensuing 16 years I leave to
this meeting to describe.

Table 2. The Miller Careers II. Selected Publications

James A. Miller	Elizabeth C. Miller
Miller, Rusch and Bauman et al.	Miller and Baumann
1941-1946	1944-1946
Assistant Professor, 1946	Assistant Professor, 1949

(1947) Miller and Miller
        The Metabolism and Carcinogenicity of p-Dimethylaminoazobenzene
        and Related Compounds in the Rat
            Cancer Research 7: 39-41

(continued)

Table 2.   (continued)

(1947)   Miller and Miller
         The Presence and Significance of Bound Aminoazodyes in the Livers
         of Rats Fed p-Dimethylaminoazobenzene
              Cancer Research 7: 468-480

(1949)   Mueller, G. C. and Miller, J. A.
         The Reductive Cleavage of 4-Dimethylaminoazobenzene by Rat Liver:
         The Intracellular Distribution of the Enzyme System and Its
         Requirement for Triphosphopyridine Nucleotide.
              J. Biol. Chem. 180: 1125-1136

(1951)   Sorof, Cohen, Miller and Miller
         Electrophoretic Studies on the Soluble Proteins from Livers of
         Rats Fed Aminoazo Dyes
              Cancer Research 11: 383-387

(1956)   Conney, Miller and Miller
         The Metabolism of Methylated Aminoazo Dyes.   V.   Evidence for
         Induction of Enzyme Synthesis in the Rat by 3-Methylcholanthrene.
              Cancer Research 16: 450-459

(1960)   Cramer, Miller and Miller
         N-Hydroxylation: A New Metabolic Reaction Observed in the Rat
         with the Carcinogen 2-Acetylaminofluorene.
              J. Biol. Chem. 235: 885-888

(1961)   Miller, Miller and Hartman
         N-Hydroxy-2-acetylaminofluorene: A Metabolite of 2-acetylamino-
         fluorene with Increased Carcinogenic Activity in the Rat.
              Cancer Research 21: 815-824

(1966)   Miller, Juhl and Miller
         Nucleic Acid Guanine: Reaction with The Carcinogen N-Acetoxy-2-
         Acetylaminofluorene.
              Science 153: 1125-1127

(1967)   Kriek, Miller, Juhl and Miller
         8 (N-2-Fluorenyl-acetamido)guanosine, an Arylamidation Product of
         Guanosine and the Carcinogen N-Acetoxy-N-2-fluorenylacetamide in
         Neutral Solution
              Biochemistry 6: 177-182

(1967)   Miller and Miller
         The Activation of Carcinogenic Aromatic Amines and Amides by N-
         Hydroxylation In Vivo. In "Carcinogenesis: A Broad Critique".
         M. Mandel (Ed). pp. 397-420.

(1969)   Miller and Miller
         Studies in the Mechanism of Activation of Aromatic Amine and Amide
         Carcinogens to Ultimate Carcinogenic Electrophilic Reactants
              Ann. N. Y. Acad. Sciences 163: 731-750

     Finally, I would like to include for the record, and with apologies
to Jim and Betty for this unsolicited publicity, a partial list of their
honors, including only those that were labeled as "Awards" (Table 3).

Table 3. The Miller Careers III. Awards

<u>Joint Awards to James and Elizabeth Miller</u>

1962   Langer-Teplitz Award for Cancer Research
                    Chicago Cancer Foundation

1965   Lucy Wortham James Award for Cancer Research
                    James Ewing Society

1971   Bertner Foundation Award
                    M. D. Anderson Hospital and Tumor Institute

1973   Wisconsin National Divisional Award
                    American Cancer Society

1975   Papanicolaou Award
                    Papanicolaou Institute for Cancer Research

1976   Rosenstiel Award for Basic Medical Research
                    Brandeis University

1977   National Award in Basic Science
                    American Cancer Society

1978   First Founder's Award
                    Chemical Industry Institute of Toxicology

1978   Bristol-Myers Award in Cancer Research
                    Bristol-Myers

1978   Gairdner Foundation Award
                    University of Toronto

1979   FASEB 3 M Life Sciences Award

1979   Louis and Bert Freedman Foundation Award in Biochemistry
                    New York Academy of Sciences

1980   Mott Award
                    General Motors Cancer Research Foundation

        Today we all applaud these well-deserved honors and join in adding
one more token of love and esteem for two people whom we all admire.

REACTIVITY AND TUMORIGENICITY OF BAY-REGION DIOL EPOXIDES DERIVED FROM

POLYCYCLIC AROMATIC HYDROCARBONS

D. M. Jerina*, J. M. Sayer*, S. K. Agarwal*, H. Yagi*, W. Levin[†], A. W. Wood[†], A. H. Conney[†], D. Pruess-Schwartz[§], W. M. Baird[§], M. A. Pigott[✦], and A. Dipple[✦]

*The National Institutes of Health, NIADDK, Bethesda, MD
[†]Roche Institute of Molecular Biology, Nutley, NJ
[§]Department of Medicinal Chemistry and Pharmacognosy, Purdue University, W. Lafayette, IN, and [✦] LBI-Basic Research Program, NCI-Frederick Cancer Research Facility, Frederick MD

INTRODUCTION

The polycyclic aromatic hydrocarbons are a widespread class of environmental contaminants which express their tumorigenic activity through metabolic transformation to chemically reactive species which covalently modify cellular macromolecules. Ten years ago[1,2] we proposed that bay-region diol epoxides are prime candidates for the ultimate carcinogenic metabolites of the carcinogenic hydrocarbons provided that these molecules contained a bay-region and were capable of being metabolically transformed to such diol epoxides. The noncarcinogen phenanthrene is the simplest hydrocarbon that contains a bay region, the sterically hindered, cup-shaped

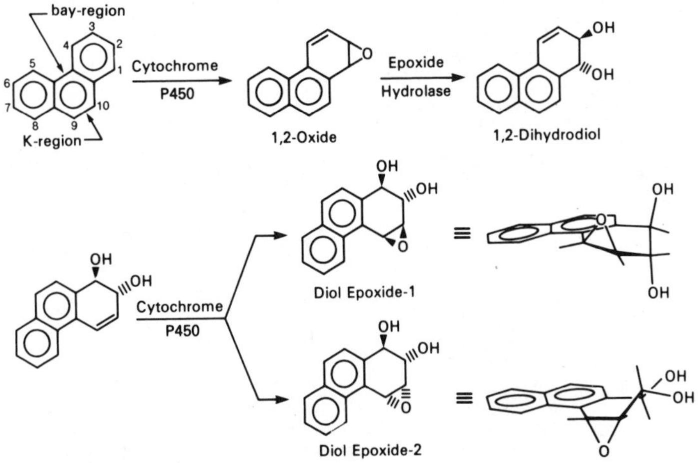

Fig. 1. Metabolism of the noncarcinogen phenanthrene to its dia-stereomeric bay-region diol epoxides. Preferred conformations of the diastereomers are shown. Relative but not absolute configuration is indicated.

area between carbons-4 and -5 of the molecule. The metabolic transformation of this hydrocarbon[3,4] into bay-region diol epoxides is illustrated in Figure 1. Initially, the hydrocarbon is oxidized by the cytochrome P450 monooxygenase system to form a 1,2-oxide that is subsequently converted in the presence of microsomal epoxide hydrolase to a *trans*-1,2-dihydrodiol by allylic attack of water. Although many arene oxides show a marked tendency to isomerize spontaneously to phenols,[5] epoxide hydrolase is often highly competitive in intercepting these reactive species. The 1,2-dihydrodiol is subject to further attack by the cytochrome P450 system to form a mixture of diastereomerically related diol epoxides in which the benzylic hydroxyl group is either cis (isomer-1 series) or trans (isomer-2 series) to the epoxide oxygen. One of the interesting features of these diol epoxides is that the isomer-1 series shows a slight preference for the conformation in which the hydroxyl groups are pseudoaxial whereas the isomer-2 series shows a marked preference for the conformation in which these groups are pseudo-equatorial (*vide* NMR).[6-8] These conformational preferences can be altered through structural variations as will be discussed later.

There are marked stereochemical preferences associated with the metabolism of polycyclic aromatic hydrocarbons to bay-region diol epoxides. These are typified by the metabolism of benzo(a)pyrene (Figure 2, reviewed in ref. 9) with cytochrome P450c and epoxide hydrolase. These preferences may be summarized as follows: i) cytochrome P450c displays a marked tendency to form benzo-ring arene oxides in which the benzylic oxirane carbon has R-absolute configuration, ii) microsomal epoxide hydrolase generally shows complete regiospecificity for attack at the allylic carbon of arene oxides such that the predominant enantiomer above is converted into the *trans*-(R,R)-dihydrodiol while a small amount of the (S,S)-dihydro-

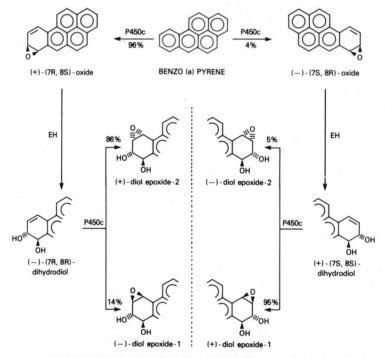

Fig. 2.   Stereoselective metabolism of benzo(a)pyrene to bay-region 7,8-diol-9,10-epoxides by cytochrome P450c and epoxide hydrolase (EH). With cytochrome P450c, the most active cytochrome P450 isozyme for hydrocarbon metabolism presently known[10], the (+)-(7R,8S)-diol-(9S,10R)-epoxide predominates[11].

diol is formed from the minor arene oxide enantiomer, and iii) cytochrome P450c exhibits a high facial preference on metabolism of the dihydrodiol enantiomers to diol epoxides such that the (R,R)-dihydrodiol is converted primarily to the diol epoxide-2 diastereomer while the minor (S,S)-dihydrodiol is metabolized mainly to the diol epoxide-1 diastereomer. Thus, although four optically active bay-region diol epoxides are metabolically possible, the (R,S,S,R)-diol epoxide-2 optical isomer predominates from benzo(a)pyrene and several other hydrocarbons studied.[12]

To date more than a dozen hydrocarbons have been studied in sufficient detail either to prove or implicate bay-region diol epoxides as their metabolically formed ultimate carcinogens (reviewed in ref. 9); these include benzo(a)pyrene, dibenzopyrenes, benzo(c)phenanthrene, chrysene and 5-methylchrysene, benz(a)anthracenes including 3-methylcholanthrene and 7,12-dimethylbenz(a)anthracene, dibenz(a,h)anthracene, several benz-acridines, 15,16-dihydro-11-methylcyclopenta(a)phenanthrene-17-one and benzo(b)fluoranthene. During the past several years, we have expended substantial effort in an attempt to identify key factors responsible for the expression of biological activity by bay-region diol epoxides. To this end, numerous diol epoxides have been synthesized, often in optically pure form, such that their mutagenic and tumorigenic activity could be compared with properties such as relative configuration (isomer-1 versus isomer-2), preferred conformation (axial versus equatorial hydroxyl groups), absolute configuration and chemical reactivity.

CONFORMATIONAL FACTORS AND TUMOR RESPONSE

The first study which conclusively established the nature of an ultimate carcinogen for any of the polycyclic aromatic hydrocarbons was done with the diastereomeric 7,8-diol-9,10-epoxides of benzo(a)pyrene.[13] With only one known exception (discussed below), high tumorigenicity, when observed, has been limited to diol epoxide-2 diastereomers, whereas the diol epoxide-1 diastereomers have been found to be weak or inactive, as illustrated by the tumor data in Table 1 for the pairs of diastereomers from benzo(a)pyrene, benz(a)anthracene, benz(c)acridine and chrysene. Although numerical comparison of tumorigenicity between diastereomers from a given hydrocarbon is not always possible due to very low activity for the diol epoxide-1 isomers, the markedly higher activity for the diol epoxide-2 diastereomers is apparent; more than 50-fold in the benz(c)acridine case.

The difference in activity between the two series of diastereomers may not be entirely due to differences in relative configuration, since the diastereomers also have different preferred conformations for their hydroxyl groups (Figure 1). In an attempt to identify the individual importance of these factors, relative configuration versus preferred conformation, in expression of tumorigenic activity, we have sought to prepare bay-region diol epoxides with altered conformations. In the absence of specific steric or electronic factors (discussed below), benzo-ring dihydrodiols have a marked preference for the conformation in which their hydroxyl groups are pseudoequatorial.[18,19] If the dihydrodiol forms part of a bay region, steric hindrance within the bay region forces the hydroxyl groups into the pseudoaxial conformation. The 9,10-dihydrodiol of benzo(e)pyrene is a case in point. Not only does the dihydrodiol prefer the pseudoaxial conformation, but also both diastereomers of the 9,10-diol-11,12-epoxide prefer this conformation.[20] As can be seen in Table 1, both diastereomers are weak or inactive as carcinogens thus suggesting that pseudoaxial hydroxyl groups may be a deterrent to tumorigenic activity. Further support for this hypothesis has been obtained with studies on 6-fluorobenzo(a)pyrene. Electronic effects of the fluorine substituent cause the 7,8-dihydrodiol[21] as well as the diastereomeric

Table 1. Comparison of tumorigenic activity for diastereomeric pairs of racemic bay-region diol epoxides in the newborn mouse

Hydrocarbon Diol Epoxide	Diastereomer (dose, μmol)	Lung Tumors per Mouse	Reference
benzo(a)pyrene	control	0.13	
	isomer-1 (0.028)	0.14	
	isomer-2 (0.028)	4.42	13
benz(a)anthracene	control	0.15	
	isomer-1 (0.280)	0.56	
	isomer-2 (0.280)	13.3	14
benz(c)acridine	control	0.15	
	isomer-1 (1.05)	0.55	
	isomer-2 (0.50)	33.4	15
chrysene	control	0.17	
	isomer-1 (1.40)	0.34	
	isomer-2 (1.40)	15.91	16
phenanthrene	control	0.17	
	isomer-1 (1.40)	0.18	
	isomer-2 (1.40)	0.18	16
benzo(e)pyrene	control	0.28	
	isomer-1 (0.70)	0.44	
	isomer-2 (0.70)	0.39	17

7,8-diol-9,10-diol epoxides[22] to prefer the conformation with pseudoaxial hydroxyl groups. The fluorinated diol epoxides had less than one-fifth the mutagenic activity of their benzo(a)pyrene counterparts toward Chinese hamster V79 cells[22] and were inactive as tumorigenic agents (unpublished data) when compared to the (+)-7,8-diol-9,10-epoxide-2 of benzo(a)pyrene at a 14 nmol dose in the newborn mouse. Triphenylene and benzo(e)pyrene are close structural analogs. Like the 9,10-diol-11,12-epoxides of benzo(e)-pyrene, both diastereomers of the 1,2-diol-3,4-epoxide of triphenylene prefer the conformation with pseudoaxial hydroxyl groups.[20] Because of the low mutagenic activity of the triphenylene diol epoxides toward Chinese hamster V79 cells,[23] a generally good predictive test for tumorigenic activity, they were assumed to be inactive and have not been tested as carcinogens.

Although the 1,2-diol-3,4-epoxide-2 of phenanthrene prefers the conformation in which its hydroxyl groups are pseudoequatorial (Figure 1), it lacks tumorigenic activity (Table 1). As will be discussed later, this is not a consequence of low intrinsic chemical reactivity compared to tumorigenic bay-region diol epoxides. Nonetheless, it is tempting to speculate that pseudoequatorial hydroxyl groups enhance tumorigenicity of bay-region diol epoxides since pseudoaxial hydroxyl groups appear to diminish activity. Thus, an example of the remaining conformational type, a diol epoxide-1 diastereomer with a preference for pseudoequatorial hydroxyl

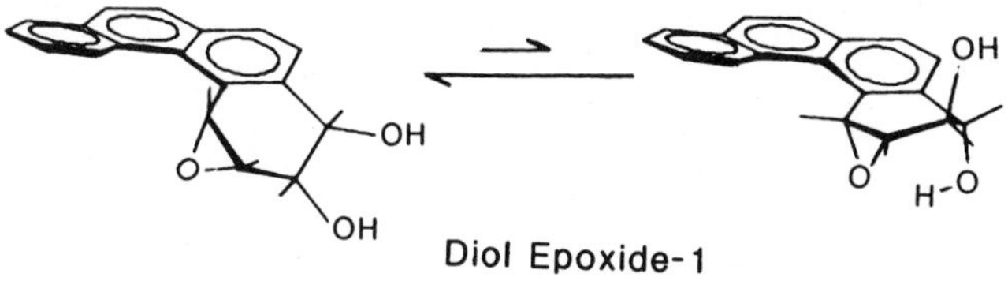

Diol Epoxide-1

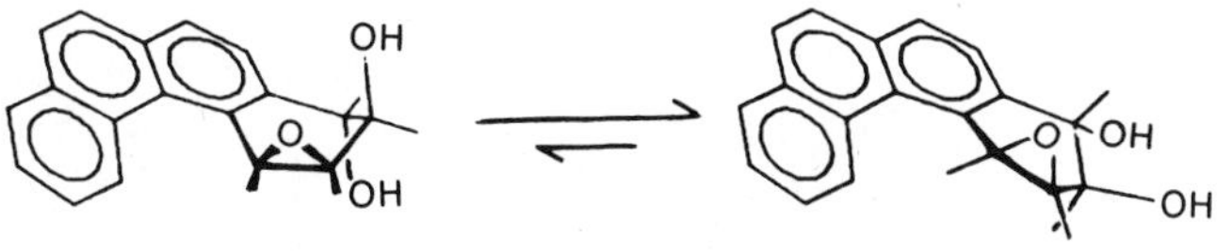

Diol Epoxide-2

Fig. 3.   Preferred conformations of the diastereomeric benzo(c)-
phenanthrene 3,4-diol-1,2-epoxides. Note that a conformation with pseudoequatorial hydroxyl groups is favored for
both diastereomers.

groups, would be of considerable interest to test for tumorigenic activity.
The 3,4-diol-1,2-epoxides of benzo(c)phenanthrene[24] provide such an example
where both diastereomers prefer a conformation with pseudoequatorial
hydroxyl groups (Figure 3). Initiation-promotion studies with these diol
epoxides on mouse skin indicated that an initiating dose of only 0.025 μmol
of both diastereomeric diol epoxides caused a >50% tumor incidence with
>1.5 papillomas per mouse.[25] Thus we not only observed the suspected high
activity for a diol epoxide-1 diastereomer with a preference for pseudoequatorial hydroxyl groups, but also identified the most tumorigenic diol
epoxides presently known. However, more recent studies in the newborn
mouse (unpublished), which parallel the experiments in Table 1, have shown
that benzo(c)phenanthrene 3,4-diol-1,2-epoxide-1 was inactive at a 0.05
μmol dose, whereas the diol epoxide-2 diastereomer produced 16 lung tumors
per mouse. This is the first example where differing conclusions can be
drawn in comparing the activity of diol epoxides in the skin versus the
newborn tumor models. In view of the difference between the models, it is
presently unclear whether pseudoequatorial hydroxyl groups aid in the
expression of tumorigenic activity or simply do not inhibit expression of
this activity. It is known, however, that the presence of hydroxyl groups
retards the ability of microsomal epoxide hydrolase to detoxify bay-region
diol epoxides to inactive tetraols[26] and also decreases the solvolytic
reactivity (under acidic conditions) of diol epoxides relative to tetrahydroepoxides[8,27] (cf. Figure 4).

SOLVOLYTIC REACTIVITY AND TUMOR RESPONSE

Diol epoxides derived from polycyclic aromatic hydrocarbons are
electrophilic molecules that undergo solvolysis in aqueous media by both
acid-catalyzed and spontaneous pathways,[8,24,27,28] according to a rate law,
$k_{obsd} = k_H a_{H^+} + k_o$. Typical pH-rate profiles for solvolysis of diol
epoxides-1 and -2 and a tetrahydroepoxide are shown in Figure 4. These
reactions have been subjected to considerable experimental scrutiny since
they provide the simplest chemical models for the postulated event in which
diol epoxides alkylate a critical macromolecule in the cell, initiating the
mutagenic or tumorigenic process. The products of the acid catalyzed
pathway for diol epoxide solvolysis (k_H), shown in Figure 5, are tetraols[29]
resulting from cis or trans attack of water at the benzylic position. In
addition to tetraols, the spontaneous reaction (k_o) can also give rise to a

15

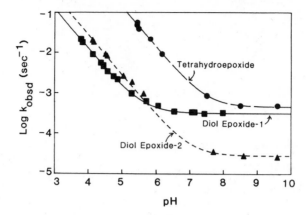

Fig. 4. Dependence on pH of the rate constant (k_{obsd}) for solvolysis of benz(a)anthracene 3,4-diol-1,2-epoxide-1, 3,4-diol-1,2-epoxide-2 and tetrahydro 1,2-epoxide, at 25 °C in 1:9 dioxane:water, ionic strength 0.1 (NaClO₄). The flat portion of the curves corresponds to k_o whereas the portion with a slope of -1 corresponds to k_H.

keto diol[29] presumably derived from hydride migration to the benzylic position; this latter product is generally observed from isomer-1 but not from isomer-2, except in the presence of unusual conformational factors.[27]

In 1976, we proposed[2] that the chemical reactivity, and hence possibly one aspect of the biological activity, of diol epoxides ought to be related to the quantum chemical parameter $\Delta E_{deloc}/\beta$, which provides a measure of the ease of formation of a resonance stabilized carbocation at the benzylic position of these molecules. The prediction was based on the expectation that the transition state for ring opening of these epoxides by water or weakly basic nucleophiles such as DNA should have substantial carbocation

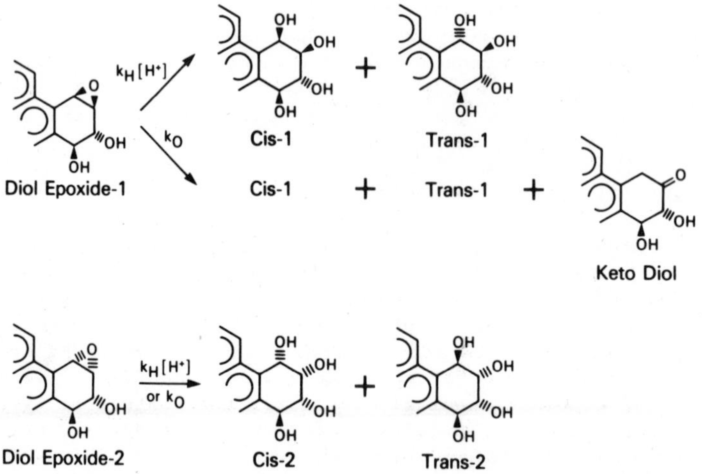

Fig. 5. Products formed from diol epoxide-1 and -2 under conditions of acid-catalyzed (k_H) and spontaneous (k_o) solvolysis.

16

Table 2. Predicted and observed solvolytic reactivity of diol epoxide-2 diastereomers compared to their ability to cause lung adenomas in the newborn mouse

Hydrocarbon Diol Epoxide	$\Delta E_{deloc}/\beta$[a]	$10^5\,k_o$[b] (sec^{-1})	$T_{1/2}$ at pH 7[c] (min)	Relative Tumorigenicity[d]
dibenzo(a,i)pyrene	0.866	120[e]	8.3	2
dibenzo(a,h)pyrene	0.845	32[e]	29	3
benzo(a)pyrene	0.794	13[f]	43	5
benz(a)anthracene	0.766	2.8[g]	210	4
benzo(e)pyrene	0.714	2.0[g]	540	nil
triphenylene	0.664	1.9[g]	550	–
phenanthrene	0.658	0.85[h]	470[h]	nil
chrysene	0.639	0.75[i]	610	1
benzo(c)phenanthrene	0.600	0.08[i]	490	6

[a]Ref. 2. [b]In 1:9 dioxane-water, ionic strength 0.1, 25°C. [c]These values reflect contributions to the rate by both k_H and k_o at this pH. [d]The numbers provide an approximate ranking of these diol epoxides with 6 representing the most tumorigenic compound. [e]Ref. 30. [f]Ref. 28. [g]Ref. 27. [h]The value of $k_o = 3.4 \times 10^{-5}$ sec^{-1} observed in water was divided by 4 to correct for a solvent effect (refs. 8, 31). [i]Ref. 24.

character. Subsequent results (Table 2) have shown that the solvolytic reactivity, as measured by k_o, is indeed well correlated over a 10^3-fold range of reactivity with $\Delta E_{deloc}/\beta$ for diol epoxides having similar conformational preferences. Two compounds in the series, the diol epoxides of benzo(e)pyrene and triphenylene, exhibit altered conformational preferences (see preceding section) which affect their solvolytic reactivity. However, for these particular compounds, the conformational effects on k_o are small. Thus even these two diol epoxides exhibit reactivities relative to other members of the series that are qualitatively consistent with their values of $\Delta E_{deloc}/\beta$. In contrast to the solvolytic reactivity, the tumorigenic activity of the diol epoxides bears little relationship to the predicted ease of epoxide ring opening via a cationic transition state. For example, phenanthrene diol epoxide, which has a low value of $\Delta E_{deloc}/\beta$ and low solvolytic reactivity,[8] is non-tumorigenic,[16,32] yet two dibenzopyrene diol epoxides, which are the most solvolytically reactive diol epoxides studied, are also less tumorigenic[33] than are diol epoxides of intermediate reactivity, such as those derived from benzo(a)pyrene and possibly benz(a)anthracene. Benzo(c)phenanthrene diol epoxide, the least solvolytically reactive diol epoxide studied,[24] is the most tumorigenic diol epoxide observed to date in the mouse skin tumor model.[25] Thus, neither high nor low solvolytic reactivity is correlated with high tumorigenic response elicited by these diol epoxides. However, relatively good correlations between $\Delta E_{deloc}/\beta$ and mutagenic activity of tetrahydroepoxides and diol epoxides have been observed.[34,35]

As illustrated in Figure 2 for benzo(a)pyrene, the combined specificity of cytochrome P450c and epoxide hydrolase results in the preferential formation of the (7R,8S)-diol-(9S,10R)-epoxide-2. The stereoselectivity associated with this pathway, initial formation of (R,R)-dihydrodiols with bay-region double bonds and subsequent oxidation to (R,S)-diol-(S,R)-

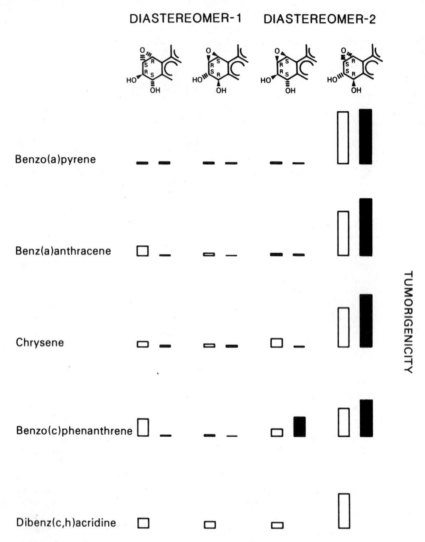

Fig. 6.   Relative tumorigenic activity of optically active bay-region diol epoxides after intraperitoneal injection into Swiss-Webster newborn mice (lung adenomas/mouse-solid bars) and after initiation-promotion on the backs of CD-1 mice (papillomas/mouse-open bars).[12] Tumor data for the diol epoxides from benzo(c)phenanthrene and dibenz(c,h) acridine are unpublished. Data are expressed as percent of total activity in a given tumor model for each isomer. Because of differences in experimental conditions, comparisons are valid only among the four configurational isomers derived from a given parent hydrocarbon, and not among diol epoxides from different hydrocarbons.

epoxide-2 isomers, has now been studied in detail for several hydrocarbons including phenanthrene, chrysene, benz(a)anthracene and 6-fluorobenzo(a)-pyrene (reviewed in refs. 36, 37). In all these cases, the (R,S)-diol-(S,R)-epoxide-2 predominates. The purpose of this section is to illustrate that this same configurational isomer of the four metabolically possible bay-region diol epoxides also has the highest tumorigenic activity. Relative tumorigenic activities for these isomers from five hydrocarbons in two tumor models are shown in Figure 6. The data are uncorrected for spontaneous tumors observed in animals treated with solvent only. For initiation-promotion on mouse skin, from 51-90% of the total activity is associated with the (R,S)-diol-(S,R)-epoxide-2. After intraperitoneal injection in the newborn mouse, this number ranges from 64-98%. Most of the other isomers have activity close enough to control values that generalizations concerning their relative activity are not warranted. The only major exception to this is found for the isomers from benzo(c)-phenanthrene, where the (4S,3R)-diol-(2S,1R)-epoxide-1 and the (4S,3R)-diol-(2R,1S)-epoxide-2 also have substantial activity in one or the other of the tumor models. Even in this case, the (4R,3S)-diol-(2S,1R)-epoxide-2 is the most active isomer in both tumor models. Perhaps the most dramatic point that can be made from Figure 6 is that absolute configuration plays a very important role in the expression of tumorigenic activity by bay-region diol epoxides.

COVALENT BINDING OF BENZO(c)PHENANTHRENE DIOL EPOXIDES TO DNA

Initial work by Borgen et al.[38] and by Sims et al.[39] pointed to the role of diol epoxides in the covalent modification of DNA by polycyclic aromatic hydrocarbons. Numerous studies (reviewed in refs. 40, 41) have since explored the nature of such binding in more detail. Because of the exceptionally high tumorigenicity of the benzo(c)phenanthrene diol epoxides, we anticipated that these compounds would represent highly interesting candidates for the study of their interactions with DNA. These diol epoxides have now been found to exhibit an exceptionally high efficiency of covalent binding, relative to hydrolysis, when allowed to react with DNA. Results of experiments designed to measure the rate and amount of covalent binding of the four possible configurational isomers of benzo(c)phenanthrene diol epoxide to DNA at pH 6.8-6.9 and 37 °C are shown in Figure 7. Reactions were initiated by adding each diol epoxide in acetone solution to a buffered solution (Tris-HCl, 10 mM) containing 0.8 mg calf thymus DNA/ml. After trapping of unreacted diol epoxide with a basic solution of mercaptoethanol,[42] aliquots of the reaction mixture were neutralized and subjected to HPLC for quantitation of the resultant thioether as well as tetraols formed in the course of the reaction. The percent recovery of tetraols in the presence of DNA was calculated by comparison with a sample of each diol epoxide that had been completely converted to tetraols by acid hydrolysis. For all four isomers, only 25-40% of the diol epoxide was recovered as tetraols after completion of reaction with DNA. Thus, 60-75% of the diol epoxide had become covalently bound to DNA. The rate of disappearance of diol epoxide-2 was markedly accelerated by DNA as indicated in the lower left panel. The half-life for diol epoxide-2 in the presence of DNA was 4-7 min, whereas the corresponding half-life in buffer alone under these conditions was ~2.3 hours. Notably, there was no large difference in either the rate of binding or the fraction of diol epoxide bound for the four configurational isomers. Under the same conditions, the disappearance of (+)-benzo(a)pyrene diol epoxide-2 also occurred at a rapid rate ($t_{1/2} < 30$ sec), but $\leq 6\%$ of the diol epoxide was bound to DNA; i.e., recovery of tetraols was nearly quantitative, within our experimental limitations. This was in accord with our expectations based on the low efficiency of binding of benzo(a)pyrene bay-region diol epoxides to DNA observed in binding studies by other workers.[43,44] Thus, the increased rate of disappearance of the benzo(c)-phenanthrene diol epoxides in the presence of DNA is due in large part to adduct formation, in contrast to the case of benzo(a)pyrene diol epoxide-2, where the pathway for the acceleration almost exclusively involves catalysis by DNA of hydrolysis[42,45,46] to the *trans*-tetraol. The non-carcinogen phenanthrene diol epoxide-2 also exhibited ≤10% binding to DNA under our conditions, and, unlike both the benzo(c)phenanthrene and benzo(a)pyrene derivatives, showed only a very small rate acceleration in the presence of 0.7 mg/ml DNA.

Because of the high efficiency of binding of the benzo(c)phenanthrene diol epoxides to DNA, these compounds provide an ideal opportunity for identification and structural study of the resultant adducts. Several products were obtained upon treatment of calf thymus DNA with each con-figurational isomer of the diol epoxide, followed by enzymatic hydrolysis

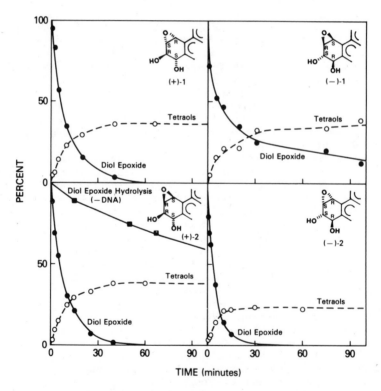

Fig. 7. Rates of diol epoxide disappearance (solid symbols) and tetraol formation (open symbols) from the four con-figurationlly isomeric benzo(c)phenanthrene diol epoxides in the presence of calf thymus DNA (0.8 mg/ml) in 1:10 acetone: water at 37 °C, pH 6.8. The amount of diol epoxide not recovered as tetraols is covalently bound to the DNA. For comparison the rate of disappearance of benzo(c)phenanthrene diol epoxide-2 under similar experimental conditions but in the absence of DNA is shown in the lower left panel.

Table 3. Chromatographic comparison of diol epoxide-deoxynucleoside adducts formed from deoxyribonucleotides and from DNA[a]

Reaction with (+)-Diol Epoxide-1 (Percent of Total Product)

Products from	(+)-DE-1/dG$_1$	(+)-DE-1/dG$_2$	(+)-DE-1/dA$_1$	(+)-DE-1/dA$_2$
dGuo	18.5	69.3		
dAdo			32.3	55.4
DNA	5.5	4.8	66.8	20.9

Reaction with (-)-Diol Epoxide-1 (Percent of Total Product)

Products from	(-)-DE-1/dG$_1$	(-)-DE-1/dG$_2$	(-)-DE-1/dA$_1$	(-)-DE-1-dA$_2$
dGuo	14.5	50.0		
dAdo			14.1	74.6
DNA	28.1	10.3	6.8	48.8

Reaction with (+)-Diol Epoxide-2 (Percent of Total Product)

Products from	(+)-DE-2/dG$_1$	(+)-DE-2/dG$_2$	(+)-DE-2/dC	(+)-DE-2/dA$_1$	(+)-DE-2/dA$_2$
dGuo	5.5	92.8			
dCyt			48.8		
dAdo				9.4	88.4
DNA	1.5	35.0	11.3	16.7	34.6

Reaction with (-)-Diol Epoxide-2 (Percent of Total Product)

Products from	(-)-DE-2/dG$_1$	(-)-DE-2/dG$_2$ [(-)-DE-2/dC]	(-)-DE-2/dA$_1$	(-)-DE-2/dA$_2$
dGuo	9.5	60.1		
dCyt		74.3		
dAdo			7.2	91.2
DNA	3.2	30.2	1.6	64.0

[a]Products are listed in order of their elution times on reverse-phase HPLC. Identification of the products from DNA was based on chromatographic and UV spectral comparison with products formed from the deoxynucleotides. The sum of adducts from individual nucleotides is not 100% since only those adducts that correspond to DNA adducts are tabulated.

to the mononucleoside adducts and reverse-phase HPLC. The base involved in each of these DNA adducts has been tentatively identified by comparison with adducts obtained from the reactions of the isomeric diol epoxides with the individual nucleotides (Table 3). To date, these identifications are based mainly on comparison of chromatographic retention times and UV

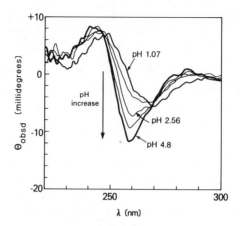

Fig. 8. Changes in the CD spectrum in 1:9 methanol:water of a
deoxyguanosine adduct of benzo(c)phenanthrene diol
epoxide-1 as a function of pH.

spectra. From a given diol epoxide, two pairs of adducts are formed by
addition to deoxyguanosine and deoxyadenosine, presumably by nucleophilic
attack of the bases at C-1 cis and trans to the epoxide oxygen (see below).
It is striking that a large proportion of the adducts formed from all four
isomers involves deoxyadenosine. Previous studies with bay-region diol
epoxides of other polycyclic aromatic hydrocarbons have generally implicated the exocyclic amino group of deoxyguanosine as a principal site of
reaction,[40,41] with only small amounts of deoxyadenosine adducts. These
deoxyadenosine adducts also appear to involve binding to the exocyclic
amino group.[47]

     Preliminary chemical studies on one of the deoxyguanosine adducts are
consistent with an adduct in which the exocyclic N^2 of deoxyguanosine has
been alkylated by the diol epoxide, analogous to previously characterized
adducts from guanine nucleosides and benzo(a)pyrene diol epoxides. Changes
in the CD spectrum[48] of this adduct, produced upon titration in 10%
methanol with acid (cf. Figure 8) or base gave pK_a values of 2.4 and 9.7,
consistent with a lack of substitution at N-7 and N-1 (O^6), and similar to
values of 1.4-2.1 and 9.1-9.8 for the guanosine adducts of benzo(a)pyrene
diol epoxides.[48,49] Methylation of the deoxyguanosine adduct with dimethyl
sulfate, followed by mild acid treatment, (pH 6, 95 °C, 1.5 hr) gave a
single chromatographic peak, which was not identical to the starting
material, and whose UV spectrum indicated the presence of both the guanine

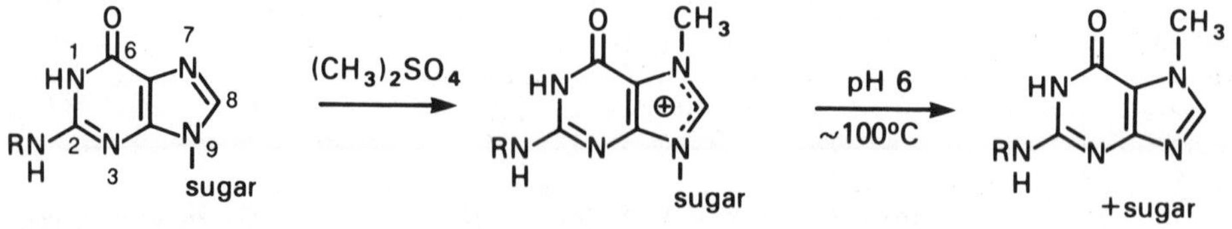

Fig. 9. Methylation and depurination reactions of a deoxyguanosine adduct of benzo(c)phenanthrene diol epoxide-1. The
hydrocarbon moiety is indicated as R.

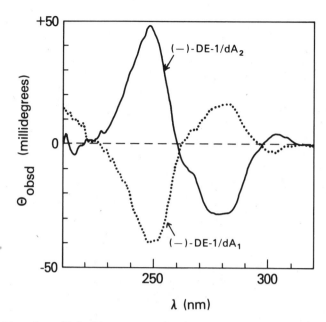

Fig. 10.   Circular dichroism spectra of two deoxyadenosine adducts
of (-)-benzo(c)phenanthrene diol epoxide-1. The spectra
are normalized to absorbance 1.0 at their absorption
maxima (252-254 nm).

and benzo(c)phenanthrene moieties, consistent with methylation at N-7 and
loss of deoxyribose (Figure 9). Thus the site of addition of the benzo(c)-
phenanthrene moiety is not on the sugar. This adduct was also shown to be
stable to 1 M potassium hydroxide at 100 °C for 2 hours, like the
corresponding benzo(a)pyrene diol epoxide adducts at N^2 of guanosine[49].

     Preliminary NMR data indicate that several of the pairs of adducts
derived from reaction of a given nucleotide with each configurational
isomer of the diol epoxide (e.g., (+)-DE-1/dG$_1$ and (+)-DE-1/dG$_2$, Table 3)
result from cis and trans attack of the nucleotide base at C-1 of the diol
epoxide.   In further support of the cis-trans relationship for these pairs
of adducts, their CD spectra are nearly mirror images, as expected for
pairs of isomers derived from the same enantiomer of the diol epoxide but
differing in configuration about C-1 (cf. refs. 47, 49). This relationship
is illustrated in Figure 10 for the adducts (-)-DE-1/dA$_1$ and (-)-DE-1/dA$_2$.

     The distribution of DNA adducts formed from the configurationally
isomeric benzo(c)phenanthrene diol epoxides is compared graphically with
their activity in two tumor models in Figure 11. The relationship between
the activity of these diol epoxides in either tumor model and the distribution of the nucleoside adducts formed from high concentrations of the diol
epoxides upon reaction with DNA *in vitro* is not obvious. For example, the
ratio of deoxyguanosine to deoxyadenosine adducts from (-)-diol epoxide-2,
which shows substantial activity in both tumor models, is 33:67, yet
(-)-diol epoxide-1, which is inactive in both tumor models, gives a very
similar ratio of deoxyguanosine to deoxyadenosine adducts of 38:56. Thus,
more subtle differences, perhaps not identifiable simply through comparison
of chemical reactivity with DNA and tumor data, must be responsible for the
differences in tumorigenic response elicited by the isomers.

     The binding of tritium-labeled benzo(c)phenanthrene to rodent embryo
cells in culture has been investigated. In a typical experiment, confluent
cultures of third-passage cells from Wistar rats were incubated for 48 hr

23

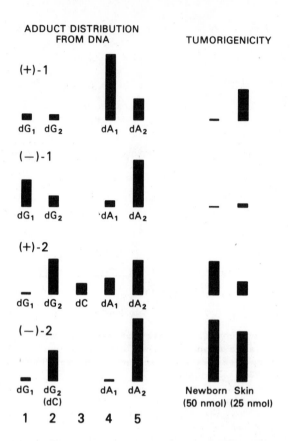

ADADUCT DISTRIBUTION FROM DNA

TUMORIGENICITY

Fig. 11. Comparison of the tumor response and adduct distribution
from calf thymus DNA for the four configurational isomers
of benzo(c)phenanthrene 3,4-diol-1,2-epoxide. The adducts
are numbered in order of their chromatographic elution
times. It is important to realize that 16 different
adducts are involved from deoxyguanosine and deoxyadenosine and the diol epoxides (cf. Table 3). Products
from deoxyguanosine and deoxycytidine with (−)-diol
epoxide-2 were co-chromatographic, and a definitive
assignment of product 2 from this isomer was also not
possible based on UV spectra. Tumor data are expressed
as percent of total activity in a given tumor model for
each isomer.

with 2.5 μM [^{3}H]-benzo(c)phenanthrene of specific activity 3.67 Ci/mmol.
Upon isolation of the DNA, covalent binding of 35 pmol of benzo(c)-
phenanthrene per mg of DNA was observed. The isolated DNA was hydrolyzed
enzymatically to the deoxyribonucleoside level, and the adducts containing
bound benzo(c)phenanthrene were analyzed by HPLC.   A preliminary
separation of adducts derived from isomer-1 (C_2-OH and C_3-OH of the
tetrahydro benzo-ring trans) and isomer-2 of the diol epoxide (C_2-OH and
C_3-OH cis) was accomplished by chromatography on an immobilized boronate
column eluted with 1 M morpholine buffer, followed by 1 M morpholine buffer

Table 4. Deoxyribonucleoside adducts of benzo(c)phenanthrene diol epoxides
identified after treatment of Wistar rat embryo cells with
[^{3}H]-benzo(c)phenanthrene[a]

Isomer-1 Adducts (26.5% of Total)	
Adduct Identification[b]	Percent of Total Radioactivity[c]
(-)-DE-1/dG$_1$	1.7
(+)-DE-1/dA$_1$	13.9
(+)-DE-1/dA$_2$	2.7

Isomer-2 Adducts (73.5% of Total)	
Adduct Identification[d]	Percent of Total Radioactivity[c]
(-)-DE-2/dG$_1$	0.6
(-)-DE-2/dG$_2$	16.0
(-)-DE-2/dA$_2$	53.5

[a]Identification was by chromatographic comparison with standards prepared
from diol epoxides of known absolute configuration and calf thymus DNA.
Only identifiable peaks are tabulated. [b]In addition to the peaks
identified, radioactivity was observed at the solvent breakthrough (4.6%)
and in two small unknown peaks (1.2%).   [c]Based on a total of 100% for
isomer-1 and isomer-2 adducts together.   [d]Less than 2% of the total
radioactivity was found in solvent breakthrough and unidentified peaks.

containing 10% sorbitol.[50,51] The isomer-1 adducts were eluted by the
morpholine buffer whereas the isomer-2 adducts were displaced from the
column only when sorbitol was added to the eluent.   The fractions
containing isomer-1 and isomer-2 adducts of the deoxyribonucleosides were
analyzed separately by reverse-phase HPLC, and the adducts were identified
by co-chromatography with standards derived from the reactions of calf
thymus DNA with the four configurational isomers of benzo(c)phenanthrene
diol epoxide (cf. Table 3). Table 4 summarizes the results. Similar results
were observed with Sencar mouse and Syrian hamster embryo cells. Notably,
the major portion (~70%) of the adducts identified arose from benzo(c)-
phenanthrene (4R,3S)-diol-(2S,1R)-epoxide-2, with a lesser but substantial
contribution from the (4S,3R)-diol-(2S,1R)-epoxide-1, in accordance with
predominant metabolic formation of the diol epoxide-2 isomer via the
(3R,4R)-dihydrodiol and formation of the minor diol epoxide-1 product from
the (3S,4S)-dihydrodiol (cf. Figure 2). For other hydrocarbons, such as
benzo(a)pyrene[52] and 7,12-dimethylbenz(a)anthracene,[50] both diol epoxide-2
and diol epoxide-1 adducts have also been observed. For the (-)-(4R,3S)-
diol-(2S,1R)-epoxide-2 adducts from benzo(c)phenanthrene, deoxyadenosine
adducts predominated, with a ratio of deoxyadenosine:deoxyguanosine adducts
of 3:1.   This result was qualitatively similar to that observed for the
reaction of (-)-benzo(c)phenanthrene (4R,3S)-diol-(2S,1R)-epoxide-2 with
calf thymus DNA in aqueous buffers (deoxyadenosine:deoxyguanosine adduct
ratio of ~2:1). Thus, the selectivity for the individual nucleotide units
in a purely chemical system parallels that at the higher level of
complexity involving a metabolically generated diol epoxide in intact
cells.   It is of interest to note that a high percentage of the adducts
formed from benzo(c)phenanthrene involves binding to deoxyadenosine. This
result parallels metabolic activation studies with 7,12-dimethylbenz(a)-
anthracene[53] and may be related to the deformation from planarity induced
by steric hindrance in the bay region of both these hydrocarbons.

SUMMARY

During the past decade substantial progress has been made in elucidating factors that determine the tumorigenic activity of bay-region diol
epoxides, major ultimate carcinogenic metabolites derived from polycyclic
aromatic hydrocarbons. Neither high nor low chemical reactivity of the diol
epoxides (as measured by rates of uncatalyzed solvolysis) is required for
high tumorigenic response. In contrast, aspects of molecular structure
such as conformation and absolute configuration strongly influence tumorigenic activity. The role of conformation is illustrated by the observation
that those diol epoxides whose hydroxyl groups are pseudoaxial are weak or
inactive as tumorigens. Absolute configuration is an important determinant
of biological activity of bay-region diol epoxides: in all cases studied to
date, the predominantly formed (R,S)-diol-(S,R)-epoxides are generally the
most tumorigenic of the four metabolically possible configurational
isomers.

In the course of investigating the effects of structural factors on
tumorigenic activity, we identified the (4R,3S)-diol-(2S,1R)-epoxide of
benzo(c)phenanthrene as the most potent tumorigen (in initiation-promotion
experiments on mouse skin) of the diol epoxides studied to date. Studies
of all four configurationally isomeric diol epoxides derived from benzo(c)-
phenanthrene led to the striking observation that these diol epoxides
exhibit an exceptionally high efficiency of covalent binding, relative to
hydrolysis, when allowed to react with calf thymus DNA in aqueous solution.
Thus, these diol epoxides should provide an excellent tool for the detailed
study of such binding. When the four isomeric benzo(c)phenanthrene diol
epoxides are compared, there appears to be no simple correlation between
tumorigenic response and either the extent of binding to DNA or the major
types of deoxyribonucleoside adducts formed. Deoxyribonucleoside adducts
of benzo(c)phenanthrene diol epoxide have also been identified from the DNA
of cultured rodent embryo cells after treatment of the cells with tritium-
labeled benzo(c)phenanthrene. The distribution of adducts is consistent
with predominant metabolic formation of the (4R,3S)-diol-(2S,1R)-epoxide;
deoxyadenosine is the major site in the cellular DNA attacked by this
epoxide, just as it is in DNA in solution. Further experiments are in
progress which we hope will identify more subtle aspects of the DNA binding
of benzo(c)phenanthrene diol epoxides that may be uniquely correlated with
their tumorigenic activity.

ACKNOWLEDGEMENT

The authors wish to express their appreciation to Mrs. Judy Thomas-
Younis for her valuable assistance in the preparation of this manuscript.

REFERENCES

1. D. M. Jerina and J. W. Daly, Oxidation at carbon, in: "Drug
   Metabolism-from Microbe to Man", D. V. Parke and R. L. Smith, eds.,
   Taylor and Francis Ltd., London, (1976) p. 13.
2. D. M. Jerina, R. E. Lehr, H. Yagi, O. Hernandez, P. M. Dansette, P. G.
   Wislocki, A. W. Wood, R. L. Chang, W. Levin, and A. H. Conney,
   Mutagenicity of benzo(a)pyrene derivatives and the description of a
   quantum mechanical model which predicts the ease of carbonium ion
   formation from diol epoxides, in: "In Vitro Metabolic Activation in
   Mutagenesis Testing", F. J. de Serres, J. R. Fouts, J. R. Bend, and
   R. M. Philpot, eds., Elsevier/North-Holland Biomedical Press,
   Amsterdam (1976) p. 159.
3. M. Nordqvist, D. R. Thakker, K. P. Vyas, H. Yagi, W. Levin, D. E. Ryan,

P. E. Thomas, A. H. Conney and D. M. Jerina, Metabolism of chrysene and phenanthrene to bay-region diol epoxides by rat liver enzymes, Mol. Pharmacol. 19:168 (1981).

4. K. P. Vyas, D. R. Thakker, W. Levin, H. Yagi, A. H. Conney, and D. M. Jerina, Stereoselective metabolism of the optical isomers of trans-1,2-dihydroxy-1,2-dihydrophenanthrene to bay-region diol epoxides by rat liver microsomes, Chem.-Biol. Interact. 38:203 (1982).

5. D. R. Boyd and D. M. Jerina, Arene oxides - oxepins, in: "Small Ring Heterocycles," 42 (Part 3), A. Hassner, ed., John Wiley and Sons, Inc., New York (1985) p. 197.

6. H. Yagi, O. Hernandez, and D. M. Jerina, Synthesis of ($\pm$)-7β,8α-dihydroxy-9β,10β-epoxy-7,8,9,10-tetrahydrobenzo(a)pyrene, a potential metabolite of the carcinogen benzo(a)pyrene with stereochemistry related to the antileukemic triptolides, J. Am. Chem. Soc. 97:6881 (1975).

7. R. E. Lehr, M. Schaefer-Ridder, and D. M. Jerina, Synthesis and reactivity of diol epoxides derived from non-K-region trans-dihydrodiols of benzo(a)anthracene, Tetrahedron Lett. 539 (1977).

8. D. L. Whalen, A. M. Ross, H. Yagi, J. M. Karle, and D. M. Jerina, Stereoelectronic factors in the solvolysis of bay region diol epoxides of polycyclic aromatic hydrocarbons, J. Am. Chem. Soc. 100:5218 (1978).

9. D. R. Thakker, H. Yagi, W. Levin, A. W. Wood, A. H. Conney, and D. M. Jerina, Polycyclic aromatic hydrocarbons: Metabolic activation to ultimate carcinogens, in: "Bioactivation of Foreign Compounds," M. W. Anders, ed., Academic Press, New York (1985) p. 177.

10. W. Levin, P. E. Thomas, L. M. Reik, A. W. Wood, and D. E. Ryan, Multiplicity and functional diversity of rat hepatic microsomal cytochrome P450 isozymes, in: "IUPHAR 9th International Congress of Pharmacology," Vol. 3, W. Paton, J. Mitchell, and P. Turner, eds., MacMillan Press, Ltd., London (1984) p. 203.

11. D. R. Thakker, H. Yagi, H. Akagi, M. Koreeda, A. Y. H. Lu, W. Levin, A. W. Wood, A. H. Conney, and D. M. Jerina, Metabolism of benzo(a)-pyrene. VI. Stereoselective metabolism of benzo(a)pyrene and benzo(a)pyrene 7,8-dihydrodiol to diol epoxides, Chem.-Biol. Interact., 16:281 (1977).

12. D. M. Jerina, H. Yagi, D. R. Thakker, J. M. Sayer, P. J. van Bladeren, R. E. Lehr, D. L. Whalen, W. Levin, R. L. Chang, A. W. Wood, and A. H. Conney, Identification of the ultimate carcinogenic metabolites of the polycyclic aromatic hydrocarbons: Bay-region (R,S)-diol-(S,R)-epoxides, in: "Foreign Compound Metabolism," J. Caldwell and G. D. Paulson, eds., Taylor and Francis Ltd, London (1984) p. 257.

13. J. Kapitulnik, P. G. Wislocki, W. Levin, H. Yagi, D. M. Jerina, and A. H. Conney, Tumorigenicity studies with diol epoxides of benzo(a)-pyrene which indicate that ($\pm$)-trans-7β,8α-dihydroxy-9α,10α-epoxy-7,8,9,10-tetrahydroenzo(a)pyrene is an ultimate carcinogen in newborn mice, Cancer Res. 38:354 (1978).

14. P. G. Wislocki, M. K. Buening, W. Levin, R. E. Lehr, D. R. Thakker, D. M. Jerina, and A. H. Conney, Tumorigenicity of the diastereomeric benz(a)anthracene 3,4-diol-1,2-epoxides and the (+)- and (-)-enantiomers of benz(a)anthracene 3,4-dihydrodiol in newborn mice, J. Natl. Cancer Inst. 63:201 (1979).

15. R. L. Chang, W. Levin, A. W. Wood, S. Kumar, H. Yagi, D. M. Jerina, R. E. Lehr, and A. H. Conney, Tumorigenicity of dihydrodiols and diol epoxides of benz(c)acridine in newborn mice, Cancer Res. 44:5161 (1984).

16. M. K. Buening, W. Levin, J. M. Karle, H. Yagi, D. M. Jerina, and A. H. Conney, Tumorigenicity of bay-region epoxides and other derivatives of chrysene and phenanthrene in newborn mice, Cancer Res. 39:5063 (1979).

17. R. L. Chang, W. Levin, A. W. Wood, R. E. Lehr, S. Kumar, H. Yagi, D. M. Jerina, and A. H. Conney, Tumorigenicity of the diastereomeric bay-region benzo(e)pyrene 9,10-diol-11,12-epoxides in newborn mice, Cancer Res. 41:915 (1981).

18. D. M. Jerina, H. Selander, H. Yagi, M. D. Wells, J. F. Davey, V. Mahadevan and D. T. Gibson, Dihydrodiols from anthracene and phenanthrene, J. Am. Chem. Soc. 98:5988 (1976).

19. K. P. Vyas, T. Shibata, R. J. Highet, H. J. Yeh, P. E. Thomas, D. E. Ryan, W. Levin, and D. M. Jerina, Metabolism of α-naphthoflavone and β-naphthoflavone by rat liver microsomes and highly purified reconstituted cytochrome P-450 systems, J. Biol. Chem. 258:5649 (1983).

20. H. Yagi, D. R. Thakker, R. E. Lehr, and D. M. Jerina, Benzo-ring diol epoxides of benzo(e)pyrene and triphenylene, J. Org. Chem. 44:3439 (1979).

21. D. R. Buhler, F. Unlu, D. R. Thakker, T. J. Slaga, A. H. Conney, A. W. Wood, R. L. Chang, W. Levin, and D. M. Jerina, Effect of a 6-fluoro substituent on the metabolism and biological activity of benzo(a)-pyrene, Cancer Res. 43:1541 (1983).

22. D. R. Thakker, H. Yagi, J. M. Sayer, U. Kapur, W. Levin, R. L. Chang, A. W. Wood, A. H. Conney, and D. M. Jerina, Effects of a 6-fluoro substituent on the metabolism of benzo(a)pyrene 7,8-dihydrodiol to bay-region diol epoxides by rat liver enzymes, J. Biol. Chem. 259:11249 (1984).

23. A. W. Wood, R. L. Chang, M.-T. Huang, W. Levin, R. E. Lehr, S. Kumar, D. R. Thakker, H. Yagi, D. M. Jerina, and A. H. Conney, Mutageni-city of benzo(e)pyrene and triphenylene tetrahydroepoxides and diol epoxides in bacterial and mammalian cells, Cancer Res. 40:1985 (1980).

24. J. M. Sayer, H. Yagi, M. Croisy-Delcey, and D. M. Jerina, Novel bay-region diol epoxides from benzo(c)phenanthrene, J. Am. Chem. Soc. 103:4970 (1981).

25. D. M. Jerina, J. M. Sayer, H. Yagi, M. Croisy-Delcey, Y. Ittah, D. R. Thakker, A. W. Wood, R. L. Chang, W. Levin, and A. H. Conney, Highly tumorigenic bay-region diol epoxides from the weak carcino-gen benzo(c)phenanthrene, in: "Adv. Exp. Med. Biol.: Biological Reactive Intermediates IIA," R. Snyder, D. V. Parke, J. J. Kocsis, D. J. Jollow, C. G. Gibson, and C. M. Witmer, eds., Plenum Publish-ing Co., New York (1982) p. 501.

26. J. M. Sayer, H. Yagi, P. J. van Bladeren, W. Levin, and D. M. Jerina, Stereoselectivity of microsomal epoxide hydrolase toward diol epoxides and tetrahydroepoxides derived from benz(a)anthracene, J. Biol. Chem. 260:1630 (1985).

27. J. M. Sayer, D. L. Whalen, S. L. Friedman, A. Paik, H. Yagi, K. P. Vyas and D. M. Jerina, Conformational effects in the hydrolyses of benzo-ring diol epoxides that have bay-region diol groups, J. Am. Chem. Soc. 106:226 (1984).

28. D. L. Whalen, J. A. Montemarano, D. R. Thakker, H. Yagi, and D. M. Jerina, Changes of mechanisms and product distributions in the hydrolysis of benzo(a)pyrene-7,8-diol 9,10-epoxide metabolites induced by changes in pH, J. Am. Chem. Soc. 99:5522 (1977).

29. H. Yagi, D. R. Thakker, O. Hernandez, M. Koreeda, and D. M. Jerina, Synthesis and reactions of the highly mutagenic 7,8-diol 9,10-epoxides of the carcinogen benzo(a)pyrene, J. Am. Chem. Soc. 99:1604 (1977).

30. A. W. Wood, R. L. Chang, W. Levin, D. E. Ryan, P. E. Thomas, R. E. Lehr, S. Kumar, D. J. Sardella, E. Boger, H. Yagi, J. M. Sayer, D. M. Jerina, and A. H. Conney, Mutagenicity of the bay-region diol-epoxides and other benzo-ring derivatives of dibenzo(a,h)-pyrene and dibenzo(a,i)pyrene, Cancer Res. 41:2589 (1981).

31. J. M. Sayer, R. E. Lehr, D. L. Whalen, H. Yagi, and D. M. Jerina,

Structure-reactivity indices for the hydrolysis of diol epoxides of polycyclic aromatic hydrocarbons, Tetrahedron Lett. 23:4431 (1982).

32. A. W. Wood, R. L. Chang, W. Levin, D. E. Ryan, P. E. Thomas, H. D. Mah, J. M. Karle, H. Yagi, D. M. Jerina, and A. H. Conney, Mutagenicity and tumorigenicity of phenanthrene and chrysene epoxides and diol epoxides, Cancer Res. 39:4069 (1979).

33. R. L. Chang, W. Levin, A. W. Wood, R. E. Lehr, S. Kumar, H. Yagi, D. M. Jerina, and A. H. Conney, Tumorigenicity of bay-region diol-epoxides and other benzo-ring derivatives of dibenzo(a,h)pyrene and dibenzo(a,i)pyrene on mouse skin and in newborn mice, Cancer Res. 42:25 (1982).

34. R. E. Lehr, S. Kumar, W. Levin, A. W. Wood, R. L. Chang, A. H. Conney, H. Yagi, J. M. Sayer, and D. M. Jerina, The bay region theory of polycyclic aromatic hydrocarbon carcinogenesis, in: "Polycyclic Hydrocarbons and Carcinogenesis", ACS Symposium Series 283, R. G. Harvey, ed., American Chemical Society, Washington, D. C. (1985) p. 63.

35. A. W. Wood, W. Levin, R. L. Chang, H. Yagi, D. R. Thakker, R. E. Lehr, D. M. Jerina, and A. H. Conney, Bay region activation of carcino-genic polycyclic hydrocarbons, in: "Polynuclear Aromatic Hydro-carbons," P. W. Jones and P. Leber, eds., Arbor Science Publishers, Inc., Ann Arbor, Michigan (1979) p. 531.

36. D. R. Thakker, W. Levin, H. Yagi, A. H. Conney, and D. M. Jerina, Regio-and stereoselectivity of hepatic cytochrome P450 toward polycyclic aromatic hydrocarbon substrates, in: "Adv. Exp. Med. Biol.: Biological Reactive Intermediates IIA," R. Snyder, D. V. Parke, J. J. Kocsis, D. J. Jollow, C. G. Gibson, and C. M. Witmer, eds., Plenum Publishing Co., New York (1982) p. 525.

37. D. M. Jerina, J. M. Sayer, H. Yagi, P. J. van Bladeren, D. R. Thakker, W. Levin, R. L. Chang, A. W. Wood, and A. H. Conney, Stereo-selective metabolism of polycyclic aromatic hydrocarbons to carcinogenic metabolites, in: "Microsomes and Drug Oxidations," A. R. Boobis, J. Caldwell, F. De Matteis, and C. R. Elcombe, eds., Taylor and Francis Ltd., London (1985) p. 310.

38. A. Borgen, H. Darvey, N. Castagnoli, T. T. Crocker, R. E. Rasmussen, and I. Y. Wang, Metabolic conversion of benzo(a)pyrene by Syrian hamster liver microsomes and binding of metabolites to deoxyribo-nucleic acid, J. Med. Chem. 16:502 (1973).

39. P. Sims, P. L. Grover, A. Swaisland, K. Pal, and A. Hewer, Metabolic activation of benzo(a)pyrene proceeds by a diol epoxide, Nature (London) 252:326 (1974).

40. R. G. Harvey, ed., "Polycyclic Hydrocarbons and Carcinogenesis", ACS Symposium Series 283, American Chemical Society, Washington, D. C. (1985).

41. A. Dipple, R. C. Moschel, and C. A. H. Bigger, Polynuclear aromatic carcinogens, in "Chemical Carcinogens," Second Edition, Vol. 2, ACS Monograph 182, C. E. Searle, ed., American Chemical Society, Washington, D. C. (1984) p. 41.

42. D. P. Michaud, S. C. Gupta, D. L. Whalen, J. M. Sayer, and D. M. Jerina, Effects of pH and salt concentration on the hydrolysis of a benzo(a)pyrene 7,8-diol-9,10-epoxide catalyzed by DNA and poly-adenylic acid. Chem.-Biol. Interact. 44:41 (1983).

43. T. Meehan and K. Straub, Double-stranded DNA stereoselectively binds benzo(a)pyrene diol epoxides, Nature (London), 277:410 (1979).

44. M. C. MacLeod and M.-s. Tang, Interactions of benzo(a)pyrene diol-epoxides with linear and supercoiled DNA, Cancer Res. 45:51 (1985).

45. A. Kootstra, B. L. Haas, and T. J. Slaga, Reactions of benzo(a)pyrene diol-epoxides with DNA and nucleosomes in aqueous solutions, Biochem. Biophys. Res. Commun. 94:1432 (1980).

46. N. E. Geacintov, H. Yoshida, V. Ibanez, and R. G. Harvey, Noncovalent binding of 7β,8α-dihydroxy-9α,10α-epoxytetrahydrobenzo(a)pyrene to deoxyribonucleic acid and its catalytic effect on the hydrolysis of the diol epoxide to tetrol, Biochemistry 21:1864 (1982).

47. A. M. Jeffrey, K. Grzeskowiak, I. B. Weinstein, K. Nakanishi, P. Roller, and R. G. Harvey, Benzo(a)pyrene-7,8-dihydrodiol 9,10-oxide adenosine and deoxyadenosine adducts: structure and stereochemistry, Science 206:1309 (1979).

48. H. Kasai, K. Nakanishi, and S. Traiman, Two micromethods for determining the linkage of adducts formed between polyaromatic hydrocarbons and nucleic acid bases, J. Chem. Soc., Chem. Commun. 798 (1978).

49. P. D. Moore, M. Koreeda, P. G. Wislocki, W. Levin, A. H. Conney, H. Yagi, and D. M. Jerina, In vitro reactions of the diastereomeric 9,10-epoxides of (+) and (-)-trans-7,8-dihydroxy-7,8-dihydro-benzo(a)pyrene with polyguanylic acid and evidence for formation of an enantiomer of each diastereomeric 9,10-epoxide from benzo(a)-pyrene in mouse skin, in: "Drug Metabolism Concepts," ACS Symposium Series 44, D. M. Jerina, ed., American Chemical Society, Washington, D. C. (1977) p. 127.

50. J. T. Sawicki, R. C. Moschel, and A. Dipple, Involvement of both syn- and anti-dihydrodiol-epoxides in the binding of 7,12-dimethyl-benz(a)anthracene to DNA in mouse embryo cell cultures, Cancer Res. 43:3212 (1983).

51. D. Pruess-Schwartz, S. M. Sebti, P. T. Gilham, and W. M. Baird, Analysis of benzo(a)pyrene:DNA adducts formed in cells in culture by immobilized boronate chromatography, Cancer Res. 44:4104 (1984).

52. M. Koreeda, P. D. Moore, P. G. Wislocki, W. Levin, A. H. Conney, H. Yagi, and D. M. Jerina, Binding of benzo(a)pyrene 7,8-diol-9,10-epoxides to DNA, RNA, and protein of mouse skin occurs with high stereoselectivity. Science 199:778 (1978).

53. A. Dipple, M. Pigott, R. C. Moschel, and N. Constantino, Evidence that binding of 7,12-dimethylbenz(a)anthracene to DNA in mouse embryo cell cultures results in extensive substitution of both adenine and guanine residues, Cancer Res. 43:4132 (1983).

CHEMISTRY OF COVALENT BINDING: STUDIES WITH

BROMOBENZENE AND THIOBENZAMIDE

Robert P. Hanzlik

Department of Medicinal Chemistry
University of Kansas
Lawrence, Kansas 66045

INTRODUCTION

The concept that chemically reactive metabolites of foreign substances could react covalently with cellular constituents to cause biological injury was introduced by J. A. Miller and E. C. Miller almost 40 years ago.[1-3] Since then considerable effort has been spent investigating the chemistry, biochemistry and biological consequences of reactive metabolic intermediates. Our laboratory has been interested in the metabolic activation, covalent binding and hepatotoxicity of two relatively simple organic compounds, bromobenzene and thiobenzamide. Our long term objective has been to determine the identity of their reactive metabolites, the chemistry of their covalent binding to cellular macromolecules and the relationship of the various types of covalent binding events to the biological changes which follow. This report summarizes some of our recent efforts toward these objectives.

STUDIES WITH BROMOBENZENE

Early work with bromobenzene in the laboratories of B. B. Brodie, and later J. R. Gillette, strongly implicated bromobenzene-3,4-oxide as an intermediate leading to several of the stable metabolites of this aryl halide, including 4-bromophenol, a trans-3,4-dihydrodiol, and a glutathione adduct.[4,5] This epoxide has never been isolated or synthesized, but by analogy to other epoxides it was expected to be quite reactive and capable of arylating protein nucleophiles. More recently, Hesse et al. observed in vitro, that covalent binding of bromobenzene to rat liver microsomes continues long after the disappearance of bromobenzene ceases, and suggested that perhaps secondary metabolites of bromobenzene (e.g. bromoquinones) could be involved in its covalent binding.[6]

Our recent efforts have been aimed at differentiating the contributions of epoxides vs. quinones to the overall covalent binding of bromobenzene. Two approaches have been taken. In one study we determined the effects of 26 chemical and enzymic probes on the in vitro metabolism and covalent binding of bromobenzene, in order to reveal the chemical properties of the reactive metabolites. In the other we measured the loss or retention of tritium relative to C-14 during the covalent binding of

bromobenzene in order to determine the "oxidation state" of the covalently bound residues. Collectively our results imply that a majority of
the covalent binding of bromobenzene metabolites to liver microsomes,
both in vitro and in vivo, arises via metabolites more highly oxidized
than a bromobenzene epoxide.

Methods

Liver microsomes were prepared from male Sprague-Dawley rats (180-
220 g), in some cases after treatment with phenobarbital (50 mg/kg, ip)
for three days. [2,4,6-^{3}H]- and [3,5-^{3}H]-bromobenzene were prepared in
our laboratory, and ^{14}C-bromobenzene was obtained from Pathfinder (St.
Louis). Incubations were conducted under air in teflon-capped culture
tubes containing 2 mg microsomal protein, 1.2 umole of bromobenzene, and
an NADPH generating system, all in 1.2 ml of phosphate buffer (pH 7.4,
0.1M). After 0-60 min a 600 ul aliquot was removed and added to 1.0 ml
20% trichloroacetic acid for determination of covalent binding. The
precipitated protein was washed exhaustively with acetone, aqueous methanol and ether, dissolved in 0.1 N NaOH, and an aliquot neutralized and
counted for radioactivity. The remaining 600 ul was made alkaline with 1
ml 1 N NaOH, extracted with hexane, and a 200 ul aliquot neutralized with
acid and counted to determine total water-soluble metabolites.

In some studies phenobarbital pretreated rats were administered
[2,4,6-^{3}H/^{14}C]-dual-labeled bromobenzene [sp. act. (mCi/mmole), ^{3}H/^{14}C =
0.132/0.030] (2.5 mmole/kg, ip) in corn oil (0.2 ml/rat). Four hours
later these animals were killed by ether anesthesia and their livers,
lungs and kidneys removed and homogenized. The homogenates were then
processed for determination of covalently bound radioactivity as
described above.

Effects of probe agents

Before adding chemical or enzymic probes to in vitro incubations
with bromobenzene, the character of its metabolism and covalent binding
were determined. With microsomes from untreated (UT) control rats total
metabolites (MET) increased nearly linearly for 60 min, to a level of 75
nmole/incubation. With microsomes from phenobarbital pretreated rats
(PB), the time course was decidedly non-linear, consisting of a rapid
burst of metabolism lasting about 5 min followed by a slower linear
phase, essentially parallel to the rate seen with UT microsomes, to a
level of 110 nmole/incubation at 60 min. Plots of covalent binding vs.
time closely paralleled those for metabolism; no lag in CUB was seen, and
the "burst effect" with PB microsomes was very obvious. After 60 min the
levels of covalently bound tritium corresponded to 7.5 nmole with PB
microsomes and 5 nmole with UT microsomes. The effects of 26 different
probe reagents were then measured after 15 min using PB microsomes.
Representative results are summarized in Table 1.

As expected, cytochrome P-450 inhibitors depressed CVB in parallel
with their inhibition of total metabolism of bromobenzene. In contrast
sulfur nucleophiles such as glutathione, N-acetylcysteine, bisulfite ion
or thiourea were markedly more selective for reducing CVB with minimal or
even no effect on total formation of water-soluble metabolites. This
result is consistent with the spontaneous chemical (as opposed to enzyme-
catalyzed) trapping of electrophilic metabolites by the strongly nucleophilic probe agents. Although detailed kinetic information about the
relative reactivity of these nucleophiles toward (arene) epoxides vs.
quinones is lacking, thiourea and bisulfite are both known as extremely
efficient quinone-trapping agents,[7,8] and overall the results with sulfur

Table 1. Effects of Chemical Probes on the Microsomal Metabolism and Covalent Binding of Bromobenzene. Data expressed as percent of control values.[a]

Agent	Conc. (mM)	METAB	CVB
n-Octylimidazole	0.1	12	4
	1.0	3	2
SKF-525A	1.0	17	10
Glutathione	0.1	100	59
	1.0	96	21
N-Acetylcysteine	0.1	100	91
	1.0	92	48
2-Mercaptoethanol	1.0	63	13
Bisulfite	1.0	83	26
	5.0	58	7
Thiourea	2.0	84	23
	10.0	77	13
Mannitol	100	100	94
Benzoate	20	86	93
Superoxide	(3 U)	100	82
dismutase	(50 U)	96	81
Catalase	(10 U)	88	79
Ascorbate	1	80	65
	10	74	40
Diaphorase	(2 U)	88	65
BHT	1.0	79	57
	2.0	74	42
DPPD	0.05	76	53
	0.5	63	46
o-Bromophenol	0.05	77	58
	0.20	55	23
m-Bromophenol	0.05	82	60
	0.20	45	19
p-Bromophenol	0.05	69	48
	0.20	44	20
UDPGA/NAG	3.0	92	85
TCPO	0.1	85	102
	1.0	69	87

[a]Microsomes from PB-pretreated rats were used with an initial bromobenzene concentration of 1.0 mM. Data expressed as percent of the following control values (mean $\pm$ SD): 39.1 $\pm$ 2.0 nmol MET/15 min/ 2 mg protein; 3.32 $\pm$ 0.48 nmol equivalents CVB/15 min/2 mg protein.

nucleophiles appear more consistent with quinones rather than epoxides as the major protein-alkylating species. The report that bromobenzene-3,4-oxide does not react with glutathione in the absence of glutathione transferase would seem to underscore this tentative conclusion.[9]

If quinones are involved, then they must be formed by secondary oxidations of bromophenol, which could conceivably be mediated by a number of agents. Therefore we also examined the effect of various reducing agents and anti-oxidants on the metabolism and covalent binding of bromobenzene. Ascorbate definitely inhibited CVB moreso than metabolism, while lesser (but still probably significant) degrees of selectivity were also shown by diaphorase, butylated hydroxytoluene (BHT) and N,N'- diphenyl-p-phenylenediamine (DPPD). A modest effect was seen with superoxide dismutase, but little if any with catalase or the hydroxyl radical scavengers mannitol or benzoate. p-Bromophenol inhibited bromobenzene metabolism surprisingly well, but appeared to have an even greater depressant effect on CVB; nearly indentical results were obtained with ortho- and meta bromophenol. Inhibition of binding by the bromophenols is most likely due to carrier pool dilution of the radiolabeled bromophenol metabolites, although a possible antioxidant effect of these phenols in blocking further oxidation can not be discounted (cf. effect of BHT, for example). However, in contrast to Hesse et al.,[6] we did not find that addition of UDPGA plus N-acetyl-galactosamine to trap phenolic metabolites as glucuronides had much effect on covalent binding. Finally, TCPO added to inhibit microsomal epoxide hydrolase depressed CVB to a lesser extent than it inhibited metabolism. This would be consistent with a role for an epoxide metabolite in protein alkylation, in that decreasing its hydration should give an epoxide a greater chance to alkylate a protein. However, by the same token, the epoxide would also have an extended opportunity to rearrange to a phenol, ultimately to undergo further oxidation to secondary metabolites capable of covalently binding to proteins. Thus the effects of TCPO are not entirely unambiguous to interpret.

In summary, the evidence from chemical and enzymic probes supports the hypothesis that secondary metabolites of bromobenzene play a significant, possibly even major role in leading to protein covalent binding. Further support for this hypothesis comes from studies with ^{3}H/^{14}C dual-labeled bromobenzenes described below.

Studies with [^{3}H/^{14}C]-Bromobenzene

These studies are based on the simple concept that if bromobenzene epoxides are formed and alkylate proteins, the covalently bound material should have the same tritium/carbon-14 (T/C) ratio as the starting bromobenzene, since no hydrogens are lost during formation of the epoxide or its reaction with nucleophiles. On the other hand, the formation of a bromobenzoquinone implies the loss of two ring hydrogens; a third would necessarily be lost during re-aromatization after nucleophilic addition to the quinone system. Thus if quinone intermediates were responsible for bromobenzene covalent binding, the bound material could be expected to contain much less tritium, relative to C-14, than the starting bromobenzene.

To test this hypothesis, two types of tritiated bromobenzenes were prepared, one in which the label was distributed only in positions meta to bromine (i.e. [3,5-^{3}H]-BB), and one in which tritium was distributed ortho and para to bromine (i.e. [2,4,6-^{3}H]-BB). Individually these were mixed with [^{14}C]-BB and incubated with microsome in vitro. The results are shown in Table 2.

As can be seen, with both UT and PB microsomes, there is extensive loss of tritium during bromobenzene activation and covalent binding. The bound residues contain only 25-35% as much tritium, relative to carbon-14, as the starting substrate, and the extent of loss is essentially the same for both forms of tritiated substrate. When N-acetylcysteine (1 mM) was added to incubations with PB microsomes, the total amount of covalent binding was reduced by ca. 65% (based on C-14), but the T/C ratio in the bound material was not altered compared to that observed in the absence of N-acetylcysteine. Finally, the [2,4,6-^{3}H/^{14}C] dual labeled bromobenzene was administered to phenobarbital pretreated rats, and 4 hr later their liver, lungs and kidneys were removed and analyzed for covalently bound radiolabel. The T/C ratios observed in whole liver homogenate, S-10 fraction or liver microsomes were nearly identical (0.33-0.37) and were nearly identical to those values observed in vitro. In contrast the T/C ratios in lung (0.53) and kidney (0.58) were very different.

Table 2. Covalent Binding of $[^3H/^{14}C]$-Bromobenzene.

Relative $^3H/^{14}C$ binding

Target	2,4,6-^{3}H	3,5-^{3}H
A. in vitro		
UT microsomes	0.26 (0.03)	0.31 (0.02)
PB microsomes	0.23 (0.04)	0.28 (0.05)
" " + NAC	0.29 (0.04)	0.37 (0.05)
B. in vivo (PB rats)		
Liver		
homogenate	0.37 (0.01)	
S-10	0.35 (0.03)	
microsomes	0.33 (0.02)	
Lung	0.53 (0.04)	
Kidney	0.58 (0.03)	

Quantitative interpretation of T/C ratios in the covalently bound
materials is impossible for numerous reasons (unknown extent of NIH
shift, variable tritium loss by exchange with water, multiplicity of
pathways, etc.). However, one thing is very clear from the data, and
that is that the extensive loss of tritium in the covalently bound material is simply not compatible with a major role for epoxide metabolites
in protein alkylation, although a minor role for epoxides can not be
excluded.

Since extended washing of the proteins did not alter the T/C ratios,
the losses of tritium must have occurred prior to the act of covalent
binding. This could have occurred in one or both of two ways. First,
tritium could be lost during ring oxidation of either bromobenzene or its
metabolites as a normal feature of the mechanisms involved. Second, it
is possible that one or more soluble bromobenzene metabolites might be
labile to tritium exchange with solvent water prior to its becoming
covalently bound. Catechols and hydroquinones might fall into this
group, but since bromophenols are not susceptible to deuterium-water
exchange,[10] they are not likely to be susceptible to tritium-water
exchange. Indeed, NIH-shift values for tritium are usually higher than
those for deuterium.[11]

According to these views the tritium lost from bound residues should
be found as tritiated water. When incubations were alkalinized, extracted with ether, and lyophilized, the recovered water contained negligible
C-14 but substantial tritium; in fact the amount of tritiated water was
several times the amount that could be accounted for by the tritium
deficit in the covalently bound material.

The T/C ratios in lung and kidney are significantly higher than
those for liver, but these too indicate a significant role for secondary,

more highly oxidized metabolites in covalent binding. In light of the differences in liver vs. lung and kidney T/C ratios in vivo, the observation of nearly identical liver T/C ratios in vitro and in vivo is striking. We feel that this is not simply a coincidence, but rather an indication that the biochemical and chemical events involved in covalent binding in the liver in vivo may be reasonably well represented by processes occurring in liver microsomes in vitro.

STUDIES WITH THIOBENZAMIDE

Thiobenzamide (TB) produces a variety of toxic effects in the liver, the expression of which appears to depend upon its metabolic activation. By far the major pathway for biotransformation of thiobenzamide involves "oxidative desulfuration," which proceeds via two sequential S-oxidations mediated by the flavin containing monooxygenase,[12] although the first of these oxidations may also be effected by certain isozymes of cytochrome P-450,[13] or even by "reactive oxygen species."[14] The first oxidation produces thiobenzamide S-oxide (TBSO). While TBSO is in some ways a "chemically reactive metabolite," notably toward reduction (see below), it is relatively easily synthesized and handled.[15] Administration of TBSO to rats elicits the same types of toxicities as TB, but TBSO is somewhat more potent and faster acting.[16] Further oxidation of TBSO leads to an S,S-dioxide or "sulfene" metabolite (TBSO$_2$).[12,15] Sulfenes are extremely reactive chemically. In aqueous solution TBSO$_2$ rapidly acylates water (i.e. it hydrolyzes forming benzamide); it can also acylate other nucleophiles such as ethanol (forming ethyl benzimidate).[16]

Substituent effects on thiobenzamide chemistry and toxicity

Given the above pattern of biotransformation and chemical reactivity, it seemed to us that reactions of TBSO$_2$ with cellular nucleophiles (i.e. covalent binding) might underlie the acute toxic affects of thiobenzamide. In an attempt to shed further light on the mechanism of TB-induced hepatotoxicity we have investigated the effect of substituents on several aspects of the chemistry, biotransformation, covalent binding and toxicity of thiobenzamide. Substituent effects on the toxicity of TB can be summarized as follows. Meta and/or para substituents modulate hepatotoxicity by means of electronic effects.[16,17] Ortho-substitution, on the other hand, abolishes hepatotoxicity; this is very likely a steric effect, since it occurs irrespective of the electronic character of the substituent.[16,17] Surprisingly, N-methyl substitution causes a dramatic shift in the site of injury from liver to lung, as well as a significant increase in overall toxic potency.[18]

Meta/para substitutions

The toxicity of meta- and para substituted thiobenzamides is strictly correlated with the electronic character of the substituent, being increased by electron-donating groups and decreased by electron-with drawing groups. Initially we expected this to be the case,[19] and observing it seemed to confirm our suspicions that oxidative biotransformation was required for expression of hepatotoxicity by TB. Later chemical studies confirmed that electron donating groups increase (and electron withdrawing groups decrease) the rate of TB oxidation by hydrogen peroxide in vitro.[15] To our surprise, however, such effects were not clearly observable for the microsomal oxidation of TB derivatives in vitro.[20] An escape from this seeming paradox was provided by the observation that TBSO undergoes non-enzymic reduction by thiols such as N-acetylcysteine, and that this reaction is strongly accelerated by

Table 3. Covalent Binding of Thiobenzamide and 2,6-Dichlorothiobenzamide to Rat Tissue Proteins In Vivo.

Covalent Binding (nmol/2 hr/mg protein)

Tissue	Thiobenzamide	2,6-Dichloro-thiobenzamide
Liver	3.61 ± 0.29	0.16 ± 0.01
Lung	0.92 ± 0.08	0.20 ± 0.02
Kidney	0.76 ± 0.08	0.14 ± 0.02

[a] [Carboxyl-^{14}C]-Thiobenzamide and [4-^{3}H]-2,6-dicholrothiobenzamide were administered ip to male Sprague Dawley rats (200–220g) at a dosage of 1.5 mmol/kg as a fine suspension in corn oil (4ml/kg).
[b] Results expressed as mean $\pm$ SD (n=3).

electron withdrawing groups on the ring.[15,17] Numerous S-oxides have been observed to undergo reduction in vivo. If the reduction of TBSO derivatives were to occur, and if it was accelerated by electron withdrawing substituents, it would adequately explain the effect of meta/para substitution on TB hepatotoxicity. The TB-TBSO interconversion would become in effect an _equilibrium_, and its displacement to the left by electron withdrawing groups would decrease the net amount of material progressing, via the second oxidation step, to the true toxic species, the sulfene metabolite.

N–Substitution

The biotransformation of N-Methylthiobenzamide (NMTB) follows the same scheme outlined for TB.[21] Our general expectation regarding the reactivity and covalent binding of sulfene metabolites appears amply borne out, at least in vitro. The product of lung (and liver) microsomal metabolism of [^{14}C]-NMTB is its S-oxide, NMTBSO, which in turn is further oxidized (via the sulfene) to N-methylbenzamide and covalently-bound ^{14}C. Since the latter are formed in a 1:1 ratio, the covalent binding of this compound is truly extensive! 1-Methyl-1-phenyl-3-benzoyl thiourea (MPBTU), which inhibits the S-oxidation and covalent binding of NMTB and NMTBSO in vitro,[21] also blocks their toxicity in vivo.[22]

Although in vivo covalent binding studies have not yet been done with NMTB, we anticipate that we will observe dose-dependent formation of lung residues, and that MPBTU will block their formation. Unpublished studies by D. W. Gottschall and D. A. Penney in our laboratory have shown that N-(t-butyl)-thiobenzamide lacks the lung damaging effects of NMTB, while N-ethylthiobenzamide is pneumotoxic but much less potent than NMTB. At least with liver microsomes in vitro, metabolism of [^{14}C]-N-(tBu)TB leads to efficient formation of its S-oxide as well as N-(t-butyl)-benzamide; the amount of covalent binding was ca. one-sixth the amount of amide, as compared to the 1:1 ratio observed with the pneumotoxic NMTB. The reason N-substitution shifts the toxicity of TB from liver to lungs is unknown, but the decrease in pneumotoxic potency with larger N-substituents may simply be due to a steric effect on the ability of reactive sulfene metabolites to acylate proteins (as opposed to acylation of water). However, additional in vivo experiments need to be done to test this hypothesis.

Ortho substitution

The abolition of the hepatotoxicity of TB by ortho-substitution has also been suggested to be caused by a steric effect on the reactivity of sulfene metabolites.[17] Recently we carried out an experiment to test this hypothesis in vivo. We prepared [^{14}C]-TB and [4-^{3}H]-2,6-dichloro-thiobenzamide (DCTB) and administered them (separately) ip as fine suspensions in corn oil to groups of 4 male Sprague Dawley rats (200–220 g) at a dosage of 1.5 mmole/kg. Rapid absorption was indicated for both compounds by the very prominent sedative-hypnotic effect common to nearly all thiobenzamide derivatives (except for their S-oxides). Two hours later the animals were killed and their liver, lungs and kidneys removed and analyzed for covalently bound radioactivity as described above for bromobenzene. The results of this study are shown in Table 3.

As expected, thiobenzamide metabolites became bound to liver proteins to an appreciable extent, while smaller but still significant amounts became bound to lung and kidney proteins. Also as expected, 2,6-disubstitution reduced the amount of covalently bound material to a very low level in all three tissues. Since DCTB is reported to be extensively metabolized to 2,6-dichlorobenzonitrile,[23] the reduced amounts of covalent binding are most likely attributable to the steric effects of ortho substituents directing the reactivity of the sulfene metabolite toward elimination of the sulfur moiety and nitrile formation, rather than electrophilic attack on nucleophiles. Thus it would appear that ortho substitution on the TB molecule blocks the offending reaction (covalent binding), and eliminates the associated hepatotoxicity, without fundamentally changing other aspects of the absorption or disposition of the compound relative to TB itself.

The significance of this observation probably lies in the way it affects our view of the relationship of covalent binding to toxicity, i.e. cause and effect, or merely parallel but independent phenomena? Parallel dose-response curves for covalent binding and toxicity do not in themselves establish a cause and effect relationship, although they are consistent with, or even hint at such a view. Evidence from the time-dependence of binding vs. toxicity can further support the hypothesis of cause and effect, particularly in single dose acute toxicity studies. The finding that simple substituent effects on chemical reactivity and covalent binding are associated with corresponding changes in toxicity among a series of congeneric molecules gives one yet another completely independent line of support for the hypothesis of a cause and effect relationship. Obviously the key word here is "simple," for in general, structural change in a molecule affects many of its properties and modes of interaction with biological systems (for a review see ref. 24). Fortunately for our purposes, most substituent-induced changes on the TB molecule do appear to be relatively "simple." Even here, though, we have found exceptions. One example is p-hydroxythiobenzamide which, based on the inductive effect of its OH group, should be extremely hepatotoxic yet is not;[16] the probable reason is that rapid glucuronidation obviates S-oxidation.

ACKNOWLEDGEMENT

I take pleasure in acknowledging the many valuable contributions to this work made by co-workers named in the references cited. It is also a pleasure to acknowledge financial support for our work generously provided by the National Institutes of Health and the University of Kansas General Research Fund.

REFERENCES

1. E. C. Miller and J. A. Miller, The presence and significance of bound aminoazo dyes in the livers of rats fed p-dimethylaminoazobenzene. Cancer Res. 7: 468-480 (1947).

2. E. C. Miller and J. A. Miller, In vivo combinations between carcinogens and tissue constituents and their possible role in carcinogenesis. Cancer Res. 12: 547-556 (1952).

3. J. A. Miller, Carcinogenesis by chemicals: an overview. Cancer Res. 30: 559-576 (1970).

4. B. B. Brodie, W. D. Reid, A. K. Cho, G. Sipes, G. Krishna and J. R. Gillette, Possible mechanism of liver necrosis caused by aromatic organic compounds. Proc. Nat. Acad. Sci. (U.S.), 68: 160-164 (1971).

5. D. J. Jollow, J. R. Mitchell, N. Zampaglione and J. R. Gillette, Bromobenzene-induced liver necrosis. Protective role of glutathione and evidence for 3,4-bromobenzene oxide as the hepatotoxic metabolite, Pharmacology. 11: 151-169 (1974).

6. S. Hesse, T. Wolf and M. Mezger, Involvement of phenolic metabolites in the irreversible protein-binding of ^{14}C-bromobenzene catalyzed by rat liver microsomes. Arch. Toxicol., Suppl. 4, 358-362 (1980).

7. J. W. Dodgson, The reaction of p-benzoquinone with sulphurous acid and with alkali. Part I. J. Chem. Soc., 105: 2435-2443 (1914).

8. K. T. Finley, The addition-substitution chemistry of quinones, in: "The chemistry of quinoid compounds, S. Patai, ed., John Wiley and Sons, New York (1974), pp 880-900.

9. T. J. Monks, L. R. Pohl, J. R. Gillette, M. Hong, R. J. Highet, J. A. Ferrretti and J. A. Hinson, Stereoselective formation of two bromobenzene-glutathione conjugates. Chem.-Biol. Interactions 41: 203-216 (1982).

10. R. P. Hanzlik, K. Hogberg and C. M. Judson, Microsomal hydroxylation of specifically deuterated monosubstituted benzenes. Evidence for direct aromatic hydroxylation. Biochemistry 23: 3048-3055 (1984).

11. J. W. Daly, D. M. Jerina and B. Witkop, Arene oxides and the NIH-shift: the metabolism, toxicity and carcinogenicity of aromatic compounds. Experientia 28: 1129-1149 (1972).

12. R. P. Hanzlik and J. R. Cashman, Micrrosomal metabolism of thiobenzamide and thiobenzamide S-oxide. Drug Metab. Disposition, 11: 201-205 (1983).

13. R. E. Tynes and E. Hodgson, Oxidation of thiobenzamide by the FAD-containing and cytochrome P-450-dependent monooxygenases of liver and lung microsomes. Biochem. Pharmacol. 33: 3419-3428 (1983).

14. M. Younes, Involvement of reactive oxygen species in the microsomal S-oxidation of thiobenzamide. Experientia, 41: 479-480 (1985).

15. J. R. Cashman and R. P. Hanzlik, Oxidation and other reactions of thiobenzamide derivatives of relevance to their hepatotoxicity. J. Org. Chem. 47: 4645-4650 (1982).

16. R. P. Hanzlik, J. R. Cashman and G. J. Traiger, Relative hepatotoxicityof substituted thiobenzamides and thiobenzamide S-oxides in the rat. Toxicol. Appl. Pharmacol. 55: 260-272 (1980).

17. J. R. Cashman, K. K. Parikh, G. J. Traiger and R. P. Hanzlik, Relative hepatotoxicity of ortho and meta monosubstituted thiobenzamides in the rat. Chem.-Biol. Interactions 45: 341-347 (1983).

18. J. R. Cashman, G. J. Traiger and R. P. Hanzlik, Pneumotoxic effects of thiobenzamide derivatives. Toxicology, 23: 85-93 (1982).

19. R. P. Hanzlik, K. P. Vyas and G. J. Traiger, Substituent effects on the hepatotoxicity of thiobenzamide derivatives in the rat. Toxicol. Appl. Pharmacol., 46: 685-694 (1978).

20. J. R. Cashman, Ph.D. Thesis, University of Kansas, 1982.

21. D. W. Gottschall, D. A. Penney, G. J. Traiger and R. P. Hanzlik, Oxidation of N-methylthiobenzamide and N-methylthiobenzamide S-oxide by liver and lung microsomes. Toxicol. Appl. Pharmacol., 78: 332-341 (1985).

22. D. A. Penney, D. W. Gottschall, R. P. Hanzlik and G. J. Traiger, The role of metabolism in N-methylthiobenzamide-induced pneumotoxicity. Toxicol. Appl. Pharmacol. 78: 323-331 (1985).

23. M. H. Griffiths, J. A. Moss, J. A. Rose, and D. E. Hathway, The comparative metabolism of 2,6-dichlorothiobenzamide (Prefix) and 2,6-dichlorothiobenzonitrile in the dog and rat. Biochem. J. 98: 770-781 (1966).

24. R. P. Hanzlik, Effects of substituents of reactivity and toxicity of chemically reactive intermediates. Drug Metab. Rev. 13:207-234 (1982).

ROLE OF THIOLS IN PROTECTION AGAINST BIOLOGICAL REACTIVE INTERMEDIATES

Pierluigi Nicotera and Sten Orrenius

Department of Toxicology, Karolinska Institutet
Box 60400, S-104 01 Stockholm, Sweden

INTRODUCTION

Thanks to the pioneering work of Elizabeth and James Miller[1] it is now well established that the cytotoxic and carcinogenic effects of a wide variety of chemicals are mediated by reactive products formed during their biotransformation in the organism. It is equally clear that there exist a number of protective systems which can trap, or inactivate, toxic metabolites and thereby prevent their accumulation within the tissues and subsequent toxic effects. Although phase I reactions, in particular those mediated by the cytochrome P-450-linked monooxygenase system, are most often responsible for the production of toxic metabolites, there are now many examples of metabolic activation via phase II reactions, despite the fact that the latter normally serve a protective function. Hence, it is obvious that the formation of toxic metabolites cannot be attributed to any single enzyme or enzyme system, and that the balance between metabolic activation and inactivation is absolutely critical in deciding whether exposure to a potentially toxic compound will result in toxicity, or not.

Glutathione plays a unique role in cellular defense. This tripeptide (γ-glu-cys-gly) is characterized by its reactive cysteinyl thiol and its γ-glutamyl bond that makes it resistant to normal peptidase digestion. Glutathione is present at high concentrations in mammalian cells and almost entirely in its reduced form (GSH), with glutathione disulfide (GSSG), mixed disulfides, and thioethers constituting minor fractions of the total glutathione pool[2].

Early work by Brodie and associates[3,4] indicated GSH depletion as one of the important consequences of toxic injury and established the critical role of GSH in protecting tissues from toxic effects of accumulating reactive intermediates. In fact, it was clearly shown that, in many cases, toxic effects are not displayed until tissue GSH has been depleted. More recently, the role of GSH and protein thiol depletion in toxic cell injury has been extensively investigated in our laboratory. Using freshly isolated hepatocytes and subcellular fractions, we have shown that GSH depletion and modification of certain protein thiols will result in a perturbation of intracellular Ca^{2+} homeostasis, alteration of hepatocyte morphology, and cell death. The disruption of intracellular Ca^{2+} homeostasis seems to play a predominant role in the development of irreversible cell damage and to be intimately related to the loss of cellular thiols. In this presentation, we will briefly discuss the mechanisms responsible for the perturbation of thiol and Ca^{2+} homeostasis following the metabolism of a model compound, menadione (2-methyl-1,4-naphthoquinone), in isolated hepatocytes.

Mechanisms of Glutathione Depletion

The rate of glutathione turnover differs markedly in different tissues and cell types[5]. The half-life of glutathione has been found to vary from less than an hour in kidney to several days in erythrocytes, nervous tissue, lung and spleen. The biosynthesis of GSH is catalyzed by the enzymes γ-glutamylcysteine synthetase and glutathione synthetase[6]. Under conditions of stimulated GSH consumption, the availability of the sulfur amino acid cysteine appears to be rate-limiting for GSH resynthesis. The degradation of glutathione is initiated by a reaction catalyzed by γ-glutamyltransferase[6]. This enzyme is not present in all cells, and in those cells where it is located, its active site faces the extracellular space[7]. Therefore, the degradation of glutathione depends on the efflux of intracellular glutathione and the transport to a site of γ-glutamyltransferase activity. The kidney, with its high activity of γ-glutamyltransferase appears to play a major role in this process.

Conjugation with GSH probably represents the most important cellular defense mechanism against the toxicity of a wide variety of reactive electrophiles. These reactions involve a nucleophilic attack by the thiol group of GSH towards electrophilic carbon, nitrogen, or sulfur atoms of the target molecule. Although GSH can react non-enzymatically with a large number of electrophilic compounds, this process is normally catalyzed by glutathione transferases[8]. In the liver there are at least seven cytosolic isoenzymes of glutathione transferase, which are all inducible and display overlapping substrate specificities. Although there are important examples of endogenous compounds that are metabolized by glutathione conjugation (e.g. the leukotrienes), most of the electrophilic compounds that react with glutathione are of exogenous origin and are formed as a result of reactions catalyzed by the cytochrome P-450-linked monooxygenase system.

The selenoprotein glutathione peroxidase, localized in the cytosol and mitochondria of mammalian cells, provides protection against H_2O_2 and a variety of organic hydroperoxides[9]. GSH serves as the electron donor, and the GSSG formed in the reaction is subsequently reduced back to GSH by glutathione reductase at the expense of NADPH[9]. Under conditions of oxidative stress, when the cell must cope with large amounts of H_2O_2 and/or organic hydroperoxides, the glutathione reductase is unable to keep up with the rate of glutathione oxidation, and GSSG accumulates. In an apparent effort to avoid the detrimental effects of increased intracellular levels of GSSG, the cell actively excretes the disulfide which can lead to a depletion of the cellular glutathione pool (see below).

It is now well established that the toxicity of many xenobiotics is preceded by depletion of cellular GSH[3]. Any process which leads to GSH consumption at a rate which exceeds the capacity of the cell to maintain its normal thiol status will lead to GSH depletion. Although depletion of GSH as a result of oxidation would appear to be a reversible process due to the presence of glutathione reductase, this is not necessarily the case. When the glutathione reductase cannot keep up with the rate of GSH oxidation, the GSSG is transported out of the cell. The increase in intracellular GSSG concentration, which seems to be a prerequisite for the efflux to occur, is most likely due to insufficient regeneration of NADPH[10].

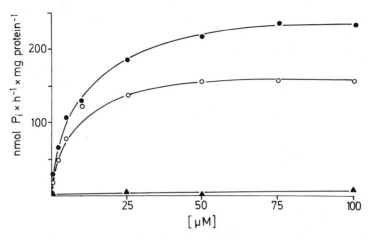

Fig. 1. Stimulation of ATP hydrolysis by GSSG and a glutathione conjugate in the plasma membrane fraction from rat hepatocytes. The plasma membrane fraction was incubated with various concentrations of GSSG (o), 2,4-dinitrophenyl glutathione (●), or 1-chloro-2,4-dinitrobenzene (▲), and samples were assayed for inorganic phosphate production. For experimental details see Nicotera et al.[14].

Studies on the mechanism of GSSG efflux have provided evidence for the existence of an ATP-dependent translocase in erythrocytes[11], and have revealed a competitive relationship between the excretion of GSSG and a glutathione conjugate in the perfused liver[12]. Studies in our laboratory have recently identified a GSSG-stimulated ATPase activity in rat liver plasma membrane fraction[13] which is also stimulated by several glutathione conjugates[14] (Fig. 1). Stimulation of ATP hydrolysis is apparent at micromolar concentrations of the substrates studied and is further enhanced by thiol-oxidizing or -alkylating agents[14]. Several enzymes are known to depend on free thiol groups for activity, including Na^+, K^+-ATPase and Ca^{2+}-ATPases. However, in contrast to these ATPases, the GS-ATPase is activated rather than inhibited by thiol-depleting agents. Although the molecular mechanism responsible for this effect needs to be further investigated, it is conceivable that activation of the GS-ATPase would result in important protective effects under both physiological and pathophysiological conditions, i.e. excretion of glutathione conjugates and a normal GSH/GSSG redox level could be maintained also under conditions of intracellular thiol depletion. The similarities between the properties of the GS-ATPase and the characteristics reported for the transport of glutathione derivatives have led us to propose that this newly discovered ATPase may function in the active extrusion of both GSSG and glutathione conjugates from hepatocytes (Fig. 2). Under conditions of oxidative stress, this activity may therefore contribute to glutathione depletion by removal of intracellular GSSG.

Mechanisms of Protein Thiol Depletion

Although the presence of free sulfhydryl groups in proteins has been recognized since the beginning of this century, interest in the role they may play as highly reactive functional groups in biological systems arose

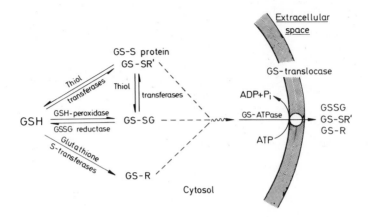

Fig. 2. Schematic illustration of intracellular reactions leading to GSH
depletion. GS-R, glutathione conjugate; GS-SR', soluble mixed
disulfides containing glutathione.

only after the discovery of glutathione by Hopkins[15]. A great number of
enzymes, catalyzing a wide variety of reactions, are now known to depend
on free sulfhydryl groups for their activity, and it is therefore not
surprising that modification of protein thiol groups can result in severe
functional damage in biological systems. The demonstration of inhibition
of glycolytic enzymes by iodoacetate was an early example of this phenomenon[16] and stimulated the interest in the role of thiols in the regulation of enzyme activity.

Thiol groups are highly reactive and participate in several different
reactions, such as alkylation, arylation, oxidation, thiol-disulfide exchange, etc. All of these reactions may be involved in the modification
of protein thiols resulting from the interaction with reactive intermediates formed during the metabolism of toxic chemicals. Thus, although the
toxicological implications of "covalent binding" (i.e. alkylation, arylation) of reactive intermediates to various proteins has often been emphasized[17], it is clear that thiol oxidation and mixed disulfide formation
may also interfere with normal protein function. As discussed below, such
modifications of protein thiols may be particularly important during
oxidative stress.

Menadione-Induced Oxidative Stress

The metabolism of quinones, e.g. menadione, by flavoenzymes can occur
by either one- or two-electron reduction routes which often differ greatly
in cytotoxicity. One-electron reduction results in formation of semiquinone
radicals which can rapidly reduce dioxygen, forming the superoxide anion
radical, O_2^-, and regenerating the quinone. Dismutation of O_2^- and production of other highly reactive species quickly lead to conditions of oxidative stress and toxicity as redox cycling of the quinone continues.
However, quinones can also undergo two-electron reduction, forming hydroquinones without production of free semiquinone intermediates. This reac-

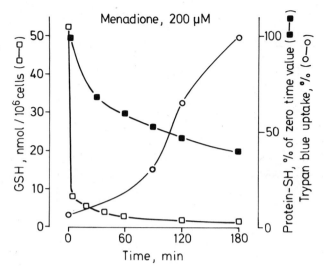

Fig. 3. Thiol depletion and cytotoxicity of menadione. Hepatocytes from phenobarbital-treated rat were incubated with 200 μM menadione and samples were assayed for GSH, protein thiols, and viability (by trypan blue uptake). For experimental details see Di Monte et al.[21].

tion is catalyzed by DT-diaphorase[18], and may serve an important protective function for the cell by competing with the single-electron pathway, since hydroquinones are often less reactive, and more easily excreted by the cell, than semiquinone radicals.

In isolated liver cells, menadione metabolism involves both one- and two-electron reduction pathways[19]. The relative contribution of the two routes depends on menadione concentration, and can easily be manipulated by selective induction of either NADPH-cytochrome P-450 reductase or DT-diaphorase. Thus, hepatocytes isolated from phenobarbital-treated rats exhibit increased redox cycling and toxicity, when exposed to menadione, whereas the reverse is true for hepatocytes from 3-methylcholanthrene-treated rats[19].

Exposure of hepatocytes to toxic concentrations of menadione is associated with extensive formation of $O_2^{\bar{}}$ and H_2O_2, and the oxidation of glutathione and pyridine nucleotides[19]. Treatments which promote one-electron reduction and redox cycling of the quinone potentiate menadione toxicity. As shown in Fig. 3, cell death is preceded by GSH depletion and a loss of protein thiols. A recent analysis[20] has shown that oxidation of GSH to GSSG, as a result of metabolism of increased amounts of H_2O_2, is responsible for most of the GSH loss during menadione metabolism by hepatocytes, whereas formation of a glutathione conjugate with menadione and of glutathione-protein mixed disulfides account for minor fractions of the GSH consumption.

As also illustrated in Fig. 3, GSH depletion occurs during the early phase of incubation of hepatocytes with toxic levels of menadione, whereas the apparent decrease in protein thiols is a later phenomenon which appears to be more closely related to the development of toxicity[21]. It

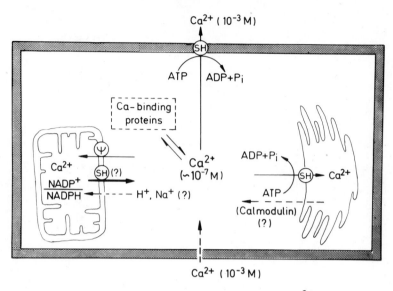

Fig. 4. Mechanisms of regulation of cytosolic Ca^{2+} concentration in
hepatocytes.

is interesting to note that menadione-induced disappearance of protein
thiols in hepatocytes is mainly due to their oxidation, and that measurable loss of protein thiols occurs only after depletion of GSH[21]. This
suggests that sulfhydryl groups in proteins critical for cell survival
are normally protected from oxidative damage by the presence of GSH. Although we observed a general decrease in protein thiols associated with
development of cytotoxicity under the experimental conditions employed
in this study, this does not imply that they are all equally important
for cell survival. The fraction of "critical thiols" is probably quite
small, and further work is required to define this fraction and to elucidate its function(s). As discussed below, maintenance of intracellular
calcium ion homeostasis represents one such function that is critically
dependent on thiol groups in specific proteins.

Disruption of Intracellular Ca^{2+} Homeostasis by Menadione Metabolism

Toxic concentrations of menadione cause the formation of numerous
small blebs on the surface of isolated hepatocytes during the early phase
of incubation[19]. Many other toxic agents, including t-butylhydroperoxide
and bromobenzene, cause similar alterations in surface structure[22]. This
type of surface blebbing can also be induced by the calcium ionophore
A23187 in the absence of extracellular Ca^{2+}, suggesting that the surface
blebbing caused by menadione and other toxins is due to a redistribution
of intracellular Ca^{2+}[22]. Further support for this idea comes from the fact
that cell Ca^{2+} is lost from hepatocytes during menadione metabolism, and
that this is preceeded by the loss of GSH[19].

As illustrated schematically in Fig, 4, a very low concentration of
Ca^{2+} in the cytosol of hepatocytes is maintained by active compartmentation processes and binding to cellular proteins, including calmodulin[23].

Mitochondrial Ca^{2+} homeostasis is regulated by a cyclic mechanism, involving Ca^{2+} uptake by an energy-dependent, ruthenium red-sensitive pathway[24], and Ca^{2+} release which is probably mediated by a Ca^{2+}/H^+ antiporter. The latter appears to be regulated by the redox level of intramitochondrial pyridine nucleotides[25], although a recent study has shown that thiols may also be important in modulating mitochondrial Ca^{2+} transport[26]. In addition, the existence of a Ca^{2+}/Na^+ antiporter may contribute to Ca^{2+} efflux also in liver mitochondria[27].

The active transport of calcium ions through the endoplasmic reticular and plasma membranes is mediated by Ca^{2+}-stimulated, Mg^{2+}-dependent ATPases which both appear to depend on free sulfhydryl groups for activity[28,29]. Whereas the microsomal Ca^{2+}-ATPase has recently been proposed to be regulated by calmodulin[30], there is no evidence for calmodulin dependence of the hepatic plasma membrane Ca^{2+}-ATPase. The recent finding that inositol 1,4,5-trisphosphate formed during α-adrenergic stimulation of the liver, causes Ca^{2+} release from the endoplasmic reticulum suggests the existence of a specific mechanism for Ca^{2+} efflux which, however, has not yet been characterized in any detail.

By developing a method by which we could measure the Ca^{2+} content of the hepatocyte mitochondrial and endoplasmic reticular compartments, we could demonstrate that menadione-induced oxidative stress causes the mobilization and loss of Ca^{2+} from both compartments[19,22]. Further studies with isolated subcellular fractions showed that menadione impairs the ability of mitochondria to take up and retain Ca^{2+} by causing the oxidation of pyridine nucleotides[32]. Menadione also inhibits Ca^{2+} uptake by rat liver microsomes by a mechanism which seems to involve oxidation/arylation of thiol group(s) critical for Ca^{2+}-ATPase activity[23]. The latter study has further revealed that arylation, oxidation, and formation of mixed disulfides with protein thiols can all affect microsomal Ca^{2+} sequestration and cause various degrees of functional impairment.

Thus menadione is able to inhibit Ca^{2+} sequestration into both the mitochondria and endoplasmic reticulum through its ability to oxidize pyridine nucleotides and intracellular thiols. The incubation of hepatocytes with menadione therefore causes the release into the cytosol of Ca^{2+} which cannot be requestered. Normally this would cause only a transient rise in the cytosolic Ca^{2+} concentration, because the plasma membrane Ca^{2+}-ATPase would remove this Ca^{2+} from the cell and the Ca^{2+} concentration would return to its usual very low level. Recent studies in our laboratory have shown, however, that the hepatic plasma membrane Ca^{2+}-ATPase is inhibited by agents which oxidize membrane protein thiol groups, including menadione[34]. Our ability to restore menadione-inhibited Ca^{2+}-ATPase activity by treatment with reducing agents, such as dithiothreitol, strongly suggests that this inhibition is due to protein thiol oxidation (Fig. 5). The fact that menadione toxicity is associated with a net loss of cell Ca^{2+} suggests that despite the inhibition of the plasma membrane Ca^{2+}-ATPase, Ca^{2+} efflux can still occur although a higher cytosolic Ca^{2+} concentration is required for stimulation of the ATPase.

The redox cycling of menadione in hepatocytes could therefore inhibit all three Ca^{2+}-translocases present in the mitochondria, endoplasmic reticulum and plasma membrane. This would undoubtedly lead to a sustained rise in cytosolic Ca^{2+}, which could cause blebbing by altering microfilament organization. Obviously, measurements of cytosolic Ca^{2+} following exposure of hepatocytes to menadione are required to substantiate this hypothesis, but this is practically very difficult. We have, however, measured the activity of phosphorylase a, an enzyme which is activated by

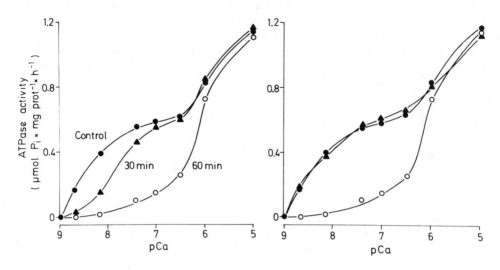

Fig. 5. Effect of menadione metabolism on Ca^{2+}-ATPase activity in hepatocyte plasma membrane fraction. Inhibition of Ca^{2+}-ATPase activity
after incubation of hepatocytes with 200 μM menadione (left panel)
and restoration of the activity by subsequent addition of 1 mM
dithiothreitol to the incubation medium (right panel). For experimental details see Nicotera et al.[34].

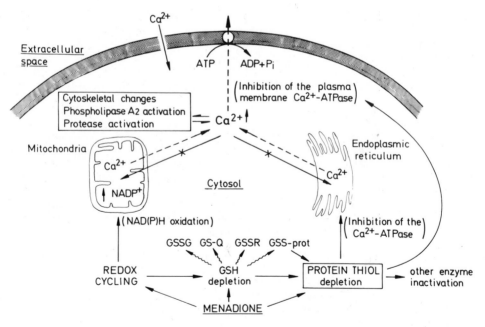

Fig. 6. Proposed scheme of cytotoxicity of menadione. GS-Q, menadione conjugate with glutathione; GSSR, soluble mixed disulfides containing
glutathione; GSS-prot, glutathione-protein mixed disulfides.

raised cytosolic Ca^{2+}, in hepatocytes subjected to oxidative stress and found that exposure of the cells to toxic levels of menadione causes prolonged activation of this enzyme[21]. Thus, menadione-induced oxidative stress depletes cellular GSH, causes a decrease in protein thiols and disrupts mechanisms which regulate Ca^{2+} homeostasis in liver cells. Our results further indicate that this leads to an uncontrollable rise in cytosolic Ca^{2+} which results in cell death. A schematic representation of these events is given in Fig. 6.

CONCLUDING REMARKS

Our studies have shown that exposure of hepatocytes to toxic levels of menadione results in extensive production of oxygen radicals, oxidation of thiols and pyridine nucleotides, and perturbation of intracellular Ca^{2+} homeostasis. The latter is associated with characteristic blebbing of the plasma membrane which appears to be an early morphological sign of cell injury. Our present work is concerned with biochemical mechanisms by which the increased cytosolic Ca^{2+} concentration may trigger cytotoxicity.

ACKNOWLEDGEMENTS

This work has been supported by grants from the Swedish Medical Research Council (proj. no. 03X-2471) and from the Swedish Council for the Planning and coordination of Research.

REFERENCES

1. J.A. Miller and E.C. Miller, The metabolic activation of carcinogenic aromatic amines and amides, Progr. Exp. Tumor Res. 11:273 (1969).
2. A. Larsson, S. Orrenius, A. Holmgren, and B. Mannervik, eds., "Functions of Glutathione: Biochemical, Physiological, Toxicological and Clinical Aspects", Raven Press, New York (1983).
3. J.R. Gillette, J.R. Mitchell, and B.B. Brodie, Biochemical basis for drug toxicity, Ann. Rev. Pharmacol. 14:271 (1974).
4. J.R. Mitchell, D.J. Jollow, W.Z. Potter, J.R. Gillette, and B.B. Brodie, Acetaminophen-induced hepatic necrosis. IV. Protective role of glutathione, J. Pharmacol. Exp. Ther. 187:211 (1973).
5. N.S. Kosower and E.M. Kosower, The glutathione status of cells, Int. Rev. Cytol. 54:109 (1978).
6. A. Meister and M. Anderson, Glutathione, Ann. Rev. Biochem. 52:711 (1983)
7. A. Meister, S.S. Tate, and L.L. Ross, Membrane bound γ-glutamyl transpeptidase, in: "The Enzymes of Biological Membranes", A. Martinosi, ed., Plenum Press, New York (1976).
8. W.B. Jakoby, The glutathione-S-transferases: a group of multifunctional detoxification proteins, Adv. Enzymol. 46:383 (1978)
9. A. Wendel, Glutathione peroxidase, in: "Enzymatic Basis of Detoxication", W.B. Jakoby, ed., Academic Press, New York (1980).
10. L. Eklöw, P. Moldéus, and S. Orrenius, Oxidation of glutathione during hydroperoxide metabolism. A study using isolated hepatocytes and the glutathione reductase inhibitor 1,3-bis(2-chloroethyl)-1-nitrosourea, Eur. J. Biochem. 138:459 (1984)
11. S.K. Srivastava and E. Beutler, The transport of oxidized glutathione from human erythrocytes, J. Biol. Chem. 244:9 (1969).
12. T.P.M. Akerboom, M. Bilzer, and H. Sies, Competition between transport of glutathione disulfide (GSSG) and glutathione-S-conjugates from perfused rat liver into the bile, FEBS Lett. 140:73 (1982).

13. P. Nicotera, M. Moore, G. Bellomo, F. Mirabelli, and S. Orrenius, Demonstration and partial characterization of glutathione disulfide-stimulated ATPase activity in the plasma membrane fraction from rat hepatocytes, J. Biol. Chem. 260:1999 (1985).
14. P. Nicotera, C. Baldi, S.-Å. Svensson, R. Larsson, G. Bellomo, and S. Orrenius, Glutathione-S-conjugates stimulate ATP hydrolysis in the plasma membrane fraction of rat hepatocytes, FEBS Lett. in press, (1985).
15. F.G. Hopkins, An autooxidable constituent of the cell, Biochem. J. 15:296 (1921).
16. E. Lundsgaard, Biochem. Z. 217:162 (1930).
17. D.J. Jollow, J.R. Mitchell, W.Z. Potter, D.C. Davis, J.R. Gillette, and B.B. Brodie, Acetaminophen-induced hepatic necrosis. II. Role of covalent binding in vivo, J. Pharmacol. Exp. Ther. 187:195 (1973).
18. L. Ernster, DT-diaphorase, Meth. Enzymol. 10:309 (1967).
19. H. Thor, M.T. Smith, P. Hartzell, G. Bellomo, S.A. Jewell, and S. Orrenius, The metabolism of menadione (2-methyl-1,4-naphthoquinone) by isolated hepatocytes. A study of the implications of oxidative stress in intact cells, J. Biol. Chem. 257:12419 (1982).
20. D. Di Monte, D. Ross, G. Bellomo, L. Eklöw, and S. Orrenius, Alterations in intracellular thiol homeostasis during the metabolism of menadione by isolated rat hepatocytes, Arch. Biochem. Biophys. 235:334 (1984).
21. D. Di Monte, G. Bellomo, H. Thor, P. Nicotera, and S. Orrenius, Menadione-induced cytotoxicity is associated with protein thiol oxidation and alteration in intracellular Ca^{2+} homeostasis. Arch. Biochem. Biophys. 235:343 (1984).
22. S.A. Jewell, G. Bellomo, H. Thor, S. Orrenius, and M.T. Smith, Bleb formation in hepatocytes during drug metabolism is caused by disturbances in thiol and calcium ion homeostasis, Science, 217:1257 (1982).
23. F.C. Bygrave, Intracellular calcium and its regulation in liver, in: "Progress in Clinical and Biological Research", F. Bronner and M. Peterlik, eds., A.R. Liss, New York.
24. E. Carafoli and M.Crompton, The regulation of intracellular Ca^{2+}, Curr. Top. Membr. Transp. 10:151 (1978).
25. A.C. Lehninger, A. Vercesi, and E.P. Bababunmi, Regulation of Ca^{2+} release from mitochondria by the oxidation-reduction state of pyridine nucleotides, Proc. Natl. Acad. Sci. USA, 75:1690 (1978).
26. A.E. Vercesi, Possible participation of membrane thiol groups on the mechanism of $NAD(P)^{+}$-stimulated Ca^{2+} efflux from mitochondria, Biochem. Biophys. Res. Commun. 119:305 (1984).
27. P. Goldstone and H. Crompton, Evidence for β-adrenergic activation of Na^{+}-dependent efflux of Ca^{2+} from isolated liver mitochondria, Biochem. J. 204:369 (1982).
28. L. Moore, T. Chen, H.R. Knapp, and E.J. Landon, Energy dependent calcium sequestration activity in rat liver microsomes, J. Biol. Chem. 250:4562 (1975).
29. P.B. Moore and N. Kraus-Friedman, Hepatic microsomal Ca^{2+}-dependent ATPase, Biochem. J. 214:69 (1983).
30. N. Kraus-Friedman, J. Biber, H. Murer, and E. Carafoli, Calcium uptake in isolated hepatic plasma membrane vesicles, Eur. J. Biochem. 129:7 (1982).
31. H.J. Berridge and R.F. Irwin, Inositol trisphosphate, a novel second messenger in cellular signal transduction, Nature, London, 312:315 (1984).
32. G. Bellomo, S.A. Jewell, and S. Orrenius, The metabolism of menadione impairs the ability of rat liver mitochondria to take up and retain calcium, J. Biol. Chem. 257:11558 (1982)

33. H. Thor, P. Hartzell, S.-Å. Svensson, S. Orrenius, F. Mirabelli, V. Marinoni, and G. Bellomo, On the role of thiol groups in the inhibitor of liver microsomal Ca^{2+} sequestration by toxic agents, Biochem. Pharmac. in press (1985).

34. P. Nicotera, M. Moore, F. Mirabelli, G. Bellomo, and S. Orrenius, Inhibition of hepatocyte plasma membrane Ca^{2+}-ATPase activity by menadione metabolism and its restoration by thiols, FEBS Lett. 181:149 (1985).

ROLE OF PARENCHYMAL VERSUS NON-PARENCHYMAL CELLS IN THE CONTROL OF BIO-LOGICALLY REACTIVE INTERMEDIATES

Franz Oesch, Mark Lafranconi, and Hans-Ruedi Glatt

Institut für Toxikologie, Universität Mainz,
Obere Zahlbacher Strasse 67, D-6500 Mainz, FRG

ABSTRACT

The non-parenchymal cells (NPC) of the liver have the potential to significantly influence the formation of reactive intermediates in the liver because of their critical location along the sinusoids where they are the first cells to encounter blood borne xenobiotics. To study the possible role of the NPC in the metabolism of xenobiotics, populations of NPC and parenchymal cells (PC) were prepared from rats and various xenobiotic metabolizing enzyme activities investigated. The specific activity of every enzyme studied was 12 to 1000 % higher in the PC than in the NPC populations and the pattern of activities between the 2 populations was remarkably different. The NPC also displayed a more dramatic response to Aroclor 1254 induction of enzyme activities than did the PC. Furthermore, the NPC were capable of forming biologically reactive intermediates which caused cyto- and genotoxicity. From these data we conclude that the NPC provide a distinct contribution to hepatic metabolism of xenobiotics.

INTRODUCTION

More than 90 % of the mass of the liver is from the parenchymal cells (PC) which also have very high concentrations of xenobiotic metabolizing enzyme activities.[1] For this reason and because these cells are easily obtained in good yield they have been widely used as a tool to investigate the formation of biologically reactive intermediates. However, other cell types in the liver, the non-parenchymal cells (NPC), also respond to the toxic effects of xenobiotics, in some instances when there is no measurable effect on the PC.[2-4] These types of responses are possible if there are significant differences in the metabolism of xenobiotics between cell types.

Furthermore, while it is unlikely that the NPC possess the same capabilities to metabolize xenobiotics as the PC, the NPC may significantly contribute to hepatic metabolism of xenobiotics because the majority of the NPC are located along the sinusoids. When a xenobiotic enters the hepatic circulation the first population of cells it encounters are the NPC. Therefore, by the time it reaches the PC population it is already partially altered by whatever enzyme activities may be present in the NPC.

While much has been written about the metabolic potential of PC there is a dearth of information about metabolism by NPC. The few reports available on xenobiotic metabolizing enzymes in the NPC have focused on a particular enzyme or enzyme system.[5][6] These reports have provided some very important information about the metabolic capability of NPC but were not designed to determine how relevant this capability is to the toxicology of the liver. Therefore, we wanted to measure the activities of some important xenobiotic metabolizing enzymes, both oxidative and post oxidative, in the NPC and compare them to those found in PC in order to better characterize enzyme distribution in the liver. Further, we wanted to study the distribution of cyto- and genotoxic effects of certain model hepatotoxicants which would help determine the toxicological significance of the enzyme activities. These data would provide a basis for understanding differences between cell types in susceptibility to hepatotoxicants.

MATERIALS AND METHODS

All enzymes, reagents and animals were purchased from local suppliers. *trans*-7,8-Dihydrobenzo[*a*]pyrene (B[*a*]P-7,8-diol) was prepared by Dr. K.L. Platt of our institute with published methods.[7][8] Benzphetamine was a generous gift from Dr. Anthony Lu, Merck Sharpe and Dohme, Rahway, New Jersey, U.S.A. and the Salmonella typhimurium tester strain TA-100 was provided by Dr. Bruce Ames, U.C. Berkley, California.

Male Sprague-Dawley rats weighing 200 g, purchased from Süddeutsche Versuchstierfarm, Tuttlingen, FRG were used for all experiments. Enzyme activities were induced with a single interperitoneal injection of Aroclor 1254 (500 mg/kg) five days before cell isolation.

Preparation of cells

Monodispersed preparations of hepatocytes were prepared using collagenase perfusion methods described by Glatt et al.[9] The PC and NPC populations were separated using a combination of methods described in the literature[10-13] and to be reported in detail elsewhere.[14]

Cell identification

Populations were identified based on their morphology at the light microscopical level, and their biochemical characteristics as described.[15][16]

Assays for cell viability

The viability of each preparation was assessed by counting the proportion of cells that excluded 0.2 % trypan blue. The methods of Bergmeyer and Bernt[17] were used to assay for lactate dehydrogenase (LDH) activity. The rate of protein synthesis was measured by determining the amount of ^{3}H-leucine incorporated into trichloroacetic acid (TCA) precipitable materials. Viability was also determined by colorimetrically measuring the amount of a tetrazolium salt, (3-(4,5-dimethylthiazol-2-yl)-2,5-diphenyl tetrazolium bromide) (MTT) reduced by the cells.[18]

Enzyme assays

In all assays reported, the amount of product formed was linear with both time and protein concentration. Benzphetamine demethylase was determined by colorimetrically following the formation of formaldehyde.[19] Ethoxyre orufin (ERR) O-deethylase activity was measured fluorimetrically

according to the method of Burke and Mayer.[20] Cytochrome P-450 content
was determined through a carbon monooxide difference spectrum as described by Omura and Sato.[21] Epoxide hydrolase (EH) activity was radiometri-
cally determined with ^{3}H-Benzo[a]pyrene (B[a]P) 4,5-oxide.[22] Glutathione
transferase (GST) and UDP glucuronononosyl transferase (UDPGT) activities
were measured spectrophotometrically according to the methods of Habig
et al.[23] and Bock et al.[24] respectively using 1 mM 1-chloro-2,4-dinitro-
benzene to assay for GST and 5 mM 1-napthol as the substrate for UDPGT.
Proteins were estimated according to the methods of Lowry et al.[25] using
a standard curve constructed from bovine serum albumin.

Cell mediated bacterial mutagenicity test

Freshly prepared PC (1 x 10^6 cells) or NPC (10 x 10^6 cells) were incubated for 1 hr at 37 degrees C in sterile glass test tubes with approximately 1.7 x 10^8 Salmonella typhimurium (TA-100) and the indicated
amount of test compound dissolved in 10 µl of acetone as described by
Glatt et al.[9] Total volume of the incubation was 1.11 ml. The contents
of each tube were then mixed with 2 ml of molten top agar, poured onto
histidine poor agar plates and incubated at 37 degrees C in the dark for
3 days, after which, the number histidine independent bacterial colonies
was counted.

Statistical analysis

All simple comparisons to control were done using the Student's
t-test. Multiple comparisons to a single control were accomplished with
Dunnett's t-test.[26]

RESULTS AND DISCUSSION

For this work we chose methods to isolate cells which provided NPC
populations that were essentially free from contamination with PC. This
was necessary because of the generally high specific activity of PC
enzymes and the relatively low activities we expected in the NPC. With
this method we routinely obtained from a single liver about 400 mg of
protein from the PC population and about 55 mg of protein from the NPC
populations. More importantly, the NPC population was more than 99 %
free from intact hepatocytes as measured by light microscopy and marker
enzymes[16] for PC (data not shown). However, it must be kept in mind that
the NPC preparation was not a pure population but contained a mixture of
many small cells contained in the liver as well as some dense cell
debris which sedimented with whole cells.

Measurement of enzyme activities

Enzyme activities were measurable in all cases both in the PC and
the NPC populations (Table 1). In most instances, the PC possessed
greater enzyme activities than did the NPC (Table 1). In preparations
from control animals, the differences were particularly large for the
benzphetamine demethylase (280 fold) and ethoxyresorufin deethylase
(14 fold) activities as well as for the total amount of cytochrome
P-450 (6 fold). The differences in specific activity between the two
cell populations were smaller (1.3 to 2.4 fold) for the post oxidative
enzymes (epoxide hydrolase, GST and UDPGT).

One striking feature of the data in this table is the low ratio of
oxidative to post oxidative enzyme activities measured in the NPC. The
NPC populations have consistantly lower cytochrome P-450 enzyme activities, 1/280 to 1/8, compared to the PC. Yet the post oxidative enzyme

Table 1. Enzyme activities in PC and NPC populations[a]

Enzyme Assay	Cell Type[b]	Control	Induced	Fold Increase
ERR-O-deethylase[c]	PC	14 + 5	2153 + 456	150*
	NPC	1 + 1	261 + 109	260*
Benzphetamine Demethylase[c]	PC	830 + 128	4175 + 722	5*
	NPC	3 + 4	538 + 174	180*
Cyt P-450[d]	PC	416 + 53	2400 + 590	6*
	NPC	66 + 11	454 + 103	7*
GST[e]	PC	656 + 64	2289 + 444	3*
	NPC	305 + 98	754 + 196	2*
UDPGT[c]	PC	3190 + 696	13043 + 1409	4*
	NPC	1290 + 196	4648 + 668	4*
EH[c]	PC	2460 + 470	8569 + 1365	3*
	NPC	1223 + 897	6543 + 1444	5*

[a] All enzyme activities were assayed in broken cell preparations in the presence of appropriate co-factors. Values are means $\pm$ SD from at least 6 experiments per group.

[b] Cells were isolated as described in Materials and Methods.

[c] Activity is expressed as pmole product formed/min/mg whole homogenate protein.

[d] Assayed spectrophotometrically in microsomal preparations. Units are pmole Cyt P-450/mg microsomal protein.

[e] Activity is expressed as nmole product formed/min/mg whole homogenate protein.

* Indicates significantly different from the corresponding control value ($p < 0.05$, Student's t-test).

activities are only 1/3 to 1/1.3 the activities of the PC. This indicates that the NPC have a relatively lower ability to oxidize xenobiotics to reactive electrophiles and a greater ability to conjugate or hydrolyze those products that may be formed. Based on these observations it is likely that xenobiotics metabolized by NPC have a lower probability of generating significant concentrations of biologically reactive intermediates than the PC.

Treatment of the animals with Aroclor 1254 consistently led to an increase in all investigated enzyme activities in both cell populations. However, the NPC appeared to be more responsive to the effects of Aroclor 1254 pretreatment than the PC. This was particularly notable for the cytochrome P-450 activities ERR-O-deethylase and benzphetamine demethylase. This is in agreement with the results obtained by Cantrell and Bresnick[5] in which they reported that the NPC displayed a significantly greater increase in benzo[a]pyrene hydroxylase activity in response to 3-methylcholanthrene pretreatment than the PC populations.

Viability studies

To determine if the enzyme activities measured in the NPC popula-

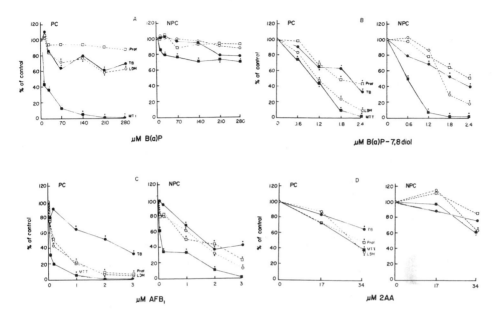

Fig. 1   Changes in 4 indices of viability in 2 hepatocyte populations
caused by metabolism of carcinogens.

All indices of cell viability were measured in cells obtained from Aroclor 1254 induced animals. These values were
determined after 1 hr incubation of cells in 1 ml microfuge
tubes with fresh buffer containing the indicated concentration
of test compound. All measurements were conducted as described
in Materials and Methods. Values represent the means of at
least 3 separate experiments and are expressed as a percent
of the control values which received no test compound. LDH
values reflect the percent of LDH remaining in the cells.
Values for trypan blue exclusion are represented in the
figure by solid circles, the reduction of MTT by solid squares,
the amount of LDH in the cells by open circles and protein
synthesis by open squares. The mean values for protein synthetic rate, % of cells excluding trypan blue, % LDH remaining in
the cell and the rate of MTT reduction in the PC preparations
were 4.4 + .4 nmole/hr/10^b cells, 61 $\pm$ 8 %, 78 $\pm$ 5 % and 189 $\pm$
35 nmole $\overline{MTT}$ reduced/hr/10^6 cells respectively. Control values
for the NPC population were; protein synthesis 3.2 $\pm$ .3 nmole/
hr/10^b cells, % of cells excluding trypan blue -93 $\pm$ -3 %, LDH
remaining in the cells -90 + -3 % and MTT reduction -35 $\pm$ -9
nmole MTT reduced/hr/10^6 cells.
* Indicates significantly different from control values,
p < 0.05 Dunnett's t-test.

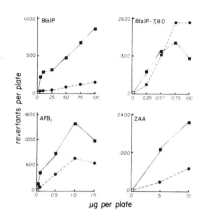

Fig. 2   PC and NPC mediated bacterial mutagenicity. One million PC and
          10^7 NPC (roughly equal protein concentration) from Aroclor
          1254 pretreated male Sprague-Dawley rats were incubated for
          1 h with bacteria (TA-100) and the indicated amount of test
          compound as described in materials and methods. The number
          of histidine independent colonies was counted after plating
          the bacteria and incubation for 3 days in the dark at 37 °C.
          Values are means from 3 experiments.
          * Indicates significantly different from control (p < 0.05
          Dunnett's t-test).

tions had any biological significance, a number of indices of viability
were monitored for dose dependent changes with carcinogens. Representative compounds were chosen from 3 major groups; the polycyclic aromatics
(B[a]P and B[a]P-7,8 diol), the aromatic amines, 2-amino anthracene (2AA)
and aflatoxins, aflatoxin B_1 (AFB$_1$).

     In all instances the PC populations appeared more sensitive to the
effects of xenobiotics than did the NPC populations. With PC, B[a]P
caused a dose dependent decrease in the reduction of MTT, the amount of
LDH remaining in the cell and the ability of the cells to exclude trypan
blue. The changes in the NPC caused by B[a]P were not as remarkable. The
only index to change was the reduction of MTT. Thus, it appears the PC
were better able to metabolize the polycyclic aromatic B[a]P to a cytotoxic species than were the NPC (Figure 2a). However, when a metabolite
of B[a]P-7,8-diol, which is the immediate precursor to a major ultimate
mutagen and carcinogen, the diol epoxide, both the NPC and the PC populations were capable of metabolizing the compound to a cytotoxic species
(Figure 2b). Likewise, both populations of cells could metabolize AFB$_1$
to cytotoxic species (Figure 2c). The aromatic amine (2AA) was not as
cytotoxic to the NPC population as the PC (Figure 2d). In all instances
the changes in viability caused by the test compounds could be attenuat-

ed by including in the incubation media 100 μM 7,8-benzoflavone, an inhibitor of the 3-methyl-cholanthrene inducible forms of cytochrome P-450 (data not shown).

Four indices of viability were used to investigate the metabolism dependent cytotoxicity caused by xenobiotics. This was done to allow for the possibility that xenobiotics metabolized to reactive intermediates might have different cellular targets and thereby demonstrate different manifestations of toxicity. That this possibility is likely is demonstrated by the different patterns of cytotoxicity observed with the various test compounds. When B[a]P was tested in PC (Figure 2a) the order of sensitivity, from most responsive to least responsive, was MTT reduction, LDH leakage, trypan blue exclusion and protein synthesis. With AFB_1 and 2AA the pattern was altered slightly with the exclusion of trypan blue displaying less change than protein synthesis.

Mutagenicity Studies

The formation of biologically reactive intermediates can have a variety of effects on cells. The above measurements were designed to monitor those events which blatantly interfered with the easily measured biochemical processes of the cell. However, other macromolecules in the cell, in particular DNA, may be damaged without resulting in such immediate effects. The damage to DNA may have profound effects on the cell leading to complex responses such as transformation or tumor formation. To measure possible genotoxic effects of metabolic activation of xenobiotics by the NPC population, the ability of these cells to mediate mutagenicity in bacteria was monitored using the Salmonella typhimurium mutagenicity assay.

These assays show that the NPC were consistently weaker mediators of bacterial mutagenicity than were the PC (Figure 3). The one exception is the mutagenicity caused by B[a]P 7,8-diol in which the PC and NPC showed roughly equal capabilities for metabolic activation. Nevertheless, the NPC seem to be able to metabolically activate a wide range of premutagens to mutagenic species.

SUMMARY

To study the potential of NPC in the liver to metabolize xenobiotics we have used techniques described in the literature to prepare fractions of hepatocytes free from PC. Some xenobiotic metabolizing enzyme activities in the NPC populations were measured and found to be consistantly lower than those activities found in the PC. Despite the normally low enzyme activities measured in the NPC populations, these cells were capable of metabolizing some xenobiotics to cytotoxic and genotoxic species which indicates that the metabolic potential of these cells, although low, is nonetheless potentially significant. Because of the distinctly different pattern of xenobiotic metabolizing enzymes in the 2 cell populations, the exact role of the NPC in the control of biologically reactive metabolites and the toxicity produced by them will depend upon the structural elements of the xenobiotic in question.

ACKNOWLEDGEMENTS

This work was supported by The Deutsche Forschungsgemeinschaft and the Alexander von Humboldt Foundation (W.M.L.). We also thank Ms. I. Böhm for typing this manuscript.

REFERENCES

1. R. Daoust, Liver function, in: "American Institute Biological Sciences" publication # 4, R. W. Brauer, ed., pp. 3-10 (1958).
2. J. L. Creech Jr., and M. N. Johnson, Angiosarcoma of the liver in the manufacture of polyvinyl chloride, J. Occup. Med. 16:150-151 (1974).
3. M. A. Bedell, J. G. Lewis, K. C. Billings, and J. A. Swenberg, Cell specificity in hepatocarcinogenesis: preferential accumulation of O^6 methylguanine in target cell DNA during continuous exposure of rats to 1,2-dimethylhydrazine, Cancer Res., 42:3079-3083 (1982).
4. H. Druckrey, "Organospecific carcinogenesis in the digestive tract", in: Topics in Chemical Carcinogenesis, W. Nakahara, S. Takayama, T. Sugimura, and S. Odashima, eds., University Park Press, Tokyo, pp. 73-101 (1972).
5. E. Cantrell, and E. Bresnick, Benzpyrene hydroxylase activity in isolated parenchymal and nonparenchymal cells of rat liver, J. Cell Biol. 52:316-321 (1972).
6. J. Morland, and H. Olsen, Metabolism of sulfadimidine, sulfanilamide, para aminobenzoic acid, and isoniazid in suspensions of parenchymal and nonparechymal rat liver cells, Drug Metab. Disp. 5:511-517 (1977).
7. D. J. McCaustland, and J. F. Engel, Metabolites of aromatic hydrocarbons. II. Synthesis of 7,8-dihydrobenzo[a]pyrene-7,8-diol and 7,8-dihydrobenzo[a]pyrene-7,8-epoxide, Tetrahedron Lett., 2059-2552 (1975).
8. P. P. Fu, and R. G. Harvey, Synthesis of the diols and the dioepoxides of carcinogenic hydrocarbons, Tetrahedron Lett., 2059-2062 (1977).
9. H. R. Glatt, R. Billings, K. L. Platt, and F. Oesch, Improvement of the correlation of bacterial mutagenicity with carcinogenicity of benzo[a]pyrene and four of its major metabolites by activation with intact liver cells instead of cell homogenate, Cancer Res. 41:270-277 (1981).
10. D. M. Mills, and D. Zucker-Franklin, Electron microscopic study of isolated Kupffer cells. Am. J. Pathol. 54:147-155 (1969).
11. J. A. Swenberg, M. A. Bedell, K. C. Billings, D. R. Umbenhauer, and A. Pegg, Cell specific differences in O^6-alkylguanine DNA repair activity during continuous exposure to carcinogen, Proc. Natl. Acad. Sci. 79:5499-5502 (1982).
12. D. P. Praaning van Dalen, and D. L. Knook, Quantitative determinations of in vivo endocytosis by rat liver Kupffer and endothelial cells facilitated by an improved cell isolation method, FEBS Lett. 141:229-232 (1982).
13. D. L. Knook, and E. Ch. Sleyster, Isolated parechymal, Kupffer and endothelial rat liver cells characterized by their lysosomal enzyme content, Biochem. Biophys. Res. Comm. 96:250-257 (1980).
14. W. M. Lafranconi, H. R. Glatt, and F. Oesch, Xenobiotic metabolizing enzymes of rat liver non-parenchymal cells, manuscript in preparation.
15. D. L. Knook, and E. Ch. Sleyster, Separation of Kupffer and endothelial cells of the rat liver by centrifugal elutriation. Exp. Cell Res. 99:444-449 (1976).
16. T. J. J. van Berkel, J. F. Koster, and W. C. Hulsmann, Distribution of L and M type pyruvate kinase between parenchymal and Kupffer cells of rat liver, Biochim. Biophys. Acta 276:425-429 (1972).
17. H. U. Bergmeyer, and E. Bernt, in: Methoden der enzymatischen Analyse, H. U. Bergmeyer, ed., Verlag Chemie, Weinheim/Bergstraße, FRG, pp. 533-538 (1970).
18. T. Mossman, Rapid colorimetric assay for cellular growth and survival: Appliction to proliferation and cytotoxicity assays, J. Immunol. Methods 65:55-63 (1983).

19. A. Y. H. Lu, A. Somogyi, S. West, R. Kutzman, and H. H. Conney, Pregnenolone-16α-carbonitrile: A new type of inducer of drug metabolizing enzymes, Arch. Biochem. Biophys. 152:457-462 (1972).
20. M. D. Burke, and R. T. Mayer, Ethoxyresurofin: Direct fluorimetric assay of a microsomal O-dealkylation which is preferentially inducible by 3-methylcholanthrene, Drug Metab. Disp. 2:583-588 (1974).
21. T. Omura, and R. Sato, The carbon monooxide binding pigment of liver microsomes. J. Biol. Chem. 239:2370-2378 (1964).
22. H. U. Schmassmann, H. R. Glatt, and F. Oesch, A rapid assay for epoxide hydratase activity with benzo[a]pyrene 4,5-(K-region)-oxide as substrate, Anal. Biochem. 74:94-104 (1976).
23. W. H. Habig, M. J. Pabst, and W. B. Jakoby, Glutathione transferases: The first step in mercapturic acid formation, J. Biol. Chem. 249:7130-7139 (1974).
24. K. W. Bock, B. Burchell, G. J. Dutton, O. Hanninen, G. J. Mulder, I. S. Owens, G. Sies, and T. R. Tephly, UDP-glucuronosyltransferase activities: Guidelines for consistent interim terminology and assay conditions, Biochem. Pharmacol. 32:953-955 (1983).
25. O. H. Lowry, J. Rosebrough, A. L. Farr, and R. J. Randall, Protein measurement with the Folin phenol reagent, J. Biol. Chem. 193:265-275 (1951).
26. C. W. Dunnett, New tables for multiple comparisons with a control Biometrics 6:482-494 (1964).

SIGNIFICANCE OF COVALENT BINDING OF CHEMICALLY REACTIVE METABOLITES OF
FOREIGN COMPOUNDS TO PROTEINS AND LIPIDS

James R. Gillette

Laboratory of Chemical Pharmacology
National Heart, Lung, and Blood Institute
Bethesda, Maryland

Shortly after the initial finding by the Millers (Miller and Miller,
1947; Miller et al., 1949) that the administration of N,N-dimethyl-4-
aminoazobenzene (DAB) to rats resulted in the covalent binding of metabolites of the azo-dye to liver proteins, there was considerable interest
in the possibility that the covalent binding to specific proteins might
initiate the series of events that led to carcinogenesis. Recognition
that tumors probably develop from the transformation of cells in which
alterations were inherited by daughter cells, however, led to the
mutagenesis hypothesis of carcinogenesis, and the emphasis of research
shifted from covalent binding of metabolites to proteins to their
covalent binding to DNA and RNA. Since that time, however, many foreign
compounds have been shown to be transformed to chemically reactive
metabolites that combine with proteins and lipids. Indeed, even
carcinogens that presumably evoke mutagenic changes in DNA almost
invariably also become covalently bound to proteins.

While studying the mechanism of the teratogenic effects of thalidomide, Dr. Bernard Brodie and I considered the possibility that drugs and
biologically active metabolites might react irreversibly with various
cellular components and thereby cause various toxicities in addition to
cancer. Since aqueous solutions of thalidomide undergo rapid hydrolysis
at pH 7.4, it seemed possible that the teratogenic effects of the drug
might be mediated by covalent binding to substances in the fetuses.
Although it was rather simple to establish that thalidomide did indeed
become covalently bound to proteins such as serum albumin (Schumacher et

63

al., 1948), we still were unable to demonstrate at that time any relationship between the covalent binding of the drug to plasma proteins and the teratogenic effect in animals.

With this experience in mind, we were faced with the need to develop a strategy by which it might be possible to determine whether a toxicity caused by a foreign compound might evoke the toxicity through the formation of a chemically reactive metabolite. Clearly some toxicities are caused by stable metabolites and some foreign compounds are converted to chemically reactive metabolites without manifesting any obvious toxicity. But when the compound evoked a toxicity and became covalently bound, it did not necessarily follow that the toxicity was caused by a chemically reactive metabolite. Such considerations as these, however, led to the idea that the covalent binding of radiolabeled drugs might be used as an indirect measure of the chemically reactive metabolites in various organs (Gillette, 1974a, 1974b, 1982). According to this concept, a toxic chemically reactive metabolite may react with several groups on a given protein by second order reactions.

BASIC FIRST ORDER AND SECOND ORDER PHARMACOKINETIC CONCEPTS

Changes in the amount of formation of an adduct of a reactive metabolite to a given group on a given nucleophilic substance within a given cell may be expressed by the difference between the rate of formation of the adduct and the rate of repair or elimination of the adduct within the cell. This relationship may be described by a rate equation.

$$d \text{ Adduct}/dt = \text{Rate of formation} - \text{Rate of elimination} \qquad (1)$$

where,

$$\text{Rate of adduct formation} = [\text{metabolite}] \times [\text{nucleophile}] \times \text{second order constant} \qquad (1a)$$

and

$$\text{Rate of elimination} = \text{adduct} \times \text{first order constant} \qquad (1b)$$

As written, the dimensions of the second order constant in equation (1a) are volume of distribution of the metabolite within the cell divided by the mathematical product of the molar concentration of the nucleophile and time; thus the term may be visualized as a second order clearance, i.e., a rate, expressed in amount per time, divided by the concentration of the chemically reactive metabolite within the cell.

Similar rate equations may be written to describe the rate of formation of other adducts between the toxic chemically reactive metabolite and other groups on the same nucleophile. Thus the total rate of formation of all the adducts of a toxic chemically reactive metabolite to a given macromolecular substance may be described by the sum of such rate equations representing the reactions between the chemically reactive metabolite and all of the reactive groups on any given macromolecular substance within a single cell. Similar sets of rate equations would be needed to describe the reactions between the toxic chemically reactive metabolite and all of the other endogenous substances in the cell that react with the reactive metabolite. Larger sets of rate equations would be needed to describe the rates of formation of adducts within each of the other cells in the organ under investigation. The total rate of formation of all adducts of the reactive metabolite within an organ is thus the sum of a large number of rate equations that describe the rates of formation of the adducts to all the endogenous substances that react with the chemically reactive metabolite in all of the cells of the organ.

Because there is no reason to believe that the rate constants that represent the rate of reaction of a given chemically reactive metabolite with different groups of different endogenous substances will be the same, the relative rates of formation of different types of adducts will depend on the chemically reactive metabolite and the nucleophile. Moreover, there is also no reason to believe that the concentrations of either the chemically reactive metabolite or the individual nucleophiles will necessarily be the same in all of the cells of the organ, and thus there is no reason to believe that the total rate of adduct formation will necessarily be the same in all of the cells of the organ.

Recognizing the potential complexity of the relationships between the rates of formation of adducts and the incidence of toxicity, there seemed little advantage of simply administering the foreign compound and observing whether the foreign compound was covalently bound to macromolecules in organs and whether they caused toxicities in the organ. Instead, another strategy was needed. It occurred to us, however, that treatments that resulted in alterations in the area under the curve of the toxic chemically reactive metabolite within cells of the organ would result in alterations of all of the reactions of reactive metabolite with the various endogenous substances in the same direction. Thus, a change in the covalent binding of the chemically reactive metabolite to protein that paralleled a change in the incidence or severity of the toxicity would provide circumstantial evidence that the chemically reactive metabolite caused a toxicity even when the target substance was not a protein or even when the toxicity did not result directly from the formation of an adduct.

In the interest of simplicity, I have usually assumed that the factor that relates the rate of elimination of the adduct within any given cell to the amount of the adduct at any given time is a first order constant. However, there is really no evidence to support this view. Even if it were true, however, many studies have established that the turnover of the various proteins within the cells vary markedly, and thus the kinetics of elimination of covalently bound material to various proteins after the toxic chemically reactive metabolite and its precursors have been eliminated from the body may appear to be complicated. It certainly is possible if not probable that a protein may be more easily hydrolyzed when it contains several adducts than when it contains only one or no adducts. Such possibilities would further complicate the interpretation of the kinetics of covalently-bound material during the terminal phase of elimination of the adducts.

The purpose of pharmacokinetic equations is not to describe all of the possible complications in estimating exactly all of the reactions that can possibly occur in the body. Instead, the purpose is to visualize the simplest model that would be consistent with the major events and to provide insights into the effects of alterations in the values of the various parameters that describe the events. Inevitably there must be certain simplifying assumptions in any pharmacokinetic approach to a biological problem. For example, if we assume that the concentrations of nearly all of the available target groups on the nucleophiles within the cells remain virtually constant during the course of the reaction, the value of the [nucleophile] x the second order constant in equation (1a) may be written as a first order constant $[k(met \rightarrow ad)]$. If we also assume that the rate of elimination of a given adduct in equation (1b) is directly proportional to the concentration of the adduct in the cell, we may integrate the equation and calculate the area under the curve of the adduct $(AUC_{ad(cell)})$ within a given cell.

$$AUC_{ad(cell)} = AUC_{met(cell)} \; k_{(met \to ad)(cell)} / k_{ad(cell)} \qquad (2)$$

Under these conditions, the area under the curve of the total amount of covalently-bound metabolite would equal the sum of the areas under the curves of each of the adducts.

When the elimination of the covalently-bound material is very slow compared with the metabolism of the chemically reactive metabolite and its precursors, the maximum amount of adduct in the cell would approach the value predicted by the following equation.

$$AUC_{ad(cell)} = AUC_{met(cell)} \; k_{(met \to ad)(cell)} \qquad (3)$$

which is the maximum amount of the parent compound that ultimately is converted to the adduct. The total maximum amount of reactive metabolite that becomes covalently bound within a cell is, therefore, the sum of a set of such equations.

Because of the difficulties in measuring the area under the curve of the adducts within an organ, it is usually easier to measure the maximum covalent binding of the reactive metabolite as an estimate of equation (3). The validity of this approach, however, depends on the relationship between the rate constants for the elimination of the precursors of the chemically reactive metabolite and the rate constant of elimination of the adduct. To illustrate these relationships even for a very simple system, we would need to know the route of administration of the precursor, the time course of the amounts of the precursors of the reactive metabolite as well as the time course of the amounts of the adduct. If we were to assume (1) that the parent compound was administered intravenously, (2) that the kinetics of elimination could be described by a single compartment system, and (3) that the rate constant for the elimination of the chemically reactive metabolite was considerably greater than the rate constants of elimination of the parent compound and the adducts, we could derive a rather simple equation for the time course of the adduct.

$$V_d[Adduct] = \left[\frac{Dose \; k_{(pc \to met)} F_{met \to ad}}{k_{ad} - k_a} \right] \left[e^{-k_a t} - e^{-k_{ad} t} \right] \qquad (4)$$

where $k_{(pc \to met)}$ is the rate constant for the conversion of the precursor to the chemically reactive metabolite, k_a is the sum of the rate constants for all of the mechanisms of elimination of the parent drug, $F_{met \to ad}$ is the fraction of the amount of chemically reactive metabolite that is converted to the adduct, and k_{ad} is the rate constant for the elimination of the adduct. With this equation it can be shown (Gillette, 1979) that the maximum amount of the adduct would occur at,

$$t_{max} = \frac{\ln (k_a/k_{ad})}{k_a - k_{ad}} \tag{5}$$

On substitution of equation (5) into equation (4), followed by a series of transformations, we may obtain,

$$V_d [\text{Adduct}]_{max} = \frac{\text{Dose } k_{pc \to met} F_{met \to ad}}{k_a} [\frac{k_a}{k_{ad}}]^{-k_{ad}/(k_a - k_{ad})} \tag{6}$$

But $k_{pc \to met}/k_a$ is the fraction of the dose that is converted to the reactive metabolite, $F_{pc \to met}$, and thus we may rewrite equation (6) as,

$$V_d [\text{Adduct}]_{max} = \text{Dose } F_{pc \to met} F_{met \to ad} [\frac{k_a}{k_{ad}}]^{-k_{ad}/(k_a - k_{ad})} \tag{6a}$$

or

$$V_d [\text{Adduct}]_{max} = \text{Dose } F_{pc \to met} F_{met \to ad} \text{ RAD} \tag{6b}$$

Thus, the maximum amount of adduct present in an organ depends on the dose, the fraction of the dose that is converted to the adduct in the organ, and the relative values of the elimination constants representing the elimination of the parent compound from the body and the elimination of the adduct. Substituting various values for the rate constants of elimination into equation (6a) reveals that when the rate constant of elimination of the parent compound in the body is twice the rate constant of elimination of the adduct, the maximum amount will be 50% of the dose times the fraction of the dose that forms the adduct. Thus, for rapidly eliminated parent compounds, measurements of the maximum amount of covalently bound material usually will provide reasonable estimates of the amount of the covalently bound material formed from the parent compound.

Such substitutions also provide predictions of the effects of induction of enzymes that catalyze the various reactions by which the parent compound is converted to the chemically reactive metabolite and to innocuous metabolites. When the rate constant of elimination of the adduct is smaller than the rate constant of elimination of the parent compound, the effects of an inducer will be greatest when the formation of the chemically reactive metabolite represents a relative minor pathway of elimination of the parent compound. If the inducer specifically increases the amount of an enzyme that catalyzes the formation of the chemically reactive metabolite, induction would not markedly affect k_a and thus would not markedly change either t_{max} (equation 5) or the value of RAD in equation (6b). The major effect would be an increase in the fraction of the dose that is converted to the reactive metabolite and thus the fraction of the dose that forms the adduct. On the other hand, a given preferential increase in the activity of an enzyme that catalyzes the formation of major innocuous metabolites would markedly increase k_a and thereby decrease the t_{max} in equation (5), increase RAD in equation (6b), but decrease the fraction of the dose that was converted to the reactive metabolite and decrease the maximum amount of adduct that would be present in the organ. By contrast, if the formation of the chemically reactive metabolite represented the sole mechanism of elimination of the parent compound, an increase in the activity of the enzyme that catalyzed the formation of the chemically reactive metabolite would not affect the fraction of the dose that was converted to the chemically reactive metabolite, although it would decrease t_{max} and increase the value of RAD. Thus, there is no mathematical rationale for the statement that the effects of a chemically reactive metabolite depend on the relative rates of formation and inactivation of the chemically reactive metabolite. The correct statement is that the area under the curve of a chemically reactive metabolite depends on the dose times the fraction of the dose of the parent compound that is converted to the chemically reactive metabolite divided by the rate constant of elimination of the reactive metabolite in the organ. Similarly, the area under the curve of the adduct within an organ may be expressed as the dose times the fraction of the dose that forms the adduct divided by the rate constant of elimination of the adduct.

APPLICATION OF FRACTION OF THE DOSE CONCEPTS TO NONLINEAR SYSTEMS

The concept of the fraction of the dose of the parent compound that is converted to an adduct in an organ may also be applied to nonlinear systems in which the rates of formation of the chemically reactive and innocuous metabolites are not exactly proportional to the concentration of their immediate precursors and the rates of elimination of the chemically reactive metabolite are not always exactly proportional to the concentration of the chemically reactive metabolite. For example, if the parent compound were eliminated from the body solely by an enzyme that catalyzed the formation of the chemically reactive metabolite, increasing the dose would not alter the fraction of the dose of the parent compound that was converted to the chemically reactive metabolite even though the initial concentration of the parent compound exceeded the K_m of the enzyme. Moreover, as long as the elimination of the chemically reactive metabolite remains directly proportional to the concentration of the chemically reactive metabolite, the area under the curve of the adduct per dose would remain unchanged, because the fraction of the dose that was converted to the adduct would remain unchanged. Indeed, the effect of high initial concentrations of the parent compound in excess of the K_m of the enzyme would be manifested by a delay in the time at which the maximum concentration of the adduct formed from the chemically reactive metabolite would occur, a smaller maximum concentration of the adduct per dose and prolonged persistence of the adduct. Moreover, similar effects would be expected if the parent compound were eliminated from the body solely by several enzymes that had the same K_m value. Thus, there is no mathematical rationale for the statement that high doses of a parent compound that lead to concentrations of the parent compound which approach saturation of an enzyme necessarily will have marked effects on the area under the curve of the covalent binding of a chemically reactive metabolite to an intracellular nucleophile or to the incidence of toxicity; it is still theoretically possible for increasing doses to approach saturation of enzymes and still have the area under the curve of the adducts and the incidence of toxicity directly proportional to the dose.

Clearly, the effects of increasing doses on the formation of the fraction of the dose of the parent compound that is converted to a chemically reactive metabolite depends on the presence of several mechanisms of elimination of the chemically reactive metabolite having markedly different K_m values. The effects will depend on the relative importance of the elimination mechanisms at low doses of the parent compound and on which of the elimination mechanisms has the lowest K_m value. The greatest effects would be expected when the formation of the chemically reactive metabolite represents a minor pathway of elimination of the parent compound. If the major enzyme that catalyzes the formation of innocuous metabolites has the lowest K_m value, then increasing the dose will increase the fraction of the dose of the parent compound that is converted to the chemically reactive metabolite. If the enzyme that catalyzes the formation of the chemically reactive metabolite has the lowest K_m value, then increasing the dose will decrease the fraction of the dose of the parent compound that is converted to the chemically reactive metabolite.

Concentration dependent effects on areas under the curves of chemically reactive metabolites and adducts can be subtle and difficult to detect by ordinary pharmacokinetic studies of the parent compound. For example, on increasing the dose of a parent compound, the highest concentration of the parent compound may still remain well below the K_m values of the enzymes that catalyze the formation of the innocuous and the chemically reactive metabolites, but increasing the dose of the parent compound may lead to concentrations of the chemically reactive metabolite that exceed the K_m value of enzymes that inactivate the chemically reactive metabolite. Thus, increasing the dose of the parent compound may not significantly alter the half-life of the parent compound, but still increase the fraction of the chemically reactive metabolite that is converted to adducts bound to target sites.

It is also well established that dose dependent (as opposed to concentration dependent kinetics) may occur. The most common of this kind of kinetics occurs in the formation of conjugates. The rates of formation of conjugates depend on the concentration of an endogenous component as well as the concentration of the foreign compound or metabolite. When the dose of the foreign compound exceeds the amount of the endogenous component present in the body, the initial rate of formation of the conjugate depends on the amount of the endogenous component initially present in the body, but as this is depleted, the concentration of the endogenous component tends to approach a steady-state, which depends on its rate of formation divided by the rate constant of elimination. For example, it now seems likely that the dose dependent kinetics of acetaminophen metabolism and covalent binding of acetaminophen metabolites is due to at least three such dose dependent reactions. At low doses of acetaminophen, the drug is eliminated from the body predominantly by the formation of glucuronide and sulfate conjugates and to a lesser extent by the formation of a chemically reactive metabolite which in turn is eliminated predominantly by the formation of its glutathionyl conjugate. In liver, the major organ of elimination of the drug in the body, the amounts of phosphoadenosyl

phosphosulfate (PAPS) and uridine diphosphoglucuronate (UDPGA), which are
the cosubstrates for the formation of sulfate and glucoronide conjugates,
respectively, are small compared with the dose of the drug (Hjelle et
al., 1985). Thus, the initial stores of these cosubstrates are consumed
within a very few seconds and the rates of formation of the conjugates
depend in part on the rate of synthesis of the cosubstrates in
hepatocytes. If the dose of acetaminophen is increased further, the rate
of synthesis of the cosubstrates may decrease as the concentration of
their precursors, such as inorganic sulfate (Galinski and Levy, 1981) and
uridine diphosphoglucose in the cells decline and thus the rate of
synthesis of the conjugates may be further decreased. Such dose
dependent kinetics as these would result in an increase in the fraction
of the dose of acetaminophen that is converted to the chemically reactive
metabolite. Since the chemically reactive metabolite is eliminated predominantly by the formation of its glutathione conjugate, however, the
rate of inactivation of the reactive metabolite depends on the
concentration of glutathione in hepatocytes, but the rate of synthesis of
glutathione is rather slow. Initially, therefore, the inactivation of
the reactive metabolite depends on the initial store of glutathione, but
eventually the inactivation by this pathway depends on the rate of
synthesis of glutathione. Thus, increasing the dose of acetaminophen
results in increases in the fraction of the dose of acetaminophen that
becomes covalently bound.

PHYSIOLOGICAL PHARMACOKINETIC MODELS

Because the half-lives of chemically reactive metabolites can vary
from milliseconds to several hours, a comprehensive model of the covalent
binding of chemically reactive metabolites to protein and lipids in given
cells of given organs that would include the entire range of half-lives
would be hopelessly complex. Since the factors that govern the kinetics
of a reactive metabolite having a half-life of a few milliseconds will
differ greatly from the dominant factors that govern the kinetics of a
metabolite having a half-live of several hours, I have found it useful to
divide chemically reactive metabolites into several different categories
according to their half-lives (Table 1). In using this system of
classification, however, investigators should be aware that a given
chemically reactive metabolite may be included in several of the
categories, depending on the situation. For example, a chemically
reactive metabolite that is inactivated predominantly by being converted

Table 1.

Functional Classification of Metabolites

1. Ultrashort-lived Metabolites

   Metabolites never leave the enzyme.

2. Short-lived Metabolites

   Metabolites never leave the cell.

3. Intermediately-lived Metabolites

   Metabolites never enter arterial blood

4. Long-lived Metabolites

   Metabolites never leave the body

5. Ultralong-lived Metabolites

   Metabolites eliminated predominantly by excretion into bile, urine or air.

to a glutathione conjugate may be in the short-lived category in cells that contain large amounts of glutathione, but may be in the intermediately-lived or the long-lived category in cells that contain small amounts of glutathione. It may also be in the short-lived category at low doses of the parent compound that do not significantly affect the concentration of glutathione in the cells, but in the intermediate or long-lived category at high doses of the parent compound that result in marked depletion of glutathione in the cells. Such changes in category may be illustrated with the chemically reactive metabolite of bromobenzene, namely bromobenzene-3,4-oxide. Studies with rat hepatocytes revealed that the glutathione conjugates of bromobenzene were formed almost entirely from the bronobenzene-3,4-oxide as it was formed in the endoplasmic reticulum but that virtually all of the other metabolites of bromobenzene-3,4-oxide were formed from the bromobenzene-3,4-oxide that had diffused out of the hepatocytes (Monks et al., 1984; Gillette et al., 1984). Thus, these findings indicated that under the conditions of the experiment bromobenzene-3,4-oxide was predominantly an intermediately-lived metabolite or a long-lived metabolite. In the living rat, however, very little bromobenzene-3,4-oxide could be detected unless high doses of bromobenzene were administered to rats pretreated with phenobarbital (Lau et al., 1984). Moreover, the area under the curve of the bromobenzene-3,4-oxide in the systemic blood was markedly increased by pretreatment of the animals with diethyl maleate, which decreases the glutathione concentration in tissues. Thus, the data suggest that bromobenzene-3,4-oxide may be classified as an intermediately-lived metabolite at low doses of bromobenzene, but may be classified as a long-lived metabolite after administration of large doses of bromobenzene to animals pretreated with diethyl maleate.

The relative importance of the various factors that may influence the pharmacokinetics of chemically reactive metabolites may differ enormously with the category of the chemically reactive metabolite. Indeed, a factor that may be dominant in the pharmacokinetics of an ultrashort-lived chemically reactive metabolite may be of trivial importance in the pharmacokinetics of an ultralong-lived metabolite. These differences may be illustrated by brief descriptions of what might be expected in each of the categories.

Ultrashort-lived metabolites. By definition this group of chemically reactive metabolites never leaves the active site of the enzyme that catalyzes their formation. Up to a point the equations that describes the kinetics for the formation of such metabolites will be similar to the Michaelis equation that describes the kinetics of any enzymatic reaction. Thus, we may visualize that the enzyme combines with the substrate to form a reversible complex. The reversible complex then undergoes rearrangement to form a transition complex of the chemically reactive metabolite and the enzyme. A portion of the transition complex then may undergo an irreversible rearrangement to form a covalently bound adduct of the chemically reactive metabolite to the enzyme; since the adduct is bound at the active site of the enzyme, the formation of the adducts usually inactivates the enzyme and thus precursors of ultrashort-lived chemically reactive metabolites have been called suicide enzyme inhibitors. More often than not, however, the chemically reactive metabolite in the transition complex also undergoes rearrangement to a stable product that dissociates from the enzyme. Indeed, the formation of the stable metabolite usually predominates and thus the rate constant of inactivation of the enzyme can be considerably smaller than the molar V_{max} of the enzyme (i.e., maximum moles of substrate metabolized per mole

of enzyme). Thus, we may illustrate the basic equation for a suicide
inhibitor by the following model:

E(active) + Substrate

$\Updownarrow$ k(p)

ES $\rightarrow$ EP(reactive) $\rightarrow$ EP(inactive)
                                  k(i)

              $\downarrow$ k(stable)
                  k

        EP(stable) $\rightarrow$ E(active) + P(stable)

The ratio of the rate constant of formation of the inactive form of the
enzyme to the rate constant of formation of the stable metabolite (i.e.,
k(i)/k(stable)) is frequently called the partition of the reactions. As
long as the conformation of the EP(reactive) complex is the same the
relative rates of formation of the inactive enzyme and the stable
metabolite will be constant regardless of the concentration of the
substrate. It follows from this concept that a given amount of enzyme
will catalyze the metabolism of a given amount of substrate regardless of
the initial concentration of the substrate; only the time required for
the enzyme to metabolize that amount of substrate will depend on the
initial concentration of the substrate.

Several kinds of _in vitro_ studies may be devised to elucidate
whether ultrashort-lived metabolites are formed. For example, as long as
the substrate concentration is maintained at a constant value, the
decrease in enzyme activity will follow pseudo-first order kinetics; that
is, a plot of the logarithm of the enzyme activity against time will give
a straight line (Walsh et al., 1978). The pseudo-first order rate
constant, however, will be rather complex, as shown below:

$$\text{Pseudo-first order rate constant} = \frac{[S]\ k(p)}{K_m + [S]} \frac{k(i)}{k(i) + k(stable)} \quad (7)$$

where K_m is the Michaelis constant for the formation of (ES) and k(p) is
the first order rate constant that describes the conversion of (ES) to
the transition chemically reactive metabolite complex. Since k(i) +
k(stable) = k(p), however, we may also write the equation as,

$$\text{Pseudo-first order rate constant} = \frac{[S]\ k(i)}{K_m + [S]} \quad (7a)$$

The slope of the line may be altered by changing the concentration of the substrate or by addition of competitive inhibitors of the enzyme. A plot of the reciprocal of the slopes versus the reciprocal of the substrate concentration thus should provide estimates of the rate constant of inactivation of the enzyme ($k(i)$) and the K_m of the enzyme. Once the inactive adduct of the enzyme is formed, however, the enzyme cannot be reactivated by addition of the competitive inhibitors. Moreover, the addition of nucleophiles that might ordinarily be expected to react nonenzymatically with the chemically reactive metabolite would probably not prevent inactivation of the enzyme, because they would not reach the metabolite on the active site of the enzyme.

The pharmacokinetics of the ultrashort-lived chemically reactive metabolites in vivo are more complicated for several reasons. The synthesis of the enzyme being inactivated may continue and thus the amount of enzyme in a given cell at any given time will depend on the relative rates at which the enzyme is being synthesized and inactivated during the course of treatment. Moreover, the parent compound may be eliminated from the body almost solely by the enzyme being inactivated or by a combination of mechanisms. Nevertheless, there may be certain phases of the reaction during which some of the parameters of the equation may be predominant and thus some simplified forms of the general equations may be useful. For example, immediately after the administration of the parent substance the rate of synthesis of the enzyme may be negligible in comparison to the rate of inactivation of the enzyme. When the substance is eliminated solely by the enzyme being inactivated, the concentration of the substance in the body, under these conditions, may be described by the usual equation for second order reactions. On the other hand, when the parent compound is being eliminated from the body predominantly by other mechanisms, the kinetics of elimination of the parent compound will be independent of the amount of the target enzyme, but the relationship between the pharmacokinetics of the parent compound and the effects on the target enzyme may still be evaluated by relatively simple relationships. A plot of the logarithm of the activity of the enzyme in a given cell versus the area under the curve of the parent substance within the cell from time zero to time t, should be a straight line, as long as the maximum concentration of the parent substance is lower than the K_m of the enzyme being inactivated; the slope of the line will be the slope of the line obtained under these conditions will be the second order rate constant, $k(i)/K_m$.

The amount of covalent binding of the chemically reactive metabolite to the target enzyme will be dependent on the amount of enzyme initially present in the cell as well as on the pharmacokinetic parameters of the parent compound that governs its area under the curve within the cell.

Short-lived metabolites. By definition, a short-lived metabolite leaves the enzyme that catalyzes its formation but does not leave the cell in which it is formed. Thus, the covalent binding will be restricted to cells which contain the enzyme that catalyzes their formation. It is, therefore, frequently possible to identify members of this category of chemically reactive metabolites by demonstrating that covalent binding occurs with several different intracellular components by chemical analysis but by showing that the covalent binding is restricted to certain types of cells within heterogeneous organs (such as the lung) by histological techniques based either on autoradiography or by the reaction of specific antibodies raised against adducts of the chemically reactive metabolites.

The basic pharmacokinetics of the covalent binding of members of this category of chemically reactive metabolites may be described by equations (1 through 6b), but the fraction of the dose of the parent compound that becomes covalently bound should be visualized for each cell that contains enzymes that catalyze the formation and inactivation of the chemically reactive metabolite. If a parent compound were to be eliminated from the body solely by one enzyme, an inducer that increased the activity of the enzyme to different extents in different cells could still affect the magnitude of the toxicity, because preferential increases in some of the cells would increase the fraction of the dose of the parent compound that was converted to the chemically reactive metabolite in those cells. Indeed, the finding that pretreatment of animals with an inducer causes preferential increases in covalently bound material in some organs but not in others frequently provides the first clue that this kind of chemically reactive metabolite is formed.

Intermediate-lived metabolites. By definition, members of this category of chemically reactive metabolites leave the cells in which they are formed but are inactivated before they reach the arterial blood. Such metabolites thus are not recirculated through the circulatory system. The factors that govern the pharmacokinetics of members of this category of chemically reactive metabolites can be rather complex, because this kind of chemically reactive metabolite leaves cells in which it is generated and may enter surrounding cells. It also may enter the blood within the organ, be swept downstream, and enter other cells distal to those in which it is formed. Thus, the distribution of the covalent binding of members of this category of chemically reactive metabolites may be affected not only by the location of cells containing enzymes that catalyze their formation but also by the blood flow rate through the organ and diffusional rates into cells. Clearly, members of this kind of chemically reactive metabolite cannot be used in histological tests to identify cells containing enzymes that catalyzed their formation. Indeed, the chemically reactive metabolites may become covalently bound to blood cells as they flow through the organ even when the metabolites cannot be detected in systemic blood. The finding of covalently bound material in blood cells in vivo that does not occur in isolated cells incubated with the substrate frequently may be used to distinguish this category from the short-lived category.

Long-lived metabolites. By definition, members of this category enter the systemic blood and are recirculated through the organ in which they are formed. They thus should be detectable in blood. Although it is not possible to provide an accurate lower limit of the half-life of such metabolites in cells of organs, a reasonably good estimate may be obtained from the fact that the blood volume within an organ is about 5% of the organ weight and the blood flow rates through the organ. For liver, therefore, we may calculate a half-life of about 5 seconds. Since the average circuit time of blood in humans is about 30 seconds and is considerably less in rats and mice, chemically reactive metabolites having half-lives in blood that are greater than a few seconds should be considered as long-lived metabolites.

Because long-lived metabolites are recirculated in the blood, the pharmacokinetics that describe the time course of covalent binding in any given cell within a given organ can be quite complex. Indeed, the pharmacokinetics equations must describe not only the covalent binding of the chemically reactive metabolite that was generated in the cell but also the covalent binding of the reactive metabolite generated in neighboring cells plus that generated in cells upstream in the same or different organs, plus that recirculated by way of the arterial blood. With the longer-lived members of this category, however, the proportion of the covalently bound metabolite within a given cell may be largely due to the recirculated metabolite and thus the complexities in the equation that describe the contribution made by covalent binding of the metabolite before it enters the systemic circulation may become negligible. Under these conditions, the cellular localization of the enzymes that generate the chemically reactive metabolite become largely unimportant, and the body may be described by a relatively few pharmacokinetic compartments. For this reason induction of the enzyme that catalyzes the formation of the chemically reactive metabolite may tend to increase the fraction of the dose that becomes converted to adducts in all cells of the body and not just those in which the metabolite is generated. Indeed, the finding that an inducer increases the amount of metabolite in the blood, or that it increases the amount of covalent binding of the metabolite in several organs to about the same extent, frequently provides the first clue that the chemically reactive metabolite might be a member of this category of chemically reactive metabolites.

Ultralong-lived metabolites. By definition, members of this class of chemically reactive metabolites are sufficiently long lived to be excreted into bile and urine. Because the concentrations of the chemically reactive metabolites are likely to be higher in these body fluids than in blood, any toxicity caused by them is likely to be predominantly in the urinary tract or in the intestinal tract. If the chemically reactive metabolite is excreted into bile, however, the metabolite may undergo enterohepatic circulation which would increase the complexity of the pharmacokinetic equations required to describe the time course of the metabolite in different organs.

FIRST PASS EFFECTS AND FIRST PASS ORGANS

A foreign compound rarely is injected directly into the arterial circulation. Thus, to enter the body by the usual routes of administration, the foreign compound must pass through the lung and frequently several other organs before it enters the arterial circulation. Those organs through which the foreign compound must pass on its way from the site of administration to the arterial circulation are frequently called first pass organs. It can be shown mathematically that the area under the curve of the parent foreign compound in the blood within a first pass organ must be at least slightly greater than the area under the curve of the parent compound in blood in the systemic arterial circulation, regardless of the organs that participate in the elimination of the parent foreign compound. Although the magnitude of the differences in the areas under the curves for blood within a first pass organ may depend on several different factors, some of the subtle aspects may be illustrated with a pharmacokinetic model of a foreign compound that is constantly infused into portal blood but is eliminated from the body solely by the kidney under steady-state conditions. Under these conditions, we may write an equation for the steady-state concentration of the parent compound in arterial blood as:

ARTERIAL BLOOD
$$D_{ss(ART)} = K_o/CL_r \tag{8}$$

Where,

$D_{ss(ART)}$ is the steady-state concentration of the foreign compound, K_o is the steady-state infusion rate, and CL_r is the renal clearance of the foreign compound.

However, the steady-state concentration of the foreign compound in the liver will depend not only on its steady-state concentration in the arterial blood as it passes into the portal and the hepatic blood entering the liver, but also on the increase in the concentration of the foreign compound due to the infusion of the foreign compound into the portal blood; the latter concentration may be calculated from the rate of the infusion and the total blood flow rate (Q_H) through the liver. Thus,

BLOOD WITHIN LIVER

$$D_{ss(H)} = K_o/Q + K_o/CL_r \tag{9}$$

The ratio of these two steady-state concentrations thus would be,

$$D_{ss(H)}/D_{ss(ART)} = 1 + CL_r/Q \tag{10}$$

It may be shown mathematically that the presence of an enzyme in the liver will not affect the ratio shown in equation (10). It also may be shown mathematically that the clearance of the foreign compound in a nonfirst pass organ served solely by arterial blood may be added to the renal clearance term in equation (10). Furthermore, it may be shown mathematically that in completely first order systems the ratio of the areas under the curves following a rapid injection of the foreign compound into the portal blood would also be described by equation (10). It is evident, therefore, that the effects of the route of administration on these values is governed by the elimination of the foreign compound in

the nonfirst pass organs and not on the activity of the enzyme in the
first pass organ. This is not to say, however, that increases in the
activity of an enzyme within the liver will not affect the fraction of
the dose that is converted to a chemically reactive metabolite, but is
rather an estimate of the difference in the effects caused by differences
in the route of administration.

It is difficult to predict the effects of the route of administration solely by measurements of the concentrations of the foreign compound
in systemic blood. Nevertheless, one can gain a perspective from the
realization that all of the increase must have taken place during the
absorption of the foreign compound from the site of administration.
Estimates of the rate constant of absorption together with estimates of
the blood flow rate through the first pass organ permits us to make
estimates of the concentration of the parent compound within the first
pass organ and therefore to estimate whether the concentration of the
foreign compound approaches the K_m values of enzymes within the organ.
Such estimates as these can provide insights into the possible effects of
nonlinear pharmacokinetics which would be difficult if not impossible to
obtain directly by experimental data. Many of the concepts summarized
here have been developed more extensively in another paper (Gillette,
1985).

REFERENCES

Galinski, R.E., and Levy, G., 1981, Dose- and time-dependent elimination
        of acetaminophen in rats: Pharmacokinetic implications of
        cosubstrate depletion, J. Pharmacol. Exp. Therap., 219:14.

Gillette, J.R., 1974a, A perspective on the role of chemically reactive
        metabolites of foreign compounds in toxicity--I. Correlation of
        changes in covalent binding of reactive metabolites with
        changes in the incidence and severity of toxicity. Biochem.
        Pharmacol. 23:2785.

Gillette, J.R., 1974b, A perspective on the role of chemically reactive
        metabolites of foreign compounds in toxicity--II. Alterations
        in the kinetics of covalent binding. Biochem. Pharmacol.
        23:2927.

Gillette, J.R., 1979, Effects of induction of cytochrome P-450 enzymes
        on the concentration of foreign compounds and their metabolites
        and on the toxicological effects of these compounds, Drug
        Metabol. Rev. 10:59.

Gillette, J.R., 1982, The problem of chemically reactive metabolites, Drug Metabol. Rev. 13:941.

Gillette, J.R., Lau, S.S., and Monks, T.J., 1984, Intra- and extra-cellular formation of metabolites from chemically reactive species, Biochem. Soc. Transactions 12:4.

Gillette, J.R., 1985, Pharmacokinetics of biological activation and inactivation, in: "Bioactivation of Foreign Compounds," M.V. Anders, ed., Academic Press, New York.

Hjelle, J.J., Hazelton, G.A., and Klaassen, C.D., 1985, Acetaminophen decreases adenosine-3'-phosphate-5'phosphosulfate and uridine diphosphoglucuronic acid in rat liver, Drug Metabol. Dispos. 13:35.

Lau, S.S., Monks, T.C., and Gillette, J.R. 1984, Detection and half-life of bromobenzene-3,4-oxide in blood, Xenobiotica 14:539.

Miller, E.C., and Miller, J.A., 1947, The presence and significance of bound aminoazo dyes in the livers of rats fed p-dimethylamino-azobenzene, Cancer Res. 7:468.

Miller, E.C., Miller, J.A., Sapp, R.W., and Weber, G.M., 1949, Studies on protein-bound aminoazo dyes formed in vivo from 4-dimethylaminoazobenzene and its C-monomethyl derivatives, Cancer Res. 9:336.

Monks, T.J., Lau, S.S., and Gillette, J.R., 1984, Diffusion of reactive metabolites out of hepatocytes: Studies with bromobenzene, J. Pharmacol. Exptl. Therap. 228:393.

Schumacher, H., Blake, D.A., and Gillette, J.R., 1968, Disposition of thalidomide in rabbits and rats, J. Pharmacol. Exptl. Therap. 160:201.

Walsh, C., Cromantie, T., Marcotte, P., and Spencer, R., 1978, Suicide substrates for flavoprotein enzymes, Methods in Enzymology 53:437.

STRUCTURE AND FUNCTION OF CYTOCHROME P-450

F. Peter Guengerich, Tsutomu Shimada, Diane R. Umbenhauer,
Martha V. Martin, Kunio S. Misono, Linda M. Distlerath, Paul
E.B. Reilly, and Thomas Wolff

Department of Biochemistry and Center in Molecular Toxicology
Vanderbilt University School of Medicine
Nashville, Tennessee 37232

INTRODUCTION

Cytochrome P-450 (P-450) is a hemoprotein that catalyzes the oxidation
of a wide variety of substrates, some of which are converted to reactive
products that can be attacked by cellular nucleophiles or solvolyzed
(Guengerich and Liebler, 1985). Much has been learned about the isozymes in
this family of hemoproteins from studies with microorganisms and experimental animals. Biochemical studies with humans were started shortly after
the discovery of P-450, and a number of catalytic activities have been
ascribed to P-450 in humans (Boobis and Davies, 1984). One of the first
efforts at the purification of human cytochrome P-450 was that of Kaschnitz
and Coon (1975), which was followed up by several other studies. Our own
laboratory has been involved in the purification of several isozymes of human
as well as rat liver P-450 (Wang et al., 1980, Wang et al., 1983; Guengerich
et al., 1982; Distlerath et al., 1985; Guengerich et al., 1985). In the
earlier efforts catalytic activity data were difficult to evaluate because
most of the assays were not specific to individual P-450 isozymes (Wang et
al., 1980, 1983). More recently we have focused our efforts on the
biochemical characterization of human P-450s which are involved in polymorphisms of drug metabolism. Such genetic polymorphisms, studied extensively by Smith and Idle and their associates (Ayesh et al., 1984), provide
an opportunity for directly relating biochemical studies to in vivo
situations and for focusing on the regulation of specific P-450 isozymes and

their catalytic activities. We report here the purification of four such human liver P-450s and studies on their catalytic specificity.

MATERIALS AND METHODS

Sources of Human Liver Microsomes

Liver samples were obtained from individuals who met accidental deaths and donated other tissues for transplant. Arrangements were handled through the Nashville Regional Organ Procurement Agency. Tissues were perfused and chilled within 15-30 min of death and frozen ($-70^\circ C$) in small pieces as described elsewhere (Wang et al., 1983).

Purification of P-450s

Microsomes were solubilized with sodium cholate and fractionated to obtain electrophoretically homogeneous $P-450_{DB}$, $P-450_{PA}$, $P-450_{MP}$, and $P-450_{NF}$ using combinations of n-octylamino Sepharose, DEAE-cellulose, hydroxylapatite, and CM-cellulose chromatography. The methods are described in detail elsewhere (Distlerath et al., 1985; Shimada et al., 1985; Guengerich et al., 1985). Purity was established using sodium dodecyl sulfate-polyacrylamide gel electrophoresis, isoelectric focusing, prosthetic group analysis, and various immunochemical criteria (Distlerath et al., 1985; Shimada et al., 1985; Guengerich et al., 1985; Guengerich et al., 1985a).

Reconstitution of Catalytic Activities

Purified P-450s were reconstituted with rabbit or rat NADPH-P-450 reductase and L-α-1,2-dilauroyl-sn-glycero-3-phosphocholine as described elsewhere (Guengerich et al., 1982). In some cases purified human liver cytochrome b_5 was added after other components were mixed (Shimada et al., 1985; Guengerich et al., 1985).

Immunochemical Studies

Antibodies were raised to purified P-450s in rabbits as described elsewhere (Guengerich et al., 1981). Specificity of antisera and estimates of concentrations of individual P-450s in microsomal samples were made using immunoblotting methods (Guengerich et al., 1982). Immunoinhibition studies were done using immunoglobulin G fractions and human liver microsomes as described (Distlerath et al., 1985; Wolff et al., 1985; Distlerath and Guengerich, 1984).

Table 1. Isolated Human Liver P-450s

P-450	Polymorphism	Homologous rat enzyme	References
$P-450_{DB}$	Debrisoquine, sparteine, etc.	$P-450_{UT-H}$	Larrey et al., 1984; Distlerath and Guengerich, 1984; Distlerath et al., 1985; Gut et al., 1984
$P-450_{PA}$	Phenacetin	Unknown	Distlerath et al., 1985
$P-450_{MP}$	Mephenytoin	$P-450_{UT-I}$	Shimada and Guengerich, 1985; Shimada et al., 1985
$P-450_{NF}$	Nifedipine	$P-450_{PCN-E}$	Guengerich et al., 1985
$P-450p$	?	$P-450_{PCN-E}$	Watkins et al., 1985
$P-450_9$	?	?	Beaune et al., 1985

RESULTS AND DISCUSSION

Human liver P-450s which have been purified and partially characterized are listed in Table 1. Those referenced from our own laboratory are involved in polymorphisms and their substrate specificity will be discussed in this article. $P-450_9$ and P-450p have no known substrates yet. Portions of human genes related to mouse P_1-450 (Jaiswal et al., 1985), rabbit P-450-1 and P-450-6 (Pendurthi et al., 1985), rat P-450e (Phillips et al., 1985), and beef adrenal $P-450_{21\alpha}$ (White et al., 1984) have been isolated and characterized but the existence of translated human liver proteins has not been demonstrated.

$P-450_{DB}$, $P-450_{PA}$, $P-450_{MP}$, and $P-450_{NF}$ were isolated from human liver samples using techniques of protein chromatography. Electrophoretograms are shown in Fig. 1. In various instances further evidence for homogeneity was provided by isoelectric focusing, prosthetic group analysis, and immuno-chemical methods. The enzymes were purified on the basis of catalytic

activity towards the drug involved in the polymorphism under consideration
and the final preparations exhibit catalytic activity. Further evidence
that the purified proteins are related to the polymorphisms under consideration comes from studies in which specific antibodies prepared to each protein
inhibit most of the respective catalytic activity in human liver microsomes.

$P-450_{DB}$ and $P-450_{PA}$ appear to be relatively minor forms of P-450 as
judged by their abundance. However, $P-450_{MP}$ and $P-450_{NF}$ each appear to
account for on the order of 20% of total microsomal P-450 in human liver,
depending upon the individual. Fetal liver contains $P-450_{NF}$ but not $P-450_{MP}$.
$P-450_{NF}$ may be inducible by dexamethasone but more extensive studies will be
necessary to confirm this hypothesis. In the case of $P-450_{NF}$ we have been
able to correlate catalytic activity with the amount of immunochemically-
detectable $P-450_{NF}$ ($r = 0.78$, $n = 32$, $p < 0.005$). In the cases of $P-450_{DB}$,
$P-450_{PA}$, and $P-450_{MP}$ we found no correlation, suggesting that perhaps minor

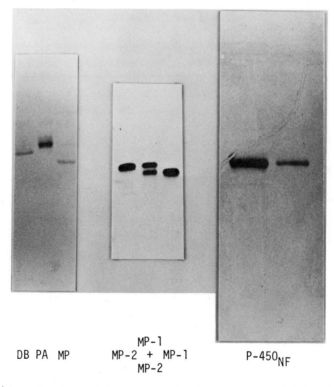

DB PA MP   MP-1
            MP-2 + MP-1 $P-450_{NF}$
                MP-2

Fig. 1.   Sodium dodecyl sulfate polyacrylamide gel electrophoresis of human
          liver P-450s. From left to right: $P-450_{DB}$ (DB), $P-450_{PA}$ (PA),
          $P-450_{MP-1}$ (MP), $P-450_{MP-2}$ (MP-2), mixture of $P-450_{MP-1}$ and
          $P-450_{MP-2}$ (MP-1 + MP-2), $P-450_{MP-1}$ (MP-1), and $P-450_{NF}$ (last two
          lanes).

alterations in the structural genes might be responsible for the "poor metabolizer" phenotypes. $P-450_{MP}$ exists as two molecular weight variants, $P-450_{MP-1}$ and $P-450_{MP-2}$ (Fig. 1). These two proteins cannot be distinguished by catalytic activity, peptide mapping, spectral properties, or any of several immunochemical techniques; however in vitro translation experiments indicate that they are translated from different mRNAs. The existence of the two proteins does not account for the polymorphic variation in catalytic activity.

Table 2 lists the catalytic activities assigned to the individual human liver P-450s, made on the basis of reconstitution, immunoinhibition, and correlation studies. Activities which have been examined and judged not to be specifically catalyzed by each of these P-450s are also listed.

Table 2. Apparent Substrate Specificity of Human P-450s

P-450	Activities	Activities shown not to be associated with each P-450
$P-450_{DB}$ [a]	Debrisoquine 4-hydroxylation	S-Mephenytoin 4-hydroxylation
	Sparteine Δ^5-oxidation	Nifedipine oxidation
	Bufuralol 1'-hydroxylation	Phenacetin O-deethylation
	Encainide O-demethylation	7-Ethoxycoumarin O-deethylation
	Propranolol 4-hydroxylation	Acetanilide 4-hydroxylation
	Lasiocarpine oxidation	Aniline 4-hydroxylation
	Monocrotaline oxidation	Diazepam N-demethylation
		Benzo(a)pyrene hydroxylation
		Benzo(a)pyrene-7,8-dihydrodiol 9,10-epoxidation
		N,N-Dimethylnitrosamine N-demethylation
		Trichloroethylene oxidation
		Vinylidene chloride oxidation
		Aflatoxin B_1 2,3-epoxidation
		Aflatoxin B_1 8-hydroxylation
		Ethylmorphine N-demethylation
		Morphine N-demethylation
		d-Benzphetamine N-demethylation
		Aminopyrine N-demethylation
		1-Naphthylamine 2-hydroxylation
		2-Naphthylamine N-hydroxylation

87

Table 2 —— Continued

P-450	Activities	Activities shown not to be associated with each P-450
		2-Naphthylamine 1-hydroxylation
		2-Naphthylamine 6-hydroxylation
		2-Aminofluorene N-hydroxylation
		2-Aminofluorene 5-hydroxylation
		Azoprocarbazine N-oxidation
P-450$_{PA}$ [a]	Phenacetin O-deethylation	Debrisoquine 4-hydroxylation
		S-Mephenytoin 4-hydroxylation
		Nifedipine oxidation
P-450$_{MP}$ [b]	S-Mephenytoin 4-hydroxylation	R-Mephenytoin 4-hydroxylation
	S-Mephenytoin N-demethylation	R-Mephenytoin N-demethylation
	S-Nirvanol 4-hydroxylation	R-Nirvanol 4-hydroxylation
	Diphenylhydantoin (mephenytoin) 4-hydroxylation	N-Methyldiphenylhydantoin 4-hydroxylation
		Diazepam N-demethylation
		Nifedipine oxidation
		d-Benzphetamine N-demethylation
		4-Nitroanisole O-demethylation
		Benzo(a)pyrene hydroxylation
		Bufuralol 1'-hydroxylation
		Phenacetin O-deethylation
P-450$_{NF}$ [c]	Nifedipine oxidation	S-Mephenytoin 4-hydroxylation
	Testosterone 6β-hydroxylation	Phenacetin O-deethylation
	Aldrin epoxidation	Bufuralol 1'-hydroxylation
	17β Estradiol 2-hydroxylation	
	17β Estradiol 4-hydroxylation	

[a]Wolff et al., 1985; Distlerath et al., 1985; Distlerath and Guengerich, 1984; Boobis et al., 1985.

[b]Shimada et al., 1985.

[c]Guengerich et al., 1985.

P-450$_{DB}$ catalyzes the oxidation of compounds which contain a basic nitrogen atom. Comparison of the substrates listed in Table 2 and the drugs ascribed to this polymorphism family in vivo has led to the development of a working model for the active site of P-450$_{DB}$ (Wolff et al., 1985). The putative perferryl oxygen species is located near the site of oxidation, which is about 5 Å away from the basic nitrogen, thought to be positioned near a carboxylic acid group on the protein (Fig. 2). A hydrophobic region is postulated to be near the heme, as many substrates are hydroxylated at benzylic carbons or contain a stretch of methylenes near the hydroxylation site.

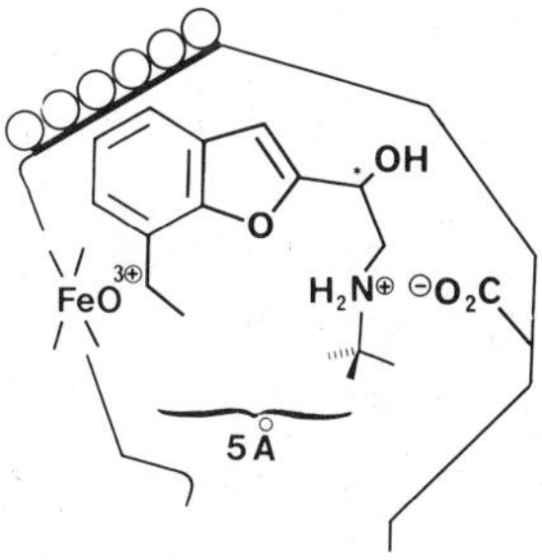

Fig. 2.   Model for the substrate binding site of P-450$_{DB}$.

To date we have not thoroughly investigated the substrate specificity of P-450$_{PA}$. Phenacetin O-deethylation and debrisoquine 4-hydroxylation are correlated in different individuals (Boobis and Davies, 1984) but biochemical studies clearly show that these activities are catalyzed by different P-450s.

Table 3. Inhibition of Catalytic Activities in Human Liver Microsomes by Anti-P-450$_{MP}$

Reaction	Maximum Percentage Inhibition[a]
R,S-Mephenytoin 4-hydroxylation	90
S-Mephenytoin 4-hydroxylation	98
R,S-Mephenytoin N-demethylation	30
S-Nirvanol 4-hydroxylation	97
d-Benzphetamine N-demethylation	15
Benzo(a)pyrene hydroxylation	12
Diphenylhydantoin 4-hydroxylation	96
(±)Bufuralol 1'-hydroxylation	5
Phenacetin O-deethylation	12
Nifedipine oxidation	10
Diazepam N-demethylation	20

[a]The amount of rabbit anti-P-450$_{MP}$ (immunoglobulin G fraction) was varied from 0 to at least 2 mg/nmol microsomal P-450.

P-450$_{MP}$ catalyzes the stereoselective hydroxylation of substituted hydantoins. Other substrates have not been identified. The catalytic activity of purified P-450$_{MP}$ is enhanced by the addition of purified rat or human cytochrome b_5 and antibodies raised to human liver cytochrome b_5 inhibit P-450$_{MP}$ activity in human liver microsomes. Results of some of the immunochemical studies with P-450$_{MP}$ are summarized in Table 3. A working model for the catalytic site of P-450$_{MP}$ is shown in Fig. 3: it involves 2 or 3 adjacent pockets. Pocket B must have groups participating in specific hydrogen bonding with sites on the hydantoin ring to explain the specificity for 4-hydroxylation of the S-isomer (left panel). The N-demethylation can be explained by one of several possibilities: a) Pocket A may have specific hydrogen bonding sites which facilitate N-demethylation. b) Hydrophobic interaction in pocket B or another pocket C may exist to facilitate placement of the hydantoin ring in pocket A and N-demethylation. Since phenytoin (diphenylhydantoin) is hydroxylated at the (pro-S) 4-position we favor a model with hydrophobic interaction in the A and C pockets (and hydrogen bonds in pocket B). This model must, however, be considered rather speculative in the absence of more information about substrate specificity.

Table 4. Correlation of Polymorphic Activities in Human Liver Microsomes

	Correlation coefficient (r) (n in parentheses)		
	Phenacetin O-deethylation	S-Mephenytoin 4-hydroxylation	Nifedipine oxidation
Debrisoquine 4-hydroxylation	0.62 (43)[a]	0.24 (20)	0.04 (22)
Phenacetin O-deethylation		0.11 (20)	0.27 (13)
S-Mephenytoin 4-hydroxylation			-0.06 (21)

[a]Statistically significant at the $p < 0.025$ confidence level. No other values are significant at the $p < 0.10$ level.

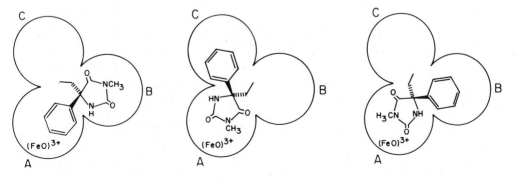

Fig. 3. Model for the substrate binding site of P-450$_{MP}$.

P-450$_{NF}$ oxidizes the dihydropyridine calcium channel blocker nifedipine. Kleinbloesem et al. (1984) reported that one-sixth of the individuals in their study were phenotypic poor metabolizers. The same studies used with P-450$_{MP}$ indicate that cytochrome $\underline{b}_5$ plays a role in the catalytic activity of P-450$_{NF}$ as well. Aldrin epoxidation was inhibited by anti-P-450$_{NF}$ in two of the three individual liver microsomal samples investigated. P-450$_{NF}$ also appears to be the major enzyme involved in the hydroxylation of two endogenous steroids, testosterone and 17β estradiol.

The reconstitution and immunochemical studies also indicated that all of the four polymorphisms were distinct from each other (Table 4). These results are in accord with the lack of correlation between catalytic activities measured in microsomes isolated from different individuals (except in the case of debrisoquine and phenacetin).

REFERENCES

Ayesh, R., Idle, J.R., Ritchie, J.C., Crothers, M.J., and Hetzel, M.R., 1984, Metabolic oxidation phenotypes as markers for susceptibility to lung cancer, Nature (London), 312:169.

Beaune, P., Flinois, J-P., Kiffel, L., Kremers, P., and Le Roux, J-P., 1985, Purification of a new cytochrome P-450 from human liver microsomes, Biochim. Biophys. Acta, in press.

Boobis, A.R., and Davies, D.S., 1984, Human cytochromes P-450, Xenobiotica, 14:151.

Boobis, A.R., Murray, S., Hampden, C.E., and Davies, D.S., 1985, Genetic polymorphism in drug oxidation: in vitro studies of human debrisoquine 4-hydroxylase and bufuralol 1'-hydroxylase activities, Biochem. Pharmacol., 34:65.

Distlerath, L.M., and Guengerich, F.P., 1984, Characterization of a human liver cytochrome P-450 involved in the oxidation of debrisoquine and other drugs by using antibodies raised to the analogous rat enzyme, Proc. Natl. Acad. Sci. U.S.A., 81:7348.

Distlerath, L.M., Reilly, P.E.B., Martin, M.V., Davis, G.G., Wilkinson, G.R., and Guengerich, F.P., 1985, Purification and characterization of the human liver cytochromes P-450 involved in debrisoquine 4-hydroxylation and phenacetin O-deethylation, two prototypes for genetic polymorphism in oxidative drug metabolism, J. Biol. Chem., in press.

Guengerich, F.P., Dannan, G.A., Wright, S.T., Martin, M.V., and Kaminsky, L.S., 1982, Purification and characterization of rat liver microsomal cytochromes P-450: electrophoretic, spectral, catalytic, and immunochemical properties and inducibility of eight isozymes isolated from rats treated with phenobarbital or β-naphthoflavone, Biochemistry, 21:6019.

Guengerich, F.P., Distlerath, L.M., Reilly, P.E.B., Wolff, T., Shimada, T., Umbenhauer, D.R., and Martin, M.V., 1985, Human liver cytochromes P-450 involved in polymorphism of drug oxidations, Xenobiotica, in press.

Guengerich, F.P., and Liebler, D.C., 1985, Drug-metabolizing enzymes and toxicity, CRC Crit. Rev. Toxicol., in press.

Guengerich, F.P., Martin, M.V., Waxman, D.J., Beaune, P.H., Kremers, P., and Wolff, T., 1985a, in preparation.

Guengerich, F.P., Wang, P., Mason, P.S., and Mitchell, M.B., 1981, Immunological comparison of rat, rabbit, and human microsomal cytochromes P-450, Biochemistry, 20:2370.

Gut, J., Gasser, R., Dayer, P., Kronbach, T., Catin, T., and Meyer, U.A., 1984, Debrisoquine-type polymorphism of drug oxidation: purification from human liver of a cytochrome P-450 isozyme with high activity for bufuralol hydroxylation, FEBS Letters, 173:287.

Jaiswal, A.K., Gonzalez, F.J., and Nebert, D.W., 1985, Human dioxin-inducible cytochrome P_1-450: complementary DNA and amino acid sequences, Science, 228:80.

Kaschnitz, R.M., and Coon, M.J., 1975, Drug and fatty acid hydroxylation by solubilized microsomal cytochrome P-450--phospholipid requirement, Biochem. Pharmacol., 24:295.

Kleinbloesem, C.H., van Brummelen, P., Faber, H., Danhof, M., Vermeulen, N.P.E., and Breimer, D.D., 1984, Variability in nifedipine pharmaco-kinetics and dynamics: a new oxidation polymorphism in man, Biochem. Pharmacol., 33:3721.

Larrey, D., Distlerath, L.M., Dannan, G.A., Wilkinson, G.R., and Guengerich, F.P., 1984, Purification and characterization of the rat liver microsomal cytochrome P-450 involved in the 4-hydroxylation of debrisoquine, a prototype for genetic variation in oxidative drug metabolism, Biochemistry, 23:2787.

Pendurthi, U.R., Saxe, D.A., Okino, S.T., and Tukey, R.H., 1985, Chromosomal localization of a human P-450 gene and characterization of cDNAs that hybridize to rabbit P-450 1 and P-450 6, Fed. Proc., 44:1446.

Phillips, I.R., Shepard, E.A., Ashworth, A., and Rabin, B.R., 1985, Isolation and sequence of a human cytochrome P-450 clone, Proc. Natl. Acad. Sci. U.S.A., 82:983.

Shimada, T., and Guengerich, F.P., 1985, Participation of a rat liver cytochrome P-450 induced by pregnenolone 16α-carbonitrile and other compounds in the 4-hydroxylation of mephenytoin, <u>Mol. Pharmacol.</u>, in press.

Shimada, T., Misono, K.S., and Guengerich, F.P., 1985, submitted for publication.

Wang, P.P., Beaune, P., Kaminsky, L.S., Dannan, G.A., Kadlubar, F.F., Larrey, D., and Guengerich, F.P., 1983, Purification and characterization of six cytochrome P-450 isozymes from human liver microsomes, <u>Biochemistry</u>, 22:5375-5383.

Wang, P., Mason, P.S., and Guengerich, F.P., 1980, Purification of human liver cytochrome P-450 and comparison to the enzyme isolated from rat liver, <u>Arch. Biochem. Biophys.</u>, 199:206.

Watkins, P.B., Maurel, P., Schuetz, E.G., Wrighton, S.A., Parker, G., and Guzelian, P.S., 1985, Identification of a human cytochrome P-450 isozyme inducible by glucocorticoids and macrolide antibodies, <u>Fed. Proc.</u>, 44:1206.

White, P.C., New, M.I., and DuPont, B., 1984, <u>HLA</u>-linked congenital adrenal hyperplasia results from a defective gene encoding a cytochrome P-450 specific for steroid 21-hydroxylation, <u>Proc. Natl. Acad. Sci. U.S.A.</u>, 81:7505.

Wolff, T., Distlerath, L.M., Worthington, M.T., Groopman, J.D., Hammons, G.J., Kadlubar, F.F., Prough, R.A., Martin, M.V., and Guengerich, F.P., 1985, Substrate specificity of human liver cytochrome P-450 debrisoquine 4-hydroxylase probed using immunochemical inhibition and chemical modeling, <u>Cancer Res.</u>, 45:2116.

ANTIBODIES AS PROBES OF CYTOCHROME P450 ISOZYMES

Paul E. Thomas, Linda M. Reik, Sarah L. Maines,
Stelvio Bandiera, Dene E. Ryan and Wayne Levin

Laboratory of Experimental Carcinogenesis and Metabolism
Roche Institute of Molecular Biology, Nutley, NJ 07110

INTRODUCTION

Cytochrome P450*, an integral membrane protein, is widely distributed in many mammalian tissues, but is present in the highest concentration in hepatic endoplasmic reticulum. It functions as the terminal oxidase of an electron transport system that is involved in the metabolism of a large number of xenobiotics as well as endogenous substrates such as steroids, bile acids, fatty acids and prostaglandins (1). Although this enzyme system formerly was believed to function principally in detoxification, it is now known to metabolically activate many compounds to reactive metabolites that initiate toxic and carcinogenic events (2). Even more interesting is the observation that only certain optical isomers of bay-region diol epoxides of polycyclic aromatic hydrocarbons are carcinogenic, and that the cytochrome P450-dependent mixed function oxidase system preferentially catalyzes the formation of the optical isomer with the highest carcinogenic activity (3).

In this paper we will explore some of the properties of 10 cytochrome P450 isozymes which we have purified from rat liver microsomes. The use of polyclonal antibodies and MAb** directed against these isozymes as probes of their structural properties as well as the use of antibodies for quantitation of these isozymes will be emphasized. Knowledge of the properties, regulation and distribution of the individual isozymes will contribute in fundamental ways to our understanding of the toxicity of xenobiotics and important physiological functions of this enzyme system.

PROPERTIES OF HEPATIC CYTOCHROMES P450

Isolation and characterization studies have shown that the seemingly infinite substrate capacity of this monooxygenase system is not only the result of a multiplicity of distinct molecular forms of this enzyme but

---

* The term cytochrome P450 is used to refer to any or all forms of liver microsomal cytochrome P450. We designate these hemoproteins in the order that they are purified, since a nomenclature for the various forms of cytochrome P450 has not been established.

** Abbreviations: MAb, monoclonal antibody/antibodies; SDS, sodium dodecyl sulfate; ELISA, enzyme-linked immunosorbent assay; TCDD, 2,3,7,8-tetrachlorodibenzo-p-dioxin.

Table 1.   Properties of Purified Rat Hepatic Cytochromes P450a-P450j

Cytochrome	Minimum Mr	λmax of Fe^{++}-CO complex (nm)	Benzo[a]pyrene 3- and 9- hydroxylation	Benzphetamine N-demethylation	Testosterone hydroxylation and product formed[1] (nmol product/min/nmol P450)								
					1α	2α	6α	6β	7α	15α	16α	16β	A[2]
P450a	48K	452	0.2	2.3	-	-	1.0	-	20.9	-	-	-	-
P450b	52K	450	0.4	133	-	-	-	-	-	-	9.1	7.2	10.8
P450c	56K	447	23.4	6.7	-	-	-	1.9	-	-	-	-	-
P450d	52K	447	0.3	3.9	-	-	-	0.7	-	-	-	-	-
P450e	52.5K	451	0.1	19.8	-	-	-	-	-	-	0.8	0.7	1.1
P450f	51K	448	< 0.1	1.3	-	-	-	-	-	-	0.8	-	-
P450g	50K	448	< 0.1	4.9	-	-	-	3.8	-	-	0.3	-	-
P450h	51K	451	1.8	52.1	-	7.3	-	-	-	0.9	7.9	-	2.8
P450i	50.5K	449	< 0.1	2.8	0.3	-	-	-	-	0.3	-	-	-
P450j	51.5K	451.5	< 0.1	5.5	-	-	-	-	-	-	-	-	-

Minimum Mr was determined in SDS-gels. Catalytic activity is expressed as nmol product formed/min/nmol of cytochrome P450 determined under conditions in which metabolism was proportional to hemoprotein concentration and time of incubation. Saturating NADPH-cytochrome c reductase and optimal dilauroylphosphatidylcholine concentrations were used in all experiments. Data listed above were from previously published reports (4-6,16).

1 Dashes indicate that no significant amount of product was observed (i.e. <0.15 nmol/min/nmol cytochrome P450)

2 "A" represents androstenedione formed from testosterone.

is also due to the broad and often overlapping substrate selectivities of the individual cytochromes. While many forms of cytochrome P450 have been purified and characterized, the total number and the diversity of their structures and functions are only beginning to be clarified. Our laboratory (4-6) has purified 10 forms of rat hepatic microsomal cytochrome P450 (P450a-P450j). Since these 10 forms of cytochrome P450 differ in their partial amino acid sequences (4,6-8), they are considered to be isozymes arising as products of distinct mRNAs. These isozymes have also been compared and contrasted with respect to their minimum Mr in SDS-gels, spectral properties, peptide maps and substrate selectivities (4-6). Selected properties of rat hepatic cytochromes P450a-P450j are given in Table 1. It is clear from the data in Table 1 that no single parameter can be used to assay unambiguously for the presence of each cytochrome P450 isozyme in liver microsomes where as many as 15 or more isozymes may exist (4-6,9-15). For example, in SDS-gels, cytochromes P450b and P450d have the same minimum Mr (52K) as do cytochromes P450f and P450h (51K). The rates of N-demethylation of benzphetamine and hydroxylation of benzo[a]pyrene differ among selected isozymes by 100-fold or more, but neither assay is a specific measure of the activity of a single isozyme (Table 1). The profile of hydroxylated metabolites of testosterone formed by the 10 isozymes is unique for several, but not all, of the isozymes (5,6,16). Moreover, there are other isozymes (10-14) whose properties overlap those listed in Table 1.

POLYCLONAL ANTIBODY SPECIFICITY

The high specificity of antibodies makes them increasingly useful as probes of the function and structure of different isozymes of cytochrome P450. However, with regard to antibody specificity, one point needs emphasis since it is often disregarded. The extent to which the specificity of an antibody can be relied upon is only as good as the extent to which the specificity has been appropriately evaluated. All too often, investigators assume that an antibody made against an apparently pure protein will react exclusively with that antigen. Two fallacies are apparent in such reasoning. The most obvious problem is protein purity; minor contaminants may elicit a disproportionately greater amount of antibody response than their relative concentration would warrant if they are significantly more antigenic than the immunogen. Secondly, one or more antigenic determinants of the protein used as immunogen may occur on other proteins in the tissue under study.

A useful approach to evaluate antibody specificity is to screen each antibody for reaction with a number of purified cytochrome P450 isozymes and other microsomal proteins by ELISA and immunodiffusion analysis as shown in Table 2. One distinct advantage of such a screen is that cross-reactions can be detected for proteins which comigrate on SDS-gels (e.g. cytochromes P450f and P450h) or resolve poorly on SDS-gels (e.g. cytochromes P450b and P450e; cytochromes P450g and P450i). In addition, cross-reactions are tested under optimal conditions in which all of the cytochrome P450 isozymes are assayed at the same concentration. This is an important approach, since there is no known liver microsomal sample that contains comparable levels of all known cytochrome P450 isozymes. Until all cytochrome P450 isozymes are purified and available, it is equally important to test antibody specificity with liver microsomes (e.g. immunoaffinity chromatography and "Western blots") to determine antibody recognition of cytochrome P450 isozymes not yet purified.

When originally prepared, antibody against cytochrome P450a was not monospecific, possibly due to either recognition of an immunochemically related protein in hepatic microsomes or the presence of a minor contaminant in the immunogen. The antibody was made monospecific by

Table 2. Collation of the Reactivity of Ten Cytochrome P450 Isozymes, NADPH-Cytochrome P450 Reductase and Epoxide Hydrolase with Eleven Polyclonal Antibodies[1]

Polyclonal Antibodies	Microsomal Enzymes											
	Cytochromes P450										Other[2]	
	a	b	c	d	e	f	g	h	i	j	Red.	EH
Anti-P450a	+++	-	-	-	-	-	-	-	-	-	-	-
Anti-P450b	-	+++	-	-	+++	+	-	-	±	-	-	-
Anti-P450c	-	-	+++	++	-	-	-	-	-	-	-	-
Anti-P450c(-d)[3]	-	-	+++	-	-	-	-	-	-	-	-	-
Anti-P450d	-	-	++	+++	-	-	-	-	-	-	-	-
Anti-P450d(-c)[3]	-	-	-	+++	-	-	-	-	-	-	-	-
Anti-P450e	-	+++	-	-	+++	+	-	-	±	-	-	-
Anti-P450f	-	+	-	-	+	+++	++	++	++	-	-	-
Anti-P450g	-	-	-	-	-	++	+++	+	++	-	-	-
Anti-Red.	-	-	-	-	-	-	-	-	-	-	+++	-
Anti-EH	-	-	-	-	-	-	-	-	-	-	-	+++

[1] The extent of cross-reaction was determined in agarose double diffusion analysis and in ELISA and is indicated with "+++" for the strongest reaction and "-" for no detectable reaction.

[2] Other microsomal enzymes tested were NADPH-cytochrome P450 reductase (Red.) and epoxide hydrolase (EH).

[3] These antibody preparations were made monospecific for the antigen by adsorption against the cytochrome P450 isozyme listed in parentheses.

absorption against a partially purified protein preparation obtained from rat liver microsomes that contained the cross-reacting protein (17). From the results in Table 2, it is clear that anti-P450a, after absorption, does not recognize the other 9 cytochrome P450 isozymes, NADPH-cytochrome P450 reductase or epoxide hydrolase. Also, when detergent-solubilized rat liver microsomes were chromatographed on an anti-P450a immunoaffinity column, only cytochrome P450a was recognized by the antibody (17). Likewise, anti-P450b, anti-P450c and anti-P450d have also been absorbed against appropriate partially purified preparations of microsomal proteins to obtain monospecific antibodies (17-19). This absorption process might obscure bona fide immunochemical relationsips by removing antibody to common determinants on other proteins. We were primarily concerned, however, with obtaining high specificity antibody for immunochemical quantitation studies and inhibition of catalysis rather than establishing immunochemical relationships. As will be discussed, MAb are inherently superior reagents for establishing immunochemical relationships among proteins although they are not necessarily any more monospecific than appropriately absorbed polyclonal antibodies.

In Ouchterlony immunodiffusion analysis, anti-P450b and anti-P450e recognize cytochromes P450b and P450e as being immunochemically identical (4). The molecular basis for this immunochemical identity is now understood since these 2 isozymes have recently been shown to differ in their amino acid sequences by only 13 out of 491 amino acids (7,20). As a consequence of this immunochemical identity, the immunoquantitation of cytochrome P450b necessarily includes cytochrome P450e (21) unless these cytochromes are first resolved. Unfortunately, only under optimal conditions can cytochromes P450b and P450e be resolved in "Western blots" of SDS-gels. Anti-P450b and anti-P450e cross-react slightly with cytochrome P450f and even less with cytochrome P450i when tested in immunodiffusion plates devoid of detergent as well as in a sensitive ELISA. In mixing experiments, we found no evidence that the cross-reaction of cytochrome P450f with anti-P450b influences the immunoquantitation of cytochromes P450b+P450e, probably because the cross-reaction is weak and is not detected in Ouchterlony plates containing detergents (5). The weak cross-reaction of anti-P450f with both cytochromes P450b and P450e provides additional evidence that one or more regions of amino acid sequence homology exist among cytochromes P450f, P450b and P450e.

In immunodiffusion studies, anti-P450f cross-reacts moderately with cytochromes P450g, P450h and P450i and weakly with cytochromes P450b and P450e giving a reaction of partial identity with cytochrome P450f (22). Cytochromes P450f, P450h and P450i react with anti-P450g to varying extents and show partial identity to cytochrome P450g in immunodiffusion plates. From these results, it is apparent that cytochromes P450f-P450i share some, but not all, epitopes. Therefore, they must have substantial regions of amino acid sequence homology. Based on the data in Table 2, we recently proposed that cytochromes P450f, P450g, P450h and P450i are members of a distinct family of immunochemically related isozymes which are more related to one another than they are to cytochromes P450a-P450e (22). Amino-terminal sequence analysis (8) has shown that there is more sequence homology in the first 15 amino acids of cytochromes P450f-P450i (47%-80%) than there is among any of the other isozymes listed in Table 2 with the exception of cytochromes P450b and P450e which have identical NH_2-terminal sequences for the first 302 amino acids (7,20). It is interesting that all 4 of these isozymes (cytochromes P450f-P450i) occur in significant amounts in untreated adult rat livers, since they have been purified from this source (5), whereas cytochromes P450b-P450e occur in very low concentrations in untreated animals (21).

Antibody prepared against cytochrome P450c and absorbed as described by Thomas et al. (21) cross-reacts with cytochrome P450d, but does not recognize any of the other 8 isozymes. Cytochrome P450d shows immunochemical partial identity with cytochrome P450c in Ouchterlony double diffusion plates when both proteins are allowed to react with anti-P450c (19,21). Similarly, cytochrome P450c shows immunochemical partial identity with cytochrome P450d when both proteins are allowed to react with anti-P450d. Both of these antibodies can be absorbed with solid phase preparations of the heterologous proteins so that they become monospecific for their respective antigens (19). The absorbed antibodies are designated "anti-P450c(-d)" and "anti-P450d(-c)" to indicate that they no longer recognize the heterologous protein (Table 2). The specificity of these absorbed antibodies has been further demonstrated by immunoaffinity chromatography which showed that no other microsomal proteins were recognized by these antibodies (19).

A thorough knowledge of antibody specificity is necessary for the correct interpretation of experimental results. For example, when anti-P450c was used to immunoprecipitate _in_ _vitro_ translation products programmed with poly(A)$^+$ mRNA from 3-methylcholanthrene-treated rats, the immunoprecipitate contained both cytochromes P450c and P450d. In contrast, when anti-P450c(-d) or anti-P450d(-c) was used in the same type of experiments, only cytochrome P450c or cytochrome P450d, respectively, was precipitated (23). Moreover, anti-P450c and anti-P450d are potent inhibitors of metabolism catalyzed by the heterologous protein, whereas after absorption they do not inhibit metabolism by the heterologous isozyme (19). These data emphasize the importance of properly absorbing and fully characterizing antibodies before they are used as specific reagents.

MONOCLONAL ANTIBODY PROBES OF CYTOCHROMES P450

We have prepared MAb against cytochromes P450c (24), P450b (25) and P450f to enable more detailed studies of the properties of these cytochromes P450 (Table 3). Since a MAb is directed against a single epitope, it is far superior to polyclonal antibodies for detailed structural mapping of the functions and homologies among the cytochrome P450 isozymes. Nine distinct MAb were made against cytochrome P450c, and each was purified from mouse ascites fluid. As anticipated from the immunochemical relationship of cytochrome P450c with cytochrome P450d, some of the MAb cross-react with cytochrome P450d (CD2, CD3 and CD5; see Table 3 legend for explanation of MAb terminology). These 3 cross-reactive MAb recognize the same epitope, or different epitopes that are not spatially distinct, since they compete with one another for binding to either cytochrome P450c or P450d. The remaining 6 MAb recognize 5 different epitopes on cytochrome P450c, and 4 of these epitopes are spatially distinct (24). Except for the epitope which cytochromes P450c and P450d share, these 6 epitopes do not occur on the other 8 isozymes tested (Table 3). Moreover, we used each MAb to probe "Western blots" of hepatic microsomes from untreated rats or rats treated with 3-methylcholanthrene and found no evidence for the recognition of poly-peptides other than cytochromes P450c and P450d in the cytochrome P450 region (Mr = 45K - 60K).

The next group of 12 MAb shown in Table 3 were prepared against rat cytochrome P450b (25). When these 12 MAb were used in a competitive ELISA for determination of spatially distinct epitopes on cytochrome P450b, we found that they recognize 6 spatially distinct epitopes. Since 3 of these epitopes could be further subdivided on the basis of differential MAb reactivity with various isozymes, there must be a total of 9 epitopes delineated by the 12 MAb. Since polyclonal anti-P450b or anti-P450e react identically with cytochromes P450b and P450e, it is not surprising that most of the MAb cross-react well with cytochrome P450e, especially since these isozymes have 97% sequence homology. In view of this high sequence homology, it is surprising that 2 of the MAb (B50 and B51) directed against spatially distinct epitopes show no detectable cross-reaction with cytochrome P450e in an ELISA. Even more remarkable is that a strain variant of cytochrome P450b from Holtzman rats (26) is not recognized by B51 but is recognized as well as cytochrome P450b by the remaining 11 MAb in an ELISA (25). From available data, it is anticipated that the Holtzman strain variant of cytochrome P450b will have no more than 1 to 2 amino acid differences compared to Long Evans cytochrome P450b. This illustrates the high discriminatory power of MAb. This example also serves to reveal a danger often overlooked when using MAb. If we had used only B51 to assay for cytochrome P450b in Holtzman rats, we would have concluded that Holtzman rats were devoid of that isozyme. Clearly, it is important to use several MAb to establish either the absence or presence of structurally homologous isozymes when evaluating strain or species differences.

Table 3. Collation of the Reactivity of Ten Cytochrome P450 Isozymes
with Twenty-three MAb in ELISA

MAb	Epitope	Cytochromes P450									
		a	b	c	d	e	f	g	h	i	j
C1	1	-	-	10	-	-	-	-	-	-	-
CD2	2	-	-	10	10	-	-	-	-	-	-
CD3		-	-	10	10	-	-	-	-	-	-
CD5		-	-	10	10	-	-	-	-	-	-
C4	4	-	-	10	-	-	-	-	-	-	-
C6	6	-	-	10	-	-	-	-	-	-	-
C7	7	-	-	10	-	-	-	-	-	-	-
C10		-	-	10	-	-	-	-	-	-	-
C8	8	-	-	10	-	-	-	-	-	-	-
BE26	26	-	10	-	-	8	-	-	-	-	-
BE32		-	10	-	-	10	-	-	-	-	-
BEA33	33	2	10	-	-	13	-	-	-	-	-
BE28	28	-	10	-	-	9	-	-	-	-	-
BE44		-	10	-	-	13	-	-	-	-	-
BE45		-	10	-	-	16	-	-	-	-	-
BEF29	29	-	10	-	-	13	9	-	-	-	-
BE46	46	-	10	-	-	5	-	-	-	-	-
B50	50	-	10	-	-	-	-	-	-	-	-
BE49	49	-	10	-	-	5	-	-	-	-	-
B51	51	-	10	-	-	-	-	-	-	-	-
BE52	52	-	10	-	-	20	-	-	-	-	-
FGHIJ57	57	-	-	-	-	-	10	10	9	10	2
FGHI62	62	-	-	-	-	-	10	2	4	1	-

Each MAb is assigned letters and a unique number. The first
letter refers to the antigen of immunization while the presence of
additional letters indicates cross-reactivity with heterologous
cytochrome(s) P450.   The ELISA value for the antigen of
immunization was arbitrarily set at a value of 10.   A dash
indicates a value of less than 10% cross-reaction. Data are
derived from the present study and previous publications (24, 25).

In our studies of polyclonal antibody specificity (Table 2) we found that anti-P450b as well as anti-P450e cross-react weakly with cytochrome P450f, and also that anti-P450f cross-reacts weakly with both cytochromes P450b and P450e. These results suggest that at least one common epitope exists among cytochromes P450b, P450e and P450f. This hypothesis is confirmed by the almost equivalent level of reactivity in ELISA of BEF29 for these 3 cytochromes P450 (Table 3). In addition, BEF29 reacts with these 3 isozymes in "Western blots". This example illustrates that results obtained with properly characterized polyclonal antibodies and MAb can be equivalent.

The 2 MAb which are directed against cytochrome P450f (Table 3) cross-react to varying extents with cytochromes P450g, P450h and P450i. Since these 2 MAb have a different specificity for cytochrome P450j and show differences in relative reactivity with cytochromes P450g, P450h and P450i, they must be directed against distinct epitopes. Clearly at least 2 epitopes are shared to varying extents by these 4 cytochrome P450 isozymes. These results further support our previous studies using polyclonal anti-P450f and anti-P450g where the substantial cross-reaction among cytochromes P450f, P450g, P450h and P450i led us to propose that these 4 isozymes are more related to one another than they are to cytochromes P450a-P450e (22). So in yet another example, the specificity of MAb need not necessarily be any different than the specificity of properly characterized polyclonal antibody.

IMMUNOCHEMICAL QUANTITATION

In 1979, we published methods for producing monospecific anti-P450b (PB-P450) and anti-P450c (MC-P448) which were used to immunoquantitate these antigens in liver microsomes from untreated rats and rats treated with 3-methylcholanthrene, phenobarbital or Aroclor 1254 (18). Since that time, we have surveyed a number of different inducers (17,21,27) and the results for a few of these appear in Table 4. Phenobarbital preferentially induces cytochromes P450b+P450e by 40-fold or more but has negligible effects on cytochromes P450c, P450d (Table 4) and cytochrome P450g (manuscript in preparation). Phenobarbital appears to modestly induce cytochrome P450a (Table 4) and P450f (manuscript in preparation). SKF-525A, a potent inhibitor of several cytochrome P450-catalyzed reactions, and trans-stilbene oxide, a potent inducer of microsomal epoxide hydrolase, are both inducers of cytochromes P450b and P450e and elicit a phenobarbital-type response even though all 3 compounds are structurally different. The pattern of isozyme induction by β-naphthoflavone and TCDD is indistinguishable from the 3-methylcholanthrene prototype, which induces cytochrome P450c 70-fold or more and, in addition, induces cytochrome P450d at least 11-fold. The latter 3 compounds are also modest inducers of cytochrome P450a. We have tested a number of compounds as inducers to see if induction of the immunorelated isozymes, cytochromes P450c and P450d, could be dissociated from each other. In a survey that included 12 structurally diverse compounds (21), 41 individual polychlorinated- or polybrominated-biphenyl isomers and congeners (27), and 18 substituted pentachlorobenzenes and tetrachlorobiphenyls (28), we were unable to find a potent inducer of either cytochrome P450c or cytochrome P450d that did not induce the immunorelated isozyme to some extent, although the relative proportions of each isozyme varied. For example, isosafrole induces greater absolute levels of cytchrome P450d compared to cytochrome P450c, whereas 3-methylcholanthrene has the opposite effect. It would appear that cytochromes P450c and P450d are coinduced by structurally diverse xenobiotics. Aroclor 1254, a commercial mixture of polychlorinated

Table 4. Immunochemical Quantitation of Hepatic Microsomal Cytochrome P450 from Immature Male Rats Treated with Inducers.

Treatment	Cytochrome			
	P450a	P450b + P450e	P450c	P450d
	induced/control			
Phenobarbital	2	40	1	1
SKF-525A	2	27	3	1
trans-Stilbene oxide	1	25	2	1
3-Methylcholanthrene	4	1	72	11
β-Napthoflavone	3	1	70	12
TCDD	6	1	76	13
Isosafrole	2	13	19	22
Phenothiazine	2	21	56	15
Aroclor 1254	4	45	50	20
2,4,5,2',4',5'-Hexachlorobiphenyl	2	73	1	1
3,4,5,3',4',5'-Hexachlorobiphenyl	8	1	43	40
2,3,4,5,3',4'-Hexachlorobiphenyl	6	47	43	10

Liver microsomes were prepared from rats administered the above compounds. Isozymes were quantitated by radial immunodiffusion with the appropriate polyclonal antibodies as described (17,21,27,28). Absolute values for controls are: 0.05-0.08 nmol cytochrome P450a/mg protein; 0.02-0.03 nmol cytochromes P450b+P450e/mg protein; 0.02-0.03 nmol cytochrome P450c/mg protein; and 0.04-0.06 nmol cytochrome P450d/mg protein.

biphenyls, is a potent inducer of cytochromes P450a-P450e. Its inducing properties are not unlike those obtained by coadministration of phenobarbital and 3-methylcholanthrene (29). The pattern of induction of cytochromes P450a-P450e by certain polychlorinated biphenyls (e.g. 2,4,5,2',4',5'-hexachlorobiphenyl) is similar to that of phenobarbital, whereas induction by other polychlorinated biphenyls (e.g. 3,4,5,3',4',5'-hexachlorbiphenyl) is similar to that of 3-methylcholanthrene. However, the regulation of these isozymes is complex, since some pure polychlorinated biphenyls (e.g. 2,3,4,5,3',4'-hexachlorobiphenyl) induce cytochromes P450a-P450e. Other compounds, such as phenothiazine and isosafrole, also induce all 5 of these isozymes.

We have used the marked isozyme specificity of the anti-P450b MAb to estimate relative levels of cytochromes P450b and P450e in "Western blots" of liver microsomes from untreated rats or rats treated with inducers. Based on the staining intensity of polypeptides corresponding to these isozymes in "Western blots" of microsomes from untreated rats, we conclude that untreated rats contain more cytochrome P450e than cytochrome P450b, although both proteins are present at very low levels (25). In "Western blots" of microsomes from phenobarbital-treated rats, it appears that there is more cytochrome P450b than cytochrome P450e. Vlasuk et al. (30), using two-dimensional isoelectric focusing-SDS gels for resolution of microsomal proteins from immature male rats treated with 5 different inducers of cytochromes P450b and P450e, estimated that the level of cytochrome P450b was consistently at least 2-fold greater than cytochrome P450e. Neither of these isozymes could be detected by protein-staining in microsomes from untreated rats (30). We have not studied the induction of cytochromes P450b and P450e using MAb, but results from immunoblots of microsomes from control rats suggest that cytochromes P450b and P450e may not be present in the same relative proportions in all rats. Guengerich et al. (10), using polyclonal antibodies that recognize both proteins, noted independent regulation of cytochromes $P450_{PB-B}$ and $P450_{PB-D}$ (corresponding to cytochromes P450b and P450e, respectively) in liver microsomes from control and phenobarbital-treated rats ($P450_{PB-B}$:$P450_{PB-D}$ ratio of 0.4 and 1.4, respectively); however, in a subsequent study (29) the corresponding ratios were 0.1 and 0.8 for control and phenobarbital-induced rats, respectively, as determined by immunoquantitation. It is possible that the determinations of cytochromes $P450_{PB-B}$ and $P450_{PB-D}$ are inaccurate since polymorphism exists at both the cytochrome P450b and P450e loci, and the Sprague-Dawley rats from the colony used in the former study are known to possess a polypeptide from each locus which are not resolved in one-dimensional isoelectric focusing gels of microsomes (31).

Clearly, the use of immunochemical methods for quantitation of cytochrome P450 isozymes has revealed considerable complexity in their regulation. These antibodies are also important probes for (a) evaluating the contribution of individual cytochrome P450 isozymes to the overall metabolism of substrates by intact microsomes, (b) determining structural relatedness among different isozymes, (c) membrane and tissue localization of the proteins, (d) immunoaffinity purification and (e) recombinant DNA technology.

ACKNOWLEDGMENT

We thank Ms. Cathy Michaud for her help in the preparation of this manuscript.

REFERENCES

1. R. Sato and T. Omura, "Cytochrome P450," Academic Press, New York (1978).
2. E.C. Miller and J.A. Miller, Searches for ultimate chemical carcinogens and their reactions with cellular macromolecules, Cancer, 47:2327-2345 (1981).
3. D.M. Jerina, J.M. Sayer, H. Yagi, P.J. van Bladeren, D.R. Thakker, W. Levin, R.L. Chang, A.W. Wood and A.H. Conney, Stereoselective metabolism of polycyclic aromatic hydrocarbons to carcinogenic metabolites, in: "Microsomes and Drug Oxidations," A.R. Boobis, J. Caldwell, F. DeMatteis, and C.R. Elcombe, eds., Taylor and Francis Ltd., London p. 310-319 (1985).

4. D.E. Ryan, P.E. Thomas, L.M. Reik and W. Levin, Purification, characterization and regulation of five rat hepatic microsomal cytochrome P450 isozymes, Xenobiotica, 12:727-744 (1982).
5. D.E. Ryan, S. Iida, A.W. Wood, P.E. Thomas, C.S. Lieber and W. Levin, Characterization of three highly purified cytochromes P450 from hepatic microsomes of adult male rats, J. Biol. Chem., 259: 1239-1250 (1984).
6. D.E. Ryan, L. Ramanathan, S. Iida, P.E. Thomas, M. Haniu, J.E. Shively, C.S. Lieber and W. Levin, Characterization of a major form of rat hepatic microsomal cytochrome P450 induced by isoniazid, J. Biol. Chem., 260: 6385-6393 (1985).
7. P-M. Yuan, D.E. Ryan, W. Levin, and J.E. Shively, Identification and localization of amino acid substitutions between two phenobarbital-inducible rat hepatic microsomal cytochromes P450 by micro sequence analyses, Proc. Natl. Acad. Sci. U.S.A, 80: 1169-1173 (1983).
8. M. Haniu, D.E. Ryan, S. Iida, C.S. Lieber, W. Levin and J.E. Shively, NH_2-terminal sequence analyses of four rat hepatic microsomal cytochromes P450, Arch. Biochem. Biophys., 235: 304-311 (1984).
9. K.-C. Cheng, and J.B. Schenkman, Purification and characterization of two constitutive forms of rat liver microsomal cytochrome P450, J. Biol.Chem., 257:2378-2385 (1982).
10. F.P. Guengerich, G.A. Dannan, S.T. Wright, M.V. Martin and L.S. Kaminsky, Purification and characterization of rat liver microsomal cytochromes P450: electrophoretic, spectral, catalytic, and immunochemical properties and inducibility of eight isozymes isolated from rats treated with phenobarbital or β-naphthoflavone, Biochemistry, 21:6019-6030 (1982).
11. P.P. Tamburini, H.A. Masson, S.K. Bains, R.J. Makowski, B. Morris and G.G. Gibson, Multiple forms of hepatic cytochrome P450: purification, characterization and comparison of a novel clofibrate-induced isozyme with other major forms of cytochrome P450, Eur. J. Biochem. 139:235-246 (1984).
12. N.A. Elshourbagy and P.S. Guzelian, Separation, purification and characterization of a novel form of hepatic cytochrome P450 from rats treated with pregnenolone-16α-carbonitrile, J. Biol. Chem., 255:1279-1285 (1980).
13. I. Jansson, J. Mole and J.B. Schenkman, Purification and characterization of a new form (RLM_2) of liver microsomal cytochrome P-450 from untreated rat, J. Biol. Chem. 260:7084-7093 (1985).
14. D. Larrey, L.M. Distlerath, G.A. Dannan, G.R. Wilkinson and F.P. Guengerich, Purification and characterization of the rat liver microsomal cytochrome P450 involved in the 4-hydroxylation of debrisoquine, a prototype for genetic variations in oxidative drug metabolism, Biochemistry 23:2787-2795 (1984).
15. T. Kamataki, K. Maeda, Y. Yamozoe, T. Nagai, and R. Kato, Sex difference of cytochrome P450 in the rat: purification, characterization, and quantitation of constitutive forms of cytochrome P450 from liver microsomes of male and female rats, Arch. Biochem. Biophys., 225:758-770 (1983).
16. A.W. Wood, D.E. Ryan, P.E. Thomas and W. Levin, Regio- and stereo-selective metabolism of two C_{19} steroids by five highly purified and reconstituted rat hepatic cytochrome P450 isozymes, J. Biol. Chem., 258:8839-8847 (1983).
17. P.E. Thomas, L.M. Reik, D.E. Ryan and W. Levin, Regulation of three forms of cytochrome P450 and epoxide hydrolase in rat liver microsomes. Effects of age, sex and induction, J.Biol. Chem. 256:1044-1052 (1981).
18. P.E. Thomas, D. Korzeniowski, D. Ryan, and W. Levin, Preparation of monospecific antibodies against two forms of rat liver cyto-

chrome P-450 and quantitation of these antigens in microsomes, Arch. Biochem. Biophys., 192:524-432 (1979).

19. L.M. Reik, W. Levin, D.E. Ryan and P.E. Thomas, Immunochemical relatedness of rat hepatic microsomal cytochromes P450c and P450d, J. Biol. Chem., 257:3950-3957 (1982).

20. Y. Fujii-Kuriyama, Y. Mizukami, K. Kawajiri, K. Sogawa and M. Muramatsu, Primary structure of a cytochrome P-450: coding nucleotide sequence of phenobarbital-inducible cytochrome P-450 cDNA from rat liver, Proc. Natl. Acad. Sci. U.S.A., 79:2793-2797 (1982).

21. P.E. Thomas, L.M. Reik, D.E. Ryan and W. Levin, Induction of two immuno-chemically related rat liver cytochrome P450 isozymes, cytochromes P450c and P450d, by structurally diverse xenobiotics, J. Biol. Chem., 258:4590-4598 (1983).

22. S. Bandiera, D.E. Ryan, W. Levin and P.E. Thomas, Evidence for a family of four immunochemically related isozymes of cytochrome P-450 purified from untreated rats, Arch. Biochem. Biophys., 240:478-482 (1985).

23. A.L. Morville, P.E. Thomas, W. Levin, L. Reik, D.E. Ryan, C. Raphael and M. Adesnik, The accumulation of distinct mRNAs for the immunochemically related cytochromes P450c and P450d in rat liver following 3-methylcholanthrene treatment, J. Biol. Chem., 258:3901-3906 (1983).

24. P.E. Thomas, L.M. Reik, D.E. Ryan and W. Levin, Characterization of nine monoclonal antibodies against rat hepatic cytochrome P450c: Delineation of five spatially distinct epitopes, J. Biol. Chem., 259:3890-3899 (1984).

25. L.M. Reik, W. Levin, D.E. Ryan, S.L. Maines and P.E. Thomas, Evaluation of monoclonal antibodies as probes to distinguish among isozymes of the cytochrome P450b subfamily, Arch. Biochem. Biophys., in press (1985).

26. D.E. Ryan, A.W. Wood, P.E. Thomas, F.G. Walz, Jr., P-M. Yuan, J.E. Shively, and W. Levin, Comparisons of highly purified hepatic microsomal cytochromes P450 from Holtzman and Long Evans rats, Biochim. Biophys. Acta, 709:273-283 (1982).

27. A.P. Parkinson, S.H. Safe, L.W. Robertson, P.E. Thomas, D.E. Ryan, L.M. Reik and W. Levin, Immunochemical quantitation of cytochrome P450 isozymes and epoxide hydrolase in liver microsomes from polychlorinated or polybrominated biphenyl-treated rats: A study of structure activity relationships, J. Biol. Chem., 258:5967-5976 (1983).

28. S.M.A. Li, M.A. Denomme, B. Leece, S. Safe, D. Dutton, A. Parkinson, P.E. Thomas, D. Ryan, S. Bandiera, L.M. Reik and W. Levin, Hexachlorobenzene and substituted pentachlorobenzenes as inducers of hepatic cytochrome P-450-dependent monooxygenases, IARC Publication Series, in press (1985).

29. G.A. Dannan, F.P. Guengerich, L.S. Kaminsky and S.D. Aust, Regulation of cytochrome P450: Immunochemical quantitation of eight isozymes in liver microsomes of rats treated with polybrominated biphenyl congeners, J. Biol. Chem., 258:1282-1288, (1983)

30. G.P. Vlasuk, D.E. Ryan, P.E. Thomas, W. Levin and F.G. Walz, Jr., Polypeptide patterns of hepatic microsomes from Long-Evans rats treated with different xenobiotics, Biochemistry, 21: 6288-6292 (1982).

31. A. Rampersaud and F.G. Walz, Jr., At least six forms of extremely homologous cytochromes P-450 in rat liver are encoded at two closely linked genetic loci, Proc. Natl. Acad. Sci. U.S.A., 80:6542 6546 (1983).

AGE AND SEX DIFFERENCES IN CONSTITUTIVE FORMS OF CYTOCHROME P-450 OF RAT LIVER MICROSOMES

John B. Schenkman, Leonard V. Favreau, and Ingela Jansson

Department of Pharmacology
University of Connecticut Health Center
Farmington, CT 06032 - U.S.A.

Knowledge of the cytochrome P-450 monooxygenase system has come a long way since the reports of its hemoprotein nature.[1,2] Initial studies revealed some drug and chemical-treated animals respond to challenge by such agents with an increase in the hepatic microsomal content of cytochrome P-450. It wasn't long, however, before it was realized that the cytochrome P-450 increase was the result of an induction of different forms of cytochrome P-450.[3,4]

Since the late 1940's the liver microsomes have been known as the site of the oxidative metabolism of drugs. The University of Pennsylvania group[5] demonstrated the involvement of cytochrome P-450 in the process as the terminal oxidase. Before long it was recognized that induction with different compounds differentially increases the V_{max} and alters the K_m for a number of substrates of the P-450 monooxygenases,[6] suggesting differential metabolism by different P-450 enzymes in microsomes. A number of induced forms of P-450 have been purified.[7] With time it became recognized that the P-450 system was involved not only in inactivation of drugs and chemicals, but could also cause activation of compounds to more toxic, reactive and even mutagenic products.[8] This led to standardized tests, such as the Ames test[9] in which liver microsomes are utilized to activate the proagent to some reactive form which can be evaluated by a bacterial assay for mutagenesis. Most frequently, the procedure used calls for microsomes of Arochlor 1254-treated animals, since these have elevated levels of cytochrome P-450.[10] However, as pointed out above, not only is the content of P-450 in the microsomes changed, but the composition of the microsomal cytochrome P-450 family is altered by this treatment. This is an important point to consider, since the many cytochromes P-450 are not all isozymes. While almost all are capable of oxidizing the same compounds, products formed from many of these differ for the different cytochromes.[11,12] Thus, in the non-induced animal, for the most part metabolites formed will be the result of different constitutive forms and may differ markedly from those of the induced forms of cytochrome P-450.

In an attempt to learn the nature of the constitutive forms of cytochrome P-450 a procedure was developed making use of liver microsomes of the adult male rat.[13,14] The procedure had to provide a high yield, as the non-induced animal possesses only about 0.5-1.0 nmol P-450 per mg microsomal protein, a level far below that of the induced animal. In

Figure 1 the typical elution profile of hemoproteins is shown, after detergent solubilization, from the lauric acid-AH Sepharose 4B column.

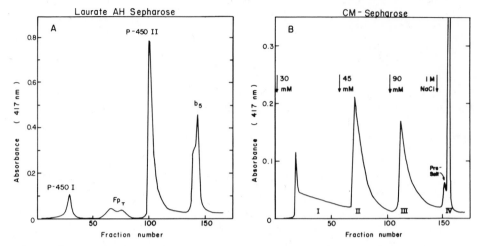

Figure 1: Isolation and separation of constitutive forms of cytochrome P-450. A. Lauric acid-AH-Sepharose chromatography of Emulgen 911-solubilized microsomes (13,14). B. Carboxymethyl sepharose CL6B chromatography of P-450II (14). Hemoprotein absorption (417nm) peaks plus pigments absorbing at this wavelength are shown for the fractionation. In B, sodium phosphate step gradients are used (arrows). P-450II contained 65% of the solubilized P-450 (14). The presalt fraction elutes shortly before the main peak when 1MNaCl in 60mM sodium phosphate is applied to the column.

Monitoring absorbance at 417 nm reveals a small amount (5%) of the cytochrome P-450 (P-450I) applied elutes immediately after the column volume. A small subsequent absorbance peak is NADPH-cytochrome P-450 reductase. The first large peak (P-450II) contains cytochrome P-450 (55% of the applied material). The last peak contains some cytochrome P-450 (ca. 10%) plus cytochrome b_5. In our purification studies to date the P-450II fraction is dialyzed and applied to a CM Sepharose CL6B column, to which it binds. It is eluted in step gradients of phosphate buffer concentrations (Figure 1B). Four fractions are obtained, CMI (30 mM), CMII (45 mM), CMIII (90 mM) and CM IV (60 mM and 1 MNaCl). The sharp spike preceding the CMI fraction contains mainly non-P-450 proteins plus

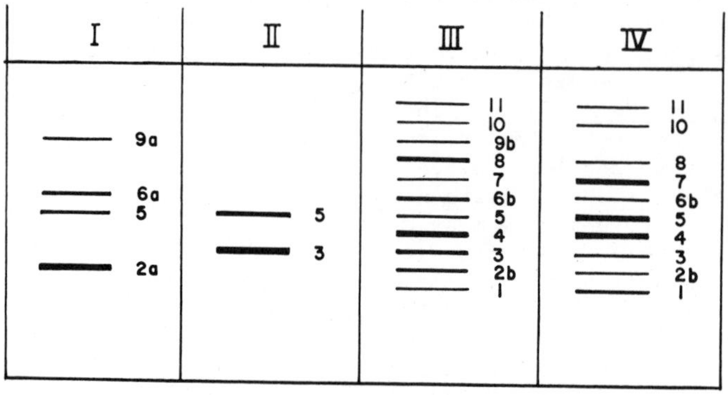

Figure 2: Sketch of protein bands seen in SDS-PAGE of fractions from CM-Sepharose CL6B. Numbers are from anode to cathode in the 45-61KDa region with permission from Schenkman et al. (14).

NADH-cytochrome b$_5$ reductase activity and has routinely been discarded.

Figure 2 shows a sketch of the proteins obtained in each CM fraction as revealed in sodium dodecylsulfate polyacrylamide gel electrophoresis SDS-PAGE, with enumeration of the bands is the 45 KDa to 60 KDa region in the direction from anode to cathode. The intensity of the bands is an indication of the amount of protein revealed for this band by Coomassie Brilliant Blue R stain. The utility of such a fingerprint is that it allows us to set a value for the potential number of cytochromes P-450 contained in the P-450IIa fraction. Undoubtedly, different P-450 forms will also be found in the P-450IIb and cytochrome b$_5$ containing fraction of the lauric acid-AH-Sepharose column. Eleven different molecular weight bands are found, three of which appear in other CM fractions separated by a step concentration (e.g. 2a and 2b). These are clearly different proteins based upon their chromatographic properties on the CM column, and put a ceiling of 14 cytochromes P-450 in the P-450IIa fraction if all of the bands are cytochromes P-450. We do not expect all of the proteins to be P-450, since there are a number of other approximately 50 KDa microsomal proteins, e.g., epoxide hydrolase[15] and UDP-glucuronic acid transferase[16].

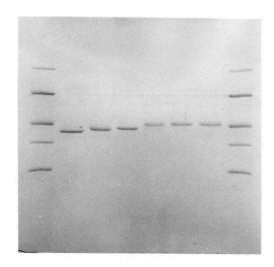

Figure 3: SDS-PAGE of constitutive forms of P-450 purified from liver microsomes of male or female rats. Tracks contain standards (1), RLM2(2), RLM3(3), fRLM4(4), RLM5(5), RLM5a(6), fRLM5a(7), standards (8). One g of each P-450 was used.

At the present time we have purified and characterized six cytochromes P-450 (Figure 3). These are, RLM2 (track 2), RLM3 (track 3), fRLM4 (track 4), RLM5 (track 5), RLM5a (track 6) and fRLM5a (track 7). The numbers represent the band number of the Rat Liver Microsomal P-450 of male and female (f) rats. RLM2 is a recently characterized protein[17] of 49 KDa, while RLM3 and RLM5 were described earlier (Cheng and Schenkman). RLM5a is a newly isolated form from male rats[18] and fRLM4 and fRLM5a are newly isolated forms from the female rat. These latter two proteins may correspond to the recently reported[19,20] P-450-15ß and DEa respectively, although DEa was called female specific.

Although rats and rat liver microsomes have been used extensively in studies on oxidative metabolism of drugs and various chemicals, a considerable number of reports exist pointing to and examining a sex difference in such metabolism in this species.[21,22] Such studies would indicate sex differences in the constitutive forms of cytochrome P-450 in liver microsomes by the rat. In fact, there have been reports of male specific and neonatally imprinted cytochrome P-450 as well as female specific forms of cytochrome P-450.[23,24]

In order to better understand the basis of the sex difference in metabolism and such reports, liver microsomes of adult male and female rats were prepared and the P-450IIa fractions were chromatographed on

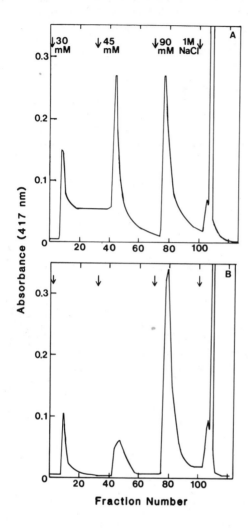

Figure 4: Comparison of carboxymethyl Sepharose separations of male and female P-450II fractions. Conditions were as in Fig. 1. A. Male P-450II fraction was used. B. Female P-450II fraction was used.

CM-Sepharose CL6B (Figure 4). The pattern observed in the male is the typical one, with the pre-CMI spike containing NADH-cytochrome b_5 reductase and other non-hemoprotein polypeptides. In the female preparation there is no CMI fraction, only the pre-CMI spike. The CMII fraction of males is usually equal to or up to 50% larger than the CMIII fraction. With females, however, the CMII peak is always very much smaller than the CMIII peak. An examination of the polypeptides in the first two CM fractions of male and female rats (Figure 5) reveals striking differences. Since the female does not have a CMI fraction the pre-CMI "spike" was compared. In both instances the NADH-cytochrome b_5 reductase is seen slightly below the lactic dehydrogenase standard (36 KDa). In the male spike a trace of RLM2 is seen (arrow). In the female preparation a sizable amount of an as yet unidentified P-450 is found, unlike in the male, with a densely staining band of migration intermediate between RLM3 and RLM5 polypeptides of male in the adjacent CMII fraction (Figure 5). The lowest molecular weight band in the male CMII fraction, RLM3, is clearly absent from the female CMII fraction.

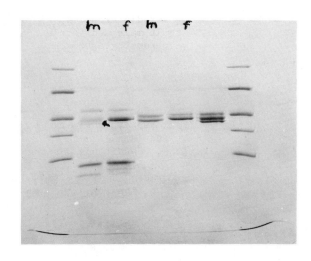

Figure 5: Comparison of protein bands seen in male and female pre-CMI fractions (tracks 2 and 3) and CMII fractions (tracks 4 and 5). Tracks 1 and 7 contain molecular weight standards and track 6 contains an equal hemoprotein (20 pmol) each mixture of RLM2, 3, 4, and 5.

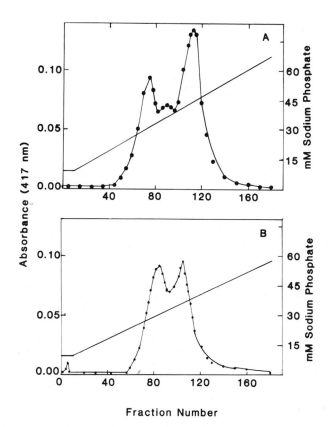

Figure 6: Hydroxylapatite separation of male (A) and female (B) CMII fractions. The fractions were eluted in sodium phosphate gradient (17). 150 nmol of male and 100 nmol of female P-450 was applied to the hydroxylapatite columns and was eluted in 10-80mM linear phosphate gradient.

Further separation of the CMII fractions of male and female rats is shown in Figure 6. The preparations are dialyzed and applied to hydroxylapatite columns on which the cytochrome P-450 binds tightly, and is eluted by a linear phosphate gradient. When the gradient is made shallow a trimodal elution is observed for the male and a bimodal elution occurs with the female cytochrome P-450. In the male, the first peak is always RLM5 and a protein elutes in the female at the corresponding ionic strength (Figure 6). Similarly, the second peak in the female eluted at the same ionic strength as the middle peak of the male. There was no corresponding protein eluting at the ionic strength of the third male peak, which is RLM3. The middle peak of the male (Figure 6) contains RLM5a and some RLM3. If the tail end of the CMII peak (Figure 4 top), where it begins to spread, is discarded, the hydroxylapatite column will separate RLM5 from RLM3 clearly into two peaks with minimal overlap. The CMII "tail" is found to be enriched in RLM5a. This form of P-450 is quite similar to RLM5 and corresponds, on chromatographic separations, to the second female peak (Figure 6) which we designate fRLM5a. As seen in SDS-PAGE (Figure 5, track 5) fRLM5 shows similar migration to RLM5, although in pure form it has an apparent minimum molecular weight of

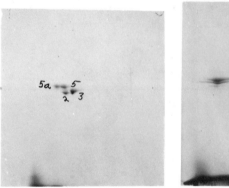 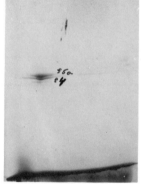

Figure 7: Two dimensional gel (IEF SDS/PAGE) of male and female proteins. Male (left) contains RLM2, 3, 5, and 5a while female (right) contains fRLM4 and fRLM5. Each protein was 25 pmol.

52,500, or about 500 greater than RLM5. The other female P-450 has a migration between RLM5 and RLM3 (figure 5, track 5 vs. track 4) and was named fRLM4. Mixing four male and female proteins together results in essentially three bands; one migrates as RLM2, one as RLM3 plus 4 and one as RLM5.

When the four male proteins were combined and submitted to two-dimensional isoelectric focusing-SDS-PAGE, a striking pattern was observed. The four separated clearly. The pI of the individual proteins and their Mr are listed in Table 1. The two proteins purified from the female CMII fraction are shown in Figure 7, right for comparison. RLM5a and fRLM5a have the same Mr and pI.

Table 1

Isoelectric Points of Different Cytochrome P-450

Form	PI
RLM3	7.1
RLM2	7.3
RLM5	7.4
RLM5a	7.6
fRLM5a	7.6
fRLM4	7.6

When the entire P-450 collected in the laurate II fraction of male and female microsomes is subjected to the two dimensional isoelectric focusing-SDS-PAGE, the pattern observed was similar to that seen with the purified proteins in the neutral pH region (Figure 8). It would appear

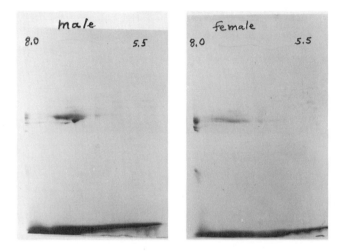

Figure 8: Two dimensional isoelectric focusing/SDS-PAGE of male and female rat liver P-450II fraction. 100 pmol of P-450 from each fraction was subjected to isoelectric focusing and then electrophoresed on SDS-PAGE in a perpendicular direction. The IEF pH showed a linear gradient from 5.5 to 8 as indicated on this Figure.

that the CMI and CMII fraction proteins recovered contain essentially all of the neutral pH proteins in the male. The RLM2, 3, 5 and 5a are clearly apparent. In contrast, the female lacks most of the neutral pI proteins, showing fRLM4 and fRLM5a, both of which smear in these gels as do the purified proteins (Figure 7 right). Several P-450's remain to be elucidated, these appearing in the basic region of the gel. The pattern for male and female of the basic proteins appears fairly similar. Thus, it would appear that the differences seen between male and female rats metabolism of drugs and other xenobiotics may reside in the neutral region P-450 proteins.

When RLM2, the most recent constitutive P-450 to be characterized,[17] was submitted to NH_2-terminal partial amino acid sequencing, it turned out to have almost the same partial sequence as P-450a[25] differing in the first 22 residues only in residue 19. Both have a Mr of about 48 KDa. However, the metabolic activity of RLM2 was so strikingly different from that reported for P-450a[12] that the two could not be the same protein. P-450a was originally found in microsomes of Arochlor 1254 treated immature male rats and it was stated it was also found in the untreated male as well. Examination of the CM Sepharose CL6B "fingerprint" of the lauric acid AH-Sepharose 4B P-450II fraction (Figure 9) revealed differences in the P-450 population between the

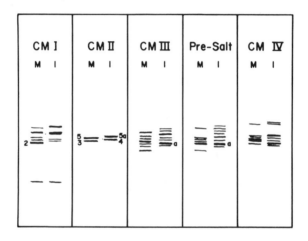

Figure 9: Sketch of protein bands observed in the male immature rat (I) compared with the mature (M) rat CM-Sepharose CL6B fractions.

immature male rat and the mature male rat. Most striking in the CMI fraction is the absence of RLM2, shown by the number for the mature males, in the sketch of an SDS-PAGE profile. In the CMII fraction both RLM3 and RLM5 are absent in the immature male; instead polypeptide bands corresponded to fRLM4 and fRLM5a are seen. The CMIII fractions of the immature male rat, as well as the pre-salt shoulder of the CMIV peak (see arrow Figure 9), possess a 48 KDa band of strong intensity P-450a while only a faint corresponding polypeptide is seen in the adult preparation. Examination of all CM fractions for testosterone 7α-hydroxylase activity, purportedly the only site of attack with P-450a,[12] revealed the greatest activities to be in the CMIII and pre-salt fractions. The data, suggest the pattern of cytochrome P-450 in the immature male rat is like that of the adult female, with female forms of P-450 in the CMII fraction and with P-450a present in abundance.

The last question our attention turned to was that of similarity of the different P-450 enzymes. In Figure 10 is shown an immunoblot of the two-dimensional isoelectric focusing SDS-PAGE of male and female rat liver microsomes. Note the similarity in appearance to the Coomassie Blue stained P-450II fractions (Figure 9). Of especial interest is the fact that goat antibody to RLM5 cross reacts with proteins in the 50 KDa region of the gel, acidic, neutral and basic proteins. The neutral

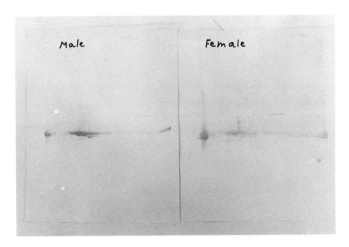

Figure 10: Immunoblot of solubilized microsomal proteins of male and female rats after SDS-PAGE. Proteins were electrophoretically transferred to nitrocellulose paper and reacted with goat antibody to RLM5. Staining was with peroxidase conjugated rabbit anti-goat IgG and diaminobenzidine H_2O_2.

proteins have already been identified and shown to be constitutive forms of P-450. Even the female forms possess antigenic similarity to RLM5. Overall, then, constitutive P-450's possess a similarity in homology sufficient to allow antigenic recognition. A further indication of the similarity of these forms is seen in the immunoblot of the purified enzymes and the CMII fraction of male and female microsomes. The data clearly indicates antigenic similarity.

Table 2

Metabolism of Testosterone by Purified Cytochromes P-450

Assay conditions are described in Methods section. The ratio of P-450 to Fp_T was 1:1. Activities are expressed as nmol product formed/min/ nmol P-450 (rounded to one decimal place).

nmol product formed/min/nmol P-450[a]

	RLM2	RLM3	RLM5	RLM5a	fRLM5a	fRLM4
Testosterone						
Total	7.3	2.4	12.1	0.5	0.2	1.3
Androstenedione	---	---	2.9	0.2	0.05	---
2α-OH Testosterone	---	---	3.6	0.1	0.05	---
6ß-OH-Testosterone	1.2	1.2	---	---	---	---
7ß-OH Testosterone	1.8	---	---	---	---	---
15α-OH Testosterone	2.7	0.4	---	---	---	1.3
16α-OH Testosterone	---	---	4.8	0.2	0.05	---

[a]Only major metabolites of testosterone are shown for each enzyme.

Despite the antigenic similarity between the different P-450 isozymes, they do not all catalyze the same reaction, as seen in Table II using a substrate capable of oxidation to a number of isomeric and epimeric products, testosterone, reveals the enzymes form different products of common substrates (Table 2). Clearly, this is something to keep in mind when assessing the ability of individual compounds to undergo metabolic activation.

In conclusion, while the use of Arochlor 1254 and other inducers will elevate the microsomal P-450 content in the S-9 fraction, they do so selectively, usually with form previously not present. And while such inducers will indicate the potential for activation of a compound to a reactive intermediate, care must be exercised to recognize that such intermediates will be formed only if enzymes capable of generating such products are present. Since the constitutive forms of cytochrome P-450 are present even after inducers are employed, it would seem beneficial that investigators also test with control microsomes or with reconstituted systems that contain constitutive P-450 isozymes for their ability to generate reactive forms with different xenobiotics.

Acknowledgements

Supported in part by United States Public Health Service Grant GM26114 from the National Institutes of Health. Leonard V. Favreau is a Stauffer Chemical Company Predoctoral Fellow.

References

1. T. Omura and R. Sato, A new cytochrome in liver microsomes, J. Biol. Chem. 237:1375-1376 (1962).
2. T. Omura and R. Sato, The carbon monoxide-binding pigment of liver microsomes. I. Evidence for its hemoprotein nature, J. Biol. Chem. 239:2370-2378 (1964).
3. S. Orrenius and L. Ernster, Phenobarbital-induced synthesis of the oxidative demethylating enzymes of rat liver microsomes, Biochem. Biophys. Res. Comm. 16:60-65 (1964).
4. N.E. Sladek and G.J. Mannering, Evidence for a new P-450 hemoprotein in hepatic microsomes from methylcholanthrene treated rats, Biochem. Biophys. Res. Comm. 24:668-674 (1966).
5. D.Y. Cooper, S. Levin, S. Narasimhulu, O. Rosenthal and R.W. Estabrook, Photochemical action spectrum of the terminal oxidase of mixed function oxidase systems, Science 147:400-402 (1965).
6. G. Powis, R. Talcott and J.B. Schenkman, Kinetic and spectral evidence for multiple species of cytochrome P-450 in liver microsomes, in: Microsomes and Drug Oxidation, V. Ullrich et al. (Eds.) pp. 127-135, Pergamon Press, NY (1977).
7. D.E. Ryan, P.E. Thomas, L.M. Reik and W. Levin, Purification, characterization and regulation of five rat hepatic microsomal cytochrome P-450 isozymes, Xenobiotica 12:727-744 (1982).
8. Biological Reactive Intermediates II. Chemical mechanism and biological effects, Part A. R. Snyder, D.V. Parke, J.J. Kocsis, D.J. Jollow, G.G. Gibson and C.M. Witmer (Eds.) Plenum Press, NY (1982).
9. B.N. Ames, J. McCann and E. Yamasaki, Methods for detecting carcinogens and mutagens with the salmonella/mammalian microsome mutagenicity test, Mutation Research 31:349-364 (1975).
10. C.L. Litterst, T.M. Farber, A.M. Baker and E.J. VanLoon, Effect of polychlorinated biphenyls on hepatic microsomal enzymes in the rat, Tox. Appl. Pharmac. 23:112-122 (1972).

11. K.-C. Cheng and J.B. Schenkman, Testosterone metabolism by cytochrome P-450 isozymes RLM3 and RLM5 and by microsomes, J. Biol. Chem. 258:11738-11744 (1983).

12. A.W. Wood, D.E. Ryan, P.E. Thomas and W. Levin, Regio-and stereoselective metabolism of two C_{19} steroids by highly purified and reconstituted rat hepatic cytochrome P-450 isozymes, J. Biol. Chem. 258:8839-8843 (1983).

13. K.-C. Cheng and J.B. Schenkman, Purification and characterization of two constitutive forms of rat liver microsomal cytochrome P-450, J. Biol. Chem. 257:2378-2385 (1982).

14. J.B. Schenkman, I. Jansson, W.L. Backes, K.-C. Cheng and C. Smith, Dissection of cytochrome P-450 isozymes (RLM) from fractions of untreated rat liver microsomal proteins, Biochem. Biophys. Res. Comm. 107:1517-1523 (1982).

15. P. Bentley, F. Oesch and A. Tsugita, Properties and amino acid composition of pure epoxide hydratase, FEBS Lett. 59:296-299 (1975).

16. K.W. Bock, D. Josting, W. Lilienblum and H. Pfeil, Purification of rat liver microsomal UDP-glucuronyltransferase. Separation of two enzyme forms inducible by 3-methylcholanthrene or phenobarbital, Eur. J. Biochem. 98:19-26 (1979).

17. I. Jansson, J. Mole and J.B. Schenkman, Purification and characterization of a new form (RLM2) of liver microsomal cytochrome P-450 from untreated rats, J. Biol. Chem. 260:7084-7093 (1985).

18. I. Jansson, P.P. Tamburini, L.V. Favreau and J.B. Schenkman, The interaction of cytochrome b_5 with four cytochromes P-450 from the untreated rat, Drug Metab. Dispo. 13:1304-1309 (1985).

19. C. MacGeoch, E.T. Morgan, J. Halpert and J.-A. Gustafsson, Purification, characterization and pituitary regulation of the sex-specific cytochrome P-450 15 ß-hydroxylase from liver microsomes of untreated female rats, J. Biol. Chem. 259:15433-15439 (1984).

20. E.T. Morgan, C. MacGeoch and J.-A. Gustafsson, Sexual differentiation of cytochrome P-450 in rat liver. Evidence for a constitutive isozyme as the male-specific 16α-hydroxylase, Molec. Pharmacol. 27:471-479 (1985).

21. R. Kato, E. Chiesara and G. Frontino, Influence of sex difference on the pharmacological action and metabolism of some drugs, Biochem. Pharmacol. 11:221-227 (1962).

22. J.B. Schenkman, I. Frey, H. Remmer and R.W. Estabrook, Sex differences in drug metabolism by rat liver microsomes, Mol. Pharmacol. 3:516-525 (1967).

23. T. Kamataki, K. Maeda, Y. Yamazoe, T. Nagai and R. Kato, Sex difference of cytochrome P-450 in the rat: Purification, characterization and quantitation of constitutive forms of cytochrome P-450 from liver microsomes of male and female rats, Arch. Biochem. Biophys. 225:758-770 (1983).

24. L.W.K. Chung and H. Chao, Neonatal imprinting and hepatic cytochrome P-450. I. Comparison of testosterone hydroxylation in a reconstituted system between neonatally imprinted and nonimprinted rats, Mol. Pharmacol. 18:543-549 (1980).

25. M. Haniu, D.E. Ryan, S. Iida, C.S. Lieber, W. Levin and J.E. Shively, NH_2-terminal sequence analyses of four rat hepatic microsomal cytochromes P-450, Arch. Biochem. Biophys. 235:304-311 (1984).

IDENTIFICATION OF INTRATISSUE SITES FOR XENOBIOTIC ACTIVATION AND

DETOXICATION

Jeffrey Baron, Jeffrey M. Voigt, Tyrone B. Whitter, Thomas
T. Kawabata, Shirley A. Knapp, F. Peter Guengerich[¶], and
William B. Jakoby[§]

The Toxicology Center, Department of Pharmacology, University
of Iowa, Iowa City, IA   52242; [¶]Center in Molecular
Toxicology, Department of Biochemistry, Vanderbilt University
Nashville, TN   37232; and [§]Laboratory of Biochemistry and
Metabolism, National Institute of Arthritis, Diabetes, and
Digestive and Kidney Diseases, National Institutes of Health
Bethesda, MD   20205

INTRODUCTION

    It is now apparent that the generation of reactive metabolites from a
multitude of xenobiotics frequently preceeds the appearance of necrosis,
mutagenesis, carcinogenesis, and other cytotoxicities (Mitchell et al.,
1976; Boyd, 1980; Wright, 1980; Miller and Miller, 1981, 1982; Boyd and
Statham, 1983). It is also evident, however, that xenobiotics that are
biotransformed into reactive metabolites usually exert relatively selective
toxic effects within most mammalian tissues, i.e., they often damage either
a specific morphological cell type or groups of morphologically similar
cells located within selected areas or regions of tissues (Mitchell et al.,
1976; Rappaport, 1979; Boyd, 1980; Baron and Kawabata, 1983; Minchin and
Boyd, 1983). Differential susceptibility of cells to toxicity that results
as a consequence of the formation of reactive metabolites undoubtedly is
related to differences in the ability of cells to both activate and
detoxicate xenobiotics (Boyd, 1980; Baron and Kawabata, 1983; Minchin and
Boyd, 1983). Primarily for this reason, the intratissue localizations and
distributions of cytochrome P-450 isozymes, NADPH-cytochrome P-450 reductase, epoxide hydrolase, glutathione S-transferases, and other enzymes that
participiate in the activation and detoxication of xenobiotics recently
have come under intensive investigation. Such investigation however, has
been hindered greatly by the heterogeneous nature of most mammalian tissues,
as well as by the fact that morphologically similar cells, hepatocytes for
example, can exhibit significant differences in their ability to metabolize
xenobiotics (Wattenberg and Leong, 1962; Gangolli and Wright, 1971; Ji et
al., 1981; Conway et al., 1982; Baron and Kawabata, 1983).

    In order to assess directly the potential of all cells found within
complex tissues to enzymatically activate and detoxicate xenobiotics, we
have made use of a combination of immunohistochemical and histochemical
procedures to investigate the intratissue localizations, distributions,
and inductions of cytochromes P-450 (Baron et al., 1978a, 1981, 1982, 1983,
1984; Redick et al., 1980; Baron and Kawabata, 1983, 1985; Kawabata et al.,

119

1984; Voigt et al., 1985), NADPH-cytochrome P-450 reductase (Baron et al.,
1978b, 1983, 1984; Taira et al., 1979, 1980a, 1980b; Redick et al., 1980;
Baron and Kawabata, 1983, 1985; Smith et al., 1983; Kawabata et al., 1984;
Voigt et al., 1985), epoxide hydrolase (Baron et al., 1980, 1983, 1984;
Redick et al., 1980; Kawabata et al., 1981, 1983, 1984; Baron and Kawabata,
1983, 1985; Ishii-Ohba et al., 1984), glutathione S-transferases (Redick
et al., 1982; Baron and Kawabata, 1983, 1985; Baron et al., 1983, 1984;
Kawabata et al., 1984; Ishii-Ohba et al., 1984), and aryl hydrocarbon
hydroxylase activity (Baron et al., 1984; Baron and Kawabata, 1985). In
the studies that are summarized here, immunohistochemical and histochemical
techniques were employed to determine where xenobiotics can be enzymatically
metabolized *in situ* within four tissues - the liver, skin, respiratory
tract, and pancreas - that are targets for the toxic actions of xenobiotics
biotransformed into reactive metabolites. The application of such methods
also enabled us to investigate the induction of selected enzymes and aryl
hydrocarbon hydroxylase activity within these tissues following *in vivo*
exposure to xenobiotics.

ENZYME IMMUNOHISTOCHEMISTRY AND HISTOCHEMISTRY

     For investigations on the intratissue localization and distribution
of xenobiotic activation and detoxication processes, antibodies were
raised against the following enzymes that had been purified to apparent
homogeneity from rat liver: cytochromes P-450 BNF-B, MC-B, PB-B, and PCN-E
[the major isozymes of cytochrome P-450 induced in rat hepatic microsomes
by β-naphthoflavone, 3-methylcholanthrene, phenobarbital, and pregnenolone-
16α-carbonitrile, respectively (Guengerich, 1978; Guengerich et al.,
1982a, 1982b)]; NADPH-cytochrome P-450 reductase (Taira et al., 1979);
epoxide hydrolase (Guengerich et al., 1979); and glutathione S-transferases
B (Habig et al., 1974), C (Habig et al., 1974), and E (Fjellstedt et al.,
1973). These antibodies were used in immunoperoxidase [unlabeled antibody
peroxidase-antiperoxidase (Sternberger et al., 1970) and/or avidin-biotin-
peroxidase (Hsu et al., 1981)] and indirect immunofluorescence staining
techniques to localize the enzymes at the light microscopic level in 4- to
7-μm-thick sections prepared from fixed, paraffin-embedded tissue (Baron
et al., 1978a, 1978b, 1980, 1981, 1982, 1983, 1984; Taira et al., 1979,
1980a, 1980b; Redick et al., 1980, 1982; Kawabata et al., 1981, 1983,
1984; Baron and Kawabata, 1983, 1985; Smith et al., 1983; Voigt et al.,
1985). Whenever antibodies raised against the nine enzymes isolated and
purified from rat liver cross-reacted with antigens in mouse and hamster
tissues, the antigens detected in these species most likely are immunochemically similar, rather than identical, to the rat enzymes. Furthermore,
immunohistochemical staining produced by these antibodies within rat
extrahepatic tissues also may indicate only the presence of antigens that
are similar, but not identical, to the rat liver enzymes. In addition,
since glutathione S-transferases C and A share a common subunit (Mannervik
and Jensson, 1982) and are immunochemically similar (Habig et al., 1974),
results obtained with antibody to transferase C must be interpreted as
indicating the presence of transferase C and/or transferase A.

     Immunoperoxidase staining gives rise to a brownish, particulate
deposit at the site of the antigen (when 3,3'-diaminobenzidine is employed
as the chromogen-substrate), whereas indirect immunofluorescence staining
results in a bright green fluorescence (when fluorescein isothiocyanate is
used as the fluorochrome). Different cell types in immunohistochemically-
stained sections were identified by examining adjacent sections that had
been stained with hematoxylin and eosin. To investigate quantitatively
the distribution and induction of enzymes across the liver lobule, the
intensities with which hepatocytes in different regions of the lobule were
stained for each enzyme were determined microfluorometrically after

completion of indirect immunofluorescence staining (Taira et al., 1979, 1980a, 1980b; Redick et al., 1980; Baron et al., 1981, 1982, 1983, 1984; Kawabata et al., 1981, 1983, 1984; Baron and Kawabata, 1983, 1984; Smith et al., 1983). Microfluorometric analyses permitted calculations of the relative extents of binding of an antibody to centrilobular, midzonal, and periportal hepatocytes, calculations which, in turn, served as indices of the relative levels of the enzyme within these cells.

Immunohistochemistry offers exquisite sensitivity and specificity for detecting and studying enzymes and other antigens, especially those whose overall tissue levels may be too low to be detected by more conventional biochemical and biophysical methodologies. Despite these advantages, this approach does not provide information regarding the biological function and/or activity of an antigen at the sites at which it is localized. The intratissue localization and distribution of certain cytochromes P-450-catalyzed xenobiotic monooxygenations, however, can be investigated <u>in situ</u> by means of histochemical methods (Wattenberg and Leong, 1962, 1970;

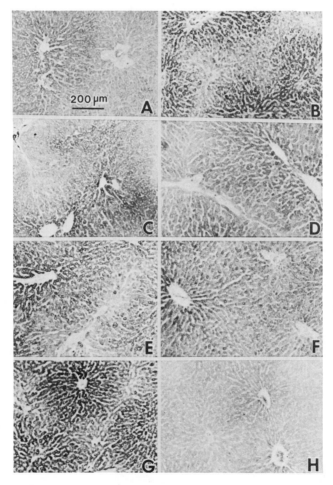

Figure 1.   Immunoperoxidase staining for NADPH-cytochrome P-450 reductase (*A*), cytochromes P-450 PB-B (*B*), MC-B (*C*), and PCN-E (*D*), epoxide hydrolase (*E*), and glutathione S-transferases B (*F*), C/A (*G*), and E (*H*) within livers of untreated rats. Reprinted from Baron et al. (1983), with permission.

Gangolli and Wright, 1971; Grasso et al. 1971; Gati et al., 1973). Thus, in order to determine more directly the sites for xenobiotic metabolism, especially those at which xenobiotics are bioactivated by cytochromes P-450, within the liver, skin, respiratory tract, and pancreas, aryl hydrocarbon hydroxylase activity was studied in unfixed cryostat sections using a modification (Baron et al., 1983) of the procedure developed by Wattenberg and Leong (1962) with which the yellowish-green fluorescence of phenolic benzo[a]pyrene metabolites can be visualized in cryostat sections.

XENOBIOTIC METABOLISM WITHIN THE LIVER

Intralobular Localization and Distribution of Xenobiotic-Metabolizing Enzymes and Aryl Hydrocarbon Hydroxylase Activity in Livers of Untreated Rats

Representative results of immunoperoxidase staining for cytochromes P-450 PB-B, MC-B, and PCN-E, NADPH-cytochrome P-450 reductase, epoxide hydrolase, and glutathione S-transferases B, C/A, and E within livers of untreated rats are presented in Fig. 1. Although not shown, indirect immunofluorescence staining yielded identical results (Baron et al., 1983). These data conclusively demonstrate that hepatocytes (parenchymal cells) throughout the liver are stained for each enzyme. There are some obvious differences, however, in the intensities with which hepatocytes in the different regions of the lobule are stained for specific enzymes: the most intense staining for each enzyme appeared to be produced within centrilobular cells, i.e., within those hepatocytes that are damaged most frequently by xenobiotics transformed into reactive metabolites (Mitchell et al., 1976; Rappaport, 1979; Baron and Kawabata, 1983.)

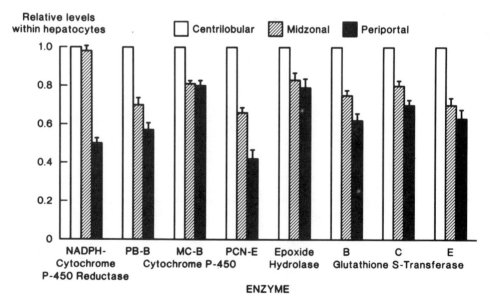

Figure 2.   Intralobular distribution of xenobiotic-metabolizing enzymes in livers of untreated rats. Microfluorometric determinations of the relative extents of binding of antibodies to hepatocytes were normalized by assigning a value of 1.0 for the extent of binding of each antibody to centrilobular cells.

Microfluorometric analysis of indirect immunofluorescence staining verified the visual findings (Fig. 2). The data demonstrate that none of the nine enzymes are distributed uniformly across the liver lobule and, furthermore, that there are significant differences in their patterns of intralobular distribution. However, except for NADPH-cytochrome P-450 reductase which is present at essentially identical levels within centrilobular and midzonal cells, centrilobular hepatocytes contain the greatest amount of each enzyme. These immunohistochemical findings suggest that centrilobular hepatocytes of untreated rats possess the greatest capacity for oxidatively metabolizing and, hence, activating xenobiotics, for hydrating and inactivating potentially toxic epoxide metabolites, and for enzymatically conjugating electrophilically-reactive metabolites with reduced glutathione. Consistent with these conclusions is the observation that the greatest benzo[a]pyrene hydroxylase activity occurs around central veins, although the xenobiotic can be seen to be hydroxylated throughout the lobule in livers of untreated rats (Fig. 3, panel *A*).

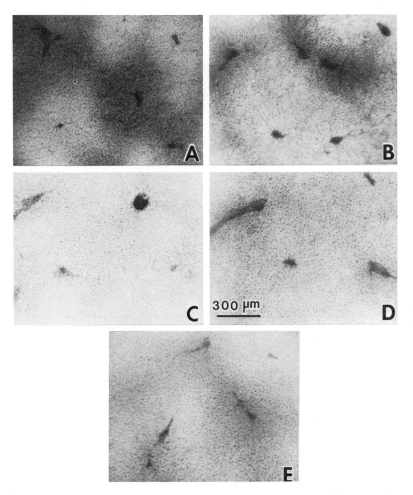

Figure 3.   Aryl hydrocarbon hydroxylase activity within livers of untreated (*A*) and phenobarbital (*B*), 3-methylcholanthrene (*C*), *trans*-stilbene oxide (*D*), and pregnenolone-16α-carbonitrile (*E*) pretreated rats. The photomicrographs show fluorescence due to benzo[a]pyrene phenols emitted from unfixed cryostat sections. Reprinted from Baron et al. (1983), with permission.

Intralobular Induction of Xenobiotic-Metabolizing Enzymes and Aryl Hydrocarbon Hydroxylase Activity

After determining the intralobular localization and distribution of xenobiotic-metabolizing enzymes and aryl hydrocarbon hydroxylase activity in livers of untreated rats, the intralobular induction of cytochromes P-450 MC-B, PB-B, and PCN-E, NADPH-cytochrome P-450 reductase, epoxide hydrolase, and benzo[a]pyrene hydroxylase activity was investigated. To assess both the location and relative extent of induction of enzymes across the liver lobule, immunohistochemical and histochemical analyses were conducted after male rats had been pretreated with phenobarbital [40 mg/kg/day, i.p. in saline for 4 and 7 days (epoxide hydrolase was studied after 7 days)], 3-methylcholanthrene (25 mg/kg/day, i.p. in corn oil for 3-4 days), pregnenolone-16α-carbonitrile (40 mg/kg, p.o. in aqueous Tween 80 every 12 hours for 4 days), *trans*-stilbene oxide (300 mg/kg/day, i.p. in corn oil for 3 days), and β-naphthoflavone (40 mg/kg/day, i.p. in corn oil for 3-4 days).

Quantitative immunohistochemical analyses demonstrated that cytochromes P-450 MC-B, PB-B, and PCN-E, NADPH-cytochrome P-450 reductase, and epoxide hydrolase usually are not induced uniformly across the lobule, with the least degree of induction often occurring within centrilobular cells (Fig. 4). Thus, these findings suggest that inducers of xenobiotic-metabolizing enzymes, especially of cytochromes P-450, can alter significantly the extents to which cells within different regions of the lobule participate in oxidative xenobiotic metabolism. The photomicrographs in Fig. 3 reinforce this conclusion and, moreover, indicate that modifications in the intralobular distributions of cytochrome P-450 isozymes are paralleled by similar alterations in the distribution of aryl hydrocarbon hydroxylase activity. As seen in panel *B* of Fig. 3, phenobarbital enhances benzo[a]-

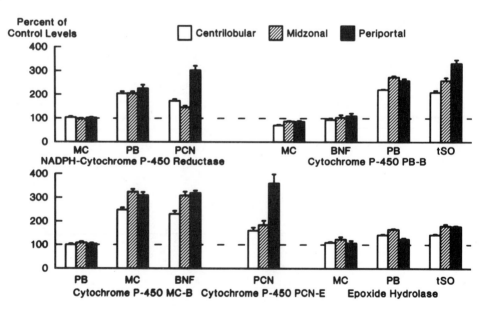

Figure 4.   Intralobular induction of enzymes by phenobarbital (PB), 3-methylcholanthrene (MC), pregnenolone-16α-carbonitrile (PCN), *trans*-stilbene oxide (tSO), and β-naphthoflavone (BNF). Percent of control levels were calculated from microfluorometric determinations of the relative extents of binding of specific antibodies to corresponding cells within livers of vehicle and xenobiotic pretreated rats.

pyrene hydroxylase activity throughout the lobule, although centrilobular cells still are considerably more active than are periportal cells. On the other hand, 3-methylcholanthrene (Fig. 3, panel *C*), *trans*-stilbene oxide (Fig. 3, panel *D*), and pregnenolone-16α-carbonitrile (Fig. 3, panel *E*) significantly reduce the difference in activity between centrilobular and periportal cells, i.e., each induces benzo[a]pyrene hydroxylase activity to a greater degree in periportal cells than in centrilobular cells. Thus, in addition to modifying the rates and extents of formation and detoxication of reactive metabolites, inducers of cytochrome P-450 isozymes also can differentially alter the extents to which different cells within a tissue such as the liver participate in the oxidative metabolism of a xenobiotic. Such alterations may explain, at least in part, why 3-methylcholanthrene and pregnenolone-16α-carbonitrile can protect against the centrilobular necrosis that results as a consequence of the bioactivation of xenobiotics such as carbon tetrachloride (Reid et al., 1971; Tuchweber et al. 1974).

XENOBIOTIC METABOLISM WITHIN THE SKIN

The skin is exposed continuously to a myriad of environmental chemicals, is a major portal of entry of these and other xenobiotics into the body, and is a target for the toxic actions of many xenobiotics, especially polycyclic aromatic hydrocarbons and other procarcinogens that are transformed into reactive metabolites (Miller and Miller, 1981; Slaga, 1984). Although often considered to act merely as a passive barrier, the skin clearly is capable of metabolizing endogenous substances as well as drugs and other exogenous chemicals (Pannatier et al., 1978; Bickers and Kappas, 1980).

The skin is a relatively heterogeneous tissue, consisting of an epidermis, a stratified, squamous, keratinizing epithelium, and a dermis, a thick, dense connective tissue layer containing blood vessels, nerves, hair follicles, and sebaceous glands. Primarily because of its complex nature, only limited information is available regarding the intracutaneous sites for xenobiotic metabolism. On the other hand, results of a number of studies (Nakai and Shubik, 1964; Wattenberg and Leong, 1970; Grasso et al., 1971; Gati et al., 1973; Vermorken et al., 1979) have shown that xenobiotics can be oxidatively metabolized within the epidermis as well as within the outer root sheath of hair follicles and sebaceous glands, both of which are derived from the epidermis. Results of our immunohistochemical and histochemical investigations are consistent with these findings.

Xenobiotic-Metabolizing Enzymes and Aryl Hydrocarbon Hydroxylase Activity within Skin of Untreated Rats

Immunoperoxidase staining produced by antibodies to rat hepatic cytochromes P-450 BNF-B, MC-B, PB-B, and PCN-E, NADPH-cytochrome P-450 reductase, epoxide hydrolase, and glutathione S-transferases B, C, and E within skin obtained from shaved, dorsal regions of female rats during the telogen, or resting phase, of the hair growth cycle (Butcher, 1934) is shown in Fig. 5. As can be seen, antibodies to cytochromes P-450 PB-B and PCN-E and the glutathione S-transferases stain epidermal cells, the outer root sheath of hair follicles, and sebaceous glands. Epidermal cells and sebaceous glands also are stained by antibodies to the cytochrome P-450 reductase and epoxide hydrolase, whereas these antibodies produce only minimal staining in hair follicles. Anti-cytochromes P-450 BNF-B and MC-B produce extremely weak staining in the epidermis and sebaceous glands, and even less in hair follicles. In agreement with results obtained with the anti-cytochromes P-450 BNF-B and MC-B, as well as with earlier reports (Wattenberg and Leong, 1970; Gati et al., 1973), virtually no benzo[a]pyrene hydroxylase activity could be detected histochemically within skin of untreated rats (Fig. 6, row *A*).

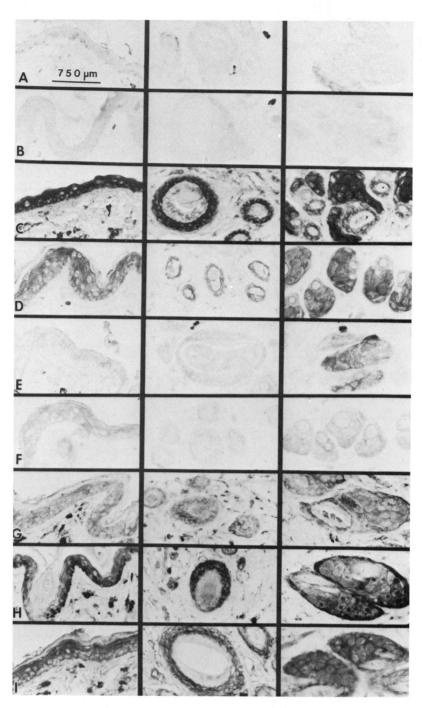

Figure 5. Immunoperoxidase staining within epidermis (left column), hair follicles (middle column), and sebaceous glands (right column) in skin of untreated rats produced by antibodies to cytochromes P-450s BNF-B (*A*), MC-B (*B*), PB-B (*C*), and PCN-E (*D*), NADPH-cytochrome P-450 reductase (*E*), epoxide hydrolase (*F*), and glutathione S-transferases B (*G*), C (*H*), and E (*I*).

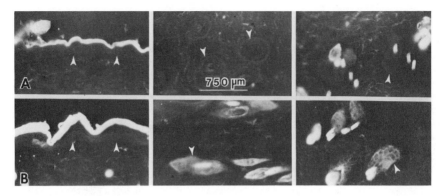

Figure 6. Benzo[a]pyrene hydroxylase activity within the epidermis (left column), hair follicles (middle column), and sebaceous glands (right column) in skin of untreated (A) and 3-methylcholanthrene pretreated (B) rats. Arrowheads indicate the basal cell layer of the epidermis, the outer root sheath of hair follicles, and sebaceous glands. Nonspecific fluorescence of keratin is seen at the edge of the epidermis, while that of hair is seen within hair follicles and/or associated with sebaceous glands.

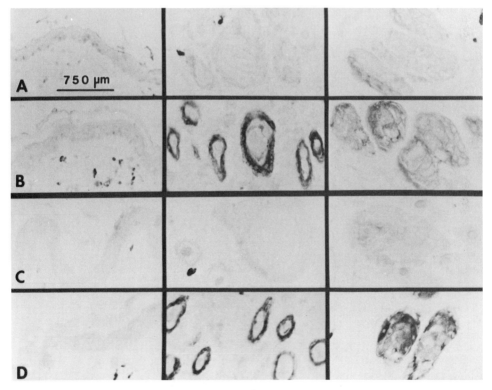

Figure 7. Immunoperoxidase staining within epidermis (left column), hair follicles (middle column), and sebaceous glands (right column) in skin of untreated (A and C) and 3-methylcholanthrene-treated (B and D) rats produced by antibodies raised against rat hepatic cytochromes P-450 BNF-B (A and B) and MC-B (C and D).

Induction of Cytochromes P-450 and Aryl Hydrocarbon Hydroxylase Activity within Rat Skin

Comparisons between the photomicrographs in rows *A* and *B* and between those in rows *C* and *D* of Fig. 7 illustrate that the topical application of 3-methylcholanthrene (4.5 μmol/rat/day, in acetone) for four consecutive days results in dramatic increases in the intensities with which hair follicles and sebaceous glands are stained by antibodies to rat hepatic microsomal cytochromes P-450 BNF-B and MC-B. Similar alterations are not apparent, however, within the epidermis. In accord with these findings, the topical application of 3-methylcholanthrene causes benzo[a]pyrene hydroxylase activity to appear within both hair follicles and sebaceous glands, while the activity remains essentially undetectable in the epidermis (Fig. 6, row *B*). These observations again serve to illustrate that, within a specific tissue, xenobiotic-metabolizing enzymes and monooxygenase activity can be affected to significantly different extents within different cell types capable of metabolizing xenobiotics.

XENOBIOTIC METABOLISM WITHIN THE RESPIRATORY TRACT

The Lung

The lung is another major portal of entry of xenobiotics into the body, is exposed continuously to chemicals that are present in the general circulation, and represents another organ in which numerous xenobiotics, including many that are transformed into reactive metabolites, selectively damage different cell types (Boyd, 1980; Baron and Kawabata, 1983; Boyd and Statham, 1983; Minchin and Boyd, 1983). Major sites within the lung at which xenobiotics can be metabolized include the bronchus, bronchiole, and alveolar wall (Wattenberg and Leong, 1962; Grasso et al., 1971; Baron and Kawabata, 1983; Boyd and Statham, 1983; Minchin and Boyd, 1983). The precise identification of those cells located within these three pulmonary structures that can metabolize xenobiotics, however, has been hampered by the complex nature of the organ: at least 40 distinctive cell types are present (Sorokin, 1970), some of which (e.g., type II pneumocytes) cannot be identified unequivocally at the light microscopic level. Nevertheless, bronchial epithelial cells (Grasso et al., 1971; Cohen and Moore, 1976; Kahng et al., 1981), nonciliated bronchiolar epithelial (Clara) cells (Boyd, 1977, 1980; Devereux et al., 1982; Boyd and Statham, 1983; Minchin and Boyd, 1983), and type II pneumocytes (Devereux et al., 1979; Teel and Douglas, 1980; Devereux and Fouts, 1981; Devereux et al., 1982) have each been shown conclusively to be capable of oxidatively metabolizing and activating xenobiotics. Our immunohistochemical and histochemical findings are consistent with the participation of these pulmonary cell types in xenobiotic metabolism and, further, suggest that xenobiotics also can be metabolized within ciliated bronchiolar epithelial cells as well as within alveolar wall cells other than type II pneumocytes.

Xenobiotic-Metabolizing Enzymes and Aryl Hydrocarbon Hydroxylase Activity within Lungs of Untreated Rats. The photomicrographs in Fig. 8 demonstrate that antigens related to rat hepatic cytochromes P-450 BNF-B, PB-B, and PCN-E, NADPH-cytochrome P-450 reductase, epoxide hydrolase, and glutathione S-transferases B, C/A, and E are present within the bronchial epithelium, both Clara and ciliated epithelial cells of the bronchiole, as well as within type II pneumocytes and possibly other cells in alveolar walls in lungs of untreated rats. Marked differences in the intensities with which bronchial, bronchiolar, and alveolar cells are stained for the various antigens are obvious, however, with the alveolar wall being stained least intensely by each antibody.

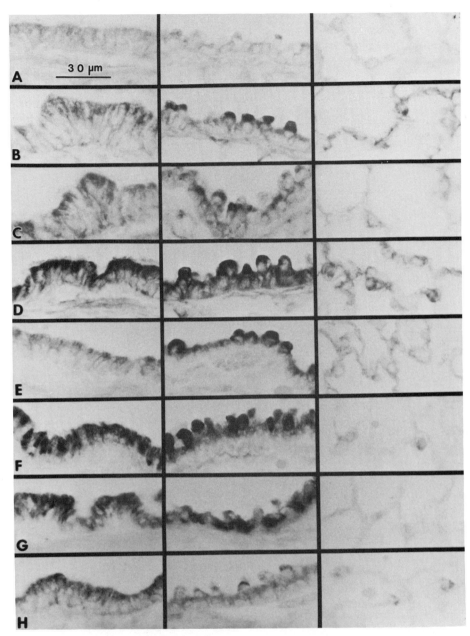

Figure 8.   Immunoperoxidase staining within bronchial (left column) and
            bronchiolar (middle column) epithelia and the alveolar wall
            (right column) in lungs of untreated rats produced by antibodies
            raised against rat hepatic cytochromes P-450s BNF-B (*A*), PB-B
            (*B*), and PCN-E (*C*), NADPH-cytochrome P-450 reductase (*D*),
            epoxide hydrolase (*E*), and glutathione S-transferases B (*F*), C
            (*G*), and E (*H*).

In agreement with the findings that antigens related to rat hepatic microsomal cytochromes P-450 BNF-B, PB-B, and PCN-E are present within the bronchial and bronchiolar epithelia and the alveolar walls, the photomicrographs in row *C* of Fig. 9 illustrate that benzo[a]pyrene hydroxylase activity also can be detected histochemically within these structures in lungs of untreated rats.

Induction of Cytochromes P-450 and Aryl Hydrocarbon Hydroxylase Activity within the Rat Lung. Although an antigen related to rat hepatic microsomal cytochrome P-450 MC-B could not be detected immunohistochemically within lungs of untreated rats, the presence of this antigen was observed in some, but not all, bronchial epithelial cells, Clara cells, as well as within the alveolar wall, especially within type II pneumocytes, 7 days

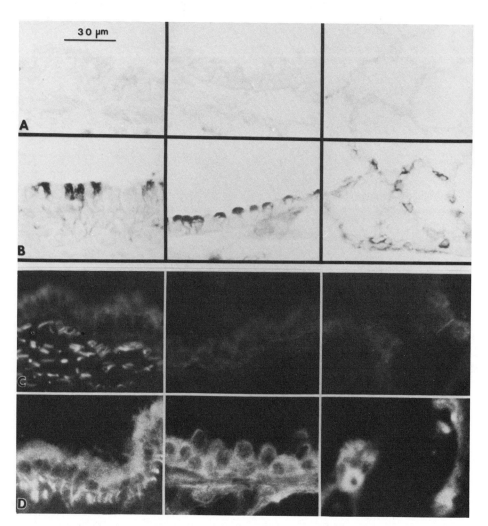

Figure 9.   Immunoperoxidase staining produced by the antibody raised against rat hepatic microsomal cytochrome P-450 MC-B (*A* and *B*) and benzo[a]pyrene hydroxylase activity (*C* and *D*) within bronchial (left column) and bronchiolar (middle column) epithelia and the alveolar wall (right column) in lungs of untreated (*A* and *C*) and Aroclor 1254 pretreated (*B* and *D*) rats.

after rats had received a single 500 mg/kg dose of Aroclor 1254 (i.p. in corn oil) (Fig. 9). Aroclor 1254 pretreatment similarly caused enhanced staining of some, but not all, bronchial and bronchiolar epithelial cells and of the alveolar wall by antibody to cytochrome P-450 BNF-B (data not shown). The photomicrographs in Fig. 9 demonstrate that the administration of Aroclor 1254 results in a dramatic increase in pulmonary benzo[a]pyrene hydroxylase activity. In this instance, however, induction can be seen to occur throughout the bronchial and bronchiolar epithelia and the alveolar wall, an observation suggesting that other cytochrome P-450 isozymes capable of catalyzing the hydroxylation of benzo[a]pyrene are induced by Aroclor 1254 throughout these structures.

The Nasal Mucosa

The nasal mucosa is another tissue exposed continuously to air-borne xenobiotics as well as to chemicals present in the general circulation. It represents another potentially important area within the respiratory tract where xenobiotics can be metabolized and frequently is damaged by xenobiotics, including many that must undergo bioactivation in order to exert their toxic effects (Hecht et el., 1980; Reznik et al., 1980; Lee et al., 1982; Brittebo et al., 1983). Virtually nothing is known, however, regarding the specific sites at which xenobiotics can be metabolized within this tissue.

The nasal mucosa is another relatively complex and heterogeneous tissue, consisting of olfactory and respiratory regions, both of which are lined by a columnar epithelium that lies above a lamina propria. In the olfactory region, the lamina propria contains acini and ducts of Bowman's glands, while seromucous gland acini and ducts are found in the lamina propria in the respiratory region. Autoradiographic analyses have demonstrated that reactive metabolites of tobacco-specific nitrosamines preferentially bind to the olfactory and respiratory epithelia and to glands in the lamina propria of both regions (Brittebo et al., 1983). These cells may then represent the primary sites for xenobiotic activation within the nasal mucosa.

Xenobiotic-Metabolizing Enzymes and Aryl Hydrocarbon Hydroxylase Activity within the Nasal Mucosa of Untreated Rats.
The photomicrographs in Fig. 10 illustrate immunoperoxidase staining produced by the antibodies that were raised against rat hepatic cytochromes P-450 BNF-B, MC-B, PB-B, and PCN-E, NADPH-cytochrome P-450 reductase, epoxide hydrolase, and gluta-thione S-transferases B, C, and E within nasal tissue of untreated rats. Antigens related to the rat hepatic enzymes are observed within olfactory and respiratory epithelial cells as well as within Bowman's and seromucous glands in the lamina propria of these regions. There are, however, a number of obvious differences in the intensities of staining produced by some antibodies between the olfactory and respiratory regions, as well as among the different cell types present within each region. Consistent with observations that antibodies to cytochromes P-450 BNF-B and MC-B produce significantly more intense staining within cells in the olfactory region, the photomicrographs in row *C* of Fig. 11 demonstrate that the olfactory epithelium and Bowman's glands exhibit greater aryl hydrocarbon hydroxylase activity than do the epithelium and seromucous glands in the respiratory region.

Induction of Cytochromes P-450 and Aryl Hydrocarbon Hydroxylase Activity within Rat Nasal Mucosa.
Seven days after rats had been treated with a single 500 mg/kg dose of Aroclor 1254 (i.p. in corn oil), dramatic increases are seen in the intensities with which the olfactory epithelium and Bowman's glands are stained by the antibody to cytochrome P-450 MC-B (Fig. 11). Considerably less pronounced alterations in staining by the

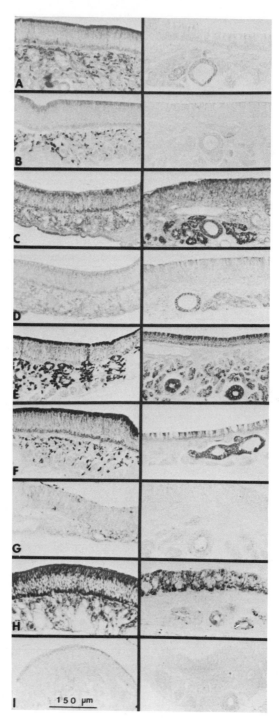

Figure 10. Immunoperoxidase staining within olfactory (left column) and respiratory (right column) regions in the nasal mucosa of untreated rats produced by antibodies raised against rat hepatic cytochromes P-450s BNF-B (*A*), MC-B (*B*), PB-B (*C*), and PCN-E (*D*), NADPH-cytochrome P-450 reductase (*E*), epoxide hydrolase (*F*), and glutathione S-transferases B (*G*), C (*H*), and E (*I*).

anti-cytochrome P-450 MC-B are apparent within the respiratory epithelium and seromucous glands. Thus, after rats have been pretreated with Aroclor 1254, the respiratory region still contains significantly lower levels of the antigen related to hepatic microsomal cytochrome P-450 MC-B than does the olfactory region. Although the data are not presented, comparable results were obtained with the antibody raised to rat hepatic microsomal cytochrome P-450 BNF-B. In contrast, although Aroclor 1254 significantly enhances aryl hydrocarbon hydroxylase activity within the olfactory epithelium, a much more pronounced increase in activity occurs within the respiratory epithelium and seromucous glands which now appear to be the major sites for the monooxygenation of benzo[a]pyrene and possibly other xenobiotics within the nasal mucosa.

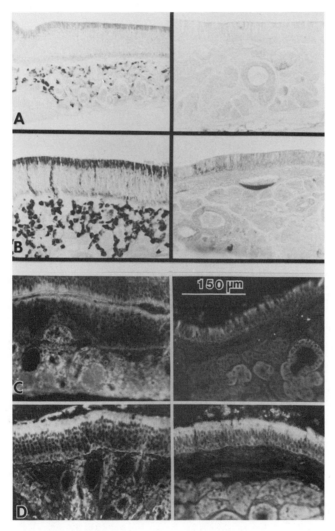

Figure 11.   Immunoperoxidase staining produced by the antibody directed against rat hepatic microsomal cytochrome P-450 MC-B (A and B) and benzo[a]pyrene hydroxylase activity (C and D) within olfactory (left column) and respiratory regions (right column) in the nasal mucosa of untreated (A and C) and Aroclor 1254 pretreated (B and D) rats.

The pancreas represents still another example of a complex, hetero-geneous tissue in which many xenobiotics that require bioactivation before producing cytotoxicity selectively damage different cell types (Lazarus and Shapiro, 1972; Flaks and Lucas, 1973; Pour at al., 1979; Pour and Wilson, 1980; Bockman, 1981; Flaks et al., 1982; Chow and Fischer, 1984). Of primary importance, the exocrine pancreas is one of the major targets for the carcinogenic actions of nitrosamines (Levitt et al., 1977; Pour et al., 1979; Pour and Wilson, 1980; Flaks et al., 1981, 1982) and polycyclic aromatic hydrocarbons (Dissin et al., 1975; Bockman et al., 1976, 1978; Bockman, 1981). There are important species differences, especially with respect to the putative cells of origin of chemically-induced exocrine pancreatic adenocarcinomas in rats and hamsters (Bockman et al., 1976, 1978; Levitt et al., 1977; Pour, 1978; Pour et al., 1979; Pour and Wilson, 1980; Bockman, 1981; Flaks et al. 1981, 1982). Such differences may be due, at least in part, to differences in the intrapancreatic distribution of xenobiotic-metabolizing enzymes in rats and hamsters. Attempts to identify the precise sites at which xenobiotics can be metabolized within this tissue, however, have failed to provide conclusive information, primarily because the exocrine pancreas contains several different cell types: acinar cells and cells associated with the pancreatic duct system, i.e., centroacinar cells and epithelial cells of intralobular, interlobular, and common ducts. Furthermore, although various xenobiotics can be oxida-tively metabolized within the pancreas (Iqbal et al. 1977; Black et al., 1980, 1981; Scarpelli et al., 1980, 1982; Black and Stoming, 1983; Wiebkin et al., 1984), and although inducers of cytochromes P-450 and cytochromes P-450-cataylzed xenobiotic monooxygenations in extrapancreatic tissues also are capable of stimulating xenobiotic metabolism in the pancreas (Iqbal et al., 1977; Black et al., 1980; Scarpelli et al., 1980, 1982; Black and Stoming, 1983; Wiebkin et al., 1984), it has not been possible to detect the presence of cytochrome P-450 in pancreatic preparations by means of conventional spectrophotometric methodology (Ichikawa and Yamano, 1967).

Xenobiotic-Metabolizing Enzymes and Aryl Hydrocarbon Hydroxylase Activity within Pancreases of Untreated Rats and Hansters

Rat Pancreas. The photomicrographs in Fig. 12 demonstrate that antibodies raised against rat hepatic NADPH-cytochrome P-450 reductase, epoxide hydrolase, and glutathione S-transferases B, C, and E stain acinar cells, the epithelia of intralobular, interlobular, and common ducts, and cells within islets in the rat pancreas. Staining for an antigen related to rat hepatic microsomal cytochrome P-450 BNF-B could not be detected readily within either rat pancreatic islets or duct epithelial cells, whereas extremely low levels of this antigen appear to be present within acinar cells (Fig. 13, column A). In contrast, the antibody directed against rat hepatic microsomal cytochrome P-450 PB-B produced appreciable staining of islet as well as acinar cells (Fig. 13, column D). Antibody to cytochrome P-450 PB-B also stained, albeit with extremely weak intensity, epithelial cells of both intralobular and interlobular ducts.

In agreement with these observations, rat pancreatic acinar cells and islets exhibit low, but detectable, benzo[a]pyrene hydroxylase activity, whereas virtually no hydroxylase activity occurs within the epithelia of either intralobular or interlobular ducts (Fig. 13, column D). Although a low level of hydroxylase activity also can be detected within the epithelium of the common duct, these results are consistent with the suggestion that the acinar cell is the site of the initial damage induced by polycyclic aromatic hydrocarbons in the rat pancreas and, therefore, may represent

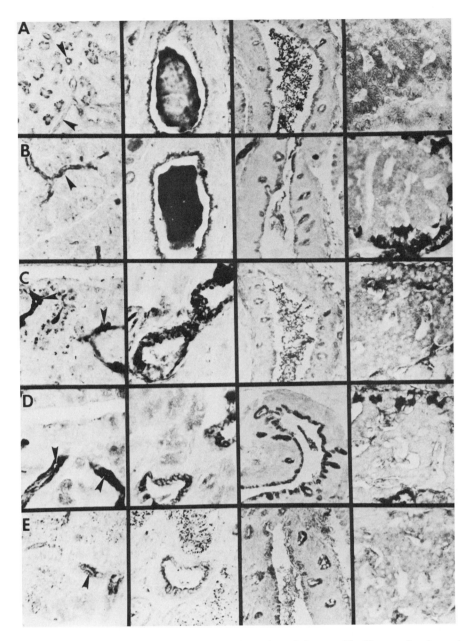

Figure 12. Immunoperoxidase staining produced by antibodies raised against
rat hepatic NADPH-cytochrome P-450 reductase (A), epoxide
hydrolase (B), and glutathione S-transferases B (C), C (D),
and E (E) within (columns from left to right) acinar cells and
intralobular ducts (arrowheads), medium and large interlobular
ducts, common ducts, and islets in pancreases of untreated rats.

the cell of origin of the exocrine pancreatic adenocarcinomas induced in
rats by these carcinogens (Bockman et al., 1976, 1978; Bockman, 1981).

Hamster Pancreas. Immunohistochemical staining produced within the
hamster pancreas by most of the antibodies raised against rat hepatic
enzymes closely resembled that seen within the rat pancreas (Kawabata et

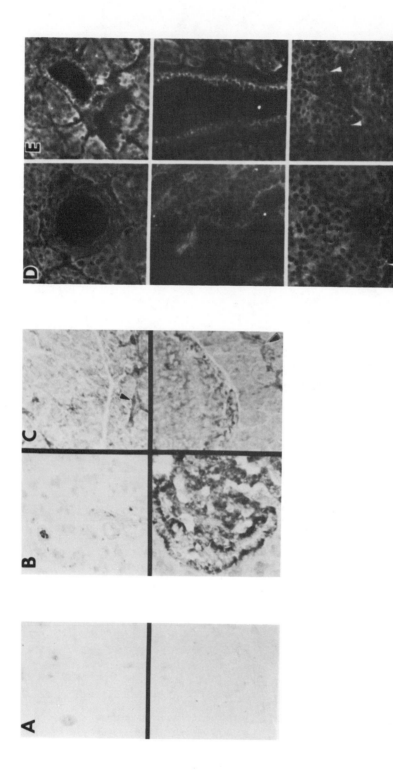

Figure 13. Immunoperoxidase staining produced by antibodies raised against rat hepatic microsomal cytochromes P-450 BNF-B (column *A*) and PB-B (columns *B* and *C*) and aryl hydrocarbon hydroxylase activity (columns *D* and *E*) within pancreases of untreated rats (columns *A*, *B*, and *D*) and hamsters (columns *C* and *E*). Arrowheads indicate intralobular ducts (column *C*) and edges of islets (columns *D* and *E*).

al., 1984; Baron and Kawabata, 1985). However, an antigen related to rat hepatic microsomal cytochrome P-450 PB-B is present within pancreatic ducts as well as acinar and islet cells in the hamster pancreas (Fig. 13, column C). Consistent with this finding, duct epithelia in the hamster pancreas exhibit significant degrees of benzo[a]pyrene hydroxylase activity: these cells, in fact, appear be more active in hydroxylating this xenobiotic than are either acinar or islet cells (Fig. 13, column E). Thus, these findings may explain why the duct epithelial cell appears to be the cell of origin of chemically-induced exocrine pancreatic adenocarcinomas in the hamster (Levitt et al., 1977; Pour et al., 1977, 1979; Takahaski et al., 1977; Pour, 1978; Pour and Wilson, 1980).

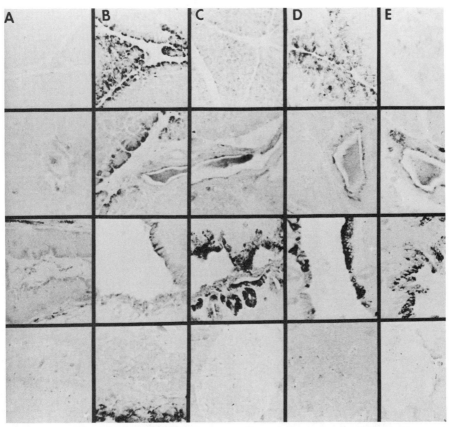

Figure 14. Effects of 3-methylcholanthrene and Aroclor 1254 pretreatments on immunoperoxidase staining produced by the antibody directed against rat hepatic microsomal cytochrome P-450 BNF-B within (top to bottom rows) rat pancreatic lobules, acinar cells and interlobular ducts, common ducts, and islets. Pretreatments: A, none; B, 3-methylcholanthrene administered i.p. (25 mg/kg/day in corn oil for 5 days); C, 3-methylcholanthrene administered p.o. (50 mg/kg/ day in corn oil for 5 days); D, Aroclor 1254 administered i.p. (a single 500 mg/kg dose in corn oil 5 days before rats were sacrificed); and E, Aroclor 1254 administered p.o. (a single 500 mg/kg dose in corn oil 5 days before rats were sacrificed).

Induction of Cytochrome P-450 and Aryl Hydrocarbon Hydroxylase Activity within the Rat Pancreas

Although alterations in staining for antigens related to rat hepatic NADPH-cytochrome P-450 reductase, epoxide hydrolase, and glutathione S-transferases B, C/A, and E were not apparent within the pancreas after rats had been pretreated with either 3-methylcholanthrene or Aroclor 1254, both xenobiotics caused an enhanced degree of staining of acinar and duct epithelial cells by the antibody to cytochrome P-450 BNF-B (Fig. 14). The photomicrographs shown in Fig. 14 illustrate, however, that the route of administration of each xenobiotic profoundly influences the intrapancreatic induction of the antigen related to rat hepatic microsomal cytochrome P-450 BNF-B. For example, when administered intraperitoneally, 3-methylcholanthrene (column *B*) and Aroclor 1254 (column *D*) induce this antigen primarily within those acinar cells that are located at or near the periphery of the pancreatic lobules, whereas oral administration of these xenobiotics results in a more uniform induction of this antigen throughout the lobule (columns *C* and *E*).

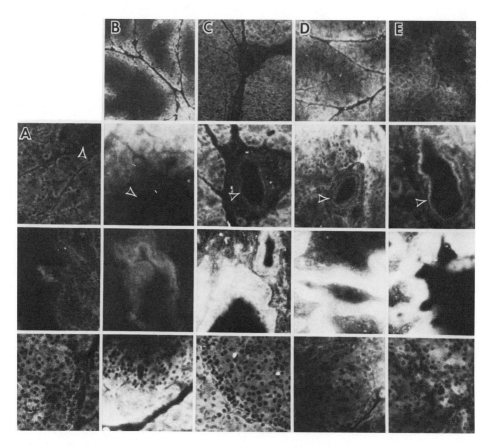

Figure 15. Effects of 3-methylcholanthrene and Aroclor 1254 on benzo[a]-pyrene hydroxylase activity within (top to bottom rows) rat pancreatic lobules, acinar cells and interlobular ducts, common ducts, and islets. Pretreatments were the same as those described for Figure 14.

Observations on the intrapancreatic induction of aryl hydrocarbon hydroxylase activity by 3-methylcholanthrene and Aroclor 1254 shown in Fig. 15 are entirely consistent with the effects of these xenobiotics on the antigen related to cytochrome P-450 BNF-B. Thus, the intraperitoneal administrations of 3-methylcholanthrene (column *B*) and Aroclor 1254 (column *D*) result in the most pronounced enhancement of hydroxylase activity at the edge of the pancreatic lobule, while hydroxylase activity is enhanced quite uniformly throughout the lobule after the oral administration of each inducer (columns *C* and *E*). Also in agreement with their effects on the antigen related to rat hepatic microsomal cytochrome P-450 BNF-B (Fig. 14), both 3-methylcholanthrene and Aroclor 1254 enhance benzo[a]pyrene hydroxylase activity within the epithelia of rat pancreatic ducts (Fig. 15). Moreover, epithelial cells of the larger interlobular ducts and the common duct seem to be the sites at which benzo[a]pyrene undergoes oxidative metabolism to the greatest extent within pancreases of rats pretreated with either 3-methylcholanthrene or Aroclor 1254.

SUMMARY AND CONCLUSIONS

Results of immunohistochemical and histochemical investigations on xenobiotic-metabolizing enzymes and aryl hydrocarbon hydroxylase activity have demonstrated that xenobiotic activation and detoxication do not occur uniformly throughout the liver, skin, respiratory tract, and pancreas, four tissues that are targets for the toxic actions of xenobiotics that are biotransformed into reactive metabolites. It has been shown that there can be significant differences in the levels and activities of xenobiotic-metabolizing enzymes among even morphologically similar cells, that an inducer can affect a specific xenobiotic-metabolizing enzyme to significantly different extents within different cells in a tissue, and that inducers of xenobiotic-metabolizing enzymes can alter differentially the extents to which different cells within a tissue participate in xenobiotic metabolism. These studies also have revealed that the route of administration of an inducer can affect significantly the induction of xenobiotic-metabolizing enzymes and aryl hydrocarbon hydroxylase activity within an organ such as the pancreas.

Some of the immunohistochemical findings reported for the cellular localizations of xenobiotic-metabolizing enzymes within specific tissues, e.g., the nasal mucosa, may not appear to be entirely consistent with the intratissue distribution of benzo[a]pyrene hydroxylase activity, especially after induction. However, it must be appreciated that other cytochrome P-450 isozymes undoubtedly are present within these tissues which, although not studied, also are capable of catalyzing aryl hydrocarbon hydroxylase activity.

ACKNOWLEDGMENTS

This research was supported in part by United States Public Health Service Grants GM33253, CA30140, ES01590, ES00267, and ES02205.

REFERENCES

Baron, J., and Kawabata, T. T., 1983, Intratissue distribution of activating and detoxicating enzymes, in: "Biological Basis of Detoxication," J. Caldwell and W. B. Jakoby, eds., p. 105, Academic Press, New York.
Baron, J., and Kawabata, T. T., 1985, Localization and distribution of carcinogen-metabolizing enzymes and benzo[a]pyrene hydroxylase activity within rat and hamster pancreas, in: "Experimental Pancreatic

Carcinogenesis," D. G. Scarpelli, J. R. Reddy, and D. S. Longnecker, eds., CRC Press, Boca Raton.

Baron, J., Redick, J. A., and Guengerich, F. P., 1978a, Immunohistochemical localizations of cytochromes P-450 in rat liver, Life Sci., 23:2627.

Baron, J., Redick, J. A., Greenspan, P., and Taira, Y., 1978b, Immunohistochemical localization of NADPH-cytochrome c reductase in rat liver, Life Sci., 22:1097.

Baron, J., Redick, J. A., and Guengerich, F. P., 1980, Immunohistochemical localization of epoxide hydratase in rat liver, Life Sci., 26:489.

Baron, J., Redick, J. A., and Guengerich, F. P., 1981, An immunohistochemical study on the localizations and distributions of phenobarbital- and 3-methylcholanthrene-inducible cytochromes P-450 within the livers of untreated rats, J. Biol. Chem., 256:5931.

Baron, J., Redick, J. A., and Guengerich, F. P., 1982, Effects of 3-methylcholanthrene, β-naphthoflavone, and phenobarbital on the 3-methylcholanthrene-inducible isozyme of cytochrome P-450 within centilobular, midzonal, and periportal hepatocytes, J. Biol. Chem., 257:953.

Baron, J., Kawabata, T. T., Redick, J. A., Knapp, S. A., Wick, D. G., Wallace, R. B., Jakoby, W. B., and Guengerich, F. P., 1983, Localization of carcinogen-metabolizing enzymes in human and animal tissues, in: "Extrahepatic Drug Metabolism and Chemical Carcinogenesis," J. Rydstrom, J. Montelius, and M. Bengtsson, eds., p. 73, Elsevier Science Publishers, Amsterdam.

Baron, J., Kawabata, T. T., Knapp, S. A., Voigt, J. M., Redick, J. A., Jakoby, W. B., and Guengerich, F. P., 1984, Intrahepatic distribution of xenobiotic-metabolizing enzymes, in: "Foreign Compound Metabolism," J. Caldwell and G. D. Paulson, eds., p. 17, Taylor and Francis, London.

Bickers, D. R., and Kappas, A., 1980, The skin as a site of chemical metabolism, in: "Extrahepatic Metabolism of Drugs and Other Foreign Chemicals," T. E. Gram, ed., p. 295, Spectrum Publications, New York.

Black, O., and Stoming, T. A., 1983, Characterization of benzopyrene metabolism in rat pancreas, Cancer Lett., 18:97.

Black, O., Murrill, E., and Pallas, F., 1980, Pancreatic metabolism of benzo(a)pyrene in vitro and in vivo in the Long-Evans rat, Res. Commun. Chem. Pathol. Pharmacol., 29:291.

Black, O., Murrill, E., and Fanska, C., 1981, Metabolism of 3-methylcholanthrene in rat pancreas, Dig. Dis. Sci., 26:358.

Bockman, D.E., 1981, Cells of origin of pancreatic cancer: experimental animal tumors related to human pancreas, Cancer, 47:1528.

Bockman, D. E., Black, O., Mills, L. R., Mainz, D. L., and Webster, P. D., 1976, Fine structure of pancreatic adenocarcinoma induced in rats by 7,12-dimethylbenz[a]anthracene, J. Natl. Cancer Inst., 57:931.

Bockman, D. E., Black, O., Mills, L. R., and Webster, P. D., 1978, Origin of tubular complexes developing during induction of pancreatic adenocarcinoma by 7,12-dimethylbenz(a)anthracene, Am. J. Pathol., 90:645.

Boyd, M. R., 1977, Evidence for the Clara cell as a site of cytochrome P-450-dependent mixed-function oxidase activity in lung, Nature (London), 269:713.

Boyd, M. R., 1980, Biochemical mechanisms in chemical-induced lung injury: roles of metabolic activation, CRC Crit. Rev. Toxicol., 7:103.

Boyd, M. R., and Statham, C. N., 1983, The effect of hepatic metabolism on the production and toxicity of reactive metabolites in extrahepatic organs, Drug Metab. Rev., 14:35.

Brittebo, E. B., Castonguay, A., Furuya, K., and Hecht, S. S., 1983, Metabolism of tobacco-specific nitrosamines by cultured rat nasal mucosa, Cancer Res., 43:4343.

Butcher, E. O., 1934, The hair cycles in the albino rat, Anat. Res., 61:5.

Chow, S. A., and Fischer, L. J., 1984, Alterations in rat pancreatic B-cell function induced by prenatal exposure to cyproheptadine, Diabetes, 33:572.

Cohen, G. M., and Moore, B. P., 1976, Metabolism of [^{3}H]benzo[a]pyrene by different portions of the respiratory tract, Biochem. Pharmacol., 25:1623.

Conway, J. G., Kauffman, F. C., Ji, S., and Thurman, R. G., 1982, Rates of sulfation and glucuronidation of 7-hydroxycoumarin in periportal and pericentral regions of the liver lobule, Mol. Pharmacol., 22:509.

Devereux, T. R., and Fouts, J. R., 1981, Xenobiotic metabolism by alveolar type II cells isolated from rabbit lung, Biochem. Pharmacol., 30:1231.

Devereux, T. R., Hook, G. E. R., and Fouts, J. R., 1979, Foreign compound metabolism by isolated cells from rabbit lung, Drug Metab. Dispos., 7:70.

Devereux, T. R., Jones, K. G., Bend, J. R., Fouts, J. R., Statham, C. N., and Boyd, M. R., 1982, In vitro metabolic activation of the pulmonary toxin, 4-ipomeanol, in nonciliated bronchiolar epithelial (Clara) and alveolar type II cells isolated from rabbit lung, J. Pharmacol. Exp. Ther., 220:223.

Dissin, J., Mills, L. R., Mains, D. L., Black, O., and Webster, P. D., 1975, Experimental induction of pancreatic adenocarcinomas in rats, J. Natl. Cancer Inst., 55:857.

Fjellstedt, T. A., Allen, R. H., Duncan, B. K., and Jakoby, W. B., 1973, Enzymatic conjugation of epoxides with glutathione, J. Biol. Chem., 248:3702.

Flaks, B., and Lucas, J., 1973, Persistent ultrastructural changes in pancreatic acinar cells induced by 2-acetylaminofluorene, Chem.-Biol. Interact., 6:91.

Flaks, B., Moore, M. A., and Flaks, A., 1981, Ultrastructural analysis of pancreatic carcinogenesis. IV. Pseudoductular transformation of acini in the hamster pancreas during N-nitroso-bis(2-hydroxypropyl)amine carcinogenesis, Carcinogenesis, 2:1241.

Flaks, B., Moore, M. A., and Flaks, A., 1982, Ultrastructural analysis of pancreatic carcinogenesis. VI. Early changes in hamster acinar cells induced by N-nitroso-bis(2-hydroxypropyl)amine, Carcinogenesis, 3:1063.

Gangolli, S., and Wright, M., 1971, The histochemical demonstration of aniline hydroxylase activity in rat liver, Histochem. J., 3:107.

Gati, E., Calop, J. Y., and Lafaverges, F., 1973, Histochemical demonstration of the benzo(a)pyrene hydroxylase activity in animal skin, Ann. Histochem., 18:311.

Grasso, P., Williams, M., Hodgson, R., Wright, M. G., and Gangolli, S. D., 1971, The histochemical distribution of aniline hydroxylase activity in rat tissues, Histochem. J., 3:117.

Guengerich, F. P., 1978, Separation and purification of multiple forms of microsomal cytochrome P-450. Partial characterization of three apparently homogeneous cytochromes P-450 prepared from livers of phenobarbital- and 3-methylcholanthrene-treated rats, J. Biol. Chem., 253:7931.

Guengerich, F. P., Wang, P., Mitchell, M. B., and Mason, P. S., 1979, Rat and human liver microsomal epoxide hydrolase. Purification and evidence for the existence of multiple forms, J. Biol. Chem., 254: 12248.

Guengerich, F. P., Dannan, G. A., Wright, S. T., Martin, M. V., and Kaminsky, L. S., 1982a, Purification and characterization of liver microsomal cytochromes P-450: electrophoretic, spectral, catalytic, and immunochemical properties and inducibility of eight isozymes isolated from rats treated with phenobarbital or β-naphthoflavone, Biochemistry, 21:6019.

Guengerich, F. P., Dannan, G. A., Wright, S. T., Martin, M. V., and Kaminsky, L. S., 1982b, Purification and characterization of microsomal cytochromes P-450, Xenobiotica, 12:701.

Habig, W. H., Pabst, M. J., and Jakoby, W. B., 1974, Glutathione S-transferases. The first enzymatic step in mercapturic acid formation, J. Biol. Chem., 249:7130.

Hecht, S. S., Chen, C. B., Ohmori, T., and Hoffman, D., 1980, Comparative carcinogenicity in F344 rats of the tobacco-specific nitrosamines, N'-nitrosonornicotine and 4-(N-methyl-N-nitrosamine)-1-(3-pyridyl)-1-butanone, Cancer Res., 40:298.

Hsu, S. M., Raine, L., and Fanger, H., 1981, Use of avidin-biotin-peroxidase complex (ABC) in immunoperoxidase techniques: a comparison between ABC and unlabeled antibody (PAP) procedures, J. Histochem. Cytochem., 29:577.

Ichikawa, Y., and Yamano, T., 1967, Electron spin resonance of microsomal cytochromes. Correlation of the amount of CO-binding species with so-called microsomal Fe_x in microsomes of normal tissues and rat liver microsomes of sudan III-treated animals, Arch. Biochem. Biophys., 121:742.

Iqbal, Z. M., Varnes, M. E., Yoshida, A., and Epstein, S. S., 1977, Metabolism of benzo(a)pyrene by guinea pig pancreatic microsomes, Cancer Res., 37:1011.

Ishii-Ohba, H., Guengerich, F. P., and Baron, J., 1984, Localization of epoxide-metabolizing enzymes in rat testis, Biochim. Biophys. Acta, 802:326.

Ji, S., Lemasters, J. J., and Thurman, R. G., 1981, A fluorometric method to measure sublobular rates of mixed-function oxidation in the hemoglobin-free perfused rat liver, Mol. Pharmacol., 19:513.

Kahng, M. W., Smith, M. W., and Trump, B. F., 1981, Aryl hydrocarbon hydroxylase in human bronchial epithelium and blood monocyte, J. Natl. Cancer Inst., 66:227.

Kawabata, T. T., Guengerich, F. P., and Baron, J., 1981, An immunohisto-chemical study on the localization and distribution of epoxide hydrolase within livers of untreated rats, Mol. Pharmacol., 20:709.

Kawabata, T. T., Guengerich, F. P., and Baron, J., 1983, Effects of phenobarbital, trans-stilbene oxide, and 3-methylcholanthrene on epoxide hydrolase within centrilobular, midzonal, and periportal regions of rat liver, J. Biol. Chem., 258:7767.

Kawabata, T. T., Wick, D. G., Guengerich, F. P., and Baron, J., 1984, Immunohistochemical localization of carcinogen-metabolizing enzymes within the rat and hamster exocrine pancreas, Cancer Res., 44:215.

Lazarus, S. S., and Shapiro, S. H., 1972, Serial morphologic changes in rabbit pancreatic islet cells after streptozotocin, Lab. Invest., 27:174.

Lee, K. P., and Trochimowicz, H. J., 1982, Induction of nasal tumors in rats exposed to hexamethylphosphoramide by inhalation, J. Natl. Cancer Inst., 68:157.

Levitt, M. H., Harris, C. C., Squire, R., Springer, S., Wenk, M., Mollelo, C., Thomas, D., Kingsbury, E., and Newkirk, C., 1977, Experimental pancreatic carcinogenesis. I. Morphogenesis of pancreatic adeno-carcinoma in the Syrian golden hamster induced by N-nitroso-bis(2-hydroxypropyl)amine, Am. J. Pathol., 88:5.

Mannervik, B., and Jensson, H., 1982, Binary combinations of four protein subunits with different catalytic specificities explain the relation-ship between six basic glutathione S-transferases in rat liver cytosol, J. Biol. Chem., 257:9909.

Miller, E. C., and Miller, J. A., 1981, Mechanisms of chemical carcinogen-esis, Cancer, 47:1055.

Miller, E. C., and Miller, J. A., 1982, Reactive metabolites as key inter-mediates in pharmacologic and toxicologic responses: examples from chemical carcinogenesis, in: "Biological Reactive Intermediates - II. Chemical Mechanisms and Biological Effects," R. Snyder, D. V. Parke, J. J. Kocsis, D. J. Jollow, G. G. Gibson, and C. M. Witmer, eds., p. 1, Plenum Press, New York.

Minchin, R. F., and Boyd, M. R., 1983, Localization of metabolic activation and deactivation systems in the lung: significance to the pulmonary toxicity of xenobiotics, Ann. Rev. Pharmacol. Toxicol., 23:217.

Mitchell, J. R., Nelson, S. D., Thorgeirsson, S. S., McMurtry, R. J., and Dybing, E., 1976, Metabolic activation: biochemical basis for many drug-induced liver injuries, in: "Progress in Liver Diseases," H. Popper and F. Schaffner, eds., Vol. V, p. 259, Grune and Stratton, New York.

Nakai, T., and Shubik, P., 1964, Autoradiographic localization of tissue-bound tritiated 7,12-dimethylbenz[a]anthracene in mouse skin 24 and 48 hours after single application, J. Natl. Cancer Inst., 33:887.

Pannatier, A., Jenner, P., Testa, B., and Etter, J. C., 1978, The skin as a drug-metabolizing organ, Drug Metab. Rev., 8:319.

Pour, P., 1978, Islet cells as a component of pancreatic ductal neoplasms. I. Experimental study: ductular cells, including islet cell precursors, as primary progenitor cells of tumors, Am. J. Pathol., 90:295.

Pour, P., and Wilson, R. B., 1980, Experimental tumors of the pancreas, in: "Tumors of the Pancreas," A. R. Moosa, ed., p. 37, Williams and Wilkins, Baltimore.

Pour, P., Althoff, J., and Takahashi, M., 1977, Early lesions of pancreatic ductal carcinoma in the hamster model, Am. J. Pathol., 88:291.

Pour, P., Salmasi, S. Z., and Runge, R. G., 1979, Ductular origin of pancreatic cancer and its multiplicity in man comparable to experimentally induced tumores. A preliminary study, Cancer Lett., 6:89.

Rappaport, A. M., 1979, Physioanatomical basis of toxic liver injury, in: "Toxic Injury of the Liver," E. Farber and M. M. Fisher, eds., p. 1, Marcel Dekker, New York.

Redick, J. A., Kawabata, T. T., Guengerich, F. P., Krieter, P. A., Shires, T. K., and Baron, J., 1980, Distributions of monooxygenase components and epoxide hydratase within the livers of untreated male rats, Life Sci., 27:2465.

Redick, J. A., Jakoby, W. B., and Baron, J., 1982, Immunohistochemical localization of glutathione S-transferases in livers of untreated rats, J. Biol. Chem., 257:15200.

Reid, W. D., Christie, B., Eichelbaum, M., and Krishner, G., 1971, 3-Methylcholanthrene blocks hepatic necrosis induced by administration of bromobenzene or carbon tetrachloride, Exp. Mol. Pathol., 15:363.

Reznik, G., Reznik-Schuller, H., Ward, J. M., and Stinson, S. F., 1980, Morphology of nasal cavity tumors in rats after chronic inhalation of 1,2-dibromo-3-chloropropane, Br. J. Cancer, 42:772.

Scarpelli, D. G., Rao, M. S., Subbarao, V., Beversluis, M., Gurka, D. P., and Hollenberg, P. F., 1980, Activation of nitrosamines to mutagens by postmitochondrial fraction of hamster pancreas, Cancer Res., 40:67.

Scarpelli, D. G., Kokkinakis, D. M., Rao, M. S., Subbarao, V., Luetteke, N., and Hollenberg, P. F., 1982, Metabolism of the pancreatic carcinogen N-nitroso-2,6-dimethylmorpholine by hamster liver and component cells of pancreas, Cancer Res., 42:5089.

Slaga, T. J., ed., 1984, "Mechanisms of Tumor Promotion. Vol. II. Tumor Promotion and Skin Carcinogenesis," CRC Press, Boca Raton.

Smith, M. T., Redick, J. A., and Baron, J., 1983, Quantitative immunohisto-chemistry: a comparison of microdensitometric analysis of unlabeled antibody peroxidase-antiperoxidase staining and of microfluorometric analysis of indirect fluorescent antibody staining for nicotinamide adenosine dinucleotide phosphate (NADPH)-cytochrome c (P-450) reductase in rat liver, J. Histochem. Cytochem., 31:1183.

Sorokin, S. P., 1970, The cells of the lung, in: "Morphology of Experimental Respiratory Carcinogenesis," P. Nettesheim, M. G. Hanna, Jr., and J. W. Deatherage, eds., p. 3, U.S. At. Energy Comm., Oak Ridge.

Sternberger, L. A., Hardy, P. H., Cuculis, J. J., and Meyer, H. G., 1970, The unlabeled antibody enzyme method of immunohistochemistry. Preparation and properties of soluble antigen-antibody complex (horseradish peroxidase-antihorseradish peroxidase) and its use in identification of spirochetes, J. Histochem. Cytochem., 18:315.

Taira, Y., Redick, J. A., Greenspan, P., and Baron, J., 1979, Immunohistochemical studies on electron transport proteins associated with
cytochromes P-450 in steroidogenic tissues. II. Microsomal NADPH-
cytochrome *c* reductase in the rat adrenal, Biochim. Biophys. Acta,
583:148.

Taira, Y., Redick, J. A., and Baron, J., 1980a, An immunohistochemical
study on the localization and distribution of NADPH-cytochrome *c*
(P-450) reductase in rat liver, Mol. Pharmacol., 17:374.

Taira, Y., Greenspan, P., Kapke, G. F., Redick, J. A., and Baron, J.,
1980b, Effects of phenobarbital, pregnenolone-16α-carbonitrile, and
3-methylcholanthrene pretreatments on the distribution of NADPH-
cytochrome *c* (P-450) reductase within the liver lobule, Mol. Pharmacol.,
18:304.

Takahashi, M., Pour, P., Althoff, J., and Donnelly, T., 1977, Sequential
alteration of the pancreas during carcinogenesis in Syrian golden
hamsters by N-nitrosobis(2-oxopropyl)amine, Cancer Res., 37:4602.

Teel, R. W., and Douglas, W. H. J., 1980, Aryl hydrocarbon hydroxylase
activity in type II alveolar lung cells, Experientia, 36:107.

Tuchweber, G., Werringloer, J., and Kourounakis, P., 1974, Effect of
phenobarbital or pregnenolone-16α-carbonitrile (PCN) pretreatment on
acute carbon tetrachloride hepatotoxicity in rats, Biochem. Pharmacol.,
23:513.

Vermorken, A. J. M., Goos, C. M. A. A., Roelofs, H. M. J., Henderson, P.
Th., and Bloemendal, H., 1979, Metabolism of benzo[*a*]pyrene in isolated
human scalp hair follicles, Toxicology, 14:109.

Voigt, J. M., Guengerich, F. P., and Baron, J., 1985, Localization of a
cytochrome P-450 isozyme (cytochrome P-450 PB-B) and NADPH-cytochrome
P-450 reductase in rat nasal tissue, Cancer Lett., 27:241.

Wattenberg, L. W., and Leong, J. L., 1962, Histochemical demonstration of
reduced pyridine nucleotide dependent polycyclic hydrocarbon
metabolizing systems, J. Histochem. Cytochem., 10:412.

Wattenberg, L. W., and Leong, J. L., 1970, Benzpyrene hydroxylase activity
in mouse skin, Proc. Am. Assoc. Cancer Res., 11:81.

Wiebkin, P., Schaeffer, B. K., Longnecker, D. S., and Curphey, T. J., 1984,
Oxidative and conjugative metabolism of xenobiotics by isolated rat
and hamster acinar cells, Drug Metab. Dispos., 12:427.

Wright, A. S., 1980, The role of metabolism in chemical mutagenesis and
chemical carcinogenesis, Mutat. Res., 75:215.

MONOCLONAL ANTIBODY-DIRECTED ANALYSIS OF CYTOCHROME P-450

Fred K. Friedman, Sang S. Park, Byung J. Song, Kuo C. Cheng,
T. Fujino and Harry V. Gelboin

National Cancer Institute
National Institutes of Health
Bethesda, Md. 20205

The cytochromes P-450 metabolize a diverse array of xenobiotic and endobiotic compounds, including carcinogens, drugs, and steroids (1-3). The different isozymes differ in their substrate and product specificities and reactivities. The types and amounts of isozymes in a tissue therefore regulate the metabolic conversion of substrates to products. Progress in understanding the role of individual P-450s in metabolism of specific substrates, and in relating P-450 phenotype to individual differences in sensitivity to drugs and carcinogens, has been hindered by the multiplicity of the P-450s. The approach we have taken to the multiplicity problem is to prepare and use monoclonal antibodies (MAbs) as specific probes to individual and epitope-specific classes of P-450s (4-6).

A monoclonal antibody specifically interacts with a single antigenic site, or epitope, on the surface of the P-450, and is therefore specific for those P-450s with this epitope. If the epitope recognized by the MAb is present on only one isozyme, the MAb is specific for that isozyme. On the other hand, if the MAb recognizes a epitope which is present on several isozymes, these are related by this common epitope and may be said to consititute an epitope-specific class of P-450s.

We have prepared panels of several hundred MAbs to ten different forms of P-450. These MAbs have been used for different purposes: preparation of immunoadsorbents which can be used to purify specific P-450s from various sources; development of immunoassays to detect P-450s; and using MAbs that inhibit enzymatic activity, to determine the contribution of specific P-450s to the total P-450 catalyzed activity of a tissue. These MAb based methods can then be applied to different tissues and individuals for deter- mination of their P-450 phenotype.

The MAbs that are prepared are initially characterized by several criteria (4-6): by their ability to bind the P-450 antigen in a radio- immunoassay, to immunoprecipitate the P-450, and to inhibit its enzymatic activity. In addition, we classify the MAbs according to epitope specifi- city, as measured by the ability of MAbs to compete with one another for binding to P-450 epitopes (7).

IMMUNOPURIFICATION

An important application of MAbs is for the simple and rapid purification of P-450 (8,9). The MAb is immobilized on a Sepharose support to yield an immunoadsorbent which binds specific P-450s from tissues, and these P-450s are subsequently eluted at pH 3.0. This is a simple, one-step procedure which is more efficient than conventional purification methods. Immunopurification from rat tissues is shown on the sodium dodecylsulfate-polyacrylamide gel (SDS-PAGE) in Fig. 1. Lanes 2, 6, and 10 show the many proteins present in liver, lung, and kidney microsomes. In control experiments with a nonspecific MAb (lanes 3, 7, and 11) the immunoadsorbent does not bind any protein from the tissue microsomes. However, with MAbs specific for MC-induced rat liver P-450, proteins were indeed bound. With MAb 1-31-2, a single protein was isolated from liver (lane 5) and from lung (lane 9). With another MAb, 1-7-1, two proteins were obtained from liver (lane 4) and one from lung (lane 8). No protein was detected from kidney using either MAb (lanes 12 and 13).

Figure 2 shows SDS-PAGE of the purified P-450s obtained from different species and tissues, using these two MAbs. There are liver P-450s from rat, C57 and DB mice, guinea pig, and hamster, and a rat lung P-450. In each case, a single- or two-step immunopurification yields an electro-phoretically homogeneous P-450. This data is summarized in Table 1, which shows the molecular weights of these P-450s. Thus 1-31-2 binds a single species of MW 57K from rat and C57 mouse liver and 1-7-1 binds both 56K and 57K proteins from these tissues. A 57K protein from rat lung is also bound by both monoclonals. 1-7-1 also binds P-450s with different MW from DB mice, guinea pig, and hamster. Such immunopurifications are useful since they are simple and rapid, and secondly, since all the isozymes purified with a given MAb contain a common epitope, this method also serves to immunochemically relate the purified P-450s.

These P-450s are quite suitable for various structural studies, such as determination of amino-terminal sequences (10, 11). The sequences we

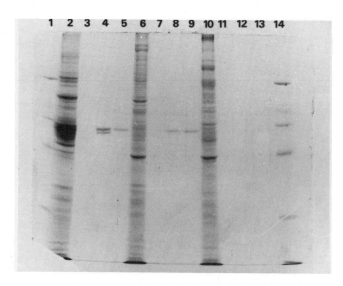

Fig. 1. SDS-PAGE of MC-rat microsomal proteins and proteins immunopurified with MAbs. Lanes 1 and 14 contain standards. Lanes 2, 6,
and 10 contain liver, lung, and kidney microsomes and the three
subsequent lanes contain the results of immunopurification with
a nonspecific MAb, 1-7-1, and 1-31-2, respectively.

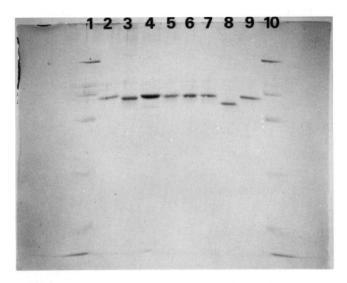

Fig. 2: SDS-PAGE analysis of immunopurified proteins from MC-treated animals. Lanes 1 and 10 contain standards. Lanes 2-9 contain proteins purified from: rat liver (lanes 2 and 3), rat lung (lane 4), $C_{57}BL/6$ mouse liver (lanes 5 and 6) DBA/2 mouse liver (lane 7), guinea pig liver (lane 8) and hamster (lane 9).

Table 1. Purified cytochromes P-450

Species, Tissue	Molecular Weight of Cytochrome P-450 Purified with Sepharose-		
	MAb 1-7-1		MAb 1-31-2
rat liver	56,000	57,000	57,000
rat lung	57,000		57,000
$C_{57}BL/6$ mouse liver	56,000	57,000	57,000
DBA/2 mouse liver	56,000		N.D.
guinea pig liver	53,000		N.D.
hamster liver	57,000		N.D.

N.D.: None detectable

obtained for six P-450s from rats, mice, and guinea pigs are shown in Fig. 3. The sequence of the 57K rat P-450 is the same as the major MC-inducible rat P-450 isozyme, as deduced from the nucleotide sequence (12). The C57 mouse sequence is the same as that of the Pl-450 isozyme as deduced from the nucleotide sequence (13). The rat 56K P-450 corresponds to the P-450d isozyme (14), which in the major isosafrole-induced form, and the mouse 56K isozymes correspond to the sequence of mouse P3-450 (13). The last three sequences of the 56K proteins are identical, and the rat and mouse 57K P-450s on the first two lines are highly homologous. The guinea pig sequence is newly derived and has more limited homology with the other P-450 sequences. The sequence data and its agreement with known sequences shows that these immunopurified P-450s are sufficiently pure for further structural studies.

Another method with which we have characterized these P-450s is peptide mapping after limited proteolysis (11). Fig. 4 shows the peptides

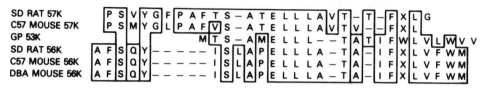

Fig. 3: Amino terminal sequences of immunopurified cytochromes P-450.

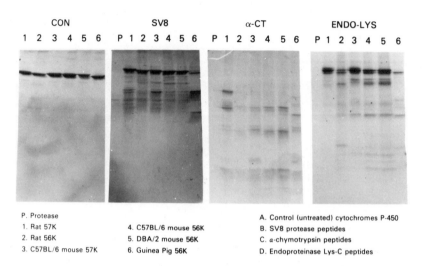

CON	SV8	α-CT	ENDO-LYS
1 2 3 4 5 6	P 1 2 3 4 5 6	P 1 2 3 4 5 6	P 1 2 3 4 5 6

P. Protease
1. Rat 57K
2. Rat 56K
3. C57BL/6 mouse 57K
4. C57BL/6 mouse 56K
5. DBA/2 mouse 56K
6. Guinea Pig 56K

A. Control (untreated) cytochromes P-450
B. SV8 protease peptides
C. α-chymotrypsin peptides
D. Endoproteinase Lys-C peptides

Fig. 4: SDS-PAGE peptide mapping of immunopurified cytochromes P-450.

generated by three proteases, SV8, chymotrypsin, and endoproteinase lysine-C. Each lane represents a different purified P-450. While the maps of the rat and mouse P-450s (lanes 1-5) have a number of similarities as well as some differences, the most distinct patterns are those of guinea pig (lanes 6). Thus, both sequence analysis and peptide mapping show that the guinea pig isozyme is least homologous to the other P-450s.

REACTION PHENOTYPING

Another very useful application of the MAbs is for reaction phenotyping of tissues. A MAb which inhibits a P-450 can determine the contribution of that P-450 to the total tissue activity for any P-450 substrate in a crude tissue preparation.

Table 2 shows the effect of MAb 1-7-1 on two P-450 dependent activities, aryl hydrocarbon hydroxylase (AHH) and ethoxycoumarin deethylase (ECD) (5). It inhibits both activities of the purified major MC-inducible form of rat liver P-450 by 92%. This MAb was used to determine the contribution of the MAb-sensitive P-450 in liver microsomes from control, phenobarbitol (PB) and MC-treated rats. It inhibits these activities in MC-treated rats by about 70%, indicating that this proportion of the total activities derives from MAb-sensitive P-450. The remaining 30% of the activities therefore depends on other P-450s, ones that are insensitive to this MAb. However, for liver microsomes from control and PB-treated rats, there is little or no inhibition, indicating that the activities in these microsomes derive entirely from the second type of P-450, which is insensitive to this MAb.

148

Table 2. Inhibition of AHH and ECD by MAb 1-7-1

Enzyme Source	AHH		ECD	
	Control	% Inhibition	Control	% Inhibition
Purified MC-P-450*	1177	92	110	92
MC-Microsomes**	1170	75	5280	71
PB-Microsomes	600	0	1266	8
Control Microsomes	367	0	815	0

*pmol/min/nmol;   **pmol/min/mg

Table 3. MAb 1-7-1 Inhibition of Hepatic AHH and ECD of Different Species and Strains

Species	Induction	AHH (pmol/min/mg)		ECD (nmol/min/mg)	
		Control	% Inhibition	Control	% Inhibition
Rat	none	229	0	1.06	0
SD	PB	141	9	2.73	0
	MC	2099	81	8.37	65
Mouse	none	186	0	2.46	15
C57BL/6	PB	855	4	10.9	8
	MC	4883	85	14.0	55
Mouse	none	328	11	4.81	8
DBA/2	PB	610	17	16.2	3
	MC	283	2	4.02	17
Guinea Pig	none	282	26	2.00	0
	PB	703	0	2.88	0
	MC	1642	51	5.31	0
Hamster	none	100	0	8.69	0
	PB	192	0	18.0	5
	MC	387	47	22.1	0

We have applied this method to reaction phenotype tissue P-450 in different species (15). Table 3 shows the results obtained with four different species exposed to MC or PB. In MC-induced C57 mice 88% of AHH and 39% of ECD is contributed by the MAb specific P-450, which means that very little of the AHH but 61% of the ECD derives from the insensitive type of P-450. Another example is guinea pig and hamster, where about half the AHH, but none of the ECD is catalyzed by MAb-specific P-450.

Reaction phenotyping of human tissues was also carried out (16, 17), as shown in Fig. 5. The MAb 1-7-1 inhibits both placenta and lymphocyte activities, and these thus derive from MAb-sensitive P-450. On the other hand, monocytes and liver activities are not inhibited and are therefore catalyzed by a different type of P-450. The placenta data show that 90% of the AHH is inhibited, which indicates that it nearly entirely derives from the MAb sensitive type of P-450. However, the ECD is only partly inhibited, by about 50%, which means that half of the ECD is contributed by P-450 which is sensitive to the MAb and the remainder is catalyzed by a different type of P-450, which is MAb-insensitive.

These results for both animal and human tissues therefore show how reaction phenotyping can be used to identify and measure the tissue P-450s that are responsible for specific metabolic reactions. This approach can be applied to the study of any P-450 substrate and we have successfully

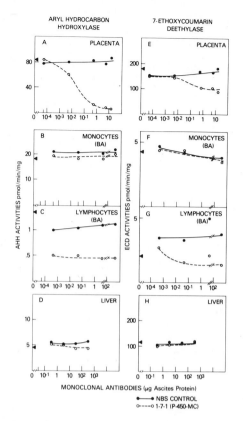

Fig. 5: Effect of MAb 1-7-1 on the AHH and ECD activities of different human tissues.

applied it in numerous collaborative studies with a variety of drugs and carcinogens as well as endogenous substrates (18). It can be used not only to measure substrate disappearance and product formation, but also consequences of P-450 catalyzed reactions such as toxicity, mutagenesis, and DNA and protein binding.

RADIOIMMUNOASSAY

We have also used the MAbs to directly measure P-450 by radioimmunoassay (RIA) (9, 19), which is advantageous since this method is independent of catalytic activity. An RIA for specific P-450 in rat liver is shown in Fig. 6. The amount of P-450 in a tissue is determined by its ability to compete with a known amount of P-450, for binding to radiolabeled MAb 1-7-1. The greater the amount of P-450 in the tissue, the greater will be its inhibition of MAb binding to the known P-450. Thus, liver microsomes from MC-treated rats contain the greatest amount of P-450, while liver from control and PB-treated rats contain less, by about 50-fold, of this P-450. These results are also consistent with those of reaction phenotyping experiments.

The RIA was further applied to additional tissues and species (9), as shown in Fig. 7. Table 4 summarizes the RIA data by listing the amount of protein needed to achieve 50% inhibition of binding by 1-7-1. The lower this parameter, the more P-450 with the MAb-specific epitope. The RIA can thus directly measure the level of specific P-450s in tissues.

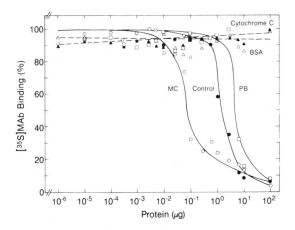

Fig. 6: Competitive RIA for cytochromes P-450 in liver microsomes of control, PB, and MC-treated rats, using [^{35}S] MAb 1-7-1

MAbs with different specificities react differently in the RIA, and yield different curves. By carrying out these RIAs with a series of monoclonals, we can therefore compare the MAb-specific P-450 content of different tissues and species (7).

A promising application of RIA is for measurement of P-450 in tissues in which they are present at low levels and difficult to otherwise detect. This is the case, for example, with human tissues such as placenta and lymphocytes. It has been established that smoking induces AHH activity in human placenta (20). We thus applied the RIA method to placenta obtained from women who smoked as well as nonsmokers (21). Figure 8 shows the RIA curves for some individual placenta. The RIA measured P-450 in all the placenta from smokers, but little or no P-450 was detected in placenta from nonsmokers, a result which is very consistent with the higher levels of AHH in placenta from smokers.

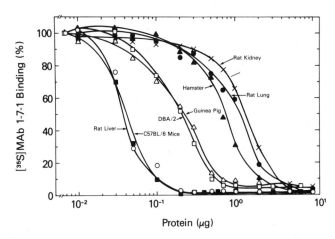

Fig. 7. Competitive RIA for microsomes from livers of MC-treated rat, C$_{57}$BL/6 and DBA/2 mice, hamster, and guinea pig; and kidney and lung from MC-treated rats.

Table 4. Competitive RIA for Cytochromes P-450 With MAb 1-7-1 Epitope

Tissue Source	I50(μg)
Rat liver	.042
C57BL mouse liver	.050
Guinea pig liver	.21
DBA/2 mouse liver	.23
Hamster liver	.78
Rat lung	1.2
Rat kidney	1.4

We assessed the reliability of the RIA and compared it to the AHH activity assay. During a 70 hour incubation at 4°C, the AHH of rat liver and placenta was considerably diminished. The tissues were stable for RIA measurements, with no loss in binding during this time. This stability suggests the utility and advantage of the RIA for screening of human tissues or other labile samples.

The RIA was also used to measure MAb 1-7-1 specific P-450 in lymphocytes obtained from different individuals (21). The P-450 was measured in control as well as lymphocytes treated with benzanthracene, an AHH inducer. An elevation of P-450 of the induced relative to the control curves was observed. The capability of RIA to detect P-450s in lymphocytes is especially promising for studies on individual human variation in P-450 phenotype since the RIA is readily adaptable to screening of large numbers of samples, and because lymphocytes are a relatively easily obtainable human tissue.

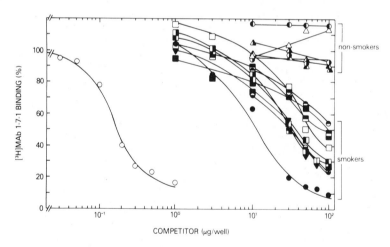

Fig. 8: Competitive RIA of placenta from smokers and nonsmokers, using [^{3}H] MAb 1-7-1.

CONCLUSION

MAbs offer a new, multidimensional approach for the analysis of P-450s. The nature and complexity of the P-450 problem renders it especially suitable for analyses with MAbs, since they offer the potential resolution required to distinguish closely related forms of P-450. When used in immunoadsorbents for isolation of specific P-450s, they provide

purified P-450s in a procedure that is carried out more readily and efficiently than conventional purification methods. These immunopurified P-450s
are suitable for subsequent structural studies. The amount of specific
P-450 in tissues can also be directly measured by radioimmunoassay. This
method is especially suitable for evaluating the P-450 content of large
numbers of samples. Additional advantages are its sensitivity, which
allows for detection of minor P-450 forms which are present in low levels,
and that it detects P-450 independent of functional catalytic activity. In
addition, MAbs that inhibit catalytic activity can determine the contribution of specific P-450s to the overall tissue metabolism of P-450
substrates.

Characterization of the P-450 phenotype by such MAb based methods will
help define the role of specific P-450s in metabolism or various drugs and
carcinogens, and may thus help to clarify the relationship of P-450 phenotype to individual differences in responsiveness to drugs and carcinogens.

REFERENCES

1. A. Y. H. Lu, and S. B. West, Multiplicity of mammalian microsomal
    cytochromes P-450, Pharmacol. Rev., 31:277 (1980).
2. H. V. Gelboin, Benzo[a]pyrene metabolism, activation, and carcino-
    genesis: Role of mixed-function oxidases and related enzymes,
    Physiol. Rev., 60:1107 (1980).
3. R. Sato, and R. Kato, "Microsomes, Drug Oxidations, and Drug Toxicity,"
    Scientific Societies Press, Japan (1982).
4. S. S. Park, S. J. Cha, H. Miller, A. V. Persson, M. J. Coon, and
    H. V. Gelboin, Monoclonal antibodies to rabbit liver cytochrome P-450
    LM2 and P-450 LM4, Mol. Pharmacol., 21:248 (1981).
5. S. S. Park, T. Fujino, D. West, F. P. Guengerich, and H. V. Gelboin,
    Monoclonal antibodies that inhibit enzyme activity of 3-methyl-
    cholanthrene induced cytochrome P-450, Cancer Res., 42:1798 (1982).
6. S. S. Park, T. Fujino, H. Miller, F. P. Guengerich, and H. V. Gelboin,
    Monoclonal antibodies to phenobarbital-induced rat liver cytochrome
    P-450, Biochem. Pharmacol., 33:2071 (1984).
7. B. J. Song, H. V. Gelboin, S. S. Park, F. K. Friedman, Epitope-
    relatedness and phenotyping of hepatic cytochrome P-450 with mono-
    clonal antibodies, Biochem. J., (in press).
8. F. K. Friedman, R. C. Robinson, S. S. Park, and H. V. Gelboin, Mono-
    clonal antibody-directed immunopurification and identification of rat
    liver cytochromes P-450, Biochem. Biophys. Res. Commun., 116:859 (1983).
9. K. C. Cheng, H. V. Gelboin, B. J. Song, S. S. Park, and F. K. Friedman,
    Detection and purification of cytochromes P-450 in animal tissues with
    monoclonal antibodies, J. Biol. Chem., 259:12279 (1984).
10. K. C. Cheng, H. C. Krutzsch, S. S. Park, P. H. Granthan, H. V. Gelboin,
    and F. K. Friedman, Amino-terminal sequence analysis of six cytochrome
    P-450 isozymes purified by monoclonal antibody directed immunopurifi-
    cation, Biochem. Biophys. Res. Commun., 123:1201 (1984).
11. K. C. Cheng, H. V. Gelboin, S. S. Park, and F. K. Friedman, Structural
    relatedness of six monoclonal antibody immunopurified cytochromes
    P-450, (submitted).
12. Y. Yabusaki, M. Shimizu, H. Murakami, K. Nakamura, K. Oeda, and
    H. Ohkawa, Nucleotide sequence of a full-length cDNA coding for
    3-methylcholanthrene-induced rat liver cytochrome P-450 MC. Nucleic
    Acid Res., 12:2929 (1984).
13. S. Kimura, F. J. Gonzalez, and D. W. Nebert, The murine Ah locus:
    comparison of the complete cytochrome P1-450 and P3-450 cDNA nucleo-
    tide and amino acid sequences. J. Biol. Chem., 259:10705 (1984).

14. L. H. Botelho, D. E. Ryan, P-M. Yuan, R. Kutny, J. E. Shirley, and W. Levin, Amino-terminal and carboxy-terminal sequence of hepatic microsomal cytochrome P-450d, a unique hemoprotein from rats treated with isosafrole, Biochemistry, 21:1152 (1982).

15. T. Fujino, D. West, S. S. Park, and H. V. Gelboin, Monoclonal antibody-directed phenotyping of cytochrome P-450-dependent aryl hydrocarbon hydroxylase and 7-ethoxycoumarin deethylase in mammalian tissues, J. Biol. Chem., 259:9044 (1984).

16. T. Fujino, S. S. Park, D. West, and H. V. Gelboin, Phenotyping of cytochrome P-450 in human tissues with monoclonal antibodies, Proc. Natl. Acad. Sci. USA, 70:3682 (1982).

17. T. Fujino, K. Gottlieb, D. K. Manchester, S. S. Park, D. West, H. L. Gurtoo, and H. V. Gelboin, Monoclonal antibody phenotyping of interindividual differences in cytochrome P-450 dependent reactions of single and twin human placenta. Cancer Res., 44:3916 (1984).

18. H. V. Gelboin, and F. K. Friedman, Monoclonal antibodies for studies on xenobiotic and endobiotic metabolism: cytochromes P-450 as paradigm. Biochem. Pharmacol., 34:2225 (1985).

19. B. J. Song, T. Fujino, S. S. Park, F. K. Friedman, and H. V. Gelboin, Monoclonal antibody-directed radioimmunoassay of specific cytochromes P-450, J. Biol. Chem., 259:1394 (1984).

20. H. V. Gelboin, Carcinogens, drugs and cytochromes P-450, N. Engl. J. Med., 309:105 (1983).

21. B. J. Song, H. V. Gelboin, S. S. Park, G. Tsokos, and F. K. Friedman, Monoclonal antibody-directed radioimmunoassay detects cytochrome P-450 in human placenta and lymphocytes, Science, 228:490 (1985).

154

DRUG INTERACTIONS WITH MACROLIDE ANTIBIOTICS : SPECIFICITY OF

PSEUDO-SUICIDE INHIBITION AND INDUCTION OF CYTOCHROME P-450

Patrick M. Dansette, Marcel Delaforge, Eric Sartori,
Philippe Beaune°, Maryse Jaouen and Daniel Mansuy

Laboratory of chemical and biochemical pharmacology and
toxicology, CNRS UA400 45 rue des St Pères 75270 Paris Cedex 06
and°Laboratory of BIochemistry, CHU Necker 156 rue de
Vaugirard 75730 Paris Cedex 15 France

ABSTRACT: Macrolide antibiotics like Erythromycin and Tri-acetyl
oleandomycin (TAO) are metabolized to nitrosoderivatives which cause
inactivation of Cytochrome P-450 by forming stable complex with the Iron of
the hemoporphyrin. Several derivatives of erythromycin having lost their
cladinose moiety are stronger inducer of liver cytochrome P-450 itself. The
major form of cytochrome P-450 induced by all these macrolides in rat liver
electrophoretically and immunogically indistiguishable from the major form
induced by pregnenolone 16 $\propto$ carbonitrile (PCN). This form is particularly
able to metabolize macrolide and to lead to the corresponding 456nm absorbing
cytochrome P-4540 complexes <u>in vivo</u> and <u>in vitro</u> .

INTRODUCTION

  Cytochrome dependent monooxygenases play a key role in pharmacology since
they control the elimination rate of drugs as well as their plasmatic level
and their therapeutic effect. They also play a key role in toxicology as they
are one priviliged site of formation of reactive metabolite inside the cells.

SCHEME I: DRUG-DRUG INTERACTIONS

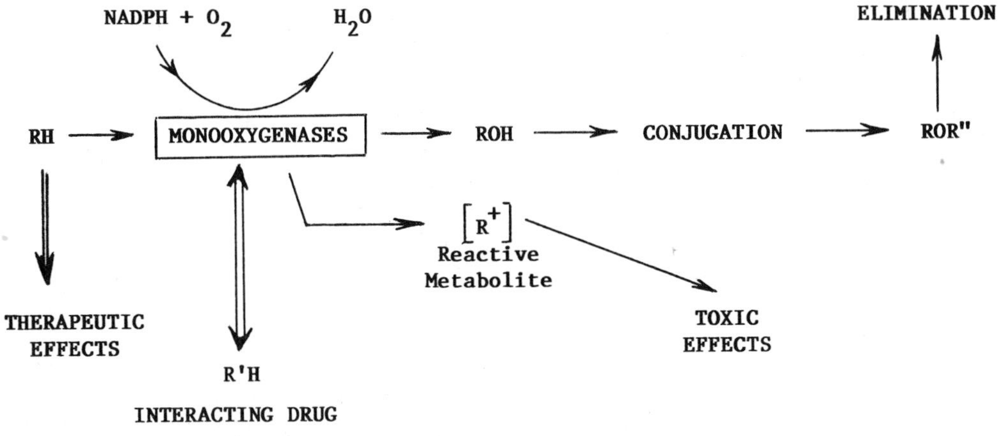

ABBREVIATIONS : TAO= Triacetyloleandomycin, PCN= Pregnenolone carbonitrile,
E= Erythromycin, EM= Erythralosamine, DAEM= Diacetylerythralosamine, MAEM=
Monoacetylerythralosamine, UT= Untreated rats.

155

Many compounds including drugs are known to greatly modify the levels and activities of these monooxygenases by acting as inducers or inhibitors. Such effects are at the origine of many drug interactions, either by increasing their elimination rate and decreasing their plasmatic half life and possibly decreasing their therapeutic effect or by decreasing their metabolism, increasing their plasma concentration and their plasmatic half life leading to overdose effects which may be sometime usefull from a therapeutic point of view, but which can also lead in many cases to clinical accidents.

In order to understand the mechanism of such drug interactions, and even to predict them, it is necessary to understand, at the molecular level, how a drug or a xenobiotic acts as an inducer or an inhibitor of cytochrome P-450, and to find out which structural features are related with the inducing or inhibiting effect.

We describe here a class of compounds which greatly affect cytochrome P-450 and lead to very stable Iron-Metabolite complexes of cytochrome P-450 in vivo. When this happens, the complexed cytochrome P-450 loses the ability to bind and thus to activate dioxygen and in that sence the starting compound acts as pseudo suicide-substrate of cytochrome P-450. Several types of complexes involving the coordination chemistry of cytochrome P-450, with Fe-O, Fe-N and Fe-C bonds are known (1).

This communication will concentrate on a class of compounds the Macrolide Antibiotics , which have been know in the past few years to cause severe interactions with other drugs (2).
They inhibit drastically the first path effect of the asociated drug, leading to a great increase of plasma concentration of this particular drug, up to ten times in the case of TAO-dihydroergotamine(2, 3).

TAO and erythromycin have been shown to affect hepatic cytochrome P-450 and to prolong the effect of drugs and worse to cause secondary toxic effects as shown in the following table (Table I).

TABLE I

TYPICAL DRUG—DRUG INTERACTIONS

DRUG A	DRUG B	METABOLISM	EFFECTS
PHENOBARBITAL	ETHYNYLESTRADIOL	↗	↘
PILOCARPINE	HEXOBARBITAL	↘	↗
CIMETIDINE	WARFARIN	↘	↗
PROPOXYPHENE	DIPHENYLHYDANTOIN	↘	↗
NORETHISTERONE	ANTIPYRINE	↘	↗
TROLEANDOMYCIN (TAO)	ERGOTAMINE	↘	↗
ERYTHROMYCIN	THEOPHYLLIN	↘	↗

We have shown in the past that TAO and erythromycin are oxydized by cytochrome P-450 to form iron-metabolite complexes which are very stable in vivo as well as in vitro (4, 5). Such complexes have been also shown in vivo in man (6).

It is very probable that these complexes involve $Iron^{II}$-nitroso alkane structure, since they have the same spectral and physicochemical properties of other numerous complexes described formely.(1, 7, 8, 9, Table II).

The nitrosoalkane would be formed upon chemical oxidation of the sugar tertiary amine function upon oxidative dealkylation followed by N-oxidation (8). It has been found that nitrosoalkane which are chemically unstable, but can be formed in situ close to the iron of the hemoporphyrin bind strongly to it to form stable Fe^{II}-RNO complex. This complex can be dissociated in vitro by oxidation with relatively powerfull oxidizers like potassium ferricyanide to Fe^{III} hemoprotein which no longer bind RNO ligands (scheme II).

SCHEME II: METABOLIC OXIDATION OF TERTIARY AMINES TO NITROSOALKANES

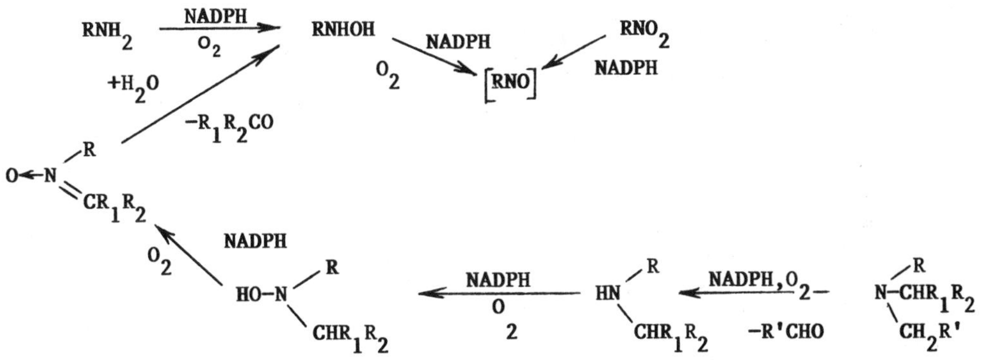

We have studied into details the effects of TAO and other related macrolides on hepatic cytochromes P-450, with the aim to help understanding the consequences of the in vivo formation of cytochrome P-450-Iron-metabolite complexes.

We have tried to answer three questions :

1) What are the structural features important for the formation of cytochrome P-450 macrolide complexes ?
2) Are induction by macrolides and complexation related ?
3) If not is the induced form a specific one ?

STRUCTURAL FEATURES RELATED TO MACROLIDE CYTOCHROME P-450 COMPLEXES

When rats are treated with TAO, an important part (70%) of cytochrome P-450 is complexed with the 455 nm spectral complex, and strong induction (four fold) of total cytochrome P-450 is obtained. When a kinetic study of in vivo complex formation is made, the result shows that a single dose of 500 mg/kg already induces cytochrome P-450 and at the same time leads to complexed cytochrome . But three dosing one day a part lead to a 4-5 time increase in total cytochrome P-450 most of it being complexed, and the half life of the complex can be estimated to about 3 days. One week after the cytochrome level has come back to normal, a reintroduction of TAO produces a more efficient induction of both total Cytochrome P-450 and complexed cytochrome P-450.

During the all experiment free cytochrome P-450 remains nearly constant (10) (Fig. 1).

TABLE II

COMPARISON OF MACROLIDES ANTIBIOTICS ABILITY TO FORM R-NO CYTOCHROME P-450

COMPLEXES IN VIVO IN RAT AND TO CAUSE DRUG INTERACTIONS IN MAN

DRUG	RAT a	MAN b	DRUG	RAT	MAN
Troleandomycin (TAO)	+	+	Josamycine base	–	?
Oleandomycin	+	+/–	Josamycine propionate	–	?
Erythromycin base	+	+	Tylosin tartrate	–	–
Erythromycin estolate	+	+	Spiramycin base	–	–
Erythromycin propionate	+	+	Spiramycin adipate	–	–
Erythromycin acetate	+	?	Platenomycin A1	–	
Erythrom.glucoheptonate	+		Platenomycin B1	–	
Methymycin	+		Carbomycin	–	–
M 4365 A2	+				
M 4365 G2	+		Rifampicin	–	+
Cirramycin A1	+		Lincomycin	–	–

a) + means: Cytochrome P-450 R-NO complex obtained in vivo or in vitro .
b) + means: known drug interaction; – : no interaction yet reported.

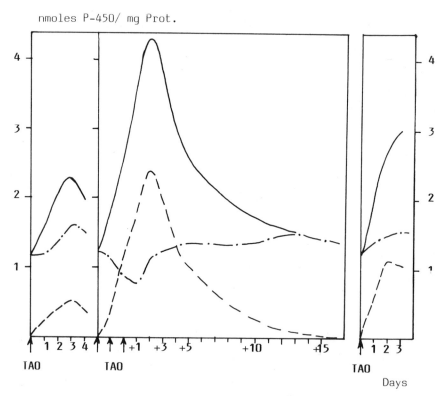

Fig. 2: In vivo modification of cytochromes P-450 in liver microsomes of rats treated with I.P. doses (500mg/kg in corn oil). ___ = total, _ _ _ = complexed, _._._ = free cytochrome P-450. A) after one single injection. B) after 3 daily injections. C) after a new injection 22 days after B.

We have examined a number of related antibiotics and found that only a small number were able to lead to such complex in vivo (11) (Fig. 2).

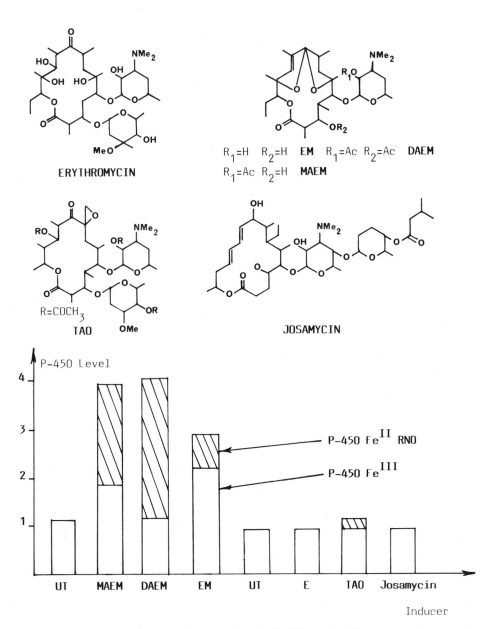

Fig. 2: Structures of macrolide analogs of TAO, and effects on the Cytochrome P-450 level in liver microsomes of rats treated with 3 daily doses of the drug (200mg/kg in corn oil).

Thus only those involving the following caracteristics are able to form significant amount of complex.

a) a readily accessible tertiary amine function.
b) a relative unhindered region near this tertiary amine.
c) a sufficient hydrophobicity of the molecule.

From the numerous compounds examined, TAO and erythromycin estolate were the most active in the formation of these complexes (11).

Some drugs which are very closely related to TAO like josamycin and spiramycin are not suited to make this type of complex and are less dangerous in respect to the potential drug interactions. Josamycin for instance which only differs from TAO by substitution of a hydrogen in β of the amino group by a sugar moiety in not able to for complex, and data from Pharmaco-toxicology survey show that it is not hepatotoxic, and after many years of use only one controversial drug interaction has been reported (12).

Moreover, we have recently found that erythromycin derivatives having lost their cladinose sugar and rearranged to more hydrophobic derivatives by internal cyclisations (DAEM and MAEM) are good substrates for complex formation in vitro , and are also better complex inducers in vivo than TAO, giving very stable RNO complexes. These derivatives are also very good inducers of cytochrome P-450, much stronger than TAO and erythromycin (13).

STRUCTURAL FEATURES RELATED TO INDUCTION OF CYTOCHROME P-450

All the compounds which are suited for giving macrolide cytochrome P-450 complexes are also good inducers. However we have found that complex formation is not required for induction of cytochrome P-450 since derivatives without the tertiary amino group, in particular desaminoerythralosamine were found to induce cytochrome P-450 and were absolutely unable to form complexes.

IS THE INDUCED CYTOCHROME P-450 A SPECIFIC ONE ?

When microsomes from TAO treated rats are analised by immunotransfer electrophoresis (Western blot), using several specific antibodies to cytochrome P-450 isozymes (either prepared by Ph. Beaune or given by Dr. F. Guengerich), we have found that the induced form is very similar if not identical to the major isozyme induced by PCN (PCN-E). Moreover the amount of complex formed in vitro with TAO by liver microsomes from rats treated with diverse inducers correlates well with the amount of immunologically quantified cytochrome P-450 (PCN-E) (13, 14) (data not shown).

The major form induced by DAEM, MAEM, TAO and PCN is very suited to form in vitro the 456 absorbing cytochrome P-450-metabolite complex in vivo or in vitro as shown in Table II.
This reaction can be considered as specific of the PCN-induced cytochrome P-450 and can be easily used for its detection.

POSSIBLE CONSEQUENCES FOR DRUG INTERACTIONS INVOLVING MACROLIDE ANTIBIOTICS RELATED TO TAO

Several antibiotics related to TAO induce in rat a cytochrome P-450 for identical or very similar to that induced by steroids like PCN or dexamethasone (13,15):
- For macrolides which have inducing capacities but are not able to form complexes such as desaminoerythromycin or rifampicin one should expect an

increase of the elimination of drugs which are metabolized by that particular form.
- For those leading to induction into mostly complexed cytochrome P-450, the net effect is a decrease of the available active cytochrome P-450, and of the first path effect of drugs metabolized by this particular form. Since the complex formed is very stable the half life of cytochrome P-450 may be increased; this may lead to accumulation in the liver and still higher induction of that particular isozyme of cytochrome P-450.
- Another effect of such macrolide could be to decrease the relative proportion of other forms of cytochrome P-450 as already found with other inducers (14, 16) and thus lead to a slower elimination of the drugs normally metabolized by these cytochromes.

Macrolide antibiotics have various effects on the liver cytochromes P-450. The consequences on drug interactions will depend of the ability of the antibiotic to induce the PCN-E cytochrome P-450 isozyme and to lead to complex formation but also upon the nature of the associated drug and of its main metabolic and elimination pathway. The induction of the same cytochrome P-450 isozyme by steroids and macrolides may result in interactions between both series. In that context further work is needed to explain at the molecular level clinical reports on associations of oral contraceptives and corticosteroids with macrolide antibiotics.

REFERENCES

1. D. Mansuy, Use of model systems in biochemical toxicology:heme models, Rev.Biochem. Toxicol., 3 ,283 (1981).
2. T.M. Ludden, Pharmacokinetic interactions of the macrolide antibiotics, Clinical Pharmacokinetics. 10 , 63 (1985).
3. A. Hayton, Precipitation of accute ergotism by Triacetyloleandomycin, New Zealand Medical Journal. 69 , 42 (1969).
4. D. Pessayre, D. Descatoire, M. Konstantinova-Mitcheva, J.C. Vandscheer, B. Cobert, R. Level, J.P. Benhamou, M. Jaouen and D. Mansuy, Self induction by Triacetyloleandomycin of its own tranformation into a metabolite forming a stable 456 nm-absorbing complex with cytochrome P-450, Biochem. Pharmacol. 30, 553 (1981).
5. D. Mansuy, M. Delaforge, E. Le Provost, J.p. Flinois, S. Columelli and Ph. Beaune, Induction of cytochrome P-450 in rat liver by the antibiotic Troleandomycin: partial purification and properties of cytochrome P-450-troleandomycin metabolite complex, Biochem. Biophys. Res. Commun. 103 , 1201 (1981)
6. D. Pessayre, D. Larrey, J. Vitaux, P. Breil, J. Belghiti and J.P. Benhamou, Formation of an inactive cytochrome P-450 Fe(II)-metabolite complex after administration of troleandomycin in human, Biochem. Pharmacol. 31 , 1699 (1982).
7. D. Mansuy, Coordination chemistry of cytochrome P-450 and iron-porphyrins: relevance to pharmacology and toxicology, Biochimie, 60 , 969 (1978).
8. D. Mansuy, Ph. Beaune, J.C. Chottard, J.F. Bartoli and P. Gans, The nature of the "455 nm absorbing complex" formed during the cytochrome P-450 dependant oxidative metabolism of amphetamine, Biochem. Pharmacol. 25 , 609 (1976).
9. B. Lindeke, U. Paulsen-Sorman, G. Hallstrom, A.H. Khutier, A.K. Cho and R.C. Kammerer, Cytochrome P-455 nm complex formation in the metabolism of phenyl-alkylamines. VI Structure-activity relationship in metabolic intermediary complex formation with a series of -substuted 2-phenylethylamines and corresponding N-hydroxylamines, Drug Metab. Dispos., 10 ,700 (1982).
10. M. Delaforge,M. Jaouen and D. Mansuy, The cytochrome P-450 metabolite complex derived from troleandomycin : Properties in vitro and stability in vivo, Chem.-Biol. Interactions, 51 , 371 (1984).

11. M. Delaforge, M. Jaouen and D. Mansuy, Dual effects of macrolide antibiotics on rat liver cytochrome P-450. Induction and formation of metabolite complexes : a structure-activity relationship, Biochem. Pharmacol., 32 , 2309 (1983).

12. J.Y. Grolleau, M. Martin, B. De La Guerande, J. Barrier and P. Peltier, Ergotisme aigu lors d'une association josamycine/tartrate d'ergotamine, Therapie, 36 , 319 (1981).

13. E. Sartori, M. Delaforge, D. Mansuy and Ph. Beaune, Some erythromycin derivatives are strong inducers in rats of a cytochrome P-450 very similar to that induced by 16 -pregnenolone carbonitrile, Biochem. Biophys. Res. Commun., 128 , 1434 (1985).

14. F.P. Guengerich, G.A. Dannan, S.T. Wright, M.V. Martin and L.S. Kaminsky, Purification and characterization of liver microsomal cytochromes P-450: Electrophoretic, spectral, catalytic and immunochemical properties and inducibility of eight isozymes isolated from rats treated with phenobarbital or -naphthoflavone, Biochemistry, 21 , 6019 (1982).

15. P.S. Guzelian, P.B. Watkins, E. Schuetz, S. Wrighton and P. Maurel, Degradation of cytochrome P-450 DEX is inhibited by troleandomycin in primary monolayer cultures of adult rat hepatocytes, Abstracts of the 6th International Symposium on Microsomes and Drug Oxidation, Brighton, 1984.

16. P. Thomas et al., This Symposium.

ISOMERIC COMPOSITION OF N-ALKYLATED PROTOPORPHYRINS PRODUCED

BY SUBSTITUTED DIHYDROPYRIDINES IN VIVO AND IN VITRO

Francesco De Matteis and Anthony H. Gibbs

MRC Toxicology Unit, M.R.C. Laboratories
Woodmansterne Road, Carshalton
Surrey SM5 4EF, U.K.

INTRODUCTION

Substituted dihydropyridines, like DDC and its 4-ethyl
analogue (4-ethyl-DDC)* give rise to N-alkylated porphyrins
when given to rats and mice in vivo, by donating their intact
4-alkyl group to one of the pyrrole nitrogens of haem in
hepatic cytochrome P-450 (De Matteis et al., 1981; Ortiz
de Montellano et al., 1981; Tephly et al., 1981). Because
of the asymmetrical arrangement of the two vinyl and propionate
side chains in protoporphyrin IX, four structural isomers
of N-monoalkylated protoporphyrins are possible, according
to which of the four pyrrole nitrogens has been substituted.
In a previous paper (De Matteis et al., 1983) we have provided
evidence that the isomeric composition of the N-ethyl protoporphyrin produced in vivo by treatment with 4-ethyl-DDC
depends on the particular cytochrome P-450 which predominates
at the time of treatment, suggesting a role for the apo-cyto-
chrome in directing alkylation preferentially on to one of
the pyrrole nitrogens. In this present paper we wish to
report additional findings in vivo which are compatible with
this interpretation. Production of N-ethyl protoporphyrin
is also found in vitro, when isolated microsomes are incubated
with 4-ethyl-DDC in presence of NADPH and, although a change
in isomeric profile is again found with microsomes containing
different cytochrome P-450 enzymes, nevertheless the picture
obtained with isolated membranes in vitro is different from
that seen in the intact cells in vivo.

METHODS

Male rats of the Porton (Wistar-derived) strain were
pretreated with inducers of cytochrome P-450 enzymes as described
(De Matteis et al., 1983) and were given 4-ethyl-DDC intraperitoneally (100 mg/kg body wt)30 min befor killing. In

*Abbreviations used: DDC, 3,5-diethoxycarbonyl-1,4-dihydro-
2,4,6-trimethylpyridine; 4-ethyl-DDC, 3,5-diethoxycarbonyl-
4-ethyl-1,4-dihydro-2,6-dimethylpyridine; h.p.l.c., high
performance liquid chromatography.

one experiment in which the recovery of N-ethylprotoporphyrin
from liver homogenates was studied, an uninduced rat was
injected intraperitoneally with 2 µCi of 5-amino[4-^{14}C]levulinate
2 h before being dosed with 4-ethyl-DDC. Alkylated porphyrins
were extracted from liver homogenates and purified as described
(De Matteis et al., 1983); they were estimated in the Sephadex
LH 20 eluate, using the Soret absorption of the dication
derivative (De Matteis et al., 1982) and were resolved into
the four structural isomers by the h.p.l.c. procedure of
Ortiz de Montellano et al. (1981).

Microsomes were obtained from liver homogenates of fed
rats, either controls or pretreated with inducers; they were
washed and stored frozen as a pellet at -70 C. Microsomal
suspensions (3-5 mg protein/ml) were incubated aerobically
at 37 C for 10 min with 4-ethyl-DDC (0.5 mM) in the presence
of NADPH (1 mM), a NADPH-generating system, phosphate buffer,
pH 7.4 (100 mM) and EDTA (1 mM) in a total volume of 3.15ml.
To stop the reaction the incubation mixture was added to
20 volumes of 5% H_2SO_4 in methanol and methylation allowed
to proceed for 20 h in the dark at 4 C. Proteins were removed
by centrifugation and the methylated products were transferred
to $CHCl_3$. After washing with water, the organic phase was
dried under N_2 and dissolved in 80 µl methanol for analysis
on h.p.l.c., using a Nucleosil 5 column (4.5 mm x 250 mm),
isocratic elution with dichloromethane/methanol/ conc.NH_3
(sp.gr. 0.88) (50:50:0.1, by vol) and a flow rate of 2 ml/min.
This procedure allowed good separation of the N-ethylated
protoporphyrin from haem but could only resolve the alkylated
porphyrin into two fractions, the first comprising the two
isomers N_A and N_B, the second the isomers N_C and N_D (the
suffixes A-D indicate the pyrrole ring that is N-ethylated
in each structural isomer, see De Matteis et al., 1983).

Male 6 weeks-old mice of the MFI (Ah-responsive) and
of the DBA-2 (Ah-nonresponsive) strains were also used. They
were pretreated with a single intraperitoneal injection of
either β-naphthoflavone (80 mg/kg body wt) in arachis oil or
oil alone 48 h before being given either DDC or 4-ethyl-DDC
(10 mg/kg body wt). They were killed 1 h thereafter. N-Alkylated porphyrins were extracted, estimated and resolved into
individual isomers as in the experiments with rats, except
that 4 to6 mice were pooled for each experiment.

RESULTS AND DISCUSSION

When rats were given a single injection of 4-ethyl-DDC
30 min before killing, the amount and isomeric composition
of the resulting N-ethyl protoporphyrin isolated from their
livers was found to depend on whether they had received a
preliminary treatment with phenobarbitone or β-naphthoflavone
(Table 1). These results confirm previous observations (De
Matteis et al., 1963) obtained in rats killed 1 h after 4-ethyl-
DDC, where too the predominant isomer was N_C in control rats,
N_A in phenobarbitone-induced rats and N_D in β-naphthoflavone-
induced rats.

Further evidence that the change in isomeric composition
due to induction was not an artifact of the isolation procedure,
but reflected the "native" isomeric composition existing
in the liver at the time of killing, was obtained as follows.

Table 1. Effect of treatment of rats with two inducers of cytochrome P-450 on amount and isomeric composition of $\underline{N}$-ethylprotoporphyrin isolated from liver after treatment with 4-ethyl-DDC.

N-Ethyl protoporphyrin isolated

Pretreatment	Total (pmol/g liver)	Isomer composition (%)			
		N_B	N_A	N_C	N_D
None (Control)	1965 ± 119	3.7 ± 0.1	26.4 ± 0.6	47.6 ± 0.7	22.4 ± 0.8
Phenobarbitone	2697 ± 90**	3.9 ± 0.1	61.5 ± 1.4**	17.0 ± 0.9**	17.5 ± 0.7*
β-Naphthoflavone	2686 ± 127*		33.3 ± 2.6+	11.8 ± 2.2**	55.0 ± 4.2**+

Values are means ± SEM of at least 4 observations. * $P < 0.005$, ** $P < 0.001$, when compared with corresponding 'control' values. + $P < 0.001$, when compared with corresponding 'phenobarbitone values'.

Table 2. Amount and specific radioactivity of three isomers of N-ethylprotoporphyrin isolated from the liver homogenates of rats.

		Control	β-Naphthoflavone	Mixture Expected	Found
Amount of isomers (nmol/g liver)	N_A	0.59	0.74	0.67	0.68
	N_C	0.91	0.27	0.57	0.54
	N_D	0.54	0.82	0.69	0.69
Specific radio-activity of isomers (dpm/nmol)	N_A	2,340	0	1,047	1,093
	N_C	2,010	0	1,550	1,450
	N_D	2,260	0	905	933

A control rat (prelabelled with 5-amino[4-^{14}C]levulinate) and a β-naphthoflavone-induced rat were both given 4-ethyl-DDC and their liver homogenates were extracted individually and as an equal mixture. Values obtained in the mixture were compared to those calculated from the individual homogenates.

Table 3. Amount and isomeric composition of N-ethyl protoporphyrin isolated from the liver of mice after 4-ethyl-DDC. Effect of pretreatment with β-naphthoflavone (BNF).

N-Ethyl protoporphyrin

Strain	Pretreat-ment	Total (pmol/g liver)	Isomer composition (%)			
			N_B	N_A	N_C	N_D
MFI	Oil	1121	2.3	26.3	48.9	23.6
	BNF	1781	1.3	18.5	17.9	62.4
DBA-2	Oil	1257	1.9	30.5	51.7	16.0
	BNF	1548	1.9	30.7	42.3	25.2

Values are means of two observations, each obtained with the combined livers of 4-6 mice.

One control and one β-naphthoflavone-induced rat were both
given 4-ethyl-DDC (the control having also received 5-amino-
[4-^{14}C]levulinate 2 h before 4-ethyl-DDC, so as to label
its N-ethyl protoporphyrin). The two liver homogenates were
then extracted individually and as an equal mixture; the
amount and specific activity of the three main isomers found
in the mixture were then compared to those expected from
the individual values of the two homogenates (Table 2).
A good agreement was found.

The production of N-ethyl protoporphyrin by treatment
with 4-ethyl-DDC was also studied in mice of the MFI and
DBA-2 strains, both with and without β-naphthoflavone pretreatment (Table 3). In control mice of either strain the isomeric
profile was quite similar to that found in control rats,
that is isomer N_C predominated. After pretreatment with
β-naphthoflavone, however, the expected marked increase in
isomer N_D was only seen in the MFI (Ah responsive) strain.
This is an agreement with the N_D isomer originating preferentially from the β-naphthoflavone-inducible cytochrome
(P-448), as the DBA-2 strain would not be expected to respond
significantly with induction to β-naphthoflavone treatment.

The effect of pretreatment with β-naphthoflavone on amount
and isomeric composition of N-methyl protoporphyrin (produced
by treatment with DDC, the 4-methyl dihydropyridine) was
also studied in both strains of mice (Table 4). The effect
of DDC was found to differ from that of 4-ethyl-DDC in two
main respects. First, as reported in rats after induction
with 3-methylcholanthrene (Coffman et al., 1982) we found
induction with β-naphthoflavone to cause a reduced yield
of N-methyl protoporphyrin, an effect only seen in the Ah-
responsive strain (MFI). Second, the major isomer of N-methyl
protoporphyrin was N_A in both control and β-naphthoflavone
treated mice of both strains, so no marked change in isomeric
profile was found after induction, even in the responsive
strain. These findings are compatible with the suggestion
(Coffman et al., 1962) that DDC, unlike 4-ethyl-DDC, is not
readily utilized by cytochrome P-448 as a suicide substrate.

Production of N-ethylprotoporphyrin could also be demonstrated in vitro when rat liver microsomes were incubated
aerobically with 4-ethyl-DDC in presence of NADPH (Table 5).
Although pretreatment with inducers increased the yield of
N-ethylprotoporphyrin (per mg of microsomal protein) and
changed significantly the isomeric profile (as compared with
that obtained with control microsomes), nevertheless the
isomeric composition obtained in vitro was different from
that seen when the pigment was produced in vivo. Attempts
to modify the isomeric profile of the N-ethylprotoporphyrin
produced by 3-methylcholanthrene microsomes by using fresh
(rather than stored) microsomes, by omitting EDTA or by adding
cytosol to the incubation mixture (2 ml of cytosol obtained
from a 50% liver homogenate per 3.15 ml total incubation)
were not successful.

Authentic synthetic N-ethylprotoporphyrin incubated
(at 1 μM concentration) with liver microsomes in presence
of NADPH was partly degraded, an effect prevented by adding
catalase to the incubation mixture. The $N_A + N_B$ mixture
of isomers were lost to a greater extent than the $N_C + N_D$

Table 4. Amount and isomeric composition of N-methyl proto-
porphyrin isolated from the liver of mice after
DDC. Effect of pretreatment with β-naphthoflavone
(BNF).

N-methyl protoporphyrin

Strain	Pretreat-ment	Total (pmol/g liver)	Isomer composition (%)			
			N_B	N_A	N_C	N_D
MFI	Oil	1056	5.8	83.2	10.0	1.1
	BNF	647	6.6	74.1	15.7	4.8
DBA-2	Oil	1576	6.3	82.2	9.3	2.3
	BNF	1431	5.4	81.8	9.5	1.6

Values are means of two observations, each obtained with
the pooled livers of 4-6 mice.

mixture (30-40% loss, as compared with 10-20%, respectively).
It is unlikely that this difference in degradation rate between
various isomers could have accounted for the isomeric profile
found in vitro, as control incubations where catalase was
also added showed production of N-ethylprotoporphyrin (from
4-ethyl-DDC) to be very marginally decreased with no significant
alteration of percent composition of isomers.

SUMMARY AND CONCLUSIONS

     The ability of inducers of different cytochrome P-450
enzymes to increase the in vivo production of N-ethylprotoporphyrin caused by 4-ethyl-DDC treatment, changing at the
same time its isomeric profile, has now been confirmed.
The main isomer accumulating in the control liver is N_C (where
pyrrole ring C is N-ethylated); after pretreatment with phenobarbitone and β-naphthoflavone the main isomers which accumulate
are N_A and N_D, respectively. Similar changes in isomeric
profile are found after β-naphthoflavone treatment in mice,
provided that a responsive strain is used, capable -that
is- of showing cytochrome P-448 induction after this inducer.
An isotopic dilution experiment indicates that the change
in isomeric profile caused by β-naphthoflavone reflects a
genuine difference in isomeric composition and does not result
from differential recoveries of the various isomers. These
results are compatible with the hypotheses that the alkylated
porphyrins originate from the haem of cytochrome P-450 and
that the various apo-cytochromes direct alkylation on different
pyrrole nitrogens; possibly by changing the orientation of
the binding site for the dihydropyridine substrate with respect
to the haem prosthetic group, so as to render different pyrrole
nitrogens more accessible to the liberated reactive ethyl
group.

Table 5. Amount and isomeric composition of N-ethylproto-porphyrin produced in vitro by isolated rat liver microsomes. Effect of pretreatment with phenobarbitone or 3-methylcholanthrene.

Pretreatment of rats in vivo	N-Ethylprotoporphyrin produced in vitro	
	Total (pmol/mg protein)	Isomeric composition (N_A+N_B expressed as % of total)
None, control	39.2 ± 3.7	34.0 ± 1.3 $(38.3)^+$
Phenobarbitone	$53.7 \pm 1.9^*$	$44.1 \pm 2.3^*$ (68.1)
3-Methyl-cholanthrene	$202.7 \pm 13.4^{**}$	$62.0 \pm 0.3^{**}$ (29.4)

Values given are means ± SEM of at least three observations.
* $P < 0.02$, when compared to corresponding control values,
** $P < 0.02$, when compared to corresponding phenobarbitone values. + The isomeric composition obtained in vivo with the same pretreatment is shown in parentheses for comparison.

Attempts to reproduce in vitro the same isomeric profile found in vivo after inducers were not successful. This may indicate that the isolated membranes differ from their native state in the intact cell with respect to the stereochemistry of the active site of their cytochromes P-450.

REFERENCES

Coffman, B.L., Ingall, G. and Tephly, T.R., 1982. The formation of N-alkylprotoporphyrin IX and destruction of cytochrome P-450 in the liver of rats after treatment with 3,5-diethoxycarbonyl-1,4-dihydrocollidine and its 4-ethyl analog. Arch.Biochem.Biophys., 218:220.

De Matteis, F., Gibbs, A.H., Farmer, P.B. and Lamb, J.H., 1981, Liver production of N-alkylated porphyrins caused in mice by treatment with substituted dihydropyridines. Evidence that the alkyl group on the pyrrole nitrogen atom originates from the drug. FEBS Lett., 129:328,

De Matteis, F., Hollands, C., Gibbs, A.H., De Sa, N. and Rizzardini, M., 1982, Inactivation of cytochrome P-450 and production of N-alkylated porphyrins caused in isolated hepatocytes by substituted dihydropyridines. Structural requirements for loss of haem and alkylation of the pyrrole nitrogen atom. FEBS Lett., 145:87.

De Matteis, F., Gibbs, A.H. and Hollands, C., 1983, N-Alkylation of the haem moiety of cytochrome P-450 caused by substituted dihydropyridines. Preferential attack of different pyrrole nitrogen atoms after induction of various cytochrome P-450 isoenzymes. Biochem.J., 211:455.

Ortiz de Montellano, P.R., Beilan, H.S. and Kunze, K.L., 1981, N-Alkylprotoporphyrin IX formation in 3,5-dicarbethoxy-1,4-dihydrocollidine-treated rats. Transfer of the alkyl group from the substrate to the porphyrin. J.Biol.Chem., 256:6708.

Tephly, T.R., Coffman, B.L., Ingall, G., Abou Zeit-Har, M.S., Goff, H.M., Tabba, H.D. and Smith, K.M., 1981, Identification of N-methylprotoporphyrin IX in livers of untreated mice and mice treated with 3,5-diethoxycarbonyl-1,4-dihydrocollidine: Source of the methyl group. Arch.Biochem. Biophys., 212:120.

UDP-GLUCURONYLTRANSFERASES AND THEIR TOXICOLOGICAL

SIGNIFICANCE

Karl Walter Bock, Barbara S. Bock-Hennig, Gösta
Fischer[*] Werner Lilienblum and Gerhard Schirmer

Departments of Pharmacology and Toxicology and of
Pathology[*] University of Göttingen, 3400 Göttingen
FRG

INTRODUCTION

In general chemicals are converted by phase-I enzymes of
drug metabolism to a variety of nucleophilic and electrophilic
metabolites (Fig. 1). The interaction of the more reactive
electrophilic metabolites with critical cellular macromolecules plays a major role in their toxicity (Miller and Miller,
1981). Therefore much interest was given to the control of
electrophilic metabolites. However the more stable and more
abundant nucleophilic metabolites can also be readily converted to reactive metabolites. For example studies on benzene
toxicity have shown that the major primary metabolite phenol
is further oxidized to quinols, catechol and hydroquinone, in
liver. These quinols are accumulating in bone marrow where
they may be further oxidized to radical intermediates (Greenlee et al., 1981). Similarly, benzo(a)pyrene (BP) quinols are
readily converted to reactive semiquinones. Moreover they
undergo toxic redox-cycles between quinones and quinols with a
continuous generation of reactive oxygen species (Lorentzen
and Ts'o, 1977; Lorentzen et al., 1979; Lilienblum et al.,

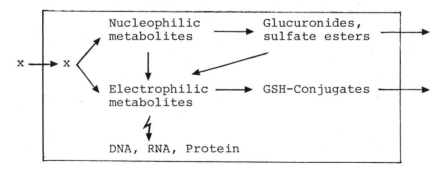

Fig. 1. Control of nucleophilic and electrophilic metabolites
in cellular drug metabolism.

Table 1. Isoenzymes of UDP-glucuronosyltransferase (GT)

Isoenzyme	Substrate	M_r $(\times 10^{-3})$	References
Phenol-GT (GT_1, GT_{MC})	1-Naphthol 4-Nitrophenol	56 54	Falany et al.,1983 Bock et al.,1979
Bilirubin-GT	Bilirubin	54 53	Scragg et al.,1985 Roy Chowdhury et al.,1985
3-OH-androgen-GT	Androsterone Etiocholanolone Lithocholic acid	52	Falany et al.,1985
17-OH-steroid-GT	Testosterone Estradiol 1-Naphthol 4-Nitrophenol	50	Falany et al.,1985
3-OH-estrogen-GT	Estrone Estradiol	?	Falany et al.,1985

1985). Therefore the control of nucleophilic metabolites by UDP-glucuronyltransferase (GT) and sulfotransferase cannot be neglected. In the case of benzene toxicity conjugation leads to the elimination of quinols. In support of this observation the time course of induction of UDP-glucuronyltransferase (GT) by 3,4,3',4'-tetrachlorobiphenyl mirrored the time course of protection against benzene toxicity in the bone marrow and lymphoid organs (Greenlee and Irons, 1981). In this context it is to be noted that some conjugates, in particular sulfate esters, are more reactive than the parent compounds (Miller and Miller, 1978).

Similar to findings with other drug metabolizing enzymes it has recently been demonstrated that GT consists of a family of isoenzymes with distinct but overlapping substrate specificity (Bock et al., 1979; Dutton, 1980; Falany et al., 1983; Roy Chowdhury et al., 1985; Scragg et al., 1985). Phenol-GT (GT_1) is inducible by 3-methylcholanthrene(MC)-type inducers and prefers planar phenols as substrates. Other isoenzymes are more specific for endogenous compounds such as bilirubin and steroids (Table 1). However the 17-hydroxy-steroid-GT also conjugates exogenous planar phenols. The existence of multiple isoenzymes has been suggested earlier in studies on differential induction of GT activities by two prototypes of inducing agents of the P-450 system, phenobarbital and MC (Bock et al., 1973; Wishart, 1978; Bock et al., 1979; Lilienblum et al., 1982; Watkins et al., 1982; Bock et al., 1984). There is a remarkable similarity between the induction of P-450 and GT. For example, coordinate induction of P_1-450 and GT_1 has been suggested in genetic studies with 'responsive', i.e. MC-inducible, and 'non-responsive' inbred strains of mice (Owens, 1977). In isolated hepatocytes the isoenzyme level has been shown to be a major determinant of glucuronide formation although other factors may be important in some instances (Ullrich and Bock, 1984).

In the following recent advances and problems in the elucidation of multiple isoenzymes are exemplified in the search of functional and molecular properties of GT_1.

FUNCTIONAL PROPERTIES OF GT_1

It is obviously difficult to characterize 'acceptor unspecific' isoenzymes. Despite its preference for planar phenols GT_1 shows regioselectivity in the conjugation of planar BP phenols. For example the conjugation of 3-hydroxy-BP is stimulated in liver microsomes of MC-treated rats, the conjugation of 9-hydroxy-BP is not (Bock et al., 1984). Evidence for a special role of GT_1 in the detoxication of polycyclic aromatic hydrocarbons stems from previous studies of BP mutagenicity using the Ames-Test (Bock et al., 1981; Bock et al., 1982a). In these studies cofactors of the GT reaction were added (UDP-glucuronic acid and UDP-N-acetylglucosamine) and microsomes from untreated controls, phenobarbital- and MC-treated rats were used as enzyme sources. Addition of the cofactors decreased BP mutagenicity but the effect was most pronounced using microsomes from MC-treated rats. These findings may illustrate a role for coordinate induction of P_1-450 and GT_1, namely to decrease the risk of P_1-450 induction.[1]

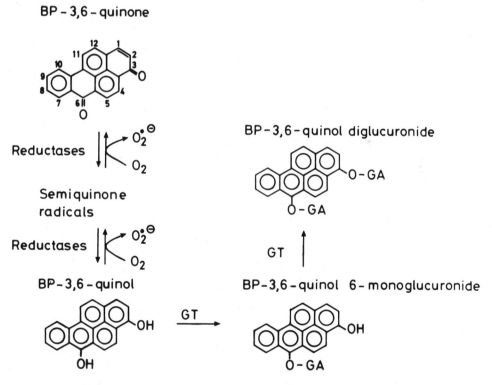

BP - 3,6 - quinone

Reductases

Semiquinone radicals

Reductases

BP-3,6-quinol

BP-3,6-quinol diglucuronide

GT

BP-3,6-quinol 6-monoglucuronide

Fig. 2 Conversion of BP-3,6-quinone to the corresponding quinol glucuronides. GA, glucuronic acid.

BP-3,6-quinol may be a most useful functional probe of GT$_1$ (Fig. 2). It is formed from the corresponding quinone by P-450 reductase and DT-diaphorase, the latter enzyme bypassing the semiquinone step. The quinol is autoxidized and then undergoes toxic redox-cycles with a continuous production of reactive oxygen species. The resulting toxicity can be prevented by conjugation. It was recently found that glucuronidation of the quinol proceeds in two steps, forming the mono- and diglucuronide (Lilienblum et al., 1985). The two conjugates can be detected simultaneously due to their different fluorescent spectra. Diglucuronide formation was low in liver microsomes from untreated controls and phenobarbital-treated rats, but it was dramatically stimulated in microsomes from MC-treated rats (Fig. 3). The time course of mono- and diglucuronide formation suggests a precursor-product relationship. From the initial rates the reactions were analysed kinetically. It was found that the increase of diglucuronide formation was not due to a change in K$_m$ but due to a 40-fold increase in V$_{max}$. This is the greatest induction factor ever found for a MC-inducible GT reaction. This finding suggests that either (a) BP-3,6-quinol 6-monoglucuronide is a more selective substrate for GT$_1$ than simple phenols such as 1-naphthol or (b) more than one MC-inducible GT exists in rat liver. Diglucuronide formation was also high with isolated hepatocytes from

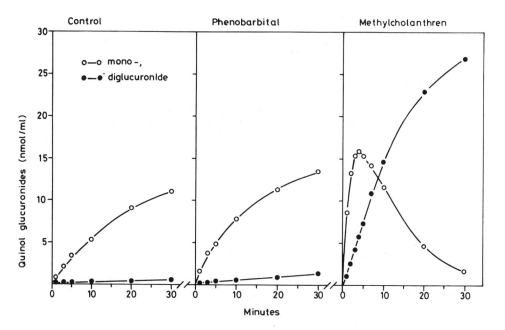

Fig. 3. Glucuronidation of BP-4,6-quinol to its mono- (o) and diglucuronide (●) with liver microsomes from untreated controls and from phenobarbital- or MC-treated rats. BP-3,6-quinone (25 μM) was reduced to the quinol by addition of 0.1 M ascorbic acid to the microsomes (0.1 mg/ml) in the assay described previously (Lilienblum et al., 1985).

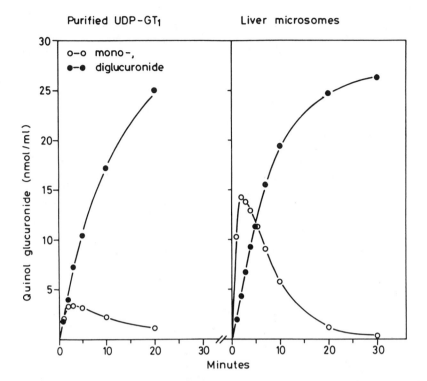

Fig. 4. Glucuronidation of BP-3,6-quinol to its mono- (o) and diglucuronide (●) with purified GT_1 and microsomes from MC-treated rats (0.02 and 0.2 mg protein, respectively).

MC-treated rats (Lilienblum et al., 1985) and with the purified GT_1 (Fig. 4). However the dramatic increase of diglucuronide formation was not observed in liver microsomes from MC-treated responsive C57BL/6 mice (unpublished results).

The toxicological significance of quinol glucuronidation was studied using H4IIE-cells, a permanent rat hepatoma cell line derived from a Reuber hepatoma (Reuber, 1961; Pitot et al., 1964). Pilot experiments showed that this cell line contained high GT_1 activity but no phenol sulfotransferase activity. The cells were grown for 24 h in the presence of various concentrations of BP-3,6-quinone. Cytotoxicity was followed measuring ³H-thymidine incorporation into cellular DNA (Fig. 5). The experiments were carried out in the absence and presence of salicylamide, an inhibitor of glucuronide formation. Under our conditions no cytotoxicity was observed in the absence of the inhibitor. However a dose dependent increase of cytotoxicity was seen in its presence. Hence glucuronidation was able to prevent cytotoxicity.

MOLECULAR PROPERTIES OF GT_1

GT_1 was purified to apparent homogeneity. The purified enzyme preparation showed a subunit molecular weight of about 54000 (Bock et al., 1979). Polyclonal antibodies were raised in rabbits (Pfeil and Bock, 1983). They were used to study

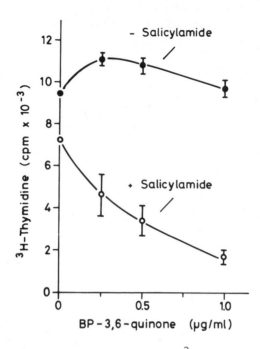

Fig. 5. Effects of BP-3,6-quinone on ^{3}H-thymidine incorporation into DNA of H4IIE-cells.
BP-3,6-quinone and ^{3}H-thymidine (0.07 μCi/ml) were
added to 1.5 x 10^5 cells. When indicated 2 mM salicylamide was also present. After 22 hours in culture the
cells were washed and lysed with 0.5% sodium dodecyl
sulfate. DNA was precipitated with 10% trichloroacetic
acid and collected on glass filters.

GT polypeptides in microsomes (Fig. 6). After electrophoretic
separation the proteins were transferred to nitrocellulose
and immunostained (Towbin et al., 1979). Anti-GT$_1$ antibodies
and peroxidase-conjugated swine anti-rabbit antibodies were
used successively. Four polypeptides at 50, 52, 54 and 56 kD
were recognized by the anti-GT$_1$ antibody. The 54 and 56 kD
polypeptides were inducible by MC-type inducers. This finding
raises the question: Are the two inducible polypeptides
different gene products or are they posttranslational modifications of a common gene product? Posttranslational modification of GT was suggested by studies on GT synthesis in
vitro (Mackenzie and Owens, 1984).

Immunochemical cross-reactivity between rat and mouse GT
has been observed in several laboratories (Burchell, 1979;
Pfeil and Bock, 1983; Mackenzie et al., 1984). Similar to the
rat system several mouse liver polypeptides were stained by
the antibody. Two polypeptides at 54 and 55 kD were inducible
by MC. Interestingly immunochemical cross-reactivity could
also be demonstrated between rat and human GT. A polypeptide
at about 54 kD was immunostained. This finding was supported
using competitive inhibition of enzyme binding in an ELISA and
by enzyme precipitation and inhibition (unpublished results).
Binding between the human enzyme and the anti-rat GT$_1$ antibody

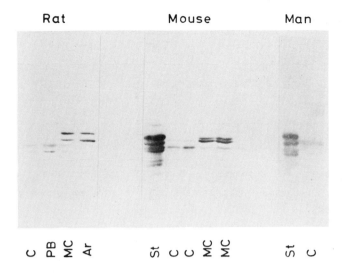

Fig. 6. Immunoblot analysis of GT polypeptides in rat, mouse and human liver microsomes.
Rat: Microsomes (2.5 µg protein) were from untreated controls (C), phenobarbital(PB)-, MC- and Aroclor 1254-treated male Wistar rats. Mouse: Microsomes (2.5 and 5 µg protein) were from untreated and MC-treated male C57BL/6 mice. Man: Microsomes (30 µg protein) were from a 5 year old boy. St, microsomes from MC-treated rats (5 µg protein) were used as molecular weight standards.

appears to be weak since it was not observed in Ouchterlony double-diffusion analysis (Pfeil and Bock, 1983).

Similar to the liver enzyme inducible GT polypeptides at 54 and 56 kD could also be demonstrated by immunoblot analysis of microsomes from rat small intestine and kidney (Koster, Schirmer and Bock, unpublished results). Aroclor 1254 (500 mg/ kg, once i.p., treatment for 7 days) was used as an inducer of GT_1 in extrahepatic tissues.

The antibodies were also used to localize GT_1 immuno-histochemically in the liver lobule. It was found that the enzyme was preferentially localized in the centrilobular region (Ullrich et al., 1984). The immunohistochemical findings were substantiated by enzyme determinations in microdissections from the centrilobular and periportal region.

'PERMANENT INDUCTION' OF GT_1 IN THE COURSE OF HEPATOCARCINO-GENESIS

Earlier studies indicated high GT_1 activities in hepa-tomas and liver nodules (Lueders et al., 1979; Bock et al., 1982b). Using the immunohistochemical method increased GT_1 could be demonstrated in early focal lesions (Fischer et al., 1983a). GT-positive focal lesions were observed after the administration of a variety of carcinogens: 2-acetylamino-

Table 2. Altered drug metabolizing enzymes in focal lesions produced by a single injection of N-nitrosomorpholine (NNM) (75 mg/kg i.p., 24 h after partial hepatectomy)

Treatment	GT_1 (µmol/min/g dry weight)	Arylhydrocarbon hydroxylase (nmol/min/g dry weight)
NNM (180 days after initiation)		
Focal tissue	21.6 ± 4.2 (12)[a]	26 ± 16 (4)
Extrafocal tissue	6.8 ± 1.6 (12)	56 ± 11 (4)
NNM (330 days after initiation)		
Focal tissue	34.0 ± 8.0 (12)	33 ± 8 (4)
Extrafocal tissue	6.2 ± 1.4 (12)	60 ± 3 (4)

[a] Data represent the means $\pm$ S.D. The number of determinations is given in brackets. For the arylhydrocarbon hydroxylase assay three microdissections were pooled, i.e., four determinations represent the enzyme activity of 12 microdissections. 1-Naphthol was used to assay GT_1 activity.

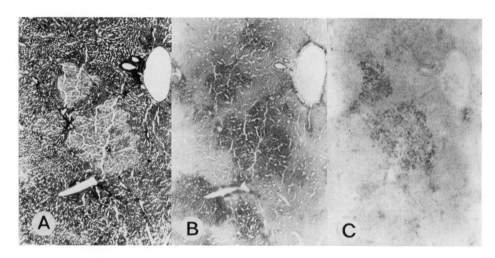

Fig. 7. Histochemical correlation between ATPase-negative (A), ɣ-glutamyltranspeptidase-positive (B) and GT-positive (C) focal lesions. They were produced by administration of methapyrilene-HCl (0.1 percent) in drinking water for 120 days. ɣ-Glutamyltranspeptidase-positive staining in zone 1, which is unrelated to carcinogenesis, is seen in B, left lower corner.

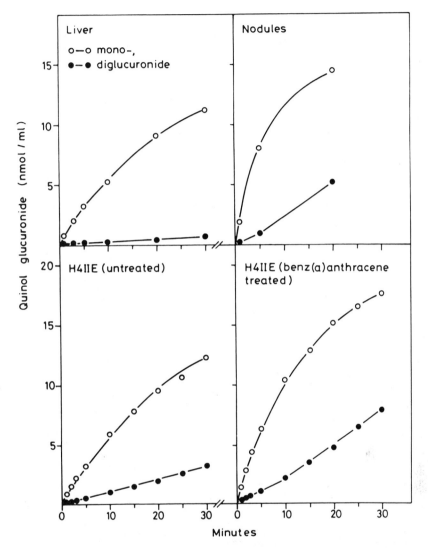

Fig. 8. Glucuronidation of BP-3,6-quinol to its mono- (o) and
diglucuronide (●) with microsomes from control liver,
pooled liver nodules (produced by feeding 0.05% 2-ace-
tylaminofluorene in the diet) and in H4IIE-cells,
grown in the absence and presence of benz(a)anthra-
cene (5 µg/ml) for 24 h.

fluorene, diethylnitrosamine, N-nitrosomorpholine (Fischer et al., 1985a) and the antihistaminic drug methapyrilene (Fischer et al., 1983b). The latter has been shown to be carcinogenic in the rat (Lijinski et al., 1980). GT-positive foci were correlated with ATPase-negative and γ-glutamyltranspeptidase-positive focal lesions (Fig. 7). Increased GT_1 activity was substantiated by enzyme determinations in microdissections obtained from focal and extrafocal tissue (Table 2). After initiation of hepatocarcinogenesis by a single injection of N-nitrosomorpholine GT_1 activity was markedly increased in focal lesions analysed 170 and 330 days after initiation. In contrast, P-450 dependent arylhydrocarbon hydroxylase activity was lower compared with extrafocal tissue. These findings are in agreement with studies in liver nodules indicating that P-450-dependent reactions, often involved in the activation of chemicals, were lower than in the surrounding liver (Cameron et al., 1976) whereas conjugating enzymes such as GT_1 and GSH-transferases were higher in liver nodules (Bock et al., 1982b; Yin et al., 1982; Aström et al., 1983). This altered pattern of drug metabolizing enzymes is consistent with increased toxin-resistance of initiated hepatocytes, for example, resistance to the mitoinhibitory effect of 2-acetyl-aminofluorene (Farber, 1984). This effect has been proposed as the basis for the Solt-Farber model of tumor promotion (Solt and Farber, 1976). Enzyme activity data were supported by pharmacokinetic studies of 2-acetylaminofluorene metabo-lism in nodule-bearing rats (Eriksson, 1985). In view of the 'resistance phenotype' of initiated hepatocytes cytotoxins may be considered as 'copromoters' since they stimulate the factors responsible for regenerative growth of liver foci.

GT_1 activities, including BP-3,6-quinol glucuronidation, were found to be high in several liver nodules and some hepa-tocellular carcinomas produced by feeding 2-acetylaminofluo-rene (Bock, Schirmer and Eriksson, unpublished results) and in hepatoma(H4IIE) cells. BP-3,6-quinol glucuronidation could be further increased (2-fold) by growing these cells in the pre-sence of benz(a)anthracene (Fig. 8). Because of the permanent increase of GT_1 activity after initiation of hepatocarcino-genesis we termed this phenomenon 'permanent induction' to be able to distinguish it from regular, reversible induction. The permanently increased GT_1 in foci and nodules could still be further and reversibly induced by phenobarbital or 3-methyl-cholanthrene (Fischer et al., 1985a). These findings are simi-lar to those reported with GSH-transferase and DT-diaphorase (Pickett et al., 1985). 'Permanent induction' probably does not represent simply an increased gene expression of certain isoenzymes but may include alterations in their regulatory and structural properties.

Recently GT-positive foci were also found in human liver. They were detected in liver surrounding focal nodular hyper-plasia in a woman after long-term use of oral contraceptives (Fischer et al., 1985b). Increased glucuronidation was also found in human squamous carcinomas of the lung and colonic adenocarcinomas (Gibby et al., 1985).

CONCLUSIONS

UDP-glucuronyltransferase represents a family of isoenzymes with distinct but overlapping substrate specificity
for nucleophilic endogenous and exogenous compounds. One isoenzyme in rat liver (GT_1) is inducible by 3-methylcholanthrene-
type inducers and prefers phenolic metabolites of planar aromatic hydrocarbons as substrates. Benzo(a)pyrene-3,6-quinol
may be a most useful functional probe of GT_1. It is conjugated to mono- and diglucuronides, and diglucuronide formation in particular is strongly induced (40-fold) in liver
microsomes from 3-methylcholanthrene-treated rats. Glucuronidation of the quinol prevents autoxidation to semiquinones and
toxic redox-cycles with the generation of reactive oxygen
species. Conjugation of phenols and quinols demonstrates that
the control of these abundant nucleophilic metabolites may be
as important in the bodies detoxication system as the control
of reactive electrophilic metabolites.

Immunoblot analysis using anti-GT_1 antibodies showed that
both 54 and 56 kD polypeptides are inducible by 3-methyl-
cholanthrene in rat liver microsomes. Immunochemical cross-
reactivity between rat, mouse and human UDP-glucuronyltransferase was also observed suggesting 'GT_1-like' enzymes in
human liver.

UDP-glucuronyltransferase, mainly GT_1, was permanently
increased following initiation of hepatocarcinogenesis in
early foci, nodules, hepatomas and some hepatocellular carcinomas whereas P-450 dependent reactions were decreased. The
altered pattern of drug metabolizing enzymes is consistent
with increased toxin-resistance of initiated hepatocytes and
may be responsible for the action of cytotoxins as 'copromoters' of carcinogenesis.

ACKNOWLEDGEMENTS

We thank the Deutsche Forschungsgemeinschaft and the
Stiftung Volkswagenwerk for financial support.

REFERENCES

Aström, A., DePierre, J.W., and Eriksson, L., 1983, Characteri-
    zation of drug-metabolizing systems in hyperplastic
    nodules from the livers of rats receiving 2-acetylamino-
    fluorene, Carcinogenesis, 4:577.
Bock, K.W., Fröhling, W., Remmer, H., and Rexer, B., 1973,
    Effects of phenobarbital and 3-methylcholanthrene on
    substrate specificity of rat liver microsomal UDP-glucu-
    ronyltransferase, Biochim. Biophys. Acta, 327:46.
Bock, K.W., Josting, D., Lilienblum, W., and Pfeil, H., 1979,
    Purification of rat liver microsomal UDP-glucuronyltrans-
    ferase, Eur. J. Biochem., 98:19.
Bock, K.W., Bock-Hennig, B.S., Lilienblum, W., and Volp, R.F.,
    1981, Release of mutagenic metabolites of benzo(a)pyrene
    from the perfused rat liver after inhibition of glucuroni-
    dation and sulfation by salicylamide. Chem.-Biol. Inter-
    actions, 36:167.

Bock, K.W., Bock-Hennig, B.S., Lilienblum, W., Pfeil, H. and
    Volp, R.F., 1982a, Roles of UDP-glucuronosyltransferase
    in the inactivation of benzo(a)pyrene, in: "Biological
    Reactive Intermediates II", R. Snyder, D.V. Parke, J.J.
    Kocsis, D.J. Jollow, C.G. Gibson, C.M. Witmer, eds.,
    p. 53, Plenum Press, New York and London.
Bock, K.W., Lilienblum, W., Pfeil, H., and Eriksson, L.C.,
    1982b, Increased UDP-glucuronyltransferase activity in
    preneoplastic liver nodules and Morris hepatomas, Cancer
    Res., 42:3747.
Bock, K.W., Lilienblum, W., Ullrich, D., and Fischer, G.,
    1984, Differential induction of UDP-glucuronosyltrans-
    ferases and their 'permanent induction' in preneoplastic
    rat liver, Biochem. Soc. Transact., 12:55.
Burchell, B., 1979; Purification of mouse liver UDPglucurono-
    syltransferase, Med. Biol., 57:265.
Cameron, R., Sweeney, G.D., Jones, K., Lee, G., and Farber,
    E., 1976, A relative deficiency of cytochrome P-450
    and arylhydrocarbon (benzo(a)pyrene) hydroxylase in
    hyperplastic nodules induced by 2-acetylaminofluorene in
    rat liver, Cancer Res., 36:3888.
Dutton, G.J., 1980, "Glucuronidation of Drugs and Other
    Compounds", CRC Press, Inc., Boca Raton, Florida.
Eriksson, L.C., 1985, Aspects of drug metabolism in hepato-
    cyte nodules in rats, in: "Advances in Glucuronide Conju-
    gation", S. Matern, K.W. Bock, W. Gerok, eds., p. 295,
    MTP Press, Lancaster.
Falany, C.N., and Tephly, T.R., 1983, Separation, purification
    and characterization of three isoenzymes of UDP-glucuro-
    nyltransferase from rat liver microsomes, Arch. Biochem.
    Biophys., 227:248.
Falany, C.N., Kirkpatrick, R.B., and Tephly, T.R., 1985,
    Comparison of rat and rabbit liver UDP-glucuronosyltrans-
    ferase activities, in: "Advances in Glucuronide Conju-
    gation", S. Matern, K.W. Bock, W. Gerok, eds., p. 41,
    MTP Press, Lancaster.
Farber, E., 1984, Chemical carcinogenesis: a current biolo-
    gical perspective, Carcinogenesis, 5:1.
Fischer, G., Ullrich, D., Katz, N., Bock, K.W., and Schauer,
    A., 1983a, Immunohistochemical and biochemical detection
    of uridine-diphosphate-glucuronyltransferase (UDP-GT)
    activity in putative preneoplastic liver foci, Virchows
    Arch., 42:193.
Fischer, G., Schauer, A., Bock, K.W., Ullrich, D., and Katz,
    N.R., 1983b, Immunohistochemical demonstration of in-
    creased UDP-glucuronyltransferase in putative preneo-
    plastic liver foci, Naturwissenschaften, 70:153.
Fischer, G., Ullrich, D., and Bock, K.W., 1985a, Effects of
    N-nitrosomorpholine and phenobarbital on UDP-glucuronyl-
    transferase in putative preneoplastic foci of rat liver,
    Carcinogenesis, 6:605.
Fischer, G., Schauer, A., Hartmann, H., and Bock, K.W.,
    1985b, Increased UDP-glucuronyltransferase in putative
    preneoplastic foci of human liver after long-term use of
    oral contraceptives, Naturwissenschaften, 72:277.
Gibby, E.M., D'Arcy Doherty, M., and Cohen, G.M., 1985,
    Alterations in glucuronic acid and sulphate ester conju-
    gation in normal and tumour tissues, in: "Advances in
    Glucuronide Conjugation", S. Matern, K.W. Bock, W. Gerok,
    eds., p. 317, MTP Press Lancaster.

Greenlee, W.F., Gross, E.A., and Irons, R.D., 1981, Relationship between benzene toxicity and the disposition of ^{14}C-labelled benzene metabolites in the rat, Chem.-Biol. Interactions, 33:285.

Greenlee, W.F., and Irons, R.D., 1981, Modulation of benzene-induced lymphoerytropenia in the rat by 2,4,5,2',4',5'-hexachlorobiphenyl and 3,4,3',4'-tetrachlorobiphenyl, Chem.-Biol. Interactions, 33:345.

Lilienblum, W., Walli, A.K., and Bock, K.W., 1982, Differential induction of rat liver microsomal UDP-glucuronosyltransferase activities by various inducing agents, Biochem. Pharmacol., 31:907.

Lilienblum, W., Bock-Hennig, B.S., and Bock, K.W., 1985, Protection against toxic redox-cycles between benzo(a)pyrene-3,6-quinone and its quinol by 3-methylcholanthrene-inducible formation of the quinol mono- and diglucuronide, Molec. Pharmacol., 27:451.

Lijinski, W., Reuber, M.D., and Blackwell, B.-N., 1980, Liver tumors induced in rats by oral administration of the antihistaminic methapyrilene hydrochloride, Science, 290:817.

Lorentzen, R.J., and Ts'o, P.O.P., 1977, Benzo(a)pyrenedione/benzo(a)pyrenediol oxidation-reduction couples and the generation of reactive reduced molecular oxygen, Biochem., 16:1473.

Lorentzen, R.J., Lesko, S.A., McDonald, K., and Ts'o, P.O.P., 1979, Toxicity of metabolic benzo(a)pyrenediones to cultured cells and the dependence upon molecular oxygen, Cancer Res., 39:3194.

Lueders, K.K., Dyer, H.M., Thompson, E.B., and Kuff, E.L., 1979, Glucuronyltransferase activity in transplantable rat hepatomas, Cancer Res., 30:274.

Mackenzie, P.I., Gonzalez, F.J., and Owens, I.S. 1984, Cloning and characterization of DNA complementary to rat liver UDP-glucuronosyltransferase mRNA, J. Biol. Chem., 259:12153.

Mackenzie, P.I., and Owens, I.S., 1984, Cleavage of nascent UDP glucuronosyltransferase from rat liver by dog pancreatic microsomes, Biochem. Biophys. Res. Comm., 122:1441.

Miller, E.C., and Miller, J.A., 1981, Searches for ultimate chemical carcinogens and their reactions with cellular macromolecules, Cancer, 47:2327.

Owens, I.S., 1977, Genetic regulation of UDP-glucuronosyltransferase induction by polycyclic aromatic compounds in mice, J. Biol. Chem., 252:2828.

Pfeil, H., and Bock, K.W., 1983, Electroimmunochemical quantification of UDP-glucuronosyltransferase in rat liver microsomes, Eur. J. Biochem., 131:619.

Pickett, C.B., Williams, J.B., Lu, A.Y.H., and Cameron, R.G., 1984, Regulation of glutathione transferase and DT-diaphorase mRNAs in persistent hepatocyte nodules during chemical hepatocarcinogenesis, Proc. Natl. Acad. Sci. USA, 81:5091.

Pitot, H.C., Peraino, C., Morse jr., P.A., and Potter, V.R., 1964, Hepatomas in tissue culture compared with adapting liver in vivo, Natl. Cancer Inst. Monogr., 13:229.

Reuber, M.D., 1961, A transplantable bile-secreting hepatocellular carcinoma in the rat, J. Natl. Cancer Inst., 26:891.

Roy Chowdhury, J., Roy Chowdhury, N., Novikoff, P.M., Novikoff, A.B., and Arias, I.M., 1985, UDP-glucuronyltransferase: problems within a biological 'family', in: "Advances in Glucuronide Conjugation", S. Matern, K.W. Bock, W. Gerok, eds., p. 33, MTP Press, Lancaster.

Scragg, I., Celier, C., and Burchell, B., 1985, Congenital jaundice in rats due to the absence of hepatic bilirubin UDP-glucuronyltransferase enzyme protein, FEBS Letters, 183:37.

Solt, D., and Farber, E., 1976, New principle for the analysis of chemical carcinogenesis, Nature, 263:701.

Towbin, H., Staehelin, T., and Gordon, J., 1979, Electrophoretic transfer of proteins from polyacrylamide gels to nitrocellulose sheets: procedure and some applications, Proc. Natl. Acad. Sci. USA, 76:4350.

Ullrich, D., and Bock, K.W., 1984, Glucuronide formation of various drugs in liver microsomes and in isolated hepatocytes from phenobarbital- and 3-methylcholanthrene-treated rats, Biochem. Pharmacol., 33:97.

Ullrich, D., Fischer, G., Katz, N., and Bock, K.W., 1984, Intralobular distribution of UDP-glucuronosyltransferase in livers from untreated, 3-methylcholanthrene- and phenobarbital-treated rats, Chem.-Biol. Interactions, 48:181.

Watkins, J.B., Gregus, Z., Thompson, T.N., and Klaassen, C.D., 1982, Induction studies on the functional heterogeneity of rat liver UDP-glucuronosyltransferase, Toxicol. Appl. Pharmacol., 64:429.

Wishart, G.J., 1978, Demonstration of functional heterogeneity of hepatic uridine diphosphate glucuronosyltransferase activities after administration of 3-methylcholanthrene and phenobarbital to rats, Biochem. J., 174:671.

Yin, Z., Sato, K., Tsuda, H., and Ito, N., 1982, Changes in activities of uridine diphosphate-glucuronyltransferase during chemical hepatocarcinogenesis, Gann, 73:239.

EXPRESSION AND SEQUENCE ANALYSIS OF RAT LIVER GLUTATHIONE S-TRANSFERASE

GENES

Cecil B. Pickett, Claudia A. Telakowski-Hopkins,
Gloria J.-F. Ding, and Victor D.-H. Ding

Department of Molecular Pharmacology and Biochemistry
Merck Sharp and Dohme Research Laboratories
Rahway, New Jersey   07065-0900

INTRODUCTION

The rat liver glutathione S-transferases are a family of enzymes which catalyze the conjugation of the reduced sulfhydryl group of glutathione with various electrophiles. In addition, the transferases bind with high affinity various exogenous hydrophobic compounds as well as potentially toxic endogenous compounds such as bilirubin and heme (1-3). The enzymes are comprised of binary combinations of at least six major subunits, Yα, Ya, Yb1, Yb2, Yc and Yn, which can be separated by one-dimensional SDS-polyacrylamide gel electrophoresis (4-6).

Our laboratory has been interested in the molecular mechanisms which lead to an elevation of glutathione S-transferases in response to xenobiotic administration. Early studies from our group indicated that the amount of functional glutathione S-transferase mRNA was elevated by phenobarbital or 3-methylcholanthrene (7-9). More recently, we have constructed cDNA clones complementary to the Ya, Yb1 and Yc mRNAs of the rat liver glutathione S-transferases in order to examine the effect of phenobarbital and 3-methylcholanthrene on the expression of the glutathione S-transferase genes (10-12). In this paper, we have summarized our data using RNA blot and in vitro transcription analysis to study the regulation of the glutathione S-transferase genes by pheno-barbital and 3-methylcholanthrene. Furthermore, we provide evidence based on DNA sequence analysis of the Ya and Yc cDNA clones that these two mRNAs are members of the same glutathione S-transferase gene family.

MATERIALS AND METHODS

Preparation of cDNA and Construction of Recombinant Plasmids

In order to prepare dS cDNA, the glutathione S-transferase mRNAs were purified by polysomal immunoabsorption techniques as described previously (10). Both the first and second strands of the cDNA were synthesized as described by Gubler and Hoffman (13) and tailed with dCTP using terminal deoxynucleotidyl transferase as described previously (10).

In Vitro Labeling of cDNAs

cDNAs were labeled _in vitro_ with ^{32}P-dCTP by Nick translation; at the 5' end with ^{32}P ATP using T_4 polynucleotide kinase; at the 3' end with ^{32}P dideoxy-ATP using terminal deoxynucleotidyl transferase.

In Vitro Transcription Assay

Nuclei were isolated from rat liver as described by Blobel and Potter (14). _In vitro_ transcription assays using isolated nuclei were performed essentially as described by McKnight and Palmiter (15).

Nucleotide Sequence Analysis

The chemical method of Maxam and Gilbert was used for DNA sequence analysis (16). Appropriate restriction fragments were 5' and 3' end labeled and subjected to DNA sequence analysis.

RNA Blots

All poly(A$^+$)-RNA was isolated from total rat liver RNA by oligo(dT)-cellulose affinity chromatography. Total rat liver RNA was isolated by the procedure of Chirgwin, et al. (17). Gel electrophoresis of RNA was carried out in 1.5% agarose gels containing 10 mM methyl mercury hydroxide. The RNA was electrophoresed according to the procedure of Bailey and Davidson (18) and transferred to DBM paper as described by Alwine et al. (19). Conditions for prehybridization and hybridization have been described previously (10).

RESULTS

Effect of Phenobarbital and 3-Methylcholanthrene on the Expression of Rat Liver Glutathione S-Transferase Genes

We have used two cDNA clones, pGTB38 and pGTA/C44, in RNA-blot hybridization experiments and in nuclear run-off experiments (_in vitro_ transcription assay) to determine the effect of phenobarbital and 3-methylcholanthrene on the expression of the transferase genes. pGTB38 contains a 900 bp insert complementary to a Ya mRNA whereas clone pGTA/C44 contains a 1100 bp insert complementary to the Ybl mRNA.

As can be seen from Table 1, at 24 hrs after phenobarbital or 3-methylcholanthrene administration the Ya mRNAs are elevated ~8-fold and 11-fold, respectively, whereas the Yb mRNAs are elevated ~3-fold. More recently we have utilized the 3' untranslated region of a Yc cDNA clone, pGTB42, in RNA slot blot analysis to determine whether the Yc mRNA was elevated by phenobarbital. The 3' untranslated regions of the Ya and Yc clones have no sequence homology. Our results demonstrate that only the Ya mRNAs are elevated significantly by phenobarbital. Similarly, we have utilized the 3' untranslated regions of a Ybl clone, pGTA/C44, and Yb2 clone, pGTA/C48, to determine the effect of phenobarbital treatment on these two mRNAs. Our data clearly demonstrate that both the Ybl and Yb2 mRNAs are elevated by phenobarbital administration.

In order to determine whether the accumulation of mRNA was due to transcriptional activation of the glutatione S-transferase genes, we utilized purified nuclei isolated from rats treated for various times with phenobarbital or 3-methylcholanthrene. As presented in Table 2 and 3, the transcriptional rates of the Ya-Yc and Yb genes were elevated 5-fold, 12 hrs and 6 hrs, respectively after phenobarbital administration. In

contrast, the transcriptional rates of the Ya-Yc genes were elevated
~8-fold at 16 hrs after 3-methylcholanthrene administration, whereas the
transcriptional rates of the Yb genes were elevated ~5-fold at 6 hrs after
3-methylcholanthrene administration. The elevation in transcriptional
activity of the glutathione S-transferase genes is sufficient to account
for the increase in glutathione S-transferase mRNA levels determined by
RNA blot hybridization.

TABLE 1

Induction of Rat Liver Glutathione S-Transferases mRNAs
by Phenobarbital and 3-Methylcholanthrene

		Relative mRNA Levels	
Treatment	Time h	Ya-Yc	Yb
Phenobarbital	0	1.0	1.0
	12	5.4	2.0
	24	7.9	2.6
3-Methylcholanthrene	0	1.0	1.0
	12	5.0	1.6
	24	11.0	3.1

The mRNA level at the 0 hr time point was arbitrarily
assigned a value of 1.

TABLE 2

Transcriptional Activity of Glutathione S-Transferase Genes
at Various Times After Phenobarbital Treatment

	Glutathione S-Transferase Gene Transcription (ppm)	
Hours After PB Administration	Ya Genes	Yb Genes
0	10	3.5
2	28	7.0
4	30	4.2
6	34	12.5
8	50	11.0
12	47	8.1
16	39	6.8

^{32}P-RNA was isolated from in vitro transcription assays and
hybridized to nitrocellulose dishes containing plasmid DNA
from pGTB38. Radioactivity absorbed to filters containing
pBR322 was subtracted from values obtained for the
hybridization of ^{32}P-RNA to the cDNA filter. Parts per
million hybridized was calculated based on the total cpm
incubated with each cDNA filter. Livers from two groups of
rats (3-4 rats/group) were utilized for the isolation of
nuclei. All transcriptional assays were determined in
duplicate or triplicate. The total input cpm of ^{32}P-RNA in
the hybridization reactions were 0.94-1.1 x 10^7 cpm.
Hybridization efficiency for the Ya clone, pGTB38, was 26.6%
whereas hybridization efficiency for the Yb clone, pGTA/C44,
was 30.3%.

TABLE 3

Transcriptional Activity of Glutathione S-Transferase Genes
at Various Times After 3-Methylcholanthrene Administration

Hours After 3MC Administration	Glutathione S-Transferase Gene Transcription (ppm)	
	Ya Genes	Yb Genes
0	6	2.5
2	12	7.7
4	15	8.2
6	8	12.0
8	23	3.5
12	32	4.1
16	43	5.8

Experimental details are given in the legend to Table 2. Total input cpm of $[^{32}P]$-RNA in the hybridizations were 1.0-2.5 x 10^7 cpm.

DNA Sequence Analysis of the Ya and Yc cDNA Clones

Since pGTB38 and pGTB42 represent nearly full length cDNA clones, they were excellent candidates for DNA sequence analysis. The entire nucleotide sequence of pGTB42, a Yc clone, and pGTB38, a Ya clone, is presented in Figure 1. Over identical regions of both clones, nucleotides -39 to 780, there is a 66% nucleotide sequence homology; whereas, in the protein coding region there is a 75% nucleotide sequence homology. Interestingly, the 3' untranslated regions of the two cDNA clones display no sequence homology.

The cDNA insert in pGTB42 has an open reading frame of 663 nucleotides encoding a Yc polypeptide comprised of 221 amino acids with a molecular weight of 25,322 (Fig. 2). The cDNA insert in pGTB38 has an open reading frame of 666 nucleotides encoding a Ya polypeptide comprised of 222 amino acids with a molecular weight of 25,547 (Fig. 2). The amino acid sequences of the Ya and Yc subunits have an overall homology of 68%.

```
        -40        -30        -20        -10        +1                20                   40
pGTB42 AGAGGGAGCAGCTTTTTAACAAGAGAACTCAAGCAATTGCTGCCATGCCGGGGAAGCCAGTCCTTCACTACTTCGATGGCAGGGG
pGTB38  -CCAC-AC-C-CGCT-GACAGTGAAGCA--G-------T---T-T-----------G------------A---C-C----

                   60                 80                 100                     120
       GAGAATGGAGCCCATCCGGTGGCTCCTGGCTGCAGCTGGAGTAGAGTTTGAAGAACAATTTCTGAAAACTCGGGATGACC
       C---------TG----------------------A-----G-----------GA-GC--A-AC-G-G--CA--A---T

                   140                160                180                   200
       TGGCCAGGCTAAGGAATGATGGGAGTTTGATGTTCCAGCAAGTGCCCATGGTGGAGATTGATGGGATGAAGCTGGTGCAG
       ---AA-A-----A---A--C----A----------TG-C-------------------C-------------CA---

                   220                240                260                   280
       ACCAGAGCCATTCTCAACTACATTGCCACCAAATACAACCTCTATGGGAAGGACATGAAGGAGAGAGCCCTCATCGACAT
       ----------------------C-----------TG----------------------------G--T-----

                   300                320                340                   360
       GTATGCAGAAGGAGTGGCGGATCTGGATGAAATAGTTCTCCATTACCCTTACATTCCCCCTGGGGAGAAAGAGGCAAGTC
       ----T----G--TA-TTTA-------AC------GA--A-----A-TGGTAATATG------A-ACC-A-G---A--C-AGA

                   380                400                420                   440
       TTGCCAAAATCAAGGACAAAGCAAGGAACCGTTACTTTCCTGCCTTTGAAAAGGTGTTGAAGAGCCATGGACAAGATTAT
       CC---TTGGCA--A----GGA-C-AA-----G-----G---------------------------C-----C--C

                   460                480                500                   520
       CTCGTTGGCAATAGGCTGAGCAGGGCTGATGTTTACCTAGTTCAAGTTCTCTACCATGTGGAAGAGCTGGACCCCAGCGC
       --T--A--T--C-------C-C---TA--CA-CC----GC-GG--C-----CT-T----T------T-T--TG-----CT

                   540                560                580                   600
       TTTGGCCAACTTCCCTCTGCTGAAGGCCCTGAGAACCAGAGTCAGCAACCTCCCCACAGTGAAGAAGTTTCTTCAGCCTG
       -C--A--TCT-----------------T-C-AG-G----A------G--------AT------------C--G-------

                   620                640                660                   680
       GCAGCCAGAGGAAGCCATTAGAGGATGAGAAATGTGTAGAATCTGCAGTTAAGATCTTCAGTTAATTCAGGCATCTATGG
       ----T-----A------GCCAT-----CA---CAA-C---GAA---AGG---G-T----AG-TT-AGC--AGCTGCACT

                   700                720                740                   760
       ATACACTGTACCCACAAAGCCAGCCTTCGAAAGCTTTGCAACAATCGCATATTTTGACTAAATGTTGACCCTACTTATTG
       G-C--A-T-CTTGTA--TCCAGGCT--GATGTTTTGCAAA---ATGAGAAGC-A---GTTGATCC--GCT-TTT-G-AATAAT

                   780                800                820
       GGAGGCCAACACGTTTTCTAATGCTTTTGTGTTAATTCATATAGACATGACTGATGAGGAT
       AA-AAAATGA--AAA-GG-
```

Figure 1. Comparison of DNA sequences of a Ya cDNA clone, pGTB38, and a
           Yc cDNA clone, pGTB42 (11). The dashed lines in the sequence of
           pGTB42. The first nucleotide of the initiation codon ATG is in-
           dicated as +1.

189

FIGURE 2

```
                            9                                    18
Yc MET Pro Gly Lys Pro Val Leu His Tyr Phe Asp Gly Arg Gly Arg Met Glu Pro
Ya MET Ser Gly Lys Pro Val Leu His Tyr Phe Asn Ala Arg Gly Arg Met Glu Cys

                           27                                    36
   Ile Arg Trp Leu Leu Ala Ala Ala Gly Val Glu Phe Glu Glu Gln Phe Leu Lys
   Ile Arg Trp Leu Leu Ala Ala Ala Gly Val Glu Phe Glu Glu Lys Leu Ile Gln

                           45                                    54
   Thr Arg Asp Asp Leu Ala Arg Leu Arg Asn Asp Gly Ser Leu Met Phe Gln Gln
   Ser Pro Glu Asp Leu Glu Lys Leu Lys Lys Asp Gly Asn Leu Met Phe Asp Gln

                           63                                    72
   Val Pro Met Val Glu Ile Asp Gly Met Lys Leu Val Gln Thr Arg Ala Ile Leu
   Val Pro Met Val Glu Ile Asp Gly Met Lys Leu Ala Gln Thr Arg Ala Ile Leu

                           81                                    90
   Asn Tyr Ile Ala Thr Lys Tyr Asn Leu Tyr Gly Lys Asp Met Lys Glu Arg Ala
   Asn Tyr Ile Ala Thr Lys Tyr Asp Leu Tyr Gly Lys Asp Met Lys Glu Arg Ala

                           99                                   108
   Leu Ile Asp Met Tyr Ala Glu Gly Val Ala Asp Leu Asp Glu Ile Val Leu His
   Leu Ile Asp Met Tyr Ser Glu Gly Ile Leu Asp Leu Thr Glu Met Ile Ile Gln

                          117                                   126
   Tyr Pro Tyr Ile Pro Pro Gly Glu Lys Glu Ala Ser Leu Ala Lys Ile Lys Asp
   Leu Val Ile Cys Pro Pro Asp Gln Arg Glu Ala Lys Thr Ala Leu Ala Lys Asp

                          135                                   144
   Lys Ala Arg Asn Arg Tyr Phe Pro Ala Phe Glu Lys Val Leu Lys Ser His Gly
   Arg Thr Lys Asn Arg Tyr Leu Pro Ala Phe Glu Lys Val Leu Lys Ser His Gly

                          153                                   162
   Gln Asp Tyr Leu Val Gly Asn Arg Leu Ser Arg Ala Asp Val Tyr Leu Val Gln
   Gln Asp Tyr Leu Val Gly Asn Arg Leu Thr Arg Val Asp Ile His Leu Leu Glu

                          171                                   180
   Val Leu Tyr His Val Glu Glu Leu Asp Pro Ser Ala Leu Ala Asn Phe Pro Leu
   Leu Leu Leu Tyr Val Glu Glu Phe Asp Ala Ser Leu Leu Thr Ser Phe Pro Leu

                          189                                   198
   Leu Lys Ala Leu Arg Thr Arg Val Ser Asn Leu Pro Thr Val Lys Lys Phe Leu
   Leu Lys Ala Phe Lys Ser Arg Ile Ser Ser Leu Pro Asn Val Lys Lys Phe Leu

                          207                                   216
   Gln Pro Gly Ser Gln Arg Lys Pro Leu Glu Asp Glu Lys Cys Val Glu Ser Ala
   Gln Pro Gly Ser Gln Arg Lys Pro Ala Met Asp Ala Lys Gln Ile Glu Glu Ala

   Val Lys Ile Phe Ser
   Arg Lys Val Phe Lys Phe
```

Comparison of the amino acid sequences of rat liver glutathione
S-transferase Ya and Yc subunits (11).

DISCUSSION

In this paper, we have used RNA blot hybridization experiments and in vitro transcription assays to demonstrate that members of the Ya-Yc gene family and the Yb gene family are transcriptionally activated by phenobarbital and 3-methylcholanthrene. The transcriptional activation of these gene families leads to an accumulation of the Ya, Yb1 and Yb2 mRNAs; however, the Yc mRNA level is not affected. These data suggest that transcriptional activation is primarily responsible for the elevation of the rat liver glutathione S-transferase isozymes when animals are treated with phenobarbital or 3-methylcholanthrene.

From our DNA sequence analysis of the Ya and Yc cDNA clones, we have been able to deduce the complete amino acid sequence of the Ya and Yc polypeptides. Although the Yc subunit migrates on SDS-polyacrylamide gels with an apparent Mr which is 2000-2500 larger than the Ya subunit, the actual molecular weight of this subunit is 25,322 which is virtually the same size obtained for the Ya subunit (Mr=25,567). Therefore the differences in the migration of the subunits on SDS-polyacrylamide gels are most likely due to differences in the amount of SDS bound to the polypeptides rather than any real difference in their molecular weight.

Based upon the nucleotide differences in the coding region of the Ya and Yc mRNAs along with the divergent 5'- and 3'-untranslated regions, it appears that the Ya and Yc subunits of the rat liver glutathione S-transferase are derived from two different genes rather than by post-transcriptional processing of a single gene. The hypothesis that the Ya subunit is generated from the Yc subunit by post-translational proteolysis (20,21) can be ruled out.

Although it appears that the Ya and Yc subunits are derived from related yet different genes, these genes most likely arose from a common ancestral gene which during the course of evolution duplicated and diverged. The divergent 5'-untranslated regions found between the Ya and Yc clones may play a role in the differential regulation of the Ya and Yc mRNAs by xenobiotics (7-9) as well as their tissue specific regulation (22).

Recently, we have completed the sequencing of a Yb1 cDNA clone, pGTA/C44 and a Yb2 cDNA clone pGTA/C48. The two Yb clones share ~85% nucleotide and amino acid sequence homology in the protein coding region whereas the 3' untranslated regions are only 35% homologous. These two clones share no sequence homology with the Ya or Yc clones. Our data suggest a minimum of two gene families in the glutathione S-transferase super family.

The construction and characterization of the Ya, Yc, Yb1 and Yb2 cDNA clones should facilitate the isolation and characterization of the structural genes as well as elucidate the mechanisms by which these genes are regulated by xenobiotics.

ACKNOWLEDGEMENT

We would like to thank Joan Kiliyanski for her assistance in the preparation of this manuscript.

REFERENCES

1. I. M. Arias, G. Fleischner, R. Kirsch, S. Mishkin and Z. Gatmaitan, in: "Glutathione: Metabolism and Function", I. M. Arias and W. B. Jakoby, eds., Raven Press, New York (1976).
2. G. Litwack, B. Ketterer and I. M. Arias, Ligandin: A hepatic protein which binds steroids, bilirubin, carcinogens and a number of exogenous organic anions, Nature (Lond.) 234:466 (1971).
3. W. B. Jakoby and W. H. Habig, in: "Enzymatic Basis of Detoxification", W. B. Jakoby, ed., Academic Press, New York (1980).
4. N. M. Bass, R. E. Kirsch, S. A. Tuff, I. Marks and S. J. Saunders, Ligandin heterogeneity: Evidence that the two non-identical subunits are the monomers of two distinct proteins, Biochim. Biophys. Acta 492:163 (1977).
5. C. C. Reddy, N.-Q. Li and C.-P.D. Tu, Identification of a new glutathione S-transferase from rat liver cytosol, Biochem. Biophys. Res. Commun. 121:1014 (1984).
6. J. D. Hayes, Purification and characterization of glutathione S-transferases P, S and N. Isolation from rat liver of Yb_1 Yn protein, the existence of which was predicted by subunit hybridization in vitro, Biochem. J. 224:839.
7. C. B. Pickett, W. Wells, A. Y. H. Lu and B. F. Hales, Induction of translationally active rat liver glutathione S-transferase B messenger RNA by phenobarbital, Biochem. Biophys. Res. Commun. 99:1002 (1981).
8. C. B. Pickett, C. A. Telakowski-Hopkins, A. M. Donohue, A. Y. H. Lu and B. F. Hales, Differential induction of rat hepatic cytochrome P-448 and glutathione S-transferase B messenger RNAs by 3-methylcholanthrene, Biochim. Biophys. Res. Commun. 104:611.
9. C. B. Pickett, A. M. Donohue, A. Y. H. Lu and B. F. Hales, Rat liver glutathione S-transferase B: The functional mRNAs specific for the Ya Yc subunits are induced differentially by phenobarbital, Arch. Biochem. Biophys. 215:539 (1982).
10. C. B. Pickett, C. A. Telakowski-Hopkins, G. J.-F. Ding, L. Argenbright and A. Y. H. Lu, Rat liver glutathione S-transferases: Complete nucleotide sequence of a glutathione S-transferase mRNA and the regulation of the Ya, Yb, and Yc mRNAs by 3-methylcholanthrene and phenobarbital, J. Biol. Chem. 259:5182 (1984).
11. C. A. Telakowski-Hopkins, J. A. Rodkey, C. D. Bennett, A. Y. H. Lu and C. B. Pickett, Rat liver glutathione S-transferases: Construction of a cDNA clone complementary to a Yc mRNA and prediction of the complete amino acid sequence of a Yc subunit, J. Biol. Chem. 260:5820 (1985).
12. G. J.-F. Ding, A. Y. H. Lu and C. B. Pickett, Rat liver glutathione S-transferases: Nucleotide sequence analysis of a Ybl cDNA clone and prediction of the complete amino acid sequence of the Ybl subunit, J. Biol. Chem. in press (1985).
13. U. Gubler and B. S. Hoffman, A simple and very efficient method for generating cDNA libraries, Gene 25:263 (1983).
14. G. Blobel and V. R. Potter, Nuclei from rat liver: Isolation method that combines purity with high yield, Science (Wash., D.C.) 154:1662 (1966).
15. G. S. McKnight and R. D. Palmiter, Transcriptional regulation of the ovalbumin and conalbumin genes by steroid hormones in chick oviduct, J. Biol. Chem. 258:8081 (1979).
16. A. M. Maxam and W. Gilbert, Sequencing end-labeled DNA with base-specific chemical cleavages, Methods Enzymol. 65:499 (1980).

17. J. M. Chirgwin, A. E. Przybyla, R. J. MacDonald and W. J. Rutter, Isolation of biologically active ribonucleic acid from sources enriched in ribonuclease, Biochem. 18:5294 (1979).
18. J. M. Bailey and N. Davidson, Methylmercury as a reversible denaturing agent for agarose gel electrophoresis, Anal. Biochem. 70:75 (1976).
19. J. C. Alwine, D. J. Kemp, B. A. Parker, J. Reiser, J. Renart, G. R. Stark and G. M. Wahl, Detection of specific RNAs in specific fragments of DNA by fractionation in gels and transfer to diazobenzyloxymethyl paper, Methods Enzymol. 68:220 (1979).
20. M. M. Bhargava, N. Ohmi, J. Listowsky and I. M. Arias, Structural, catalytic, binding, and immunological properties associated with each of the two subunits of rat liver ligandin, J. Biol. Chem. 255:718 (1980).
21. N. C. Scully and T. J. Mantle, Tissue distribution and subunit structures of the multiple forms of glutathione S-transferase in the rat, Biochem. J. 193:367 (1981).
22. C.P.-D. Tu, M. J. Weiss, N. Li and C. C. Reddy, Tissue-specific expression of the rat glutathione S-transferases, J. Biol. Chem. 258:4659 (1983).

RAT CYTOSOLIC EPOXIDE HYDROLASE

Franz Oesch, Ludwig Schladt, Renate Hartmann,
Christopher Timms, and Walter Wörner

Institut für Toxikologie, Universität Mainz
Obere Zahlbacher Strasse 67, D-65 Mainz, FRG

ABSTRACT

Rat liver microsomal and cytosolic epoxide hydrolase may be distinguished through differences in substrate specificity: styrene 7,8-oxide is preferentially hydrolyzed by the microsomal form, while *trans*-stilbene oxide is the prefered substrate for cytosolic epoxide hydrolase. Large interindividual differences in the specific activity of Sprague-Dawley (outbred strain) liver cytosolic epoxide hydrolase were observed, varying from 2 to 77 pmol/min x mg protein. Interindividual variations were much lower for microsomal epoxide hydrolase. The specific activity of Fischer F-344 (inbred strain) liver cytosolic epoxide hydrolase varied only by a factor of 2. The specific activity of cytosolic epoxide hydrolase using *trans*-stilbene oxide as the substrate was highest in kidney and heart, followed by liver, brain, lung, testis, and spleen. For microsomal epoxide hydrolase, the specific activity was much lower in extrahepatic tissues than in liver. None of the commonly used inducers of xenobiotic metabolizing enzymes caused significant changes in rat liver cytosolic epoxide hydrolase. However, peroxisome proliferating drugs were found to drastically increase cytosolic epoxide hydrolase activity. Treatment for one week with a diet containing clofibrate (0.25 %), tiadenol (0.5 %) or acetylsalicylic acid (1 %) caused a 8, 13 and 5 fold increase in cytosolic epoxide hydrolase activity respectively in the liver which parallelled the induction of peroxisomal β-oxidation activity (13, 19 and 5 fold, respectively).

INTRODUCTION

Epoxides are generated by the oxidation of an aliphatic or aromatic double bond primarily through the action of cytochrome P-450 monooxygenases.[1] This reaction is usually the first step in transforming lipophilic xenobiotic compounds into metabolites of higher water solubility which, in turn, enhances excretion. For the next step there are several alternative pathways[2][3]: cytochrome P-450-mediated reduction to the parent compound; non-enzymatic rearrangement to phenols (arene oxides), to aldehydes or ketones (alkene oxides); conjugation with glutathione (enzymatically or non-enzymatically); addition of water by epoxide hydrolases leading to *trans*-dihydrodiols.

In rat liver at least three different forms of epoxide hydrolases exist, which can be distinguished by their substrate specificities.[4] Most work on epoxide hydrolase has been carried out with a microsomal form (mEH$_b$), which shows a broad spectrum substrate specificity for epoxides from a vast range of structurally diverse compounds including benzo(a)pyrene 4,5-oxide and styrene-7,8-oxide.[2 5] Whereas microsomal epoxide hydrolase is strongly inhibited by 3,3,3-trichloropropene 1,2-oxide (TCPO) this is not true for the microsomal cholesterol oxide hydrolase, which catalyzes the hydrolysis of cholesterol-5,6-oxide.[6 7] In contrast to the two microsomal epoxide hydrolases, the third form is localized in the cytosolic fraction. Properties of cytosolic epoxide hydrolase differ from that of the other two forms.[4 8] Using *trans*-stilbene oxide as a diagnostic substrate for cytosolic epoxide hydrolase, activity was detected in all rat tissues investigated. However, there were high interindividual variations in Sprague-Dawley liver epoxide hydrolase activity. Furthermore, peroxisome-proliferating drugs were shown to be very potent inducers of rat cytosolic epoxide hydrolase. Where microsomal epoxide hydrolase is compared with the cytosolic enzyme in this study, mEH$_b$ is referred to.

MATERIALS AND METHODS

Materials

[^{3}H]styrene oxide ([^{3}H]STO), 11.7 GBq/mmol and [^{3}H]*trans*-stilbene oxide ([^{3}H]TSO), 0.41 GBq/mmol were synthesized as described in references.[9 10]

All other chemicals used were of analytical grade or the purest grade commercially available. Male Sprague-Dawley rats (180-220 g) were obtained from Süddeutsche Versuchstieranstalt, Tuttlingen, FRG; male Fischer F-344 rats (150-160 g) were obtained from Charles River Wiga GmbH, Sulzfeld, FRG. Animals were kept at constant temperature, under a constant light-dark cylce and had free access to water and a defined diet (Altromin).

Methods

Subcellular fractions. Rats were killed by cervical dislocation. Livers were perfused with ice-cold homogenization buffer (10 mM Tris-HCl, pH 7.4, containing 0.25 M sucrose). Homogenization of livers or other tissues was done in homogenization buffer using an Ultra-Turrax tissue homogenizer to give a 25 % (w/v) homogenate. The homogenate was centrifuged for 10 min at 600 g and the resulting supernatant for 15 min at 10,000 g. The pellet was discarded and the supernatant was centrifuged for 60 min at 100,000 g to yield 2 fractions; the microsomal pellet and the cytosol. The pellet was resuspended in homogenization buffer at a final protein concentration of approximately 15 mg/ml.

Enzyme assays. Cytosolic epoxide hydrolase activity was assayed in glass centrifuge tubes containing 5 nmol [^{3}H]TSO, 0.125 M Tris-HCl, pH 7.4 and 1.25 mM 1-chloro-2,4-dinitrobenzene (CDNB) in a total volume of 200 µl. The CDNB was added in order to reduce substrate depletion caused by glutathione conjugation. The reaction was stopped by addition of 3 ml of light petroleum ether and 250 µl dimethylsulfoxide. The unhydrolyzed substrate was extracted by shaking the tubes for 2-3 min followed by a brief centrifugation at 1000 x g to resolve the phases. After discarding the ether phase, the aqueous phase was washed again with 3 ml of light petroleum ether. The diol was then extracted into 1 ml of ethyl acetate and an aliquot was counted.

The assay for microsomal epoxide hydrolase was performed as described in reference 9. Activity of cyanide-insensitive peroxisomal β-oxidation was determined in the 600 $\underline{g}$ supernatant according to Bieri et al.[11]

RESULTS AND DISCUSSION

Interindividual Differences

Using *trans*-stilbene oxide as substrate, large interindividual differences in the specific activity of Sprague-Dawley rat liver cytosolic epoxide hydrolase were observed (Table 1). The variation of the specific activity was between 2 to 77 pmol/min x mg protein. With *trans*-ethyl styrene oxide as substrate for cytosolic epoxide hydrolase qualitatively identical results were obtained.

The large interindividual variation in cytoslic epoxide hydrolase is consistent with previous reports. We had observed that the specific activity of cytosolic epoxide hydrolase from human liver biopsies showed a 539-fold interindividual variation.[12] In addition, the specific activity of soluble epoxide hydrolase from human leukocytes varied 5-fold.[13] By contrast, the activities of microsomal epoxide hydrolase are relatively consistent between individual animals, routinely varying by less than 1.5-fold.

The reason for the variation of the specific activity of the cytosolic epoxide hydrolase in human samples is not known. The Sprague-Dawley rat strain used was outbred. With inbred Fischer F-344 rats the variation in the specific activity of cytosolic epoxide hydrolase was much lower. The specific activity varied only by a factor of two, from 24 to 48 pmol/min x mg protein. The mean specific activity is nearly the same in both rat strains. Therefore, the main reason for the large interindividual differences observed with Sprague-Dawley rats may result from their genetic constitution. As in Sprague-Dawley rats specific activity of microsomal epoxide hydrolase in Fischer F-344 rats showed low interindividual differences, but the specific activity was lower than in Sprague-Dawley rats. This is in agreement with our earlier obserations.[14]

Organ distribution of cytosolic epoxide hydrolase activity

In earlier experiments we had determined the activity of microsomal epoxide hydrolalse in several tissues of Sprague-Dawley rats.[15] In order to compare the organ distribution of the two forms of epoxide hydrolase

Table 1. Interindividual variation of cytosolic and microsomal epoxide hydrolase activity in the liver cytosol of different rat strains

Rat strain	Mode of breeding	cytosolic epoxide hydrolase activity[a]	microsomal epoxide hydrolase activity[b]
Sprague-Dawley	outbred	38 + 18 (n = 88)	11.6 + 1.3 (n = 14)
Fischer F-344	inbred	36 $\overline{+}$ 6 (n = 24)	4.2 $\overline{+}$ 0.8 (n = 24)

Male Fischer F-344 (~ 150 g) or Sprague-Dawley rats (~ 200 g) were used.
a Activities are given in pmol/min x mg protein. TSO was used as substrate.
b Activities are given in nmol/min x mg protein. STO was used as substrate.
The values are means + S.D.

Table 2   Organ distribution of Sprague-Dawley cytosolic and microsomal
           epoxide hydrolases

Organ	cytosolic epoxide hydrolase[a],[b]	microsomal epoxide hydrolase[a],[c]
kidney	1.56	0.11
heart	1.44	0.003
liver	1.00	1.00
brain	0.37	0.02
lung	0.17	0.06
testis	0.10	0.23
spleen	0.09	0.02

[a] Activities of liver epoxide hydrolases are defined as 1.00.
[b] Activity of cEH was determined using TSO as substrate. Specific activity of liver was 34.4 pmol/min x mg protein.
[c] Data are taken from reference 15. Specific activity of liver was 6.4 nmol/min x mg protein using benzo[*a*]pyrene 4,5-oxide as substrate.

within the same species, the specific activity of cytosolic epoxide hydrolase was measured in seven tissues. The specific activity of cytosolic epoxide hydrolase using *trans*-stilbene oxide as substrate was highest in kidney and heart, followed by liver, brain, lung, testis, and spleen (Table 2). In contrast to cytosolic epoxide hydrolase activity microsomal epoxide hydrolase activity was highest in liver and much lower in extrahepatic tissues.

In Fischer F-344 rat kidneys the activity of cytosolic epoxide hydrolase was twofold higher than in liver. In contrast to these two rat strains, mice and rabbit specific activity was highest in liver followed by kidney and other tissues.[8]

Induction of cytosolic epoxide hydrolase

None of the compounds listed in Table 3 caused significant changes of specific activity of Sprague-Dawley rat liver cytosolic epoxide hydrolase. Hammock and Ota[16] observed no significant induction of cytosolic epoxide hydrolase activity after i.p. injection of phenobarbitone, 3-methylcholanthrene, Aroclor 1254, *trans*- and *cis*-stilbene oxides, pregnenolone-16α-carbonitrile and chalcone in male C57BL/6 mice. From these data it is clear that the commonly used inducers of xenobiotic metabolizing enzymes

Table 3   Compounds which did not act as inducers of male Sprague-Dawley
           rat liver cytosolic epoxide hydrolase

(1) Aroclor 1254   (5) Pentacene
(2) 3,3',4,4'-Tetrachlorobiphenyl   (6) Phenobarbitone
(3) Benzo[*c*]chrysene   (7) ß-Naphthoflavone
(4) Picene   (8) *trans*-Stilbene oxide

Animals were treated with the indicated compounds by i.p. injection of 27.8 mg/kg (3,4 and 5), 80 mg/kg (6), 25 mg/kg (7), 400 mg/kg (8) on three consecutive days, whereas compounds (1) and (2) were given twice (500 mg/kg and 43.8 mg/kg, respectively).

fail to induce cytosolic epoxide hydrolase in rat and mouse liver.

However, it was reported that the hypolipidemic drugs clofibrate, di(2-ethylhexyl)phthalate[15] and nafenopin[17] induced cytosolic epoxide hydrolase of mice by a factor of about 2. With the exception of C3H mice, microsomal epoxide hydrolase was also induced by these drugs in all mouse strains tested.

For our investigation we chose Fischer F-344 rats, because this strain showed less interindividual differences in the activity of cytosolic epoxide hydrolase, and because of their very low specific activity of cytosolic epoxide hydrolase compared to mouse. The low basal activity in the Fischer F-344 rats might make them a more sensitive model for the detection of induction. Animals were treated with three structurally unrelated hypolipidemic compounds: clofibrate, tiadenol and acetylsalicylic acid. All of them have been reported to have hypolipidemic activity, and to cause hepatic peroxisome proliferation and induction of some mostly peroxisomal enzymes.[18]

Recently, it has been shown that triglyceride-lowering drugs are tumourigenic in rats and mice, but they are negative as mutagens in the Ames test.[19] Table 4 shows that treatment of Fischer rats for one week with a diet containing the peroxisome proliferating drugs clofibrate (0.25 %), tiadenol (0.5 %) or acetylsalicylic acid (1 %) caused a drastic increase in cytosolic epoxide hydrolase activity of rat liver (8, 13 and 5 fold), which parallelled the induction of peroxisomal β-oxidation activity (13, 19 and 5 fold). The activity of microsomal epoxide hydrolase was only slightly increased (< 1.5 fold). Clofibrate and tiadenol, but not acetylsalicylic acid, caused a marked increase in the liver/body weight ratio. Tiadenol treatment doubled the liver weight compared to control animals.

Table 4   Effect of clofibrate, tiadenol and acetylsalicylic acid on liver/body weight ratio, activity of peroxisomal β-oxidation system and cytosolic and microsomal epoxide hydrolase activity of Fischer F-344 liver

Compound	liver/body weight ratio %	peroxisomal β-oxidation[a]	cytosolic epoxide hydrolase[b]	microsomal epoxide hydrolase[c]
control	4.1 + 0.3	5.3 + 0.3	38 + 8	4.0 + 0.5
clofibrate	6.3 + 0.5	70 + 7	312 + 45	5.8 + 0.6
tiadenol	8.5 + 0.6	98 + 7	505 + 34	5.7 + 0.6
acetyl-salicylic acid	4.2 + 0.1	25 + 3	178 + 36	5.8 + 0.6

4 animals received a diet containing 0.25 % clofibrate, 0.5 % tiadenol or 1 % acetylsalicylic acid for one week. The mean dose were 205, 548 and 630 mg/kg and day. 12 control animals were used. All assays were done in duplicate for each individual. The values are means + S.D.

[a] Activities are given in nmol NAD$^+$ reduced/min x mg protein.
[b] Activities are given in pmol/min x mg protein. TSO was used as substrate.
[c] Activities are given in nmol/min x mg protein. STO was used as substrate.

Peroxisome proliferating drugs are known to be potent inducers of enzymes involved in the metabolism of fatty acids, e.g. peroxisomal β-oxidation sysem, acyl-CoA hydrolase and acyl-carnitine transferases.[20] Induction of cytosolic epoxide hydrolase might, therefore, indicate a possible participation of cytosolic epoxide hydrolase in the metabolism of fatty acids.

Acknowledgment: This work was supported by the Deutsche Forschungsgemeinschaft (SFB 302).

REFERENCES

1. R. Synyder, D. V. Parke, J. J. Kosicis, D. J. Jollow, C. G. Gibson and C. G. Witmer, Advances in Experimental Medicine and Biology. Biological Reactive Intermediates-II, Chemical Mechanism and Biological Effects. Plenum Press, New York (1982).
2. F. Oesch, Mammalian epoxide hydrases. Inducible enzymes catalysing the inactivation of carcinogenic and cytotoxic metabolites derived from aromatic and olefinic compound, Xenobiotica 3: 305 (1973).
3. W. B. Jakoby and W. H. Habig, Glutathione transferases, in: "Enzymatic Basis of Detoxication", Vol. 2, W. B. Jakoby, ed., pp. 63-94. Academic Press, New York (1980).
4. C. Timms, F. Oesch, L. Schladt and W. Wörner, Multiple forms of epoxide hydrolase, in: "Proceedings of the 9th International Congress of Pharmacology", J.F. Mitchell, W. Paton, P. Turner, eds., Macmillan Press, Londen (1984).
5. A. Y. H. Lu and G. T. Miwa, Molecular properties and biological functions of microsomal epoxide hydrase, Ann. Rev. Pharmacol. Toxicol. 20: 513 (1980).
6. F. Oesch, C. W. Timms, C. H. Walker, T. M. Guenthner, A. J. Sparrow, T. Watabe and C. R. Wolf, Existence of multiple forms of microsomal epoxide hydrolases with radically different substrate specifities, Carcinogenesis 5: 7 (1984).
7. W. Levin, D. P. Michaud, P. E. Thomas and D. M. Jerina, Distinct rat hepatic microsomal epoxide hydrolases catalyze the hydration of cholesterol 5,6α-oxide and certain xenobiotic alkene and arene oxides, Archs. Biochem. Biophys. 220: 485 (1983).
8. S. S. Gill and B. D. Hammock, Distribution and properties of a mammalian soluble epoxide hydrase, Biochem. Pharmacol. 29: 389 (1980).
9. F. Oesch, D. M. Jerina, J. W. Daly, A radiometric assay for hepatic epoxide hydrase activity with 7-^{3}H-styrene oxide, Biochim. Biophys. Acta 227: 685 (1971).
10. F. Oesch, A. J. Sparrow and K. L. Platt, Radioactively labelled epoxides part II. (1) Tritium labelled cyclohexene oxide, trans-stilbene oxide and phenanthrene 9,10-oxide, J. Labelled Compound Radiopharm. 17: 93 (1980).
11. F. Bieri, P. Bentley, F. Waechter and W. Stäubli, Use of primary cultures of adult rat hepatocytes to investigate mechanism of action of nafenopin, a hepatocarcinogenic peroxisome proliferator, Carcinogenesis 5: 1033 (1984).
12. I. Mertes, R. Fleischmann, H. R. Glatt and F. Oesch, Interindividual variations in the activities of cytosolic and microsomal epoxide hydrolase in human liver, Carcinogenesis 6: 219 (1985).
13. J. Seidegård, J. W. DePierre and R. W. Pero, Measurement and characterisation of membrane-bound and soluble epoxide hydrolase activities in resting mononuclear leukocytes from human blood, Cancer Res. 44: 3654 (1984).
14. F. Oesch, A. Zimmer and H. R. Glatt, Microsomal epoxide hydrolase in

different rat strains, <u>Biochem. Pharmacol</u>. 32: 1763 (1983).

15. F. Oesch, H. R. Glatt and H. Schmassmann, The apparent ubiquity of epoxide hydratase in rat organs, <u>Biochem. Pharmacol.</u> 26: 603 (1977).

16. B. D. Hammock and K. Ota, Differential induction of cytosolic epoxide hydrolase, microsomal epoxide hydrolase, and glutathione S-transferase activities, <u>Toxicol. Appl. Pharmacol.</u> 71: 254 (1983).

17. F. Waechter, F. Bieri, W. Stäubli and P. Bentley, Induction of cytosolic and microsomal epoxide hydrolases by the hypolipidaemic compound nafenopin in the mouse liver, <u>Biochem. Pharmacol.</u> 33: 31 (1984).

18. A. J. Cohen and P. Grasso, Review of the hepatic response to hypolipidaemic drugs in rodents and assessment of its toxicological significance to man, <u>Fd. Cosmet. Toxicol.</u> 19: 585 (1981).

19. J. K. Reddy, D. L. Azarnoff and C.E. Hignite, Hypolipidaemic hepatic peroxisome proliferators form a novel class of chemical carcinogens, <u>Nature</u> 283: 397 (1980).

20. J. K. Reddy and N. D. Lalwani, Carcinogenesis by hepatic peroxisome proliferators: Evaluation of the risk of hypolipidemic drugs and industrial plasticizers to humans, <u>CRC Critical Reviews in Toxicology</u> 12 (1): 1 (1983).

MULTIPLICITY OF CYTOCHROME P-450 IN MORRIS HEPATOMA

Minro Watanabe, Tetsuo Ohmachi, and Ikuko Sagami

Research Institute for Tuberculosis and Cancer
Tohoku University
Sendai, 980, Japan

INTRODUCTION

In many hepatic tumors the reduced activities of benzo[a]pyrene (BP) hydroxylase or drug-metabolizing enzymes were observed, when compared to the enzyme activities in normal adult liver, from which hepatic tumor was originated (Adamson and Fouts, 1961). Conney et al. (1957) described the remarkable and now widely recognized induction of BP hydroxylase activity in rat liver by the prior administration of BP itself, or 3-methylcholanthrene (MC), or other polycyclic hydrocarbons. An apparent induction of the drug-metabolizing enzymes by an inducer, such as MC or phenobarbital (PB), was also demonstrated in some lines of "slow-growing" Morris hepatoma (Conney and Burns, 1963; Hart et al., 1965; Watanabe et al., 1970, 1975a, 1975b), but not in Morris hepatoma 7777 (Miyake et al., 1974), Novikoff hepatoma (Hart et al., 1965), and some lines of Yoshida ascites hepatoma (Sugimura et al., 1966), all of which were considered as lines of rapidly growing tumors (Morris, 1972). It was then observed that the activity and the inducibility by inducers of the drug-metabolizing enzymes were roughly correlated with the growth rate of the tumor in the host. On the other hand, very low activities of some of the drug-metabolizing enzymes were detected in the liver from newborn and fetal animals, compared with those in the liver from the corresponding adult animals (Hart et al., 1962; Watanabe et al., 1970). The inducibility of BP hydroxylation enzyme sometimes appears to be a determinant factor among those which affect the sensitivity of animals to chemical carcinogens (Watanabe et al., 1975; Nebert et al., 1978).

The microsomal monooxygenase system containing cytochrome P-450 (P-450) and NADPH-cytochrome P-450 reductase (NADPH-P-450 reductase) plays a critical role in the metabolic activation and detoxification of various kinds of xenobiotics, such as chemical carcinogens, drugs, insecticides (Miller and Miller, 1966; Gelboin, 1969; Weisburger and Weisburger, 1973), and in the metabolism of endogenous substrates, such as prostaglandins, fatty acids and steroid hormones (Ryan et al., 1979; Kusunose et al., 1981; Vatsis et al., 1982; Wada et al., 1984). The elevated BP hydroxylation activities in liver microsomes of mice and rats treated with MC are attributed to the induction of molecular forms of P-450 which have high catalytic activities for BP hydroxylation in

---

Abbreviations used; Benzo(a)pyrene, BP; 3-Methylcholanthrene, MC; Untreated, UT; Phenobarbital, PB; Cytochrome P-450, P-450; NADPH-Cytochrome P-450 (Cytochrome c) reductase, NADPH-P-450 reductase; Potassium phosphate, K-phosphate

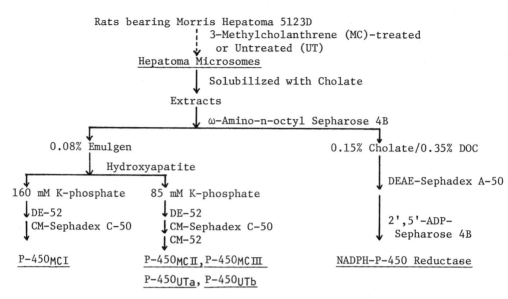

Fig. 1. Purification procedures of cytochrome P-450 and NADPH-cytochrome P-450 reductase from hepatoma

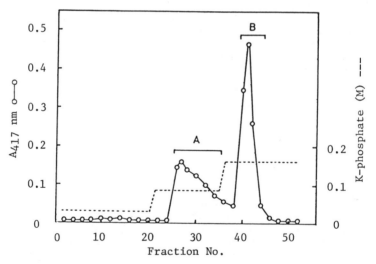

Fig. 2. Elution profiles of cytochrome P-450$_{MC}$ on hydroxyapatite column. The cytochrome P-450 fraction from the Sepharose 4B column was applied, and eluted stepwise with 85 mM (Fraction A) and 160 mM (Fraction B) potassium phosphate buffer (pH 7.25).

the reconstituted systems, and multiple forms of P-450 have been purified and extensively characterized to express their specific catalytic activities (Guengerich, 1978; Ryan et al., 1979; Negishi and Nebert, 1979).

We reported previously that in Morris hepatoma 5123D BP hydroxylation activity was very low, but apparently increased by the administration of MC, as demonstrated in the host liver, small intestine, and lung from the rats bearing this tumor and that the properties of microsomal BP hydroxylase were somewhat different between the tumor and the host liver (Watanabe et al., 1975a, 1975b). Therefore, we tried to separate and purify the molecular forms of P-450 from Morris hepatoma 5123D and to characterize their molecular properties and catalytic specificities in different kinds of hydroxylation reaction.

SEPARATION AND PURIFICATION OF CYTOCHROME P-450 FROM HEPATOMA

Hepatoma P-450$_{MC}$ in MC-treated Tumor-bearing Rats

Saine and Strobel (1976) reported that P-450 was purified from Morris hepatoma 5123tc(H) of the tumor-bearing rats treated with phenobarbital and that the specific content of P-450 was 1.0 nmol per mg protein, indicating of apparently lower specific content, compared with the values reported for hepatic P-450.

Morris hepatoma 5123D, supplied by Dr. H. P. Morris, Howard University, Washington, D.C. in 1980, has been maintained by serial transplantation in this institute. A piece of solid tumor was inoculated intramuscularly into the leg of BUF/MA strain of rat   Approximately one month later the tumor-bearing rats were treated intraperitoneally with MC (30 mg/kg body weight/day) for 3 days and sacrificed 24 hr after. Purification procedures were

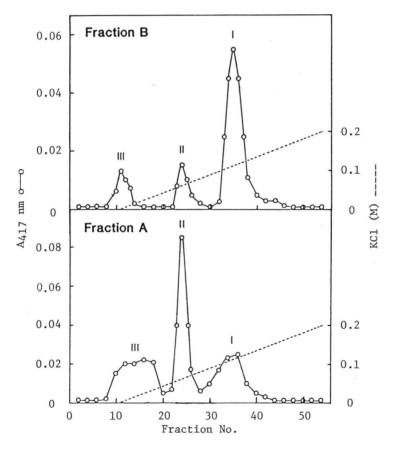

Fig. 3. Elution profiles on DE-52 columns of the fractions eluted with 160 mM (B) and 85 mM (A) potassium phosphate from hydroxyapatite column. Elutions were performed with a linear gradient from 0 to 0.2 M KCl in the buffer at room temperature.

Table 1. Purification of Cytochromes P-450$_{MC}$ from Hepatoma Microsomes of the Tumor-bearing Rats Treated with 3-Methylcholanthrene

Fractions	Protein (mg)	Cytochrome P-450$_{MC}$		
		Total content (nmol)	Specific content (nmol/mg protein)	Yield (%)
Microsome	642	136	0.218	100
Cholate Extract	282	107	0.319	78.6
ω-Amino-n-octyl Sepharose 4B Eluate	43.5	48.4	1.11	35.6
P-450$_{MCI}$				
Hydroxyapatite 160 mM Eluate	7.0	13.1	1.87	9.6
DE-52 Eluate	0.44	6.51	14.9	4.8
CM-Sephadex C-50 Eluate	0.15	2.56	16.8	1.9
P-450$_{MCII}$				
Hydroxyapatite 85 mM Eluate	10.0	14.7	1.47	10.9
DE-52 Eluate	0.55	4.51	8.20	3.3
CM-Sephadex C-50 Eluate	0.30	2.50	8.25	1.8

schematically shown in Fig. 1. All the procedures for the purification except for DEAE-cellulose (DE-52) colum chromatography were carried out at 0-4 °C. Hepatoma microsomes were solubilized with 0.6% sodium cholate in 0.1 M potassium phosphate buffer (pH 7.25) containing 1 mM EDTA, 1 mM DTT, 1 μg/ml leupeptin, 1 μg/ml pepstatin, and 20% glycerol. The supernatant fluid, obtained after centrifugation at 105,000 g for 60 min, was applied to an ω-amino-n-octyl Sepharose 4B. The fractions containing P-450$_{MC}$ were eluted with the buffer containing 0.4% sodium cholate and 0.08% Emulgen 913 (Ohmachi et al., 1985). The eluates from Sepharose 4B column were chromatographed on a hydroxyapatite column in the presence of 0.2% Emulgen 913 and 20% glycerol. After washing with 30 mM potassium phosphate buffer (pH 7.25), 85 and 160 mM potassium phosphate buffer (pH 7.25) were used to obtain the two main fractions containing P-450, as shown in Fig. 2. The two eluates were dialyzed separately and each dialyzed solution was applied to a DEAE-cellulose (DE-52) column. The column was eluted with a linear gradient of 0 to 0.2 M KCl in the buffer containing 0.2% Emulgen 913, 0.5% sodium cholate, and 20% glycerol. Fig. 3 shows that each P-450 fraction is divided into three fractions, respectively. The two main fractions of P-450 were prepared from the fractions eluted from the 85 and 160 mM potassium phosphate buffer, and peak I was mainly obtained from the latter fraction and peak II, from the former fraction. After the further purification the purified P-450$_{MCI}$ and P-450$_{MCII}$ were obtained respectively from the peak I and II fractions eluted from DE-52 column. The results of a typical purification of hepatoma P-450$_{MCI}$ and P-450$_{MCII}$ are shown in Table 1. The specific contents of purified hepatoma P-450$_{MCI}$ and P-450$_{MCII}$ were 16.8 and 8.25 nmol per mg protein, respectively, with the recovery of 1.9 and 1.8 %, compared to the contents in microsomes.

Hepatoma P-450$_{UT}$ in Untreated Tumor-bearing Rats

Cytochrome P-450$_{UT}$ was prepared from hepatoma microsomes of untreated tumor-bearing rats. The purification procedures were almost similar to those for hepatoma P-450$_{MC}$. The fractions eluted from the ω-amino-n-octyl Sepharose 4B column were applied to a hydroxyapatite column. A main peak fraction eluted with 85 mM potassium phosphate buffer was obtained, but any major

Table 2. Purification of Cytochromes P-450$_{UT}$ from Hepatoma Microsomes of Untreated Tumor-bearing Rats

Fractions	Protein (mg)	Cytochrome P-450$_{UT}$		
		Total content (nmol)	Specific content (nmol/mg protein)	Yield (%)
Microsome	795	110	0.138	100
Cholate Extract	488	103	0.211	93.6
ω-Amino-n-octyl Sepharose 4B Eluate	65	71.4	1.10	64.9
Hydroxyapatite 85 mM Eluate	17.7	33.1	1.87	30.1
P-450$_{UTa}$				
DE-52 Eluate	1.71	6.5	3.80	5.8
CM-52 Eluate	1.06	7.5	7.07	6.8
P-450$_{UTb}$				
DE-52 Eluate	1.71	8.9	5.20	8.0
CM-Sephadex C-50 Eluate	0.66	4.9	7.42	4.0

Table 3. Purification of NADPH-Cytochrome P-450 Reductase from Hepatoma of the Tumor-bearing Rats Treated with 3-Methylcholanthrene

Fractions	Protein (mg)	Total activity (μmol/min)	Specific activity (μmol/min/mg protein)	Yield (%)
Microsome	642	61.0	0.095	100
Cholate Extract	282	61.5	0.218	100
ω-Amino-n-octyl Sepharose 4B Eluate	13.3	34.9	2.63	57.2
DEAE-Sephadex A-50 Eluate	1.68	21.8	12.9	35.7
2',5'-ADP-Sepharose 4B Eluate	0.25	9.6	38.5	15.6

fraction containing P-450 was not obtained even in the case of 160 mM potassium phosphate solution as a elution buffer. These chromatographic profiles of P-450$_{UT}$ seem to be apparently different from those of P-450$_{MC}$ which were previously reported. The results of a typical purification of hepatoma P-450$_{UT}$ are shown in Table 2. The specific contents of the purified P-450$_{UTa}$ and P-450$_{UTb}$ were 7.07 and 7.42 nmol per mg protein, respectively.

PURIFICATION OF NADPH-CYTOCHROME P-450 REDUCTASE FROM HEPATOMA

The preparation of highly purified NADPH-P-450 reductase from Morris hepatoma 5123tc(H) microsomes of PB-treated tumor-bearing rats was reported by Fennell and Strobel (1982), and they stated that NADPH-P-450 reductase of hepatoma appeared identical to that of liver by peptide analysis of protease digest, by Ouchterlony double-diffusion analysis and by inhibition profile of the enzyme activity following the antibody.

In this experiment we tried to purify hepatoma NADPH-P-450 reductase from MC-treated tumor-bearing rats according to the modified procedure of Imai (1976). As shown in Fig. 1, the NADPH-P-450 reductase fraction was eluted with a solution of 0.15% sodium cholate and 0.35% sodium deoxycholate from

Table 4. Catalytic Activities of Hepatoma P-450$_{MCI}$ and P-450$_{MCII}$ in the Reconstituted System[a]

Reactions	Hepatoma 5123D		Liver P-450$_c$
	P-450$_{MCI}$	P-450$_{MCII}$	
Benzo[a]pyrene 3-hydroxylation	19.2 [b]	0.50	23.7
7-Ethoxycoumarin O-deethylation	34.6	N.D.[c]	53.4
Benzphetamine N-demethylation	0.77	0.34	0.76
Aniline p-hydroxylation	1.65	0.44	1.04
Aminopyrine N-demethylation	2.47	N.D.	0.42
p-Nitroanisole O-demethylation	2.15	0.39	1.68

[a]The reconstituted system contained NADPH-cytochrome P-450 reductase and dilauroylphosphatidylcholine.
[b]The catalytic activity is expressed as nmol of product formed/min/nmol of cytochrome P-450.
[c]N.D.; Not detectable

the ω-amino-n-octyl Sepharose 4B column, and further purified by using consecutively DEAE-Sephadex A-50 and 2',5'-ADP-Sepharose 4B columns. The result of a typical purification of hepatoma NADPH-P-450 reductase is shown in Table 3, and the specific activity of the purified form was 38.5 μmol cytochrome c reduced per min per mg protein with 15.6 percent of recovery, compared to the activity in microsomes.

CATALYTIC ACTIVITY IN MICROSOMES AND IN THE RECONSTITUTED SYSTEMS

The activities of BP 3-hydroxylation in microsomes of Morris hepatomas were very low, even in the slow-growing hepatoma, compared with those in rat liver, but apparently increased the activity by a single MC-administration into the rat, whereas in the rapidly growing hepatoma the increase of the activity was not observed (Watanabe et al., 1970, 1975a, 1975b; Strobel et al., 1978). In our experiment the molecular activity of BP 3-hydroxylation was 6.47 nmol formed per min per nmol P-450 in the hepatoma microsomes of MC-treated tumor-bearing rats, whereas the activity in hepatoma microsomes of untreated tumor-bearing rats was 1.69, showing an apparent induction of the activity by MC.

The catalytic activities toward various kinds of xenobiotics were assayed in the reconstituted systems containing NADPH-P-450 reductase and hepatoma P-450$_{MCI}$ or P-450$_{MCII}$, and compared with the activities of liver P-450$_c$, a major form of P-450$_{MC}$ in MC-treated rats. As shown in Table 4, hepatoma P-450$_{MCI}$ preferentially catalyzed BP 3-hydroxylation and 7-ethoxycoumarin O-deethylation, the activities of which were slightly lower than those of rat liver P-450$_c$. It was well observed that the liver P-450$_c$ had low catalytic activities for benzphetamine and aminopyrine N-demethylation compared to the activities in liver P-450$_b$ from PB-treated rats (Ryan et al., 1979). Hepatoma P-450$_{MCI}$ also had low catalytic activities toward the two substrates above mentioned. On the other hand, the molecular activities of hepatoma P-450$_{MCII}$ toward all substrates used were very low. At present the substrate having a high affinity in the reconstituted systems containing hepatoma P-450$_{MCII}$ was not yet confirmed. As previously reported by Fennell and Strobel (1982), it was confirmed that NADPH-P-450 reductase purified from Morris hepatoma 5123D of MC-treated rats was similar to that from liver of PB-treated rats, and that in the reconstituted system for BP 3-hydroxylation hepatoma P-450$_{MCI}$ and NADPH-P-450 reductase were mutually interchangeable with liver P-450$_c$ and NADPH-P-450 reductase.

Table 5. Comparison in Properties of Cytochromes P-450MC and P-450UT in Morris Hepatoma 5123D

	Hepatoma			
	P-450UTa	P-450UTb	P-450MCI	P-450MCII
Total P-450 content in microsomes (nmol/mg protein)	0.13		0.22	
Specific content of purified forms (nmol/mg protein)	7.07	7.42	16.8	8.25
Maximum of CO-reduced complex (nm)	446.5	450.5	446.5	450.5
Spin state of heme iron[a]	low	low	low	low
Molecular weight on SDS-PAGE (Mr.×10³)	52	50	56	50
Catalytic activity in reconstituted system (mol formed/min/mol P-450)				
Benzo[a]pyrene 3-hydroxylation	N.D.[b]	0.72	19.2	0.50
7-Ethoxycoumarin O-deethylation	N.D.	N.D.	34.6	N.D.

[a]The spin state of ferric hemoprotein was suspected from the spectrophoto-metric profile in the oxidized form of the purified cytochrome P-450 after the elimination of Emulgen.
[b]Not detectable.

COMPARISON IN PROPERTIES OF CYTOCHROMES P-450$_{MC}$ AND P-450$_{UT}$ IN MORRIS HEPATOMA 5123D

Properties of the purified hepatoma P-450s were compared, as shown in Table 5. The specific contents of the purified P-450 prepared were more than 7.0 nmol per mg protein, but it was observed that some of them still contained contaminating minor protein band(s) on SDS-PAGE. Soret bands of hepatoma P-450UTa and P-450UTb under CO-reduced state were 446.5 and 450.5 nm, respec-tively, and in the oxidized form a main peak with about 415 nm was observed in both P-450UTa and P-450UTb, indicating that the both P-450UT have low-spin state of hemoprotein. By spectrophotometrical analysis, by analysis of molecular weight and of peptide composition after partial proteolysis, and by catalytic properties in the reconstituted system, no differences in the characteristics between P-450UTb and P-450MCII were observed, indi-cating that hepatoma P-450MCII might be a constituted form of P-450 in hepatoma. It was, however, confirmed that the fraction of hepatoma P-450MCI could not be detected in hepatoma microsomes of untreated tumor-bearing rats, showing that hepatoma P-450MCI was an induced form of P-450 in hepatoma. The molecular activities of hepatoma P-450UTa toward BP 3-hydroxylation and 7-ethoxycoumarin O-deethylation were not detectable, indicating that hepatoma P-450UTa was apparently different from the other three hepatoma P-450s used. In conclusion we have at least three purified forms of P-450 in Morris hepatoma 5123D, which were different each other in their characteristics of P-450, and a further purification and a precise characterization will be requested.

PRESENCE OF SIMILAR CYTOCHROME P-450 IN LIVER, LUNG AND HEPATOMA 5123D FROM 3-METHYLCHOLANTHRENE-TREATED RATS

It is of interest to know whether a similar molecular form of P-450 presents in liver, lung and hepatoma from MC-treated rats. Table 6 summarized the characters of the purified P-450s in the three different tissues. As reported previously by Sagami and Watanabe (1983) and Watanabe et al. (1985), the minimum molecular weight and the peptide patterns on partial proteolysis of rat pulmonary P-450$_{MC}$ were apparently different from those of hepatic

Table 6. Similarity in Properties of Cytochromes P-450 in Liver, Lung and Hepatoma from 3-Methylcholanthrene-treated Rats

	Lung[a] P-450$_{MC}$	Liver P-450$_C$	Hepatoma P-450$_{MCI}$
Total P-450 content in microsomes (nmol/mg protein)	0.04	1.0	0.22
Specific content of purified forms (nmol/mg protein)	12.5	20.0	16.8
Molecular weight on SDS-PAGE (Mr.$\times 10^3$)	54	56	56
Maximum of CO-reduced complex (nm)	447.5	447.5	447
Spin state of heme iron	low	low	low
Catalytic activity in reconstituted system (mol formed/min/mol P-450)			
Benzo[a]pyrene 3-hydroxylation	16.9	23.7	19.2
7-Ethoxycoumarin O-deethylation	31.4	53.4	34.6
Benzphetamine N-demethylation	2.1	2.93	2.66
Antigenicity in Ouchterlony double diffusion analysis toward antibody against hepatic P-450$_C$[b]	+	+	+

[a] Pulmonary P-450$_{MC}$ and hepatic P-450$_C$ were purified by the methods of Sagami and Watanabe (1983) and of Harada and Omura (1981).
[b] The antibody was prepared following the injection of the purified rat liver P-450$_C$ into female New Zealand White rabbits.

P-450$_{MC}$ (P-450$_C$), but no differences in catalytic activity in the reconstituted system and in immunological properties in Ouchterlony double diffusion analysis were observed between liver P-450$_C$ and lung P-450$_{MC}$. Goldstein and Links (1984) and Domin et al. (1984) reported that by the analysis of Western blots of hepatic and pulmonary microsomal fractions prepared from untreated and tetrachlorodibenzo-p-dioxin-treated rats, no difference in the molecular weight of the P-450, recognized by antibody against a form 6 of rabbit P-450, was observed. The discrepancy in the similarity of the molecular weight of hepatic and pulmonary P-450$_{MC}$ may be derived from the difference of the rat strain used in each experiment, because we used BUF/MA strain and the others, Sprague-Dawley, Fisher or Long-Evans strain of rats, but it remains to be clear more precisely.

Ohmachi et al. (1985) demonstrated that hepatoma P-450$_{MC}$ could not be distinguishable from liver P-450$_{MC}$ (P-450$_C$), because no differences in the properties, that is, minimum molecular weight, peptide composition after partial proteolysis, catalytic activities in the reconstituted system and Ouchterlony double diffusion anaylsis, were detected. It is of importance to note that there are similar catalytic activities of P-450$_{MC}$ in the reconstituted systems among liver, lung and hepatoma from BUF/MA strain of rats treated with MC, and that the primary structure of pulmonary P-450$_{MC}$ is clearly different from that of hepatic or hepatoma P-450$_{MC}$.

ACKNOWLEDGEMENTS

We wish to express our hearty gratitude to Dr. H. Fujii, Mr. K. Kikuchi, Miss M. Hirata and Mrs. Y. Gunji for their help in performing these experiments and in preparation of the manuscript. This work was supported in part by a Grant-in-Aid for Cancer Research from the Ministry of Education, Science and Culture, Japan and by a grant from the Japan Tobacco and Salt Public Corporation.

REFERENCES

Adamson, R. H., and Fouts, J. R., 1961, The metabolism of drugs by hepatic tumors, Cancer Res., 21:667.

Conney, A. H., and Burns, J. J., 1963, Induced synthesis of oxidative enzymes in liver microsomes by polycyclic hydrocarbons and drugs, Adv. Enzyme Regulation, 1:189.

Conney, A. H., Miller, E. C., and Miller, J. A., 1957, Substrate induced synthesis and other properties of benzpyrene hydroxylase in rat liver, J. Biol. Chem., 228:753.

Domin, B. A., Serabjit-Singh, C. J., Vanderslice, R. R., Devereux, T. R., Fouts, J. R., Bend, J. R., and Philpot, R. M., 1984, Tissue and cellular differences in the expression of cytochrome P-450 isozymes, in: "Proceeding of 9th International Congress of Pharmacology," W. Paton, J. Mitchell and P. Turner, eds., Vol. 3, p. 219, Macmillan Press, London.

Fennell, P. M., and Strobel, H. V., 1982, Preparation of homogeneous NADPH-cytochrome P-450 reductase from rat hepatoma, Biochem. Biophys. Acta, 709:173.

Gelboin, H. V., 1969, A microsome-dependent binding of benzo[a]pyrene to DNA, Cancer Res., 29:1272.

Goldstein, J. A., and Linko, P., 1984, Differential induction of two 2,3,7,8-tetrachlorodibenzo-p-dioxin-inducible forms of cytochrome P-450 in extrahepatic versus hepatic tissues, Mol. Pharmacol., 25:185.

Guengerich, F. P., 1978, Separation and purification of multiple forms of microsomal cytochromes P-450. Partial characterization of three apparently homogeneous cytochromes P-450 isolated from liver microsomes of phenobarbital and 3-methylcholanthrene-treated rats, J. Biol. Chem., 253:7931.

Harada, N., and Omura, T., 1981, Selective induction of two different molecular species of cytochrome P-450 by phenobarbital and 3-methylcholanthrene, J. Biochem., 89:237.

Hart, L. G., Adamson, R. H., Dixon, R. C., and Fouts, J. R., 1962, Stimulation of hepatic microsomal drug metabolism in the newborn and fetal rabbit, J. Pharmacol. Exptl. Therap., 137:103.

Hart, L. G., Adamson, R. H., Morris, H. P., and Fouts, J. R., 1965, The stimulation of drug metabolism in various rat hepatomas, J. Pharmacol. Exptl. Therap., 149:7.

Imai, Y., 1976, The use of 8-aminooctyl Sepharose for the separation of some components of the hepatic microsomal electron transfer system, J. Biochem., 80:365.

Kusunose, E., Ogita, K., Ichihara, K., and Kusunose, M., 1981, Effect of cytochrome b₅ on fatty acid ω- and (ω-1)-hydroxylation catalyzed by partially purified cytochrome P-450 from rabbit kidney cortex microsomes, J. Biochem., 90:1069.

Miller, E. C., and Miller, J. A., 1966, Mechanisms of chemical carcinogenesis: Nature of proximate carcinogens and interactions with macromolecules, Pharmacol. Rev., 18:805.

Miyake, Y., Gaylor, J. L., and Morris, H. P., 1974, Abnormal microsomal cytochromes and electron transport in Morris hepatoma, J. Biol. Chem., 249:1980.

Morris, H. P., 1972, Isozymes in selected hepatomas and some biological characteristics of a spectrum of transplantable hepatomas, in "Isozymes and Enzyme Regulation in Cancer," S. Weinhouse, and T. Ono, eds., p. 95, Univ. Tokyo Press, Tokyo.

Nebert, D. W., Atlas, S. A., Guenthner, T. M., and Kouri, R. E., 1978, The Ah locus: genetic regulation of the enzymes which metabolize polycyclic hydrocarbons and the risk for cancer, in: "Polycyclic Hydrocarbons and Cancer," H. V. Gelboin and P.O.P. Ts'o, eds., Vol. 2, p. 345, Academic Press, New York.

Negishi, M., and Nebert, D. W., 1979, Structural gene products of the "Ah" locus: genetic and immunochemical evidence for two forms of mouse liver cytochrome P-450 induced by 3-methylcholanthrene, J. Biol. Chem., 254: 11015.

Ohmachi, T., Sagami, I., Fujii, H., and Watanabe, M., 1985, Microsomal monooxygenase system in Morris hepatoma: purification and characterization of cytochrome P-450 from Morris hepatoma 5123D of 3-methylcholanthrene-treated rats, Arch. Biochem. Biophys., 236:176.

Ryan, D. E., Thomas, P. E., Korzeniowski, D., and Levin, W., 1979, Separation and characterization of highly purified forms of liver microsomal cytochrome P-450 from rats treated with polychlorinated biphenyls, phenobarbital, and 3-methylcholanthrene, J. Biol. Chem., 254:1365.

Sagami, I., and Watanabe, M., 1983, Purification and characterization of pulmonary cytochrome P-450 from 3-methylcholanthrene-treated rats, J. Biochem., 93:1499.

Saine, S. E., and Strobel, H. W., 1976, Drug metabolism in liver and tumors. Resolution of components and reconstitution of activity, Mol. Pharmacol., 12:649.

Strobel, H. W., Digman, J. D., Saine, S. E., Fang, W.-F., and Fennell, P. M., 1978, The drug metabolism systems of liver and liver tumors: A comparison of activities and characteristics, Mol. Cell. Biochem., 22:79.

Sugimura, T., Ikeda, K., Hirota, K., Hozumi, M., and Morris, H. P., 1966, Chemical, enzymatic, and cytochrome assays of microsomal fraction of hepatoma with different growth rates, Cancer Res., 26:1711.

Vatsis, K. P., Theoharides, A. D., Kupfer, D., and Coon, M. J., 1982, Hydroxylation of prostaglandins by inducible isozymes of rabbit liver microsomal cytochrome P-450. Participation of cytochrome b_5, J. Biol. Chem., 257:11221.

Wada, A., Okamoto, M., Nonaka, Y., and Yamano, T., 1984, Aldosterone biosynthesis by a reconstituted cytochrome P-450$_{11\beta}$ system, Biochem. Biophys. Res. Commun., 119:365.

Watanabe, M., Konno, K., and Sato, H., 1975a, Aryl hydrocarbon hydroxylase in Morris hepatoma 5123D, Gann, Jpn. J. Cancer Res., 66:499.

Watanabe, M., Konno, K., and Sato, H., 1975b, Properties of aryl hydrocarbon hydroxylase in microsomes of Morris hepatoma 5123D and the host liver, Gann, Jpn. J. Cancer Res., 66:505.

Watanabe, M., Potter, V. R., and Morris, H. P., 1970, Benzpyrene hydroxylase activity and its induction by methylcholanthrene in Morris hepatomas, in host livers, in adult livers, and in rat liver during development, Cancer Res., 30:263.

Watanabe, M., Sagami, I., Ohmachi, T., and Fujii, H., 1985, Characteristics of purified cytochrome P-450s in microsomes of rat lung and Morris hepatoma 5123D, in: "P-450 and Chemical Carcinogenesis," Y. Tagashira and T. Omura, eds., p. 19, Plenum Press, New York.

Watanabe, M., Watanabe, K., Konno, K., and Sato, H., 1975, Genetic differences in the induction of aryl hydrocarbon hydroxylase and benzo[a]pyrene carcinogenesis in C3H/He and DBA/2 strains of mice, Gann, Jpn. J. Cancer Res., 66:217.

Weisburger, J. H., and Weisburger, E. K., 1973, Biochemical formation and pharmacological, toxicological, and pathological properties of hydroxylamines and hydroxamic acids, Pharmacol. Rev., 25:1.

THE METABOLISM OF BENZENE AND PHENOL BY A RECONSTITUTED PURIFIED PHENOBARBITAL-INDUCED RAT LIVER MIXED FUNCTION OXIDASE SYSTEM

James C. Griffiths, George F. Kalf and Robert Snyder

Joint Graduate Program in Toxicology, Rutgers University and UMDNJ, Piscataway, N.J. 08854 and the Department of Biochemistry, Thomas Jefferson University, Philadelphia, PA 19107.

INTRODUCTION

During its long history of extensive industrial use, chronic exposure of humans to benzene has been associated with blood disorders, such as aplastic anemia and leukemia. It has been our aim to study the link between the metabolism of benzene and the mechanism by which it produces bone marrow toxicity (Snyder et al., 1967, 1977, 1982). The mixed function oxidases, a family of hemoprotein cytochrome P-450 enzymes located in the smooth endoplasmic reticulum of liver as well as most other tissues, was shown by Gonasun et al. (1973) to play a key role in the metabolism of benzene. Studies by Jerina and Daly (1974) and Tunek et al. (1978) strongly supported the concept that the formation of benzene oxide is the principal first step in benzene metabolism. However, Ingelman-Sundberg and Hagbjork (1982) have suggested that the hydroxylation may occur via the insertion of a hydroxyl free radical, postulated to be generated from an "iron-catalyzed cytochrome P-450-dependent Haber Weiss reaction." Gorsky and Coon (1984) have demonstrated that the pathway by which benzene is metabolized is directly related to the concentration utilized in the in vitro reaction, i.e. with very low concentrations the reaction is mediated by a free hydroxyl-radical mechanism and at higher concentrations of benzene, in the range of its K_M, by direct cytochrome P-450-mediated oxidation. These two concepts could be reconciled if (1) there were two (or more) cytochromes P-450 involved in the hydroxylation, one with the ability to form an epoxide intermediate and another generating free radicals through a Haber Weiss mechanism, or (2) the cytochrome P-450 responsible for epoxidation also functioned as a NADPH oxidase, thereby, producing H_2O_2 which subsequently generated the hydroxyl radicals. Post and Snyder (1983) recently reported evidence which suggested that there are at least two different rat liver mixed function oxidases active in the hydroxylation of benzene. Using a reconstituted purified mixed function oxidase system from phenobarbital-induced rat liver, we have studied the initial hydroxylation step in benzene metabolism leading to phenol formation and the subsquent conversion of phenol to polyhydroxylated metabolites.

MATERIALS AND METHODS

Animals and treatment

Sprague-Dawley male rats, weighing 200-250 grams and supplied by Perfection Breeders, Douglassville, Pennsylvania, were fed Purina Lab Chow and water ad libitum. They were given sodium phenobarbital (80 mg/kg), dissolved in 0.9 % NaCl, intraperitoneally, 72, 48 and 24 hours before sacrifice.

Chemicals

All chemicals and reagents were of the highest grade available commercially. The [^{14}C]benzene was purchased from New England Nuclear (Boston, MA) and the [^{14}C]phenol from Amersham (Arlington Heights, IL). SKF 525A (β-diethylaminoethyl-diphenyl propylacetate) was a gift from Smith, Kline and French Pharmaceutical Company (Philadelphia, PA) and desferrioxamine (Desferal mesylate[*]) was a gift from the CIBA-Geigy Company (Summit, NJ).

Microsomal preparation from phenobarbital-induced animals

The animals were sacrificed by decapitation. The livers were perfused in situ with ice cold 0.9 % NaCl via the portal vein, homogenized (33 % w/v) in 0.05 M Tris buffer, pH 7.5 , 4° C, containing 1.15 % KCl, and centrifuged at 10,000 x g for 20 min at 4° C. The resulting supernatant was centrifuged at 105,000 x g for 60 min; to collect the microsomes which were resuspended to the same volume in 10 mM EDTA buffer, pH 7.5, containing 1.15 % KCl and centrifuged at 105,000 x g for 60 min. The microsomes were suspended by homogenization in 0.25 M sucrose to a volume approximately one half the original wet liver weight and stored at -70° C until used for enzyme purification.

Isolation and assay of NADPH-cytochrome c reductase and cytochrome P-450

NADPH-cytochrome c reductase (NADPH-cytochrome P-450 reductase) was prepared from phenobarbital-treated rats by a method based on the published procedures of Dignam and Strobel (1975) and Yasukochi and Masters (1976). The phenobarbital-inducible form of cytochrome P-450 was prepared by the method of West et al. (1979). Protein concentration was determined by the method of Lowry et al. (1951) as modified by the procedure of Polacheck and Cabib (1981). The cytochrome P-450 activity was measured using the method of Omura and Sato. (1964). Benzphetamine N-demthylation was measured by the method of Thomas et al. (1976). NADPH-cytochrome P-450 reductase activity was assayed by its ability to reduce cytochrome c by the method of Phillips and Langdon (1962). One unit of reductase activity is defined as the amount of enzyme catalyzing the reduction of cytochrome c at an initial rate of 1 nmol/ min at room temperature.

SDS polyacrylamide gel electrophoresis

The determination of the molecular weights of the purified enzymes was carried out with the established procedure of Laemmli (1970).

Metabolism of benzene by the reconstituted mixed function oxidase system

The reconstituted mixed function oxidase system consisting of 0.1 nmol purified cytochrome P-450, 250 units of purified NADPH-cytochrome c reductase and 25 μg of the lipid, dilauroylglyceryl-3-phosphorylcholine was added to an incubation vessel containing an NADPH-generating system made up of 2.0 mM NADP, 0.2 mM NAD, 8 mM glucose 6-phosphate, and 2 units glucose 6-phosphate dehydrogenase, to yield a total volume of 2.5 ml. [^{14}C]Benzene (4 mM, 200 dpm/ nmol) was added to start the reaction; incubation was for 30 min at 37° C with shaking. The reaction was stopped with the addition of trichloroacetic acid (final concentration 5 %). 100 ml of 10 mM ascorbic acid was added at the same time to prevent oxidation of the metabolites. The solution was alkalinized with NaOH (10 N) and extracted twice with 10 ml of toluene to remove any unreacted benzene. HCl (6 N) was added and the samples were extracted twice with 5 ml of toluene to remove the [^{14}C]phenol (Post and Snyder, 1983). Radioactivity in the toluene layer was determined by liquid scintillation counting. An aliquot of the remaining aqueous layer was added to toluene-Triton X-100 fluor (Benson, 1966) and the radioactivity was determined.

Metabolism of phenol by the reconstituted mixed function oxidase system

The reconstituted system in this case consisted of 0.2 nmol cytochrome P-450, 200 units of the NADPH-cytochrome c reductase and 50 μg of the dilauroylglyceryl-3-phosphorylcholine. The same NADPH-generating system was employed. [^{14}C]Phenol (0.2 mM, 200 dpm/ nmol) was added to start the reaction. Trichloroacetic acid stopped the reaction and the unreacted [^{14}C]phenol was extracted with toluene and the remaining radioactivity in the aqueous layer counted as described above.

Determination of benzene metabolites by high performance liquid chromatography

High-performance liquid chromatography of benzene metabolites was performed on an Altex Partisil ODS C-18 reverse-phase column. A standard solution containing phenol, catechol, hydroquinone, 1,2,4-benzenetriol, and ascorbic acid (added to retard the oxidation of the phenols) was added to the sample following the removal of the precipitated proteins. An aliquot was applied to the column, and the column was eluted with a solution of 2 % methanol in deionized water pumped through at a rate of 1 ml/ min. The absorbance of the added standards was monitored at 280 nm, and the radioactivity in each fraction was used to determine the concentration of metabolites formed during the incubation. An electrochemical detector was also utilized in the characterization of benzene metabolites, using a mobile phase consisting of 2 % acetonitrile in 0.1 M ammonium acetate, pH 4.0. The detector potential was set at +1.0 V (Lunte and Kissinger, 1983).

RESULTS

Post and Snyder (1983) found that in liver microsomes prepared from phenobarbital- and benzene-treated rats, benzene metabolism was found to have different K_M and pH optima, suggesting the involvement of several cytochrome P-450 species in benzene metabolism. We elected to study benzene metabolism in a mixed function oxidase system reconstituted from purified cytochrome P-450 and NADPH-cytochrome c reductase prepared from phenobarbital-induced rats. The NADPH-cytochrome c reductase was purified 84-fold to a specific content of 24,395 units/ mg protein and the cytochrome P-450 was purified 3-fold to 7.1 nmol/ mg protein. SDS-PAGE electrophoresis revealed that both proteins had been purified to a single homogeneous bands of 78 Kd for the reductase and 56 Kd for the P-450. Optimal assay conditions were established with respect to the concentrations of the enzymes and dilauroylglyceryl-3-phosphorylcholine.

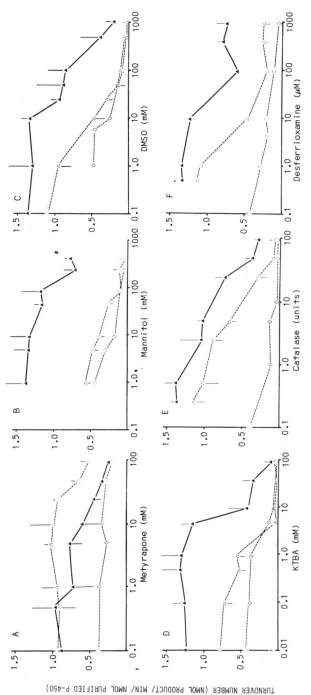

TURNOVER NUMBER (NMOL PRODUCT/ MIN /NMOL PURIFIED P-450)

Fig. 1. Inhibition of benzene & phenol metabolism by added compounds in a reconstituted M. F. O. system from phenobarbital-induced rat liver.

1A.) Metyrapone (mM); 1B.) Mannitol (mM); 1C.) DMSO (mM); 1D.) KTBA (mM); 1E.) Catalase (units*); 1F.) Desferrioxamine (μM).
* One unit is the amount of catalase able to decompose 1.0 mole of H_2O_2 per min at pH 7.0 at 25° C.
▲——▲ Phenol recovered from [^{14}C]benzene (4.0 mM) added as substrate (control: 1.26 nmol/ min/ nmol P450).
△-------△ Polyhydroxylated metabolites recovered from [^{14}C]benzene (4.0 mM) added as substrate (control: 1.21 nmol/ min/ nmol P450).
◇-------◇ Polyhydroxylated metabolites recovered from [^{14}C]phenol (0.2 mM) added as substrate (control: 0.36 nmol/ min/ nmol P450).

Under these conditions, the turnover number (nmol of product formation/ min/ nmol cytochrome P-450) for the N-demethylation of benzphetamine was 94.6 and for benzene metabolism, 3.4. (Griffiths et al., 1984a).

A Lineweaver-Burke double reciprocal plot for the determination of the parameters of benzene kinetics reveals a K_M for benzene of 13.3 mM and a Vmax of 38.5 nmol/ 30 min incubation. Mannitol (500 mM) a free radical trapping agent, reduced the Vmax to 14.7, leaving the K_M about the same. Mannitol appeared to act as a non-competitive inhibitor. (Griffiths et al., 1984b).

The effect of cytochrome P-450 inhibitors on the production of benzene metabolites was studied. In the absence of inhibitors 1.26 nmol of phenol and 1.21 nmol of polyhydroxylated compounds were produced/ min/ nmol cytochrome P-450 from 4.0 mM $[^{14}C]$benzene. When $[^{14}C]$phenol was used as the starting substrate, 0.36 nmol of polyhydroxylated metabolites were produced/ min/ nmol cytochrome P-450. The phenol was extracted from the acidified supernatant from the incubation mixture; the radioactivity remaining in the aqueous fraction was identified as the polyhydroxylated metabolites. For the remaining graphs, the abcissa displays the concentration of the added agent on a logarhythmic scale, while the ordinate represents the turnover number (nmol of products formed/ min/ nmol of purified cytochrome P-450) of the metabolite measured. Metyrapone inhibits cytochrome P-450 (Netter, 1969; Roots and Hildebrandt, 1973). At a concentration of 1.0 mM, metyrapone inhibited the production of phenol from benzene by the reconstituted mixed function oxidase system. The formation of di- and trihydroxylated metabolites was not inhibited until a concentration of inhibitor in excess of 50 mM was reached. Little phenol was recovered because it was converted to the polyhydroxylated metabolites. Metyrapone had virtually no effect on the recovery of the polyhydroxylated metabolites when radiolabeled phenol was the starting compound. (Fig. 1A).

The effects of mannitol, DMSO and KTBA, which are hydroxyl radical trapping agents (Winston and Cederbaum, 1982; Ingelman-Sundberg and Hagbjork, 1982; Cederbaum, 1983) on benzene metabolism were studied to determine the possible role free radicals play in the formation of benzene metabolites. Fig. 1B shows that the major effect of mannitol added to the benzene reaction at a concentration below 100 mM was to reduce the formation of the polyhydroxylated metabolites. Failure to metabolize phenol to the polyhydroxylated compounds resulted in recovery of large amounts of phenol. When the starting compound was phenol, increasing the concentration of mannitol decreased the recovery of the polyhydroxylated metabolites. The addition of DMSO gave similar results. (Fig. 1C). The recovery of the polyhydroxylated metabolites was inhibited, at a concentration of 1 mM when either benzene or phenol was used as a starting substrate, while the production of phenol from the radiolabeled benzene remained unchanged from that in the control, until a concentration in excess of 25 mM was used. KTBA, also depressed polyhydroxylated metabolite formation at the 1 mM mark in both reactions (Fig. 1D). In benzene metabolism, at 5 mM, when the first evidence of an effect on phenol recovery could be seen, the secondary metabolites had already dropped to negligble levels.

If the generation of free radicals is due to the presence of hydrogen peroxide during mixed function oxidase-mediated oxidations (Gillette et al., 1957), the addition of catalase (Nicholls and Schonbaum, 1963; Cederbaum and Qureshi, 1982) to the reaction would serve to break down the H_2O_2 to oxygen and water, thus short-circuiting the production of OH·. To determine whether H_2O_2 was involved in the formation of benzene metabolites, catalase was added to the incubation;

the recovery of the polyhydroxylated metabolites was depressed, beginning
with a concentration of 0.5 unit of catalase, irrespective of which
labeled compound was used to start the reaction, whereas the formation
of phenol was not affected until 50 units of enzyme had been added.
(Fig. 1E). Thus H_2O_2 may have played a role in the metabolism of phenol
but not of benzene.

Since free iron may play a catalytic role in the generation of
active oxygen species in vitro, the effect of the iron-chelator,
desferrioxamine (Westlin, 1971; Cederbaum and Dicker, 1983) was studied.
It was found to depress the recovery of the polyhydroxylated metabolites
at concentrations between 1 and 10 μM in both assays. Phenol recovery
was not reduced until concentrations between 10 and 100 μM were used.
(Fig. 1F).

DISCUSSION

Gonasun et al. (1973) demonstrated that benzene was hydroxylated by
the hepatic mixed function oxidase. Post and Snyder (1983) reported that
enzyme induction by benzene or phenobarbital generated at least two
distinct mixed function oxidases active in benzene hydroxylation in rat
liver microsomes. Accordingly we decided to isolate, reconstitute and
study this enzyme system using benzene as the substrate. The studies
reported here, have concentrated on the phenobarbital-induced enzyme and
have shown that the isolated reconstituted enzyme displays benzene
hydroxylating capacity similar to that found in intact microsomes thereby
substantiating the original findings of Gonasun et al. (1973).

The mechanism of benzene hydroxylation has been the subject of
considerable recent discussion. Figure 2 shows that the initial
hydroxylation of benzene to its primary metabolite, phenol, may proceed
via either of two pathways, (1) by enzymatic oxidation to form an epoxide
which subsequently rearranges to phenol (Jerina and Daley, 1974; Tunek
and Oesch, 1979), or (2) by direct insertion of a hydroxyl free radical
(Ingelman-Sundberg and Hagbjork, 1982). Further metabolism of phenol may
proceed via either pathway. Our data suggests that the hydroxylation of
benzene by the phenobarbital-induced reconstituted mixed function oxidase
at the substrate concentration reported here, operates predominantly by
epoxidation because compounds capable of trapping free radicals had
relatively little effect on the formation of phenol. (Griffiths et al.,
1985; Snyder et al., 1985)

The relative roles of epoxidation and hydroxy radical insertion in
aromatic hydrocarbon metabolism has also been debated. The "bay region"
theory of carcinogenesis in large measure has been related to sequential
epoxidations of polycyclic aromatic hydrocarbons separated by the
conversion of the first epoxide to a dihydrodiol (Jerina et al., 1977).
Cavalieri et al., (1983) have argued that free radical insertion may
represent an important alternative to epoxidation and that the ionization
potential of the hydrocarbon is an important indicator of the likelihood
for radical insertion. Gorsky and Coon (1984) used a reconstituted mixed
function oxidase system containing rabbit phenobarbital-inducible
$P-450_{LM2}$ to show that very low concentrations of benzene are metabolized
by a free hydroxyl radical mechanism and at higher concentrations, in the
range of the K_M of benzene (105 mM), by direct cytochrome P-450-mediated
oxidation. This concentration contrasts the amount used by
Ingelman-Sundberg and coworkers (Ingelman-Sundberg et al., 1982;
Ingelman-Sundberg and Hagbjork, 1982; Johansson and Ingelman-Sundberg,
1983) who conducted their experiments with 17 μM benzene. Possibly, in

218

the presence of very low concentrations of benzene, the free radical-mediated formation of phenol is the predominate pathway, while at the higher concentrations of benzene, closer to the K_M for the isozyme 2, the direct oxidation by P-450 is quantitatively of much greater importance. Gorsky and Coon did not find evidence for a superoxide radical-independent, hydrogen peroxide-dependent oxidation of benzene in the presence of NADPH. Anaerobically, H_2O_2 and NADPH did cause a low rate of phenol formation but the activity is much too low to contribute significantly to the rate observed with NADPH under aerobic conditions. An analysis of these arguments when applied to benzene involves first the appreciation that although benzene is a carcinogen, it is unlikely that it forms a diol epoxide. However, the fact that radical insertion did not appear to be a favored pathway is in agreement with the mechanism proposed by Cavalieri's group who argue that the ionization potential of the single benzene ring is not conducive to free radical insertion.

The formation of the di- and trihydroxylated secondary metabolites was much more sensitive to the action of the radical trapping agents, mannitol, DMSO and KTBA, and to desferrioxamine and catalase, both of which can inhibit the generation of hydroxy free radicals, than was phenol production. These data suggest that formation of the secondary metabolites was much more dependent on the presence of free radicals than was phenol formation. Trush et al. (1984) reported that the mixed function oxidase system can be a potent generator of active oxygen moieties, such as singlet oxygen, hydrogen peroxide, superoxide, and hydroxyl radical. Furthermore, the ionization potentials of phenolic compounds are more likely to be conducive to radical insertion (Cavalieri et al., 1983). These data suggest that the reconstituted phenobarbital-induced mixed function oxidase system, while acting to form benzene oxide, may generate free radicals for the further metabolism of phenol.

ACKNOWLEDGMENTS

The authors would like to thank Lois Argenbright and Regina Wang for their technical expertise in the preparation of the purified components of the mixed function oxidase system, and to Dr. Anthony Lu for his advice and interest in the project and the use of his laboratory for the initial production of the enzymes. This research project was supported by N.I.E.H.S. grant ES 02931.

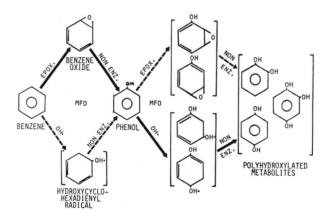

Fig. 2. Proposed mechanism of benzene metabolism in a reconstituted purified M.F.O. system from phenobarbital-induced rat liver.

REFERENCES

Benson, R., 1966, Limitiations of tritium measurements by scintillation counting emulsions, <u>Analyt. Chem.</u>, 38:1353.

Cavalieri, E. L., Rogan, E.G., Roth, R.W., Saugier, R. K. and Hakam, A., 1983, The relationship between ionization potential and horseradish peroxidase/ hydrogen peroxide-catalyzed binding of aromatic hydrocarbons to DNA, <u>Chem.-Biol. Interactions</u>, 47:87.

Cederbaum, A. I. and Qureshi, A., 1982, Role of catalase and hydroxyl radicals in the oxidation of methanol by rat liver microsomes, <u>Biochem. Pharmacol.</u>, 31:329

Cederbaum, A. I., 1983, Organic hydroperoxide-dependent oxidation of ethanol by microsomes: Lack of a role for free hydroxyl radicals, <u>Arch. Biochem. Biophys.</u>, 227:329

Cederbaum, A. I. and Dicker, E., 1983, Inhibition of microsomal oxidation of alcohols and of hydroxyl-radical-scavenging agents by the iron-chelating agent desferrioxamine, <u>Biochem. J.</u>, 210:107.

Dignam, J. D. AND STROBEL, H. W., 1975, Preparation of homogeneous NADPH-Cytochrome P-450 reductase from rat liver, <u>Biochem. Biophys. Res. Commun.</u>, 63:845.

Gillette, J. R., Brodie, B. B. and La Du, B. N., 1957, The oxidation of drugs by liver microsomes: On the role of TPNH and oxygen, <u>J. Pharmacol. Exp. Ther.</u>, 119:532.

Gonasun, L. M., Witmer, C. M., Kocsis, J. J. and Snyder, R., 1973, Benzene metabolism in mouse liver microsomes, <u>Toxicol. Appl. Pharmacol.</u>, 26:398.

Gorsky, L. D. and Coon, M. J., 1984, Evaluation of the role of free radicals in the cytochrome P-450-catalyzed oxidation of benzene and cyclohexane, <u>Drug Metab. Disp.</u>, 13:169.

Griffiths, J. C., Kalf, G. F. and Snyder, R., 1984a, The metabolism of benzene by a reconstituted purified phenobarbital(pb)-induced rat liver mixed function oxidase system, <u>Fed. Proc.</u>, 43:934.

Griffiths, J. C., Kalf, G. F. and Snyder, R., 1984b, The inhibition of benzene metabolism in a reconstituted purified phenobarbital(pb)-induced rat liver mixed function oxidase system by mannitol, <u>Proc. of the Ninth Int. Congress of Pharmacol. London</u>, 513.

Griffiths, J. C., Kalf, G. F. and Snyder, R., 1985, The metabolism of phenol by a reconstituted purified phenobarbital-induced rat liver mixed function oxidase system, <u>Fed. Proc.</u>, 44:516.

Ingelman-Sundberg, M. and Hagbjork, A. L., 1982, On the significance of the cytochrome P-450-dependent hydroxyl radical-mediated oxygenation mechanism, <u>Xenobiotica</u>, 12:673.

Ingelman-Sundberg, M., Hagbjork, A.L., Ekstrom, G., Terelius, Y. and Johansson, I., 1982, The cytochrome P-450 dependent hydroxyl radical-mediated oxygenation mechanism: implications in pharmacology and toxicology, <u>in</u>: "Cytochrome P-450 Biochemistry, Biophysics and Environmental Implications," E. Heitanen, M. Laitinen, and O. Hanninen, eds., Elsevier Biomedical Press, Amsterdam

Jerina, D. and Daly, J., 1974, Arene oxides: A new aspect of drug metabolism, _Science_, 185:573.

Jerina, D. M., Yagi, H. and Hernandez, O., 1977, Steroeselective synthesis and reactions of a diol-epoxide from benzo(a)pyrene, _in_: "Biological Reactive Intermediates," D. J. Jollow, J. J. Kocsis, R. Snyder and H. Vainio, eds., Plenum Press, New York.

Johansson, I. and Ingelman-Sundberg, M., 1983, Hydroxyl radical-mediated, cytochrome P-450-dependent metabolic activation of benzene in microsomes and reconstituted enzyme systems from rabbit liver, _J. Biol. Chem._, 258:7311.

Laemmli, U. K., 1970, Cleavage of structural proteins during assembly of the head of bacteriophage T4, _Nature_, 227:680.

Lowry, O. H., Rosenbrough, N. J., Farr, A. L. and Randall, R. J., 1951, Protein measurement with the folin phenol reagent, _J. Biol. Chem._, 193:265.

Lunte, S. M. and Kissinger, P. T., 1983, Detection and identification of sulfhydryl conjugates of _p_-benzoquinone in microsomal incubations of benzene and phenol, _Chem.-Biol. Interact._, 47:195.

Netter, K., Kahl, G. and Magnussen, M., 1969, Kinetic experiments on the binding of metyrapone to liver microsomes, _Naunyn-Schmiedeberg's Arch. Pharmakol._, 265:202.

Nicholls, P. and Schonbaum, G. R., 1963, Catalases, _in_: "The Enzymes," P. D. Boyer, H. Lardy, and K. Myrback, eds., Academic Press, New York.

Omura, T. and Sato, R., 1964, The carbon monoxide binding pigment of liver microsomes. I. Evidence for its hemoprotein nature, _J. Biol. Chem._, 239:2370.

Phillips, A. H. and Langdon, R. G., 1962, Hepatic triphosphopyridine nucleotide-cytochrome c reductase : isolation, characterization and kinetic studies, _J. Biol. Chem._, 237:2653.

Polachek, I. and Cabib, E., 1981, A simple procedure for protein determination by the Lowry method in dilute solutions and in the presence of interphering substances, _Analyt. Biochem._, 117:311.

Post, G. B. and Snyder, R., 1983, Effects of enzyme induction on microsomal benzene metabolism, _J. Toxicol. Environ. Health_, 11:811.

Roots, I. and Hildebrandt A., 1973, Non-competitive and competitive inhibition of mixed function oxidase in rat liver microsomes by metyrapone, _Naunyn-Schmeideberg's Arch. Pharmakol._, 277:27.

Snyder, R., Uzuki, R., Gonasun, L., Broomfield, E. and Wells, A., 1967, The metabolism of benzene _in vivo_, _Toxicol. Appl. Pharmacol._, 11:346.

Snyder, R., Andrews, L. S., Lee, E. W., Witmer, C. M., Reilly, M. and Kocsis, J. J., 1977, Benzene metabolism and toxicity, _in_: "Biological Reactive Intermediates," D. J. Jollow, J. J. Kocsis, R. Snyder and H. Vainio, eds., Plenum Press, New York.

Snyder, R., Longacre, S. L., Witmer, C. M. and Kocsis, J. J., 1982, Metabolic correlates of benzene toxicity, _in_: "Biological Reactive

Intermediates II," R. Snyder, D. V. Parke, J. J. Kocsis, D. J. Jollow, G. G. Gibson and C. M. Witmer, eds., Plenum Press, New York.

Snyder, R., Griffiths, J. C. and Kalf, G. F., 1985, Benzene metabolism by a reconstituted purified phenobarbital-induced rat liver mixed function oxidase system, Fed. Proc., 44:516.

Thomas, P. E., Lu, A. Y. H., Ryan, D., West, S. B., Kawalek, J. and Levin, W., 1976, Multiple forms of rat liver cytochrome P-450. Immunological evidence with antibody against cytochrome P-448, J. Biol. Chem., 251:1385.

Trush, M. A., Reasor, M. J., Wilson, M. E. and VanDyke, K., 1984, Oxidant-mediated electronic excitation of imipramine, Biochem. Pharmacol., 33:1401.

Tunek, A., Platt, K. L., Bentley, P. and Oesch, F., 1978, Microsomal metabolism of benzene to species irreversibly binding to microsomal protein and effects of modifications of this metabolism, Mol. Pharmacol., 14:920.

Tunek, A. and Oesch, F., 1979, Unique behavior of benzene monooxygenase: Activation by detergent and different properties of benzene and phenobarbital-induced monooxygenase activities, Biochem. Pharmacol., 28:3425.

West, S. B., Huang, M. T., Miwa, G. T. and Lu, A. Y. H., 1979, A simple and rapid procedure for the purification of phenobarbital-inducible cytochrome P-450 from rat liver microsomes, Arch. Biochem. Biophys., 193:42.

Westlin, W. F., 1971, Desferoxamiane as a chelating agent, Clin. Toxicol., 4:597.

Winston, G. W. and Cederbaum, A. I., 1982, NADPH-dependent production of oxy radicals by purified components of the rat liver mixed function oxidase system. I. Oxidation of hydroxyl radical scavenging agents, J. Biol. Chem., 258:1508.

Yasukochi, Y. and Masters, B. S. S., 1976, Some properties of a detergent-solubilized NADPH-cytochrome c (cytochrome P-450) reductase purified by biospecific affinity chromatography, J. Biol. Chem., 251:5337.

COMPARISON OF THE METABOLISM OF BENZENE AND ITS METABOLITE PHENOL

IN RAT LIVER MICROSOMES

Susan K. Gilmour*, George F. Kalf**, and Robert Snyder*

*Joint Graduate Program in Toxicology
Rutgers-The State University of New Jersey and the
University of Medicine and Dentistry of New Jersey/
Rutgers Medical School
Piscataway, NJ 08854

**Department of Biochemistry
Jefferson Medical College, Thomas Jefferson University
Philadelphia, PA 19107

INTRODUCTION

It has been known for many years that chronic exposure to benzene leads to bone marrow depression and aplastic anemia and in recent years it has become apparent that benzene can also be leukemogenic (Snyder, 1984). Benzene-induced bone marrow depression is caused by one or more metabolites of benzene (Snyder, et al., 1981). Cytochrome P-450 mediates the first step in benzene metabolism (Gonasun et al., 1973). The initial metabolite formed in the liver is thought to be benzene oxide (Jerina and Daly, 1974) which rearranges to form phenol. Johansson and Ingelmann-Sundberg (1983) suggested that benzene hydroxylation may occur as a result of hydroxyl radical formation during the partially uncoupled mixed function oxidase-mediated metabolism of benzene in which hydrogen peroxide is generated. Nevertheless, it is clear that the major metabolite of benzene _in vivo_ (Parke and Williams, 1953) and _in vitro_ (Gonasun et al., 1973) is phenol. Because phenol, can be further hydroxylated it is both a product and a substrate in this system. The metabolism of a substrate, either _in vivo_ or _in vitro_ is in part controlled by the concentration at which it encounters the enzyme. In the metabolism of xenobiotic compounds another controlling factor is the type of cytochrome P-450 which metabolizes the compound and hence the importance of enzyme induction. The concentration of an intermediary metabolite such as phenol is a product of the rate at which it is produced and the rate of further metabolism. The issue can be further complicated if both the initial substrate and its metabolite undergo similar reactions and the two compete at the active site of the enzyme. Given this model, we report on some aspects of the metabolism of benzene and phenol.

MATERIALS AND METHODS

Chemicals and Reagents

Benzene (thiophene-free), phenol, hydroquinone, catechol, and toluene were purchased from Fisher Scientific Company (King of Prussia, PA) and were of the highest grade available. Crystalline [U-^{14}C]phenol was purchased from Amersham-Searle (Arlington Heights, IL) and diluted with phenol to a specific acitivity of 220 to 900 dpm/nmole. [U-^{3}H]Benzene and [U-^{14}C]benzene (99% pure) were purchased from New England Nuclear Corp. (Boston, MA); [^{3}H]benzene was diluted with benzene to a specific activity of 1200 to 4700 dpm/nmole, and [^{14}C]benzene was diluted with benzene to a specific activity of 100 to 1585 dpm/nmole. Scintillation cocktails were prepared as described by Benson (1966). Glucose 6-phosphate, NADP, NAD, glucose 6-phosphate dehydrogenase, and metyrapone were products of Sigma Chemical Company (St. Louis, MO). PB[5] was purchased from J.T. Baker Chemical Company (Brick Town, NJ), and BNF and 1,2,4-benzenetriol were products of Aldrich Chemical Company (Milwaukee, WI). All other chemicals were of the highest grade available commercially.

Animals

Male Sprague-Dawley rats (175–225 g, Perfection Breeders, Douglassville, PA.) were used. They were allowed food (Purina Lab Chow) and water ad libitum. Rats were treated with PB (80 mg/kg, ip, once daily for 3 days) BNF[5] (in corn oil, 80 mg/kg, i.p. once daily for 3 days), or benzene (in corn oil, 1100 mg/kg, s.c. 24 h and 16 h before killing). Control rats were injected with the vehicle only. Rat liver microsomes were isolated as previously described (Snyder et al., 1967). Protein was determined using the biuret method (Layne, 1975) with bovine serum albumin as a standard.

Incubation Protocols

The incubation mixtures contained: 4 mg microsomal protein, 0.1M sodium phosphate-0.15 M KCl buffer (pH 7.4), and an NADPH-generating system consisting of 2 mM NADP, 0.2 mM NAD, 8 mM glucose 6-phosphate, and 2 units glucose 6-phosphate dehydrogenase. Reactions were initiated by the addition of the substrates, [^{3}H] or [^{14}C]benzene and/or [^{14}C]phenol in a total volume of 2.5 ml. The incubations were carried out in one-half ounce bottles (French squares) with teflon-lined screw caps for either 10 or 20 min at 37°C in a Dubnoff shaker. The reactions were stopped by the addition of trichloroacetic acid (5% final concentration).

Carbon monoxide inhibition studies were performed in 15 ml Warburg flasks with no center well and one side arm stoppered with a rubber diaphragm. Microsomes diluted in phosphate buffer were placed in the main compartment of the flasks with the side arm of the flask containing the NADPH-generating system and [^{14}C]phenol. After attachment to a manometer containing Brody's solution, the flasks were inserted into a Warburg-type water bath at 25°C and gassed for 5 min with the appropriate gas mixture. All gas mixtures contained 4% O_2 and either 0, 0.1, 2, 4, 8, or 32% carbon monoxide, with nitrogen as the inert carrier gas. The reaction was started by mixing the contents of the sidearm and the main compartment, and the incubation was continued with shaking at 120 oscillations/min, for an additional 10 min in the dark.

Determination of Phenol and Water Soluble Metabolites

The acidified samples were extracted twice with 5 ml toluene. Subsequent HPLC analysis revealed that only phenol was found in these toluene extracts. A one ml sample of the remaining aqueous layer containing [^{14}C]phenol metabolites was assayed for radioactivity. If benzene was used as the substrate, then the samples were alkalinized by the addition of NaOH and extracted twice with 10 ml toluene to remove unreacted benzene. [Exposure to alkaline pH degrades hydroquinone and catechol to unidentified polar compounds, and this can explain the unknown polar metabolite of benzene which has been previously reported from this laboratory (Post and Snyder, 1983). However, less than 0.2% of phenol is oxidized when subjected to these alkaline conditions for 45 min. and the phenol produced was then separated by acidification with hydrochloric acid followed by two extractions with toluene. The radioactivity in the combined toluene extracts was measured after the addition of Liquifluor POPOP-PPO toluene concentrate and one ml of the remaining aqueous layer was counted separately.]

Reverse-Phase High Performance Liquid Chromatography

Ascorbic acid (1 mM) was added as an antioxidant to samples from microsomal incubations when the reaction was stopped with TCA. The precipitated protein was removed, the pH adjusted to 3-4, and a freshly prepared solution containing phenol (0.1 mg/ml), catechol (0.1 mg/ml), and hydroquinone (0.05 mg/ml) as markers was added. The HPLC conditions described by Andrews et al. (1979) and Greenlee et al. (1981) for the analysis of benzene metabolites were used with some modifications. A 100 ul aliquot of each sample was injected into a Beckman Ultrasil ODS C-18 reverse-phase column, and the column was eluted at a flow rate of 1 ml/min with a solution of 10% methanol in water acifidied to pH 3.2 with formic acid. The UV absorption of the standards was monitored at a wavelength of 280 nm. Fractions were collected from the column, and the radioactivity was determined.

Spectral Studies

Washed microsomes from perfused livers of either benzene-treated or untreated rats were prepared and diluted with 0.1 M phosphate-KCl buffer, pH 7.4, to a final concentration of 3.5 nmoles cytochrome P-450/ml (2.7 to 3.5 mg protein/ml) as measured bycording to the procedure of Omura and Sato (1964). The binding spectra were determined by the method of Schenkman et al. (1967) using a Hitachi Model 220A spectrophotometer. Additions of either 10 ul of phenol diluted in phosphate buffer or 0.2 ul of benzene were added to the sample cuvette. The change in absorbance from peak to trough was measured for each substrate addition as an index of binding to cytochrome P-450 (Remmer, et al., 1966). The spectral dissociation constant (K_s) was determined by linear regression analysis of the double reciprocal plot of the absorbance increment versus the substrate concentrations (Schenkman, 1970).

RESULTS

Preliminary reports on this work have been published (Gilmour et al., 1983a). At about the same time a report was published by Sawahata and Neal (1983) on the metabolism of phenol and the results of both laboratories agreed that (1) phenol metabolism in rat liver microsomes required NADPH and oxygen, (2) hydroquinone was the major metabolite obtained by reversed phase HPLC and smaller amounts of catechol were formed, and (3) the reaction was inhibited by metyrapone and SKF 525A.

Table 1
Kinetic Parameters of the Microsomal Metabolism of Phenol[a]

Treatment	Km (mM)	Vmax (nmole/mg protein/min)	Vmax (nmole/nmole P-450/min)
Control[b]	0.10 ± 0.04	1.13 ± 0.29	1.56 ± 0.37
PB[b]	0.11 ± 0.03	2.04 ± 0.26[d]	1.46 ± 0.15
Benzene[b]	0.29 ± 0.15[d]	6.08 ± 2.89[d]	6.21 ± 1.72[d]
BNF[c]	0.09 ± 0.05	1.68 ± 0.40	0.99 ± 0.35

[a] Various concentrations of [^{14}C] phenol (0.005 to 0.4 mM) were incubated with liver microsomes (4 mg protein/ml) prepared from untreated rats or rats treated with either PB, benzene, or BNF. The experimental procedures are described in Materials and Methods. The values are expressed as the mean ± S.D. and are calculated from a Lineweaver-Burk plot with each data point on the plot representing the mean of duplicate samples. Student's t-test was used at a significance level of $p < 0.05$.
[b] Values calculated from 4 experiments each consisting of duplicate samples.
[c] Values calculated from 2 experiments each consisting of duplicate samples.
[d] Significantly different from the control ($p < 0.05$).

Table 2
Effect of Enzyme Induction on the Distribution of
[^{14}C]Phenol Microsomal Metabolites[a]

Enzyme Induction	nmoles/mg Protein/10 Min				% of Total Metabolites	
	Hydroquinone	Catechol	Total Phenol Metabolites	Unreacted Phenol	Hydroquinone	Catechol
Control[b]	1.8 ± 0.9	0.5 ± 0.1	2.3	8.8 ± 0.4	78	22
Phenobarbital[c]	2.7 ± 0.5	1.6 ± 0.4	4.3	5.2 ± 0.3	63	37
Benzene[c]	6.0 ± 0.3	1.3 ± -	7.3	3.4 ± 0.3	82	18

Enzyme Induction	nmoles/nmole Cytochrome P-450/10 Min		
	Hydroquinone	Catechol	Total Phenol Metabolites
Control[b]	2.4 ± 0.6	0.8 ± 0.3	3.2
Phenobarbital[c]	2.5 ± 0.5	1.5 ± 0.4	4.0
Benzene[c]	12.0 ± 0.3	2.6 ± -	14.6

[a] [^{14}C]Phenol (30 uM) with a specific activity of 905 dpm/nmole was incubated for 10 min at 37°C with 4 mg microsomal protein and a NADPH-generating system as described previously under Methods. After removal of TCA-precipitated protein by centrifugation, the [^{14}C]phenol metabolites present in the supernatants of each sample were analyzed using a reverse phase ODS-C18 HPLC column as described under Methods.
[b] Values are given as mean ± S.D. for triplicate samples taken from two separate incubations.
[c] Values expressed as mean ± range for duplicate samples taken from two separate incubations.

We report here that the metabolism of phenol by this enzyme system follows first order kinetics. Fig. 1 shows a plot of the metabolism of phenol in microsomes as a function of time and log phenol concentration. The resulting straight lines at each concentration indicate first order kinetics and, therefore, a direct relationship between phenol concentration and the rate of its metabolism.

The kinetics of phenol metabolism were studied using control microsomes and microsomes from animals which had been induced using PB, benzene and BNF (Table 1). Double reciprocal plots of the data were drawn and the kinetic constants were determined. The Km for phenol appears to be similar in control phenobarbital, beta-naphthoflavone and benzene-induced microsomes. However, whether the data is viewed as metabolism/mg microsomal protein or as metabolism/nmole cytochrome P-450, the benzene-induced preparation was most active.

Table 2 shows the formation of hydroquinone and catechol from phenol by microsomes from PB or benzene-induced animals. Although hydroquinone was the predominant metabolite formed by each microsomal preparation, the data suggest that the PB-induced preparations formed less hydroquinone as a percent of total metabolite and more catechol. The increased formation of metabolites after benzene and PB induction is reflected in the recovery of less unreacted phenol. When the data are expressed on the basis of cytochrome P-450, PB induction does not appear to have increased the turnover number for phenol metabolism whereas benzene induction has increased it 4-fold.

Sawahata and Neal (1983) reported that a 1:1 mixture of air:CO inhibited phenol metabolism with the result that hydroquinone formation was decreased 62%, catechol formation 41% and covalent binding to protein was 28%. We have explored the effect of CO on phenol metabolism somewhat more extensively (Fig. 2). Since the degree of inhibition of mixed function oxidase reactions is dependent upon the CO/O_2 ratio, rather than on the absolute concentration of CO (Cooper et al.,1965) the microsomal metabolism of phenol was observed over a range of CO/O_2 ratios. The oxygen concentration in the reaction vessels was kept at 4% and the ratio was varied from 0.5-8. Phenol metabolism was inhibited by 50% at a CO/O_2 ratio of about 1.3, which is within the range commonly observed for other mixed function oxidase reactions. (Rosenthal and Cooper, 1968). The inset in Fig. 2 shows that the values for the Warburg partition coefficient (K) were not constant, unlike reports for other cytochrome P-450 substrates (Cooper et al., 1979).

Substrates of cytochrome P-450 when added to microsomes, cause difference spectra which may be considered characteristic of the particular enzyme-substrate complex. Fig. 3 presents a comparison of the spectral changes produced by benzene and phenol. In both cases the difference spectra exhibit a peak at 385 nm, a trough at 420 nm and a crossover point at about 405 nm. Control microsomes were saturated after adding 11.7 mM benzene whereas concentrations of phenol above 3) demonstrated a Type I spectral change when benzene was added to microsomes from benzene-induced animals. We observed the same phenomenon when phenol was added to microsomes from benzene-induced animals. Fig. 4 presents a double reciprocal plot prepared using the technique of Schenkman (1970). The absorbance difference between 385 and 420 nm was plotted versus the phenol concentration. Whereas the intercept on the abcissa (sometimes called the Ks) was unaltered by induction, the value for Amax in the control microsomes (1.26×10^{-3}) was increased (1.64×10^{-3}) by 30%.

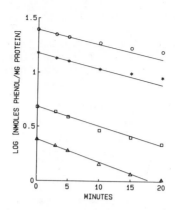

Figure 1. Disappearance of [14C]phenol in rat liver microsomal incubations. Varying concentrations of [14C]Phenol (0.005 mM, 0.01 mM, 0.03 mM, 0.05 mM) were incubated with microsomes in the presence of a NADPH-generating system for the indicated time periods. The residual [14C] phenol was determined after separation of hydroquinone and catechol metabolites. Each data point represents the mean of two experiments each consisting of duplicate samples. The lines were drawn on the basis of linear regression analysis.

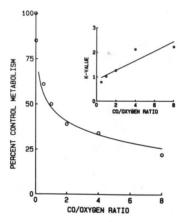

Figure 2. Carbon monoxide inhibition of microsomal phenol metabolism. [14C]phenol (0.1 mM) was incubated for 10 min at 25°C with hepatic microsomes from benzene-treated rats in the presence of a NADPH-generating system and 4% O_2 (v/v). CO concentration ranged from 0 to 32% (v/v) with the balance in all gas mixtures made up with nitrogen. Data are expressed as the percentage of the control metabolism versus increasing ratios of CO/O_2. Control metabolism is the total formation of hydroquinone and catechol from phenol with 4% O_2 and no CO and was 23.4 nmol metabolites/10 min. Data are expressed as the mean of two experiments consisting of triplicate samples. The curve was drawn after logarithmic regression analysis of the data points. The inset plots the Warburg partition coefficient (K) for each data point versus increasing ratios of CO/O_2. The partition coefficient, K, is defined as

$$K = (n)/(1-n) \,/\, (CO)/(O_2)$$

where n = rate or metabolism in the presence of CO divided by the rate without CO.

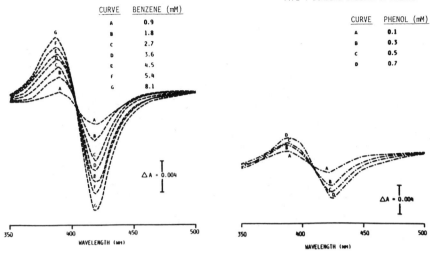

Figure 3. Binding spectra produced by the addition of benzene or phenol to control microsomes.

Curve	Benzene (mM)	Phenol (mM)
A	0.9	0.1
B	1.8	0.3
C	2.7	0.5
D	3.6	0.7
E	4.5	
F	5.4	
G	8.1	

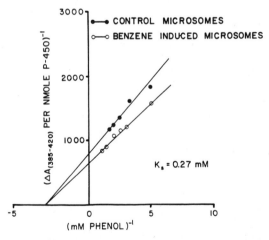

Figure 4. Double reciprocal plot of spectral change produced by phenol. Binding spectra were determined in liver microsomes from untreated rats (●) and from benzene-treated rats (o).

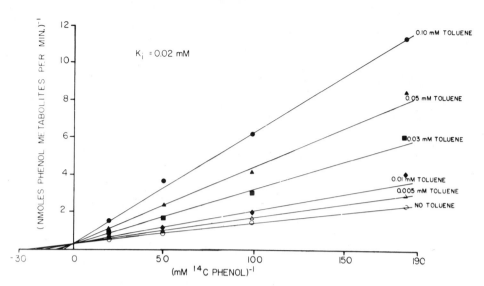

Figure 5. Lineweaver-Burk plot of toluene inhibition of [14C]phenol metabolism. [14C]Phenol was incubated with rat hepatic microsomes and varying concentrations of toluene (none, 0.005 mM, 0.01 mM, 0.03 mM, 0.05 mM, 0.10 mM). The slope and K_I for each line were determined after linear regression analysis of the data points. The data are taken from two experiments consisting of triplicate samples.

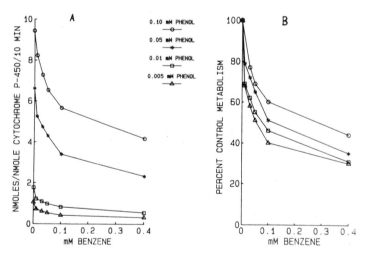

Figure 6. Inhibition of the the microsomal metabolism of [14C]phenol by benzene. Symbols represent data obtained when different concentrations of [14C]phenol (0.10 mM, 0.05 mM, 0.01 mM, 0.005 mM) were incubated with varying concentrations of non-radiolabeled benzene and hepatic microsomes from untreated rats. Metabolism is expressed in (A) nmoles/nmole cytochrome P-450/10 min. or in (B) percent control metabolism where control metabolism is phenol metabolism in the absence of added benzene. Total formation of hydroquinone and catechol is shown. Each data point represents the mean of two experiments each consisting of triplicate samples.

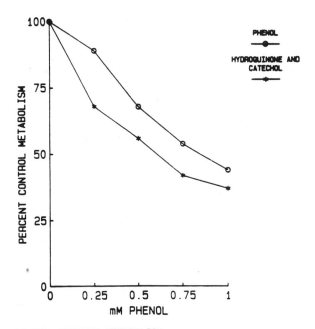

CONTROL BENZENE METABOLISM =
14.1 NMOLES PHENOL +
1.5 NMOLES HYDROQUINONE AND CATECHOL

Figure 7. Inhibition of the microsomal metabolism of [^{14}C]benzene by phenol.

Toluene, a competitive inhibitor of benzene metabolism (Andrews et al., 1977) also competitively inhibits phenol metabolism with an identical K_I of 0.02 mM (Fig. 5). Benzene also inhibited phenol metabolism (Fig. 6) but it was not po sible· to characterize the inhibition because the data could not be linearized using conventional transformations. Phenol also inhibited microsomal benzene metabolism (Fig. 7). Although the addition of equimolar concentrations of phenol inhibited benzene metabolism by 50%, the ratio of hydroquinone and catechol to phenol formed from benzene remained constant.

The formation of phenol and its subsequent oxidation to hydroquinone should follow one of two pathways: In the first, phenol resulting from either benzene oxide formation or hydroxyl radical insertion, dissociates from the enzyme and then reassociates before further hydroxylation occurs; alternatively, hydroxylation occurs while the phenol remains bound to the enzyme. Since free phenol is observed, there must be some dissociation which reflects the relative rates of subsequent hydroxylation versus the rate of dissociation. This question can be studied by adding a pool of phenol to the incubation mixture during the metabolism of labeled benzene. We have selected data from Fig. 7 to explore this relationship. The starting concentration of [^{14}C]benzene was 0.8mM (2000 nmoles). In the absence of added phenol, 14.1 nmoles of phenol and 1.5 nmoles of hydroquinone and catechol were formed. In another incubation which contained added phenol (0.25mM) [^{14}C]benzene (204 dpm/nmole) was converted to 13.6 nmoles [^{14}C]phenol. The addition of 0.25mM phenol (625 nmoles) should result in an isotope dilution of the [^{14}C]phenol derived from [^{14}C]benzene. Thus, the new specific activity of [^{14}C]phenol would be (13.6/638.6) × 204 dpm/nmole = 4.3 dpm/nmole. Since the total observed radioactivity associated with the aqueous layer after acidic toluene extractions was 208 [^{14}C]dpm, the calculated amount of hydroquinone and catechol from benzene-derived phenol would be a total of 48 nmoles (208/4.3). The fact that the incubation of [^{14}C]benzene without added phenol yielded only 1.5 nmoles hydroquinone and catechol metabolites indicates that the addition of phenol did not result in an isotope dilution. A possible explanation could be that some phenol derived from benzene does not dissociate from the enzyme and is subject to a second hydroxylation, resulting in hydroquinone and catechol formation. In this case, added phenol would not be metabolized in the same proportion as phenol derived from benzene, and an isotope dilution effect would not be observed. To answer this question, 0.8 mM [^{3}H]benzene was incubated for five minutes with 0.225 mM [^{14}C]phenol. While total benzene metabolism was decreased by 23%, none of the [^{14}C]phenol was metabolized; [^{14}C]Phenol, incubated 5 min with microsomes in the absence of benzene, yielded 64 nmoles metabolites.

DISCUSSION

The significance of these observations lies in the suggestion that locally produced phenol is the precursor to the reactive intermediate in benzene toxicity and that hydroquinone or p-benzoquinone appear to be reasonable candidates for the toxic metabolite (Tunek et al.,1980; Greenlee et al.,1981 ; Lunte and Kissinger, 1983; Rushmore et al., 1984). While these metabolites are readily detected in liver microsomal preparations, Gollmer et al. (1984) found only phenol and covalently bound metabolites in bone marrow microsomes.

The microsomal metabolism of benzene and phenol possess some distinctive similarities largely because they are cytochrome P-450-mediated. Thus, both are inhibited by metyrapone, SKF 525A, toluene, and CO, and each can inhibit the metabolism of the other. The metabolism of both substrates is best stimulated after benzene pretreatment. Although PB or BNF treatment increase the total liver cytochrome P-450 content, they do not increase the specific enzyme activity for phenol. Similar results have also been obtained using benzene as a substrate with no stimulation observed following pretreatment with either BNF (Post et al., 1983) or benzo[a]pyrene (Gonasun et al., 1973). Furthermore, though PB treatment increases benzene microsomal metabolism, there is no significant difference when benzene metabolism is compared on a per nmole cytochrome P-450 basis (Post et al., 1983). These observations suggest that the isozyme(s) of cytochrome P-450 that is mainly responsible for benzene and phenol metabolism is not the same form that is induced by either BNF or PB.

Besides benzene and phenol hydroxylation, very few other cytochrome P-450-mediated reactions are stimulated following benzene treatment. Recently, benzene treatment of rabbits has been shown to increase the microsomal oxidation of ethanol (Ingelman-Sundberg and Hagbjork, 1982). However, neither the demethylation of ethylmorphine or codeine nor the hydroxylation of benzo[a]pyrene are increased by benzene treatment (Post et al., 1983). Studies using a 9000xg supernatant showed that the oxidative metabolism of both hexobarbital and chlorpromazine are unaffected by benzene treatment while the reductive metabolism of p-nitrobenzoate and neoprontosil is increased (Saito et al., 1973). Moreover, since the first step in the cytochrome P-450 reaction cycle is the binding of the substrate to the enzyme, it is significant that the magnitude of the binding spectra of both benzene and phenol is increased in microsomes from benzene-treated animals. It is probable that the larger spectral change produced by benzene is due to the more hydrophobic nature of benzene which can more easily interact with this membrane-bound enzyme.

Carbon monoxide inhibits the microsomal oxidation of phenol, which is in agreement with the data reported by Sawahata and Neal (1983). However, our data provide evidence for the competition of carbon monoxide with oxygen for the active center of the enzyme by demonstrating increasing inhibition of phenol microsomal oxidation in the presence of a constant 4% oxygen but with varying carbon monoxide/oxygen ratio. Although 50% inhibition of phenol microsomal metabolism occurred at a CO/O_2 ratio of 1.3, which is within the range commonly found for other mixed function oxidase reactions (Rosenthal and Cooper, 1967), the Warburg partition values were not constant. While previous studies of the CO inhibition of the microsomal metabolism of endogenous steroids, acetanilide, codeine, and monomethyl-4-aminopyrine yielded constant K values over a range of CO/O_2 ratios (Cooper et al., 1965), variable K values with changing CO/O_2 have been reported for the CO inhibition of the demethylation of 3-ethylmorphine and benzphetamine (Cooper et al., 1979) and the hydroxylation of benzo[a]pyrene (Cooper et al., 1977). This variability in K values may be attributed to the derivation of this partition coefficient which does not take into account what is generally agreed to be the rate-limiting step in cytochrome P-450 reactions, the reduction of cytochrome P-450 (Fe^{+2}) to the ferrous form which can then interact with CO (Cooper et al., 1979).

The failure to see an isotope dilution effect during incubations of benzene with microsomes in the presence of excess of phenol suggests that phenol produced from benzene is preferentially metabolized over exogenously added phenol. A likely explanation is that successive hydroxylations of benzene to phenol to dihydroxy metabolites occur without the substrate leaving the binding site of the enzyme. Since free phenol is also formed from benzene, some phenol is released withoout further hydroxylation. However, these results do not demonstrate whether the unlabeled phenol is also being metabolized by other enzyme pathway(s) present in the preparation. Recently, Smart and Zannoni (1985) have suggested that an intermediate metabolite of benzene which is formed before phenol, is responsible for a large portion of covalent binding when benzene is incubated with microsomes.

There are several similarities in the metabolism of benzene and phenol which include 1) the metabolism and Type I binding spectra of each substrate is increased by benzene treatment; 2) both are competitively inhibited by toluene; and 3) each inhibits the other's metabolism suggesting that the same enzyme may metabolize both substrates. However, the possibility of two different benzene-inducible cytochrome P-450 isozymes, one of which can metabolize benzene and the other phenol, and for which each substrate can act as an inhibitor of the other cannot be ruled out from these data.

ACKNOWLEDGMENTS

We thank Drs. David Cooper and Heinz Schleyer for very valuable discussions on the interpretation of K in the CO inhibition studies and for the use of their laboratory at the University of Pennsylvania during the carbon monoxide inhibition studies.

REFERENCES

Andrews, L.S., Lee, E.W., Witmer, C.M., Kocsis, J.J., and Snyder, R. (1977). Effects of toluene on metabolism, disposition, and hematopoietic toxicity of (^{3}H) benzene. Biochem. Pharmacol. 26:293-300.

Andrews, L.S., Sasame, H.A., and Gillette, J.R. (1979). ^{3}H-Benzene metabolism in rabbit bone marrow. Life Sciences 25, 567-572.

Cooper, D.Y., Levin, S., Narasimhulu, S., and Rosenthal, O. (1965). Photochemical action spectrum of the terminal oxidase of mixed function oxidase systems. Science 147, 400-402.

Cooper, D.Y., Schleyer, H., Rosenthal, O., Levin, S., Lu, A.Y., Kuntzman, R., and Conney, A.H. (1977). Inhibition by CO of hepatic benzo[a]pyrene hydroxylation and its reversal by monochromatic light. Eur. J. Biochem. 74, 69-75.

Cooper, D.Y., Schleyer, H., Leviin, S., Eisenhardt, R.H., Novack, B., and Rosenthal, O. (1979). The reevaluation of cytochrome p-450 as the terminal oxidase in hepatic microsomal mixed function oxidase catalyzed reactions. Drug Metab. Rev. 10, 153-185.

Gilmour, S. and Snyder, R. (1983a). Metabolism of benzene and phenol in rat liver microsomes. Fed. Proc.42, 1136.

Gilmour S. and Snyder, R.(1983b). Similarities in the microsomal metabolism of benzene and its metabolite phenol. The Pharmacologist 25, 210.

Golmer, L., Graf, H, and Ullrich, V. (1984). Characterization of the benzene monooxygenase in rabbit bone marrow. Biochem. Pharmacol. 33, 3597-3602.

Gonasun, L.M., Witmer, C.M., Kocsis, J.J., and Snyder, R. (1973). Benzene metabolism in mouse liver microsomes. Toxicol. Appl. Pharmacol. 26, 398-406.

Greenlee, W.F., Chism, J.P., and Rickert, D.E. (1981a). A novel method for the separation and quantitation of benzene metabolites using high pressure liquid chromatography. Anal. Biochem. 112, 367-370.

Greenlee, W.F., Sun, J.S. and Bus, J.S. (1981b). A proposed mechanism of benzene toxicity: Formation of reactive intermediates from polyphenol metabolites. Toxicol. Appl. Pharmacol.59, 187-195.

Ingelman-Sundberg, M. and Hagbjork, A.L. (1982). On the significance of the cytochrome P-450-dependent hydroxyl radical-mediated oxygenation mechanism. Xenobiotica 12, 673-686.

Jerina, D. and Daly, J.W.(1974). Arene oxides: A new aspect of drug metabolism, Science, 185, 573-82.

Johansson, I. and Ingelman-Sundberg, M. (1983). Hydroxyl radical-mediated, cytochrome P-450-dependent metabolic activation of benzene in microsomes and reconstituted enzyme systems from rabbit liver.J. Biol. Chem. 258, 7311-7316.

Lunte, S.M. and Kissenger, P.T. (1983). Detection and identification of sulfhydryl conjugates of p-benzoquinone in microsomal incubations of benzene and phenol. Chem.-Biol. Interactions 47,195-212.

Omura, T. and Sato, R. (1964). The carbon monoxide-binding pigment of liver microsomes. I. Evidence for its hemoprotein nature. J. Biol. Chem. 239, 2370-2378.

234

Parke, D.V. and Williams, R.T. (1953). Studies in detoxication. The metabolism of benzene containing ^{14}C benzene. Biochem. J. 54, 231-238.

Post, G. and Snyder, R. (1983). Effects of enzyme induction on microsomal benzene metabolism. J. Toxicol. Environ. Health 11,811-825.

Remmer, H., Schenkman, J.B., Estabrook, R.W., Sasame, H., Gillette, J.R., Narasimhulu, S., Cooper, D.Y. and Rosenthal, O. (1966). Drug interaction with hepatic microsomal cytochrome. Mol. Pharmacol. 2, 187-190.

Rosenthal, O. and Cooper, D.Y. (1967) Methods of determining the photochemical action spectrum. In Methods of Enzymology, Vol. X, R.W. Estabrook and M.E. Pullman, eds., Colowick and Kaplan, series eds., pp. 616-628, Academic press, New York.

Rushmore, T., Snyder, R., and Kalf, G.F. (1984). Covalent binding of benzene and its metabolites to DNA in rabbit bone marrow mitochondria in vitro. Chem.-Biol. Interactions 49, 133-154.

Saito, F., Kocsis, J.J., and Snyder, R. (1973). Effect of benzene on hepatic metabolism and ultrastructure. Toxicol. Appl. Pharmacol. 26, 209-217.

Sawahata, T. and Neal, R.A. (1983). Biotransformation of phenol to hydroquinone and catechol by rat liver microsomes. Mol. Pharmacol.23, 453-460.

Schenkman, J.B. (1970). Studies on the nature of the Type I and Type II spectral changes in liver microsomes. Biochemistry 9, 2081-2091.

Schenkman, J.B., Remmer, H., and Estabrook, R.W. (1967). Spectral studies of drug interaction with hepatic microsomal cytochrome. Mol. Pharmacol. 3,113-123.

Smart, R.C. and Zannoni, V.G. (1985). Effect of ascorbate on covalent binding of benzene and phenol metabolites to isolated tissue preparation. Toxicol. Appl. Pharmacol. 77, 334-343.

Snyder, R., Longacre, S.L., Witmer, C.M., Kocsis, J.J., Andrews, L.S., and Lee, E.W. (1981). Biochemical toxicology of benzene. In: Reviews in Biochemical Toxicology, Elsevier/North Holland Publishing Co., New York, New York, 3:123-53.

Snyder, R. (1984). The benzene problem in historical perspective. Fund. Appl. Toxicol. 4, 692-699.

Tunek, A., Platt, K.L., Przyblyski, M. and Oesch, F. (1980). Multi-step metabolic activation of benzene. Effect of superoxide dismutase on covalent binding to microsomal macromolecules, and indentification of glutathione conjugates using high pressure liquid chromatography and field desorption mass spectrometry. Chem.-Biol. Interactions 331, 1-7.

FURTHER EVIDENCE FOR A ROLE OF ISOZYMES OF P450 IN THE METABOLISM

AND TOXICITY OF CARBON DISULFIDE (CS_2)

R.J. Rubin and R. Kroll

The Johns Hopkins University
School of Hygiene and Public Health
Baltimore, Maryland 21205

INTRODUCTION

CS_2 has been shown to undergo sequential desulfuration in the liver to form CO_2. Previous work by Neal and his colleagues[1,2] has revealed that the initial desulfuration is catalyzed by a cytochrome P450-dependent enzyme system which results in the formation of an electrophilic sulfur atom and carbonyl sulfide (COS). They have further shown[3] that the second desulfuration step is catalyzed primarily by a soluble, acetazoleamide-inhibitable carbonic anhydrase which results in the formation of CO_2 and hydrogen sulfide (H_2S) (Figure 1).

It has been suggested[2] that the covalent binding of the electrophilic sulfur released by the microsomal P450-mediated metabolism of CS_2 to hepatic macromolecules is the cause of CS_2-induced hepatotoxicity.

We have recently reported[4] that the kinetics of metabolism of CS_2 to COS by rat liver microsomes is consistent with the existence of at least 2 forms of P450, one a high affinity/low capacity (HA/LC) enzyme and the second a low affinity/high capacity (LA/HC) enzyme. We have further shown[4] that it is the HA/LC form of the enzyme that is more sensitive to either induction by the in vivo administration of alcohols such as ethanol and isopropanol (18 hour pretreatment) or to direct inhibition by these alcohols. With this initial evidence for a role of isozymes of P450 in the metabolism of CS_2, we set out in the present study to investigate the effects of isopropanol on the hepatotoxicity of CS_2 under conditions in which the HA/LC form of the enzyme is either induced or inhibited. The objective was to correlate in vivo toxic responses to in vitro changes in enzyme activities.

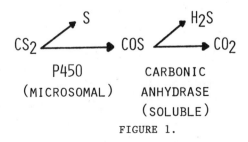

P450
(MICROSOMAL)

CARBONIC
ANHYDRASE
(SOLUBLE)

FIGURE 1.

TABLE 1. Effect of 18-hr Isopropanol Pretreatment on the Hepatic Aniline Hydroxylase Response to In Vivo CS_2

Change in Aniline Hydroxylase Activity[a]
ΔUnits of Activity

CS_2 (mg/kg i.p.)	N	Control Pretreatment	N	Isopropanol Pretreatment (2.8 gms/kg p.o.) 18 hrs
		3 hrs Post CS_2		
0	18	--- (17±1)	18	--- (35±1)
1	7	+2±2	3	-20±3*§
5	7	-3±2	4	-22±2*§
10	7	-6±2*	4	-26±2*§
50	5	-11±2*	-	---
100	6	-9±2*	6	-30±2*§
		24 hrs Post CS_2		
0	18	--- (17±1)	18	--- (35±1)
100	6	+3±2	6	-17±2*§
500	3	-8±3*	3	-20±3*§
625	6	-9±2*	6	-27±2*§
1250	2	-13±3*	1	-31

a) Values in parentheses at 0 mg/kg CS_2 represent the actual levels of aniline hydroxylase activity. All other values represent the mean change in enzyme activity (± S$\bar{x}$) from this level. A unit of activity is defined as one nmole of p-aminophenol formed per min per mg protein x 10^2 for a 10,000 g supernatant fraction.

*) p<.05 for t test relative to respective 0 mg/kg CS_2.

§) p<.05 for t test comparing isopropanol pretreatment relative to control pretreatment for a given dose of CS_2.

RESULTS

We have previously shown in rats that an 18-hour pretreatment with an oral dose of isopropanol (2.8 gms/kg) results in an approximately 2-fold increase in the Vmax for the HA/LC component of the CS_2 microsomal metabolism[4]. In Table 1 is shown the effect of the same dosing regimen for isopropanol on a hepatotoxic response to CS_2. In this case the hepatotoxic response being assessed is the ability of i.v. administered CS_2 to decrease a P450-mediated function of the liver i.e. aniline hydroxylase activity. It is a measure of the ability of CS_2 to inactivate cytochrome P450[2]. The dose-response to CS_2 at both 3 hrs and 24 hrs is shown in the Table. It should be noted that under either control or isopropanol pretreatment conditions, the 3-hour response is a more sensitive indicator of CS_2-induced damage than is the 24 hour response. For example, under conditions of control pretreatment a 100 mg/kg dose of CS_2 results in no significant alteration in aniline hydroxylase activity at 24 hours while at 3 hours there is significant inhibition at doses as low as 10 mg/kg. Likewise, under conditions of isopropanol pretreatment, a 100 mg/kg dose of CS_2 results in a greater decrement in aniline hydroxylase activity at 3 hours than at 24 hours. These temporal relationships appear to be related to the

TABLE 2.   Effect of 18-hr Isopropanol Pretreatment on Plasma
Transaminase Response to In Vivo CS_2

Plasma Glutamic-Pyruvic Transaminase Activity[a]

CS_2 mg/kg i.p.	SF Units/ml			
	N	Control Pretreatment	N	Isopropanol Pretreatment
0	18	24±1	18	30±2
100	6	27±4	6	29±4
500	3	22±2	3	37±3*§
625	6	27±3	6	59±5*§

a) $\bar{x} \pm S\bar{x}$ 24 hours post CS_2 administration.

*) p<.05 for t test relative to 0 mg/kg CS_2.

§) p<.05 for t test for isopropanol pretreatment relative to control
pretreatment for a given dose of CS_2.

ability of the inactivated P450 to be regenerated by 24 hours.

More importantly, the data indicate that for each tested dose of CS_2,
pretreatment with isopropanol resulted in a significantly greater decrease
in aniline hydroxylase activity than in the absence of isopropanol. For
example, at 100 mg/kg CS_2 at 24 hours there was no significant change in
aniline hydroxylase activity in the absence of isopropanol pretreatment,
while in the presence of the alcohol the same dose of CS_2 resulted in a
decrease of approximately 18×10^{-2} units of activity. Similarly, a dose of
1 mg/kg of CS_2 resulted in no significant change in enzyme activity at 3
hours in the absence of isopropanol, while in its presence the same dose of
CS_2 resulted in a decrease of approximately 22×10^{-2} units of activity.
These data clearly indicate the ability of isopropanol to enhance the inhibitory effect of CS_2 on a P450-mediated function of the liver.

In Table 2 are shown the results of the effect of isopropanol pretreatment on another hepatotoxic parameter of CS_2, i.e. the induced elevation of
plasma transaminase activity. It can be seen that at doses of CS_2 up to
625 mg/kg there was no statistically significant alteration in plasma GPT
activity in the absence of isopropanol. In the presence of isopropanol
statistically significant increases in GPT activity were seen only at doses
of 500 mg/kg and greater. However, at 100 mg/kg CS_2, a dose at which there
is already a marked isopropanol-induced enhancement of the inhibitory effect
of CS_2 on aniline hydroxylase activity (Table 1), there was no apparent increase in hepatoxicity as assessed by changes in plasma GPT activity.

In the previous experiments isopropanol was given 18 hours prior to
the administration of CS_2. At that time point the great majority of the
isopropanol has been cleared from the body[5] and any effects observed for the
interaction of the alcohol and CS_2 must be related to induced changes in the
metabolism of CS_2. In the next set of experiments, isopropanol was administered for periods of time ranging from simultaneous administration to 2
hours pretreatment. Under these conditions the enzyme(s) involved in the
metabolism of CS_2 to COS have not yet been induced and presumably both
CS_2 and isopropanol are present in the liver simultaneously. This experimental design was selected to evaluate the ability of the presence of isopropanol to prevent hepatotoxicity by inhibiting CS_2 activation.

TABLE 3. Effect of the Presence of Isopropanol on the Hepatic Aniline
Hydroxylase Response to In Vivo CS_2

	Aniline Hydroxylase Activity[a]			
Pretreatment	Control Pretreatment		Isopropanol Pretreatment	
Time (mins.)	$-CS_2$	$+CS_2$[b]	$-CS_2$	$+CS_2$
0	18±1.0[c]	5±0.4*	17±0.8	5±0.5*
30	18±0.4	5±0.9*	19±0.6	5±0.2*
60	16±0.6	6±0.5*	17±0.3	5±0.4*
120	18±0.8	5±0.3*	17±1.3	3±0.3*

a) nmoles p-aminophenol formed per min per mg protein x 10^2 for a 10,000 g
   supernatant fraction.

b) CS_2 dose = 1250 mg/kg i.p. Animals were sacrificed 3 hrs post treatment.

c) $\bar{x} \pm S\bar{x}$; N=6 for all groups.

*) p<.05 for t test relative to respective control ($-CS_2$) at each time
   point.

In Table 3 it can be seen that CS_2 at a dose of 1250 mg/kg i.p. significantly decreased aniline hydroxylase activity in the absence of isopropanol (control pretreatment). Isopropanol pretreatment, for periods
ranging from simultaneous administration to 120 min. pretreatment, did not
significantly alter this CS_2-induced damage to the P450-dependent enzyme
system.

Table 4 shows the effect of an identical experimental protocol on the
CS_2-induced elevation of plasma GPT activity. CS_2 (1250 mg/kg i.p. - 3
hrs) produced approximately a five-fold increase in plasma GPT activity in
the absence of isopropanol. In the presence of isopropanol, the CS_2-induced
elevation was not significantly affected when the alcohol was administered
either simultaneously or 30 minutes prior to the CS_2. However, when the
pretreatment time was lengthened to 60 and 120 minutes, a marked protection
against the effect of CS_2 on plasma GPT (approximately 75%) was seen. Presumably, this 60-120 minute delay was required for absorption and significant blood and liver levels of isopropanol to occur following oral administration of the alcohol.

DISCUSSION AND CONCLUSION

The release of a reactive sulfur atom following the microsomal metabolism of CS_2 to COS has been suggested to be the cause of the hepatotoxicity
associated with CS_2 exposure[2]. The results in this paper support this
suggestion in that in vivo pretreatment with isopropanol 18 hours prior to
CS_2, a protocol which we previously had shown significantly elevated the
in vitro rate of hepatic metabolism of CS_2 to COS, also enhances the destruction of cytochrome P450 induced by CS_2 in vivo. Interestingly enough,
this enhancement of the in vivo toxicity of CS_2 by isopropanol was not
accompanied by an associated increase in the release of GPT from liver into
plasma. Thus, the two parameters of hepatotoxicity were effectively
dissociated from each other and one could see an enhanced destruction of
P450 in the absence of an alteration in membrane leakage of the hepatocyte.

240

TABLE 4. Effect of the Presence of Isopropanol on the Plasma Transaminase Response to In Vivo CS_2

Pretreatment Time (mins.)	Plasma Glutamic-Pyruvic TRansaminase Activity (SF Units/ml)			
	Control Preatment		Isopropanol Pretreatment	
	$-CS_2$	$+CS_2^a$	$-CS_2$	$+CS_2^a$
0	19 ± 1^b	97±13*	16±2	83±9*
30	23±3	98±19*	21±1	82±17*
60	26±2	104±12*	20±2	43±3*§
120	22±2	93±17*	22±3	42±2*§

a) CS_2 dose = 1250 mg/kg i.p. Animals were sacrificed 3 hrs post treatment.

b) $\bar{x} \pm S\bar{x}$; N=6 for all groups.

*) p<.05 for t test relative to respective control ($-CS_2$) at each time point.

§) p<.05 for t test comparing the $+CS_2$ response in control vs. isopropanol pretreatment groups.

The results with the "near simultaneous" administration of isopropanol also supports the contention that it is the metabolism of CS_2 which generates a hepatotoxic species. We had previously shown[4] that the in vitro addition of isopropanol to a microsomal preparation inhibits the rate of metabolism of CS_2 to COS and presumably a reactive sulfur atom. In the current experiments the administration of isopropanol, under conditions which assure significant levels of the alcohol to be present in the liver at the same time that hepatic exposure to CS_2 occurs, significantly decreased the hepatotoxicity of CS_2 as assessed by plasma GPT levels. Once again, a dissociation of hepatotoxic parameters was noted; no effect on the CS_2-induced destruction of P450 was noted under these conditions.

Thus, in addition to our earlier evidence for the existence of at least two different isozymes of P450 that metabolize CS_2[4], we have interpreted the present data to indicate the P450-mediated formation of two different reactive intermediates of CS_2 in vivo. Each intermediate would appear to be responsible for mediating a different toxic response, with the formation of each being differentially affected by the exposure protocol. It remains possible that each intermediate is formed in vivo by a different isozyme of P450, or conversely that the two reactive intermediates are sequentially formed from an initial intermediate metabolite with the relative balance of the 2 reactive metabolites dependent upon the rate of formation of the initial intermediate.

In addition to these mechanistic interpretations, it should be pointed out that the marked interaction seen with 18-hour isopropanol pretreatment and CS_2 (Table 1) results in the lowest dose (1 mg/kg) that we have found reported in the literature to cause an alteration in hepatic function. This strong interactive effect might be the explanation for the increasing incidence of hepatic malfunction recently reported[6] to occur in humans exposed in their occupational setting to both CS_2 and a variety of other organic solvents, including alcohols.

REFERENCES

1. R.R. Dalvi, R.E. Poore and R.A. Neal, Studies of the Metabolism of Carbon Disulfide by Rat Liver Microsomes, *Life Sciences*, 14:1785 (1974).
2. R.R. Dalvi, A.L. Hunter and R.A. Neal, Toxicological Implications of the Mixed Function Oxidase-Catalyzed Metabolism of CS_2, *Chem.-Biol. Interactions*, 10:347 (1975).
3. C.P. Chengelis and R.A. Neal, Hepatic Carbonyl Sulfide Metabolism, *Biochem. Biophys. Res. Comm.*, 90:993 (1979).
4. R.J. Rubin, B. Taffe and P. Egner, Interaction of Alcohols with Carbon Disulfide Metabolism, *Toxicologist*, Abstract 602:151 (1984).
5. G.L. Plaa, W.R. Hewitt, P. duSouich, G. Caille and S. Lock, Isopropanol and Acetone Potentiation of Carbon Tetrachloride-Induced Hepatotoxicity: Single versus Repetitive Pretreatment in Rats, *J. Tox. Env. Health*, 235 (1982).
6. M. Dossing and L. Ranek, Isolated Liver Damage in Chemical Workers, *Brit. J. Ind. Med.*, 41:142 (1984).

EFFECT OF A SUICIDE SUBSTRATE ON THE METABOLISM OF STEROIDS AND XENOBIOTICS

AND ON CYTOCHROME P-450 APOPROTEINS

T.R. Tephly*, K.A. Black*, M.D. Green*, B.L. Coffman*, G.A. Dannan
and F.P. Guengerich

*The Toxicology Center, University of Iowa, Iowa City, IA 52242
and Department of Biochemistry and Center in Molecular Toxicology
Vanderbilt University, Nashville, TN 37232

INTRODUCTION

Xenobiotic administration can significantly alter forms of hepatic microsomal cytochrome P-450. Individual forms of cytochrome P-450 are known to be increased by agents, such as phenobarbital, 3-methyl-cholanthrene, β-naphthoflavone and pregnenolone-16α-carbonitrile[1-3]. Certain chemicals decrease the levels of these isoenzymes by modulating the heme biosynthetic pathway and, thereby, affect the availability of heme and the amount of hepatic cytochrome P-450, e.g. cobaltous chloride[4,5]. Some agents can act as suicide substrates (i.e., mechanism-based inhibitors) for one or more cytochrome P-450 forms and, thereby, promote catabolic destruction of these cytochrome P-450 species[6,7]. Derivatives of 3,5-diethoxycarbonyl-1,4-dihydrocollidine (DDC) interact rapidly with cytochrome P-450, promote the rapid destruction of hepatic microsomal cytochrome P-450 and lead to the generation in vivo of N-alkyl protoporphyrins which, in turn, inhibit the activity of mitochondrial ferrochelatase, the enzyme catalyzing the last step in heme biosynthesis[8-11]. The result is a decreased total hepatic microsomal cytochrome P-450 level and a marked hepatic porphyria[12,13].

The current study was designed to investigate the consequences of administration of the 4-ethyl derivative of DDC, 3,5-diethoxycarbonyl-2,6-dimethyl-4-ethyl-1,4-dihydropyridine (DDEP), an active suicide substrate for cytochrome P-450 in rats. The metabolism of several drugs in vivo, xenobiotic and steroid metabolism in vitro, and quantitative, specific antibody-directed analysis of individual forms of rat hepatic microsomal cytochrome P-450 species were performed.

MATERIALS AND METHODS

Radiochemicals

[Dimethylamine-^{14}C]aminopyrine (114 mCi/mmol), [^{14}C]formaldehyde (14 mCi/mmol) and [N-methyl-^{14}C]morphine hydrochloride (58 mCi/mmol) were obtained from Amersham Corporation (Arlington Heights, IL). Sodium [^{14}C]formate (40-60 mCi/mmol) and [4-^{14}C]testosterone (52 mCi/mmol) were obtained from New England Nuclear (Boston, MA).

Other Chemicals and Reagents

Diethoxycarbonyl-2,6-dimethyl-4-ethyl-1,4-dihydropyridine (DDEP) was synthesized from propionaldehyde and ethylacetoacetate using a method described by DeMatteis and Prior[14]. The purity was verified by NMR spectroscopy.

Animals

Male, Sprague-Dawley rats (200 g) were obtained from Bio-Labs (St. Paul, MN) and housed in stainless steel, wire-bottomed cages. The rats were maintained on a 12-hr light:dark cycle, and environmental temperature and humidity were rigidly controlled. Purina Rodent Chow #5001 and water were provided ad libitum.

Animal Treatments

Rats received 3-methylcholanthrene dissolved in peanut oil (40 mg/kg, intraperitoneally) daily for 3 days. Control animals received oil alone. Phenobarbital-treated rats received sodium phenobarbital dissolved in saline (80 mg/kg, intraperitoneally) daily for 4 days. DDEP was suspended in peanut oil and injected intraperitoneally (100 mg/kg).

Liver Preparations

Hepatic microsomes were isolated from a 25% wet weight/volume liver homogenate prepared in 1.15% KCl. The homogenate was centrifuged at 9,000 x g for 20 min, and the resulting supernatant was centrifuged at 135,000 x g for 30 min to sediment the microsomal pellet. The pellet was homogenized in 10 mM Tris acetate buffer (pH 7.4), containing 20% glycerol and 1 mM sodium EDTA.

Enzyme Assays

Cytochrome P-450 content was determined by the method of Omura and Sato[15]. Aniline hydroxylation was measured by the method of Imai et al.[16] and 7-ethoxyresorufin deethylation was determined by the method of Prough et al.[17]. NADPH-cytochrome c reductase activity and p-nitrophenol glucuronidation were determined as described by the methods of Baron and Tephly[18] and Bock et al.[19], respectively. Testosterone hydroxylation was determined by procedures described by Waxman et al.[20]. In this method, the major metabolic products separated and quantified were 2α-, 7α-, 16α- and 16β-hydroxytestosterone. Protein was determined by the method of Bradford[21].

In Vivo Studies

Rats received [dimethylamine-^{14}C]aminopyrine at a dose of 40 μmoles per kg (5 μCi/kg) or [^{14}C-methyl]morphine (morphine sulfate equivalent to 17.5 μmoles of morphine base/kg, 10 μCi/kg) by intraperitoneal administration. The vehicle was saline. DDEP was administered (100 mg/kg) 2 hr prior to the administration of isotope. Immediately following aminopyrine or morphine injection, rats were placed in glass metabolic chambers, one rat per chamber; and $^{14}CO_2$ excreted in the breath was collected and quantified as described previously[22]. In certain studies, aminopyrine metabolism was followed as described by Black et al.[23]. ^{14}C-Formaldehyde and ^{14}C-formate oxidation in vivo were studied as reported previously[24]. Hepatic folates were determined by the method of McMartin et al.[25].

Quantitation of Cytochrome P-450 Apoproteins

Individual forms of cytochrome P-450 were purified from rat hepatic microsomes as reported previously by Guengerich et al.[3]. Antibodies to individual purified cytochrome P-450s were raised in rabbits and IgG fraction subsequently prepared using standard methods[26]. Quantitative estimates of individual cytochrome P-450 isozyme levels in liver microsomal fractions were performed by "Western blotting" as described by Guengerich et al.[27]. Immunoquantitations were generally performed using microsomal samples prepared from individual animals and are expressed as nmol P-450 isozyme/mg microsomal protein (mean ± S.D. for 3 to 5 individuals).

RESULTS

Effects of DDEP Pretreatment on Drug Metabolism In Vivo

Table 1 shows that DDEP administration (100 mg/kg) 2 hours prior to the administration of $[^{14}C]$-aminopyrine resulted in a 95% inhibition of the rate of pulmonary $^{14}CO_2$ excretion. Following DDEP pretreatment, aminopyrine blood levels remained elevated for prolonged time periods in agreement with a profound inhibition of N-demethylation. Marked inhibition of the rate of N-demethylation was also found when [N-methyl-^{14}C]morphine metabolism to $^{14}CO_2$ was studied (Table 1).

The inhibition of aminopyrine metabolism was not due to an interference with one-carbon metabolism by DDEP. The rate of ^{14}C-formaldehyde oxidation to $^{14}CO_2$ in control rats (1.41 ± 0.16 % of dose/min, mean ± SE,) was not significantly different compared with DDEP-pretreated animals (1.18 ± 0.18% of dose/min). Likewise, ^{14}C-formate oxidation to $^{14}CO_2$ was not affected by DDEP pretreatment (control 0.268 ± 0.05 vs. DDEP-pretreated 0.231 ± 0.015 % dose/min). Total hepatic folate, tetrahydrofolate, 5-methyltetrahydrofolate and 10-formyltetrahydrofolate levels were unaffected by DDEP pretreatment.

Table 1. Effect of DDEP on Metabolism of $[^{14}C$-Methyl]morphine and [Dimethyl-^{14}C]aminopyrine to $^{14}CO_2$ In Vivo

	Peak Rate of $^{14}CO_2$ Exhalation (% of dose/min)		Total $^{14}CO_2$ Exhaled (% of dose)	
	Control	DDEP	Control	DDEP
$[^{14}C$-methyl] morphine	0.22 ± 0.04	0.004 ± 0.002*	a)15 ± 2.7	0.5 ± 0.1*
[dimethyl-^{14}C] aminopyrine	0.91 ± 0.15	0.017 ± 0.002*	b)45 ± 3.5	2.0 ± 0.3*

*Value is significantly different from that for oil-treated rats (P < 0.01).
a) $^{14}CO_2$ collected for 4 hrs. b) $^{14}CO_2$ collected for 2 hrs.
DDEP (100 mg/kg) or peanut oil (5 ml/kg) was administered intraperitoneally 2 hr prior to the administration of $[^{14}C$-methyl]-morphine or [dimethyl-^{14}C] aminopyrine and initiation of $^{14}CO_2$ collection. Each value is the mean ± S.E. obtained from 4 rats.

Effect of DDEP on the In Vitro Metabolism of Xenobiotics and Steroids

Previous studies in this laboratory and others[12,13] have shown that, after DDEP administration, there was a marked accumulation of N-ethylproto-porphyrin IX and a significant decrease in total microsomal cytochrome P-450. Ethoxyresorufin deethylation was studied because it is largely dependent on the 3-methylcholanthrene or β-naphtoflavone-inducible $P-450_{\beta NF-B}$ isozyme[3]. 3-Methylcholanthrene pretreatment greatly increased the in vitro deethylation of ethoxyresorufin (2.62 ± 0.29 nmol/min/mg) compared to untreated (0.042 ± 0.001 nmol/min/mg) rat liver microsomes, a 62-fold increase. Pretreatment of animals with phenobarbital had only a slight effect (2-fold increase) on the deethylation of ethoxyresorufin (0.106 ± 0.003 nmol/min/mg). After DDEP treatments there was a rapid decrease in ethoxyresorufin deethylation in all groups which reached a nadir within 2-4 hr and remained low for 24 hr (Figure 1).

Testosterone hydroxylations at the 16α, 2α, 7α and 16β positions may be used as indicators of the isozymes $P-450_{UT-A}$, $P-450_{UT-F}$ and $P-450_{PB-B}$[3,20]. $P-450_{UT-A}$, a major isozyme in microsomes from untreated rats, catalyzes the 16α- and 2α-hydroxylation of testosterone. Phenobarbital induces $P-450_{PB-B}$, the major isozyme in microsomes from PB-induced rats. $P-450_{PB-B}$ catalyzes both 16α-and 16β-hydroxylation of testosterone. The 7α-hydroxylation is catalyzed by $P-450_{UT-F}$. $P-450_{PB-C}$, $P-450_{PCN-E}$, and $P-450_{UT-H}$ have little or no hydrolase activity towards testosterone[20,28].

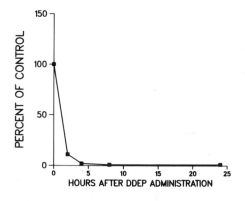

Fig. 1   Inhibition of 7-ethoxyresorufin deethylation by DDEP in microsomes from 3-methylcholanthrene-treated rats. Hepatic 7-ethoxyresorufin deethylation activity was determined at various times after DDEP (100 mg/kg) administration. Values represent means of 4-6 animals.

Figure 2 shows that 16α-, 16β- and 2α-hydroxylations are rapidly decreased in microsomes from untreated and 3-methylcholanthrene-treated rats after treatment with DDEP, while these activities remain relatively constant in microsomes from phenobarbital-treated rats. The 7α-hydroxylation activity remains high for at least 8 hours after DDEP in microsomes from all groups of rats before declining. These data suggest that there is a differential effect of DDEP on individual P-450 isozymes.

Effects of DDEP on Protein Moieties of Individual Cytochrome P-450 Forms Determined by Immunochemical Quantification

DDEP produces a decrease in the immunoreactive protein moieties of cytochrome P-450 forms, P-450$_{UT-A}$, P-450$_{PCN-E}$ and P-450$_{PB-C}$ in hepatic microsomes from all treatment groups. P-450$_{\beta NF-B}$ in microsomes from

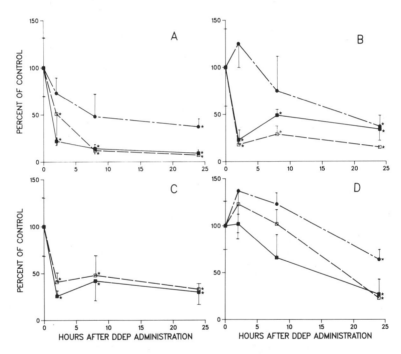

Fig. 2. Effect of the administration of DDEP on the production of various testosterone hydroxylation products in rat liver microsomes. Following administration of DDEP (100 mg/kg) the hydroxylation activities towards testosterone at the 16α (A), 16β (B), 2α (C) and 7α (D) positions were determined. Control rates for uninduced (■———■) rat liver microsomes were; 2α - 0.26 ± 0.08 nmol/min/mg protein, 16α - 0.33 ± 0.13, 16β - 0.09 ± 0.02 and 7α - 0.21 ± 0.02. Control rates for 3-methylcholanthrene (□----□) treated rats were 2α - 0.25 ± 0.09, 16α - 0.19 ± 0.06, 16β - 0.27 ± 0.16 and 7α - 0.43 ± 0.12. Control rates for phenobarbital (●—·—●) treated rats were 16α - 0.35 ± 0.13, 16β - 0.24 ± 0.10 and 7α - 0.16 ± 0.05. Values represent means ± SD obtained with three animals. All values marked with an asterisk are significantly lower than the corresponding value measured at time zero at $p < 0.05$.

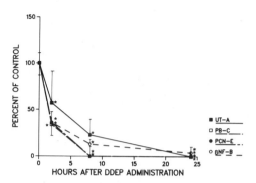

Fig. 3   Hepatic microsomal cytochrome P-450 isozymes affected strongly by
DDEP treatment. Cytochrome P-450 isozymes were quantitated in rat
liver microsomes at various times after DDEP (100 mg/kg) administration to PB- 3-methylcholanthrene-induced rats. Values
represent means ± SD of measurements made with three animals. All
values marked with an asterisk are significantly lower than the
corresponding values measured at time zero at $p < 0.05$.

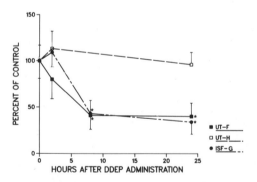

Fig. 4   Hepatic microsomal cytochrome P-450 isozymes less affected by DDEP
treatment. Cytochrome P-450 isozymes were quantitated in rat
liver microsomes at various times after DDEP administration (100
mg/kg) to 3-methylcholanthrene-induced rats. Values represent
means ± SD of measurements made with three animals. All values
marked with an asterisk are significantly lower than the corresponding values measured at time zero at $p < 0.05$.

248

3-methylcholanthrene-treated rats is also decreased (figure 3). In contrast, $P-450_{UT-H}$ appears to be unaffected by DDEP and the concentration of form $P-450_{UT-F}$ is decreased only at the later time periods (figure 4). The $P-450_{ISF-G}$ and $P-450_{PB-B}$ concentrations in microsomes from 3-methyl-cholanthrene- and phenobarbital- treated rats, respectively, are also lower only at the later time periods ($P-450_{PB-B}$ data not shown).

DISCUSSION

Treatment of rats with DDEP leads to a series of events which include reaction of the dihydropyridine with hepatic microsomal cytochrome P-450, a transfer of the 4-ethyl groups to the prosthetic heme group and a catabolic destruction of the hemoprotein[13]. This results in a rapid decrease in spectrally quantifiable total hepatic microsomal cytochrome P-450 and the formation of N-ethylprotoporphyrin IX. The latter metabolite inhibits mitochondrial ferrochelatase activity, the enzyme mediating the last step in heme biosynthesis, the result being an induced porphyria.

As a consequence of the destruction of the hemoprotein, a profound inhibition of enzymatic activities towards substrates whose metabolism is catalyzed by several cytochrome P-450 isozymes takes place both in vivo and in vitro, following DDEP treatment. The N-demethylation in vivo of amino-pyrine and morphine and the hydroxylation of aniline in vitro (data not shown) are rapidly decreased following DDEP treatment.

Ethoxyresorufin deethylation, which is linked to $P-450_{BNF-B}$, is also rapidly diminished in vitro in microsomes from DDEP-treated rats; however, the 7α-hydroxylation of testosterone is not altered significantly until 24 hours following DDEP treatment. Other hydroxylations of testosterone, such as 16α-, 16β- and 2α-hydroxylations are more markedly affected in untreated and 3-methylcholanthrene induced rats, compared to the 7α-hydroxylation. These activities are less altered in phenobarbital induced rats. Other hepatic microsomal activities, such as NADPH-cytochrome c reductase activity and p-nitrophenol glucuronidation are not changed (data not shown) over the time course of these experiments (24 hours after DDEP). Also, DDEP did not display general hepatoxicity as determined by examination of several transaminase levels at doses used in this experiment, and histological examination of the livers showed no abnormalities following DDEP doses of 100mg/kg.

The DDEP treatment leads to a decrease in immunoreactive protein moieties of a number of cytochrome P-450 isozymes. The apoprotein of $P-450_{\beta NF-B}$ is particularly affected and markedly decreased as likewise are the protein moieties of $P-450_{UTA}$, $P-450_{PCN-E}$ and $P-450_{PB-C}$. In contrast, $P-450_{UT-H}$ apoprotein is not affected at all. Interestingly there is an apparent correlation between the effect of DDEP on 7α-hydroxylation of testosterone which is mediated by $P-450_{UT-F}$ and the effect on the P-450 apoprotein. Also, in microsomes from phenobarbital treated rats the effect of DDEP on $P-450_{PB-B}$ appears to correlate with the effect on the 16α- and 16β-hyroxylations of testosterone.

The mechanism of destruction of specific cytochrome P-450 apoproteins cannot be derived from these experiments. The decrease in apoprotein follows the decreases in enzymatic activities and spectrally quantifiable cytochrome P-450, suggesting that the loss of the heme prosthetic group promotes catabolic destruction of the apoprotein or that the protein moieties are directly attacked by metabolites of DDEP. The generation of ethyl radicals after interaction of DDEP with hepatic microsomes has been demonstrated by Augusto et al.[29]. This type of agent may be involved in the destruction of the protein moieties of cytochrome P-450 isozymes.

DDEP does not affect certain forms of cytochrome P-450 to the degree it perturbs others. This might be because DDEP is not a substrate for those forms not altered, or that these isozymes are so situated in the endoplasmic membrane, that protection against DDEP is afforded by endogenous components in the membrane. Finally, one cannot rule out selective endogenous substrates protection as a potential protective mechanism for certain isozymes.

Supported by NIH Grants GM 12675, ES 01590 and ES 00267.

REFERENCES

1. F. P. Guengerich, Separation and purification of multiple forms of microsomal cytochrome P-450, J. Biol. Chem. 252:3970 (1977).
2. D. E. Ryan, P. E. Thomas, D. Korzeniowski and W. Levin, Separation and characterization of highly purified forms of liver microsomal cytochrome P-450 from rats treated with polychlorinated biphenyls, phenobarbital, and 3-methylcholanthrene, J. Biol. Chem. 254:1365 (1979).
3. F. P. Guengerich, G. A Dannan, S. T. Wright, M. V. Martin and L. S. Kaminsky, Purification and characterization of liver microsomal cytochromes P-450: electrophoretic, spectral, catalytic, and immunochemical properties and inducibility of eight isozymes isolated from rats treated with phenobarbital or β-naphthoflavone, Biochemistry 21:6019 (1982).
4. T. R. Tephly and P. Hibbeln, The effect of cobalt chloride administration on the synthesis of hepatic microsomal cytochrome P-450, Biochem. Biophys. Res. Commun. 42:589 (1971).
5. T. R. Tephly, C. R. Webb, P. Trussler, F. Kniffen, E. Hasegawa and W. Piper, The regulation of heme synthesis related to drug metabolism, Drug Metab. Dispos. 1:259 (1973).
6. P. R. Ortiz de Montellano and B. A. Mico, Destruction of cytochrome P-450 by allylisopropylacetamide is a suicidal process, Arch. Biochem. Biophys. 206:43 (1981).
7. H. H. Liem, E. F. Johnson and U. Müller-Eberhard, The effect in vivo and in vitro of allylisopropylacetamide on the content of hepatic microsomal cytochrome P-450 2 of phenobarbital treated rabbits, Biochem. Biophys. Res. Commun. 111:926 (1983).
8. F. DeMatteis, A. H. Gibbs and T. R. Tephly, Inhibition of protohaem ferrolyase in experimental porphyria. Isolation and partial characterization of a modified porphyrin inhibitor, Biochem. J. 188:145 (1980).
9. T. R. Tephly, A. H. Gibbs, G. Ingall and F. DeMatteis, Studies on the mechanism of experimental porphyria and ferrochelatase inhibition produced by 3,5-diethoxycarbonyl-1,4-dihydrocollidine, J. Int. Biochem. 12:993 (1980).
10. P. R. Ortiz de Montellano, H. S. Beilan and K. L. Kunze, N-methyl-protoporphyrin IX: chemical synthesis and indentification as the green pigment produced by 3,5-diethoxycarbonyl-1,4-dihydrocollidine treatment, Proc. Natl. Acad. Sci. USA 78:1490 (1981).
11. T. R. Tephly, B. L. Coffman, G. Ingall, A. Zeit-Har, H. M. Goff, H. D. Tabba and K. M. Smith, Identification of N-methylprotoporphyrin IX in livers of untreated mice and mice treated with 3,5-diethoxy-carbonyl-1,4-dihydrocollidine (DDC): Source of the methyl group, Arch. Biochem. Biophys. 212:120 (1981).
12. F. DeMatteis, A. H. Gibbs, P. B. Farmer and J. H. Lamb, Liver production of N-alkylated porphyrins caused in mice by treatment with substituted dihydropyridines: evidence that the alkyl group on the pyrrole nitrogen originates from the drug, FEBS Letters, 129:328 (1981).

13. B. L. Coffman, G. Ingall and T. R. Tephly, The formation of N-alkylprotoporphyrin IX and destruction cytochrome P-450 in the liver of rats after treatment with 3,5-diethoxycarbonyl-1,4-dihydrocollidine and its 4-ethyl analogue, Arch. Biochem. Biophys. 218:220 (1982).

14. F. DeMatteis and B. E. Prior, Experimental hepatic porphyria caused by feeding 3,5-diethoxycarbonyl-1,4-dihydro-2,46,-trimethylpyridine, Biochem. J. 83:1(1962).

15. T. Omura and R. Sato, The carbon monoxide-binding pigment of liver microsomes II solubilization, purification, and properties, J. Biol. Chem. 239:2370 (1964).

16. Y. Imai, A. Ito and R. Sato, Evidence for biochemically different types of vesicles in the hepatic microsomal fraction, J. Biochem. 60:417 (1966).

17. R. A. Prough, M. D. Burke and R. T. Mayer, Direct fluorometric methods for measuring mixed function oxidase activity, Methods Enzymol. 52:399 (1978).

18. J. Baron and T. R. Tephly, Effect of 3-amino-1,2,4-triazole on the induction of rat hepatic microsomal oxidases, cytochrome P-450 and NADPH-cytochrome c reductase by phenobarbital, Mol. Pharmacol. 5:10 (1969).

19. K. W. Bock, B. Burchell, G. J. Dutton, O. Hanninen, G. J. Mulder, I. S. Owens, G. Siest and T. R. Tephly, UDP-glucuronyltransferase activities. Guidelines for consistent interim terminology and assay conditions. Biochem. Pharmacol. 32:953 (1983).

20. D. J. Waxman, A. Ko and C. Walsh, Regioselectivity and stereoselectivity of androgen hydroxylations catalysed by cytochrome P-450 isozymes purified from phenobarbital-induced rat Liver, J. Biol. Chem. 258:11937 (1983).

21. M. M. Bradford, A rapid and sensitive method for quantitation of microgram quantities of protein utilizing the principle of protein dye binding, Analyt. Biochem. 72:248 (1976).

22. K. E. McMartin, G. Martin-Amat, A. B. Makar and T. R. Tephly, Methanol poisoning. V. Role of formate metabolism in the monkey, J. Pharmacol. Exp. Ther. 201:564 (1977).

23. K. A. Black, V. Virayotha and T. R. Tephly, Reduction of hepatic tetrahydrofolate and inhibition of exhalation of $^{14}CO_2$ formed from [dimethylamino-^{14}C]-aminopyrine in nitrous oxide-treated rats, Hepatology 4:871 (1984).

24. J. T. Eells, A. B. Makar, P. E. Noker and T. R. Tephly, Methanol poisoning and formate oxidation in nitrous oxide treated rats, J. Pharmacol. Exp. Ther. 217:57 (1981).

25. K. E. McMartin, V. Virayotha and T. R. Tephly, High-pressure liquid chromatography separation and determination of rat liver folates, Arch. Biochem. Biophys. 209:127 (1981).

26. L. S. Kaminsky, M. J. Fasco and F. P. Guengerich, Production and application of antibodies to rat liver cytochrome P-450, Methods Enzymol. 74: 262 (1981).

27. F. P. Guengerich, P. Wang and N. K. Davidson, Estimation of isozymes of microsomal cytochrome P-450 in rats, rabbits and humans using immunochemical staining coupled with sodium dodecyl sulfate-polyacrylamide gel electrophoresis, Biochemistry 21:1698 (1982).

28. D. Larrey, L. M. Distlerath, G. A. Dannan, G. R. Wilkinson and F. P. Guengerich, F.P. Purification and characterization of the rat liver microsomal cytochrome P-450 involved in the 4-hydroxytion of debrisoquine, a prototype for genetic variation in oxidative drug metabolism, Biochemistry 23:2787 (1984).

29. O. Augusto, H. S. Beilan and P. R. Ortiz de Montellano, The catalytic mechanism of cytochrome P-450. Spin-trapping evidence for one-electron substrate oxidation, J. Biol. Chem. 257:11288 (1982).

ACTIVATION OF CYTOCHROME P-450 HEME IN VIVO[1]

Helen W. Davies, Paul E. Thomas[*] and Lance R. Pohl

Laboratory of Chemical Pharmacology, National Heart,
Lung, and Blood Institute, National Institutes of Health
Bethesda, Maryland 20892. [*]Laboratory of Experimental
Carcinogenesis and Metabolism, Roche Institute of Molecular
Biology, Nutley, New Jersey 07110

INTRODUCTION

It is well known that the administration of CCl_4 produces substantial decreases in the activities of cytochromes P-450 in the liver as well as in other organs.[1-5] The inactivation is most likely due to the observed destruction of cytochrome P-450 heme,[1] but neither the mechanism nor the products of that destruction have been elucidated. It has been found in the present study that the pathway of the CCl_4 inactivation of rat liver cytochromes P-450 in vivo includes the irreversible binding of heme-derived products to microsomal proteins and that a large fraction of the irreversibly bound heme-derived products are attached specifically to cytochromes P-450. These findings suggest that CCl_4 destroys cytochromes P-450, at least in part, by producing activated heme moieties which subsequently bind irreversibly to cytochrome P-450 protein. In addition, the levels of two cytochrome P-450 isozymes in microsomes are decreased following CCl_4 treatment. It may be that the irreversible binding of heme fragments functions as a signal for proteolytic degradation of cytochrome P-450 proteins.

MATERIALS AND METHODS

Chemicals

Chemicals were obtained from the following sources: $[^{14}C]NaHCO_3$ (54.0 mCi/mmole) and $[3,5-^3H]ALA$ (1.8 Ci/mmole) from New England Nuclear (Boston, MA); CCl_4 from Fisher Scientific Co. (Fairlawn, NJ); DEAE Affi-Gel Blue and SDS-PAGE molecular weight standards (carbonic anhydrase, 31,000 daltons; ovalbumin, 45,000 daltons; bovine serum albumin, 66,200 daltons; phosphorylase B, 92,500 daltons; beta-galactosidase, 116,250 daltons; and myosin, 200,000 daltons) from Bio-Rad (Richmond, CA); normal rabbit IgG and agarose from Miles Laboratories (Naperville, IL); GelBond Film from FMC Corporation (Rockland, ME); Maxifluor scintillation counting solution from J.T. Baker Chemical Co. (Philadelphia, PA).

---

[1]A preliminary communication describing a portion of this work has been recently published in Biochem. Pharmacol. 34:3203 (1985).

Preparation of Rat Liver Microsomes containing [3H] Labeled Heme and [14C] Labeled Proteins

Male Sprague Dawley rats (70-100 g, Taconic Farms, Germantown, NY) were pretreated with phenobarbital (80 mg/kg in saline, ip) for 4 days and then administered [14C]NaHCO3 (11.4 mCi/kg, 210 umole/kg) and [3,5-3H]ALA (8.40 mCi/kg, 4.67 umole/kg) to radiolabel microsomal protein and heme, respectively, following essentially the procedure of Parkinson et al.[6] Two hr and 4 hr after the administration of [14C]NaHCO3 and [3,5-3H]ALA respectively, two rats were injected with either CCl4 (26 mmole/kg, ip, as a 50% solution in sesame oil) or sesame oil (control rats). Animals were killed by decapitation one hr later. Liver microsomes were prepared as previously described.[7] They were resuspended in 10mM Tris-acetate (pH 7.4) containing 20% (v/v) glycerol and 1 mM EDTA (to give a protein concentration of 20-25 mg/ml) and stored at -80°C until used. The activities (dpm/mg protein; average of two values) of the microsomes from the control and CCl4 treated rats were as follows: control group, [3H]label 280,000, [14C]label 35,000; CCl4 group, [3H]label 213,000, [14C]label 32,000.

Irreversible Binding of Heme Radiolabel to Microsomal Protein

Microsomes were precipitated in 5 volumes of acetone containing 0.5 M HCl. The pellets were washed with the acetone-HCl mixture until the counts in the supernatants were background. This usually required two washes. The resulting protein pellets were dried under vacuum, dissolved in 1.0 ml of 1.0 N NaOH, and aliquots were counted by scintillation spectrophotometry.

Radioelectrophoretograms of Microsomal Proteins

Microsomal proteins were separated by SDS-PAGE essentially by the method of Laemmli[8] in a separating gel (9% acrylamide) 1.5 mm thick and 12 cm long. The gels were cut into 2 mm sections, solubilized in 0.5 ml of a 19:1 mixture (v/v) of 30% H_2O_2 and concentrated NH_4OH, and after the addition of 30 ul of glacial acetic acid, were counted.

Purification of Two Forms of Cytochrome P-450 from Phenobarbital Treated Rats

Two forms of cytochrome P-450, of molecular weights 52,000 and 54,000 daltons (P-450-52 kD and P-450-54 kD), were purified from phenobarbital treated rats by modifications of the methods of Imai and Sato[9] and Guengerich et al..[10] The procedure has been described in detail elsewhere.[11] In brief, cytochromes P-450 were isolated from solubilized microsomes by chromatography on an octylamine Sepharose column. Separation of P-450-54 kD and P-450-52 kD was achieved on a hydroxylapatite column. P-450-54 kD was eluted from that column in buffer containing 90 mM potassium phosphate, whereas P-450-52 kD was eluted in 150 mM potassium phosphate. Each was further purified to electrophoretic homogeneity as determined by SDS-PAGE on a column of DE-52 anion exchange resin and concentrated on a hydroxylapatite column. The final preparation of P-450-54 kD had a specific content of 17.0 nmole/mg protein and, based on its physical properties and the way it was purified, most likely corresponds to cytochromes PB-B of Guengerich et al.,[10,12] P-450b of Ryan et al.,[13] and PB-4 of Waxman and Walsh.[14] The 52 kD cytochrome P-450 had a specific content of 8.8 nmole/mg, an absolute oxidized spectrum with a Soret maximum at 417.7 nm, an absolute reduced spectrum with a Soret maximum at 413.5 nm, a carbon monoxide-reduced difference spectrum with a Soret maximum at 449.3 nm, and a metyrapone binding spectrum to ferrous cytochrome P-450 with a maximum at 444.8 nm. This enzyme does not appear to correspond to any form purified by other investigators.

Antisera to purified 52 kD and 54 kD cytochromes P-450 were produced in female New Zealand White rabbits essentially by the method of Kamataki et al..[15] The IgG fraction was purified from the serum by DEAE Affi-Gel Blue chromatography as described by the Bio-Rad Laboratories and concentrated by ultrafiltration to a protein concentration of 10 mg/ml. While antibody to 54 kD cytochrome P-450 is known to be reactive with at least one other form of cytochrone P-450 (P-450e),[16] no reaction was seen on Ouchterlony double-diffusion plates between anti-P-450 52 kD and P-450-54 kD or between anti-P-450-54 kD and P-450-52 kD. In addition, each antibody gave only a single immunoprecipitin band against solubilized microsomes on Ouchterlony plates.

Immunochemical Analysis of Cytochrome P-450-52 kD

Cytochrome P-450-52 kD did not react on Ouchterlony double-diffusion plates with antibodies against cytochromes P450a-P450e;[16] but it did react moderately with anti-P450f and weakly with anti-P450g. The reaction of cytochrome P-450-52 kD with anti-P450f on Ouchterlony plates gave an immunoprecipitin band of partial identity with cytochrome P450f indicating that it was immunochemically related to cytochrome P450f. Similarly, in the same test the immunoprecipitation band formed by cytochrome P-450-52 kD reacting with anti-P450g showed partial identity with cytochrome P450g. Consequently, cytochrome P-450-52 kD is immunochemically related to both cytochromes P450f and P450g and can be considered as another member of the cytochromes P450f, P450g, P450h and P450i family which has been recently described.[17]

Immunochemical Identification of P-450-54 kD and P-450-52 kD [^{3}H]Heme-Derived Adducts

Microsomes were solubilized to a final concentration of 8 mg/ml in 50 mM potassium phosphate (pH 7.4) containing 20% (v/v) glycerol, 0.1 mM EDTA, 0.1 M KCl, 1% (w/v) sodium cholate, and 0.2% (v/v) Emulgen 911 (buffer A) and an aliquot (25-30 ug) was mixed with anti-P-450-54 kD IgG or anti-P-450-52 kD IgG to precipitate immunoreactive P-450-54 kD or P-450-52 kD, respectively, according to the method of Newman et al..[18] After the immunoprecipitates were solubilized in 60 mM Tris-HCl (pH 6.8) containing 2.3% (w/v) SDS, 10% (v/v) glycerol, and 5% (v/v) mercaptoethanol, samples were analyzed for irreversibly bound [^{3}H]heme-derived label by the acetone-HCl precipitation procedure and radioelectrophoretography as described above. All experiments were repeated with normal IgG; the non-specific binding was subtracted from the above results to give corrected levels of irreversibly bound [^{3}H]heme-derived label.

Radial Immunodiffusion

Immunodiffusion plates were prepared from gel containing 0.9% (w/v) agarose, 1 M glycine, 0.1 mM EDTA, 0.9% (w/v) NaCl, 0.2% (w/v) NaN$_3$, and 1% (v/v) rabbit antiserum. The mixture of all components except antiserum was heated to boiling to dissolve the agarose. When the solution had cooled to 60°C the antiserum was added and the gel was cast on GelBond Film on 10x10 cm glass plates (15 ml agarose solution per plate). Solutions of microsomes and standard solutions of the purified cytochrome P-450 antigens were prepared in buffer A and 8 ul portions were placed into wells of 4 mm diameter. Plates were kept at 4°C in sealed humid boxes until diffusion was complete, generally in about 2 weeks. Plates were then washed thoroughly with several changes of phosphate buffered saline, followed by several changes of distilled water, to remove unreacted proteins, and dried in air. The dried gels were stained with 0.5% (w/v) Coomassie blue R-250 in a mixture of isopropanol, acetic acid and water (45:10:45) for 90 min, destained in the mixture of isopropanol, acetic acid and water for 10 min, rinsed with distilled water, and dried in air.

Other Assays

Cytochromes b_5 and P-450 were determined by the method of Omura and Sato[19] and protein was measured by the method of Lowry et al.[20] with BSA as the standard.

All experimental results represent the average of single determinations from two rats.

RESULTS

Administration of CCl_4 to phenobarbital treated rats caused extensive loss of spectrally determined cytochromes P-450 (Table 1). After one hr 64% of the cytochromes P-450 were destroyed. As has been reported by other investigators,[1,2] cytochrome b_5 was not affected by CCl_4 treatment (results not shown).

Precipitation of the microsomal proteins with acetone-HCl, which will extract noncovalently bound heme from hemeproteins,[21] showed that the CCl_4 treatment caused a substantial fraction (28%) of the microsomal [^{3}H]heme radiolabel to become irreversibly bound to microsomal proteins (Table 1). Radioelectrophoretograms of the microsomes (Figure 1) revealed that the irreversibly bound [^{3}H] label was confined to proteins in the region of cytochromes P-450 (47 kD to 56 kD). Only a trace of the heme-derived label was observed in the radioelectrophoretogram of the control microsomes because noncovalently bound heme dissociates from apoprotein during SDS-PAGE and migrates on the gel with or in front of the tracking dye.[6]

Immunoprecipitation of the microsomes with antibodies to two major forms of liver microsomal cytochrome P-450 purified from phenobarbital induced rats (P-450-52 kD and P-450-54 kD) established conclusively (Table 2) that there was heme-derived radiolabel irreversibly bound to cytochromes P-450. The two antibodies precipitated 61% and 100% of the total irreversibly bound [^{3}H] label in the microsomes from the CCl_4-treated and the untreated control rats respectively. Radioelectrophoretograms of the immunoprecipitates confirmed that the antibodies precipitated mainly [^{14}C] label associated with immunoreactive P-450-52 kD or P-450-54 kD from both control and CCl_4-treated rats (results not shown).

Table 1. Destruction of cytochromes P-450 and irreversible binding to protein of [^{3}H]heme radiolabel in rat liver microsomes 1 hour after CCl_4 treatment.

Treatment	% Loss of cyto-chromes P-450	Irreversible binding of [^{3}H] label to protein
		(% of total microsomal [^{3}H]label)
Control	0	4
CCl_4	64	28

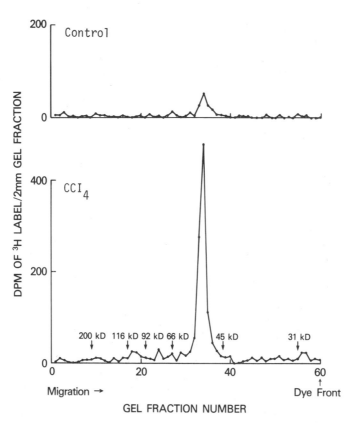

Fig. 1.   SDS-PAGE radioelectrophoretograms of 40 ug samples
of liver microsomes from control and CCl_4-treated
rats.

Antibodies to the two forms of cytochrome P-450 were used in radial
immunodiffusion assays to measure the levels of P-450-54 kD and P-450-52
kD proteins in microsomes from control and CCl_4-treated rats (Figure 2).
From standard curves of precipitin ring areas vs. antigen concentrations,
it was determined that in control microsomes the concentration of
proteins immunoreactive with anti-P-450-54 kD was 1.15 nmole/mg protein,
while the concentration of proteins reactive with anti-P450-52 kD was
0.65 nmole/mg protein.   Since total spectrally determined cytochromes
P-450 in control microsomes was 2.30 nmole/mg protein, P-450-54 kD
accounted for up to 50% of the cytochromes P-450 in phenobarbital induced
microsomes, while P-450-52 kD accounted for up to 28%.   It must be
recognized that these figures represent upper limits, because both
P-450-54 kD and P-450-52 kD have been found to be immunochemically
related to other forms of cytochrome P-450.   In the microsomes prepared
from rats which had been treated with CCl_4, the concentrations of
proteins immunoreactive with anti-P-450-54 kD and with anti-P-450-52 kD
were 0.77 nmole/mg protein and 0.33 nmole/mg protein, respectively.

DISCUSSION

It has now been shown that the destruction of rat liver cytochromes
P-450 by CCl_4 is due, at least in part, to the irreversible binding of
cytochrome P-450 heme-derived product or products to microsomal
proteins.   The heme of cytochrome b_5, the other major heme component in

Table 2. Immunoprecipitation of irreversibly bound heme-
derived [3H] radiolabel from microsomes of control
and CCl$_4$-treated rats, using antibodies against
cytochromes P-450-52 kD and P-450-54 kD.

Treatment	% of irreversibly bound [3H] label precipitated by	
	Anti-P-450-52 kD	Anti-P-450-54 kD
Control	25	75
CCl$_4$	22	39

liver microsomes,[19] is not affected. Furthermore, immunoprecipitation of microsomal protein with antibodies against two major forms of cytochrome P-450 established definitively that the heme degradation products were bound mainly to cytochrome P-450 protein (Table 2). Sixty-one percent of the irreversibly bound heme-derived radiolabel could be precipitated by the two antibodies. Whether the remaining heme fragments were bound to other cytochromes P-450 remains to be determined.

Although it is not yet clear how CCl$_4$ causes the formation of heme-derived protein adducts, the mechanism likely involves, at least in part, the activation of the heme moiety into an electrophilic product or products by reactive CCl$_4$ metabolites such as CCl$_3$., CCl$_3$OO., CCl$_3$OOH, or electrophilic chlorine.[22,23] Active CCl$_4$ metabolites may be directly involved in the destruction of cytochrome P-450 heme,[24-26] or they may act indirectly by promoting the formation of lipid hydroperoxides which destroy cytochrome P-450.[27-29] In any event, it appears that heme-derived species may be produced at the site of cytochrome P-450 and bind before they diffuse to other proteins.

In addition to destroying the heme moiety of cytochrome P-450, CCl$_4$ has been reported to produce an apparent loss of liver microsomal proteins of molecular weights of approximately 52-58 kD as determined by SDS-PAGE.[2,3,30,31] It has been suggested, but not proven, that the lost proteins are cytochromes P-450.[2,3,30] The results of the radial immuno-diffusion assays indicated that cytochrome P-450 proteins are indeed lost following CCl$_4$ treatment (Figure 2). Assuming that protein immuno-reactivity was not changed by CCl$_4$ treatment, the immunoreactive 54 kD and 52 kD cytochromes P-450 were decreased by 33% and 49%, respectively.

The fact that there was significant irreversible binding of heme fragments to each of these isozymes may provide important information concerning the mechanism of their apparent losses from microsomes. It may be that the irreversible binding of heme fragments makes them more susceptible to proteolytic degradation by tissue proteases. If the 'tagging' of cytochrome P-450 proteins by heme fragments is a general phenomenon, it may explain, at least in part, how other compounds which destroy the heme of cytochrome P-450 also produce apparent losses of cytochrome P-450 apoprotein.[32-35] Moreover, the small amount of heme-derived material found irreversibly bound to cytochromes P-450 in control animals may represent a similar process that occurs slowly under normal physiological conditions.

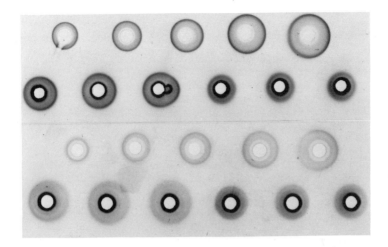

Fig. 2. Radial immunodiffusion gels for assay of the concentrations of cytochromes P-450-54 kD and P-450-52 kD in microsomes from control and CCl$_4$-treated rats. The top row contained P-450-54 kD at concentrations of (from left to right) 0.51, 0.77, 1.03, 1.54, and 2.06 nmole/ml. The second row contained microsomes (1mg/ml) from control (3 wells at left) and CCl$_4$-treated (3 wells at right) rats. The third row contained P-450-52 kD at concentrations of (from left to right) 0.28, 0.42. 0.56, 0.84, and 1.14 nmole/ml. The fourth row contained microsomes (2mg/ml) from control (3 wells at left) and CCl$_4$-treated (3 wells at right) rats.

REFERENCES

1. W. Levin, M. Jacobson, and R. Kuntzman, Incorporation of radioactive delta-aminolevulinic acid into microsomal cytochrome P-450: Selective breakdown of the hemoprotein by allylisopropylacetamide and carbon tetrachloride, Arch. Biochem. Biophys. 148:262 (1972).
2. B. Head, D. E. Moody, C. H. Woo, and E. A. Smuckler, Alterations of specific forms of cytochrome P-450 in rat liver during acute carbon tetrachloride intoxication, Toxicol. App. Pharmacol. 61:286 (1981).
3. T. Noguchi, K. L. Fong, E. K. Lai, L. Olson, and P. B. McCay, Selective early loss of polypeptides in liver microsomes of CCl$_4$-treated rats, Biochem. Pharmacol. 31:609 (1982).
4. E. G. D. de Toranzo, M. C. Villarruel, and J. A. Castro, Early destruction of cytochrome P-450 in testis of carbon tetrachloride poisoned rats, Toxicol. 10:39 (1978).
5. W. C. Brogan, III, P. I. Eacho, D. E. Hinton, and H. D. Colby, Effects of carbon tetrachloride on adrenocortical structure and function in guinea pigs, Toxicol. Appl. Pharmacol. 75:118 (1984).
6. A. Parkinson, P. E. Thomas, D. E. Ryan, and W. Levin, The in vivo turnover of rat liver microsomal epoxide hydrolase and both the apoprotein and heme moieties of specific cytochrome P-450 isozymes, Arch. Biochem. Biophys. 225:216 (1983).
7. L. R. Pohl, R. V. Branchflower, R. J. Highet, J. L. Martin, D. S. Nunn, T. J. Monks, J. W. George, and J. A. Hinson, The formation

of diglutathionyl dithiocarbonate as a metabolite of chloroform, bromotrichloromethane, and carbon tetrachloride, Drug Metab. Dispos. 9:334 (1981).

8. U. K. Laemmli, Cleavage of structural proteins during the assembly of the head of bacteriophage T4, Nature (Lond.) 227:680 (1970).

9. Y. Imai and R. Sato, A gel-electrophoretically homogenous preparation of cytochrome P-450 from liver microsomes of phenobarbital pre-treated rabbits, Biochem. Biophys. Res. Commun. 60:8 (1974).

10. F. P. Guengerich, G. A. Dannan, S. T. Wright, M. V. Martin, and L. S. Kaminski, Purification and characterization of liver microsomal cytochromes P-450: Electrophoretic, spectral, catalytic, and immunochemical properties and inducibility of eight isozymes iso-lated from rats treated with phenobarbital or beta-naphthoflavone, Biochemistry 21:6019 (1982).

11. H. Satoh, J. R. Gillette, H. W. Davies, R. D. Schulick, and L. R. Pohl, Immunochemical evidence of trifluoroacetylated cytochrome P-450 in the liver of halothane treated rats, Mol. Pharmacol. (in press).

12. A. R. Steward, G. A. Dannan, P. S. Guzelian, and F. P. Guengerich, Changes in the concentration of seven forms of cytochrome P-450 in primary cultures of adult rat hepatocytes, Mol. Pharmacol. 27:125 (1985).

13. D. E. Ryan, P. E. Thomas, D. Korzeniowski, and W. Levin, Separation and characterization of highly purified forms of liver microsomal cytochrome P-450 from rats treated with polychlorinated biphenyls, phenobarbital, and 3-methylcholanthrene, J. Biol. Chem. 254:1365 (1979).

14. D. J. Waxman and C. Walsh, Phenobarbital-induced rat liver cytochrome P-450. Purification and characterization of two closely related isozymic forms, J. Biol. Chem. 257:10446 (1982).

15. T. Kamataki, D. H. Belcher, and R. A. Neal, Studies of the metabolism of diethyl p-nitrophenyl phosphorothionate (parathione) and benz-phetamine using an apparently homogeneous preparation of rat liver cytochrome P-450: Effect of a cytochrome P-450 antibody prepara-tion, Mol. Pharmacol. 12:921 (1976).

16. D. E. Ryan, P. E. Thomas, and W. Levin, Purification and characteri-zation of a minor form of hepatic microsomal cytochrome P-450 from rats treated with polychlorinated biphenyls, Arch. Biochem. Biophys. 216:272 (1982).

17. S. Bandiera, D. E. Ryan, W. Levin, and P. E. Thomas, Evidence for a family of four immunochemically related isozymes of cytochrome P-450 purified from untreated rats, Arch. Biochem. Biophys. 240:478 (1985).

18. S. L. Newman, J. L. Barwick, N. A. Elshourbagy, and P. S. Guzelian, Measurement of the metabolism of cytochrome P-450 in cultured hepatocytes by a quantitative and specific immunochemical method, Biochem. J. 204:281 (1982).

19. T. Omura and R. Sato, The carbon monoxide-binding pigment of liver microsomes. I. Evidence for its hemoprotein nature, J. Biol. Chem. 239:2370 (1964).

20. O. H. Lowry, N. J. Rosebrough, A. L. Farr, and R. J. Randall, Protein measurement with the Folin phenol reagent, J. Biol. Chem. 193:265 (1951).

21. M. D. Maines and M. W. Anders, Characterization of the heme of cyto-chrome P-450 using gas chromatography/mass spectrometry, Arch. Biochem. Biophys. 159:201 (1973).

22. B. A. Mico and L. R. Pohl, Reductive oxygenation of carbon tetra-chloride: Trichloromethylperoxyl radical as a possible inter-mediate in the conversion of carbon tetrachloride to electrophilic chlorine, Arch. Biochem. Biophys. 225:596 (1983).

23. L. R. Pohl and B. A. Mico, Electrophilic halogens as potentially

toxic metabolites of halogenated compounds, <u>Trends Pharmacol. Sci.</u> 5:61 (1984).

24. H. De Groot and W. Haas, O$_2$-Independent damage of cytochrome P-450 by CCl$_4$ metabolites in hepatic microsomes, <u>FEBS Lett.</u> 115:253 (1980).

25. H. De Groot and W. Haas, Self-catalysed, O$_2$-independent inactivation of NADPH- or dithionite-reduced microsomal cytochrome P-450 by carbon tetrachloride, <u>Biochem. Pharmacol.</u> 30:2343 (1981).

26. G. Fernandez, M. C. Villarruel, E. G. D. de Toranzo, and J. A. Castro, Covalent binding of carbon tetrachloride metabolites to the heme moiety of cytochrome P-450 and its degradation products, <u>Res. Commun. Chem. Pathol. Pharmacol.</u> 35:283 (1982).

27. E. G. Hrycay and P. J. O'Brien, Cytochrome P-450 as a microsomal peroxidase utilizing a lipid peroxide substrate, <u>Arch. Biochem. Biophys.</u> 147:14 (1971).

28. E. Jeffery, A. Kotake, R. El Azhary, and G. J. Mannering, Effects of linoleic acid hydroperoxide on the hepatic monooxygenase systems of microsomes from untreated, phenobarbital-treated, and 3-methylcholanthrene-treated rats, <u>Mol. Pharmacol.</u> 13:415 (1977).

29. T. D. Lindstrom and S. D. Aust, Studies on cytochrome P-450-dependent lipid hydroperoxide reduction, <u>Arch. Biochem. Biophys.</u> 233:80 (1984).

30. T. Noguchi, K. L. Fong, E. K. Lai, S. S. Alexander, M. M. King, L. Olson, J. L. Poyer, and P. B. McCay, Specificity of a phenobarbital-induced cytochrome P-450 for metabolism of carbon tetrachloride to the trichloromethyl radical, <u>Biochem. Pharmacol.</u> 31:615 (1982).

31. P. A. Krieter and R. A. Van Dyke, Cytochrome P-450 and halothane metabolism. Decrease in rat liver microsomal P-450 <u>in vitro</u>, <u>Chem. Biol. Interact.</u> 44:219 (1983).

32. H. H. Liem, E. F. Johnson, and U. Muller-Eberhard, The effect <u>in vivo</u> and <u>in vitro</u> of allylisopropylacetamide on the content of hepatic microsomal cytochrome P-450 2 of phenobarbital treated rabbits, <u>Biochem. Biophys. Res. Commun.</u> 111:926 (1983).

33. H. L. Bonkovsky, J. F. Sinclair, J. F. Healey, P. R. Sinclair, and E. L. Smith, Formation of cytochrome P-450 containing haem or cobalt-protoporphyrin in liver homogenates of rats treated with phenobarbital and allyisopropylacetamide, <u>Biochem. J.</u> 222:453 (1984).

34. L. M. Bornheim, A. N. Kotake, and M. A. Correia, Differential haemin-mediated restoration of cytochrome P-450 N-demethylases after inactivation by allylisopropylacetamide, <u>Biochem. J.</u> 227:277 (1985).

35. T. R. Tephly, K. A. Black, M. D. Green, B. L. Coffman, G. A. Dannan, and F. P. Guengerich, Effect of the suicide substrate, 3,5-diethoxycarbonyl-2,6-dimethyl-4-ethyl-1,4-dihydropyridine (DDEP), on the metabolism of xenobiotics and on cytochrome P-450 apoproteins, Abstract No. 32, Third International Symposium on Biological Reactive Intermediates, June 6-8, 1985.

THE FAD-CONTAINING MONOOXYGENASE OF LUNG AND LIVER TISSUE

FROM RABBIT, MOUSE AND PIG: SPECIES AND TISSUE DIFFERENCES

Patrick J. Sabourin, R. E. Tynes,* B. P. Smyser*
and E. Hodgson*
Lovelace Inhalation Toxicology Research Institute
P. O. Box 5890, Albuquerque, NM 87185 and
*Interdepartmental Toxicology Program
North Carolina State University, Box 7613
Raleigh, NC 27650-7613

INTRODUCTION

The FAD-containing monooxygenase (EC 1.14.13.8) is a microsomal flavoprotein capable of oxidizing a wide range of nitrogen, sulfur and phosphorus
containing compounds. These include secondary and tertiary (but not primary) aliphatic amines, arylamines, hydrazines, sulfides, thiols, thioamides
and thioureas.[1-6] The pig liver FAD-containing monooxygenase (FMO) has been
purified and extensively characterized,[7-9] however, only recently has this
enzyme been studied in other species and tissues.[10-14,17] It is apparent
from these studies that there are species and tissue differences in the
flavin-containing monooxygenase. This paper describes some of the more
recent advances towards characterizing this enzyme in different species and
tissues and also extends the large list of compounds which are substrates.

PURIFICATION OF THE FAD-CONTAINING MONOOXYGENASES

Mouse and pig liver microsomal FMO have been purified to homogeneity
using affinity chromatographic methods (Table 1) as previously described.[10]
The FMO activity is followed during the purification by monitoring oxidation
of thiobenzamide.[15] The behavior of these two FMO's during purification indicates species differences. The pig liver FMO will not bind to the 2',5'-
ADP-agarose column using conditions identical to those used for the mouse
liver FMO purification, however, the mouse lung and liver FMO, rabbit lung
and liver FMO and rat liver FMO[11] all bind to this column.

Table 1. Purification of Mouse and Pig Liver
FAD–Containing Monooxygenase

Fraction	Specific Activity nmole/min/mg	% Recovery
Mouse		
Microsomes	10	(100)
Solubilized microsomes	13	81
Cibacron blue sepharose	100	48
Procion red agarose	250	35
2',5'–ADP-agarose	1200	11
Pig		
Microsomes	16	(100)
Solubilized microsomes	21	82
Cibacron blue sepharose	436	64
Procion red agarose	954	44

The FMO has also been purified from rat liver,[11] mouse lung and rabbit lung and liver.[13] Sodium dodecyl sulfate polyacrylamide gel electrophoresis (SDS-PAGE) of the purified FMO's from mouse, rabbit and pig liver or lung are shown in Figure 1. All of the purified FMO's migrate with an apparent molecular weight between 55–58,000. Rabbit liver and lung FMO clearly migrate as different molecular weight proteins. Western blotting of rabbit liver and lung FMO also reveals immunoreactive bands at these differing molecular weights. This molecular weight difference has not been observed in other species examined. The purified mouse liver FMO migrates on SDS-polyacrylamide gel electrophoresis (PAGE) as two closely spaced protein bands when identified by silver or immunoreactive staining. However, western blotting of mouse liver microsomes indicates a single immunoreactive protein band. This discrepancy may be due to artifacts produced during purification or during PAGE of the purified mouse liver FMO. The mouse liver and lung FMO migrate similarly on SDS-PAGE but at a slightly higher apparent molecular weight than the pig liver FMO. However, the molecular weight of the FMO in the liver vs. the lung of mice, pigs or rats are not different. When microsomes are analyzed by western blotting, only a single immunoreactive band is seen for liver or lung of rats and mice and liver of rabbits. Curiously, the rabbit lung FMO appears to consist of three very closely spaced MW protein bands whether purified FMO or microsomes are used.

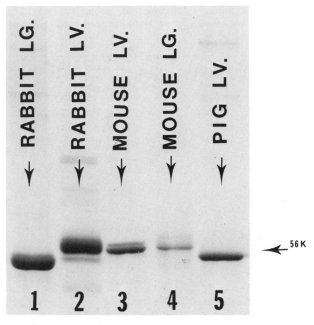

PURIFIED FMO

Fig. 1.   SDS-polyacrylamide gel electrophoresis of purified FAD-containing
monooxygenases from the lung and liver of several species.   Gels
were silver stained.   Protein amounts are as follows: rabbit lung
FMO, 2.0 µg; rabbit liver FMO, 0.5 µg; mouse liver FMO, 0.5 µg,
mouse lung FMO, 2.0 µg; and pig liver FMO, 0.5 µg.

KINETIC AND IMMUNOLOGICAL DIFFERENCES IN FMO BETWEEN SPECIES

Kinetic analysis of the mouse and pig liver FMO with a large number of
nitrogen- and sulfur-containing compounds indicates that purified FMO from
both species oxidizes the same compounds.[12,23]   Table 2 illustrates the
kinetic constants of selected compounds from major classes of compounds oxidized by FMO.   In some cases, the mouse and pig liver FMO have quite different kinetic constants (Km).   For example, the pig liver FMO has a much lower
Km for N-methylaniline and N,N-dimethylaniline than the mouse liver FMO.
Interestingly, cysteamine, a proposed physiological substrate for the FMO,[16]
has similar Km values for mouse or pig liver FMO.

Benzphetamine, a classical substrate of the cytochrome P-450 dependent
monooxygenase, is oxidized, albeit poorly, by the FMO.   Thiobenzamide, which
is a very good FMO substrate, can also be oxidized by the cytochrome P-450

Table 2. Kinetic Constants of Mouse Vs. Pig Liver
FAD-Containing Monooxygenases

	Km (mM)	
	Mouse	Pig
Trimethylamine	2,340	617
Triethylamine	2,890	1,090
N-Methylaniline	1,060	343
N,N-Dimethylaniline	105	11
Imipramine	27	7
Benzphetamine	1,670	74
Benzylmethylsulfide	2	2
Dimethylsulfide	34	11
Cysteamine	65	59
Thiobenzamide	8	3
Phorate	32	12

dependent monooxygenase system.[21] The rate of oxidation by the FMO, at saturating substrate concentrations, is the same for all substrates. However,
in most cases, the amount of substrate presented to the enzyme in vivo will
be at less than saturating concentrations. Therefore, the relative contribution of the FMO and cytochrome P-450 dependent monooxygenases to oxidations
of foreign compounds depends not only on the relative amounts of the two enzymes, but also on the substrate affinity.

Although, in general, the lung and liver FMO catalyze oxidation of the
same classes of compounds, there are some noticeable exceptions. Imipramine
is a very good substrate for the mouse liver and lung and pig liver FMO[12,17]
(see Table 2), however, it (and also chlorpromazine) are poor substrates for
the rabbit lung enzyme.[14,17]

Using an antiserum prepared in rabbits against pig liver FMO, mouse and
pig liver microsomes and purified FMO's were compared by Ouchterlony analysis for immunological cross-reactivity.[12] Equivalent immunoprecipitin
bands were seen between the anti-pig liver FMO antibody and either pig liver
microsomes or purified pig liver FMO, however, only a faint band was elicited with mouse liver microsomes or purified mouse liver FMO. This faint
band formed a spur with the band elicited by the pig liver FMO, indicating
that it was due to a different epitope. The fact that antibody to pig liver
FMO can be used to identify FMO's from pig liver and rat, mouse and rabbit
liver and lung FMO's by western blotting (a much more sensitive immunological technique) indicates that these FMO's all share some common antigenic
determinants.[24]

The lung and liver forms of the FMO in rabbit and mouse are also immunologically distinct. Antibody raised in guinea pigs to rabbit lung FMO does not cross-react in Ouchterlony analysis with rabbit liver FMO or solubilized rabbit liver microsomes.[14] Also, antisera raised to rabbit or mouse liver FMO's reacts only slightly with the lung FMO from the same species.[14] These observations indicate that lung and liver isozymes of the FMO in mice and rabbits have different antigenic determinants.

Two independent research groups have recently provided compelling evidence that rabbit lung microsomal FMO is distinctly different from the liver FMO.[13-14] Tynes et al.[13] extended these studies to mouse liver and lung FMO. Although mouse liver and lung FMO migrate similarly on SDS-PAGE, the lung enzyme has catalytic properties different from the liver enzyme but quite similar to rabbit lung FMO.

Primary aliphatic amines are not substrates for the mouse[12] or pig liver FMO,[1,12] however, long chain aliphatic amines, such as N-octylamine, may activate these enzymes. Recently, it was found that the mouse and rabbit lung FMO oxidize NADPH at a high rate in the presence of long chain aliphatic primary amines.[13]

In the absence of substrate, the FMO catalyzes the oxidation of NADPH at a low rate. This oxidation is due to breakdown of the hydroperoxyflavin-enzyme intermediate to the reduced enzyme and H_2O_2 creating a futile cycle in which NADPH is oxidized (Fig. 2). This "uncoupled" NADPH oxidation is stimulated slightly by long chain primary amines with pig, mouse and rabbit liver FMO. The large stimulation of NADPH oxidation by these primary amines with rabbit and mouse lung FMO led to the hypothesis that the lung FMO (at least in these 2 species) actually catalyzes the oxidation of these primary alkylamines. These primary amines act as substrates having a maximal rate of NADPH oxidation (Vmax) similar to other substrates. A similar Vmax for all substrates is expected from the catalytic mechanism since the breakdown of the hydroxyflavin (NADP-E-OH) is considered to be the rate limiting step.[8] Although this mechanism has only been studied using the purified pig liver FMO, data in other species is consistent with this mechanism. Longer chain alkylamines such as decyl- and dodecylamine appear to be the best substrates giving maximal catalytic rates at 1 mM concentrations. The compound is no longer a substrate if a bulky group (tertoctylamine) or a second amine group (1,8-diaminooctane) is added. Oxidation of the primary aliphatic amine by the lung FMO should result in the production of the appropriate hydroxylamine. Hydroxylamine formation was detected when rabbit lung FMO was

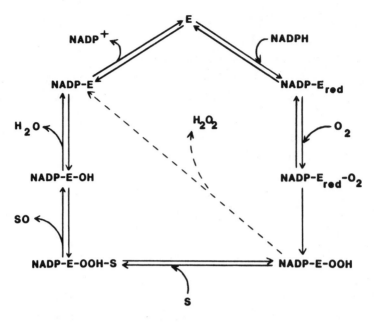

Fig. 2. Mechanism of the pig liver FAD-containing monooxygenase. This mechanism was developed by Poulsen and Ziegler.[7] E represents the oxidized FAD-containing monooxygenase flavoprotein and E_{red} the reduced flavoprotein. S represents the substrate.

incubated with n-octylamine or n-decylamine, although the production of H_2O_2 from the hydroperoxyflavin accounts for greater than half of the NADPH oxidation.[13] No hydroxylamine is detected when the pig liver FMO is incubated with 10 mM n-octylamine.

OXIDATION OF PHOSPHINES BY THE FAD-CONTAINING MONOOXYGENASE

A variety of organophosphate pesticides are oxidized by the FAD-containing monooxygenase.[2] Whereas, the cytochrome P-450 dependent monooxygenases generally oxidize phosphorothioates at the thiono sulfur, the thioether sulfur atom is the site of oxidation by the FMO as with phorate being converted to the sulfoxide (Fig. 3). However, in compounds such as fonofos (O-ethyl S-phenyl ethylphenylphosphine)[18] and diethylphenylphosphine sulfide[19] (Figs. 3 and 4) there is no available thioether sulfur and the thiono sulfur group is oxidized leading to the formation of the oxon.

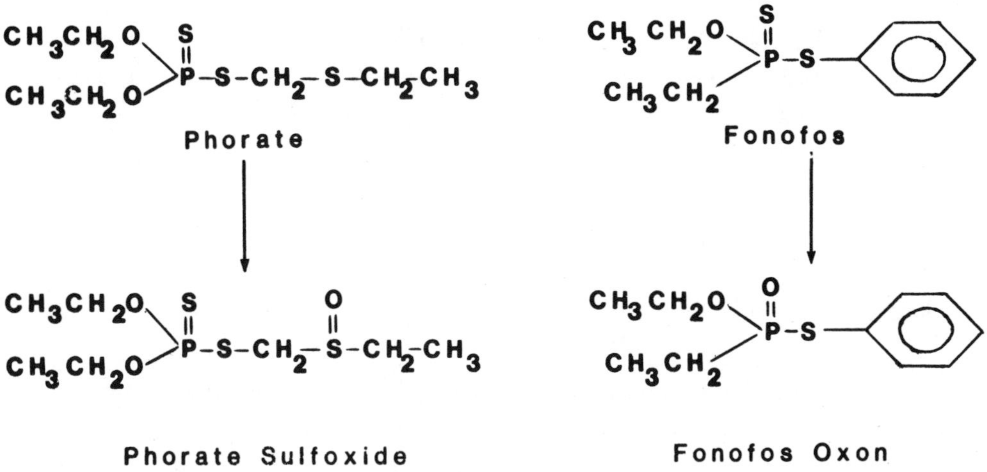

Fig. 3. Oxidation of organophosphorus pesticides by the FAD-containing monooxygenase.

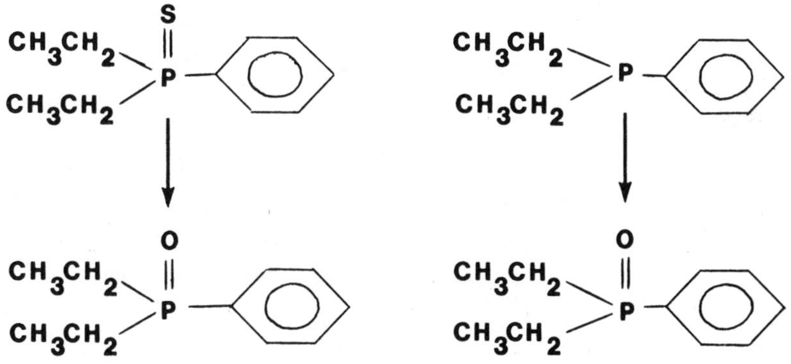

Fig. 4. Oxidation of diethylphenylphosphine sulfide and diethylphenylphosphine by the FAD-containing monooxygenase.

269

Recently it was demonstrated that diethylphenylphosphine, which contains neither a nitrogen nor a sulfur atom, stimulates NADPH oxidation by the pig liver FMO.[19] The Km for this reaction was less than 3 μM indicating that it is a very good substrate of the FMO. Diethylphenylphosphine oxide was identified as the product of this reaction as well as the product of diethylphenylphosphine sulfide oxidation (Fig. 4). Therefore, the FMO can oxidize on the phosphorus atom as well as a nitrogen or sulfur atom. It is unclear at this time if the attack of the oxygen atom on compounds such as fonofos by the FMO is at the S atom or at the P atom. Clearly, at least in phosphines, the FMO can oxidize a phosphorus atom.

CONCLUSIONS

It is clear that there are significant species and tissue differences in the FAD-containing monooxygenase (FMO). These differences, both in substrate specificity and/or kinetic properties, will affect the role of the FMO in the metabolism of foreign compounds. The relative contribution of the FMO and the cytochrome P-450 dependent monooxygenase system to the initial oxidation of a compound may affect the types and/or amounts of oxidation product(s) produced as well as the stereochemistry of the product.[22] This in turn may ultimately affect the deactivation, toxicity or carcinogenicity of the compound. Therefore, the qualitative as well as quantitative properties of the FMO are important when considering how a compound will be metabolized in different species and even in different organs of the same species.

ACKNOWLEDGEMENTS

This investigation was supported in part by PHS Grants No. ES-00044 and ES-07046 from the National Institute of Environmental Health Science and in part by U. S. Department of Energy Contract No. DE-AC04-76EV01013.

REFERENCES

1.   D. M. Ziegler, Microsomal flavin-containing monooxygenase: oxygenation of nucleophilic nitrogen and sulfur compounds, Enzymatic Basis of Detoxification 1:201-227 (1980).

2.   N. P. Hajjar and E. Hodgson, Flavin adenine dinucleotide-dependent monooxygenase: its role in the sulfoxidation of pesticides in mammals, Science 209:1134–1136 (1980).

3.   L. L. Poulsen, Organic sulfur substrates for the microsomal flavin-containing monooxygenase, Rev. Biochem. Toxicol. 3:33–49 (1981).

4.   R. A. Prough, P. C. Freeman, and R. N. Hines, The oxidation of hydra-zine derivatives catalyzed by the purified liver microsomal FAD-containing monooxygenase, J. Biol. Chem. 256:4178–4184 (1981).

5.   M. W. Kloss, J. Cavagnaro, G. M. Rosen, and E. J. Raukman, Involvement of FAD-containing monooxygenase in cocain-induced hepatotoxicity, Toxicol. Appl. Pharmacol. 64:88–93 (1982).

6.   D. M. Ziegler, E. M. McKee, and L. L. Poulsen. Microsomal flavopro-tein-catalyzed N-oxidation of arylamines, Drug Metab. Dispos. 1:314–321 (1973).

7.   L. L. Poulsen and D. M. Ziegler, The liver microsomal FAD-containing monooxygenase, spectral characterization and kinetic studies, J. Biol. Chem. 254:6449–6455 (1979).

8.   N. B. Beaty and D. P. Ballou, The oxidative half-reaction of liver microsomal FAD-containing monooxygenase, J. Biol. Chem. 256: 4619–4625 (1981).

9.   N. B. Beaty and D. P. Ballou, The reductive half-reaction of liver microsomal FAD-containing monooxygenase, J. Biol. Chem. 256:4611–4618 (1981).

10. P. J. Sabourin, B. P. Smyser, and E. Hodgson, Purification of the flavin-containing monooxygenase from mouse and pig liver microsomes, Int. J. Biochem. 16:713–720 (1984).

11. T. Kimura, M. Kodama, and C. Nagata, Purification of mixed-function amine oxidase from rat liver microsomes, Biochem. Biophys. Res. Commun. 110:640–645 (1983).

12. P. J. Sabourin and E. Hodgson, Characterization of the purified micro-somal FAD-containing monooxygenase from mouse and pig liver, Chem.-Biol. Interactions 51:125–139 (1984).

13. R. E. Tynes, P. J. Sabourin, and E. Hodgson, Identification of distinct hepatic and pulmonary forms of microsomal flavin-containing mono-oxygenase in the mouse and rabbit, Biochem. Biophys. Res. Commun. 126:1069–1075 (1985).

14. D. E. Williams, D. M. Ziegler, D. J. Nordin, S. E. Hale, and B. S. S. Masters, Rabbit lung flavin-containing monooxygenase is immuno-chemically and catalytically distinct from the liver enzyme, Biochem. Biophys. Res. Commun. 125:116–122 (1984).

15. J. R. Cashman and R. P. Hanzlik, Microsomal oxidation of thiobenzamide. A photometric assay for the flavin-containing monooxygenase, Biochem. Biophys. Res. Commun. 98:147-153 (1981).

16. D. M. Ziegler, L. L. Poulsen, and B. M. York, Role of the flavin-containing monooxygenase, in: "Maintaining Cellular Thiol: Disulfide Balance in Functions of Glutathione: Biochemical, Physiological, Toxicological, and Clinical Aspects," A. Larsson et al., eds., Raven Press, NY, pp. 297-305 (1983).

17. R. E. Tynes and E. Hodgson, Catalytic activity and substrate specificity of the flavin-containing monooxygenase in microsomal systems: characterization of the hepatic, pulmonary and renal enzymes of the mouse, rabbit and rat, Arch. Biochem. Biophys. 240: 77-93 (1985).

18. N. P. Hajjar and E. Hodgson, The microsomal FAD-dependent monooxygenase as an activating enzyme: fonofos metabolism, Biological Reactive Intermediates-II Part B:1245-1253 (1982).

19. B. P. Smyser and E. Hodgson, Metabolism of phosphorus-containing compounds by pig liver microsomal FAD-containing monooxygenase, Biochem. Pharmacol. 34:1145-1150 (1985).

20. C. J. Serabjit-Singh, P. W. Albro, I. G. C. Robertson, and R. M. Philpot, Interactions between xenobiotics that increase or decrease the levels of cytochrome P-450 isozymes in rabbit lung and liver, J. Biol. Chem. 258:12827-12834 (1983).

21. P. E. Levi, R. E. Tynes, P. J. Sabourin, and E. Hodgson, Is thiobenzamide a specific substrate for the microsomal FAD-containing monooxygenase?, Biochem. Biophys. Res. Commun. 107:1314-1317 (1982).

22. D. J. Waxman, D. R. Light, and C. Walsh, Chiral sulfoxidations catalyzed by rat liver cytochromes P-450, Biochem. 21:2499-2507 (1982).

23. R. E. Tynes and E. Hodgson, Magnitude of involvement of the mammalian flavin-containing monooxygenase in the microsomal oxidation of pesticides, J. Agric. Food Chem. 33:471-479 (1985).

23. G. A. Dannan and F. P. Guengerich, Immunochemical comparison and quantitation of microsomal flavin-containing monooxygenases in various hog, mouse, rat, rabbit, dog, Mol. Pharmacol. 22:787 (1982).

OXYGEN RADICAL FORMATION DURING REDOX CYCLING OF BLEOMYCIN-Fe(III) CAT-

ALYZED BY NADPH-CYTOCHROME P-450 REDUCTASE OF LIVER MICROSOMES AND NUCLEI

Hermann Kappus and Ismail Mahmutoglu

Free University of Berlin, FB 3, WE 15, Rudolf-Virchow-
Clinic, Augustenburger Platz 1, D-1000 Berlin 65, FRG

INTRODUCTION

Bleomycin, a glycopeptide antibiotic is successfully used in the chemotherapy of various tumors. It has been suggested that its anticancer activity is due to oxygen radicals formed by a reduced complex of bleo-mycin and iron ions (for review see Burger et al., 1981). A bleomycin-Fe(II)-complex damages DNA resulting in the formation of strand breaks and the release of free bases and malondialdehyde, the latter originating from the deoxyribose moiety of DNA (Giloni et al., 1981; Gutteridge et al., 1981). The bleomycin-Fe-complex can be reduced chemically or by xanthine oxidase (Sausville et al., 1978). Furthermore, increases in DNA damage have been observed when microsomes of different organs and NADPH were incubated with bleomycin and iron ions (Bickers et al., 1984; Trush et al., 1982; Yamanaka et al., 1978). We found that isolated liver micro-somal NADPH-cytochrome P-450 reductase is able to catalyze bleomycin-related DNA strand breaks and base and malondialdehyde release in the presence of ferric ions (Scheulen et al., 1981; Scheulen and Kappus, 1984). This has been confirmed and extended recently (Kilkuskie et al., 1984). Therefore, it is obvious that this enzyme is responsible for the effects observed in microsomes.

However, bleomycin has also a high affinity to DNA which leads to site-specific cleavage of DNA (Mirabelli et al., 1985; Umezawa et al., 1984). Therefore, we were interested whether a nuclear enzyme reduces the bleomycin-Fe-complex which activates oxygen in close proximity to DNA. We found that NADPH- and oxygen-consumption considerably increased in liver cell nuclei in the presence of bleomycin and $FeCl_3$ indicating that NADPH-cytochrome P-450 reductase reduces the bleomycin-Fe(III)-complex in nuclei. Further studies with this enzyme isolated from liver microsomes suggest that redox cycling of the bleomycin-Fe-complex is associated with the formation of hydroxyl radicals.

METHODS

Materials

All experiments were carried out with the clinically used Bleo-mycinum Mack[R] containing 55 - 70 % bleomycin A_2 and 25 - 32 % bleomycin B_2

(Mack, Illertissen, FRG). All chemicals and biochemicals of the purest grade available were purchased either from Merck (Darmstadt, FRG), Sigma (München, FRG), Serva (Heidelberg, FRG), or Boehringer (Mannheim, FRG). All gases used were obtained from Linde (Berlin, FRG).

Experiments with cell nuclei

Male Wistar rats of about 200 g starved overnight were decapitated, the livers removed, perfused with 0.9 % NaCl solution and homogenized in 0.33 M sucrose. The homogenate was centrifuged at 960g for 10 min and the pellet suspended and rehomogenized in 1.93 M sucrose containing 1 mM $MgCl_2$. The next centrifugation step was 70 min at 69,500g. The remaining pellet was suspended in 2.2 M sucrose containing 3 mM $MgCl_2$ and centrifuged at 48,900g for 60 min. This centrifugation step was repeated once. The resulting nuclear pellet was suspended in 1 M sucrose containing 1 mM $MgCl_2$ and centrifuged at 960g for 10 min. The pellet was washed with 0.33 M sucrose containing 1 mM $MgCl_2$. The isolated nuclei were suspended in 25 mM TES-buffer (N-tris(hydroxymethyl)methyl-2-aminomethanesulfonic acid-Na-salt) containing 1 mM $MgCl_2$. The purity of the isolated nuclei was checked by microscopy as well as by measuring the microsomal glucose-6-phosphatase (Swanson, 1955) and the mitochondrial succinate dehydrogenase (Fleischer and Fleischer, 1967).

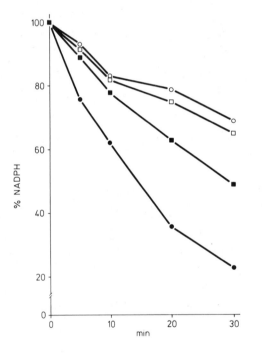

Fig. 1. NADPH-consumption in rat liver nuclei (1 mg protein/ ml) in presence of bleomycin (100 µg/ml) and $FeCl_3$ (0.1 mM). Initial NADPH concentration 0.05 mM. ●—● complete system, ■—■ minus $FeCl_3$, ○—○ minus bleomycin, o—o minus $FeCl_3$, minus bleomycin.

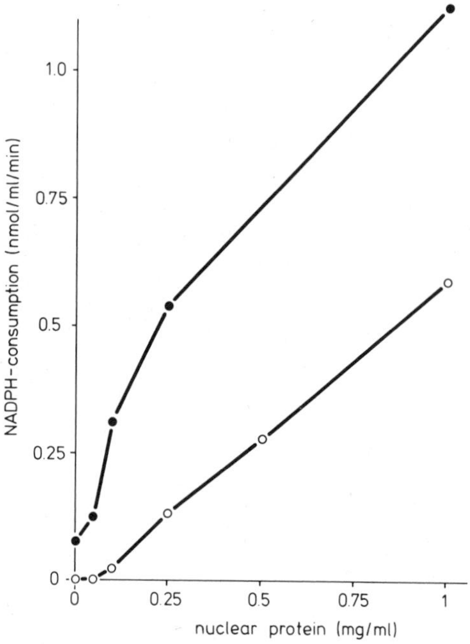

Fig. 2. Effect of rat liver nuclear protein concentration on NADPH-consumption in presence of bleomycin (100 µg/ ml) and $FeCl_3$ (0.1 mM). Initial NADPH concentration 0.05 mM. ●—● complete system, o—o minus $FeCl_3$, minus bleomycin.

274

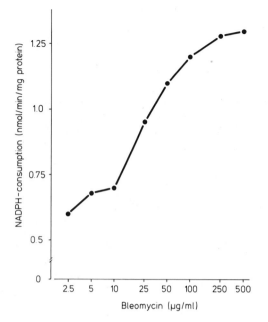

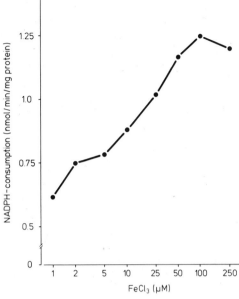

Fig. 3. Effect of bleomycin on NADPH-consumption in rat liver nuclei (1 mg protein/ml) in presence of FeCl$_3$ (0.1 mM). Initial NADPH concentration 0.05 mM.

Fig. 4. Effect of FeCl$_3$ on NADPH-consumption in rat liver nuclei (1 mg protein/ml) in presence of bleomycin (100 µg/ml). Initial NADPH concentration 0.05 mM.

In general, liver cell nuclei (1 mg nuclear protein/ml) were aerobically incubated at 37 °C in 25 mM TES with 2.5 mM MgCl$_2$, 0.05 mM NADPH, 100 µg/ml bleomycin and 0.1 mM FeCl$_3$ under shaking. At various times aliquots were removed, diluted with water and the NADPH content measured using a fluorimetric method with excitation at 350 nm and emission at 465 nm. Oxygen consumption was measured directly in the nuclear suspension using a Clark-oxygen-electrode.

Experiments with isolated NADPH-cytochrome P-450 reductase

Liver microsomes obtained from male Wistar rats were used to isolate NADPH-cytochrome P-450 reductase according to standard procedures as previously described (Scheulen et al., 1981). The enzyme activity was determined with cytochrome c as substrate.

In general, the isolated reductase (0.2 U/ml) was aerobically incubated in 20 mM phosphate buffer pH 7.5 at 37 °C with 0.5 mM NADPH, 100 µg/ml bleomycin, 0.1 mM FeCl$_3$ and 1 mM methional (freshly destilled) under shaking. Sealed flasks were used as previously described (Kappus and Muliawan, 1982). Gas samples were removed from the head space of the flask (Kappus and Muliawan, 1982) and ethene analyzed by a gas chromatographic method originally developed to measure ethane (Filser et al., 1983).

RESULTS

Figure 1 demonstrates that liver cell nuclei incubated under the conditions described consume NADPH. In the presence of FeCl$_3$ no further increase could be observed whereas with bleomycin (without $FeCl_3$) NADPH-consumption almost doubled. But in the presence of bleomycin and $FeCl_3$ NADPH-consumption was much more pronounced. The increase in NADPH-consumption was almost exclusively due to the liver cell nuclei (Figure 2). Figures 3 and 4 show that maximal NADPH-consumption depended on the bleomycin as well on the $FeCl_3$ concentrations. Besides NADPH oxygen was consumed by the nuclei incubated (Figure 5). Oxygen was not consumed in the absence of NADPH. Oxygen consumption was maximal in the presence of NADPH, bleomycin and $FeCl_3$ and depended on the amount of nuclei (Figure 5), although with $FeCl_3$ or bleomycin alone some oxygen was consumed. When both $FeCl_3$ and bleomycin were omitted a significant amount of oxygen was only consumed with the higher concentration of nuclei (Figure 5).

Figure 6 which shows an experiment with isolated liver microsomal NADPH-cytochrome P-450 reductase incubated with NADPH, bleomycin, $FeCl_3$ and methional demonstrates that the conversion of methional to ethene (ethylene), a measure for hydroxyl radicals (Yamazaki, 1977), is depended on the presence of oxygen. Hydroxyl radicals occurred only in the presence of the enzyme, NADPH and bleomycin (Table 1). Without the addition of $FeCl_3$ ethene formation was considerably lower (Table 1). Compounds known as hydroxyl radical trapping agents partially inhibited the formation of ethene (Table 1).

DNA which is the target macromolecule for the oxygen radicals formed by a reduced bleomycin-Fe(II)-complex inhibited the conversion of methional to ethene (Figure 7).

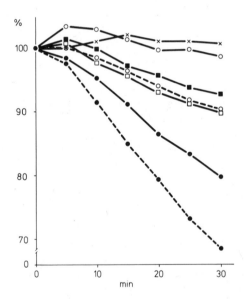

Fig. 5. Oxygen-consumption in rat liver nuclei (--- 1 mg protein/ml; —— 0.5 mg protein/ml) in presence of bleomycin (100 μg/ml), $FeCl_3$ (0.1 mM) and NADPH (0.05 mM). The initial oxygen concentration of the incubation mixture equilibrated with air is set to 100 %. ●--● complete system, o—o minus bleomycin, ■—■ minus $FeCl_3$, x—x minus NADPH, 8—8 minus bleomycin, minus $FeCl_3$.

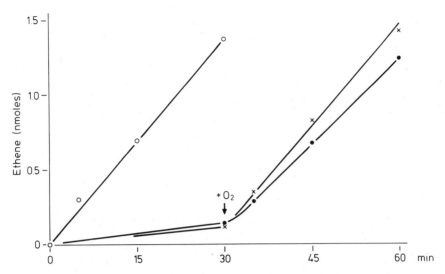

Fig. 6. Ethene formation from methional induced by bleomycin (50 µg/ml), FeCl$_3$ (0.1 mM), NADPH (0.5 mM) and isolated NADPH-cytochrome P-450 reductase (0.2 U/ml). o—o aerobic incubation, •—• anaerobic incubation (N$_2$) until 30 min, then aerobic incubation, x—x anaerobic incubation (He) until 30 min then aerobic incubation. Total ethene formed by 1 ml incubation.

Table 1. Ethene formation from methional formed during 30 min aerobic incubation. Influence of NADPH-cytochrome P-450 reductase, NADPH, bleomycin, FeCl$_3$ and various inhibitors.

	-% Ethene (mean ± SD, n = 3)
Complete system (100 µg/ml bleomycin, 0.1 mM FeCl$_3$, 0.5 mM NADPH, 0.2 U/ml isolated NADPH-cytochrome P-450 reductase, 1 mM methional)	100
- Reductase	0
- NADPH	0
- Bleomycin	2 ± 3
- FeCl$_3$	25 ± 5
+ DMSO (10 mM)	65 ± 5
+ Mannitol (50 mM)	62 ± 8
+ Glycerol (50 mM)	61 ± 9
+ GSH (10 mM)	21 ± 4

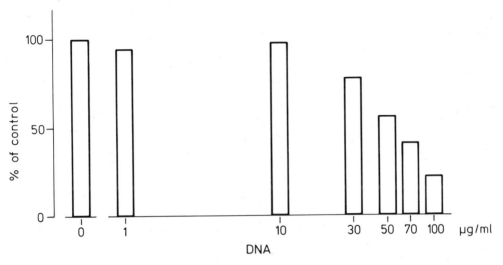

Fig. 7. Effect of DNA (salmon testes) on ethene formation from methional
induced by bleomycin (50 µg/ml), $FeCl_3$ (0.1 mM), NADPH (0.5 mM)
and isolated NADPH-cytochrome P-450 reductase (0.2 U/ml). Total
ethene formed by 1 ml incubation within 30 min.

DISCUSSION

     The isolation procedure described here yields liver cell nuclei with-
out any mitochondrial contamination as measured by the marker enzyme succinate dehydrogenase (data not shown). On the other hand, based on protein
content glucose-6-phosphatase activity, a marker enzyme of microsomes, was
in all preparations about 10 % of the activity measurable in microsomes
(data not shown). A persistent part of this enzyme activity has been
observed in all liver nuclear preparations yet published and has been
ascribed to the identity of nuclear and microsomal membranes (Arion et al.,
1983). Therefore, we suggest that NADPH- and oxygen-consumption measured
here are due to an enzyme present in the intact nucleus and do not originate from other cell particles attached to the nuclei after isolation.
This was also confirmed by microscopic examination (data not shown).

     That both NADPH- and oxygen-consumption increased considerably in the
presence of bleomycin and $FeCl_3$ is consistent with the involvement of an
NADPH-dependent nuclear enzyme. This enzyme may reduce the bleomycin-
Fe(III)-complex to bleomycin-Fe(II), a complex which has been shown to
activate molecular oxygen to reactive species which damage nuclear DNA
(for review see Burger et al., 1981). We do not have direct evidence that
nuclear DNA is damaged in our incubation system. But we observed that the
nuclear organisation was destroyed and malondialdehyde was released during
aerobic incubation with NADPH, bleomycin and $FeCl_3$ (data not shown). That
NADPH- and oxygen-consumption also increased - but to a lesser extent - in
the presence of only bleomycin (without $FeCl_3$ added) suggests that metal
ions, e.g. iron ions, are present in the isolated nuclei. The reagents
used including TES-buffer did not contain metals. This finding is very
important in relation to the cytotoxic activity of bleomycin in intact
cells, because it offers the possibility that bleomycin complexes metals
present in the cell nucleus and that these complexes can be reduced in
close proximity to DNA by an NADPH-dependent enzyme, probably NADPH-cytochrome P-450 reductase. This enzyme has been found in a number of nuclear
preparations and is involved in the mixed function oxidation of drugs

observed in liver cell nuclei (for review see Romano et al., 1983) as well as in the nuclear redox cycling of several quinone antibiotics (Bachur et al., 1982; Kennedy et al., 1982; Sinha et al., 1984). The reduced bleomycin-metal-complex formed by this nuclear enzyme binds and activates oxygen.

That liver microsomal NADPH-cytochrome P-450 reductase is able to redox cycle a bleomycin-Fe-complex is now well established (Scheulen et al., 1981; Scheulen and Kappus, 1984). During this cycle DNA added is cleaved as indicated by the formation of strand breaks (Kilkuskie et al., 1984; Scheulen et al., 1981) and the release of bases and malondialdehyde (Scheulen and Kappus, 1984). The latter has been observed by the chemically reduced bleomycin-Fe-complex and suggests that hydroxyl radicals are involved (Giloni et al., 1981; Grollman et al., 1985). In the present study we show that during incubation of NADPH-cytochrome P-450 reductase, bleomycin, $FeCl_3$, NADPH and oxygen methional is converted to ethene, an indication for the formation of hydroxyl radicals (Yamazaki, 1977). The inhibitory effect of hydroxyl radical trapping agents are in favour of this reactive oxygen species. DNA which inhibited ethene formation might compete with methional for these hydroxyl radicals. But it cannot be excluded that this effect is due to binding of bleomycin to DNA. Although it has long been known that a reduced bleomycin-Fe-complex is able to activate molecular oxygen to reactive species including hydroxyl radicals which seriously damage DNA, it has not been shown that such a reduced complex can be formed and recycled in an intact cell.

In conclusion our data indicate that the reduction of the bleomycin-Fe-complex is catalyzed in liver cell nuclei by an NADPH-dependent enzyme, probably NADPH-cytochrome P-450 reductase, and that hydroxyl radicals are formed during redox cycling of bleomycin-Fe(III/II) with this enzyme. If this process also occurs in tumor cells the therapeutic efficacy of the anti-cancer drug bleomycin may depend on the presence and activity of this enzyme in the nuclei.

ACKNOWLEDGEMENT

This study was supported by the Deutsche Forschungsgemeinschaft, Bonn, FRG.

REFERENCES

Arion, W. J., Schulz, L. O., Lange, A. J., Telford, J. N., and Walls, H. E., 1983, The characteristics of liver glucose-6-phosphatase in the envelope of isolated nuclei and microsomes are identical, J. Biol. Chem., 258:12661.
Bachur, N. R., Gee, M. V., and Friedman, R. D., 1982, Nuclear catalyzed antibiotic free radical formation, Cancer Res., 42:1078.
Bickers, D. R., Dixit, R., and Mukhtar, H., 1984, Enhancemant of bleomycin-mediated DNA damage by epidermal microsomal enzymes, Biochim. Biophys. Acta, 781:265.
Burger, R. M., Peisach, J., and Horwitz, S. B., 1981, Mechanism of bleomycin action: in vitro studies, Life Sci., 28:715.
Filser, J. G., Bolt, H. M., Muliawan, H., and Kappus, H., 1983, Quantitative evaluation of ethane and n-pentane as indicators of lipid peroxidation in vivo, Arch. Toxicol., 52:135.
Fleischer, S., and Fleischer, B., 1967, Removal and binding of polar lipids in mitochondria and other membrane systems, Meth. Enzymol., 10:406.
Giloni, L., Takeshita, M., Johnson, F., Iden, Ch., and Grollman, A. P., 1981, Bleomycin-induced strand-scission of DNA, J. Biol. Chem., 256:8608.

Grollman, A. P., Takeshita, M., Pillai, K. M. R., and Johnson, F., 1985, Origin and cytotoxic properties of base propenals derived from DNA, Cancer Res., 45:1127.

Gutteridge, J. M. C., Rowley, D. A., and Halliwell, B., 1981, Superoxide-dependent formation of hydroxyl radicals in the presence of iron salts, Biochem. J., 199:263.

Kappus, H., and Muliawan, H., 1982, Alkane formation during liver microsomal lipid peroxidation, Biochem. Pharmacol., 31:597.

Kennedy, K. A., Sligar, S. G., Polomski, L., and Sartorelli, A. C., 1982, Metabolic activation of mitomycin c by liver microsomes and nuclei, Biochem. Pharmacol., 31:2011.

Kilkuskie, R. E., Macdonald, T. L., and Hecht, S. M., 1984, Bleomycin may be activated for DNA cleavage by NADPH-cytochrome P-450 reductase, Biochemistry USA, 23:6165.

Mirabelli, Ch. K., Huang, Ch.-H., Fenwick, R. G., and Crooke, S. T., 1985, Quantitative measurement of single- and double-strand breakage of DNA in Escherichia coli by the antitumor antibiotics bleomycin and talisomycin, Antimicrob. Agents Chemother., 27:460.

Romano, M., Facchinetti, T., and Salmona, M., 1983, Is there a role for nuclei in the metabolism of xenobiotica? Drug Metab. Rev., 14:803.

Sausville, E. A., Peisach, J., and Horwitz, S. B., 1978, Effect of chelating agents and metal ions on the degradation of DNA by bleomycin, Biochemistry, 17:2740.

Scheulen, M. E., and Kappus, H., 1984, The activation of oxygen by bleomycin is catalyzed by NADPH-cytochrome P-450 reductase in the presence of iron ions and NADPH, in: "Oxygen Radicals in Chemistry and Biology," W. Bors, M. Saran, D. Tait, eds., Walter de Gruyter & Co., Berlin, New York.

Scheulen, M. E., Kappus, H., Thyssen D., and Schmidt, C. G., 1981, Redox cycling of Fe(III)-bleomycin by NADPH-cytochrome P-450 reductase, Biochem. Pharmacol., 30:3385.

Sinha, B. K., Trush, M. A., Kennedy, K. A., and Mimnaugh, E. G., 1984, Enzymatic activation and binding of adriamycin to nuclear DNA, Cancer Res., 44:2892.

Swanson, M. A., 1955, Glucose-6-phosphatase from liver, Meth. Enzymol., 2:541.

Trush, M. A., Mimnaugh, E. G., Ginsburg, E., and Gram, T. E., 1982, Studies on the interaction of bleomycin A_2 with rat lung microsomes. II. Involvement of adventitious iron and reactive oxygen in bleomycin-mediated DNA chain breakage, J. Pharmacol. Exp. Ther., 221:159.

Umezawa, H., Takita, T., Sugiura, Y., Otsuka, M., Kobayashi, S. and Ohno, M., 1984, DNA-bleomycin interaction - Nucleotide sequence-specific binding and cleavage of DNA by bleomycin, Tetrahedron, 40:501.

Yamanaka, N., Kato, T., Nishida, K., and Ota, K., 1978, Enhancement of DNA chain breakage by bleomycin A_2 in the presence of microsomes and reduced nicotinamide adenine dinucleotide phosphate, Cancer Res., 38:3900.

Yamazaki, I., 1977, Free radicals in enzyme-substrate reactions, in: "Free Radicals in Biology, III," W. A. Pryor, ed., Academic Press, New York, San Francisco, London.

SUPEROXIDE DISMUTASE MODIFICATION AND GENOTOXICITY OF

TRANSITION-METAL ION CHELATORS

T. P. Coogan, D. G. Stump, D. A. Barsotti and
I. Y. Rosenblum

Department of Pharmacology and Toxicology
Philadelphia College of Pharmacy and Science
Philadelphia, PA 19104

INTRODUCTION

Superoxide radicals are considered to be of major
significance in the mechanism of oxygen-mediated cell toxicity.
Superoxide and/or superoxide-derived free radical species have
been implicated in damage to cells in vivo (Fridovich, 1978) and
in vitro (Holland et al., 1982). Specifically, damage to DNA
(Emerit et al., 1982; Cunningham and Lokesh, 1983), proteins
(Lavelle et al., 1973) and lipids (Chance et al., 1979) has been
reported in a number of experimental systems. Superoxide
dismutase (SOD), a widely distributed enzyme in aerobic tissues,
is recognized for its important scavenging function in the
primary defense of cells (Fridovich, 1978). SOD protects the
cell against the deleterious effects of superoxide by catalyzing
its dismutation to molecular oxygen and hydrogen peroxide. In
eucaryotic cells, two forms of SOD are found; a cytoplasmic
enzyme containing copper and zinc and a mitochondrial enzyme
containing manganese (Fridovich, 1978). The importance of the
metalloprotein complex in the catalytic function of SOD has been
firmly established (Fridovich, 1978).

Differential sensitivity of testicular tissue to the
cytotoxic and mutagenic effects of chemotherapeutic drugs and
other environmental toxicants has been reported (Meistrich,
1984). The presence of various cell types, e.g. germinal and
interstitial, in testicular tissue may explain some of these
differences. Germinal cells are known to be an important target
for both oxidative (Holland and Storey, 1981;) and drug-induced
(Meistrich, 1984) damage. However, little information exists
regarding the distribution and role of SOD in spermatogenic
cells. In the present investigation, the occurrence of SOD
activity in isolated, premeiotic male germ cells (MGC) was
established. The effects of three transition metal ion
chelators - diethyldithiocarbamate (DDC), bathocuproine
disulfonate (BC), and deferoxamine mesylate (DM) - on cellular
SOD activity were assessed and compared to the effects of
chelator treatment in a noncellular system, using purified

bovine erythrocyte SOD. The ability of each chelator to induce
single-strand DNA damage was also determined.

MATERIALS AND METHODS

Chemicals

All chemicals were purchased from Sigma Chemical Co. (St.
Louis, MO) except as indicated. Trypsin was purchased from Difco
Laboratories (Detroit, MI). Proteinase-K, tetraethylammonium
hydroxide and 3,5-diaminobenzoic acid dihydrochloride were
obtained from E. Merck (Darmstadt, F.R.G.), RSA Laboratories
(Ardsley, NY) and Aldrich Chemicals (Milwaukee, WI),
respectively. Deferoxamine mesylate (Desferal) was obtained
from CIBA Pharmaceutical Co. (Summit, NJ).

Isolation of Male Germ Cells and Exposure Protocol

Immature, Sprague-Dawley rats were obtained from Taconic
Farms (Germantown, NY). Three to four week old animals were
sacrificed by cervical dislocation and their testes removed.
Seminiferous tubules were separated and carefully teased apart,
and male germ cells isolated using a modification of the
procedure of Romrell et al. (1976).

Approximately one gram of tubules were transferred to 20 ml
of 0.01 M phosphate buffered saline (pH 7.2), containing 0.1%
glucose (PBSG). To remove connective tissue and interstitial
cells, collagenase (1 mg/ml) and trypsin (2.5 mg/ml) were added
to the tissue. Following incubation for 18 min at 33°C in a
shaker water bath (150 cycles/min), the suspension was allowed
to sediment and the supernatant decanted. The pellet was washed
twice in 20 ml PBSG and incubated a second time in 20 ml of the
above mixture with the addition of DNase (1.0 ug/ml). After
incubating for 15 min (33°C, 150 cycles/min), the tissue was
disrupted by gently pipetting and layered over 10 ml of PBSG
containing 0.5% BSA and 1.0 ug/ml DNase. The cells were
pelleted by centrifugation (2000 x g, 10 min) at 4°C and
resuspended in 30 ml PBSG. Spermatogenic cells were collected
following filtration of this suspension through coarse (40 mesh)
and fine (100 mesh) stainless steel filters (A. H. Thomas,
Swedesboro, NJ). Cell densities were estimated using a
hemacytometer, and cell viability was determined using the
method of trypan blue exclusion. Exposures to the various test
compounds were conducted by incubation of isolated MGC
suspensions (1 x 10^{6} cells/ml) for one hour at 37°C in a
humidified atmosphere of 6% carbon dioxide. DNA single-strand
breakage and SOD activity were assessed following incubation.

Determination of SOD Activity

SOD activity was measured using the xanthine oxidase –
cytochrome c reduction method (McCord and Fridovich, 1969).
Protein content was determined using the method of Lowry et al.
(1951). Results are expressed as means $\pm$ S.D. Significant
differences due to various treatments (control vs. treatments)
were analyzed using Dunnett's procedure with $p < 0.05$ (Steele and
Torrie, 1960).

Non-Cellular System. The effects of varying concentrations
of each chelating agent on purified bovine erythrocyte SOD were
determined. Prior to the addition of xanthine oxidase, either
water (blank) or 330ng of SOD was added to each cuvette
containing the indicated concentration of test compound. Zero
percent inhibition was defined as the difference between rates
at 0 and 330 ng SOD. Rate differences were compared to control
values to determine percent inhibition. All samples were
analyzed in triplicate and each assay result represents the mean
of three rate determinations.

   Cellular System. After exposure, MGC (2×10^7 cells) were
pelleted, resuspended in 0.5 ml PBSG and lysed by repeated (x3)
freeze-thaw cycles. Cellular debris was pelleted and three 50
ul aliquots of supernatant were assayed for SOD activity.
Samples were assayed in triplicate and data were expressed as
units of SOD per mg protein.

Determination of DNA Damage

   DNA damage was determined using a modification of the
alkaline elution technique of Kohn, et al. (1981a) and Bradley
(1982). After incubation, cells were gently mixed and 2 ml
aliquots (2×10^6 cells) were diluted in 10 ml of ice-cold PBSG
and loaded onto elution filters (2 um pore size, 25 mm,
polyvinyl chloride filters; Nucleopore Corp., Pleasanton, CA).
Cells were disrupted by adding 1.5 ml of a lysis solution
containing proteinase-K (0.5 mg/ml), 2% sodium dodecyl sulfate
(SDS) and 25 mM disodium EDTA (pH 9.7). After 30 min, the
solution was filtered by gravity and filters were washed with
0.5 ml of lysis solution without proteinase-K. DNA was eluted
in the dark (40 ul/min) using a solution containing 40 mM EDTA
and 1% SDS, adjusted to pH 12.1 with tetrapropylammonium
hydroxide. Fractions were collected at 3 h intervals for 12 h.
DNA content of each fraction and filter was determined using a
microfluorometric DNA assay (Kissane and Robins, 1958) as
modified by Erickson et al.(1980) and Bradley (1982).

RESULTS

   Three transition-metal ion chelating agents - DDC, BC and
DM - were assessed for their ability to inhibit SOD in a
cellular and non-cellular system. Each compound was screened
initially using purified bovine erythrocyte SOD. As shown in
Table 1, all of the compounds inhibited SOD activity; DDC was
the most potent. At 0.5 mM DDC, 64% inhibition of SOD activity
was demonstrated. At the same concentration, neither BC nor DM
was effective. However, 86% inhibition was observed using a
twenty fold higher concentration of DM (10.0 mM). DM exposure
resulted in a concentration-dependent inhibition of SOD. In
this system, exposure to BC was limited by solubility to 2.3 mM
and the maximum inhibition was determined to be 27%. Although
the maximum inhibition reported for DDC was 64%, exposure to
concentrations >1.0 mM resulted in total inhibition of the blank
reaction.

   Typically, one gram of seminiferous tubules, obtained from
the testes of prepubertal rats, yielded 2×10^8 cells.
Following the isolation procedure outlined above, a single cell

TABLE 1

Effects of Transition-Metal Ion Chelators on Bovine Eythrocyte SOD

Treatment	% Inhibition
10.0 mM DDC	-[a]
1.0 mM DDC	64
0.5 mM DDC	64
0.1 mM DDC	18
2.3 mM BC[b]	27
1.0 mM BC	27
0.5 mM BC	9
0.1 mM BC	18
10.0 mM DM	86
5.0 mM DM	57
1.0 mM DM	14
0.5 mM DM	0

[a] Blank reaction inhibited.
[b] Insoluble at higher concentrations

suspension of MGC (>90%) was obtained that consisted of spermatogonia and premeiotic spermatocytes. Viability, as determined by trypan blue exclusion, was >90% before specific treatments and >80% following exposure to the chelators for one hour at 37°C.

The effects of transition-metal ion chelators on SOD activity in MGC is presented in Table 2. The SOD content of isolated, immature MGC was determined to be 5.53 ± 1.49 units per mg protein. After a one hour exposure to each of the chelators, only DDC was able to significantly inhibit SOD activity. A 54% reduction in activity was obtained following exposure to 2.5 mM DDC. Upon exposure to higher concentrations of DDC, SOD activity was inhibited to levels below the standard assay range (66 - 330 ng).

TABLE 2

Effects of Transition-Metal Ion Chelators on SOD Activty in Male Germ Cells.

	n	U SOD[a]/mg protein[b]	% inhibition[c]
PBS	17	5.53 ± 1.49	-
10.0 mM DDC	3	0 [d,e]	100
7.5 mM DDC	3	0 [d,e]	100
5.0 mM DDC	3	0 [d,e]	100
2.5 mM DDC	3	2.56 ± 0.41 [d]	54
1.0 mM DDC	3	4.38 ± 0.56	21
0.1 mM DDC	3	3.89 ± 0.65	30
10.0 mM DM	3	6.39 ± 0.71	0
1.0 mM DM	5	4.60 ± 1.19	17
0.1 mM DM	3	7.33 ± 0.95	0
10.0 mM BC	6	5.54 ± 0.98	0
1.0 mM BC	5	4.68 ± 2.23	15
0.5 mM BC	6	5.78 ± 2.68	0

[a] One unit = SOD required to inhibit cytochrome c reduction by 50%.
[b] Values are expressed as the mean $\pm$ S.D.
[c] Inhibition of SOD activity compared to PBS control.
[d] Value significantly different from control, $p < 0.05$
[e] Value below standard assay range.

The capacity of each of the chelating agents to induce single-strand DNA breaks (SSBs) was determined and the resultant data are presented in Figure 1. Although 10 mM DDC induced a frequency of SSBs no different from that of the PBS control, exposure to 0.1 and 1.0 mM DDC resulted in an increase in single-strand DNA damage. Although BC did not induce SSBs, an inverse concentration-dependent elution profile was observed. The frequency of SSBs after exposure to DM was no different compared to controls.

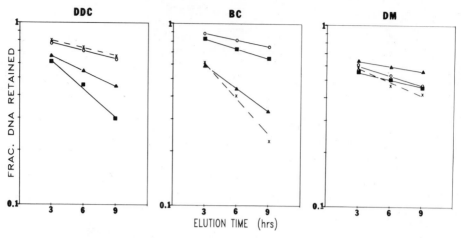

Fig. 1.   Single-strand DNA damage following 1 hr exposure
to chelator. Fractions were collected at 3, 6 and
9 hrs. Concentrations tested were: x - - x, PBS
Control; o————o, 10 mM; ■———■ , 1 mM; ▲———▲ ,
0.1 mM.

DISCUSSION

    Male germ cells are an important target for several drugs as
well as many environmental toxicants. Recently, an in vitro
model using isolated, premeiotic rat MGC to probe genotoxicant-
induced DNA damage was described (Coogan et al., 1985).   In
these studies, the response of MGC to known genotoxic agents was
evaluated, and the applicability of the model to the study of
genotoxic mechanisms was demonstrated. Since DNA damage
following toxic insult can be mediated by superoxide radicals
(Emerit et al., 1982; Cunningham and Lokesh, 1983), it was of
interest to determine the occurrence of SOD in male germ cells.
If the inhibition of SOD resulted in enhanced DNA damage, then a
role for superoxide radicals would be indicated. Transition-
metal ion chelators have been reported to inhibit SOD activity
(Heikkila et al., 1978). In the present investigation, the
inhibitory effects of DDC, BC and DM on SOD activity were
determined. The capacity of these compounds to induce
single-strand DNA damage was also assessed.

The current experiments establish that premeiotic male germ cells contain SOD. Previously, SOD has been reported to occur in whole testicular homogenates (Carstensen et al., 1976) and spermatozoa (Mennella and Jones, 1980). Since the testes contain various cell types, including germ cells at several stages of differentiation, the localization of SOD within this tissue represents important new information.

Diethyldithiocarbamate and bathocuproine disulfonate are known copper chelating agents. DDC inhibited male germ cell and purified bovine erythrocyte SOD activities. These data are consistent with previous studies demonstrating the ability of DDC to inhibit SOD in a non-cellular system, as well as in homogenates of brain and liver (Heikkila et al., 1976). In vivo exposure to DDC has been reported to decrease SOD activity in brain (Heikkila et al., 1976; Puglia and Loeb, 1984), liver (Heikkila et al., 1976) and lung (Frank et al., 1978). In contrast to DDC, BC did not inhibit cellular SOD activity, and was only marginally effective in the inhibition of purified SOD.

Heikkila et al. (1976) proposed that the inhibition of SOD by DDC may be due to copper chelation. The present studies demonstrate that BC, although not as potent as DDC, was inhibitory in the non-cellular system. Perhaps this effect was due to greater accessability of DDC to the active site of SOD, as suggested by Puglia and Loeb (1984) for other copper chelators.

DM, an iron-specific chelating agent, had no effect on male germ cell SOD activity, but resulted in a concentration-dependent inhibition of purified bovine SOD. One possible interpretation of these data is that DM does not cross the plasma membrane. The results of recently conducted studies in this laboratory (Coogan et al., 1985) suggest that DM does enter MGC; however, the intracellular concentration was not determined. Another possibility is that DM resulted in non-specific chelation of copper, leading to the inhibition of bovine SOD. The high concentration of DM required to inhibit SOD activity supports this explanation.

The ability of the chelating agents to induce single-strand DNA damage was addressed to determine their potential use as molecular probes in genotoxicity studies. The results indicate that DDC induced single-strand DNA damage at low (0.1 and 1.0 mM), but not high (10 mM), concentrations. One hypothesis is that DDC at low concentrations binds to DNA and single-strand breaks are generated during excision repair. This mechanism of DNA strand scission has been suggested by other investigators (Hanawalt et al., 1979; Kohn, 1981b). DDC is known to affect enzymatic processes in addition to SOD (Eneanya et al., 1981; Puglia and Loeb, 1984). It is possible that 10 mM DDC inhibits enzymatic repair; therefore, strand scission does not occur and the DNA in treated cells elutes at a rate similar to controls. MGC pretreated for one hour at 37°C with 10 mM DDC, followed by a one hour incubation in DDC-naive PBSG resulted in a significant increase in single-strand DNA damage (unpublished data). The resumption of repair that occurs following replacement of the incubation media would account for the observed increase in single-strand scission. Further

experiments are needed to assess the effects of DDC on DNA
repair.

Exposure to BC resulted in a decreased frequency of
single-strand breaks that was inversely related to
concentration. This relationship may reflect DNA interstrand
crosslinking. Another possibility is that single-strand breaks
in control cells were generated during incubation by a
copper-mediated process. Therefore, the decrease in
single-strand damage may be due to copper chelation. Further
experimentation is required to clarify this observation.

In contrast to the copper chelators, DM had no effect on the
frequency of single-strand breaks. DM is used clinically in the
treatment of acute iron intoxication, chronic iron overload, and
chronic iron storage diseases. Desferal was found to be
non-cytotoxic and non-genotoxic in male germ cells following
exposure for one hour at high concentrations.

Of the three transition-metal ion chelators studied, only
DDC significantly altered SOD activity in isolated, premeiotic
MGC. Although the paradoxical genotoxic effects of this
compound precludes its use as a molecular probe in genotoxicity
studies, this finding is of significant clinical value. DDC is
a metabolite of disulfiram, a compound used in the treatment of
alcoholism (Eneanya et al., 1981). The demonstration of
DDC-induced genotoxicity in male germ cells also has important
toxicologic implications. Following the administration of
disulfiram, this lipophilic drug has been found widely
distributed in several tissues, including the testes (Eneanya et
al., 1981). Therefore, the current findings add a previously
unreported dimension to the toxicologic profiles of disulfiram
and DDC.

ACKNOWLEDGEMENT

This investigation was supported by a grant from the
Pfeiffer Research Foundation.

REFERENCES

Bradley, M.O., Dysart, G., Fitzsimmons, K., Harbach, P., Lewin,
J. and Wolf, G. 1982, Measurements by filter elution of DNA
single- and double-strand breaks in rat hepatocytes:
effects of nitrosamines and gamma-irradiation, Cancer Res.,
42:2592-2597.

Carstensen, H., Marklund, S., Damber, J.E., Nasman, B. and
Lindgren, S., 1976, No effect of oxygen in vivo on plasma or
testis testosterone in rats and no induction of testicular
superoxide dismutase. J. Steroid. Biochem., 7:465-467.

Chance, B., Sies, H. and Boveris, A., 1979, Hydroperoxide
metabolism in mammalian organs. Physiol. Rev., 59:527-605.

Coogan, T.P., Rosenblum, I.Y., Barsotti, D.A., and Heyner, S.,
1985, Genotoxicity in male germ cells: a mechanistic
approach. The Toxicologist, 5:736

Cunningham, M.L. and Lokesh, B., 1983, Superoxide anion generated by potassium superoxide is cytotoxic and mutagenic to Chinese hamster ovary cells. Mutation Res., 121:299-304.

Emerit, I., Keck, M., Levy, A., Feingold, J. and Michelson, A.M. 1982, Activated oxygen species at the origin of chromosome breakage and sister-chromatid exchanges. Mutation Res., 103:165-172.

Eneanya, D.I., Bianchine, J.R., Duran, D.O. and Andresen, B.D., 1981, The action and metabolic fate of disulfiram. Ann. Rev. Pharmacol. Toxicol., 21:575-596

Erickson, L.C., Osieka, R., Sharkey, N.A. and Kohn, K.W., 1980, Measurement of DNA damage in unlabeled mammalian cells analyzed by alkaline elution and a fluorometric DNA assay. Anal. Biochem., 106:169-174

Frank, L., Wood, D.L. and Roberts, R.J., 1978, Effects of diethyldithiocarbamate on oxygen toxicity and lung enzyme activity in immature and adult rats. Biochem. Pharmacol., 27:251-254

Fridovich, I., 1978, The biology of oxygen radicals. Science 201:875-880.

Hanawalt, P.C., Cooper, P.K., Ganesan, A.K. and Smith, C.A., 1979, DNA repair in bacteria and mammalian cells. Ann. Rev. Biochem., 48:783-836

Heikkila, R.E., Cabbat, F.S. and Cohen, G., 1976, In vivo inhibition of superoxide dismutase in mice by diethyl-dithiocarbamate. J. Biol. Chem., 251:2182-2185

Heikkila, R.E., Cabbat, F.S. and Cohen, G., 1978, Inactivation of superoxide dismutase by several thiocarbamic acid derivatives. Experientia 34:1553-1554.

Holland, M.K. and Storey, B.T., 1981, Oxygen metabolism in mammalian spermatozoa. Generation of hydrogen peroxide by rabbit epedidymal spermatozoa. Biochem. J., 198:273-280.

Holland, M.K., Alvarez, J.G. and Storey, B.T., 1982, Production of superoxide and activity of superoxide dismutase in rabbit epedidymal spermatozoa. Biol. Reprod. 27:1109-1118.

Kissane, J.M. and Robins, E., 1958, The fluorometric measurement of deoxyribonucleic acid in animal tissues with special refererence to the central nervous system. J. Biol. Chem., 233:184-188.

Kohn, K.W., Ewig, R.A.G., Erickson, L.C. and Zwelling, L.A., 1981a, Measurement of strand breaks and cross-links by alkaline elution, in: "DNA Repair, A Laboratory Manual of Research Procedures," E. Friedberg and P. Hanawalt, eds., Marcel Dekker Inc., NY.

Kohn, K.W., 1981b, DNA damage in mammalian cells. Bioscience, 31:593-597.

Lavelle, F., Michelson, A.M. and Dimitrijevic, L., 1973, Biological protection by superoxide dismutase. *Biochem. Biophys. Res. Commun.*, 55:350-357.

Lowry, O.H., Rosenbrough, N.J., Farr, A.L. and Randall, R.J., 1951, Protein measurement with the folin phenol reagent. *J. Biol. Chem.*, 193:265-275

McCord, J.M. and Fridovich, I., 1969, Superoxide dismutase: an enzymatic function for erythrocuprein (hemocuprein). *J. Biol. Chem.*, 244:6049-6055.

Meistrich, M.L., 1984, Stage-specific sensitivity of spermatogonia to different chemotherapeutic drugs. *Biomedicine & Pharmacotherapy*, 38:137-142.

Mennella, M.R.F. and Jones, R., 1980, Properties of spermatozoal superoxide dismutase and lack of involvement of superoxides in metal-ion catalysed lipid peroxidation reactions in semen. *Biochem. J.*, 191:289-297.

Puglia, C.D. and Loeb, G.A., 1984, Influence of rat brain superoxide dismutase inhibition by diethyldithiocarbamate upon the development of central nervous system oxygen toxicity. *Toxicol. Appl. Pharmacol.*, 75:258-264.

Romrell, L.J., Bellve, A.R.B. and Fawcett, D.W., 1976, Separation of mouse spermatogenic cells by sedimentation velocity. *Dev. Biol.*, 49:119-131.

Steele, R.G.D. and Torrie, J.H., 1960, "Principles and Procedures of Statistics," McGraw-Hill, NY.

TOXICOLOGIC IMPLICATIONS OF THE IRON-DEPENDENT ACTIVATION OF BLEOMYCIN A_2 BY MOUSE LUNG MICROSOMES

Michael A. Trush[1] and Edward G. Mimnaugh[2]

[1]Dept. Environmental Health Sciences, Johns Hopkins University Baltimore, MD 21205 and [2]National Cancer Institute, NIH Bethesda, MD 20205

INTRODUCTION

The bleomycins are a family of glycopeptide antiobiotics currently being utilized in the chemotherapy of human neoplastic diseases (1). Unlike many antineoplastic agents, bleomycin does not elicit significant myelosuppressive activity, which is advantageous not only when considering bleomycin as a single chemotherapeutic agent but also when incorporating it in multiple agent modalities. The administration of bleomycin is accompanied however by the development of dose-limiting pulmonary toxicity which can manifest as life-threatening interstitial fibrosis. Understanding the underlying chemico-biological interactions by which bleomycin initiates alterations in cellular function is important from the standpoint of developing rational chemoprotective strategies.

Recent studies in cell-free chemical systems have provided insight into the actions of bleomycin at the molecular level (2-4) and in particular have provided evidence supporting the concept that a reactive intermediate of bleomycin mediates DNA deoxyribose cleavage, yielding toxic base propenals (5). Such observations are relevant to the interaction of bleomycin with cellular systems since alterations in DNA are believed to account for bleo-mycin's cytotoxic effect on neoplastic cells (6). In this regard, bleomycin is preferentially toxic to aerobic cells (7) while simultaneous exposure to non-toxic levels of oxygen potentiate bleomycin-induced pulmonary toxicity (8). In addition to the degree of cellular oxygenation, differences in the tissue content of bleomycin hydrolase, a cytosolic aminopeptidase, have been proposed as the basis for the sensitivity or resistance of tumors to the cytotoxic actions of bleomycin (9). Moreover, normal host tissues, such as the lung or skin, which are relatively deficient in this enzyme are more susceptible to bleomycin-induced toxicity (9). The importance of bleomycin hydrolase in cells is that is hydrolyzes the terminal amino group from the molecule (see Table 3) thereby preventing the proper binding of iron, an obligatory cofactor in order for bleomycin to damage DNA in cell-free chemical systems (10). In addition to bleomycin hydrolase, Matsuda et al. (11) have recently demonstrated that E. coli having increased levels of detoxi-fying systems for reactive oxygen exhibited a resistance to the cytotoxic action of bleomycin A_2. Based on these observations it would appear that the mechanism of bleomycin-induced toxicity in intact cells is iron-, oxygen- and reactive oxygen-dependent. Thus, any biological system proposed to account for the activation of bleomycin and its cytotoxicity in lung cells should be modulated by these factors.

METHODS

Microsomal Preparation. Lungs from male CDF_1 mice were homogenized in 2 volumes of 150 mM KCl-50 mM Tris-HCl buffer (pH 7.4) in a Potter-type glass homogenizer with a motor driven Teflon pestle. After homogenization, tissues were diluted to 25% w/v, centrifuged at 9000 x g for 20 min and the supernatants were centrifuged at 100,000 x g for 1 hr. The resulting microsomal fractions were washed by resuspending in KCl-Tris-HCl buffer and centrifuging a second time at 100,000 x g for 1 hr. Unless otherwise stated, the washed microsomes were resuspended in KCl-Tris-HCl-buffer that had been bubbled with O_2. Microsomal protein was determined by the method of Lowry et al. (12) using bovine serum albumin as the standard. Microsomes were diluted to 3.5 mg of protein per ml in O_2 saturated KCl-Tris-HCl buffer.

Incubation Conditions for Determining Bleomycin-Mediated Cleavage of DNA Deoxyribose. Kuo and Haidle (13) initially demonstrated that bleomycin-mediated chain breakage was accompanied by the generation of a product which reacts with thiobarbituric acid (TBA), yielding a chromophore which can be measured spectrophotometrically at 533 nm. This reaction with TBA is due to aldehyde groups liberated from the ring opening of deoxyribose derived from those bases released after bleomycin-mediated chain breakage. Accordingly, microsomes (final concentration, 0.5 mg of microsomal protein per ml) were incubated at 37°C in the absence or presence of bleomycin A_2 (100 µM) and/or calf thymus DNA (0.25 mg/ml) with an NADPH-generating system in glass vials (1.8 cm diameter x 4 cm high) in a total volume of 1.75 ml. The NADPH-generating system consisted of NADP (1.9 mM), glucose-6-phosphate (20 mM), glucose-6-phosphate dehydrogenase (1.1 I.U./ml) and magnesium chloride (9mM). Incubations were conducted in a Dubnoff metabolic shaker with a covered hood connected to an O_2 atmosphere (flow rate, 5.0 liters/min). Reactions were stopped by the addition of 0.75 ml of a 2.0 M trichloroacetic acid-1.7 N HCl solution and the samples were centrifuged at 1000 x g for 10 min at room temperature. One-half milliliter aliquots of the clear supernatants were then reacted with 2 ml of 1% TBA and the chromophore was developed at 90°C for 10 min. After the samples were cooled, the absorbance at 533 nm was determined. Bleomycin-mediated DNA deoxyribose cleavage is expressed as nanomoles of TBA reacting material per milligram protein per 60 min using an extinction coefficient of $1.53 \times 10^{-5} M^{-1} cm^{-1}$ at 533 nm. Note that the absorption at 533 nm from incubations conducted in the absence of DNA is attributed to lipid peroxidation. Zero time blanks containing all of the components of the incubation mixture were used throughout.

RESULTS

Incubation of bleomycin A_2 with mouse lung microsomes and a NADPH-generating system in the presence of DNA but not its absence, resulted in significant generation of thiobarbituric acid reactive products (TBAR) (Table 1). The generation of TBAR is attributed to the cleavage of DNA deoxyribose by an activated bleomycin intermediate (2,3). In the absence of the NADPH-generating system or in the presence of the NADPH-generating system and heat-inactivated microsomes, there was negligible bleomycin-mediated cleavage of DNA deoxyribose. These results indicate that the activation of bleomycin was dependent on a source of reducing equivalents and catalytically active microsomal enzyme(s). In addition to the NADPH-generating system, NADPH or NADH could serve as the source of reducing equivalents with NADH giving approximately 50% of the activity obtained with an equivalent concentration of NADPH (data not presented). Bleomycin-mediated cleavage of DNA deoxyribose was also dependent on aerobic conditions as indicated by the significant reduction in TBAR when incubations were conducted under a nitrogen atmosphere.

Table 1. Interaction of Bleomycin A_2 with Mouse Lung Microsomes:
Enhancement of DNA Damage but not Lipid Peroxidation

Addition(s) to Mouse Lung Microsomes	TBAR nmoles/mg protein/60 min	Primary reaction yielding TBAR
NADPH-generating System (GS)	5.6 ± 0.3	Lipid peroxidation
GS, DNA	1.9 ± 0.4[a]	Lipid peroxidation
GS, Blm A_2	0.6 ± 0.2[a]	Lipid peroxidation
GS, DNA, Blm A_2	41.4 ± 4.5[a]	Deoxyribose cleavage
DNA, Blm A_2	0.4 ± 0.2[b]	---
GS, DNA, Blm A_2 and heat-inactivated microsomes	0.3 ± 0.2[b]	---
GS, DNA, Blm A_2 and nitrogen atmosphere	3.6 ± 1.2[b]	Deoxyribose cleavage

Mouse lung microsomes (0.5 mg protein per ml) were incubated under O_2 for
60 min and as indicated in the presence of an NADPH-generating system, 100
μM bleomycin A_2 (Blm A_2) and calf thymus DNA (0.25 mg/ml). Data are mean
$\pm$ S.D. (N=3-6).

[a]$p < 0.05$ from addition of NADPH-generating system.

[b]$p < 0.05$ from GS, DNA, Blm A_2.

Table 2. Effect of Various Additions which Modulate the Availability of
Iron or Reactive Oxygen on Bleomycin A_2-Mediated DNA Deoxyribose
Cleavage Catalyzed by Mouse Lung Microsomes.

Addition	DNA deoxyribose cleavage nmoles TBAR/mg protein/60 min
None[a]	43.7 ± 5.3
EDTA (10^{-4}M)	2.5 ± 1.2[b]
Fe^{3+} (10^{-4}M)	97.4 ± 10.6[b,c]
Ascorbic acid (10^{-4}M)	76.0 ± 7.3[b,c]
GSH (10^{-3}M)	34.0 ± 2.1[b]
SOD (10^{-5}M)	26.9 ± 3.6[b]
Redox cycling chemicals	
Mitomycin C (5×10^{-4}M)	82.7 ± 3.5[b,c]
Paraquat (5×10^{-4}M)	83.6 ± 7.2[b,c]

[a]Mouse lung microsomes (0.5 mg protein per ml) were incubated under O_2 for
60 min in the presence of an NADPH-generating system, 10^{-4}M Blm A_2 and
0.25 mg/ml calf thymus DNA. Data are mean $\pm$ S.D. (N=3-6).

[b]$p < 0.05$ from None.

[c]The addition of ascorbic acid, Fe^{3+}, mitomycin C or paraquat in the absence
of bleomycin did not cleave DNA deoxyribose.

[d]Bleomycin inhibited the lipid peroxidation of microsomes alone or that
enhanced by the addition of mitomycin C or paraquat.

As shown in Table 1, incubation of bleomycin A_2 with mouse lung microsomes in the presence of an NADPH-generating system and under O_2 significantly
inhibited the generation of malondialdehyde originating from lipid peroxidation. This inhibiting effect of bleomycin is in contrast to the stimulation
of mouse lung microsomal lipid peroxidation observed with other pulmonary
toxins (14,15) and is probably due to bleomycins' ability to chelate metals
and in this particular situation, iron (16). It is of note that under these
incubation conditions the bleomycin molecule subsequently loses its
ability to both cleave DNA deoxyribose and to inhibit lipid peroxidation
(16-19).

Studies with cell-free chemical systems have emphasized that the activation of bleomycin to a DNA-damaging intermediate is highly dependent on
the availability of iron, the ability of bleomycin to bind iron and the subsequent reduction of ferric iron to the ferrous form (20). The ascorbic
acid radical and the superoxide anion both have the capability to reduce
ferric iron (21,22). Accordingly, addition of EDTA or SOD significantly
inhibited bleomycin-mediated DNA deoxyribose cleavage whereas addition of
Fe^{3+}, ascorbic acid or superoxide generating compounds, namely mitomycin C
and paraquat, dramatically enhanced this reaction (Table 2). These results
with mitomycin C and paraquat are in contrast with those of Schulen and
Kappus (23) who observed a reduction in bleomycin-mediated generation of
TBAR in the presence of mitomycin C and paraquat with NADPH cytochrome P-450
reductase and NADPH. However, as previously discussed, if the concentration
of NADPH becomes limiting, then chemical redox cycling will not be maintained
and reactions mediated by reactive oxygen may not proceed and for that matter
may artifactually appear to be inhibited (14). In fact, these authors did
note a substantial increase in NADPH consumption in the presence of these
redox cycling chemicals. Nonetheless, both our results (18,24) and that of
Schulen and Kappus (23) are consistent with the involvement of iron and
reactive oxygen in the cleavage of DNA deoxyribose by bleomycin.

Bleomycin is a complex molecule in that it has moieties which recognize
and bind to specific regions of DNA and a separate and distinct metal binding
portion (Table 3). Metabolism of bleomycin by bleomycin hydrolase yields
desamido-bleomycin (des-Blm) which lacks the fifth axial nitrogen required
for proper iron coordination and subsequent drug activation (10). In contrast to bleomycin A_2, des-Blm A_2 failed to elicit significant TBAR from
DNA with mouse lung microsomes even in the presence of exogenous ferric iron
(Table 3). On the other hand, two bleomycin analogs, namely Blm-BAPP and
Blm-PEP, modified at the terminal substituent but not the iron binding sites
were quite effective in mediating deoxyribose cleavage. Blm-BAPP and Blm-PEP
are pulmonary toxins (25) whereas subcutaneous injection of mice with des-Blm
A_2 does not result in pulmonary fibrosis (26).

DISCUSSION

One of the possible results of the interaction of a drug with enzyme
systems is its metabolism to a biologically reactive intermediate (27).
Covalent binding to protein or nucleic acid and/or alterations to membrane
phospholipids or nucleic acids have been used as indications that activation
of a chemical to a reactive intermediate has occurred. Previous studies with
cell-free chemical systems have indicated that the antineoplastic drug bleomycin is "activated" to an intermediate which cleaves DNA deoxyribose (2-4),
yielding toxic base propenals (5). The results of this study demonstrate
that the interaction of bleomycin A_2 with mouse lung microsomes results in
drug activation as evidenced by the significant cleavage of DNA deoxyribose.
Thus, bioactivation of bleomycin by both tumor microsomes (25) and lung
microsomes may be the mechanism of bleomycin's chemotherapeutic effect and
its selective toxicity to the lung and skin. Bleomycin activated by either
the microsomal or nuclear membrane mixed-function oxidase system is capable

Table 3. Comparison of DNA Deoxyribose Cleavage Mediated by Several Bleomycin Analogs

*Denotes proposed iron-binding sites

Bleomycin	Terminal Substituent (X)	Drug-mediated DNA Deoxyribose Cleavage nmoles TBAR/mg protein/60 min
A_2	$-NH-(CH_2)_3-S^+-(CH_3)_2$	37.2 ± 2.9[a]
A_2, Fe^{3+}		80.2 ± 8.0
BAPP	$-NH(CH_2)_3-NH-(CH_2)_3-NH-(CH_2)_3CH_3$	44.5 ± 4.3
BAPP, $F3^{3+}$		81.9 ± 3.3
PEP	$-NH-(CH_2)_3-NH-C \text{(H)(CH}_3)-\bigcirc-H_2SO_4$	38.2 ± 1.3
PEP, Fe^{3+}		84.8 ± 3.5
Desamido Blm A_2	$-NH-(CH_2)_3-S^+-(CH_3)_2$	1.8 ± 0.4
Desamido Blm A_2, Fe^{3+}		6.3 ± 1.3

[a]Incubations were conducted as described in Table 1. Data are mean $\pm$ S.D. (N=3-6).

of damaging DNA in nuclei from rabbit lung or the AH66 hepatoma (28,29). On the other hand, the interaction of bleomycin with lung microsomes does not result in lipid peroxidation (Table 1 and 16-19,29).

While bioactivation may be a requisite process in order for bleomycin to initiate a cytotoxic response, it appears that other bleomycin-subcellular interactions could exert considerable influence on the extent to which bleomycin interacts with and is activated by the endoplasmic reticulum or the nucleus (Figure 1). For example, bleomycin exists as a copper complex in the plasma and following penetration into cells the copper is removed by a sulphydryl-dependent process (30). Iron can not directly displace this copper (18,31) and consequently the rate of copper removal would have a modulating influence on the iron-dependent bioactivation of bleomycin since interaction of the bleomycin-copper complex with lung microsomes does not result in deoxyribose cleavage (18). However, once the copper is removed the bleomycin molecule can bind iron, possibly from non-heme membrane proteins (32), or alternatively be metabolized by bleomycin hydrolase (33). Metal-bound bleomycin is resistant to the actions of this enzyme. Bleomycin hydrolase is a cytosolic aminopeptidase which hydrolyzes the terminal amino group from the blactaminopyrimidyl moiety of the molecule yielding a metabolite, desamido-bleomycin, that can not properly bind iron and thus is not very effectively activated in cell-free chemical system (10). Interestingly, subcutaneous administration of desamido-bleomycin does not elicit pulmonary toxicity in mice (26). The inability of this metabolite to undergo iron-dependent mixed-function oxidase catalyzed activation could very well account for why there is a correlation between host tissue or tumor cell bleomycin hydrolase levels and their relative resistance or susceptibility toward the cytotoxic actions of bleomycin (9). While there appears to be very little difference between bleomycin analogs insofar as their ability to be activated by the mixed-function oxidase system (Table 3), there do appear to be differences in their rate of metabolism by bleomycin hydrolase (33). In this regard, the B_2 isomer is rapidly metabolized by this enzyme and its intratracheal administration elicits much less pulmonary toxicity than does administration of the A_2 isomer (34).

Studies on the mechanism by which bleomycin alters DNA demonstrated that this reaction was dependent on oxygen and on the presence of oxidizing or reducing agents such as O_2^-/H_2O_2, ascorbic acid or dithiothreitol (20). Addition of EDTA or anaerobiosis inhibited bleomycin-mediated DNA damage. It was proposed that the interaction of bleomycin with Fe^{2+} resulted in the generation of reactive oxygen metabolites and that these metabolites were directly responsible for the DNA damage (20). Subsequent studies have not supported this concept (35), but suggest rather that the oxidation/reduction of iron bound by bleomycin is critical for the activation of bleomycin. Like these cell-free chemical systems, it appears that iron and molecular oxygen are important in order for bleomycin A_2 to be activated by the mixed-function oxidase system. The importance of iron in the activation of bleomycin A_2 was illustrated by the inhibition of bleomycin-mediated DNA deoxyribose cleavage by metal chelators, such as EDTA (Table 2) and desferroxamine (data not presented), and by the enhancement of this reaction by the addition of exogenous Fe^{3+}. Moreover, desamido-bleomycin A_2, which lacks the fifth axial nitrogen required for proper iron coordination and subsequent drug activation, elicited minimal damage to DNA even in the presence of exogenous Fe^{3+}. The form of iron required as a cofactor in the activation of bleomycin is Fe^{2+}. Thus, it appears that mixed-function oxidase systems catalyze bleomycin-mediated DNA damage by maintaining Fe^{3+} bound to bleomycin as Fe^{2+}. If so, factors which enhance this reaction should similarly enhance the activation of bleomycin and as a consequence the cleavage of DNA deoxyribose. O_2^- can reduce Fe^{3+} to Fe^{2+} (22) and as such is a likely biochemical basis for the stimulatory actions of mitomycin C and paraquat since their enzyme-catalyzed redox cycling results in O_2^- generation. Oxidation

PROCESSES INVOLVED IN THE CELLULAR PHARMACODYNAMICS OF BLEOMYCIN

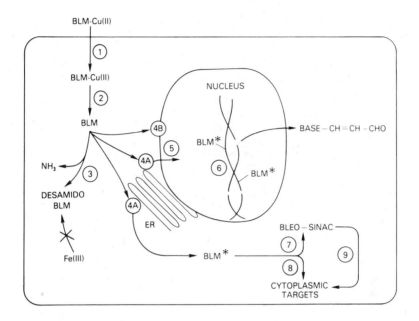

FIGURE 1.

1. Cellular uptake of the bleomycin-copper complex (BLM-Cu(II)).

2. Removal of copper from bleomycin by cytoplasmic components.

3. Metabolism of bleomycin by bleomycin hydrolase.

4. Acquisition of iron (Fe^{3+}) to form a bleomycin-iron complex, interaction with and subsequent activation to an activated intermediate (BLM*) by endoplasmic reticulum (A) or nuclear membrane-associated (B) mixed-function oxidase enzymes.

5. Diffusion of activated BLM (BLM*) from the endoplasmic reticulum (ER) into the nucleus.

6. BLM*-mediated DNA chain breakage.

7. Self-inactivation of BLM to a non-DNA damaging products (BLEO-SINAC).

8. Possible interaction of BLM* with cytoplasmic targets.

9. Possible interaction of BLEO-SINAC with cytoplasmic targets.

of ascorbic acid by molecular oxygen also results in the generation of O_2^- (36) and thus may account for its enhancing effect on bleomycin-mediated deoxyribose cleavage as well. Ascorbic acid radical has also been shown to reduce iron (21). Thus, one explanation for the inhibitory effect of SOD is that it is competing with and preventing the O_2^--mediated reduction of Fe^{3+} bound to bleomycin. Such a role for the superoxide anion has been proposed in a cell-free system containing bleomycin (37). It is also possible that the labile, activated intermediate of bleomycin may result directly from reactive oxygen attack.

Although the microsomal mixed-function oxidase enzymes are an efficient system by which bleomycin can be activated to a DNA damaging intermediate, the interaction of bleomycin with microsomes and NADPH in the absence of DNA results in a modification of the bleomycin molecule so that its ability to subsequently mediate DNA damage is considerably reduced (16-19). This same process has also been noted in cell-free chemical systems (2,20). While this reaction can be perceived as a detoxification type of metabolism, it is possible that the bleomycin molecule is altered in such a fashion as to yield a fragment(s) capable of interacting with cellular targets other than DNA. Interestingly, the studies of Raisfeld et al. (38) demonstrate that moieties found in the terminal substituent of bleomycin are capable of damaging the lung. Thus, the mixed-function oxidase catalyzed activation-inactivation of bleomycin offers a mechanism by which such moieties could arise in cells.

While the data presented in this study emphasize that the binding of iron in a required coordination and its subsequent redox cycling, mediated by O_2^- or H_2O_2, are important in the microsome catalyzed metabolic activation of bleomycin A_2 to a DNA damaging intermediate, the chemical nature of this intermediate remains unclear. Recent studies have provided evidence that an electronically excited state intermediate of bleomycin is generated in the presence of Fe^{2+} and H_2O_2 (4). Antholine et al. (39) have also shown that the decomposition of photoexcited bleomycin results in a bleomycin methyl radical and a radical localized on the bithiazole moiety. Interestingly, the bithiazole moiety is involved in DNA binding. Both electronically excited states and radical intermediates could alter biomolecules through hydrogen abstraction (40). As such, it is important to assess whether similar kinds of intermediates are generated by biological systems and if so, if their generation is iron- and reactive oxygen-dependent.

ACKNOWLEDGEMENTS

We wish to thank Dr. John S. Lazo, Department of Pharmacology, Yale University School of Medicine for the generous gift of desamido-bleomycin A_2. Marletta Regner is recognized for her expert preparation of this manuscript. Support from BRSG is gratefully acknowledged.

REFERENCES

1. S.K. Carter and R.H. Blum, Current status of American studies with bleomycin, Prog. Biochem. Pharmacol. 11:158 (1976).
2. R.M. Burger, J. Peisach and S.B. Horwitz, Activated bleomycin a transient complex of drug, iron and oxygen that degrades DNA, J. Biol. Chem. 256:11634 (1981).
3. L. Giloni, M. Takeshita, F. Johnson, C. Iden and A.P. Grollman, Bleomycin-induced strand scission of DNA, mechanism of deoxyribose cleavage, J. Biol. Chem. 256:8608 (1981).
4. M.A. Trush, E.G. Mimnaugh, Z.H. Siddik and T.E. Gram, Bleomycin-metal interaction: ferrous iron-initiated chemiluminescence, Biochem. Biophys. Res. Commun. 112:378 (1983).

5. A.P. Grollman, M. Takeshita, K.M.R. Pillar and F. Johnson, Origin and cytotoxic properties of base propenals derived from DNA, Cancer Res. 45:1127 (1985).
6. Z.M. Igbal, K.W. Kohn, R.A.G. Ewig and A.J. Fornace, Single strand scission and repair of DNA in mammalian cells by bleomycin, Cancer Res. 36:3834 (1976).
7. B.A. Teicher, J.S. Lazo and A.C. Sartorelli, Classification of anti-neoplastic agents by their selective toxicities toward oxygenated and hypoxic tumor cells, Cancer Res. 41:73 (1981).
8. A.F. Tryka, W.A. Skornik, J.J. Godelski and J.D. Brain, Potentiation of bleomycin-induced lung injury by exposure to 70% oxygen, Amer. Rev. Resp. Dis. 126:1074 (1982).
9. O. Yoshioka, N. Amano, K. Takahashi, A. Matsuda and H. Umezawa, in: "Bleomycin: Current Status and New Developments", S.K. Carter, S.T. Crooke and H. Umezawa, eds., Academic Press, New York (1978).
10. Y. Sugiura, Bleomycin-iron complexes. Electron spin resonance study, ligand effect and implication for action mechanism, J. Amer. Chem. Soc. 102:5208 (1980).
11. Y. Matsuda, M. Kitahare, K. Maeda and H. Umezawa, Correlation between level of defense against active oxygen in Esherchia coli K12 and resistance to bleomycin, J. Antibiotics (Tokyo), 35:931 (1982).
12. O.H. Lowry, N.J. Rosenbrough, A.L. Farr and R.J. Randall, Protein measurement with Folin phenol reagent, J. Biol. Chem. 193:265 (1951).
13. M.T. Kuo and C.W. Haidle, Characterization of chain breakage in DNA induced by bleomycin, Biochim. Biophys. Acta, 335:109 (1973).
14. M.A. Trush, E.G. Mimnaugh, E. Ginsburg and T.E. Gram, In vitro stimu-lation by paraquat of reactive oxygen-mediated lipid peroxidation in rat lung microsomes, Toxicol. Appl. Pharmacol. 60:279 (1981).
15. M.A. Trush, E.G. Mimnaugh, E. Ginsburg and T.E. Gram, Studies on the in vitro interaction of mitomycin c, nitrofuran and paraquat with pulmonary microsomes: stimulation of reactive oxygen-dependent lipid peroxidation, Biochem. Pharmacol. 31:805 (1982).
16. M.A. Trush, Demonstration that the temporary sequestering of adventitious iron accounts for the inhibition of microsomal lipid peroxidation by bleomycin A_2, Res. Commun. Chem. Pathol. Pharmacol. 37:21 (1982).
17. M.A. Trush, E.G. Mimnaugh, E. Ginsburg and T.E. Gram, Studies on the interaction of bleomycin A_2 with rat lung microsomes. I. Characteri-zation of factors which influence bleomycin-mediated DNA chain breakage, J. Pharmacol. Exp. Ther. 221:152 (1982).
18. M.A. Trush, Studies on the interaction of bleomycin A_2 with rat lung microsomes. III. Effect of exogenous iron on bleomycin-mediated DNA chain breakage, Chem.-Biol. Interactions, 45:65 (1983).
19. M.A. Trush and E.G. Mimnaugh, Different roles for superoxide anion in the toxic actions of bleomycin and paraquat, in: "Oxy Radicals and Their Scavenger Systems. Vol. II. Cellular and Medical Aspects", R.A. Greenwald and G. Cohen, eds., Elsevier Biomedical, New York (1983).
20. R.M. Burger, J. Peisach and S.B. Horwitz, Mechanism of bleomycin action: in vitro studies, Life Sci. 28:715 (1981).
21. CC. Winterbourn, Comparison of superoxide with reducing agents in the biological production of hydroxyl radicals, Biochem. J. 182:625 (1979).
22. K.-L. Fong, P.B. McCay, J.L. Poyer, H.P. Misra and B.B. Keele, Evidence for superoxide-dependent reduction of Fe^{3+} and its roles in enzyme-generated hydroxyl radical formation, Chem.-Biol. Interactions, 15: 77 (1976).
23. M.E. Scheulen and H. Kappus, The activation of oxygen by bleomycin is catalyzed by NADPH P-450 reductase in the presence of iron ions and NADPH, in: "Oxygen Radicals in Chemistry and Biology", W. Bors, M. Saran and D. Tait, eds., Walter deGruyter and Co., Berlin (1984).
24. M.A. Trush, E.G. Mimnaugh, E. Ginsburg and T.E. Gram, Studies on the interaction of bleomycin A_2 with rat lung microsomes. II. Involvement

of adventitious iron and reactive oxygen in bleomycin-mediated DNA chain breakage, J. Pharmacol. Exp. Ther. 221:159 (1982).

25. E. Ginsburg, T.E. Gram and M.A. Trush, A comparison of the pulmonary toxicity and chemotherapeutic activity of bleomycin-BAPP to bleomycin and pepleomycin, Cancer Chemother. Pharmacol. 12:111 (1984).

26. J.S. Lazo and C.J. Humphreys, Lack of metabolism as the biochemical basis of bleomycin-induced pulmonary toxicity, Proc. Natl. Acad. Sci. (USA), 80:3064 (1983).

27. M.R. Boyd, Metabolic activation and chemical-induced lung disease: implications for the cancer field, in: "Organ-directed Toxicity, Chemical Indices and Mechanisms", S. Brown and D.S. Davies, eds., Pergamon Press Ltd., New York (1981).

28. M.A. Trush, K.A. Kennedy, B.K. Sinha and E.G. Mimnaugh, Bleomycin-mediated deoxyribose cleavage in rabbit lung nuclei, Pharmacologist, 25:225 (1983).

29. N. Yamanaka, M. Fukushima, K. Koizumi, K. Nishida, T. Kato and K. Ota, Enhancement of DNA chain breakage by bleomycin and biological free radical producing systems, in: "Tocopherol, Oxygen and Biomembranes", C. DeDuve and O. Hagaishi, eds., Elsevier, Amsterdam (1978).

30. H. Umezawa, Bleomycin: Discovery, chemistry and action, Gann Monograph Cancer Res. 19:3 (1976).

31. J.H. Freedman, S.B. Horwitz and J. Peisach, Reduction of copper(II)-bleomycin: A model for in vivo drug activity, Biochem. 21:2203 (1982).

32. W.C. Mackellar and F.L. Crane, Iron and copper in plasma membranes, J. Bioenerg. Biomem. 14:241 (1982).

33. H. Umezawa, T. Takeuchi, S. Hori, F. Sawa and M. Ishizuka, Studies on the mechanism of antitumor effect of bleomycin on squamous cell carcinoma, J. Antibiotics (Tokyo), 25:409 (1972).

34. I.H. Raisfeld, Pulmonary toxicity of bleomycin analogs, Toxicol. Appl. Pharmacol. 56:326 (1980).

35. L.O. Rodriguez and S.M. Hecht, Iron (II)-bleomycin. Biochemical and spectral properties in the presence of radical scavengers, Biochem. Biophys. Res. Commun. 104:1470 (1982).

36. M. Scarpa, R. Stevanato, P. Viglino and A. Rigo, Superoxide ion as active intermediate in the autoxidation of ascorbate by molecular oxygen, J. Biol. Chem. 258:6695 (1983).

37. Y. Sugiura, T. Suzuki, J. Kuwahara and H. Tanaka, On the mechanism of hydrogen peroxide-, superoxide-, and ultraviolet light-induced DNA cleavages of inactive bleomycin-iron (III) complex, Biochem. Biophys. Res. Commun. 105:1511 (1982).

38. I.H. Raisfeld, J.P. Chovan and S. Frost, Bleomycin pulmonary toxicity: Production of fibrosis by bithiazole-terminal amine and terminal amine moieties of bleomycin A_2, Life Sci. 30:1391 (1982).

39. W.E. Antholine, T. Sarna, R.C. Sealy, B. Kalyanaraman, G.D. Shields and D.H. Petering, Free radicals from the photodecomposition of bleomycin, Photochem. Photobiol. 41:393 (1985).

40. M.A. Trush, E.G. Mimnaugh and T.E. Gram, Activation of pharmacologic agents to radical intermediates: implications for the role of free radicals in drug action and toxicity, Biochem. Pharmacol. 31:3335 (1982).

THE PEROXIDATIVE ACTIVATION OF BUTYLATED HYDROXYTOLUENE TO

BHT-QUINONE METHIDE AND STILBENEQUINONE

David C. Thompson[1], Young Nam Cha[2] and Michael A. Trush[1]

[1]Dept. Environmental Health Sciences, Johns Hopkins University
Baltimore, MD 21205 and [2]Dept. Pharmacology, Yonsei University
College of Medicine, Seoul, Korea

Butylated hydroxytoluene (BHT, 2,6-di-tert-butyl-4-methyl-phenol) is a commonly used antioxidant allowed in foods in amounts up to 0.02% of the weight of fat present. BHT helps prevent undesirable oxidation reactions from occurring by acting as a free radical scavenger. BHT is also used as a stabilizer in pesticides, gasolines and lubricants, soaps and cosmetics, and as an antiskinning agent in paints and inks (1). BHT has been shown to have a protective effect against the toxicity and carcinogenicity of a wide variety of chemicals (2). However, several recent animal studies have questioned the presumed safety of this antioxidant. For example, BHT has been shown to cause lung damage in mice (3,4), hemorrhagic death in rats (5) and can act as a tumor promoter in both mice and rats (6,7). One of the best characterized toxic effects of BHT is the destruction of type I alveolar and pulmonary endothelial cells (8) in the mouse lung. This lung damage is thought to arise from the biotransformation of BHT into BHT-quinone methide (2,6-di-tert-butyl-4-methylene-2,5-cyclohexadienone) (9,10), a highly reactive compound (see Figure 1). BHT has been demonstrated to be metabolized to BHT-quinone methide in vivo in the mouse (10) and rat (11). This reaction is presumably catalyzed by a cytochrome P-450 related enzyme (12,13). As a class of chemical compounds, quinone methides have been shown to react with cellular nucleophiles including amines, carbohydrates, alcohols, thiols, and olefins (14).

Peroxidase enzymes have recently been shown to catalyze the activation of a wide variety of xenobiotic compounds to reactive intermediates and these enzymes, particularly prostaglandin H synthase, have been suggested to play a role in the extrahepatic toxicity and carcinogenicity of several compounds (15,16). Since antioxidants are good electron donors we investigated whether BHT might be metabolically activated to BHT-quinone methide by peroxidase enzymes. This study reports on the metabolic activation of BHT by two peroxidase enzymes: horseradish peroxidase and prostaglandin H synthase.

METHODS

Materials: (Ring U-^{14}C) BHT (20 mCi/mmole) was purchased from Amersham. Arachidonic acid was obtained from Nu Chek Prep (Elysian, MN). BHT, BHA and other test compounds were obtained from Sigma or Aldrich. Horseradish peroxidase (type II) was obtained from Sigma and prostaglandin H synthase was prepared from ram seminal vesicles obtained from Dr. L. Marnett, Wayne State University. Microsomes were prepared as a 25% homogenate in 0.15 M KCl adjusted to pH 7.8. The homogenate was centrifuged twice at 9,000 g for

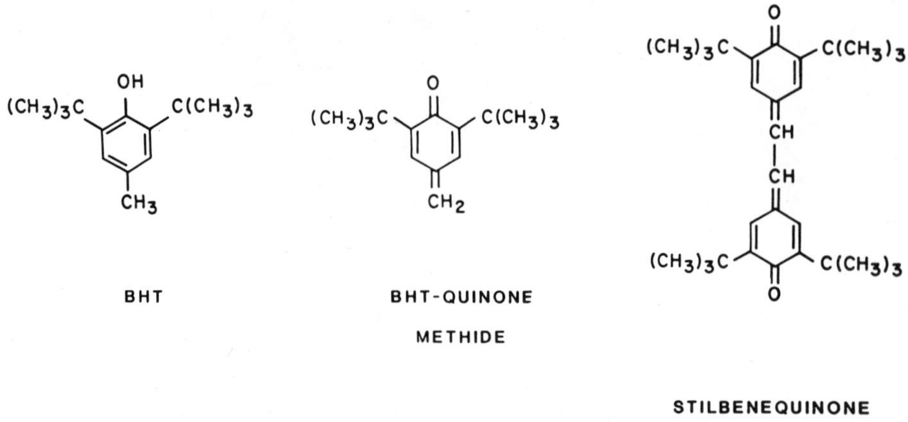

Figure 1.   Structures of BHT, BHT-quinone methide and stilbenequinone.

20 minutes.   The combined supernatants were filtered through two layers of cheesecloth and then centrifuged at 100,000 g for 1 hour.   The final micro-somal pellet was resuspended in 0.15 M KCl buffer at a protein concentration of approximately 10 mg/ml and then rapidly frozen in a methanol/dry ice bath and stored at -80°.

Covalent Binding of BHT Metabolite(s) to Microsomal Protein.   Reactions were initiated by the addition of arachidonic acid (330 μM) and allowed to proceed for 10 minutes at room temperature.   Each tube contained 0.5 μCi BHT (diluted to the appropriate concentration with cold BHT), approximately 1 mg ram semi-nal vesicle microsomes (specific activity of prostaglandin H synthase 0.52 μmoles arachidonic acid metabolized/minute/mg) in a total of 1 ml of 0.1 M Tris buffer, pH 8.0.   BHT and other test compounds were dissolved in DMSO, water, acetone or methanol and added to the reaction in volumes not exceeding 10 μl.   Reactions were stopped with 4 ml methanol and the protein pellets were repeatedly extracted with 2 ml methanol or methanol/ether (3:1) until no further radioactivity could be extracted (generally 12-15 washes).   Pellets were dissolved in 1 ml of 1 N NaOH and the radioactivity of an aliquot was counted in 10 ml of scintillation fluid.   Protein was determined using the Lowry method (17).

Formation of BHT-Quinone Methide and Stilbenequinone.   Reactions contained 15 units horseradish peroxidase, 0.9 mM hydrogen peroxide, 200 μM BHT and 100 to 500 μM of the various test compounds in 1 ml of 0.01 M phosphate buffer, pH 7.0.   BHT-quinone methide and stilbenequinone (3,5,3',5'-tetra-tert-butyl-stilbene-4,4'-quinone) were determined to have aqueous absorption maxima of 300 nm and 460 nm respectively and were detected spectrophotometrically. Authentic samples of BHT-quinone methide (18) and stilbenequinone (19) were synthesized and compared to products obtained from these reactions to confirm the identity of these spectral peaks.

RESULTS

We investigated the metabolic activation of BHT to a reactive inter-mediate(s) by two procedures: (1) assessing the covalent binding of this metabolite to microsomal protein; and (2) monitoring the formation of BHT-quinone methide and its subsequent dimerization product, stilbenequinone (see Figure 1).   The concentration dependent covalent binding of BHT to microsomal protein catalyzed by prostaglandin H synthase in the presence of arachidonic acid is presented in Table 1.   We have observed that

TABLE 1

Covalent Binding of BHT to Protein Catalyzed by Prostaglandin H Synthase

BHT Concentration	nmoles bound/mg protein[1]
10 μM	0.36 ± .04
50 μM	1.62 ± .19
100 μM	2.84 ± .43
200 μM	4.46 ± .60
500 μM	8.15 ± .78

[1]Values represent mean ± standard error of triplicate determinations.

TABLE 2

Modification of BHT Covalent Binding by Various Compounds[1]

Reaction	% of Control	Type of Radical Formed from Added Compound
Complete system[2]	100[3]	---
+ Glutathione (100μM)	57	---
+ Glutathione (1 mM)	7	---
+ Phenylbutazone	46	Peroxy
+ Diphenylisobenzofuran	84	Peroxy
+ Phenidone	265	N-cation
+ Tetramethylphenylenediamine	223	N-cation
+ BHA	439	Phenoxy
+ Phenol	167	Phenoxy
+ Acetaminophen	151	Phenoxy

[1]Concentration of all test agents was 100 μM except where noted.

[2]Complete system contained 50 μM BHT (0.5 μCi), 1 mg RSV microsomes and
330 μM arachidonic acid in 1 ml 0.01 M phosphate buffer, pH 7.0.

[3]Complete system (100%) = 1.65 ± .09 nmoles BHT bound/mg protein in a
10 minute incubation. Values represent means of triplicate determinations.

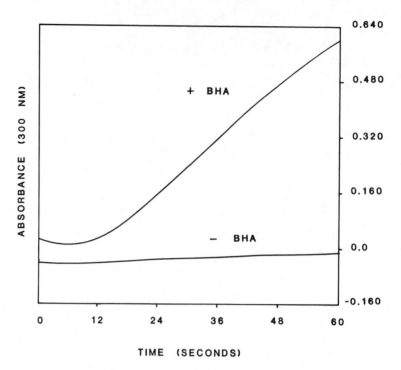

Figure 2. Formation of BHT-quinone methide from horseradish peroxidase-catalyzed oxidation of 200 μM BHT in the presence or absence of 100 μM BHA.

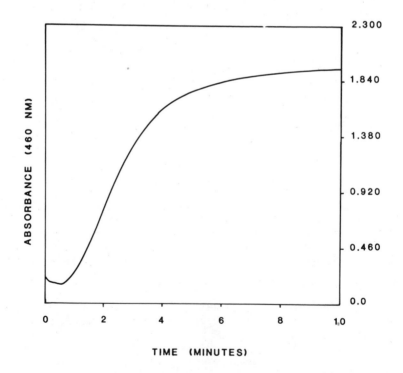

Figure 3. Formation of stilbenequinone from horseradish peroxidase-catalyzed oxidation of 200 μM BHT in the presence of 100 μM BHA.

hydrogen peroxide can also stimulate BHT covalent binding in this system indicating the peroxidative nature of this activation reaction (not shown). The ability of various pharmacologic agents to modify the covalent binding of BHT is shown in Table 2. Glutathione was an effective inhibitor of BHT binding. In the presence of butylated hydroxyanisole (BHA), however, the covalent binding of BHT was greatly enhanced and BHT-quinone methide was detected as a metabolic intermediate (see Figure 2). Several other compounds which are known to be cooxidation substrates for prostaglandin H synthase were also tested for their ability to influence the covalent binding of BHT. These included phenylbutazone and diphenylisobenzofuran which form peroxy radicals, phenidone and tetramethylphenylenediamine which form nitrogen-centered cation radicals, and phenol and acetaminophen which form phenoxy radicals when activated by peroxidase enzymes. The results in Table 2 demonstrate that several of these compounds stimulated BHT metabolism and that this stimulation may be related to the type of radical intermediate formed by the compound.

The formation of BHT-quinone methide from the horseradish peroxidase-catalyzed oxidation of BHT is shown in Figure 2. A small amount of BHT-quinone methide (1.18 nmoles/minute using an extinction coefficient of 27,000 $M^{-1}cm^{-1}$ from reference 18) was detected during the metabolism of 200 µM BHT in the absence of any activators. In the presence of 100 µM BHA the formation of BHT-quinone methide was greatly enhanced (26.6 nmoles/minute). The increased formation of this metabolite may thus be responsible for the increased covalent binding of BHT that was seen in the presence of BHA (see Table 2). In the presence of BHA, we also observed the formation of stilbenequinone. The formation of stilbenequinone was not observed from BHT alone, even at 1 mM BHT. The formation of stilbenequinone was preceeded by a lag period of about 1 minute after the start of the reaction (see Figure 3) supporting the contention that BHT-quinone methide was formed initially and subsequently dimerized to form the stilbenequinone. In addition to the compounds reported here, we have also observed that several other compounds can enhance the covalent binding of BHT. These include diethylstilbestrol, estradiol, methimazole, guaiacol, methyl paraben and eugenol. Compounds which enhanced BHT binding were also observed to enhance the formation of BHT-quinone methide and stilbenequinone whereas compounds which inhibited BHT binding did not enhance BHT-quinone methide or stilbenequinone formation (not shown).

The effect of glutathione on the horseradish peroxidase-catalyzed formation of BHT-quinone methide from 200 µM BHT in the presence of 100 µM BHA is shown in Figure 4. 100 µM glutathione inhibited BHT-quinone methide formation by approximately 50% while 500 µM glutathione completely inhibited the formation of BHT-quinone methide. The formation of stilbenequinone was similarly inhibited. If 100 µM glutathione was added to the reaction mixture after BHT-quinone methide had already been formed, the peak representing BHT-quinone methide rapidly disappeared suggesting the possible formation of a BHT-glutathione conjugate. During such reactions glutathione may also be oxidized to GSSG with the concommitent conversion of the BHT metabolite back to BHT.

DISCUSSION

This report demonstrates that the peroxidase-mediated activation of BHT results in the formation of a reactive intermediate(s) capable of covalently binding to microsomal protein. Since only a small amount of BHT-quinone methide was detected spectrophotometrically from the metabolism of BHT alone, perhaps BHT-quinone methide is not the only BHT metabolite responsible for this binding, as has been suggested in the literature (13). Peroxidase enzymes utilize one electron donors as cofactors and thus BHT should form a radical (a one electron oxidation product) prior to forming BHT-quinone

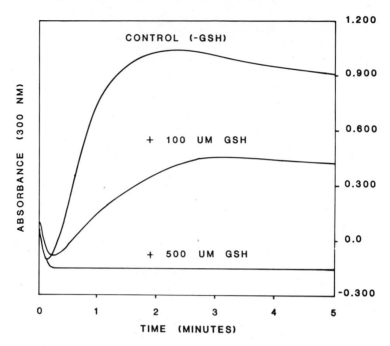

Figure 4. Effect of glutathione (GSH) on the formation of BHT-quinone methide from horseradish peroxidase-catalyzed oxidation of 200 μM BHT in the presence of 100 μM BHA.

methide (a two electron oxidation product). This BHT radical might be re-sponsible for the covalent binding of BHT in the absence of any activators. In the presence of activators such as BHA, the increase in the formation of BHT-quinone methide may account for the increase in covalent binding. Thus, both a free radical of BHT and BHT-quinone methide may be involved in the covalent binding of BHT to tissue macromolecules and hence its toxicity.

In addition to BHA, several other compounds were capable of enhancing BHT binding and BHT-quinone methide formation. It appears that the ability of a compound to enhance BHT binding depends on whether it can form a radical, what type of radical intermediate it forms, and whether BHT interacts with the radical by donating an electron to it. A possible mechanism for the en-hancement of BHT binding by these compounds is suggested in Figure 5. As illustrated in the schematic, a xenobiotic compound (XH) is preferentially oxidized by the peroxidase to a radical (X˙). In the presence of BHT, this xenobiotic-derived radical would be recycled back to the parent molecule by accepting an electron from BHT. The end result is the regeneration of the parent xenobiotic molecule and the enhanced conversion of BHT to a radical first, then BHT⁻quinone methide and ultimately stilbenequinone. A similar mechanism has been proposed by Kurechi and Kato (20) for a non-enzymatic system utilizing tert-butyl hydroperoxide as the oxidizing agent. In their system, BHA enhanced the metabolism of BHT to BHT-quinone methide. As a consequence, the parent BHA molecule was regenerated and thus accounted for an increased antioxidant capacity observed in the presence of both BHT and BHA.

Most chemicals which have been approved for use in food and cosmetic products have been tested for their toxicological effects on an individual basis. Toxic effects resulting from the interaction of two or more compounds are more difficult to predict or assess, yet nevertheless may be very im-portant. Recently Reed et al. (21) observed that phenylbutazone enhanced the prostaglandin H synthase catalyzed epoxidation of 7,8-dihydroxy-7,8-

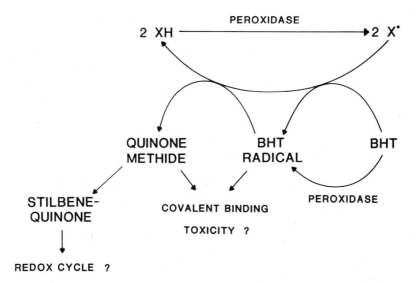

Figure 5. Possible mechanism for the formation of BHT-quinone methide and stilbenequinone from the peroxidase-catalyzed oxidation of BHT in the presence of a xenobiotic activator (XH).

dihydrobenzo[a]pyrene. The primary event was the formation of a peroxy radical from phenylbutazone which appeared to be the epoxidizing agent. The same type of chemical-chemical interaction is proposed here with the end result being the enhanced formation of the toxic BHT-quinone methide.

The formation of stilbenequinone may also be of toxicologic significance since this compound may be a substrate for cellular enzymes such as cytochrome P-450 reductase and thus redox cycle and generate reactive oxygen species. These reactive oxygen species may further contribute to BHT-induced toxicity.

We have clearly demonstrated that peroxidase enzymes are capable of metabolizing BHT to BHT-quinone methide in the presence of several activator compounds. This observation suggests that the involvement of peroxidase enzymes in the mechanism of toxicity of BHT should be examined. Since there are apparently a number of compounds which can serve as activators of BHT metabolism in vitro, perhaps there are endogenous compounds which can serve as activators in vivo as well. There are many phenolic compounds naturally present in foods (and thus ingested along with BHT) as well as many endogenous phenols which might serve as such an activator. The presence of an endogenous activator in a tissue with substantial peroxidase activity might render that tissue susceptible to BHT toxicity.

ACKNOWLEDGEMENTS

We gratefully acknowledge the financial support of the National Institute of Occupational Safety and Health OH01833-02 and National Institutes of Health ES07141. We also thank Marletta Regner for typing the mansucript.

REFERENCES

1. Clinical Toxicology of Commercial Products, 5th Ed., R.E. Gosselin, R.P. Smith and H.C. Hodge, eds., Williams and Wilkins, Baltimore (1984).
2. H. Babich, Butylated hydroxytoluene (BHT): A review, Environ. Res. 29:1 (1982).
3. A.A. Marino and J.T. Mitchell, Lung damage in mice following intraperitoneal injection of butylated hydroxytoluene, Proc. Soc. Exp. Biol. Med. 140:122 (1972).
4. H. Witschi and W. Saheb, Stimulation of DNA synthesis in mouse lung following intraperitoneal injection of butylated hydroxytoluene, Proc. Soc. Exp. Biol. Med. 147:690 (1974).
5. G. Takahashi and K. Kiraga, Dose-response study of hemorrhagic death by dietary butylated hydroxytoluene (BHT) in male rats, Tox. Appl. Pharmacol. 43:399 (1978).
6. H.P. Witschi, Enhancement of lung tumor formation in mice, in: "Carcinogenesis", M.J. Mass, et al., eds., Raven Press, New York (1985).
7. K. Imaida, S. Fukushima, T. Shirai, T. Masui, T. Ogiso and N. Ito, Promoting activities of butylated hydroxyanisole, butylated hydroxytoluene and sodium L-ascorbate on forestomach and urinary bladder carcinogenesis initiated with methylnitrosourea in F344 male rats, Gann 75:769 (1984).
8. L.J. Smith, The effect of methylprednisolone on lung injury in mice, J. Lab. Clin. Med. 101:629 (1983).
9. T. Mizutani, I. Ishida, K. Yamamoto and K. Tajima, Pulmonary toxicity of butylated hydroxytoluene and related alkyl-phenols: Structural requirements for toxic potency in mice, Tox. Appl. Pharmacol. 62:273 (1982).
10. T. Mizutani, K. Yamamoto and K. Tajima, Isotope effects on the metabolism and pulmonary toxicity of butylated hydroxytoluene in mice by deuteration of the 4-methyl group, Tox. Appl. Pharmacol. 69:283 (1983).
11. O. Takahashi and K. Hiraga, 2,6-di-tert-butyl-4-methylene-2,5-cyclohexadienone: A hepatic metabolite of butylated hydroxytoluene in rats, Fd. Cosmet. Toxicol. 17:451 (1979).
12. J.P. Kehrer and H. Witschi, Effects of drug metabolism inhibitors on butylated hydroxytoluene-induced pulmonary toxicity in mice, Tox. Appl. Pharmacol. 53:333 (1980).
13. K. Tajima, K. Yamamoto and T. Mizutani, Formation of a glutathione conjugate from butylated hydroxytoluene by rat liver microsomes, Biochem. Pharmacol. 34:2109 (1985).
14. A.B. Turner, Quinone methides, Quart. Rev. 18:347 (1964).
15. L.J. Marnett and T.E. Eling, Cooxidation during prostaglandin biosynthesis: A pathway for the metabolic activation of xenobiotics, in: "Reviews in Biochemical Toxicology, 5,", E. Hodgson and J.R. Bond, eds., Elsevier Biomedical, New York (1983).
16. S.M. Cohen, T.V. Zenser, G. Muraski, S. Fukushima, M.B. Mattammal, N.S. Rapp and B.B. Davis, Aspirin inhibition of N-(4-[5-nitro-2-furyl]-2-thiazolyl)formamide-induced lesions of the urinary bladder correlated with inhibition of metabolism by bladder prostaglandin endoperoxide synthetase, Cancer Res. 41:3355 (1981).
17. O.H. Lowry, N.J. Rosebrough, A.L. Farr and R.J. Randall, Protein measurement with the Folin phenol reagent, J. Biol. Chem. 193:265 (1951).
18. H. Becker, Quinone dehydrogenation. I. The oxidation of monohydric phenols, J. Org. Chem. 30:982 (1965).
19. C.D. Cook, N.G. Nash and H.R. Flanagan, Oxidation of hindered phenols. III. The rearrangement of the 2,6-di-t-butyl-4-methylphenoxy radical, J. Amer. Chem. Soc. 77:1783 (1955).

20. T. Kurechi and T. Kato, Studies on the antioxidants. XX. The effect of butylated hydroxytoluene on tert-butylhydroperoxide-induced oxidation of butylated hydroxyanisole, Chem. Pharm. Bull. 31:1772 (1983).
21. G.A. Reed, E.A. Brooks and T.E. Eling, Phenylbutazone-dependent epoxidation of 7,8-dihydroxy-7,8-dihydrobenzo[a]pyrene, J. Biol. Chem. 259: 5591 (1984).

ACTIVATION OF XENOBIOTICS BY HUMAN POLYMORPHONUCLEAR LEUKOCYTES VIA

REACTIVE OXYGEN-DEPENDENT REACTIONS

Michael A. Trush[1], Thomas W. Kensler[1] and John L. Seed[2]

[1]Department of Environmental Health Sciences and [2]Immunology
and Infectious Diseases, The Johns Hopkins University, School
of Hygiene and Public Health, Baltimore, Maryland 21205

INTRODUCTION

Reactive oxygen intermediates have become widely implicated in various
pathologic states, chemical-induced tissue injury and chemical carcinogenesis
(1-5). While much of the interest in the role of oxy-radicals in these processes has centered on the direct interaction of reactive oxygen metabolites
with biomolecules, it is becoming increasingly apparent that molecular
oxygen-derived oxidants can also participate in the metabolic activation of
chemicals (6,7). It has been hypothesized that polymorphonuclear leukocytes
(PMNs) may be a useful cellular model to study the interaction and possible
activation of compounds by reactive oxygen species (Figure 1)(8,9). Resting
PMNs release measurable quantities of superoxide (O_2^{-}), hydrogen peroxide
(H_2O_2) and hydroxyl radical ($\cdot OH$), while activation of their redox metabolism by both particulate and soluble stimulants results in an increased
rate in the generation of these molecular oxygen-derived oxidants (10,11).
The utilization of H_2O_2 by the PMN enzyme myeloperoxidase (MPO) results in
the formation of hypochlorous acid and an O_2 metabolite or complex with
singlet oxygen ($^1O_2$)-like reactivity (11). The data presented in this
study demonstrate that bleomycin A_2 and benzo[a]pyrene-7,8-dihydrodiol
(BP-7,8-dihydrodiol) are activated to genotoxic derivatives as a result of
their interaction with PMN-derived oxidants. Such an activation mechanism
could provide an explanation as to how neoplasms often develop at sites of
ongoing inflammation (12).

METHODS

Cell Isolation. Blood was drawn from normal healthy volunteers into heparinized tubes and then centrifuged at 150 x g for 10 min after which the
plasma and buffy coat containing mononuclear cells were discarded. The remaining leukocytes and erythrocytes were mixed with an equal volume of 6%
dextran and incubated in inverted syringes for 45 min at 37°. The upper
leukocyte-rich fractions were ejected through a 16 gauge needle (90° bend),
pooled and spun at 150 x g for 5 min at 4°. Contaminating erythrocytes
were lysed by adding cold 0.155 M NH_4Cl, 0.01 M $KHCO_3$ and 0.1 mM EDTA
buffer (pH 7.4). PMNs were washed, resuspended in PBS and counted on a
hemacytometer. This procedure yielded a preparation of PMNs that was 95%
pure.

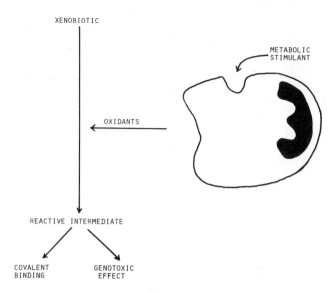

XENOBIOTIC

METABOLIC
STIMULANT

OXIDANTS

REACTIVE INTERMEDIATE

COVALENT   GENOTOXIC
BINDING   EFFECT

Fig. 1.

Measurement of Chemiluminescence (CL) Responses. CL responses were monitored using an ambient temperature liquid scintillation spectrometer (Model 3003, Packard Instrument Co.) operated in the out-of-coincidence mode. The counter was set as follows: gain, 100%; window A to infinity with discriminators set at 0 to 1000 and input selector 1 + 2. Experiments were begun by incubating 7×10^6 PMNs in 3 ml of PBS for 10 min at 37° in dark adapted polyethylene vials. After the background CL of each vial was determined, the PAHs (3 μM) or vehicle (DMSO, 0.1%) was added, any response noted and the reactions subsequently initiated by addition of 12-0-tetradecanoyl-phorbol-13-acetate (TPA) (100 ng/ml). CL was monitored for 0.2 min at 75-sec intervals and vials were maintained at 37° between counting. All additions to the vials as well as the CL counting procedure were performed in a darkened room. Results are expressed at counts/unit time minus background. Data are presented as peak (maximum) responses and temporal response curves.

Covalent Binding to DNA. Trans[³H]BP-7,8-dihydrodiol (3 μM, 367 mCi/mmole), obtained from the NCI Chemical Carcinogen Repository, was incubated with PMNs (1×10^7) and calf thymus DNA (1.5 mg) at 37° for 60 min in a total volume of 3 ml of PBS. After 60 min, the PMNs were removed by centrifugation (150 x g) and the supernatant extracted by shaking with an equal volume of phenolic reagent (100 phenol:8 hydroxyquinoline:11 iso-amyl alcohol) for 20 min. The phases were separated by centrifugation at 10,000 x g for 20 min and the phenolic layer removed and discarded. The aqueous layer was re-extracted with 3 ml $CHCl_3$ reagent (24 $CHCl_3$:1 iso-amyl alcohol) until the washes contained background levels of radioactivity (generally 5 washes). The aqueous layer was then washed with 3 ml of water saturated ether to remove residual solvents. The aqueous layer containing DNA was precipitated with cold ethanol overnight at -20°. Precipitated DNA was dried under N_2 and redissolved in 0.1 M potassium phosphate buffer and radioactivity bound to DNA determined. The concentration of DNA was determined by the diphenylamine reaction. Data are expressed as pmol of [³H]BP-7,8-dihydrodiol bound per mg of DNA.

312

<u>Mutagenesis Assay</u>. Mutagenesis assays were conducted as described by Ames et al. with minor modifications. <u>Salmonella</u> <u>typhimurium</u> strain TA100 was cultured overnight in 5 ml of Vogel-Bonner Medium E, containing 2% glucose, 1% nutrient broth and 0.85 mM histidine and biotin, then centrifuged at 1000 x g and the bacterial pellet resuspended in an equal volume of PBS. This washing step was repeated three times in order to remove PMN-stimulating substances from the bacterial broth. <u>Salmonella</u> (300 μl) were then incubated with PMNs ($2x10^6$), the indicated PAH ($10\mu M$) with or without TPA (100 ng/ml) in a total volume of 3 ml of PBS. After 1 hour of incubation at 37° in the dark, 1 ml of the incubation medium was transferred to 1 ml of 1.5% minimal agar containing .05 mM histidine and biotin and then poured on a minimal agar plate. Revertant colonies were determined after 72 hours of incubation at 37°.

<u>Sister Chromatid Exchange</u>. Sister chromatid exchanges were done in Chinese hamster V-79 cells. PMNs ($2x10^6$ ml), PAH ($3 \mu M$) and TPA (100 ng/ml) were coincubated with the V-79 cells for 1 hours, after which they were replaced with fresh media. The V-79 cells were then cultured in Eagles MEM containing 6 μg/ml bromodeoxyuridine for 28 hours (2 cell doublings). Colcemid was added (10^7 M) during the last 4 hours. The cells were then harvested, stained and SCE's scored.

<u>Bleomycin-mediated DNA Deoxyribose Cleavage</u>. Cleavage of DNA deoxyribose yields a product which reacts with thiobarbituric acid (TBA) yielding a chromophore which can be measured at 533 nm (13). This assay was utilized to assess BLM-mediated DNA deoxyribose cleavage and the incubation conditions are described in the legend to Table 1. Bleomycin-mediated DNA deoxyribose cleavage is expressed as nanomoles of TBA reacting product (TBAR) per 10^7 PMNs per 60 min using an extinction coefficient of $1.53 \times 10^{-5}M^{-1}$ cm^{-1}. Zero-time blanks were used throughout.

RESULTS

Bleomycin A_2 is a glycopeptide antibiotic which exhibits antiprolifera-tive activity probably via alterations in DNA (14). Both release of nucleic acid bases and cleavage of DNA deoxyribose are observed in the presence of bleomycin (14). These reactions are mediated by an "activated" bleomycin intermediate (15). This activation process is facilitated by O_2^- genera-ting systems (16,17), possibly by the following reaction:

$$Blm - Fe^{3+} \xrightarrow{O_2^-} Blm - Fe^{2+} - O_2 \longrightarrow \text{"Activated" Bleomycin Intermediate}$$

The data in Table 1 demonstrate that the interaction of bleomycin A_2 with PMNs resulted in deoxyribose cleavage of exogenously added calf thymus DNA. Bleomycin-mediated DNA deoxyribose cleavage was observed with resting PMNs while stimulation of their redox metabolism by the phorbol diester TPA resulted in a 6-fold increase in this response. No significant thiobarbi-turic acid reactive products could be detected with TPA-stimulated PMNs in the absence of bleomycin. Previous studies have demonstrated that DNA damage occurs in PMNs following TPA stimulation(18). PMNs have also been shown to elicit reactive oxygen-dependent genetic lesions when co-incubated with bacteria or mammalian cells (19).

Activated bleomycin A_2 is generated by autoxidation of Fe^{2+}, xanthine oxidase, microsomal mixed-function oxidase enzyme and NADPH cytochrome P450 reductase (20-23). Addition of superoxide dismutase (SOD) inhibits bleomy-cin-mediated DNA deoxyribose cleavage in each of these systems. As shown in Table 1, SOD inhibited bleomycin-mediated DNA deoxyribose cleavage catalyzed by both resting and TPA-stimulated PMNs, verifying the involve-ment of reactive oxygen in this process.

TABLE 1. Bleomycin-Mediated DNA Deoxyribose Cleavage in the Presence of
Resting and TPA-Stimulated Human PMNs: Inhibition by SOD

Redox Metabolic State of PMNs	Additions to PMNs	DNA Deoxyribose Cleavage nmol TBAR/10^7 PMNs/60 min
Resting	None	0.1 ± 0.1[a]
	DNA	0.2 ± 0.1
	BLM A_2	0.2 ± 0.1
	BLM A_2, DNA	1.3 ± 6.2[b]
	BLM A_2, DNA, SOD	0.4 ± 0.2[c]
Stimulated	TPA	0.2 ± 0.2
	TPA, BLM A_2	0.1 ± 0.1
	TPA, DNA	0.3 ± 0.1
	TPA, BLM A_2, DNA	6.5 ± 0.6[d,e]
	TPA, BLM A_2, DNA, SOD	1.3 ± 0.5[f]

a. PMNs (1×10^7) were incubated in the absence or presence of bleomycin A_2
(BLM A_2, 100 μM), calf thymus DNA (0.25 mg/ml) and TPA (5 ng/ml) under
an O_2 atmosphere (5 1/min) at 37° for 60 min. Total incubation volume
was 1.75 ml. After 60 min, 0.75 ml of 2.0 M trichloroacetic acid -1.7N
HCl was added and then a 0.5 ml aliquot of the supernatant was assayed
for thiobarbituric acid reactive products (TBAR)(22). Values are mean
$\pm$ S.D. (N=3).

b. $p < 0.01$ from all other values for resting PMNs.

c. $p < 0.01$ BLM A_2, DNA. SOD concentration was 10 μg/ml. Heat-inactivated
SOD (10 min, 90°C) was not inhibitory.

d. $p < 0.01$ from BLM A_2, DNA.

e. $p < 0.01$ from all other values for stimulated PMNs.

f. $p < 0.01$ from TPA, BLM A_2, DNA.

g. No significant increase in TBAR was observed under an N_2 atmosphere.

Various oxidants have also been shown to oxygenate BP-7,8-dihydrodiol.
For example, interaction of BP-7,8-dihydrodiol with $^1O_2$-generating systems
results in the generation of chemiluminescence (CL) resulting from a 9,10-
dioxetane intermediate (24). Peroxy radicals derived from lipid peroxidation reactions (25) or phenylbutazone (26) can oxidize BP-7,8-dihydrodiol
to the carcinogenic and mutagenic BP-7,8-diol-9,10-epoxide. The data presented in Table 2 illustrate that the addition of BP-7,8-dihydrodiol to
TPA-stimulated PMNs resulted in a significant enhancement in CL, probably
due to the generation of the 9,10-dioxetane intermediate (26). The generation of CL by PMNs is dependent upon their ability to generate $O_2^{\overline{\cdot}}$, H_2O_2
and an MPO-derived oxidant which exhibits chemical reactivity similar to
$^1O_2$. Concordantly, CL responses of PMNs are inhibited by SOD or azide, a
MPO inhibitor (10,11). Addition of CuDIPS (0.5 μM), a biomimetic superoxide
dismutase (27,28), or azide (1.0 mM) inhibited the CL from BP-7,8-dihydro-
diol 90 and 98%, respectively. The antioxidant BHA (0.1 mM) was inhibitory
(97%). On the other hand, SOD (10 μg/ml) did not inhibit BP-7,8-dihydro-
diol-derived CL but did inhibit (90%) the CL from PMNs in the absence of
this polycyclic aromatic hydrocarbon (PAH). The biochemical basis for the
apparent discrepancy in the actions of CuDIPS and SOD has been discussed

TABLE 2. Chemiluminescent and Genotoxic Reactions Resulting from the Interaction of BP-7,8-Dihydrodiol with Human PMNs

Additions to PMNs	Chemiluminescence (peak counts/0.2 min)x10^{-4}	Covalent Binding pmol bound/mg DNA	Mutagenesis Histidine Revertants	Sister Chromatid Exchange Exchange/Chromosome
None	---	---	79	0.4
TPA	3.2	---	70	0.3
BP-7,8-Dihydrodiol	3.4	0.8	95	0.4
TPA, BP-7,8-Dihydrodiol	27.6	7.0	249	1.6
TPA, BP-7,8-Dihydrodiol, CuDIPS	2.5	2.8	---	0.4
TPA, BP-7,8-Dihydrodiol, Azide	1.0	1.2	---	0.4

previously (30). CuDIPS, azide and BHA (29) also inhibited TPA-stimulated PMN CL in the absence of BP-7,8-dihydrodiol.

Covalent binding of chemicals to macromolecules is used as an indicator of the generation of a biologically reactive intermediate(s)(31). Significant covalent binding of [^{3}H]BP-7,8-dihydrodiol to exogenously added calf thymus DNA occurred in the presence of PMNs (Table 2). A small degree of covalent binding was observed with resting PMNs. Stimulation of their metabolism by TPA resulted in a 10-fold increase in the binding of BP-7,8-dihydrodiol to DNA which could be significantly inhibited by CuDIPS or azide. Thus, both the generation of CL from BP-7,8-dihydrodiol and its activation to a genotoxic species appear to be oxidant-dependent and more particularly, MPO-dependent. Purified MPO has been shown to activate several other carcinogenic compounds (32,33).

As a biological reflection of its activation to a DNA-binding intermediate, the interaction of BP-7,8-dihydrodiol with TPA-stimulated PMNs elicited mutagenesis in <u>Salmonella typhimurium</u> strain TA100 and sister chromatid exchanges (SCE) in Chinese hamster V-79 cells. CuDIPS and azide, but not SOD (data not presented), inhibited the induction of SCE's by BP-7,8-dihydrodiol. In contrast to the observations with BP-7,8-dihydrodiol, the interaction of benzo[a]pyrene with TPA stimulated PMNs did not yield enhanced CL or elicit mutagenesis (30). Moreover, PAH derivatives lacking a double bond at the 9,10 position were not chemiluminescent substrates and did not elicit genotoxic responses in bacteria or V-79 cells (Table 3), indicating a site-specificity to this PMN-mediated activation process. The rank order for both the CL and genotoxic response was the same, namely 7,8-dihydro-BP><u>trans</u> BP-7,8-dihydrodiol><u>cis</u> BP-7,8-dihydrodiol (Table 3 and 30). Similarly, those derivatives which were not genotoxic were not chemiluminescent. These relationships further substantiate the proposal that the generation of CL is indicative of a PAH's ability to undergo metabolic activation at the 9,10 position.

TABLE 3. Chemiluminescence and Genotoxic Responses of Various Polycyclic Aromatic Hydrocarbons in the Presence of TPA-Stimulated Human PMNs

Compound	Chemiluminescence	Mutagenesis	Sister-Chromatid Exchange
<u>trans</u> BP-7,8-Dihydrodiol	+	+	+
<u>trans</u> BP-7,8-Dihydrodiol-9,10-Dihydro	-	-	-
7,8-Dihydro-BP	+	+	+
9,10-Dihydro-BP	-	-	-

DISCUSSION

It has become increasingly evident over the last decade that the metabolic activation of xenobiotics to reactive metabolites and the subsequent interaction of this metabolite with cellular macromolecules account for many of the acute, and probably, chronic toxicities caused by exogenous chemicals (31). Thus, understanding the mechanisms by which chemicals are metabolically activated is important from the standpoint of developing strategies to prevent this process and the subsequent development of toxicity. Several studies have demonstrated that metabolically stimulated PMNs are capable of activating xenobiotics (34-36); however, these studies have not examined whether biological consequences occur as a result of PMN-mediated chemical activation. The results of this study clearly demonstrate that this activation mechanism can result in the generation of chemical intermediates which are genotoxic.

The interaction of bleomycin A_2 with TPA-stimulated human PMNs resulted in DNA deoxyribose cleavage, a reaction mediated by an "activated" bleomycin intermediate. The chemical identity of this intermediate is presently unclear although a bleomycin derived electronically exicted state and radicals have been implicated (37,38). SOD inhibited the activation of bleomycin by TPA-stimulated PMNs indicating a dependency on O_2^-, which is consistent with the other systems which activate this drug (20-23). In addition to its selective lung and skin toxicity, bleomycin induces SOD-inhibitable transformation of hamster embryo cells (39).

Like bleomycin A_2, BP-7,8-dihydrodiol was activated by PMNs and this reaction resulted in both the generation of a chemiluminescent intermediate and one which covalently binds to DNA. CuDIPS, a biomimetic SOD, and azide, a MPO inhibitor, inhibited both of these reactions. These inhibitor studies demonstrated a dependency on both an oxygen-derived oxidant and the catalytic actions of MPO in this hydrocarbon activation process. One of the characteristics of oxidation reactions that are highly dependent on MPO, including those yielding CL, is that they exhibit a sensitivity to inhibition by azide but not to SOD (11,36). Addition of azide inhibited CL, DNA covalent binding and induction of SCE's by BP-7,8-dihydrodiol whereas SOD was ineffective. The seemingly paradoxical inhibition of these reactions by CuDIPS, a biomimetic SOD, may be due to the interaction of such copper-complexes to interact with oxygen, possibly in a superoxide state, bound to MPO (40).

From the various benzo[a]pyrene congeners investigated, it appears that the activation of these PAHs to genotoxic derivatives occurred at the 9,10 double bond, the position at which BP-7,8-dihydrodiol is oxidized to a bay region epoxide by the cytochrome P-448 system (41), during lipid peroxidation (25) and by an oxidant originating from the peroxidase component of PGS (42). Battista et al. (43) have suggested that a peroxy radical was responsible for the oxidation of BP-7,8-dihydrodiol observed during lipid peroxidation, whereas a cytochrome P450-oxo complex was implicated in this action by the mixed-function oxidase system. Interestingly, a chloro-peroxy-dependent mechanism has been implicated in MPO-catalyzed dioxygenations (11).

Recently Guthrie et al. (44) concluded that the PGS system was more selective than the cytochrome P450 system in the activation of PAHs to mutagenic derivatives; only dihydrodiols with adjacent double bonds in the bay region were activated by the PGS system. Considering that PMNs have several features in common with the PGS system, including the ability to generate CL, involvement of peroxidase and the capability to oxidize organic molecules (42), it is possible that PMNs exhibit the same type of selectivity toward PAHs. In fact, preliminary experiments have indicated that chrysene-1,2-dihydrodiol and benzo[a]anthracene-3,4-dihydrodiol, dihydrodiols with adjacent double bonds in the bay region, are chemiluminescent substrates with TPA-stimulated PMNs but that chrysene-3,4-dihydrodiol and benzo[a]anthracene-1,2-dihydrodiol are not.

Like PGS, PMNs did not activate the parent BP to a derivative which was mutagenic to <u>Salmonella typhimurium</u> TA100. Quinones, rather than an epoxide, are the principal metabolites which arise from the interaction of BP with the PGS system (45). As a result of their enzyme-catalyzed redox cycling and accompanying reactive oxygen generation these quinones can elicit mutagenesis in the oxidant sensitive <u>Salmonella</u> tester strain TA104 (46). The data in this study also suggests that oxy radicals originating from these quinone metabolites could serve to activate other BP metabolites such as the 7,8-dihydrodiol. Whether quinones are formed from the interaction of BP with PMNs was not determined, but remains a possibility. Another intriguing possibility is that the electronically excited 9,10 dioxetane intermediate could have biological reactivity distinct from that of other reactive metabolites, such as epoxides or quinones. For example, chemical intermediates in an electronically excited state have been implicated in DNA damage (47,48). In addition, the ring opened dialdehyde resulting from the oxygenolytic cleavage of the dioxetane intermediate could react with tissue constituents in a manner similar to the reactive aldehyde metabolite to toxic methyl furans (49). 1,3-Diphenylisobenzofuran is oxidized by both the PGS system (42) and PMNs (50). Therefore, depending on the type of oxidant(s) generated by a particular biological system, it is possible that different spectrums of reactive products from BP-7,8-dihydrodiol could be observed.

For most xenobiotics, metabolic activation to genotoxic derivatives is a prerequisite step in the initiation of carcinogenesis (51). Recently, Hennings et al. (52) demonstrated that the repetitive treatment of mouse epidermis with phorbol diester was not particularly effective in eliciting the progression of benign papillomas to malignant carcinomas but that treatment of the papillomas with initiating agents dramatically increased this progression to malignancy. It appears then that metabolic activation of procarcinogens could play an important role not only in the initiation of multistage carcinogenesis, but in the progression to malignancy as well. This study clearly demonstrates that PMNs can metabolically activate carcinogens, including aflatoxin B$_1$ and N-OH-AFF (data not presented), to genotoxic derivatives via a reactive oxgyen-dependent, cytochrome P450 independent reaction. Thus, under conditions where there is an accumulation of metabolically stimulated PMNs, such as exists at sites of inflammation, it is conceivable that these PMNs could serve as a primary or secondary source of carcinogen activation. Investigations are in progress to evaluate the importance of this process <u>in vivo</u>. These studies may provide a molecular basis for the observed association between the development of some malignancies and sites of ongoing inflammation (12) as well as to identify the role of chronic inflammatory states as risk factors for neoplasia.

ACKNOWLEDGEMENTS

We gratefully acknowledge financial support for this research from the National Institutes of Health ES 03760 and ES 02300, BRSG and the American Cancer Society SIG-3.

REFERENCES

1. B.N. Ames, Dietary carcinogens and anticarcinogens: oxygen radicals and degenerative diseases, <u>Science</u>, 221:1256 (1983).
2. M.A. Trush, E.G. Mimnaugh and T.E. Gram, Activation of pharmacologic agents to radical intermediates: implications for the role of free radicals in drug action and toxicity, <u>Biochem. Pharmacol.</u>, 31:3335 (1982).
3. B. Freeman and J. Crapo, Free radicals and tissue injury, <u>Lab. Invest.</u>, 47:412 (1982).
4. T.W. Kensler and M.A. Trush, Role of oxygen radicals in tumor promotion, <u>Env. Mutagenesis</u>, 6:593 (1984).
5. T.W. Kensler and M.A. Trush, Oxygen free radicals in chemical carcinogenesis, <u>in</u>: "Superoxide Dismutase, Vol. III: Pathological States", L.W. Oberley, ed., CRC Press, Boca Raton, FL, in press (1985).

6. I. Johansson and M. Ingleman Sundberg, Hydroxyl radical-mediated cyto-
    chrome P-450-dependent metabolic activation of benzene in microsomes
    and reconstituted enzyme systems from rabbit liver, J. Biol. Chem.,
    258:7311 (1983).
7. E. Dybing, S.D. Nelson, J.R. Mitchell, H.A. Sasame and J.R. Gillette,
    Oxidation of α-methyldopa and other catechols by cytochrome P-450-
    generated superoxide anion: possible mechanism of methyldopa hepatitis,
    Mol. Pharmacol., 12:911 (1976).
8. M.A. Trush, M.E. Wilson and K. VanDyke, The generation of chemilumine-
    scence by phagocytic cells, in: "Methods Enzymol", 57, M. DeLuca, ed.,
    Academic Press, NY (1978).
9. A.A. Roman-Franco, Non-enzymatic extramicrosomal bioactivation of chemi-
    cal carcinogens by phagocytes: A proposed new pathway, J. Theor. Biol.,
    97:543 (1982).
10. S.J. Klebanoff, Oxygen metabolism and the toxic properties of phagocytes,
    Ann. Inter. Med., 93:480 (1980).
11. R.C. Allen, Biochemiexcitation: chemiluminescence and the study of bio-
    logical oxygenation reactions, in: "Chemical and Biological Generation
    of Excited States", W. Adam and G. Cilento, eds., Academic Press, NY,
    (1982).
12. H.B. Demopoulos, D.D. Pietronigro and M.L. Seligman, The development of
    secondary pathology with free radical reactions as a threshold mechan-
    ism, J. Amer. Coll. Toxicol., 2:173 (1983).
13. M.T. Kuo and C.W. Haidle, Characterization of chain breakage of DNA in-
    duced by bleomycin, Biochim. Biophys. Acta, 335:109 (1973).
14. H. Umezawa, Bleomycin: discovery, chemistry and action, Gann Monogr.
    Cancer Res., 19:3 (1976).
15. R.M. Burger, J. Peisach and S.B. Horwitz, Activated bleomycin - a tran-
    sient complex of drug, iron and oxygen that degrades DNA, J. Biol.
    Chem., 256:11636 (1981).
16. E.A. Sausville, J. Peisach and S.B. Horwitz, Effect of chelating agents
    and metal ions on the degradation of DNA by bleomycin, Biochem., 17:
    2740 (1978).
17. M.A. Trush and E.G. Mimnaugh, Different roles for superoxide anion in
    the toxic actions of bleomycin and paraquat, in: "Oxy Radicals and
    Their Scavenger Systems: Cellular and Medical Aspects, Vol. 2", R.
    Greenwald and G. Cohen, eds., Elsevier/North Holland, New York (1983).
18. H.W. Birnboim, DNA strand breaks in human leukocytes exposed to a tumor
    promoter, phorbol myristate acetate, Science, 215:1247 (1982).
19. A.B. Weitberg, S.A. Weitzman, M.Destrempes, S.A. Latt and T.R. Stossel,
    Stimulated human phagocytes produce cytogenic changes in cultured
    mammalian cells, New Eng. J. Med., 308:26 (1983).
20. J.W. Lown and S-K. Sim, The mechanism of the bleomycin-induced cleavage
    of DNA, Biochem. Biophys. Res. Commun., 77:1150 (1977).
21. R. Ishida and T. Takahashi, Increased DNA chain breakage by combined
    action of bleomycin and superoxide radical, Biochem. Biophys. Res.
    Commun., 66:1432 (1975).
22. M.A. Trush, E.G. Mimnaugh, E. Ginsburg and T.E. Gram, Studies on the
    interaction of bleomycin A_2 with rat lung microsomes. II. Involvement
    of adventitious iron and reactive oxygen in bleomycin-mediated DNA
    chain breakage, J. Pharmacol. Exp. Therap., 221:159 (1982).
23. M.E. Scheulen, H. Kappus, D. Thyssen and C.G. Schmidt, Reduction cycling
    of Fe(III)-bleomycin by NADPH-cytochrome P-450-reductase, Biochem.
    Pharmacol., 30:3385 (1981).
24. H.H. Seliger, A. Thompson, J.P. Hamman and G.H. Posner, Chemilumi-
    nescence of benzo[a]pyrene-7,8-diol, Photochem. Photobiol., 36:359
    (1982).
25. T.A. Dix and L.J. Marnett, Metabolism of polyclic aromatic hydrocarbon
    derivatives to ultimate carcinogens during lipid peroxidation,
    Science, 221:77 (1983).

26. G.A. Reed, E.A. Brooks and T.E. Eling, Phenylbutazone-dependent epoxidation of 7,8-dihydroxy-7,8-dihydrobenzo[a]pyrene, J. Biol. Chem. 259:5591 (1984).

27. T.W. Kensler, D.M. Bush and W.J. Kozumbo, Inhibition of tumor promotion by a biomimetic superoxide dismutase, Science, 221:75 (1983).

28. T.W. Kensler and M.A. Trush, Inhibition of oxygen radical metabolism in phorbol ester-activated polymorphonuclear leukocytes by an antitumor promoting copper complex with superoxide dismutase-mimetic activity, Biochem. Pharmacol., 32:3485 (1983).

29. W.J. Kozumbo, M.A. Trush and T.W. Kensler, Are free radicals involved in tumor promotion? Chem.-Biol. Interactions, in press (1985).

30. M.A. Trush, J.L. Seed and T.W. Kensler, Oxidant-dependent metabolic activation of polycyclic aromatic hydrocarbons by phorbol ester-stimulated human polymorphonuclear leukocytes: possible link between inflammation and cancer, Proc. Natl. Acad. Sci. USA, in press (1985).

31. M.R. Boyd, Biochemical mechanisms in chemical induced lung injury: roles of metabolic activation, CRC Crit. Rev. Toxicol., 7:163 (1980).

32. K. Takanaka, P.J. O'Brien, Y. Tsuruta and A.H. Rahimtula, Tumor promoter stimulated irreversible binding of N-methylaminobenzene to polymorphonuclear leukocytes, Cancer Letters, 15:311 (1982).

33. Y. Tsuruta, V.V. Subrahmanyam, W. Marshall and P.J. O'Brien, Peroxidase-mediated irreversible binding of arylamine carcinogens to DNA in intact polymorphonuclear leukocytes activated by a tumor promoter, Chem.-Biol. Interactions, 53:25 (1985).

34. S.J. Klebanoff, Estrogen binding by leukocytes during phagocytosis, J. Exp. Med., 145:983 (1977).

35. T.W. Kensler and M.A. Trush, Inhibition of phorbol ester-stimulated chemiluminescence in human polymorphonuclear leukocytes by retinoic acid and 5,6-epoxyretinoic acid, Cancer Res., 41:216 (1981).

36. M.A. Trush, M.J. Reasor, M.E. Wilson and K. VanDyke, Oxidant-mediated electronic excitation of imipramine, Biochem. Pharmacol., 33:1401 (1984).

37. M.A. Trush, E.G. Mimnaugh, Z.H. Siddik and T.E. Gram, Bleomycin-metal interaction: ferrous iron-initiated chemiluminescence, Biochem. Biophys. Res. Commun., 112:378 (1983).

38. W.E. Antholine, T. Sarna, R.C. Sealy, B. Kalyanaraman, G.D. Shields and D.H. Petering, Free radicals from the photodecomposition of bleomycin, Photochem. Photobiol., 41:393 (1985).

39. C. Borek and W. Troll, Modifiers of free radicals inhibit in vitro the oncogenic actions of x-rays, bleomycin and the tumor promoter 12-0-tetradecanoyl phorbol-13-acetate, Proc. Natl. Acad. Sci. USA, 80:1304 (1983).

40. C. Auclair, H. Gautero and P. Boivin, Effects of salicylate-copper complex on the metabolic activation in phagocytizing granulocytes, Biochem. Pharmacol., 29:3105 (1980).

41. A.H. Conney, Induction of microsomal enzymes by foreign chemicals and carcinogenesis by polycyclic aromatic hydrocarbons: G.H.A. Clowes Memorial Lecture, Cancer Res., 42:4875 (1982).

42. L.J. Marnett, Polycyclic aromatic hydrocarbon oxidation during prostaglandin biosynthesis, Life Sci., 29:531 (1981).

43. J.R. Battista, T.A. Dix and L.J. Marnett, The mechanism of hydroperoxide-dependent epoxidation of 7,8-dihydroxy-7,8-dihydrobenzo[a]pyrene by rat liver microsomes, Proc. Amer. Assoc. Cancer Res., 25:114 (1984).

44. J. Guthrie, I.G.C. Robertson, E. Zeiger, J.A. Boyd and T.E. Eling, Selective activation of some dihydrodiols of several polycyclic aromatic hydrocarbons to mutagenic products by prostaglandin synthetase, Cancer Res., 42:1620 (1982).

45. L.J. Marnett, G.A. Reed and J.T. Johnson, Prostaglandin synthetase dependent benzo[a]pyrene oxidation: products of the oxidation and inhibition of their formation by antioxidants, Biochem. Biophys. Res. Commun., 79:569 (1977).

46. P.L. Chesis, D.E. Levin, M.T. Smith, L. Ernster and B.N. Ames, Mutagenicity of quinones: pathways of metabolic activation and detoxification, Proc. Natl. Acad. Sci. USA, 81:1696 (1984).

47. A. Faljoni, M. Haun, M.E. Hoffman, R. Meneghini, N. Duran and G. Cilento, Photochemical-like effects in DNA caused by enzymically energized triplet carbonyl compounds, Biochem. Biophys. Res. Commun., 80:490 (1978).

48. S.A. Toledo, A. Zaha and N. Duran, DNA strand scission in E. coli by electronically excited state molecules generated by enzymatic systems, Biochem. Biophys. Res. Commun., 104:990 (1982).

49. V. Ravindranath, L.T. Burka and M.R. Boyd, Reactive metabolites from the bioactivation of toxic methylfurans, Science, 224:884 (1984).

50. H. Rosen and S.J. Klebanoff, Formation of singlet oxygen by the myeloperoxidase-mediated antimicrobial system, J. Biol. Chem., 252:4803 (1977).

51. H.C. Pitot, Biological and enzymatic events in chemical carcinogenesis, Ann. Rev. Med., 30:25 (1979).

52. H. Hennings, R. Shores, M.L. Wenk, E.F. Spangler, R. Tarone and S.H. Yuspa, Malignant conversion of mouse skin is increased by tumor initiators and unaffected by tumor promoters, Nature, 289:353 (1981).

METABOLISM OF DIEHTYLSTILBESTROL BY HORSERADISH PEROXIDASE AND PROSTAGLANDIN

SYNTHASE - EVIDENCE FOR A FREE RADICAL INTERMEDIATE

David Ross[1,4], Rolf J. Mehlhorn[2], Peter Moldeus[3], and Martyn
T. Smith[1]

[1]Department of Biomedical and Environmental Health Sciences
School of Public Health, and [2]Lawrence Berkeley Laboratory
University of California, Berkeley, CA 94720; [3]Department of
Toxicology, Karolinska Institute, Stockholm, Sweden
[4]present address: School of Pharmacy, University of Colorado
Boulder, CO 80309

INTRODUCTION

Diethylstilbestrol (DES) is carcinogenic in animals (1) and its use as
an abortifacient in humans has been linked to the appearance of reproductive
tract tumours in both the male and female offspring of mothers who were
treated with the drug (2-4). The mechanisms underlying DES carcinogenicity
are as yet unresolved but have been thought to involve the estrogenic
activity of the compound (1). Recent data, however, demonstrates that
hormonal activity alone is insufficient to explain the carcinogenicity of
DES (5-7).

DES is metabolized in both animals and man (8,9) and has been shown to
be extensively oxidized to reactive intermediates by peroxidases in utero
(10). One of these intermediates, DES quinone, covalently binds to DNA (11)
and is thus a possible cause of the carcinogenic activity of DES. A recent
study showed, however, that horseradish peroxidase (HRP) catalyzed
metabolism of ^{14}C-DES derivatives which could not form quinones but which
had at least one free hydroxyl group resulted in DNA binding (12). These
authors therefore proposed the phenoxy free radical derived from DES was a
possible genotoxic species. In this study we have examined whether or not
free radicals are produced during the metabolism of DES by HRP and
prostaglandin synthase (PGS), a peroxidase that is present in organs
susceptible to DES-induced carcinogenesis (13,14).

MEHTODS

Chemicals. DES and its dipropionate analog, glutathione (GSH),
hydrogen peroxide, HRP Type VI, and hematin were obtained from Sigma
Chemical Co., Mo. 5,5-dimethyl-1-pyrroloine-N-oxide (DMPO) was obtained
from Aldrich Chemical Co., Wi. Purified PGS was a generous gift from Prof.
L. Marnett, Department of Chemistry, Wayne State University, Detroit, Mi.
p-Phenetidine-HCl was synthesized as described previously (15).

Incubations with HRP. The following conditions were used: hydrogen
peroxide (0.1mM), HRP (3 g/ml), DES (0.1mM), DMPO (100mM), and in some

cases GSH (5mM) in potassium phosphate buffer (0.1M, pH=7.4).

Incubations with PGS. The following conditions were used: hydrogen peroxide (0.05mM), purified PGS (equiv to 12.3μg protein), DES or DES dipropionate (0.1mM), hematin (1μM) and DMPO (100mM).

EPR Spectroscopy. This was performed on a Varian E109E spectrometer at room temperature. Instrument conditions: microwave power (10mM), modulation amplitude (1.25G), time constant (0.064s), receiver gain (2.5 x 10^4) and scan time (one minute).

RESULTS AND DISCUSSION

In the presence of DMPO a six line paramagnetic signal (a^N = 14.9G, a^H = 18.3G) could be detected during the HRP-catalyzed oxidation of DES (Fig 1A). Control reactions without DES, in the presence of boiled enzyme, without enzyme, without hydrogen peroxide or without DMPO produced signals indistinguishable from background noise (Fig 1B). The EPR signal was short lived and decayed appreciably within three successive one minute scans. When the spin trap was added fifteen minutes after initiation of the reaction, no signal could be detected, confirming the transient nature of the radical.

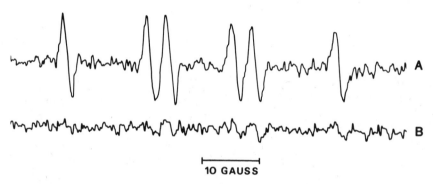

10 GAUSS

Fig 1 EPR spectra observed during HRP-catalyzed oxidation of DES in the presence of DMPO
a) DES as a DMSO solution, HRP, hydrogen peroxide and DMPO;
b) as in a but with DES omitted from the DMSO.

When GSH was included in the reaction mixture, an EPR signal consistent with the generation of a glutathionyl radical – DMPO adduct (16–18) was observed (Fig 2A). This signal was enzyme and DES-dependent (Fig 2B,C) and was not produced if GSH was either omitted from the reaction or added after fifteen minutes (Fig 2D). These data show that the radical produced during HRP-catalyzed metabolism of DES can interact with GSH to form a thiyl radical. To show that similar reactions also occurred in the presence of a biologically relevant peroxidase, the reactions were repeated using PGS in place of HRP. The same 6 line EPR signal obtained during HRP-catalyzed metabolism of DES was also obtained during PGS-catalyzed oxidation when either arachidonic acid (data not shown) or hydrogen peroxide (Fig 3A) was used as substrate for the enzyme. This shows that it is the hydroperoxidase component of PGS which is responsible for the production of the radical species which gives rise to the EPR signal. Controls without DES (Fig 3B), enzyme (Fig 3C) or hydrogen peroxide (Fig 3D), or containing boiled enzyme (Fig 3E) produced signals of much lower intensity than those obtained in the complete incubation system. PGS-catalyzed oxidation of DES dipropionate (Fig 3F) or p-phenetidine HCl (Fig 3G) did not induce the formation of EPR signals above control levels, showing that a free hydroxy group is needed for DES derivatives to cause radical formation and that enzyme turnover induced by very efficient co-substrates of PGS, such as p-phenetidine, is not sufficient to generate this signal.

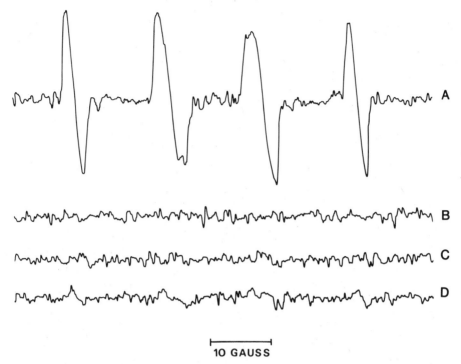

10 GAUSS

Fig 2   EPR spectra observed during HRP-catalyzed oxidation of DES in the presence of DMPO and GSH
a)   as Fig 1a but plus GSH (5mM);   b)   as Fig 3a but minus enzyme
c)   DES omitted from the DMPO;   d)   GSH added after incubating a GSH-free reaction mixture for 15 min.

Our experiments therefore show that a free radical is produced, which can be trapped using DMPO, during either HRP or PGS-catalyzed oxidation of DES. The one electron nature of the peroxidatic oxidation of DES by HRP was further underlined by the generation of a glutathionyl radical when GSH was included in the system. This shows that GSH can reduce the radical and in the process be oxidized to a thiyl radical. The identity of the 6 line EPR signal cannot be deduced from our data: the hyperfine coupling constants cannot be correlated with available published constants (19) although the signal is similar to the DMPO-formate radical adduct (20). Thus this data is not conclusive proof that the 6 line EPR signal obtained during HRP or PGS-catalyzed oxidation of DES in the presence of DMPO is indicative of a DMPO-DES adduct. Indeed a small EPR signal with identical coupling constants could be generated during turnover of PGS in the absence of DES (Fig 3B). This small signal was presumably produced by metabolism of some component of the control reaction mixture to a radical which could be trapped by DMPO. Attempts to characterize the putative DMPO-DES adduct by mass spectrometry are in progress.

Whatever the source of the control signal shown in Fig 3B, the observation of identical EPR signals during metabolism of DES by both HRP and PGS and the generation of a glutathionyl radical during HRP-catalyzed metabolism of DES in the presence of glutathione shows that peroxidatic oxidation of DES occurs via a one electron mechanism. Since peroxidase-catalyzed oxidation of DES occurs in all _in vitro_ systems in which DES has been found to be genotoxic (12,21) and in organs susceptible to DES-induced carcinogenicity (22), the generation of DES-derived free radicals is therefore a possible determinant of DES-induced genotoxicity and carcinogenicity.

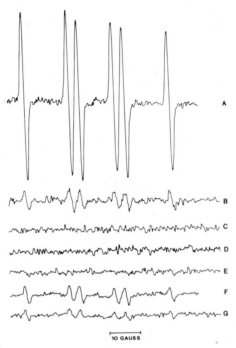

Fig 3   EPR spectra observed during hydrogen peroxide dependent PGS catalyzed oxidation of DES in the presence of DMPO.
a) H_2O_2, PGS, hematin, DES, DMPO; b) DES omitted from DMSO;
c) minus enzyme; d) minus H_2O_2; e) boiled enzyme; f) with DES dipropionate (0.1mM) in place of DES; g) p-phenetidine HCl (0.5mM) in place of DES.

ACKNOWLEDGEMENT

Supported by the National Foundation for Cancer Research. The authors would like to thank Michael Murphy for secretarial assistance.

REFERENCES

1.   IARC Monographs on the Evaluation of the Carcinogenic Risk of Chemicals to Humans, Vol 21, Sex Hormones (II); IARC, Lyon (1979).

2.   P. Greenwald, J.J. Barlow, P.C. Nasca and W.S. Burnett. Vaginal cancer after maternal treatment with synthetic oestrogens. N. Engl. J. Med. 285:390 (1971).

3.   A.L. Herbst, H. Ulfelder and D.C. Poskanzer. Adenocarcinoma of the vagina. Association of maternal stilbestrol therapy with tumor appearance in young women. N. Eng. J. Med. 284:878 (1971).

4. W.B. Gill, G.F.B. Schumacher, M.M. Hubby and R.R. Blough. Male genital tract changes in humans following intrauterine exposure to diethylstibestrol in Developmental Effects of Diethylstibestrol in Pregnancy (A.L. Herbst and H. A. Bern, eds.) Thieme-Stratton Inc., New York (1981).

5. J.A. McLachlan, A. Wong, G.H. Degen and J.C. Barrett. Morphological and neoplastic transformation of Syrian hamster embryo fibroblasts by diethylstibestrol and its analogs. Cancer Res. 42:3040 (1982).

6. J.G. Liehr. 2 Fluoroestradiol. Separation of estrogenicity from carcinogenicity. Mol. Pharmacol. 23:278 (1983).

7. J.J. Li, S.A. Li , J.K. Klicka, J.A. Parson and L.K.T. Lam. Relative carcinogenic activity of various synthetic and natural estrogens in the Syrian hamster kidney. Cancer Res. 43:5200 (1983).

8. M. Metzler. The metabolism of diethylstibestrol. CRC Crit. Rev. Biochem. 10:171 (1981).

9. M. Metzler and J.A. McLachlan. Oxidative metabolites of diethylstibestrol in the fetal, neonatal and adult mouse. Biochem. Pharmacol. 27:1087 (1978).

10. M. Metzler and J.A. McLachlan. Peroxidase mediated oxidation--a possible pathway for metabolic activation of diethylstibestrol. Biochem. Biophys. Res. Commun. 85:874 (1978).

11. J.G. Liehr, B.B. Dague, A.M. Ballatore and J. Henkin. Diethylstilbestrol (DES) quinone--a reactive intermediate in DES metabolism. Biochem. Pharmacol. 32:3711 (1983).

12. M. Metzler and B. Epe. Peroxidase-mediated binding of diethylstilbestrol analogs to DNA in vitro: a possible role for a phenoxy radical. Chem. Biol. Interactions 50:351 (1984).

13. B.B. Davis, M.B. Mattamal and T.V. Zenser. Renal metabolism of drugs and xenobiotics. Nephron 27:187 (1981).

14. M.H. Abel and D.T. Baird. The effect of 17 beta estradiol and progesterone on prostaglandin production by human endometrium maintained in organ culture. Endocrinology 106:1599 (1980).

15. B. Anderson, R. Larsson, A. Rahimtula and P. Moldeus. Hydroperoxidase-dependent activation of p-phenetidine catalyzed by prostaglandin synthase and other peroxidases. Biochem. Pharmacol. 32:1045 (1983).

16. G. Saez, P.J. Thornalley, H.A.O. Hill, R. Hems and J.V. Bannister. The production of free radicals during the autooxidation of cysteine and their effect in isolated rat hepatocytes. Biochim. Biophys. Acta. 419:24 (1982).

17. L.S. Harman, C. Mottley and R.P. Mason. Free radical metabolites of L cysteine oxidation. J. Bio. Chem. 259:5600 (1984).

18. D. Ross, E. Albano, U. Nilsson and P. Moldeus. Thiyl radicals formation during peroxidase catalyzed metabolism of acetaminophen in the presence of thiols. Biochem. Biophys. Res. Commun. 125:109 (1984).

19. E.G. Janzen and J. I-Ping-Liu. Radical addition reactions of 5,5-
     dimethyl-1-pyrroline-N-oxide-ESR spin trapping with a cyclic nitrone.
     J. Mag. Resonance 9:510 (1973).

20. E. Finkelstein, G.M. Rosen and E.J. Rauckman. Spin trapping of
     superoxide and hydroxyl radical: Practical Aspects. Arch. Biochem.
     Biophys. 200:1 (1980).

21. M. Metzler. Diethylstilbestrol: reactive metabolites derived from a
     hormonally active compound, in Biochemical Basis of Chemical
     Carcinogenesis (H. Greim, R. Jung, M. Kramer, H. Marquardt and F.
     Oesch, eds.). Raven Press, New York (1984).

22. R. Maydl, R.R. Newbold, M. Metzler and J.A. McLachlan.
     Diethylstibestrol metabolism by the fetal genital tract. Endocrinology
     113:46 (1983).

THIYL RADICALS--THEIR GENERATION AND FURTHER REACTIONS

David Ross[1] and Peter Moldeus

Department of Toxicology, Karolinska Institute
Stockholm, Sweden
[1]present address: School of Pharmacy
University of Colorado, Boulder, CO 80309

INTRODUCTION

Amongst the functions of glutathione (GSH) is the protection of cells against reactive electrophilic species. Many xenobiotics can generate reactive substrate-derived radicals during metabolism (1) and other free radicals such as superoxide anion radical (O_2^-) can be generated during normal cellular function (2). Thus the interaction of GSH with free radicals is of both physiological and toxicological significance. The reduction of free radicals by GSH, however, generates another radical--the glutathionyl radical (GS·), the fate of which in biological systems is poorly understood.

In this study we describe the generation of thiyl radicals during horseradish peroxidase (HRP) catalyzed oxidation of acetaminophen and p-phenetidine in the presence of GSH and the interaction of O_2^- with GSH. Furthermore, using the HRP system as a model system in which to generate glutathionyl radicals, we have described the further reactions of GS· by following the kinetics of oxygen uptake and oxidized glutathione (GSSG) formation, two reactions of GS· which are known to occur from radiolysis studies in chemical systems (3-7).

METHODS

EPR spectroscopy was performed using a Varian E9 spectrometer at 25°C.

Oxygen uptake was measured at 25°C using a Clarke electrode.

GSH/GSSG was measured by HPLC according to (8). All data has been corrected for the oxidation of GSH by hydrogen peroxide in the absence of co-substrate.

Xanthine/xanthine oxidase: incubations contained xanthine (1.5mM), xanthine oxidase (30mU/ml), GSH (5-20mM), catalase (40μg/ml) to remove hydrogen peroxide, DETAPAC (1mM) to remove impurities in oxygen-saturated potassium phosphate buffer 0.1M (pH=7).

HRP system: incubations were performed in potassium phosphate buffer

0.1M (pH=8) containing EDTA (1mM) at 25°C. Reactions were terminated by the addition of catalase (0.05 ml, 1500 u/ml)

Phenetidine removal and quantitation of p-phenetidine metabolites was measured as in (9). Regeneration of p-phenetidine was calculated as the difference between the removal of p-phenetidine in the presence and absence of GSH.

Experiments under nitrogen were performed using nitrogen-saturated buffers in sealed flasks flushed with nitrogen. Additions were performed through rubber seals using Hamilton syringes.

RESULTS AND DISCUSSION

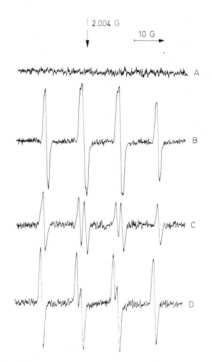

Fig 1.   DMPO-thiyl radical adducts generated during HRP catalyzed oxidation of p-phenetidine in the presence of thiols and DMPO Reaction contained; A, p-phenetidine (0.5mM), HRP (0.2µg/ml), hydrogen peroxide (0.25mM) and DMPO (0.1M); B, as A but plus GSH (5mM); C, as A but plus cysteine (5mM); D, as A but plus N-acetyl cysteine (5mM).

HRP catalyzed oxidation of either p-phenetidine or acetaminophen in the presence of thiols and the spin trap DMPO, leads to the generation of EPR signals consistent with the generation of the appropriate DMPO-thiyl radical adduct (10,11, Fig 1). Such EPR signals were not observed in the absence of co-substrate, HRP, hydrogen peroxide or thiol. This data shows that acetaminophen or p-phenetidine derived radicals can interact with thiols to form thiyl radicals which can be observed using EPR spectroscopy in conjunction with the spin trap DMPO.

The interaction of O_2^- with GSH is of physiological importance and we studied this reaction using the xanthine/xanthine oxidase system as a model system to generate O_2^-. It has previously been reported that the interaction of GSH and O_2^- generates singlet oxygen (12) but the production of a glutathionyl radical, although proposed as an intermediate in this reaction, has never been shown. Fig 2A shows the EPR signal observed during xanthine oxidase catalyzed metabolism of xanthine in the presence of DMPO and this signal is consistent with the generation of an adduct of superoxide and DMPO (13, DMPO-OOH). When GSH (5-20mM) was included in the reaction mixture, some change in the EPR signal, relative to that detected in the absence of thiol, was observed (Fig 2B). This mixed spectrum implied that another paramagnetic species was formed in the presence of GSH. A reference spectrum of a DMPO-glutathionyl radical adduct is shown in Fig 2C and it can be seen that the spectrum observed in Fig 2B could be considered a composite of a DMPO-OOH adduct (Fig 2A) and a DMPO-glutathionyl radical adduct (Fig 2C). Unequivocal identification of the additional radical formed in the presence of GSH as a DMPO-glutathionyl radical adduct however is not possible from these spectra. That the DMPO-glutathionyl radical adduct cannot be observed clearly is probably a function of the rapid further reactions of the glutathionyl radical which will compete with DMPO-adduct formation.

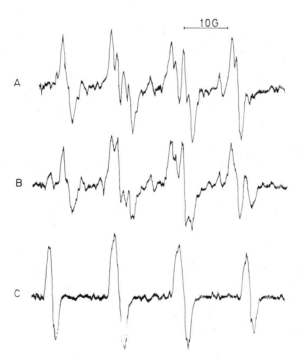

Fig 2.   EPR spectra observed during xanthine oxidase catalyzed metabolism of xanthine in the presence of DMPO (0.1M)
A - complete system; B - plus GSH (20mM); C - reference spectrum of DMPO-glutathionyl radical adduct.

We have investigated these further reactions of the glutathionyl radical by using the HRP-catalyzed oxidation of acetaminophen or p-phenetidine in the presence of GSH as a model system in which to generate glutathionyl radicals. The glutathionyl radical may either dimerize, interact with oxygen or react with GSH (3-7). These reactions and their rate constants are shown in Fig 3. The readily measurable indices of these reactions are oxygen uptake and GSSG formation and these both occur during HRP-catalyzed oxidation of acetaminophen and p-phenetidine (data not shown). The reduction of substrate-derived radicals by GSH should lead to a regeneration of substrate and this was indeed found to be the case during HRP-catalyzed oxidation of p-phenetidine, the amount of regeneration proportional to GSH concentration (Fig 4). The amounts of GSSG generated and oxygen uptake observed however increased with increasing concentrations of GSH even after maximal regeneration of p-phenetidine had ocurred (Fig 4).

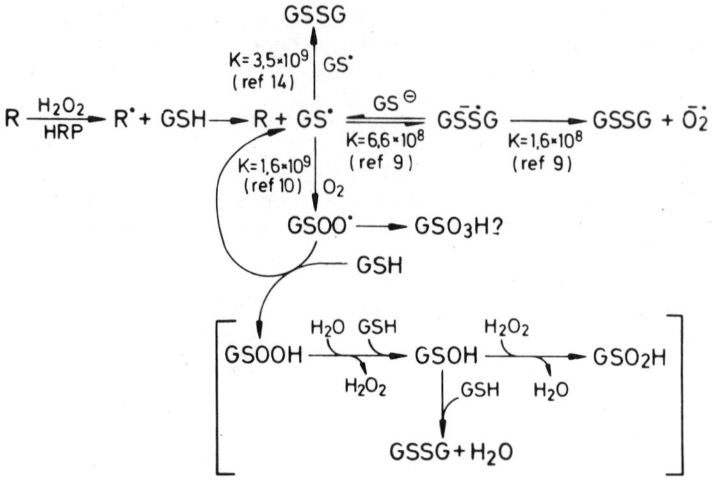

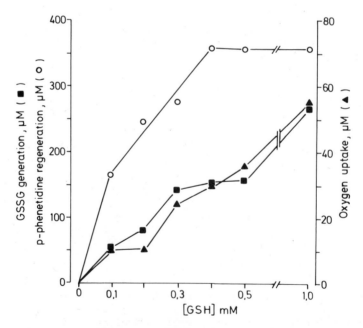

Fig 3.   Proposed scheme for the interaction of radicals with GSH and the subsequent fate of the products.

Fig 4.   Oxygen consumption (▲), GSSG generation (■) and p-phenetidine regeneration (o) during HRP-catalyzed oxidation of p-phenetidine (HRP 0.2 μg/ml, p-phenetidine 0.5mM, hydrogen peroxide 0.2mM)

These data indicate that oxygen uptake and GSSG formation occur at high thiol concentration via a process independent of the initial reaction leading to substrate regeneration. This secondary process was however still dependent on GSH concentration and this probably reflects the operation of one or a mixture of two pathways. The thiyl radical can react with $G\overline{S}$ to form the glutathione anion radical which in turn can interact with oxygen forming GSSG. Alternatively the peroxysulphenyl radical formed by addition of oxygen to the glutathionyl radical may abstract a hydrogen atom from the excess GSH present forming glutathione sulphenyl hydroperoxide (GSOOH) which can undergo various reactions including the generation of GSSG (see Fig 3).

Dimerization of gluathione radicals is one of the possible mechanisms of GSSG generation during these reactions (Fig 3). When HRP-catalyzed oxidation of p-phenetidine or acetaminophen (paracetamol) in the presence of GSH was performed under an atmosphere of nitrogen, substantial decreases in the amount of GSSG generated, relative to the quantities formed under aerobic conditions, were observed (Fig 5). Nitrogen saturation conditions did not affect HRP activity as measured by p-phenetidine removal and appearance of metabolites of p-phenetidine (data not shown). These data indicate that the presence of oxygen is essential for maximal thiol oxidation and that thiyl radical dimerization, which would not be affected by nitrogen saturation conditions, appears to play a minor role in GSSG generation despite the high rate constant for this reaction calculated from radiolysis experiments.

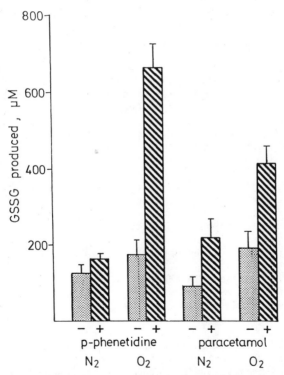

Fig 5.   GSSG generation during HRP catalyzed oxidation of acetaminophen (paracetamol) and p-phenetidine in the presence of GSH under aerobic and nitrogen saturation conditions. Mean ± s.d of 3 observations (p-phenetidine 0.5mM, HRP 2μg/ml, H_2O_2 0.25mM; acetaminophen 0.5mM, HRP 25μg/ml, H_2O_2 0.25mM; GSH = 5mM in both cases.)

In summary, our data suggest that free radicals of widely differing structure, e.g., acetaminophen–, p–phenetidine–, and oxygen–derived radicals, can react with GSH and other thiols to form a common intermediate--the thiyl radical. These data also confirm the observations that glutathionyl radicals participate in reactions leading to oxygen uptake and GSSG generation. More importantly, however, they show that the major route of GSSG formation is not via thiyl radical dimerization but is via mechanism(s) dependent on oxygen concentration such as via the glutathione anion radical (GSSG⁻) or the glutathione peroxysulphenyl radical (GSOO⁻)

ACKNOWLEDGEMENT

Supported by the Royal Society of Great Britain and the Swedish Medical Research Council. The authors would also like to thank Michael Murphy for secretarial assistance.

REFERENCES

1. R.P. Mason. Free radical intermediates in the metabolism of toxic chemicals, in Free Radicals in Biology vol. 5 (ed. Pryor, W.A.), pp. 161–222, Academic Press, New York (1982).

2. I. Fridovich. The biology of oxygen radicals. Science 201:875 (1978).

3. M. Quintiliani, R. Badillo, M. Tamba and G. Gorin. Radiation chemical basis for the role of glutathione in cellular radiation sensitivity. In, Modification of Radiosensitivity of Biological Systems, pp. 29–37, IAEA, Vienna (1976).

4. A. Al Thannon, J.P. Barton, J.E. Pucker, R. Sims, C.W. Trumborne and R.V. Winchester. The radiolysis of aqueous solutions in the presence of oxygen. Int. J. Radiat. Phys. Chem. 6:233 (1974).

5. M. Quintiliani, R. Badiello, M. Tamba, A. Esfandi and G. Gorin. Radiolysis of glutathione in oxygen containing solutions of pH=7. Int. J. Radiat. Biol. 32:195 (1977).

6. M. Lal. ^{60}Co–radiolysis of reduced glutathione in aerated solutions at pH values between 1–7. Can J. Chem 54:1092 (1976).

7. M.Z. Hoffman and E. Hayon. One electron reduction of the disulfide linkage in aqueous solution. Formation, protonation and decay kinetics of the RSSR radical. J. Am Chem. Soc. 94:7950 (1972).

8. D.J. Reed, J.R. Babson, P.W. Beatty, A.E. Brodie, W.W. Ellis and D.W. Potter. HPLC analysis of nanomole levels of glutathione, glutathione disulfide and related thiols and disulfides. Anal. Biochem. 106:55 (1980).

9. D. Ross, R. Larsson, B. Andersson, U. Nilsson, T. Lindquist, B. Lindeke and P. Moldeus. The oxidation of p–phenetidine by horseradish peroxidase and prostaglandin synthase and the fate of glutathione during such oxidations. Biochem. Pharmacol. 34:343 (1985).

10. G. Saez, D.J. Thornalley, H.A.O. Hill, R. Hems and J.V. Bannister. The production of free radicals during the autoxidation of cysteine and their effect in isolated rat hepatocytes. Biochim. Biophys. Acta. 719:24 (1982).

11. L.S. Harman, M. Mottley and R.P. Mason. Free radical metabolites of L-cysteine oxidation. J. Biol. Chem. 259:5606 (1984).

12. H. Wefers and H. Sies. Oxidation of glutathione by the superoxide radical to the disulfide and the sulfonate yielding singlet oxygen. Eur. J. Biochem. 137:29 (1983).

13. E. Finkelstein, G.M. Rosen and E.J. Rauckman. Spin trapping of superoxide and hydroxyl radical: Practical aspects. Arch. Biochem. Biophys. 200:1 (1980).

MECHANISM OF THE CYTOCHROME P-450

CATALYZED ISOMERIZATION OF HYDROPEROXIDES

Michael D. Wand and John A. Thompson

Pharmaceutical Sciences Division
School of Pharmacy
University of Colorado
Boulder, Colorado 80309

INTRODUCTION

The interactions of hydroperoxides with cytochrome P-450 have been studied extensively (O'Brien, 1982; Sligar et al., 1984). It is well known that these compounds can donate an oxygen to the ferric form of P-450. The resulting activated form of the enzyme can oxidize many substrates with results similar to those of NADPH/O_2-supported P-450 oxidations. Mechanistic details of the O-O bond cleavage step and the nature of the oxidant remain unsettled issues. Homolytic cleavage would produce an alkoxy radical and an iron-coordinated hydroxyl radical similar to peroxidase Compound II (equation 1, where Fe represents the heme iron of P-450). Heterolytic cleavage of the O-O bond would produce the alcohol

$$(1) \quad ROOH + Fe^{3+} \longrightarrow RO\cdot + (FeOH)^{3+}$$

$$(2) \quad ROOH + Fe^{3+} \longrightarrow ROH + (FeO)^{3+}$$

and an iron-oxo species analogous to peroxidase Compound I (equation 2). Much of the early work in this area suggests that homolysis is the principal result when hydroperoxides interact with P-450. For example, Griffin (1980) successfully trapped a methyl radical from the microsomal decomposition of cumene hydroperoxide (COOH). This species is produced when the cumyloxy radical undergoes β-scission. Blake and Coon (1981) invoked homolysis to explain the results of structure-activity relationships on the benzylic hydroxylation of substituted toluenes by rabbit liver P-450$_{LM2}$ that were supported by analogs of COOH. According to the mechanism proposed, the cumyloxy radical is the oxidant that abstracts H$\cdot$ from the substrate. The resulting carbon-centered radical then interacts with $(FeOH)^{3+}$ forming the benzylic alcohol and the native enzyme. More recent work with P-450 and peroxyphenylacetic acid demonstrated that peroxy compounds can undergo both homolytic and heterolytic scission of the O-O bond. McCarthy and White (1983) provided evidence that hydroxylations of cosubstrates in the P-450-peroxyacid system were a consequence of heterolysis, and that $(FeO)^{3+}$ was the oxidant. Recently, it was reported that P-450 catalyzes both types of O-O bond cleavage, and that the actual mechanism is dependent on the structure of the peroxy compound (Lee and Bruice, 1985).

We are studying the products of hydroperoxide transformation in order to elucidate further the nature of hydroperoxide-P-450 interactions (Thompson and Wand, 1985). It was determined that hydroperoxides can be isomerized to diols in a caged sequence involving heterolytic cleavage of the O-O bond and oxidation of the resulting alcohol at the active site of P-450. The formation of other products that are radical-derived, show the ability of the enzyme to catalyze both homolytic and heterolytic processes with the same substrate. The results described in this report provide additional data concerning the interactions of BOOH and COOH (Figure 1) with P-450. BOOH is produced from the antioxidant BHT during

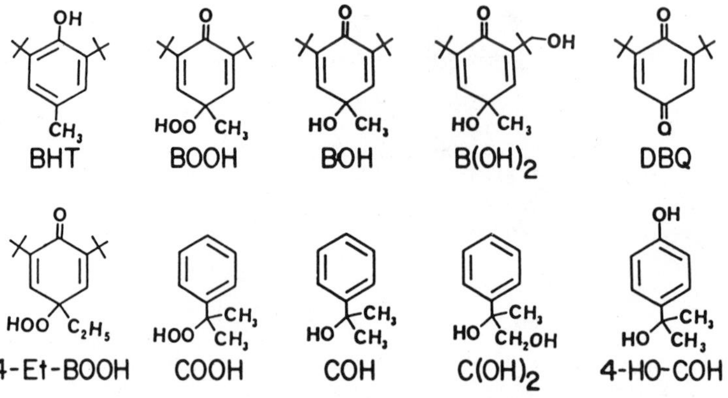

Figure 1. Structures of the Compounds Involved in this Study.

turnover of the microsomal monooxygenase system, and also can be isomerized by P-450 to $B(OH)_2$. COOH, on the other hand, is isomerized to a 1:1 mixture of the 2 diols $C(OH)_2$ and 4-HO-COH. Both hydroperoxides also give rise to products resulting from homolysis and expulsion of a methyl radical. The hydroperoxide 4-Et-BOOH was employed to delineate further the differences between homolytic and heterolytic pathways.

MATERIALS AND METHODS

Chemicals

BHT, DBQ, 3,5-di-t-butyl-4-hydroxybenzylalcohol, 3,5-di-t-butyl-4-hydroxybenzaldehyde, COOH, COH, $C(OH)_2$, acetophenone, 4-hydroxyacetophenone, t-butylhydroperoxide and methylmagnesium iodide were purchased from Aldrich. BOH, $4-CD_3-BOH$, BOOH, and BOOH were synthesized as described (Kharasch and Joshi, 1975; Nishinaga et al., 1975; Thompson and Wand, 1985). 4-HO-COH was prepared from methylmagnesium iodide and 4'-hydroxy-acetophenone, and CD_3-COH was synthesized from trideuteriomethylmagnesium iodide (prepared from CD_3I) and acetophenone. The 4-ethyl analog of BHT was synthesized by adding methylmagnesium iodide to 3,5-di-t-butyl-4-hydroxybenzaldehyde; the resulting alcohol was reduced to the 4-ethyl compound by hydrogenation in ethanol catalyzed by palladium on charcoal. This compound was converted into 4-Et-BOOH as described for the preparation of BOOH. All products were purified by flash chromatography on silica gel and characterized by standard methods, including UV, ^{1}H NMR and MS. Hemoglobin, hematin, horseradish peroxidase Type VI, NADP[+], NADPH, glucose 6-phosphate and glucose 6-phosphate dehydrogenase were obtained from Sigma.

Incubations

Microsomes and purified cytochrome P-450 were prepared from the livers of phenobarbital-treated male Sprague-Dawley rats (140-180 g) as described (Thompson et al., 1984; Thompson and Wand, 1985). Microsomal incubations were normally conducted at either 25° or 37°C in 50 mM phosphate buffer (pH 7.4) containing 1 mg/ml of protein and 0.5 mM substrate (added in 5-7 μl of $(CH_3)_2SO$ per ml of incubate). In some cases, an NADPH-generating system consisting of $NADP^+$, glucose 6-phosphate and glucose 6-phosphate dehydrogenase, and 5 mM $MgCl_2$ were added also. Reactions were terminated by cooling to 4°C, saturating with NaCl and shaking with cold ether. The substrates BHT, BOOH and COOH were purified before use by semi-preparative HPLC with a 10 x 250 mm RP-18 Hibar column (Merck) and mixtures of acetonitrile-water as the mobile phase.

Analyses

Incubate extracts in ether were evaporated carefully and redissolved in acetonitrile. Samples were analyzed by HPLC with a 4.6 x 250 mm Altex Ultrasphere ODS column and eluted with 1 ml/min of acetonitrile-water (60:40, changed to 80:20 after 20 min for BHT and BOOH incubates; 40:60 for CHP incubates). Quantitative work was conducted by constructing calibration curves with internal standards. GC/MS analyses were performed either in the electron or chemical ionization mode with a Hewlett Packard 5984A instrument. The GC column (2 mm x 1.9 m) was packed with 3% OV-22 on Supelcoport (Supelco), and temperature programming was employed over the range 120-220°C. Trimethylsilyl (TMS) derivatives were prepared by heating $B(OH)_2$ in 100 μl of dichloromethane and 20 μl of N-methyl-N-(trimethylsilyl)-trifluoroacetamide (Aldrich) at 60°C, or with 100 μl of the silylation reagent and pyridine at 90°C for $C(OH)_2$ and 4-HO-COH.

RESULTS

Formation and Destruction of BOOH

Small amounts of BOOH were observed by HPLC in incubates conducted with highly purified BHT, liver microsomes and an NADPH-generating system at 15°C (Figure 2). The hydroperoxide was also rapidly destroyed in

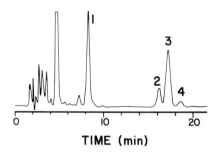

TIME (min)

Figure 2. Chromatogram of Metabolites Formed from BHT in Microsomes. Analysis was by HPLC with UV detection at 225 nm. Metabolites are (1) 3,5-di-t-butyl-4-hydroxybenzyl alcohol, (2) BOH, (3) 3,5-di-t-butyl-4-hydroxybenzaldehyde and (4) BOOH.

microsomes (see below). For this reason, BOOH formation was investigated by measuring its reduction product BOH (peak 2). The data in Table 1 show that BOH (and, therefore, BOOH) was produced in microsomes, and that its formation is dependent on NADPH. Incubates were conducted at low temperature, but substantially more BOH was produced at 37°C. These results indicate that BHT undergoes 1-electron oxidation as a result (directly or indirectly) of turnover of the monooxygenase system. The resulting delocalized phenoxy radical could then combine with O_2 at the 4-position of the ring leading to BOOH.

Table 1. Conversion of BHT to BOH by Liver Microsomes

Incubation Time[a] (min)	BOH Formation (nmol/mg protein)	
	−NADPH	+NADPH
0	0.13	0.28
6	0.15	1.21, 1.15

[a] Purified BHT incubated with rat liver microsomes (1 mg/ml) at 15°C with or without NADPH.

Characteristics of the interaction between BOOH and microsomal cytochrome P-450 were examined. The results in Table 2 show that this hydroperoxide destroys P-450, but not as effectively as COOH. Data in Table 3 show that BOOH can support the P-450-catalyzed N-demethylation of N,N-dimethylaniline, however, BOOH is also less effective here than COOH.

Table 2. Destruction of Cytochrome P-450 by Hydroperoxides

Hydroperoxide[a]	P-450 Concentration (nmol/mg protein)	% Destruction
none	2.81 ± 0.02	
BOOH	1.99 ± 0.06	29
COOH	1.16 ± 0.01	59

[a] Rat liver microsomes (1.7 mg protein/ml) were incubated with the hydroperoxide (0.5 mM) or the vehicle alone at 37°C for 10 min.

The concentration of BOOH decreased sharply during the first minute when incubated with microsomes alone, and this was accompanied by the rapid formation of BOH (Figure 3). A second product, the diol $B(OH)_2$, was formed nearly linearly with time at 15°C, but the rate of formation at 25°C was distinctly biphasic. When NADPH was added at the 16 min time

Table 3. Hydroperoxide-Supported Demethylation of N,N-Dimethylaniline

Hydroperoxide[a]	Formaldehyde formation (nmol/nmol P-450)
BOOH	32.6 ± 1.1
COOH	66.8 ± 2.2
t-Butylhydroperoxide	20.0 ± 1.6

[a] N,N-Dimethylaniline (2 mM) was incubated at 37°C for 10 min with liver microsomes (2 mg/ml) and hydroperoxides (1.0 mM) were added in ethanol. Results are the means ± SE, N=3.

point, the sudden change in concentrations suggests that this cofactor supported the conversion of BOH to $B(OH)_2$. The ability of microsomes to hydroxylate BOH was confirmed; incubates conducted with BOH as the substrate in the presence of NADPH produced $B(OH)_2$ as the major product, along with minor amounts of a second product tentatively identified as the glycol from hydroxylation of the 4-methyl group of BOH.

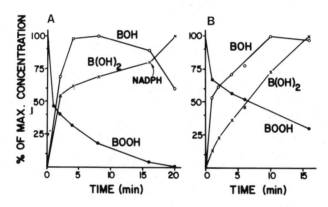

Figure 3. Time Courses of BOOH Destruction and Product Formation. Incubations were conducted with BOOH and liver microsomes at (A) 25°C and (B) 15°C. NADPH was absent at the beginning of each incubation, but was added at the 16 min time point of the 25°C reaction. Each compound is plotted as a percentage of its maximum concentration.

Mechanism of $B(OH)_2$ Formation

Microsomal incubates conducted at 25°C with a 1:1 mixture of BOOH and the 4-trideuteriomethyl analog of BOH ($4-CD_3-BOH$) led to a 95:5 mixture of nondeuterated to deuterated $B(OH)_2$ after 4 min, and an 89:11 mixture after 16 min. When di-^{18}O-labeled hydroperoxide (BOOH) was incubated, the product diol was fully labeled with ^{18}O in both hydroxyl groups. The incubation of a 54:44 mixture of BOOH and BOOH that contained about 2% of mono-^{18}O-labeled BOOH gave the mass spectral results shown in Figure 4 after 1 min at 25°C. The pseudomolecular ions appeared at m/z 397, 399 and 401 for the unlabeled, mono-^{18}O-labeled and di-^{18}O-labeled diol. After correcting for contributions due to naturally occurring heavy isotopes, the amounts of these three species matched those of

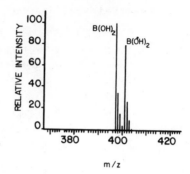

Figure 4. Mass Spectral Data of Diols from the BOOH/BÖÖH Experiment.
The diols were analyzed as their trimethylsilyl derivatives
by chemical ionization GC/MS.

the original mixture of hydroperoxides. The results presented here indicate that during the early (rapid) phase of the reaction, BOOH is converted to BOH by heterolytic cleavage of the 0-0 bond and $B(OH)_2$ is then
formed by attack of the reactive iron-oxo species of P-450 before the
alcohol migrates from the active site of the enzyme. Thus, there is

$$BOOH \xrightarrow{\text{P-450}} BOH + (FeO)^{3+} \longrightarrow B(OH)_2 + Fe^{3+}$$

no transfer of oxygen from one molecule of hydroperoxide to another molecule of hydroperoxide or alcohol. At longer incubation times, however,
BOH that does escape from the active site, can reenter and undergo hydroxylation in the second (slow) phase of diol formation.

Quantitative data for the conversion of BOOH and COOH to diols are
presented in Table 4. When BOH was incubated with horseradish peroxidase, hemoglobin and hematin, BOH was produced but $B(OH)_2$ was not.

Table 4. Hydroperoxide Isomerization by Cytochrome P-450

Incubate[a]	Diol formation (nmol/nmol P-450)		
	$B(OH)_2$	$C(OH)_2$	4-HO-COH
Microsomes, PB-treated	7.9 ± 0.3	26 ± 1	29 ± 1
Boiled microsomes	nd[b]	nd	nd
Purified P-450	11[c]	18 ± 1	11 ± 1

[a] Incubated with 0.5 mM BOOH at 25°C or COOH at 37°C for
10 min and either microsomes (1 mg protein/ml) or purified
P-450 (0.2 nmol/ml). Results are the means ± SE of 3 determinations.
[b] None detected.
[c] Average of 2 determinations.

Conversion of COOH to Diols

    COOH was stable when incubated in buffer but was almost completely destroyed when incubated with microsomes at 37°C for 10 min (Figure 5).

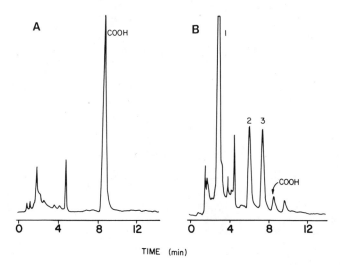

TIME (min)

Figure 5. Chromatographic Analysis of COOH and Its Transformation Products. The hydroperoxide was incubated with (A) buffer alone or (B) rat liver microsomes at 37°C for 10 min. The incubates were analyzed by HPLC with UV detection at 254 nm. Compounds were identified by comparing chromatographic and spectral properties with authentic standards: (1) $C(OH)_2$ and 4-HO-COH; (2) COH and (3) acetophenone.

Four products were identified, including the 2 diols $C(OH)_2$ and 4-HO-COH which coeluted in peak 1. The compounds were isolated from the HPLC column and analyzed by GC/MS methods. The data shown in Table 4 demonstrates that the diols were formed in approximately a 1:1 ratio. When COH was incubated under the same conditions with microsomes and NADPH, the diols also were produced in the same ratio, but the concentrations were 10-fold less. Incubation of COOH with an equal concentration of

Table 5. Caged Isomerization of COOH

Incubation[a]	Unlabeled diol formed as a percentage of total diol	
	$C(OH)_2$	4-HO-COH
COOH + CD_3-COH	98.7	97.4
COOH + BÖOH**	94.8	99.8

[a] Incubations were conducted with microsomes at 37°C for 16 min. The hydroperoxides and CD_3-COH were present at 0.5 mM, and the isotope content of products was determined by GC/MS.

trideuteriomethyl COH (CD$_3$-COH) led to only traces of deuterium labeled diols (Table 5). Incubation of COOH with an equal concentration of BOOH** also led to only small amounts of ^{18}O-containing diols. These results indicate that COOH underwent heterolytic O-O bond cleavage by P-450 and that attack by the iron-oxo species on the intermediate COH occurred

$$\text{COOH} \xrightarrow{\text{P-450}} \text{COH} + (\text{FeO})^{3+} \longrightarrow \text{C(OH)}_2 + 4\text{-HO-COH}$$

before the alcohol escaped from the enzyme. It also appears that less COH migrated back into the active site to undergo hydroxylation than occurred with BOH during the conversion of BOOH to B(OH)$_2$.

Homolytic Cleavage of the O-O Bond

About 1% of the COOH added to microsomes was converted to acetophenone, presumably by homolysis of the O-O bond and β-scission. Similarly, BOOH was converted to the quinone DBQ. Replacing the 4-methyl substituent of BOOH with an ethyl group would not be expected to have a large effect on products of heterolysis (the diol), but should affect substantially the amount of quinone formed. This is due to the fact that once formed, the alkoxy radical should undergo β-scission more readily to give an ethyl radical compared to a methyl radical. The results (Figure 6) show a small decrease in diol formation, but a 10-fold increase in DBQ formation. These results provide additional evidence that B(OH)$_2$ and DBQ arise from BOOH by different initial events, i.e. heterolysis vs. homolysis.

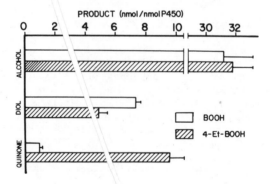

Figure 6. Comparison of Products Formed from BOOH and 4-Et-BOOH. Incubations were conducted with microsomes at 25°C for 10 min.

DISCUSSION

BHT is an important antioxidant used widely as a preservative of foods, drugs and other consumer products. As such, its various routes of biotransformation should be thoroughly understood. Several years ago, Chen and Shaw (1974) reported that BHT is converted to BOOH in rat liver microsomal incubates containing an NADPH-generating system. Their results indicated that relatively large amounts of the hydroperoxide were formed, that BOOH persisted at high concentrations throughout a 40 min incubation period, and that BOOH was converted to 3,5-di-t-butyl-4-hydroxybenzyl alcohol. They reported also that BOH was converted to BOOH when incubated with microsomes in the presence of NADPH. Due to these unexpected results, we reinvestigated the formation and destruction of BOOH. This hydroperoxide was rapidly destroyed in microsomes, and was converted mainly to BOH, and to several lesser products including $B(OH)_2$ and DBQ. The benzyl alcohol metabolite reported by Chen and Shaw was not detected in the presence of or absence of NADPH. The major metabolite of BOH was $B(OH)_2$, with only traces of one other metabolite which was identified tentatively as the 4-hydroxymethyl analog of BOH. We were able to confirm, however, that BHT is partially converted to BOOH in microsomes with NADPH. Small amounts of the hydroperoxide were observed at low incubation temperatures and identified by co-chromatography with an authentic sample of BOOH on both reverse and normal phase (silica) HPLC, as well as by its UV properties. Hydroperoxide formation was investigated further by quantitating the major transformation product, BOH. Turnover of the microsomal monooxygenase system apparently causes a 1-electron oxidation of BHT, that could be a direct consequence of P-450 oxidation as has been reported for other compounds with low oxidation potentials (Augusto et al., 1982; Cavalieri and Rogan, 1984). Alternatively, oxidation could occur as a consequence of an interaction of BHT with another reactive oxidant generated in the microsomal system. The delocalized phenoxy radical from BHT could then combine with triplet O_2 and abstract a hydrogen atom to produce the hydroperoxide.

Our results show that BOOH has properties similar to those of other hydroperoxides (Nordblom et al., 1976) with regard to its interactions with cytochrome P-450; it supports the oxidation of cosubstrates and partially destroys the enzyme. Detailed studies concerning the conversion of BOOH to $B(OH)_2$ and COOH to diols have provided information on the mechanistic aspects of P-450-hydroperoxide interactions. The formation of 2 diols from 20α-hydroperoxycholesterol with bovine adrenocortical mitochondria has been reported (Van Lier and Smith, 1970). We have found that the conversion of BOOH and COOH to diols by rat liver P-450 is an isomerization process, which occurs principally within a cage (via the intermediate alcohol) at the active site of the enzyme. It seems most likely that these conversions are initiated by heterolysis of the peroxy O-O bond. The resulting iron-oxo species $(FeO)^{3+}$ is a sufficiently strong oxidant to abstract H• from a t-butyl group of BOH or a methyl group of COH (Groves et al., 1978). In addition, attack at the aromatic ring of COH to produce 4-HO-COH indicates that $(FeO)^{3+}$ is the oxidant.

Both hydroperoxides are converted to products derived from their alkoxy radicals followed by β-scission. Thus, BOOH and COOH undergo homolysis in microsomes to produce small quantities of DBQ and acetophenone, respectively. Once formed, the alkoxy radicals can certainly react by pathways other than β-scission (e.g., H• abstraction). When the β-scission pathway was facilitated by substituting a 4-ethyl group for the 4-methyl of BOOH (since expulsion of CH_3CH_2• is more favored energetically than expulsion of CH_3•), the formation of DBQ increased 10-fold. This result, and the fact that diol formation from 4-Et-BOOH was only slightly affected compared to DBQ production, indicates that isomerization products and β-cleavage products are produced by different routes of O-O bond cleavage.

The results presented here provide additional mechanistic evidence that hydroperoxides are processed by both heterolytic and homolytic pathways by cytochrome P-450.

ACKNOWLEDGEMENTS

This work was supported by NIH Grant CA33497. We thank Ms. Susan Mastovich for her excellent technical assistance.

345

REFERENCES

Augusto, O., Beilan, H.S. and Ortiz de Montellano, P.R., 1982, The catalytic mechanism of cytochrome P-450: Spin-trapping evidence for one-electron substrate oxidation, J. Biol. Chem., 257:11288.

Blake, R.C. and Coon, M.J., 1981, On the mechanism of action of cytochrome P-450: Evaluation of homolytic and heterolytic mechanisms of oxygen-oxygen bond cleavage during substrate hydroxylation by peroxides, J. Biol. Chem., 256:12127.

Cavalieri, E.L. and Rogan, E.G., 1984, One-electron and two-electron oxidation in aromatic hydrocarbon carcinogenesis, in: "Free Radicals in Biology, Volume VI," W.A. Pryor, ed., Academic Press, New York.

Chen, C. and Shaw, Y-S., 1974, Cyclic metabolic pathway of a butylated hydroxytoluene by rat liver microsomal fractions, Biochem. J., 144:497.

Griffin, B.W., 1980, Detection of free radical species derived from cumene hydroperoxide in model hemeprotein systems and in rat liver microsomes by spin-trapping techniques, in: "Microsomes, Drug Oxidations and Chemical Carcinogenesis," M.J. Coon, A.H. Conney, R.W. Estabrook, H.V. Gelboin, J.R. Gillette and P.J. O'Brien, eds., Academic Press, New York.

Groves, J.T., McClusky, G.A., White, R.E. and Coon, M.J., 1978, Aliphatic hydroxylation by highly purified liver microsomal cytochrome P-450. Evidence for a carbon radical intermediate, Biochem. Biophys. Res. Commun., 81:154.

Kharasch, M.S. and Joshi, B.S., 1957, Reactions of hindered pheonols. II. Base-catalyzed oxidations of hindered phenols, J. Org. Chem., 22:1439.

Lee, W.A. and Bruice, T.C., 1985, Homolytic and heterolytic oxygen-oxygen bond scissions accompanying oxygen transfer to iron (III) porphyrins by percarboxylic acids and hydroperoxides. A mechanistic criterion for peroxidase and cytochrome P-450, J. Am. Chem. Soc., 107:513.

McCarthy, M-B. and White, R.E., 1983, Competing modes of peroxyacid flux through cytochrome P-450, J. Biol. Chem., 258:11610.

Nishinaga, A., Itahara, T. and Matsuura, T., 1975, Base-catalyzed oxygenation of 2,6-di-t-butylphenols. A convenient method for preparation of p-quinols, Bull. Chem. Soc. Japan, 48:1683.

Nordblom, G.D., White, R.E. and Coon, M.J., 1976, Studies on hydroperoxide-dependent substrate hydroxylation by purified liver microsomal cytochrome P-450, Arch. Biochem. Biophys., 175:524.

O'Brien, P.J., 1982, Hydroperoxides and superoxides in microsomal oxidations, in: "Hepatic Cytochrome P-450 Monooxygenase System," J.B. Schenkman and D. Kupfer, eds., Pergamon Press, New York, p. 567.

Sligar, S.G., Gelb, M.H. and Heimbrook, D.C., 1984, Bio-organic chemistry and cytochrome P-450-dependent catalysis, Xenobiotica, 14:63.

Thompson, J.A., Ho, B. and Mastovich, S.L., 1984, Reductive metabolism of 1,1,1,2-tetrachloroethane and related chloroethanes by rat liver microsomes, Chem.-Biol. Interact., 51:321.

Thompson, J.A. and Wand, M.D., 1985, Interaction of cytochrome P-450 with a hydroperoxide derived from butylated hydroxytoluene: Mechanism of isomerization, J. Biol. Chem., in press.

Van Lier, J.E. and Smith, L.L., 1970, Sterol metabolism VIII. Conversion of cholesterol 20α-hydroperoxide to 20α,21- and 20α,22R-dihydroxycholesterol by adrenal cortex mitochondria, Biochem. Biophys. Acta, 210:153.

INDOLE-3-CARBINOL INHIBITS LIPID

PEROXIDATION IN CELL-FREE SYSTEMS

Howard G. Shertzer, Michael P. Niemi and M. Wilson Tabor

Kettering Laboratory
Department of Environmental Health
University of Cincinnati Medical Center
Cincinnati, OH 45267-0056

SUMMARY

Free radicals mediate toxicological and carcinogenic responses of tissues to many chemicals. Cellular defenses against radical mediated damage utilize endogenous substances such as tocopherol, ascorbate and GSH. Here we report a new antioxidant, indole-3-carbinol (I-3-C), a natural constituent of human diet. In chlorobenzene containing soy phospholipids, lipid oxidation was initiated with azobisisobutyronitrile; I-3-C inhibited formation of thiobarbituric acid-reactive material in a dose-dependent manner. Similar results were obtained in an aqueous system containing phospholipid vesicles initiated by Fe/ascorbate. For both systems I-3-C was less effective than tocopherol or BHT as antioxidant. To assess these antioxidant effects in vivo, mice were treated with I-3-C by gavage. A hepatic post-mitochondrial supernatant fraction isolated 2 hours after treatment showed dose-dependent decreases in NADPH-mediated lipid oxidation which correlated with decreases in ^{14}C-nitrosodimethylamine covalent binding to protein. Although hepatotoxicity may not involve lipid oxidation per se, it does indicate that free radical damage had occurred. Inhibition of damage by I-3-C suggests that this dietary component has the potential to ameliorate radical mediated chemical toxicity.

INTRODUCTION

Free radicals mediate toxicological and carcinogenic responses to many chemicals (Demopoulos et al., 1980; Floyd, 1982; Slater, 1984). Potential target molecules for such radicals include thiol groups, enzymes, amino acids, nucleotides, and unsaturated fatty acids. In the presence of oxygen, free radical attack on unsaturated fatty acids results in lipid peroxidation which is easily quantified. The toxicological consequences of lipid oxidation per se are uncertain (Slater, 1984). However, the generation of highly reactive free radical species may be reflected by the formation of thiobarbiturate-reactive metabolites which are products of oxidative damage. In this paper we report the inhibition of lipid peroxidation by indole-3-carbinol (I-3-C), which is a natural dietary constituent in humans.

MATERIALS AND METHODS

Semipurified asolectin (soybean phospholipids) was obtained from Associated Concentrates, Woodside, L.I., N.Y., which was further purified by precipitation from a chloroform solution with acetone. The procedure was repeated 3 times to devoid the phospholipids of contaminating tocopherols. The purified phospholipid was dried in a desiccator; chloroform solutions of the phospholipid were stored at -20°C under argon. I-3-C, NDMA and MDA were obtained from Aldrich Chemical Co., Milwaukee, WI, whereas TBA, BHT, indole, alpha-tocopherol and ascorbic acid were purchased from Sigma Chemical Co., St. Louis, MO. Ascorbic acid was purified by repeated extractions with MeOH, followed by an acetone extraction; it was dried under vacuum. AIBN was from Eastman Kodak Co., Rochester, NY. Scinti Verse® and Scinti Gest® were from Fisher Scientific Co., Pittsburgh, PA. [14C]-NDMA was obtained from Amersham/Searle, Arlington Heights, IL. All other chemicals were of reagent grade or better and were used without further purification.

Male ICR Swiss mice, about 30g, were allowed water and Purina mouse chow ad lib. A light-dark photocycle of 12-hr was maintained, and the bedding consisted of hardwood chips. Animals were treated by gavage (feeding needle PS20) with corn oil alone or I-3-C in corn oil vehicle; doses of about 1.5 ul/g body wt of corn oil were used. Prior to use, the corn oil was rendered peroxide free by passage through a ferrous Dowex AG 1-X8 column (Shertzer and Tabor, 1985). After 2 hr, the mice were sacrificed and 10% (v/v) whole liver homogenates were prepared at 0°C in 0.1 M KCl, 50 mM HEPES-KOH (pH 7.4). A 14S fraction was prepared as described previously (Shertzer, 1984).

For studies o' the in vivo effects of I-3-C on in vitro parameters of hepatic metabolism, reactions were performed as described in the legend of Figure 5 and aliquots of this reaction mixture were used for the following assays. For lipid peroxidation, 0.6 ml 1% (w/v) TBA in 0.28% (w/v) NaOH, was mixed with 1 ml of the reaction mixture/TCA/BHT, and heated at 100°C for 30 min. MDA standards were run simultaneously. After cooling, samples were extracted with 3 ml methylene chloride. The difference in absorbance (532 nm minus 600 nm) was used to calculate MDA equivalents of lipid oxidation. For NDMA demethylation, TCA precipitated protein from the reaction mixtures was pelleted by centrifugation, and a 0.8 ml aliquot of the supernatant solution was mixed with 0.8 ml modified Nash reagent (300 g ammonium acetate, 4 ml 2,4-pentanedione, 6 ml glacial acetic acid in 1 liter aqueous solution). The reaction was heated at 60°C for 20 min and read at 412 nm against formaldehyde standards (Nash, 1953). The precipitated protein containing bound [14C] was washed 4X with 5% TCA in 80% MeOH, or until no [14C] was present in the supernatant solution. The pellet was dissolved in 1 ml Scinti Gest® and counted with Scinti Verse® cocktail, in a Packard Tri-Carb 460 CD Liquid Scintillation system, which was programmed to convert cpm to dpm on the basis of known quenched standards.

For the cell-free lipid oxidation assay, an aliquot of chloroform containing phospholipid was mixed with 40 vol acetone. Following centrifugation, the supernatant solution was decanted, and the phospholipid dried under vacuum. For the hydrophobic lipid oxidation system (Fig. 1-3), lipid was dissolved in chlorobenzene. To 1 ml of the reaction mixture/BHT mixtures, described in the legend to Figure 1, was added 1 ml 5% TCA and 0.6 ml TBA reagent. The mixture was heated to 100°C for 30 min. The upper aqueous phase (containing TBA reactive material) was extracted with 4 vol dichloromethane. For the Fe/ascorbate aqueous lipid oxidation system, 6.25 ml 0.1 M KP$_i$ (pH 7.4) was added to 12.5 mg dried phospholipids. After flushing with argon for 2 min, the suspension was sealed with Parafilm® and sonicated in a bath sonicator (Laboratory Supplies Co., Hicksville, N.Y.) at full power for about 60 sec or until the suspension was translucent. The resulting phospholipid vesicles were added to the reaction mixture described in the legend to Figure 4. After stopping the reaction by the addition of BHT, 0.1 vol 50% (w/v) TCA and 0.6 vol 1% (w/v) TBA in 0.28% (w/v) NaOH were added, and the mixture was incubated at 100°C for 30 min. Absorbances were measured at 532 nm and 600 nm; the resulting difference in absorbance at these two wavelengths was compared to that observed for MDA standards.

RESULTS

Inhibition of lipid oxidation in the non-enzymatic systems. In order to validate our test systems for the study of lipid oxidation, we utilized

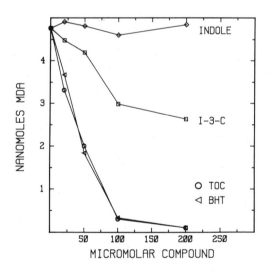

Fig. 1. Effects of various compounds on AIBN initiated lipid peroxidation. The reaction mixture contained 200 ug/ml phospholipid, and the various compounds as indicated in the Figure (toc:alpha-tocopherol), in 2.5 ml chlorobenzene. The reaction was initiated with 10 mM AIBN and performed at 37°C under white fluorescent lighting. At 2 hr the reactions were terminated with 10 ul 0.5M BHT in DMSO. TBA reactive material was determined as described in the Methods section. The results shown are the average values for 2 experiments.

the conventional polyunsaturated fatty acid chain-breaking antioxidants, alpha-tocopherol and BHT, as positive controls. In the chlorobenzene system, these compounds produced the same degree of inhibition of lipid oxidation, determined as TBA reactive substances (Fig. 1). Since there are many products of lipid peroxidation, and TBA reacts with many compounds other than malondialdehyde (Gutteridge, 1984), we calculated our results as TBA reactive malondialdehyde equivalents. The % inhibition at lower tocopherol and BHT concentrations was about 1.2% per uM inhibitor. I-3-C also inhibited the formation of TBA reactive substances at about 0.37% inhibition per uM I-3-C. Therefore, under these assay conditions, I-3-C is about 30% as effective an antioxidant as tocopherol or BHT. In contrast, indole (Fig. 1) and anthracene (data not shown) did not inhibit lipid oxidation even at concentrations of 200 uM.

The efficiency of I-3-C in preventing free radical mediated lipid peroxidation was found to be inversely related to the rate of generation of radical species. In the AIBN-initiated system, the rate of free radical generation by homolytic cleavage (thermal decomposition) of AIBN is proportional to AIBN concentration at any given temperature (Burton and Ingold, 1981). By altering AIBN concentration, the efficiency of I-3-C as an antioxidant could be modified (Fig. 2); the lower the rate of chain initiation, the greater the protectivity by I-3-C. These data, Figure 2, also reveal two components to the I-3-C effect. At less than 10 uM I-3-C,

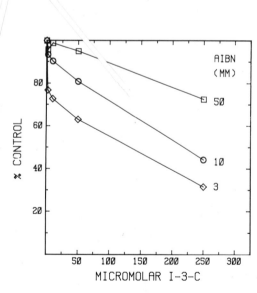

Fig. 2   Effect of AIBN concentration on I-3-C inhibition of lipid peroxidation. The reaction conditions are described in the legend to Fig. 1. Results shown are the average values for 2 experiments.

peroxidation was exquisitely sensitive to the rate of radical generation. At 50 mM AIBN there was little (if any) I-3-C effect, while at 3 mM AIBN, lipid oxidation was inhibited 24% by 2 uM I-3-C. Above 10 uM I-3-C, the rate of radical generation did not appear to affect the antioxidant efficacy of I-3-C (parallel lines in Fig. 2).

When AIBN initiated lipid oxidation was monitored over time, it was observed (Fig. 3) that I-3-C was inhibitory at early time points, but lost efficiency as the peroxidation assay proceeded. These data would be consistent with the consumption of I-3-C in the course of its acting as an antioxidant, similar to the antioxidant action of tocopherol and ascorbate.

In order to examine the potential antioxidation by I-3-C in an aqueous system, ferrous ion was used in conjunction with micromolar concentrations of ascorbate (Fig. 4). At these concentrations, lipid peroxidation was directly related to ascorbate concentration. At higher (millimolar) ascorbate concentrations, ascorbate may inhibit peroxidation (data not shown, and Shaefer et al., 1975); as this would complicate the interpretation of I-3-C effects, ascorbate concentrations were maintained below 400 uM. Within this ascorbate concentration range, an I-3-C dose dependent decrease in Fe/ascorbate mediated lipid oxidation was observed (Fig. 4). This effect was especially evident at 10 to 100 uM ascorbate. At higher ascorbate concentrations, antioxidation was evident only at 250 uM I-3-C.

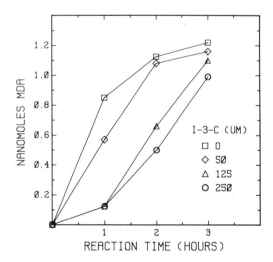

Fig. 3. Time course of AIBN-initiated lipid peroxidation at different I-3-C concentrations. The assay conditions are described in the legend to Fig. 2. The average values from 2 experiments are shown.

It should be noted that in the two nonenzymatic assay systems, it was necesssary to extract tocopherol from the phospholipids with acetone; without this process, peroxidation of lipids did not occur.

<u>Inhibition of lipid oxidation in hepatic cell fraction</u>. In order to determine the potential biological relevance of the antioxidation properties of I-3-C, mice were treated with I-3-C by gavage and the hepatic post-mitochondrial cell fractions (14S) were examined after 2 hrs. The data in Figure 5 show that although the rate of NDMA demethylation remained unchanged, I-3-C pretreatment produced a decrease in covalent binding of [^{14}C]-NDMA metabolites to protein concomitant with a decrease in lipid peroxidation. The relationship between covalent binding of NDMA and lipid peroxidation is shown by the data in Figure 6. The r^2 value of the correlation is 0.82, indicating that 82% of the variance in covalent

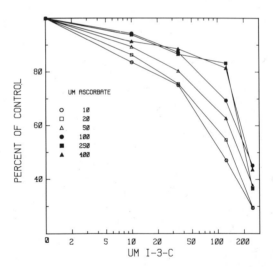

Fig. 4. Effect of ascorbate concentration on ferrous initiated lipid peroxidation. The reaction mixture for the oxidation of phospholipid vesicles contained 200 ug/ml phospholipid in 0.1 M KPi (pH 7.4). The reaction was initiated by the addition of ascorbic acid as shown, plus 10 uM Fe(NH$_4$)$_2$(SO$_4$)$_2$. The reaction continued at 37°C in a shaking water bath for 30 min, and was stopped by adding 10 ul 0.5M BHT in DMSO. TBA reactive material was determined as described in the <u>Methods</u> section. The average values for 2 experiments are shown.

binding is explained by, or directly related to, lipid peroxidation. However, it is not clear whether this relationship is causal or casual.

DISCUSSION

Free radicals are involved in the mediation of various types of cellular and tissue injuries, such as cell necrosis, radiation damage, transition metal overload, inflammation and chemical carcinogenesis

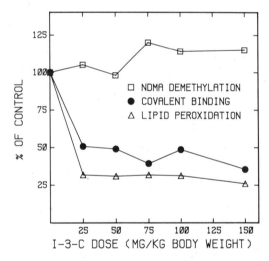

Fig. 5. Effect of in vivo I-3-C dosing on in vitro parameters of hepatic metabolism. Animals were treated as described in the Methods section. The reaction mixture contained 0.25 ml 14S fraction, 1 mM NADP$^+$, 10 mM glucose 6-phosphate, 1 U/ml glucose 6-phosphate dehydrogenase, 10 mM MgCl$_2$, 1 mM semicarbazide, 10 mM glucose, 5 mM NDMA containing 10^5 dpm/umole [^{14}C]-NDMA, and 100 mM HEPES-KOH (pH 7.4) in a final volume of 2 ml. The reaction was started with NDMA, incubated 20 min with shaking at 37°C, and stopped with 0.1 vol 50% (w/v) TCA (for demethylation and covalent binding), or TCA plus 10 ul 0.5 M BHT in DMSO (for lipid peroxidation). In all cases, aliquots were also taken at zero time for determining blank values. These were subtracted from the values obtained after 20 min. The results shown are the average value for 2 experiments.

(Floyd, 1982; Slater, 1984).   Various synthetic antioxidants have been shown to intervene in these toxicological processes, and thus, afford protection from a variety of insulting sources (Kahl, 1984).   Among the types of protectivity afforded by antioxidants is the amelioration of hepatotoxicity from NDMA precursors by phenolic antioxidants (Astill and Mulligan, 1977).   Although these authors imply that protectivity results from the prevention of NDMA formation, this was not demonstrated.   In previous reports from our laboratory, treatment of animals by gavage with I-3-C was shown to protect against the subsequent covalent binding of benzo(a)pyrene and NDMA to mouse hepatic DNA and protein, both in vivo and in vitro (Shertzer, 1983; Shertzer, 1984).   Additionally, N-nitrosamines are thought to exert their toxic effects primarily via a reactive electrophile that results from metabolic activation (Krueger, 1972); however, there is some evidence to suggest (Duran and Faljoni, 1978; Nagata et al., 1973) that free radicals and reactive oxygen species may be produced in the course of N-nitroso compound metabolism.   For example, N-nitrosodipropylamine is metabolized by horseradish peroxidase,

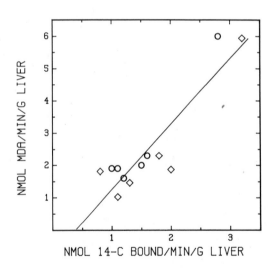

Fig. 6. Correlation between lipid peroxidation and covalent binding.   The individual data (n=12) from Fig. 5 were replotted.   A linear least squares regression analysis was performed and the line of best fit is shown. ($y=2.3x - 0.77$).   The r^2 value = 0.82, and the standard error of the regression estimate = 0.75.   The circles and the diamonds are the results from the first and second sets of experiments, respectively.

producing singlet oxygen mediated chemiluminescence that can be inhibited by singlet oxygen traps. A major radical species that may be involved in the toxicological activation of certain N-nitroso compounds is the nitroso radical. The direct acting mutagen N-methyl-N'-nitro-N-nitrosoguanidine and other related compounds produce a characteristic nitroso radical esr signal when mixed with 105,000 x g supernatant from mammalian tissue [28]. Generation of the signal required iron and sulfhydryl groups. Although N-nitrosamines did not produce this esr signal, this would not be expected, because these compounds require metabolic activation, presumably by cytochrome P-450 mediated mixed function oxidation. These microsomal enzymes would be absent from the 105,000 X g supernatant fraction. Thus it has not been resolved whether part of the toxicity and carcinogenicity associated with N-nitroamines is mediated by reactive oxygen or radical species.

When tissue is homogenized in the absence of divalent metal chelators such as EDTA, NADPH reducing equivalents may be used by the cytochrome P-450 mixed function oxidase enzyme system to produce reactive chemical species which in turn can produce secondary effects such as lipid peroxidation (McCay and Poyer, 1976; Kornbrust and Mavis, 1980). The relationship between lipid peroxidation and cellular toxicity is not clear. We have used lipid peroxidation as an indicator that free radicals had been generated and had persisted long enough to react with other chemical or cellular components. The ability of I-3-C to quench a free radical mediated reaction in three different systems suggests that I-3-C may intervene in the toxicity of agents that operate by free radical mechanisms. Since hepatic GSH, ascorbate levels, and GSH transferase activities were unaltered by I-3-C pretreatment, we suggest that the I-3-C effect in inhibiting lipid peroxidation is mediated by I-3-C itself in the two cell free reaction systems. In the enzymatic NADPH mediated reaction system, we propose that either I-3-C, or an enzymatic or chemical reaction product of I-3-C is responsible for inhibiting lipid oxidation. We are currently evaluating these possibilities.

ACKNOWLEDGEMENTS

Supported by NIH grants ES-03373 and CA-38277.

REFERENCES

Astill, B. D. and Mulligan, L. T., 1977, Phenolic antioxidants and the inhibition of hepatotoxicity from N-dimethylnitrosamine formed in situ in the rat stomach, Fd. Cosmet. Toxicol., 15:167.

Burton, G. W. and Ingold, K. U., 1981, Autoxidation of biological molecules. 1. The antioxidant activity of vitamin E and related chain-breaking phenolic antioxidants in vitro, J. Am. Chem. Soc., 103:6472.

Demopoulos, H. B., Pietronigro, D. D., Flamm, E. S. and Seligman, M. L., 1980, The possible role of free radical reactions in carcinogenesis, in: "Cancer and the Environment," H. B. Demopoulos and M. A. Mehlman eds., Pathotox Publishers, Park Forest South, Illinois.

Duran, N. and Faljoni, A., 1978, Singlet oxygen formation during peroxidase catalyzed degradation of carcinogenic N-nitrosamine, Biochem. Biophys. Res. Commun., 83:287.

Floyd, R. A., 1982, "Free Radicals and Cancer," Marcel Dekker, New York.

Gutteridge, J. M. C., 1984, Reactivity of hydroxyl and hydroxyl-like radicals discriminated by release of thiobarbituric acid-reactive material from deoxy sugars, nucleosides and benzoate, Biochem. J., 224:761.

Kahl, R., 1984, Synthetic antioxidants: biochemical actions and interference with radiation, toxic compounds, chemical mutagens and chemical carcinogens, Toxicology, 33:185.

Kornbrust, D. J. and Mavis, R. D., 1980, Microsomal lipid peroxidation. 1. Characterization of the role of iron and NADPH, Mol. Pharmacol., 17:400.

McCay, P. B. and Poyer, J. L., 1976, Enzyme-generated free radicals as initiators of lipid peroxidation in biological membranes, in: "The Enzymes of Biological Membranes," A. Martonosi, ed., V. 4, Plenum Press, New York.

Nagata, C., Ioki, Y., Kodama, M. and Tagashira, Y., 1973, Free radical induced in rat liver by a chemical carcinogen, N-methyl-N'-nitro-N-nitrosoguanidine, Ann. N.Y. Acad. Sci., 222:1031.

Nash, J., 1953, Estimation of formaldehyde, Biochem. J., 55:416.

Schaefer, A., Komlos, M. and Seregi, A., 1975, Lipid peroxidation as the cause of the ascorbic acid induced decrease of adenosine triphosphatase activities of rat brain microsomes and its inhibition by biogenic amines and psychotropic drugs, Biochem. Pharmacol., 24:1781.

Shertzer, H. G., 1983, Protection by indole-3-carbinol against covalent binding of benzo(a)pyrene metabolites to mouse liver DNA and protein, Fd. Chem. Toxicol., 21:31.

Shertzer, H. G., 1984, Indole-3-carbinol protects against covalent binding of benzo[a]pyrene and N-nitrosodimethylamine metabolites to mouse liver macromolecules, Chem.-Biol. Interact., 48:81.

Shertzer, H. G. and Tabor, M. W., 1985, Peroxide removal from organic solvents and vegetable oils, J. Environ. Sci. Hlth., 20: in press.

Slater, T.F., 1984, Free radical mechanisms in tissue injury, Biochem. J., 222:1.

GLUTATHIONE PEROXIDASE AND REACTIVE OXYGEN

SPECIES IN TCDD-INDUCED LIPID PEROXIDATION

S.J. Stohs, Z.F. Al-Bayati, M.Q. Hassan, W.J. Murray and
H.A. Mohammadpour

Department of Biomedical Chemistry
University of Nebraska Medical Center
Omaha, NE 68105

SUMMARY

Previous studies have shown that high doses of TCDD induce hepatic
lipid peroxidation and inhibit selenium dependent glutathione peroxidase
(GSH-Px) activity. The dose dependent effects of TCDD on hepatic lipid
peroxidation (malondialdehyde content) and GSH-Px activity were determined. A dose as low as 1 µg/kg induced hepatic lipid peroxidation and
inhibited GSH-Px. Based on the use of scavengers of reactive oxygen
species, lipid peroxidation (malondialdehyde formation) by hepatic
microsomes from both control and TCDD-treated rats appears to be due
primarily to H_2O_2. The results indicate that superoxide, hydroxyl
radical and singlet oxygen are also involved. The differences in the
reactive oxygen species involved in microsomal lipid peroxidation between
control and TCDD treated animals appear to be quantitative rather than
qualitative. A 5.9-fold greater rate of malondialdehyde (MDA) formation
by microsomes from TCDD treated animals occurred as compared to controls,
while livers of TCDD rats had an MDA content that was 5.0-fold greater
than the controls. These differences may be due in part to an enhanced
production of H_2O_2 as well as a decrease in the activity of selenium
dependent glutathione peroxidase which metabolizes H_2O_2.

INTRODUCTION

Peroxidation of membrane lipids is of recent research interest due
to its prominent role in biochemical toxicology[1] and the aging process[2].
Both microsomal[1,3,4] and mitochondrial[5-7] lipid peroxidation have been
studied. NADPH-induced lipid peroxidation as determined by malondialdehyde (MDA) formation has been shown to require iron as a cofactor[3,4] and
to be mediated by the enzyme NADPH cytochrome c (P-450) reductase[8].

2,3,7,8-Tetrachlorodibenzo-p-dioxin (TCDD) is one of the most toxic
trace impurities formed during the commercial synthesis of the herbicide
2,4-D[9,10]. The molecular mechanism for the toxicity of TCDD is not
known[9,10]. Previous investigations have shown that hepatic microsomal[11,12]
MDA formation is induced as a result of TCDD administration to rats[11,12]
and guinea pigs[12]. The toxicity of TCDD may be due in part to enhanced

membrane lipid peroxidation[11,12]. TCDD also inhibits the activity of the enzyme selenium dependent glutathione peroxidase (GSH-Px)[12], which removes H_2O_2 and hydroperoxides from cells[13].

Since initial studies on TCDD induced MDA formation and inhibition of GSH-Px activity utilized high doses of the xenobiotic[11,12], the dose dependent effects were determined. Furthermore, in an attempt to determine the relative contributions of various oxygen species, we have examined the types of reactive oxygen species involved in MDA formation by microsomes from control and TCDD-treated rats, employing scavengers of reactive oxygen species.

MATERIALS AND METHODS

TCDD was obtained from Dow Chemical Company, Midland, MI. Superoxide dismutase (2.75 U/μg), catalase (0.78 U/μg), mannitol, sodium benzoate, ethanol, α-tocopherol, and Triton X-100 were purchased from Sigma Chemical Company, St. Louis, MO. Histidine HCl was obtained from Nutritional Biochemicals Corp., Cleveland, OH, and dimethylfuran was procured from Aldrich Chemical Company, Milwaukee, WI.

Female Sprague-Dawley rats (140-160 gm; Sasco Inc., Omaha, NE) were allowed free access to water and lab chow. They were maintained at a temperature of $21°C$, with lighting from 6:00 a.m. to 6:00 p.m. daily for 3-5 days prior to use. The animals were given corn oil intragastrically or TCDD dissolved in corn oil. Animals were killed by decapitation.

Livers were homogenized in 0.05 M Tris HCl-1.15% KCl buffer (pH 7.4), and microsomes were prepared by differential centrifugation. The microsomes were resuspended in 0.10 M phosphate buffer, pH 7.4, at a final concentration of 1-2 mg protein/ml with or without the presence of 0.2% (v/v) Triton X-100[14], a non-ionic detergent. The Triton X-100 was employed to provide dispersion of the microsomal membranes[14]. Microsomal protein was determined by the method of Lowry et al.[15].

Malondialdehyde (MDA) formation was estimated using the thiobarbituric acid method as described by Miles et al.[16]. MDA content of whole liver was determined by the thiobarbituric acid method of Uchiyama and Mihara[14]. GSH-Px activity in the cytosol was determined as previously described[12].

RESULTS

The results in Figure 1 demonstrated that the ability of TCDD to induce hepatic lipid peroxidation as determined by MDA content was dose dependent. All animals were killed 6 days after treatment with TCDD or the vehicle. A dose as low as 1 μg TCDD/kg produced a significant increase in hepatic MDA content. The MDA contents resulting from 25, 50 and 100 μg TCDD/kg were not significantly different from one another.

The ability of TCDD to inhibit hepatic GSH-Px activity is presented in Figure 2. A dose as low as 1 μg TCDD/kg was also able to significantly inhibit GSH-Px activity 6 days post-treatment. Similar to the results for MDA formation, GSH-Px activity following doses of 25, 50 and 100 μg TCDD/kg were not significantly different.

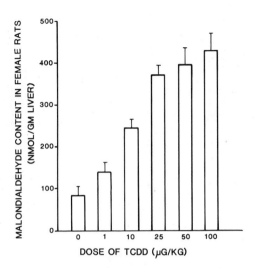

Fig. 1. Female rats received 0, 1, 10, 25, 50 or 100 μg TCDD/kg in corn oil P.O. and were killed 6 days post-treatment. Lipid peroxidation was assessed by determining MDA content of whole liver. Each value is the mean of 4-6 rats with the S.D.

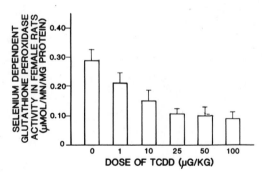

Fig. 2. Rats were treated with TCDD as described in Fig. 1 Selenium dependent GSH-Px activity was determined on liver cytosol. Each value is the mean of 4-6 rats with the S.D.

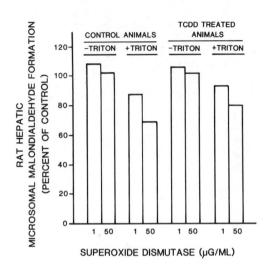

Fig. 3. Effect of superoxide dismutase on MDA formation by microsomes from control and TCDD-treated rats in the presence and absence of 0.2% Triton X-100. Rats were treated with 40 μg TCDD/kg for 3 days and killed 6 days post-treatment. Data is expressed as percent of MDA formation without the addition of superoxide dismutase. Each value is the mean of 4-8 experiments.

The effects of various scavengers of reactive oxygen species on in vitro lipid peroxidation as determined by MDA formation with Triton X-100 treated and untreated hepatic microsomes from control and TCDD-treated rats are presented in Figures 3-6. Mean values for MDA formation for microsomes from control and TCDD-treated animals were 1.52 and 8.90 nmol/min/mg protein, respectively. Each value is the mean of 4-8 separate experiments, and the results are expressed as percent of values without scavengers. In the presence of Triton X-100, microsomes from control and TCDD-treated animals formed 20-40% less MDA.

Superoxide dismutase at two concentrations in the absence of Triton did not inhibit MDA formation with microsomes from either control or TCDD-treated rats (Figure 3). A small increase in lipid peroxidation was actually observed with microsomes from both control and TCDD-treated animals. However, in the presence of Triton, superoxide dismutase produced a concentration dependent, partial inhibition of lipid peroxidation with microsomes from control and TCDD-treated animals.

Catalase had no effect on microsomal lipid peroxidation at 1 μg/ml (Figure 4). However, at 50 and 500 μg/ml, catalase produced a concentration-dependent inhibition of MDA formation with microsomes from both control and TCDD-treated rats in the absence of Triton. In the presence of Triton, MDA formation was inhibited by all three concentrations of catalase. Complete inhibition was observed with the microsomal preparations at a concentration of 500 μg catalase/ml in the presence of Triton.

At 5 and 50 mM concentrations, the hydroxyl radical scavengers ethanol, mannitol and sodium benzoate had little effect on MDA formation by microsomes from control animals, and stimulated (20-30%) MDA formation by microsomes from TCDD-treated rats (data not shown). However, the presence of Triton facilitated the inhibition of lipid peroxidation by

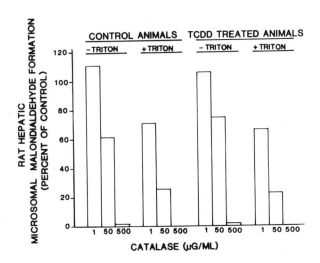

Fig. 4. Effect of catalase on MDA formation by microsomes from control and
TCDD-treated rats in the presence and absence of 0.2% Triton X–100.

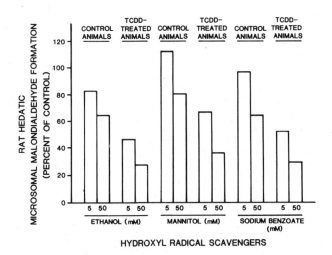

Fig. 5. Effect of the hydroxyl radical scavengers ethanol, mannitol and
sodium benzoate on MDA formation by microsomes from control and TCDD-treated
rats. Microsomes were dispersed in 0.2% Triton X–100.

all three hydroxyl radical scavengers, particularly at a 50 mM concentration with microsomes from both control and TCDD-treated animals (Figure 5). Greater inhibition of MDA formation by hydroxyl radical scavengers at both concentrations was observed in the presence of Triton with microsomes from TCDD-treated animals as compared to microsomes from control animals.

The effects of the singlet oxygen scavengers histidine and dimethylfuran were examined (Figure 6). In the absence of Triton, histidine inhibited MDA formation associated with microsomes from both control and treated rats. Greatest inhibition was observed for histidine with control microsomes. In the presence of Triton, histidine had no effect on lipid peroxidation with control microsomes but facilitated greater inhibition of lipid peroxidation by microsomes from TCDD-treated rats. Dimethylfuran inhibited lipid peroxidation by control microsomes at both 5 and 50 mM in the absence of Triton. At the same concentrations, dimethylfuran had little effect on lipid peroxidation by microsomes from TCDD-treated animals in the absence of Triton. In the presence of Triton, dimethylfuran had little effect on MDA formation by microsomes from control animals but facilitated lipid peroxidation by microsomes from treated animals.

The non-specific free radical scavenger vitamin E (d-α-tocopherol) inhibited lipid peroxidation associated with microsomes from both control TCDD-treated rats (data not shown). Complete inhibition of MDA formation occurred at 10 mM vitamin E with microsomes from both control and TCDD-treated animals in the presence of Triton X-100.

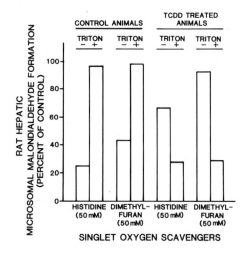

Fig. 6. Effect of the singlet oxygen quenchers histidine and dimethylfuran on MDA formation by microsomes from control and TCDD-treated animals in the presence and absence of 0.2% Triton X-100.

DISCUSSION

The ability of TCDD to enhance hepatic MDA content (Figure 1) and inhibit GSH-Px activity (Figure 2) is dose dependent, with a dose as low as 1 μg/kg being effective. An inverse correlation between the effects of TCDD on MDA content and GSH-Px activity appears to exist. The enhanced MDA content may be due in part to accumulation of H_2O_2 due to a decrease in the activity of GSH-Px. The mechanism whereby TCDD inhibits GSH-Px activity is not known, but may be due to altered selenium metabolism.

With respect to the involvement of H_2O_2 and various free radicals in lipid peroxidation by microsomes from control and TCDD-treated animals, the results suggest that quantitative differences rather than qualitative differences exist. Microsomes from TCDD treated rats formed MDA at a rate that was 5.9-fold greater/mg protein than control microsomes. Based on the results observed with the addition of catalase, a primary role for H_2O_2 is indicated in MDA formation by microsomes (Figure 4) from control and TCDD-treated rats. The formation of H_2O_2 by cytochrome P-450 was first reported by Gillette et al.[18], and its role in xenobiotic metabolism has been recently investigated[19-22]. H_2O_2 formation by the mixed function oxidase system mainly results from dismutation of superoxide anion radicals which are released from the oxycomplex of cytochrome P-450[22].

Superoxide dismutase had little effect on lipid peroxidation by microsomes from both control and TCDD-treated animals. Dixit et al.[23] have observed little effect of superoxide dismutase on lipid peroxidation by epidermal microsomes, and concluded that superoxide is not produced or is not responsible. However, with Triton dispersed microsomes (Figure 3), superoxide dismutase produced partial inhibition of lipid peroxidation. The results indicate that the physical state of the microsomes must be taken into account when employing inhibitors to gain information concerning free radical reaction systems. Based on these observations, superoxide is involved in MDA formation by microsomes from both control and TCDD-treated animals.

Caution should be exercised in interpreting results involving the use of free radical scavengers since the specificity may not be as great as expected. In the absence of Triton X-100, the hydroxyl radical scavengers ethanol, mannitol, and benzoate were ineffective inhibitors of lipid peroxidation with hepatic microsomes (Figure 5) from both control and TCDD-treated rats. The reason for the increase in MDA formation by these scavengers at low concentrations is not known. Similar observations have been reported by other investigators using free radical scavengers with microsomes from epidermis[23] and erythrocyte membranes[14]. In the presence of Triton dispersed microsomes, all three hydroxyl radical scavengers at a 50 mM concentration produced partial inhibition of MDA formation. Ethanol, mannitol and benzoate at this concentration were approximately 50% more effective as inhibitors of lipid peroxidation by Triton dispersed microsomes from TCDD-treated animals as compared to microsomes from control animals. Hydroxyl radical may ·play a more prominent role in the enhanced MDA formation associated with microsomes from TCDD-treated rats. The reaction of H_2O_2 and superoxide will result in the generation of more highly reactive oxygen species, namely, hydroxy radicals[24]. Furthermore, this reaction is catalyzed by metals, especially iron[24]. The effect of TCDD on iron metabolism has not been studied in detail. However, preliminary studies in our laboratories indicate that the increase in microsomal lipid peroxidation following the administration of TCDD to rats is not due to an increase in hepatic iron content.

The singlet oxygen quenchers histidine and dimethylfuran produced partial inhibition of lipid peroxidation by microsomes from control animals in the absence of Triton X-100 (Figure 6), but had little inhibitory effect in the presence of Triton. With microsomes from TCDD-treated animals, these two scavengers were much more effective as inhibitors of lipid peroxidation in the presence of Triton. The reason for these differences is not known. The results suggest that singlet oxygen may be involved in microsomal lipid peroxidation in both control and TCDD-treated animals. Differences in the lipid solubility of the two scavengers and in the relative abilities to permeate the microsomal membranes may partially account for the observed effects. Singlet oxygen may be produced by cytochrome P-450 enzymes[25].

The results suggest that H_2O_2 may be the more important contributor to lipid peroxidation involving microsomes from both control and TCDD-treated rats. However, the use of free radical scavengers indicates that other reactive oxygen species including superoxide, hydroxyl radical and singlet oxygen are also involved. The marked increase in lipid peroxidation by microsomes from TCDD-treated animals[11,12] may be due to an enhanced microsomal production of H_2O_2 and other reactive oxygen species in conjunction with a decrease in the activity of selenium dependent GSH-Px.

ACKNOWLEDGEMENTS

These studies were supported in part by a grant from the March of Dimes. The authors thank Ms. Anne Bailey for technical assistance.

REFERENCES

1. J. S. Bus, and J. E. Gibson, Lipid peroxidation and its role in toxicology, Rev. Biochem. Toxicol. 1:125 (1979).
2. A. L. Tappel, B. Fletcher, and D. W. Deamer, Effects of anti-oxidants and nutrients on lipid peroxidation fluorescent products and aging parameters in the mouse, J. Gerontol. 28:415 (1973).
3. P. Hochstein, and L. Ernster, ADP-activated lipid peroxidation coupled to the TPNH oxidase system of microsomes, Biochem. Biophys. Res. Commun. 12:388 (1963).
4. D. J. Kornbrust, and R. D. Mavis, Microsomal lipid peroxidation, Mol. Pharmacol. 17:408 (1980).
5. E. D. Wills, Mechanism of lipid peroxide formation in animal tissues, Biochem. J. 99:667 (1966).
6. T. J. Player, D. J. Mills, and A. A. Horton, Age-dependent changes in rat liver microsomal and mitochondrial NADPH-dependent lipid peroxidation, Biochem. Biophys. Res. Comm. 78:1397 (1977).
7. Y. A. Vlademirov, V. I. Olenev, T. B. Suslova, and Z. P. Cheremisina, Lipid peroxidation in mitochondrial membrane, Adv. Lipid Res. 17:173 (1980).
8. T. C. Pederson, and S. D. Aust, NADPH-dependent lipid peroxidation catalyzed by purified NADPH cytochrome c reductase from rat liver microsomes, Biochem. Biophys. Res. Commun. 48:789 (1972).
9. R. J. Kociba, and B. A. Schwetz, Toxicity of 2,3,7,8-tetra-chlorodibenzo-p-dioxin (TCDD), Drug. Met. Rev. 13:387 (1982).
10. A. Poland, and J. C. Knutson, 2,3,7,8-Tetrachlorodibenzo-p-dioxin and related halogenated aromatic hydrocarbons: Examination of the mechanism of toxicity, Ann. Rev. Pharmacol. Toxicol. 22:517 (1982).

11. S. J. Stohs, M. Q. Hassan, and W. J. Murray, Lipid peroxidation as a possible cause of TCDD toxicity, Biochem. Biophys. Res. Commun. 11:854 (1983).

12. M. Q. Hassan, S. J. Stohs, and W. J. Murray, Comparative ability of TCDD to induce lipid peroxidation in rats, guinea pigs and Syrian golden hamsters, Bull. Environ. Contam. Toxicol. 31:649 (1983).

13. R. A. Sunde, and W. G. Hoekstra, Structure, synthesis and function of glutathione peroxidase, Nutr. Rev. 38:265 (1980).

14. A. W. Girotti, and J. P. Thomas, Superoxide and hydrogen peroxide-dependent lipid peroxidation in intact and Triton-dispersed erythrocyte membranes, Biochem. Biophys. Res. Comm. 118:474 (1984).

15. O. H. Lowry, A. L. Rosebrough, A. L. Farr, and R. J. Randall, Protein measurement with the folin-phenol reagent, J. Biol. Chem. 193:265 (1951).

16. P. R. Miles, J. R. Wright, L. Bowman, and H. D. Colby, Inhibition of hepatic microsomal lipid peroxidation by drug substrates without drug metabolism, Biochem. Pharmacol. 29:565 (1980).

17. M. Uchiyama, and M. Mihara, Determination of malondialdehyde precursor in tissues by thiobarbituric acid test, Anal. Biochem. 86:271 (1978).

18. J. R. Gillette, B. B. Brodie, and B. N. LaDu, The oxidation of drugs by liver microsomes: on the role of TPNH and oxygen, J. Pharmacol. Exptl. Ther. 111:532 (1957).

19. R. W. Estabrook, C. Martin-Wixtrom, Y. Saehi, R. Renneberg, A. Hildebrandt, and J. Werringloer, The peroxidatic function of liver microsomal cytochrome P-450: comparison of hydrogen peroxide and NADPH-catalyzed N-demethylation reactions, Xenobiotica 14:87 (1984).

20. Y. Ohta, I. Ishiguro, J. Maito, and R. Shinohara, Effect of superoxide dismutase on hydroxylase activity and hydrogen peroxide formation in anthranilamide hydroxylation by a rat liver microsomal monooxygenase system, Biochem. Int. 8:617 (1984).

21. H. Kuthan and V. Ullrich, Oxidase and oxygenase function of the microsomal cytochrome P-450 monooxygenase system, Eur. J. Biochem. 126:583 (1982).

22. A. Bast, and G. R. M. M. Haenen, Cytochrome P-450 and glutathione: what is the significance of their interrelationship in lipid peroxidation?, Trends Biol. Sci. 9:510 (1984).

23. R. Dixit, H., Mukhtar, and D. R. Bickers, Evidence that lipid peroxidation in microsomal membranes of epidermis is associated with generation of hydrogen peroxide and singlet oxygen, Biochem. Biophys. Res. Commun. 105:546 (1982).

24. C. E. Thomas, L. A. Morehouse, and S. D. Aust, Ferritin and super-oxide-dependent lipid peroxidation, J. Biol. Chem. 260:3275 (1985).

25. G. H. Posner, J. R. Lever, K. Miura, C. Lisek, H. H. Seliger, and A. Thompson, A chemiluminescent probe specific for singlet oxygen, Biochem. Biophys. Res. Commun. 123:869 (1984).

ACTIVATION OF NITROSAMINES TO MUTAGENS BY RAT AND RABBIT NASAL, LUNG

AND LIVER S-9 HOMOGENATES

Alan R. Dahl

Lovelace Inhalation Toxicology Research Institute
Albuquerque, NM

Over 30 nitrosamines have been reported to cause cancer in the nasal
cavities of laboratory animals (Dahl, 1985). A majority of the observed
tumors followed systemic administration of the nitrosamine, although in a
few experiments inhalation was the route of exposure. Nitrosamine-caused
nasal tumors and the binding of nitrosamine-introduced radioactivity in the
nasal cavity (Brittebo et al., 1981; Löfberg et al., 1982) may result from
the activity of the high concentrations of xenobiotic metabolizing enzymes
in the nasal cavity (Dahl et al., 1982; Bond 1983; Hadley and Dahl, 1983).

Because nitrosamines are common nasal carcinogens in the rat and other
species, it was the purpose of this study to determine whether rat nasal S-9
homogenates could activate certain nitrosamines to mutagens and to compare
the results from the nasal homogenates with those from liver and lung homogenates. The rat nasal cavity, however, yields only about 20 mg of S-9 protein compared to 200 mg for rabbits. Therefore, rabbit nasal tissue S-9
homogenate was tested as a source of nasal enzyme for mutagenicity tests and
to determine the nitrosamine concentrations needed for adequate mutagenic
response.

MATERIALS AND METHODS

Diethanolnitrosamine (NDELA) was purchased from Columbia Organic Chemical Co., Cassatt, SC. All other nitrosamines were purchased from Sigma,
St. Louis, MO. The three dialkylnitrosamines tested were N,N-dimethylnitro-
samine (NDMA), N,N-diethylnitrosamine (NDEA), and N,N-dipropylnitrosamine
(NDPA). The two cyclic nitrosamines tested were N-nitrosopyrrolidine (NPYR)
and N-nitrosopiperidine (NPIP). All other chemicals used were of the highest purity readily available.

Specific pathogen-free, male F344/N rats, 15-20 weeks of age, reared at
the Institute, were used in these studies.

Rabbits used in these studies were New Zealand white male rabbits, 20
to 26 weeks old (Bell Breeding Ranch, Clovis, NM).

Animals were killed by carbon dioxide asphyxiation. The tissues were
removed and S-9 homogenates were prepared as previously described (Hadley
and Dahl, 1983). To minimize bacterial contamination, the S-9 homogenates

from each tissue were filtered through an Acrodisc® disposable 25 mm
filter assembly (Gelman Sciences, Inc., Ann Arbor, MI, product number 4184).
After filtration, the filtered S-9 homogenate was analyzed for protein by
the method of Lowry.

Prior to the mutagenicity assay, cytotoxicity assays were carried out
with the nitrosamines without S-9. At the highest concentration (100
μmoles/plate) used for the mutagenicity assays there was little cytotoxicity demonstrated in either TA-100 or TA-98 strains.

Nasal tissue, liver and lung S-9 homogenates were tested for their ability to activate nitrosamines to mutagens using the Ames _Salmonella_/mammalian
microsome assay. Rabbit tissues were assayed for bioactivation capacity in
tester strains TA-98 and TA-100 at 4 concentrations of nitrosamines (0.1, 1,
10, and 100 micromoles per plate). Tissue S-9 protein for these assays was
1 mg protein per plate for the nose and liver, and 2 mg protein per plate
for the lung. Because only the 100 micromoles per plate nitrosamine concentration was clearly mutagenic using rabbit tissue homogenates and only in
TA-100, the scarcer rat tissue homogenates were tested only at that concentration and only in that strain. Only NPIP and NPYR were mutagenic and required activation and so were tested with graded concentrations (0.5-3.0 mgm
protein per plate) of rat and rabbit tissue S-9 homogenate. Positive controls (benzo(a)pyrene, 2-aminoanthracene) confirmed the activitating capability of all S-9 fractions.

RESULTS

Diethanolnitrosamine (NDELA) was the only nitrosamine tested that
showed a positive response in TA-98. At 100 μmoles/plate, NDELA gave a
positive response 4-5 times background in TA-98 and 2-3 times background in
TA-100. The responses appeared independent of whether or not S-9 homogenates were present. Lower concentrations of NDELA did not give positive
responses.

None of the nitrosamines tested, except NDELA discussed above and possibly NPIP at 100 μmoles/plate in TA-98, gave positive responses at any
concentration in either TA-98 or TA-100 in the absence of S-9 homogenate.
Only the cyclic nitrosamines NPIP and NPYR were substantively mutagenic and
required S-9 activation and only in TA-100.

NPIP was mutagenic in TA-100 with all tissue homogenates. NPYR was mutagenic with nasal tissue of both species and with rabbit liver tissue.
These two cyclic nitrosamines were tested with stepped concentrations of S-9
protein. Results are graphed in Figures 1-4. The most pronounced feature
occurs in Figures 1 and 2 where the rat nasal S-9 homogenate is decidedly
more activating at all protein concentrations and toward both NPYR and NPIP
than either lung or liver S-9. Rabbit nasal, lung and liver tissue S-9 homogenates had similar activating capabilities towards NPIP at all protein concentrations (Fig. 3); whereas, with NPYR at lower protein concentrations,
the liver S-9 was a better activator than either lung or nasal S-9 (Fig. 4).

DISCUSSION

All of the test substrates that were studied, except NDELA, have been
reported to be rat nasal carcinogens, and all except NDELA and NPYR are rat
lung carcinogens. All cause rat liver carcinomas (IARC, 1978). Only NDMA
and NDEA have been tested for carcinogenicity in the rabbit. Both caused
liver and lung tumors (IARC, 1978). The observation that rat nasal S-9 homogenate also activates some nitrosamines, that are rat nasal carcinogens

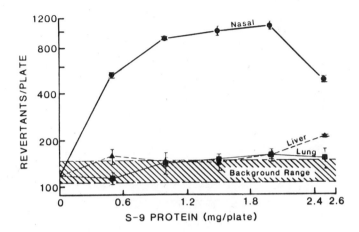

Fig. 1. Mutagenicity of 100 micromoles NPIP per plate in <u>Salmonella</u> TA-100 with rat tissue S-9. Error bars are standard errors (N = 3).

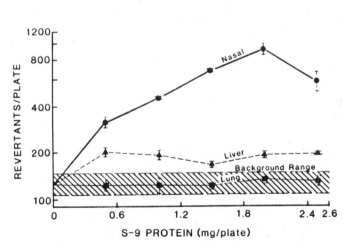

Fig. 2. Mutagenicity of 100 micromoles NPYR per plate in <u>Salmonella</u> TA-100 with rat tissue S-9. Error bars are standard errors (N = 3).

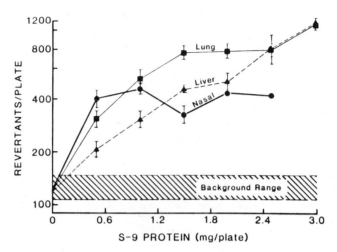

Fig. 3. Mutagenicity of 100 micromoles NPIP per plate in <u>Salmonella</u> TA-100
with rabbit tissue S-9. Error bars are standard errors (N = 3).

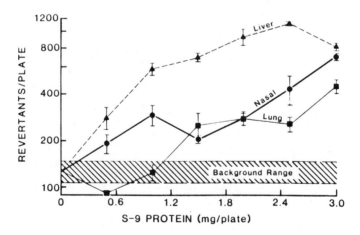

Fig. 4. Mutagenicity of 100 micromoles NPYR per plate in <u>Salmonella</u> TA-100
with rabbit tissue S-9. Error bars are standard errors (N = 3).

lends evidence to the importance of nasal enzymes in the causation of nasal cancer.

The types of cytochromes P-450 in the rabbit or rat nasal cavities are not known with certainty. However, antibodies against rabbit P-450 types LM2 and LM4 were ineffective inhibitors of rabbit nasal monooxygenase activities (Dahl et al., 1983) whereas antibody against type LM3a was effective (Ding et al., 1985). The LM3a isozyme is reported to catalyze nitrosamine dealkylation, the step that activates nitrosamines to potent alkylating agents (Yang et al., 1985). High concentrations of this or similar isozymes in the nasal cavity relative to isozymes that metabolize nitrosamines to less toxic products (e.g., beta-carbon hydroxylation) may explain why rat nasal S-9 is a better activator of nitrosamines than is liver or lung S-9.

Interesting comparisons (and possibly correlations) can be made among three observations with NPYR in rats. First, the mutagenicity reported here (Fig. 1) was high with nasal S-9 and absent with lung S-9. Second, carbon-14 labeled NPYR injected intravenously in rats binds extensively to the nasal mucosa but not to the bronchi (Brittebo et al., 1981). Third, NPYR induced an olfactory carcinoma in one of 29 rats given NPYR in the drinking water (IARC, 1978). (The concentration of NPYR reaching the olfactory tissue was likely to be low by the oral route. An inhalation exposure might have yielded much higher olfactory tumor incidences.) It is reasonable to conjecture that these three observations are related to metabolic activation of NPYR by nasal tissue enzymes.

One of the purposes for the experiments reported was to determine if plentiful rabbit nasal tissue could substitute in the Ames test for scarcer nasal tissues of animals commonly used in carcinogenicity tests (e.g., mouse, Syrian hamster, rat). This seemed reasonable because the xenobiotic metabolizing capabilities of the nasal tissues from common laboratory animals, including all of the above, are similar (Hadley and Dahl, 1983). The results reported here indicate that, at least for screening compounds, the rabbit nasal tissue may substitute for the scarcer nasal tissues. More testing is required to increase confidence in this procedure.

In conclusion, despite the relative difficulty of obtaining it, nasal S-9, especially rat nasal S-9, may be worth using in addition to liver S-9 in the Ames Salmonella test because it is a substantially better activator for certain compounds. In particular, suspect nasal carcinogens and compounds administered by inhalation probably should be tested with nasal S-9 in the Ames test.

ACKNOWLEDGMENTS

I thank A. Brooks for his valuable comments and N. Stephens for conducting the Ames tests. Research supported by the U. S. Department of Energy under Contract Number DE-AC04-76EV01013 and was conducted in facilities fully accredited by the American Association for the Accreditation of Laboratory Animal Care.

REFERENCES

Bond, J. A., 1983, Some biotransformation enzymes responsible for polycyclic aromatic hydrocarbon metabolism in rat nasal turbinates: effects on enzyme activities of in vitro modifiers and intraperitoneal and inhalation exposure of rats to inducing agents, Cancer Res., 43:4805-4811.

Brittebo, E. B., Löfberg, B., and Tjälve, H., 1981, Extrahepatic sites of metabolism of n-nitrosopyrrolidine in mice and rats, Xenobiotica, 11: 619-625.

Dahl, A. R., 1985, Activation of carcinogens and other xenobiotics by nasal cytochromes P-450, in: "Microsomes and Drug Oxidations," A. R. Boobis, J. Caldwell, F. DeMatteis, and C. R. Elcombe, eds., Taylor Francis, Philadelphia, pp. 299-309.

Dahl, A. R., Hadley, W. M., Hahn, F. F., Benson, J. M., and McClellan, R. O., 1982, Cytochrome P-450-dependent monooxygenases in olfactory epithelium in dogs: possible role in tumorigenicity, Science, 216: 57-59.

Dahl, A. R., Hall, L., and Hadley, W. M., 1983, Characterization and partial purification of rabbit nasal cytochrome P-450, The Toxicologist, 3:19.

Ding, X., Koop, D. R., Coon, M. J., 1985, Extrahepatic identification of ethanal-inducible rabbit cytochrom P-450 isozyme 3a, Fed Proceedings, 44:1449.

Hadley, W. M. and Dahl, A. R., 1983, Cytochrome P-450-dependent monooxygenase activity in nasal membranes of six species, Drug Metab. Dispos., 11:275-276.

IARC, 1978, IARC monographs on the evaluation of the carcinogenic risk of chemicals to humans: some N-nitroso compounds, WHO Publications, Albany, NY.

Löfberg, B., Brittebo, E. B., and Tjälve, H., 1982, Localization and binding of N'-nitrosonornicotine metabolites in the nasal region and in some other tissues of Sprague-Dawley rats, Cancer Res., 42:2877-2883.

Yang, C. S., Koop, D. R., Wang, T., and Coon, M. J., 1985, Immunochemical studies on the metabolism of nitrosomines by ethanol-inducible cytochrome P-450. Biochem. Biophys. Res. Comm., 128:1007-1013.

STRUCTURE OF THE ADDUCT OF GLUTATHIONE

AND ACTIVATED 3-MATHYLINDOLE*

Garold S. Yost[1], Mark R. Nocerini[2], James R. Carlson[2], and Daniel J. Liberato[3]

[1]College of Pharmacy and [2]Department of Animal Sciences, Washington State University, Pullman WA; and [3]National Institutes of Health Bethesda, MD.

INTRODUCTION

3-Methylindole (3MI) is a pneumotoxic metabolite of tryptophan fermentation in ruminants.[1] The toxicity of 3MI is organ, cell, and species specific. In ruminants and horses, the most susceptible species, the lung is the target organ. Type I alveolar epithelial and nonciliated bronchiolar epithelial (Clara) cells are the most susceptible targets within the lungs of ruminants,[2,3] while Clara cells only are primarily damaged by 3MI in horses.[4] Exposure of man to 3MI is through intestinal absorption and by cigarette smoke although the toxic manifestations of this exposure have not yet been assessed.[5-7] Cytochrome P-450 monooxygenases are responsible for the bioactivation of 3MI to a toxic intermediate.[8] Several studies have demonstrated that the P-450 system is a necessary component in the metabolic intoxication process.[9-13] Although oxidation of 3MI is required for tissue injury to occur, the precise nature of the toxic intermediate has not been established. We have demonstrated that 3-methyloxindole and indole-3-carbinol, major metabolites of 3MI in ruminants, are not toxic to goats.[14] Thus, the toxic compound is most likely an intermediate between 3MI and these major metabolites, or is formed by another metabolic pathway which does not produce either of the two major metabolites.

Glutathione (GSH) status is important in the modulation of 3MI pneumotoxicity.[10] We have recently demonstrated that a GSH adduct of activated 3MI is formed with goat lung microsomes,[15] and that the rate of formation of the GSH-3MI adduct in microsomes can be correlated to the organ-selective toxicity of 3MI.[16]

In order to establish the nature of the reactive 3MI intermediate, we have isolated and determined the structure of the GSH-3MI adduct using thermospray LC/MS, and UV and NMR spectrometry. This work provides evidence that the reactive intermediate is an imine methide, formed by hydrogen abstraction by cytochrome P-450 from the methyl position of 3MI.

Supported by grant HL13645 from the US PHS.

MATERIALS AND METHODS

Chemicals

[Methyl-^{14}C]-3-Methylindole (specific activity: 3.41 mCi/mmol) was prepared by custom synthesis by New England Nuclear Corporation (Boston, MA). The radiochemical and chemical purity was greater than 95%. GSH, NADP, glucose-6-phosphate, and glucose-6-phosphate dehydrogenase were purchased from Sigma Chemical Company (St. Louis, MO). Nanograde solvents were used for preparative HPLC; all other reagents were of analytical grade. HPLC solvents for LC/MS were HPLC grade methanol (Burdick and Jackson, Muskeson, MI) and water from a purifier (Hydro Services, Rockville, MD). Ammonium acetate (HPLC grade) was obtained from J.T. Baker (Phillipsburg, NJ). The mobile phase was filtered (0.45 µm) and degassed before use.

Microsomal Incubations

Goats obtained from the Department of Animal Sciences (Washington State University, Pullman) were killed by intravenous injection of sodium pentobarbital. Lungs were removed, rinsed twice with 0.01 M NaH_2PO_4 buffer (pH 7.4) containing 1.15% KCl, and placed on ice. Tissue samples were homogenized in 3 parts phosphate- KCl buffer with a Polytron (Brinkmann Instruments, Inc., Westbury, NY), and microsomes were isolated by centrifugation as described previously.[16] In a total volume of 2.0 ml of 0.1 M NaH_2PO_4 buffer (pH 7.4), the following incubation components were added to tubes: lung microsomes (10 mg), 1.0 mM ^{14}C-3MI, 8.0 mM GSH, and an NADPH-generating system consisting of 2.5 mM NADP, 7.5 mM glucose-6-phosphate, 5.0 mM $MgCl_2$, and 2 I.U. of glucose-6-phosphate dehydrogenase. The components were incubated at 37°C in a water bath with shaking for 1 hr. Incubations were stopped by adding 4.0 ml ice-cold methanol. Samples were centrifuged for 20 min (4°C) at 10,000g, and supernatant fractions were transferred to vials and stored at -20°C.

Isolation of the GSH-3MI Adduct

The adduct was isolated from incubation supernatants by HPLC. A Spectra-Physics SP8000B liquid chromatograph (San Jose, CA) equipped with a SF770 UV detector (Shoeffel Instrument Corp., Westwood, NJ) and a 25 x 0.46 cm reverse-phase C_8 10µ LiChrosorb column (Spectra-Physics) was used. Metabolites were separated using a gradient system which consisted of 0.005% (v/v) acetic acid in water adjusted to pH 6 with 1 N NaOH and methanol (0-100% methanol over 23 min). The flow rate was 2.0 ml/min and column temperature was maintained at 40°C. Under these conditions, the GSH-3MI adduct eluted at 12.5 min, and was detected by UV absorbance and radioprofile as previously reported.[15] Approximately 130 aliquots of incubation supernatants (100 µl each) were chromatographed using an automated sample injection/fraction collection system. Pooled samples of GSH-3MI adduct were lyophilized, solubilized in 0.8 ml water, and rechromatographed under the same HPLC conditions. Collected eluate fractions which contained the adduct were again pooled and lyophilized. Approximately 30% of the purified sample was used for mass spectrometry. The remaining portion was analyzed by NMR.

Liquid Chromatography/Mass Spectrometry

An LKB Model 2150 HPLC pump (LKB, Bromma, Sweden) was used along with an Altex Ultrasphere C_8 column (15 cm x 4.6 mm) (Altex, Berkeley, CA). Samples were introduced with a microsyringe through a Rheodyne

Model 7125 sample injection valve (Rheodyne, Cotaoi, CA).

The design and operation of the thermospray interface to the mass spectrometer were as described previously.[17,18] Solvent (75% 0.1 M ammonium acetate (pH 6.7) and 25% methanol (v/v) is pumped at conventional flow rates (1 ml/min) through the interface without splitting. Spectra were obtained with a Biospect quadrupole mass spectrometer (Scientifc Research Instruments, Baltimore, MD). The scan functions were controlled by an Incos data system (Finnigan/MAT, San Jose, CA). Mass calibration was accomplished using a solution consisting of 0.02 M ammonium acetate, 0.9 µM tetrabutylammonium iodide and 0.09 µM benzyltriphenylphosphonium chloride. The optimum interface temperatures were determined by making repeated direct injections of 3MI while varying the transfer line and vaporizer temperatures. The objective was to generate both protonated molecular and fragment ions with good overall sensitivity.

^{1}H-NMR

A Nicolet NT-200 WB 200 MHz FT NMR was used. The GSH-3MI adduct was solubilized in D_2O. Chemical shifts were relative to the chemical shift of HOD (4.74 ppm). Fourier transform of 12,128 free induction decays yielded the spectrum of the GSH-3MI adduct. 3MI was solubilized

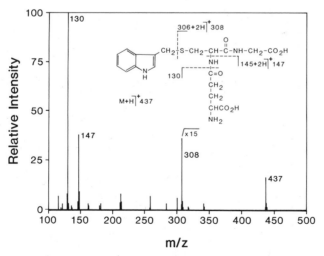

FIG. 1.   Positive ion thermospray mass spectrum of the GSH-3MI adduct. Data reproduced with permission of ASPET from Nocerini, Yost, Carlson, Liberato and Breeze, Drug Metabolism and Disposition, Vol. 13, No. 6 (1985).

in d$_4$-methanol. Chemical shifts of 3MI were relative to TMS=0.
Fourier transform of 224 free induction decays yielded the spectrum of
3MI.

RESULTS

The structure of the GSH-3MI adduct was determined by UV and NMR
spectrometry and by thermospray LC/MS. The UV spectrum of the adduct
is not shown but the absorbance maxima of the purified adduct were
determined to be at 216 and 280 nm. 3MI exhibits the same maxima; thus
the indole chromophore was assumed to be unchanged in the production of
the GSH adduct.

Repeated initial attempts to determine the molecular ion of the
adduct by fast atom bombardment mass spectrometry were unsuccessful.
The use of LC/MS was, however, immediately successful. The positive
ion thermospray LC/MS technique provided a molecular ion of m/z 437
corresponding to m + 1 for the addition of GSH to 3MI without simultaneous incorporation of oxygen into the molecule. Fig. 1 illustrates
the mass spectrum of the GSH-3MI adduct. Fragments of the adduct
appear at 147 and 130 which correspond to successive cleavage of amino

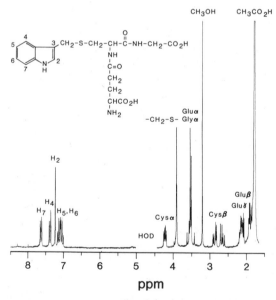

FIG. 2.   Proton NMR spectrum of the GSH-3MI adduct in D$_2$O. Data repro-
duced with permission of ASPET from Nocerini, Yost, Carlson,
Liberato and Breeze, Drug Metabolism and Disposition, Vol. 13,
No. 6 (1985).

acid fragments from the tripeptide, as shown in fig. 1. In addition, a m/z 308 ion is shown which corresponds to the molecular weight of GSH. Thus, the adduct was shown to have a molecular ion of 436 and fragments corresponding to the presence of GSH.

The most valuable information concerning the structure of the adduct was obtained by proton NMR. Fig. 2 is a full NMR spectrum of the GSH-3MI adduct. The NMR spectrum for 3MI is shown in fig. 3. Solvent impurities in the GSH-3MI spectrum were noted between 4.5 - 5 ppm for HOD, at 3.2 ppm for methanol, and at 1.8 ppm corresponding to acetic acid. The methanol impurity was confirmed by spiking the adduct sample with this solvent. The peak assigned to acetic acid was eliminated by passing the adduct sample through a C_{18} Sep-Pak cartridge before NMR analysis.

The most striking change in the portion of the adduct spectrum arising from the 3MI moiety was the lack of a singlet for the methyl group at 2.3 ppm. In addition, a broad singlet at 6.9 ppm corresponding to the H2 proton on 3MI was shifted downfield, by approximately 0.4 ppm, to 7.3 ppm. There was an additional singlet at 3.9 ppm which was not part of either the GSH or 3MI spectra. This singlet corresponded to 2 protons. These results, when considered together, indicate that conjugation of GSH occurs at the methyl group of 3MI. Transformation of the methyl into a methylene would explain the resonance at

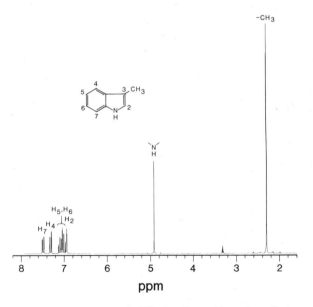

FIG. 3. Proton NMR spectrum of 3MI in d_4-methanol. Data reproduced with permission of ASPET from Nocerini, Yost, Carlson, Liberato, and Breeze, <u>Drug Metabolism and Disposition</u>, Vol. <u>13</u>, No. 6 (1985).

3.9 ppm, which is shifted downfield due to the electronegativity of sulfur attachment. Also, the H2 proton of 3MI is shifted downfield by 0.4 ppm, again due to the electronegativity of sulfur substitution on the former methyl group. The resonances for the GSH protons are as noted in fig. 2. These values are typical for GSH adducts from other electrophilic xenobiotics.[19-21] The glutamate α and glycine α assignments at 3.5 ppm were the only resonances that could not be differentiated with absolute certainty. The glutamate α resonance was assigned on the basis of irradiation of the glutamate β resonance at 1.9 ppm, whereupon the multiple at 3.5 ppm collapsed to 2 singlets at 3.51 and 3.52 ppm, corresponding to the non-coupled glycine α and glutamate α protons, respectively.[19-21] Other irradiation experiments confirmed the assignment of the cysteine protons as well as the glutamate protons, as indicated in fig. 2.

The aromatic region of the GSH-3MI adduct was identical to the aromatic region of 3MI (fig. 3) with the exception as noted above that the H2 proton was shifted downfield in the spectrum of the GSH-3MI adduct. Of particular note are the chemical shifts and coupling constants for all of the other aromatic protons in the GSH-3MI adduct. Since the shifts and coupling constants are unchanged from the original compound, it is highly unlikely that any substitution has occurred on the aromatic ring of 3MI or on the carbon next to the nitrogen of the indole moiety.

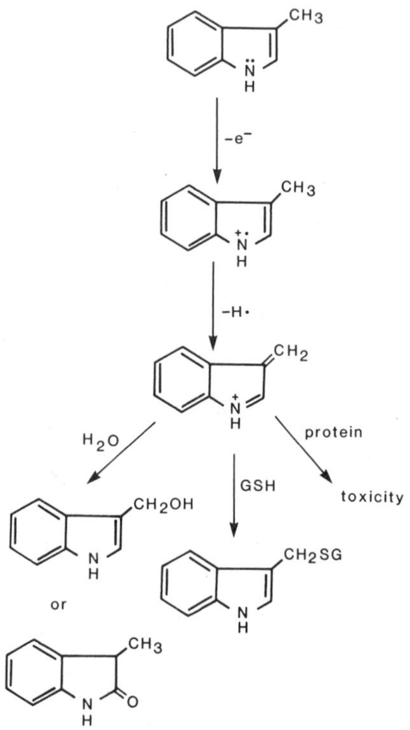

FIG. 4.   Proposed mechanism of 3MI pneumotoxic bioactivation.

DISCUSSION

The structure of the GSH-3MI adduct has been shown to be 3-[(glutathion-S-yl)-methyl]indole. We have unequivocally demonstrated that the site of attachment of GSH to the 3MI moiety occurs at the methyl position and thus an intermediate of 3MI metabolism formed by oxidation at the methyl position is a logical chemical entity for an electrophilic species.

As shown in fig. 4, we propose that the GSH-3MI adduct is formed via the oxidation of 3MI to a imine methide intermediate which then traps GSH and regenerates the aromatic indole nucleus. It is highly feasible also that the imine methide can trap water as shown in fig. 4 followed by tautomerization to 3-methyloxindole, the major urinary metabolite of 3MI in ruminants.[22] Water could also be trapped by the imine methide at the exocyclic methyl position to form indole-3-carbinol, another significant metabolite of 3MI.[22] The imine methide metabolite then accounts for the major metabolic products of 3MI as well as accounting for the fact that neither of the two major metabolites is toxic even though bioactivation, presumably by cytochrome P-450, results in a toxic species.

Oxidation of 3MI via the successive one electron oxidation of the indole nitrogen and then the methyl C-H, represents a reasonable mechanism in light of recent mechanisms of one electron heteroatom oxidations,[23] and reports of trapping of nitrogen-centered free radicals during 3MI metabolism.[24] Thus, production of the electrophilic imine methide and alkylation of vital cellular proteins provide an intriguing mechanism of 3MI pneumotoxicity.

REFERENCES

1. J.R. Carlson, M.T. Yokoyama, and E.O. Dickinson: Induction of pulmonary edema and emphysema in cattle and goats with 3-methylindole. Science 176, 298-299 (1972).
2. T.W. Huang, J.R. Carlson, T.M. Bray, and B.J. Bradley: 3-Methylindole-induced pulmonary injury in goats. Am. J. Pathol. 87, 647-666 (1977).
3. B.J. Bradley and J.R. Carlson: Ultrastructural pulmonary changes induced by intravenously administered 3-methylindole in goats. Am. J. Pathol. 99, 551-560 (1980).
4. M.A. Turk, R.G. Breeze, and A.M. Gallina: Pathologic changes in 3-methylindole-induced equine bronchiolitis. Am. J. Pathol. 110, 209-218 (1983).
5. J.S. Fordtran, W.B. Scroggie, and D.E. Potter: Colonic absorption of tryptophan metabolites in man. J. Lab. Clin. Med. 64, 125-132 (1964).
6. M.T. Yokoyama and J.R. Carlson: Microbial metabolites of tryptophan in the intestinal tract with special reference to skatole. Am. J. Clin. Nutr. 32, 173-178 (1979).
7. D. Hoffmann and G. Rathkamp: Quantitative determination of 1-alkylindoles in cigarette smoke. Anal. Chem. 42, 366-370 (1970).
8. J.R. Carlson, M.R. Nocerini, and R.G. Breeze: The role of metabolism in 3-methylindole-induced acute lung injury. In "Progress in Tryptophan and Serotonin Research" (H.G. Schlossberger, W. Kochen, B. Linzen, and H. Steinhart, eds.), pp. 483-489. Walter de Gruyter and Co., Berlin, 1984.
9. T.M. Bray and J.R. Carlson: Role of mixed-function oxidase in 3-methylindole-induced acute pulmonary edema in goats. Am. J. Vet. Res. 40, 1268-1272 (1979).

10. M.R. Nocerini, J.R. Carlson, and R.G. Breeze: Effect of gluta-
    thione status on covalent binding and pneumotoxicity of 3-methyl-
    indole in goats. Life Sci. 32, 449-458 (1983).
11. T.M. Bray, J.R. Carlson, and M.R. Nocerini: In vitro covalent
    binding of 3-[^{14}C]methylindole metabolites in goat tissues. Proc.
    Soc. Exp. Biol. Med. 176, 48-53 (1984).
12. R.G. Breeze, W.W. Laegreid, and B.M. Olcott: Role of metabolism
    in the immediate effects and pneumotoxicity of 3-methylindole in
    goats. Br. J. Pharmacol. 82, 809-815 (1984).
13. M.S.M. Hanafy and J.A. Bogan: Pharmacological modulation of the
    pneumotoxicity of 3-methylindole. Biochem. Pharmacol. 31, 1765-
    1771 (1982).
14. M.J. Potchoiba, J.R. Carlson, and R.G. Breeze: Metabolism and
    pneumotoxicity of 3-methyloxindole, indole-3-carbinol, and
    3-methylindole in goats. Am. J. Vet. Res. 43, 1418-1423 (1982).
15. M.R. Nocerini, J.R. Carlson, and G.S. Yost: Electrophilic meta-
    bolites of 3-methylindole as toxic intermediates in pulmonary
    oedema. Xenobiotica 14, 561-564 (1984).
16. M.R. Nocerini, J.R. Carlson, and G.S. Yost: Glutathione adduct
    formation with microsomally activated metabolites of the pulmonary
    alkylating and cytotoxic agent, 3-methylindole. Toxicol. Appl.
    Pharmacol. In press (1985).
17. D.J. Liberato, C.C. Fenselau, M.L. Vestal, and A.L. Yergey:
    Characterization of glucuronides with a thermospray liquid
    chromatography/mass spectrometry interface. Anal. Chem. 55,
    1741-1744 (1983).
18. A.L. Yergey, D.J. Liberato, and D.S. Millington: Thermospray
    liquid chromatography/mass spectrometry for the analysis of
    L-carnitine and its short-chain acyl derivatives. Anal. Biochem.
    139, 278-283 (1984).
19. J.A. Hinson, T.J. Monks, M. Hong, R.J. Highet, and L.R. Pohl: 3-
    (Glutathion-S-yl)acetaminophen: a biliary metabolite of aceta-
    minophen. Drug Metab. Dispos. 10, 47-50 (1982).
20. T.J. Monks, L.R. Pohl, J.R. Gillette, M. Hong, R.J. Highet, J.A.
    Ferretti, and J.A. Hinson: Stereoselective formation of bromo-
    benzene glutathione conjugates. Chem.-Biol. Interact. 41,
    203-216 (1982).
21. E.J. Moss, D.J. Judah, M. Przybylski, and G.E. Neal: Some
    mass-spectral and n.m.r. analytical studies of a glutathione
    conjugate of aflatoxin B$_1$. Biochem. J. 210, 227-233 (1983).
22. A.C. Hammond, J.R. Carlson, and J.D. Willett: The metabolism and
    disposition of 3-methylindole in goats. Life Sci. 25, 1301-1306
    (1979).
23. F.P. Guengerich and T.L. Macdonald: Chemical mechanisms of
    catalysis by cytochromes P-450: A unified view. Acc. Chem. Res.
    17, 9-16 (1984).
24. S. Kubow, E.G. Janzen, and T.M. Bray: Spin-trapping of free
    radicals formed during in vitro and in vivo metabolism of 3-
    methylindole. J. Biol. Chem. 259, 4447-4451 (1984).

STUDIES ON THE MECHANISM OF S-CYSTEINE CONJUGATE METABOLISM AND TOXICITY

IN RAT LIVER, KIDNEY, AND A CELL CULTURE MODEL

James L. Stevens, Patrick Hayden and Gail Taylor

Division of Biochemistry and Biophysics
Center for Drugs and Biologics, FDA
Bethesda, MD 20205

INTRODUCTION

Cysteine conjugate β-lyases are enzymes which may be involved in the nephrotoxicity caused by a variety of S-cysteine conjugates which are formed from halogenated hydrocarbons.[1] The enzymes catalyze the cleavage of S-cysteine conjugates to ammonia, pyruvate and a thiol whose sulfur is derived from cysteine (eq. 1).[2,3]

$$NH_2-CH-COOH \xrightarrow{\;H_2O\;} CH_3-\overset{O}{\overset{\|}{C}}-COOH \;+\; :NH_3 \;+\; RSH \qquad (eq.\ 1)$$
$$|$$
$$CH_2-SR$$

The thiol containing cleavage fragments may be the species which are responsible for the nephrotoxicity of these compounds. This hypothesis is based on the observations of Schultze and coworkers who investigated the toxicity of S-1,2-dichlorovinyl-L-cysteine (DCVC)[4-6]. These workers found that DCVC was cleaved by mammalian and bacterial enzymes to pyruvate and ammonia, as well as an unidentified fragment which covalently bound to cellular macromolecules. However, there is no direct evidence linking the metabolism of S-cysteine conjugates and the toxicity of reactive electrophilic metabolites in vivo. In addition, it is not clear why the kidney is the target organ. We have investigated the properties of mammalian cysteine conjugate β-lyases in liver and kidney and developed a model culture system which responds to the toxic effects of a variety of S-cysteine conjugates. With these systems, we hope to establish a mechanism of toxicity for S-cysteine conjugates which are nephrotoxic.

Rat liver and kidney contain the highest activities for cysteine conjugate β-lyase in this species.[7] Our studies have concentrated on elucidating the nature of the enzymes in rat kidney and liver in order to determine; a) the cofactor requirements of the enzymes, b) the differences or similarities between the enzymes in the two organs, and c) whether the enzymes use endogenous compounds as substrates.

Our work in cell culture models employs two continuous cell lines, the first, LLC-PK1, has differentiated functions similar to the proximal tubule of kidney,[8] and the second, rabbit lung fibroblasts, is of mesynchymal origin, presumably a nontarget tissue in vivo. These two lines are used to address a) the nature of the characteristics of specific cells which may relate to the organ specificity for S-cysteine conjugate toxicity, and b) the relationship between metabolism, toxicity, and the production of reactive cleavage fragments. We will employ DCVC as a model in these studies in order to establish a mechanism to which we can compare data from other toxic conjugates.

Methods

S-1,2-Dichlorovinyl-L-cysteine and glutathione conjugates were synthesized by the method of McKinney et al.[9] The cysteine and glutathione conjugates of hexachlorobutadiene, hexafluoropropylene and chorotrifluoroethylene were synthesized by the method of van Bladderen et al.[10] using cysteine or glutathione, instead of N-acetyl-L-cysteine, as the starting material. All compounds were characterized by NMR spectroscopy and sometimes by FAB-mass spectrometry and elemental analysis. Physical data on the compounds will appear elsewhere.

Hepatic cysteine conjugate β-lyase was purified as reported.[11] Antibody was raised against purified cysteine conjugate β-lyase as described[12] and linked to Activated CH-Sepharose® as described by the manufacturer (Pharmacia). Antibody precipitation from rat liver and kidney cortex cytosol and the cysteine conjugate β-lyase assays for the immunologic experiments were done as previously reported.[13]

Both LLC-PK1 cells and rabbit lung fibroblasts were grown in 75cm^2 flasks in DMEM containing 10% fetal calf serum and 2 mM glutamine. Toxicity experiments were done using cells seeded into 24 well culture dishes. Cell death was assayed 14 hr later by measuring the release of lactate dehydrogenase (LDH) into the culture media by standard techniques. The percent cell death was calculated by comparing the release of LDH due to toxin to the amount released by Triton X-100 lysis of the cells.

Metabolism of DCVC was measured by one of two methods. Disappearance of substrate was measured by reversed phase HPLC analysis of deproteinized incubation mixtures. Pyruvate formation was measured by extracting ^{14}C-pyruvate produced from ^{14}C-DCVC into ethyl acetate at pH 1 (66% efficiency). Dinitrophenylhydrazones of pyruvate were prepared by standard techniques and analyzed by reversed phase HPLC. Details of the HPLC methods will appear elsewhere. Pyruvate (10 mM) was included in the incubations, as well as the incubations for the binding experiments described below, in order to stimulate the metabolism of DCVC. The mechanism of the stimulation will be discussed elsewhere, but we believe it is due to a minor transamination reaction which regulates the activity of the enzyme. This type of regulation has been discussed.[14]

Binding of metabolites to cellular macromolecules was measured by washing trichloroacetic acid insoluble material from incubation mixtures two times with methanol and two times with water. The washed trichloroacetic acid insoluble material was solubilized in 2N NaOH, neutralized and counted to determine the amount of radioactivity bound.

Studies on Rat Liver and Kidney Cysteine Conjugate β-lyase

The β-elimination reaction catalyzed by cysteine conjugate β-lyases (eq 1) is similar to β-elimination reactions catalyzed by pyridoxal 5'-phosphate (PLP) dependent enzymes.[15] We have purified the rat hepatic cysteine conjugate β-lyase and identified its cofactor and substrate requirements.[12] In addition, we have compared the kidney and liver cysteine conjugate

$$\text{(2-}NH_2\text{-}C_6H_4)\text{-}\overset{\overset{\displaystyle O}{\|}}{C}\text{-}CH_2\text{-}\underset{\underset{\displaystyle NH_2}{|}}{CH}\text{-}COOH \xrightarrow{\ H_2O\ } CH_3\text{-}\underset{\underset{\displaystyle NH_2}{|}}{CH}\text{-}COOH \ + \ (2\text{-}NH_2\text{-}C_6H_4)\text{-}COOH \qquad \text{(eq. 2)}$$

β-lyases immunologically.[13]

Rat hepatic cysteine conjugate β-lyases. Rat hepatic cysteine conjugate β-lyase was first described by Anderson and Schultze[4] and Tateishi et al.[7] Subsequently, the rat liver cytosolic enzyme was purified to homogeneity.[12] We have found that the rat liver enzyme contains PLP as a cofactor and is identical to rat liver kynureninase, an enzyme which cleaves the naturally occurring substrate kynurenine to alanine and anthranilic acid (eq. 2).[13,16] This reaction proceeds through the formation of an aminoacrylate-PLP complex formed by a reaction equivalent to a hydrolysis at the γ-carbonyl of kynurenine (eq. 3). This intermediate has carbanion character at the β-carbon and appears to be stabilized by resonance involving the Schiff base. Protonation of the carbanion yields alanine, an observed reaction product.

The hydrolytic cleavage of kynurenine differs from the β-elimination reactions catalyzed by PLP dependent enzymes (eq. 4). The aminoacrylate intermediate resulting from β-elimination has carbonium ion character at the β-carbon.[15] In this case the aminoacrylate is released into solution as an iminoalanine and hydrolyzes nonenzymatically to pyruvate and ammonia, the observed reaction products. Enzymes which contain PLP and perform β-eliminaton reactions are sometimes subject to inactivation during the catalytic cycle.[15] Inactivation by substrates which β-eliminate has been observed with bacterial kynureninase and with rat liver cysteine conjugate β-lyase.[12,17] Two mechanisms have been proposed for the inactivation, a) the Michaels addition of an enzyme bound nucleophile to the β-carbon of the aminoacrylic acid which results from the β-elimination[15] or b) an aldol type condensation between the aminoacrylate released at the active site, and the aldimine of the pyridoxal phosphate-enzyme complex.[18] Either mechanism could explain the inactivation of hepatic cysteine conjugate β-lyase, however, further experiments are necessary to determine if either mechanism is operative.

Immunological comparison of the rat liver and kidney cytosolic enzymes. Antibody was prepared against the rat hepatic cytosolic cysteine conjugate

(eq. 3)[a]

[a] the equation represents the enzyme bound pyridoxal phosphate cofactor showing the carbonyl carbon (*CH) in Schiff base with kynurenine. The kynurenine is shown as a hydrate to illustrate the proposed nature of the cleavage.

$$RS-CH_2-\overset{\displaystyle\curvearrowleft}{\underset{\|}{C}}-COOH \quad\longrightarrow\quad RS^- \quad\longrightarrow\quad CH_2=C-COOH \qquad (eq.\ 4)^b$$

(Left structure) RS–CH₂–C–COOH with +NH, *CH, (P) attached to a pyridine ring with OH, CH₃, N–H

(Right structure) CH2=C-COOH with +NH, *CH, (P) attached to pyridinium ring with OH, CH₃, +N–H

b   the equation represent the enzyme and pyridoxal phosphate as
    described in equation 3. In this case the β-elimination of a
    thiol from an S-cysteine conjugate is shown.

β-lyase and used to investigate the enzyme(s) present in kidney cytosol.
Anti-rat hepatic cysteine conjugate β-lyase antibody bound to Sepharose 4B
(anti-CBL Sepharose) was mixed with liver and kidney cytosol, and the
antibody reactive protein removed along with the anti-CBL Sepharose by
centrifugation. Assay of the supernatant for cysteine conjugate β-lyase
activity revealed that activity with DCVC and S-2-benzothiazolyl-L-cysteine
(BZC) was removed from liver cytosol, but not from kidney cytosol (Table 1).
However, activity with kynurenine was removed from the cytosol of both
organs. There is residual cysteine conjugate β-lyase activity remaining in
rat liver cytosol following removal of anti-CBL Sepharose reactive protein,
suggesting that another enzyme may also be present in rat liver cytosol.
However, the major form of cysteine conjugate β-lyase in cytosol from the two
organs are immunologically distinct.

Studies on the mechanism of toxicity in model culture systems

      The toxicity of halogenated hydrocarbons which are thought to exert
their toxic effects via the formation of S-cysteine conjugates shows marked
specificity for the proximal tubule of the kidney.[1] We investigated both the
biochemical characteristics which might make the proximal tubule a target,
and the role in toxicity of S-cysteine conjugate metabolism using two culture
models. The first, LLC-PK1 cells, has γ-glutamyl transpeptidase activity
($30\ nmol\cdot min^{-1}\cdot mg^{-1}$) and developes differentiated function of kidney in
culture.[8] The second, rabbit lung fibroblasts, has very little γ-glutamyl

Table 1.   Immunological comparison of cysteine conjugate β-lyase
           activities in rat liver and kidney cytosol.

Substrate	anti-CBL Sepharose (mg)	Activity $nmol\cdot mg^{-1}\cdot min^{-1}$	
		liver cytosol	kidney cytosol
DCVC	0	1.3 ± 0.1	3.1 ± 0.2
	400	0.4 ± 0.1	3.1 ± 0.2
BZC	0	1.3 ± 0.1	0.9 ± 0.1
	400	0.3 ± 0.1	0.9 ± 0.2
Kynurenine	0	1.3 ± 0.1	0.2 ± 0.01
	400	<0.01	<0.01

a   activity represents the specific activity remaining after the addition
    of the specified amount of anti-CBL Sepharose to 1 ml of cytosol.

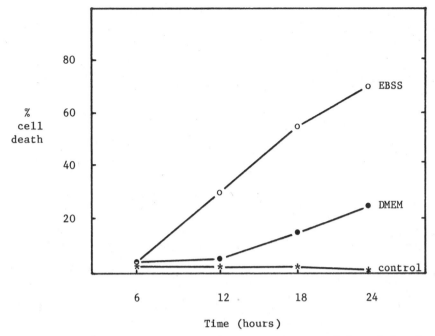

Figure 1. The effect of different media on DCVC toxicity in
LLC-PK1 cells.

DCVC (0.2 mM) was added to cultures in 24 well dishes and aliquots
(50 μl) withdrawn and assayed for LDH release at the times shown.
o=Earle's Balanced Salt Solution (EBSS); ●= Dulbecco's Modified Eagle's
Medium (DMEM); *= controls in either DMEM or EBSS to which no DCVC was
added.

transpeptidase activity (1 nmol·min·$^{-1}$·mg^{-1}). We used this difference and
the inhibitor of γ-glutamyl transpeptidase, AT-125, to determine if
mercapturic acid biosynthetic enzymes can play a role in producing toxic
S-cysteine conjugates. The role of metabolism of S-cysteine conjugates in
toxicity was investigated in the LLC-PK1 cells.

The effect of culture conditions on the toxicity of DCVC. Two different
media were tested using LLC-PK1 cells; Earle's Balanced Salt Solution (EBSS)
and Dulbecco's Modified Eagle's Medium (DMEM), both without added fetal calf
serum. The data show that DCVC was more toxic in the EBSS than in DMEM (Fig.
1). DMEM contains a variety of essential and nonessential amino acids as well
as vitamins and other additives. Though initial experiments were done in
DMEM, much of the mechanistic work we will present was done in the EBSS.
This system was used in order to avoid interference by amino acids and
vitamins at either uptake or activation steps. Indeed it is not clear at
this point whether the decreased toxicity in DMEM is due to the presence of
nutrients necessary for cellular repair of DCVC damage, or interference at
event prior to and including activation of DCVC to toxic species.

The role of mercapturic acid biosynthesis in S-cysteine conjugate
toxicity. Mercapturic acid biosynthesis is the major physiological source of
S-cysteine conjugates.[2] The mercapturate pathway is initiated by the
formation of an S-glutathione conjugate which is then degraded to an
S-cysteine conjugate in the kidney and liver. In order to determine if the
presence of enzymes which degrade S-glutathione conjugates to S-cysteine

Table 2. Comparison of the toxicity of DCVC and DCVG in LLC-PK1 cells and rabbit lung fibroblasts (RLF) in DMEM.

Conjugate	mM	Percent Cell Death	
		LLC-PK1	RLF
none	−	4 ± 1	6 ± 0
S-1,2-dichlorovinyl-L-glutathione	1	59 ± 1	10 ± 1
S 1,2-dichlorovinyl-L-cysteine	1	62 ± 5	48 ± 11

conjugates is involved in the organ specificity of S-cysteine conjugate toxicity, we compared the toxicity of DCVC and its glutathione analogue S-1,2-dichlorovinyl-L-glutathione (DCVG) in LLC-PK1 cells and rabbit lung fibroblasts. DCVC was toxic to both LLC-PK1 cells and rabbit lung fibroblasts, but DCVG was less toxic in the RLF cells (Table 2).

If the observed difference in the toxicity of DCVG between the fibroblasts and LLC-PK1 cells is due to the difference in the amount of γ-glutamyl transpeptidase, and consequently the ability to degrade DCVG to DCVC, then inhibition of γ-glutamyl transpeptidase activity in LLC-PK1 cells should produce a decrease in toxicity. The inhibitor of γ-glutamyl transpeptidase, AT-125, reduced γ-glutamyl transpeptidase activity 95% in LLC-PK1 cells when included in cultures at a concentration of 0.5 mM. This treatment protected the cells from the toxicity of DCVG, but had no effect on the toxicity of DCVC (Table 3).

The role of DCVC metabolism by LLC-PK1 cells in toxicity. The data suggested that S-glutathione conjugates must be metabolized to their corresponding S-cysteine conjugates for toxicity to occur. However, it was not yet determined if metabolism of DCVC played a role in cell death or if the parent compound could be toxic in LLC-PK1 cells. We approached this problem in two ways. First, we compared the metabolism and toxicity of S-1,2-dichlorovinyl-D-cysteine (D-DCVC) to DCVC. If the parent compound itself is toxic, then D-DCVC should be toxic, however, if metabolism is required, then only one isomer might be toxic. Second, we used an inhibitor of pyridoxal phosphate enzymes, aminoxyacetic acid (AOA), to block the metabolism of DCVC to toxic species.[19,20]

Stereospecificity of DCVC metabolism and toxicity in LLC-PK1 cells. When we compared the metabolism and toxicity of D-and L-DCVC we found that D-DCVC was not toxic(Table 4). In addition, D-DCVC was not metabolized by homogenates of LLC-PK1 cells, but L-DCVC is metabolized. However, when both

Table 3. The effect of AT-125 on the toxicity of DCVC and DCVG in LLC-PK1 cells in DMEM.

AT-125 (0.5 mM)	Percent Cell Death[a]	
	DCVC (1 mM)	DCVG (1 mM)
+	65 ± 7	11 ± 1
−	67 ± 4	64 ± 2

[a] cell death values for control cultures to which no DCVC was added were 4 ± 1 in the presence or absence of AT-125 (0.5 mM).

Table 4. Comparison of the metabolism and toxicity of
L-and D-DCVC in LLC-PK1 cells in EBSS.

DCVC Isomer	Metabolism nmol 60 min^{-1}·mg^{-1}	Cell Death percent
D-DCVC	<0.01	12 ± 4
L-DCVC	3.7 ± 0.1	66 ± 4
D-DCVC + L-DCVC	n.d	31 ± 6

Metabolism experiments were done using 50 nmol of DCVC, 4 mg of
LLC-PK1 cell homogenate, 10 mM pyruvate and 50 mM Tris·HCl in a 2 ml
volume DCVC. disappearance from incubations was assayed by HPLC
analysis as described in the Methods. Toxicity was determined using
50 μM L-DCVC and 1 mM D-DCVC in Earle's Balanced Salt Solution.
Control LDH values are 4 ± 2. n.d.= not determined, n=3

D-DCVC and L-DCVC were included in cultures simultaneously, D-DCVC inhibited
the toxicity of L-DCVC. Since D-amino acids are transported by LLC-PK1
cells [21] and renal proximal tubule [22], we conclude that both D- and L-DCVC
enter the cells, but only L-DCVC is metabolized. The two isomers interact
at some point in the uptake or activation, hence the inhibition by D-DCVC.

The metabolism of DCVC by β-elimination in LLC-PK1 cells and the role of
reactive species production in cell death. Though, the results suggested that
DCVC is metabolized by LLC-PK1 cells, the mechanism had not been clearly
established. We synthesized both ^{14}C-and ^{35}S-labelled DCVC from ^{14}C-and
[35]S-labelled cysteine and developed an assay which exploits the extraction of
pyruvate into ethyl acetate at pH 1 (see Methods). Using this assay,
extractable counts from both ^{35}S-and ^{14}C-DCVC appeared in the ethyl acetate
extract (Table 5). However, the recovery of the two labels differed
suggesting that the cysteine moiety of DCVC had been cleaved and the sulfur
and carbons seperated. Dinitrophenylhydrazone derivatives were prepared from
incubation which contained either ^{35}S-or ^{14}C-labelled DCVC and the hydrazones
analyzed by HPLC as described in the Methods. Greater than 90% of the
[14]C-product cochromatographed with the authentic dinitrophenylhydrazone of
pyruvate. There were no ^{35}S-counts recovered in the same fractions. These
data show that DCVC is metabolized in LLC-PK1 cells, by β-elimination, to
pyruvate and a cleavage fragment, presumably dichlorovinylthiol. Ammonia was
not measured because of the relatively low specific activity in these cells;
0.1 nmol·mg^{-1}·min^{-1} and 3.1 nmol.mg^{-1}·min^{-1} for LLC-PK1 homogenate and rat
kidney homogenate, respectively.

To determine if the cleavage fragment was a reactive species which could
covalently bind to cellular macromolecules, as reported by Schultze and
coworkers,[4-6] we measured the binding of [14]C- and ^{35}S-label from [14]C- or
[35]S-labelled DCVC to trichloroacetic acid insoluble material from cell
homogenates. The data show that only the ^{35}S-label becomes bound to cellular
macromolecules and that the binding can be prevented by the addition of
glutathione to the incubation mixtures (Table 6). Therefore, it appears that
in LLC-PK1 cells, DCVC is metabolized to an electrophilic species, containing
sulfur which reacts with nucleophiles on cellular macromolecules.

Table 5. The metabolism of [14]C- and ^{35}S-labelled DCVC
to extractable products in LLC-PK1 homogenates.[a]

Label	Product extracted nmol.mg^{-1}.30 min^{-1}
[14]C-DCVC	7.2 ± 0.3
^{35}S-DCVC	3.9 ± 0.6

[a]Incubations contained 1mM DCVC, 10 mM pyruvate, approximately 1 mg of protein and 50 mM Tris·HCl, pH 8, in a 0.25 ml volume.

Having established that DCVC was metabolized to a reactive species in LLC-PK1 cells and that the metabolism appeared to be involved in the toxicity, we treated the cells with an inhibitor of pryidoxal phosphate dependent enzymes, aminoxyacetic acid (AOA). We then measured the effect of AOA on the toxicity of DCVC, covalent binding of ^{35}S-metabolites to cellular macromolecules and the ability of cells treated with AOA to metabolized ^{14}C-DCVC to pyruvate. Cells in 24 well culture dishes were treated with ^{35}S-or unlabelled DCVC in the presence or absence of 0.1 mM AOA.   In addition, three 15cm culture dishes of LLC-PK1 cells were treated with 0.1 mM AOA.   At 14 hours the metabolism of DCVC to pyruvate was measured in homogenates of the AOA treated cells from the 15cm dishes and compared to the metabolism of DCVC by homogenates from untreated cells; neither group was treated with DCVC. AOA was not included in the homogenization buffer to exclude the possibility that the enzyme might not be inhibited in the cells, but might be inhibited when they were homogenized in the presence of AOA. The data show that AOA protected the cells from DCVC toxicity, inhibited the covalent binding and the ability of the cells to metabolize DCVC.

We also compared the toxicity of several S-cysteine conjugates in the LLC-PK1 cells. Table 8 shows that all the conjugates are toxic, but the potencies may differ.

Conclusions

The data presented here and elsewhere[11],[13] show that the major form of hepatic cysteine conjugate β-lyase is kynureninase, but that the kidney

Table 6. The binding of DCVC metabolites to trichloroacetic acid insoluble materials in LLC-PK1 homogenates.[a]

Label	Product bound nmol.mg^{-1}·30 min^{-1}
[14]C-DCVC	0.3 ± 0.1
^{35}S-DCVC	5.0 ± 0.1
^{35}S-DCVC + GSH	0.4 ± 0.1

[a] LLC-PK1 homogenates were incubated with 1mM ^{35}S-DCVC and the binding determined as described in the Methods.

Table 7. The effect of aminoxyacetic acid on the metabolism, toxicity and covalent binding of DCVC and DCVC metabolites in EBSS.[a]

Treatment	Cell Death	Covalent Binding nmol·well	Metabolism nmol.mg-1.min-1
35S-DCVC	55 ± 2	0.5 ± 0.1	-
35S-DCVC + AOA	16 ± 2	<0.1	-
control	4 ± 4	-	4.3 ± 0.2
control + AOA	11 ± 1	-	1.2 ± 0.1

[a] Metabolism was measured in homogenates of cells which had been treated with AOA (0.1 mM) but no DCVC. Covalent binding and toxicity were measured with DCVC (50 μM) in triplicate cultures from two seperate experiments; a well contains approximately 0.4 mg of cellular protein, or 1 X 10^6 cells.

cortex cytosolic enzyme is immunologically different. It is possible that this difference contributes to the organ specificity of S-cysteine conjugate toxicity. However, other factors such as subcellular localization, substrate specificity and specific transport systems are probably also very important.

The data from the cell culture experiments show that the toxicity of S-glutathione conjugates depends on the processing to the S-cysteine conjugate. In addition, the inhibition of covalent binding, toxicity, and metabolism by AOA in LLC-PK1 cells suggests that reactive intermediate production correlates with cytotoxicity. We must be careful in interpretting these results, however. Without proof that the reactive intermediate is causing the damage, an alternate interpretations is that the covalent binding is merely a measure of metabolism in intact cells. Since, LLC-PK1 cells respond to the toxicity of a variety of toxic S-cysteine and S-glutathione conjugates, they appear to be a good model for comparing data concerning the mechanism of metabolism, toxicity and leaving group reactivity for a number of toxic S-cysteine conjugates.

Table 8. Comparison of the toxicity of S-cysteine conjugates in LLC-PK1 cells in EBSS.

conjugate	mM	% cell death
S-1,2-dichlorovinyl-L-cysteine	0.1	46 ± 10
S-1,1,2-trifluoro-2-chloroethyl -L-cysteine	0.1	71 ± 4
S-1,2,3,4,4-hexachloro-1,3 -butadienyl-L-cysteine	0.2	29 ± 3
S-1,1,2,3,3,3-hexafluoropropyl -L-cysteine	0.4	28 ± 7

References

1. A.A. Elfarra and M.W. Anders, Renal processing of glutathione conjugates.

Role in nephrotoxicity, Biochem. Pharmacol. 33:3729(1984).

2. W.B. Jakoby, J. Stevens, M.W. Duffel and R.A. Weisiger, The terminal enzymes of mercapturate formation and the thiomethyl shunt, Rev. Biochem. Toxicol. 6:97(1984).

3. J.J. Rafter, J. Bakke, G. Larsen, B. Gustafsson and J.-A. Gustafsson, Role of the intestinal microflora in the formation of sulfur containing conjugates of xenobiotics. Rev. Biochem. Toxicol. 5:387(1983).

4. P.M. Anderson and M.O., Schultze, Cleavage of S-(1,2-dichlorovinyl)-L-cysteine by an enzyme of bovine origin, Arch. Biochem. Biophys. 111: 593(1965).

5. R.F. Derr and M.O. Schultze, The metabolism of [35]S-(1,2-dichlorovinyl)-L-cysteine in the rat, Biochem. Pharmacol. 12:465(1963).

6. R.K. Bhattacharya and M.O. Schultze, Arch. Biochem, Biophys. Properties of DNA treated with S-1,2-dichlorovinyl-L-cysteine and a lyase. 153:105(1972).

7. M. Tateishi and H. Shimizu, Cysteine conjugate β-lyase, in "Enzymatic basis of detoxication," W.B. Jakoby,ed., Academic Press, New York (1980).

8. F.V. Sepulveda and J.P. Pearson, Neutral amino acid transport in cultured kidney tubule cells, in, "Tissue Culture of Epithelial Cells," Mary Taub, ed., Plenum Press, New York (1985).

9. L.L. McKinney, J.C. Picken, F.B. Weakley, A.C. Eldridge, R.E. Campbell, J.C. Cowen and H.E. Beister, Possible toxic factor of trichloroethylene extracted soybean meal, J. Am. Chem. Soc. 81:909(1959).

10. P.J. van Bladeren, W. Buys, D.D. Breimer and A. van der Gen, The synthesis of mercapturic acids and their esters. Eur. J. Med. Chem. 15: 495(1980).

11. J.L. Stevens, Isolation and characterization of an enzyme with both cysteine conjugate β-lyase activity and kynureninase activity, J. Biol. Chem. 260: 7945(1985).

12. J.L. Stevens and W.B. Jakoby, Cysteine conjugate β-lyase, Mol. Phramacol. 23:761(1983).

13. J.L. Stevens, Cysteine conjugate β-lyase in rat kidney cortex: Subcellular localization and relationship of the hepatic enzyme, Biochem. Biophys. Res. Comm. 129:499(1985).

14. E.W. Miles, Special aspects of transaminases, in "Transaminases," P. Christen and D. Metzler eds., John Wiley and Sons, New York(1985).

15. C. Walsh, "Enzymatic reaction mechanisms," W.H. Freeman and Co., San Fransisco (1979).

16. G.S. Bild and J.C. Morris, Detection of β-carbanion formation during kynurenine hydrolysis catalyzed by Pseudomonas marginalis kynureninase, Arch. Biochem. Biophys. 235:41(1984).

17. G.M. Kishore, Mechanism-based inactivation of bacterial kynureninase by β-substituted amino acids, J. Biol. Chem. 259:10669(1984).

18. H. Ueno and D. Metzler, Chemistry of the inactivation of cytosolic aspartate aminotransferase by serine O-sulfate, Biochemistry, 21:4387 (1982).

19. M.W. Anders, et al., this volume.

20. T, Beeler and J.E. Churchich, Reactivity of the phosphopyridoxal groups of cystathionase, J. Biol. Chem. 251:5267(1976).

21. C.A. Rabito and M.V. Karish, Polarized amino acid transport by an epithelial cell line of renal origin. J. Biol. Chem. 258, 2543(1983).

22. B. Sacktor, Na$^+$ gradient dependent transport systems in renal proximal tubule brush border membrane vesicles. in, Membranes and Transport, M. Taub, ed., Plenum Press, New York (1982).

FORMATION AND IDENTIFICATION OF NAPHTHOQUINONE GLUTATHIONE CONJUGATES

FOLLOWING MICROSOMAL METABOLISM OF 1-NAPHTHOL

Marion G. Miller, John Powell* and Gerald M. Cohen

The Toxicology Unit and *Department of Pharmaceutical
Chemistry, The School of Pharmacy, University of London
29/39 Brunswick Square, London, WC1N 1AX, UK

INTRODUCTION

Previous studies have indicated that 1-naphthol is metabolised by the cytochrome P-450 mixed function oxidase enzyme system to form a reactive species capable of covalently binding to microsomal protein (1,2). Furthermore, Hesse and Mezger (1) suggested the involvement of quinones and/or semiquinones in 1-naphthol-dependent covalent binding. In more recent studies, 1,4-naphthoquinone formed from 1-naphthol has been directly measured with both rat liver microsomes (2,3) and a reconstituted cytochrome P-450 enzyme system (4). From this, it was hypothesised that the toxicity of 1-naphthol, previously reported in isolated hepatocytes (5), may be mediated by naphthoquinone metabolites of 1-naphthol.

In the present studies, we have further investigated the identity of the reactive species responsible for 1-naphthol covalent binding. Theoretically, the oxidative metabolism of 1-naphthol could result in formation of either 1,4- or 1,2-naphthoquinone both of which have the potential ability to interact with cellular nucleophiles including glutathione (2) (Fig. 1). In agreement, addition of glutathione markedly inhibited 1-naphthol covalent binding with a concomitant increase in the amount of polar metabolites derived from 1-naphthol (2,4). In order to ascertain whether these metabolites were glutathione conjugates of the proposed naphthoquinone metabolites, naphthoquinone glutathione conjugates were synthesised, identified by Nuclear Magnetic Resonance and Mass Spectrometry and separated by HPLC and TLC. Evidence is presented which strongly suggests that 1,4-naphthoquinone is the major species responsible for the microsomal covalent binding of 1-naphthol.

MATERIALS AND METHODS

Materials
[1-^{14}C]-1-Naphthol (56 μCi/ mole) was purchased from Amersham International plc., UK. 1,2-Naphthoquinone (1,2-NQ) and 1,4-naphthoquinone (1,4-NQ) were obtained from Fluka (Switzerland) and the latter was purified by sublimation.

Synthesis of Naphthoquinone Glutathione Conjugates
As described by Nickerson et al (6) 1 mmole of naphthoquinone was dissolved in 95% ethanol and to this was added 1 mmole of reduced

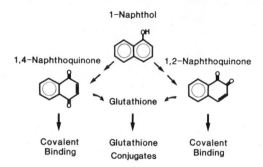

Figure 1. Proposed Metabolic Activation of 1-Naphthol.

glutathione. After cooling to $4°$ C, any crystals which appeared in the
solution were filtered and washed with 95% ethanol.

Nuclear Magnetic Resonance (NMR) Spectroscopy
    Proton NMR was carried out using a Bruker Model WP 80 SY instrument.
The synthesised samples were dissolved in D_2O which had been acidified by
addition of DC1. Glutathione was dissolved in D_2O.

Mass Spectrometry
    Mass spectral studies used a VG Analytical ZAB-1F instrument equipped
for either fast atom bombardment (FAB) or electron impact (EI) methods of
ionisation. For FAB analysis, compounds were dissolved in acidified acetonitrile, mixed with glycerol and inserted by direct probe. For EI
analysis, trimethyl silyl derivatives were synthesised by mixing with Tri-
Sil/BSA and were subsequently inserted by direct probe.

Separation of Naphthoquinone Glutathione Conjugates
    A. TLC. Two TLC systems were employed. Separation by reverse phase
TLC used Analtech RPS Uniplates and a 0.05 M phosphate buffer pH 4.5 :
methanol (70 : 30 v/v) solvent mixture. Normal phase TLC separation
utilised Merck aluminium backed silica plates and a solvent mixture of
propan-2-ol : water : acetic acid (70 : 20 : 10 v/v).
    B. HPLC. Chromatography was carried out using an Altex ODS Ultrsphere
column (25 cm). Samples were injected on to the column and eluted at a
flow rate of 1 ml/min with an isocratic solvent system of 25% methanol :
water (80 : 20 v/v)(Solvent A) and 75% 100 mM sodium acetate, pH 5.0
(Solvent B) for 25 min followed by a 15 min linear gradient to 100% Solvent
A which was maintained for a further 5 min.

Rat Liver Microsomes
    Washed liver microsomes were prepared from male Sprague-Dawley rats
(200-300 g) according to the method of Ernster et al (7) as previously
described (2). Freshly prepared microsomes were used in all experiments.

Metabolism of $[1-^{14}C]$-1-Naphthol
    $[1-^{14}C]$-1-Naphthol (20 µM) was incubated with rat liver microsomes (1
mg/ml) in 0.12 M Tris-HCl buffer, pH 7.4 at $37°$ C for up to 60 min in the

392

presence of an NADPH generating system as previously described (2). Following incubation with or without glutathione (1 mM), the microsomal protein was precipitated with 2 volumes of ice-cold methanol. The protein was removed by centrifugation and subsequently subjected to analysis for covalent binding. The supernatant fraction was analysed for 1-naphthol and its metabolites by the TLC and HPLC procedures described above.

Determination of Covalent Binding

[1-^{14}C]-1-Naphthol covalent binding to microsomal protein was measured by the method of Kappus and Remmer (8) as previously described (2).

RESULTS

Synthesis of Naphthoquinone Glutathione Conjugates

Two yields of crystalline solid were obtained from the reaction of 1,4-NQ and glutathione. An orange-brown solid rapidly crystallised out of the cooled reaction mixture and was filtered and washed with 95% ethanol. A second yellow solid crystallised from the filtrate. Both compounds were subjected to analysis by NMR and Mass Spectrometry.

The reaction of 1,2-NQ and glutathione initially yielded a brown-coloured solution which further reacted to form a tarry black compound. No crystals were obtained from this solution.

Identification of Naphthoquinone Glutathione Conjugates

A. NMR. The proton NMR spectrum of reduced glutathione is illustrated in Fig. 2A. The resonances representing the protons of the glutathione molecule were interpreted from their chemical shifts (ppm) and the ratios of the number of protons in each chemical grouping were established by integration. Similarly, Fig. 2B illustrates the proton NMR spectrum obtained for the yellow crystallisation product from the reaction of 1,4-NQ with glutathione. The spectrum suggests the presence of 2-glutathionyl-1,4-NQ since resonance 7 represents the single unsubstituted proton at position 3 of the quinone ring. Moreover, resonance 8 exhibits the chemical shift usually associated with aromatic protons and which, after integration, indicated that the ratio of aromatic protons to glutathione protons was 1 : 1 suggesting that this molecule was the monosubstituted glutathione conjugate of 1,4-NQ. Fig. 2C depicts the spectrum obtained for the orange-brown crystallisation product from the reaction of 1,4-NQ with glutathione. The marked decrease in resonance 7, associated with the proton on the naphthoquinone ring suggested that this compound was 2,3-diglutathionyl-1,4-NQ. Also, from integration of resonance 8, the ratio of aromatic protons to glutathionyl protons was 1 : 2 indicating formation of a disubstituted glutathione conjugate. This spectrum also suggested contamination of 2,3-diglutathionyl-1,4-NQ with approximately 30% 2-glutathionyl-1,4-NQ, which was confirmed using HPLC and TLC.

The reaction mixture obtained after mixing of 1,2-NQ and glutathione was also examined using proton NMR. The spectra were not readily interpretable.

B. Mass Spectrometry. Mass spectral analysis, using the FAB ionisation technique, of the compound tentatively identified as 2-glutathionyl-1,4-NQ yielded a base peak with a mass of 465 corresponding to a (M + 2) ion. Similarly, for 2,3-diglutathionyl-1,4-NQ a (M + 2) ion of 770 was identified. This (M + 2) base peak has been documented for other quinonoid compounds when these have been subjected to FAB mass spectrometry (9). For EI mass spectrometry, the compounds were first derivatised by silylation and, as illustrated in Fig. 3 the compound thought to be 2-glutathionyl-1,4-NQ exhibited a large number of fragmentation ions consistent with its

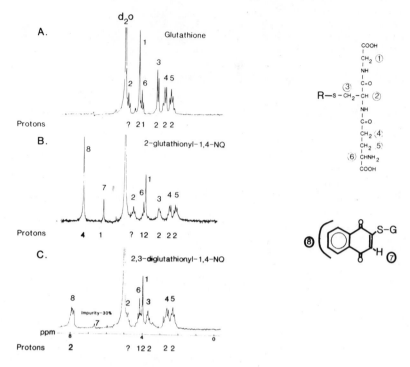

Figure 2. Proton NMR spectra of - A, Glutathione; B, 2-glutathionyl-
1,4-NQ and C, 2,3-diglutathionyl-1,4-NQ. Each resonance is assigned
to the chemical groupings indicated. The ratio of protons in each
molecule are shown.

	m/z
M+·	679
M − TMS	606
M − GLY TMS	533
M−GLY TMS −TMS	460
GSH − GLY TMS	304
GSH−GLYTMS−TMS−S	271
1,4−NQ CYS	258
ɤ GLU 2TMS	201
1,4−NQSH	190
1,4−NQ	157
GLY TMS	147
ɤ GLU	131

2-glutathionyl −1,4−NQ

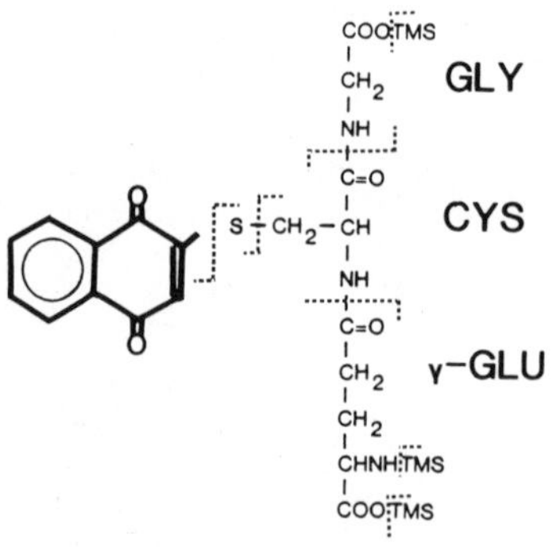

Figure 3. Fragmentation ions obtained from EI mass spectrometry of the trimethylsilyl derivative of 2-glutathionyl1-
1,4-NQ.

suggested identity. However, all of the fragmentation ions could not be
assigned molecular structures. The 2,3-diglutathionyl-1,4-NQ derivative
showed an EI fragmentation pattern similar to that observed with
2-glutathionyl-1,4-NQ although no parent ion could be detected.

Overall, these data are consistent with the proposed formation of
the mono- and di-substituted glutathione conjugates of 1,4-NQ.

Effect of Glutathione on [1-^{14}C]-Naphthol Covalent Binding to Rat Liver Microsomes

The amount of radioactivity irreversibly bound to microsomal protein
was determined after various times of incubation with [1-^{14}C]-1-naphthol
(20 µM) in the presence and absence of glutathione (1 mM). In the absence
of glutathione, greater than 50% of the 1-naphthol was covalently bound
after 60 min incubation (Fig. 4 and Ref. 2). However, in the presence of
glutathione there was a marked inhibition of covalent binding (Fig. 4).

Effect of Glutathione on the Metabolism of [1-^{14}C]-1-Naphthol

In the presence of glutathione, a marked increase in the formation of
methanol-soluble metabolites was observed (Fig. 5) in agreement with our
earlier studies (2). Analysis of these metabolites by HPLC and TLC
revealed that most of the radioactivity cochromatographed with either
2-glutathionyl-1,4-NQ or 2,3-diglutathionyl-1,4-NQ. The time course of
formation of the metabolites is depicted in Fig. 6. Immediate analysis of
the samples indicated that 2-glutathionyl-1,4-NQ was the predominant
metabolite. However, after storage overnight (at -20° C), the majority of
the radioactivity cochromatographed with 2,3-diglutathionyl-1,4-NQ
suggesting a further reaction of 2-glutathionyl-1,4-NQ in the presence of
excess glutathione (Results not shown).

DISCUSSION

In rat liver microsomes, [1-^{14}C]-1-naphthol was metabolised in the
absence of glutathione to covalently bound products (Fig. 4) and in the
presence of glutathione to compounds proposed to be naphthoquinone
glutathione conjugates (Fig. 5). The methanol-soluble products of the
microsomal metabolism of 1-naphthol in the presence of glutathione co-
chromatographed (Fig. 6) with synthetic 2-glutathionyl-1,4-NQ and 2,3-
diglutathionyl-1,4-NQ (Figs 2 and 3). From the data, the amount of
1-naphthol covalent binding inhibited by addition of glutathione
approximately equalled the amount of 1,4-NQ glutathione conjugates formed.
Hence, we conclude that a large component of the microsomal covalent
binding of 1-naphthol can be accounted for by formation of 1,4-NQ or by a
metabolite derived from this naphthoquinone.

Glutathione conjugates of 1,2-NQ were not positively identified due
to an apparent lack of stability. However, since the reaction products of
1,2-NQ and glutathione were separated from 1,4-NQ glutathione conjugates
and little or no radioactivity was associated with the former compounds,
the data provides no evidence for formation of significant amounts of
1,2-NQ following the microsomal metabolism of 1-naphthol. Nevertheless, a
component of 1-naphthol covalent binding was not inhibited by glutathione
(Fig. 4) and this could represent 1,2-NQ binding to protein in close
proximity to its site of generation. Alternatively, a proportion of
1,4-NQ may be bound to protein as well as reacted with glutathione.
Another possibility is that the residual glutathione-insensitive covalent
binding of 1-naphthol could represent formation of a metabolite capable of
covalent binding but which is not readily conjugated with glutathione.
Previous studies have also provided evidence that 1,4-NQ and not 1,2-NQ is
responsible for 1-naphthol covalent binding (2,4). In particular,
ethylene diamine, which reacts specifically with o-quinones (10) including

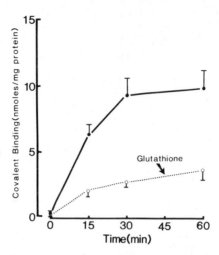

Figure 4. Effect of glutathione on $[1-^{14}C]$-1-naphthol covalent binding in rat liver microsomes. Microsomes were incubated with 1-naphthol (20 μM) in the absence (●) and presence (o) of gluta-thione (1 mM). Covalent binding was measured at various incubation times as described in "Materials and Methods". Data are mean ± SEM, N = 4.

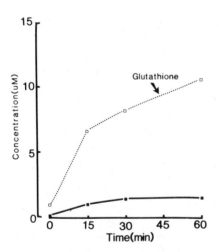

Figure 5. Effect of glutathione on formation of methanol-soluble metabolites of 1-naphthol in rat liver microsomes. Microsomes were incubated with 1-naphthol (20 μM) in the absence (■) and presence (□) of glutathione (1 mM). 1-Naphthol and its metabolites were separated as described in "Materials and Methods". Results are from one experiment typical of three.

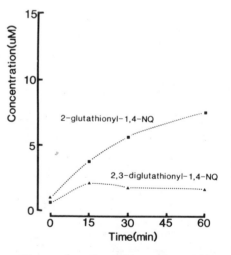

Figure 6. Formation of naphtho-quinone-glutathione conjugates following microsomal metabolism of 1-naphthol. Rat liver micro-somes were incubated with 1-naph-thol (20 μM) in the presence of glutathione (1 mM). 1-Naphthol and naphthoquinone-glutathione conjugates were separated as described in "Materials and Methods". Results are from one experiment typical of three.

1,2-NQ (2), had no effect on 1-naphthol covalent binding (2).

1-Naphthol and both 1,2-NQ and 1,4-NQ are known to be toxic to isolated
hepatocytes (5). These compounds also caused a dose-dependent depletion
of hepatocellular glutathione which occurred prior to the onset of cell
death (5). Furthermore a concentration and time-dependent covalent
binding of $[1-^{14}C]$-1-naphthol in isolated hepatocytes has been noted
(data not shown). From these data, we have hypothesised that the toxicity
of 1-naphthol may be mediated via naphthoquinone formation (2,4,5). Naph-
thoquinones may be toxic either by redox cycling with consequent generation of reactive semiquinone intermediates and active oxygen species or by
covalent binding to cellular macromolecules (11,12). Further studies are
in progress to determine whether naphthoquinone glutathione conjugates are
formed as metabolites of 1-naphthol in isolated hepatocytes.

ACKNOWLEDGEMENTS: This work was supported in part by a grant from the
Cancer Research Campaign. MGM thanks the Sterling Winthrop Foundation for
partial support. We thank Mrs Mary Fagg for typing the manuscript.

REFERENCES

1.   Hesse, S. and Mezger, M., 1979. Involvement of phenolic metabolites
      in the irreversible protein-binding of aromatic hydrocarbons:
      reactive metabolites of $[^{14}C]$ naphthalene and $[^{14}C]$-1-naphthol
      formed by rat liver microsomes. Mol. Pharmacol., 16: 667.
2.   d'Arcy Doherty, M. and Cohen, G.M., 1984. Metabolic activation of
      1-naphthol by rat liver microsomes to 1,4-naphthoquinone and
      covalent binding species. Biochem. Pharmacol. 33: 3201.
3.   Fluck, D.S., Rappaport, S.M., Eastmond, D.A. and Smith, M.T. 1984.
      Conversion of 1-naphthol to naphthoquinone metabolites by rat liver
      microsomes: Demonstration by high performance liquid chromatography
      with reductive electrochemical detection. Arch. Biochem. Biophys.
      235: 351.
4.   d'Arcy Doherty, M., Makowski, R., Gibson, G.G. and Cohen, G.M. 1985.
      Cytochrome P-450 dependent metabolic activation of 1-naphthol to
      naphthoquinones and covalent binding species. Biochem. Pharmacol.
      34: 2261.
5.   d'Arcy Doherty, M., Cohen, G.M. and Smith, M.T. 1984. Mechanisms
      of toxic injury to isolated hepatocytes by 1-naphthol. Biochem.
      Pharmacol. 33: 543.
6.   Nickerson, W.J., Falcone, G. and Strauss, G. 1963. Studies on
      quinone-thioethers. I. Mechanism of formation and properties of
      thiodione. Biochemistry. 3: 537.
7.   Ernster, L., Siekevitz, P. and Palade, G.E. 1962. Enzyme
      structure relationships in the endoplasmic reticulum of rat liver:
      a morphological and biochemical study. J. Cell Biol. 15: 541.
8.   Kappus, H. and Remmer, H. 1975. Metabolic activation of
      norethisterone to an irreversibly protein-bound derivative by rat
      liver microsomes. Drug Metab. Disp. 3: 338.
9.   Ross, D. Larsson, R., Norbeck, K., Ryhage, R. and Moldeus, P. 1985.
      Characterization and mechanism of formation of reactive products
      formed during peroxidase-catalyzed oxidation of p-phenetidine.
      Mol. Pharmacol., 27: 277.
10.   Jellinck, P.H. and Irwin, L. 1963. Interaction of oestrogen
      quinones with ethylenediamine. Biochem. Biophys. Acta. 78: 778.
11.   Kappus, H. and Sies, H. 1981. Toxic drug effects associated with
      oxygen metabolism: redox cycling and lipid peroxidation.
      Experientia. 37: 1233.
12.   Thor, H., Smith, M.T., Hartzell, P., Bellomo, G., Jewell, S.A. and
      Orrenius, S. 1982. The metabolism of menadione (2-methyl-1,4-
      naphthoquinone) by isolated hepatocytes. J. Biol. Chem. 257: 12419.

397

THE FORMATION OF N-GLUCURONIDES CATALYZED BY PURIFIED HEPATIC

17β-HYDROXYSTEROID AND 3α-HYDROXYSTEROID UDP-GLUCURONYLTRANSFERASES

Mitchell D. Green, Yacoub Irshaid and Thomas R. Tephly

Toxicology Center
Department of Pharmacology
University of Iowa
Iowa City, IA   52242

INTRODUCTION

Conjugation of many endogenous and exogenous chemicals with glucuronic acid, catalyzed by UDP-glucuronyltransferases (EC 2.4.1.17), is an important step in their detoxification and elimination from the body. We have previously reported on the purification of three different UDP-glucuronyltransferases (UDPGTs) from hepatic microsomes of female Sprague-Dawley rats[1]. A 17β-hydroxysteroid UDPGT catalyzes the glucuronidation of the 17β-hydroxy position of steroids, such as testosterone and 17β-estradiol, and xenobiotics such as p-nitrophenol and 1-naphthol. A 3α-hydroxysteroid UDPGT has been purified which is specific for the glucuronidation of the 3α-position of steroids and bile acids[2,3]. A 3-methylcholanthrene-inducible form of p-nitrophenol UDPGT catalyzes the conjugation of various xenobiotic substrates, such as p-nitrophenol, 4-methylumbelliferone and 1-naphthol, with glucuronic acid[1].

One of the major metabolic pathways for aromatic amines is N-glucuronidation[4,5]. Arylamine N-glucuronides are labile and easily hydrolyzed to the parent amine under weakly acidic conditions, which can exist in the urinary bladder[6]. This may result in exposure of the urinary bladder to relatively lipophilic and potentially toxic substances. This paper presents some of our initial work on the identification of UDPGT(s) responsible for the N-glucuronidation of arylamines. Specifically the N-glucuronidation of α-naphthylamine (α-NA) and 4-aminobiphenyl (4-ABP) were investigated. The results indicate that α-NA is a substrate for the 17β-hydroxysteroid UDPGT, whereas, the 3α-hydroxysteroid UDPGT catalyzes the glucuronidation of 4-ABP.

METHODS

Female Sprague-Dawley rats (180–220g, BioLabs, St. Paul, MN) were used in this study. Hepatic microsomes were prepared in ice cold 1.15% KCl and stored at −70°C until used. All buffers used in this study contained 20% glycerol, 0.1 mM dithiothreitol and 0.05% Emulgen 911.

Glucuronidation of p-nitrophenol, testosterone, androsterone and estrone were determined as described previously[1,7]. α–NA, and 4–ABP N-glucuronidation assays were performed using the spectrofluorometric method described by Lilienblum and Bock[8]. All reactions were conducted at 37°C with 5 mM UDP-glucuronic acid (2 mM for column fractions) and 100 μg of phosphatidylcholine.

Chromatofocusing of solubilized microsomal protein was conducted using a pH 9 to pH 7 gradient[2]. Column fractions were assayed for 3α-hydroxysteroid and 17β-hydroxysteroid UDPGT activities using androsterone and testosterone, respectively, as substrates. Affinity chromatography, using UDP-hexanolamine bound to Sepharose 4B, was performed as described previously[1]. Kinetic parameters were determined by regression analysis of double reciprocal plots of velocity versus substrate concentration. When kinetic experiments were performed, the assays were initiated by the addition of enzyme to the reaction mixture.

RESULTS

The hepatic microsomal glucuronidation rates for α–NA, 4–ABP and other selected substrates are shown in Table 1. The microsomal rate of α–NA N-glucuronidation is relatively high and comparable to that for

Table 1. Glucuronidation Rates of Various Substrates in Solubilized Hepatic Microsomes from Sprague-Dawley Rats

Substrate	Enzymatic Activity[*]
p-Nitrophenol	89.4 ± 3.7
Testosterone	5.5 ± 0.3
Androsterone	12.5 ± 1.2
Estrone	0.4 ± 0.1
α-Naphthylamine	60.3 ± 3.3
4-Aminobiphenyl	3.4 ± 0.5

[*]Rates are expressed as nmol substrate conjugated/min/mg protein. Each value represents mean ± S.E. (n = 4).

p-nitrophenol. In contrast, the rate of glucuronidation for 4-ABP is approximately 20-fold lower than that of α-NA. These results agree with those obtained by Lilienblum and Bock[8].

Chromatofocusing chromatography of solubilized hepatic microsomes produced different patterns of elution for α-NA and 4-ABP UDPGT activities (Figure 1). N-Glucuronidation activity for α-NA co-eluted with

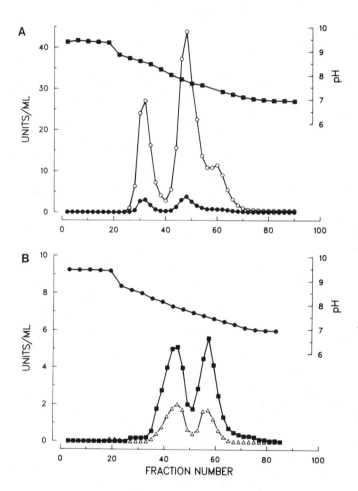

Fig. 1 Chromatofocusing profiles of solubilized
Sprague-Dawley rat liver microsomes using
a pH 9 to pH 7 gradient. In panel A
glucuronidation rates for testosterone
(closed circle) and α-naphthylamine (open
circles) in chromatofocusing column fractions are shown (pH gradient, solid
squares). Panel B shows the co-elution of
androsterone (solid squares) and 4-amino-
biphenyl (open triangle) UDPGT activities
(pH gradient, closed circles). One unit
of UDPGT activity represents 1 nmol of
substrate conjugated per minute.

testosterone (a substrate for 17β-hydroxysteroid UDPGT) activity (Figure 1A). Major peaks of α-NA and testosterone glucuronidation activities were observed at pH 8.5 and pH 7.8. Figure 1B shows that 4-ABP N-glucuronidation activity co-eluted with androsterone UDPGT activity. Two peaks of androsterone and 4-ABP UDPGT activities were observed at pH 7.8 and pH 7.5. No 4-ABP UDPGT activity was detected in pH 8.5 chromatofocusing fractions.

17β-Hydroxysteroid and 3α-hydroxysteroid UDPGT were further purified using affinity chromatography[1]. Testosterone and α-NA UDPGT activities co-eluted from the affinity column when 0.1 mM UDP-glucuronic acid was applied (Figure 2). The purified 17β-hydroxysteroid UDPGT catalyzes the N-glucuronidation of α-NA, but not of 4-ABP. The affinity column elution profile for 4-ABP and androsterone glucuronidation activities were similar. These UDPGT activities were specifically eluted with 0.1 mM UDP-glucuronic acid. 3α-Hydroxysteroid UDPGT purified from rat liver microsomes catalyzes the N-glucuronidation of 4-ABP.

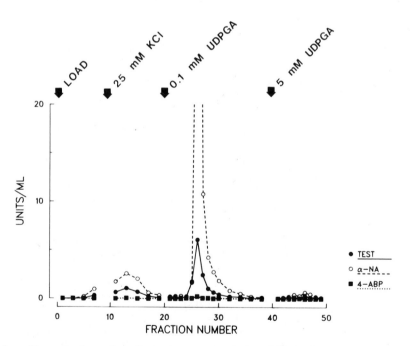

Fig. 2. Chromatofocusing fractions containing 17β-hydroxysteroid UDPGT activity (pH 8.5 peak) were pooled and applied to the affinity column and eluted as indicated. One unit of UDPGT activity represents one nmole of substrate conjugated per minute. The activity for α-NA in the peak fraction was 64 units/ml.

Even though α-NA and 4-ABP were glucuronidated by separate UDPGTs, both arylamines are able to inhibit the glucuronidation catalyzed by other UDPGTs (Table 2). α-NA and 4-ABP are competitive inhibitors of both 3α-hydroxysteroid and 17β-hydroxysteroid UDPGTs. These data suggest that both aromatic amines are interacting similarly with the active sites of the different purified UDPGTs, even though they may not be glucuronidated by that particular isoenzyme.

DISCUSSION

Many studies have demonstrated that N-glucuronidation of aromatic amines occurs in hepatic microsomes[8,10]. Our studies show that N-glucuronidation of arylamine xenobiotics is catalyzed by specific purified rat liver UDPGTs. Surprisingly, the enzymes which conjugate the aromatic amines α-NA and 4-ABP are those which also catalyze steroid O-glucuronidation; specifically, the 17β-hydroxysteroid and 3α-hydroxysteroid UDPGTs. It should be noted that 4-ABP represents the first xenobiotic identified as a substrate for the 3α-hydroxysteroid UDPGT; previously, only 3α-hydroxy steroids and bile acids have been reported to react with this isoenzyme[2].

Further support for the role of 3α-hydroxysteroid UDPGT in the N-glucuronidation of 4-ABP has been obtained in experiments using Wistar rat liver microsomes. Matsui et al.[11,12] originally demonstrated that about 50% of the Wistar rat population were deficient in their ability to

Table 2. Inhibition of Purified 17β-Hydroxysteroid and 3α-Hydroxysteroid UDP-Glucuronyltransferase Activities by Arylamines

Inhibitor	17β-Hydroxysteroid UDPGT	3α-Hydroxysteroid UDPGT
α-NA	+ (C)	+ (C)
4-ABP	+ (C)	+ (C)

A plus (+) indicates the alternate substrate is inhibitory at a concentration of 1 mM. The type of inhibition, as determined from double reciprical plots of reaction velocity vs substrate concentration, is given in parenthesis. C = competitive inhibition. Testosterone and androsterone were used as the varied substrates for 17β- and 3α-hydroxysteroid UDPGTs, respectively.

glucuronidate androsterone. We have subsequently shown that this deficiency is due, in large part, to reduced amounts of 3α-hydroxysteroid UDPGT
in these animals[13]. Preliminary studies have shown that Wistar rat liver
microsomes with low amounts of 3α-hydroxysteroid UDPGT are also deficient
in their ability to conjugate 4-ABP. The N-glucuronidation of 4-ABP by a
constitutive rat liver UDPGT is in agreement with an induction study by
Lilienblum and Bock[8] which showed that microsomal 4-ABP UDPGT activity was
not induced by phenobarbital or 3-methylcholanthrene treatments.

N-Glucuronidation of α-NA is catalyzed by the 17β-hydroxysteroid
UDPGT. Preliminary experiments also indicate that β-naphthylamine (β-NA)
is also a substrate for the 17β-hydroxysteroid UDPGT. The current results
have toxicological significance. Various species have been shown to
exhibit differential susceptibility to tumor development induced by α-NA
and β-NA[14-16]. It is possible that the levels of 17β-hydroxysteroid UDPGT,
or other UDPGTs) for which α-NA and β-NA are substrates, in different
species may correlate with the relative potency of these compounds.
Another important possibility is the presence or absence of this isoenzyme
in various extrahepatic tissues. Experiments are currently in progress to
determine whether α-NA and β-NA are substrates for the 3-methylcholan-
threne-inducible UDPGT.

ACKNOWLEDGMENT

This study was supported by NIH research grant GM 26221.

REFERENCES

1. C. N. Falany and T. R. Tephly, Separation, purification and characterization of three isoenzymes of UDP-glucuronyltransferase from rat
   liver microsomes, Arch. Biochem. Biophys. 227: 248 (1983).
2. R. B. Kirkpatrick, C. N. Falany and T. R. Tephly, Glucuronidation of
   bile acids by rat liver 3-OH androgen UDP-glucuronyltransferase,
   J. Biol. Chem. 259: 6176 (1984).
3. C. N. Falany, R. B. Kirkpatrick and T. R. Tephly, Comparison of rat
   and rabbit liver UDP-glucuronyltransferase activities, in:
   "Advances in Glucuronide Conjugation," S. Matern, K. W. Bock and
   W. Gerok, eds., MTP Press, Lancaster (1985).
4. E. Boyland, D. Manson and S. F. D. Orr, The biochemistry of aromatic
   amines. 2. The conversion of arylamines into arylsulphamic acids
   and arylamine-N-glucosiduronic acids, Biochem. J. 65: 417 (1957).
5. I. M. Arias, Ethereal and N-linked glucuronide formation by normal
   and Gunn rats in vitro and in vivo, Biochem. Biophys. Res. Commun.
   6:81 (1961).
6. F. F. Kadlubar, L. E. Unruh, T. J. Flammang, D. Sparks, R. K. Mitchum
   and G. J. Mulder, Alteration of urinary levels of the carcinogen,
   N-hydroxy-2-naphthylamine, and its N-glucuronide in the rat by
   control of urinary pH, inhibition of metabolic sulfation and
   changes in biliary excretion, Chem. - Biol. Interact. 33:129
   (1981).

7. R. H. Tukey, R. E. Billings and T. R. Tephly, Separation of oestrone UDP-glucuronyltransferase and p-nitrophenol UDP-glucuronyltransferase activities, Biochem. J. 171:659 (1978).

8. W. Lilienblum and K. W. Bock, N-Glucuronide formation of carcinogenic aromatic amines in rat and human liver microsomes, Biochem. Pharmacol. 33:2041 (1984).

9. R. H. Tukey and T. R. Tephly, Purification and properties of rabbit liver estrone and p-nitrophenol UDP-glucuronyltransferases, Arch. Biochem. Biophys. 209:565 (1981).

10. C. Y. Wang, K. Zukowski and M. S. Lee, Glucuronidation of carcinogenic arylamine metabolites by rat liver microsomes, Biochem. Pharmacol. 34:837 (1985).

11. M. Matsui, F. Nagai and S. Aoyagi, Strain differences in rat liver UDP-glucuronyltransferase activity towards androsterone, Biochem. J. 179:483 (1979).

12. M. Matsui and M. Hakozaki, Discontinuous variation in hepatic uridine diphosphate glucuronyltransferase toward androsterone in Wistar rats. A regulatory factor for in vivo metabolism of androsterone, Biochem. Pharmacol. 28:411 (1979).

13. M. D. Green, C. N. Falany, R. B. Kirkpatrick and T. R. Tephly, Strain differences in purified rat hepatic 3α-hydroxysteroid UDP-glucuronosyltransferase, Biochem. J. In press (1985).

14. G. M. Bonser, D. B. Clayson, J. W. Jull and L. N. Pyrah, The carcinogenic properties of 2-amino-1-naphthol hydrochloride and its parent amine 2-naphthylamine, Br. J. Cancer 6:412 (1952).

15. D. B. Clayson and R. C. Garner, in: "Chemical Carcinogens," C.E. Searle, ed., American Chemical Society, Washington, D.C. (1976).

16. R. M. Hicks, R. Wright and J. St. J. Wakefield, The induction of rat bladder cancer by 2-naphthylamine, Br. J. Cancer 46:646 (1982).

405

IMPROVED METHOD FOR DETERMINATION OF CELLULAR THIOLS, DISULFIDES AND PROTEIN MIXED DISULFIDES USING HPLC WITH ELECTROCHEMICAL DETECTION

Charles B. Jensen, Scott J. Grossman and David J. Jollow

Department of Pharmacology
Medical University of South Carolina
Charleston, SC 29425

INTRODUCTION

The broad interest in GSH[*] as a cellular mediator in the defense against toxicological injury has prompted the development of a variety of methods for measuring cellular glutathione, comprised predominantly of GSH and its disulfides, GSSG and GSSProt. As investigations have increasingly focused on the biochemical mechanisms underlying the disposition of glutathione, the need has arisen for methods providing greater analytical capabilities. The earliest methods developed for thiol determination utilized the stoichiometric reaction of thiols with chemical agents to produce a colored (Ellman, 1959) or fluorescent (Hissin and Hilf, 1976) product. These methods were convenient, but they did not provide a specific determination of GSH. While GSH constitutes the major non-protein thiol in most cell types, inclusion of other thiols or non-thiol compounds (Benson et al., 1975) in the measurement led to overestimation of GSH levels. In addition, chemical methods lacked adequate sensitivity to measure low levels of GSH (Jensen et al., 1985). Despite these difficulties, a difference method has been developed (Habeeb, 1973) and utilized (Isaacs & Binkley, 1977) to measure hepatic GSH, GSSG and GSSProt. Specificity for GSH and sensitivity have been increased by combining enzymatic recycling with a colorimetric determination (Tietze, 1969), although this method did not differentiate between GSH and GSSG. While GSH could be removed from the reaction using N-ethylmaleimide for GSSG determination, this thiol-blocking reagent, as well as protein precipitants and denaturants, can alter the kinetics of the reaction (Griffith, 1980; Brigelius et al., 1983) leading to erroneous results. Recently, rapid chromatographic analysis using HPLC has been applied to GSH determinations (Reed et al., 1980). While specificity was insured based on column retention time and the sensitivity was adequate for the determination of hepatic GSSProt (Brigelius et al., 1983), lengthy pre-column derivitization was required. In the present report, a new method that utilizes HPLC separation and thiol-specific electrochemical detection for thiol determination is demonstrated. This analytical system provides some advantages lacking in previous methods. Specificity for GSH and other thiols is based on column retention time and does not require thiol

---

[*]GSH, reduced glutathione; GSSG, glutathione disulfide, oxidized glutathione; GSSProt, glutathione-protein mixed disulfide(s); HPLC-EC, high performance liquid chromatography with electrochemical detection.

derivitization. The sensitivity of the method allows quantitation of GSH in the picomole range, which is adequate for the determination of hepatic GSSProt. Finally, the presence of endogenous biological materials or reagents used in sample preparation has been found not to interfere with thiol detection. The applicability of this method to biological systems is illustrated by the determination of the glutathione status in rat liver and erythrocytes.

METHODS

The procedure used to prepare rat livers for assay is outlined in Figure 1 and is described in greater detail by Jensen et al. (1985). Recoveries of standard amounts of GSH, GSSG and GSSProt added to biological samples were 93.0 ± 1.4% (n=5), 93.8 ± 1.8% (n=5) and 108% (n=2) respectively. The chromatographic system consisted of a reverse phase C_{18} column with a mobile phase of 5 mM heptanesulfonic acid, 50 mM KH_2PO_4 in an 8% methanol/water mixture, pH 2.5. The detector electrode was a Au/Hg amalgam with an applied voltage of -100mV. Figure 2 includes typical chromatograms of (A) an acid soluble supernate from rat liver, and (B) sodium borohydride releasable material from a washed protein precipitate. In both cases, GSH and cysteine have been identified by their retention times relative to standards. The limit of detection for GSH is 0.5 pmoles.

HEPATIC GLUTATHIONE STATUS

The toxicities of some xenobiotics have been shown to be inversely related to hepatic GSH levels, which demonstrates a uniform diurnal cycling (Jaeger et al., 1973). Further, Isaacs and Binkley (1977) reported that the diurnal variations in GSH correlated with reciprocal formation of GSSProt. Because non-specific methods were used to quantify

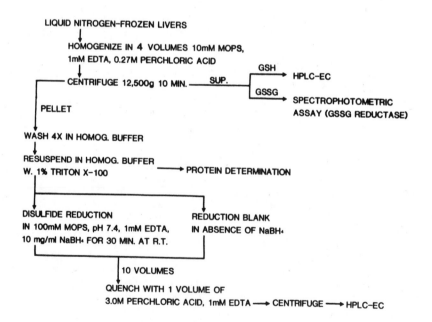

Figure 1. Outline of sample preparation for GSH, GSSG and GSSProt determinations.

the hepatic glutathione status in these studies, re-examination of the phenomenon was undertaken using HPLC-EC. The advantages of specificity and selectivity were particularly important for the direct measurement of GSSProt levels. The diurnal variation of GSH previously reported (Jaeger et al., 1973; Isaacs and Binkley, 1977) and based on a colorimetric assay was confirmed using the HPLC-EC method (Fig. 3A). GSSProt levels (Fig. 3B) were not related to GSH levels in either magnitude or direction. Based on these observations, it is concluded that diurnal changes in GSH levels do not result from changes in the formation of GSSProt, but, by implication, from changes in total cellular glutathione equivalents.

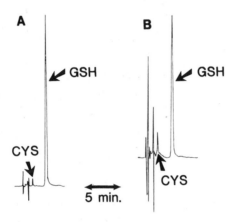

Figure 2. Typical HPLC-EC chromatograms of (A) acid-soluble supernate, and (B) reduced protein precipitate from rat liver.

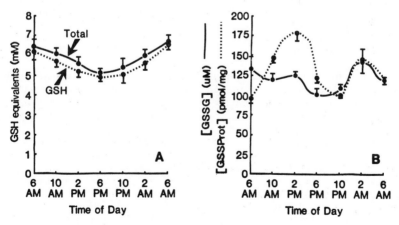

Figure 3. Diurnal variations in rat hepatic glutathione status. (A) GSH and total glutathione equivalents. In (B) GSSG is represented by the solid line, GSSProt by the dashed line.

Table 1. Effects of acetaminophen (700 mg/kg, i.p., three hours prior to sacrifice) and fasting (24 hours) on hepatic glutathione status.

TREATMENT	[GSH] (mM)	[GSSG] (uM)	[GSSProt] (pmol/mg)
FED, NORMAL	6.22 ± 0.51	243 ± 35	165 ± 12
FED, ACET.	2.47 ± 0.21	62.3 ± 6.4	174 ± 8
FASTED, NORMAL	3.39 ± 0.31	112 ± 16	238 ± 10
FASTED, ACET.	0.833 ± 0.102	31.3 ± 7.1	239 ±15

To determine whether GSSProt levels could be perturbed by more dramatic changes in rat hepatic GSH levels, the effects of a 24-hour fast and/or a toxic dose of acetaminophen (700 mg/kg) on the glutathione status were determined (Table 1). Acetaminophen, fasting and a combination of both produced decreases in the levels of both GSH and GSSG. However, acetaminophen had no effect on GSSProt levels in either the fed or fasted states. Fasting produced a slight elevation in GSSProt levels. Thus, the hepatic levels of GSSProt appear to be unrelated to hepatic GSH levels.

ERYTHROCYTIC GLUTATHIONE STATUS

Arylhydroxylamines have been demonstrated to produce hemolytic anemia (Harrison and Jollow, 1985). It is generally thought that the hemolytic agent produces an oxidative stress on erythrocyte metabolism, which is manifested as an oxidation of GSH to GSSG. This idea was tested by determining the glutathione status of rat erythrocytes following in vitro exposure to the toxic N-hydroxy metabolite of dapsone. Dapsone is known to induce hemolytic anemia in man (Degowin, 1966). N-hydroxydapsone (DDS-NOH)produced a rapid decrease in GSH levels (Fig. 4); however, no concomitant increase of GSSG was observed. Instead, a reciprocal increase in GSSProt levels occurred that fully accounted for the loss of GSH. Clearly, a simple oxidative stress model cannot explain these observations in the rat, but it is possible that a reactive thiol intermediate or a rapid thiol exchange reaction between GSSG and protein thiol groups can account for the formation of GSSProt.

In a comparative study, the effect of DDS-NOH on human erythrocytic glutathione status was investigated. Figure 5 demonstrates that, in the human erythrocyte, GSH appeared to be oxidized to GSSG in response to the same concentration of DDS-NOH that was used in the rat erythrocyte studies. However, little change was observed in GSSProt levels. Clearly, the effects of DDS-NOH on rat and human erythrocytic glutathione status are different. Because dapsone has been demonstrated to be hemolytic in both rat (Grossman, 1985) and man (Degowin, 1966), this difference strongly suggests that there may be significant differences in the mechanisms underlying the hemolytic response to dapsone in these two susceptible species.

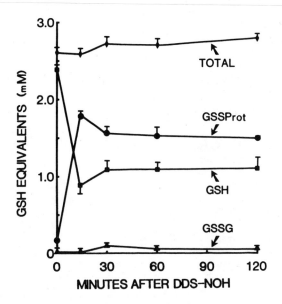

Figure 4. Effect of N-hydroxydapsone (120 μM) on rat erythrocytic glutathione status in vitro.

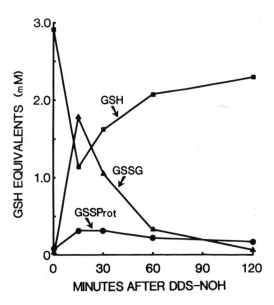

Figure 5. Effect of N-hydroxydapsone (120 μM) on human erythrocytic glutathione status in vitro.

SUMMARY

The role of glutathione in cellular responses to toxic insult is well
established in the literature and is largely based on chemical determinations of GSH. Although these methods have proved to be adequate to demonstrate the relationships between cellular levels of GSH and xenobiotic
toxicity, they may be inadequate for careful investigations regarding the
regulation of cellular concentrations of GSH, GSSG and GSSProt. Enzymatic
and chromatographic methods provide increased specificity and sensitivity,
but the problems of variable accuracy or tedious derivitization steps
remain. The HPLC-EC method constitutes a significant improvement over
previous methods by satisfying the requirements for selectivity, sensitivity, accuracy and convenience.

The utility of the HPLC-EC method has been described in two systems
of toxicological interest, particularly with regard to the determination
of GSSProt. In further studies, the HPLC-EC method has been useful in
investigating the mechanism(s) of alternate nucleophiles, such as cysteamine. The capability to determine various thiols simultaneously in a
given biological sample greatly facilitates these investigations. In
view of its power as an analytical tool, the HPLC-EC method has the potential for widespread application to the study of cellular thiols.

ACKNOWLEDGEMENTS

The technical skill and advice of Jennifer Schulte were invaluable
during these studies. The work was supported by grants from the U.S.
Public Health Service (GM 30546 and HL 30038).

References

Benson, J. and Hare, P., 1975, o-phthalaldehyde: fluorogenic detection of
primary amines in the picomole range. Comparison with fluorescamine and
ninhydrin, Proc. Nat. Acad. Sci., 72: 619-622.

Brigelius, R., Muckel, C., Akerboom, T. and Sies, H., 1983, Identification and quantitation of glutathione in hepatic protein mixed disulfides
and its relationship to glutathione disulfide, Biochem. Pharmacol., 32:
2529-2534.

Degowin, R., Eppes, R., Powell, R. and Carson, P., 1966, The haemolytic
effects of diaphenylsulfone (DDS) in normal subjects and in those with
glucose-6-phosphate dehydrogenase deficiency, Bull. Wld. Hlth. Org., 35:
165-179.

Ellman, G.L., 1959, Tissue sulfhydryl groups, Arch. Biochem. Biophys., 82:
70-77.

Griffith, O.W., 1980, Determination of glutathione and glutathione disulfide using glutathione reductase and 2-vinyl-pyridine, Analyt. Biochem.,
106: 207-212.

Grossman, S.J., 1985, Studies on the mechanism of dapsone-induced hemolytic anemia, Ph.D. thesis, Med. Univ. S. Carolina.

Habeeb, A.T., 1973, Reaction of protein sulfhydryl groups with Ellman's
reagent, Meth. Enz., 25: 457-464.

Harrison, J. and Jollow, D., 1985, Contribution of aniline metabolites to aniline-induced methemoglobinemia, J. Pharm. Exp. Ther., accepted for publication.

Hissin, P.J. and Hilf, R., 1976, A fluorometric method for determination of oxidized and reduced glutathione in tissues, Analyt. Biochem. 74: 214-226.

Isaacs, J. and Binkley, T., 1977, Glutathione dependent control of protein disulfide-sulfhydryl content by subcellular fractions of hepatic tissue, Biochim. Biophys. Acta, 497: 192-204.

Jaeger, R., Connolly, R. and Murphy, S., 1973, Diurnal variation of hepatic glutathione concentration and its correlation with 1,1-dichloroethylene inhalation toxicity in rats, Res. Commun. Chem. Path. Pharm., 6: 465-471.

Jensen, C., Grossman, S. and Jollow, D., 1985, Role of glutathione-protein mixed disulfides in the diurnal variation of the hepatic glutathione status, Biochem. Pharm., submitted for publication.

Reed, D., Babson, J., Beatty, P., Brodie, A., Ellis, W. and Potter, D., 1980, High performance liquid chromatography analysis of nanomole levels of glutathione, glutathione disulfide, and related thiols and disulfides, Analyt. Biochem., 106: 55-62.

Tietze, F., 1969, Enzymatic method for quantitative determination of nanogram amounts of total and oxidized glutathione: applications to mammalian blood and other tissues, Analyt. Biochem., 27: 502-522.

ARYLSULFOTRANSFERASE IV CATALYZED SULFATION OF 1-NAPHTHALENEMETHANOL

Michael W. Duffel and Maria N. Janss

Division of Medicinal Chemistry and Natural Products
College of Pharmacy
University of Iowa
Iowa City, Iowa 52242

INTRODUCTION

A variety of drugs, carcinogens, and other xenobiotics either contain benzylic alcohol functional groups or are metabolized by hydroxylation at a benzylic carbon. Several studies indicate that the route for formation of reactive metabolites from molecules containing benzylic hydroxyls may proceed via sulfation.[1-7] Benzylic sulfate esters have been reported as metabolites of such diverse compounds as 1-methylamino-2-phenylpropan-2-ol (an isomer of ephedrine)[1], 1´-hydroxysafrole,[2,5] 7,12-dimethylbenz-anthracene,[3,4] and various mono- and di-nitrotoluenes.[6,7] Benzylic sulfates are electrophilic metabolites and can react with cellular nucleophiles due to the ease with which the sulfate acts as a leaving group, and the ability of the aromatic ring to stabilize a positive charge at the benzylic carbon. The ability of these reactive sulfate esters to bind covalently to cellular macromolecules has led to an interest in their role in carcinogenesis and other toxic responses.

Numerous in vivo[2,5-7] and in vitro[1-5] studies have established the role of sulfation in conversion of benzylic alcohols into reactive metabolites. In order to identify the sulfotransferase(s) in hepatic cytosol responsible for these sulfation reactions, we have examined purified preparations of rat hepatic arylsulfotransferase. Although purification was carried out with 2-naphthol as sulfate acceptor, 1-naphthalenemethanol was used as a model substrate to determine benzylic alcohol sulfation. A homogeneous preparation of arylsulfotransferase IV, also categorized as tyrosine methylester sulfotransferase (EC 2.8.2.9), catalyzed the following reaction:

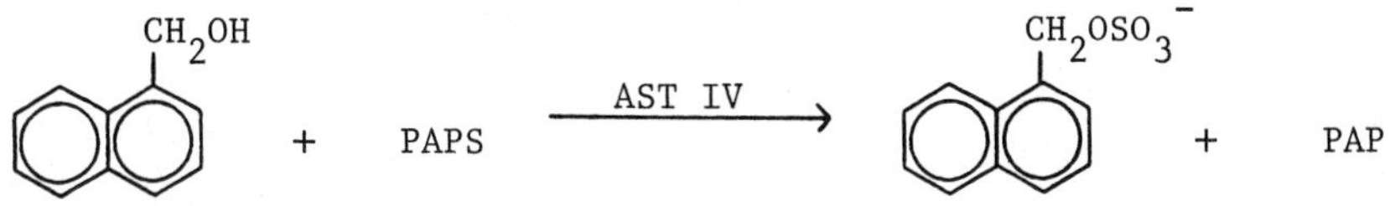

415

Arylsulfotransferase IV (AST IV) is one of at least four isoenzymes of phenol sulfotransferase purified from male rat liver.[8-11] It is distinguished from arylsulfotransferases I and II by its physical properties, pH optimum, and ability to catalyze sulfation of arylhydroxamic acids and peptides containing N-terminal tyrosines.[8] The substrate specificity of another isoenzyme, AST III, has not been as fully characterized, although it can be readily separated from AST IV by isoelectric focusing. A fifth isoenzyme, with some similarities to AST III and AST IV, has been purified from rat liver.[10] This enzyme has properties different from AST I, II, and IV, but its possible identity with AST III has been neither confirmed or refuted. An N-hydroxy-2-acetylaminofluorene sulfotransferase which is active with phenols has also been purified to homogeneity from rat liver.[11] The enzyme shares many of the physical and catalytic properties of AST IV.[8,11]

In this paper we present evidence for the 3´-phosphoadenylylsulfate (PAPS) dependent sulfation of 1-naphthalenemethanol catalyzed by one isoenzyme of arylsulfotransferase, AST IV, and report the pH optimum for this reaction. A quantitative assay for 1-naphthalene methanol sulfation is also presented. This method is a modification of a methylene blue ion-pair extraction assay for steroid sulfation[12] which has also been used to determine phenol sulfation.[9] Evidence is also presented for the participation of 1-naphthalenemethanol in the AST IV catalyzed exchange reaction, wherein SO_3^- is transferred from 4-nitrophenyl sulfate to an acceptor in the presence of a catalytic amount of adenosine 3´,5´-bisphosphate (PAP).

MATERIALS AND METHODS

Hepatic arylsulfotransferase IV was purified from male Sprague-Dawley rats (BioLabs) using the following modifications of a previously published procedure:[8] 10% (v/v) glycerol and 1.0 mM dithiothreitol were added to all purification buffers, 2-mercaptoethanol was deleted from buffer mixtures, and an ammonium sulfate fractionation (30-50%) was substituted for the affigel blue chromatography step. Protein concentrations were determined[13] using bovine serum albumin as standard, and sulfotransferase assays were carried out with 2-naphthol as sulfate acceptor.[8] PAPS was prepared as described previously.[9] All other assay and buffer components were obtained from commercial sources.

Assays for determination of 1-naphthalenemethanol sulfation were conducted by modification of a methylene blue ion-pair extraction method described for 2-naphthol sulfation.[9] Reaction mixtures contained 0.25 mM 1-naphthalenemethanol, 0.2 mM PAPS, 5 mM 2-mercaptoethanol, arylsulfotransferase IV, and potassium phosphate (0.1 M, pH 7.0) in a total volume of 0.4 ml at 37°. Reactions were started by addition of enzyme and incubations were carried out for 10 min. Reactions were terminated by addition of 0.5 ml of methylene blue reagent[12] followed by 2 ml of chloroform. After mixing well, and centrifugation to separate the layers, the chloroform layer was removed, dried with 50-100 mg anhydrous sodium sulfate, and the absorbance determined at 651 nm.

Assay mixtures for the AST IV catalyzed exchange reaction contained 0.2 M potassium phosphate (pH 7.4), 0.4 μM adenosine 3´,5´-bisphosphate (PAP), 0.2 mM 4-nitrophenyl sulfate, 0.24 mg AST IV, and either 0.5 mM 1-naphthalenemethanol or 5 μM 2-naphthol in a final volume of 1.05 ml at 37°. 4-Nitrophenol formation was monitored at 400 nm.

1-Naphthalenemethyl sulfate was prepared by a published method,[14] and the chemical structure was confirmed by ^{13}C NMR spectroscopy.

RESULTS

The methylene blue ion-pair extraction assay for 2-naphthol sulfation was evaluated for determination of 1-naphthalenemethanol sulfation. A linear standard curve (r=0.999) described the dependence of the final absorbance of the ion-pair at 651 nm on the amount of 1-naphthalenemethyl sulfate (2.0 to 32 nmol) added to 0.4 ml water and carried through the extraction procedure described in MATERIALS AND METHODS. When 10 nmol of 1-naphthalenemethyl sulfate were carried through the extraction procedure, an absorbance of 0.29 was obtained. Efficiency of extraction of either 2 nmol or 10 nmol of 1-naphthalenemethyl sulfate added to standard assay mixtures is seen in Table 1. Although recovery of 1-naphthalenemethyl sulfate was acceptable with purified preparations of AST IV, extraction was not quantitative with rat liver 100,000xg supernatant in the assay.

The methylene blue assay procedure was used to determine the rate of sulfation of 1-naphthalenemethanol under a variety of conditions. The sulfation of 1-naphthalenemethanol was completely dependent on the presence of PAPS, 1-naphthalenemethanol, and AST IV in the reaction mixture, and deletion of any one of these resulted in no product formation. As seen in Table 2, the reaction was also inhibited by pentachlorophenol, an effective inhibitor of phenol sulfotransferases.[15]

Table 1. Recovery of 1-Naphthalenemethyl sulfate[a]

Assay Mixture	%Recovery	
	10 nmol	2 nmol
No Enzyme	102 ± 6	93 ± 4
Purified AST IV	96 ± 2	91 ± 9
Rat Liver Supernatant	81 ± 3	21 ± 9

[a]Assay mixtures were as described in MATERIALS AND METHODS, except that either 2 or 10 nmoles of 1-naphthalenemethyl sulfate was substituted for 1-naphthalenemethanol. AST IV was purified through the ATP agarose chromatography step, and 15 µg was added to assay mixtures. Rat liver 100,000xg supernatant (0.4 mg protein per assay was added as indicated. Values are the mean ± standard deviation of three determinations.

Table 2. Pentachlorophenol Inhibition of 1-Naphthalenemethanol
          Sulfation

Assay Mixture[a]	Sulfation Rate (nmol/min/mg protein)
Complete	19.4
+10 μM Pentachlorophenol	1.4

[a]Assays were conducted at pH 7.0 and 37° as described in
MATERIALS AND METHODS.

Since previous studies[8] on the specificity of AST IV indicated that
the pH optimum varies with the sulfate acceptor used, the effect of pH on
the sulfation of 1-naphthalenemethanol was compared to pH effects with
2-naphthol as sulfate acceptor. As seen in Figure 1, the pH optimum for
1-naphthalenemethanol sulfation was pH 6.5 - 7.0. Optimal AST IV activity

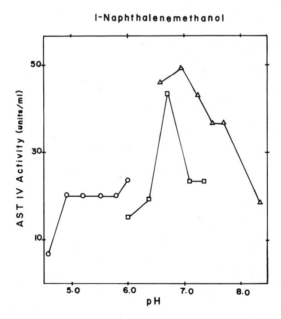

Fig. 1. Optimum pH for 1-Naphthalenemethanol
         Sulfation. Assays were conducted as
         described in MATERIALS AND METHODS with
         the following buffer modifications:
         O——O , 0.1 M sodium acetate;
         □——□ , 0.1 M sodium succinate;
         △——△ , 0.1 M sodium phosphate.

418

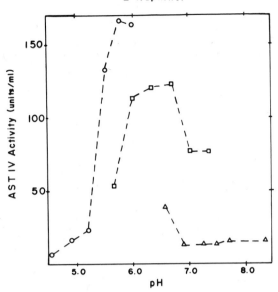

Fig. 2. Optimum pH for 2-Naphthol sulfation.
Assays were conducted as described in
MATERIALS AND METHODS with the following
buffer modifications: O - - O , 0.1 M
sodium acetate; □- - -□ , 0.1 M sodium
succinate; △ - - - △ , 0.1 M sodium phosphate.

with 2-naphthol as sulfate acceptor was obtained at pH 5.7 in sodium acetate,
and pH 6.0 - 6.7 in sodium phosphate (Figure 2).

A homogeneous preparation of AST IV was used to compare the rate of
sulfation of 1-naphthalenemethanol with the rate of 2-naphthol sulfation at
pH 5.5 and 7.0 (Table 3). While the sulfation rate is comparable at pH 7.0
with the two acceptors, there is a difference of over 100-fold in the rates
of reaction at pH 5.5.

1-Naphthalenemethanol sulfation rate was also compared to 2-naphthol
sulfation at intermediate and final steps of AST IV purification. As seen
in Table 4, catalytic activity with 1-naphthalenemethanol as sulfate acceptor
increased in approximate proportion to the increase (6-7 fold) in activity
with 2-naphthol as sulfate acceptor during purification. DEAE-cellulose
chromatography is the purification step in which AST I and II are removed;
ATP agarose chromatography is the final step in purification of AST IV.

In addition to serving as a sulfate acceptor in the AST IV catalyzed
reaction requiring PAPS, 1-naphthalenemethanol also participated in a
sulfate exchange reaction similar to that originally described by Gregory
and Lipmann using 4-nitrophenyl sulfate as donor and phenol as acceptor.[16]
The reaction requires a catalytic amount of PAP and is easily followed

Table 3. Sulfation of 1-Naphthalenemethanol and 2-Naphthol
Catalyzed by Homogeneous Arylsulfotransferase IV

Sulfate Acceptor[a]	Sulfation Rate[b]	
	pH 5.5	pH 7.0
1-Naphthalenemethanol	4.7	34
2-Naphthol	610	46

[a]Reaction mixtures contained 0.25 mM sulfate acceptor, 0.2 mM
PAPS, 5 mM 2-mercaptoethanol, 20 µg AST IV, and either sodium
acetate (0.25 M, pH 5.5) or potassium phosphate (0.1 M, pH 7.0)
in a total volume of 0.4 ml at 37°, as described in MATERIALS
AND METHODS.

[b]Rates are expressed as nmol product formed per min per mg
protein.

by observing the absorbance of 4-nitrophenol at 400 nm. The reaction
involving 1-naphthalenemethanol is as follows:

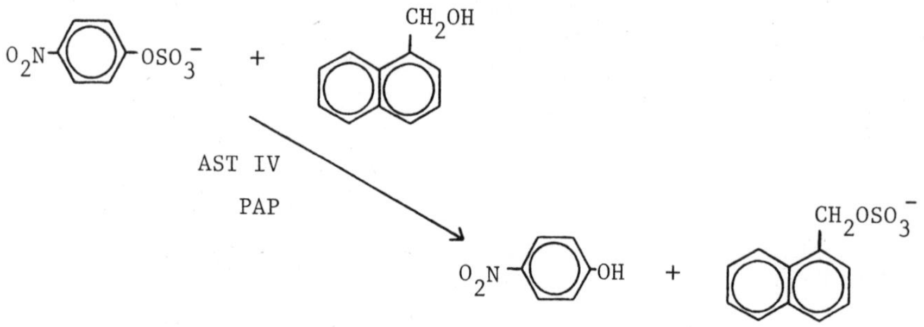

Using 1-naphthalenemethanol as acceptor, with reaction conditions as
described in MATERIALS AND METHODS, the rate of the AST IV catalyzed
exchange reaction was 0.45 ± 0.02 (n=3) nmol 4-nitrophenol formed per

Table 4. Sulfation of 2-Naphthol and 1-Naphthalenemethanol After
Intermediate and Final AST IV Purification Steps

Purification Step	Sulfate Acceptor	
	2-Naphthol pH 5.5	1-Naphthalenemethanol pH 7.0
DEAE-Cellulose	90	6
ATP-Agarose	612	34

[a]Values represent rates of sulfation (nmol product/min/mg protein)
of the given substrate at each pH using assay conditions given in
MATERIALS AND METHODS.

min per mg of AST IV. The corresponding rate using the same AST IV preparation and 2-naphthol as acceptor was 2.8 nmol 4-nitrophenol formed per min per mg of AST IV.

DISCUSSION

Arylsulfotransferase IV, purified to homogeneity from rat liver, catalyzed the sulfation of 1-naphthalenemethanol in a reaction dependent on PAPS, enzyme, and 1-naphthalenemethanol, and inhibited by pentachlorophenol. The reaction was monitored using a methylene blue ion-pair extraction assay which was quantitative for partially purified and highly purified enzyme preparations. However, this assay procedure did not permit quantitative determination of 1-naphthalenemethyl sulfate when used with 100,000xg rat liver supernatant fractions. Whether this was due to binding of the 1-naphthalenemethyl sulfate to cytosol protein, glutathione transferase catalyzed conjugation of the 1-naphthalenemethyl sulfate, or to other causes for poor extraction, remains to be determined.

The pH optimum for AST IV catalyzed sulfation of 1-naphthalenemethanol was 6.5 - 7.0. This is similar to the pH optimum (pH 6.4 in sodium phosphate) displayed by the N-hydroxy-2-acetylaminofluorene sulfotransferase,[11] but is at a higher pH than the optimum for 2-naphthol sulfation obtained in this and previous[8] studies on AST IV.

Finally, 1-naphthalene methanol participates in the SO_3^- exchange reaction catalyzed by AST IV and PAP. Based on the previously described mechanism of the exchange reaction,[17] the reaction involving 1-naphthalene-methanol most likely proceeds via the following two steps:

1.) O_2N—⬡—OSO_3^- + PAP ⇌ O_2N—⬡—OH + PAPS

2.) PAPS + [naphthalene-CH_2OH] ⇌ PAP + [naphthalene-$CH_2OSO_3^-$]

The catalytic concentration of PAP is regenerated in the second reaction. While this reaction might prove useful as a qualitative assay, its use in quantitation of AST IV may be very difficult due to the same inhibition effects observed when phenols are used as acceptors.[17] Thus, the relative rates of 2-naphthol and 1-naphthelenemethanol sulfation determined with the methylene blue procedure (Table 4) are more informative than differences detected in the exchange reaction.

ACKNOWLEDGEMENTS

We gratefully acknowledge the excellent technical assistance of Mr. Brian Anderson. This research was supported by National Institutes of Health grants RR09222 and CA38683.

REFERENCES

1. U. Bicker and W. Fischer, Enzymatic aziridine synthesis from β-amino alcohols - a new example of endogenous carcinogen formation, Nature 249:344 (1974).

2. P. G. Wislocki, P. Borchert, J. A. Miller, and E. C. Miller, The metabolic activation of the carcinogen 1′-hydroxysafrole in vivo and in vitro and the electrophilic reactivities of possible ultimate carcinogens, <u>Cancer Res.</u> 36:1686 (1976).

3. T. Watabe, T. Ishizuka, M. Isobe, and N. Ozawa, A 7-hydroxymethyl sulfate ester as an active metabolite of 7,12-dimethylbenz[a]-anthracene, <u>Science</u> 215:403 (1982).

4. T. Watabe, T. Ishizuka, M. Isobe, and N. Ozawa, Conjugation of 7-hydroxymethylbenz[a]anthracene (7-HMBA) with glutathione via a sulphate ester in hepatic cytosol, <u>Biochem. Pharmacol.</u> 31:2542 (1982).

5. E. W. Boberg, E. C. Miller, J. A. Miller, A. Poland, and A. Liem, Strong evidence from studies with brachymorphic mice and penta-chlorophenol that 1′-sulfooxysafrole is the major ultimate electro-philic and carcinogenic metabolite of 1′-hydroxysafrole in mouse liver, <u>Cancer Res.</u> 43:5163 (1983).

6. G. L. Kedderis, M. C. Dyroff, and D. E. Rickert, Hepatic macromolecular covalent binding of the hepatocarcinogen 2,6-dinitrotoluene and its 2,4-isomer <u>in vivo</u>: modulation by the sulfotransferase inhibitors pentachlorophenol and 2,6-dichloro-4-nitrophenol, <u>Carcinogenesis</u> 5:1199 (1984).

7. D. E. Rickert, R. M. Long, M. C. Dyroff, and G. L. Kedderis, Hepatic macromolecular covalent binding of mononitrotoluenes in Fischer-344 rats, <u>Chem. Biol. Interact.</u> 52:131 (1984).

8. R. D. Sekura and W. B. Jakoby, Aryl sulfotransferase IV from rat liver, <u>Arch. Biochem. Biophys.</u> 211:352 (1981).

9. R. D. Sekura and W. B. Jakoby, Phenol sulfotransferases, <u>J. Biol. Chem.</u> 254:5658 (1979).

10. R. T. Borchardt and C. S. Schasteen, Phenol sulfotransferase. I. Purification of a rat liver enzyme by affinity chromatography, <u>Biochim. Biophys. Acta</u> 708:272 (1982).

11. S.-C. G. Wu and K. D. Straub, Purification and characterization of N-hydroxy-2-acetylaminofluorene sulfotransferase from rat liver, <u>J. Biol. Chem.</u> 251:6529 (1976).

12. Y. Nose and F. Lipmann, Separation of steroid sulfokinases, <u>J. Biol. Chem.</u> 233:1348 (1958).

13. A. Bensadoun and D. Weinstein, Assay of proteins in the presence of interfering materials, <u>Analyt. Biochem.</u> 70:241 (1976).

14. J. J. Clapp and L. Young, Formation of mercapturic acids in rats after the administration of aralkyl esters, <u>Biochem. J.</u> 118:765 (1970).

15. G. J. Mulder and E. Scholtens, Phenol Sulphotransferase and Uridine Diphosphate Glucuronyltransferase from Rat Liver <u>in vivo</u> and <u>in vitro</u>, <u>Biochem. J.</u> 165:553 (1977).

16. J. D. Gregory and F. Lipmann, The transfer of sulfate among phenolic compounds with 3′,5′-diphosphoadenosine as coenzyme, <u>J. Biol. Chem.</u> 229:1081 (1957).

17. M. W. Duffel and W. B. Jakoby, On the mechanism of aryl sulfotrans-ferase, <u>J. Biol. Chem.</u> 256:11123 (1981).

ACTIVATED PHASE II METABOLITES: COMPARISON OF ALKYLATION BY 1-O-ACYL
GLUCURONIDES AND ACYL SULFATES

Richard B. van Breemen, Catherine C. Fenselau, and
Deanne M. Dulik

Department of Pharmacology, School of Medicine
Johns Hopkins University, Baltimore, MD 21205

ABSTRACT

1-O-acyl glucuronides are reactive Phase II metabolites which can
alkylate chemical nucleophiles. Industrial sulfate ester mixed anhydrides
have been reported to be active acylating agents. This study was undertaken
in order to establish that sulfate ester mixed anhydrides of clinically
useful drugs could be synthesized, purified, and characterized as reactive
chemical species. Their ability to alkylate 4-(p-nitrobenzyl)pyridine (NBP)
and their stability in aqueous solution was compared with 1-O-acyl
glucuronide conjugates of the same drugs. Synthesis of the 1-O-acyl
glucuronides of the hypolipidemic agent, clofibric acid, and the
nonsteroidal antiiflammatory drugs flufenamic acid and indomethacin were
catalyzed by immobilized microsomal rabbit liver UDP-glucuronyltransferase.
Potassium salts of the sulfate ester mixed anhydrides of these drugs were
synthesized chemically by temperature-controlled reaction with
chlorosulfonic acid in anhydrous pyridine. Half-lives at pH 2.0, 6.0, 7.4,
and 10.0 were determined for each compound. The reactivity of the acyl
glucuronides and sulfate ester mixed anhydrides towards the standard
chemical nucleophile, 4-(p-nitrobenzyl pyridine (NBP), was measured using a
spectrophotometric assay at several substrate concentrations. Acyl sulfate
ester mixed anhydrides were shown to be 3-20 times more reactive towards NBP
than their corresponding 1-O-acyl glucuronides. For both glucuronides and
sulfate esters, relative reactivitiy towards NBP was: clofibric acid >
indomethacin > flufenamic acid. This behavior paralleled the hydrolytic
instability of the compounds.

INTRODUCTION

Many xenobiotic and endogenous carboxylic acids undergo Phase II
conjugation to form 1-O-acyl glucuronides. These metabolites are reactive
intermediates which have been shown to alkylate both biological (1,2) and
chemical (3) nucleophiles. This reaction proceeds via nucleophilic attack
on the acyl carbon of the ester bond with subsequent loss of glucuronic
acid. The resulting alkylated biopolymer formed in vivo by this mechanism
could have altered biological activity and potential toxicity.

Carboxylic acids may also be postulated to form sulfate ester mixed
anhydrides as a product of Phase II metabolism. The compounds have been
shown to be powerful acylating agents (4) in industrial applications.
However, their existence in biological matrices has not been documented,

423

perhaps due to their hydrolytic instability during routine metabolite isolation procedures. Nevertheless, these relatively unstable compounds would be expected to undergo alkylation reactions analogously to acyl glucuronides in a mechanistic sense, since sulfate anion serves as an excellent leaving group.

The nonsteroidal antiinflammatory drugs, flufenamic acid and indomethacin, and the hypolipidemic agent, clofibric acid, were chosen for a comparative study of the alkylating ability of acyl glucuronides and acyl sulfate ester mixed anhydrides. These drugs, which are shown in Figure 1, were chosen because their major human metabolite is the acyl glucuronide.

Both the pH stability and chemical reactivity of these compounds were evaluated and compared. pH stability of the compounds was studied under acidic, basic and physiological conditions. Chemical reactivity was determined by reaction of each compound with a standard chemical nucleophile, 4-(p-nitrobenzyl)pyridine (NBP). NBP has been shown to react with alkylating agents to give a quaternary pyridinium salt. In the presence of base, this intermediate forms a blue color which may be quantitated spectrophotometrically (5,6).

MATERIALS AND METHODS

Chemicals. Flufenamic acid, indomethacin, and uridine-5'-diphospho-glucuronic acid (sodium salt) were supplied by Sigma Chemical Company (St. Louis, MO). Clofibric acid, 4-(p-nitrobenzyl)pyridine, triethylamine, chlorosulfonic acid, and potassium hydrogen carbonate were obtained from Aldrich Chemical Company (Milwaukee, WI). Ethylene glycol was purchased from J.T. Baker Chemical Company (Phillipsburg, NJ). Tetrahydrofuran from Burdick and Jackson (Muskegon, MI) was distilled from lithium aluminum hydride before use.

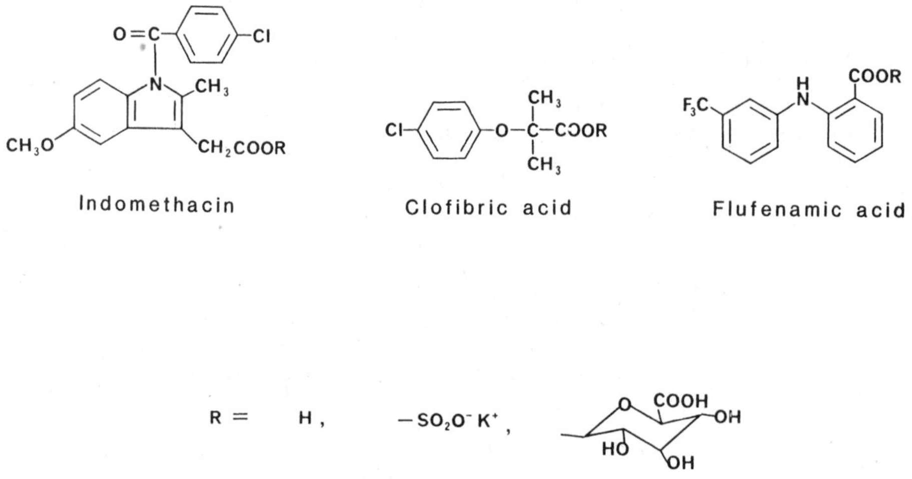

Indomethacin

Clofibric acid

Flufenamic acid

$$R = \quad H, \quad -SO_2O^-\ K^+,$$

Fig. 1. Drugs Used in This Study

Preparation of glucuronides and sulfate ester mixed anhydrides.
1-0-acyl glucuronides of clofibric acid, flufenamic acid, and indomethacin were prepared by reaction with uridine-5'-diphosphoglucuronic acid (UDPGA) which was catalyzed by partially purified rabbit liver UDP-glucuronyltransferase (7,8). Products were purified by reversed-phase Sep-Pak (Waters Associates Inc., Milford, MA) extraction followed by reversed-phase HPLC (3). Structural confirmation was obtained by TLC and FAB mass spectrometry. Yields as determined with radiolabelled aglycones were: clofibric acid glucuronide, 3%; flufenamic acid glucuronide, 7%; indomethacin glucuronide, 28%.

The corresponding sulfate ester mixed anhydrides were chemically synthesized by a modification of. the method of Feigenbaum and Neuberg (9,10). For example, to a flame-dried 3-necked round bottom flask equipped with nitrogen inlet and magnetic stir bar was added anhydrous pyridine (10 ml). After cooling to -70°C in a dry ice-isopropanol bath, chlorosulfonic acid (0.26 g, 2.3 mmol) was added via syringe to produce pyridine-SO_3 as a white solid. A solution of indomethacin (0.5 g, 1.4 mmol) in anhydrous pyridine (5 ml) was then added and the reaction allowed to proceed from -70°C to -30°C over 2.5 hr. The reaction was quenched with potassium hydrogen carbonate (0.3 g, 3.0 mmol) in H_2O (1 ml) with evolution of CO_2 and formation of the corresponding potassium salt of the sulfate ester. The reaction mixture was evaporated to dryness, then taken up in methanol (10ml), centrifuged to remove KCl, and purified by normal phase Sep-Pak extraction. The hygroscopic product was dried under vacuum for several hours at room temperature. Quantitated sulfate esters were stored in buffered aqueous solution in the freezer to retard autohydrolysis (11). Yields were: clofibric acid, 1%; flufenamic acid, 4%; indomethacin, 2%. Purity was determined by TLC, reversed-phase HPLC, and FAB mass spectrometry.

Determination of pH stability and half-lives.
Aliquots (20 ul) of the original buffered solution of sulfate ester were diluted to 1.0 ml using the appropriate pH phosphate buffer (pH 2.0, 6.0, 7.4, and 10.0). Each solution was placed in a 37°C water bath, and aliquots (20 ul) were taken at various time intervals depending on the substrate and analyzed by reversed-phase HPLC. HPLC conditions for sulfate esters were: ODS-Ultrasphere column, (Beckman), 150 x 4.6mm, 5um; isocratic mobile phase, $H_2O:CH_3CN$ (80:20); flow rate 1.0 ml/min, and UV detection at 254 nm. Retention time for sulfate esters was about 8 min, and for the corresponding free acid, about 24 min. Loss of conjugate was measured using peak height for the glucuronides and peak area for the sulfate esters (due to poorer peak shape in this case). pH stability determinations were done in duplicate, and half-lives were calculated by linear regression analysis.

NBP Alkylaton studies.
The method of van Breemen and Fenselau (5) was employed. Standard solutions of glucuronide or sulfate ester mixed anhydride (0.93mM to 1.55 mM) in phosphate buffer (0.067 M, pH 6.0) were used. Reactions were done using amber vials (4 ml) with teflon-lined screw caps under argon atmosphere. Standard reaction mixtures contained aqueous potassium phosphate (0.067 M, pH 6.0, 250 ul), NBP (2% w/v in ethylene glycol, 500 ul), and tetrahydrofuran (250 ul). The glucuronide standard solution was prepared in tetrahydrofuran and the sulfate ester standard solution in aqueous phosphate buffer. Reaction volumes were reduced by a factor of 10 for the glucuronides. Vials were vortexed for 1 min, equilibrated to 60°C in a dark oven for 2 min, and then incubated at 60°C for 0, 20, 40 or 60 min. The reaction was quenched by freezing the sample vial in a dry ice-acetone bath. After thawing (15 min at room temp),

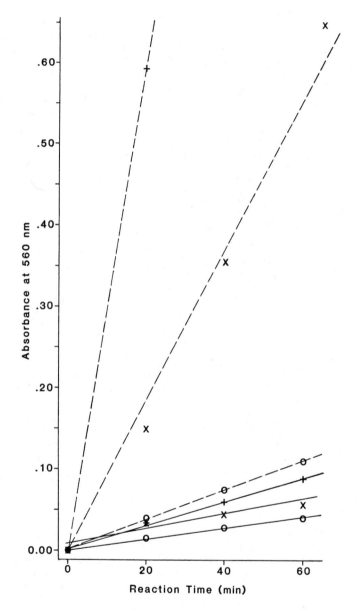

Fig. 2.   Reaction of NBP with Indomethacin (**x**), Clofibric (+) or

Flufenamic (**O**) Glucuronide (——) or Sulfate (— —) 1.55mM triethylamine (1:1 v/v in acetone, 1000 ul) was added to each vial under nitrogen with subsequent development of blue color due to alkylated NBP. Spectrophotometric absorbance at 560 nm was determined. Control NBP experiments were done using the free carboxylic acids. No alkylation was observed to occur for any of these controls.

RESULTS

The half-lives of 1-O-acyl glucuronides and sulfate ester mixed anhydrides of flufenamic acid, indomethacin, and clofibric acid are shown in Table 1. Acyl glucuronides were in general much more stable in buffered aqueous solution than the corresponding sulfate esters. The glucuronides were found to be stable in acidic media (pH 2 and 6) and very labile under basic conditions, which is consistent with literature observations (12). For both its glucuronides and sulfates, flufenamic acid was found to be the most stable conjugate towards hydrolysis. Clofibric acid sulfate ester was by far the least stable, with a half-life of less than one hour at physiological pH, compared to 21 hours for flufenamic acid sulfate ester.

The relative rate constants for NBP alkyation of these compounds are given in Table 2. The rate determined for chlorambucil, a highly reactive nitrogen mustard antineoplastic agent, is shown for purposes of comparison. The sulfate ester mixed anhydrides were found to be 3-20 times more reactive towards NBP than the corresponding 1-O-acyl glucuronides. The most reactive of the sulfates, clofibric acid, was only seven times less reactive than chlorambucil. For the sulfate esters, the order of the NBP reactivity paralleled their instability in aqueous solution: clofibric acid > indomethacin > flufenamic acid. The same relative order of NBP reactivity was observed for the glucuronides. The rate of reactivity towards NBP for these compounds is illustrated in Figure 2.

DISCUSSION

Sulfate ester mixed anhydrides have not been reported as drug metabolites to date. However, known literature methods for their preparation could be employed, even though they were found to be unstable in acidic, alkaline, or unbuffered systems. Because of autohydrolysis these sulfate esters could not be prepared as free hydrogen sulfates. Therefore, they were stabilized and isolated as potassium salts. Analysis by fast atom bombardment (FAB) mass spectrometry produced protonated molecular ions $[MK+H]^+$.

The sulfate ester mixed anhydrides of these three drugs were shown to be much more reactive towards NBP than the corresponding glucuronides. This increased reactivity could be explained in two ways: (a) the sulfate anion is a better stabilized leaving group than is glucuronic acid, and (b) the carbonyl carbon of the anhydride bond is much more electron deficient and therefore more susceptible to nucleophilic attack. It is important to note that the chemical reactivity of both the glucuronides and sulfates is conferred by these functional groups, since the free carboxylic acids show no reaction with NBP.

Based on this study of the chemical, chromatographic, and spectroscopic characteristics of these sulfate ester mixed anhydrides, efforts are now under way to test their enzymatic formation in vitro and in vivo.

ACKNOWLEDGEMENTS

We gratefully acknowledge the support of the U.S. Public Health Service Grant GM-21248 and the National Institutes of Health Grant CA09243.

Table 1. Half Lives of 1-0-Acyl Glucuronides and Sulfate Ester Mixed
Anhydrides in Aqueous Solution at 37°C

Drug	pH	1-0-Acyl Glucuronide	Sulfate Ester
Flufenamic	2.0	1109.0 hr	< 1. min
Acid	6.0	70.7 hr	13.4 hr
	7.4	7.0 hr	21.2 hr
	10.0	4.8 min	14.6 min
Indomethacin	2.0	71.1 hr	< 1. min
	6.0	21.0 hr	5.4 hr
	7.4	1.4 hr	7.7 hr
	10.0	< 1 min	< 1 min
Clofibric	2.0	124.0 hr	< 1 min
Acid	6.0	48.4 hr	54 min
	7.4	7.3 hr	41 min
	10.0	< 2 min	< 1 min

Table 2. Relative Rate Constants for Alkylation of NBP.

	min^{-1}.
Flufenamic 1-0-acyl glucuronide	0.436 ± 0.135
Indomethacin 1-0-acyl glucuronide	0.709 ± 0.081
Clofibric 1-0-acyl glucuonide	1.08 ± 0.23
Flufenamic acid sulfate ester mixed anhydride	1.15 ± 0.22
Indomethacin sulfate ester mixed anhydride	4.81 ± 0.15
Clofibric acid sulfate ester mixed anhydride	19.10 ± 3.3
Chlorambucil	137 ± 6

REFERENCES

1. van Breemen, R.B., and Fenselau, C., 1985, Acylation of Albumin by 1-O-Acyl Glucuruonides, Drug Metab. Disp., 13:318.

2. Boudinot, F.D., Homon, C.A., Jusko, W.J. and Ruelius, H.W., 1985, Protein Binding of Oxazepam and its Glucuronide Conjugates to Human Albumin, Biochem. Pharmacol., 34: 2115.

3. Stogniew, M., and Fenselau, C. 1982, Electrophilic Reactions of Acyl-Linked Glucuronides, Drug. Metab. Disp., 10:609.

4. Overberger, C.G. and Sarlo, E., 1963, Mixed Sulfonic-Carboxylic Anhydrides, J. Am. Chem. Soc., 85: 2446.

5. van Breemen, R.B., and Fenselau, C., Reaction of 1-O-Acyl Glucuronides with 4-(p-Nitrobenzyl)pyridine (NBP), submitted.

6. Epstein, J., Rosenthal, R.W., and Ess, R.J., 1955, Use of γ--(4-Nitrobenzyl) pyridine as an Analytical Reagent for Ethyleneimines and Alkylating Agents, Anal. Chem., 27: 1435.

7. Parikh, I. MacGlashan, D., and Fenselau, C., 1976, Immobilized Glucuronosyltransferase for the Synthesis of Conjugates, J. Med. Chem., 19:296.

8. Fenselau, C., Pallante, S., and Parikh, I., 1976, Solid-Phase Synthesis of Drug Glucuronides by Immobilized Glucuronosyltransferase, J. Med. Chem., 19:679.

9. Feigenbaum, J.; and Neuberg, C.A., 1941, Simplified Method for the Preparation of Aromatic Sulfuric Acid Esters, J. Am. Chem. Soc., 63: 3529.

10. Dulik, D.M., 1984, The Phase II Metabolic Fate of 1,2,3-Benzenetriol (Pyrogallol), Diss. Abstr. Int. B., 45: 1202.

11. Benkovic, S.J. and Dunikoski, L.K., 1970, Intramolecular catalysis of sulfate ester hydrolysis, Biochemistry, 9:1390.

12. Mulder, G.J., 1981, "Sulfation of Drugs and Related Compounds," CRC Press, Boca Raton, FLA.

13. Suter, C.M., 1944, "The Organic Chemistry of Sulfur," John Wiley and Sons, Inc., New York, NY.

REACTIONS OF OXAPROZIN-1-0-ACYL GLUCURONIDE IN SOLUTIONS OF HUMAN PLASMA AND ALBUMIN

H.W. Ruelius[1], S.K. Kirkman, E.M. Young, and F.W. Janssen[2]

Research Division, Wyeth Laboratories Inc.

INTRODUCTION

Acyl glucuronides are major metabolites of many carboxylic acids including drugs, various chemicals, as well as endogenously formed compounds such as bilirubin and benzoic acid. Because of their susceptibility to nucleophilic attack acyl glucuronides are generally less stable than other glucuronides (1). Two reactions, hydrolysis and rearrangement (isomerization by acyl migration), contribute to this instability. Both reactions can occur at physiologic pH values. The extent to which they occur depends on the reactivity of the acyl glucuronide which in turn depends on the structure of its aglycone (2,3).

The reactivity of unstable acyl glucuronides in the two reactions at pH values near neutrality suggests that both reactions may contribute to the in vivo disposition of the parent drug. Indeed, in vivo hydrolysis of the glucuronide of clofibric acid has been proposed as the explanation of the prolonged half-life of this drug in renal patients (4,5) whereas different susceptibilities to rearrangement have been found to contribute to the dissimilar fate of two antiinflammatory agents in the rat (3).

Acyl glucuronides also react with albumin resulting in the covalent binding of the aglycone. This reaction has been reported for zomepirac acyl glucuronide (6), for conjugated bilirubin (7,8) and for the acyl glucuronides of benoxaprofen, indomethacin, flufenamic acid, and clofibric acid (9). Thus, acyl glucuronides are potentially reactive metabolites.

In the present study we examined the reactions of oxaprozin-1-0-acyl glucuronide and its C-2 isomer in human plasma and solutions of human serum albumin. Although this glucuronide has not been detected in the plasma of any species following administration of oxaprozin, it was chosen for these model reactions because it has intermediate reactivity with respect to

[1] Present address: Johns Hopkins University, School of Medicine, Dept. of Pharmacology, Biophysics Bldg. B-7, 725 North Wolfe Street, Baltimore, MD 21205.
[2] Send reprint requests to: Mr. Frank Janssen, Research Division, Wyeth Laboratories, Inc., P.O. Box 8299, Philadelphia, PA 19101.

hydrolysis and rearrangement (3). The structure of oxaprozin glucuronide is shown below:

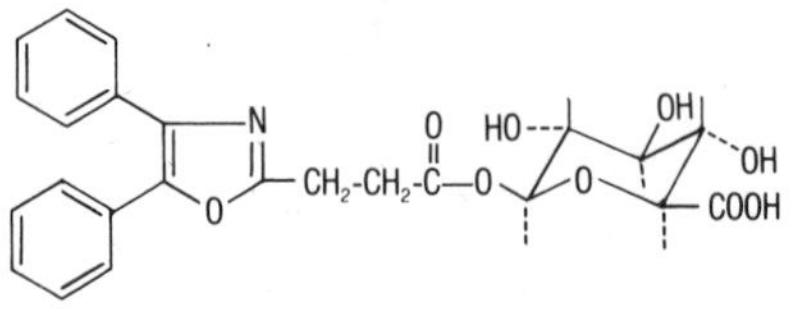

This glucuronide is a major urinary excretion product in man and rhesus monkey (10,11) and practically the only drug related entity in the bile of the dog (12).

MATERIALS AND METHODS

Materials. ^{14}C-oxaprozin (s.a. 0.116 µCi/mmol) and ^{14}C-UDPGA (uridine diphosphoglucuronic acid; s.a.254 µCi/mmol) were purchased from ICN Corp (Irvine, CA); UDPGA and various human albumins from Sigma Chemical Co. (St. Louis, MO); naproxen was obtained from Syntex Laboratories, Inc. (Palo Alto, CA); aspirin from Mallinckrodt Inc. (Paris, KY); liquid scintillation fluid, Hydrofluor®, from National Diagnostics (Somerville, NJ); and normal human plasma from the heparinized blood of healthy volunteers. Acetylated albumin was prepared using p-nitrophenyl acetate (13,14). It was subjected to ultrafiltration and dialysis against 0.02M acetic acid to remove p-nitrophenol. Its concentration in reactions was the same as albumin.

Preparation of oxaprozin glucuronide and its C-2 isomer. Oxaprozin 1-0-acyl glucuronide, labeled in either the oxaprozin or the glucuronide moiety was biosynthesized by an enzyme reaction. A crude glucuronyl transferase solution was made from the liver of a male rhesus monkey by centrifuging the 10% homogenate prepared in 1.15% KCl containing 5 mM Tris, pH 7.4, at 9000 x g. The reaction was carried out as follows in a final volume of 4 ml. To 1.5 ml of 0.1 M potassium phosphate buffer, pH 6.0, was added 0.4 ml distilled water, 0.2 ml MgCl$_2$ (0.1 M), 0.1 ml oxaprozin solution (2.3 mg/ml 0.1 M Tris buffer, pH 8.5), and 1.6 ml enzyme solution (4 mg protein/ml). After pre-incubation with shaking at 37°C for 5 min, 0.2 ml UDPGA solution (34.8 mg/ml in distilled water) was added. Following a 1 hour reaction, 8 ml of acidic CH$_3$CN (1% H$_3$PO$_4$ in CH$_3$CN) was added and after 5 min the mixture was centrifuged. The supernate was concentrated under N$_2$ to about 5 ml, centrifuged, and the glucuronide was isolated from the mixture by HPLC essentially as described for Wy-18,251 glucuronide (2). The yield of glucuronide was about 7.5% of the added oxaproxin, i.e. 27 µg/reaction. The characterization and proof of structure of the glucuronide has been described (11). The C-2 isomer of oxaprozin glucuronide was prepared by rearrangement of the 1-0-acyl glucuronide in pH 9.5 buffer for 100 min, and the isomer was isolated by HPLC as described for the glucuronide.

Reactions of oxaprozin glucuronide and its C-2 isomer with plasma or albumin. In vitro reactions were carried out in 6x50 mm test tubes at 37°C. The ingredients were added in the order: buffer (0.1M or 0.2M) potassium phosphate of pHs 6.0, 6.5, 7.0, or 7.5, suitably buffered plasma or albumin solution (fatty acid free, fraction V, final conc. 6.7-27 mg/ml), and glucuronide or C-2 isomer (final conc 16 µg/ml). A concentration of 6.7 mg/ml albumin resulted in a 2.8 fold molar excess of albumin/substrate and

432

this was used in all experiments except where noted. The final reaction volume varied from 30-375 μl. Solutions were sampled in duplicate at zero time and at various times up to 1 hr. One aliquot was precipitated with 5 vols acidic acetone for determination of covalent binding and the other was treated with the acidic acetonitrile solution (5 volumes) for HPLC assay. All samples were centrifuged. The precipitates for ^{14}C determination were washed 3 times with acidic acetone, redissolved in water and assayed in 5 ml Hydrofluor. For the HPLC assays the supernates were diluted with 1.5 volumes of 0.05M NaH$_2$PO$_4$ before injection onto the column. All test samples were compared to suitable controls containing only buffer and glucuronide or C-2 isomer. In the inhibition studies the inhibitor was added just before the albumin solution.

Analyses. Radioactivity was determined in a Packard Tri-Carb Liquid Scintillation Spectrometer Model 3390 at efficiencies near 75%. HPLC analyses were done on a series of three 0S-032, RP-8, Spheri-5 cartridges (Brownlee Labs, Santa Clara, CA) at a flow rate of 0.35 ml/min at room temperature using an LDC constametric III G pump. The mobile phase was acetonitrile:0.05 M sodium dihydrogen phosphate buffer (26:74) and detection was at 280 mμ, by an LDC UVIII monitor. Automatic injection was accomplished with an IBM LC/9505 sample handler. Detector output was integrated with a Hewlett Packard 3390A reporting integrator. Retention times were 6.3, 6.6, 7.5 and 18.5 min for the glucuronide, C-3 isomer, C-2 isomer and oxaprozin, respectively.

RESULTS

Hydrolytic activity of human plasma and serum albumin. The hydrolysis of oxaprozin glucuronide in plasma adjusted to various pH values is shown in Fig. 1. At pH 6.0 about 10% of the added glucuronide is hydrolyzed within 1 hour; at pH 7.4 hydrolysis amounts to 34%, more than a 3 fold increase. In buffer solutions spanning the same range of pH values hydrolysis increases from nil at pH 6.0 to less than 5% at pH 7.4. Thus, hydrolytic cleavage of oxaprozin glucuronide is about 7 times faster in plasma than in buffer of pH 7.4.

The hydrolysis of oxaprozin glucuronide and one of its rearrangement products (C-2) in albumin solutions of pH 7.0 is shown in Fig. 2. The two substances are hydrolyzed at about the same rate. Also indicated in Fig. 2 is the extent of hydrolysis of these two substrates in plasma. These data indicate that the hydrolytic activity in plasma is fully accounted for by the

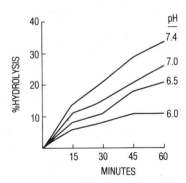

Fig. 1.   Hydrolysis of oxaprozin glucuronide in human plasma buffered 1:10. Hydrolysis (oxaprozin formed) is corrected for that observed in 0.1 M buffer alone of the same pH.

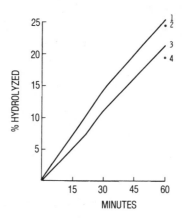

Fig. 2.   Hydrolysis of oxaprozin glucuronide and its C-2 isomer by
          human albumin (final conc. 27 mg/ml) and plasma at pH 7.0.
          The hydrolysis (oxaprozin formed) is corrected for that
          observed in 0.1 M buffer alone. Curve 1 = C-2 isomer in
          albumin. Curve 3 = oxaprozin glucuronide in albumin. Point
          2 = C-2 isomer in plasma. Point 4 = oxaprozin glucuronide in
          plasma.

albumin. Subsequent experiments show that reaction mixtures containing as
little as 5 mg/ml of albumin are about as effective as 40 mg/ml, the
approximate concentration of albumin in plasma (data not shown).

    Rearrangement (isomerization) of oxaprozin glucuronide in albumin
solutions. Table 1 shows the effect of pH in buffer and albumin solutions on
the rearrangement of oxaprozin glucuronide as a function of time. Comparison
of two solutions of the same pH indicates the extent of enhancement by
albumin, i.e. the net catalytic effect of albumin on the rearrangement
reaction (last column).

    In both buffer and albumin solutions rearrangement increases with time
and hydroxyl ion concentration but at different rates. Thus, the reaction
attributable to pH alone increases from negligible values at pH 6 to over
one-half of the total at pH 7.5. As a result the net catalytic effect of
albumin, i.e. the rearrangement in excess of that observed in buffer of the
same pH, is less pronounced at higher pH values.

    Covalent binding of oxaprozin to albumin. Fig. 3 presents the time
course of covalent binding of oxaprozin observed when oxaprozin glucuronide is
incubated with albumin solutions of various pH values. Generally, covalent
binding increases with time and pH, but at pH 7.5 it reaches a plateau after
45 minutes.

    Table 2 lists the percentage of radioactivity covalently bound when
various ^{14}C-labeled substrates are incubated with albumin. Extensive
incorporation of radioactivity occurs when the ^{14}C-label is in the oxaprozin
moiety of oxaprozin glucuronide. In contrast, when the glucuronic acid moiety
is labeled little uptake of radioactivity is observed, indicating that it is
the oxaprozin moiety which is covalently bound from oxaprozin glucuronide.
Little radioactivity is incorporated from labeled unconjugated oxaprozin or
the C-2 isomer of oxaprozin glucuronide.

434

Table 1. pH Dependent Time Course of Rearrangement of Oxaprozin Glucuronide in Buffer and Albumin Solutions[*]

pH	Sample Time Min	Buffer (0.2 M)	Albumin	Net
6.0	15	<2	3	< 2
	30	<2	8	< 7
	45	<2	11	<10
	60	2	14	12
6.5	15	<2	8	< 7
	30	2	13	11
	45	3	15	12
	60	6	21	15
7.0	15	4	12	8
	30	6	22	16
	45	9	34	25
	60	13	36	23
7.5	15	7	19	12
	30	12	34	22
	45	18	40	22
	60	24	40	16

[*] Percent of substrate rearranged as measured by respective areas.

Reactions of oxaprozin glucuronide with albumin. The time course of the three reactions (hydrolysis, rearrangement and covalent binding) of oxaprozin glucuronide with human serum albumin at pH 7.0 is shown in Fig. 4. At the end of the 60 minute incubation period, 20% of the added glucuronide is hydrolyzed, 36% is rearranged and about 22% is covalently bound as oxaprozin. Less than 25% is still present as unreacted glucuronide. In the absence of albumin, 13% of added glucuronide is rearranged (curve 4) while less than 5% is hydrolyzed (data not shown). The curves for rearrangement (curve 1) and covalent binding (curve 2) appear to plateau between 45 and 60 min whereas that for hydrolysis (curve 3) does not; it may even become a little steeper. This is plausible since the rearrangement products (the isomeric oxaprozin-

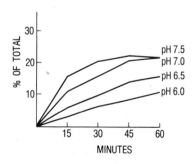

Fig. 3. Time course of covalent binding of oxaprozin (from its glucuronide) to human albumin at various pHs.

Table 2. Covalent Binding[*] of Oxaprozin to Human Albumin

Substrate	% Radioactivity Bound
^{14}C-Oxaprozin Glucuronide	22
Oxaprozin ^{14}C-Glucuronide	0.6
^{14}C-Oxaprozin C-2 Isomer	2.1
^{14}C-Oxaprozin	1.0

[*] Binding determined after 1 hr at pH 7.0.

glucuronic acid esters) are also subject to albumin catalyzed hydrolysis as shown above for the C-2 isomer.

Inhibition of covalent binding. The compounds that were tested for their ability to inhibit covalent binding of oxaprozin (from oxaprozin glucuronide) are listed in Table 3. Besides oxaprozin itself, naproxen, another non-steroidal antiinflammatory drug, almost completely inhibited covalent binding. On the other hand, aspirin had virtually no effect on this reaction. Decanoic acid is also a strong inhibitor.

The effect of these inhibitors is not restricted to the binding reaction. In the last column of Table 3 the percentages of oxaprozin glucuronide remaining at the end of the incubation period are listed . In the presence of strong inhibitors (oxaprozin, naproxen, decanoic acid) about 90% of the glucuronide is still present, in other words very little of any reaction (hydrolysis, rearrangement or covalent binding) has taken place.

Acetylation of albumin results in considerable reduction of all three reactions. Hydrolysis and rearrangement are reduced to about one-third and covalent binding to about one-half of the values obtained with untreated albumin (data not shown).

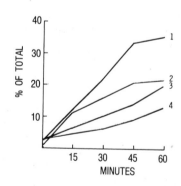

Fig. 4. Time course of hydrolysis and rearrangement of oxaprozin glucuronide and covalent binding of oxaprozin (from its glucuronide) on incubation with human albumin at pH 7.0. Curve 1 = Rearrangement of glucuronide in albumin. Curve 2 = Covalent binding of oxaprozin. Curve 3 = Hydrolysis of glucuronide in albumin. Curve 4 = Rearrangement of glucuronide in buffer. Hydrolysis in buffer was less than 5% throughout.

Table 3.   Inhibition of Covalent Binding of Oxaprozin (from
Oxaprozin Glucuronide) to Human Albumin

Inhibitor	Molar Ratio Inhibitor/Albumin	% Inhibition of Binding[a]	% Glucuronide Remaining[b]
None	-	-	22
Oxaprozin	1.2	94	91
Naproxen	1.5	96	94
Decanoic Acid	2.0	92	88
Aspirin	1.9	6	21

[a] Binding in the presence of inhibitors vs binding with
oxaprozin glucuronide alone, 1 hr, pH 7.0.
[b] % of 0.2M buffer control, 1 hr, pH 7.0

DISCUSSION

Three reactions between oxaprozin glucuronide and human serum albumin
have been investigated in the present study, namely hydrolysis, rearrangement
and covalent binding. The first two reactions occur also in aqueous
solutions in the absence of albumin but the rates are much slower at
identical pH values. Thus, the effect of albumin is catalytic. In the third
reaction, covalent binding, albumin becomes chemically modified. This
interaction has been recently described by other investigators for several
acyl glucuronides (6-9). It results in the acylation of albumin by the
aglycone of the respective glucuronides and, presumably, the liberation of
glucuronic acid.

The accelerated hydrolysis of oxaprozin glucuronides in human plasma
(Fig. 1) is entirely due to the albumin content of plasma i.e. there is no
esterase involved. The C-2 isomer of oxaprozin glucuronide is hydrolyzed at
about the same rate as the glucuronide (Fig. 2) however, the ethyl ester of
oxaprozin is not cleaved[1]. This suggests a certain specificity of the
catalytic site (as opposed to unspecific esterase-like activity). The
hydrolytic effect described here is different from the so-called esterase-
like activity of human serum albumin (15,16) which results in the acylation
of albumin by reactive esters i.e. in chemical modification of albumin and
which is therefore not a catalytic effect.

Rearrangement is quantitatively the most important reaction that takes
place within 1 hour (Fig. 4 and Table 1). However, the shape of curves 1
and 3 (Fig. 4) suggests that hydrolysis (oxaprozin formation) may become more
important as time goes on because the rearrangement products are also
hydrolyzed to oxaprozin while the substrate for rearrangement (the acyl
glucuronide) is being diminished. The faster degradation of zomepirac
glucuronide in blood and plasma (as compared to buffer) is probably due to
enhanced rearrangement in the presence of albumin (17).

Covalent binding occurs to a significant extent only with oxaprozin
glucuronide, not with its isomer or aglycone. Incorporation of radioactivity
from ^{14}C-oxaprozin glucuronide (but not from oxaprozin ^{14}C-glucuronide) is
evidence that only the oxaprozin moiety is covalently bound to albumin (Table
2). Covalent binding of oxaprozin increases with time and pH value and
reaches a maximum of about 22% of the added glucuronide after 45 min at pH
7.5 or 60 min at 7.0 (Fig. 3). Very little substrate is left at the end of

---

[1] F.W. Janssen, et al. unpublished observations.

the incubation time. It is therefore conceivable that an even greater extent
of acylation can be achieved by increasing the glucuronide to albumin ratio.
The extent of covalent binding to human serum albumin varies considerably
among various acyl glucuronides. For example, with triclomisole (Wy-18,251)
glucuronide it is much lower than with oxaprozin glucuronide[I], probably
because the former is more extensively hydrolyzed and rearranged. For
zomepirac glucuronide 3% binding has been reported at pH 7.4 (6).

Despite obvious differences, the three reactions between oxaprozin
glucuronide and albumin have certain features in common. They are pH
dependent, albeit not to the same extent. They can be considered to be
transacylation reactions in which 1) the acyl residue is transferred to the
hydroxyl ion (hydrolysis) regenerating the parent compound (the aglycone), 2)
the acyl residue migrates intramolecularly to an adjacent hydroxy group of
the glucuronic acid moiety (rearrangement) or 3) the acyl residue is
transferred to a reactive group on the albumin molecule resulting in covalent
binding. Most important, all three reactions are inhibited by the same
compounds (Table 3) suggesting a common "reactive site" on the albumin
molecule. Naproxen (and several other non-steroidal antiinflammatory agents)
as well as decanoic acid (and other short chain fatty acids) are known to
bind reversibly to a relatively specific site of human serum albumin (13,
18-20) which has been named the indole, and benzodiazepine binding site (21,
22). A highly reactive tyrosine residue has been shown to be an essential
part of this site (23-25). Chemical modification of this residue by several
reagents diminishes the binding capacity of albumin for ligands specifically
binding to the benzodiazepine site and, vice-versa, the presence of these
ligands inhibits the reactions acylating tyrosine (13, 23-25).

We therefore propose that the first step in the interactions between
oxaprozin glucuronide and human serum albumin is <u>reversible</u> binding of the
intact glucuronide to the benzodiazepine site (see scheme below). It has
been recently demonstrated that glucuronides can reversibly bind to this site
(14). Prevention of this first step by compounds that either reversibly bind
to the benzodiazepine site (e.g. naproxen, decanoic acid) or selectively
acetylate it (13) greatly reduces all three reactions. Aspirin which does
not significantly inhibit the reactions of oxaprozin glucuronide (Table 4)
binds only weakly to the benzodiazepine site (18,21). Non-defatted human
serum albumin fraction V reacts to a lesser extent with oxaprozin glucuronide
than fatty acid free fraction V (data not shown).

Scheme 1. Interactions Between Oxaprozin Glucuronide and Human Serum Albumin

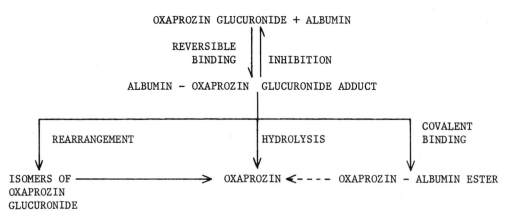

The scheme presents a comprehensive picture of all observed

438

reactions. The reaction leading to covalent binding of oxaprozin is analoguous to the acylation of albumin by reactive esters (13,15,16). For this reason an ester structure is assumed for covalently bound oxaprozin. We have not observed significant deacylation of the oxaprozin-albumin ester under physiological conditions. The corresponding zomepirac ester was found to persist in the systemic circulation for at least 48 hours. It was cleaved in vitro by strong alkali (6).

Oxaprozin glucuronide has never been detected in the plasma of any species and neither rearrangement products nor covalently bound oxaprozin was present in plasma of man, rhesus monkey or dog (10-12). Moreover, the covalent binding reaction, as well as the hydrolysis and rearrangement, is almost completely inhibited by low concentrations of oxaprozin. It is therefore unlikely that the reactions of oxaprozin glucuronide described above will occur in vivo or that these reactions have biological implications. The situation may be different with acyl glucuronides which are present in plasma. Thus far, in vivo acylation of albumin has been observed with zomepirac glucuronide leading to a product which is only slowly cleared from plasma (6). One of the few other acyl glucuronides that have been detected in plasma is clofibric acid glucuronide. As mentioned earlier, its in vivo hydrolysis is the probable cause of the prolonged half-life of this drug in patients with renal dysfunction (4,5). Rowe and Meffin (26) were able to increase the clearance of clofibric acid in rabbits by administering diisopropylfluorophosphate. They attribute the increased clearance to the inhibition of esterases assumed to be the agents responsible for the regeneration of clofibric acid from its glucuronide. The present study indicates that the regenerating agent may be albumin rather than an esterase. Diisopropylfluorophosphate reacts with the tyrosine residue of the albumins of several species (including the rabbit) similarily to p-nitrophenylacetate (24). It is therefore conceivable that the ability to hydrolyze acyl glucuronides is greatly diminished in diisopropylphosphorylated albumin as it is in acylated albumin.

At present we have only limited data on the reaction between albumin and acyl glucuronides other than oxaprozin. There are considerable differences in reactivity among acyl glucuronides indicating the need for further study.

SUMMARY AND CONCLUSIONS

Hydrolysis and rearrangement (isomerization by acyl migration) of oxaprozin glucuronide are greatly accelerated by plasma and human serum albumin. Albumin accounts for all the hydrolytic activity in plasma and no esterase is involved. The isomeric esters formed by rearrangement are also good substrates for the hydrolysis reaction. Another reaction between oxaprozin glucuronide and albumin leads to covalent binding of the aglycone. Similar reactions leading to covalent binding have been described for other acyl glucuronides by several investigators. In the case of oxaprozin, there is little or no potential for biological significance of covalent binding because the reaction is almost entirely inhibited by low concentrations of the drug. All three reactions are pH dependent but not to the same extent. They can be considered to be transacylations to the hydroxyl ion (hydrolysis), to a different OH-group of the glucuronic acid moiety (rearrangement) or to a nucleophilic group on the albumin molecule (covalent binding). All three reactions are greatly inhibited by the same compounds suggesting a common reaction site. This site has certain features in common with the indole or benzodiazepine binding site of human serum albumin. A scheme is proposed in which the first step is reversible binding of the acyl glucuronide to this site in analogy to the known reversible binding of reactive esters (such as p-nitrophenyl acetate) to the same site. All three reactions are inhibited by compounds such as naproxen and decanoic acid which are known to also inhibit the acylation of albumin by reactive esters and the reversible binding of benzodiazepines.

REFERENCES

1. E.M. Faed, Properties of acyl glucuronides: implications for studies of the pharmacokinetics and metabolism of acidic drugs, Drug Metab Rev 15:1213-1249 (1984).
2. F.W. Janssen, S.K. Kirkman, C. Fenselau, M. Stogniew, B.R. Hofmann, E.M. Young, and H.W. Ruelius, Metabolic formation of N- and O-glucuronides of 3-(p-chlorophenyl)thiazolo-[3,2-a] benzimidazole-2-acetic acid, Drug Metab Dispos 10:599-604 (1982).
3. H.W. Ruelius, E.M. Young, S.K. Kirkman, R.T. Schillings, S.F. Sisenwine, and F.W. Janssen, Biological fate of acyl glucuronides in the rat, Biochem Pharmacol 34:451-452 (1985).
4. R. Gugler, The effect of disease on the response to drugs, in "Proceedings of the 7th International Congress of Pharmacology", P. Duchêne-Mairwaz, ed., Pergamon Press, Oxford (1979).
5. E.M. Faed and E.G. McQueen, Plasma half-life of clofibric acid in renal failure, Br J Clin Pharmacol 7:407-410 (1979).
6. P.C. Smith, A.F. McDonagh, and L.Z. Benet, Covalent binding of zomepirac acyl glucuronide to albumin in healthy human volunteers, Hepatology 4:1059 (1984).
7. A. Gautam, H. Seligson, E.R. Gordon, D. Seligson, and J.L. Boyer, Irreversible binding of conjugated bilirubin to albumin in cholestatic rats, J Clin Invest 73:873-877 (1984).
8. A.F. McDonagh, L.A. Palma, J.J. Lauff, and T-W. Wu, Origin of mammalian biliprotein and rearrangement of bilirubin glucuronide in vivo in the rat, J Clin Invest 74:763-770 (1984).
9. R.B. van Breemen and C. Fenselau, Acylation of albumin by 1-O-acyl glucuronides, Drug Metab Dispos 13:318-320 (1985).
10. F.W. Janssen, W.J. Jusko, S.T. Chiang, S.K. Kirkman, P.J. Southgate, A.J. Coleman and H.W. Ruelius, Metabolism and kinetics of oxaprozin in normal subjects, Clin Pharmacol Ther 27:352-362 (1980).
11. F.W. Janssen, S.K. Kirman, J.A. Knowles, and H.W. Ruelius, Disposition of 4,5-diphenyl-2-oxazolepropionic acid (oxaprozin) in beagle dogs and rhesus monkeys, Drug Metab Dispos 6:465-475 (1978).
12. A.J. Lewis, R.P. Carlson, J. Chang, S.C. Gilman, S. Nielson, M.E. Rosenthale, F.W. Janssen, and H.W. Ruelius, The pharmacological profile of oxaprozin, an antiinflammatory and analgesic agent with low gastrointestinal toxicity, Curr Ther Res 34:777-794 (1983).
13. G.E. Means and M.L. Bender, Acetylation of human serum albumin by p-nitrophenyl acetate, Biochem 14:4989-4994 (1975).
14. F.D. Boudinot, C.A. Homon, W.J. Jusko, and H.W. Ruelius, Protein binding of oxazepam and its glucuronide conjugates to human albumin, Biochem Pharmacol 34:2115-2121 (1985).
15. N. Ohta, Y. Kurono, and K. Ikeda, Esterase-like activity of human serum albumin II:reaction with N-trans-cinnamoylimidazoles, J Pharm Sci 72: 385-388 (1983).
16. Y. Kurono, T. Kondo, and K. Ikeda, Esterase-like activity of human serum albumin III: enantioselectivity in the burst

phase of reaction with p-nitrophenyl α-methoxyphenyl acetate, Arch Biochem Biophys 227:339-341 (1983).

17. P.C. Smith, J. Hasegawa, P.N.J. Langendijk, and L.Z. Benet, Stability of acyl glucuronides in blood, plasma and urine: studies with zomepirac, Drug Metab Dispos 13:110-112 (1985).

18. I. Sjöholm, B. Ekman, A. Kober, I. Ljungstedt-Pahlman, B. Seiving, and T. Sjodin, Binding of drugs to human serum albumin. XI. The specificity of three binding sites as studied with albumin immobilized in microparticles, Mol Pharmacol 16:767-777 (1979).

19. S-W.M. Koh and G.E. Means, Characterization of a small apolar anion binding site of human serum albumin, Arch Biochem Biophys 192:73-79(1979)

20. N.P. Sollenne and G.E. Means, Characterization of a specific drug binding site of human serum albumin, Mol Pharmacol 15:754-757 (1979)

21. K.J. Fehske, W.E. Müller, U. Schlafer, and U. Wollert, Characterization of two important drug binding sites on human serum albumin, Prog Drug Protein Binding, Proc Lect Symp, 2nd 5-15 (1981)

22. W.E. Müller and U. Wollert, Benzodiazepines: specific competitors for the binding of L-tryptophan to human serum albumin, Naunyn-Schmiedeberg's Arch Pharmacol 288:17-27 (1975).

23. K.J. Fehske, W.E. Müller, and U. Wollert, A highly reactive tyrosine residue as part of the indole and benzodiazepine binding site of human serum albumin, Biochem Biophys Acta 577:346-359 (1979).

24. G.E. Means and H-L. Wu, The reactive tyrosine residue of human serum albumin: characterization of its reaction with diisopropylfluorophosphate, Arch Biochem Biophys 194:526-530 (1979).

25. K.J. Fehske, W.E. Müller, and U. Wollert, Direct demonstration of the highly reactive tyrosine residue of human serum albumin located in fragment 299-585, Arch Biochem Biophys 205:217-221 (1980)

26. B.J. Rowe and P.J. Meffin, Diisopropylfluorophosphate increases clofibric acid clearance: supporting evidence for a futile cycle, J Pharmacol Exp Ther 230:237-241 (1984).

NEPHROTOXIC AMINO ACID AND GLUTATHIONE S-CONJUGATES: FORMATION AND

RENAL ACTIVATION

M.W. Anders, Lawrence H. Lash, and Adnan A. Elfarra

Department of Pharmacology
University of Rochester School of Medicine and Dentistry
Rochester, New York 14642

INTRODUCTION

The mechanisms by which chemicals produce tissue damage include, for example, biotransformation to reactive, electrophilic metabolites, initiation of lipid peroxidation, and formation of toxic, reduced oxygen metabolites. The liver, because of its abundant capacity for biotransformation, is frequently the target organ for chemicals whose toxicity is associated with bioactivation. Extrahepatic target organs are well known, and this toxicity may also be associated with target organ bioactivation.

Many nephrotoxic chemicals have been identified, but the mechanisms by which nephrotoxins produce kidney damage are incompletely understood (Hook et al., 1979; Rush et al., 1984). Chloroform (Branchflower et al., 1984), cephaloridine (Kuo et al., 1982), and 4-ipomeanol (Boyd and Dutcher, 1981), for example, probably undergo bioactivation in the kidney. In contrast, the nephrotoxicity of hexachloro-1,3-butadiene is not associated with cytochrome P-450-dependent bioactivation (Rush et al., 1984). Chlorotrifluoroethene is a potent nephrotoxin, but this toxicity is not attributable to the formation of inorganic fluoride (Buckley et al., 1982), as would be expected if cytochrome P-450-dependent metabolism was involved.

These observations have led us to seek an alternative explanation for the toxicity of nephrotoxic halogenated alkanes and alkenes. An important clue was provided by the findings that the cysteine conjugate S-(1,2-dichlorovinyl)-L-cysteine is nephrotoxic in all species studied (Terracini and Parker, 1965) and that S-(2-chloroethyl)-DL-cysteine, a putative metabolite of the nephrotoxin 1,2-dichloroethane, is nephrotoxic (Elfarra et al., 1985). These results showed that cysteine S-conjugates are nephrotoxic and suggested the hypothesis that glutathione conjugation followed by renal processing of the glutathione S-conjugates to the corresponding cysteine S-conjugates, which may be direct-acting nephrotoxins or may undergo renal bioactivation, may account for the nephrotoxicity of certain halogenated alkanes and alkenes. This hypothesis and the supporting evidence has been described in detail in a recent review (Elfarra and Anders, 1984) and is illustrated in Fig. 1.

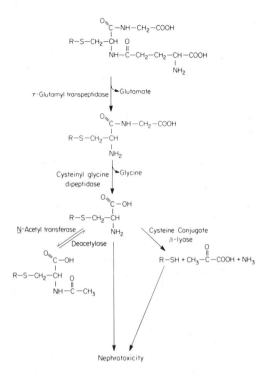

Fig. 1. Metabolism of S-substituted glutathione conjugates.
(Reproduced with permission from Elfarra and Anders, 1984).

The objective of this review is to summarize the available data
about the biosynthesis and renal processing of nephrotoxic glutathione
and cysteine S-conjugates and related compounds.

BIOSYNTHESIS OF NEPHROTOXIC GLUTATHIONE S-CONJUGATES

The hypothesis described above to explain the nephrotoxicity of
certain halogenated alkanes and alkenes requires the initial formation of
a glutathione S-conjugate. Cytosolic and microsomal glutathione S-
transferases catalyze the formation of glutathione S-conjugates (Jakoby,
1978; Morgenstern and DePierre, 1983); the glutathione S-transferase
activity ascribed to liver mitochondria appears to be due to adventitious
cytosolic transferase activity (Ryle and Mantle, 1984).

The glutathione-dependent conjugation of chlorotrifluoroethene
(CTFE) has been studied in detail (Dohn et al., 1985). The product of
the reaction is S-(2-chloro-1,1,2-trifluoroethyl)glutathione (Fig. 2),
which was identified by SIMS and by ^{1}H and ^{19}F NMR. Moreover, the
hepatic microsomal glutathione S-transferases were more efficient than
the hepatic cytosolic or the renal microsomal or cytosolic transferases
in catalyzing the reaction. The reaction, when catalyzed by the hepatic
microsomal transferases, appears to be under tight stereochemical
control, and a chiral center is introduced into the product. When the
glutathione conjugate was synthesized chemically or enzymatically with
the hepatic cytosolic transferases as the catalysts, ^{19}F NMR studies
showed the presence of both enantiomers; in contrast, when the hepatic
microsomal transferases were the catalysts, one enantiomer predominated.

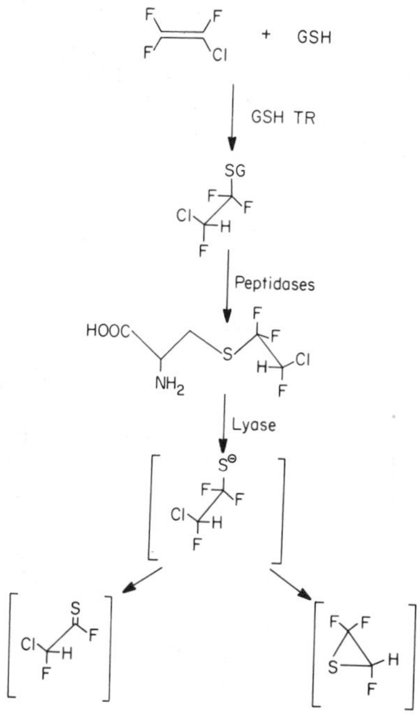

Fig. 2. Proposed metabolic pathway for the bioactivation of CTFE. GSH = glutathione; GSH TR = cytosolic or microsomal glutathione S-transferases; peptidases = γ-glutamyl transpeptidase and cysteinylglycine dipeptidase; lyase = cysteine conjugate β-lyase. Proposed structures are shown in brackets. (Reproduced with permission from Dohn et al., 1985).

Studies in other laboratories also support the concept that the hepatic microsomal glutathione S-transferases catalyze the formation of nephrotoxic glutathione S-conjugates. Hexachloro-1,3-butadiene is metabolized by hepatic microsomal transferases to S-(1,1,2,3,4-pentachloro-1,3-butadienyl)glutathione (Nash et al., 1984; Wolf et al., 1984), which is nephrotoxic. Similarly, tetrafluoroethene is metabolized to S-(1,1,2,2-tetrafluoroethyl)glutathione, and the hepatic microsomal glutathione S-transferases are the most efficient catalysts for the reaction; this glutathione conjugate is further metabolized to S-(1,1,2,2-tetrafluoroethyl)-L-cysteine, which is nephrotoxic (Odum and Green, 1984).

Thus, although a definitive biological function has not been attributed to the microsomal glutathione S-transferases, these enzymes do catalyze the initial step in the bioactivation of nephrotoxic halogenated alkanes and alkenes.

NEPHROTOXICITY OF AMINO ACID AND GLUTATHIONE S-CONJUGATES

S-Haloalkyl Conjugates, which are Direct-Acting Nephrotoxins

S-(2-Chloroethyl)cysteine (CEC), the cysteine conjugate of 1,2-dichloroethane, is a representative example of a direct-acting nephrotoxic conjugate. Although the biological formation of CEC from

1,2-dichloroethane has not been established, cysteine conjugates that could arise from CEC have been isolated (Nachtomi et al., 1966; Yllner, 1971). Treatment of rats with CEC at doses greater than 50 mg/kg causes a marked renal dysfunction accompanied by histopathological changes in the kidney (Elfarra et al., 1985).

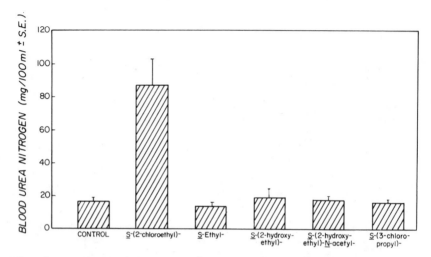

Fig. 3. Nephrotoxicity of S-(2-chloroethyl)-DL-cysteine and analogues. Rats were given saline or the compounds (0.45 mmol/kg, i.p.) dissolved in saline; the rats were killed 36 h after treatment, and blood urea nitrogen concentrations were measured. (Data are from Elfarra et al., 1985).

CEC treatment did not cause histopathologically detectable hepatic lesions, and SGPT activities were only elevated slightly. In vitro, CEC is a direct acting mutagen (Rannug et al., 1978) and is toxic to isolated rat hepatocytes (Webb et al., 1985). CEC-induced cytotoxicity was associated with a rapid depletion of glutathione, inhibition of Ca^{2+}-ATPase, and initiation of lipid peroxidation (Webb et al., 1985). CEC is not a substrate for cysteine conjugate β-lyase, and analogues of CEC in which the chlorine atom is replaced by a hydrogen atom (S-ethyl-L-cysteine), a hydroxyl group [S-(2-hydroxyethyl)-DL-cysteine; S-(2-hydroxyethyl)-N-acetyl-DL-cysteine], or chloromethylene group [S-(3-chloropropyl)-DL-cysteine] are not toxic (Fig. 3) (Elfarra et al., 1985; Webb et al., 1985). These results show that with CEC, internal displacement of the chlorine atom by the sulfur atom to form an electrophilic episulfonium ion may be responsible for the toxicity associated with CEC. In the case of S-ethyl-L-cysteine, the formation of an episulfonium ion cannot occur and with S-(2-hydroxyethyl)-DL-cysteine and S-(2-hydroxyethyl)-DL-N-acetyl-DL-cysteine, episulfonium ion formation is unlikely under physiological conditions, because the chlorine atom is a much better leaving group than the hydroxyl group. S-(3-Chloropropyl)-DL-cysteine was used to examine the possibility that direct displacement (S_N2) of the chlorine atom by nucleophiles rather than the formation an electrophilic sulfonium ion is responsible for CEC-induced toxicity. The formation of a 4-membered cyclic sulfonium ion is thermodynamically and kinetically less favored than the formation of a episulfonium ion. Thus, the finding that S-(3-chloropropyl)-DL-cysteine was not toxic strongly implicates episulfonium ion formation in CEC-induced toxicity, as shown in Fig. 4.

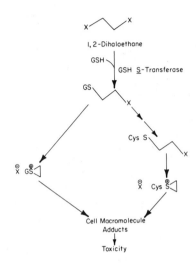

Fig. 4. Postulated mechanism of glutathione-dependent toxicity of 1,2-dihaloethanes. GSH, glutathione; Cys, cysteine; x, chlorine. (Reproduced with permission from Elfarra et al., 1985).

The organ selectivity of CEC and the protection against CEC-induced nephrotoxicity by probenecid suggest that active transport by renal tubular cells results in the accumulation of CEC in renal tissue in toxic concentrations.

S-Haloalkyl Conjugates, which Undergo Renal Bioactivtion

The glutathione and cysteine S-conjugates of chlorotrifluoroethene (CTFE), S-(2-chloro,1,1,2-trifluoroethyl)glutathione (CTFG) and S-(2-chloro-1,1,2-trifluoroethyl)-L-cysteine (CTFC), also require bioactivation for the expression of toxicity. In vivo studies have shown that both conjugates are potent nephrotoxins (D.R. Dohn et al., unpublished observations). The onset of toxicity is rapid, and distinctive lesions involving the proximal tubule are observed in the kidneys of male rats as early as 1.5 h after intravenous administration of 100 μmol/kg CTFG or CTFC. At later times after administration of the conjugates (6-24 h), proteinic casts are observed in the proximal tubules and in the collecting ducts. Some evidence of tubular regeneration is observed 72 h after CTFG administration. Lesions are generally found in the proximal convoluted tubules and in the cortical portion of the loop of Henle. These lesions are similar to those observed after CTFE inhalation (Potter et al., 1981; Buckley et al., 1982).

Renal function is markedly affected by CTFG and CTFC. Administration of 100 μmol/kg of CTFG or CTFC causes a 100-fold increase in urine glucose excretion rates, a 10-fold increase in urine protein excretion rates, and a 4-fold increase in blood urea nitrogen concentrations 24 h after exposure to the conjugates. These effects are consistent with the morphological damage observed. The dose-response curve for CTFG is steep; doses of 50 and 100 µmol/kg cause similar lesions and similar changes in renal function parameters. In contrast, a dose of 10 µmol/kg produces no apparent changes in renal morphology or renal function.

CTFG and CTFC are also toxic to isolated rat renal proximal tubular cells (Lash et al., 1985). Cell viability, as assessed by trypan blue exclusion and lactate dehydrogenase leakage, declines during a 4 h incubation from approximately 90% to 42% in the presence of 0.1 mM CTFG and declines to 30% in the presence of 0.1 mM CTFC (Fig. 5A). Dose-dependent toxicity is also observed; concentrations of the conjugates as low as 0.01 mM cause a significant loss in cell viability. The γ-glutamyltransferase inactivator AT-125 (L-(αS,5S)-α-amino-3-chloro-4,5-dihydro-5-isoxazoleacetic acid) (Reed et al., 1980) prevents CTFG toxicity (Fig. 5A), indicating a role for this enzyme in the bioactivation of CTFG. AOAA protects isolated kidney cells against CTFC toxicity (Fig. 5B), indicating that metabolism of CTFC by the β-lyase is responsible for its nephrotoxicity. This conclusion is supported by the lack of toxicity of the α-methyl analogue of CTFC $\underline{S}$-(2-chloro-1,1,2-trifluoroethyl)-DL-α-methylcysteine, which cannot be metabolized by the β-lyase (L.H. Lash and M.W. Anders, unpublished observations). These results are consistent with the hypothesis that CTFE nephrotoxicity is due to formation of a glutathione conjugate, which is then metabolized by γ-glutamyltransferase, cysteinylglycine dipeptidase, and cysteine conjugate β-lyase to the ultimate toxin.

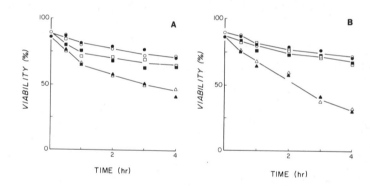

Fig. 5. Effect of CTFG and CTFC on viability of isolated kidney cells. Cells (1 x 10^6/ml) were incubated with the indicated additions of 37°C, and viability was measured by trypan blue exclusion (open symbols) or lactate dehydrogenase leakage (filled symbols). Results are the mean of 3-4 cell preparations; the relative standard deviations were less than 5%. A (left). Effect of CTFG on cell viability: Control cells (O,●); 0.1 mM CTFG (Δ,▲); 0.1 mM CTFG + 0.25 mM AT-125 (□,■). B (right). Effect of CTFC on cell viability: Control cells (O,●); 0.1 mM CTFC (Δ,▲); 0.1 mM CTFC + 0.1 mM aminooxyacetic acid (□,■).

Hassall et al. (1984) have demonstrated metabolism of CTFE to CTFG and CTFC in isolated rabbit renal tubules. Exposure of renal tubules to CTFE, CTFG, or CTFC inhibits active organic anion transport. Inhibition of γ-glutamyltransferase with AT-125 or serine-borate protects against CTFG toxicity in isolated renal tubules, supporting the role for γ-glutamyltransferase in CTFG bioactivation.

The nephrotoxic cysteine conjugate of tetrafluoroethene S-(1,1,2,2-tetrafluoroethyl)-L-cysteine also requires renal bioactivation. Odum and Green (1984) have shown that exposure of rats to tetrafluoroethene produces similar renal lesions and similar alterations in renal function parameters as those found after administration of CTFG and CTFC. Liver cytosol and microsomes catalyze the conversion of tetrafluoroethene to the glutathione conjugate. The corresponding cysteine conjugate was identified in rat bile and is further metabolized by renal cysteine conjugate β-lyase. Thus, as for CTFE, tetrafluoroethene requires bioactivation by formation of a glutathione conjugate and subsequent metabolism by renal enzymes to the ultimate nephrotoxin.

TABLE 1. The effect of AT-125 on S-(1,2-dichlorovinyl)

glutathione-induced nephrotoxicity*

Treatment**	Blood Urea Nitrogen (mg %)		Urine Glucose (mg/24 hr)	
	24 hr	48 hr	0-24 hr	24-48 hr
S-(1,2-Dichlorovinyl)glutathione (0.23 mmol/kg)	36 ± 3[†]	59 ± 10[†]	699 ± 146[†]	423 ± 207[†]
S-(1,2-Dichlorovinyl)glutathione (0.23 mmol/kg) + AT-125 (10 mg/kg)	35 ± 5	43 ± 10	225 ± 72[‡]	309 ± 38
AT-125 only (10 mg/kg)	17 ± 1	18 ± 1	6 ± 1	4 ± 1
Saline	13 ± 2	16 ± 2	8 ± 3	11 ± 4

* Data are from Elfarra et al. (1985a).

**Male Fischer 344 rats (175-225 g) were given S-(1,2-dichlorovinyl)glutathione in isotonic saline or saline alone. Some rats were given AT-125 [Acivicin, L-(αS,5S)-α-amino-3-chloro-4,5-dihydro-5-isoxazoleacetic acid] 1 hr before treatment. Values are means ± S.D. of at least four rats per group.

[†] Significantly different ($p < .05$) from animals given saline alone at the same time.

[‡] Significantly different ($p < 0.05$) from animals given 0.23 mmol/kg S-(1,2-dichlorovinyl)glutathione at the same time.

S-Haloalkenyl Conjugates, which Undergo Renal Bioactivation

S-(1,2-Dichlorovinyl)-L-cysteine (DCVC), which was isolated from soya bean meal that had been extracted with trichloroethene, is a potent nephrotoxin (Elfarra et al., 1985a). In vitro, DCVC is cleaved by cysteine conjugate β-lyase to produce pyruvic acid, ammonia, and sulfur-containing reactive metabolites that combine with proteins and nucleic acids (Bhattacharya and Schultze, 1967; 1973). The glutathione conjugate of trichloroethene S-(1,2-dichlorovinyl)glutathione (DCVG) is a potent nephrotoxin in vivo, and its nephrotoxicity is blocked by the γ-glutamyltransferase inactivator AT-125 (Table 1). These results show that renal metabolism of DCVG to DCVC is required for the expression of DCVG-induced nephrotoxicity (Elfarra et al., 1985a).

To examine the role of cysteine conjugate β-lyase in DCVC-induced renal damage, the nephrotoxicity of S-(1,2-dichlorovinyl)-DL-α-methylcysteine (DCVMC), which cannot be cleaved by β-lyase, was evaluated (Fig. 6). DCVMC was not nephrotoxic, which supports a role for β-lyase in DCVC-induced nephrotoxicity. Moreover, aminooxyacetic acid (AOAA), an inhibitor of pyridoxal phosphate-dependent enzymes (Wallach 1960; Beeler and Churchich, 1976) and of cysteine conjugate β-lyase (Elfarra et al., 1985a), is a potent inhibitor of renal mitochondrial β-lyase both in vivo and in vitro (Elfarra et al., 1985a). Only about 10% of the total β-lyase activity in homogenates as well as in mitochondrial fractions of the kidney was detected 1 hr after giving AOAA (0.5 mmol/kg). With kidney mitochondrial β-lyase, the inhibition constant K_i for AOAA was 0.1 μM. These results and the finding that AQAA (0.5 mmol/kg) blocks the nephrotoxicity of DCVC (Fig. 6) clearly demonstrate a role for renal β-lyase in DCVC-induced nephrotoxicity. Moreover, the finding that probenecid inhibits the nephrotoxicity of DCVC shows a role for the renal anion transport system in DCVC-induced kidney damage (Fig. 6).

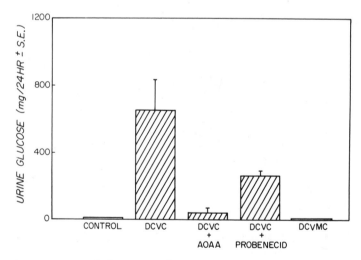

Fig. 6. Nephrotoxicity of S-(1,2-dichlorovinyl)-L-cysteine (DCVC) and S-(1,2-dichlorovinyl)-DL-α-methylcysteine (DCVMC). Rats were given DCVC (0.12 mmol/kg, i.p.) or DCVMC (0.23 mmol/kg, i.p.) in saline or saline alone; urine was collected for 24 h and analyzed for glucose content. Some rats were given aminooxyacetic acid (AOAA; 0.5 mmol/kg, i.p.) or probenecid (0.175 mmol/kg, i.p.) 1 h before giving DCVC. (Data are from Elfarra et al., 1985).

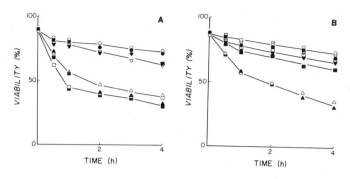

Fig. 7. Effect of DCVG and DCVC on viability of isolated kidney
cells. Cells (1 x 10^6/ml) were incubated with the indicated
additions at 37°C, and viability was measured by trypan blue
exclusion (open symbols) or lactate dehydrogenase leakage (filled
symbols). Results are the mean of 3-4 cell preparations; the
relative standard deviations were less than 5%. A (left). Effect
of DCVG on cell viability: Control cells (O,●); 1 mM DCVG (△,▲); 1
mM DCVG + 0.25 mM AT-125 (▽,▼); 1 mM DCVG + 1 mM glycylglycine
(◲,◼). B (right). Effect of DCVC on cell viability: Control cells
(O,●); 1 mM DCVC (△,▲); 1 mM DCVC + 0.1 mM aminooxyacetic acid
(▽,▼); 1 mM DCVC + 0.1 mM probenecid (☐,◼).

Isolated renal proximal tubular cells have been employed to study
the in vitro nephrotoxicity of DCVG and DCVC (Lash et al., 1985). Cell
viability, as assessed by trypan blue exclusion and lactate dehydrogenase
leakage, decreases in a time-dependent manner when cells are incubated
with 1 mM DCVC (Fig. 7B). During a 4 h incubation, viability declines
from approximately 90% to 35% in the presence of either DCVG or DCVC.
Inhibition of γ-glutamyltransferase by addition of AT-125 prevents DCVG
toxicity, and stimulation of the enzyme by addition of the γ-glutamyl
acceptor glycylglycine potentiates DCVG-induced toxicity (Fig. 7A),
indicating that formation of the cysteine conjugate from DCVG is a
required step in the bioactivation process. Inhibition of cysteine
conjugate β-lyase by addition of AOAA or inhibition of organic anion
transport by addition of probenecid protects the cells from DCVC toxicity
(Fig. 7B). These results, as well as the in vivo studies described
above, show that the formation of DCVC from DCVG, probenecid-sensitive
transport into the cell, and metabolism by the cysteine conjugate β-lyase
to the ultimate nephrotoxin are involved.

Renal cysteine conjugate β-lyase activity, which is believed to
generate the ultimate nephrotoxin from DCVC and certain other cysteine
conjugates, is found in the cytosolic and mitochondrial fractions of rat
kidney homogenates (Dohn and Anders, 1982; Elfarra et al., 1985; L.H.
Lash and M.W. Anders, unpublished observations). Studies with rat kidney
slices (Parker, 1965) and rat liver mitochondria (Stonard and Parker,
1971a,b; Stonard, 1973) showed that DCVC is a potent inhibitor of
respiration and indicated that mitochondria may be the target site at
which DCVC exerts its toxic effects in the cell. The findings that DCVC
is a potent inhibitor of mitochondrial respiration in isolated kidney
cells and that the onset of this inhibition is more rapid than the onset
of cell death (L.H. Lash and M.W. Anders, unpublished observations)
support this hypothesis. Thus, mitochondrial cysteine conjugate β-lyase
may play a more important role in DCVC bioactivation than the cytosolic
activity.

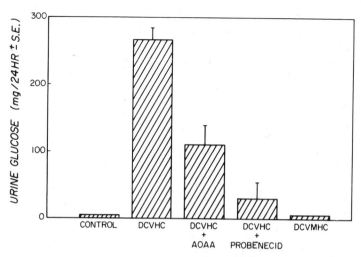

Fig. 8 Nephrotoxicity of S-(1,2-dichlorovinyl)-L-homocysteine (DCVHC) and S-(1,2-dichlorovinyl)-DL-α-methylhomocysteine (DCVMHC). Rats were given DCVHC (5 umol/kg, i.p.) or DCVMHC (230 umol/kg, i.p.) in saline or saline alone; urine was collected for 24 h and analyzed for glucose content. Some rats were given aminooxyacetic acid (AOAA; 0.5 mmol/kg,i.p.) or probenecid (0.175 mmol/kg, i.p.) 1 h before giving DCVHC. (Data are from Elfarra and Anders, 1985).

The metabolism and nephrotoxicity of homocysteine S-conjugates, which are analogues of the nephrotoxin DCVC, were also studied to test the hypothesis that homocysteine S-conjugates, if cleaved by cystathionine γ-lyase (Carroll et al., 1949), should yield sulfur-containing reactive intermediates similar to those produced by the action of β-lyase on the corresponding cysteine S-conjugates (Fig. 8). The homocysteine conjugate of trichloroethene S-(1,2-dichlorovinyl)-L-homocysteine (DCVHC) is 50 to 100 times more nephrotoxic than DCVC (Elfarra and Anders, 1985). S-(1,2-Dichlorovinyl)-DL-α-methylhomocysteine (DCVMHC) was not nephrotoxic (Elfarra and Anders, unpublished results), and treatment of rats with the cystathionine γ-lyase inhibitors DL-propargylgylcine and AOAA and the inhibitor of the organic anion transport system probenecid protected against DCVHC-induced nephrotoxicity (Elfarra and Anders, 1985). These results are consistent with a bioactivation mechanism of DCVHC that involves the renal metabolism of DCVHC by cystathionine γ-lyase or by enzymatic deamination followed by nonenzymatic β-elimination to yield the reactive S-1,2-dichlorovinylmercaptan, which is responsible for DCVHC-induced nephrotoxicity. The rate of DCVC N-acetylation is four-fold greater than the rate measured with DCVHC as the substrate (Elfarra and Anders, unpublished results). Thus, the remarkable nephrotoxic potency of DCVHC compared with DCVC may be partially due to differences in rates of detoxication by N-acetylation of DCVC and DCVHC.

CONCLUSIONS

The data presented validate the hypothesis that the nephro-toxicity of certain halogenated alkanes and alkenes is associated with hepatic glutathione S-conjugate formation followed by renal metabolism to the corresponding cysteine S-conjugates, which may be direct-acting nephrotoxins, e.g. CEC, or which may undergo activation by renal cysteine conjugate β-lyase to the penultimate or ultimate nephrotoxic species, e.g. DCVC and CTFC. The target organ selectivity is determined by renal γ-glutamyltransferase and cysteinylglycine dipeptidase activities; this was proven by studies with AT-125, which inhibits irreversibly γ-glutamyltransferase and blocks the toxicity of nephrotoxic glutathione S-conjugates. Furthermore, renal tubular anion transport systems also play a role, because the toxicity of both direct-acting cysteine conjugates and cysteine conjugates that require metabolic activation is inhibited by probenecid.

Finally, although most studies have dealt with halogenated alkanes and alkenes, recent results indicate that this mechanism can be generalized to other classes of compounds. For example, the nephrotoxicity of bromohydroquinone may involve formation of a diglutathionyl conjugate, which undergoes renal bioactivation (Monks et al., 1985).

Future studies to identify the reactive intermediates produced and to clarify the enzymology of renal bioactivation are warranted.

ACKNOWLEDGEMENTS

The studies conducted in the authors' laboratory were supported by NIEHS grant ES03127. The authors thank Ms. Lori Mittelstaedt for preparation of the manuscript.

REFERENCES

Beeler, T., and Churchich, J.E., 1976, Reactivity of the phospho-pyridoxal groups of cystathionine, J. Biol. Chem., 251:5267.

Bhattacharya, R.K., and Schultze, M.O., 1967, Enzymes from bovine and turkey kidneys which cleave S-(1,2-dichlorovinyl)-L-cysteine, Comp. Biochem. Physiol., 22:723.

Bhattacharya, R.K., and Schultze, M.O., 1973, Modification of polynucleotides by a fragment produced by enzymatic cleavage of S-(1,2-dichlorovinyl)-L-cysteine, Biochem. Biophys. Res. Commun., 53:172.

Boyd, M.R., and Dutcher, J.S., 1981, Renal toxicity due to reactive metabolites formed in situ in the kidney: Investigations with 4-ipomeanol in the mouse, J. Pharmacol. Exp. Ther., 216:640.

Branchflower, R.V., Nunn, D.S., Highet, R.J., Smith, J.H., Hook, J.B., and Pohl, L.R., 1984, Nephrotoxicity of chloroform: Metabolism to phosgene by the mouse kidney, Toxicol. Appl. Pharmacol., 72:159.

Buckley, L.A., Clayton, J.W., Nagle, R.B. and Gandolfi, A.J., 1982, Chlorotrifluoroethylene nephrotoxicity in rats: A subacute study, Fund. Appl. Toxicol., 2:181.

Carroll, W.R., Stacy, G.W., and du Vigneaud, V., 1949, α-Ketobutyric acid as a product in the enzymatic cleavage of cystathionine, J. Biol. Chem., 180:375.

Dohn, D.R., and Anders, M.W., 1982, Assay of cysteine conjugate β-1yase activity with S-(2-benzothiazolyl)cysteine as the substrate, Anal. Biochem., 120:379.

Dohn, D.R., Quebbemann, A.J., Borch, R.F., and Anders, M.W., 1985, Enzymatic reaction of chlorotrifluoroethene with glutathione: 19F NMR evidence for stereochemical control of the reaction, Biochemistry, in press.

Elfarra, A.A., and Anders, M.W., 1984, Renal processing of glutathione conjugates. Role in nephrotoxicity, Biochem. Pharmacol., 33:3729.

Elfarra, A.A., and Anders, M.W., 1985, S-(1,2-Dichlorovinyl)-L-homocysteine (DCVHC), an analogue of the renal toxin S-(1,2-dichlorovinyl)-L-cysteine (DCVC), is a potent nephrotoxin, Fed. Proc., 44:1624.

Elfarra, A.A., Baggs, R.B., and Anders, M.W., 1985, Structure-nephrotoxicity relationships of S-(2-chloroethyl)-DL-cysteine and analogs: Role for an episulfonium ion, J. Pharmacol. Exp. Ther., 233:512.

Elfarra, A.A., Jakobson, I., and Anders, M.W., 1985a, Mechanism of S-(1,2-dichlorovinyl)glutathione-induced nephrotoxicity, Biochem. Pharmacol., in press.

Hassall, C.D., Gandolfi, A.J., Duhamel, R.C., and Brendel, K., 1984, The formation and biotransformation of cysteine conjugates of halogenated ethylenes by rabbit renal tubules, Chem.-Biol. Interact., 49:283.

Hook, J.B., McCormack, K.M., and Kluwe, W.M., 1979, Biochemical mechanisms of nephrotoxicity, Rev. Biochem. Toxicol., 1:53.

Jakoby, W.B., 1978, The glutathione transferases in detoxification, in: "Functions of Glutathione in Liver and Kidney," H. Sies and A. Wendel, eds., Springer-Verlag, New York, p. 157.

Kuo, C-H., Braselton, W.E., and Hook, J.B., 1982, Effect of phenobarbital on cephaloridine toxicity and accumulation in rabbit and rat kidneys, Toxicol. Appl. Pharmacol., 64:244.

Lash, L.H., Dohn, D.R., Elfarra, A.A., and Anders, M.W., 1985, Nephrotoxicity of glutathione and cysteine conjugates in isolated rat kidney cells, Pharmacologist, 27:227.

Monks, T.J., Lau, S.S., Highet, R.J., and Gillette, J.R., 1985, Glutathione conjugates of 2-bromohydroquinone are nephrotoxic, Drug Metab. Disp., 13:in press.

Morgenstern, R., and DePierre, J.W., 1983, Microsomal glutathione transferase, Eur. J. Biochem., 134:591.

Nachtomi, E., Alumot, E., and Bondi, A., 1966, The metabolism of ethylene dibromide in the rat. I. Identification of detoxification products in urine. Israel J. Chem., 4:239.

Nash, J.A., King, L.J., Lock, E.A., and Green, T., 1984, The metabolism and disposition of hexachloro-1:3-butadiene in the rat and its relevance to nephrotoxicity, Toxicol. Appl. Pharmacol., 73:124.

Odum, J., and Green, T., 1984, The metabolism and nephrotoxicity of tetrafluoroethylene in the rat, Toxicol. Appl. Pharmacol., 76:306.

Parker, V.H., 1965, A biochemical study of the toxicity of S-dichlorovinyl-L-cysteine, Food Cosmet. Toxicol., 3:75.

Potter, C.L., Gandolfi, A.J., Nagle, R.B., and Clayton, J.W., 1981, Effects of inhaled chlorotrifluoroethylene and hexafluoropropene on the rat kidney, Toxicol. Appl. Pharmacol., 59:431.

Rannug, U., Sundvall, A., and Ramel, C., 1978, The mutagenic effect of 1,2-dichloroethane on Salmonella typhimurium. I. Activation through conjugation with glutathione in vitro. Chem.-Biol. Interact., 20:1.

Reed, D.J., Ellis, W.W., and Meck, R.A., 1980, The inhibition of γ-glutamyltrannspeptidase and glutathione metabolism of isolated rat kidney cells by L-(αS,5S)-α-amino-3-chloro-4,5-dihydro-5-isoxazoleacetic acid (AT-125; NSC-163501), Biochem. Biophys. Res. Commun., 94:1273.

Rush, G.F., Smith, J.H., Newton, J.F., and Hook, J.B., 1984, Chemically induced nephrotoxicity: Role of metabolic activation. CRC Crit. Rev. Toxicol., 13:99.

Ryle, C.M., and Mantle, T.J., 1984, Studies on the glutathione S-transferase activity associated with rat liver mitochondria, Biochem. J., 222:553.

Stonard, M.D., 1973, Further studies on the site and mechanism of action of S-(1,2-dichlorovinyl)-L-cysteine and S-(1,2-dichlorovinyl)-3-mercaptopropionic acid in rat liver, Biochem. Pharmacol., 22:1329.

Stonard, M.D., and Parker, V.H., 1971a, 2-Oxoacid dehydrogenases of rat liver mitochondria as the site of action of S-(1,2-dichlorovinyl)-L-cysteine and S-(1,2-dichlorovinyl)-3-mercaptopropionic acid, Biochem. Pharmacol., 20:2417.

Stonard, M.D., and Parker, V.H., 1971b, The metabolism of S-(1,2-dichlorovinyl)-L-cysteine by rat liver mitochondria, Biochem. Pharmacol., 20:2429.

Terracini, B., and Parker, V.H., 1965, A pathological study on the toxicity of S-dichlorovinyl-L-cysteine, Food Cosmet. Toxicol., 3:67.

Wallach, D.P., 1960, The inhibition of gamma aminobutyric-alpha-ketoglutaric acid transaminase in vitro and in vivo by amino-oxyacetic acid, Biochem. Pharmacol., 5:166.

Webb, W., Elfarra, A., Thom, R., and Anders, M., 1985, S-(2-Chloroethyl)-DL-cysteine (CEC)-induced cytotoxicity: A role for the episulfonium ion, Pharmacologist, 27:228.

Wolf, C.R., Berry, P.N., Nash, J.A., Green, T., and Lock, E.A., 1984, Role of microsomal and cytosolic glutathione S-transferases in the conjugation of hexachloro-1:3-butadiene and its possible relevance to toxicity, J. Pharmacol. Exp. Ther., 228:202.

Yllner, S., 1971, Metabolism of 1,2-dichloroethane-^{14}C in the mouse. Acta Pharmacol. Toxicol., 30:257.

THE ROLE OF GLUTATHIONE IN THE TOXICITY
OF XENOBIOTIC COMPOUNDS: METABOLIC ACTIVATION
OF 1,2-DIBROMOETHANE BY GLUTATHIONE

I. Glenn Sipes, David A. Wiersma,
and David J. Armstrong

Dept. of Pharmacology and Toxicology
College of Pharmacy, Univ. of Arizona
Tucson, AZ   85721

INTRODUCTION

Glutathione (GSH) is a ubiquitous tripeptide involved in cellular
defense mechanisms and the metabolism of xenobiotic compounds.
Detoxification of free radicals and activated oxygen species by direct
reaction, as well as that mediated by the enzymic activity of
glutathione peroxidase, is a well known and described biochemical
function. Equally well known and described is the role of GSH in
conjugation. Conjugation with GSH occurs nonenzymatically and through
the action of GSH S-transferases. Leukotrienes, active autacoids
thought to be involved in inflammatory processes, are formed by the
conjugation of GSH with fatty acids derived from arachidonic acid. One
important function of GSH S-transferases is the conjugation of GSH with
certain xenobiotic compounds, thus enhancing excretion of the
xenobiotic. Furthermore, activated metabolites produced from xenobiotic
compounds by the action of mixed function oxygenase may also be
conjugated with GSH. This occurs both directly and enzymatically. The
net result of conjugation with active metabolites is detoxification of
these reactive chemical species. In certain cases, the GSH conjugate
may ultimately result in toxic reactions. Stepwise, enzymatic
degradation of certain GSH conjugates may result in reactive
intermediates that result in tissue injury. Some glutathione
conjugates, for example, may be enzymatically toxified by the sequential
actions of gamma-glutamyl transferase and cysteine conjugate beta-lyase
(Dohn and Anders, 1982). Other GSH conjugates may undergo
intramolecular rearrangement to unstable intermediates that also result
in toxic reactions. For example, the conjugation of vicinal-
dihaloalkanes may result in an intramolecular rearrangement to a
reactive, transitory episulfonium ion (van Bladeren et al., 1980;
Schasteen and Reed, 1981; Livesay and Anders, 1982).

Stable and Unstable Metabolites

Formation of a stable GSH conjugate results in the detoxification
of many environmentally-occurring xenobiotic compounds or their reactive
metabolites produced by mixed function oxygenases. The reaction of an
active metabolite of bromobenzene is an example of a detoxification

process highly dependent on the activity of GSH S-transferase conjugation enzymes (Brodie et al., 1971; Jollow et al., 1974). Bromobenzene, in liver, is activated by the cytochrome P-450 system to a reactive epoxide. In the presence of GSH and GSH S-transferase this reactive epoxide is readily conjugated with GSH, preventing reaction of the epoxide with adjacent cellular macromolecules, such as protein, lipids, or nucleic acids. Prevention of these reactions protects the cell from potential damage. Thus, GSH conjugation "detoxifies" bromobenzene. Furthermore, such conjugation greatly enhances the water solubility of the bromobenzene molecule allowing excretion of the conjugate into the bile for elimination. Similar detoxification occurs with the conjugation of GSH with reactive metabolites of chloroform, aflatoxin B_1, vinyl chloride, acetaminophen and a host of other environmental contaminants including the pesticide methyl parathion (Clark et al., 1976).

Not only are these GSH conjugates excreted in the bile, but they may also undergo further enzymic processing to produce mercapturic acids. The enzymatic actions of gamma-glutamyltranspeptidase and non-specific peptidase remove the glutamate and glycine residues, leaving a thioether cysteine conjugate. This conjugate undergoes N-acetylation to yield the mercapturic acid. Mercapturic acids are readily excreted in the urine. However, formation of mercapturic acids is not without hazard in certain cases. For example, chlorotrifluoroethylene, a metabolite of the anesthetic halothane, may be conjugated with GSH and subsequently processed to form the corresponding halogenated vinyl cysteine (Dohn and Anders, 1982; Gandolfi et al., 1981). This vinyl cysteine has been shown to be nephrotoxic, probably by activation of the cysteinyl sulfur group through the action of the enzyme, cysteine conjugate beta-lyase. Thus, for certain xenobiotic compounds formation of a stable GSH conjugate may be the first step toward production of a toxic metabolite.

The formation of stable conjugates, with certain exceptions, leads to detoxification of a reactive xenobiotic compound or its metabolites. In contrast, with some xenobiotic compounds conjugation with GSH results in the formation of an unstable metabolite. These initial conjugates undergo internal molecular rearrangements forming transitory, highly reactive metabolites. Often, it is thought, this may involve the formation of a three-membered ring with a positive charge known as an episulfonium ion. Similar reactive species are thought to form during the conjugation of GSH with dihalogenated alkanes (van Bladeren et al., 1980a; 1981), dihalogenated cyclohexanes (van Bladeren et al., 1979), and certain halogenated ethylenes (Dohn and Anders, 1982).

THE ROLE OF GSH IN THE TOXICITY OF 1,2-DIBROMOETHANE

The role of GSH in the metabolic activation of a xenobiotic compound has been extensively examined for 1,2-dibromoethane (DBE), also known as ethylene dibromide. This paper will examine several aspects of the toxicity and biotransformation of this dihalogenated alkane which we have studied in order to determine the role of GSH conjugation in the genotoxicity of DBE.

Toxicity of DBE

DBE has a number of widely diverse and economically important uses. Besides its use as a chemical intermediate in organic syntheses, DBE has been used extensively as an antiknock additive in gasoline, a preservative and as a pesticide. The latter use has led to contamination of the environment through use as a soil, fruit and grain fumigant. Pollution of water supplies and recognition of its toxic nature have led to curtailment and banning of its commercial uses.

Acute exposure of humans to intoxicating levels of DBE results in metabolic acidosis, renal and hepatic failure as well as the necrosis of muscle and other organs (Letz et al., 1984). In animals as well, pulmonary, hepatic and renal damage results from intoxicating exposure (Rowe et al., 1952). Estimates of the oral LD_{50} in experimental animals range from 55 - 420 mg/kg (NIOSH, 1977).

Sublethal or chronic exposure of animals to DBE appears to damage extrahepatic organs rather than liver. In particular, DBE exposure results in testicular atrophy and spermatogenic toxicity (Edwards et al., 1970; Short et al., 1979). Renal damage has also been observed. Long-term exposure of rats to DBE has produced tumors of the forestomach, spleen, liver, kidney and nasal cavity (Olson et al., 1973; Wong et al., 1982). Similar results have been observed in mice exposed to DBE (Olson et al., 1973; Van Duuren et al., 1979). Site of the tumors depends on the route of DBE exposure.

The carcinogenic effects of DBE are correlated with the results of mutagenicity assays (van Bladeren et al., 1980b). In a typical Ames assay, van Bladeren showed that DBE itself is slightly mutagenic. The addition of microsomes and co-factors for mixed function oxygenase activity did not alter DBE toxicity. However, addition of a liver cytosolic fraction containing GSH S-transferases and GSH greatly enhanced the mutagenicity of DBE toward S. typhimurium TA100. These results suggested that conjugation with GSH may produce an unstable reactive intermediate which can interact with nucleic acids.

Metabolism of DBE

That DBE is metabolized by two enzyme systems was established by Nactomi (1970) and Hill et al., (1978). The microsomal cytochrome P-450 enzymes convert DBE to bromoacetaldehyde (see Figure 1) which in turn is conjugated with GSH to form S-2-hydroxyethylglutathione. This metabolite is processed further to the mercapturic acid, which is

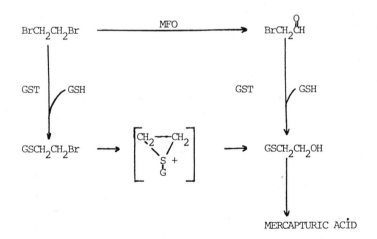

Figure 1. DBE Metabolic Pathways. MFO, mixed function oxidase; GST, GSH S-transferase.

459

excreted in the urine. The GSH S-transferase pathway conjugates DBE with GSH. The S-2-bromoethylglutathione formed may spontaneously rearrange to form the transitory reactive episulfonium ion. Upon hydration this reactive chemical species forms S-2-hydroxyethyl glutathione, the same metabolite resulting from the cytochrome P-450 pathway. Further processing produces the mercapturic acid, S-2-hydroxyethyl-N-acetylcysteine, the major urinary metabolite of DBE.

The toxic manifestations of DBE metabolism are thought to result from the interaction of reactive metabolites with cellular macromolecules. In each pathway reactive metabolites are produced; cytochrome P-450 activity produces bromoacetaldehyde while GSH S-transferase conjugation results in the episulfonium ion. In the absence of detoxification mechanisms these reactive metabolites covalently bind to nucleic acids, proteins, or lipids. This binding alters these macromolecules, in ways not well understood, to produce toxic effects.

Competing Pathways of Metabolism in DBE Toxicity

Since the toxicity of DBE results from its metabolism and that metabolism proceeds by two pathways, each of which can produce active metabolites, one focus of our study of DBE toxicity has been to examine the role that each of the two pathways of DBE metabolism may play in the observed toxicity. For example, the ability of cytosolic GSH S-transferase to enhance the mutagenicity of DBE, while cytochrome P-450 does not (van Bladeren et al., 1980b) suggests that the GSH-derived episulfonium ion may preferentially interact with DNA. Furthermore, experiments using subcellular preparations of rat (Shih and Hill, 1981) and human (Wiersma and Sipes, 1983) liver have shown that metabolism of DBE by cytochrome P-450 produces adducts which bind preferentially to proteins. Thus, metabolism by these two pathways may produce different toxicities.

Interaction of DBE Metabolites with DNA

In experiments using subcellular fractions of rat liver and isolated hepatocytes, Sundheimer, et al. (1982), examined the ability of these two activation systems to promote covalent binding of ^{14}C-DBE metabolites to DNA. Cytochrome P-450 activity produced only minimal binding to exogenous DNA. Cytosol-induced binding, catalyzed by the GSH S-transferase, was 5-10 times greater. In the intact hepatocyte experiments, where both systems were functioning, only 2.4% of the radioactivity bound to nucleic acids was bound to DNA. Prior incubation of the cells with diethylmaleate which reduced the intracellular GSH content to 40% of control decreased the amount of DNA-bound radioactivity to 50% of control. Treatment of the cells with SKF-525A to inhibit the cytochrome P-450 system did not alter the amount of covalently-bound adducts formed with nucleic acids. Using isolated GSH S-transferases and DBE, Ozawa and Guengerich (1983) have shown that the DNA-bound adduct contains equal amounts of DBE and GSH-derived equivalents. Following desulfuration and isolation of modified DNA bases, they have identified N[7]-ethyl-guanine as the predominate adduct and have concluded that the original adduct was S-[2-(N[7]-guanyl)ethyl] gluthathione. These results demonstrate that the unstable GSH S-transferase metabolite binds to DNA and suggests that it may be responsible for the DBE-induced mutagenic effects. Further inference suggests that it also may be involved in the carcinogenicity of DBE.

White et al. (1981, 1983, 1984) further examined the role of the two pathways to actually produce hepatic DNA damage. They utilized the alkaline elution technique of Kohn et al. (1976) which quantifies the number of single-strand breaks produced in the DNA. Modifications of this procedure can also identify xenobiotic induced DNA-DNA and DNA-protein crosslinks. In early experiments it was established that exposure of rats or mice to DBE results in the formation of alkali-labile sites in hepatic DNA, which is dose dependent. These are converted to single-strand breaks during the alkaline elution procedure (White et al., 1981).

To relate the pathway of DBE metabolism to the production of strand breaks, DBE metabolism was modulated by use of tetradeutero-DBE (d_4DBE) (White et al., 1983; 1984). Due to the different mechanisms by which the two enzyme systems attack the DBE molecule, the rate of cytochrome P-450 metabolism of d_4DBE is reduced compared to that of DBE, while the rate of direct conjugation with GSH is unaltered. Bromide release during metabolism of d_4DBE with microsomal incubations was reduced by more than 70% as compared to incubations with DBE (Table 1). On the other hand, bromide release during cytosolic incubations with d_4DBE was equal to that with DBE. When d_4DBE was given to mice, plasma bromide concentration was less than that produced by d_4DBE. Similarly, in studies with isolated hepatocytes, less bromide was liberated from d_4DBE.

These results suggest that DNA damage in vivo should decrease if it were due to a cytochrome P-450 derived metabolite. DNA damage would remain unaltered or be increased if it resulted from the GSH S-transferase metabolite. In the mouse, administration of d_4DBE (50 mg/kg, i.p.) did not affect the amount of hepatic DNA damage compared to that due to DBE for up to three hours after administration. At later time points greater DNA damage was observed in hepatic DNA from mice treated with d_4DBE (Table 2). Additionally, in isolated cells from rat liver, 5-10 min incubations with d_4DBE or DBE showed equivalent DNA damage (White et al. 1984). The conclusion drawn from these experiments was that the genotoxic effects of DBE observed in mouse liver and rat hepatocytes were due to GSH S-transferase mediated conjugation of DBE with GSH.

Since the greater hepatic DNA damage due to d_4DBE at the later time points in the mouse study could be explained by a difference in the tissue distribution and disposition of d_4DBE, Armstrong, et al. (1985), compared the effect of time on the disposition and tissue distribution of d_4DBE and DBE. Swiss-Webster mice were placed in all glass metabolism chambers after receiving an intraperitoneal dose of DBE or d_4DBE (50 mg/kg). Cumulative expiration and tissue concentrations were determined over a 24 hour period (Figure 2). Mice receiving d_4DBE expired about 50% of the dose which was seven times more than those mice receiving DBE. In the liver and testis of these animals, the concentration of d_4DBE was about twice that in the mice receiving DBE. These results suggest that elimination of DBE by the normal route(s) was severely hampered in mice given d_4DBE. Since previous evidence suggests that the major elimination route for DBE is metabolism by cytochrome P-450, it appears that in d_4DBE exposed animals this pathway is less available. Therefore, in the d_4DBE exposed animals expiration becomes the major route of elimination as the remaining metabolic route via the GSH S-transferases becomes saturated. The continued availability of d_4DBE for metabolism via the GSH S-transferase is the likely cause of the greater hepatic DNA damage observed at later time points.

Interaction of DBE Metabolites with Proteins

The contrasting roles of the two activating systems have also been

461

examined in vitro using subcellular fractions from human liver (Wiersma et al., 1983). Cytochrome P-450 activity produced a high amount of protein-bound adducts. When GSH was added to the incubation, either alone or in the the presence of cytochrome P-450, no binding to protein was observed. A very small amount of binding to DNA was observed to be due to cytochrome P-450 activity, but that due to microsomal GSH S-transferase was over twice as great. Cytosolic GSH S-transferase-induced DNA binding was even greater. Thus, it appears that toxic effects not due to DNA damage may result from the reactive metabolite produced by the cytochrome P-450 system. Furthermore, the detoxification role of GSH in this pathway is evident as its presence inhibited the binding to protein.

DBE Toxicity to Extrahepatic Organs

Despite the high concentration of both toxifying and detoxifying enzymes in liver, the primary sites of DBE toxicity following low or chronic exposure are extrahepatic organs. In particular, the carcinogenic effects of DBE have been found in the forestomach, spleen, adrenal gland and kidney of experimental animals (Olson et al., 1973; Van Duuren et al., 1979; Wong et al., 1982). Since the metabolite which prefers to bind to DNA arises from the GSH S-transferase pathway, one explanation for this selective toxicity may be that GSH conjugation is the predominant pathway of DBE metabolism in these organs.

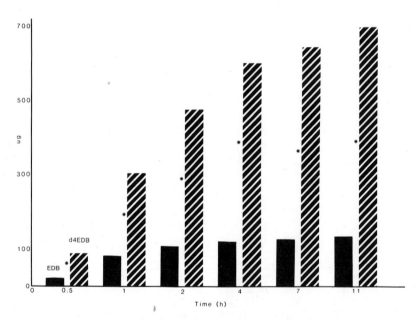

Fig. 2. The cumulative expiration of DBE or d_4DBE by mice. Male, Swiss-Webster mice were given DBE or d_4DBE (50 mg/kg, i.p.) and placed in all glass metabolism cages. Air was drawn through the cage, the DBE or d_4DBE trapped in 1,2-propanediol and quantified by gas-liquid chromatography using flame ionization detection. An asterisk indicates an expiration of d_4DBE greater than that from DBE (Student's t-test, P 0.05).

Table 1. The effect of deuterium substitution on
the metabolism of DBE by mouse liver
microsomal and cytosolic preparations.[a]

Time (min)	Bromide released from d_4DBE (% of that from DBE)	
	Cytosol	Microsomes
0.5	79 $\pm$ 6	–
1.0	113 $\pm$ 10	–
5.0	133 $\pm$ 10	18 $\pm$ 12[b]
10.	100 $\pm$ 10	19 $\pm$ 4[b]
20.		23 $\pm$ 1[b]

[a]   Microsomes and cytosol were prepared from mouse
liver using common methods and incubated with
either DBE or d_4DBE (1 mM) at a protein
concentration of 1 mg/ml for the indicated
times.   NADPH was added to microsomal
incubations while GSH (1 mM) was included in
the cytosolic incubations.   Bromide ion
released from the substrates was methylated and
quantitated by gas liquid chromatography.
Results are the mean $\pm$ S.E.M. of 3 animals.

[b]   Significantly less bromide released   from d_4DBE
than DBE; Student's t-test ($P < 0.05$).

Table 2. The effect of deuterium substitution on
DBE-induced hepatic DNA damage.[a]

Time (hr)	Elution Rate Constant (% of that by DBE)
3	84 $\pm$ 5
8	197 $\pm$ 36[b]
24	220 $\pm$ 50[b]
72	194 $\pm$ 45[b]

[a]   Mice received DBE or d_4DBE (50 mg/kg, i.p.) and
were killed at the indicated times.   Hepatic
nuclei were prepared and the alkaline elution
procedure performed (White, et al.   1981).
Elution rate constants were calculated and that
in the d_4DBE mice expressed as a percentage of
that in mice treated with DBE (mean $\pm$ S.E.M. of
3-5 mice).

[b]   Significantly greater than that due to DBE;
Student's t-test ($P < 0.05$).

A role for GSH in the metabolism of DBE by extrahepatic organs has been implicated by a number of studies. Administration of DBE to rats or mice causes a reduction in organ nonprotein sulfhydryl concentration within two hours of exposure (Kluwe et al., 1981; Wiersma and Sipes, 1983a) in liver, forestomach and kidney. Use of d_4DBE to reduce metabolism by the cytochrome P-450 pathway reduced the extent of depletion only in liver, suggesting that GSH S-transferase was the major pathway in the extrahepatic organs (Wiersma and Sipes, unpublished observations). An examination of the ability of subcellular fractions from extrahepatic tissues of the rat to metabolize DBE showed significant GSH dependent metabolism but very little cytochrome P-450 activity (MacFarland et al., 1984; Wiersma and Sipes, unpublished observations). Thus, it appears that in the extrahepatic organs sufficient protection is available for detoxification of the active metabolite produced by cytochrome P-450. On the other hand, the major route of metabolism in these organs may be through the GSH pathway.

However, even though metabolism of DBE may occur primarily through the GSH conjugation pathway in the extrahepatic organs, this may not be sufficient cause to explain the susceptibility of these organs to DBE-induced damage. Comparison of DBE-induced toxicity to forestomach and glandular stomach provides a case in point. Each of these organs contains a far greater GSH S-transferase activity than cytochrome P-450 activity. Additionally, the specific activity of the GSH S-transferase enzymes in both sections of the stomach is equal (Wiersma and Sipes, 1983b), as is the GSH content (Wiersma and Sipes, 1983a). In contrast to these similarities, Kowalski, et al. (1985), have shown DBE metabolites selectively accumulate in the forestomach and not in the glandular stomach. This finding correlates with the fact that forestomach is susceptible to DBE-induced toxicity whereas the glandular stomach is not (Olson et al., 1973). This paradox has yet to be resolved, but suggests that exposure to DBE, initial GSH content and GSH S-transferase activity are not the only factors involved in this toxicity. Perhaps the glandular stomach and certain other organs have means not present in the forestomach to protect their cells from DBE-induced toxicity.

SUMMARY

Unstable metabolites may arise during the metabolism of xenobiotic compounds with enzyme systems other than the cytochrome P-450 system. This depends on the enzyme system involved and the structure of the xenobiotic compound being metabolized. Normally detoxifying pathways may transform selected chemicals into toxic metabolites.

In our laboratory we have demonstrated that DBE is metabolized by both cytochrome P-450 and GSH S-transferases. Although the cytochrome P-450 metabolite is reactive and will covalently bind to protein and nucleic acid to some extent, the GSH S-transferase system conjugates it and under conditions of low DBE exposure is able to detoxify it. In contrast, GSH S-transferase catalyzes the direct conjugation of GSH with DBE. This can result in formation of a reactive intermediate that preferentially binds to nucleic acids and is responsible for the DNA damage observed following DBE exposure. The selective toxicity of this xenobiotic compound may be due to the preponderance of activating GSH conjugating enzymes in the extrahepatic organs. However, this difference alone does not appear sufficient to explain the selection of extrahepatic organs as sites of DBE-induced toxicity.

464

ACKNOWLEDGEMENTS

Some of the studies reported in this paper have been supported by the NIEHS and Hoffman-LaRoche, Inc. We thank Leslie Auerbach for her help with preparation of the manuscript.

REFERENCES

Armstrong, D.J., Kutob, R. and Sipes, I.G. Deuterium isotope effect on the expiration and tissue distribution of 1,2-dibromoethane. The Toxicologist 5:174, 1985.

Brodie, B.B., Reid, W., Cho, A.K., Sipes, G., Krisha, G. and Gillette, J.R. Possible mechanism of liver nicrosis caused by aromatic organic compounds. Proc. Nat'l. Acad. Sci. U.S.A. 68:160-164, 1971.

Clark, H.G., Cropp, P.L., Smith, J.N., Speir, T.W. and Tan, B.J. Photometric determination of methyl parathion GSH S-methyltransferase. Pestic. Biochem. Physiol. 6:126-131, 1976.

Dohn, D.R. and Anders, M.W. The enzymatic reaction of chlorotrifluoroethylene with glutathione. Biochem. Biophys. Res. Comm. 109:1339-1345, 1982.

Edwards, K., Jackson, H., and Jones, A.R. Studies with alkylating esters-II. A chemical interpretation through metabolic studies of the antifertility effects of ethylene dimethanesulphonate and ethylene dibromide. Biochem. Pharmacol. 19:1783-1789, 1970.

Gandolfi, A.J., Nagle, R.B., Soltis, J.J. and Plescia, F.H. Nephrotoxicity of halogenated vinyl cysteine compounds. Res. Comm. Chem. Path. Pharmacol. 33:249-261, 1981.

Hill, D.L., Shih, T.-W., Johnston, T.P. and Struck, R.F. Macromolecular binding and metabolism of the carcinogen 1,2-dibromoethane. Cancer Res. 38:2438-2442, 1978.

Jollow, D.J., Mitchell, J.R., Zampaglione, N. and Gillette, J.R. Bromobenzene induced liver necrosis. Protective role of glutathione and evidence for 3,4-bromobenzene oxide as the hepatotoxic metabolite. Pharmacology 11:151-169, 1974.

Kluwe, W.M., McNish, R., Smithson, K. and Hook, J.B. Depletion by 1,2-dibromoethane, 1,2-dibromo-3-chloropropane, tris (2,3-dibromopropyl) phosphate and hexachloro-1,3-butadiene of reduced nonprotein sulfhydryl groups in target and non-target organs. Biochem. Pharmacol. 30:2265-71, 1981.

Kohn, K.W., Erickson, L.C., Ewig, R.A. and Friedman, C.A. Fractionation of DNA from mammalian cells by alkaline elution. Biochem. 15:4629-4637, 1976.

Kowalski, B., Brittebo, E.B. and Brandt, I. Epithelial binding of 1,2-dibromoethane (EDB) in the respiratory and upper alimentary tracts of mice and rats. Cancer Res. 45:2616-2625, 1985.

Letz, G.A., Pond, S.M., Osterloh, J.D., Wade, R.L. and Becker, C.E. Two fatalities after acute occupational exposure to ethylene dibromide. J. Amer. Med. Assoc. 252:2428-2431, 1984.

MacFarland, R.T., Gandolfi, A.J. and Sipes, I.G. Extra-hepatic GSH-dependent metabolism of 1,2-dibromoethane (DBE) and 1,2-dibromo-3-chloropropane (DBCP) in the rat and mouse. Drug Chem. Toxicol. 7:213-227, 1984.

Nachtomi, E. The metabolism of ethylene dibromide in the rat. The enzymic reaction with glutathione in vitro and in vivo. Biochem. Pharmacol. 19:2853-2860, 1970.

NIOSH, National Institute for Occupational Safety and Health, Criteria for a recommended standard: occupational exposure to ethylene dibromide. DHEW (NIOSH) Publication No. 77-221. National Institute for Occupational Safety and Health, 1977.

Olson, W., Habermann, R., Weisburger, E., Ward, J., and Weisburger, J. Brief communication: induction of stomach cancer in rats and mice by halogenated aliphatic fumigants. J. Natl. Cancer Inst. 51 (6):1933, 1973.

Ozawa, N. and Guengerich, F.P. Evidence for formation of an S-(2-(N[7]-guanyl)ethyl) glutathione adduct in glutathione-mediated binding of the carcinogen 1,2-dibromoethane to DNA. Proc. Nat'l. Acad. Sci. U.S.A. 80:5266-5270, 1983.

Rowe, V., Spencer, H., McCollister, D., Hollingsworth, R., and Adams, E. Toxicity of ethylene dibromide determined on experimental animals. A.J.A. Arch. Ind. Hyg. Occupational Med. 6:158-73, 1952.

Schasteen, C.S. nd Reed, D.J. The mechanism of degradation by hydrolysis of S-(2-haloethyl)-cysteine analogs. The Pharmacologist 23:176, 1981.

Shih, T.-W. and Hill, D.L. Metabolic activation of 1,2-dibromoethane by glutathione transferase and by microsomal mixed function oxidase: further evidence for formation of two reactive metabolites. Res. Comm. Chem. Pathol. Pharmacol. 33:449-461, 1981.

Short, R.D., Winston, J.M., Hong, C.-B., Minor, J.L., Lee, C.-C. and Seifter, J. Effects of ethylene dibromide on reproduction in male and female rats. Toxicol. Appl. Pharmacol. 49:97-105, 1979.

Sundheimer, D.W., White, R.D., Brendel, K. and Sipes, I.G. The bioactivation of 1,2-dibromoethane in rat hepatocytes: Covalent binding to nucleic acid. Carcinogenesis 3:1129-1133, 1982.

van Bladeren, P.J., Breimer, D.D., Rotteveel-Smijs, G.M.T. and Mohn, G.R. Mutagenic activation of dibromoethane and diiodomethane by mammalian microsomes and glutathione S-transferases. Mut. Res. 24:341-346, 1980a.

van Bladeren, P.J., Breimer, D.D., Rotteveel-Smijs, G.M.T., de Jong, R.A.W., Buijs, W., van der Gen, A. and Mohr, G.R. The role of glutathione conjugation in the mutagenicity of 1,2-dibromoethane. Biochem. Pharmacol. 29:2975-2982, 1980b.

van Bladeren, P.J., van der Gen, A., Breimer, D.D. and Mohn, G.R. Stereoselective activation of vicinal dihalogen compounds to mutagens by glutathione conjugation. Biochem. Pharmacol. 28:2521-2524, 1979.

van Bladeren, P.J., Breimer, D.D., Rotteveel-Smijs, G.M.T., de Knijff, P., Mohn, G.R., Buis, W., van Meeteren-Wachli, B. and van der Gen, A. The relation between the structure of vicinal dihalogen compounds and their mutagenic activation via conjugation to glutathione. Carcinogenesis 2:499-503, 1981.

Van Duuren, B., Goldschmidt, B., Loewengart, G., Smith, A., Melchionne, S., Seidman, I., and Roth, D. Carcinogenicity of halogenated olefinic and aliphatic hydrocarbons in mice. J. Natl. Cancer Inst. 63(6):1433-1438, 1979.

White, R.D., Petry, T.W. and Sipes, I.G. The bioactivation of 1,2-dibromoethane in rat hepatocytes: deuterium isotope effect. Ghem.-Biol. Interact 49:225-233, 1984.

White, R.D., Gandolfi, A.J., Bowden, G.T. and Sipes, I.G. Deuterium isotope effect on the metabolism and toxicity of 1,2-dibromoethane. Toxicol. Appl. Pharmacol. 69:170-178, 1983.

White, R.D., Sipes, I.G., Gandolfi, A.J. and Bowden, G.T. Characterization of the hepatic DNA damage caused by 1,2-dibromoethane using the alkaline elution technique. Carcinogenesis 2:839-844, 1981.

Wiersma, D.A. and Sipes, I.G. Effect of dibromoethane and dichloroethane on the nonprotein sulfhydryl content of rat organs. The Toxicologist 3:97, 1983a.

Wiersma, D.A. and Sipes, I.G. Metabolism of 1,2-dibromoethane by cytosol of rat liver, forestomach and glandular stomach. Tox. Lett. 18 (Suppl 1):155, 1983b.

Wiersma, D.A., Schnellmann, R.G. and Sipes, I.G. Human liver microsomal and cytosolic metabolic activation of 1,2-dibromoethane. The Pharmacologist 25:104, 1983.

Wong, L.C.K., Winston, J.M., Hong, C.B. and Plotnick, H. Carcinogenicity and toxicity of 1,2-dibromoethane in the rat. Toxicol. Appl. Pharmacol. 63:155-165, 1982.

A COMPARISON OF THE ALKYLATING CAPABILITIES OF THE CYSTEINYL AND

GLUTATHIONYL CONJUGATES OF 1,2-DICHLOROETHANE

D.J. Reed and G.L. Foureman

Biochemistry and Biophysics Department
Oregon State University
Corvallis, Oregon

INTRODUCTION

Metabolic activation by glutathione conjugation of the dihaloethanes ethylene dichloride (EDC) and ethylene dibromide (EDB), continues to be important as these compounds are used commercially in enormous quantities and the mechanisms by which they exert toxicity and mutagenicity are not yet fully understood. The mutagenicity of EDC has been demonstrated to be dependent on the presence of cytosol and glutathione (Rannug et al., 1978). In line with this, is the observation of covalent association of EDC-derived radioactivity with polynucleotides in incubation mixtures containing cytosol, glutathione (GSH), and EDC (Guengerich et al., 1980). In the case of EDB, an alkylated purine has been isolated from similar incubation mixtures (Ozawa and Guengerich, 1983). These and other results with cofactors and inhibitors indicate that the GSH-transferases are initially responsible for the formation of important reactive intermediate(s) from these compounds.

The reactive moiety thought responsible for covalent interactions of dihaloethanes is the three membered sulfur containing ring, the episulfonium ion, generated subsequent to conjugation of EDC or EDB with GSH. Considerable evidence has been accumulated to implicate the episulfonium ion as the reactive species. Marchand and Abdel-Monem (1985) were able to isolate intact a cyclic 5-membered sulfur containing ring (a tetrahydrothiophene) formed in vitro from the glutathione S-transferase catalyzed reaction between 1,4-diiodobutane and GSH. However, on their metabolic pathway to mercapturic acids, glutathione conjugates are biotransformed through a series of intermediates such as the corresponding cysteine and N-acetyl cysteine conjugates. Work by Elfarra and Anders (1985) has implicated the episulfonium ion in the toxicity of S-(2-chloroethyl)-L-cysteine, the cysteine conjugate of EDC. Thus, there exists a series of compounds any one or all of which could theoretically form an episulfonium ion which, in turn, may exhibit differences in reactivity. In this study we report on the relative reactivities of S-(2-chloroethyl)-L-cysteine (CEC) and S-(2-chloroethyl)-GSH (CEG) the authentic cysteine and glutathione conjugates of EDC and present evidence for the existence of a reactive species which does not appear to be the episulfonium ion.

MATERIALS AND METHODS

S-(2-chloroethyl)-L-cysteine (CEC) and S-(2-chloroethyl)-glutathione (CEG) were synthesized by a liquid ammonia/sodium procedure (de Vigneaud and Patterson, 1936) and furthur purified by preparative HPLC on a Whatman M-9 ODS-II column in 2% methanol, 0.1% acetic acid (CEC), or in a 2% to 8% methanol, 0.1% acetic acid gradient stepped at 15 min (CEG). CEC is pure (>99%) by this method and CEG is greater than 90% pure, the main contaminant being the corresponding S-(2-hydroxyethyl)-glutathione.

Plasmid pBR322 was grown, amplified, radiolabeled, and purified according to established procedures (Norgard et al, 1979; Clewell, 1972; Radloff et al., 1967). The various forms of DNA were separated electrophoretically on 0.9% agarose gels and either visualized with ethidium dibromide or excised and counted in a liquid scintillation counter. Alkylation of 4-(p-nitrobenzyl)-pyridine (NBP) was measured according to Friedman and Boger (1961). Hydrolysis parameters were determined with an automatic pH actuated titrimeter. Nuclear magnetic resonance (NMR) was performed on a Bruker 400 MHz instrument under the following data acquisition conditions: pulse width, 30°; sweep width, 6K; filter width 7.6K; data size 16K; number of scans, 16 or 32 generated in 30 to 45 s.

RESULTS

CEC is a more efficient alkylating agent CEG toward NBP by a factor of 2.5 at 1 hr and by a factor of about 5 at the end of 24 hr (Fig. 1). In the case of CEG, alkylating activity is completed at around 1 hr; the level of alkylation products remained unchanged between 1 and 24 hr. This observation is in agreement with the limit of activity of CEG based on its half-life in aqueous solution at this pH and temperature, about 8 min. CEC, unlike CEG, did not display similar rates of alkylation of NBP and hydrolysis. Even though the half-life of CEC under these conditions is equivalent to that of CEG (Table 1), alkylation products continued to be formed after 1 hr. Thirty percent of the total 0-24 hr alkylation products were accumulated after 1 hr. Thus, CEC differs markedly from CEG in its alkylating capacity towards NPB although both compounds are alkylating agents. The differences exhibited between CEC and CEG are quantitative in the amount of product alkylated and qualitative with respect to the duration of alkylating capacity.

Schasteen and Reed (1983) present evidence that the presence of a free amino group in CEC affects the hydrolysis rate of CEC in a pH dependent manner such that when the amino group is blocked, the pH dependency of the hydrolysis rate is abolished. It is possible that this nucleophilic assistance between the free amino group and the halogen-containing carbon could also affect the alkylating capacity of CEC. Table 1 shows the pH dependency of the first order rate constant for hydrolysis for CEC and the absence of pH dependency for the N-acetylated analog of CEC. As CEG also exhibits no appreciable change in its rate constant at pH 6 or pH 8, it is concluded the free amino group provides an important nucleophilic assistance in both the hydrolysis of and alkylation by CEC. Therefore, we think that CEG and N-acetyl CEC are very similar in their lack of nucleophilic assistance during these reactions.

Table 2 shows the affect either CEC or CEG has on relaxing the supercoiled form of plasmid DNA. At equivalent concentrations CEG was ineffectual in manifesting any changes in supercoiled DNA while CEC nearly eliminated this topological form. As the relaxation or unwinding of supercoiled DNA is due to covalent modification of the DNA (Shooter, 1975), it is apparent that CEC is a more potent alkylator than CEG at the level of polynucleotides.

470

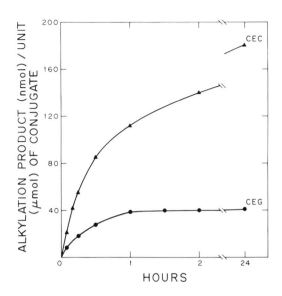

Fig. 1. Alkylation of 4-(nitrobenzyl)-pyridine (NBP) by S-(2-chloro-ethyl)-L-cysteine (CEC), and S-(2-chloroethyl)-glutathione (CEG). Incubations were carried out at 37°C with 2mM CEG or CEC present.

Table 1. First Order Rate Constants for the Hydrolysis of S-(2-chloroethyl)-L-cysteine (CEC), S-(2-chloroethyl)-glutathione (CEG), and S-(2-chloroethyl)-N-acetyl-cysteine (CENAC) at pH 6.0 or pH 8.0 and 30°C.

	$k(min^{-1})$a	
	pH 6.0	pH 8.0
CEC	0.03 ± 0.00 (n = 3)	0.11 ± 0.02 (n = 3)
CEG	0.03 ± 0.01 (n = 3)	0.04 ± 0.01 (n = 3)
CENAC	0.09 ± 0.01 (n = 4)	0.09 ± 0.02 (n = 3)

a Data derived from linear plots of the natural log of the remaining haloethyl concentration and time (min). Only plots with $R^2 \geq 0.99$ were used.

The generation of S-(2-hydroxyethyl)-L-cysteine (OHEC) during the 3 h hydrolysis of CEC is shown in Figure 2, scan B, the hydroxy-derived signal being shifted upfield from 3.0 ppm to 2.8 ppm. However, new and unexpected signals are also observed at 2.9, 3.25, and 4.1 ppm in scan B. None of these signals were explained by OHEC (scan C). When similar experiments were performed with CEG, no extraneous signals were observed (data not shown).

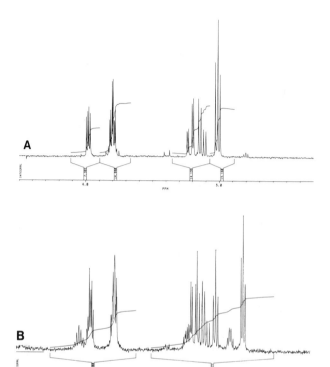

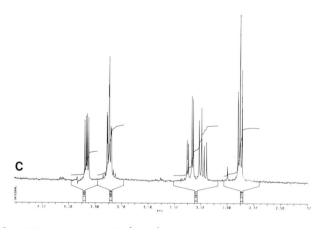

Fig. 2. NMR spectra of A) S-(2-chloroethyl)-L-cysteine at time zero and B) at the end of 3 hours after the addition of buffered D_2O (0.1 M phosphate, pH 6.0. The expected hydrolysis product, S-(2-hydroxyethyl)-L-cysteine is shown for comparison (C).

Table 2. Effect of S-(2-chloroethyl)-glutathione (CEG) and
S-(2-chloroethyl)-L-cysteine (CEC) or the Super-
helicity of pBR322.

Time (min)	% Loss of Superhelicity	
	CEC^a	CEG^a
0	0	0
30	47	3
60	70	2
120	89	4

[a] 10 mM concentration, values are normalized to 100%. Quantitation was by liquid scintillation counting of bands
excised from agarose gels. Supercoiled DNA is expressed
as the percent of total DNA present in the gel channel.
The percentage of total DNA as supercoiled DNA at zero time
was 78.7% ± 3.2. Data from Vadi et al. (1985).

DISCUSSION

     We have shown that not only that CEC can alkylate the model nucleo-
phile NBP, but that it differs markedly from CEG in the extent of its
alkylating capability. This difference may be due to quantitative differences in the reactive species, i.e., the episulfonium ion, modulated by
the cysteinyl or gluthionyl moiety. Alternatively, this difference may be
due to a qualitatively different reactive intermediate, of a non-
episulfonium nature. The data presented here suggest the latter case to
be a distinct possibility. As the predicted rate of hydrolysis of CEC
limits its existence to about 1 hour, that portion of the alkylation
occurring after this time (Figure 1) cannot be explained by evoking the
episulfonium ion. Likewise the NMR scans in Figure 2 definitely show the
presence of some hydrolysis product other than OHEC, the expected breakdown product of the episulfonium ion.

     The fact that, under identical conditions, CEG does not relax super-
coiled DNA while CEC does could also indicate subtle quantitative and
qualitative differences between the reactive intermediates of CEG and CEC
(Vadi et al., 1985). For example, Gamper et al. (1977) reported that in
the case of the benzo(a)pyrene diol epoxides, strand scission of DNA
(measured by relaxation of supercoiled DNA) accounted for only about 1% of
the total DNA alkylation. CEG, then, could be alkylating the DNA extensively but not in the same manner as CEC and not inducing strand breakage.
Another explanation as to why CEC does and CEG does not relax supercoiled
DNA may be related to the mechanism of DNA nicking. Electrophilic attack
on the nucleophilic bases followed by loss of the modified base and subsequent rearrangement of basic sites resulting in strand scission is the
expected basis for relaxation of supercoiled DNA (Singer, 1975). This
relatively slow process is contrasted by the rapid strand scission due to
formation of labile phosphotriesters (Walles and Ehrenberg, 1968). Thus,
the rapid relaxation of supercoiled DNA by CEC and not by CEG (Table 2)
could be interpreted as the reactive intermediate of CEC reacting with
phosphate groups while the reactive intermediate of CEG does not cause
this type of modification.

     A factor which may be causative of the observed differences between
CEG and CEC could be the presence of the free amino group of CEC; in CEG
the amino group is apparently shielded from lending nucleophilic assistance to the haloethyl group (Table 1).

A manner in which the free amino group of CEC may contribute to the prolonged alkylating ability observed for CEC (Figure 1) may be in promoting the formation of a reactive species other than the episulfonium ion. Although we advance no structure for this intermediate whose existence is implied in Figures 1 and 2, it is intriguing to note that Marchand and Abdel-Monem (1985) have observed the dismutation of a GSH conjugate of diiodobutane in which the postulated 5-membered sulfonium ring is actually isolated. In an analagous dismutation of haloethyl conjugates, a three-membered sulfur containing ring, ethylene sulfide, would be released, and could possibly give rise to the extraneous signals noted in Figure 2B. However, analysis of appropriate NMR spectra for this compound and, in the case of CEC, the concomitantly formed pyruvate, failed to show the presence of either compound. It is possible that ethylene sulfide or pyruvate may be oxidized under the conditions in which the NMR were taken and that these products may be elsewhere in the spectra. Alternatively, the extraneous signals in Figure 2B could be due to some cylic intermediate with the phosphate anions present in the buffer. Whatever their origin, these signals are observed during the hydrolysis of CEC and not during the hydrolysis of CEG.

The main point of this report is that the alkylating ability of CEC and CEG differ appreciably toward NBP. Also, evidence is presented which suggests that CEC is able to generate a non-episulfonium ion intermediate which is capable of alkylation; CEG is apparently unable to generate any comparable species.

REFERENCES

Clewell, D.B., 1972, Nature of ColE1 plasmid replication in the presence of chloramphenicol, J. Bacteriol., 110:667.

Elfarra, A.A., Baggs, R.B., and Anders, M.W., 1985, Structure-nephrotoxicity relationships of S-(2-chloroethyl)-DL-cysteine and analogs: role for an episulfonium ion, J. Pharmacol. Exp. Ther., 233:512.

Friedman, O.M., and Boger, E., 1961, Colorimetric estimation of nitrogen mustards in aqueous media, Anal. Chem., 33:906.

Gamper, H.B., Tung, A.S.C., Straub, K., Bartholomew, J.C., and Calvin, M., 1977, DNA strand scission by benzo(a)pyrene diol epoxide, Science, 197:671.

Guengerich, F.P., Crawford, W.M., Domoradzki, J.Y., MacDonald, T.L., and Watanabe, P.G., 1980, In vitro activation of 1,2-dichloroethane by microsomal and cytosolic enzymes, Toxicol. Appl. Pharmacol., 55:303.

Marchand, D.H., and Abdel-Monem, M.M., 1985, Glutathione S-transferase catalyzed conjugation of 1,4-disubstituted butanes with glutathione in vitro, Biochem. Biophys. Res. Commun., 128:360.

Norgard, M.V., Emigholtz, K., and Monahan, J.J., 1979, Increased amplification of pBR322 plasmic deoxyribonucleic acid in E. coli K-12 strains RR1 and X1776 grown in the presence of high concentrations of nucleoside. J. Bacteriol., 138:270.

Ozawa, N., and Guengerich, F.P., 1983, Evidence for formation of an S-[2-(N^7-guanyl)ethyl]glutathione adduct in glutathione-mediated binding of the carcinogen 1,2-dibromoethane to DNA, Proc. Natl. Acad. Sci. USA, 80:5266.

Radloff, R. Bauer, W., and Vinograd, J., 1967, A dye-buoyant density method for the detection and isolation of closed circular duplex DNA: the closed circular DNA in Hela cells. Proc. Natl. Acad. Sci. USA, 57:1514.

Rannug, U., Sundrall, A., and Ramel, C., 1978, The mutagenic effect of 1,2-dichloroethane or Salmonella typhimurium. 1: activation through conjugation with glutathione in vitro, Chem.-Biol. Interact., 20:1.

Schasteen, C.S., and Reed, D.J., 1983, The hydrolysis and alkylation activation of S-(2-haloethyl)-L-cysteine analogs-evidence for extended half-life, Toxicol. Appl. Pharmacol., 70:423.

Shooter, K.V., 1975, Assays for phosphotriester formation in the reaction of bacteriophage R17 with a group of alkylating agents. Chem.-Biol. Interact., 11:575.

Singer, B., and Fraenkel-Conrat, H., 1975, The specificity of different classes of ethylating agents toward various sites in RNA, Biochemistry, 14:772.

Vadi, H.V., Schasteen, C.S., and Reed, D.J., 1985, Interactions of S-(2-haloethyl)-mercapturic acid analogs with plasmid DNA, Toxicol. Appl. Pharmacol., in press.

duVigneaud, V., Patterson, W.I., 1936, The synthesis of djenkolic acid, J. Biol. Chem., 114:533.

Walles, S., and Ehrenburg, L., 1968, Effects of β-hydroxyethylation and β-methoxyethylation on DNA in vitro, Acta Chem. Scand., 22:2727.

COVALENT REACTIONS IN THE TOXICITY OF SO_2 AND SULFITE

Daniel B. Menzel, DOuglas A. Keller, and Kwan-Hang Leung

Depts. of Pharmacology and Medicine, Comprehensive
Cancer Center, Duke Univ. Med Ctr, Durham, Nc. 27710

INTRODUCTION

Sulfur dioxide (SO_2) is a major urban air pollutant, resulting from combustion of sulfur containing fossil fuels. It readily dissolves in water forming sulfurous acid, which dissociates to form bisulfite and sulfite ions (collectively referred to as sulfite), in a ratio depending on the pH of the solution (Petering and Shih, 1975):

$$SO_2 + H_2O \rightleftharpoons H^+ + HSO_3^- \rightleftharpoons 2H^+ + SO_3^{2-}$$

Sulfur dioxide and sulfite are also commonly used as antimicrobial and antioxidant agents in the preservation of foods and beverages (Chichester and Tanner, 1972).

The toxic effects of both SO_2 and sulfite have been extensively reviewed (Shapiro, 1977; Gunnison, 1981). Urban SO_2 exposure of humans has been correlated with increased mortality, bronchitis, asthma and other respiratory diseases. SO_2 also produces bronchoconstriction in animals and human subjects during controlled exposures (Amdur, 1969; Koenig et al., 1983; Linn et al., 1983). Asthmatic patients are thought to be particularly sensitive to SO_2. The correlation between SO_2 exposure and the mortality rate due to breast cancer, stomach cancer, and heart disease has been particularly disturbing (Shapiro, 1977). Simultaneous exposure to SO_2 and polycyclic aromatic hydrocarbons such as benzo(a)pyrene (Laskin et al. 1976) and dibenz(a,h)anthracene (Pott and Stober, 1983) increase the lung tumor and cancer incidence in rats. However, the molecular mechanisms of the toxic effects of SO_2 and sulfite have not been elucidated. One approach to the study of this problem is to investigate the reactions of SO_2, sulfites or their metabolic intermediates, with cellular components. Our emphasis has been particularly on covalent reaction products in an attempt to find a unifying concept for the apparently divergent toxic effects of sulfite and SO_2.

Sulfite reacts with all major classes of cellular molecules both <u>in vitro</u> and <u>in vivo</u> (Gunnison, 1981). One of the most well documented reactions of sulfite is the lysis of disulfide bonds by a nucleophilic displacement to form thiols and S-sulfonates:

$$R-S-S-R' + HSO_3^- \longrightarrow RSH + RS-SO_3^-$$

Sulfitolysis of protein disulfide bonds results in the formation of cysteine S-sulfonate (Gunnison and Benton, 1971), while sulfitolysis of glutathione disulfide (GSSG) produces glutathione S-sulfonate ($GSSO_3H$) (Waley, 1959). To explore the importance of the sulfitolysis reaction in the toxicity of sulfite and SO_2, we have developed methods for the detection of both cysteine S-sulfonate formed from proteins and $GSSO_3H$ formed from GSSG. The potential metabolic consequences of $GSSO_3H$ on GSH-dependent reactions have been studied using the glutathione S-transferase activity. In order to place these reactions in proper perspective, the destruction of $GSSO_3H$ by GSSG reductase has also been investigated.

MATERIALS AND METHODS

<u>Culture of Human A549 Cells Under Conditions of Constant GSH Content</u>

A549 cells were grown in Dulbecco's Modified Eagle's Medium (DMEM) with dialyzed fetal calf serum (FCS) and kanamycin sulfate (100 ug/mL). Insulin (5 mg/mL), transferrin (5 mg/mL) and selenous acid (5 ng/mL) were added as ITS Premix (Collaborative Research). Cells were

478

determined to be mycoplasma free, and were cultured in monolayer in 5% CO_2 and a humid atmosphere at 37°C. For experiments in 2% or 10% FCS, cells were grown prior to passage in media containing the same FCS concentration to be used in the study. For studies in serum-free media, cells were grown in 2% FCS, and then replated in media lacking FCS.

For determination of cell number and GSH content, cells adhering to the culture dish were washed with saline and removed with trypsin. A hemocytometer was used for cell counting. GSH analysis was by high performance liquid chromatography (Reed, et al., 1980). Student's t test was used to determine statistical significance.

Formation of Glutathione S-Sulfonate in Lung Cells.

A549 cells were grown in DMEM with 10% FCS. Cells were exposed to 0, 2, 20, or 40 mM sulfite in Hepes buffer for 45 min. at 37°C. Cells were then detached from the plate. Cell extracts were prepared by treatment with perchloric acid, and the GSH content determined by anion exchange HPLC. Compounds were detected by post-column derivatization with o-phthalaldehyde to form a fluorescent product (Keller and Menzel, 1985).

Sulfite-Binding Proteins in Rat and Human Cells.

Rat lungs (Sprague Dawley, 200-250 g) were perfused with ice-cold saline to remove blood, homogenized, and cytosol prepared by centrifugation at 100,000g. A549 cell cytosol was prepared by freeze-thawing the cells, followed by centrifugation. Lung cytosol, A549 cell cytosol, and rat plasma were treated with 0.1 mM $Na_2{}^{35}SO_3$ for 2 hours at 37°C. The samples were dialyzed to remove free sulfite, then separated by gel filtration on Sephadex G-200. The ^{35}S content was determined by liquid scintillation counting using an LKB Rackbeta scintillation counter.

Metabolism of Gultathione S-Sulfonate

$GSSO_3H$ was prepared by reaction of glutathione and sodium tetrathionate as described by Eriksson and Rundfelt (1968). Glutathione S-transferase (GST) (E.C. 2.5.1.18) activity was determined spectrophotometrically at 340nm at 25° using 1-chloro-2,4-dinitrobenzene as the substrate (Habig, et al., 1974). In experiments where the effects of GSSO$_3$H were studied, 1 - 100 uM GSSO$_3$H was included.

The enzyme activities mediating the reduction of GSSG (GSSG reductase) and GSSO$_3$H (GSSO$_3$H reductase) were determined spectrophotometrically by measuring the oxidation of NADPH at 340 nm (Carlberg and Mannervik, 1975), except that the assay was carried out at pH 7.0. In experiments where the effect of GSH on the reduction of GSSO$_3$H was studied, the cytosol of A549 cells was dialyzed against three changes of phosphate buffered saline, pH 7.2, at 4°C for eight hours each prior to use, to remove endogenous GSH. The sulfite concentration in the reaction mixture was determined by the p-rosaniline method (West and Gaeke, 1956).

RESULTS AND DISCUSSION

Variation of Glutathione Content in Cultured Cells

Table 1 shows both cell number and glutathione content for A549 cells grown in 0, 2, or 10 % FCS, with and without ITS (taken from Post et al., 1983). At each time point, GSH content was greater as the serum concentration was increased. With media containing 2 or 10% FCS, GSH levels increased between 0 and 24 hours after passage, and decreased at each time point thereafter. In 10 % FCS, a 195 % increase in GSH occurred in the first 24 hours, while in 2% FCS, a 31% increase was observed. The ITS mixture affected cell growth only in the 0 % FCS cells, by increasing plating efficiency from 16 to 31%. Growth rates and time dependence of GSH levels did not change with ITS present. The question of whether the time-dependent decrease in GSH is due to a depletion of GSH precursors in the media was investigated by plating cells in 10 % FCS and replacing the media with fresh media every 24 hours. Figure 1 illustrates that changing the media daily did not prevent the time-dependent decrease in GSH.

Table 1

Effects of Time Since Passage, Serum Concentration, and ITS on GSH Levels in A549 Cells

Media Used:	0% FCS		0% FCS+ITS		2% FCS		2% FCS+ITS		10% FCS		10% FCS+ITS	
	Cells (x10⁶)	GSH[a]	Cells (x10⁶)	GSH	Cells (x10⁶)	GSH	Cells (x10⁶)	GSH	Cells (x10⁶)	GSH	Cells (x10⁶)	GSH
Hours After Passage:												
0	0.91± 0.01	15.8± 5.3	0.91± 0.01	15.8± 5.3	2.70± 0.12	18.5± 1.8[d,e]	3.30± 0.14	7.4± 2.1[d]	0.83± 0.02	41.0± 5.8[d]	1.13± 0.22	42.0± 1.3[d]
24	0.15± 0.07	4.5± 1.1[b,c]	0.28± 0.10	5.9± 1.5[b,c]	2.25± 0.09	24.2± 3.6[e]	1.93± 0.35	19.6± 4.3[e]	0.54± 0.12	120.8± 44.2	0.93± 0.28	92.2± 12.0
48	0.32± 0.07	8.5± 0.4[c]	0.34± 0.15	13.5± 5.3	3.02± 0.59	16.9± 3.7[d]	3.36± 0.63	15.6± 2.6	N.A.[f]	N.A.	N.A.	N.A.
72	—g—	——	—g—	——	4.00± 0.45	15.3± 2.7[d,e]	3.55± 0.16	12.7± 1.4[d,e]	1.01± 0.15	51.6± 8.0[d]	1.93± 0.74	31.4± 1.6[d]
96	——	——	——	——	6.54± 0.76	4.7± 1.2[d,e]	4.86± 0.68	7.8± 0.8[d]	1.60± 0.55	37.2± 0.9[d]	3.08± 0.94	6.8± 0.9[d]

[a]GSH expressed in nmoles/10⁶ cells
[b]p<0.05 when compared to 0 hours in same media
[c]p<0.05 when compared to 2% FCS at same time after passage
[d]p<0.05 when compared to 24 hours in same media
[e]p<0.05 when compared to 10% FCS at same time after passage
[f]N.A. = Not analyzed
[g]Experiment terminated because cells lost attachment

(From Post et al., 1983. Reproduced with permission.)

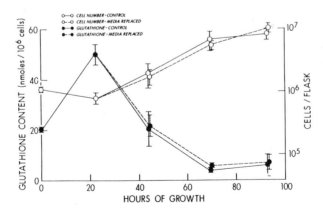

Figure 1. Effects of changing media daily on growth and GSH content in A549 cells. Cells were grown in media containing 10 % FCS, and media was removed and replaced with fresh media at 24, 48, and 72 hours. Values are mean ± S.D. for 3 flasks of cells. (From Post et al., 1983. Reproduced with permission.)

Fluctuations in non-protein sulfhydryl content in maintained cell lines has been reported previously (Harris and Patt, 1969; Ohara and Terasima, 1969). Harris and Teng (1972) suggested that the rapid increase in GSH after passage was due to an increase in the proportion of proliferating cells, which synthesize protein at a higher rate than plateau phase cells. Our results indicate otherwise, since the decrease in GSH content observed after 24 hours occurred during log-linear growth.

Large differences in GSH content occurred with time during log-phase growth. Thus, when reporting GSH levels, it is important to mention the time after passage at which measurements were made. Serum concentration appears to be highly significant in determining GSH content. Since the metabolism and toxicity of xenobiotic compounds can be mediated by the availability of GSH, knowledge of intracellular GSH levels is important for interpreting studies of these compounds. Standardization of culture conditions is necessary when studying the effects of xenobiotic compounds in cultured cells.

These studies also raise the question of the role of serum proteins in regulating intracellular GSH content. The dependence of A549 cells on serum in the presence of adequate levels of extracellular precursors of GSH suggests that regulation of GSH content in vivo may be highly complex. Given the central role of GSH in detoxication reactions, especially those involving covalent reactions, the serum effect may be central to toxicity and carcinogensity of xenobiotic compounds.

Formation of Glutathione S-Sulfonate in Lung Cells

Glutathione S-sulfonate was detected in A549 cells, on day 1 of growth, following exposure to 2, 20, and 40mM sulfite. A dose-dependent increase in $GSSO_3H$ was evident, while the ratio of $GSH/GSSO_3H$ decreased in a dose-dependent manner. In cells exposed to 20 mM sulfite, the $GSSO_3H$ was 0.474 nmole/mg protein, with a $GSH/GSSO_3H$ ratio of 56. Thus the exposure of lung cells to sulfite results in the formation of $GSSO_3H$. The effects of $GSSO_3H$ formation on carcinogen detoxification is discussed below.

Sulfite-Binding Proteins in Rat and Human Cells

Rat plasma contains two proteins with sulfite-reactive disulfide bonds, fibronectin and albumin (Gregory and Gunnison, 1984). Rat lung cytosol (Fig. 2) contains at least one other protein with sulfite binding capabilities. It has a molecular weight of greater than 200,000 daltons. A species with similar characteristics is present in A549 cell cytosol. Treatment with mercaptoethanol eliminates the radiolabel from the protein.

Although the identity of these intracellular proteins is not known, these proteins may represent specific sites of interaction of sulfite or inhaled SO_2 with cellular biopolymers. Since integrity of disulfide bonds may be crucial to protein function, the existence of protein S-sulfonates may result in altered cellular activities.

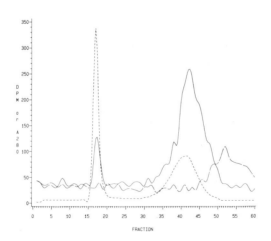

Figure 2. Rat lung cytosol, chromatographed on Sephadex G-200, after exposure to $Na_2{}^{35}SO_3$. Fractions of 0.5 mL were counted for ^{35}S (————), and analyzed for absorbance at 280 nm (----). ^{35}S content of fractions after addition of 2-mercaptoethanol to the sample prior to chromatography was also determined (— —). No chamge in absorbance at 280 nm was seen after treatment with 2-mercaptoethanol. (Absorbance at 280 nm is x 1000.)

Glutathione S-Transferase Activities in Rat Liver and Lung
and A549 Cells and Its Inhibition by Glutathione S-Sulfonate

The kinetic constants of the cytosolic and microsomal
GST with 1-chloro-2,4-dinitrobenzene as the substrate in the
rat liver and lung and human A549 cells are shown in Table
2. The cytosolic GST activity in the rat liver was
approximately 10 times higher than those in the rat lung and
A549 cells, while the K_m of GSH for enzymes from all three
sources were similar. Microsomal GST activities were less
than 3% that of the cytosolic GST in each case.

GST activities catalyzing the conjugation reaction of
GSH and benzo(a)pyrene-4,5-oxide (BP-4,5-oxide), r-7,t-8-
dihydroxy-t-9,10-epoxy-7,8,9,10-tetrahydrobenzo(a)pyrene
(anti-BPDE), and r-7,t-8-dihydroxy-c-9,10-epoxy-7,8,9,10-
tetrahydrobenzo(a)pyrene (syn-BPDE) were also determined
(Leung and Menzel, 1985a). Briefly, GSH-BP-4,5-oxide
conjugate was formed most efficiently with V_{max}'s at 2.38,
2.85, and 1.16 umole conjugate formed per min. per mg
protein from rat liver, lung, and A549 cell, respectively.
Conjugates of anti- and syn-BPDE were formed at rates of
approximately 30 and 10 %, that of BP 4,5-oxide.

Figure 3 represents a typical Lineweaver-Burk plot of
the inhibition of rat lung GST by $GSSO_3H$ using 1-chloro-2,4-
dinitrobenzene as the substrate. This study clearly
indicates that $GSSO_3H$ is a potent competitive inhibitor of
GST activities. The K_i values of $GSSO_3H$ for GST from
various enzyme sources are shown in Table 3. With the
exception of rat lung microsomal GST, the K_i of $GSSO_3H$
toward GST in each case was approximately one order of
magnitude lower than the K_m of GSH. Cysteine S-sulfonate, a
product formed following sulfitolysis of cystine, did not
inhibit the GST activity at a concentration as high as 1 mM.

SO_2 and many airborne carcinogens such as polycyclic
aromatic hydrocarbons (PAH) coexist in polluted urban air.
Reports by Laskin et al. (1976), and Pott and Stober (1983)
suggested that SO_2 may potentiate the carcinogenicity of
PAHs. However, the molecular basis of the cocarcinogenic
effect of SO_2 is uncertain. Sulfite inhibits DNA synthesis,

Table 2

Kinetic Constants of Glutathione S-Transferases in the Rat
Liver and Lung, and Human A549 Cells

Enzyme source	V_{max} (nmol product min^{-1} mg protein^{-1})	K_m (GSH) (μM)
Rat liver		
Cytosol	1600	80
Microsomes	40	160
Rat lung		
Cytosol	130	93
Microsomes	4	48
A549 cell		
Cytosol	170	130
Microsomes	5	500

(From Leung et al., 1985. Reproduced with permission.)

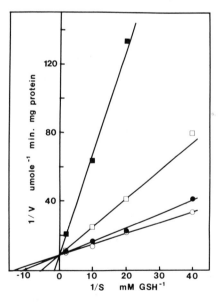

Figure 3.   Lineweaver-Burk plot of the glutathione S-transferase activities in rat lung cytosol mediating the reaction of glutathione and 1-chloro-2,4-dinitrobenzene in the presence of various amounts of glutathione S-sulfonate. (O), 0; (●), 1; (□), 10; and (■), 100 uM.   (From Leung et al., 1985. Reproduced with permission.)

Table 3

K_i Values of Glutathione S-Sulfonate for Glutathione S-Transferases in the Rat Liver and Lung, and Human A549 Cells

Enzyme source	K_i (μM)
Rat liver	
Cytosol	14
Microsomes	21
Rat lung	
Cytosol	9
Microsomes	24
A549 cell	
Cytosol	4
Microsomes	60

(From Leung et al., 1985. Reproduced with permission.)

disrupts DNA structure, and is a mutagen (Shapiro, 1977). Thus, it is possible that sulfite might cause genetic lesions directly which could add to those promoted by PAHs. How such reactions might the observed cocarcinogenic effects is not clear.

On the other hand, GST-mediated reactions represent essential detoxification pathways for PAH such as BP (Hesse et al., 1982; Jernstrom et al., 1984) and are of particular importance in the lung (Ball et al., 1979; Smith et al., 1980) where epoxides of BP are predominantly conjugated with GSH, while sulfation and glucuronidation are relatively inactive. In this study, we have shown that GSSO$_3$H is a strong competetive inhibitor of cytosolic and microsomal GST in the rat liver, lung and human lung A549 cells using both chromogenic and BP epoxide substrates (Leung and Menzel, 1985). For the chromogenic substrate, the K$_i$ of GSSO$_3$H is considerably lower than the K$_m$ of GSH, suggesting a higher affinity of the enzyme for GSSO$_3$H than GSH. The formation of GSSO$_3$H in the presence of low concentrations of GSH could lead to decreased conjugation of BP epoxides and other reactive intermediates of PAHs. Transient concentrations of PAH reactive intermediates could thus occur on simultaneous inhalation of SO$_2$ and PAHs leading to the increased tumor rates observed by Laskin et al. (1976) and Pott and Stober (1983).

Table 4

Kinetic Constants of GSSG and GSSO₃H Reductases in the
Cytosol of Rat Liver and Lung, and Human A549 Cells

Enzyme source	GSSG reductase			GSSO₃H reductase		
	V_{max} (nmol min^{-1} mg protein^{-1})	K_m (GSSG) (μM)	K_m (NADPH) (μM)	V_{max} (nmol min^{-1} mg protein^{-1})	K_m (GSSO₃H) (μM)	K_m (NADPH) (μM)
Liver	110	25	8	110	313	13
Lung	55	25	11	55	200	20
A549 cells	185	38	6	185	400	ND*

* ND, Not determined.

(From Leung et al., 1985. Reproduced with permission.)

Metabolic Fate of GSSO₃H

GSSO₃H can be degraded _in vitro_ in the presence of
NADPH and cytosolic protein from rat liver and lung and A549
cells. Under identical conditions, degradation of cysteine
S-sulfonate was not detected.

Some properties of GSSG and GSSO₃H reductases have been
compared. Both reductases were cytosolic enzymes with
identical V_{max} values, suggesting that they are the same
enzyme (Table 4). The K_m of GSSO₃H was approximately ten
times higher than those of GSSG in all three enzyme sources
examined. The relatively high substrate K_m suggestes that
GSSO₃H, when formed from exgenous sulfite may not be readily
reduced. The transient level of GSSO₃H may be significant
in inhibiting GST activity, as has been discussed in the
previous section. Two further pieces of evidence showed
that both GSSG and GSSO₃H reductases are the same protein.
Purified yeast GSSG reductase, which contained a single band
in SDS-polyacrylamide gel electrophoresis, showed identical
V_{max} values for the reduction of GSSG and GSSO₃H. In
addition, bis-(2-chloroethyl)nitrosourea, which is a known
GSSG reductase inhibitor (Babson and Reed, 1979) also
inhibits GSSO₃H reductase activity.

The nature of GSSO₃H reduction has been examined
previously. Winnell and Mannervik (1969) have proposed that
the reduction of GSSO₃H is due to an initial reaction of

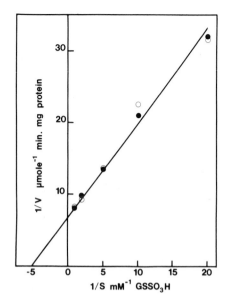

Figure 4. Lineweaver-Burk plot of GSSO₃H reductase activities in the cytosol of A549 cells which had either been dialyzed (O) or not dialyzed (●). (From Leung et al., 1985. Reproduced with permission.)

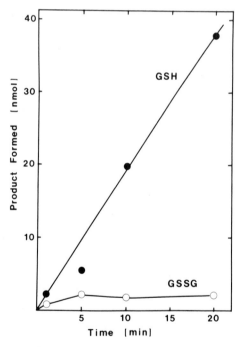

Figure 5. Formation of GSH and GSSG at various times following the enzymatic reduction of GSSO₃H. Incubation mixtures consisted of 0.1 M phosphate buffer, pH 7.0, 0.5 mM EDTA, 0.1 mM NADPH, 0.5 mM GSSO₃H, and 65 ug dialyzed cytosolic protein of A549 cells in a final volume of 1.0 mL. (From Leung et al., 1985. Reproduced with permission.)

GSSO$_3$H with GSH to form GSSG, which is then reduced by NADPH catalyzed by GSSG reductase. We have found that GSH was not required for the GSSO$_3$H reductase activity. A549 cell cytosol which had been dialyzed (containing 0.08 uM GSH per assay) exhibited identical reaction kinetics as those which had not been dialyzed (30 uM GSH per assay) (Fig. 4). Furthermore, purified yeast GSSG reductase, which was shown not to contain GSH, contained GSSO$_3$H reductase activity.

GSH and GSSG were formed as products following enzymatic reduction of GSSO$_3$H (Fig. 5). It remains unclear at present whether GSH or GSSG is the immediate product of the reduction. Free sulfite was not detected as a reaction product, suggesting that it is either bound, presumably to the enzyme, or further reacted to other oxidation states. Though the thiol-transferase pathway suggested by Winnell and Mannervik (1969) cannot be excluded, since this reaction can be initiated by the GSH produced during the reduction of GSSO$_3$H, we propose an alternative mechanism, based on our findings, that GSSO$_3$H acts as a co-substrate for GSSG reductase, and is directly reduced by the enzyme to form GSH with the transfer of the sulfite moiety to the enzyme protein.

SUMMARY

Toxic effects of SO$_2$ and sulfite such as bronchitis and bronchoconstriction have been well documented. SO$_2$ has also been suggested to potentiate carcinogenic effects of PAH. However, the molecular basis of these toxic effects is unclear. We have examined the covalent reaction of SO$_2$ and sulfite with cellular proteinacious and nonproteinaceous sulfhydryl compounds using rat liver, and lung and human lung derived A549 cells. Reactions of sulfite and protein in rat and human lung cells reveals at least three proteins with sulfite-reactive disulfide bonds. Besides fibronectin and serum albumin, which had been reported to contain sulfonated products following exposure to sulfite, we have found one other protein with sulfite-binding capabilities. Since the integrety of disulfide bonds is crucial to the tertiary structure and thus protein function, the disruption of protein structure by sulfitolysis may result in altered cellular activities leading to biochemical lesions. Using carefully controlled conditions, reproducible GSH contents can be found in cultured cells and used as an experimental basis for stuyding alterations in the GSH and GSSG content of cells. Sulfitolysis of GSSG results in the formation of $GSSO_3H$ in A549 cells, and possibly in the lung. $GSSO_3H$ can be reduced enzymatically by GSSG reductase. However, the K_m of $GSSO_3H$ is high compared to that of GSSG, suggesting the existence of a transient concentration of $GSSO_3H$ once it is formed. Cysteine S-sulfonate is, however, not reduced by cytosolic extracts in the presence of NADPH and would have to be eliminated from the cell by other means. $GSSO_3H$ is a strong competitive inhibitor of GST in rat liver and lung and A549 cells, using 1-chloro-2,4-dinitrobenzene as a substrate. It also inhibits the formation of GSH conjugates of BP 4,5-oxide, anti and syn BPDE, but to a lesser extent. These results suggest that SO_2 may affect the detoxification of xenobiotic compounds by inhibiting, via formation of $GSSO_3H$, the enzymatic conjugation of GSH and reactive electrophiles. Since GSH conjugation represents the major pathway of elimination of BP epoxides in the lung, our results offer a possible explanation for the cocarcinogenicity of SO_2 with PAHs.

These data suggest that the sulfitolysis reaction of sulfite is the common reaction mechanism mediating the underlying biochemical reactions leading to both the toxic and cocarcinogenic properties of SO_2. Quantitation of sulfitolysis products and their interaction·with cellular processes should provide a coherent scheme relating SO_2 and sulfite toxicity among animal species and humans.

REFERENCES

Amdur, M. O., 1969. Toxicological appraisal of particulate matter, oxides of sulfur and sulfuric acid. J. Air Pollution Control Assoc. 19, 638-646.
Babson, J. R. and D. J. Reed, 1978. Inactivation of glutathione reductase by 2-chloroethyl nitrosourea-derived isocyanates. Biochem. Biophys. Res. Comm. 83, 754-762.
Ball, L. M., J. L. Plummer, B. R. Smith and J. R. Bend, 1979. Benzo(a)pyrene oxidation, conjugation and

disposition in the isolated perfused rabbit lung: role of glutathione transferases. Med. Biol. 57, 298-305.

Carlberg, I. and B. Mannervik, 1975. Purification and characterization of the flavoenzyme glutathione reductase from rat liver. J. Biol. Chem. 250, 5475-5480.

Chichester, D. F. and F. W. Tanner, Jr., 1972. Antimicrobial food additives. In Handbook of Food Additives (T. E. Furia ed.), 2nd ed., pp. 115-184. CRC Press, Cleveland.

Eriksson, B. and M. Rundfelt, 1968. Reductive decomposition of S-sulfoglutathione in rat liver. Acta Chem. Scand. 22, 562-570.

Gregory, R. E. and A. F. Gunnison, 1984. Identification of plasma proteins containing sulfite-reactive disulfide bonds. Chem.-Biol. Interact. 49, 55-69.

Gunnison, A. F., 1981. Sulfite toxicity: A critical review of in vitro and in vivo data. Fd. Cosmet. Toxicol. 19, 667-682.

Gunnison, A. F. and A. W. Benton, 1971. Sulfur dioxide:sulfite, interactions with mammalian serum and plasma. Arch. Environ. Health 22, 381-388.

Habig, W. H., M. J. Pabst and W. B. Jakoby, 1974. Glutathione S-transferases: The first enzymatic step in mercapturic acid formation. J. Biol. Chem. 249, 7130-7139.

Harris, J. W. and H. M. Patt, 1969. Non-protein sulfhydryl content and cell cycle dynamics of Erlich ascites tumor. Exp. Cell Res. 56, 134-141.

Harris, J. W. and S. S. Teng, 1972. Sulfhydryl groups during the S phase: Comparison of cells from G_1, plateau phase G_1, and G_0. J. Cell Physiol. 81, 91-96.

Hesse, S., B. Jernstrom, M. Martinez, P. Moldeus, L. Christoulides and B. Ketterer, 1982. Inactivation of DNA binding metabolites of benzo(a)pyrene and benzo(a)pyrene-7,8-dihydrodiol by glutathione and glutathione S-transferases. Carcinogenesis 3, 757-760.

Jernstrom, B., L. Dock and M. Martinez, 1984. Metabolic activation of benzo(a)pyrene-7,8-dihydrodiol and benzo(a)pyrene-7,8-dihydrodiol-9,10,epoxide to protein binding products and the inhibition effect of glutathione and cysteine. Carcinogenesis 5, 199-204.

Keller, D. A. and D. B. Menzel, 1985. Picomole analysis of glutathione, glutathione disulfide, glutathione S-sulfonate, and cysteine S-sulfonate by HPLC. Analytical Biochemistry (submitted).

Koenig, J. Q., W. E. Paison, M. Horike and R. Frank, 1983. A comparison of the pulmonary effects of 0.5 ppm versus 1.0 ppm sulfur dioxide plus sodium chloride droplets in asthmatic adolescents. J. Tox. Environ. Health 11, 129-139.

Laskin, S., M. Kuschner, A. Sellakuman and G. V. Katz, 1976. Combined carcinogen-irritant animal inhalation studies. In Air Pollution and the Lung (E. F. Aharoson, A. Ben-David and M. A. Klingberg, eds.), pp. 190-213. Wiley, New York.

Leung, K.-H., G. B. Post and D. B. Menzel, 1985. Glutathione S-sulfonate, a sulfur dioxide metabolite, as a competitive inhibitor of glutathione S-transferase, and its reduction by glutathione reductase. Toxicol. Appl. Pharm. 77, 388-394.

Leung, K.-H. and D. B. Menzel, 1985a. Formation of glutathione conjugate of benzo(a)pyrene epoxides *in vitro* and its inhibition by glutathione S-sulfonate. Toxicol. Appl. Pharm. (submitted).

Linn, W. S., T. G. Venet, D. A. Shamoo, L. M. Valencia, U. T. Anzai, C. E. Spier, and J. D. Hackney, 1983. Respiratory effects of sulfur dioxide in heavily exercising asthmatics. Am. Rev. Respir. Dis. 127,287-283.

Ohara, H. and T. Terasima, 1969. Variations of cellular sulfhydryl content during cell cycle of HeLa cells and its correlation to cyclic change of X-ray sensitivity. Exp. Cell Res. 58, 182-185.

Petering, D. F. and N. T. Shih, 1975. Biochemistry of bisulfite - sulfur dioxide. Environmental Res. 9, 55-65.

Pott, F. and W. Stober, 1983. Carcinogenicity of airborne combustion products observed in subcutaneous tissue and lungs of laboratory rodents. Environ. Health Perspect. 47, 293-303.

Post, G. B., D. A. Keller, K. A. Conner and D. B. Menzel, 1983. Effects of culture conditions on glutathione content in A549 cells. Biochem. Biophys. Res. Comm. 114, 737-742.

Reed, D. J., J. R. Babson, P. W. Beatty, A. E. Brodie, W. W. Ellis and D. W. Potter, 1980. High-performance liquid chromatograph analysis of nanomole levels of glutathione, glutathione disulfide, and related thiols and disulfides. Anal. Biochem. 106, 55-62.

Shapiro, R., 1977. Genetic effects of bisulfite (sulfur dioxide). Mutation Res. 39, 149-176.

Smith, B. R., J. L. Plummer, L. M. Ball and J. R. Bend, 1980. Characterization of pulmonary arene oxide biotransformation using the perfused rabbit lung. Cancer Res. 40, 101-106.

Waley, S. G., 1959. Acidic peptides of the lens. 5. S-sulfoglutathione. Biochem. J. 71, 132-137.

Winnell, M. and B. Mannervik, 1969. The nature of the enzymatic reduction of S-sulfoglutathione in liver and peas. Biochem. Biophys. Acta 184, 374-380.

West, P. W. and G. C. Gaeke, 1956. fixation of sulfur dioxide as disulfitomercurate (II) and subsequent colorimetric estimation. Anal. Chem. 28, 1816-1819.

ONE- AND TWO-ELECTRON OXIDATION OF REDUCED GLUTATHIONE BY PEROXIDASES

Ronald P. Mason

Laboratory of Molecular Biophysics
National Institute of Environmental Health Sciences
Research Triangle Park, NC 27709

The oxidation of glutathione by horseradish peroxidase or lactoperoxidase forms a thiyl free radical, as demonstrated with the spin-trapping ESR technique. Reactions of this thiyl free radical result in oxygen consumption, which is inhibited by the radical trap 5,5-dimethyl-1-pyrroline-N-oxide. In contrast to L-cysteine oxidation, glutathione oxidation is highly hydrogen peroxide-dependent. The oxidation of glutathione by glutathione peroxidase forms GSSG without forming a thiyl radical intermediate except in the presence of the thiyl radical-generating horseradish peroxidase.

INTRODUCTION

The sulfhydryl group of L-cysteine plays many important roles in both the structure and function of proteins. These roles are modulated by the oxidation of L-cysteine. The ease of L-cysteine oxidation is also responsible for the radioprotection of intracellular GSH[1], the bactericidal effect of cysteine[2], and the general toxicity of cysteine. For this reason N-acetylcysteine, and not cysteamine, is used to treat cases of acetaminophen overdose[3]. The oxidation of thiol compounds, including L-cysteine, by metal ions and radiation has been studied for a number of years. In fact, in the event of World War III it will be sulfhydryl compounds one takes to try to prevent radiation poisoning[4]. Among the stable products of oxidation are L-cystine, L-cysteine sulfinic acid and L-cysteine sulfonic acid.

$$ RSH \xrightarrow{-2e^-} RSSR \xrightarrow{-2e^-} RSO_2H \xrightarrow{-2e^-} RSO_3H $$

These diamagnetic products of oxidation by metal ions or irradiation form via free radical intermediates, with the L-cysteine thiyl radical either dimerizing to form L-cystine or reacting with molecular oxygen or hydrogen peroxide to form the oxygen-containing products.

$$ RSH \xrightarrow{-e^-} RS^{\cdot} \xrightarrow{-e^-} RSOH \xrightarrow{-e^-} R\dot{S}O \xrightarrow{-e^-} RSO_2H \xrightarrow{-e^-} $$

$$ R\dot{S}O_2 \xrightarrow{-e^-} RSO_3H $$

Although the importance of free radical in the radiolytic and metal ion oxidation of L-cysteine is clear, no direct evidence of a role for cysteine-derived free radicals in an enzymatic reaction had been reported until the completion of a spin-trapping study of the free radical metabolites formed via the oxidation of cysteine and L-cysteine sulfinic acid by the peroxidase prototype, horseradish peroxidase[5].

Since much of the biochemistry of the tripeptide GSH is dominated by cysteine, it is natural to extend our spin-trapping investigations to GSH. Spin trapping is a technique in which a diamagnetic molecule (or spin trap) reacts with a free radical to produce a more stable radical (or spin adduct) which is readily detectable by ESR. Spin adducts are substituted nitroxide free radicals, which, for free radicals, are relatively stable.

RESULTS AND DISCUSSION

The ESR spectrum of the DMPO-glutathione thiyl radical adduct is obtained in an incubation of 10 mM GSH, 50 µM hydrogen peroxide and 100 µg/ml of horseradish, implying that the oxidation of GSH by horseradish peroxidase and hydrogen peroxide forms a thiyl free radical which reacts with the spin trap DMPO to form the DMPO-glutathione thiyl radical adduct (Fig. 1).

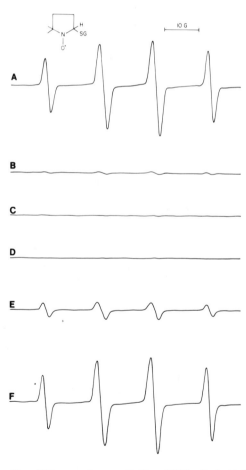

Fig. 1.   The ESR spectrum of the DMPO-glutathione thiyl radical adduct produced in a system of horseradish peroxidase and hydrogen peroxide under aerobic conditions[6]. A. Incubation containing

10 mM glutathione, 50 μM hydrogen peroxide, 90 mM DMPO and 0.1 mg/ml horseradish peroxidase in Tris/HCl buffer (1 mM DTPA), pH 8.0. B. Same as in A, but without addition of hydrogen peroxide. C. Same as in A, but without addition of horseradish peroxidase. D. Same as in A, but heat-denatured enzyme. E. Same as in A, but with 1000 units/ml catalase. F. Same as in A, but with 40 μg/ml superoxide dismutase. Instrumental conditions: microwave power, 20 mW; modulation amplitude, 0.67 G; time constant, 0.25 sec; scan rate, 40 G/min.

This radical adduct has a distinctive ESR spectrum. As expected, the omission of hydrogen peroxide or horseradish peroxidase, or the use of heat-denatured horseradish peroxidase results in negligible signal. In accordance with the dependence on hydrogen peroxide, 1000 units/ml of catalase strongly inhibited formation of the GS-adduct. Superoxide dismutase had no effect even at 40 μg/ml. These spectra were recorded as 2 min scans of the magnetic field. We noted that repetitive scans showed radical decay; therefore, we set the magnetic field at the position of maximum signal amplitude and proceeded to make measurements of the time course of the radical's concentration. At 1.0 mg of horseradish peroxidase per ml, the radical decays after a rapid formation phase (Fig. 2).

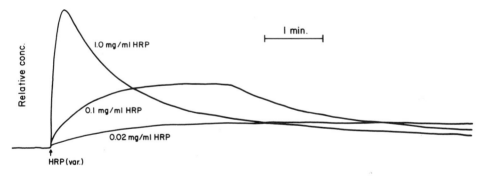

Fig. 2. Time course showing the effect of variable concentrations of horseradish peroxidase on the formation of the DMPO-glutathione thiyl radical horseradish peroxidase in Tris/HCl buffer (1 mM DTPA), pH 8.0[6]. Instrumental conditions: microwave power, 5 mW; modulation amplitude, 0.67 G; time constant, 0.25 sec; scan time 8 min.

The rate of radical formation appears to be linear in enzyme concentration, with a prolonged steady state at the lowest concentration. The incubation with 0.2 mg/ml HRP clearly appears to abruptly run out of something. When we varied the hydrogen peroxide concentration, it was immediately clear that hydrogen peroxide was limiting (Fig. 3).

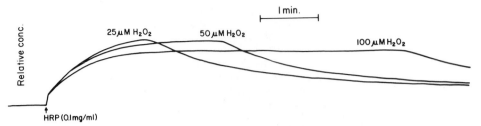

Fig. 3. Time course showing the effect of variable concentrations of hydrogen peroxide on the formation of the DMPO-glutathione thiyl radical adduct[6]. Incubations contained 10 mM glutathione, 0.1 mg/ml horseradish peroxidase and hydrogen peroxide in Tris/HCl buffer (1 mM DTPA), pH 8.0. Instrumental conditions: microwave power, 5 mW; modulation amplitude, 0.67 G; time constant, 0.25 sec; scan time, 8 min.

When we initiated the formation of the GS-adduct in the presence of 5 units/ml of catalase, we saw an induced break in the steady-state radical concentration which occurred at earlier and earlier time points as the concentration of catalase was increased (Fig. 4A).

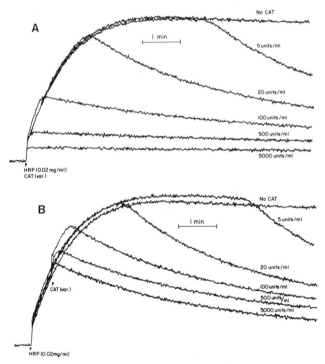

Fig. 4. Time course showing the effect of variable concentrations of catalase on the formation of the DMPO-glutathione thiyl radical adduct[6]. A. Variable concentrations of catalase added simultaneously with horse-radish peroxidase in incubations consisting of 10 mM glutathione, 50 μM hydrogen peroxide, and 0.02 mg/ml horseradish peroxidase in Tris/HCl buffer (1 mM DTPA),

pH 8.0. B. Variable concentrations of catalase added 0.5 min after initiation with horseradish peroxidase using incubation conditions described in A. Instrumental conditions: microwave power, 5 mW; modulation amplitude, 0.67 G; time constant, 0.25 sec; scan time, 8 min.

It is interesting that even at 5000 units/ml of catalase a substantial concentration of GS-adduct is formed. If catalase is added 30 sec after initiation, the same general effects are observed (Fig. 4B).

Since free radicals usually react with oxygen, it is not surprising that this system consumes oxygen, as determined with a Clark oxygen electrode (Fig. 5).

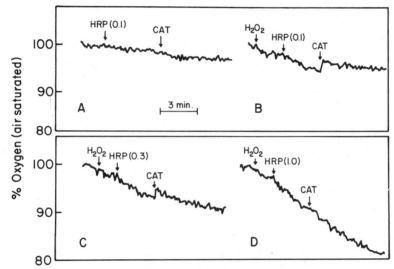

Fig. 5.   Oxygen uptake curves showing the effect of catalase on the glutathione/hydrogen peroxide system containing variable concentrations of horseradish peroxidase[6]. Tris/HCl buffer (1 mM DTPA), pH 8.0, containing 10 mM glutathione was placed in the sample chamber and other components added at the points indicated. Final concentrations were: 10 mM glutathione, 50 μM hydrogen peroxide, 5000 units/ml catalase, 0.1-1.0 mg/ml horseradish peroxidase.

The dependence of radical formation on hydrogen peroxide is also seen in the oxygen consumption. Little oxygen uptake occurs in the absence of hydrogen peroxide (Fig. 5A). In the presence of hydrogen peroxide, not only does oxygen uptake occur, but catalase inhibits the oxygen uptake and even causes an oxygen rebound due to the formation of oxygen via the disproportionation of hydrogen peroxide (Fig. 5B). As the horseradish peroxidase concentration is increased, it appears to out-compete catalase for the hydrogen peroxide, and at 1.0 mg of horseradish peroxidase per ml, even 5,000 units of catalase has no effect (Fig. 5D). It is known that horseradish peroxide has a much lower K_m for hydrogen peroxide than does catalase.

The consumption of oxygen by the horseradish peroxidase-catalyzed oxidation of reduced glutathione can be explained on the basis of known reactions of the thiyl radical. The thiyl free radical reacts with thiol anion to form the glutathione disulfide anion free radical, which is air-oxidized to oxidized glutathione, forming superoxide. Even in the absence of superoxide dismutase, superoxide disproportionates rapidly to form hydrogen peroxide, which will drive the horseradish peroxidase-catalyzed reaction.

$$GS^{\cdot} + GS^{-} \rightarrow GS\dot{\bar{S}}G$$

$$GS\dot{\bar{S}}G + O_2 \rightarrow GSSG + \dot{O}_2^{-}$$

$$2\dot{O}_2^{-} + 2H^{+} \rightarrow H_2O_2 + O_2$$

These chemical reactions, including superoxide formation, are well established. In addition, glutathione-derived peroxides could result from the known addition reaction of thiyl radicals with oxygen.

$$GS^{\cdot} + O_2 \rightarrow GSOO^{\cdot}$$

At this point, the GSH/horseradish peroxidase/hydrogen peroxide system appears to be well understood, with the glutathione thiyl radical formation being hydrogen peroxide-dependent and following the classic one-electron oxidation pathway of horseradish peroxidase substrates via compound I and II.

$$HRP + H_2O_2 \rightarrow HRP\text{-compound I(green)} + H_2O$$

$$HRP\text{-compound I} + GSH(GS^{-}) \rightarrow HRP\text{-compound II(red)} + GS^{\cdot}$$

$$HRP\text{-compound II} + GSH(GS^{-}) \rightarrow HRP + GS^{\cdot}$$

Since horseradish peroxidase is of plant origin, some may question the physiological relevance of work done with this enzyme. On the other hand, saliva, tears, and milk contain lactoperoxidase, which is thought to have an antimicrobial function in vivo. This mammalian peroxidase has a soret optical spectrum which is very similar to that of thyroid peroxidase, intestine peroxidase, uterus peroxidase and eosinophil peroxidase'. As such, lactoperoxidase appears to be a useful prototype for most mammalian hemoprotein peroxidases. Glutathione thiyl free radical formation with lactoperoxidase is hydrogen peroxide-dependent and, in all other respects, is similar to the horseradish-peroxidase system.

At this point the question of the mechanism of glutathione peroxidase naturally arises. Glutathione peroxidase is thought to reduce hydrogen peroxide (as well as organic hydroperoxides) and to oxidize GSH to GSSG without forming a glutathione thiyl free radical, but this has never been tested. The proposed mechanism involves two-electron oxidation of the selenium prosthetic group to selenol without the formation of a glutathione-derived radical.

In contrast to horseradish peroxidase (Fig. 6A), glutathione peroxidase not only does not support oxygen uptake in the presence of hydrogen peroxide (Fig. 6C), but actually inhibits the horseraidsh peroxidase-catalyzed reaction.

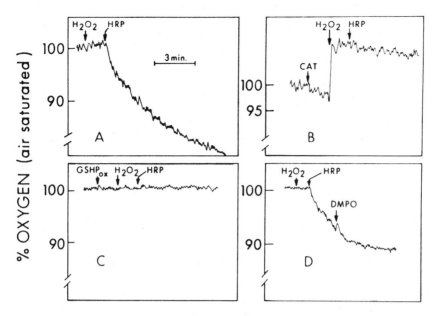

Fig. 6. Oxygen uptake curves showing the effect of glutathione peroxidase, catalase, and DMPO (90 mM) on
the glutathione/hydrogen peroxide/horseradish
peroxidase system[6]. Tris/HCl buffer (1 mM DTPA),
pH 8.0, containing 10 mM glutathione was placed in
the sample chamber and other components added at
the points indicated. Final concentrations were:
10 mM glutathione, 50 μM hydrogen peroxide, 1.0
mg/ml horseradish peroxidase, 5000 units/ml catalase, 0.25 units/ml glutathione peroxidase, 10
μl/ml DMPO.

This result is apparently due to the complete reduction of hydrogen
peroxide by glutathione peroxidase without radical formation. Analogous
inhibition of oxygen uptake can also be caused by catalase with the
destruction of hydrogen peroxide indicated by the release of oxygen (Fig.
6B). DMPO, through the formation of the glutathione thiyl adduct, also
inhibits oxygen uptake, as expected (Fig. 6D).

When we examined the inhibition of thiyl free radical formation by glutathione peroxidase (Fig. 7A, 7B), we found that the expected inhibition
was preceeded by a stimulation phase lasting a few seconds.

499

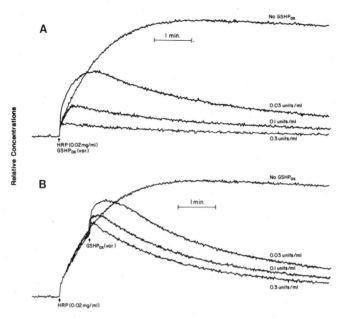

Fig. 7. Time course showing the effect of
variable concentrations of glutathione
peroxidase on the formation of the
DMPO-glutathione thiyl radical adduct[6].
A. Variable concentrations of glutathione peroxidase added simultaneously
with horseradish peroxidase in incubations consisting of 10 mM glutathione,
50 μM hydrogen peroxide, and 0.02 mg/ml
horseradish peroxidase in Tris/HCl
buffer, pH 8.0, with DTPA. B. Variable
concentrations of glutathione peroxidase
added 0.8 min after initiating with horseradish peroxidase using the incubation
conditions described in A. Instrumental
conditions: microwave power, 5 mW;
modulation amplitude, 0.67 G; time
constant, 0.25 sec; scan time, 8 min.

Apparently some species other than hydrogen peroxide formed by horseradish peroxidase and GSH is responsible for this stimulation effect.
Nevertheless, we have not been able to detect thiyl free radical formation
with glutathione peroxidase and hydrogen peroxide or even with organic
hydroperoxides such as cumene hydroperoxide or t-butyl-hydroperoxide. In
short, we have observed glutathione thiyl free radical formation by glutathione peroxidase only in the presence of horseradish peroxidase.

Clearly, hydrogen peroxide alone is not responsible for this stimulation phase. I propose that the glutathione thiyl radical oxidizes the Se
of glutathione peroxidase and transforms it into a free radical-producing
peroxidase which forms the thiyl free radical, as does horseradish
peroxidase. This is one possible scheme for such a transformation.

Damage by GS·

$$GS· + GSHP_{ox}\text{-SeOH} \rightarrow GSHP_{ox}\text{-}\overset{\cdot}{S}eO$$

$$H_2O_2 + GSHP_{ox}\text{-}\overset{\cdot}{S}eO \rightarrow GSHP_{ox}\text{-}\overset{\cdot}{S}eO_2$$

Peroxidase GS· Formation

$$GSH + GSHP_{ox}\text{-}\overset{\cdot}{S}eO_2 \rightarrow GSHP_{ox}\text{-SeO}_2H + GS·$$

$$GSH + GSHP_{ox}\text{-SeO}_2H \rightarrow GSHP_{ox}\text{-}\overset{\cdot}{S}eO + GS·$$

Repair by GSH

$$GSH + GSHP_{ox}\text{-}\overset{\cdot}{S}eO \rightarrow GSHP_{ox}\text{-SeOH} + GS·$$

Analogous investigations of the effects of glutathione peroxidase using oxygen consumption led to the same conclusions. Although, on the time scale of minutes, glutathione peroxidase is the best inhibitor of thiyl radical formation we have found, on the time scale of seconds, it is a powerful catalyst of thiyl radical formation. Both the rate of formation and the total concentration of thiyl radical increase with the horseradish peroxidase concentration. Glutathione peroxidase is not merely destroying a GSH-derived inhibitor of horseradish peroxidase, because the glutathione peroxidase-dependent reaction is faster than the initial rate of the horseradish peroxidase-dependent reaction. Clearly this effect is not simply a release of product inhibition. This stimulation effect is not duplicated with bovine serum albumin, glutathione transferase B, or GSSG. As the glutathione peroxidase concentration is lowered, both the stimulation and the inhibition of oxygen uptake are decreased with distinguishable effects even at 300-fold dilution. Thiyl radical damage of glutathione peroxidase followed by a burst of thiyl free radical formation and eventual repair of the glutathione peroxidase by glutathione accounts for all of the data, but is not the only possible explanation.

The reaction of the glutathione thiyl free radical with itself will form GSSG.

$$GS· + GS· \rightarrow GSSG$$

The reaction of the glutathione thiyl free radical with GS⁻ will be followed by air oxidation of the glutathione disulfide anion free radical which also leads to GSSG formation.

$$GS· + GS^- \rightarrow G\overset{\cdot\,-}{S}SG + H^+$$

$$G\overset{\cdot\,-}{S}SG + O_2 \rightarrow GSSG + \overset{\cdot}{O}_2^-$$

Therefore, we have assayed for oxidized glutathione formation using glutathione reductase-linked NADPH oxidation (Table 1).

Table 1. GSSG Formation in the Horseradish and Glutathione Peroxidase
Systems

Incubation Components	[GSSG] (μM)[a] (% Complete System)
Horseradish peroxidase + H_2O_2	124.3 ± 8.0 (100)
Horseradish peroxidase + H_2O_2 + nitrogen	59.6 ± 2.1 (48)
Horseradish peroxidase + H_2O_2 + DMPO	15.6 ± 0.5 (13)
Horseradish peroxidase + H_2O_2 + catalase	61.3 ± 1.4 (49)
Horseradish peroxidase + H_2O_2 + glutathione peroxidase	56.3 ± 2.8 (45)
Glutathione peroxidase + H_2O_2	53.0 ± 0.9 (43)
Horseradish peroxidase	41.3 ± 3.7 (33)
H_2O_2	27.6 ± 2.9 (22)
---	17.6 ± 3.3 (14)

[a]Stirring for 30 min in air or N_2 with 10 mM GSH. The incubations also
contained horseradish peroxidase (0.3 mg/ml), catalase (5000 units/μl),
glutathione peroxidase (0.2 units/ml), hydrogen peroxide (50 μM), and
DMPO (90 mM) as indicated[6].

Horseradish peroxidase and 50 μM hydrogen peroxide form 124 μM GSSG, which
is over twice the initial hydrogen peroxide concentration. Clearly, oxidizing equivalents are primarily provided by oxygen, as is consistent with
the oxygen uptake. Nitrogen strongly inhibits GSSG formation and gives
the expected one-to-one stoichiometry of GSSG with hydrogen peroxide. The
DMPO radical trap, catalase, and glutathione peroxidase all inhibit oxidized glutathione formation. In contrast to horseradish peroxidase, 53 μM
GSSG is formed from 50 μM hydrogen peroxide by glutathione peroxidase,
giving the expected one-to-one stoichiometry. We detected some GSSG with
horseradish peroxidase alone in accord with the small amount of GS thiyl
radical and oxygen uptake detected without hydrogen peroxide. Hydrogen
peroxide itself is known to oxidize GSH to oxidized glutathione, but the
reaction is not complete even within 30 minutes. Lastly, some autoxidation of GSH also occurs.

In summary, the oxidation of L-cysteine by horseradish peroxidase
forms the cysteine thiyl free radical[5], which reacts with oxygen resulting
in an oxygen-dependent enzymatic chain reaction. In addition, the oxidation of cysteine sulfinic acid forms both carbon- and sulfur-centered free
radicals[5]. In addition, the oxidation of reduced glutathione by horseradish peroxide forms the GS· thiyl free radical and oxidized glutathione in a hydrogen peroxide-dependent manner[6]. The oxidation of
glutathione by glutathione peroxidase forms oxidized glutathione without
forming the GS· thiyl free radical when hydrogen peroxide, cumene hydroperoxide, or t-butyl-hydroperoxide is reduced. Nevertheless, glutathione
peroxidase forms the GS thiyl radical in the presence of horseradish
peroxidase[6].

References
1. M.Z. Baker, R. Badiello, M. Tamba, M. Quintiliani, and G. Gorin,
Pulse radiolytic study of hydrogen transfer from glutathione to
organic radicals, Int. J. Radiat. Biol. 41:595 (1982).
2. G.K. Nyberg, G.P.D. Granberg, and J. Carlsson, Bovine superoxide
dismutase and copper ions potentiate the bactericidal effect of
autoxidizing cysteine, Appl. Environ. Microbiol. 38:29 (1979).
3. L.F. Prescott, J. Park, A. Ballantyne, P. Adriaenssens, and A.T.
Proudfoot, Treatment of paracetamol (acetaminophen) poisoning with
N-acetylcysteine, The Lancet II:432 (1977).

4. E.P. McGovern, N.F. Swynnerton, P.D. Steele, and D.J. Mangold, HPLC assay for 2-(3-aminopropylamino)ethanethiol (WR-1065) in plasma, Int. J. Radiation Oncology Biol. Phys. 10:1517 (1984).
5. L.S. Harman, C. Mottley, and R.P. Mason, Free radical metabolites of L-cysteine oxidation, J. Biol. Chem. 259:5606 (1984).
6. L.S. Harman, D.K. Carver, J. Schreiber, and R.P. Mason, One- and two-electron oxidation of reduced glutathione by peroxidases, J. Biol. Chem. (in press).
7. S. Ohtaki, H. Nakagawa, S. Nakamura, M. Nakamura, and I. Yamazaki, Characterization of hog thyroid peroxidase, J. Biol. Chem. 260: 441 (1985).

REACTIVE OXYGEN SPECIES FORMED IN VITRO AND IN CELLS: ROLE OF THIOLS(GSH).

MODEL STUDIES WITH XANTHINE OXIDASE AND HORSERADISH PEROXIDASE

Heribert Wefers and Helmut Sies

Institut für Physiologische Chemie I
Universität Düsseldorf
Düsseldorf, West Germany

INTRODUCTION

Reactive oxygen species are being widely studied with respect to their biological significance. When a disturbance occurs in the delicate balance between prooxidants and antioxidants in favor of the former, there is a biological response thought to be the basis of a number of physiological and pathophysiological phenomena. This condition is being referred to as oxidative stress (see (1)).

Among several biologically important types of reactive oxygen species, we have been interested in recent years in those of photoemissive nature. The exposure of cells to oxidative conditions of diverse nature can be accompanied by an elevated production of free radicals which, in turn, is expressed as an enhanced generation of electronically excited states leading to the production of low-level chemiluminescence(see (2)).

Studying the process of quinone redox cycling in intact cells and using low-level chemiluminescence as a parameter, we made the observation that, in contrast to our initial expectation, the level of photoemission elicited with a quinone, menadione, was decreased when cellular thiols were depleted rather than increased (3). This initial observation led to the study of the role of thiols and thiyl radicals in terms of metabolically generated reactive intermediates. This new area complements long-standing interests in the field of radiation biochemistry, where thiols have been studied in a number of respects.

The present work describes recent studies on thiol reactions in model reactions of enzymatic nature, xanthine oxidase and horseradish peroxidase.

THE START: DECREASED PHOTOEMISSION IN INTACT CELLS WITH GSH DEPLETION

In the intact hemoglobin-free perfused rat liver, the infusion of menadione into the portal vein leads to the production of photoemissive species. The spectral distribution of this low-level chemiluminescence is predominantly in the red region of the spectrum. From this and other evidence it may be concluded that singlet oxygen is being generated as a consequence of quinone redox cycling.

In livers taken from rats pretreated with phorone, GSH was depleted to very low levels (less than 5 % of controls). Interestingly, the photoemission observed with a given concentration of quinone was largely decreased in these GSH-depleted livers as compared to controls(Fig.1), whereas the reverse was seen when a hydroperoxide, t-butyl hydroperoxide, was employed instead of the quinone. The increase in the level of photo-emissive species in the state of GSH depletion during hydroperoxide meta-bolism is readily plausible by the loss of GSH as the cosubstrate in the GSH peroxidase reaction.

To resolve the apparent paradox observed with the quinone, we syn-thesized the menadione GSH conjugate, 2-methyl-3-glutathionyl-1,4-naphtho-quinone, also called thiodione(4). It was found(3) that the glutathione conjugate was capable of redox cycling similar to menadione itself, so that the conjugation of menadione with GSH is not a detoxication reaction in itself.

Using the menadiol GSH conjugate and studying its autoxidation, we found that photoemission attributable to singlet oxygen was observed only when GSH was also present in the cuvette(3,5). This indicated that GSH, as a free thiol, was involved in generating photoemissive species, and a further study using another source of oxygen radicals was prompted, em-ploying the xanthine oxidase reaction.

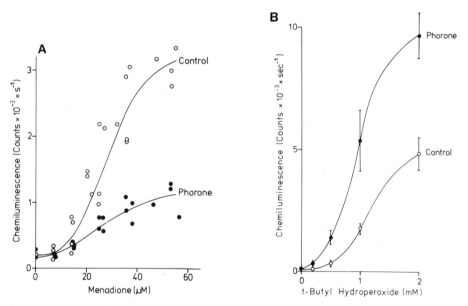

Fig. 1. Dependence of low-level chemiluminescence emitted from perfused rat liver on menadione(A) or t-butyl hydroperoxide(B) concen-tration. Data from controls are compared to those depleted in GSH by phorone pretreatment. Chemiluminescence was from wave-lengths greater than 620 nm, using a Kodak cutoff filter. Points indicate plateau levels. From Ref.(3).

MODEL STUDIES WITH XANTHINE OXIDASE

Xanthine oxidase with xanthine or acetaldehyde as substrates yields low-level chemiluminescence(6). This photoemission, however, is mainly in the nonred part of the visible spectrum(5) and therefore may be attributed to excited carbonyls. Singlet oxygen in the dimol emission reaction emits at 634 nm and 703 nm(7) and thus cannot contribute to a large extent to the xanthine oxidase-dependent chemiluminescence. This is in agreement with the lack of singlet oxygen formation observed for the xanthine oxida-se reaction using chemical traps for this excited state of oxygen(8).

However, as shown in Fig. 2, the addition of GSH to the xanthine

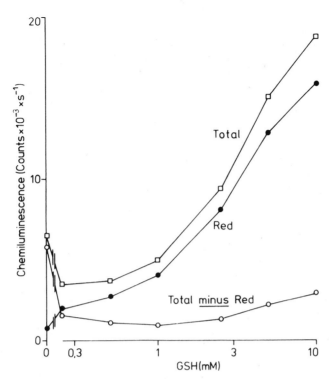

Fig. 2. Glutathione-dependent generation of red chemiluminescence in the xanthine oxidase/acetaldehyde system. Total: Photoemission observed without filters; red: photoemission observed with a cut-off filter(beyond 620 nm). Xanthine oxidase: 30 mU/ml, acetaldehyde, 5 mM. From Ref.(9).

oxidase system leads to a substantial emission of red light(beond 620 nm). Based on the criteria available, this photoemissive species generated by the superoxide anion radical and the thiol is singlet molecular oxygen.

Thus, like in the autoxidation of the quinol, the xanthine oxidase as a superoxide-generating system leads to a photoemissive species in the reaction with GSH. In further analyzing the reactions involved, we found that in addition to disulfide formation, there is the formation of further oxidation products, e.g. glutathione sulfonate.

Formation of GSSG and glutathione sulfonate

As shown in Fig. 3, parallel to the oxidation of xanthine as measured by the formation of urate, the disulfide and the sulfonate were detected as oxidation products. Like the photoemission, the production of these compounds was suppressed when superoxide dismutase was added. GSH addition led to an extra oxygen uptake, also abolished in the presence of SOD, and a balance sheet of oxygen uptake accounted for the formation of GSSG and the sulfonate(5). Sulfonate formation was about 6-15 % of GSSG formation. For example, when GSSG increased from 40 μM(already present in the initial incubation using 5 mM GSH) to 160 μM, the glutathione sulfonate increased from less than 1 μM to 18 μM in the incubation. Thus, the sulfonate is a minor product; however, this oxygen-containing metabolite is of interest regarding the nature of the reaction pathways involved(see below).

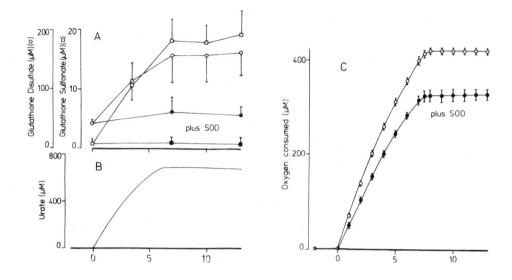

Fig. 3.   Formation of glutathione disulfide and glutathione sulfonate in the
xanthine oxidase/xanthine system.   Time courses of (A)GSSG and sulfonate formation without(open symbols) and with(full symbols)
superoxide dismutase; (B) urate formation; (C) oxygen consumption.
Incubations in oxygen-saturated phosphate buffer at pH 7.0 and
37°C.   Xanthine, 700 μM, GSH, 5 mM, and catalase, 40 μg/ml.
Xanthine oxidase was added at zero time, 30 mU/ml.   From Ref.(5).

The reaction sequence proposed(5) is an initial reaction of GSH with
the radical, either as O_2^- or as the protonated form, $HO_2^{\cdot}$, to yield the thiyl
radical, $GS^{\cdot}$,

$$O_2^- + H^+ + GSH \longrightarrow GS^{\cdot} + H_2O_2$$

$$R^{\cdot} + GSH \longrightarrow GS^{\cdot} + RH$$

similar to the reaction of GSH with radiation-induced free radicals.   The
further fate of the thiyl radical depends on the conditions for reaction,
and the major reactions include the addition of oxygen, the reaction with
GSH to form the glutathione disulfide anion radical, and the self-reaction
to form GSSG:

$$GS^{\cdot} + O_2 \longrightarrow GSOO^{\cdot}$$

$$GS^{\cdot} + GSH \longrightarrow GSSG^- + H^+$$

$$GS^{\cdot} + GS^{\cdot} \longrightarrow GSSG$$

$GSOO^{\cdot}$ is regarded as one intermediate in the formation of stable end products such as the sulfonate.   As these reactions mentioned above do not
apparently depend on the nature of the generating reaction, further work
was done with horseradish peroxidase as another source of $GS^{\cdot}$.

The oxidation of compounds by horseradish peroxidase(HRP) at the expense of hydrogen peroxide involves the formation of Compound I and, in the case of a two-step reduction of the enzyme, Compound II(10,11). HRP-mediated oxidation of cysteine as electron donor has been shown to proceed via the thiyl radical, using esr techniques and spin traps(12). The enzymatic cycle carrying out this oxidation is shown by the three following reactions:

$$HRP + H_2O_2 \longrightarrow Compound\ I + 2\ H_2O$$

$$Compound\ I + RSH \longrightarrow Compound\ II + RS^\cdot$$

$$Compound\ II + RSH \longrightarrow HRP + RS^\cdot$$

The hydrogen peroxide may be added to the system or, under aerobic conditions, can be generated by autoxidation of the thiol.

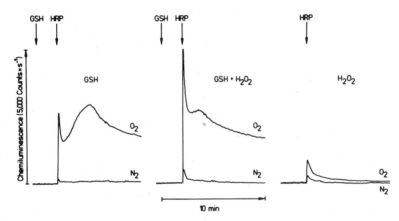

Fig. 4. Red photoemission in the reaction between GSH and HRP. The incubation buffer was gassed with O_2 or N_2 before and during the experiment. GSH, 5 mM; H_2O_2, 1 µM; HRP, 100 µg/ml. From Ref.(13).

As shown in Fig. 4, red photoemissive species are formed in the reaction of GSH with HRP, and these are dependent on the presence of oxygen. There are two phases in photoemission. The first one(Phase I) is associated with H_2O_2, whereas the second phase(Phase II) is not. This is further established by the actions of various enzymatic or nonenzymatic scavengers as shown in Fig. 5. Catalase abolished Phase I without affecting Phase II. The selenoenzyme, GSH peroxidase, abolished both Phase I and Phase II whereas the non-selenium GSH peroxidase activity exhibited by GSH transferase was capable of inhibiting mainly Phase II. These observations led to the suggestion that Phase II is related to an organic hydroperoxide; we proposed GSOOH as being responsible for Phase II photoemission(13).

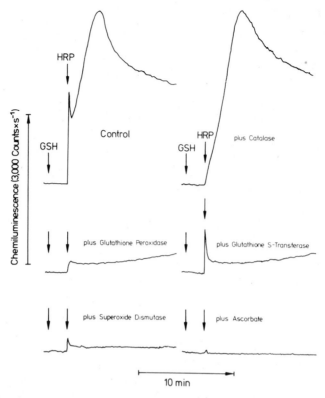

Fig. 5. Effects of enzymatic and nonenzymatic antioxidants on the GSH plus
         HRP dependent generation of red photoemissive species. For details,
         see Ref.(13).

Formation of GSSG and glutathione sulfonate

The analysis of HRP—mediated production of these oxidation products of
GSH yielded values comparable to those observed in the xanthine oxidase
reaction mentioned above. However, the relation of sulfonate to GSSG was
slightly lower with HRP than with xanthine oxidase. For example, when GSSG
production was 61 μM, that of the sulfonate was 2.4 μM(13).

Nature of photoemissive species

As in the xanthine oxidase reaction, HRP—catalyzed formation of photoemissive species in the presence of GSH was selectively in the red spectral region(13). As shown in Fig. 6, replacement of H_2O by D_2O in the incubation buffer led to a significant increase in both Phase II and Phase I
photoemission. These and other criteria again are in agreement with the
formation of singlet molecular oxygen in the reaction. Pathways for the
generation of singlet oxygen from GSOO˙ have been suggested in Ref(5);
however, the chemistry of GSOO˙ requires further study before final conclusions can be made; the proposed reactions are the reaction of GSOO˙
with GSH or with itself.

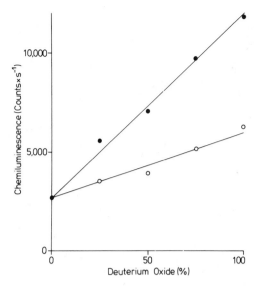

Fig. 6. Phase I(o) and Phase II (●) chemiluminescence of the HRP/GSH
reaction as a function of the D_2O concentration. From Ref.(13).

BIOLOGICAL SIGNIFICANCE

The in vitro detection of excited species by reactions involving the
GS˙ radical under aerobic conditions points to a novel aspect regarding
GSH as a radical scavenger. The two model reactions, xanthine oxidase and
horseradish peroxidase, provided evidence for the formation of photoemissive species from GSH during the production of radicals. Thus, in intact
cells the protective role of GSH or other thiols may depend on further
mechanisms to handle the thiyl radical and/or the oxygen addition products
such as RSOO˙. These further mechanisms may include enzymatic or nonenzymatic steps. A new role for glutathione S-transferases may reside in the
suppression of the GS˙-mediated generation of excited species(see Fig.5).
Also, ascorbate or other nonenzymatic scavengers are of importance in this
respect.
        So far, the stable end product of glutathione oxidation, glutathione
sulfonate, has not yet been shown to be formed in intact cells. Therefore,
it seems that powerful antioxidant capacity exists in cells in order to
quench the thiyl radicals effectively. However, as this area of research
is just emerging, it seems worthwhile to pursue the potential of the thiyl
radical as generated during enzyme catalysis.

ACKNOWLEDGEMENTS

These studies were supported by Deutsche Forschungsgemeinschaft,
Schwerpunktsprogramm "Mechanismen toxischer Wirkungen von Fremdstoffen",
and by the National Foundation for Cancer Research, Washington.

REFERENCES

1. H. Sies, Oxidative Stress: Introductory Remarks, in: "Oxidative Stress",
    H. Sies, ed., p.1, Academic Press, Orlando (1985)
2. H. Sies and E. Cadenas, Oxidative stress: damage to intact cells and
    organs, Phil.Trans.R.Soc.Lond.B., in press (1985)
3. H. Wefers and H. Sies, Hepatic low-level chemiluminescence during redox
    cycling of menadione and the menadione glutathione conjugate. Re-
    lation to glutathione and NAD(P)H:quinone reductase(DT diaphorase)
    activity, Arch.Biochem.Biophys. 224,568 (1983)
4. W.J. Nickerson, G. Falcone and G. Strauss, Studies on quinone-thioethers.
    I. Mechanism of formation and properties of thiodione.
    Biochemistry 2, 537 (1963)
5. H. Wefers and H. Sies, Oxidation of glutathione by the superoxide radi-
    cal to the disulfide and the sulfonate yielding singlet oxygen,
    Eur.J.Biochem. 137, 29 (1983)
6. R.M. Arneson, Substrate-induced chemiluminescence of xanthine oxidase
    and aldehyde oxidase, Arch.Biochem.Biophys. 136, 352 (1970)
7. A.U. Khan and M. Kasha, Chemiluminescence arising from simultaneous
    transitions in pairs of singlet oxygen molecules,
    J.Amer.Chem.Soc. 92, 3293 (1970)
8. T. Nagano and I. Fridovich, Does the aerobic xanthine oxidase reaction
    generate singlet oxygen? Photochem.Photobiol. 41, 33 (1985)
9. H. Wefers and H. Sies, Formation of excited species during glutathione
    oxidation, in:"4th International Conference on Superoxide and
    Superoxide Dismutase", Rome (J.Geelen, ed.) Elsvier Science Publ.
    B.V., Amsterdam, in press (1985)
10. I. Yamazaki, Peroxidase, in: "Molecular mechanisms of oxygen activation",
    O. Hayaishi, ed., p.535, Academic Press, New York (1974)
11. I. Yamazaki, M. Tamura and R. Nakajima, Horseradish peroxidase C,
    Mol.Cell.Biochem. 40, 143 (1981)
12. L.S. Harman, C. Mottley and R.P. Mason, Free radical metabolites in L-
    cysteine oxidation, J.Biol.Chem. 259, 5606 (1984)
13. H. Wefers, E. Riechmann and H. Sies, Excited species generation in
    horseradish peroxidase-catalyzed oxidation of glutathione.
    J.Free Radicals Biol.Med. 1, in press (1985)

ACTIVE OXYGEN AND TOXICITY

Steven D. Aust, Craig E. Thomas, Lee A. Morehouse,
Morio Saito, and John R. Bucher

Center for the Study of Active Oxygen in Biology and
and Medicine
Michigan State University
East Lansing, MI 48824-1319

Active forms of oxygen are becoming increasingly implicated in the etiology of numerous disease states and the toxicities of various drugs and chemicals. Among the former are the initiation and promotion of tumors (Petkau, 1980) and rheumatoid arthritis (Rowley et al., 1984), while the latter includes toxicities such as that associated with anthracycline antibiotics (Goodman and Hochstein, 1977), and paraquat (Bus et al., 1974). The generation of active oxygen species during normal cellular metabolism such as prostaglandin and leukotriene biosynthesis (Kalyanaraman and Sivarajah, 1984) or by stimulated polymorphonuclear leukocytes or macrophages (Babior and Peters, 1981) is now widely recognized.

The term active oxygen refers to a rather broad category of oxidants which includes certain oxygen radicals or other species such as iron-oxygen complexes; the common denominator being the ability to catalyze the inappropriate oxidation of biological material such as membrane phospholipids, proteins, and DNA. Molecular oxygen is not generally considered to be "active" because its direct reaction with most organic compounds is spin forbidden. Partially reduced species of dioxygen such as O_2^- and H_2O_2, while not particularly reactive with organic compounds, are still frequently classified as active species of oxygen. Production of these species is known to occur during normal cellular metabolism. Superoxide is produced by a number of enzyme systems including xanthine oxidase, aldehyde oxidase, and flavin dehydrogenases or perhaps by autoxidation of compounds such as ubiquinone, epinephrine and flavoproteins (for review see Freeman and Crapo, 1982). Likewise, H_2O_2 is produced directly by enzymes such as glycolate oxidase and D-amino acid oxidase or simply by non-enzymatic or enzymatic dismutation of O_2^-.

Most cells possess mechanisms for the adequate removal of these species of oxygen under normal conditions. Removal of O_2^- is catalyzed by superoxide dismutase (SOD), an enzyme found in all aerobic organisms. One product of this reaction is H_2O_2 which, in turn, is reduced to H_2O by the action of catalase and/or the seleno-enzyme glutathione peroxidase (GSH-Px). Thus, cells are generally capable of removing or detoxifying these reduced oxygen species. However, it is often proposed that the toxicities of many drugs and chemicals result from excessive generation of these "oxy radicals", perhaps by exceeding the capacity of cellular enzyme systems to efficiently remove them.

It must be considered however that in aqueous systems neither O_2^- nor H_2O_2 are particularly reactive with organic compounds (Fridovich, 1983). Therefore their proposed deleterious effects may be the result of their participation in reactions leading to other more reactive species. In general the formation of more reactive radicals requires the presence of transition metals such as iron and copper. The most widely proposed mechanism for the generation of a strong oxidant involving both O_2^- and H_2O_2 is the iron-catalyzed Haber-Weiss reaction which produces the highly reactive hydroxyl radical ($\cdot OH$) (McCord and Day, 1978).

$$O_2^- + Fe^{3+} ---\rightarrow O_2 + Fe^{2+}$$

$$2O_2^- + 2H^+ ---\rightarrow O_2 + H_2O_2$$

$$Fe^{2+} + H_2O_2 ---\rightarrow Fe^{3+} + OH^- + \cdot OH \text{ (Fenton's Reaction)}$$

While the $\cdot OH$ is clearly of sufficient reactivity to oxidize nearly all biomolecules it is conceivable that its highly reactive nature may not provide the selectivity necessary to allow $\cdot OH$ access to membrane unsaturated phospholipids. Thus, a general concensus as to the in vivo feasibility of $\cdot OH$ as an initiator of lipid peroxidation has not been reached and several alternative initiators have been proposed. However, it is clear that, irrespective of the mechanism, a requirement for transition metals such as iron is implicit in virtually all proposals. Work in our laboratory has focused on developing an understanding of the various means by which iron may be involved in the initiation and propagation of lipid peroxidation.

Initial studies confirmed that Fenton's reagent was capable of initiating liposomal peroxidation in a ·OH-dependent fashion as peroxidation was inhibitable by ·OH traps and catalase (Tien et al., 1981). However, a mixture of EDTA-Fe^{2+} and H$_2$O$_2$ was unable to promote lipid peroxidation while ADP-Fe^{2+} and H$_2$O$_2$ initiated peroxidation in a reaction that was not affected by ·OH traps (Figure 1). These findings led to the realization that peroxidation dependent upon nucleotide-chelated iron may involve a species other than ·OH.

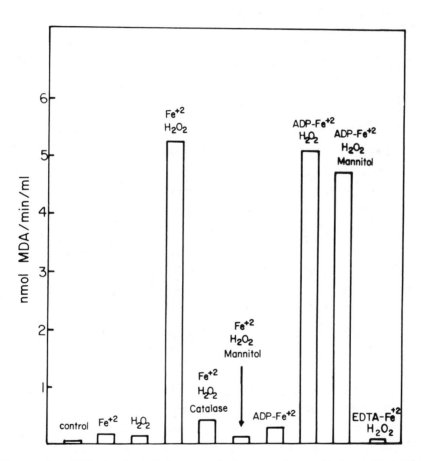

Figure 1. Effect of chelators, catalase, and mannitol on Fenton's reagent-dependent peroxidation of phospholipid liposomes.

Reaction mixtures contained liposomes (1.0 mM lipid phosphate) in 0.04 M NaCl, pH 7.5 at 37°C. Additions were 0.2 mM FeCl$_2$, 0.1 mM H$_2$O$_2$, ADP-Fe^{2+} (1.0 mM :0.2 mM), EDTA-Fe^{2+} (0.22 mM:0.2 mM), 10 mM mannitol and 1.0 U catalase/ml. Aliquots of 1 ml were removed at specific intervals and assayed for malondialdehyde (Data from Tien et al., 1981).

A critical role for O_2^- in the iron-catalyzed Haber-Weiss reaction and in other proposed mechanisms is the reduction of ferric iron. Using xanthine oxidase, low rates of O_2^--dependent peroxidation of liposomes occurred with EDTA-Fe^{3+} but when EDTA-Fe^{3+} was combined with ADP-Fe^{3+} a very marked stimulation was observed (Tien et al., 1982a). Since EDTA-Fe^{3+} could promote O_2^--dependent peroxidation of fatty acids dispersed with detergent we reasoned that the inability of EDTA-Fe^{2+} generated ·OH to initiate liposomal peroxidation was due to a steric inaccessibility to polyunsaturated fatty acids when they are in a phospholipid liposomal configuration.

Physiological concentrations of cellular reducing agents are such that it is thought that the extent of iron reduction via O_2^- produced within cells could not approach that due to other intracellular reducing processes (Winterbourn, 1979). Controversy also exists concerning direct ferric iron reduction by thiols, or indirect reduction by O_2^- produced by thiol autoxidation (Rowley and Halliwell, 1982). Therefore, Tien et al. investigated the abilities of thiols such as GSH, cysteine, and DTT to initiate liposomal peroxidation (Tien et al., 1982b). In general, the thiols appeared to act as prooxidants or antioxidants, depending upon their concentration. Similar results were obtained when microsomes were used (Figure 2). Iron reduction appeared to be direct as SOD had no effect on thiol-dependent peroxidation or cytochrome c reduction. From this work it was apparent that peroxidation was highly sensitive to the oxidation state of the iron. In the absence of, or at low concentration of thiols, insufficient iron would be reduced to support lipid peroxidation. But at high thiol concentrations all of the iron would be reduced and little peroxidation was observed. For most thiols tested there appeared to be a critical concentration where lipid peroxidation was maximal and at which we would expect a mixture of both ferrous and ferric iron.

In an effort to further characterize the initiating species, and support our hypothesis that both ferrous and ferric are required for lipid peroxidation, we used a variety of ferrous iron chelates, thereby eliminating the requirement for a reductant (Bucher et al., 1983b). While the mechanisms of autoxidation are likely to differ depending upon the chelator, it is likely that production of the partially reduced oxygen species (O_2^-, H_2O_2, and ·OH) or their iron-bound equivalents will

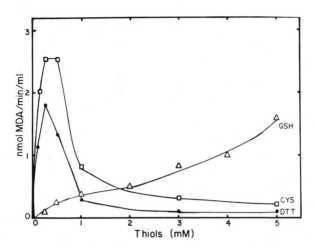

Figure 2. Effect of varying thiol concentration on thiol-dependent
microsomal lipid peroxidation.

Reaction mixtures (5 ml) contained microsomes (0.5 mg protein/ml),
ADP-Fe^{3+} (0.5 mM:0.1 mM), and varying concentrations of thiols in 30 mM
NaCl, pH 7.0 at $37^{\circ}C$. Aliquots of 1 ml were removed at specified
intervals and assayed for malondialdehyde. GSH = reduced glutathione,
CYS = cysteine, DTT = dithiothreitol.

occur (Weiss, 1953). The participation of O_2^{-} and H_2O_2 in
ferrous-dependent lipid peroxidation was assessed by adding SOD and
catalase to the system. It was found that peroxidation of
detergent-dispersed linoleate dependent upon autoxidation of free ferrous
iron, EDTA-Fe^{2+} or DTPA-Fe^{2+} was inhibited by SOD or catalase. Other
ferrous iron chelates tested (AMP, ADP, phosphate) were capable of
initiating peroxidation but activity was not inhibited by catalase and
was, in fact, slightly stimulated by the addition of SOD (Table 1).

Table 1. Effect of Chelate, SOD and Catalase on Ferrous-Dependent
Peroxidation of Detergent-Dispersed Linoleate

Chelate		$\Delta234nm/min$ Control	% of Control +SOD (50U/ml)	+Catalase (100U/ml)
Fe^{2+} alone	(100.0)	0.029	34	79
EDTA-Fe^{2+}	(7.5)	0.057	81	42
DTPA-Fe^{2+}	(12.5)	0.099	58	64
ADP-Fe^{2+}	(15.0)	0.052	123	112
AMP-Fe^{2+}	(100.0)	0.007	114	114
Phosphate-Fe^{2+}	(25.5)	0.060	140	105

Reaction mixtures contained ferrous chelates at 5:1 chelate to Fe^{2+}
except for DTPA (diethylenetriamine penta-acetic acid) which was 1.1:1.0.
Concentrations of Fe^{2+} in uM are given in parentheses. Chelates were
incubated with linoleate (0.16 mg/ml) and Lubrol Px (1%) in 30 mM NaCl,
pH 7.0. Reactions was continuously monitored at 234 nm for diene
conjugation. The rate for Fe^{2+} alone was not linear so rates were
calculated for first 3 min while the rate for ADP-Fe^{2+} is for the first
minute.

The conclusions drawn from these studies were that ferrous-
dependent peroxidation could be initiated via ·OH (in the case of EDTA
and DTPA) or an iron-bound oxidant of similar oxidizing capability
(phosphate containing chelators). However, more careful evaluation and
further study of peroxidation initiated by nucleotide chelated ferrous
iron led us to conclude that the picture was even more complicated.
When ADP-Fe^{2+} was added to phospholipid liposomes or micelles of
arachidonate a lag of 5 to 10 min was observed prior to evidence of
peroxidation (Bucher et al., 1983a). It was assumed that the lag period
was required to allow for autoxidation of some of the iron to form an
initiator, however SOD, catalase, and ·OH traps again had no effect on
the lag period. The only product of ferrous autoxidation which had been
ignored was ferric iron. It was subsequently demonstrated that the
addition of ferric iron eliminated the lag period, indicating the
immediate formation of an initiator (Figure 3). ADP-chelated ferric
iron was found to have no effect on the autoxidation of ADP-Fe^{2+}
therefore we concluded that initiation was dependent on a
ferrous-dioxygen-ferric complex, a modification of Hochstein's perferryl
hypothesis (Hochstein, 1981). Accordingly, SOD, catalase and ·OH
scavengers had no effect on lipid peroxidation initiated by a combination

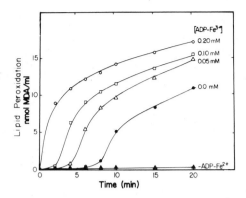

Figure 3. The effect of ferric iron on ferrous iron-dependent peroxidation of phospholipid liposomes.

Reaction mixtures (5 ml) contained liposomes (1 μmol lipid phosphate/ml), ADP-Fe^{2+} (1.7 mM ADP, 0.1 mM FeCl$_2$), and ADP-Fe^{3+} (17:1, ADP to iron) as indicated in 30 mM NaCl, pH 7.0 at 37°C. Aliquots of 1 ml were removed at specified intervals and assayed for malondialdehyde (Data from Bucher et al., 1983a).

of ferrous and ferric iron (Figure 4). The ability of such a complex to initiate peroxidation was further confirmed by using AMP-Fe^{2+} at pH 6.0, where its autoxidation is very slow, and demonstrating that peroxidation could be initiated immediately upon the addition of the ferric chelate. Further evidence was provided by the experiment in which H$_2$O$_2$ was added to liposomes containing ADP-Fe^{2+}. Hydrogen peroxide promotes the oxidation of ferrous iron (Fenton's Reaction) and accordingly, the lag

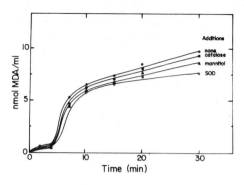

Figure 4. Effect of SOD, catalase, and mannitol on
ferrous-ferric-dependent peroxidation of phospholipid liposomes.

Reaction mixtures (5 ml) contained liposomes (1 μmol lipid phosphate/ml)
ADP-Fe^{2+} (0.85 mM ADP, 0.05 mM FeCl$_2$), ADP-Fe^{3+} (0.85 mM ADP:0.05 mM),
and where indicated, catalase (1 U/ml), mannitol (10 mM) or SOD (1 U/ml)
in 30 mM NaCl, pH 7.0 at 37°C. Aliquots of 1 ml were removed at
specified intervals and assayed for malondialdehyde.

period could be eliminated (Figure 5). Again the inability of ·OH traps
to affect the lag period in this experiment demonstrated that the
required product of autoxidation was most likely ferric iron.

These studies not only provided evidence for a novel initiating
species but importantly, suggested a means by which cells may protect
themselves against oxidative damage. Rather than relying on the enzyme
systems discussed earlier, control of cellular lipid peroxidation may be
governed by carefully regulating the oxidation state of the iron. For
example, maintenance of all cellular iron in the reduced state by
cellular reductants such as GSH or ascorbate would be expected to prevent
formation of the initiating species. Or, more simply, it may be adequate
to prevent interactions between reduced iron chelates and oxidized iron
chelates within the cellular millieu.

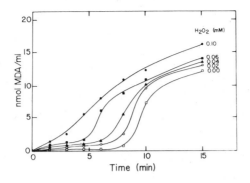

Figure 5.   Effect of H_2O_2 on liposomal peroxidation initiated by
ADP-Fe^{2+}:

Reaction mixtures (5 ml) contained liposomes (1 μmol lipid phosphate/ml),
ADP-Fe^{2+} (1.7 mM:0.1 mM) and the specified H_2O_2 concentrations in 30 mM
NaCl, pH 7.0 at 37°C. Aliquots of 1 ml were removed at specified
intervals and assayed for malondialdehyde.

It became apparent that further characterization of cellular lipid
peroxidation and mechanisms for its control would require a study of
physiological iron chelates. The transport and storage forms of iron are
well known to be transferrin and ferritin (Aisen and Listowsky, 1980),
respectively, however the potential for these proteins to provide iron
for participation in redox reactions is not well known. It has been
demonstrated that reduction and oxidation of the iron is required for the
release and uptake of iron from these proteins, respectively (Aisen and
Listowsky, 1980). It is often proposed that cells contain low molecular
weight iron chelates which function to transport iron between these
proteins (Jacobs, 1977) and iron containing enzymes and it is this iron
pool which is suggested to provide the iron necessary for formation of an
initiator of lipid peroxidation. While evidence for these low molecular
weight iron complexes in reticulocytes has been provided (Bartlett,
1976), definitive identification in tissues is lacking.

The majority of body iron is stored within ferritin as a ferric hydroxide micelle (Harrison, 1977). We chose to study the ability of ferritin to provide iron for initiation of peroxidation not only because of its abundance but also because release of iron from ferritin requires reduction and is enhanced by chelation; conditions under which lipid peroxidation would also likely be favored. Cellular reductants of ferritin iron have not been conclusively identified but it is likely that the spherical nature of the protein shell and the narrow channels through which the core is accessible impose size constraints on the reductants (Jones et al., 1978). We felt that $O_2^{\bar{}}$ would be a likely candidate for release of ferritin iron because of its reduction potential and small size. Using a phospholipid liposomal system consisting of purified rat liver ferritin and xanthine oxidase (as a $O_2^{\bar{}}$ generating system) we wereable to demonstrate low rates of lipid peroxidation (Thomas et al., 1985) (Figure 6). However, as enzyme activity was increased rates of MDA formation declined. Remniscent of our realization that autoxidation of ferrous iron produces ferric iron we came to the realization that increasing xanthine oxidase activity increases not only $O_2^{\bar{}}$ production but increases H_2O_2 generation as well, which would cause the oxidation of the released, reduced iron. Therefore catalase was added to the system. A dramatic increase in lipid peroxidation was observed, and importantly, maximum rates of lipid peroxidation occurred at a higher xanthine oxidase activity. The addition of increasing concentrations of catalase could concomitantly increase both the rate of MDA formation and the activity of xanthine oxidase at which maximal peroxidation was observed. ·

Iron release studies indicated that iron was released from ferritin in the reduced state and that the H_2O_2 produced in this system oxidized the iron. Thus, once again, lipid peroxidation appeared to be highly sensitive to the amount of ferrous and ferric iron present. In the absence of catalase all of the iron is rapidly oxidized by H_2O_2 and accordingly, rates of lipid peroxidation were quite low. Removal of H_2O_2 by catalase prevented oxidation and allowed formation of the initiator. However, it was apparent that when the catalatic capacity of catalase was exceeded, by increasing xanthine oxidase activity, peroxidation decreased rapidly. This phenomenon was shown not to be unique to ferritin as catalase also stimulated xanthine oxidase-dependent liposomal peroxidation using low concentrations of iron (Thomas et al., 1985). We had not previously made this observation as in vitro lipid peroxidation

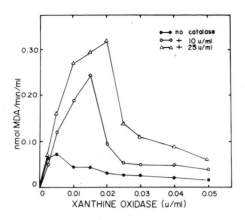

Figure 6. Effect of varying xanthine oxidase concentration on
ferritin-dependent liposomal peroxidation.

Reaction mixtures (5 ml) contained liposomes (1 μmol lipid phosphate/ml),
rat liver ferritin (200 uM Fe), ADP (1 mM), xanthine (0.33 mM), and
varying concentrations of xanthine oxidase and catalase as indicated in
50 mM NaCl, pH 7.0, at $37^{\circ}C$. Aliquots of 1 ml were removed at specified
intervals and assayed for malondialdehyde (Data from Thomas et al.,
1985).

experiments generally utilize much higher concentrations of iron so that
only a portion of the iron is reduced and formation of an initiating
species, presumably $Fe^{2+}-O_2-Fe^{3+}$, is dependent on the rate of iron
reduction since ferric is always present in excess.

All of these studies have led to the conclusions that the term
active oxygen not only encompasses the partially reduced oxygen species
but that iron-oxygen complexes such as $Fe^{2+}-O_2-Fe^{3+}$ are intimately
involved in the initiation of the oxidation of biological compounds. To
date we have been unable to demonstrate that $Fe^{2+}-O_2-Fe^{3+}$ can oxidize
substrates such as ethanol therefore it appears to be a weaker oxidant

than •OH, however it is clearly capable of catalyzing the abstraction of methylene hydrogens from unsaturated lipids. It is possible that these characteristics are what imparts a degree of selectivity to the oxidant such that it appears to have access to membrane unsaturated phospholipids presented in a liposomal configuration whereas the •OH does not.

It is important to note that not only have these various studies in several model systems shed light on an initiating species but have suggested a heretofore largely ignored mechanism for cellular control of lipid peroxidation; that being the redox state of the cell (Figure 7).

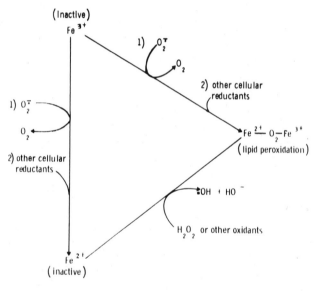

Figure 7. Possible mechanisms for the formation of an initiator of lipid peroxidation.

As mentioned, cells may 1) maintain iron all in one form (reduced or oxidized) or 2) minimize interaction between the two forms, thereby preventing formation of an initiating species. The chelation of iron by

524

proteins such as ferritin and transferrin may also be a means of maintaining the iron in a redox inactive state. Disruption of the mechanisms which function to govern the redox balance of the cell may then have severe consequences. For example, it is intriguing to consider that lipid peroxidation associated with redox active agents or GSH depletors may be a result of their interfering with the cells ability to maintain all available iron in the ferrous state. Alternatively, it is conceivable that O_2^- generated during autoxidation of reduced chemicals may release ferrous iron from its normal ferric storage form in ferritin, perhaps exceeding the capacity of the cell to maintain the iron in one form or the other. In agreement, we have recently demonstrated the ability of paraquat and purified NADPH cytochrome P450 reductase to release ferritin iron and promote lipid peroxidation (Saito et al., In Press). Thus, while the importance of enzymatic defense systems such as SOD, catalase, and GSH-Px cannot be ignored, we suggest that control of cellular redox balance and accordingly, of the iron within the cell, may be an important means by which cells can limit active oxygen-related toxicities.

ACKNOWLEDGEMENTS

The authors would like to acknowledge the secretarial assistance of Ms. Cathy M. Custer in the preparation of this manuscript. Supported in part by grants from the National Science Foundation, PCM-8302974 and the National Institute of Health, GM-33443.

REFERENCES

Aisen, P., and Listowsky, I., 1980, Iron transport and storage proteins, Ann. Rev. Biochem., 49:357.

Babior, B.M., and Peters, W.A., 1981, The O_2^- producing enzyme of human neutrophils. Further properties, J. Biol. Chem., 256:2321.

Bartlett, G.R., 1976, Phosphate compounds in rat erythrocytes and reticulocytes, Biochem. Biophys. Res. Comm., 70:1055.

Bucher, J.R., Tien, M., and Aust, S.D., 1983a, The requirement for ferric in the initiation of lipid peroxidation by chelated ferrous iron, Biochem. Biophys. Res. Comm., 111:777.

Bucher, J.R., Tien, M., Morehouse, L.A., and Aust, S.D., 1983b, Infuence of superoxide dismutase and catalase on strong oxidant formation during autoxidation of ferrous chelates, in: "Oxy Radicals and Their Scavenger Systems. Volume I: Molecular Aspects", G. Cohen and R.A. Greenwald, eds., Elsevier Science Publishing Co., New York.

Bus, J.S., Aust, S.D., and Gibson, J.E., 1974, Superoxide- and singlet oxygen-catalyzed lipid peroxidation as a possible mechanism for paraquat (methyl viologen) toxicty, Biochem. Biophys. Res. Comm., 58:749.

Freeman, B.A., and Crapo, J.D., 1982, Biology of disease. Free radicals and tissue injury, Lab. Invest., 47:412.

Fridovich, I., 1983, Superoxide radical: an endogenous toxicant, Ann. Rev. Pharmacol. Toxicol., 23:239.

Goodman, J., and Hochstein, P., 1977, Generation of free radicals and lipid peroxidation by redox cycling of adriamycin and daunomycin, Biochem. Biophys. Res. Comm., 77:797.

Harrison, P.M., 1977, Ferritin: An iron storage molecule, Sem. Hematol., 14:55.

Hochstein, P., 1981, Nucleotide-iron complexes and lipid peroxidation: mechanisms and biological significance, Israel J. Chem., 21:52.

Jacobs, A., 1977, Low molecular weight intracellular iron transport compounds, Blood, 50:433.

Jones, T., Spencer, R., and Walsh, C., 1978, Mechanism and kinetics of iron release from ferritin by dihydroflavins and dihydroflavin analogues, Biochemistry, 17:4011.

Kalyanaraman, B., and Sivarajah, K., 1984, The electron spin resonance study of free radicals formed during the arachidonic acid cascade and cooxidation of xenobiotics by prostaglandin synthase, in: "Free Radicals in Biology, Volume 6", W.A. Pryor, ed., Academic Press, New York.

McCord, J.M., and Day, E.D., 1978, Superoxide-dependent production of hydroxyl radical catalyzed by the iron-EDTA complex, FEBS Lett., 86:139.

Petkau, A., 1980, Radiation carcinogenesis from a membrane perspective, Acta Physiol. Scand., Supplemental, 492:81.

Rowley, D.A., and Halliwell, B., 1982, Superoxide-dependent formation of of hydroxyl radicals in the presence of thiol compounds, FEBS Lett., 138:33.

Rowley, D., Gutteridge, J.M.C., Blake, D., Farr, M., and Halliwell, B., 1984, Lipid peroxidation in rheumatoid arthritis: thiobarbituric acid-reactive material and catalytic iron salts in synovial fluid from rheumatoid patients, Clin. Sci., 66:691.

Saito, M., Thomas, C.E., and Aust, S.D., In Press, Paraquat and ferritin-dependent lipid peroxidation, J. Free Rad. Biol. Med.

Thomas, C.E., Morehouse, L.A., and Aust, S.D., 1985, Ferritin and superoxide-dependent lipid peroxidation, J. Biol. Chem., 260:3275.

Tien, M., Svingen, B.A., and Aust, S.D., 1981, Initiation of lipid peroxidation by perferryl complexes, in: "Oxygen and Oxy-Radicals in Chemistry and Biology", M.A.J. Rodgers and E.L. Powers, eds., Academic Press, New York.

Tien, M., Svingen, B.A., and Aust, S.D., 1982a, An investigation into the role of hydroxyl radical in xanthine oxidase-dependent lipid peroxidation, Arch. Biochem. Biophys., 216:142.

Tien, M., Bucher, J.R., and Aust, S.D., 1982b, Thiol-dependent lipid peroxidation, Biochem. Biophys. Res. Comm., 107:279.

Weiss, J., 1953, The autoxidation of ferrous ions in aqueous solution, Experientia, 9:61.

Winterbourn, C.C., 1979, Comparison of superoxide with other reducing agents in the biological production of hydroxyl radicals, Biochem. J., 182:625.

STUDIES ON THE MECHANISM OF ACTIVATION AND THE MUTAGENICITY OF RONIDAZOLE, A 5-NITROIMIDAZOLE

Gerald T. Miwa, Peter Wislocki, Edward Bagan, Regina Wang, John S. Walsh and Anthony Y.H. Lu

Department of Animal Drug Metabolism
Merck Sharp and Dohme Research Laboratories
Rahway, New Jersey 07065, U.S.A.

INTRODUCTION

5-Nitroimidazoles are used for treating protozoal and bacterial diseases in man and in animals and for potentiating the effects of chemo- and radiation therapy (Lossick, 1982; Goldman, 1980; Muller, 1981; Chessin, et al.,1978). They possess the unique property of requiring low oxygen tension in order for productive enzymatic nitro reduction to occur. This property is especially advantageous for the expression of therapeutic activity under anaerobic or hypoxic conditions such as against anaerobes and solid tumors. The presence of oxygen causes the reoxidation of a one electron reduced product, the nitroanion radical, back to the parent nitroimidazole and, consequently, results in the redox cycling of the drug and the generation of superoxide anion under hypoxic conditions. Since nitro reduction is an obligatory step for the biological activities of these drugs, the redox cycling may represent either a futile pathway in the activation of this drug or a mechanism for potentiating the activity of the nitroimidazole if a product of the superoxide anion underlies these activities.

Although nitroimidazoles enjoy a reputation as highly effective agents in their therapeutic arena, they have also earned a reputation as mutagens (Muller, 1981; Voogd, et al., 1979). Previous studies in our laboratory have elucidated many of the details of the metabolic activation of ronidazole which were responsible for the alkylation of proteins (West, et al., 1982a,b; Miwa, et al., 1982; Miwa, et al., 1984; Wislocki, et al., 1984a). The present studies were directed at evaluating if similar factors were also responsible for the mutagenic activity of this drug.

RESULTS AND DISCUSSION

The Metabolic Activation of Ronidazole and Protein Alkylation

The enzymatic generation of reactive, electrophilic intermediates from a chemical frequently results in somatic or genetic damage to the organism (Miller and Miller, 1982). The metabolism of ronidazole in turkeys can largely be accounted for in terms of imidazole ring fragmentation products such as oxalic acid, methylamine and acetamide (Rosenblum, et al., 1972). A minor pathway results in the descarbamoyl product, the only known intact imidazole metabolite. However, recent evidence indicates that reductive metabolism, catalyzed by hepatic and perhaps gut bacterial

Table 1. Structure of Ronidazole and Position of Radiolabels

Compound	Position of Label
Ia	$N\text{-}^{14}CH_3$
Ib	$4,5\text{-}^{14}C$
Ic	$2\text{-}^{14}CH_2\text{-}$
Id	$\text{-}^{14}CONH_2$
Ie	$N\text{-}C^3H_3$
If	4-^3H

enzymes, results in the formation of reactive intermediates capable of alkylating protein cysteine thiols (West, *et al.*, 1982a,b; Miwa, *et al.*, 1982; Miwa, *et al.*, 1986). In an effort to characterize the reactive intermediate, initial studies were directed at characterizing the protein-bound product since this represented a readily detectable, stabilized adduct of this intermediate.

Studies to determine the structural features of the parent nitroimidazole, retained after protein alkylation, employed several ^{14}C isotopically labeled forms of ronidazole (Compounds **Ia - Id**, Table 1). Table 2 summarizes the protein alkylation data obtained from microsomal incubations and from experiments in intact rats. The *in vitro* experiments (compounds **Ia** through **Ic**) revealed that most of the imidazole framework was retained in the protein-bound product but that alkylation had occurred with loss of the carbamate radiolabel (**Id**). Similar studies carried out in intact rats demonstrated that, although the protein-bound products were about two orders of magnitude lower than that observed *in vitro*, the N-methyl, 4,5-carbons and 2-methylene constituents of the imidazole framework were retained, in agreement with results obtained with microsomal incubations.

Structure-activity studies provided indirect evidence for the addition of protein nucleophiles to the 2-methylene carbon of ronidazole (Table 3). Compounds not containing the carbamate moiety (**VII** and **X**) or having the carbamate beta to the imidazole (**VIII**) and, therefore, inappropriately positioned for elimination facilitated by the imidazole π-electrons, gave much less protein-binding than the parent drug (**I**), suggesting that alkylation occurred primarily through loss of the carbamate. However, substitution at C_4 (**VI**) also reduced protein-binding. Moreover, the reduction in protein-binding of **VI** relative to **I** could not be completely explained

Table 2. Alkylation of Hepatic Proteins by Specifically
Labeled Ronidazole Substrates

Compound	in vitro (nmol/mg)	in vivo (nmol/mg)
Ia	1.76	0.012
Ib	1.91	0.012
Ic	1.73	0.016
Id	0.21	--

Table 3. Protein-binding and Mutagenicity Structure-Activity Relationship

Compound	R_1	R_2	binding[a]	mutagenicity[b]
			(activity expressed as % of I)	
I	$-CH_2OCONH_2$	-H	100	100
VI	$-CH_2OCONH_2$	$-CH_3$	15	2
VII	$-CH_2OH$	-H	14	18
VIII	$-CH_2CH_2OCONH_2$	-H	19	75
IX	$-CH_2CH_2OCONH_2$	$-CH_3$	--	3
X	$-CH_3$	-H	17	20
XI	$-CH_3$	$-CH_3$	12	<0.5
XII	$-CH_3$	-F	--	<0.5

[a] Protein-binding was assayed in microsomal incubations.
[b] Mutagenicity was determined in Ames strain TA100.

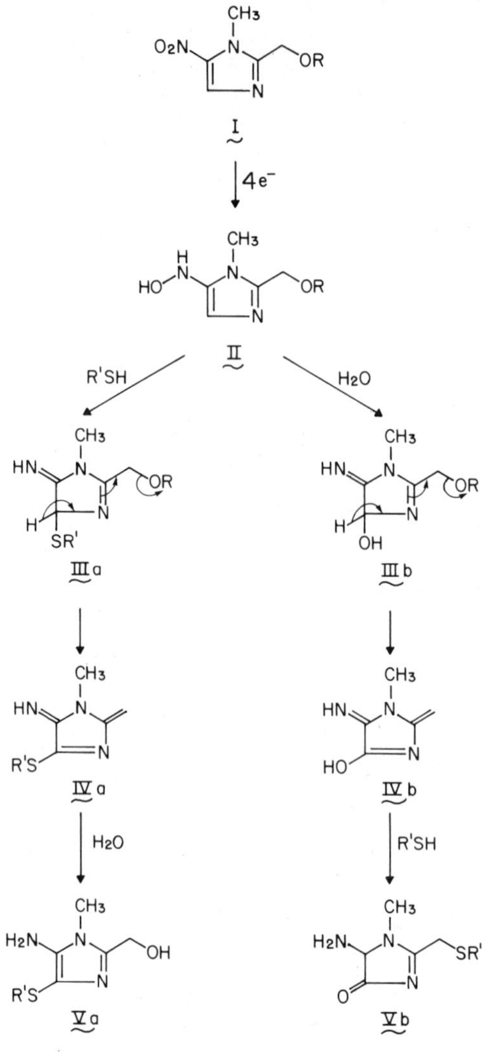

Figure 1. Postulated Activation Mechanism for Ronidazole

in terms of a lower rate of metabolism[1] but could be rationalized in terms of a two-step activation mechanism requiring the obligatory nucleophilic attack at C_4 before carbamate elimination (Fig. 1).

In this mechanism, the major pathway for protein alkylation is envisioned to occur by the initial four electron reduction of the nitro substituent generating a hydroxylamine intermediate (II) susceptible to nucleophilic attack by solvent water.[2] The alternative pathway of nucleophilic addition to II to give a protein-bound product such a Va is thought to be a minor pathway for reasons that will be discussed. Subsequent loss of the C_4 proton generates a Michael-like accepter (IVb),

---

[1] Data not shown

[2] An alternative pathway of reduction to the nitrosoimidazole, nucleophilic attack and subsequent reduction has not been ruled out.

	^{14}C (nmol/mg)	^{3}H (nmol/mg)	^{3}H-release (%)[a]
in vitro[b]			
lyophilisate[c]	0.0	22.2	
			90.1
protein-bound	1.31	0.13	
in vivo[d]			
lyophilisate[c]	0.0	4.22	
			84.3
protein-bound	0.268	0.042	

a ^{3}H released = (^{14}C-bound - ^{3}H-bound)/^{14}C-bound.

b Rat liver microsomes were anaerobically incubated with a mixture (1.0 mM) of 4,5-[^{14}C]- and 4-[^{3}H]-ronidazole and an NADPH-generating system for 30 min at 37 ^{0}C. Following the incubation, the reaction was quenched with TCA and lyophilized to recover the ^{3}H$_2$O released. The quantity of each label that was protein bound was also determined.

c Radioactivity in the lyophilisate is expressed as nmol ronidazole equivalence/mg of protein in the original sample.

d Two rats were dosed, by gavage, with a mixture of 2-[^{14}CH$_2$]- and 4-[^{3}H]-ronidazole (10 mg/kg). Six hrs after dosing, the rats were sacrificed and the livers homogenized. An aliquot of the homogenate was lyophilized to determine the quantity of ^{3}H released while a second aliquot was used to determine the quantity of each radiolabel that was protein-bound.

which is attacked by protein sulfhydryls at the 2-methylene carbon, yielding an immobilized aminoimidazalone (Vb).

This mechanism is supported by the fact that protein-alkylation occurs with loss of the C$_4$ proton both *in vitro* and *in vivo* (Table 4). In these studies, a mixture of ronidazole containing the 2-^{14}CH$_2$- (Ic) and 4-^{3}H (If) labels were incubated with liver microsomes or dosed to rats and the extent of protein alkylation calculated from the ^{14}C bound while the amount of 4-^{3}H released was determined following the lyophilization of the samples. Almost all of the protein-bound product did not contain the 4-^{3}H, consistent with extensive nucleophilic attack at C$_4$. In addition, *in vitro* studies rule out alternative explanations of proton exchange from ronidazole or one of its nitro reduced metabolites since loss of the C$_4$-proton required enzymatic activity with no proton loss occurring from the parent nitroimidazole or from an aminoimidazole analogue (data not shown).

Following acid hydrolysis of the protein-bound metabolite, several ring fragmentation products were identified. Methylamine, arising from the hydrolytic cleavage of N_1-C_2 and N_1-C_5 bonds, was quantitatively liberated (Miwa, *et al.*, 1984). Carboxymethylcysteine, formed from cleavage of the N_1-C_2 and C_2-C_3 bonds has also been recently identified[3] and provided the first unequivocal evidence for cysteine thiol addition to the 2-methylene carbon of ronidazole.

Thus, the major pathway for protein alkylation is envisioned to occur by the stepwise nucleophilic addition of water to C_4 followed by loss of the carbamate and addition of protein cysteine thiol to the 2-methylene carbon. Critical to elimination of the carbamate and subsequent protein alkylation is the addition of water to C_4, explaining the low protein-binding observed for VI relative to I.

Relationship Between Protein Alkylation and the Mutagenicity of Ronidazole
　　The potential role of the C_4 position in the mutagenicity observed in strain TA100 was investigated with 4-substituted nitroimidazoles. Mutagenic activities (Table 3) generally followed a pattern similar to protein-binding, suggesting a role for the C_4 carbon in the expression of this activity. Thus, a substituent at C_4 (VI compared to I; IX compared to VIII or XI and XII compared to X) results in dramatic reductions in mutagenic activities. The carbamate appears also to contribute to the mutagenicity (VII compared to I). However, closer examination revealed that, although both an unsubstituted C_4 and carbamate alpha to the imidazole gave the highest protein-binding and mutagenicity (I), the position of the carbamate was much less important for mutagenicity. Compound VIII was relatively more mutagenic than expected based on its protein-binding activity. These data suggest that loss of the carbamate, *per se*, although implicated in the protein-alkylation by I, is not important in the expression of mutagenicity. In contrast, mutagenicity is almost entirely dependent on an unsubstituted C_4 position, suggesting a possible role of nucleophilic addition at this carbon for the expression of mutagenicity.

The Mutagenicity of Soluble and Protein-Bound Metabolites of Ronidazole
　　The identical results obtained *in vitro* and *in vivo* in terms of: (1) protein-binding of the imidazole ring framework, (2) loss of the carbamate, and (3) C_4 proton loss provided strong evidence for the relevance of microsomal metabolic studies to the intact animal. Since conditions could be employed to substantially increase the degree of metabolism and the quantity of protein-binding, relative to that observed *in vivo*, advantage was taken of microsomal incubations to evaluate the mutagenicity of the soluble and protein-bound metabolites of I.

Prolonged microsomal incubations (46 hrs at 37 °C) were performed to maximize conversion of ronidazole to its metabolites. An aliquot of the soluble fraction from this incubation was analyzed for residual, unmetabolized drug by HPLC reverse isotope dilution analysis (Table 5, RIDA) while a second aliquot was examined for mutagenicity. The mutagenicity of this aliquot was converted to the corresponding concentration of ronidazole by comparison to a standard curve relating mutagenicity to the concentration of drug. RIDA analysis revealed that about 82% of the drug had been metabolized under these conditions. The concentration of residual drug was comparable to the concentration of drug estimated by mutagenicity indicating that all of the mutagenicity could be accounted for in terms of the residual drug.

---

[3] P. Wislocki R. Alvaro, and G. Miwa, unpublished observation

Table 5. Mutagenicity of Soluble Ronidazole Metabolites

| | | Ronidazole (mM) Assay | |
Treatment	Ronidazole[a] Metabolized (%)	RIDA	Mutagenicity
No Addition	82	1.36	1.40
+ Cysteine (50 mM)	97	0.19	0.20

[a] Ronidazole (10 mM) was anaerobically incubated for 46 hr at 37 °C with liver microsomes and an NADPH-generating system.

When cysteine was added to an identical incubation, the extent of metabolism could be increased (Table 5), presumably because the cysteine aids in preserving the activity of the activating enzyme. Even with the higher conversion condition achieved, the low mutagenicity of the supernatant could be accounted for in terms of residual drug. Thus, the evidence demonstrates the absence of mutagenic properties of these soluble metabolites and suggests that the mutagenic property of this nitroimidazole derives from short lived intermediate(s) formed prior to the formation of the stable products tested. This conclusion is supported by the absence of mutagenicity of metabolically formed reactive intermediates which have been trapped by exogenous cysteine (Wislocki. *et al.*, 1984a,b)

Although the assessment of the mutagenic activity of the soluble metabolites was readily accomplished, assessing the mutagenicity of the protein-bound metabolites of ronidazole posed a considerably more difficult task since the metabolites were immobilized to protein macromolecules. Hydrolytic conditions permitting high recovery of low molecular weight products were required for testing in the Ames assay. Proteolysis was essential since the alternative, acid hydrolysis, was known to result in imidazole ring destruction. After successive treatments of the protein-bound metabolites by several proteases, a low molecular weight fraction (<1000 mw), representing approximately 50% of the total radioactivity tested, was prepared. The composition of this sample was not characterized but was undoubtedly composed of a mixture of small peptide and cysteine adducts.

The addition of quantities of protein hydrolysates containing bound metabolites of ronidazole sufficient for detection of mutagenicity caused an attenuation in the response of the TA100 strain to ronidazole. No mutagenicity greater than that observed with a control sample, which did not contain protein-drug hydrolysate, was observed in samples containing 0.025 mM of residual ronidazole (Table 6). In order to assess the mutagenicity of the hydrolyzed samples, known quantities of drug were added to aliquots of the hydrolysates to determine the detection limit and response factor for this assay under these conditions. The results

Table 6. Mutagenicity of Protein-Bound Ronidazole Metabolites

Sample[a]	Bound Products (mM)	Added Drug (mM)	Total Drug (mM)	TA 100 Revertants
Hydrolysates of Untreated Microsomes	0	0	0	277
Hydrolysates of Microsomes + Ronidazole[b]	4.0	0	0.025	259
	4.0	0.05	0.075	566
	4.0	0.125	0.150	936

[a] Proteolysis gave products with mw < 1000

[b] Samples with a total of 200 nmol equivalents of ronidazole products also contained 1.25 nmol of residual unbound drug

demonstrate that the detection limit (concentration of drug giving a 2-fold increase in revertants over background) was approximately 0.075 mM of total drug. Since radioactivity equivalent to 4.0 mM of protein-bound metabolite was present in these samples, an upper limit for the relative mutagenicity of these products could be estimated from the ratio of this concentration and the detection limit. Estimated in this fashion, the mutagenicity of the hydrolyzed products was determined to be no greater than 2% of the parent nitroimidazole.

SUMMARY

Substantial evidence implicates the obligatory nucleophilic attack by water at C_4 for the elimination of the carbamate and subsequent immobilization by electrophilic attack on protein thiols. Consequently, the strong correlation between the structural requirements for protein alkylation and for mutagenicity in TA100 suggests a possible role of nucleophilic addition at C_4 or at the 2-methylene carbon for the expression of mutagenicity. Further studies directed at evaluating this possibility are currently in progress.

REFERENCES

Chessin, H., McLaughlin, T., Mroezkowski, Z., Rupp, W.D., and Low, K.B., 1978, Radiosensitization, Mutagenicity, and Toxicity of Escherichia coli by Several Nitrofurans and Nitroimidazoles, Radiation Research, 75: 424.

Goldman, P., 1980, Metronidazole: Proven Benefits and Potential Risks, John Hopkins Med. J., 147: 1.

Lossick, J.G., 1982, Treatment of Trichomonas vaginalis Infections, Revs. Infect. Dis., 4: S801.

Miller, E.C. and Miller, J.A., 1982, Reactive Metabolites as Key Intermediates in Pharmacologic and Toxicologic Responses: Examples from Chemical

Carcinogenesis, in: "Biological Reactive Intermediates II. Chemical Mechanisms and Biological Effects," R. Snyder, D.V. Parke, J.J. Kocsis, D.J. Jollow, C.G. Gibson, and C.M. Witmer, ed., p.1, Plenum Press, New York.

Miwa, G.T., West, S.B., Walsh, J.S., Wolf, F.J., and Lu, A.Y.H., 1982, Drug Residue Formation from Ronidazole, A 5-Nitroimidazole. III. Studies on the Mechanism of Protein Alkylation *In Vitro*, Chem.-Biol. Interactions, **41**: 297.

Miwa, G.T., Alvaro, R.F., Walsh, J.S., Wang, R. and Lu, A.Y.H., 1984, Drug Residue Formation from Ronidazole, A 5-Nitroimidazole. VII. Comparison of Protein-Bound Products formed *In Vitro* and *In Vivo*, Chem.-Biol. Interactions, **50**: 189.

Miwa, G.T., Wang, R., Alvaro, R., Walsh, J.S., and Lu, A.Y.H., 1986, The Metabolic Activation of Ronidazole [(1-Methyl-5-Nitroimidazole-2-yl)-methyl carbamate] to Reactive Metabolites by Mammalian, Cecal Bacterial and *T. foetus* Enzymes, Biochem. Pharmacol., in press.

Muller, M., 1981, Action of Clinically Utilized 5-Nitroimidazoles on Microorganisms, Scand. J. Infect. Dis., Suppl. 26: 31.

Rosenblum, C., Trenner, N.R., Buhs, R.P., Hiremath, C.B., Koniuszy, F.R. and Wolf, D.E., 1972, Metabolism of Ronidazole (1-Methyl-5-Nitroimidazol-2-ylmethyl Carbamate), J. Agr. Food Chem., **20**: 360.

Voogd, C.E., Van der Stel, J.J., and Jacobs, J.J.J.A.A., 1979, The Mutagenic Action of Nitroimidazoles and Some Imidazoles, Mutation Research, **66**: 207.

West, S.B., Wislocki, P.G., Fiorentini, K.M., Alvaro, R., Wolf, F.J. and Lu, A.Y.H., 1982, Drug Residue Formation from Ronidazole, A 5-Nitroimidazole. I. Characterization of *In Vitro* Protein Alkylation, Chem.-Biol. Interactions, **41**: 265.

West, S.B., Wislocki, P.G. and Lu, A.Y.H., 1982, Drug Residue Formation from Ronidazole, A 5-Nitroimidazole. II. Involvement of Microsomal NADPH-Cytochrome P-450 Reductase in Protein Alkylation *In Vitro*, Chem.-Biol. Interactions, **41**: 281.

Wislocki, P.G., Bagan, E.S., VandenHeuval, W.J.A., Walker, R.W., Arison, B.H., Lu, A.Y.H. and Wolf, F.J., 1984, Drug Residue Formation from Ronidazole, A 5-Nitroimidazole. V. Cysteine Adducts Formed Upon Reduction of Ronidazole by Dithionite or Rat Liver Enzymes in the Presence of Cysteine, Chem.-Biol. Interactions, **49**: 13.

Wislocki, P.G., Bagan, E.S., Cook, M.M., Bradley, M.O., Wolf, F.J. and Lu, A.Y.H., 1984, Drug Residue Formation from Ronidazole, A 5-Nitroimidazole. VI. Lack of Mutagenic Activity of Reduced Metabolites and Derivatives of Ronidazole. Chem.-Biol. Interactions, **49**: 27.

CARCINOGEN-DNA ADDUCT FORMATION AS A PREDICTOR OF METABOLIC ACTIVATION

PATHWAYS AND REACTIVE INTERMEDIATES IN BENZIDINE CARCINOGENESIS

Fred F. Kadlubar, Yasushi Yamazoe, Nicholas P. Lang*,
David Z. J. Chu*, and Frederick A. Beland

National Center for Toxicological Research, Jefferson, AR 72079
*Veterans Administration Medical Center, Little Rock, AR 72205

INTRODUCTION

Benzidine (BZ) is a strong hepatocarcinogen in rats, mice and hamsters and induces mammary tumors in female rats (1-4). However, BZ has been primarily implicated as a human urinary bladder carcinogen as a result of occupational exposure in the dye and rubber industries (5). Limited carcinogenicity experiments in dogs confirm BZ's ability to induce bladder carcinomas, although its potency in this species appears to be less than that of 4-aminobiphenyl and 2-naphthylamine (6,7). Over the last few years, we have examined carcinogen-DNA adduct formation from BZ in relation to specific metabolic activation pathways and have suggested the involvement of different biologically reactive intermediates leading to liver and urinary bladder carcinogenesis (8-15). In this paper, we will review the findings which resulted in these conclusions. In addition, we will present data on the mechanisms of BZ peroxidation and on the ability of human bladder and colon to catalyze the peroxidative activation of BZ to DNA-bound adducts.

MECHANISMS OF BZ-INDUCED HEPATOCARCINOGENESIS

When we began our studies, BZ had been extensively tested as a hepatocarcinogen in rodents and Morton et al. (16) had demonstrated that rat hepatic cytosol catalyzed the rapid, sequential N-acetylation of BZ to N-acetylbenzidine (ABZ) and N,N'-diacetylbenzidine (DABZ). Furthermore, they had shown (16,17) that DABZ was N-hydroxylated to N-hydroxy-N-N'-diacetylbenzidine (N-OH-DABZ) by rat liver microsomal monooxygenases and that N-OH-DABZ could be converted to reactive electrophiles by hepatic N,O-acyltransferase(s) and sulfotransferase(s). Thus, in collaboration with Dr. C.N. Martin from the University of York, we sought to identify the carcinogen-DNA adducts formed in rodent liver in vivo and to evaluate the role of these pathways in the formation of the ultimate carcinogenic metabolite(s).

In our initial experiment (8), rats were given a single i.p. injection (25 mg/kg) of [ring-^{3}H]BZ, [ring-^{14}C]DABZ, or ABZ radiolabeled with either ring-^{14}C or with acetyl-^{3}H. After 24 hrs, the animals were sacrificed and the liver DNA was isolated and analyzed for covalently bound ^{3}H or ^{14}C. The data clearly indicated that the levels of binding of [^{3}H]BZ, [ring-^{14}C]ABZ, and [acetyl-^{3}H]ABZ to hepatic DNA were comparable (15-30 adducts/10^6 nucleotides), while binding of [ring-^{14}C]DABZ could not be detected (<1 adduct/10^6 nucleotides). Similar results were subsequently obtained using [acetyl-^{3}H]ABZ and [acetyl-^{3}H]DABZ in both rats and hamsters (9). Enzymatic hydrolysis of the hepatic DNA from BZ- and ABZ-treated animals and subsequent analysis by high pressure liquid chromatography (hplc) further indicated the presence of a single putative carcinogen-deoxyribonucleoside adduct which accounted for >90% of the total bound ^{3}H or ^{14}C.

Though somewhat fortuitous, these findings provided a critical insight into the metabolic activation pathways that could lead to carcinogen-DNA adduct formation. First of all, the lack of binding observed after DABZ treatment indicated that a pathway involving N-hydroxylation of DABZ did not play a major role in carcinogen-DNA binding. Secondly, the comparable binding levels obtained with [ring-^{3}H] or [acetyl-^{3}H]ABZ indicated that the carcinogen-DNA adduct that was isolated had retained at least one acetyl group which had not undergone metabolic N-acetyltransfer (with concomitant dilution of the radiolabel).

In order to identify the acetylated BZ-DNA adduct found in vivo, the syntheses of four different nucleoside adducts were carried out (Fig. 1). N-Acetoxy-DABZ, an electrophilic synthetic ester of N-OH-DABZ (17), was reacted with deoxyguanosine (dG) to yield N-(deoxyguanosin-8-yl)-DABZ (N-dG-DABZ). This product was purified, characterized by mass and ^{1}H-NMR spectral analyses, and then subjected to selective deacetylation reactions. Incubation of N-dG-DABZ with carboxyl esterase for 4 hrs at 37°C yielded a single product which was purified and characterized as N-(deoxyguanosin-8-yl)-N-acetylbenzidine (N-dG-N-ABZ), while treatment of N-dG-DABZ with methanolic ammonia at 37°C for 4 hrs (18) resulted in its quantitative conversion to N-(deoxyguanosin-8-yl)-N'-acetylbenzidine (N-dG-N'-ABZ). This adduct could also be conveniently prepared by reaction of dG or DNA with synthetic N-hydroxy-N'-acetylbenzidine (N'-OH-ABZ) under mildly acidic conditions, a finding which suggested that this N-hydroxy arylamine might serve in vivo as an ultimate or proximate carcinogenic metabolite (vide infra). For chromatographic comparisons, a completely deacetylated N-(deoxyguanosin-8-yl)benzidine (N-dG-BZ) adduct was also prepared by carboxyl esterase treatment of N-dG-N'-ABZ.

With the availability of these adduct standards, hepatic DNA was obtained from rats, mice, and hamsters treated with radiolabeled BZ under various regimens. Upon hydrolysis of the DNA samples and subsequent hplc, a single radioactive component was eluted which consistently cochromatographed with authentic N-dG-N'-ABZ and was separated from the other adduct standards (Fig. 2). Further identification of the in vivo adduct as N-dG-N'-ABZ was achieved by analysis of its pH/solvent partitioning characteristics and by its conversion to N-dG-BZ upon carboxyl esterase treatment. Thus, the metabolic activation of BZ in the three rodent species sensitive to hepatocarcinogenesis each resulted in formation of N-dG-N'-ABZ as the major DNA adduct in the carcinogen-target tissue.

The structure of N-dG-N'-ABZ was consistent with studies reported for at least ten other carcinogenic arylamines, where C8-substituted dG adducts are formed by reaction of DNA with electrophilic arylnitrenium ions generated by heterolytic decomposition of protonated N-hydroxy arylamines or their O-esters (reviewed in 19,20). Consequently, a metabolic

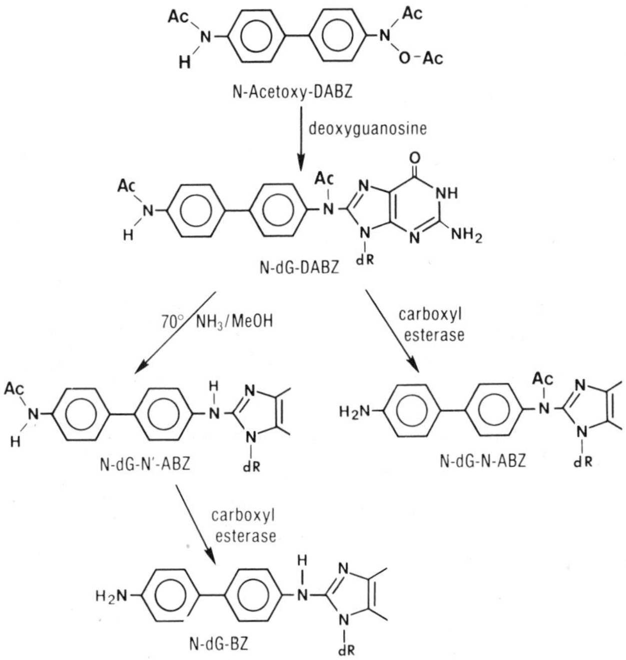

Fig. 1. Synthetic routes for the preparation of N-dG-DABZ, N-dG-N-ABZ, N-dG-N'-ABZ and N-dG-BZ. Ac = acetyl. For experimental details, see ref. 8.

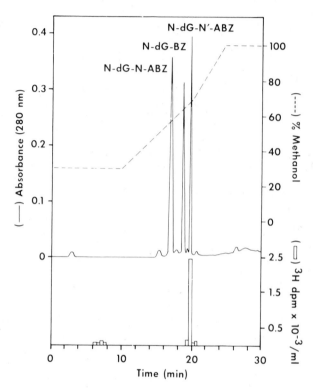

Fig. 2. Hplc profile of a mouse liver DNA hydrolysate obtained at 24 hrs
after treatment with 80 ppm [ring-^{3}H]BZ in the drinking water
for 1 week (from ref. 8). The histogram shows the radioactivity
associated with the in vivo BZ-modified DNA and the absorbance
at 280 nm indicates elution of added synthetic adducts.
N-dG-DABZ, which elutes near 18 min, was not included as a
marker in this chromatogram. The separation was done on a
Waters analytical uBondapak-C$_{18}$ column with a flow rate of 2
ml/min and a water/methanol gradient as indicated.

activation pathway involving formation of N'-OH-ABZ or N-OH-DABZ was proposed (Fig. 3). Since N'-OH-ABZ was shown to react directly with DNA to yield N-dG-N'-ABZ (vide supra), a mechanism involving protonation of the N-hydroxy group, loss of water, and arylnitrenium ion formation appeared likely. Alternatively, N'-OH-ABZ might be esterified by hepatic sulfotransferases or O-acetyltransferases to an unstable O-ester which would be expected to decompose to the same reactive electrophile. Another pathway could involve formation of N-OH-DABZ, possibly by metabolic N-acetylation of N'-OH-ABZ, and its subsequent conversion by hepatic N,O-acyltransferase to an unstable N-acetoxy-N'-acetylbenzidine derivative, which would yield the same arylnitrenium ion and an N-dG-N'-ABZ adduct.

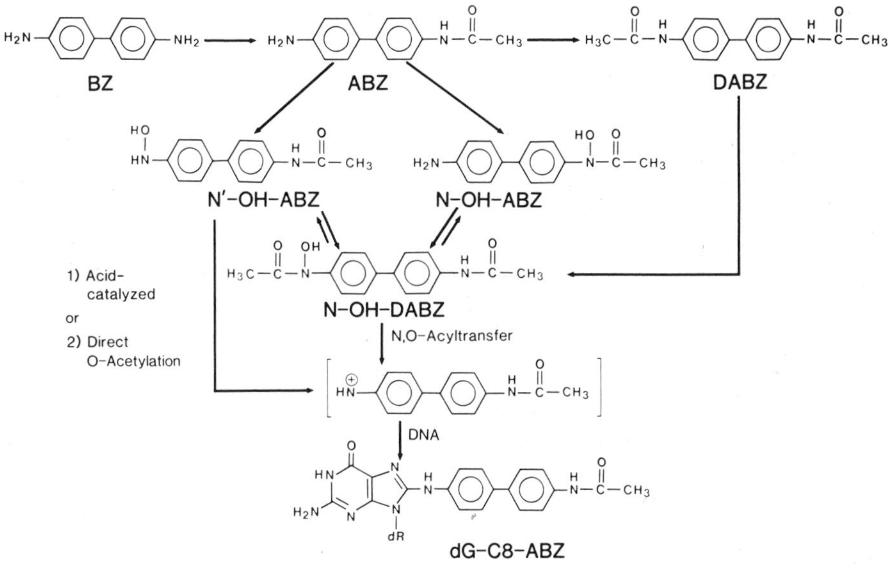

Fig. 3.   Proposed metabolic pathways responsible for the covalent binding of BZ to hepatic DNA (taken from ref. 11).   dG-C8-ABZ = N-dG-N'-ABZ.

     In order to assess the role of these pathways in the metabolic activation of BZ, a study of the hepatic N-oxidation, acetyl-transfer, DNA-binding of acetylated BZ metabolites was undertaken in vitro using rat and mouse tissue preparations (11).

     In the rat (Table 1), N-oxidation of ABZ and DABZ was catalyzed by NADPH- and NADH-fortified hepatic microsomes; but the rate of total N-oxidation at both nitrogens of ABZ was about 50 times faster than the rate of DABZ N-oxidation.   The conversion of ABZ to N-OH-ABZ and N'-OH-ABZ were nearly equivalent.   N-Acetylation of BZ to ABZ and of ABZ to DABZ by acetyl CoA and hepatic cytosol was quite rapid and the initial rates were 2 to 3-times faster than the rate of ABZ N-oxidation.   In

Table 1. Metabolism of BZ and Its Acetylated Derivatives by
Hepatic Microsomes and Cytosols of Rats and Mice[a]

Reaction	Rates of Conversion $\pm$ S.D.; n = 3	
	Rat	Mouse
ABZ -> N-OH-ABZ[b]	0.11 $\pm$ 0.01	0.19 $\pm$ 0.04
ABZ -> N'-OH-ABZ[b]	0.13 $\pm$ 0.03	0.92 $\pm$ 0.08
DABZ -> N-OH-DABZ[b]	0.005 $\pm$ 0.001	0.26 $\pm$ 0.01
BZ -> ABZ[b]	0.82 $\pm$ 0.02	2.76 $\pm$ 0.40
ABZ -> DABZ[b]	0.66 $\pm$ 0.03	2.85 $\pm$ 0.01
DABZ -> ABZ[b]	0.016 $\pm$ 0.001	2.93 $\pm$ 0.12
ABZ -> BZ[b]	0.011 $\pm$ 0.001	2.71 $\pm$ 0.33
N-OH-ABZ -> N-OH-DABZ[b]	0.49 $\pm$ 0.03	1.87 $\pm$ 0.08
N'-OH-ABZ -> N-OH-DABZ[b]	0.53 $\pm$ 0.02	1.91 $\pm$ 0.14
N-OH-DABZ -> N-OH-ABZ[b]	0.19 $\pm$ 0.03	5.96 $\pm$ 0.38
N-OH-DABZ -> N'-OH-ABZ[b]	0.90 $\pm$ 0.05	1.40 $\pm$ 0.14
N'-OH-ABZ + Acetyl CoA -> DNA Binding[c]	63 $\pm$ 8	21 $\pm$ 2
N'-OH-DABZ -> DNA Binding[c]	117 $\pm$ 16	2 $\pm$ 1

[a] For experimental details, see ref. 11.
[b] Rates are expressed as nmoles product formed/min/mg protein.
[c] DNA binding levels are expressed as pmoles substrate bound/mg DNA/15 min.

contrast, the microsomal N-deacetylation of ABZ and DABZ was very slow (<3% of the rate of N-acetylation). For the N-hydroxylated metabolites, hepatic cytosolic N-acetylation of both N-OH-ABZ and N'-OH-ABZ to form N-OH-DABZ occurred rapidly; while microsomal N-deacetylation of N-OH-DABZ to N'-OH-ABZ was 4 to 5-times faster than N-OH-ABZ formation. These data were consistent with the negligible binding of DABZ to hepatic DNA in vivo and indicated an activation pathway involving conversion of BZ to ABZ. Furthermore, although oxidation of ABZ occurred at both nitrogens, the high rates of N-acetylation of both N-OH-ABZ and N'-OH-ABZ and the selective N-deacetylation of N-OH-DABZ to N'-OH-ABZ suggested that, at steady-state, only N-OH-DABZ and N'-OH-ABZ would be available for further esterification reactions. In this regard, N'-OH-ABZ appeared to be readily activated to a DNA-binding metabolite by the acetyl CoA-dependent, cytosolic O-acetyltransferase, but not sulfotransferase; while N-OH-DABZ was rapidly converted to a DNA-bound product by hepatic cytosolic N,O-acyltransferase, as originally indicated by Morton et al. (16). In addition, hydrolysis and hplc analysis of the DNA modified in vitro by N-OH-DABZ in the presence of partially purified rat hepatic N,O-acyltransferase showed the same N-dG-N'-ABZ observed in vivo from BZ. Thus, both N'-OH-ABZ and N-OH-DABZ appear to serve as proximate carcinogenic metabolites in the rat and they undergo conversion by acetyltransferases to an electrophilic arylnitrenium ion, forming a single N-dG-N'-ABZ adduct (Fig. 3).

In the mouse (Table 1), hepatic microsomes also catalyzed the N-oxidation of both ABZ and DABZ with total ABZ N-oxidation being 4-fold greater than the N-oxidation of DABZ. However, in contrast to the rat, the conversion of ABZ to N'-OH-ABZ was 5-fold higher than its conversion to N-OH-ABZ. The N-acetylation of BZ to ABZ and of ABZ to DABZ occurred rapidly, but so did the N-deacetylation of these substrates. In the case of the N-hydroxylated derivatives, N'-OH-ABZ and N-OH-ABZ were rapidly N-acetylated to N-OH-DABZ; while N-OH-DABZ deacetylation to N-OH-ABZ was preferred over deacetylation to N'-OH-ABZ. These data indicated that N-acetylation and N-deacetylation of BZ and its N-acetylated and N-hydroxylated metabolites proceed reversibly in the mouse and that, at steady-state, all three N-hydroxy metabolites should be present. However, when mouse hepatic cytosol was assayed for N-OH-DABZ N,O-acyl-transferase and N-OH-ABZ or N'-OH-ABZ sulfotransferase, little or no activity could be detected. Only the acetyl CoA-dependent binding of N'-OH-ABZ to DNA was observed. Therefore, N'-OH-ABZ likely serves as the proximate carcinogenic metabolite in the mouse and formation of N-dG-N'-ABZ could arise upon O-acetylation of N'-OH-ABZ or by its direct reaction (21) with nuclear DNA (Fig. 3).

MECHANISMS OF BZ-INDUCED URINARY BLADDER CARCINOGENESIS

Our interest in the metabolic activation of BZ in relation to urinary bladder carcinogenesis largely arose from the realization that the urothelial DNA adducts and their mechanism of formation were likely to be completely different than those found for BZ-induced hepatocarcino-genesis. The carcinogenic activity of BZ is targeted to the urinary bladder of both humans and dogs, and the latter species is known to be unable to carry out the metabolic N-acetylation of BZ or other carcino-genic arylamines (22,23). In humans, N-acetylation of arylamines is known to exhibit a genetic polymorphic distribution with both slow and fast acetylator phenotypes; and, in cases of documented exposure to car-cinogenic arylamines, humans with the slow acetylator phenotype were found to be at much higher risk for the development of bladder cancer (reviewed in 24). In preliminary experiments with dogs (25), we showed the BZ-DNA adduct(s) formed in the urothelium had chromatographic and partitioning properties that were clearly different from those of the acetylated BZ-nucleoside derivatives previously studied. Thus, it was evident that metabolic activation of BZ for urinary bladder carcino-genesis was unlikely to involve any acetylation reactions. At the same time, it had been reported by Zenser and coworkers that prostaglandin endoperoxide synthetase (now called prostaglandin H synthase; PHS) from renal medullary or seminal vesical microsomes was able to catalyze the arachidonic acid-dependent peroxidative activation of BZ to nucleic acid- and protein-bound derivatives (26,27). PHS, which is responsible for the initial reactions in prostaglandin biosynthesis, was also known to be widely distributed in extrahepatic tissues and had been shown to mediate the co-oxidation of certain drugs and carcinogens, such as acetaminophen, nitrofurans, and polycyclic hydrocarbons (reviewed in 28). Thus, in collaboration with Dr. T.V. Zenser at the V.A. Medical Center in St. Louis, we sought to determine the role of PHS in the metabolic activation of BZ and to elucidate the mechanism of formation and the structural identity of the BZ-DNA adduct(s).

In our initial experiment (12), we examined the in vitro, arachidonic acid-dependent activation of BZ and several carcinogenic arylamines to form covalent adducts with DNA, using a solubilized PHS preparation from ram seminal vesicle microsomes. Of the substrates tested, BZ was the most extensively metabolized and gave the highest levels of covalent DNA binding. Activation of arylamines by PHS was subsequently confirmed by Morton et al. (29) who also showed that ABZ and DABZ were not substrates for this activation pathway. However, in order to suggest the involvement of PHS in carcinogen activation, it was obvious that the metabolism of BZ by PHS should be demonstrated with a carcinogen-target tissue. Consequently, microsomal enzyme preparations from dog bladder epithelium, kidney, and liver were prepared and both prostaglandin biosynthesis and the PHS-mediated activation of BZ (and 2-naphthylamine) were determined (13). As expected, no activity was detected in the liver or the renal cortex, but renal inner and outer medullary microsomes contained PHS and could metabolize BZ. However, the most striking result was obtained with bladder epithelial microsomes. This preparation was found to contain very high levels of PHS (i.e., 10 times higher than the dog renal inner medulla) and could rapidly catalyze the metabolism of BZ and other arylamines to DNA-, RNA- and protein-bound adducts. Thus, the presence of PHS and its activation of aromatic amines was demonstrated in a carcinogen-target tissue.

Recently, we have obtained preliminary data which suggest that PHS is also present at high levels in the epithelium of the human bladder and colon. As shown in Table 2, arachidonic acid-dependent binding of BZ to DNA was mediated by both bladder and colon microsomal preparations; and the activity was inhibited from 50-90% by addition of indomethacin, a selective inhibitor of PHS. Further studies are in progress to characterize PHS in human tissues and to establish its substrate specificity toward arylamine carcinogens.

Table 2. Arachidonic Acid-Dependent Activation of [ring-^{3}H]BZ by Human Tissue Microsomes[a]

Tissue	DNA Binding (pmol bound/mg DNA/mg protein)
Bladder epithelium	
No. 1	661
No. 2	457
Colon epithelium	
No. 1	294
No. 2	633
No. 3	244
Liver	<10

[a] Incubations were conducted at 37°C for 5 min and contained 50 mM potassium phosphate buffer (pH 7.4) 2.5 mg/ml DNA, 20 uM [ring-^{3}H]BZ (121 mCi/mmol), and 0.025-0.100 mg microsomal protein/ml. The DNA was isolated by multiple solvent extractions and precipitations and the level of bound [^{3}H]BZ was determined as described previously (12,14). The results represent the average of duplicate determinations.

In view of the possible involvement of PHS in human carcinogenesis, the mechanism of BZ peroxidation and the nature of its binding to DNA was further investigated. Several studies (30-33) had indicated that BZ is peroxidized, perhaps sequentially, to both one-electron (BZ radical cation) and two-electron (benzidine diimine; BZDI) intermediates; and both of these potentially electrophilic species have been proposed to be responsible for BZ binding to macromolecules. Recently, we have synthesized [2,2'-^{3}H] BZDI and have compared its binding to DNA (cf. ref. 32) under several reaction conditions (14). [^{3}H]BZDI (20-50 uM) was highly reactive and incubation with DNA (2.5 mg/ml) resulted in covalent binding of 30-45% of the added BZDI. Since the reactivity was relatively unaffected by pH (5.0 vs 7.0), oxygen (air vs argon atmospheres), or by addition of free radical traps or simple nucleophiles, a mechanism involving formation of N-hydroxy-benzidine or the BZ radical cation seemed unlikely. Upon enzymatic hydrolysis of the BZDI-modified DNA and hplc analysis, a major adduct accounting for 70-80% of the total binding was obtained (Fig. 4). The results of mass and ^{1}H-NMR spectral studies

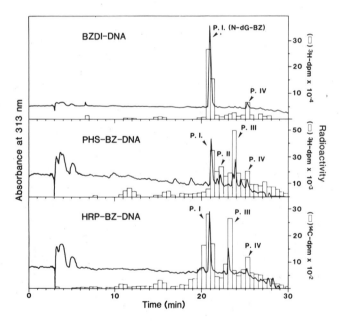

Fig. 4. Hplc profiles of BZ-modified DNA hydrolysates obtained by reaction of DNA with [ring-^{3}H] BZDI, [ring-^{3}H]BZ/PHS/arachidonic acid, or [ring-^{14}C]BZ/horseradish peroxidase (HRP)/H$_2$O$_2$. The solid line shows the absorbance at 313 nm and the radioactivity is plotted as histograms. The enzymatic hydrolysis procedures and the hplc conditions are the same as those described in ref. 14.

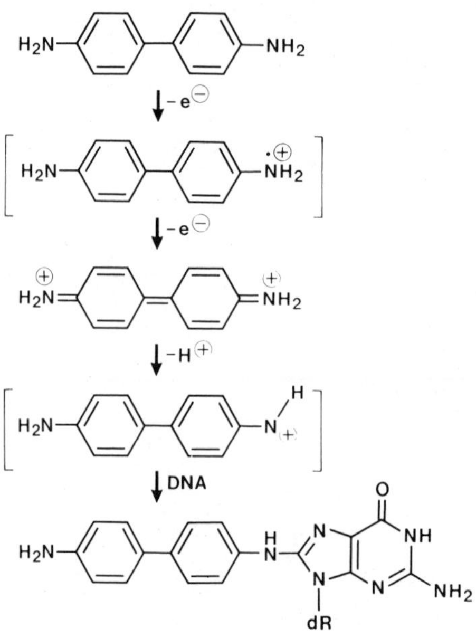

Fig. 5.   Proposed mechanism for the peroxidative activation of BZ leading to N–dG–BZ formation in the DNA.

indicated its identity as N-(deoxyguanosin-8-yl)benzidine, which strongly suggests the participation of a non-acetylated BZ nitrenium ion as the biologically reactive intermediate (vide supra). Remarkably, this was the same N-dG-BZ adduct that had been prepared earlier by deacetylation of N-dG-N'-ABZ. Similar experiments with BZ-modified DNA, obtained from in vitro incubations with PHS in the presence of arachidonic acid, and with horseradish peroxidase and H_2O_2, indicated that N-dG-BZ was also formed as a major DNA adduct during the enzymatic peroxidation of BZ (Fig. 4). Several other adducts are also evident and their identification and mechanism of formation is currently under investigation. In addition, preliminary data indicate the same BZ-DNA adducts observed in vitro are also present in the dog urothelium after oral administration of BZ.

In conclusion, a major pathway for the metabolic activation of BZ appears to involve its peroxidative metabolism by urothelial PHS to a reactive BZDI. From the reaction characteristics of synthetic BZDI, a mechanism involving a simple deprotonation to an electrophilic aryl-nitrenium ion is suggested to result in BZ-DNA adduct formation in the urinary bladder (Fig. 5).

ACKNOWLEDGEMENT

The authors wish to acknowledge the essential contributions of our collaborators and coworkers throughout the course of these studies, especially Drs. C. N. Martin, T. V. Zenser and C. B. Frederick.

REFERENCES

1. T. J. Haley, Benzidine revisited: a review of the literature and problems associated with the use of benzidine and its congeners, Clin. Toxicol. 8:13-42 (1975).
2. S. D. Vesselinovitch, K. V. N. Rao, and N. Mihailovich, Factors modulating benzidine carcinogenicity bioassay, Cancer Res. 35:2814-2819 (1975).
3. C. J. Nelson, K. P. Baetcke, C. H. Frith, R. L. Kodell, and G. Schieferstein, The influence of sex, dose, time, and cross on neoplasia in mice given benzidine dihydrochloride, Toxicol. Appl. Pharmacol. 64:171-186 (1982).
4. K. C. Morton, C. Y. Wang, C. D. Garner, and T. Shirai, Carcinogenicity of benzidine, N,N'-diacetylbenzidine, and N-hydroxy-N,N'-diacetylbenzidine for female CD rats, Carcinogenesis 2:747-752 (1981).
5. IARC, "IARC Monographs on the Evaluation of the Carcinogenic Risk of Chemicals to Humans," Vol. 29, Lyon, pp. 149-183 (1982).
6. J. L. Radomski, The primary aromatic amines: their biological structure-activity relationships, Annu. Rev. Pharmacol. Toxicol. 19:129-157 (1979).
7. S. Spitz, W. H., Maguigan, and K. Dobriner, The carcinogenic action of benzidine, Cancer (Phila.) 3:789-804 (1950).
8. C. N. Martin, F. A. Beland, R. W. Roth, and F. F. Kadlubar, Covalent binding of benzidine and N-acetylbenzidine to DNA at the C-8 atom of deoxyguanosine in vivo and in vitro, Cancer Res. 42:2678-2686 (1982).
9. J. C. Kennelly, F. A. Beland, F. F. Kadlubar, and C. N. Martin, Binding of N-acetylbenzidine and N,N'-diacetylbenzidine to hepatic DNA of rat and hamster in vivo and in vitro, Carcinogenesis 5:407-412 (1984).
10. C. N. Martin, F. A. Beland, J. C. Kennelly, and F. F. Kadlubar,

Binding of benzidine, N-acetylbenzidine, N,N'-diacetylbenzidine
and Direct Blue 6 to rat liver DNA, Environ. Health Persp.
49:101-106 (1983).

11. C. B. Frederick, C. C. Weis, T. J. Flammang, C. N. Martin, and F. F.
Kadlubar, Hepatic N-oxidation, acetyl-transfer and DNA-binding
of the acetylated metabolites of the carcinogen, benzidine,
Carcinogenesis 6:959-965 (1985).

12. F. F. Kadlubar, C. B. Frederick, C. C. Weis, and T. V. Zenser,
Prostaglandin endoperoxide synthetase-mediated metabolism of
carcinogenic aromatic amines and their binding to DNA and
protein, Biochem. Biophys. Res. Commun. 108:253-258 (1982).

13. R. W. Wise, T. V. Zenser, F. F. Kadlubar, and B. B. Davis, Metabolic
activation of carcinogenic aromatic amines by dog bladder and
kidney prostaglandin H synthase, Cancer Res. 44:1893-1897 (1984).

14. Y. Yamazoe, R. W. Roth, and F. F. Kadlubar, Reactivity of benzidine
diimine with DNA to form N-(deoxyguanosin-8-yl)-benzidine,
Carcinogenesis, in press (1985).

15. Y. Yamazoe, F. A. Beland, and F. A. Kadlubar, Evidence for benzidine
diimine as a reactive intermediate in the peroxidase-mediated
binding of benzidine to DNA, Proc. Amer. Assoc. Cancer Res.
26:85 (1985).

16. K. C. Morton, C. M. King, and K. P. Baetcke, Metabolism of benzidine
to N-hydroxy-N,N'-diacetylbenzidine and subsequent nucleic acid
binding and mutagenicity, Cancer Res. 39:3107-3113 (1979).

17. K. C. Morton, F. A. Beland, F. E. Evans, N. F. Fullerton, and F. F.
Kadlubar, Metabolic activation of N-hydroxy-N,N'-diacetyl-
benzidine by hepatic sulfotransferase, Cancer Res. 40:751-757
(1980).

18. E. Kriek and J. G. Westra, Structural identification of the
pyrimidine derivatives formed from N-(deoxyguanosin-8-yl)-2-
aminofluorene in aqueous solution at alkaline pH, Carcinogenesis
1:459-468 (1980).

19. F. F. Kadlubar and F. A. Beland, Chemical properties of ultimate
carcinogenic metabolites of arylamines and arylamides. In:
Polycyclic Hydrocarbons and Carcinogenesis, ACS Symposium Series
183, Harvey, R.G., ed., American Chemical Society, Washington,
DC, pp 341-370 (1985).

20. F. A. Beland and F. F. Kadlubar, The formation and persistence of
arylamine-DNA adducts in vivo, Environ. Health Persp. 62, in
press (1985).

21. C. B. Frederick, J. B. Mays, D. M. Ziegler, F. P. Guengerich, and
F. F. Kadlubar, Cytochrome P-450- and flavin-containing monooxygenase-catalyzed formation of the carcinogen N-hydroxy-2-amino-
fluorene and its covalent binding to nuclear DNA, Cancer Res.
42:2671-2677 (1982).

22. L. A. Poirier, J. A. Miller, and E. C. Miller, The N- and ring-
hydroxylation of 2-acetylaminofluorene and the failure to detect
N-acetylation of 2-aminofluorene in the dog, Cancer Res.
23:790-800 (1963).

23. G. M. Lower, Jr., and G. T. Bryan, Enzymatic deacetylation of
carcinogenic arylacetamides by tissue microsomes of the dog and
other species, J. Toxicol. Environ. Health 1:421-432 (1976).

24. R. A. Cartwright, Historical and modern epidemiological studies on
populations exposed to N-substituted aryl compounds, Environ.
Health Persp. 49:13-19 (1983)

25. F. A. Beland, D. T. Beranek, K. L. Dooley, R. H. Heflich, and F. F.
Kadlubar, Arylamine-DNA adducts in vitro and in vivo: their role
in bacterial mutagenesis and urinary bladder carcinogenesis,
Environ. Health Persp. 49:125-134 (1983).

26. T. V. Zenser, M. B. Mattammal, and B. B. Davis, Cooxidation of

benzidine by renal medullary prostaglandin cyclooxygenase, J. Pharmacol. Exp. Ther. 211:460-464 (1979).

27. T. V. Zenser, M. B. Mattammal, H. J. Armbrecht, and B. B. Davis, Benzidine binding to nucleic acids mediated by the peroxidative activity of prostaglandin endoperoxide synthetase, Cancer Res. 40:2839-2845 (1980).

28. L. J. Marnett and T. E. Eling, Cooxidation during prostaglandin biosynthesis: a pathway for the metabolic activation of xenobiotics, In: Reviews in Biochemical Toxicology, Hodgson, E., Bend, J.R., and Philpot, R.M., eds, Elsevier, Amsterdam, pp. 135-172 (1983).

29. K. C. Morton, C. M. King, J. B. Vaught, C. Y. Wang, M.-S. Lee and L. J. Marnett, Prostaglandin H synthase-mediated reaction of carcinogenic arylamines with tRNA and homopolyribonucleotides, Biochem. Biophys. Res. Commun. 111:96-103 (1983).

30. R. W. Wise, T. V. Zenser, and B. B. Davis, Prostaglandin H synthase metabolism of the urinary bladder carcinogens benzidine and ANFT, Carcinogenesis 4:285-289 (1983).

31. P. D. Josephy, T. Eling, and R. P. Mason, Co-oxidation of benzidine by prostaglandin synthetase and comparison with the activation of horseradish peroxidase, J. Biol. Chem. 258:5561-5569 (1983).

32. R. W. Wise, T. V. Zenser, and B. B. Davis, Characterization of benzidinediimine: a product of peroxidase metabolism of benzidine, Carcinogenesis 5:1499-1503 (1984).

33. Y. Tsuruta, P. D. Josephy, A. D. Rahimtula, and P. J. O'Brien, Peroxidase-catalyzed benzidine binding to DNA and other macromolecules, Chem.-Biol. Interactions 54:143-158 (1985).

PURIFICATION, PROPERTIES AND FUNCTION OF N-HYDROXYARYLAMINE

O- ACETYLTRANSFERASE

Ryuichi Kato, Kazuki Saito, Atsuko Shinohara and
Tetsuya Kamataki

Department of Pharmacology, School of Medicine, Keio
University, Tokyo, 160 Japan

INTRODUCTION

Among numbers of environmental or chemical carcinogens, those which
are contained in daily foods can be regarded as the most causative factors
of cancers. Thus, the discovery of mutagens/carcinogens in the pyrolysis
products of amino acids and proteins was a striking matter. Some of the
promutagens isolated so far induce remarkable numbers of revertants when
incubated with Salmonella typhimurium TA98 bacterial cells in the presence
of 9,000 xg supernatant fraction of liver homogenates of PCB-treated rats
with cofactors. The specific mutagenicity of some of promutagens
represented as the number of revertants per μg of a promutagen, reaches to
levels 1000 times as high as benzo(a)pyrene.

Our studies on the mechanisms of activation of these promutagens have
clarified that IQ (2-amino-3-methylimidazo[4,5-f]quinoline), Trp-P-1 (3-
amino-1,4-dimethyl-5H-pyrido[4,3-b]indole), Trp-P-2 (3-amino-1-methyl-5H-
pyrido[4,3-b]indole), Glu-P-1 (2-amino-6-methydipyrido[1,2-a:3',2'-d]-
imidazole) and Glu-P-2 (2-aminodipyrido[1,2-a:3',2'-d]imidazole) were
hydroxylated at NH_2-moiety as a common initial process by a cytochrome P-
450-containing monooxygenase enzyme system in liver microsomes.[1,2] Among
forms of cytochrome P-450 examined, a high spin form of cytochrome P-450,
namely P-448-H which was later identified as P-450d, was characteristic
for its high activity in the N-hydroxylation of these promutagens.[1,3,4]
The N-hydroxylated products per se were mutagenic to bacterial cells, but
were capable of binding to DNA or tRNA to extents much less than expected,
especially in the presence of reducing agents.[1,5]

The amounts bound to DNA of radioactivity derived from ^{3}H-[ring
labeled] N-hydroxy-Trp-P-2 (N-OH-Trp-P-2) increased to considerable
extents when seryl-tRNA synthetase purified from yeast was added. An
enzyme, probably prolyl-tRNA synthetase rather than seryl-tRNA synthetase,
was found to function as an activating enzyme in the cytosol of rat
livers.[6] These results suggest the possibility that the N-hydroxylated
compounds can be further activated to DNA-binding species in any organs,
since prolyl-tRNA synthetase must be distributed in all organs.[7] Despite
the significance of prolyl-tRNA synthetase, the activity of this enzyme in
the activation of mutagens was not so high as the level we expected. In
fact, in our preliminary studies, we had discovered that an enzyme(s) in
liver cytosol activated N-OH-Trp-P-2, to a DNA-binding species in the
presence of acetyl-CoA.[6]

A NEW ENZYME, O-ACETYLTRANSFERASE, IN THE CYTOSOL OF SALMONELLA
TYPHIMURIUM TA98

O-Acetyltransferase in S. typhimurium TA98 as an Activating Enzyme

As mentioned above, only a small amount of N-OH-Trp-P-2 bound to DNA
without further metabolic activation, whereas N-OH-Trp-P-2 induced
mutation of bacteria without the addition of an activation enzyme system.
Thus, it was assumed that an enzyme(s) in the S. typhimurium cells was
responsible for further activation of the N-hydroxylated product.
However, no information was available on the biotransformation of any such
compounds to reactive species by a bacterial enzyme(s).

S. typhimurium strain TA98/1,8-DNP$_6$ is a mutant isolated as resistant
to the mutagenicity of 1,8-dinitropyrene. It was also shown recently that
this strain was non-responsive to certain nitroso- and N-hydroxy-
aminoarenes, probably indicating that an enzyme(s) other than
nitroreductase was deficient in this particular strain bacteria.[8]

The number of revertants induced by N-hydroxy-Glu-P-1 (N-OH-Glu-P-1)
was compared with S. typhimurium TA98 and TA98/1,8-DNP$_6$ strains. N-OH-
Glu-P-1 produced about 170 His[+] revertants/plate in the TA98 strain, but
only 41 His[+] revertants in the TA98/1,8-DNP$_6$ strain. After examination of
factors involved in the enhancement of mutation, which must be present in
TA98 but at least to much less amounts in TA98/1,8-DNP$_6$, we found that an
enzyme(s), existing in the cytosol of the TA98 strain of bacteria,
increased the DNA binding of N-OH-Glu-P-1 in the presence of acetyl-CoA as
shown in Fig. 1.[9] The cytosol fractions from the TA98 cells activated N-
OH-Glu-P-1 and N-OH-Trp-P-2, although the extent of the activation of N-
OH-Trp-P-2 was much lower in comparison with the case of N-OH-Glu-P-1.
From these and other lines of evidence, we tentatively termed this enzyme
acetyl-CoA: N-hydroxyarylamine O-acetyltransferase.

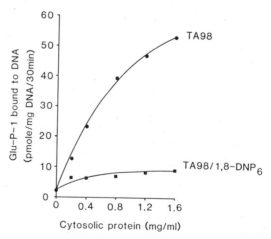

Fig. 1. Comparison of the ability of cytosol from S. typhimurium TA98
         with TA98/1,8-DNP$_6$ to activate N-OH-Glu-P-1 in the presence of
         acetyl-CoA.

Partial Purification of O-Acetyltransferase from S. typhimurium TA98

To further characterize the O-acetyltransferase, we partially purified this enzyme from the cytosol of S. typhimurium TA98 by means of streptomycin sulfate treatment, ammonium sulfate fractionation and column chromatography on DEAE-cellulose and Sephadex G-150.[10] The molecular weight of bacterial O-acetyltransferase was estimated to be 48,000 by gel filtration. The O-acetyltransferase activity, as measured by the amounts of binding of Glu-P-1 moiety to tRNA using N-[^{3}H]-OH-Glu-P-1 as a substrate, was purified by about 370-fold with a recovery of 28%. The partially purified preparation of O-acetyltransferase showed some substrate specificities. The specificity for acyl donors is shown in Fig. 2.

Acetyl-CoA was the best acyl donor, followed by propionyl-CoA, butyryl-CoA and hexanoyl-CoA. N-OH-AAF did not serve as an acyl donor for the enzymatic binding of N-OH-Glu-P-1 to tRNA. The Km value for acetyl-

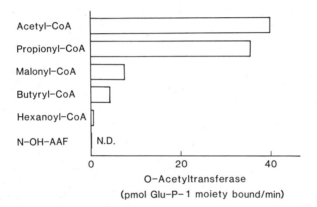

Fig. 2. Acyl donor specificity of O-acetyltransferase partially purified from S. typhimurium TA98.

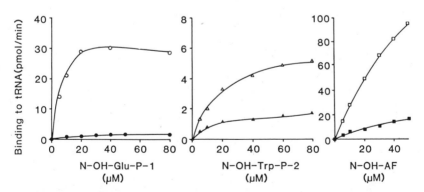

Fig. 3. Substrate specificity of O-acetyltransferase. tRNA-binding of N-OH-Glu-P-1, N-OH-Trp-P-2, and N-OH-AF was assayed in the presence of 2.5 μg/ml of enzyme (○), or in the absence of enzyme (●).

CoA was calculated to be 3.3 μM.[10] This low Km value indicates the efficient activation of the N-hydroxyarylamine in the bacterial cells. The partially purified O-acetyltansferase was capable of activating N-OH-Glu-P-1, N-OH-Trp-P-2 and N-hydroxyaminofluorene (N-OH-AF), although the efficiency varied with the substrate used (Fig. 3). The activity was inhibited by sulfhydryl reagents such as p-chloromercuribenzoate, indicating that N-hydroxyarylamine O-acetyltransferase possesses a sulfhydryl group(s) as a functional group. In addition, the O-acetyltransferase was inhibited by pentachlorophenol, a well-known inhibitor of sulfotransferase, while 2,6-dichloro-4-nitrophenol, another representative inhibitor of sulfotransferase, did not inhibit the activity. Paraoxon, a potent inhibitor of deacetylase, did not affect the O-acetyltransferase appreciably.

To roughly estimate the contribution of the N-hydroxyarylamine O-acetyltransferase to the bacterial mutation caused by N-OH-Glu-P-1 and N-OH-AF, the mutagenic activities of these N-hydroxyarylamines were measured in the presence of pentachlorophenol. As can be seen in Fig. 4, the numbers of revertants in TA98 caused by the mutagens were decreased by pentachlorophenol to levels close to those in TA98/1,8-DNP[6], lending support to the idea that the O-acetyltransferase plays major roles in the activation processes[10]. The reactive species is assumed to be very unstable and the access to bacterial DNA is limited, since the addition of the purified enzyme to a mixture containing TA98 and other necessary components resulted in a reduced number of revertants [10]. The inability of the externally added O-acetyltransferase to increase the rate of mutation may be accounted for by the assumption that the reactive metabolite(s) bound spontaneously to surrounding nucleophiles such as dithiothreitol or free coenzyme A or the bacterial cell surface.

MAMMALIAN N-HYDROXYARYLAMINE O-ACETYLTRANSFERASE

N-Hydroxyarylamine O-Acetyltransferase in Various Mammalian Species

It was of interest to determine whether or not N-hydroxyarylamine O-acetyltransferase was present in mammalian tissues. We assumed that the

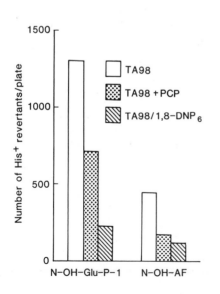

Fig. 4.
Effects of pentachlorophenol (PCP) on the mutagenicity of N-OH-Glu-P-1 and N-OH-AF.

acetyl–CoA-dependent DNA-adduct formation that we had reported previously [6] was catalyzed by the corresponding O-acetyltransferase. The activity determined as the formation of the covalent adducts varied depending on animal species and organs.[11,12] Unlike the bacterial enzyme, the mammalian enzyme could utilize N-OH-AAF as well as acetyl-CoA for the O-acetylation of N-OH-Glu-P-1. The highest activity was seen in cytosol of livers in animal species tested. Marked species differences were observed in the activity of acetyl-CoA dependent covalent binding of N-hydroxyarylamines by liver cytosol.[11,12]

Hamsters showed the highest capacity in the activation of N-OH-Glu-P-1 and N-OH-Trp-P-2 followed by rats. Rapid acetylator rabbits, which showed high activities in the N-acetylation of aromatic amines, showed the highest activities in the acetyl-CoA dependent activation of N-OH-AF but the activation of N-OH-Glu-P-1 was relatively low and that of N-OH-Trp-P-2 was undetectable. A slow acetylator rabbit, mice and guinea-pigs showed intermediate activities with N-OH-AF, but showed very low or non-detectable activities with N-OH-Glu-P-1 and N-OH-Trp-P-2. In a dog liver, the activities with these substrates were undetectable. These results clearly show the presence of marked species differences linked with substrate specificity.

The general pattern of species differences in the activity of acetyl-CoA dependent activation of N-hydroxyarylamines was similar to that observed with arylhydroxamic acid N,O-acetyltransferase activity, although there are some differences in several aspects.

In accordance with our results, the presence of acetyl–CoA dependent activation of N-hydroxyarylamine in mammalian tissues has recently been reported by Kadlubar's group.[13,14]

Purification and Properties of Rat Liver N-Hydroxyarylamine O-Acetyltransferase

The N-hydroxyarylamine O-acetyltransferase in mammalian tissues was first partially purified from rat liver cytosol. For purification of the enzyme, there was a need to develp a new and simple method to detect the O-acetyltransferase activity. We postulated that N-acetoxyarylamine formed by the O-acetyltransferase could react with sulfhydryl group-containing compounds to produce a parent arylamine as shown in Fig. 5.

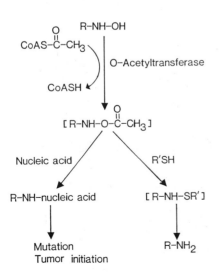

Fig. 5.
Proposed mechanism of activation of N-hydroxy-arylamines and of reduction to parent amines.

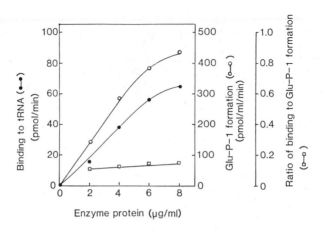

Fig. 6. The covalent binding of N-OH-Glu-P-1 to tRNA and reduction to Glu-P-1 mediated by bacterial O-acetyltransferase.

The reaction could be monitored fluorometrically when N-OH-Glu-P-1 was used as the substrate. The reaction proceeded linearly with time of incubation and amounts of bacterial enzyme, in parallel with amounts of binding to tRNA (Fig. 6).

Rat liver N-hydroxyarylamine O-acetyltransferase was partially purified by means of ammonium sulfate fractionation, DE-52 and Sephadex G-150 chromatography. The enzyme was not homogeneous as judged by sodium dodecyl sulfate (SDS)-polyacrylamide gel electrophoresis, but the molecular weight was estimated to be 28,000 daltons by gel filtration. It was noted that the N-OH-Glu-P-1 O-acetyltransferase activity was not separated from Glu-P-1 N-acetylase activity. The ratios of arylamine N-acetylase activity to N-hydroxyarylamine O-acetyltransferase activity in all the fractions of ammonium sulfate were nearly identical (Fig. 7). Among acyl donors, acetyl-CoA showed the highest activity, followed by propionyl-CoA and others. This acyl donor specificity was similar to that seen with bacterial N-hydroxyarylamine O-acetyltranferase except that N-OH-AAF was also an acyl donor in the rat O-acetyltransferase. The partially purified preparation after gel filtration also retained both activities (Fig. 8). Recently, Allaben and King[15] have purified rat liver arylhydroxamic acid N,O-acetyltransferase. Their purification procedure and some characteristics of the purified enzymes are similar to ours. These results suggest that one enzyme can support three different catalytic functions: arylamine N-acetylation, arylhydroxamic acid N,O-acetyltransfer and N-hydroxyarylamine O-acetyltransfer.

Purification and Properties of Hamster Liver O-Acetyltransferase

To verify our hypothesis, we purified hamster liver N-hydroxyarylamine O-acetyltansferase to an electrophoretical homogeneity by essentially the same methods adopted for the purification of the rat O-acetyltransferase, except that high performance KB-hydroxylapatite chromatography was added at a final step of purification. Throughout the purification procedure, activities of N-OH-Glu-P-1 O-acetyltransferase, N-OH-AAF N,O-acetyltransferase and 2-aminofluorene (2-AF) N-acetylase were copurified. The O-acetyltransferase eluted from KB-hydroxylapatite column

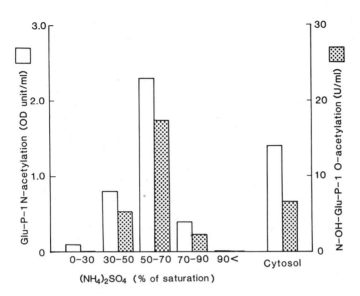

Fig. 7. Ammonium sulfate fractionation of rat liver cytosol: co-
purification of Glu-P-1 N-acetylase and N-OH-Glu-P-1 O-
acetyltransferase.

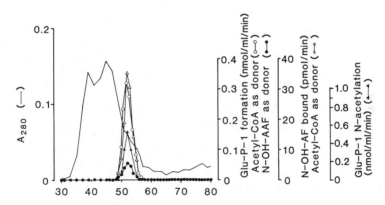

Fig. 8. The separation of rat hepatic N-hydroxyarylamine O-acetyl
transferase by Sephacryl S-200 chromatography. The amounts of
enzyme solution added to the incubation mixture are as follows;
15 μl per 300 μl for the assay of Glu-P-1 formation,
50 μl per 500 μl for tRNA-binding of N-hydroxy-2-aminofluorene,
10 μl per 100 μl for Glu-P-1 N-acetylation.

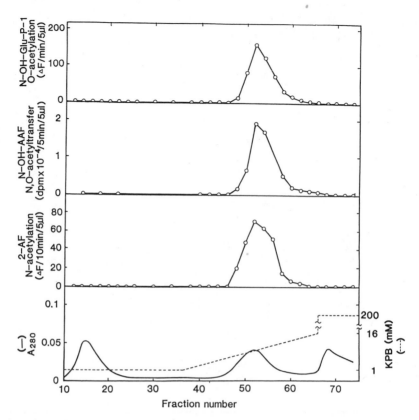

Fig. 9. High performance KB-hydroxylapatite chromatography of the
acetyltransferase.

was homogeneous as judged by SDS-polyarylamide gel electrophoresis; the 2-
AF N-acetylase and N-OH-AAF N,O-acetyltransferase activities were eluted
exactly together with N-OH-Glu-P-1 O-acetyltransferase activity (Fig. 9).
The molecular weights estimated by gel filtration and SDS-polyacrylamide
gel electrophoresis were 33,000, indicating that the O-acetyltransferase
was active in a monomeric molecular state. The activities of the purified
enzyme were inhibited by iodoacetamide, pentachlorophenol, 1-nitro-2-
naphthol and 2-AF, but not by thiolactomycin which inhibited the bacterial
O-acetyltransferase (Table 1). The inhibition pattern was quite similar
to the rat O-acetyltransferase. Furthermore, these inhibitors inhibited
not only N-hydroxyarylamine O-acetyltransferase activity but
arylhydroxamic acid N,O-acetyltransferase and arylamine N-acetylase
activities. Thiolactomycin affected neither of these activities.

Are N-Hydroxyarylamine O-Acetyltransfer, Arylhydroxamic Acid N,O-Acetyltransfer and Arylamine N-Acetylation Catalyzed by One Enzyme?

In 1974, Glowinski et al.[16] showed that arylhydroxamic acid N,O-
acetyltransferase and arylamine N-acetylase in rabbit liver cytosol were
probably the same enzyme.

Our present results support a new concept that N-hydroxyarylamine O-
acetyltransfer, arylhydroxamic acid N,O-acetyltransfer and arylamine N-
acetyltransfer are catalyzed by one enzyme on the basis of the following
evidence:

Table 1. Effects of Inhibitors on the Activities of Acetyltransferase Purified from Hamster Liver Cytosol

Inhibitor	Concn. (μM)	N-OH-Glu-P-1 O-acetylation	N-OH-AAF N,O-acetyltransfer	2-AF N-acetylation
		(% of control)		
None	–	100[a]	100[b]	100[c]
Iodoacetamide	50	2.1	0.0	0.0
Pentachlorophenol	50	24.1	42.3	72.2
1-Nitro-2-naphtol	50	25.0	64.9	73.3
2-AF	10	0.8	23.2	–
Thiolactomycin	100	131	109	89.9

[a]158 pmol Glu-P-1 formed/min
[b]16.0 pmol aminofluorene bound to tRNA/min
[c]1.20 nmol acetylaminofluorene formed/min

1) During purification, the three activities were copurified and could not be separated from each other.
2) The inhibitors of N-hydroxyarylamine O-acetyltransferase, such as pentachlorophenol and 1-nitro-2-naphthol, similarly inhibited arylhydroxamic acid N,O-acetyltransferase and arylamine N-acetylase.
3) Arylhydroxamic acid can serve as an acetyl donor for N-hydroxyarylamine O-acetyltransferase and arylamine N-acetylase.

The mechanism of acetyltransfer is given in Fig. 10. When the acetyl donor is acetyl-CoA and the acceptor is arylamine, the reaction is called arylamine N-acetylation, and when the acceptor is N-hydroxyarylamine, the reaction is called N-hydroxyarylamine O-acetyltransferase. On the other hand, when the acetyl donor is arylhydroxamic acid, the reaction is called arylhydroxamic acid N,O-acetyltransferase. However, we observed some disproportions among these three enzymatic activities during the purification processes and dissimilarities among three enzymatic activities in the species and genetic differences with different substrates used. These results indicate that one enzyme can catalyze

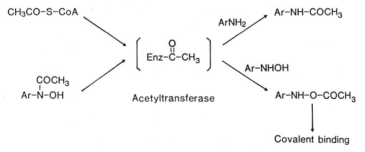

Fig. 10. Mechanism of hamster acetyltransferase in catalyzing arylamine N-acetylation, N-hydroxyarylamine O-acetyltransfer and arylhydroxamic acid N,O-acetyltransfer.

three different reactions, but there are some isozymes which may be different in their substrate specificity and in their stability during the purification process. Further studies are needed to elucidate these problems.

REFERENCE

1. Y. Yamazoe, M. Shimada, T. Kamataki, and R. Kato, Microsomal activation of 2-amino-3-methylimidazo[4,5-f]quinoline, a pyrolysate of sardine and beef extracts, to a mutagenic intermediate, Cancer Res., 43:5768 (1983).
2. R. Kato, T. Kamataki, and Y. Yamazoe, N-Hydroxylation of carcinogenic and mutagenic aromatic amines, Environ. Health Prerspec., 49:21 (1983).
3. T. Kamataki, K. Maeda, Y. Yamazoe, N. Matsuda, K. Ishii, and R. Kato, A high-spin form of cytochrome P-450 highly purified from polychlorinated biphenyl-treated rats, Molec. Pharmacol., 24:146 (1983).
4. R. Kato, T. Kamataki, and Y. Yamazoe, High spin cytochrome P-448 with high activity for mutagenic activation of aromatic amines, in "Developments in the Science and Practice of Toxicology", A. W. Hayes, R. C. Schnell, T. S. Miya, ed., Elsevier Science Publishers B.V., Amsterdam (1983).
5. S. Mita, K. Ishii, Y. Yamazoe, T. Kamataki, R. Kato, and T. Sugimura, Evidence for the involvement of N-hydroxylation of 3-amino-1-methyl-5H-pyrido[4,3-b]indole by cytochrome P-450 in the covalent binding to DNA, Cancer Res., 41:3610, (1981).
6. Y. Yamazoe, M. Shimada, T. Kamataki, and R. Kato, Covalent binding of N-hydroxy-Trp-P-2 to DNA by cytosolic proline-dependent system, Biochem. Biophys. Res. Commun., 107:165 (1982).
7. Y. Yamazoe, M. Shimada, A. Shinohara, K. Saito, T. Kamataki, and R. Kato, Catalysis of the covalent binding of 3-hydroxyamino-1-methyl-5H-pyrido[4,3-b]indole to DNA by a L-proline- adenosine triphosphate-dependent enzyme in rat hepatic cytosol, Cancer Res., 45:2495 (1985).
8. E. C. McCoy, G. D. McCoy, and H. S. Rosenkranz, Esterification of arylhydroxylamine: Evidence for a specific gene product in mutagenesis, Biochem. Biophys. Res. Commun., 108:1362 (1982).
9. K. Saito, Y. Yamazoe, T. Kamataki, and R. Kato, Mechanism of activation of proximate mutagens in Ames' tester strains: The acetyl-CoA dependent enzyme in Salmonella typhimurium TA98 deficient in TA98/1,8-DNP$_6$ catalyzes DNA-binding as the cause of mutagenicity, Biochem. Biophys. Res. Commun., 116:141 (1983).
10. K. Saito, A. Shinohara, T. Kamataki, and R. Kato, Metabolic activation of mutagenic N-hydroxyarylamines by O-acetyltransferase in Salmonella typhimurium TA98, Arch. Biochem. Biophys., 239:286 (1985).
11. A. Shinohara, K. Saito, Y. Yamazoe, T. Kamataki, and R. Kato, DNA binding of N-hydroxy-Trp-p-2 and N-hydroxy-Glu-P-1 by acetyl-CoA dependent enzyme in mammalian liver cytosol, Carcinogenesis, 6:305 (1985).
12. A. Shinohara, K. Saito, T. Kamataki, and R. Kato, Acetyl coenzyme A-dependent activation of carcinogenic N-hydroxyarylamines: Mechanism of activation, species difference, tissue distribution and acetyl donor, Cancer Res., Submitted for purblication (1985).
13. T. J. Flammang, and F. F. Kadlubar, Acetyl coenzyme A-dependent metabolic activation of carcinogenic N-hydroxyarylamines, Proc. Am. Assoc. Cancer Res., 25:120 (1984).
14. T. J. Flammang, J. G. Westra, F. F. Kadlubar, and F. A. Beland, DNA adducts formed from the probable proximate carcinogen, N-hydroxy-3,2'-dimethyl-4-aminobiphenyl, by acid catalysis or S-acetyl coenzyme A-dependent enzymatic esterification, Carcinogenesis, 6:251 (1985).

15. W. T. Allaben, and C. M. King, The purification of rat liver arylhydroxamic acid N,O-acyltransferase, <u>J. Biol. Chem.</u>, 259:12128 (1984).
16. I. B. Glowinski, and W. W. Weber, Evidence that arylhydroxamic acid N,O-acyltransferase and the genetically polymorphic N-acetyltransferase are properties of the same enzyme in rabbit liver, <u>J. Biol. Chem.</u>, 255:7883 (1980).

METABOLISM AND CARCINOGENICITY OF NITROTOLUENES

Douglas E. Rickert, John P. Chism and Gregory L. Kedderis[1]

Department of Biochemical Toxicology and Pathobiology
Chemical Industry Institute of Toxicology
P. O. Box 12137
Research Triangle Park, NC 27709

INTRODUCTION

Technical grade dinitrotoluene (DNT) is a mixture of isomers (76.4% 2,4-DNT, 18.8% 2,6-DNT and 4.7% other isomers) of demonstrated hepatocarcinogenicity in Fischer-344 rats. The incidence of neoplasia is higher in males than in females and is primarily associated with 2,6-DNT. The sex and isomer specificity of DNT is reflected in an in vivo-in vitro hepatocyte unscheduled DNA synthesis assay for potential genotoxicity, as less DNA synthesis is seen in hepatocytes from female rats than those from male rats treated with DNT, and 2,6-DNT is approximately 10-fold more active than 2,4-DNT. Detailed studies of the metabolism of 2,4- and 2,6-DNT indicate that each isomer is converted to a dinitrobenzyl alcohol in a cytochrome P-450-dependent process. The alcohols are then conjugated with glucuronic acid and excreted in the bile and urine. Male rats excrete more of the glucuronide conjugates in the bile than do female rats. Intestinal microflora hydrolyze the glucuronide and reduce at least one nitro group of the aglycone. The resulting aminonitrobenzyl alcohols are reabsorbed and converted to active metabolites in the liver (reviewed in Rickert et al., 1984a). Based on the ability of sulfotransferase inhibitors to decrease the covalent binding of 2,6-DNT to hepatic DNA, we postulated that 2-amino-6-nitrobenzyl alcohol was N-hydroxylated to yield 2-hydroxylamino-6-nitrobenzyl alcohol and conjugated with sulfate to yield an unstable N,O-sulfate which decomposed to an electrophilic nitrenium ion capable of reacting with hepatic DNA (Kedderis et al., 1984).

No carcinogenicity studies have been performed using the mononitrotoluene (NT) isomers, but sex and isomer differences in the in vivo-in vitro Fischer-344 rat hepatocyte unscheduled DNA synthesis assay suggest similarities in the activation pathways for the mono- and dinitrotoluenes. 2-NT is active in that assay while 3- and 4-NT are not. Less unscheduled DNA synthesis is seen in hepatocytes from female rats than in those from

1 Present address: Department of Animal Drug Metabolism
                    Merck Sharp & Dohme Research Laboratories
                    P. O. Box 2000
                    Rahway, NJ 07065-0900

male rats treated with 2-NT, and intestinal microflora are required for the genotoxic effects of 2-NT (Doolittle et al., 1983). We have recently demonstrated that 2-aminobenzyl alcohol and 2-amino-6-nitrobenzyl alcohol are N-hydroxylated by rat hepatic microsomes (Kedderis and Rickert, 1985a) and that 2-NT is activated in a manner similar to 2,6-DNT (Rickert et al., 1984b; Chism and Rickert, 1985). Those studies are reviewed here.

METABOLISM AND EXCRETION OF MONONITROTOLUENES

Previous work has shown that all three mononitrotoluene isomers are metabolized to nitrobenzyl alcohols in a cytochrome P-450-dependent process (deBethizy and Rickert, 1984). If the activation of 2-NT proceeds like that of 2,6-DNT, then conjugation with glucuronic acid and biliary excretion should be important to the covalent binding of 2-NT-related material to hepatic macromolecules. In order to assess the role of biliary excretion, nine Fischer-344 rats of each sex, 80-90 days old, were anesthetized with methoxyflurane and the common bile duct was cannulated. The cannula was connected to a glass receptacle with a side arm. The receptacle was implanted in the abdomen and the side arm was exteriorized to allow periodic removal of bile. The animals were allowed to recover for 30 min after regaining the righting reflex, and they were then given 1 ml saline. Thirty min after receiving the saline, three animals of each sex were given 2-, 3- or 4-nitro[U-ring ^{14}C]toluene (20 µCi/rat; 200 mg/kg) and placed in metabolism cages. An additional group of 9 rats of each sex were subjected to anesthesia and glass receptacle implantation, but the bile duct was not cannulated. These animals will be referred to as sham-operated controls. Saline and radiolabeled 2-, 3- or 4-nitrotoluene were administered as above. A third group of animals (controls) underwent no surgical procedure or anesthesia. They were given a dose of radiolabeled 2-, 3- or 4-nitrotoluene as above and placed in metabolism cages. Twelve hours after administration of the nitrotoluene dose, the animals were killed and their livers removed and stored frozen (-40°C) until analyzed for total and covalently bound radioactivity. Covalently bound radioactivity was measured by the exhaustive extraction method described by Sun and Dent (1980). Urine was collected over dry ice and analyzed for nitrotoluene metabolites by HPLC. Bile was removed from the receptacle at 1, 2, 3, 4, 6, 9 and 12 hours after the dose and analyzed for nitrotoluene metabolites by HPLC.

Urine was a major route of excretion of nitrotoluene-related material, accounting for about 40% of the dose of 3- and 4-NT and 75-80% of the dose of 2-NT (Table 1). Sham operation decreased the urinary excretion of 2-NT but not of 3- or 4-NT. Bile duct cannulation decreased the urinary excretion of each isomer compared to that in sham operated controls, suggesting that at least a portion of the dose excreted in the bile was subject to enterohepatic recycling. Biliary excretion of each isomer was greater in male rats than in female rats and accounted for nearly 30% of the dose of 2-NT in males.

While a variety of metabolites were detected and identified in bile, most of the radioactivity in bile could be accounted for by the isomeric acetamidobenzoic acids, nitrobenzoic acids, S-(nitrobenzyl)-glutathiones, S-(nitrobenzyl)-N-acetylcysteines and nitrobenzylglucuronides (Table 2).

The majority of the 2-NT-derived material in bile of both male and female rats was due to 2-nitrobenzylglucuronide. Males excreted nearly 3 times as much of this metabolite in bile as did females. A similar sex difference was found in the biliary excretion of 2,6-DNT by isolated perfused livers from male and female rats. When the initial perfusate

Table 1. Excretion of mononitrotoluenes by male or female
control, sham operated or bile duct cannulated rats*

2-Nitrotoluene

	Control		Sham Operated		Bile Duct Cannulated	
	male	female	male	female	male	female
Urine	74.9±0.9	79.9	52.0±4.6	63.3±1.6	36.2±6.2	32.9±2.8
Bile	ND	ND	ND	ND	28.6±3.3	9.6±1.8

3-Nitrotoluene

	Control		Sham Operated		Bile Duct Cannulated	
	male	female	male	female	male	female
Urine	38.0±3.0	38.5±7.8	44.5±0.8	46.3±4.0	23.0±2.1	33.3±0.9
Bile	ND	ND	ND	ND	10.8±3.1	4.3±0.8

4-Nitrotoluene

	Control		Sham Operated		Bile Duct Cannulated	
	male	female	male	female	male	female
Urine	40.0±2.3	38.6±5.5	49.9±2.0	42.2±5.5	27.7±0.4	23.4±6.2
Bile	ND	ND	ND	ND	9.8±1.0	1.3±0.3

*Values are mean (±S. E.) per cent dose excreted in 12 hours. N=3 in
each case, except for control females which received 2-NT; only two of
the animals urinated between 6 and 12 hr. ND=not determined.

Table 2. Biliary excretion of nitrotoluenes
in male and female rats*

	2-Nitrotoluene		3-Nitrotoluene		4-Nitrotoluene	
	male	female	male	female	male	female
ActBA	ND	ND	1.2±0.5	0.4±0.1	0.7±0.2	0.1±0.0
NBA	ND	ND	3.4±1.7	1.7±0.4	2.8±0.4	0.8±1.1
S(NB)GSH	4.9±0.8	0.4±0.1	1.0±0.1	0.1±0.0	2.8±0.1	0.1±0.0
S(NB)NAC	1.0±0.2	0.4±0.1	0.5±0.0	0.1±0.0	0.7±0.1	<0.1
NBGl	22.0±2.3	8.3±1.5	2.8±0.6	0.7±0.2	0.9±0.2	0.1±0.0

*Values are mean percent dose (±S. E.) excreted in bile of male or female
Fischer-344 rats 12 hr following 2-, 3- or 4-NT administration. Only
metabolites whose total excretion in 12 hr was greater than 1.0% of the
dose in at least one sex are included in this table. Abbreviations used
are: ActBA = acetamidobenzoic acid, NBA = nitrobenzoic acid, S(NB)GSH =
S-(nitrobenzyl)glutathione, S(NB)NAC = S-(nitrobenzyl)-N-acetylcysteine,
NBGl = nitrobenzylglucuronide and ND = not detected.

concentration of 2,6-DNT was 20 μM, livers from male rats excreted 3 times as much 2,6-dinitrobenzylglucuronide in the bile in 2 hr as livers from female rats. At an initial perfusate concentration of 70 μM livers from male rats excreted nearly 9 times as much in the bile as did livers from female rats (Long and Rickert, 1982). The majority of the biliary metabolites of 3- or 4-NT were comprised of glutathione conjugates and the isomeric nitrobenzoic acids and isomeric acetamidobenzoic acids. In contrast to 2-NT, less than one-third of the of the 3- or 4-NT-derived material excreted in the bile was nitrobenzylglucuronide.

Less 2-NT-derived than 3- or 4-NT-derived material remained in the livers of rats given a dose 12 hr earlier (Table 3), but 3-5 times as much radioactivity was covalently bound in male rats which had received radiolabeled 2-NT as compared to those which had received 4- or 3NT, respectively. Approximately equal concentrations of 2-, 3- and 4-NT-derived material were covalently bound in livers of female rats.

Sham operation decreased concentrations of 2-, 3- and 4-NT-derived material covalently bound in livers of male rats (Table 3). Sham operation had no effect on the concentration of 3- or 4-NT-derived material bound to macromolecules of livers from female rats, but it decreased by 30% the concentration of 2-NT-derived material bound. Bile duct cannulation markedly reduced the concentration of covalently bound material derived from 2-NT in livers from male rats to 7% of that found in livers of sham operated controls. Smaller, but substantial, decreases were observed in the concentration of 2-NT-derived material bound in livers from female rats and in the concentrations of 3- or 4-NT-derived material bound in livers from male and female rats.

These data are consistent with the hypothesis that enterohepatic recycling plays a role in the activation of the mononitrotoluenes to compounds capable of covalently binding to hepatic macromolecules. They further suggest that the nitrobenzylglucuronides are precursors to the active metabolites. The most marked differences are seen when enterohepatic recycling of 2-NT-derived material in males is interrupted; smaller differences are seen in females. The biliary excretion of 2-nitrobenzylglucuronide in male rats is three times that in females, and it is at least 10 times as great for 2-NT as for 3- or 4-NT. The decrease in covalent binding brought about by sham operation is not inconsistent with the hypothesis that biliary excretion of 2-nitrobenzylglucuronide is important to the activation of 2-NT. Watkins and Klaassen (1983) have shown that methoxyflurane and several other anesthetics temporarily reduce hepatic concentrations of UDPGA. When hepatic concentrations of UDPGA are reduced, the excretion of the glucuronides of bilirubin, diethylstilbestrol, and valproic acid is significantly decreased (Gregus et al., 1983). It is possible that the biliary excretion of 2-nitrobenzylglucuronide was decreased in the animals exposed to methoxyflurane, and that this resulted in the decrease in hepatic macromolecular covalent binding of 2-NT-derived material in the sham operated animals compared to control animals.

The residual covalent binding of NT-derived material in bile duct cannulated animals could be due to nitrobenzylglucuronides reaching the intestinal microflora via the circulation rather than the bile, but it seems more likely that other pathways of activation exist. The identities of the alternate pathways are not known, but they are apparently not capable of producing sufficient DNA excision repair to be detected in the in vivo-in vitro rat hepatocyte unscheduled DNA synthesis assay (Doolittle et al., 1983).

Table 3. Hepatic macromolecular covalent binding of nitrotoluenes in control, sham operated or bile duct cannulated male and female rats*

2-Nitrotoluene

	Control		Sham operated		Bile duct cannulated	
	male	female	male	female	male	female
Total	260+16	74+12	254+40	109+7	55+1	67+17
CVB	36.6+2.6	11.2+1.1	12.9+3.4	7.9+0.6	0.9+0.1	1.7+0.1
%Control	100	100	35	71	2	15
%Sham			100	100	7	22

3-Nitrotoluene

	Control		Sham operated		Bile duct cannulated	
	male	female	male	female	male	female
Total	548+46	300+20	135+40	260+46	326+51	255+69
CVB	6.9+0.5	7.9+0.4	3.2+0.1	8.4+0.7	1.7+0.2	4.4+0.7
%Control	100	100	46	106	25	56
%Sham			100	100	53	52

4-Nitrotoluene

	Control		Sham operated		Bile duct cannulated	
	male	female	male	female	male	female
Total	647+126	430+50	241+5	421+59	468+55	261+40
CVB	10.7+0.7	8.5+0.6	5.8+0.3	8.7+1.7	2.2+0.3	4.7+0.4
%Control	100	100	57	102	22	55
%Sham			100	100	38	54

*Values are mean nmoles nitrotoluene equivalents/gram liver ($\pm$S.E.) for 3 rats 12 hr following a dose of radiolabeled 2-, 3- or 4-NT. Total = total radioactivity, both bound and unbound; CVB = covalently bound material. %Control = mean CVB expressed as a percentage of the control value. %Sham = mean CVB expressed as a percentage of the value for sham operated controls.

N-HYDROXYLATION OF 2-AMINO-6-NITROBENZYL ALCOHOL AND 2-AMINOBENZYL ALCOHOL

If the hypothesis outlined in the Introduction is correct, then 2-amino-6-nitrobenzyl alcohol and 2-aminobenzyl alcohol should be substrates for microsomal N-hydroxylation. In order to test this, 2-amino-6-nitrobenzyl alcohol was synthesized (Kedderis and Rickert, 1985a) and incubated (0.5 mM final concentration) with rat liver microsomes (2.78 mg) in the presence of NADPH (1.0 mM) in a total volume of 3.0 ml (0.1 M phosphate buffer, pH 7.7). The reaction was stopped after 3 min at 37°C by the addition of 3 ml ethyl acetate. 2-, 3- And 4-aminobenzyl alcohol were incubated under the same conditions for 30 min. Each substrate yielded metabolites capable of reducing Fe^{3+} to yield a ferrous-bathophenanthroline complex which could be quantitated at 535 nm (Tsen, 1961). Such a result is consistent with the presence of aminophenols and/or hydroxylamines. The rate of formation of reducing equivalents from 2-amino-6-nitrobenzyl alcohol (1.65$\pm$0.05 nmol reducing equivalents/min/mg microsomal protein) was greater than the rate of formation of reducing equivalents from 2-, 3-, or 4-aminobenzyl alcohol (0.29±0.01, 0.18±0.00, or 0.04±0.00 nmol reducing equivalents/min/mg microsomal protein, respectively).

HPLC analysis of ethyl acetate extracts of reaction mixtures containing 2-amino-6-nitrobenzyl alcohol or 2-aminobenzyl alcohol revealed the formation of 2 metabolites from each substrate which were capable of reducing ferric iron. 2-Hydroxylamino-6-nitrobenzyl alcohol and 2-hydroxylaminobenzyl alcohol were synthesized by zinc dust–ammonium chloride reduction of 2,6-dinitrobenzyl alcohol and 2-nitrobenzyl alcohol, respectively. Comparison of the retention times of the synthesized standards with those of the products of microsomal metabolism showed that one metabolite from each substrate co-eluted with the hydroxylamine. The second metabolite of 2-aminobenzyl alcohol was not identified, but a combination of proton NMR and UV-visible spectral analysis allowed a tentative assignment of 2-amino-5-hydroxy-6-nitrobenzyl alcohol to the second metabolite of 2-amino-6-nitrobenzyl alcohol (Kedderis and Rickert, 1985a). It seems reasonable that the second metabolite of 2-aminobenzyl alcohol is also a ring hydroxylated aminobenzyl alcohol.

In order to quantitate each metabolite formed from 2-amino-6-nitrobenzyl alcohol, standard curves were generated for the HPLC analysis (UV detection at 254 nm) of 2-hydroxylamino-6-nitrobenzyl alcohol using the synthesized standard. Total reducing equivalents were determined from colorimetric assay of an ethyl acetate extract of each reaction mixture, reducing equivalents due to the hydroxylamine were determined by HPLC and reducing equivalents due to the phenolic metabolite of 2-amino-6-nitrobenzyl alcohol were determined by the difference between total reducing equivalents and those due to the hydroxylamine. Formation of total reducing equivalents (1.22±0.02 nmol/min/mg microsomal protein in control incubations) was decreased by addition of 0.5 mM SKF-525A (39% of control), metyrapone (31% of control) or octylamine (51% of control), suggesting that the reaction was dependent upon cytochrome P-450. Methimazole, a high affinity substrate for microsomal flavin-containing monooxygenase (Poulsen and Ziegler, 1979), also inhibited the reaction at a concentration of 0.5 mM, but heat inactivation of microsomal flavin-containing monooxygenase had no effect on the reaction. This result is consistent with other data from this laboratory which indicates that the flavin-containing monooxygenase-catalyzed metabolism of methimazole results in the formation of a compound which causes a loss of active cytochrome P-450 (Kedderis and Rickert, 1985b).

Pretreatment of rats with phenobarbital significantly increased the rate of formation of both 2-hydroxylamino-6-nitrobenzyl alcohol and 2-amino-5-hydroxy-6-nitrobenzyl alcohol, while pretreatment of rats with β-naphthoflavone increased the rate of aminophenol formation but did not affect hydroxylamine formation (Table 4).

Hepatic microsomes isolated from male rats N-hydroxylated 2-amino-6-nitrobenzyl alcohol 3 times as rapidly as microsomes isolated from female rats. The formation of 2-amino-5-hydroxy-6-nitrobenzyl alcohol from 2-amino-6-nitrobenzyl alcohol was also more rapid in incubations containing microsomes from male than in those containing microsomes from female rats (Table 5).

DISCUSSION

The data presented here are consistent with our proposed mechanism for the activation of the genotoxic nitrotoluenes. Both the carcinogen 2,6-DNT and the genotoxicant 2-NT are converted to nitrobenzylglucu-

Table 4. Effect of Phenobarbital or β-Naphthoflavone Pretreatment
on the Microsomal Metabolism of 2-Amino-6-Nitrobenzyl Alcohol*

Pretreatment	Initial rate	
	nmol hydroxylamine formed/min/nmole cytochrome P-450	nmole aminophenol formed/min/nmole cytochrome P-450
None	0.75+0.11 (100%)	2.35+0.05 (100%)
Phenobarbital	1.27+0.16**(169%)	3.49+0.15**(148%)
b-Naphthoflavone	0.63+0.09 (84%)	2.76+0.16**(117%)

*Values are mean (+S. E.) for 3 determinations. The numbers in paren-
theses are percent of control (no treatment).
**Significantly different from control value (no treatment), p<0.05.

ronides which are excreted in the bile. In both cases biliary excretion
in male rats is greater than in female rats, a difference which could
explain the sex differences seen in covalent binding to hepatic macromolecules, genotoxicity in the in vivo-in vitro hepatocyte unscheduled
DNA repair assay and, ultimately, in the carcinogenicity of 2,6-DNT.
The action of intestinal microflora can convert the glucuronides into
2-aminobenzyl alcohol and 2-amino-6-nitrobenzyl alcohol, which are
substrates for hepatic cytochrome P-450-dependent N-hydroxylation.
Since we have previously shown that administration of sulfotransferase
inhibitors decreases covalent binding of 2,6-DNT- and 2-NT-derived
material to DNA by over 95% (Kedderis et al., 1984; Rickert et al.,
1984b), it seems possible that the ultimate reactive metabolites of 2-NT
and 2,6-DNT are unstable N,O-sulfates which decompose to electrophilic
nitrenium ions.

An alternate pathway which can not be conclusively excluded at
present is suggested by the work of Watabe et al. (1982) and Boberg et
al. (1983). It is possible that sulfotransferase converts 2-amino-6-
nitrobenzyl alcohol and 2-aminobenzyl alcohol to unstable benzylsulfates
which decompose to yield carbonium ions. We do not favor this pathway
because preliminary experiments have shown that incubation of either

Table 5. Sex Difference in the Microsomal Metabolism
of 2-Amino-6-Nitrobenzyl Alcohol*

Sex	Initial rate	
	nmol hydroxylamine formed/min/nmol cytochrome P-450	nmole aminophenol formed/min/nmol cytochrome P-450
Male	0.75+0.11	2.35+0.05
Female	0.23+0.03**	1.75+0.07**

*Values are mean (+S. E.) for 3 determinations.
**Significantly different from value for microsomes from male rats;
p<0.05.

2-aminobenzyl alcohol or 2-amino-6-nitrobenzyl alcohol with rat hepatic cytosol and a PAPS generating system fails to produce metabolites capable of covalently binding to protein or calf thymus DNA (Chism, J. P. and Rickert, D. E., unpublished observations).

Whether the aminophenol metabolite of 2,6-DNT has a role in the carcinogenicity of that compound is presently unclear. C-hydroxylation of 2-acetylaminofluorene is usually considered to be a detoxification reaction (Thorgeirsson et al., 1983), but ring hydroxylated metabolites of 2-acetylaminofluorene can be oxidized in vitro by the cytochrome c-cytochrome oxidase system to reactive iminoquinones which covalently bind to protein (King et al., 1963; Nagasawa and Osteraas, 1964). The aminophenol metabolite of 2,6-DNT must be synthesized and subjected to similar in vitro experiments to determine whether it, too, can be oxidized to a reactive iminoquinone.

REFERENCES

Boberg, W. W., Miller, E. C., Miller, J. A., Poland, A., and Liem, A., 1983, Strong evidence from studies with brachymorphic mice and pentachlorophenol that 1'-sulfooxysafrole is the major ultimate electrophilic and carcinogenic metabolite of 1'-hydroxysafrole in mouse liver, Cancer Res., 43:5163.

Chism, J. P. and Rickert, D. E., 1985, Isomer and sex specific bioactivation of mononitrotoluenes: role of enterohepatic circulation, Drug Metab. Dispos. (in press).

deBethizy, J. D. and Rickert, D. E., 1984, Metabolism of nitrotoluenes by freshly isolated Fischer 344 rat hepatocytes, Drug Metab. Dispos., 12:45.

Doolittle, D. J., Sherrill, J. M. and Butterworth, B. E., 1983, The influence of intestinal bacteria, sex of the animal, and position of the nitro group on the hepatic genotoxicity of nitrotoluene isomers in vivo, Cancer Res., 43:2836.

Gregus, Z., Watkins, J. B., Thompson, T. N. and Klaassen, C. D., 1983, Depletion of hepatic uridine diphosphoglucuronic acid decreases the biliary excretion of drugs, J. Pharmacol. Exp. Ther., 225:256.

Kedderis, G. L. and Rickert, D. E., 1985a, Characterization of the oxidation of amine metabolites of nitrotoluenes by rat hepatic microsomes: N- and C-hydroxylation, Mol. Pharmacol., 28: (in press).

Kedderis, G. L. and Rickert, D. E., 1985b, Loss of rat liver microsomal cytochrome P-450 during methimazole metabolism: role of flavin-containing monooxygenase, Drug Metab. Dispos., 13:58.

Kedderis, G. L., Dyroff, M. C. and Rickert, D. E., 1984, Hepatic macromolecular covalent binding of the hepatocarcinogen 2,6-dinitrotoluene and its 2,4 isomer in vivo: modulation by the sulfotransferase inhibitors pentachlorophenol and 2,6-dichloro-4-nitrophenol, Carcinogenesis, 5:1199.

King, C. M., Gutmann, H. R. and Chang, S. F., 1963, The oxidation of o-aminophenols by cytochrome c and cytochrome oxidase. IV. Interaction of 2-imino-1,2-fluorenoquinone and of 2-imino-2,3-fluorenoquinone with bovine serum albumin, J. Biol. Chem., 238:2199.

Long, R. M. and Rickert, D. E., 1982, Metabolism and excretion of 2,6-dinitro[^{14}C]toluene in vivo and in isolated perfused rat livers, Drug Metab. Dispos., 10:455.

Nagasawa, H. T. and Osteraas, A. J., 1964, The biological arylation of proteins in vitro by a metabolite of the carcinogen N-2-fluorenylacetamide, Biochem. Pharmacol., 13:713.

Poulsen, L. L. and Ziegler, D. M., 1979, The liver microsomal FAD-containing monooxygenase: spectral characterization and kinetic studies, J. Biol. Chem., 254:6449.

Rickert, D. E., Butterworth, B. E. and Popp, J. A., 1984a, Dinitrotoluene: Acute toxicity, oncogenicity, genotoxicity, and metabolism, CRC Crit. Rev. Toxicol., 13:217.

Rickert, D. E., Long, R. M., Dyroff, M. C. and Kedderis, G. L., 1984b, Hepatic macromolecular covalent binding of mononitrotoluenes in Fischer-344 rats, Chem.-Biol. Interact., 52:131.

Sun, J. D. and Dent, J. G., 1980, A new method for measuring covalent binding of chemicals to cellular macromolecules, Chem.-Biol. Interact., 32:41.

Thorgeirsson, S. S., Glowinski, I. B. and McManus, M. E., 1983, Metabolism, mutagenicity and carcinogenicity of aromatic amines, Rev. Biochem. Toxicol., 5:349.

Tsen, C. C., 1961, An improved spectrophotometric method for the determination of tocopherols using 4,7-diphenyl-1,10-phenanthroline, Anal. Chem. 33:849.

Watkins, J. B. and Klaassen, C. D., 1983, Chemically-induced alteration of UDP-glucuronic acid concentration in rat liver, Drug Metab. Dispos., 11:37.

Watabe, T., Ishizuka, T., Isobe, M. and Ozawa, N., 1982, A 7-hydroxymethyl sulfate ester as an active metabolite of 7,12-dimethylbenz[a]anthracene, Science, 215:403.

HYDROXYLAMINES AND HEMOLYTIC ANEMIA

David J. Jollow, Scott J. Grossman and James H. Harrison

Department of Pharmacology
Medical University of South Carolina
Charleston, SC 29425

INTRODUCTION

Hemolytic anemia, the uncompensated loss of red blood cells from the circulation, has been recognized as a side effect of drugs and other chemicals for over 50 years (Muelens, 1926; Cordes, 1926). This response is commonly associated with the aminoquinoline drugs, pamaquine and primaquine (Beutler, 1959); indeed the extensive studies carried out with these drugs in the 1940's and 50's by Alving, Carson, Beutler, and others still serve as the basis for most of our knowledge in the clinical sequelae of drug-induced hemolysis.

Drugs known to induce a hemolytic response exhibit a broad structural diversity and display a wide range of biological and chemical activities (Beutler, 1959). While there does not appear to be any common link among their pharmacological activities which predisposes them to be hemolytic agents, many of them possess an arylamine nucleus, or like chloramphenicol, may be converted to an aromatic amine by reductive metabolism in vivo. Aniline, chemically the simplest member of the series, is both hemolytic and methemoglobinemic (Gosselin et al., 1976).

The hemolytic response to these compounds is seen routinely in normal individuals exposed to high concentrations of the agent (Degowin et al., 1966; Kellermeyer et al., 1962). Thus, workers in the chemical industry and patients taking high doses of hemolytic drugs such as dapsone are at greatest risk among the normal population. However, a smaller subpopulation of indviduals have been shown to be exceptionally sensitive to these compounds (Beutler, 1959; Carson, 1960). This trait, glucose-6-phosphate dehydrogenase (G6P-D) deficiency is an X-linked recessive trait (McKusick and Ruddle, 1977) estimated to affect approximately 100 million people worldwide (Carson, 1960). While many varients of this trait are known, the A$^-$ type (affecting primarily Negro males), the Mediterranian, and the Canton types constitute the majority of the individuals affected (Beutler, 1972).

Clinically, the acute hemolytic phase in A$^-$ individuals given primaquine is characterized by a fall in erythrocytic GSH, and the appearance of methemoglobinemia and intra-erythrocytic inclusion bodies known as Heinz bodies (Dern et al., 1954; Degowin et al., 1966). Two to four days after initiation of therapy, the patients experience reductions in hematocrit, hemoglobin levels, and red cell count. Individuals whose red blood cells had been labelled with radioactive chromium showed enhanced rates of disappearance of blood radioactivity associated with splenic accumulation of the label (Salvidio et al., 1967). Reticulocytosis, increased serum bilirubin, and darkened urine appear concurrently with the onset of anemia. A recovery phase typically begins after 7-10 days, even if the drug is continued, since the younger erythrocytes are relatively resistant to the drug. However, the hemolytic response can be provoked by a stepwise increase in dosage, indicating the relative nature of the younger cells' resistance. Red blood cells of normal individuals are similarly susceptible provided that the dose of drugs such as primaquine or dapsone is sufficiently high (Tarlov et al., 1962; Kellermeyer et al., 1962; Degowan, 1967).

Although the mechanism(s) by which the red cell is damaged and subsequently removed from the circulation is still unknown, a great deal of evidence has accumulated which suggests that oxidation of critical sites within the erythrocyte plays a crucial role (Tarlov et al., 1962; Cohen and Hochstein, 1964; Miller and Smith, 1970; Carrell et al., 1975). Clinical evidence suggesting an oxidative attack includes: 1) increased sensitivity to in vivo hemolysis is associated with decreased ability to produce reducing equivalents (NADPH) via the pentose shunt (i.e., G6P-D deficiency), 2) oxidation of hemoglobin to methemoglobin frequently precedes hemolytic anemia, and 3) loss of GSH from erythrocytes is characteristic of the hemolytic response in vivo. Major support for the concept has been obtained with in vitro studies using direct acting redox compounds such as acetylphenylhydrazine, which induce in the red cell in vitro the characteristic phenomena associated with the in vivo hemolytic response (hemoglobin oxidation, Heinz body formation, GSH depletion) (Beutler, 1959, 1971). Collectively, the "oxidative stress" concept has successfully unified a majority of the information available, and in particular provided a rational basis for the increased susceptibility of G6P-D deficient individuals.

However, hemolytic drugs such as primaquine and dapsone cannot participate directly in cyclic redox reactions and do not cause observable changes in erythrocytes in vitro at concentrations associated with in vivo hemolysis (Frazer and Vesell, 1968; Scott and Rasbridge, 1973; Glader and Conrad, 1973). The concept has thus arisen that the parent drug molecules are themselves inactive, but are metabolized in vivo to more reactive species (Tarlov et al., 1962; Beutler, 1971) which are capable of exerting oxidative stress on the red cell. Such reactive metabolites of hemolytic drugs have not as yet been identified, and hence direct test of the postulate of oxidative stress as the cause of drug-induced hemolysis has not been possible. The purpose of this communication is to review some recent studies which indicate that the reactive metabolites mediating aniline- and dapsone-induced hemolytic anemia in rats are respectively, phenylhydroxylamine (PHA) and N-hydroxydapsone (DDS-NOH)/N-hydroxymonoacetyldapsone (MADDS-NOH).

ANILINE-INDUCED HEMOLYTIC ANEMIA

Administration of a single dose of aniline (2.25 mmoles/kg) to rats whose erythrocytes had previously been tagged with radioactive chromium resulted in a marked increase in the rate of removal of ^{51}Cr-tagged red cells from the circulation (Fig. 1) (Harrison and Jollow, 1985). The enhanced disappearance of tagged cells was accompanied by decrease in hematocrit, increase in spleen weight, and the appearance of the radiolabel almost exclusively in the spleen. The decrease in the time necessary for the removal to 50% of the radiolabel present in the circulation (^{51}Cr-T_{50}) when aniline was administered was dose-dependent (Harrison and Jollow, 1985).

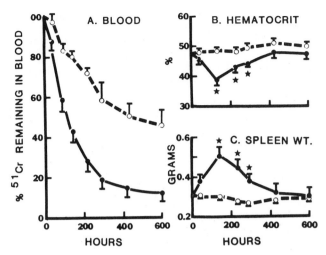

Fig. 1.   Effect of aniline on the survival of ^{51}Cr-tagged RBC in rats.
Groups of six rats received labeled red cells 24 hr prior to
aniline (2.25 mmole/kg) (closed circles) or saline vehicle (open
circles).   Blood samples were obtained from the orbital sinus
immediately prior to (T_0) and at intervals after aniline or
vehicle.   Blood radioactivity values are expressed as a % of
the T_0 sample. Hematocrits and spleen weights were determined
on additional animals in each group in a parallel experiment.

The metabolic clearance of aniline has been extensively studied
(Parke, 1960; Kao et al., 1978). The major primary pathways (Fig. 2) involve ring oxidation to yield various phenols; 2-, 3-, and 4-aminophenol,
and after acetylation, 4-hydroxyacetanilide.   In addition, N-oxidation
yields phenylhydroxylamine (PHA) which in the red cell reacts with oxyhemoglobin to yield nitrosobenzene and methemoglobin (Kiese, 1974).
Comparison of the hemolytic capacities of these metabolites (Table 1)
indicated that while the amino phenols were inactive as hemolytic agents
at the maximum doses which could be tested (compatible with survival of
the animals for the 20 day observation period), the N-hydroxy derivative
was about ten times more hemolytic than aniline.

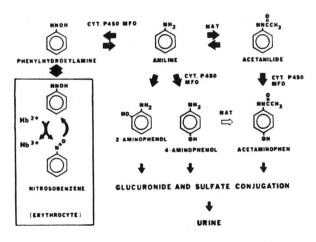

Fig. 2.   Pathways of aniline metabolism.

TABLE 1. COMPARISON OF THE ABILITY OF ANILINE AND ANILINE METABOLITES TO REDUCE THE T_{50} Cr VALUES OF RATS[a]

	Aniline	2-Amino-Phenol	3-Amino-Phenol	4-Amino Phenol	Acetanilide	PHA
			mmoles/kg			
ED_{50}	1.95	N.O.[b]	N.O.[b]	N.O.[b]	2.13	0.18

[a] Rats received ^{51}Cr-tagged RBC two days prior to test compound (day 0). Decline in blood radioactivity was monitored for 600 hrs after administration of test compound and the time needed for 50% reduction in radiolabel determined graphically. Plots of dose vs T_{50} Cr values were used to generate ED_{50} values for the hemolytic effect (Harrison and Jollow, 1985).

[b] N.O. No observed reduction in T_{50} Cr as compared with saline controls at the maximum doses (Ca 1 - 1.5 mmole/kg) that could be administered compatible with survival of the test animals for the observation period.

To test the ability of PHA to act as a direct acting hemolytic agent, isologous ^{51}Cr-tagged red cells were incubated in vitro for two hours with and without PHA, washed and then administered intravenously to rats (Table 2). As compared with saline controls, aniline did not exert hemolytic activity, confirming its lack of direct hemolytic effect. In contrast, PHA elicited a marked hemolytic response.

TABLE 2. EFFECT OF INCUBATION IN VITRO OF RED CELLS WITH ANILINE AND PHA ON THE T_{50} Cr VALUES FOLLOWING ADMINISTRATION TO RATS.

	Saline	Aniline	PHA
	-	1500 µM	200 µM
T_{50} Cr (hrs)	550 ± 140	485 ± 50	190 ± 10

Washed isologous ^{51}Cr-tagged RBC were incubated for 2 hrs at 37° with saline, aniline, or PHA at the concentrations indicated, then washed and administered intravenously to rats. Decline in blood radioactivity was monitored for 20 days and the T_{50} Cr values determined graphically.

To determine if PHA was formed in vivo from aniline in sufficient amounts to account for the hemolytic activity of aniline in rats, PHA levels in blood were measured after administration of aniline (Harrison and Jollow, 1983) and the area under the concentration-time curves (AUC) determined. Administration of PHA to additional rats in doses which yielded the same blood AUCs for PHA as was seen after the hemotoxic doses of aniline were found to induce similar hemolytic responses (Harrison and Jollow, 1985). Thus the data indicated that the hemolytic activity of the parent compound could be largely if not entirely accounted for by the hemotoxic activity of its metabolite, PHA.

DAPSONE-INDUCED HEMOLYTIC ANEMIA

Dapsone is known to be metabolized primarily to its N-hydroxy (DDS-NOH) and monoacetyl-N-hydroxy derivatives (MADDS-NOH) (Fig. 3). Determination of the hemolytic activity of dapsone (400 μmoles/kg) and its two major metabolites, DDS-NOH (200 μmole/kg) and MADDS-NOH (200 μmole/kg) (Fig. 4) indicated that all three compounds are hemolytic in the rat and that the two hydroxylamines have approximately equal hemotoxicity (Grossman and Jollow, 1985).

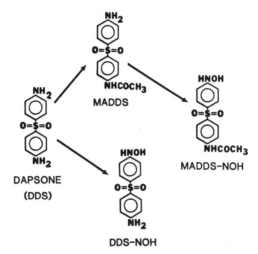

Fig. 3.   Pathways of dapsone metabolism.

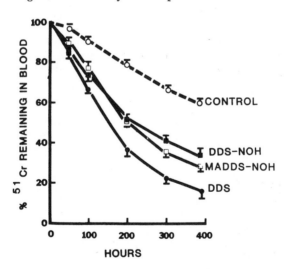

Fig. 4.   Effect of dapsone and dapsone metabolites on the survival of ^{51}Cr-tagged RBC in rats. Dapsone (DDS, 400 μmole/kg), N-hydroxy dapsone (DDS-NOH, 200 μmole/kg), monoacetyl-N-hydroxydapsone (MADDS-NOH, 200 μmoles/kg) or saline vehicle were administered to groups of five rats 48 hrs after ^{51}Cr-tagged RBC had been given intravenously. Blood samples were obtained from the orbital sinus at the times indicated and blood radioactivity determined.

Incubation of chromium tagged red cells with dapsone, DDS-NOH, or saline in vitro for 2 hrs at 37° followed by washing and administration to rats resulted in marked decrease in the T_{50} Cr value for DDS-NOH treated cells as compared with the saline treated controls. The EC_{50} for this effect was about 120 μM. As with aniline, dapsone lacked direct hemolytic activity (Grossman and Jollow, 1985).

To determine if the N-hydroxylamines of dapsone are sufficiently potent to account for the hemotoxicity of the parent drug, comparison studies analogous to those described above for aniline and PHA were performed. Dapsone was administered to rats and the AUC for DDS-NOH and MADDS-NOH was determined. Additional rats received a dose of DDS-NOH to match the combined AUC for DDS-NOH and MADDS-NOH observed after dapsone. The hemolytic response was measured in both groups of animals. Similar hemolytic activity was seen, indicating that the hydroxylamine metabolites of dapsone are sufficiently potent to account for dapsone-induced hemolytic anemia in rats.

MECHANISM STUDIES

As noted above, in spite of 30 years of research, the mechanism by which the red cell is damaged and prematurely removed from the circulation is unknown. While even a brief review of the voluminous literature available on this subject is beyond the scope of this communication, several highly pertinent points may be noted. Thus Cohen and Hochstein (1964), Goldberg and Steen (1976a,b), and Misra and Friedovich (1976) have explored the roles of active oxygen species in the process and have described roles for the various cellular defense mechanisms (glutathione peroxidase, catalase, and superoxide dismutase). Jandl and colleagues (Allen and Jandl, 1961; Jacob and Jandl, 1962a,b) and more recently Kosower et al. (1982) have clearly established that membrane free sulfhydryls have important role(s) in the hemolytic process. Of particular interest, depletion of erythrocytic GSH by diamide and other compounds have been observed to cause loss of membrane sulfhydryl groups accompanied by a marked rise in high molecular weight aggregate ($> 10^6$ daltons) of membrane protein (Kosower et al., 1982).

Experimentally, we have observed that red cells treated with DDS-NOH in vitro undergo marked deformation as seen by scanning electron-microscopy. Morphologically, the altered cells appear as extreme forms of echinocytes. The extent of the morphological change in red cell incubates with DDS-NOH is concentration-dependent and roughly correlates with hemotoxicity. Since the red cell cytoskeleton is thought to play an important role in the shape of the erythrocyte, the effect of DDS-NOH on cytoskeleton protein was examined.

As illustrated in Figure 5, DDS-NOH causes marked alteration in the SDS-PAGE pattern of the membrane protein. Major changes include loss of band 4.2, decrease in height and a broadening of bands 1, 2.1, and 3. Band 5 became obscured by the appearance of a protein with slightly higher MW. Of note, new bands appeared corresponding to MWs of 27,000 and 40,000. Membrane associated hemoglobin was indicated by the marked increase in the band at 17,000. These membrane changes were reversible by dithiothreitol which caused the release of protein corresponding to hemoglobin monomers (Grossman and Jollow, 1985). Of importance, the alterations in membrane protein patterns induced by DDS-NOH are distinctly different from those caused by the direct acting redox compounds such as diamide (Kosower et al., 1982).

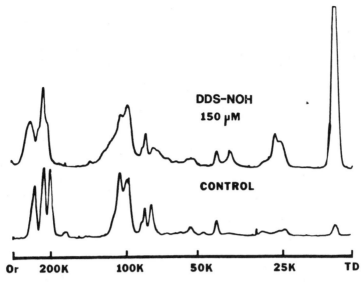

Fig. 5. Effect of DDS-NOH on the SDS-PAGE pattern of erythrocyte membrane protein. Red cells were incubated at 37° for 60 min with saline or DDS-NOH (150 μM). Ghosts were prepared, washed, solubilized, and subjected to electrophoresis as described by Fairbanks et al. (1971).

CONCLUDING REMARKS

The results of this study indicate that the hemolytic activity seen in rats after administration of aniline or the antileprosy drug, dapsone, is due largely if not entirely, to the action of their arylhydroxylamine metabolites, PHA, and DDS-NOH plus MADDS-NOH, respectively. These hydroxylamine metabolites are direct acting and are formed from the parent compounds in amounts adequate to account for the toxicity of the parent compounds. However, perhaps the most important outcome of the study is that the availability of the reactive metabolites now permits direct examination of the mechanism underlying drug-induced hemolytic anemia, as distinct from that caused by model compounds such as acetylphenylhydrazine, diamide and organic peroxides. Initial studies on the effects of the arylhydroxylamines on membrane proteins indicate that the membrane proteins are drastically altered and that the alterations occur by the addition of hemoglobin via the formation of disulfide links.

While a complete description of the molecular events underlying arylhydroxylamine-induced hemolytic anemia cannot yet be given, the following scenario may be helpful in integrating our present understanding (Fig. 6): upon crossing the cell membrane, DDS-NOH interacts with oxyhemoglobin causing the formation of methemoglobin, nitroso-dapsone (DDS-NO), and a partially reduced oxygen species. The DDS-NO may be reduced back to DDS-NOH, allowing a cyclic generation of methemoglobin and active oxygen species (Kiese, 1974). The active oxygen species may be detoxified by catalase, superoxide dismutase and glutathione peroxidase. If active oxygen species are produced in excess, depletion of erythrocytic GSH would occur via peroxidase activity especially if NADPH production was suboptimal as in G6P-D deficient erythrocytes. In addition, the active oxygen species are likely to react directly with glutathione and/or protein sulfhydryls to generate active thiol species. Addition and exchange reactions between activated sulfhydryl groups in glutathione, hemoglobin, and membrane proteins could be expected to result in the formation of a variety of mixed disulfides including the membrane protein-hemoglobin

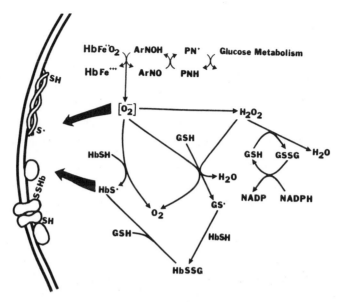

Fig. 6.   Hypothetical schema of events occurring within the red cell
           after challenge with DDS-NOH.   Active oxygen is represented as
           superoxide anion.   ArNOH and ArNO designate respectively the
           hydroxylamine and nitroso analogs of the arylamine; PN⁺ and PNH
           represent the oxidized and reduced forms of pyridine nucleotide;
           and Hb indicates hemoglobin.

adducts seen by SDS-PAGE (Fig. 5).   It seems reasonable that the alteration in the membrane protein skeleton may result in the deformation of
erythrocyte structure into the severe form of echinocyte seen by scanning
electron microscopy, and that this deformation may promote splenic sequestration.   However, it should be emphasized that direct evidence to
support this latter conjecture is not presently available and that alternate final scenarios, including roles for lipid peroxidation and/or immune
mechanisms, are still equally plausible.

ACKNOWLEDGEMENTS

     This work was supported by a grant from the United States Public
Health Service (HL 30038).   The authors thank Jennifer Schulte and Marie
Meadowcroft for their excellent assistance in the preparation of this
manuscript.

REFERENCES

Allen, D.W. and J.H. Jandl: Oxidative hemolysis and precipitation of
hemoglobin. II.   Role of thiols in oxidant drug action.   J. Clin. Invest.
40: 454 (1961).

Beutler, E.:   The hemolytic effect of primaquine and related compounds:
A review.   J. Hematol. 14: 103 (1959).

Beutler, E.:   Abnormalities of the hexose monophosphate shunt.   Sem.
Hematol. 8: 311 (1971).

Beutler, E.: Glucose-6-phosphate dehydrogenase deficiency. In The Metabolic Basis of Inherited Disease, 3rd Edition: (J.B. Stanbury, J.B. Wyngaarden and D.S. Frederickson, eds.); New York: McGraw Hill (1972) p.1358.

Carrell, R.W., C.C. Winterbourn and E.A. Rachmilewitz: Activated oxygen and hemolysis. Brit. J. Haematol. 30: 259 (1975).

Carson, P.E.: Glucose-6-phosphate dehydrogenase deficiency in hemolytic anemia. Fed. Proc. 19: 995 (1960).

Coetzer, T., and S. Zail: Membrane protein complexes in GSH depleted red cells. Blood 56: 159 (1980).

Cohen, G. and P. Hochstein: Generation of hydrogen peroxide in erythrocytes by hemolytic agents. Biochemistry 3: 895 (1964).

Cordes, W.: Experiences with plasmochin in malaria. United Fruit Co. (Med. Dept.) 15th Annual Report: 66 (1926).

DeGowan, R.L.: A review of the therapeutic and hemolytic effects of dapsone. Arch. Intern. Med. 120: 242 (1967).

Degowin, R.L., R.B. Eppes, R.D. Powell and P.E. Carson: The haemolytic effects of diaphenylsulfone (DDS) in normal subjects and in those with glucose-6-phosphate dehydrogenase deficiency. Bull. W.H.O. 35: 165 (1966).

Dern, R.J., E. Beutler and A.S. Alving: The hemolytic effect of primaquine: II. The natural course of the hemolytic anemia and the mechanism of its self-limiting character. J. Lab. Clin. Med. 44: 171 (1954).

Fairbanks, G., T.L. Steck, and D.F.H. Wallach: Electrophoretic analysis of the major polypeptides of the human erythrocyte membrane. Biochemistry 10: 2606 (1971).

Frazer, I.M. and E.S. Vesell: Effects of drugs and drug metabolites on erythrocytes from normal and G-6-PD deficient individuals. Ann. N.Y. Acad. Sci. 151: 777 (1968).

Glader, B.E. and M.E. Conrad: Hemolysis by diphenylsulfones: Comparative effects of DDS and hydroxylamine-DDS. J. Lab. Clin. Med. 81: 267 (1973).

Goldberg, B. and A. Stern: Production of superoxide anion during the oxidation of hemoglobin by menadione. Biochim. Biophys. Acta 437: 628 (1976a)

Goldberg, B. and A. Stern: The mechanisms of superoxide anion generation by the interaction of phenylhydrazine with hemoglobin. J. Biol. Chem. 251: 3045 (1976b).

Gosselin, R.E., H.C. Hodge, R.P. Smith and M.N. Gleason: Clinical Toxicology of Commercial Products. Williams and Wilkins Co., Baltimore, 4th edition, pp 29-35, 1976.

Grossman, S.J. and D.J. Jollow, in preparation, 1985.

Habeeb, A.F.S.A.: Reaction of protein sulfhydryl groups with Ellman's reagent Met. Enzymol. Vol 25B: 457 (1972).

Harrison, J.H. and D.J. Jollow: Rapid and sensitive method for the microassay of nitrosobenzene plus phenylhydroxylamine in blood. J. Chromatog. 277: 173 (1983).

Harrison, J.H. and D.J. Jollow: Role of phenylhydroxylamine in aniline-induced hemolysis. J. Pharmacol. Exp. Ther. Accepted (1985).

Jacob, H.S. and J.H. Jandl: Effects of sulfhydryl inhibition on red blood cells. I. Mechanism of hemolysis. J. Clin. Invest. 41: 779 (1962a).

Jacob, H.S. and J.H. Jandl: Effects of sulfhydryl inhibition on red blood cells. II. Studies in vivo. J. Clin. Invest. 41: 1514 (1962b).

Kao, J., J. Faulkner, and J.W. Bridges: Metabolism of aniline in rats, pigs and sheep. Drug Metab. Dispos. 6: 549 (1978).

Kellermeyer, R.W., A.R. Tarlov, C.J. Brewer, P.E. Carson and A.S. Alving: Hemolytic effect of therapeutic drugs: Clinical considerations of the primaquine-type hemolysis. J.A.M.A. 180: 388 (1962).

Kiese, M.: Methemoglobinemia: A comprehensive treatise. CRC Press, Cleveland, Ohio, 1974.

Kosower, N.S., Y. Zipser and Z. Faltin: Membrane thiol-disulfide status in glucose-6-phosphate dehydrogenase deficient red cells. Relationship to cellular glutathione. Biochim. Biophys. Acta 691: 345 (1982).

McKusick, V.A. and F.H. Ruddle: The status of the gene map of the human chromosomes. Science 196: 390 (1977).

Miller, A. and H.C. Smith: The intracellular and membrane effects of oxidant agents on normal red cells. Brit. J. Haematol. 19: 417 (1970).

Misra, H.P. and I. Fridovich: The oxidation of phenylhydrazine: Superoxide and mechanism. Biochemistry 15: 681 (1976)

Muehlens, P.: Die behandlung der naturichen menschlichen Malaria-infektion mit Plasmochin. Nature 14: 1162 (1926).

Parke, D.V.: Studies in detoxification. 84. The fate of aniline in the rabbit and other animals. Biochem. J. 77: 483 (1960).

Salvidio, E., I. Pannacciulli, A. Tizianello and F. Aimer: Nature of hemolytic crisis and the fate of G-6-PD deficient, drug damaged erythrocytes in Sardinians. New Eng. J. Med. 276: 1339 (1967).

Scott, G.L. and M.R. Rasbridge: The in vitro action of dapsone and its derivatives on normal and G6PD deficient red cells. Brit. J. Haematol. 24: 307 (1973).

Tarlov, A.R., G.J. Brewer, P.E. Carson and A.S. Alving: Primaquine sensitivity. Archives of Internal Medicine 109: 137 (1962).

ELECTROPHILIC SULFURIC ACID ESTER METABOLITES AS ULTIMATE CARCINOGENS

James A. Miller and Elizabeth C. Miller

McArdle Laboratory for Cancer Research
Medical School
University of Wisconsin
Madison, WI 53706, USA

INTRODUCTION

It is a great privilege to be the honorees of the Third International Symposium on Biological Reactive Intermediates. We wish to thank the Co-chairmen and the Organizing Committee for this signal honor and for the opportunity to contribute to this Symposium.

We encountered biological reactive intermediates indirectly in the 1940's in some of our studies on the metabolism of the carcinogenic aminoazo dyes (E.C. Miller and J.A. Miller, 1947; J.A. Miller and E.C. Miller, 1953). The new phenonomenon of the covalent binding of these dyes to the liver protein of rats in vivo and similar findings made later with other carcinogens led us on a long search for intermediate metabolites that would explain these results. Eventually, studies in our laboratory and those by many other investigators demonstrated that the majority of chemical carcinogens require metabolic activation to form electrophiles that combine covalently in non-enzymatic reactions with various nucleophiles in cellular macromolecules during the initiation of carcinogenesis (E.C. Miller and J.A. Miller, 1981). Much evidence now strongly supports the concept that chemical carcinogenesis originates in the reaction of carcinogenic electrophiles with bases in cellular DNA. Reaction with DNA appears to be the first step in the initiation stage of the multistage process of carcinogenesis that can be induced in many mammalian cells by chemical carcinogens (Fig. 1) (Weinstein, 1981; Weinstein et al., 1984). The completion of the initiation stage appears to involve heritable changes in the base sequence of cellular DNA. The subsequent stage of promotion apparently requires the clonal expansion of initiated cells in which the altered DNA is expressed in changes in growth control (see also Alexander, 1985). Recent studies strongly suggest that an important genetic target of carcinogenic electrophiles is the ras family of proto-oncogenes (Bishop, 1982; Weinberg, 1982; Marshall et al., 1984; Zarbl et al., 1985; and references therein). In several cases single point mutations within a ras coding sequence appear sufficient to cause the malignant transformation of appropriate indicator cells in culture. These changes also are found in the ras oncogene in some to many primary tumors induced in animals by chemical carcinogens. The fraction of the tumors in which these changes are found differs with the experimental conditions. The roles of proto-oncogenes in the initiation and promotion stages is not clear (Balmain, 1985). How-

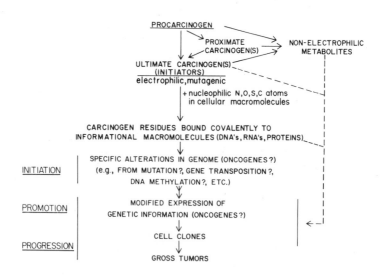

PROCARCINOGEN

PROXIMATE CARCINOGEN(S) → NON-ELECTROPHILIC METABOLITES

ULTIMATE CARCINOGEN(S) (INITIATORS)
electrophilic, mutagenic

+ nucleophilic N,O,S,C atoms in cellular macromolecules

CARCINOGEN RESIDUES BOUND COVALENTLY TO INFORMATIONAL MACROMOLECULES (DNA's, RNA's, PROTEINS)

INITIATION — SPECIFIC ALTERATIONS IN GENOME (ONCOGENES?) (e.g., FROM MUTATION?, GENE TRANSPOSITION?, DNA METHYLATION?, ETC.)

PROMOTION — MODIFIED EXPRESSION OF GENETIC INFORMATION (ONCOGENES?)

CELL CLONES

PROGRESSION — GROSS TUMORS

Fig. 1. A general outline of the metabolic activation of the majority of chemical carcinogens and the origin of tumors in multistage carcinogenesis. The dashed lines indicate possible modulations of the promotion stage by metabolites of chemical carcinogens.

ever, changes in more than one proto-oncogene may be required in the malignant transformation of certain cells in culture (Land et al., 1983; Ruley, 1983), and ras and/or other oncogenes may also play a role in the ability of tumor cells to metastasize (Grieg et al., 1985).

The identification of electrophilic ultimate carcinogens in the initiation of chemical carcinogenesis has generally involved comparisons of the DNA adducts formed in target tissues by chemical carcinogens with the adducts formed non-enzymatically by candidate synthetic electrophiles with DNA or its constituents. In most cases the quantitation of specific adducts in vivo and tumor induction have not been studied under the same conditions. In particular, tumor induction protocols have often employed repetitive doses of carcinogen, so that there was no separation of the initiation and promotion stages.

We wish to report here recent studies by our group (Dr. K. Barry Delclos, Dr. Timothy R. Fennell, Eric W. Boberg, Chen-Ching Lai, Amy Liem, and Roger W. Wiseman) on the identification of ultimate carcinogenic metabolites formed in vivo in mouse liver from members of several classes of chemical carcinogens in relation to DNA adduct and hepatoma formation. For these studies we have used 12-day-old male B6C3F$_1$ mice given a single intraperitoneal injection of carcinogen. Without further treatment these animals develop multiple gross hepatomas by about 10 months. In this system the promotion of the initiated cells appears to occur physiologically during the growth of the male liver. We have determined the structures of the DNA adducts formed in these livers, the levels of the adducts 9 hours after injection of carcinogen, and the extent of hepatoma formation at 10 months. The latter two endpoints were examined in response to two conditions that limit the formation of sulfuric acid esters in vivo. One of these conditions is the inhibition of hepatic sulfotransferase activities by the administration of pentachlorophenol (Mulder and Meerman, 1978; Meerman et al., 1980, 1981). The other condition is the congenital defi-

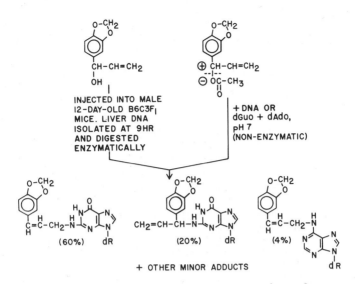

Fig. 2. Structures of adducts formed in mouse liver DNA on administration of 1'-hydroxysafrole or formed nonenzymatically <u>in vitro</u> by reaction at pH 7 of 1'-acetoxysafrole with deoxynucleosides.

ciency of the physiological sulfo group donor 3'-phosphoadenosine-5'-phosphosulfate (PAPS) found in brachymorphic mice (Sugahara and Schwartz, 1979, 1982).

SAFROLE AND 1'-HYDROXYSAFROLE

Safrole (1-allyl-3,4-methylenedioxybenzene) is one of the many closely related alkenylbenzenes that occur naturally in a variety of plant species that are sources of essential oils and spices (Leung, 1980). Safrole is a major component of oil of sassafras and a minor constituent of several other essential oils and spices. It is a weak to moderately active hepatocarcinogen in rats and mice (J.A. Miller et al., 1982). The related spice components estragole (1-allyl-4-methoxybenzene) and methyleugenol (1-allyl-3,4-dimethoxybenzene) have carcinogenic activities similar to that of safrole, but several more extensively substituted alkenylbenzenes have so far shown little or no carcinogenic activity (E.C. Miller et al., 1983).

Safrole and estragole are hydroxylated at the 1'-position of the allyl side chains by microsomal cytochrome P-450 activities in mouse and rat liver (Swanson et al., 1981). The 1'-hydroxy metabolites are more potent carcinogens than the parent alkenylbenzenes (Borchert et al., 1973; Drinkwater et al., 1976; E.C. Miller et al., 1983). The 1'-hydroxy metabolites are further metabolized by liver enzymes to three kinds of electrophiles: 1'-esters of sulfuric acid (1'-sulfooxy derivatives), 2',3'-epoxides, and 1'-oxo derivatives (Oswald et al., 1971; Wislocki et al., 1976, 1977; Fennell et al., 1984). Of these electrophiles, only the esters appear to form adducts in DNA <u>in vivo</u> (Phillips et al., 1981a,b). Fig. 2 shows the structures of the three principal DNA adducts found in mouse liver after administration of 1'-hydroxysafrole (Phillips et al., 1981b). The same adducts are formed <u>in vitro</u> upon reaction of DNA or dGuo and dAdo with the synthetic electrophilic ester 1'-acetoxysafrole. Entirely analogous results were obtained with 1'-hydroxyestragole and 1'-acetoxy-

585

estragole (Phillips et al., 1981a). The minor DNA adducts include deoxy-guanosine residues substituted at the N-7 and C-8 positions with isosafrolyl or isoestragolyl groups (Wiseman et al., 1985).

In view of the presence of sulfotransferase activity for 1'-hydroxy-safrole in rat and mouse liver (Wislocki et al., 1976), the two conditions described above that limit sulfuric acid ester formation in vivo were used to probe the role of 1'-sulfoöxysafrole generated in vivo in the formation of DNA adducts and hepatomas in mouse liver (Boberg et al., 1983). Table 1 shows that the levels of DNA adducts in the livers of male 12-day-old B6C3F$_1$ mice given a dose of $[2',3'-^3H]$1'-hydroxysafrole 45 min after a single dose of 0.04 μmol of pentachlorophenol per gram of body weight were reduced to 15% of those observed for mice not pretreated with this inhibitor. The same pretreatment with pentachlorophenol also reduced the average number of hepatomas per mouse at 10 months to less than 10% of that observed in the absence of the sulfotransferase inhibitor. However, no effect of pretreatment with pentachlorophenol was noted on hepatoma formation by diethylnitrosamine, a carcinogen which is not acti-vated by esterification (Preussmann and Stewart, 1984). Similar strong inhibition of hepatic tumor formation by pentachlorophenol was observed when a single dose of estragole or 1'-hydroxyestragole was administered to 12-day-old male B6C3F$_1$ mice (Wiseman, Miller, Miller, and Liem, unpub-lished data). Likewise, continuous dietary administration of pentachloro-phenol strongly inhibited hepatoma formation by dietary safrole or 1'-hydroxysafrole in adult female CD-1 mice (Boberg et al., 1983). A similar dependence of DNA adduct and tumor formation in mouse liver upon 1'-sulfo-öxysafrole formation was noted in the PAPS-deficient brachymorphic mice. When these mice were given 1'-hydroxysafrole (Table 2), the level of DNA adducts in the livers of 12-day-old male B6C3F$_2$ brachymorphic mice was only 15% as high as in the livers of their phenotypically normal littermates given the same treatment. Similarly, the average number of hepatomas in the brachymorphic mice was no more than 10% of that found in the normal littermates. The average number of hepatomas in brachymorphic mice given a single injection of diethylnitrosamine was only 40% as great as in their phenotypically normal littermates (Lai et al., 1985). The basis of this reduction is not known; the microsomal oxidative deethylation of diethyl-nitrosamine is as active in the livers of brachymorphic mice as in their normal littermates (Kirkheide, Fennell, Miller, and Miller, unpublished data). In any case, the difference between the very low tumorigenic re-sponse to 1'-hydroxysafrole in the brachymorphic mice, as compared to the 40% response to diethylnitrosamine, was large. Thus, these data strongly support the concept that 1'-sulfoöxysafrole is a critical metabolite

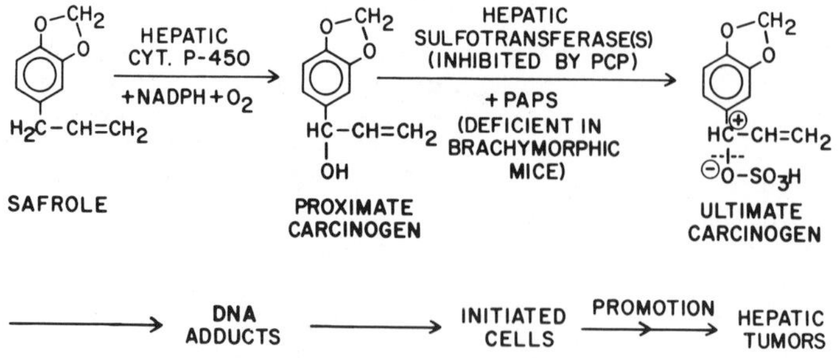

Fig. 3. The hepatic metabolism of safrole to the electrophile 1'-sulfoöxysafrole, the initiation of carcinogenesis in preweanling male mouse liver, and the promotion of the initiated cells to yield gross liver tumors in adulthood. PCP = pentachlorophenol.

TABLE 1. Effect of pretreatment with pentachlorophenol on hepatic DNA
          adduct formation and hepatoma initiation by 1'-hydroxy-
          safrole[a]

| Carcinogen (μmol/g body weight) | Penta-chloro-phenol | DNA adducts (pmol/mg) | Hepatomas | |
			Incidence (%)	Average number/liver
1'-Hydroxy-safrole (0.2)	−	190	97	4.4
" (0.2)	+	24	10	0.1
" (0.1)	−	68	86	2.2
" (0.1)	+	9	12	0.2
Diethylnitros-amine (0.01)	−		100	12
	+		100	14
None	− or +		12	0.1

[a] 12-day-old male B6C3F$_1$ mice were injected intraperitoneally with 0.04
μmol of pentachlorophenol per gram body weight 45 min before the intraperitoneal injection of carcinogen. The DNA adducts were assayed at 9
hr, and the hepatomas were enumerated at 10 mo.

TABLE 2. The formation of hepatic DNA adducts and of hepatomas on
          administration of 1'-hydroxysafrole to brachymorphic and
          phenotypically normal B6C3F$_2$ mice[a]

| Carcinogen (μmol/g body weight) | Phenotype | DNA adducts (pmol/mg) | Hepatomas | |
			Incidence (%)	Average number/liver
1'-Hydroxy-safrole (0.2)	Normal	110	43	1.2
" (0.2)	Brachymorphic	16	6	0.1
" (0.15)	Normal		56	1.9
" (0.15)	Brachymorphic		0	0.0
Diethylnitros-amine (0.02)	Normal		80	11.6
" (0.02)	Brachymorphic		52	5.0
None	Normal		11	0.1
	Brachymorphic		2	0.02

[a] 12-day-old male B6C3F$_2$ mice were injected intraperitoneally with tri-
octanoin solutions of 1'-hydroxysafrole or diethylnitrosamine. The DNA
adducts were assayed at 9 hr, and the hepatomas were enumerated at 15
mo (1'-hydroxysafrole) or 9 mo (diethylnitrosamine).

of 1'-hydroxysafrole for the formation of DNA adducts and hepatomas in the male B6C3 mice (Fig. 3). The complete data that support this conclusion have been published (Boberg et al., 1983).

1'-HYDROXY-2',3'-DEHYDROESTRAGOLE

In our studies on the metabolism of 1'-hydroxyestragole it was labeled in the allyl group at the 2',3'-carbons by the partial reductive tritiation of 1'-hydroxy-2',3'-dehydroestragole (1'-hydroxy-DHE) (Phillips et al., 1981a). Carcinogenicity testing of this synthetic acetylenic intermediate showed that it was approximately 10 times more active in the mouse liver than 1'-hydroxyestragole (E.C. Miller et al., 1983; Fennell et al., 1985). 1'-Hydroxy-DHE labeled at the 1'-carbon with tritium formed approximately 3 times as many covalently bound adducts in mouse liver DNA as were obtained with 1'-hydroxyestragole (Fennell et al., 1985). The acetic acid ester of 1'-hydroxy-DHE was electrophilic, and the NMR spectra of the adducts formed from it showed that the triple bond remained at its original position and that only the 1'-carbon became attached to the•nucleophilic bases of deoxyguanosine and deoxyadenosine. Deoxyguanosine formed two adducts: N^2-(2',3'-dehydroestragol-1'-yl)deoxyguanosine and N-7-(2',3'-dehydro-estragol-1'-yl)guanine. The latter adduct evidently formed by rapid depurination of a labile N-7-(2',3'-dehydroestragol-1'-yl)deoxyguanosine adduct. Deoxyadenosine also yielded two adducts: N^6-(2',3'-dehydroestragol-1'-yl)deoxyadenosine and an as yet uncharacterized adduct. No adducts were detected in attempts to react 1'-acetoxy-2'3'-dehydroestragole with deoxycytidine or thymidine. Surprisingly, only one adduct, N^2-(2',3'-dehydro-estragol-1'-yl)deoxyguanosine, was found in the hepatic DNA of mice administered [1'-^{3}H]1'-hydroxy-DHE. No significant loss of this adduct was noted in the hepatic DNA by 21 days after the injection of a single carcinogenic dose of labeled 1'-hydroxy-DHE.

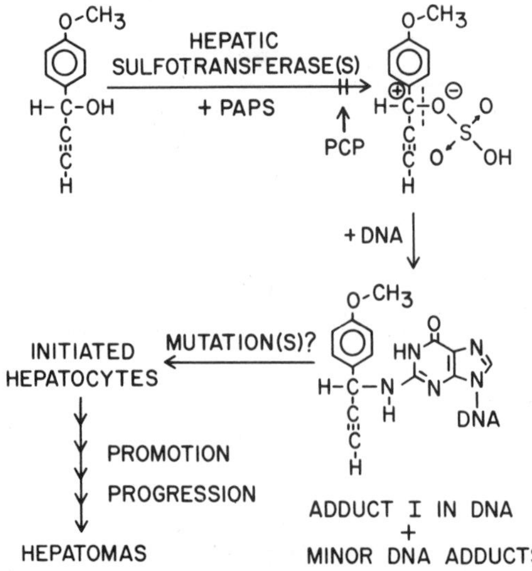

Fig. 4. The hepatic metabolism of 1'-hydroxy-2', 3'-dehydroestragole in the initiation of hepatocarcinogenesis in preweanling male B6C3F$_1$ mice. PCP = pentachlorophenol.

No cytosolic coenzyme A-dependent acetyltransferase activity for 1'-hydroxy-DHE was detected in the livers of male B6C3F$_1$ mice. However, sulfotransferase activity was found in these cytosols and it was strongly inhibited by pentachlorophenol. Pretreatment of 12-day-old male B6C3F$_1$ mice by pentachlorophenol lowered by 87% the levels of DNA adducts formed on administration of 1'-hydroxy-DHE. Similarly, the formation of hepatomas at 10 months in comparably treated mice was reduced by 94% by the pretreatment with pentachlorophenol. Thus, as noted above for 1'-hydroxy-safrole, these data strongly indicate that 1'-sulfoöxy-2',3'-dehydroestragole is the major electrophilic and carcinogenic metabolite of 1'-hydroxy-DHE in the livers of the infant male B6C3F$_1$ mice (Fig. 4).

N-HYDROXY-2-ACETYLAMINOFLUORENE

The metabolic activation of 2-acetylaminofluorene (AAF) has been studied extensively in the rat by many investigators (E.C. Miller and J.A. Miller, 1981). The conversion of this versatile carcinogen to the proximate carcinogenic metabolite N-hydroxy-AAF by hepatic cytochrome P-450 activity appears to be an obligatory reaction for carcinogenesis by this aromatic amide. N-Hydroxy-AAF is in turn metabolized by liver preparations to several electrophiles: N-sulfoöxy-AAF, N-acetoxy-AAF, 2-nitroso-fluorene, N-acetoxy-2-aminofluorene, and N-hydroxy-2-aminofluorene (electrophilic at acid pH). N-Sulfoöxy-AAF and N-acetoxy-AAF react with deoxyguanosine to form the adducts N-(deoxyguanosin-8-yl)AAF and 3-(deoxyguanosin-N^2-yl)AAF. N-Acetoxy-2-aminofluorene and N-hydroxy-2-aminofluorene (at acid pH) react with deoxyguanosine to form the adduct N-(deoxyguanosin-8-yl)-2-aminofluorene. The DNA adducts found in the liver of the rat after administration of AAF or N-hydroxy-AAF consist of N-(deoxyguanosin-8-yl)-2-aminofluorene (about 60% of the total adducts), and the two acetylated

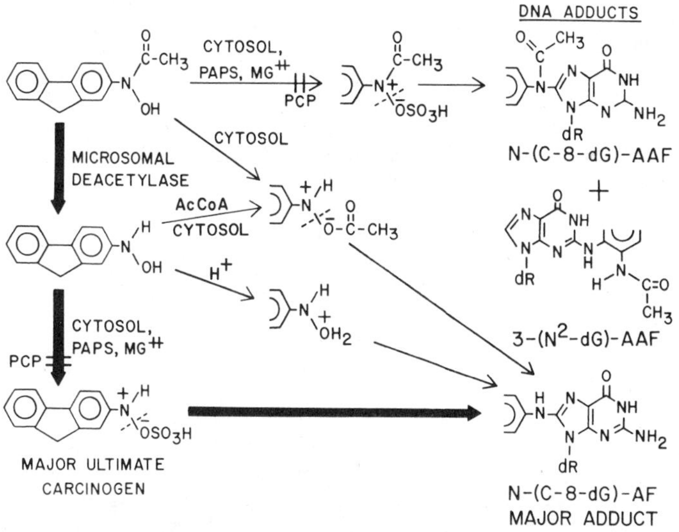

Fig. 5. The enzymatic activation of N-hydroxy-AAF in the infant male B6C3F$_1$ mouse liver to yield N-sulfo-öxy-2-aminofluorene as the major electrophilic precursor of N-(deoxyguanosin-8-yl)-2-aminofluorene [N-(C-8-dG)-AF] adducts in the DNA. PCP = pentachlorophenol, AcCoA = acetyl coenzyme A.

adducts N-(deoxyguanosin-8-yl)AAF and 3-(deoxyguanosin-N^2-yl)AAF comprise
the remainder (Meerman et al., 1981). Rats given N-hydroxy-AAF and pentachlorophenol contain reduced amounts of the latter two acetylated adducts
in their hepatic DNA, with no reduction of the non-acetylated adduct (Meerman et al., 1981). The latter adduct is frequently regarded as the adduct
probably responsible for the initiation of liver carcinogenesis in the
rat by AAF or N-hydroxy-AAF.

Recently it was noted that the metabolism of N-hydroxy-AAF in the
mouse differs in several respects from that described above for the rat.
In the male 12-day-old B6C3F$_1$ mouse N-(deoxyguanosin-8-yl)-2-amino-fluorene
accounted for at least 90% of the adducts in the liver DNA after administration of a carcinogenic dose of N-hydroxy-AAF (Lai et al., 1985). Furthermore, the hepatic cytosols of these mice contained sulfotransferase activity
for N-hydroxy-2-aminofluorene, a previously unrecognized activity. This
activity and the much greater deacetylase activity for N-hydroxy-AAF in
mouse liver than in rat liver (Schut et al., 1978; Lai et al., 1985) suggested that DNA adduct and hepatoma formation by N-hydroxy-AAF in the
mouse were dependent on the formation of N-sulfoöxy-2-aminofluorene. Data
on B6C3F$_1$ mice given a single injection of 0.06 μmol of N-hydroxy-AAF per
gram body weight supported this concept. Under these conditions pretreatment of the mice with 0.04 μmol of pentachlorophenol per gram body weight
reduced the level of N-(deoxyguanosin-8-yl)-2-aminofluorene adducts in
the hepatic DNA at 9 hr by about 90% (from 2.9 to 0.3 pmol/mg DNA) and
also reduced the average number of hepatomas per liver at 10 months by
about 90% (from 10 to 1 hepatomas per liver). In brachymorphic mice,
which are deficient in the synthesis of PAPS, the administration of N-
hydroxy-AAF yielded levels of DNA adducts only 25% of those found in the
liver DNA of comparably treated phenotypically normal littermates, and
the average number of hepatomas in the brachymorphic mice was only 10% as
high as in the normal littermates.

The liver cytosols of 12-day-old B6C3F$_1$ male mice contained acyltransferase activity and coenzyme A-dependent acetyltransferase activity for
N-hydroxy-AAF as well as the newly recognized sulfotransferase activity
for N-hydroxy-2-aminofluorene. Thus, any of these enzyme activities could
have formed esters of N-hydroxy-2-aminofluorene that would react with DNA
to form N-(deoxyguanosin-8-yl)-2-aminofluorene adducts. However, the
acyltransferase and acetyltransferase activities were very much less sensitive to inhibition by pentachlorophenol than was the sulfotransferase
activity. Thus, the deacetylation of N-hydroxy-AAF to N-hydroxy-2-amino-
fluorene and the subsequent esterification of this hydroxylamine to form
N-sulfoöxy-2-aminofluorene appears to be the major metabolic pathway in
the livers of infant male B6C3F$_1$ mice for the metabolic activation of
this carcinogen for reaction with DNA and the induction of hepatomas (Fig.
5).

4-AMINOAZOBENZENE

It has long been recognized that 4-aminoazobenzene has little or no
hepatocarcinogenicity in the rat while its N-methyl and N,N-dimethyl derivatives are strong hepatocarcinogens in this species (J.A. Miller and
E.C. Miller, 1953). Recently, it has been found that these three dyes
are equally strong hepatocarcinogens in B6C3F$_1$ mice given a single intraperitoneal injection of dye at 12 days of age (Delclos et al., 1984).
The administration of the N-methyl and N,N-dimethyl derivatives of 4-
aminoazobenzene leads to the formation of three DNA adducts in the livers
of these mice: N-(deoxyguanosin-8-yl)-N-methyl-4-aminoazobenzene, 3-
(deoxyguanosin-N^2-yl)-N-methyl-4-aminoazobenzene, and N-(deoxyguanosin-8-
yl)-4-aminoazobenzene. In contrast, 4-aminoazobenzene gives rise to only

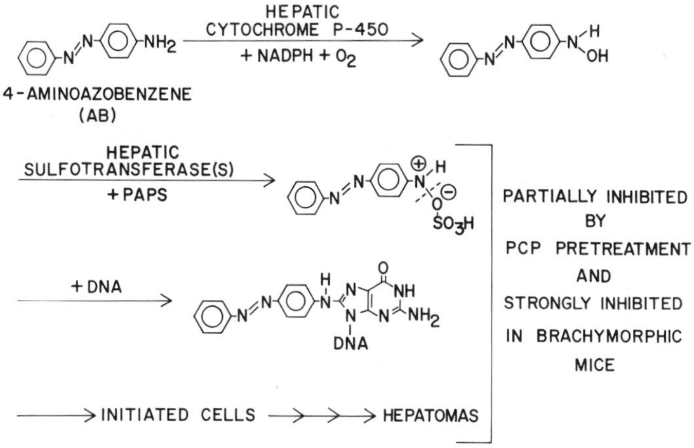

Fig. 6. The metabolic activation of 4-aminoazobenzene in
the livers of 12-day-old male B6C3F$_1$ mice and the
subsequent steps in hepatocarcinogenesis initiated
by this aminoazo dye. PCP = pentachlorophenol.

one DNA adduct, N-(deoxyguanosin-8-yl)-4-aminoazobenzene (Delclos et al.,
1984). This adduct is also formed non-enzymatically upon the incubation
of N-hydroxy-4-aminoazobenzene and an excess of acetic anhydride with DNA
or deoxyguanosine at pH 7; N-acetoxy-4-aminoazobenzene is presumed to be
the reactive intermediate in this reaction. No acetyl coenzyme A-dependent
acetyltransferase activity for N-hydroxy-4-aminoazobenzene was detected in
the liver cytosols of B6C3F$_1$ mice. However, PAPS-dependent sulfotransferase activity in these cytosols mediated the formation of N-(deoxyguano-
sin-8-yl)-4-aminoazobenzene from N-hydroxy-4-aminoazobenzene and deoxyguanosine. Unlike the sulfotransferase activities discussed previously in
this report, the sulfotransferase activity for N-hydroxy-4-aminoazobenzene
is only partially inhibited by pentachlorophenol. In this respect it
resembles human platelet sulfotransferase activity for dopamine (Rein et
al., 1982). Pentachlorophenol, at levels of 10 and 100 µM, inhibited the
sulfotransferase activity for N-hydroxy-4-aminoazobenzene by only 20 and
70%, respectively (Delclos, Miller, Miller, and Liem, submitted for publication). Likewise, 0.04 µmol of pentachlorophenol per gram of body weight
administered to infant male B6C3F$_1$ mice 45 minutes before the injection
of 0.05 or 0.10 µmol of 4-aminoazobenzene per gram body weight inhibited
the formation of N-(deoxyguanosin-8-yl)-4-aminoazobenzene adducts in the
hepatic DNA by only 30-40%. Hepatoma formation in comparably treated
mice was inhibited only about 50%.

The level of adducts in the liver DNA of male brachymorphic mice
administered 0.15 µmol of 4-aminoazobenzene per gram body weight was only
12% of that found in the livers of similarly treated phenotypically normal
male littermates; hepatoma formation in comparably treated brachymorphic
mice was only 10% of that found in the normal littermates. This marked
inhibition of the formation of DNA adducts and hepatomas from the administration of 4-aminoazobenzene in the livers of mice deficient in the ability
to synthesize PAPS is strong evidence that N-sulfoöxy-4-aminoazobenzene
is the critical electrophilic metabolite for both adduct and hepatoma
formation. The limited but parallel inhibitions of these activities by
pentachlorophenol pretreatment of B6C3F$_1$ mice given 4-aminoazobenzene are
further evidence for this conclusion. These findings are outlined in
Fig. 6.

CONCLUDING COMMENTS

The data in this report are the first <u>in vivo</u> demonstrations that sulfuric acid esters of certain carcinogens or their proximate carcinogenic metabolites are electrophiles responsible for the formation of DNA adducts and the initiation of carcinogenesis. These data also form some of the most direct evidence that correlate the formation of a specific electrophilic metabolite with the formation of DNA adducts and tumor formation in a target tissue under identical conditions. The use of a system of carcinogenesis in which a single dose of carcinogen initiates carcinogenesis through reactions that can be specifically inhibited to limit in parallel both DNA adduct formation and the subsequent appearance of tumors was critical to these results.

Sulfuric acid esters may be important in the metabolic activation of other xenobiotics, including non-carcinogens as well as carcinogens. For example, the oxidations of 7-methylbenz(a)anthracene, 7,12-dimethylbenz(a)anthracene, and 6-methylbenzo(a)pyrene to hydroxymethyl derivatives which are esterified to form electrophilic sulfuric acid esters may be a pathway of metabolic activation <u>in vivo</u> of these hydrocarbons (Cavalieri et al., 1979; Watabe et al., 1982; Watabe, 1983). While some of these esters have electrophilic, mutagenic, or carcinogenic activities or can be formed in PAPS-dependent reactions catalyzed by liver cytosols, there are as yet no data that show whether or not these esters are critical to carcinogenic processes induced by these hydrocarbons. Tests with sulfotransferase inhibitors and brachymorphic mice may help to decide this issue.

The known sulfotransferase activities for xenobiotics, steroids, and other low molecular weight compounds (Roy, 1981; this report) appear to occur predominantly in the cytosols of cells. This raises the question of how highly reactive electrophilic sulfuric acid esters generated in this cellular fraction reach the nuclear DNA, the presumed reaction site critical in the initiation of carcinogenesis. A corollary question is the possible occurrence of sulfotransferase activities in cell nuclei. Sulfotransferase activities for cerebrosides appear to be localized in membranes of microsomes and the Golgi apparatus (Farrell and McKhann, 1971; Fleischer and Smigel, 1978). Stöhrer and his colleagues (1979) reported preliminary observations on the apparent occurrence of sulfotransferase activity for N-hydroxy-AAF and purine-N-oxides in rat liver nuclei. Preliminary studies in our laboratory (Boberg, Miller and Miller, unpublished data) indicate that the activity for PAPS-dependent esterification of 1'-hydroxysafrole by mouse liver nuclei is very low. Further study is needed.

The chemical synthesis of the reactive sulfuric acid esters of carcinogens has been accomplished in only two cases (Maher et al., 1968; Kriek and Hengeveld, 1978; Beland et al., 1983). The availability of more of these reactants in pure form would make it possible to study many aspects of their reactions with cellular nucleophiles (see, for example, Gutmann et al., 1985; Smith et al., 1985).

ACKNOWLEDGMENT

This research was supported by Public Health Service Grants CA-07175, CA-09135, CA-09020, and CA-22484 from the National Cancer Institute, U.S. Department of Health and Human Services.

REFERENCES

Alexander, P., 1985, Do cancers arise from a single transformed cell or is monoclonality of tumors a late event in carcinogenesis?, Brit. J. Cancer, 51:453.

Balmain, A., 1985, Transforming ras oncogenes and multistage carcinogenesis, Brit. J. Cancer, 51:1.

Beland, F. A., Miller, D. W., and Mitchum, R. K., 1983, Synthesis of the ultimate hepatocarcinogen, 2-acetylaminofluorene N-sulphate, J. Chem. Soc. Chem. Commun., 30.

Bishop, J. M., 1982, Retroviruses and cancer genes, Adv. Cancer Res., 37:1.

Boberg, E. W., Miller, E. C., Miller, J. A., Poland, A., and Liem, A., 1983, Strong evidence from studies with brachymorphic mice and pentachlorophenol that 1'-sulfoöxysafrole is the major ultimate electrophilic and carcinogenic metabolite of 1'-hydroxysafrole in mouse liver, Cancer Res., 43:5163.

Borchert, P., Wislocki, P. G., Miller, J. A., and Miller, E. C., 1973, The metabolism of the naturally occurring hepatocarcinogen safrole to 1'-hydroxysafrole and the electrophilic reactivity of 1'-acetoxysafrole, Cancer Res., 33:575.

Cavalieri, E., Roth, R., and Rogan, E., 1979, Hydroxylation and conjugation at the benzylic carbon atom: a possible mechanism of carcinogenic activation for some methyl-substituted aromatic hydrocarbons, in: "Polynuclear Aromatic Hydrocarbons," P. W. Jones and P. Leder, eds., Ann Arbor Science, Ann Arbor.

Delclos, K. B., Tarpley, W. G., Miller, E. C., and Miller, J. A., 1984, 4-Aminoazobenzene and N,N-dimethyl-4-aminoazobenzene as equipotent hepatic carcinogens in male C57BL/6 x C3H/He F_1 mice and characterization of N-(deoxyguanosin-8-yl)-4-aminoazobenzene as the major persistent hepatic DNA-bound dye in these mice, Cancer Res., 44:2540.

Drinkwater, N. R., Miller, E. C., Miller, J. A., and Pitot, H. C., 1976, The hepatocarcinogenicity of estragole (1-allyl-4-methoxybenzene) and 1'-hydroxyestragole in the mouse and the mutagenicity of 1'-acetoxyestragole in bacteria, J. Natl. Cancer Inst., 57:1323.

Farrell, D. F., and McKhann, G. M., 1971, Characterization of cerebroside sulfotransferase from rat brain, J. Biol. Chem., 246:4694.

Fennell, T. R., Miller, J. A., and Miller, E. C., 1984, Characterization of the biliary and urinary glutathione and N-acetylcysteine metabolites of the hepatic carcinogen 1'-hydroxysafrole and its 1'-oxo metabolite in rats and mice, Cancer Res., 44:3231.

Fennell, T. R., Wiseman, R. W., Miller, J. A., and Miller, E. C., 1985, The major role of hepatic sulfotransferase activity in the metabolic activation, DNA adduct formation, and carcinogenicity of 1'-hydroxy-2',3'-dehydroestragole in infant male $B6C3F_1$ mice, Cancer Res., in press.

Fleischer, B., and Smigel, M., 1978, Solubilization and properties of galactosyl-transferase and sulfotransferase activities of Golgi membranes in Triton X-100, J. Biol. Chem., 253:1632.

Grieg, R. G., Koestler, T. P., Trainer, D. L., Corwin, S. P., Miles, L., Kline, T. P., Sweet, R., Yokoyama, S., and Poste, G., 1985, Tumorigenic and metastatic properties of "normal" and ras-transfected NIH/3T3 cells, Proc. Natl. Acad. Sci. USA, 82:3698.

Gutmann, H. R., Smith, B. A., and Springfield, J. R., 1985, Interaction of the ultimate carcinogenic metabolites of N-hydroxy-2-acetylaminofluorene with nucleophiles, Proc. Amer. Assoc. Cancer Res., 26:115.

Kriek, E., and Hengeveld, G. M., 1978, Reaction products of the carcinogen N-hydroxy-4-acetylamino-4'-fluorobiphenyl with DNA in liver and kidney of the rat, Chem.-Biol. Interact., 21:179.

Lai, C.-C., Miller, J. A., Miller, E. C., and Liem, A., 1985, N-Sulfoöxy-2-aminofluorene is the major ultimate electrophilic and carcinogenic metabolite of N-hydroxy-2-acetylaminofluorene in the livers of infant C57BL/6J x C3H/HeJ F$_1$ (B6C3F$_1$) mice, Carcinogenesis, 6:1037.

Land, H., Parada, L. F., and Weinberg, R. A., 1983, Tumorigenic conversion of primary embryo fibroblasts requires at least two cooperating oncogenes, Nature (London), 304:596.

Leung, A. Y., 1980, Encyclopedia of Common Natural Ingredients Used in Food, Drugs, and Cosmetics," John Wiley & Sons, New York.

Maher, V. M., Miller, E. C., Miller, J. A., and Szybalski, W., 1968, Mutations and decreases in density of transforming DNA produced by derivatives of the carcinogens 2-acetylaminofluorene and N-methyl-4-aminoazobenzene, Mol. Pharmacol., 4:411.

Marshall, C. J., Vousden, K. H., and Phillips, D. H., 1984, Activation of c-Ha-ras-1 proto-oncogene by in vitro modification with a chemical carcinogen, benzo(a)pyrene diol-epoxide, Nature (London), 310:586.

Meerman, J. H., Beland, F. A., and Mulder, G. J., 1981, Role of sulfate in the formation of DNA adducts from N-hydroxy-2-acetylaminofluorene in rat liver in vivo. Inhibition of N-acetylated aminofluorene adduct formation by pentachlorophenol, Carcinogenesis, 2:413.

Meerman, J. H., van Doorn, A. B. D., and Mulder, G. J., 1980, Inhibition of sulfate conjugation of N-hydroxy-2-acetylaminofluorene in isolated perfused rat liver and in the rat in vivo by pentachlorophenol and low sulfate, Cancer Res., 40:3772.

Miller, E. C., and Miller, J. A., 1947, The presence and significance of bound aminoazo dyes in the livers of rats fed p-dimethylaminoazobenzene, Cancer Res., 7:468.

Miller, E. C., and Miller, J. A., 1981, Searches for ultimate chemical carcinogens and their reactions with cellular macromolecules, Cancer, 47:2327.

Miller, E. C., Swanson, A. B., Phillips, D. H., Fletcher, T. L., Liem, A., and Miller, J. A., 1983, Structure-activity studies of the carcinogenicities in the mouse and rat of some naturally occurring and synthetic alkenylbenzene derivatives related to safrole and estragole, Cancer Res., 43:1124.

Miller, J. A., and Miller, E. C., 1953, The carcinogenic aminoazo dyes. Adv. Cancer Res., 1:339.

Miller, J. A., Miller, E. C., and Phillips, D. H., 1982, The metabolic activation and carcinogenicity of alkenylbenzenes that occur naturally in many spices, in: "Carcinogens and Mutagens in the Environment," Vol. 1, Food Products," H. F. Stich, ed., CRC Press, Boca Raton, Florida.

Mulder, G. J., and Meerman, J. H. N., 1978, Glucuronidation and sulphation in vivo and in vitro; selective inhibition of sulphation by drugs and deficiency of inorganic sulphate, in: A. Aito, ed., "Conjugation Reactions in Drug Biotransformation," Elsevier/North Holland Biomedical Press, Amsterdam.

Oswald, E. O., Fishbein, L., Corbett, B. J., and Walker, M. P., 1971, Identification of tertiary aminomethylenedioxypropiophenones as urinary metabolites of safrole in the rat and guinea pig, Biochim. Biophys. Acta, 230:237.

Phillips, D. H., Miller, J. A., Miller, E. C., and Adams, B., 1981a, Structures of the DNA adducts formed in mouse liver after administration of the proximate hepatocarcinogen 1'-hydroxyestragole, Cancer Res., 41:176.

Phillips, D. H., Miller, J. A., Miller, E. C., and Adams, B., 1981b, The N^2-atom of guanine and the N^6-atom of adenine residues as sites for covalent binding of metabolically activated 1'-hydroxysafrole to mouse-liver DNA in vivo, Cancer Res., 41:2664.

Preussmann, R. and Stewart, B. W., 1984, N-Nitroso Compounds, in: "Chemical Carcinogens," 2nd edit., ACS Monograph 182, C. E. Searle, ed., American Chemical Society, Washington, D.C.

Rein, G., Glover, V., and Sandler, M., 1982, Multiple forms of phenolsul-

photransferase in human tissues. Selective inhibition by dichloro-
nitrophenol, Biochem. Pharmacol., 31:1893.

Roy, A. B., 1981, Sulfotransferases, in: "Sulfation of Drugs and Related
Compounds," G. J. Mulder, ed., CRC Press, Inc., Boca Raton, Florida.

Ruley, H. E., 1983, Adenovirus early region 1A enables viral and cellular
transforming genes to transform primary cells in culture, Nature
(London), 304:602.

Smith, B. A., Gutmann, H. R., and Springfield, J. R., 1985, Interaction
of nucleophiles with the enzymatically-activated carcinogen, N-hydroxy-
2-acetylaminofluorene, and with the model ester, N-acetoxy-2-acetyl-
aminofluorene, Carcinogenesis, 6:271.

Stöhrer, G., Harmonay, L. A., and Brown, G. B., 1979, Sulfotransferase in
rat liver nuclei, Proc. Am. Assoc. Cancer Res., 20:285.

Sugahara, K., and Schwartz, N. B., 1979, Defect in 3'-phosphoadenosine
5'-phosphosulfate formation in brachymorphic mice, Proc. Natl. Acad.
Sci. USA, 76:6615.

Sugahara, K., and Schwartz, N. B., 1982, Defect in 3'-phosphoadenosine
5'-phosphosulfate synthesis in brachymorphic mice, Arch. Biochem.
Biophys., 214:602.

Swanson, A. B., Miller, E. C., and Miller, J. A., 1981, The side-chain
epoxidation and hydroxylation of the hepatocarcinogens safrole and
estragole and some related compounds by rat and mouse liver microsomes, Biochim. Biophys. Acta, 673:504.

Watabe, T., 1983, Metabolic activation of 7,12-dimethylbenz[a]anthracene
(DMBA) and 7-methylbenz[a]anthracene (7-MBA) by rat liver P-450 and
sulfotransferase. J. Toxicological Sciences, 8:119.

Watabe, T., Ishizuka, T., Isobe, M., and Ozawa, N., 1982, A 7-hydroxy-
methyl sulfate ester as an active metabolite of 7,12-dimethylbenz[a]an-
thracene, Science, 215:403.

Weinberg, R. A., 1982, Oncogenes of spontaneous and chemically induced
tumors, Adv. Cancer Res., 36:149.

Weinstein, I. B., 1981, Current concepts and controversies in chemical
carcinogenesis, J. Supramol. Struct. Cell. Biochem., 17:99.

Weinstein, I. B., Gattoni-Celli, S., Kirschmeier, P., Hsiao, W., Horowitz,
A., and Jeffrey, A., 1984, Cellular targets and host genes in multistage carcinogenesis, Federation Proc., 43:2287.

Wiseman, R. W., Fennell, T. R., Miller, J. A., and Miller, E. C., 1985,
Further characterization of the DNA adducts formed by electrophilic
esters of the hepatocarcinogens 1'-hydroxysafrole and 1'-hydroxyestra-
gole in vitro and in mouse liver in vivo, including new adducts at
C-8 and N-7 of guanine residues, Cancer Res., 45:3096.

Wislocki, P. G., Borchert, P., Miller, J. A., and Miller, E. C., 1976,
The metabolic activation of the carcinogen 1'-hydroxysafrole in vivo
and in vitro and the electrophilic reactivities of possible ultimate
carcinogens, Cancer Res., 36:1686.

Wislocki, P. G., Miller, E. C., Miller, J. A., McCoy, E. C., and Rosenkranz, H. S., 1977, Carcinogenic and mutagenic activities of safrole,
1'-hydroxysafrole, and some known or possible metabolites, Cancer
Res., 37:1883.

Zarbl, H., Sukumar, S., Arthur, A. V., Martin-Zanca, D., and Barbacid,
M., 1985, Direct mutagenesis of Ha-ras-1 oncogenes by N-nitroso-N-
methylurea during initiation of mammary carcinogenesis in rats, Nature
(London), 315:382.

BIOLOGICAL REACTIVE INTERMEDIATES OF MYCOTOXINS

Dennis P. H. Hsieh

Department of Environmental Toxicology
University of California
Davis, CA 95616

INTRODUCTION

Mycotoxins are a variety of highly toxic small molecules produced by many fungi growing in nature or under laboratory conditions. In a handbook compiled by Cole and Cox (1981), more than 200 of these toxic fungal products and their metabolites were described.

Some of these compounds have been associated with a number of well documented human and animal deseases (Rodricks et al., 1977). Others are potential foodborne toxicants in view of their occurrence in foodstuffs and their significant toxicity to bioassay systems. The alleged use of Fusarium toxins as warfare agents in recent years has escalated the significance of the mycotoxin problem to a new, high level (Watson et al., 1984). At present, not only the mycotoxins that occur naturally in foodstuffs are of concern, but those that can be produced only under controlled conditions are also a potential threat to human and animal health.

Among the more notorious mycotoxins that have been associated with human and animal mycotoxicoses are those listed in Table 1. The chemical structures of these prototype mycotoxins are shown in Fig. 1.

Aflatoxins such as aflatoxins B_1, G_1, and M_1 are potent hepatotoxins and hepatocarcinogens (Newberne and Rogers, 1981). The large class of trichothecene mycotoxins such as T2 toxin, nivalenol, and deoxynivalenol are potent dermatotoxins and immunotoxins that have caused very serious human mycotoxicosis problems in Russia known as alimentary toxic aleukia (Ueno, 1983). This class of mycotoxins are the ones claimed to be the active ingredients of the warfare agent, "yellow rain". Zearalenone is a potent estrogen that has caused estrogenic syndromes such as false heat and infertility in swine and other domestic animals (Mirocha et al., 1977). Ochratoxin A is a significant nephrotoxin that has caused chronic kidney damage to humans and animals (Krogh, 1977).

Table 1. Some mycotoxins of human and animal health significance.

Generic class	Representative compounds	Associated disease
Aflatoxins	Aflatoxins B_1, G_1, M_1	Liver cancer in humans and fish
Trichothecenes	T2 toxin, nivalenol, deoxynivalenol	Alimentary toxic aleukia "yellow rain" syndrome
Zearalenone	Zearalenone	"False heat" and intertility in swine
Ochratoxins	Ochratoxin A	Renal damage in humans and animals

Although mycotoxins are produced by fungi of many taxonomical classifications, most of the better known mycotoxins are produced by three genera: Aspergillus, Penicillium, and Fusarium. It is obvious from Fig. 1 that mycotoxins are diversified xenobiotics with different pharmacologies. Some of them require metabolic activation for their toxicity while others are direct acting toxins. Among the direct acting mycotoxins are trichothecenes and ochratoxins.

In this communication, only the mycotoxins that require metabolic activation to form reactive intermediates are of interest. In view of the limited volume, detail elaboration will only be focused on the relatively well-studied aflatoxin B_1 (AFB_1) to illustrate the formation, the biochemical reactions, and the toxicologic implications of the biological reactive intermediates of mycotoxins.

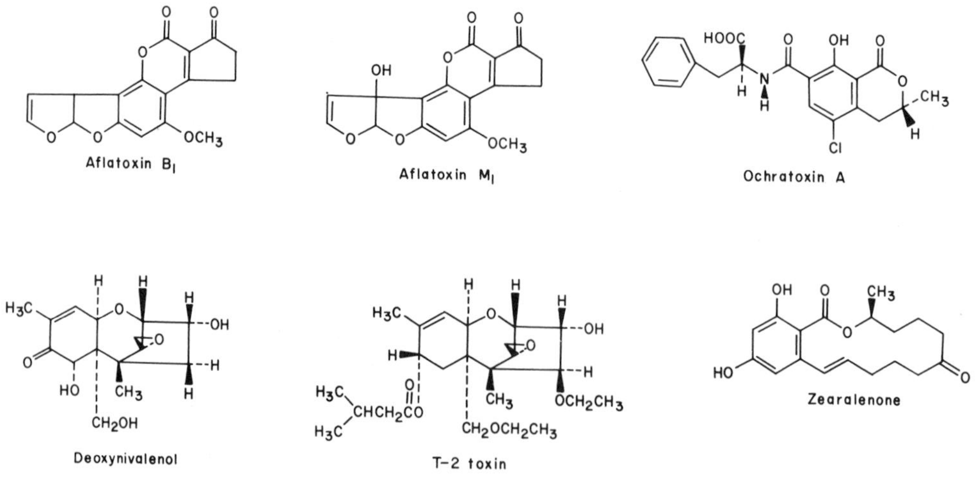

Fig. 1. Chemical structures of some proto-type mycotoxins.

AFB$_1$ is the most abundant and the most potent member of the aflatoxin family produced by <u>Aspergillus flavus</u> and <u>Aspergillus parasiticus</u>. It has been widely detected in the samples of corn, peanuts, cottonseeds and other grains, oilseeds, and tree nuts (Stoloff, 1977). It is highly hepatotoxic (Newberene and Rogers, 1981), immunotoxic (Pier et al., 1979), and has induced tumor formation in the liver of all the species of laboratory animals tested (Wogan, 1973). It is one of the most potent hepatocarcinogens known for the rat and rainbow trout.

FORMATION OF REACTIVE INTERMEDIATES

Formation of the biological reactive intermediates of mycotoxins involves primarily the phase 1 drug-metabolizing reactions: oxidation, reduction, and hydrolysis. These reactions provide the necessary chemical structures for phase 2 reactions, which are generally conjugations. The phase 1 metabolic transformations for AFB$_1$ are summarized in Fig. 2.

AFB$_1$ undergoes oxidative hydroxylation to form aflatoxins, M$_1$ and Q$_1$, O-demethylation to form aflatoxin P$_1$, and epoxidation to form the putative 8,9-oxide of AFB$_1$, or also known as AFB$_1$ epoxide (Hsieh et al., 1977). AFB$_1$ can also be reduced to aflatoxicol (Patterson and Roberts, 1971; Wong and Hsieh, 1978).

Except for the 8,9-oxide of AFB$_1$, all the other metabolites have been isolated and characterized. These metabolites do not appear to be the active forms of AFB$_1$, because in the Ames mutagenicity assay, they all require metabolic activation to be mutagenic, and the potency is considerably lower than that of the parent compound, AFB$_1$ (Wong and Hsieh, 1976).

The putative 8,9-oxide of AFB$_1$ (Swenson et al., 1974; 1975) is generally accepted as the active electrophilic form of AFB$_1$ which may attack nucleophilic nitrogen, oxygen, and sulfur heteroatoms in cellular constituents. Aflatoxicol, which is readily convertible back to AFB$_1$, is possibly serving as a reservoir of AFB$_1$ since its formation may prolong cellular exposure to the carcinogen (Patterson and Roberts, 1971, Patterson, 1973; Salhab and Edwards, 1977). Therefore, reduction of AFB$_1$ to aflatoxicol is not considered a detoxification process. This is in keeping with the relatively high mutagenic potency of aflatoxicol among the primary AFB$_1$ metabolites.

Aflatoxin M$_1$ is considerably less potent than AFB$_1$, though definitely mutagenic and carcinogenic (Green et al., 1982; Hsieh et al.; 1984). It is widely present in the excreta of animals exposed to AFB$_1$ (Stoloff, 1980). Its biotransmission to the milk of AFB$_1$-exposed dairy cattle has been a food safety problem of great concern.

Aflatoxin Q$_1$ possesses even lower mutagenic and carcinogenic potency (Wong and Hsieh, 1976; Hendricks et al., 1980a). The enzyme that converts B$_1$ to Q$_1$ is present abundantly in the primate livers (Masri et al., 1974; Hsieh et al., 1974; Buchi et al., 1974). However, Q$_1$ seems to be produced only under <u>in vitro</u> conditions and has not been found in any appreciable concentratinos in the excreta of AFB$_1$-exposed animals (Wong and Hsieh, 1980). Recently, Q$_1$ has been found to be as immunotoxic as AFB$_1$ (Pier, et al., 1985). In view of its relatively abundant formation in primates, its immunotoxicity warrants further examinations.

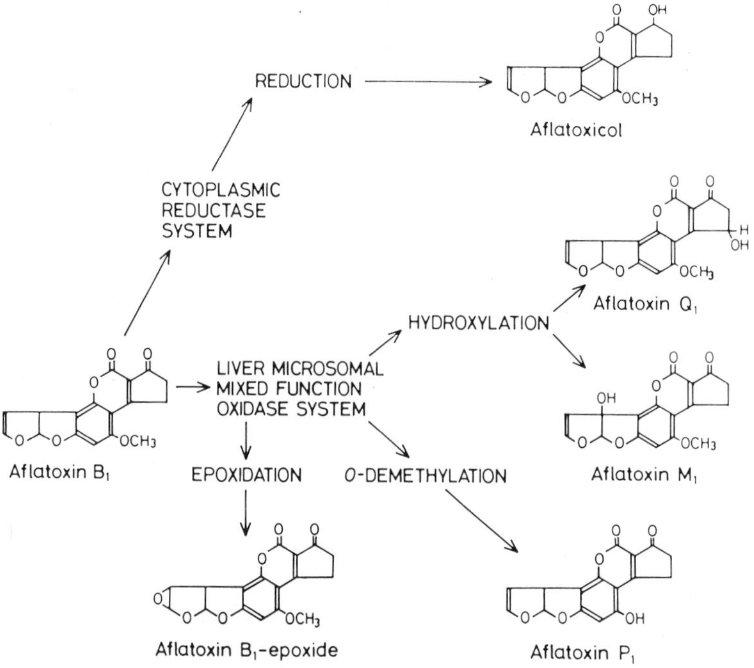

Fig. 2. The phase 1 metabolic transformations of AFB_1.

Aflatoxin P_1 is almost non-toxic and its occurrence in the excreta of animals orally administered AFB_1 is also at very low levels (Dalezios et al., 1973).

It should be noted that the 8,9-double bond in the bisfuran ring system of the AFB_1 molecule, which can form the electrophilic site responsible for the toxicity, is preserved in the primary metabolites of AFB_1 and yet their mutagenic and carcinogenic potencies are reduced markedly from that of the parent compound. Apparently, the substituted coumarin moiety of the AFB_1 molecule is an optimal structure for its toxicity; any modification at this moiety has resulted in a reduction of potency. These structural changes probably alter the bioavailability of the toxin at the receptor site and also its affinity to the receptor. This may explain why a lower mutagenic and carcinogenic potency is also observed of sterigmatocystin and versicolorin A (Wong et al., 1977; Hendricks et al., 1980b), both are bisfuranoid mycotoxins with fused xanthone and anthraquinone moieties, respectively.

The enzyme systems involved in these primary metabolic transformations of AFB_1 have been largely characterized. The activation of AFB_1, or the formation of its 8,9-epoxide, is mediated by a specific microsomal and nuclear cytochrome P-450 associated monooxygenase (Yoshizawa et al., 1981; Ueno et al., 1983a) Aflatoxins Q_1 and P_1 are formed by the action of different types of nonspecific cytochrome P-450 associated monooxygenases, whereas aflatoxin M_1 is formed by a specific cytochrome-P448 associated monooxygenase (Raina et al., 1983). The various monooxygenase enzymes are associated with the microsomal fraction of the liver cell. On the other hand, both the reductase and the dehydrogenase involved in the reversible transformation of AFB_1 to aflatoxicol are localized in the cytosol fraction (Salahab and Edwards, 1977).

Other mycotoxins are metabolized via similar phase 1 reactions to reactive metabolites, For example, emodin, a highly mutagenic monoanthraquinone toxin produced by some species of <u>Penicillium</u> and <u>Aspergillus,</u> is activated to the mutagenic 2-hydroxyemodin by a specific hepatic microsomal cytochrome P-450 monooxygenase (Wells et al., 1975). Two other hydroxylated metabolites, 5-hydroxy- and 4-hydroxyemodin, are much less potent promutagens than the parent compound. The structures of emodin and its metabolites are shown in Fig. 3.

On the other hand, zearalenone, the estrogenic mycotoxin produced by <u>Fusarium roseum</u> and other <u>Fusarium</u> species primarily in maize (Mirocha, et al., 1977), can be stereospecifically reduced by different forms of reductases to α- and β-zearalenol (Tashiro et al., 1980; Ueno et al., 1983b). The α-zearalenol, but not β-zearalenol, is ten times more estrogenic than the parent compound and appears to be the active form of zearalenone. The zearalenone reductases found in various tissues are mostly microsome associated (Tashiro et al., 1983). The partially purified enzymes from the rat hepatic microsomes are distinctly different from those of 3-hydroxysteroid dehydrogenase and other ketone reductases. The structures of zearalenone and its metabolites are shown in Fig. 4.

FORMATION OF BIOCHEMICAL LESIONS

The reactive intermediates of mycotoxins may react with molecular receptors to result in the formation of biochemical lesions. The common molecular receptors are cellular critical molecules such as DNA, RNA, functional proteins, enzyme cofactors, and membrane constituents. The reactions between mycotoxins and their molecular receptors may be covalent-irreversible or noncovalent-reversible.

<u>Noncovalently Bound Complex Formation</u>

Noncovalent, reversible interactions between proteins and mycotoxins have been observed in numerous examples of competitive inhibition of metabolic enzymes by various mycotoxins and the complex formation between mycotoxins and cellular constituents (Hsieh, 1979).

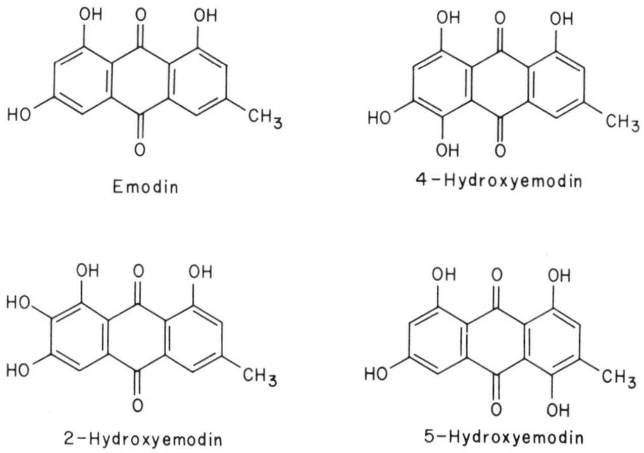

Fig. 3. Emodin and its metabolites.

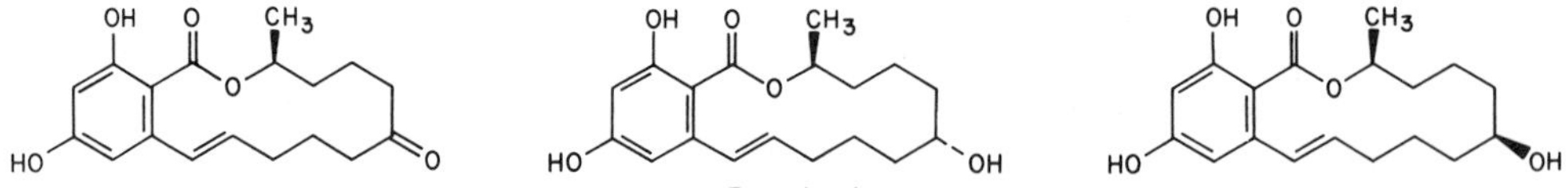

Fig. 4.   Zearalenone and its metabolites.

Aflatoxins B_1, G_1, and M_1 inhibit electron transport between cytochromes b and c or c1 (Site II) in rat liver mitochondria (Doherty and Campbell, 1972; 1973). In avian liver mitochondria, the inhibition occurs at Site I (between the substrate and cytochrome b) as well as Site II and is much more severe than in the rat mitochondria (Obidoa and Siddiqui, 1978). The biochemical effects of AFB_1 on liver mitochondria appear not to require metabolic activation.

Zearalenone possesses a high affinity for cytoplasm estrogen receptors (Tashiro et al., 1980). In rat, zearalenone is carried into the brain through the blood-brain barrier and binds with the estrogen receptors of both the hypothalamus and hypophysis (Kitagawa et al., 1982), suggesting that zearalenone affects the estrogen feedback system through the estrogen receptor of the rat brain.

Certain proteins such as plasma albumin which bind mycotoxins reversibly and noncovalently may also serve as carriers to transport reactive metabolites or as reservoirs of the toxins to stabilize and prolong cellular exposure to the toxins.

Covalently Bound Adduct Formation

Proteins are the most common receptors in the cell vulnerable to covalently bound adduct formation with reactive intermediates of mycotoxins because of their abundance as cellular constituents and the presence of nucleophilic N, O, and S heteroatoms in their functional groups.

Nonspecific-irreversible-covalent bindings to proteins by mycotoxins may cause conformational changes which denature or block the active sites for binding endogenous substrates and result in damage to various cellular structures and functions. Binding to heteroatoms away from the active site or to inert proteins, on the other hand, may represent a means by which the toxins are sequestered and deactivated. Therefore, covalent binding to proteins is not as directly correlated with toxicity as covalent binding to nucleic acids, especially DNA (Swenson et al., 1977; Rice and Hsieh, 1982).

As mentioned earlier, the AFB_1 epoxide is a reactive intermediate that may exert an electrophilic attack on the N-7 position of the guanyl residue in nucleic acids to result in the formation of DNA and RNA adducts of AFB_1 (Lin et al., 1977; Essigmann et al., 1977; Croy et al., 1978; Groopman et al., 1981). The AFB_1-guanine adduct at the N-7 position is so far the only identified adduct between this carcinogen and the nucleic acids. The structure of this adduct is shown in Fig. 5.

Another reactive intermediate of AFB_1 which is of relatively minor significance is the AFB_1 dihydrodiol, or 8,9-dihydro-8,9-dihydroxy AFB_1. The diol is formed by enzymatic or spontaneous hydrolysis of the AFB_1 epoxide (Neal and Colley, 1979; Decad et al., 1979; Ch'ih et al., 1983). Once formed, the dihydrodiol may undergo a structural rearrangement to form a putative dialdehyde phenolate intermediate which

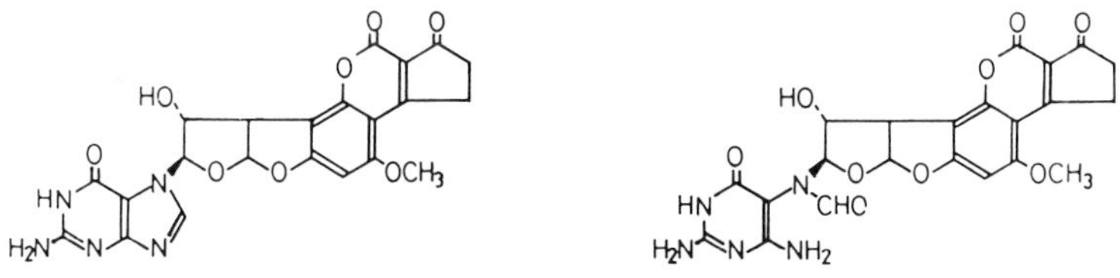

8,9-Dihydro-8-(N^7-guanyl)-
9-hydroxy-aflatoxin B_1

8,9-Dihydro-8-(N^5-formyl-
2',5',6'-triamino-4'-oxo-
N^5-pyrimidyl)-9-hydroxy-
aflatoxin B_1

Fig. 5.   Major guanyl adducts of AFB_1 derived from AFB_1-modified DNA.

can condense with the primary amino groups of proteins and other
cellular constituents to form the Schiff base (Neal and Colley, 1979),
as shown in Fig. 6.

The Schiff base formation represents an important mode of covalent
binding of AFB_1 to proteins. It has also been demonstrated that the
dihydrodiol binds to DNA in vitro and is a direct acting mutagen (Coles
et al., 1980). In addition, certain model esters of the dihydrodiol at
the 8-hydroxy position have been shown to form adducts with DNA, similar
to the action of the AFB_1 epoxide. This suggests that esterification at
the 8-hydroxyl of the dihydrodiol may be another mechanism of metabolic
activation for AFB_1 (Coles et al., 1980).

The relatively minor role of the dihydrodiol and its esters in the
modification of nucleic acids is probably due to the high reactivity of
the dihydrodiol which chemically reacts with proteins at the sites of
its formation, rendering it unavailable for further reactions (Neal and
Colley, 1979).

In addition to the Schiff bases formed between amino groups and the
dialdehyde intermediate of the dihydrodiol of AFB_1, lesions in proteins
may also arise from conjugation of protein sulfhydryl groups with the
electrophilic AFB_1 epoxide, as shown in Fig. 7.

The reactive intermediates of other mycotoxins presumably also form
covalently bound adducts with nucleic acids and proteins to result in
specific biochemical lesions that underlie their biological effects.
The biochemical lesions of other mycotoxins have not been characterized
in as much detail as are those of aflatoxins.

DETOXIFICATION OF REACTIVE INTERMEDIATES

Several types of Phase 2 reactions that lead to the detoxification
of mycotoxins involve conjugation to glucuronic acid, sulfate,
glutathione, and non-target proteins. The main detoxification mechanism
of AFB_1 is conjugation of the reactive metabolite, AFB_1 epoxide, with
glutathione as mediated by the glutathione-S-transferases (Degen and
Neumann, 1978; 1981), as shown in Fig. 8. Pretreatment with
phenobarbital and butylated hydroxytoluene, agents that induce liver

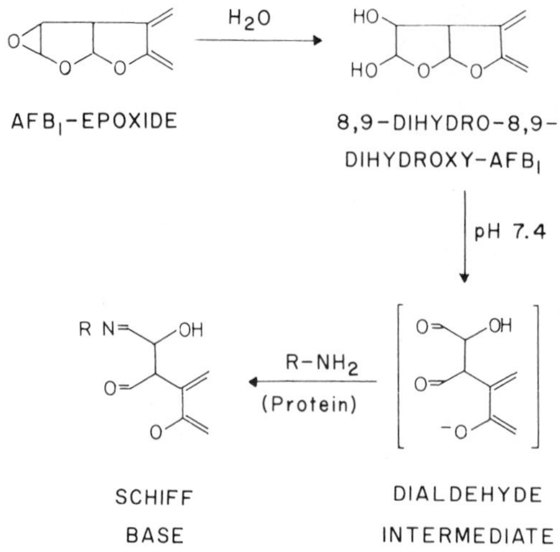

Fig. 6. Hydrolysis of the AFB$_1$ epoxide
and Schiff base formation.

glutathione-S-transferases, inhibits the ability of AFB$_1$ to bind to
hepatic nuclear DNA (Swenson et al., 1975; Salocks et al., 1981;
Fukayama and Hsieh, 1984) and hence reduces the carcinogenicity of
AFB$_1$. The glutathione conjugate of AFB$_1$ is primarily excreted through
the bile.

Another possible detoxification pathway for the AFB$_1$ epoxide is its
hydration to the dihydrodiol, followed by conjugation at one of the
hydroxyl groups with glucuronide and sulfate.

Alternatively, the active forms of AFB$_1$ may react with non-target
proteins and be sequestered and deactivated. Both the Schiff base
formation and conjugation of the protein sulfhydryl groups with the AFB$_1$
epoxide are conceivably significant detoxification mechanisms.

FATE AND IMPLICATIONS OF BIOCHEMICAL LESIONS

The biochemical lesions of mycotoxins can be removed from the cell
by a number of processes including spontaneous decomposition of the
complexes and adducts formed, receptor turnover, and specific repair
processes.

If lesions occurring on functional proteins, such as enzymes,
affect only their activity but not their biosynthesis, the effects can
be reversed as soon as the injured proteins are replaced by newly
synthesized proteins. Therefore, enzymes involved in the biosynthesis

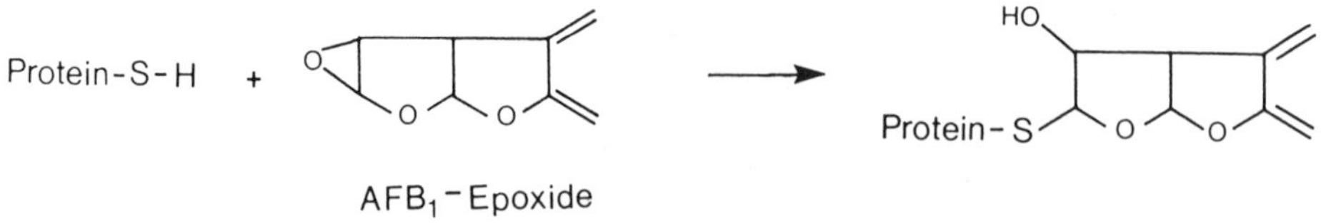

AFB$_1$-Epoxide

Fig. 7. Conjugation of the aflatoxin B$_1$ epoxide with glutathione.

604

of protein and nucleic acids are of particular significance as receptors of mycotoxins. Other crucial protein receptors have been found related to ATP production, immune mechanisms, neurotransmissions, hormone functions, and membrane transport. Lesions in these protein receptors may lead to cellular death or alterations of immune, hormonal, and other vital activities of affected animals.

Lesions occurring in DNA and RNA are of particular significance because alterations in the precise structures of DNA and RNA molecules by adduct formation will impair the template activity of these informational molecules and result either in inhibition of the synthesis of DNA, RNA, and proteins or in point mutations leading to the synthesis of erroneous macromolecules (Edwards and Wogan, 1970; Yu, 1977). The consequences of adduct formation in nucleic acids in vivo may be cellular death or cellular transformation to cancerous conditions.

AFB_1 has been extensively used as a model hepatocarcinogen to elucidate the mechanism of the initiation process of carcinogenesis in the liver. Studies of the specific adduction of the epoxide of AFB_1 at the N-7 position of the guanyl residue in DNA and the consequences of this chemical lesion have shed some light on the mechanism of mutagenesis and carcinogenesis. As mentioned earlier, the principle adduct of DNA and AFB_1, upon acid hydrolysis, is 8,9-dihydro-9-hydroxy-8-(N7-guanyl)-AFB_1(Fig. 5). AFB_1-guanyl residue in DNA undergoes two fate processess: a large portion of the N-7 adducts are readily removed spontaneously or enzymatically to give rise to apurinic sites in the DNA, whereas the remaining portion is transformed to a persistent formamidopyrimidine derivative, which upon acid hydrolysis yields 8,9-dihydro-8-(N5-formyl-2',5',6' triamino-4, oxo-N5-pyrimidyl)-9-hydroxy AFB_1 (Fig. 5) (Croy and Wogan, 1981a; 1981b). The former process can lead to mutation while the latter process may arrest DNA replication or also lead to mutation.

The two lesions differ in their mechanisms of action. For apurinic sites, it has been established that there is a strong preference for adenine insertion opposite an apurinic site in double-stranded DNA (Foster et al., 1983). Therefore, depurination at guanine residues could lead specifically to a G-C to T-A transversion during replication of the modified DNA (Schaaper et al., 1983) and will result in a base-pair substitution mutation of the DNA. This type of mutation has been observed as a mechanism for the activation of cellular proto-oncogenes. The frameshift mutagenicity of AFB_1 is less pronounced than its base-pair substitution activity (Stark, 1980). Single-strand breaks occurring at aguaninic sites are believed to be responsible for the frameshift mutagenicity of AFB_1 (D'Andrea and Haseltine, 1978).

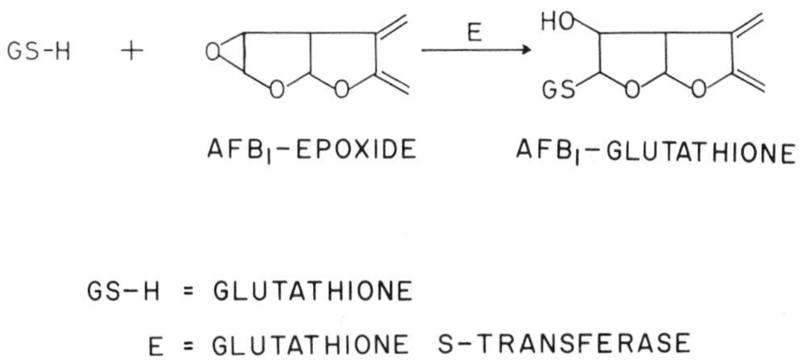

AFB₁-EPOXIDE   AFB₁-GLUTATHIONE

GS-H = GLUTATHIONE

E = GLUTATHIONE S-TRANSFERASE

Fig. 8. Conjugation of aflatoxin B₁ with protein
        sulhydryl groups.

On the other hand, the formation of the formamidopyrimidine derivative as a repair-resistant lesion in the DNA may result in point mutation or arrest of DNA synthesis.

In addition to the ability of these genetic changes to initiate the carcinogenesis in the cell, the various biochemical lesions formed by the reactive intermediates of AFB_1 can produce cytotoxicity and immunotoxicity, and alter the hormonal activities in animals. These toxicities are consistent with the promotional activity of an epigenetip carcinogen. All these activities of the reactive intermediates of AFB_1 makes it an extremely potent and complete hepatocarcinogen for certain animal models.

Zearalenone, on the other hand, exhibits its physiological activity by binding with cytosolic estrogen-receptor protein in a fashion similar to the action of steroid hormone. It possesses a high affinity for cytoplasm estrogen receptors (Tashiro et al., 1980). Upon translocation to the nuclei, the zearalenone-estrogen-receptor complex is believed to stimulate protein biosynthesis, especially the synthesis of specific proteins (Kawabata et al., 1982) and nuclear RNA polymerase I and II. In mice, zearalenone increases uterine permeability to uridine and amino acids, and such changes were followed by an increase of protein and nucleic acid synthesis, in a manner similar to the natural estrogen E2 (Ueno and Yagasaki, 1970).

SUMMARY AND CONCLUSIONS

In summary, the various reactions involved in the formation and actions of biological reactive intermediates of AFB_1 are undoubtedly common to the wide variety of mycotoxins. The structural diversity of mycotoxins and their widespread occurrence in nature has made them a significant class of model environmental toxicants for mode of action studies. Identification of the reactive intermediates of prototype mycotoxins and elucidation of their reactions with molecular receptors will provide information useful for the design of antidotes of other remedial measures and for a better understanding of the mechanisms of action of comparable environmental toxicants. This type of studies will also permit the more precise biochemical analyses to be applied to the assessment of health risk due to exposure to mycotoxins.

REFERENCES

Buchi, G. H., Muller, P. M., Roebuck, B. D., and Wogan, G. N., 1974, Aflatoxin Q1: a major metabolite of aflatoxin B1 produced by human liver, Res. Commun. Chem. Pathol. Pharmacol., 8:585.

Ch'ih, J. J., Lin T. and Devlin, T. M., 1983, Activation and deactivation of aflatoxin B1 in isolated rat hepatocytes, Biochem. Biophys. Res. Comm., 110:668.

Cole, R. J. and Cox, R. H., 1981, "Handbook of Toxic Fungal Metabolites", Academic Press, New York.

Coles, B. F., Welch, A. M., Hertzog, P. J., Lindsay Smith, J. R. and Garner, R. C., 1980, Biological and chemical studies on 8,9-dihdroxy-8,9-dihydro aflatoxin B1 and some of its esters, Carcinogene., 1:79.

Croy, R. G. and Wogan, G. N., 1981a, Temporal patterns of covalent DNA adducts in rat liver after single and multiple doses of aflatoxin B1, Cancer Res., 41:197.

Croy, R. G. and Wogan, G. N., 1981b, Quantitative comparison of covalent aflatoxin-DNA adducts formed in rat and mouse livers and kidneys, J. Natl. Canc. Inst., 66:761.

Croy, R. G., Essigman, J. M., Reinhold, V. N. and Wogan, G. N., 1978, Identification of the principal aflatoxin B1-DNA adduct formed in vitro in rat liver, Proc. Natl. Acad. Sci., USA, 75:1745.

Dalezios, J. I., Hsieh, D. P. H. and Wogan, G. N., 1973, Excretion and metabolism of orally administered aflatoxin B1 by rhesus monkeys, Fd Cosmet. Toxicol., 11:605.

D'Andrea, A. D. and Haseltine, W. H., 1978, Modification of DNA by AFB1 creates alkali-labile lesions in DNA at positions of guanine and adenine, Proc. Natl. Acad. Sci. USA, 75:4120.

Decad, G. M., Dougherty, K. K., Hsieh, D. P. H. and Byard, J. L., 1979, Metabolism of aflatoxin B1 in cultured mouse hepatocytes: Comparison with rat and effects of cyclohexene oxide and diethyl maleate, Toxicol. Appl. Pharmacol., 50:429.

Degen, G. H. and Neumann, H.-G., 1978, The major metabolite of aflatoxin B1 in the rat is a glutathione conjugate, Chem-Biol. Interact., 22:239.

Degen, G. H. and Neumann, H.-G., 1981, Differences in aflatoxin B1-susceptibility of rat and mouse are correlated with the capability in vitro to inactivate aflatoxin B1-epoxide, Carcinogene., 2:299.

Doherty, W. P. and Campbell, T. C., 1972, Inhibition of rat liver mitochondria electron transport flow by aflatoxin B1, Res. Commun. Chem. Pathol. Pharmacol., 3:601.

Doherty, W. P. and Campbell, T. C., 1973, Aflatoxin inhibition of rat liver mitochondria, Chem-Biol Interact., 7:63.

Edwards, G. S. and Wogan, G. N., 1970, Aflatoxin inhibition of template activity of rat liver chromation, Biochem. Biophys. Acta, 224:597.

Essigmann, J. M., Croy, R. J., Nadzan, A. M., Busby, W. F., Reinhold, Jr., V. N., Buchi, G. and Wogan, G. N., 1977, Structural identification of the major DNA adduct formed by aflatoxin B1 in vitro, Proc. Natl. Acad. Sci., USA, 74:1870.

Foster, P. L., Eisenstadt, E., and Miller, J. H., 1983, Base substitution mutations induced by metabolically activated AFB1, Proc. Natl. Acad. Sci. U. S., 80:2695.

Fukayama, M. Y. and Hsieh, D. P. H. 1984, Effects of butylated hydroxytoluene on the in vitro metabolism, DNA binding and mutagenicity of aflatoxin B1 in the rat, Fd. Chem. Toxicol., 22:355.

Green, C. E., Rice, D. W., Hsieh, D. P. H. and Byard, J. L., 1982, The comparative metabolisim and toxic potency of aflatoxin B1 and aflatoxin M1 in primary cultures of adult-rat hepatocytes, Fd. Chem. Toxicol., 20:53.

Groopman, J. D., Croy, R. G. and Wogan, G. N., 1981, In vitro reactions of aflatoxin B1-adducted DNA, Proc. Natl. Acad. Sci., USA, 78:5445.

Hendricks, J. D., Sinnhuber, R. O., Wales, J. H., Stack, M. E. and Hsieh, D. P. H., 1980a, Carcinogenic response of rainbow trout (Salmo gairdineri) to aflatoxin Q1 and synergistic effect of cyclopropenoid fatty acids, J. Natl. Canc. Inst., 64:1503.

Hendricks, J. D., Sinnhuber, R. O., Wales, J. H., Stack, M. E., and Hsieh, D. P. H., 1980b, Hepatocarcinogenicity of sterigmatocystin and versicolorin A to rainbow trout (Salmo gairdneri) embryos, J. Natl. Canc. Inst., 64:1503.

Hsieh, D. P. H., 1979, Basic metabolic effects of Mycotoxins, in: "Interactions of Mycotoxins in Animal Production", National Academy of Sciences, Washington, D. C.

Hsieh, D. P. H., Dalezios, J. I., Krieger, R. I., Masri, M. S. and Haddon, W. F., 1974, Use of monkey liver microsomes in production of aflatoxin Q1, J. Agric. Fd. Chem., 22:515.

Hsieh, D. P. H., Wong, Z. A., Wong, J. J., Michas, C. and Ruebner, B. H., 1977, Comparative metabolism of aflatoxin, in: "Mycotoxins in Human and Animal Health", J. V. Rodricks, C. W. Hesseltine and M.

A. Mehlman, eds., Pathotox, Park Forest South, IL.

Hsieh, D. P. H., Cullen, J. M., and Ruebner, B. H., 1984, Comparative hepatocarcinogenicity of aflatoxins B1 and M1 in the rat, Fd. Chem. Toxicol., 22:1027.

Kawabata, Y. Tashiro, F. and Ueno, Y., 1982, Synthesis of a specific protein induced by zearalenone and its derivatives in rat uterus, J. Biochem. (Tokyo), 91:801.

Kitagawa, M., Tashiro, F. and Ueno. Y., 1982, Interaction between zearalenone, an estrogenic mycotoxin, and the estrogen receptor of the rat brain, Proc. Jpn. Assoc. Mycotoxicol., 15:28.

Krogh, P., 1977, Ochratoxins, in: "Mycotoxins in Human and Animal Health", J. V. Rodricks, C. W. Hesseltine and M. A. Mehlman, eds., Pathotox, Park Forest South, IL.

Lin, J., Miller, J. A. and Miller, E. C., 1977, 2,3-Dihydro-2-(guan-7-yl)-3-hydroxy-aflatoxin B1, a major acid hydrolysis product of aflatoxin B1-DNA or ribosomal RNA adducts formed in hepatic microsome-mediated reactions and in rat liver in vivo, Cancer Res., 37:4430.

Masri, M. S., Haddon, W. F., Lundin, R. E. and Hsieh, D. P. H., 1974, Aflatoxin Q1. A newly identified major metabolite of aflatoxin B1 in monkey liver, J. Agric. Fd. Chem., 22:512.

Mirocha, C. J., Pathre, S. V., Christensen, C. M., 1977, Zearalenone, in: "Mycotoxins in Human and Animal Health", J. V. Rodricks, C. W. Hesseltine and M. A. Mehlman, eds., Pathotox, Park Forest South, IL.

Neal, G. E. and Colley, P. J., 1979, The formation of 2,3-dihydro-2,3-dihydroxy aflatoxin B1 by the metabolism of aflatoxin B1 in vitro by rat liver microsomes, FEBS Lett., 101:382.

Newberne, P. M. and Rogers, A. E., 1981, Animal toxicity of major environmental mycotoxins, in: "Mycotoxins and N-Nitroso Compounds: Environmental Risk", R. C. Shank, ed., CRC Press, Boca Raton, Florida.

Obidoa, O. and Siddiqui, H. T., 1978, Aflatoxin inhibition of avian hepatic mitochondria, Biochem. Pharmacol., 27:547.

Patterson, D. S. P., 1973, Metabolism as a factor in determining the toxic action of the aflatoxins in different animal species, Fd. Cosmet. Toxicol., 11:287.

Patterson, D. S. P. and Roberts, B. A., 1971, The in vitro reduction of aflatoxins B1 and B2 by soluble avian liver enzymes, Fd. Cosmet. Toxicol., 9:829.

Pier, A. C., Richard, J. L., and Thurston, J. R., 1979, The influences of mycotoxins on resistance and immunity, in: "Interactions of Mycotoxins in Animal Production", National Academy of Sciences, Washington, D. C.

Pier, A. C., Varman, M. J., and Belden, E. L., 1985, Aflatoxin suppression of cell mediated immune responses and interaction with T-2 toxin, in: "6th International Symposium on Mycotoxins and Phytotoxins", Abstract, S 27, IUPAC, Pretoria, Republic of South Africa.

Raina, V., Williams, C. J. and Gurtoo, H. L., 1983, Genetic expression of aflatoxin metabolism. Effects of 3-methylcholanthrene and β-naphtoflavone on hepatic microsomal metabolism and mutagenic activities of aflatoxins, Biochem. Pharmacol., 32:3755.

Rice, D. W. and Hsieh, D. P. H., 1982, Aflatoxin M1: In vitro preparation and comparative in vitro metabolism versus aflatoxin B1 in the rat and mouse, Res. Comm. Chem. Path. Pharmacol., 35:467.

Rodricks, J. V., Hesseltine, C. W., and Mehlman, M. A., 1977, "Mycotoxins in Human and Animal Health", Pathotox, Park Forest

South, Illinois.

Salhab, A. S. and Edwards, G. S., 1977, Comparative _in vitro_ metabolism of aflatoxicol by liver preparations from animals and humans, Cancer Res., 37:1016.

Salocks, C. B., Hsieh, D. P. H. and Byard, J. L., 1981, Butylated hydroxytoluene pretreatment protects against cytotoxicity and reduces covalent binding of aflatoxin B1 in primary hepatocyte cultures, Toxicol. Apl. Pharmacol., 59:331.

Schaaper, R. M., Kunkel, T. A. and Loeb, L. A., 1983, Infidelity of DNA synthesis associated with bypass of apurinic sites, Proc. Natl. Acad. Sci., USA, 80:487.

Stark, A. A., 1980, Mutagenicity and Carcinogenicity of mycotoxins: DNA binding as a possible mode of action, Ann. Rev. Microbiol., 34:235.

Stoloff, L., 1977, Aflatoxins-An overview, in: "Mycotoxins in Human and Animal Health", J. V. Rodricks, C. W. Hesseltine and M. A. Mehlman, eds., Pathotox, Park Forest South, IL.

Stoloff, L., 1980, Aflatoxin M1 in perspective, J. Fd. Prot., 43:226.

Swenson, D. H., Miller, J. A. and Miller, E. C., 1974, Aflatoxin B1-2,3-oxide: Evidence for its formation in rat liver _in vivo_ and by human liver microsomes _in vitro_., Biochem. and Biophys., Res. Comm., 60:1036.

Swenson, D. H., Miller, J. A. and Miller, E. C., 1975, The reactivity and carcinogenicity of aflatoxin B1-2,3-dichloride, a model for the putative 2,3-oxide metabolite of aflatoxin B1, Cancer Res., 35:3811.

Swenson, D. H., Lin, J., Miller, E. C., and Miller, J. A., 1977, Aflatoxin 2,3-oxide as a probable intermediate in the covalent binding of AFB1 and AFB2 to rat liver DNA and ribosomal RNA _in vivo_, Cancer Res., 37:172.

Tashiro, F., Kawabata, Y., Naoi, M. and Ueno, Y., 1980, Zearalenone estrogen receptor interaction and RNA synthesis in rat uterus. in: "Medical Mycology", H. L. Preusser, ed., Gustav Fisher Verlag, New York.

Tashiro, F., Shibata, A., Nishimura, N. and Ueno, Y., 1983, Zearalenone reductase from rat liver. J. Biochem. (Tokyo), 93:1557.

Ueno, Y., 1983, "Trichothecenes: Chemical, Biological and Toxicological Aspects", Elsevier, New York.

Ueno, Y., Ishii, K., Omata, Y., Kamataki, T. and Kato, R., 1983a, Specificity of hepatic cytochrome P-450 isoenzymes from PCB-treated rats and participation of cytochrome b_5 in the activation of aflatoxin B1, Carcinogene., 4:1071.

Ueno, Y., Tashiro, F. and Kobayashi, T., 1983b, Species difference on zearalenone-reductase activity, Fd. Chem. Toxicol., 21:167.

Watson, S. A., Mirocha, C. J. and Hayes, A. W., 1984, Analysis for trichothecenes in samples from Southeast Asia associated with "yellow rain", Fund. Appl. Toxicol., 4:700.

Wells, J. M., Cole, R. J. and Kirskey, J. W., 1975, Emodin, a toxic metabolite of Aspergillus wentii isolated from weevil-damaged chestnuts, Appl. Microbiol., 30:26.

Wogan, G. N., 1973, Aflatoxin carcinogenesis, in: "Methods in Cancer Research", H. Busch, ed., Academic Press, New York.

Wong, J. J. and Hsieh, D. P. H., 1976, Mutagenicity of aflatoxins related to their metabolism and carcinogenic potential, Proc. Natl. Acad. Sci., USA, 73:2241.

Wong, J. J., Singh, R. and Hsieh, D. P. H., 1977, Mutagenicity of fungal metabolites related to aflatoxin biosynthesis, Mutat. Res., 44:447.

Wong, Z. A. and Hsieh, D. P. H., 1978, Aflatoxicol: major aflatoxin B1 metabolite in rat plasma, Science, 200:325.

Wong, Z. A. and Hsieh, D. P. H., 1980, The comparative toxicokinetics of aflatoxin B1 in the monkey, rat and mouse, Toxicol. Appl. Pharmacol., 55:115.

Yoshizawa, H., Uchimaru, R. and Ueno, Y., 1981, Metabolism of aflatoxin B1 in the nuclei isolated from rat liver, J. Biochem., Tokyo, 89:443.

Yu, F.-L., 1977, Mechanism of aflatoxin B1 inhibition of rat hepatic nuclear RNA synthesis. J. Biol. Chem., 252:3245.

BIOLOGICAL REACTIVE INTERMEDIATES OF PYRROLIZIDINE ALKALOIDS

Donald R. Buhler and Bogdan Kedzierski

Department of Agricultural Chemistry and
Environmental Health Sciences Center
Oregon State University
Corvallis, Oregon

INTRODUCTION

The hepatotoxic and carcinogenic pyrrolizidine alkaloids (PAs) present in Senecio and a number of other plant genera are responsible for the poisoning of humans and animals in many parts of the world (Hirono, 1981; Huxtable, 1979; McLean, 1970). Fatalities or severe cases of poisoning in humans have resulted from the consumption of grain contaminated by PA containing seeds or by drinking supposedly therapeutic herbal teas that contain PAs (Hirono, 1981; Huxtable, 1980; McLean, 1970). There is also increasing evidence that the PAs are excreted in toxic form in the milk of cows and goats consuming PA containing plants such as tansy ragwort (Senecio jacobaea) (Dickinson et al., 1976; Luthy et al., 1983; Miranda et al., 1981; Deinzer et al., 1982).

Over 200 PAs have been isolated and their structures identified. Many of the more toxic PAs are alicyclic esters of the necine base retronecine (I) or its enantiomer heliotridine (II). Toxicities of the PAs vary according to the nature of the esterified necic acids. PAs can be mono-esters, diesters or closed ring diesters, the latter being generally the most hepatotoxic and carcinogenic of the PAs. Senecionine (III) is a typical example of the closed ring macrocyclic ester type of PA.

The PAs are not toxic per se but are biotransformed in vivo, primarily by the liver, to highly reactive metabolites that are responsible for the tissue damage. Culvenor and his associates have suggested (1962) that toxicity of PAs resulted from the alkylation of cellular constituents. Mattocks (1968, 1972a, 1972b) then showed a correlation between pyrrole formation and PA toxicity. The PAs are now known to be metabolized by cytochrome P-450 monooxygenases through dehydrogenation of the pyrrolizidine nucleus to form highly reactive pyrrole type metabolites (Guengerich and Liebler, 1985; Hirono, 1981; Peterson and Culvenor, 1983). The alkaloids are thought to be initially converted to unstable dehydroalkaloids (PA pyrroles), intermediates that are potent electrophiles capable of either alkylating cellular nucleophiles or undergoing rapid hydrolysis to the more stable necine pyrrole and the corresponding nececic acids. Dehydroretronecine [(R)-6,7-dihydro-7-hydroxy-1-hydroxymethyl-5H-pyrrolizine; (R)-DHP], the alleged necine pyrrole product of the hydrolysis of retronecine-based dehydroalkaloids, has also been shown to be a potent electrophile capable of covalently binding to tissue constituents, including DNA (Fennell et al., 1985; Robertson, 1982; Wickramanayaki et al., 1985). Contrary to the previously accepted belief that configuration of the necine pyrrole was retained, we have recently found that the major microsomal metabolite of the PAs is the racemic DHP (IV) rather than its R or S enantiomers (Kedzierski and Buhler, 1985). These results indicate that hydrolysis of the putative dehydroalkaloid proceeds with fission of the C7-O bond and formation of a carbonium ion at C7 according to a S_N1 mechanism. Other important pathways for PA metabolism include N-oxidation and hydrolysis to the corresponding necine bases (Peterson and Culvenor, 1983). The major pathways for metabolism of retronecine-type PAs such as senecionine (III) are summarized schematically in Figure 1.

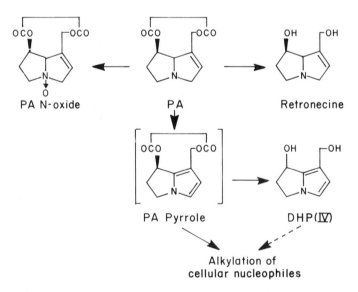

Figure 1.   Major metabolism pathways for macrocyclic retronecine-type pyrrolizidine alkaloids.

In this study we report the development of a new high performance liquid chromatography (HPLC) system for determination of the PAs and their major microsomal metabolites. This system was then utilized to study the in vitro metabolism of eight different PAs and to investigate the mechanism for DHP formation.

MATERIALS AND METHODS

The PAs senecionine (III; Sn), seneciphylline, jacobine, jacoline and retroresine were isolated from Senecio jacobaea extracts by preparative reversed-phase HPLC on a RP-8 column eluted with potassium phosphate buffer (pH 6.1) and methanol. Monocrotaline was obtained commercially from the Trans World Chemical Co., Washington, DC; lasiocarpine and heliotrine were generously provided by C.C.J. Culvenor, Parkville, Australia. Standards of the PA N-oxides and (R)-DHP were prepared from the various PAs and monocrotaline, respectively (Culvenor et al., 1970; Mattocks, 1969). ^{18}O-Water (97.92%) was purchased from Prochem/Isotopes, Summit, NJ.

Livers were obtained from untreated male Sprague-Dawley rats or males pretreated with phenobarbital (PB) or β-naphthoflavone (BNF), 80 mg/kg i.p. for three days or from untreated male and female Swiss-Webster mice; the livers were homogenized in 10 mM potassium phosphate buffer, pH 7.4 containing 1.15% KCl; and microsomes were then prepared. Incubation mixtures contained in a final volume of 1 ml: 100 mM phosphate buffer (pH 7.6), 1 mM PA, 1 mM NADP$^+$, 5 mM glucose-6-phosphate, 5 mM MgCl$_2$, 1 unit glucose-6-phosphate dehydrogenase, and 0.5 mg microsomal protein. Incubations, initiated by addition of microsomes, were for 15 min at 37°C. The chilled reaction mixtures were then centrifuged either at 105,000 g or, more conveniently, at 46,000 g for 30 min. Aliquots of the resulting supernatants were next injected directly onto a PRP-1 reversed-phase styrene-divinylbenzene column and eluted with an acetonitrile--0.1 M NH$_4$OH gradient (Ramsdell and Buhler, 1981). Detection was by uv absorption at 220 nm.

Electron impact mass spectra were obtained with a Finnigan Model 4023 quadrupole mass spectrometer. Temperature of the ion source was 190°C, and samples were introduced by the probe inlet. Circular dichroism (CD) spectra were measured on a Jasco Model 41A instrument as described previously (Kedzierski and Buhler, 1985).

RESULTS

PAs and their major in vitro metabolites could be readily separated by HPLC on a PRP-1 column eluted with an acetonitrile--0.1 M NH$_4$OH gradient. Chromatograms of supernatants from blank incubations performed without added PAs, showed no uv absorbing peaks at retention times greater than 10 min, therefore, the PAs and their metabolites could be readily detected. Analysis of a typical incubation mixture of Sn with hepatic microsomes from PB pretreated rats is shown in Figure 2. Two major metabolite peaks labeled DHP and SnNO co-chromatographed with authentic (R)-DHP and senecionine N-oxide (SnNO), respectively. The mass spectra of the two Sn metabolites isolated following HPLC separation were also identical to those of authentic (R)-DHP and SnNO. The four smaller uv absorbing peaks shown in Figure 2 (marked with astericks) are as yet unidentified metabolites. Comparable results were found when Sn was incubated with male or female mouse liver microsomes. Microsomal incubation mixtures from other PAs gave similar results, showing prominent DHP peaks and varying amounts of the corresponding N-oxides and/or other unidentified metabolites.

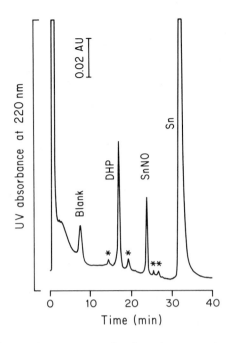

Figure 2. HPLC analysis of Sn metabolites
after the in vitro incubation
of Sn with rat liver microsomes
from PB induced rats.

With standard curves prepared from authentic (R)-DHP and the corresponding PA N-oxides, the HPLC assay was used to quantitate the microsomal
metabolism of eight different PAs. A marked difference was observed in the
metabolism to DHP by the various PAs, with lasiocarpine exhibiting the
highest rates of pyrrole formation and monocrotaline the lowest (Table 1).
DHP levels were highest with microsomes from PB induced rats whereas
microsomes from BNF pretreated animals caused only a slight increase in DHP
production.

Table 1. Microsomal Metabolism of Pyrrolizidine Alkaloids. DHP Formation.

| Alkaloid | DHP (nmol/min/mg protein)[a] | | |
	Control	PB	BNF
Lasiocarpine	7.05 ± 1.23	25.9 ± 3.5	9.45 ± 3.61
Senecionine	3.90 ± 0.90	16.1 ± 2.0	4.60 ± 1.92
Jacobine	2.80 ± 0.37	14.2 ± 1.2	3.40 ± 0.71
Retrorsine	2.90 ± 0.70	9.60 ± 1.05	3.35 ± 0.82
Seneciphylline	2.35 ± 0.57	10.7 ± 1.5	3.60 ± 2.20
Jacoline	0.60 ± 0.16	3.55 ± 0.41	0.75 ± 0.19
Heliotrine	0.35 ± 0.10	1.75 ± 0.10	0.50 ± 0.20
Monocrotaline	0.20[b]	1.15 ± 0.10	0.20[b]

[a] Means of incubations from 4 animals ± S.D.
[b] Detection limits of the method.

Formation of PA N-oxides was seen with only three PAs (Table 2), apparently because of limitations in the detection limits of the HPLC method used. Somewhat greater quantities of PA N-oxides (Table 2) than of DHP (Table 1) were formed in the oxidation of Sn, retrorsine and seneciphylline by control microsomes. Treatment of the rats with either PB or BNF, however, significantly reduced conversion of the PAs to their N-oxide metabolites.

Table 2. Microsomal Metabolism of Pyrrolizidine Alkaloids.
PA N-Oxide Formation.

Alkaloid	PA N-oxide (nmol/min/mg protein)[a]		
	Control	PB	BNF
Senecionine	8.05 ± 1.86	6.50 ± 0.90	3.20 ± 1.57
Retrorsine	3.65 ± 0.55	3.40 ± 0.28	1.65 ± 0.41
Seneciphylline	2.95 ± 0.55	4.75 ± 0.52	2.15 ± 0.87

[a] Means of incubations from 4 animals ± S.D.

Metabolism of PAs to DHP and PA N-oxides was significantly inhibited by SKF-525A or metyrapone, suggesting that rat cytochrome P-450 isozymes may be involved in both oxidations (Table 3). To consider the possibility that the microsomal flavin-containing monooxygenase (FMO) (Ziegler, 1984) may be involved in the N-oxidation of PAs, additional studies were performed. No clear cut evidence was obtained, however, for the involvement of the FMO system in the metabolism of PAs to their N-oxides since both DHP and SnNO formation were influenced similarly by added n-octylamine, higher pH, or heat pretreatment of microsomes.

Table 3. Effects of Inhibitors and Temperature Pretreatment on the Microsomal Oxidation of Senecionine

Treatment/Inhibitor	% of Control Activity			
	pH 7.6		pH 8.4	
	DHP	SnNO	DHP	SnNO
SKF 525A (1 mM)	5.7	13.3	8.5	21.7
Metyrapone (1 mM)	4.3	43.9	4.9	47.4
N-Octylamine (1 mM)	44.3	41.3	46.3	38.3
Temperature pretreatment[a]	63.6	65.2	83.3	70.7

[a] Microsomes pretreated at 50°C for 1 min.

Following the in vitro metabolism of Sn by rat hepatic microsomes performed in the presence of $H_2^{18}O$, the major pyrrolic metabolite DHP was isolated by preparative HPLC and examined by mass spectrometry. Both the C7 and C9 oxygens of DHP were found to come from the solvent (molecular ion of DHP m/z 157 for the $H_2^{18}O$ incubation as compared to m/z 153 for the $H_2^{16}O$ experiment). These results indicate that hydrolysis of the proposed intermediate dehydrosenecionine is accomplished by alkyl-oxygen fission of both the C7-O and C9-O bonds. To prove that DHP already formed did not exchange hydroxyl groups with the solvent during microsomal incubations and subsequent isolation procedures, in vitro incubations were conducted using synthetic (R)-DHP as the substrate. The (R)-DHP, isolated after the incubation, showed a Cotton effect in the CD spectrum, as described earlier (Kedzierski and Buhler, 1985), confirming the configurational stability (hence the lack of exchange) of DHP to both incubation and isolation conditions.

DISCUSSION

Using the PRP-1 HPLC method described, Sn was shown to be metabolized by rat and mouse liver microsomes to both DHP (IV) and SnNO. DHP was also a major microsomal metabolite of the seven other PAs studied. Mattocks and Bird (1983) using a colorimetric assay also found that Sn was oxidized by microsomes to both the necine pyrrole (DHP) and SnNO, although the necine pyrrole was the predominate metabolite in their experiments. (R)-DHP or (S)-DHP were also frequently reported metabolites in many other in vivo and in vitro studies (Hirono, 1981; Huxtable, 1979; Peterson and Culvenor, 1983). However, in recent studies by Segall and co-workers (Eastman and Segall, 1982; Segall et al., 1984) that also employed an improved HPLC method for the examination of in vitro PA metabolism mixtures, no traces of DHP were seen in the microsomal metabolism of Sn. In fact, the only pyrroles detected after the oxidation of Sn by mouse hepatic microsomes were small quantities of 7-methoxy DHP and an aldehyde formed from DHP through oxidation at the C9' position (Segall et al., 1984). The reasons for this discrepancy are not readily apparent.

As observed previously by others (Mattocks and Bird, 1983), microsomal conversion of PAs to pyrolic metabolites varied appreciably (by more than 30-fold) between the PAs tested (Table 1). Formation of DHP from lasio-carpine was the highest while monocrotaline showed the lowest conversion to DHP. There was a generally good correlation between the acute toxicities of the various PAs in rats (Bull et al., 1968) and microsomal DHP produc-tion (Table 1), giving further evidence for the central role of pyrroles in PA toxicity. A similar relationship between acute hepatotoxicity of PAs and the quantities of pyrrolic metabolites found in 16 livers of rats given the alkaloids has been noted by Mattocks (1972b).

The marked variation in DHP concentrations seen in Table 1 could result from differences in the rates of formation of the dehydroalkaloid intermediates or they could be a reflection of a differential stability of the different dehydroalkaloids. That the latter is not the case may be concluded from the studies of Karchesy and Deinzer (1981). These investi-gators, studying the kinetics of alkylation of 4-p-nitrobenzylpyridine by various chemically prepared dehydroalkaloids, found an order of reactivity completely different from the relative rates of DHP formation shown in Table 1. Thus, in their studies dehydrojacobine was the most active alkylating agent and would undergo hydrolysis to DHP most readily while dehydrosenecionine showed a much lower reactivity. It, therefore, seems likely that the differences in microsomal oxidation of PAs to DHP summarized in Table 1 reflect the relative substrate specificities of the PAs for the cytochrome P-450 monooxygenases.

While BNF pretreatment had little effect on the microsomal metabolism of various PAs to DHP, this conversion was significantly enhanced by PB induction (Table 1). This oxidation was, therefore, likely catalyzed by a P-450 type of cytochrome P-450 isozyme rather than a P-448 type of isozyme. Supporting evidence for this conclusion has been obtained from studies conducted with reconstituted rat hepatic cytochrome P-450 systems (Ramsdell and Buhler, 1985). These latter experiments have shown that the major PB induced rat cytochrome P-450 isozyme (P-450$_{PB-B}$) had high activity for the oxidation of PAs to DHP while the principle BNF inducible rat cytochrome P-450 (P-450$_{BNF-B}$) was devoid of such activity.

It is clear, however, that the microsomal metabolism of PAs to their corresponding N-oxides is catalyzed by enzymes systems different from those responsible for pyrrole formation. Formation of PA N-oxides by microsomes from PB pretreated rats was either decreased or unchanged from that found with control microsomes while BNF induction caused a significant decrease in N-oxide production (Table 2). PB and BNF, however, influenced DHP formation in quite a different manner (Table 1), indicating the probable involvement of different cytochrome P-450 isozymes in the two pathways. Inhibitor studies (Table 3) and preliminary studies with antibodies against NADPH-cytochrome P-450 reductase (Buhler et al., 1985) also support involvement of cytochrome P-450 monooxygenases in the oxidation of PAs to both pyrrole and N-oxides. It seems likely, however, that the microsomal FMO system may also be involved in conversion of PAs to their N-oxides since anti-reductase antibodies do not completely inhibit microsomal N-oxide formation, even with a 20-fold (w/w) excess of antibody to microsomal protein (Buhler et al., 1985). Nevertheless, addition of n-octylamine or incubation at higher pH, both of which would be expected to enhance FMO activity (Ziegler, 1984), did not increase N-oxide formation (Table 3). Similarly, heat pretreatment of microsomes to inactivate the more temperature-sensitive FMO enzyme, had similar effects on both DHP and N-oxide production. Studies currently in progress in our laboratory with purified cytochrome P-450 isozymes and FMO, however, will likely determine the relative roles of both enzymes to PA oxidation.

Results of the ^{18}O study indicate that hydrolysis of the dehydroalkaloid, the proposed initial product in the metabolic transformation of PAs to DHP, is accomplished by alkyl-oxygen fission of both the C7-O and C9-O bonds. We have demonstrated previously (Kedzierski and Buhler, 1985) that the strong stabilizing effect of the pyrrole ring on the carbonium ion that forms at C7 results in substitutions at that position occurring by a S$_N$1 mechanism. A similar stabilizing effects on the carbonium ion formed at C9 may also result in a S$_N$1 mechanism of nucleophilic substitution at this position. A suggested sequence of carbonium ion formation and subsequent nucleophilic substitution reactions are shown in Figure 3.

The highly electrophilic carbonium ion intermediates can either react with various cellular nucleophiles or with water to form DHP (IV), the latter being detected as a major in vitro metabolite of both retronecine (I) and heliotridine (II) type PAs. Although not yet demonstrated conclusively, it is likely that DHP is also a major in vivo metabolite of the PAs. Since the enantiomer (R)-DHP is capable of alkylating DNA and other cellular nucleophiles (Fennell et al., 1985; Robertson, 1982; Wickramanayaki et al., 1985), it is probable that the actual in vivo or in vitro metabolite, the racemic (R,S)-DHP will act in a similar manner.

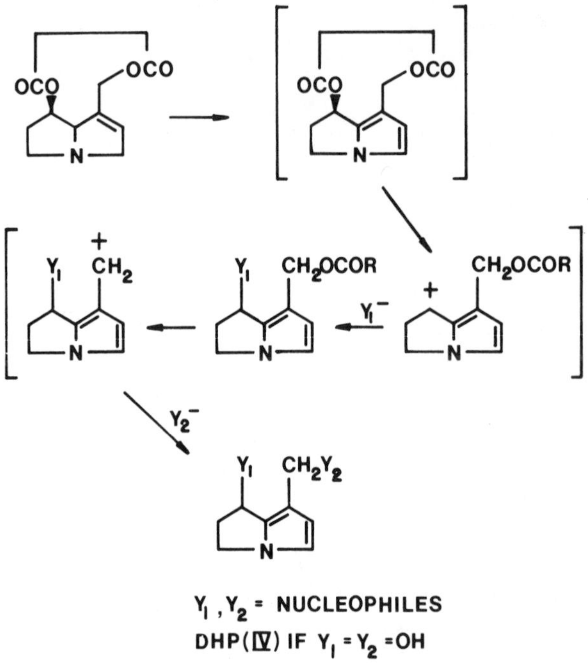

Y_1, Y_2 = NUCLEOPHILES

DHP($\underline{IV}$) IF $Y_1 = Y_2 =$ OH

Figure 3. Proposed sequential carbonium ion formation and nucleophilic substitution
reactions that results in the conversion
of dehydroalkaloids (PA pyrroles) to DHP
or other nucleophilic addition products.

We believe, however, that the carbonium ion intermediates (Figure 3)
formed from the dehydroalkaloids play a decisive role in the toxicity of
PAs. These highly reactive electrophiles are capable of reacting at both
the C7 and C9 positions (Figure 3), consistent with the demonstrated
ability of PAs to cross-link DNA in vivo (Petry et al., 1984). It is
likely that these more potent carbonium ion intermediates rather than DHP
are responsible for much of the alkylation of tissue macromolecules,
including DNA, that is associated with PA toxicity.

REFERENCES

Buhler, D. R., Williams, D. E., and Kedzierski, B., 1985, to be published.
Bull, L. B., Culvenor, C. C. J., and Dick, A. T., 1968, "The Pyrrolizidine
    Alkaloids," John Wiley and Sons, New York.
Culvenor, C. C. J., Dann, A. T., and Dick, A. T., 1962, Alkylation as the
    mechanism by which the hepatotoxic pyrrolizidine alkaloids act on cell
    nuclei. Nature, 195:570.
Culvenor, C. C. J., Edgar, J. A., Smith, L. W. and Tweeddale, H. J., 1970,
    Dihydropyrrolizines. III. Preparation and reaction of derivatives
    related to pyrrolizidine alkaloids. Aust. J. Chem., 23:1853.
Deinzer, M. L., Arbogast, B. L., Buhler, D. R., and Cheeke, P. R., 1982,
    Gas chromatographic determination of pyrrolizidine alkaloids in goat's
    milk. Anal. Chem., 54:1811.
Dickinson, J. O., Cooke, M. P., King, R. R., and Mohamed, P. A., 1976, Milk
    transfer of pyrrolizidine alkaloids in cattle. J. Am. Vet. Med.
    Assoc., 169:1192.

Eastmann, D. F., and Segall, H. J., 1982, A new pyrrolizidine alkaloid metabolite, 19-hydroxysenecionine isolated from mouse hepatic microsomes in vitro. Drug Metab. Dispos., 10:696.

Fennell, T. R., Robertson, K. A., Miller, J. A., Miller, E. C., and Stewart, B. C., 1985, The hepatocarcinogenicity in mice of dehydroretronecine (DR) and its reaction with deoxyguanonine (dGUO) to yield unstable adducts. Proc. Am. Assoc. Cancer Res., 26:83.

Guengerich, F. P., and Liebler, D. C., 1985, Enzymatic activation of chemicals to toxic metabolites. C.R.C. Crit. Rev. Toxicol., 14:259.

Hirono, I., 1981, Natural carcinogenic products of plant origin. C.R.C. Crit. Rev. Toxicol., 8:235.

Huxtable, R. J., 1979, New aspects of the toxicity and pharmacology of pyrrolizidine alkaloids. Gen. Pharmacol., 10:159.

Huxtable, R. J., 1980, Herbal teas and toxins: Novel aspects of pyrrolizidine poisoning in the United States. Perspect. Biol. Med., 24:1.

Karchesy, J. J., and Deinzer, M. L., 1981, Kinetics of alkylation reactions of pyrrolizidine alkaloid derivatives. Hetrocyclics, 16:631.

Kedzierski, B., and Buhler, D. R., 1985, Configuration of necine pyrroles--Toxic metabolites of pyrrolizidine alkaloids. Toxicol. Lett., 25:115.

Luthy, J., Heim, T., and Schlatter, C., 1983, Transfer of [^{3}H]pyrrolizidine alkaloids from Senecio vulgaris L. and metabolites into rat milk and tissues. Toxicol. Lett., 17:283.

Mattocks, A. R., 1968, Toxicity of pyrrolizidine alkaloids. Nature, 217:723.

Mattocks, A. R., 1969, Dihydropyrrolizine derivatives from unsaturated pyrrolizidine alkaloids. J. Chem. Soc., (C) 1155.

Mattocks, A. R., 1972a, Toxicity and metabolism of Senecio alkaloids, in: "Phytochemical Ecology," J. B. Harborne, ed., Academic Press, New York.

Mattocks, A. R., 1972b, Acute hepatotoxicity and pyrrolic metabolites in rats dosed with pyrrolizidine alkaloids. Chem.-Biol. Interact., 5:227.

Mattocks, A. R., and Bird, I., 1983, Pyrrolic and N-oxide metabolites formed from pyrrolizidine alkaloids by hepatic microsomes in vitro: Relevance to in vivo hepatotoxicity. Chem.-Biol. Interact., 43:209.

McLean, E. K., 1970, The toxic actions of pyrrolizidine (Senecio) alkaloids. Pharmacol. Rev., 22:429.

Miranda, C. L., Cheeke, P. R., Goeger, D. E., and Buhler, D. R., 1981, Effect of consumption of milk from goats fed Senecio jacobaea on hepatic drug metabolizing enzyme activities in rats. Toxicol. Lett., 8:343.

Peterson, J. E., and Culvenor, C. C. J., 1983, Hepatotoxic pyrrolizidine alkaloids, in: "Pytochemical Ecology,", J. B. Harborne, ed., Academic Press, New York.

Petry. T. W., Bowden, G. T., Huxtable, R. J., and Sipes, I. G., 1984, Characterization of hepatic DNA damage induced in rats by the pyrrolizidine alkaloid monocrotaline. Cancer Res., 44:1505.

Ramsdell, H. S., and Buhler, D. R., 1985, to be published.

Ramsdell, H.S., and Buhler, D.R., 1981, High performance liquid chromatographic analysis of pyrrolizidine (Senecio) alkaloids using a reversed-phase styrene-divinylbenzene resin column. J. Chromatogr., 210:154.

Robertson, K. A., 1982, Alkylation of N^2 in deoxyguanosine by dehydroretronecine, a carcinogenic metabolite of the pyrrolizidine alkaloid monocrotaline. Cancer Res., 42:8.

Segall, H. J., Dallas, J. L., and Haddon, W. F., 1984, Two dihydropyrrolizine alkaloid metabolites isolated from mouse hepatic microsomes in vitro. Drug Metab. Dispos., 12:68.

Wickramanayake, P. P., Arbogast, B. L., Buhler, D. R., Deinzer, M. L., and
    Burlingame, A. L., 1985, Alkylation of nucleosides and nucleotides by
    dehydroretronecine: Characterization of adducts by liquid secondary
    ion mass spectrometry. J. Am. Chem. Soc., 107:2485.
Ziegler, D. M., 1984, Metabolic oxygenation of organic nitrogen and sulfur
    compounds, in: "Drug Metabolism and Drug Toxicity," J. R. Mitchell
    and M. G. Horning, eds., Raven Press, New York.

MUTAGENS AND CARCINOGENS FORMED DURING COOKING

John H. Weisburger, Takuji Tanaka, William S. Barnes,*
and Gary M. Williams

Naylor Dana Institute, American Health Foundation
Valhalla, New York 10595-1599; *Clarion Univ. of Penn.
Clarion, PA 16214

INTRODUCTION

In the late 1970s, Sugimura et al. (1977; 1983; 1986) noted that powerful mutagens were produced during flame or charcoal broiling of fish or
meat. This observation started a new field of research in nutritional
carcinogenesis. The mutagens were eventually identified as heterocyclic
aromatic compounds with an exocyclic amino group and often an ortho-methyl
group. The structure of the food mutagen 2-amino-3-methylimidazo[4,5-f]-
quinoline (IQ) (I) mimicks that of the synthetic chemical 3,2'-dimethyl-4-
aminobiphenyl (DMAB) (II), a carcinogen for the colon, breast, and pros-
tate (Figure 1). IQ contains both the quinoline ring system and the primary amine with an ortho-methyl substituent, which characterizes DMAB.
Recent tests (Sugimura, 1985) show that the mutagens in this newly discovered class of mutagens, including IQ itself along with other IQ-type compounds, are potent carcinogens and deserve consideration as candidates for
the genotoxic carcinogens implicated as the possible causative factors of
the nutritionally-linked cancers (Weisburger and Horn, 1982). The key
elements underlying this area will be described.

IQ is not only a powerful mutagen in the Ames Salmonella typhimurium
test system but also can induce unscheduled DNA synthesis (UDS) in the
primary culture rat hepatocyte (HPC) test of Williams (Barnes et al.,
1985). Any compound active as a mutagen in the Ames test and capable of
inducing DNA repair in the Williams HPC test is likely to be a carcinogen
(Weisburger and Williams, 1984). Several bioassays now confirm this
association (Ohgaki et al., 1984; Takayama et al., 1984; Tanaka et al.,
1985). In comparison with an established carcinogen for the rat mammary
gland, 4-aminobiphenyl (4-AB), acting as a positive control, IQ was highly
active for this target organ (Tanaka et al., 1985). Also, IQ induced
Zymbal's gland tumors and to a lesser extent, neoplastic nodules, carcinomas and hemangioendotheliomas in the liver; pancreatic neoplasms, and
renal pelvis and urinary bladder neoplasms. Furthermore, preneoplastic
lesions were seen in liver, pancreas, and adrenal cortex. Thus, IQ is a
multipotent carcinogen for the rat. These data together with those of
Takayama et al. (1984) and Ohgaki et al. (1984) suggest that specific
modes of cooking lead to the formation of compounds that are carcinogens
for mice and rats of several strains. In part, the target organs affected
are those corresponding to the nutritionally-linked cancers prevalent
throughout the Western World: intestinal tract, pancreas, and breast.

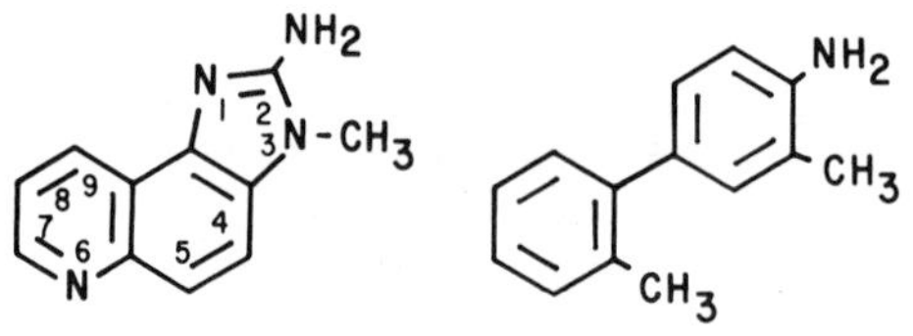

Figure 1. Comparative structures of I, 2-amino-3-methylimidazo[4,5-f]-quinoline, IQ and II, 3,2'-dimethyl-4-aminobiphenyl, DMAB.

MATERIALS AND METHODS

Chemicals

DMAB was purchased as the hydrochloride salt from Ash-Stevens (Detroit, MI). The free base was obtained by extraction from a saturated solution of sodium bicarbonate into diethyl ether. Extracted DMAB was eluted through a Clin Elut disposable column (Fisher Scientific, Pittsburgh, PA), and solvent was removed by rotary evaporation. No impurities were detectable by HPLC; the correct mass spectrum completed structural identification. IQ was acquired from Dr. David Dime, Toronto Research Chemicals, Toronto, Canada. No impurities were detectable by HPLC; the compound had the correct mass spectrum (Kasai et al., 1980). 4-AB was purchased from Aldrich Chemicals, Milwaukee, WI. For the carcinogenicity tests, IQ and 4-AB were converted to the hydrochloride salt by using an excess of gaseous HCl in methanol. The product was then filtered, washed with cold ethanol and ether, and stored at 4°C until used.

Assay for Bacterial Mutagenicity

Media for bacterial mutagenicity assays were from Difco Laboratories (Detroit, MI). NADP and glucose-6-phosphate were from Sigma Chemical Co. (St. Louis, MO). Salmonella typhimurium tester strains TA98 and TA100 were obtained from Dr. B.N. Ames (Ames et al., 1975; Maron and Ames, 1983). The S9 fraction of liver was prepared from male Sprague-Dawley rats induced with Aroclor 1254 (Maron and Ames, 1983). Each assay was routinely performed as described (Barnes et al., 1985). Controls included vehicle control, positive control (2-aminoanthracene for both TA98 and TA100), and sterility controls containing soft agar + test agent, soft agar + S9 mix, and soft agar alone. Spontaneous revertant frequencies in this series of experiments were 40 ± 9/plate with S9 for TA98, 140 ± 15/plate with S9 for TA100. Revertant/nmol of compound and revertants/micro g of compound were calculated by linear regression analysis of the data, without subtracting spontaneous revertant frequencies.

Carcinogen bioassay in female rats

Weanling female Sprague-Dawley rats (Sprague Dawley, Madison, Wi.) were maintained in quarantine for 2 weeks, and routine health tests were performed. At six weeks of age, a total of 100 rats found suitable for bioassay were selected and transferred to holding rooms of the Research Animal Facility, randomized into groups of 32 each experimental rats, 27 vehicle control rats, and 9 untreated control rats, and were kept three to a polycarbonate cage with hardwood bedding. The rooms were maintained at $23 \pm 2°C$ and 40-60% humidity with 8-10 air changes per hour and with a 12-hr light/dark cycle. Drinking water (tap) was provided by an automatic distribution system.

During the quarantine period, all rats were fed the standard NIH-07 diet ad libitum. At 6 weeks of age, beginning with the chemical treatment period, all rats, including controls, were switched to the standard NIH-07 diet but with 15% corn oil added for a total fat content of 20% by weight.

The hydrochloride salts of IQ and 4-AB were dissolved in a solution of 5% Emulphor (EL 620, GAF Corp., New York) in distilled water at a concentration such that 0.4 mmole/kg body weight of IQ or 0.22 mmole/kg of 4-AB were present in 0.25 ml solutions. These solutions were administered by gavage on the following regimens. The first doses were given to rats when they were 6 weeks old and continued 3 times per week for weeks 1-4, at a time when the female Sprague-Dawley rat is maximally sensitive to the induction of mammary tumors. However, because of beginning toxicity, the 3x/week treatment for 4-AB was administered only from weeks 1-3. For IQ, the gavage was continued at 2x/week from weeks 5-8, and for 4-AB, this protocol was used from weeks 4-8. For the next 23 weeks, the dosage was administered once a week. After that time, all of the animals continued without further chemical treatment on the control diet. One group of rats was given the vehicle control, 0.25 ml 5% Emulphor solution in distilled water on the same schedule as the IQ group. An additional group of 9 rats served as untreated controls.

Beginning eight weeks after the first gavage treatment, the animals were carefully palpated once a week to detect mammary tumors (Huggins, 1979). All rats were killed 52 weeks after the first dose. They were necropsied, and any abnormalities were noted, including the multiplicity of tumors. The tissues were fixed in 10% buffered formalin and sections were routinely prepared and stained for microscopic diagnosis. Histologically, mammary tumors were diagnosed according to the classification of Young and Hallows (1973). Liver neoplasms were diagnosed as neoplastic nodules, hepatocellular carcinomas, or hemangioendotheliomas according to the criteria described by Stewart et al. (1980).

RESULTS

Carcinogen Bioassay in Female Sprague Dawley Rats

At the beginning of the test, the average weight of all rats was 126 g. At week 10, group 1 (untreated) weighed an average of 270 g; group 2 (vehicle), 274 g; group 3 (IQ), 256 g; and group 4 (4-AB), 247 g. At week 31, at the end of the treatment, the weights were 371, 389, 343, and 347 g, respectively. Thus, the dosages administered were consistent with proper toxicologic practice. In the early weeks of the test, the rats given 4-AB, but not those on IQ, turned blue for some hours after each gavage, suggesting an effect on the hematopoietic system. This observation was made even though the dose rate for 4-AB (0.22 mmole/kg) was about one-half that of IQ (0.4 mmole/kg). Survival in all groups was good--

Table 1. Incidence of Tumors in Female Sprague Dawley Rats Treated with IQ or 4-AB

Group No. Treatment	No. of Rats	No. of rats with tumors	No. of animals with:									
			Mammary tumors				Liver tumors[a]				Ear Duct tumors	Other tumors
			Total	Carcinoma	Fibro-adenoma	Others	Total	NN	HC	Others		
1 Control	9	0	0	0	0	0	0	0	0	0	0	0
2 Vehicle Control	27	2	2 (3)	0	2 (3)	0	0	0	0	0	0	0
3 IQ	32	23[b]	14[b] (22)	14[b] (21)	0	1[c] (1)	6[b] (9)	3 (5)	2 (2)	2[c] (2)	11[b] (12)	5[d] (6)
4 4-AB	32	20[b]	19[b] (35)	18[b] (32)	2 (2)	1[e] (1)	0	0	0	0	0	0

Numbers in parentheses are the total number of tumors.
[a]NN: Neoplastic nodule, HC: Hepatocellular carcinoma; [c]Hemangioendothelioma; [d]Granulocytic leukemia, 2 rats: pancreatic acinar cell adenoma, 1 rat: islet cell adenoma, 1 rat: transitional cell papillomas of pelvis and urinary bladder, 1 rat: [e]fibrosarcoma.
[b]Significantly different from Group 2 by Fisher's exact probability test ($P < 0.05$).
From Tanaka et al., 1985.

Table 2. The Incidence of Preneoplastic Lesions In Female
Sprague Dawley Rats.

Group and treatment	1 Control	2 Vehicle Control	3 IQ	4 4-AB
Organ		Number of rats		
Liver: Altered liver cell foci	0	0	17[b] (53)[a]	5[b] (16)
Pancreas: Atypical hyperplastic acinar cell lesions	0	0	19[b] (59)	7[b] (22)
Adrenal cortex: Altered proliferative foci.	0	0	5[b] (16)	2 (6)

[a] Numbers in parentheses are % of lesion-bearing rats.
[b] Significantly different from Group 2 by Fisher's exact
    probability test ($P < 0.05$).
Adapted from Tanaka et al., 1985

97/100 rats were alive at 6 months. However, 3 to 5 rats in the groups
receiving IQ or 4-AB died or had to be killed each subsequent month because of the occurrence of cancers, especially in the mammary gland or ear
duct. In all groups, 78 of the starting 100 rats were alive when the test
was terminated at 52 weeks.

Neoplasms in the mammary gland. The major finding in this experiment
was the induction of cancer in the mammary gland. With IQ, the first palpable tumor was noted during the 12th experimental week and with the positive control carcinogen 4-AB during week 16. Progressively more mammary
tumors were palpated during the test. No regression of palpable tumors
was noted. At the end of the 52-week study period, 14 of the 32 rats giv-
en IQ exhibited 21 mammary carcinomas and one a hemangioendothelioma
(Table 1) (Tanaka et al., 1985). In the group of 32 rats given 4-AB, 18
had 32 mammary carcinomas, 2 had fibroadenomas, and one a fibrosarcoma.
In the 27 control rats intubated with vehicle, 2 had three mammary fibroadenomas. None of the nine untreated controls had palpable or microscopic
mammary gland tumors.

Zymbal's Gland Neoplasms. Eleven of the 32 rats given IQ had 12 keratinizing epidermoid carcinomas in the ear duct (Table 1). The earliest
growth appeared at 40 weeks. One animal had bilateral tumors. The right
ear duct was involved in 6/10 tumors. This lesion was absent in rats giv-
en 4-AB or either of the two control groups.

Miscellaneous Preneoplastic or Neoplastic Lesions. In the IQ group,
17/32 rats had altered liver cell foci (Table 2), 3 had 5 neoplastic nodules, 2 had hepatocellular carcinomas, and 2 more had hemangioendotheliomas in the liver (Table 1). With 4-AB, 5/32 had altered liver cell foci
but no liver neoplasms (Tables 1 and 2). The controls had normal livers.

In the IQ group, 19/32 had atypical hyperplastic acinar cell lesions
in the pancreas, 1 had an acinar cell adenoma, and 1 had an islet cell
adenoma of the pancreas (Tables 1 and 2). With 4-AB, only 7/32 had atypical hyperplastic acinar cell lesions but failed to show neoplasia in the
pancreas.

With IQ, 5 altered proliferative foci in the adrenal cortex were seen and there were 2 in the group given 4-AB (Table 2). In addition, in the group of rats given IQ, 2 had leukemia and 1 had a papilloma of the pelvis and urinary bladder (Table 1). None of the control or vehicle control rats had abnormalities in these organs.

DISCUSSION

In agreement with the positive results from the battery of in vitro tests described previously (Caderni et al., 1983; Coresi and Dolara, 1983; Terada et al., 1983; Bird and Bruce, 1984; Minkler and Carrano, 1984; Barnes et al., 1985; Sugimura, 1985; 1986), IQ can be classified as genotoxic and has been found to be a multipotential, powerful carcinogen in the rat. The chemicals were administered by gavage to obtain quantitative information of dosage and also to use the expensive chemical IQ most economically. The total dosage administered per rat was 4.4 mmoles (871 mg) of IQ and 2.4 mmoles (400 mg) of 4-AB, the positive control carcinogen. The overall duration of the test was one year. Thus, a high yield of mammary cancer was induced with intermittent, limited administration of IQ. In addition, other neoplasms were found, including those in the ear duct, and to a lesser extent, in the liver, pancreas, and urinary tract. Compared with the positive control, the human carcinogen 4-AB, IQ appears to have potency of the same order as 4-AB, allowing for the 2-fold lower level of 4-AB administered. Nonetheless, IQ demonstrated a greater versatility, and yet specificity, by inducing cancer mainly in the mammary gland and ear duct. The induction of mammary gland neoplasms may have been potentiated by dietary fat, as has been noted with other mammary carcinogens, especially the closely related homocyclic 3,2'-dimethyl-4-aminobiphenyl (Reddy et al., 1980). In addition, Takayama et al. (1984) induced cancer in the intestinal tract, including small and large intestines as well as lesions in the pancreas and ear duct, by feeding 300 ppm IQ in the diet to Fischer strain rats.

There have been no detailed studies on the mode of action of these newly discovered carcinogens. With the homocyclic arylamines and arylamides, N-hydroxylation is the key first activation step (see Miller and Miller, 1981; Kadlubar, this volume), and this was observed with the heterocyclic amines upon in vitro metabolism with subcellular fraction of liver (Nagao et al., 1983; Alldrick et al., 1985; Kato, this volume), and in preliminary studies in vivo (Sjödin and Jägerstad, 1984; Bergman, 1985; Barnes and Weisburger, 1985; Sato, 1986). The combined data obtained, therefore, suggest that IQ and related heterocyclic mutagens and carcinogens most likely are converted to proximate and ultimate carcinogenic forms by N-oxidation on the exocyclic amino group. The results of the mutagenicity test of typical chemicals in this series point in the same direction. Furthermore, the metabolism experiments indicate that N-hydroxylation is an essential activation reaction (Nagao et al., 1983; Kato, this volume).

IQ at high dosages induced cancer at several target organs. In mice, neoplasms were found in the forestomach, liver, and bladder. In rats, the organs affected were small and large intestine, breast, pancreas, ear duct, liver, and urinary bladder. However, the amounts usually consumed are much smaller, although the intake begins at a young age and is fairly continuous. There are no data, as yet, on the effect of lower dosages in animal models, especially as regards shifts in target organs as a function of dosage. Also, of relevance is the fact that for organs such as colon, breast, and pancreas, nutritional parameters such as dietary fat levels may yield potentiation through promotion. Thus, in the human setting, those target organs would be preferentially affected, whereas insufficient carcinogen would be available to express carcinogenesis in the liver, for example.

Essentially, observations so far show that IQ is a potent carcinogen in mice and in rats of two strains at target organs such as intestinal tract, breast, and pancreas, that represent the main nutritionally-linked cancers in the Western World. It is essential to validate these findings and establish whether these agents in fried food are actually the genotoxic carcinogens associated with these major cancers seen in the Western World (Weisburger, 1977), as well as in Japan where the incidence of these cancers is increasing as the Japanese people progressively assume Western dietary habits (Wynder and Hirayama, 1977; Segi et al., 1981).

REFERENCES

Alldrick, A.J., Rowland, I.R., and Coutts, T.M., 1985, Interspecies differences in the activation of MeIQ and IQ to active mutagens by hepatic S9-fractions, Abstract, in: "Fourth International Conference on Environmental Mutagens," Nordic Environmental Mutagen Society, Stockholm, p. 321.

Ames, B.N., McCann, J., and Yamasaki Y., 1975, Methods for detecting carcinogens and mutagens with the Salmonella/mammalian microsome mutagenicity test, Mutat. Res., 31:347.

Barnes, W. and Weisburger J.H., 1985, Fate of the food mutagen 2-amino-3-imidazole[4,5-f]quinoline (IQ) in Sprague Dawley rats. I. Mutagens in the urine, Mutat. Res., 156:83.

Barnes, W.S., Lovelette, C.A., Tong, C., Williams, G.M., and Weisburger, J.H., 1985, Genotoxicity of the food mutagen, 2-amino-3-methylimidazo[4,5-f]quinoline (IQ) and analogs, Carcinogenesis, 6:441.

Bergman, K., 1985, Autoradiographic distribution of ^{14}C-labeled 3H-imidazo[4,5-f]quinoline-2-amines in mice, Cancer Res., 45, 1351.

Bird, R.P., and Bruce, W.R., 1984, Damaging effect of dietary components to colon epithelial cells in vivo: Effect of mutagenic heterocyclic amines, J. Natl. Cancer Inst., 73:237.

Caderni, G., Kreamer, B.L., and Dolara, P., 1983, DNA damage of mammalian cells by the beef extract mutagen 2-amino-3-methylimidazo[4-5f]quinoline, Food Chem. Toxicol. 21:641.

Coresi, E. and Dolara, P., 1983, Neoplastic transformation of BALB 3T3 mouse embryo fibroblasts by the beef extract mutagen 2-amino-3-methylimidazo[4-5d]quinoline, Cancer Letts., 20:43.

Huggins, C.B., 1979, "Experimental Leukemia and Mammary Cancer Induction, Prevention, Cure. University of Chicago Press, Chicago, Illinois.

Kadlubar, F.F., The role of arylamine-DNA adducts in benzidine carcinogenesis and mutagenesis, this volume.

Kasai, H., Yamaizumi, Z., Wakabayashi, K., Nagao, M., Sugimura, T., Yokoyama, S., Miyazawa, T., Spingarn, N.E., and Weisburger, J.H, 1980, Potent novel mutagens produced by broiling fish under normal conditions, Proc. Jpn. Acad., 56B:278.

Kato, R., Saito, K., Shinohara, A., and Kamataki, T., 1985, Purification properties and function of N-hydroxy-arylamine o-acetyltransferases, this volume.

Maron, D.M. and Ames, B.N., 1983, Revised methods for the Salmonella mutagenicity test, Mutat. Res., 113:173.

Miller, E.C. and Miller, J.A., 1981, Mechanisms of chemical carcinogenesis, Cancer, 47:1055; see also this volume.

Minkler, J.L. and Carrano, A.V., 1984, In vivo cytogenetic effects of the cooked-food-related mutagens Trp-P-2 and IQ in mouse bone marrow, Mutat. Res., 140:49.

Nagao, M. Fujita, Y., Wakabayashi, K., and Sugimura, T., 1983, Ultimate
    forms of mutagenic and carcinogenic heterocyclic amines produced
    by pyrolysis, Biochem. Biophys. Res., Commun., 114:626.
Ohgaki, H., Kusama, K., Matsukura, N., Morino, K., Hasegawa, H., Sato,
    S., Takayama, S., and Sugimura, T, 1984, Carcinogenicity in mice
    of a mutagenic compound, 2-amino-3-methylimidazo[4,5-f]quinoline,
    from broiled sardine, cooked beef and beef extract, Carcinogene-
    sis, 5:921.
Reddy, B.S., Cohen, L.A., McCoy, G.D., Hill, P., Weisburger, J.H., and
    Wynder, E.L., 1980, Nutrition and its relationship to cancer,
    Adv. Cancer Res., 30:237.
Sato, S., 1986, Symposium on Formation of Mutagens During Cooking and Heat
    Processing of Foods, Environ. Health Perspect., 00:000.
Segi, M., Tominaga, S., Aoki, K., and Fujimoto, I., Eds., 1981, "Cancer
    Mortality and Morbidity Statistics," Gann Monogr. on Cancer Res.
    26:1.
Sjödin, P. and Jägerstad, M., 1984, A balance study of ^{14}C-labelled 3H-
    imidazo[4,5-f]quinolin-2-amines (IQ and MeIQ) in rats, Fd. Chem.
    Toxic., 22:207.
Stewart, H.L., Williams, G.M., Keysser, C.H., Lombard, L.S., and Montali,
    R.J., 1980, Histologic typing of liver tumors of the rat, J.
    Natl. Cancer Inst., 64:177.
Sugimura, T., 1985, Carcinogenicity of mutagenic heterocyclic amines form-
    ed during the cooking process, Mutat. Res., 150:33.
Sugimura, T., 1986, Symposium on Formation of Mutagens During Cooking and
    Heat Processing of Foods, 1986, Environ. Hlth Perspect., 00:000.
Sugimura, T., Nagao, M., Kawachi, T., Honda, M., Yahagi, T., Seino,
    Y, Sato, S., Matsukura, N., Matsushima, T., Shirai, A., Sawamura,
    M., and Matsumoto, H., 1977, Mutagen-carcinogens in food, with
    special reference to highly mutagenic pyrolytic products in broil-
    ed foods, in: "Origins of Human Cancer," H.H. Hiatt, J.D. Watson,
    and J.D. Winsten, Eds., Cold Spring Harbor Laboratories, Cold
    Spring Harbor, New York, p. 1561.
Sugimura, T. Sato, S., and Takayama, S., 1983, New mutagenic hetero-
    cyclic amines found in amino acid and protein pyrolysates and in
    cooked food, in: "Environmental Aspects of Cancer: The Role of
    Macro and Micro Components of Foods," E. L. Wynder, G.A. Leveille,
    J.H. Weisburger, and G.E. Livingston, Eds., Food and Nutrition
    Press, Westport, Conn., p. 167.
Takayama, S. Nakatsuru, Y., Masuda, M., Ohgaki, H., Sato, S., and
    Sugimura, T, 1984, Demonstration of carcinogenicity in F344 rats
    of 2-amino-3-methylimidazo[4,5-f]quinoline from broiled sardine,
    fried beef and beef extract, Gann, 75:467.
Tanaka, T., Barnes, W.S., Weisburger, J.H. and Williams, G.M., 1985,
    Multipotential carcinogenicity of the fried food mutagen 2-amino-
    3-methylimidazo[4,5-f]quinoline (IQ) in rats, Jap. J. Cancer
    Res. (Gann), 76:570.
Terada, M., Nakayasu, M., Sakamoto, H., Nakasato, F., and Sugimura, T.
    1983, Induction of diphtheria toxin-resitant cells by mutagens-
    carcinogens, in: "ADP-Ribosylation, DNA Repair and Cancer," M.
    Miwa et al., Eds., Japan Sci. Soc. Press, Tokyo, p. 277.
Weisburger, J.H., 1977, Current views on mechanisms concerned with the
    etiology of cancers in the digestive tract, in: "Pathophysiology
    of Carcinogenesis in Digestive Organs," E. Farber, T. Kawachi, T.
    Nagayo, H. Sugano, T. Sugimura, and J.H. Weisburger, Eds., Uni-
    versity Park Press, Baltimore, Maryland, p. 1.
Weisburger, J.H. and Horn, C.L., 1982, Nutrition and cancer: Mechanisms of
    genotoxic and epigenetic carcinogens in nutritional carcinogene-
    sis, Bull. N.Y. Acad. Med., 58:296.

Weisburger, J.H. and Williams, G.M., 1984, Bioassay of Carcinogens: in vitro and in vivo tests, in: "Chemical Carcinogens," ACS Monogr. 182, Vol. 2, C.E. Searle, Ed., American Chemical Society, Wash. DC, p. 1323.

Wynder, E.L. and Hirayama, T., 1977, Comparative epidemiology of cancers of the United States and Japan, Prev. Med. 6, 567.

Young, S. and Hollowes, R.C., 1973, Tumours of the mammary gland, in: "Tumours of the Rats, Pathology of Tumours in Laboratory Animals," Vol. 1, Part 1, V.S. Turusov, Ed., International Agency for Research on Cancer, Lyon, France, p. 31.

Acknowledgements: These investigations were supported by PHS Grants Numbers CA-24217 (Large Bowel Cancer Project) and CA-29602 awarded by the National Cancer Institute, DHHS. The authors acknowledge the excellent technical assistance of Jane Maher, Karin Gilbert, and Bruce Griffith; C. Choi, M. Reddy. J. Reinhardt, and A.M. Keizer in the Research Animal Facility, and Ms. Clara Horn for editorial assistance.

QUINONE IMINES AS BIOLOGICAL REACTIVE INTERMEDIATES

D. Porubek, M. Rundgren, R. Larsson, E. Albano, D. Ross,
S.D. Nelson and P. Moldéus

Department of Toxicology, Karolinska Institutet
S-104 01 Stockholm, Sweden

INTRODUCTION

Quinone imines represent a class of biological reactive intermediates that has been the focus of intensive toxicological research in recent years. Two representative quinone imines that have been subjects of research in this laboratory and others are shown in Fig. 1. These quinone imines are of toxicological significance since N-acetyl-p-benzoquinone imine is a postulated reactive intermediate formed during metabolism of the analgesic drug acetaminophen[1] while 4-(ethoxyphenyl)-p-benzoquinone imine is a possible reactive intermediate formed during metabolism of the analgesic phenacetin[2]. While acetaminophen is still widely used, phenacetin has largely been discontinued.

Quinone imines are electrophilic reactive intermediates that can potentially engage in a number of biochemical interactions when formed metabolically. These interactions may include reactions with cellular constituents such as reduced glutathione[3-5], macromolecules including structural proteins[6] and enzymes[7], and nucleic acids[8]. In addition, quinone imines may undergo redox cycling which may lead to the formation of activated oxygen species such as superoxide, hydrogen peroxide and hydroxyl radical[9] which could in turn react with lipids.

The quinone imines shown in Fig. 1 are convenient examples of this class of reactive intermediates since their formation illustrate alternative mechanisms by which they may be metabolically generated. Additionally, these quinone imines display typical biochemical effects that are believed to play critical roles in the manifestation of toxicity of acetaminophen and phenacetin. In turn, the postulated mechanisms of metabolic formation

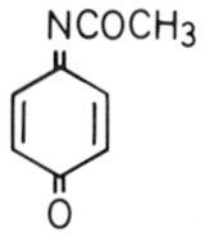

NCOCH3

N-acetyl-p-benzoquinone imine

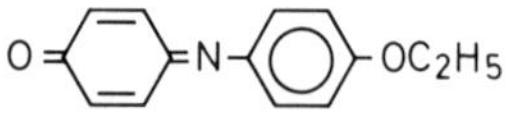

$O=$ ⟨⟩ $=N-$ ⟨O⟩ $-OC_2H_5$

4-(ethoxyphenyl)-p-benzoquinone imine

Fig. 1. Two representative quinone imines.

and biochemical effects of N-acetyl-p-benzoquinone imine and 4-(ethoxy-phenyl)-p-benzoquinone imine will be discussed.

N-Acetyl-p-Benzoquinone Imine (NAPQI)

As a postulated reactive intermediate of acetaminophen, NAPQI has become a central figure in investigations concerning the toxicity of this drug. Generally considered safe when used at therapeutic doses, acetaminophen taken in very large doses causes severe hepatic necrosis[10] which is sometimes accompanied by renal damage[11].

NAPQI was postulated to be the reactive intermediate involved in acetaminophen toxicity as early as 1973[1]. At that time the envisaged mechanism whereby acetaminophen was biotransformed into NAPQI involved initial N-hydroxylation followed by spontaneous loss of water. Since this time the intermediacy of N-hydroxy acetaminophen in the generation of NAPQI has fallen out of favor because studies with synthetic N-hydroxy acetaminophen have revealed the rate of its dehydration to NAPQI to be quite slow[12]. In light of this finding and the inability to detect N-hydroxy acetaminophen as a metabolite of acetaminophen[13] the currently favored mechanism of NAPQI formation in liver involves direct dehydrogenation of acetaminophen by cytochrome P-450. Direct dehydrogenation can be envisaged to occur by successive one electron oxidations thus generating intermediate semiquinone radicals[14]. Alternatively, the reaction could occur by a concerted two electron oxidation involving the perferryl form of cytochrome P-450. Studies that demonstrate no significant loss of oxygen-18 in the glutathione conjugate when $|p-^{18}O|$-acetaminophen is bioactivated support a direct dehydrogenation by cytochrome P-450.

Since the time that NAPQI was initially proposed as the reactive intermediate involved in acetaminophen toxicity it has been possible to generate NAPQI chemically in solutions[16-18] and finally prepare crystalline samples in 1982[19]. The availability of pure preparations of NAPQI have greatly accelerated investigations concerning the role of this quinone imine in acetaminophen toxicity.

When allowed to react with reduced glutathione (GSH), NAPQI can function as an oxidant to generate glutathione disulfide (GSSG) or function as an electrophile to form a glutathione conjugate. The relative amounts of GSSG and glutathione conjugate that are formed are in part governed by the relative amounts of NAPOI and GSH initially present (Fig. 2). When added to isolated rat hepatocytes NAPQI also reacts very rapidly with intracellular GSH to form GSSG (Fig. 3a). In cells with functioning glutathione reductase, the rapid increase in GSSG is transient since the GSSG is rapidly reduced back to GSH. However, inhibition of glutathione reductase by treatment of hepatocytes with 1,3-bis(2-chloroethyl)-1-nitrosourea (BCNU) allowed this increased level of GSSG to be maintained[20].

The time course of formation of the glutathione conjugate is somewhat delayed when compared to the almost immediate depletion of GSH and instantaneous formation of GSSG following exposure to NAPQI (Fig. 3b). Glutathione conjugate formation increases for the first 5 min of incubation. In

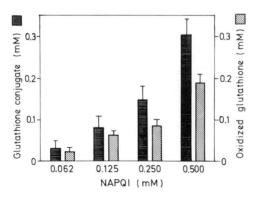

Fig. 2. Reaction of increasing concentrations of NAPQI with GSH (0.5 mM) in 0.05 M Tris-HCl buffer, pH 7.4. Glutathione conjugate (▤) and GSSG (▨).

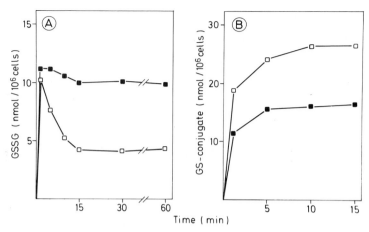

Fig. 3. Effect of NAPQI (250 μM) on glutathione redox state in isolated
hepatocytes. Concentrations of GSSG (A) and glutathione conjugate
(B) determined for cells preincubated with (filled symbols) or
without (open symbols) BCNU.

cells treated with BCNU considerably less glutathione conjugate is formed
apparently because less GSH is available as a result of reductase inactivation.

NAPQI also reacts with protein thiol groups in isolated hepatocytes[20].
The reactions are rapid and appear to be due to oxidation and arylation
but since most of the thiol groups can be restored by treatment with
dithiothreitol (DTT) the predominant reaction appears to be oxidation
(Fig. 4a). Furthermore, the oxidation of protein thiols appears to be a
major event in cytotoxicity since DTT treatment also resulted in protection against cytotoxicity caused by NAPQI (Fig. 4b). Although the loss of
protein thiol groups appears to be of major importance for the cytotoxicity caused by NAPQI the ultimate cause of cell damage remains unclear.
However, recent findings indicate that NAPQI could exert at least some of
its toxic effects via perturbation of Ca^{2+} homeostasis[7]. Therein it was
shown that NAPQI can cause elevation of cytosolic Ca^{2+}. The increase appeared to be due to release of mitochondrial Ca^{2+} and inhibition of the
high affinity Ca^{2+}-ATPase activity of the plasma membrane caused primarily
by the oxidation of protein thiol groups.

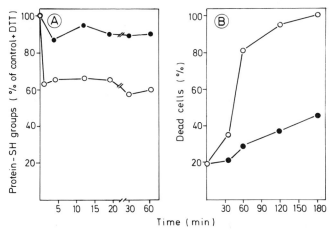

Fig. 4. Effect on DTT (10 mM) on protein thiols (A) and cytotoxicity (B)
of hepatocytes exposed to NAPQI (400 μM). DTT was added 4 min
after NAPQI (filled symbols) or was left out (open symbols).

From the studies described here it is evident that if the putative
reactive metabolite of acetaminophen, NAPQI, is added as such to isolated
hepatocytes the oxidation and arylation of soluble and protein thiols
occurs. However, if NAPQI is generated from acetaminophen in isolated
hepatocytes no GSSG formation can be observed even in cells treated with
BCNU (Fig. 5). A possible explanation for these observations is that under
conditions of continuous production from acetaminophen, NAPQI may preferentially undergo enzyme mediated conjugation rather than reduction by
GSH. The conjugation of NAPQI with GSH has been shown to be fascilitated
by the action of purified glutathione transferases[21].

Although acetaminophen does not cause detectable oxidation of GSH
the oxidation of protein thiol groups may still be of importance for its
toxicity. That oxidation rather than arylation of protein thiol groups
may indeed be the cause of acetaminophen toxicity is indicated by results
obtained with the acetaminophen analogue 3,5-dimethylacetaminophen. The
analogue has been shown to be as hepatotoxic as acetaminophen in vivo[22].
The oxidation product of this analogue, 3.5-dimethyl-NAPQI, does not form
a conjugate when reacted with GSH. Instead GSSG is formed stoichiometri-
cally[23]. Upon addition of 3,5-dimethyl-NAPQI to isolated hepatocytes an
immediate GSH depletion due to GSSG formation results since all the GSH

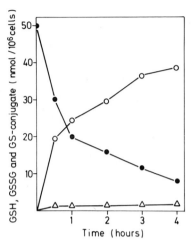

Fig. 5. Effect of acetaminophen (2 mM) on glutathione redox status in isolated hepatocytes pretreated with BCNU. Glutathione conjugate (o), GSH (•) and GSSG (Δ).

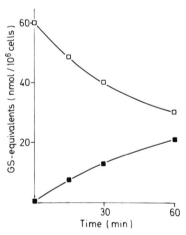

Fig. 6. Effect of 3,5-dimethylacetaminophen (5 mM) on GSH (open symbols) and GSSG (closed symbols) in isolated hepatocytes pretreated with BCNU.

was recoverable by DTT reduction[24]. In the presence of NAPQI only part of
the GSH depletion is recovered by DTT. Interestingly, when 3,5-dimethyl
NAPQI is supposedly generated intracellularly from 3,5-dimethylacetamino-
phen the depletion of cellular GSH is nearly all accounted for by GSSG
formation (Fig. 6). Thus it appears likely that an oxidation mechanism,
specifically oxidation of protein thiol groups, may be a major determinant
in the hepatotoxicity of 3,5-dimethylacetaminophen and possibly acetaminophen.

In summary, the proposed reactive intermediate of acetaminophen NAPQI,
reacts with GSH to form GSSG and a glutathione conjugate when added to
isolated hepatocytes (Fig. 7). In addition, NAPQI reacts with cellular
protein thiols resulting in oxidation and arylation. The consequences of
these protein alterations may include enzyme inactivation. Such a process
has been shown for enzymes involved in Ca^{2+} homeostasis. The perturbation of Ca^{2+} homeostasis may be a critical event in the mechanism of
NAPQI and thus acetaminophen toxicity.

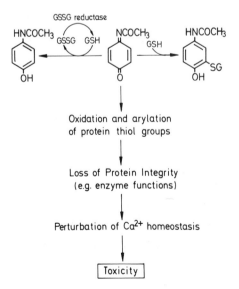

Fig. 7. Scheme of proposed mechanism for the hepatotoxicity of acetaminophen.

4-(Ethoxyphenyl)-p-benzoquinone imine (4-EPPBQI)

As a possible reactive intermediate of phenacetin, 4-EPPBQI has been much studied in this laboratory as of late. Unlike acetaminophen, phenacetin is highly nephrotoxic with damage occurring after prolonged use of therapeutic doses[25]. Reports of different kinds of kidney damage exist including cases of severe renal papillary necrosis[26]. An increased rate of cancer in the kidney and urinary tract is also associated with long-term phenacetin exposure[27,28].

The metabolism of phenacetin in vivo is quite complex but major primary metabolites include acetaminophen and p-phenetidine[29]. It has been suggested that acetaminophen is responsible for the therapeutic effects of phenacetin but it is not likely that acetaminophen is responsible for the nephrotoxic effects since it is less toxic than phenacetin in this regard. The nephrotoxicity of phenacetin has also been suggested to be caused by one of its major metabolites, N-hydroxyphenacetin[30]. It is of interest that the acetate and sulfate esters of this metabolite have been found to bind to nucleic acids and were mutagenic[31]. Another minor metabolite of phenacetin, p-aminophenol, is a known nephrotoxin and it is possible that it is at least partially responsible for the nephrotoxicity of phenacetin.

Alternatively, the kidney is an organ which contains high levels of the enzyme prostaglandin synthetase[32] (PGS) and the activity is particularly high in the inner medulla. PGS is an enzyme complex that possesses both cyclooxygenase and hydroperoxidase activities with the hydroperoxidase component requiring the availability of two reducing equivalents per cleavage reaction[32]. These reducing equivalents may be acquired by the oxidation of cosubstrate molecules[33]. Although phenacetin cannot serve as a cosubstrate in this reaction many of its metabolites can including the major metabolites acetaminophen and p-phenetidine[8].

As previously discussed the nephrotoxicity of phenacetin is probably not due to its metabolism to acetaminophen since acetaminophen is much less nephrotoxic. Being a unique metabolite of phenacetin, p-phenetidine might instead be playing a critical role. When the PGS dependent oxidation of p-phenetidine was studied using ram seminal vesicle microsomes (RSVM) or horse radish peroxidase (HRP) as enzyme source several intensely colored products were formed which could be separated by thin layer chromatography[35].

The formation of colored products during peroxidase-catalyzed oxidation of aromatic amines and phenols is well established and is indicative

of free radical intermediates[36]. That a free radical species of p-phenetidine is generated during HRP catalyzed oxidation has been shown in experiments where the spin probe OXANOH was used to detect the unstable p-
phenetidine radical[35].

The major metabolite generated by either RSVM or HRP oxidation of
p-phenetidine has been isolated by preparative thin layer chromatography
and characterized by mass spectral analysis[37]. The proposed mechanism of
formation of this quinone imine, 4-(ethoxyphenyl)-p-benzoquinone imine
(4-EPPBQI) involves the initial production of a diimine dimer formed by
head-to-tail coupling of two-phenetidine radicals with the concomitant
production of ethanol (Fig. 8). Since imines are quite unstable to hydrolysis only the quinone-imine can be isolated but the detection of an
equivalent amount of generated ammonia is support for the proposed mechanism[37]. In fact, the unstable diimine has been trapped isolated and characterized as its BHA adduct[37] as has been done with the diimine product
formed during peroxidase catalyzed oxidation of benzidine.

As indicated by its structure 4-EPPBQI is an electrophilic intermediate and if allowed to react with GSH the formation of GSSG and mono-
and bisconjugates results (unpublished observations). As with NAPQI the
relative amount of GSSG formed increases as the initial amount of GSH is
increased. Following reduction of 4-EPPBQI by GSH the resultant 4-(ethoxyphenyl)-p-phenol can be oxidized by oxygen thus generating active oxygen

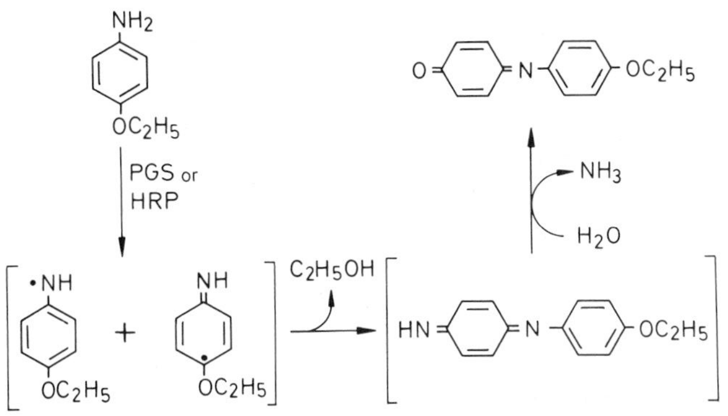

Fig. 8. Proposed mechanism of formation of 4-EPPBQI.

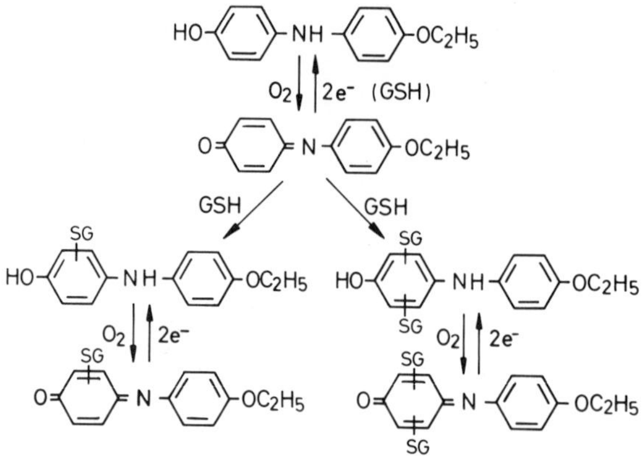

Fig. 9. Possible mechanisms of redox cycling by 4-EPPBQI and its mono- and bisglutathione conjugates.

species (Fig. 9). Similarly, the mono- and bisconjugates can engage in such redox reactions with GSH and oxygen. Whether this redox cycling has any toxicological relevance in vivo remains to be determined.

CONCLUSION

While the formation of NAPQI is believed to occur by direct oxidation of acetaminophen by cytochrome P-450, 4-EPPBQI has been postulated to arise from radical recombination following oxidation of p-phenetidine (a metabolite of phenacetin) by a peroxidase such as PGS. Thus, these examples represent two mechanisms whereby quinone imines could be metabolically formed.

The bioactivation of acetaminophen and phenacetin to reactive quinone imines may be responsible for the toxicities associated with these xenobiotics. While definitive proof for the involvement of quinone imines in the toxicity of acetaminophen and phenacetin is still unavailable studies do indicate that many of the biochemical effects of NAPQI go a long way in explaining the toxicity of acetaminophen. Nevertheless, identification of the ultimate toxicological lesions that are responsible for the toxicities of these compounds may be slow in coming. This, of course, is a direct result of the rich biochemistry that would be associated with intracellular generation of quinone imines. Only when all the potential biochemical interactions of quinone imines are addressed can their importance in acetaminophen and phenacetin toxicity be properly assessed.

REFERENCES

1. D.J. Jollow, J.R. Mitchell, W.Z. Potter, D.C. Davis, J.R. Gillette, and B.B. Brodie, Acetaminophen-induced hepatic necrosis. II. Role of covalent binding in vivo, J. Pharmacol. Exp. Ther. 187:195 (1973).

2. R. Larsson, D. Ross, M. Nordenskjöld, B. Lindeke, L.-I. Olsson, and P. Moldéus, Reactive products formed by peroxidase catalyzed oxidation of p-phenetidine, Chem.-Biol. Interactions, 52:1 (1984).

3. J.R. Mitchell, D.J. Jollow, W.Z. Potter, J.R. Gillette, and B.B. Brodie, Acetaminophen-induced hepatic mecrosis. IV. Protective role of glutathione, J. Pharmacol. Exp. Ther. 187:211 (1973).

4. J.R. Rice and P.T. Kissinger, Cooxidation of benzidine by horse radish peroxidase and subsequent formation of possible thioether conjugates of benzidine, Biochem. Biophys. Res. Comm. 104:1312 (1982).

5. B. Andersson, R. Larsson, A. Ramimtula, and P. Moldéus, Prostaglandin synthase and horse radish peroxidase catalyzed DNA binding of p-phenetidine, Carcinogenesis, 5:161 (1984).

6. A.G. Streeter, D.C. Dahlin, S.D. Nelson, and T.A. Baillie, The covalent binding of acetaminophen to protein, evidence for cysteine residues as major sites of arylation in vitro, Chem.-Biol. Interactions, 48: 349 (1984).

7. M. Moore, H. Thor, G. Moore, S. Nelson, P. Moldéus, and S. Orrenius, The toxicity of acetaminophen and N-acetyl-p-benzoquinone imine (NAPQI) in isolated hepatocytes is associated with thiol depletion and increased cytosolic Ca^{2+}. J. Biol. Chem. in press, (1985).

8. B. Andersson, M. Nordenskjöld, A. Rahimtula, and P. Moldéus, Prosta-glandin syntehtase-catalyzed activation of phenacetin metabolites to genotoxic products, Molec. Pharmacol. 22:479 (1982).

9. G.M. Rosen, W.V. Singletary, Jr., E.J. Rauckman, and P.G. Killenberg, Acetaminophen hepatotoxicity: An alternative mechanism, Biochem. Pharmacol. 32:2053 (1983).

10. L.F. Prescott, W. Wright, P. Roscoe, and S.S. Brown, Plasma paraceta-mol half-life and hepatic necrosis in patients with paracetamol overdosage, Lancet, 1:519 (1971).

11. G.A. Mudge, M.W. Gemboys, and G.G. Duggin, Covalent binding of metabo-lites of acetaminophen to kidney protein and depletion of renal glutathione, J. Pharmacol, Exp. Ther. 206:218 (1978).

12. M.W. Gemboys, G.W. Gribble, and G.A. Mudge, Synthesis of N-hydroxy-acetaminophen, a postulated toxic metabolite of acetaminophen, and its phenolic sulfate conjugate, J. Med. Chem. 21:649 (1978).

13. J.A. Hinson, L.R. Pohl, and J.R. Gillette, N-Hydroxyacetaminophen: A microsomal metabolite of N-hydroxyphenacetin but apparently not of acetaminophen, Life Sci. 24:2133 (1979).

14. S.D. Nelson, D.C. Dahlin, E.J. Rauckman, and G.M. Rosen, Peroxidase-mediated formation of reactive metabolites of acetaminophen, Molec. Pharmacol. 20:195 (1981).

15. J.A. Hinson, S.D. Nelson, and J.R. Gillette, Metabolism of $|p-^{18}O|$-phenacetin: The mechanism of activation of phenacetin to reactive metabolites in hamsters, Molec. Pharmacol. 15:419 (1979).

16. I.C. Calder, M.J. Creek, and P.J. Williams, N-Hydroxyphenacetin as a precursor of 3-substituted 4-hydroxyacetamide metabolites of phenacetin, Chem. Biol. Interact. 8:87 (1974).

17. D.J. Miner and P.T. Kissinger, Evidence for the involvement of N-acetyl-p-quinone imine in acetaminophen metabolism, Biochem. Pharmacol. 28:3285 (1979).

18. I.A. Blair, A.R. Boobis, and D.S. Davies, Paracetamol oxidation: Synthesis and reactivity of N-acetyl-p-benzoquinone imine, Tetr. Letters, 21:4947 (1980).

19. D.C. Dahlin and S.D. Nelsom, Synthesis decomposition kinetics and preliminary toxicological studies of pure N-acetyl-p-benzoquinone imine, a proposed toxic metabolite of acetaminophen, J. Med. Chem. 25:885 (1982).

20. E. Albano, M. Rundgren, P.J. Harvison, S.D. Nelson, and P. Moldéus, Mechanisms of N-acetyl-p-benzoquinone imine (NAPQI) cytotoxicity, Molec. Pharmacol. in press, (1985).

21. J.A. Hinson, B. Coles, S.D. Nelson, and B. Ketterer, Glutathione transferase catalyzed conjugation of the reactive metabolite of acetaminophen (Abstract), IUPHAR 9th International Congress of Pharmacology, London (1984).

22. C.R. Fernando, I.C. Calder, and K.N. Ham, Studies on the mechanism of toxicity of acetaminophen. Synthesis and reactions of N-acetyl-2,6-dimethyl- and N-acetyl-3,5-dimethyl-p-benzoquinone imines, J. Med. Chem. 23:1153 (1980).

23. G.M. Rosen, E.J. Rauckman, S.P. Ellington, D.C. Dahlin, J.C. Christie, and S.D. Nelson, Reduction and glutathione conjugation reactions of N-acetyl-p-benzoquinone imine and two dimethylated anaogues, Molec. Pharmacol.25:151 (1984).

24. D.J. Porubek, M. Rundgren, S.D. Nelson, and P. Moldéus, Investigation of acetaminophen toxicity with an acetaminophen analogue, 3,5-dimethylacetaminophen (in preparation).

25. O. Spühler and H.U. Zollinger, Die Chronische-Interstitille Nephritis, Z. Klin. Med. 151:1 (1953).

26. U. Bengtsson, A comparative study of chronic non-obstructive pyelo-nephritis and renal papillary necrosis, Acta Med. Scand. (Suppl.) 388:5 (1962).

27. N. Hultengren, C. Lagergren, and A. Ljungqvist, Carcinoma of the renal pelvis in renal papillary necrosis, Acta Clin. Scand. 130:314 (1965).

28. S. Johansson and L. Wahlqvist, Tumors of urinary bladder and ureter associated with abuse of phenacetin-containing analgesics, Acta Pathol. Microbiol. Scand. Sect. A. 85:768 (1977).

29. R.L. Smith and J.A. Timbell, Factors affecting the metabolism of phenacetin. I. Influence of dose, chronic dosage route of administration and species on the metabolism of ($1-^{14}$C-acetyl) phenacetin, Xenobiotica, 4:489 (1974).

30. I.C. Calder, D.E. Goss, P.J. Williams, C.C. Funder, C.R. Green, K.N. Ham, and J.D. Tange, Neoplasia in the rat induced by N-hydroxy-phenacetin, a metabolite of phenacetin, Pathology, 8:1 (1976).

31. J.B. Vaught, P.B. Mc Garvey, M.-S. Lee, C.D. Garner, C.Y. Wang, E.M. Linsmaier-Bednar, and C.M. King, Activation of N-hydroxyphenacetin to mutagenic and nucleic acid binding metabolites by acyl transfer deacylation and sulfate conjugation, Cancer Res. 41:3424 (1981).

32. B. Samuelsson, M. Goldyne, E. Granström, M. Hamburg, S. Hammarström, and C. Malmsten, Prostaglandins and thromboxanes, Ann. Rev. Biochem. 47:997 (1978).

33. L.J. Marnett and T.E. Eling, Cooxidation during prostglandin biosynthesis: A pathway for the metabolic activation of xenobiotics, in: "Reviews in Biochemical Toxicology 5", E. Hodgson, J.R. Bend, and R.M. Philpot, eds., Elsevier Biomedical, New York (1983).

34. D. Ross, R. Larsson, B. Andersson, U. Nilsson, T. Lindqvist, B. Lindeke, and P. Moldéus, The oxidation of p-phenetidine by horse radish peroxidase and prostaglandin synthetase and the fate of glutathione during such reactions, Biochem. Pharmacol. 34:343 (1985)

35. B.C. Saunders, Peroxidases and catalases, in: "Inorganic Biochemistry 2", G.L. Eichhorn, ed., Elsevier, New York (1973).

36. D. Ross, R. Larsson, K. Norbeck, R. Ryhage, and P. Moldéus, Charac-

terization and mechanism of formation of reactive products formed during peroxidase-catalyzed oxidation of p-phenetidine, <u>Molec. Pharmacol</u>. 27:277 (1985).

THE ROLE OF REACTIVE INTERMEDIATES IN SULFHYDRYL-DEPENDENT IMMUNO-
TOXICITY: INTERFERENCE WITH MICROTUBULE ASSEMBLY AND MICROTUBULE-
DEPENDENT CELL FUNCTION

Richard D. Irons

Chemical Industry Institute of Toxicology
Research Triangle Park, NC 27709

INTRODUCTION

The critical role of protein and low molecular weight thiols in
cell homeostasis has long been appreciated. Sulfhydryl(SH) reagents
have been shown to interfere with numerous biochemical processes critical
for cell integrity including numerous SH-dependent enzymes, protein, RNA
and DNA synthesis and in the maintenance of normal intracellular redox
potential. On a different level, the lymphocyte appears to be unusually
sensitive to certain SH-reagents at concentrations that do not result in
non-specific cytotoxicity or cell death. SH reagents, for example, are
potent suppressors of normal immune cell function via mechanisms that
apparently are distinct from cytotoxic events frequently associated with
the exposure to SH-reactive intermediates. Both surface and intracellular SH groups have been shown to be involved in lymphocyte activation
and maintenance of normal cell function, however, this discussion will
be restricted largely to membrane penetrating SH-reagents.

Our interest in SH-dependent lymphocyte function evolved from the
study of the quinone metabolites of benzene, which apart from inducing
cytotoxicity at concentrations comparable to the quinoneimines previously discussed by Dr. Moldeus, suppress normal lymphocyte blastogenesis
and antibody secretion at concentrations that do not result in cell
death (Irons et al., 1981; Pfeifer and Irons, 1981). Subsequently we
have extended this observation to include a variety of membrane penetrating SH-alkylating reagents, in addition to quinones, that contain
α,β-unsaturated diketone structures and therefore carbons that act as
Michael acceptors (Irons et al., 1981; Pfeifer and Irons, 1981). These

prototype compounds are N-ethylmaleimide (NEM); p-benzoquinone (pBQ), and cytochalasin A.

Many compounds will suppress or inhibit lymphocyte mitogenic response. Most interfere with energy production, protein, RNA or DNA synthesis, or some critical function necessary for cell integrity, and consequently inhibition of blastogenic response cannot be distinguished from cell death. A particular distinction of the membrane penetrating SH-reactive compounds is their ability to interfere with intracellular thiols and block a variety of cell functions, including blastogenesis, antibody secretion, phagocytosis and lymphocytic cytotoxic responses at concentrations that do not result in a loss of plasma membrane integrity (Chaplin and Wedner, 1978; Irons et al., 1981), alter receptor binding (Berlin and Ukena, 1972; Sachs et al. 1973; Green, et al., 1976) or reduce intracellular [GSH] or [ATP] (Mazur and Williamson, 1977; Chakravarty and Echetebu, 1978; Chaplin and Wedner, 1978; Lagunoff and Wan, 1979; Irons et al., 1981; Pfeifer and Irons 1981, 1982). Examination of both the SH- sensitivity and the time dependence using a variety of biochemical and whole cell systems suggests the most sensitive intracellular thiol groups implicated in non-cytotoxic suppression of lymphocyte blastogenesis are associated with the cytoskeleton, more specifically microtubules (Irons et al., 1981; Pfeifer and Irons 1981, 1982, 1983). The peculiar sensitivity of microtubules to SH-alkylating reagents may be related to their role as a specific target for endogenous molecules, such as soluble immune response suppressor (SIRS), involved in regulating the immune response (Irons et al., 1984).

Lectin-Induced Lymphocyte Blastogenesis

Nowell first discovered that certain plant lectins, such as phytohemagglutinin (PHA), bind specifically to the lymphocyte plasma membrane and induce DNA synthesis and division in normal lymphocytes (1960). The binding of macromolecules such as lectins or antibodies to the lymphocyte cell surface triggers blastogenesis and is therefore a useful tool for studying membrane associated regulatory mechanisms involved in lymphoid proliferation. A large number of biochemical changes have been described following lectin-binding to the cell, however, none have been conclusively shown to be the absolute signal. Regardless of the absolute mechanism of signal transduction, definitive temporal associations can be made between these various processes and deterministic events resulting in lymphocyte blastogensis. The initial triggering event

appears to be confined to the plasma membrane and involves the binding
of a cross-linking ligand, such as a plant lectin, to specific sac-
charide groups on surface glycoproteins. This initial binding presumably leads to structural and/or functional alterations in the surface
receptor molecule that is necessary for signal transduction. There are
structural changes that immediately follow lectin binding. These
include surface receptor redistribution (patching and capping) and
cell-cell agglutination. Although the relationship of these events to
blastogenesis is not known, they can be detected minutes after the
addition of mitogenic divalent ligands to the cells and are associated
with blastogenesis. The cytoskeletal dependence of receptor redistribution and cell agglutination is well established (Edelman, 1973; Wang et
al., 1975; Gunther et al., 1976; McClain and Edelman, 1980).

During the first two hours following receptor activation, there is
a transient depression in basal protein synthesis that is accompanied by
a redistribution of existing mRNA associated with ribosomal units (Jagus
and Kay, 1979). At this time there is no measurable RNA synthesis.
This is followed by increased protein synthesis and then RNA synthesis.
Protein synthesis is an absolute requirement for blastogenesis (Milner,
1978; Kay, 1980).

RESULTS AND DISCUSSION

Pretreatment of resting lymphocytes with SH-alkylating reagents for
as little as 15 to 30 min is sufficient to achieve complete suppression
of lectin-induced blastogenesis (Pfeifer and Irons, 1981, 1982, 1983).
These effects are protected against by addition to the medium of low
molecular weight thiols, such as dithiothreitol (DTT) or cysteine (CYS)
but not other nucleophiles. Membrane impermeable or poorly penetrating
SH- reagents, such as 5,5'-dithiobis-2-nitrobenzoic acid (DTNB), do not
suppress blastogenesis at similar concentrations.

In order to explore further the effects of membrane SH-reactive
compounds on microtubule dependent cell function we examined the potency
and inhibition kinetics of these reagents in a variety of independent
biochemical and whole cell systems. The ability of cytochalasin A, pBQ
and NEM to suppress normal PHA blastogenic response is illustrated in
Fig. 1. As previously indicated, no alterations in cell membrane
integrity, [ATP], or [GSH] are observed at these concentrations, how-

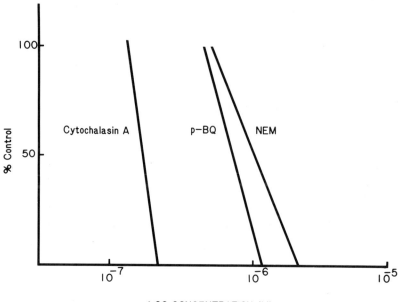

Fig. 1.　Inhibition of PHA-stimulated lymphocyte proliferative
response by membrane penetrating sulfhydryl reagents.
Regression lines calculated for ^{3}H-thymidine incorporation into PHA stimulated murine lymphocytes expressed
as a percent of control.　Conditions same as described
(Pfeifer and Irons, 1983).

ever, normal lymphocyte agglutination is blocked (Pfeifer and Irons,
1981).　As mentioned in the introduction, protein synthesis is an absolute requirement for lymphocyte blastogenesis. Therefore, suppression of
blastogenesis could conceivably result from interference with protein
synthesis.　However, an examination of the concentration dependence of
suppression of blastogenesis and inhibition of basal protein synthesis
in resting lymphocytes indicates that these compounds effectively suppress DNA synthesis at concentrations that do not interfere with protein
synthesis (Fig. 2).　This is contrasted by the effects of anisomycin, a
selective inhibitor of peptidyl transferase, for which no distinction
can be made between the concentration that blocks blastogenesis and that
inhibiting basal protein synthesis (Fig. 3).

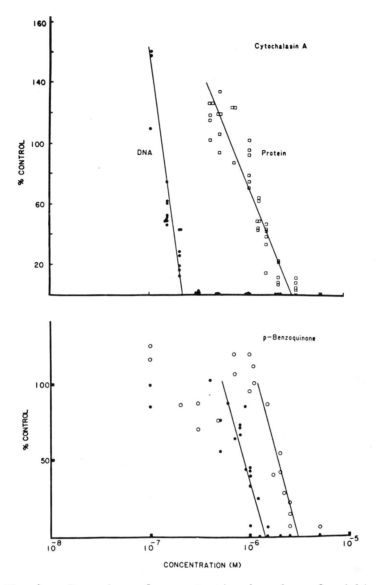

Fig. 2. Comparison of concentration-dependence for inhibition of PHA stimulated lymphocyte blastogenesis (DNA) and basal protein synthesis (Protein)
for Cytochalasin A and pBQ. Values expressed as
percent of control and represent the mean of 4
separate determinations as measured by the incorporation of ^{3}H-thymidine into PHA stimulated
murine splenic lymphocytes (DNA) or or ^{3}H-
leucine into resting lymphocytes (Protein) following 30 min. pretreatment with Cytochalasin A
or pBQ.

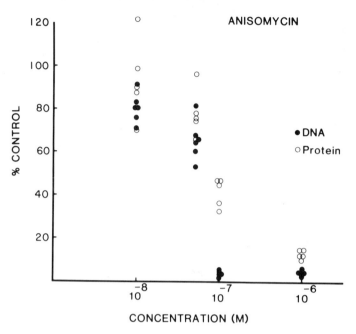

Fig. 3. Comparison of the concentration-dependent inhibition of PHA stimulated lymphocyte blastogenesis
(DNA) and basal protein synthesis (Protein) for
anisomycin. Conditions identical to those described in Fig. 2.

Because lymphocyte activation and agglutination correlate temporally
and functionally with surface receptor redistribution, the effects of
these reagents on capping was examined. The integrity of microfilaments
and active cellular metabolism are required for the redistribution of
surface receptors after binding of a divalent lectin or ligand (Bourguignon and Singer, 1977), therefore, capping is inhibited by agents
that interfere with microfilaments or energy production (Loor, 1977).
Extensive cross-linking of receptors by tetravalent ligands such as con
A restrict surface receptor movement, prevent cap formation and block
blastogenesis (Loor, 1977; McClain and Edelman, 1976). This inhibition
of receptor movement is microtubule dependent. Microtubule disrupting
agents facilitate cap formation in the presence of excess con A. This
system can therefore be utilized as a means to discriminate among cytoskeletal disruptive agents with respect to their functional specificity.

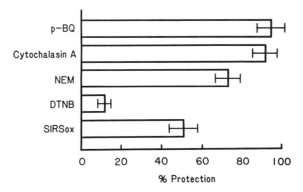

Fig. 4. Facilitation of lymphocyte capping by sulfhydryl re-
         agents and SIRSox in the presence of inhibitory con-
         centrations of concanavalin A. Results are expressed
         as the mean ± SD for three separate determinations.
         Conditions essentially as previously described (Irons
         et al., 1981, 1984).

Agents that inhibit energy production or disrupt microfilaments inhibit
capping, whereas agents that selectively disrupt microtubules at concentrations that do not interfere with energy production or microfilaments,
overcome the inhibition of capping by con A (Fig. 4).

Intact microtubules represent a dynamic process of equilibrium
between assembly and disassembly of tubulin dimers. Therefore, any
agent interfering directly with microtubule assembly should disrupt
intact microtubules in resting cells. Cytochalasin A, p-BQ and NEM
effectively disrupt the cytoplasmic array of microtubules in intact
fibroblasts at micromolar concentrations (Irons et al., 1984). Microtubules are in intimate association with calmodulin, and the dynamic
equilibrium of microtubule assembly in whole cells is governed by a
complex set of interactions involving calcium and calmodulin (Watanabe
and West, 1982). As previously indicated by Dr. Moldeus, alterations in
calcium homeostasis associated with cell death, such as those observed
following treatment of cells with quinoneimines (~ 250 µM), result in a

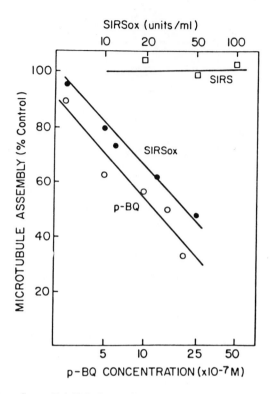

Fig. 5.   Inhibition of microtubule assembly by
pBQ and SIRSox.   Porcine brain tubulin (1.7 mg/ml)
was incubated with pBQ, SIRS or SIRSox for 1 min.
at 37°C before addition of 2.5 mM GTP.   Microtubule
assembly was measured spectrophotometrically at
350 nm.   Reproduced with permission from (Irons et
al., 1984).

disruption of the normal cytoplasmic array of microtubules.   Nevertheless, examination of the effects of SH-alkylating reagents on calcium-
calmodulin-independent microtubule assembly in vitro (Fig. 5) reveals
that these reagents directly interfere with MT assembly at micromolar or
lower concentrations independent of their effects on calcium metabolism
observed at higher concentrations.   These reagents interfere with MT
assembly at a ratio betweeen 0.1-2 to 1, depending on the source of
tubulin, with IC50's for porcine brain MT assembly ranging between 10^{-7}
and 10^{-6}M (Irons and Neptun, 1980; Pfeifer and Irons, 1982; Irons et al.,
1981, 1984).

The molecular mechanism of ·SH-dependent disruption of MT assembly appears to involve interference with GTP binding to the tubulin dimer (Irons and Neptun, 1980; Irons et al., 1981). The sensitivity of this site to SH-alkylating reagents lends well to the hypothesis that GTP binding sites on tubulin serve as a target for endogenous molecules involved in the regulation of cell function and growth. Soluble immune response suppressor (SIRS) is a protein synthesized by activated T suppressor cells that, when converted to its active form (SIRSox) by macrophages or H_2O_2, suppresses immune response in vitro and inhibits division in normal and neoplastic cells (Aune and Pierce, 1981; Aune et al., 1981; Aune et al., 1983). The effects of SIRSox on blastogenesis, surface receptor redistribution, disruption of intact microtubules and inhibition of microtubule assembly in vitro suggest a pattern of reactivity similar to SH-dependent cytoskeletal disrupting agents (Irons et al., 1984). The potency of SIRSox deserves comment. Conservative estimates of the IC50 for MT assembly by SIRSox range between ~ 10^{-16} and 10^{-12}M. Although H_2O_2 is necessary for the activation of SIRS to SIRSox, it is not consumed in reaction of SIRSox with SH groups, the chemistry of which remains obscure (Aune and Pierce, 1981a; Aune et al., 1983; Irons et al., 1984). The partial characterization of the mechanism of action of this novel protein suggests that membrane penetrating SH-reactive compounds, such as quinones and α,β-unsaturated diketones, may exert their potent effects at sub-cytotoxic concentrations by mimicking the activity of endogenous regulatory molecules.

REFERENCES

Aune, T. M., Webb, D. R. and Pierce, C. W., 1983, Purification and initial characterization of the lymphokine soluble immune response suppressor, J. Immunol., 131:2848.

Aune, T. M., and Pierce, C. W., 1981a, Mechanism of action of macrophage-derived suppressor factor produced by soluble immune response suppressor-treated macrophages, J. Immunol., 127:368.

Aune, T. M., and Pierce, C. W., 1981b, Identification and and initial characterization of a nonspecific suppressor factor produced by soluble immune response-treated macrophages, J. Immunol., 127:1828.

Berlin, R. D., and Ukena, T. E., 1972, Effect of colchicine and vinblastine on the agglutination of polymorphonuclear leukocytes by concanavalin A, Nature New Biol., 238:120.

Bourguignon, L. Y. W., and Singer, S. K., 1977, Transmembrane interactions and the mechanism of capping of surface receptors by their specific ligands, Proc. Natl. Acad. Sci. USA, 74:5031.

Chakavarty, N., and Echetebu, Z., 1978, Plasma membrane adenosine triphosphatases in rat peritoneal mast cells and macrophages- the relation of the mast cell enzyme to histamine release, Biochem. Pharmacol., 27:1561.

Chaplin, D. D., and Wedner, H. J., 1978, Inhibition of lectin-induced lymphocyte activation by diamide and other sulfhydryl reagents, Cell. Immunol., 36:303.

Edelman, G. M., 1973, Surface alterations and mitogenesis in lymphocytes, in: "Control of Proliferation in Animal Cells," Cold Spring Harbor Laboratories, NY.

Greene, W. C., Parker, C. M., and Parker, C. W., 1976, Colchicine sensitive structures and lymphocyte activation, J. Immunol., 117:1015.

Gunther, G. R., Wang, J. L. and Edelman, G. M., 1976, Kinetics of colchicine inhibition of mitogenesis in individual lymphocytes, Exptl. Cell Res., 98:15.

Irons, R. D. and Neptun, D. A., 1980, Effects of the principal hydroxy-metabolites of benzene on microtubule polymerization, Arch. Toxicol., 45:297.

Irons, R. D., Neptun, D. A., and Pfeifer, R. W., 1981, Inhibition of lymphocyte transformation and microtubule assembly by quinone metabolites of benzene: Evidence for a common mechanism, J. Reticuloendothelal Soc., 30:359.

Irons, R. D., Pfeifer, R. W., Aune, T. M. and Pierce, C. W., 1984, Soluble immune response suppressor (SIRS) inhibits microtubule function in vivo and microtubule assembly in vitro, J. Immunol., 133:2032.

Jagus, R. and Kay, J. E., 1979, Distribution of lymphocyte RNA during stimulation by PHA, Eur. J. Biochem., 100:503.

Kay, J. E., 1980, Protein synthesis during activation of lymphocytes by mitogens, Biochem. Soc. Transact., 288.

Lagunoff, D., and Wan, H., 1979, Inhibition of histamine release from rat mast cells by cytochalasin A and other sulfhydryl reagents, Biochem. Pharmacol., 28:1765.

Loor, F., 1977, Structure and dynamics of the lymphocyte surface in relation to differentiation, recognition and activation, Prog. Allergy, 23:1.

Mazur, M. T. and Williamson, J. R., 1977, Macrophage deformability and phagocytosis, J. Cell Biol, 75:185.

McClain, D. A., and Edelman, G. M., 1980, Density dependent stimulation and inhibition of cell growth by agents that disrupt microtubules, Proc. Natl. Acad. Sci. USA, 77:2748.

Milner, J., 1978, Is protein synthesis necessary for the commitment of lymphocytes to transformation?, Nature, 272:628.

Nowell, P. C., 1960, Phytohemagglutinin: An initiator of mitosis in cultures of normal human leukocytes, Cancer Res., 20:462.

Pfeifer, R. W. and Irons, R. D., 1981, Inhibition of lectin-stimulated lymphocyte agglutination and mitogenesis by hydroquinone: Reactivity with intracellular sulfhydryl groups, Exptl. Mol. Pathol., 35:189.

Pfeifer, R. W., and Irons, R. D., 1982, Effect of benzene metabolites on phytohemagglutinin-stimulated lymphopoiesis in rat bone marrow, J. Reticuloendothelial Soc., 31:155.

Pfeifer, R. W., and Irons, R. D., 1983, Alteration of lymphocyte function by quinones through sulfhydryl-dependent disruption of microtubule assembly, Intl. J. Immunopharmacol., 5:463.

Wang, J. L., Gunther, G. R. and Edelman, G. M., 1975, Inhibition by colchicine of the mitogenic stimulation of lymphocytes prior to S phase, J. Cell Biol., 66: 128.

Watanabe, K. and West, W.L., 1982, Calmodulin, activated nucleotide phosphodiesterase, microtubules and vinca alkaloids, Fed. Proc., 41:2292.

INVESTIGATION OF THE IMMUNOLOGICAL BASIS

OF HALOTHANE-INDUCED HEPATOTOXICITY

Hiroko Satoh,   James R. Gillette, Tamiko Takemura,[+] Victor
J. Ferrans,[+] Sandra E. Jelenich,[#] John G. Kenna, James
Neuberger, and Lance R. Pohl

Laboratory of Chemical Pharmacology, National Heart, Lung
and Blood Institute,[+]Pathology Branch, Ultrastructure
Section, National Heart, Lung, and Blood Institute
and [#]Department of Anesthesiology, Clinical Center
National Institutes of Health, Bethesda, Maryland, 20892
The Liver Unit, King's College Hospital and School
of Medicine and Dentistry, Denmark Hill, London SE5 9RS
U.K.

INTRODUCTION

It is well established that halothane ($CF_3CHClBr$), an inhalation
anesthetic, causes both a mild and a severe form of hepatotoxicity in
patients.[1] The milder form of hepatotoxicity is characterized by minor
elevations in serum transaminase levels and has been reported in about
20% of patients anesthetized with halothane.[2,3] The severe form of
hepatotoxicity, however, is often fatal and is much rarer.[4,5] Most of
the patients with the severe disease have high serum transaminase values
and massive hepatic necrosis. The necrosis is often centrilobular,[6]
although other histologic lesions have been reported.[7,8]

Reactive Metabolites of Halothane

The metabolism of halothane has been extensively studied in order to
determine whether a metabolite or metabolites may be involved in these
toxicities. It is now clearly recognized that liver microsomal cytochrome P-450 metabolizes halothane into two reactive products. Under
anaerobic conditions, $CF_3CHClBr$ is reduced by cytochrome P-450 to produce
the reactive radical intermediate, 1-chloro-2,2,2-trifluoroethyl radical
($CF_3CHCl.$).[9-12] This product can react with liver microsomal protein,
lipid, and presumably other unidentified target substances in the liver,
to form adducts or can abstract a hydrogen atom to produce 1-chloro-
2,2,2,-trifluoroethane (CF_3CH_2Cl). The radical can also be reduced
further by cytochrome P-450 to form 1-chloro-2,2,2-trifluoroethyl
carbanion ($CF_3CHCl:^{-1}$), which can eliminate fluoride to produce 1-chloro-
2,2-difluoroethylene (CF_2CHCl). In contrast, liver microsomes in air
catalyze the oxidation of halothane to a trifluoroacetyl halide (CF_3COX)
intermediate that either acylates tissue molecules to form trifluoroacetylated (TFA, CF_3CO-) adducts or reacts with water to form trifluoro-
acetic acid (CF_3COOH).[9,13-17]

Mild Form of Hepatotoxicity in Animals

The mild form of hepatotoxicity appears to have been produced in animals. Most of the studies with rats indicate that the toxicity is due to the reductive radical metabolite. For example, it was found that substitution of a deuterium atom in place of a hydrogen atom in halothane ($CF_3CDClBr$) slowed the rate of the oxidative pathway of metabolism without decreasing the extent of hepatotoxicity.[16,18] Other studies have shown that treatment of rats with halothane under moderate hypoxia (14% O_2), enhanced both the rate of the reductive pathway of metabolism and the extent of the hepatotoxicity.[10] Similarly, females rats were found to metabolize halothane more slowly by the reductive pathway than males and to be less susceptible than males to the hepatotoxic effect of halo- thane.[19] In addition, treatment of rats with cimetidine selectively inhibited the reductive pathway of halothane metabolism and provided partial protection against its hepatotoxic effect.[20] In contrast to these results, recent studies with a guinea-pig model of halothane-associated hepatotoxicity suggested that either the oxidative or reductive metabo- lites may produce the hepatotoxicity.[21]

Evidence for an Immune Basis of the Fulminant Form of Hepatotoxicity

Progress in understanding the basis of the fulminant form of hepatotoxicity has been considerably slower than that of the mild form of toxicity, mainly because no animal model has been developed to study this disease. Nevertheless, recent findings have indicated that this rare toxicity may have an immune basis.

It had been suggested several years ago that the fulminant form of halothane-induced hepatotoxicity might have an immune basis, because most of the patients with this toxicity had received halothane on previous occasions[6,22-24] and because its clinical features, such as eosinophilia, fever, rash, and serum liver-kidney microsomal autoanti- bodies were similar to those found in idiosyncratic drug sensitization reactions.[8,25] This hypothesis has been strengthened by several lines of evidence. For example, a cell migration test revealed that leucocytes from 8 of 12 patients with unexplained fulminant hepatic failure after halothane were sensitized to a cell-subfraction of liver homogenate from halothane-treated rabbits.[26] Similar evidence for cellular sensiti- zation to an antigen formed during halothane anesthesia was obtained by a direct lymphocyte cytotoxicity assay.[27] Furthermore, specific circu- lating antibodies were found only in the sera of patients with fulminant hepatic failure after several episodes of anesthesia.[28] These antibodies were shown by indirect immunofluorescence to react with the cell surface of hepatocytes from halothane-treated rabbits. Moreover, they rendered the hepatocytes susceptible to antibody-dependent cell-mediated cyto- toxicity (ADCC). These results suggested that halothane-induced fulmin- ant hepatotoxicity may be initiated by a reactive metabolite that alters the surface structure of hepatocytes and in susceptible individuals, induces an immune response, which in turn leads to fulminant hepato- toxicity.

Recent investigations have suggested that it is the oxidative metabolite, CF_3COX, and not the reductive radical metabolite, CF_3CHCl., of halothane that alters the surface of the rabbit hepatocytes so that they are recognized by the antibodies from the halothane hepatitis patients.[29] For example, only hepatocytes from rabbits administered halothane at oxygen tensions that promoted its oxidative and not its reductive metabolism were susceptible to ADCC induced by the human antibodies. Subsequently, sensitive peroxidase enzyme-linked immuno- sorbent and immunofluorescence antibody staining methods for identifying trifluoroacetylated (TFA) hepatocytes were developed;[30] these techniques were based upon the development of an antiserum that specifically recognized bound TFA groups. It was observed that liver sections prepared from rats at 4 hr after halothane administration were stained preferentially in the centrilobular regions with anti-TFA serum. The methods were shown to be specific for the TFA-moiety, because the staining was completely inhibited by preincubation of the anti-TFA serum with N-epsilon-TFA-L-lysine and because treatment of rats with deuterium labeled halothane (halothane-d), which was oxidatively metabolized more slowly than was halothane,[16] resulted in significantly less staining than that observed with halothane. It was additionally found that rat hepatocytes isolated 24 hr after the administration of halothane contained TFA adducts on their surface membranes.[30] This finding indicated that CF_3COX either reacted directly with constituents of the plasma membrane or with other cellular components which became incorporated into the plasma membrane.

Present Investigation

In the present investigation we have extended these immunochemical studies. A major target of CF_3COX in hepatocytes from halothane treated rats has been identified as a 54 Kd form of cytochrome P-450, which was detected in both the microsomal and plasma membrane fractions of the cell. The enzyme, however, did not appear to be inactivated by CF_3COX. Similar TFA adducts were also detected in liver homogenates of patients that had received halothane. Moreover, 2 of 6 patients that had been diagnosed as having the fulminant form of halothane hepatitis appeared to contain anti-TFA antibodies in their serum. Immunochemical examination of isolated hepatocytes and hepatocyte plasma membranes from rats not treated with halothane revealed that the plasma membrane contained cytochromes P-450 and cytochrome P-450 reductase. These findings suggest that CF_3COX or reactive metabolites of other hepatotoxic drugs or environmental chemicals may be produced in the plasma membrane where they might cause direct cellular damage or initiate an immune response by forming an immunogenic adduct.

MATERIALS AND METHODS

Materials

Chemicals were obtained from the following sources: microtiter plates (Immulon 1) from Dynatech (Alexandria, VA); goat anti-rabbit IgG-horseradish peroxidase (HRP) conjugate, goat anti-human IgG-HRP conjugate, and 4-chloro-1-naphthol from Bio-Rad (Richmond, CA); peroxidase rabbit anti-peroxidase complex (PAP) from Miles Laboratories (Naperville, IL); 2,2'-azino-di-(3-ethylbenzthiazoline sulfate) diammonium salt and 3,3'-diaminobenzidine tetrahydrochloride from Sigma Biochemicals (St. Louis, MO); trifluoracetylated rabbit serum albumin (TFA-RSA) and rabbit anti-TFA serum were prepared as previously described;[30] cytochromes P-450 (52 Kd and 54 Kd forms) and anti-serum were prepared as described elsewhere;[31] cytochrome P-450 reductase was purified by the method of Yasukochi and Masters;[32] halothane was distilled before it was used and halothane-d was prepared as described previously.[16]

Rat and Human Tissue Samples

Rat liver. Male Sprague-Dawley rats (150-250g) were obtained from Taconic Farms (Germantown, NY) and maintained on Purina rat chow and water ad libitum. Animals were treated daily with sodium phenobarbital (80 mg/kg in normal saline, ip) for 3 days. At 24 hr after the last treatment, the rats were administered halothane, halothane-d (10 mmol/kg in 50% sesame oil solution, v/v, ip) or sesame oil (ip) or were not further treated. After 4hr, the animals were killed by decapitation. Their livers were immediately perfused with approximately 30 mls of ice cooled phosphate buffered saline (PBS), removed, and placed on ice.

Human liver. Liver biopsy samples (approximately 1 g portion) were taken from the following four advised and consenting patients undergoing major abdominal surgery at the NIH Clinical Center: Patient 1, a 24 year old man; patient 2, a 15 year old woman; patient 3, a 62 year old woman; and patient 4, a 26 year old woman. Liver samples were taken intra-operatively after anesthetic induction with pentothal, nitrous oxide, narcotic and muscle relaxant and 2 to 4 hr after the beginning of halothane administration. The samples were placed on ice and within 1 hr rapidly frozen in a dry-ice acetone bath and stored at -80°C until used.

Human serum samples. Serum samples were taken from 9 advised and consenting patients. Patients 1-6 were halothane fulminant hepatitis patients, whereas 7 and 8 were paracetamol overdose hepatitis patients, and 9 was a control patient. Patients 1-8 and 9 were admitted to King's College Hospital and the NIH Clinical Center respectively. Patient 1, a 43 year old woman, had been administered halothane on 4 occasions. Serum was obtained 26 days after the last administration of halothane, at which time the patient died. Patient 2, a 53 year old woman, had been administered halothane on 2 occasions. Serum was obtained 14 days after the last administration of halothane. The patient died on day 21. Patient 3, a 48 year old woman, had been administered halothane on 2 occasions. Serum was obtained 20 days after the last administration of halothane. The patient died on day 23. Patient 4, a 14 year old woman, had been administered halothane on 3 occasions. Serum was obtained 14 days after the last administration of halothane. The patient survived. Patient 5, a 46 year old man, had been administered halothane on 3 occasions. Serum was obtained 7 days after the last administration of halothane. The patient died on day 11. Patient 6, a 57 year old woman, had been administered halothane on 2 occasions. Serum was obtained 42 days after the last administration of halothane. The patient died on day 49. Patient 7, a 40 year old woman, took an overdose of paracetamol. Serum was obtained 6 days after paracetamol was ingested. The patient survived. Patient 8, a 36 year old man, took an overdose of paracet-amol. Serum was obtained 4 days after paracetamol was ingested. The patient died on day 6. A serum sample was taken from patient 9, a 31 year old woman, before halothane was administered for the first time.

Preparation of Subcellular Fractions of Liver

Rat liver. The total homogenate, cytosol, and microsomal fractions of rat liver were isolated in the following manner. Livers were homogenized in 3 volumes of ice-cooled 20 mM Tris-HCl (pH 7.4) containing 1.15% (w/v) KCl. A portion of this total homogenate was saved and the remainder was centrifuged at 9,000g for 20 min at 4°C. The supernatant was collected and centrifuged at 105,000g for 80 min at 4°C. The resultant supernatant (cytosol fraction) was saved and the microsomal pellet was resuspended and recentrifuged at 105,000g for 80 min. The microsomal pellet was resuspended in 10 mM Tris-acetate (pH 7.4) containing 20% (v/v) glycerol and 1 mM EDTA to a protein concentration of approximately 20 mg/ml (microsomal fraction).

A plasma membrane fraction was isolated by a modification of the two phase polymer procedure of Lesko et al.[33] A 30 g portion of liver was homogenized in 400 ml of ice-cooled 0.5 mM calcium chloride containing 1 mM sodium bicarbonate, pH 7.5,(buffer A) with a Dounce

Homogenizer (25 excursions with a loosely fitting pestle). The homogenate was diluted to 3000 ml with buffer A, allowed to stand for 5 min with occassional stirring, and passed through 4 layers of cheese cloth. The filtrate was then centrifuged in a swinging bucket rotor (SBR) at 1350g for 30 min at 4°C. The pellet was resuspended in buffer A, gently homogenized (5 excursions), diluted to 1500 ml, and recentrifuged at 1000g for 15 min at 4°C. After homogenizing and resuspending the pellet an additional time in a total volume of 750 ml of buffer A, it was collected after centrifugation (SBR) and suspended in 260 ml of the top phase of the polymer mixture. Portions (20 ml) of the suspension were mixed gently with equal volumes of the bottom phase of the polymer mixture and centrifuged (SBR) at 1100g for 15 min at 4°C. The plasma membrane fraction, which was located at the interface of the two polymer solutions, was collected, resuspended in 85 ml of the top phase, mixed with 85 ml of the bottom phase, and recentrifuged (SBR) at 1100g for 15 min at 4°C. The plasma membrane fraction was further purified by repeating the latter isolation procedure once more. It was then resuspended in 100 ml of 50 mM Tris-HCl (pH 7.4) and centrifuged (SBR) at 2400g for 10 min at 4°C. This washing procedure was repeated an additional two times. The washed pellet was resuspended in a small volume of 10 mM Tris-acetate (pH 7.4) containing 20% (v/v) glycerol and 1 mM EDTA.

Human liver. Homogenates (25%, w/v) of the biopsy samples were prepared in 50 mM Tris-HCl buffered saline (pH 7.4, TBS). They were centrifuged at 9400g for 15 min at 4°C in a microfuge centrifuge. The supernatants (200 ul) were collected and centrifuged at 20 psi for 15 min at 4°C in an airfuge. The resulting microsomal pellets were washed with 150 ul of TBS, recentrifuged in the airfuge, and resuspended in 30 ul of TBS by sonicating it for 20 seconds.

All isolated cellular fractions were rapidly frozen in a dry-ice acetone bath and stored at -80°C until used.

Immunoblotting of Components of the Liver Subcellular Fractions with Anti-TFA, anti-Cytochrome P-450 (52 Kd and 54 Kd), and Anti-Cytochrome P-450 Reductase Serum

The protein constituents of the subcellular fractions were separated by sodium dodecyl sulfate polyacrylamide electrophoresis (SDS/PAGE), electrophoretically transferred to nitrocellulose sheets (SDS/PAGE blot), and stained by the immunoperoxidase method as described in detail elsewhere.[31]

Immunoelectron microscopic Examination of Hepatocytes with Anti-Cytochrome P-450-54 Kd IgG

Anti-cytochrome P-450-54 Kd IgG was purified from antiserum on an affinity column of cytochrome P-450-54 Kd by the method of Thomas et al..[34] The anti-cytochrome P-450-54 Kd IgG and normal rabbit IgG were conjugated to HRP with glutaraldehyde by the procedure of Avrameas and Ternynck.[35] Single cell suspensions of hepatocytes from phenobarbital pretreated rats were prepared as described previously.[30] Portions (100 ul) of the cell suspension (10^7 cells/ml) were diluted with PBS containing 0.1% (w/v) BSA (PBS-BSA) and collected by centrifugation at 10g for 3 min at 4°C. The cells were incubated with PBS-BSA containing 5% (v/v) normal rabbit serum for 20 min at 4°C and recollected by centrifugation. Anti-cytochrome P-450-54 Kd IgG-HRP conjugate or normal rabbit IgG-HRP conjugate (270 ug/ml in PBS-BSA) was reacted with the cells for 30 min at 4°C followed by their washing 3 times with PBS-BSA (15 min per wash). After being fixed with 1.25 % (v/v) glutaraldehyde in PBS for 15 min at $4^{\circ}C$, the cells were washed with PBS 3 times, preincubated with 0.03% (w/v) 3,3'-diaminobenzidine tetrahydrochloride in 50 mM Tris-HCl, pH 7.6, (DAB) for 30 min at room temperature, washed again, and then incubated with DAB containing 0.01% (v/v) hydrogen peroxide for 5 min at room temperature. The cells were washed 3 times with PBS, postfixed with 1% (w/v) osmium tetroxide in PBS for 45 min at room temperature, dehydrated in graded alcohols, and embedded in Poly/Bed 812. Ultrathin sections were observed with a JEOL 100B transmission electron microscope at an accelerating voltage of 60 kV.

Enzyme Linked Immunosorbent Assay (ELISA) for Anti-TFA Antibodies in Human Serum

The ELISA used was a modification of a previously described method.[30] Microtiter plates were coated with 200 ul of TFA-RSA (15 ug/ml in 0.05 M $NaHCO_3$-Na_2CO_3, pH 9.6) and incubated overnight at $4^{\circ}C$. The plates were washed 3 times (10 min per wash) with TBS containing 0.5% (w/v) casein (TBS-casein). TBS-casein containing 0.01% (w/v) thimerosal was added to block unbound sites on the surface of the microtiter wells. After 5 hr at room temperature, human serum (100 ul of a 1:200, 1:400, and 1:800 dilution in PBS containing 3% (w/v) RSA and 0.01% (w/v) thimerosal) was added to the plates, which were incubated overnight at room temperature and washed with TBS-casein. Goat anti-human IgG-HRP conjugate (100 ul of a 1:300 dilution in TBS-casein) was added to the plates. After 1 hr at room temperature, the plates were washed with TBS-casein, and incubated with the peroxidase substrate reaction mixture for 1 hr at room temperature. The titers were expressed as an optical density at 414 nm and were corrected by subtracting the optical density obtained from the corresponding human serum sample from wells that were not coated with TFA-RSA.

Statistical analysis.

All statistical comparisons were made with the Student's t test.

RESULTS

Immunoperoxidase Detection of TFA Adducts in Subcellular Fractions of Rat and Human Liver

Four hours after halothane administration to phenobarbital pretreated rats, TFA adducts were detected in various subcellular fractions of the liver by immunoperoxidase staining with anti-TFA serum of a SDS/PAGE blot (Fig. 1). Although the total cellular homogenate contained several TFA components, one major product was present with an apparent molecular weight of 54 Kd. This TFA adduct was localized in the microsomal fraction and not in the cytosolic fraction of the cell. A TFA component with identical electrophoretic properties was also detected in the plasma membrane fraction of the cell.

The total homogenate from rats treated with halothane-d contained significantly smaller amounts of TFA adducts than did the homogenates of halothane treated rats (results not shown). This finding confirmed that the staining was predominantly due to the presence of bound TFA groups and not reductive metabolites, since substitution of the hydrogen atom of halothane by a deuterium atom decreases the rate of oxidation, but does not alter the rate of reduction of halothane.[16,18]

The 54 Kd anti-TFA staining component in microsomes was subsequently found to be a 54 Kd phenobarbital inducible form of cytochrome P-450.[31]

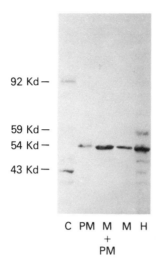

C PM M   M   H
         +
        PM

Fig. 1.   Immunoperoxidase staining with anti-TFA serum of a
           SDS/PAGE blot of liver subcellular fractions from
           phenobarbital pretreated rats 4 hr after halothane
           treatment. The SDS/PAGE and immunoblot staining
           procedures are described in detail elsewhere.[31]
           The following amounts of subcellular fractions were
           applied to the wells of the gel: M, microsomes
           (27 ug); C, cytosol (160 ug); PM, plasma membrane
           (27 ug); M + PM (27 ug each); total homogenate (100
           ug). The primary anti-TFA serum was diluted 1:1000
           and the secondary goat anti-rabbit IgG and peroxidase
           rabbit antiperoxidase complex were diluted 1:100 and
           1:1000 respectively.

This was established by passing detergent solubilized microsomes through
an affinity column of anti-cytochrome P-450-54 Kd IgG and showing that
this treatment simultaneously removed the 54 Kd protein that reacted with
anti-TFA and anti-cytochrome P-450-54 Kd. The binding of the TFA moiety
to cytochrome P-450-54 Kd did not appear to decrease the catalytic
activity of this enzyme since treatment of rats with halothane-d resulted
in less formation of the TFA adduct but in the same amount of inhibition
of benzphetamine demethylation and loss of cytochrome P-450 as that
produced by halothane (Table 1). Therefore, the inactivation of cytochrome P-450 appears mainly to be due to the reactive radical metabolite
of halothane as suggested by the _in vitro_ studies of other investigators.[36,37]

     Trifluoroacetylated protein adducts were also detected in the liver
microsomes of 4 patients, approximately two to four hours after the
beginning of halothane anesthesia (Fig. 2). As found with rats, the
adduct formation was very selective. Moreover, the TFA proteins were
all localized in the 51 Kd to 56 Kd region of the gel, which is
consistent with them being cytochromes P-450.[40]

Immunoperoxidase Detection of Cytochrome P-450, and Cytochrome P-450
Reductase in Rat Liver Plasma Membrane

     The finding that the TFA adduct in the plasma membrane had the same
apparent molecular weight as microsomal TFA cytochrome P-450-54 Kd
(Fig.1), suggested that the plasma membrane protein was also cytochrome

Table 1. Comparative effect of halothane and halothane-d on cytochrome P-450 levels and benzphetamine demethylation activity[a]

Treatment	Benzphetamine Demethylation (nmole HCHO/mg/min)	Cytochrome P-450 (nmole/mg)
Control	19.86	1.92
Halothane	13.07[b]	1.67[b]
Halothane-d	12.33[b]	1.62[b]

[a]Microsomes were prepared 4 hr after the administration of halothane, halothane-d (10 mmole/kg, ip, as a sesame oil solution) or sesame oil (control group) to phenobarbital pretreated rats. Benzphetamine demethylation activity[38] and cytochrome P-450[39] were measured by published procedures. The results are expressed as the mean of single determinations from 3 rats.
[b]Significantly different from control values (p<0.02).

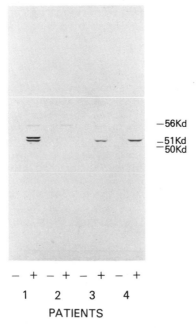

Fig. 2.   Immunoperoxidase staining with anti-TFA serum of a SDS/PAGE blot of liver microsomes from four patients approximately 2 to 4 hr after halothane anesthesia. Approximately 75 ug of each of the microsomes were applied to the wells of the gel. The primary anti-TFA serum and secondary goat anti-rabbit IgG-HRP conjugate were diluted 1:500 and 1:2000 respective.

P-450-54 Kd. If this were the case, it would indicate that either
an endoplasmic cytochrome P-450 TFA adduct had been incorporated into the
plasma membrane of the cell or that CF_3COX had been formed by cytochrome
P-450 in the plasma membrane and had reacted covalently with it. In
order to investigate the latter possibility, an SDS/PAGE blot of plasma
membranes and microsomes isolated from phenobarbital treated rats was
stained, using the immunoperoxidase method, with a mixture of anti-
cytochrome P-450 (52 Kd and 54 Kd) and anti-cytochrome P-450 reductase
serum (Fig. 3). Both cellular fractions contained nearly identical
relative amounts of immunoreactive proteins that were presumably cytochrome P-450 reductase (78 Kd) and the 52 Kd and 54 Kd forms of cytochrome P-450. The faint staining fraction above the 54 Kd protein was
likely due to the cross reaction of the anti-cytochrome P-450-54 Kd serum
with another form of cytochrome P-450 (likely cytochrome P-450e)[41] that
is immunochemically related to cytochrome P-450-54 Kd.

Since it is difficult, if not impossible, to rule out the possibility that results obtained with isolated plasma membranes (Fig. 3)
were due to contamination of this cellular fraction with smooth endoplasmic reticulum,[33] an attempt to detect cytochrome P-450-54 Kd in the
plasma membrane was repeated with freshly isolated intact hepatocytes
from phenobarbital treated rats. As seen in Fig. 4, linear regions of
immunoperoxidase staining with anti-cytochrome P-450-54 Kd IgG-HRP
conjugate of hepatocyte plasma membrane were clearly observed. The
stained vesicles on the surface of the hepatocytes were likely fragments
of endoplasmic reticulum of other hepatocytes that were disrupted during
the preparation of the cells. No staining was observed when the normal
IgG-HRP conjugate was used in place of the anti-cytochrome P-450-54 Kd
IgG-HRP conjugate (results not shown).

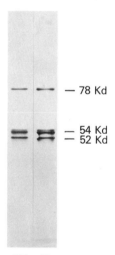

— 78 Kd

— 54 Kd
— 52 Kd

PM   M

Fig. 3.   Immunoperoxidase staining with a mixture of anti-
           cytochrome P-450 (52 Kd and 54 Kd) and anti-cyto-
           chrome P-450 reductase serum of a SDS/PAGE blot of
           liver plasma membrane and microsomes from pheno-
           barbital treated rats . Approximately 10 ug of
           plasma membrane and 3 ug microsomes were applied to
           the wells of the gel. The primary anti-cytochrome
           P-450-52 Kd, anti-cytochrome P-450-54 Kd, and anti-
           cytochrome P-450 reductase sera and the secondary
           goat anti-rabbit IgG-HRP conjugate were diluted
           1:3000, 1:10000, 1:5000, and 1:1000 respectively.

665

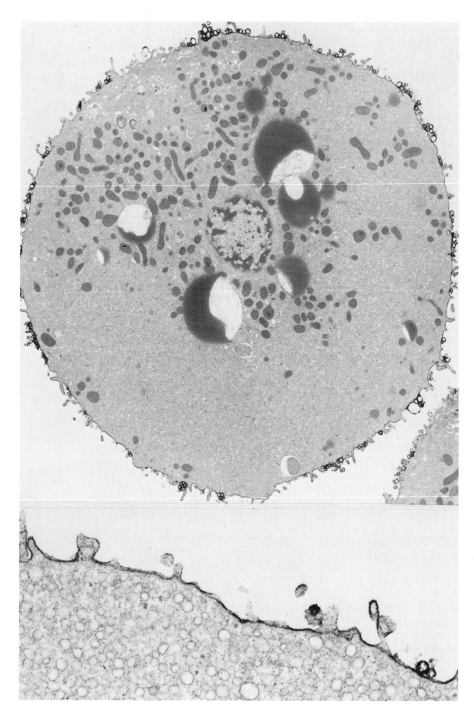

Fig. 4. Low (top) and high (bottom) magnification views showing the
immunoelectron microscopic localization of cytochrome P-450-
54 Kd in the plasma membrane of isolated hepatocytes of
phenobarbital treated rats. The cells were isolated and
immunoperoxidase stained with anti-cytochrome P-450-54 Kd
IgG-HRP conjugate as described in MATERIALS AND METHODS. The
magnification of the upper and lower panels were x 9000 and
x 27000 respectively.

Detection of Anti-TFA Antibodies in the Sera of Halothane Hepatitis Patients

The results of the ELISA indicated that two of the six patients (patients 2 and 3) diagnosed as having halothane-associated massive liver cell necrosis contained anti-TFA antibodies in their sera; the amount of their ELISA product was significantly ($p < 0.05$) higher than the nonspecific levels measured in the sera of two paracetamol hepatitis patients and one control NIH patient (Fig. 5). These results were supported by the finding that the incubation of the microtiter wells with rabbit anti-TFA serum prior to the addition of the human serum samples inhibited only the ELISA results of patients 2 and 3 (results not shown).

DISCUSSION

In this investigation, we have described specific immunoperoxidase staining methods to detect protein adducts of the reactive oxidative metabolite of halothane, CF_3COX, in subcellular fractions of rat (Fig. 1) and human liver (Fig. 2). The results indicate that CF_3COX reacts covalently with a relatively small number of proteins in the cell. In phenobarbital pretreated rats, cytochrome P-450-54 Kd, which is apparently equivalent to cytochromes PB-B of Guengerich et al.,[42] P-450b of Ryan et al.,[43] and PB-4 of Waxman and Walsh,[44] is either identical or closely related immunochemically to the protein in the endoplasmic reticulum of the cell that reacts predominantly with this metabolite.

The specificity of the reaction is likely due to the high reactivity of CF_3COX. Consequently, most of this species is probably trapped by the molecule of cytochrome P-450 that has produced it before it has had time to diffuse away and react with other proteins. It follows that the major TFA proteins detected in the human microsomes after halothane treatment (Fig. 2) and the 54 Kd protein detected in the plasma membrane fraction isolated from livers of halothane treated rats (Fig. 1) were also cytochromes P-450. Moreover, since the isolated rat liver plasma membrane from phenobarbital pretreated rats apparently contained a 54 Kd protein immunochemically related to the liver microsomal cytochrome P-450-54 Kd as well as other proteins that were similar if not identical to the microsomal 52 Kd form of cytochrome P-450 and cytochrome P-450 reductase (Fig. 3), CF_3COX may have been formed by cytochrome P-450 in the plasma membrane and covalently bound to it.

It has been controversial, however, whether cytochrome P-450 is[45-49] or is not[50-57] a constituent of the plasma membrane of cells. Although the reasons for this dichotomy are not clearly understood, they are probably due in part to a combination of factors including the methods used to isolate the plasma membrane and the sensitivities of the spectral, catalytic, or immunochemical procedures used for identifying and quantitating cytochrome P-450. One related major problem in working with isolated plasma membranes is proving that they are not significantly contaminated with smooth endoplasmic reticulum and its complement of cytochromes P-450 and other electron transport proteins. This is due to the fact that the morphometric and the glucose 6-phosphatase marker enzyme methods of purity determination have their limitations and may yield conflicting results.[46,58-60] A way to circumvent this problem is to identify cytochrome P-450 in the plasma membrane of intact hepatocytes instead of in isolated membranes. Indeed, when freshly isolated hepatocytes from phenobarbital pretreated rats were stained with anti-cytochrome P-450-54 Kd IgG-HRP conjugate, electron microscopy revealed that the plasma membrane contained a component that was immunochemically related to microsomal cytochrome P-450-54 Kd (Fig. 4).

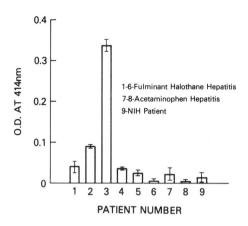

Fig. 5. Detection of anti-TFA antibodies in the sera of halothane hepatitis patients. The details of the ELISA procedure are described in MATERIALS AND METHODS. In brief, TFA-RSA was absorbed to a well of a microtiter plate, followed by the sequential addition of human serum, goat anti-human IgG-HRP conjugate, and HRP substrate reaction mixture.

It remains to be determined whether this enzyme is functionally active and accounts for the TFA adduct detected in the isolated plasma membrane (Fig. 1) and on the surface of isolated hepatocytes[30] from halothane treated rats. This idea has important implications because it suggests that toxic reactive metabolites of other compounds may be formed in the plasma membrane and produce acute damage[61] or initiate immunotoxicity[62] by altering the antigenicity of the plasma membrane.

There are several possible explanations for failing to detect anti-TFA antibodies in the sera of some of the halothane hepatitis patients studied (Fig. 5). It is conceivable that the immune mechanism of toxicity in patients 1, 4, 5, and 6 (Fig. 5) was not an antibody-dependent process, but instead was due to a specific cell-mediated immune reaction.[63] On the other hand, if the mechanism was dependent on the presence of anti-TFA antibodies, it is possible that the titer of the antibodies may have declined in the sera of these patients at the time of sampling, possibly because they became bound to TFA hepatocytes. The results may also suggest that other mechanisms of sensitization may be involved in halothane hepatitis. For example, although it was recently found that repeated administration of halothane to rabbits resulted in the formation of anti-TFA antibodies, they did not appear to promote liver damage.[64] Accordingly, an investigation with a more general ELISA procedure than the specific anti-TFA ELISA method described in this study revealed that the sera of 16 of 24 patients with halothane associated liver failure contained antibodies that bound to unidentified halothane-induced determinants in the liver microsomes of halothane treated rabbits.[65] When the constituent polypeptides in the microsomes were resolved by SDS-PAGE and the nitrocellulose blots were indirectly stained by an immunoperoxidase procedure with serum from halothane hepatitis patients, three distinct polypeptides (29 Kd, 76 Kd, and 100 Kd) were labeled, although no serum from any one patient reacted with all of the polypeptides.[66]

Clearly, the immunological basis of halothane hepatotoxicity is
quite complex. The determinants involved in the tissue sensitization are
probably not solely due to bound TFA adducts. Other potential determinants include bound reductive metabolites of halothane or altered
autoantigens produced by either CF_3COX or CF_3CHCl.. The identification of
the antigens involved in halothane hepatotoxicity, however, is no longer
an insurmountable problem because of the enormous recent advances in
immunochemical techniques, many of which have been described in the
present paper. Once these antigens have been identified, they can be
used in several important ways such as for the development of an ELISA to
detect halothane-sensitized individuals and the development of an animal
model by immunizing animals with the antigen or passively immunizing them
with an antibody to the antigen prior to the administration of halothane. Such studies may serve as models for investigating the immunological basis of other drug-induced cytotoxicities and ultimately
should lead to the design of safer, new agents that will not produce
tissue sensitization.

REFERENCES

1. L. R. Pohl and J. R. Gillette, A perspective on halothane-induced
    hepatotoxicity, **Anesth. Analg.** 61:809 (1982).
2. R. Wright, O. E. Eade, M. Chisholm, M. Hawksley, B. Lloyd,
    T. M. Moles, J. C. Edwards, and M. J. Gardner, Controlled
    prospective study of the effect on liver function of multiple
    exposure to halothane, **Lancet** 1:817 (1975).
3. J. Trowell, R. Peto, and A. C. Smith, Controlled trial of repeated
    halothane anesthetics in patients with carcinoma of the uterine
    cervix treated with radium, **Lancet** 1:821 (1975).
4. W. W. Mushin, M. Rosen, and E. V. Jones, Post-halothane jaundice in
    relation to previous administration of halothane, **Br. Med. J.** 3:18
    (1971).
5. H. T. Wark, Postoperative jaundice in children, the influence of
    halothane, **Anesthesia** 38:237 (1983).
6. R. L. Peters, H. A. Edmondson, T. B. Reynolds, J. C. Meister, and
    T. J. Curphey, Hepatic necrosis associated with halothane-induced
    hepatotoxicity, **Am. J. Med.** 47:748 (1969).
7. E. A. Gall, Report of the pathology panel. National halothane study,
    Anesthesiology 29:233 (1968).
8. D. J. Miller, J. Dwyer, and G. Klatskin, Halothane hepatitis: Benign
    resolution of severe lesion, **Ann. Intern. Med.** 89:212 (1978).
9. A. J. Gandolfi, R. D. White, I. G. Sipes, and L. R. Pohl, Bioacti-
    vation and covalent binding of halothane invitro: Studies with [^{3}H]-
    and [^{14}C]-halothane, **J. Pharmacol.Exp. Ther.**, 214:721 (1980).
10. G. K. Gourlay, J. F. Adams, M. J. Cousins, and P. Hall, Genetic
    differences in reductive metabolism and hepatotoxicity of halothane
    in three rat strains, **Anesthesiology** 55:90 (1981).
11. J. R. Trudell, B. Bosterling, and A. J. Trevor, Reductive metabolism
    of halothane by human and rabbit cytochrome P-450. Binding of
    1-chloro-2,2,2-trifluoroethyl radical to phospholipids,
    Mol. Pharmacol. 21:710 (1982).
12. H. J. Ahr, L. J. King, W. Nastainczyk, and V. Ullrich, The mechanism
    of reductive dehalogenation of halothane by liver cytochrome
    P-450, **Biochem. Pharmacol.** 31:383 (1982).
13. E. N. Cohen, J. R. Trudell, H. N. Edmunds, and E. Watson, Urinary
    metabolites of halothane in man, **Anesthesiology** 43:392 (1975).
14. D. Karashima, Y. Hirokata, A. Shigematsu, and T. Furukawa, The in
    vitro metabolism of halothane (2-bromo-2-chloro-1,1,1-trifluoro-

ethane) by hepatic microsomal cytochrome P-450, J. Pharmacol. Exp. Ther. 203:409 (1977).

15. L. P. McCarty, R. S. Malek, and E. R. Larsen, The effects of deuteration on the metabolism of halogenated anesthetics in the rat, Anesthesiology 51:106 (1979).

16. I. G. Sipes, A. J. Gandolfi, L. R. Pohl, G. Krishna, and B. R. Brown, Comparison of the biotransformation and hepatotoxicity of halothane and deuterated halothane, J. Pharmacol. Exp. Ther. 214:716 (1980).

17. R. Müller and A. Stier, Modification of liver microsomal lipids by halothane metabolites: A multi nuclear NMR spectroscopic study. Naunyn-Schmiedeberg's Arch. Pharmacol. 321:234 (1982).

18. L. R. Pohl and J. R. Gillette, Determination of toxic pathways of metabolism by deuterium substitution, Drug Metab. Reviews 15:1335 (1985).

19. J. L. Plummer, P. Hall, M. A. Jenner, and M. J. Cousins, Sex differences in halothane metabolism and hepatotoxicity in a rat model, Anesth. Analg. 64:563 (1985).

20. J. L. Plummer, S. Wanwimolruk, M. A. Jenner, P. Hall, and M. J. Cousins, Effects of cimetidine and ranitidine on halothane metabolism and hepatotoxicity in an animal model, Drug Metab. Disp. 12:106 (1984).

21. C. A. Lunam, M. J. Cousins, and P. Hall, Guinea-Pig model of halothane-associated hepatotoxicity in the absence of enzyme induction and hypoxia, J. Pharmacol. Exp. Ther. 232:802 (1985).

22. S. Belfrage, I. Ahlgren, and S. Axelson, Halothane hepatitis in an anesthetist, Lancet 2:1466 (1966).

23. G. Klatskin and D. V. Kimberg, Recurrent hepatitis attributable to halothane sensitization in an anesthetist, N. Engl. J. Med. 280:515 (1969).

24. F. M. T. Carney and R. A. Van Dyke, Halothane hepatitis: A critical review, Anesth. Analg. 51:135 (1972).

25. B. Walton, B. R. Simpson, D. Doniach, J. Perrin, and A. J. Appleyard, Unexplained hepatitis following anesthesia, Br. Med. J. 1:1171 (1976).

26. D. Vergani, D. Tsantoulas, A. L. W. F. Eddleston, M. Davis, and R. Williams, Sensitization to halothane-altered liver components in severe hepatic necrosis after halothane anesthesia, Lancet 2:801 (1978).

27. G. Mieli-Vergani, D. Vergani, J. M. Tredger, A. L. W. F. Eddleston, M. Davis, and R. Williams, Lymphocyte cytotoxicity to halothane altered hepatocytes in patients with severe hepatic necrosis following halothane anesthesia, J. Clin. Lab. Immunol. 4:49 (1980).

28. D. Vergani, G. Mieli-Vergani, A. Alberti, J. Neuberger, A. L. W. F. Eddleston, M. Davis, and R. Williams, Antibodies to the surface of halothane-altered rabbit hepatocytes in patients with severe halothane-associated hepatitis, N. Engl. J. Med. 303:66 (1980).

29. J. Neuberger, G. Mieli-Vergani, J. M. Tredger, M. Davis, and R. Williams, Oxidative metabolism of halothane in the production of altered hepatocyte membrane antigens in acute halothane-induced necrosis, Gut 22:669 (1981).

30. H. Satoh, Y. Fukuda, D. K. Anderson, V. J. Ferrans, J. R. Gillette, and Lance R. Pohl, Immunological studies on the mechanism of halothane-induced hepatotoxicity: Immunohistochemical evidence of trifluoroacetylated hepatocytes, J. Pharmacol. Exp. Ther. 233:857 (1985).

31. H. Satoh, J. R. Gillette, H. W. Davies, R. D. Schulick, and L. R. Pohl, Immunochemical evidence of trifluoroacetylated cytochromes P-450 in the liver of halothane treated rats, Mol. Pharmacol. (in press).

32. Y. Yasukochi and B. S. S. Masters, Some properties of a detergent-solubilized NADPH cytochrome c (cytochrome P-450) reductase purified by biospecific affinity chromatography, *J. Biol. Chem.* 251:5337 (1976).

33. L. Lesko, M. Donlon, G. V. Marinetti, and J. D. Hare, A rapid method for the isolation of rat liver plasma membranes using an aqueous two-phase polymer system, *Biochem. Biophys. Acta* 311:173 (1973).

34. P. E. Thomas, D. Korzeniowski, D. Ryan, and W. Levin, Preparation of monospecific antibodies against two forms of rat liver cytochrome P-450 and quantitation of these antigens in microsomes, *Arch. Biochem. Biophys.* 192:524 (1979).

35. S. Avrameas and T. Ternynck, Peroxidase labelled antibody and Fab conjugate with enhanced intracellular penetration, *Immunochemistry* 8:1175 (1971).

36. H. De Groot, U. Harnisch, and T. Noll, Suicidal inactivation of microsomal cytochrome P-450 by halothane under hypoxic conditions, *Biochem. Biophys. Res. Commun.* 107:885 (1982).

37. P. A. Krieter and R. A. Van Dyke, Cytochrome P-450 and halothane metabolism. Decrease in rat liver microsomal P-450 in vitro, *Chem. Biol. Interact.* 44:219 (1983).

38. R. V. Branchflower, R. D. Schulick, J. W. George, and L. R. Pohl, Comparison of the effects of methyl-n-butyl ketone and phenobarbital on rat liver cytochromes P-450 and the metabolism of chloroform to phosgene, *Toxicol. Appl. Pharmacol.* 71:414 (1983).

39. T. Omura and R. Sato, The carbon monoxide binding pigment of liver microsomes. I. Evidence for its hemoprotein nature, *J. Biol. Chem.* 239:2370 (1964).

40. P. P. Wang, P. Beaune, L. S. Kaminsky, G. A. Dannan, F. F. Kadlubar, D. Larrey, and F. P. Guengerich, Purification and characterization of six cytochrome P-450 isozymes from human liver microsomes, *Biochemistry* 22:5375 (1983).

41. D. E. Ryan, P. E. Thomas, and W. Levin, Purification and characterization of a minor form of hepatic microsomal cytochrome P-450 from rats treated with polychlorinated biphenyls, *Arch. Biochem. Biophys.* 216:272 (1982).

42. F. P. Guengerich, G. A. Dannan, S. T. Wright, M. V. Martin, and L. S. Kaminsky, Purification and characterization of liver microsomal cytochromes P-450: Electrophoretic, spectral, catalytic, and immunochemical properties and inducibility of eight isozymes isolated from rats treated with phenobarbital or beta-naphtho-flavone, *Biochemistry* 21:6019 (1982).

43. D. E. Ryan, P. E. Thomas, D. Korzeniowski, and W. Levin, Separation and characterization of highly purified forms of liver microsomal cytochrome P-450 from rats treated with polychlorinated biphenyls, phenobarbital, and 3-methycholanthrene, *J. Biol. Chem.* 254:1365 (1979).

44. D. J. Waxman, and C. Walsh, Phenobarbital-induced rat liver cytochrome P-450. Purification and characterization of two closely related isozymic forms, *J. Biol. Chem.* 257:10446 (1982).

45. D. E. Hultquist, D. W. Reed, P. G. Passon, and W. E. Andrews, Purification and properties of S-protein (hemoprotein$_{559}$) from human erythrocytes, *Biochim. Biophys. Acta* 229:33 (1971).

46. H. Glaumann and J. A. Gustafsson, Subcellular localization of steroid hormone metabolism in rat liver, *Exp. Mol. Pathol.* 27:221 (1977).

47. G. Bruder, A. Fink, and E. D. Jarasch, The b-type cytochrome in endoplasmic reticulum of mammary gland epithelium and milk fat globule membranes consists of two components, cytochrome b_5 and cytochrome P-420, *Exp. Cell Res.* 117:207 (1978).

48. E. D. Jarasch, J. Kartenbeck, G. Bruder, A. Fink., D. J. Morre, and W. W. Franke, B-type cytochromes in plasma membranes isolated from

rat liver, in comparison with those of endomembranes, <u>J. Cell Biol.</u> 80:37 (1979).

49. P. Stasiecki, F. Oesch, G. Bruder, E. D. Jarasch, and W. W. Franke, Distribution of enzymes involved in metabolism of polycyclic aromatic hydrocarbons among rat liver endomembranes and plasma membranes, <u>Eur. J. Cell Biol.</u> 21:79 (1980).

50. I. M. Vassiletz, E. F. Derkatchev, and S. A. Neifakh, The electron transfer chain in liver cell plasma membrane, <u>Exp. Cell Res.</u> 46:419 (1967).

51. Y. Ichikawa and T. Yamano, Cytochrome b_5 and CO-binding cytochromes in the golgi membranes of mammalian livers, <u>Biochem. Biophys. Res. Commun.</u> 40:297 (1970).

52. S. Fleischer, B. Fleischer, A. Azzi, and B. Chance, Cytochrome b_5 and P-450 in liver cell fractions, <u>Biochim. Biophys. Acta</u> 225:194 (1971).

53. P. Emmelot and C. J. Bos, Studies on plasma membranes. XVII. On the chemical composition of plasma membranes prepared from rat and mouse liver and hepatoma, <u>J. Membrane Biol.</u> 9:83 (1972).

54. P. Emmelot, C. J. Bos, R. P. van Hoeven, and W. J. van Blitterswijk, Isolation of plasma membranes from rat and mouse livers and hepatomas, <u>in</u>: "Methods in Enzymology," S. Fleischer and L. Packer, eds., Vol. 31, p. 75, Academic Press, New York (1974).

55. C. von Bahr, E. Hietanen, and H. Glaumann, Oxidation and glucuronidation of certain drugs in various subcellular fractions of rat liver: Binding of desmethylimipramine and hexobarbital to cytochrome P-450 and oxidation and glucuronidation of desmethylimipramine, aminopyrine, p-nitrophenol and 1-naphthol, <u>Acta Pharmacol. Toxicol.</u> 31:107 (1972).

56. F. C. Charalampous, N. K. Gonatas, and A. D. Melbourne, Isolation and properties of the plasma membrane of KB cells, <u>J. Cell Biol.</u> 59:421 (1973).

57. S. Matsuura, Y. Fujii-Kuriyama, and Y. Tashiro, Immunoelectron microscope localization of cytochrome P-450 on microsomes and other membrane structures of rat hepatocytes, <u>J. Cell Biol.</u> 78:503 (1978).

58. P. Emmelot and C. J. Bos, Studies on plasma membranes. XI. Inorganic pyrophosphatase, PP_1-glucose phosphotransferase and glucose-6-phosphatase in plasma membranes and microsomes isolated from rat and mouse livers and hepatomas, <u>Biochim. Biophys. Acta</u> 211:169 (1970).

59. S. Fleischer and M. Kervina, Subcellular fractionation of rat liver, <u>in</u>: "Methods in Enzymology," S. Fleischer and L. Packer, eds., Vol. 31, p. 6, Academic Press, New York (1974).

60. K. E. Howell, A. Ito, and G. E. Palade, Endoplasmic reticulum marker enzymes in golgi fractions-What does this mean?, <u>J. Cell Biol.</u> 79:581 (1978).

61. G. L. Ginsberg and S. D. Cohen, Plasma membrane alterations and covalent binding to organelles after an hepatotoxic dose of paracetamol, <u>The Toxicologist</u> 5:154 (1985).

62. K. B. Taylor and H. C. Thomas, Gastrointestinal and liver diseases, <u>in</u>: "Basic and Clinical Immunology," D. P. Stites, J. D. Stobo, H. H. Fudenberg, and J. V. Wells, eds., p. 518, Lange Medical Publications, Los Altos (1982).

63. J. V. Wells, Immune mechanisms in tissue damage, <u>in</u>: "Basic and Clinical Immunology," D. P. Stites, J. D. Stobo, H. H. Fudenberg, and J. V. Wells, eds., p. 136, Lange Medical Publications, Los Altos (1982).

64. A. H. Callis, S. D. Brooks, S. J. Waters, A. J. Gandolfi, D. O. Lucas, L. R. Pohl, H. Satoh, and I. G. Sipes, Evidence for a role of the immune system in the pathogenesis of halothane hepatitis, <u>in</u>: "Molecular Mechanisms of Anesthesia, Progress in Anesthesiology," S. H. Roth, ed., Vol. 3, Raven Press (in press).

65. J. G. Kenna, J. Neuberger, and R. Williams, An enzyme-linked immunosorbent assay for detection of antibodies against halothane-altered hepatocyte antigens, _J. Immunol. Method_ 75:3 (1984).

66. J. G. Kenna, J. Neuberger, and R. Williams, Characterization of halothane-induced antigens by immunoblotting, _Biochem. Soc. Trans._ 13:910 (1985).

INHIBITION OF RNA SYNTHESIS IN MOUSE MACROPHAGES AND LYMPHOCYTES BY BENZENE AND ITS METABOLITES

George F. Kalf, Robert Snyder[*], and Gloria B. Post

Dept. Biochemistry, Thomas Jefferson University,
Philadelphia, PA 19107 and [*]Dept. Pharmacology and
Toxicology, Rutgers University, Piscataway, NJ 08854

INTRODUCTION

Chronic exposure of laboratory animals and man to benzene leads to progressive degeneration of bone marrow resulting in aplastic anemia and leukemia (Snyder et al., 1977; Cronkite et al., 1984). The mechanism by which benzene produces myelotoxicity is unknown; substantial evidence indicates that toxicity results from its metabolism to one or more compounds (Snyder et al., 1982) which accumulate in bone marrow (Andrews et al., 1977); these have been shown to be phenol, catechol, and hydroquinone (Rickert et al., 1979) which can further oxidize to p-benzoquinone.

Hemopoiesis results from an interaction of stem cells with the marrow stroma which provides a microenvironment for the regulated proliferation and differentiation of progenitor cells (Tavasolli and Friedenstein, 1983). Benzene has been demonstrated to be selectively cytotoxic toward progenitor cells of the various hemopoietic lineages of an intermediate level of differentiation (Lee et al., 1974; Tunek et al., 1981; Wierda and Irons, 1982) as well as to the stromal microenvironment (Frash et al, 1976) and its in vitro equivalent, the marrow adherent layer (Harigaya et al., 1981). Indeed, Gaido and Wierda (1984) have recently reported that hydroquinone and its oxidation product, p-benzoquinone, inhibit the ability of stromal cells to support granulocyte/macrophage colony formation in vitro.

675

The macrophage is an essential cell of the marrow stromal micro-environment and is an important source of protein factors which promote the survival, proliferation, and differentiation of progenitor cells (Moore, 1978). T-lymphocytes, both in vivo and in vitro are also a potent source of factors which regulate hemopoiesis, (Hesketh et al., 1984; Verma et al., 1984; Prystowsky et al., 1984) and lymphocytopenia is a distinctive feature of benzene toxicity in animals (Irons et al., 1979; Irons and Moore, 1980; Green et al., 1981) and humans (Goldstein, 1984).

Since both the macrophage and the T-lymphocyte synthesize protein factors required by maturing marrow cells, it is conceivable that benzene, by metabolism in these cell types to a reactive intermediate, inhibits the production of growth factors by interfering with mRNA formation. In this connection we have previously observed an inhibition of mRNA synthesis by benzene metabolites in bone marrow nuclei (Post et al., 1984). We report here that benzene and its metabolites can inhibit RNA synthesis in mouse macrophages and lymphocytes at concentrations that have no effect on viability and that the putative toxic metabolite, p-benzoquinone, can inhibit the production of the lymphokine, inter-leukin-2 as well.

METHODS

Preparation of Peritoneal Macrophages and Spleen Lymphocytes

Male Swiss mice (20-25g) were used in all experiments. Peritoneal macrophages were elicited by the injection of 1 ml of a sterile solution of 10% proteose-peptone in PBS*, pH 7.2. After 4 days the animals were killed by cervical dislocation and the peritoneal cavity exposed. A period of four days was chosen as the harvest time for the peritoneal cells because it represents the peak of macrophage response and the nadir in neutrophils present. Since neutrophils also adhere to glass or plastic, this minimizes the problems of neutrophil contamination and

---

Abbreviations used: PBS, phosphate buffered saline; CFU-G/M, colony forming unit-granulocyte/macrophage; FCS, fetal calf serum; IC$_{50}$, molar concentration causing 50 percent inhibition; CTLL, cytotoxic lymphocyte line; Con A, concanavalin A; TCA, trichloroacetic acid; IL-2, interleukin-2.

allows for a highly purified macrophage population (Kurland, 1984). Cells were collected by rinsing the peritoneal cavity 3 times with 2 ml of cold RPMI 1640 containing 10 U/ml heparin. After collection of the cells, the red blood cells were lysed by suspension in a hypotonic solution consisting of NH_4Cl (155 mM), EDTA (0.1 mM) and $KHCO_3$ (10 mM). The cells were washed in complete medium (RPMI 1640 containing 10% FCS, HEPES, 10 mM; penicillin, 50 μg/ml; streptomycin, 50 μg/ml; L-glutamine, 2 mM; and 2-mercaptoethanol, 50 μM). The cells were counted using a hemocytometer after staining with trypan blue dye. Viability was routinely greater than 95%. The yield of cells was typically $5-6 \times 10^7$ cells from 6 mice.

Macrophages were purified by active adherence by plating 1 to 2×10^6 peritoneal exudate cells in 1 ml complete medium in a 35mm plastic or glass petri dish and allowing adherence to take place for 2 hr at $37^{\circ}C$ in a 5% CO_2 atmosphere. The medium was then aspirated and the adherent cell layer washed twice with warm medium or PBS.

Lymphocytes were prepared by forcing the spleen through a 50 mesh stainless steel screen into RPMI 1640 medium to obtain a single cell suspension. The cells were collected by centrifugation and resuspended in hypotonic NH_4Cl to lyse erythrocytes. Cells were washed and resuspended in RPMI 1640 complete medium. Macrophages were removed by allowing them to adhere to a plastic petri dish. Cells were counted and viability was determined in a hemocytometer after staining with trypan blue dye.

Exposure of Cells to Benzene and Metabolites

Metabolite solutions were prepared in dark bottles immediately before use. Macrophages, derived from 1 to 2×10^6 peritoneal exudate cells in 1 ml PBS or complete medium were incubated with metabolite for 1 hr in 35mm petri dishes at $37^{\circ}C$ in a 5% CO_2 atmosphere prior to addition of [^{3}H]uridine. Lymphocytes (2×10^6/ml) were exposed to metabolites in PBS for 30 min at 37° in 5% CO_2 atmosphere and were washed free of metabolite, counted, and diluted to 1×10^6 cells/ml in RPMI 1640 complete. RNA synthesis was assayed as described below.

For exposure to benzene, adherent macrophages were removed from a glass petri dish into 1 ml of complete medium by scraping with a rubber policeman and transferred to a 2 ml glass stoppered tube. One ml of a benzene solution composed of benzene in complete medium at twice the desired final concentration was added, the tubes were sealed and incubated at $37^{\circ}C$ for 2 hr after which [^{3}H]uridine was added and RNA synthesis allowed to proceed as described below. Lymphocytes (1×10^6/ml) were exposed to benzene in complete medium which was previously equilibrated with 5% CO_2. Incubation was carried out for 1 hr at 37° in a tightly closed screw cap bottle.

Assay for RNA Synthesis

Adherent macrophages, derived from $1-2\times10^6$ peritoneal exudate cells were incubated at 37° in 5% CO_2 atmosphere in 35mm petri dishes with [^{3}H]uridine (5 μCi in 10 μl PBS) in complete medium or PBS for 1 to 3 hr depending on the experiment. At the end of the incubation, adherent cells were scraped from the dishes with a rubber policeman, the incubation medium containing the cells was removed to a centrifuge tube and the petri dish rinsed with an additional 1 ml of PBS. Lymphocytes suspended in RPMI 1640 complete (2×10^5 cells in 0.2 ml) were incubated with [^{3}H]uridine (2 μCi in 10 μl PBS) for 4 hr at 37° in 5% CO_2. Cells were collected by centrifugation, suspended in 0.2 ml water and lysed by vortexing . The lysate (0.1 ml) was applied to a paper filter disc, and the incorporation of [^{3}H]uridine into trichloroacetic acid TCA-precipitable material was assayed as routinely carried out in this laboratory (D'Agostino et al., 1975). All incubations were conducted in triplicateand all experiments were done at least two times.

Assay for Interleukin-2

Cells (4×10^6/ml) were incubated in PBS with or without p-benzoquinone (5×10^{-6}M) for 30 min, washed, and resuspended in RPMI 1640 complete. Conditioned media were obtained by incubating the cells (1×10^6/ml) for 24 hr with or without Con A (5 μg/ml) and removing the cells by centrifugation.

The conditioned media were assayed for stimulation of proliferation of the IL-2 dependent cell lines, CTLL-CH and CTLL-DP. Cells (4×10^3)

were incubated with 0.1 ml conditioned medium in a total volume of 0.2 ml. After 44 hr, the cells were allowed to incorporate [3H]thymidine (1 μCi) for 4 hr. The cells were collected on a filter paper with a PHD cell harvester (Cambridge Technologies, Inc.) and radioactivity was determined by liquid scintillation counting. The cell lines, CTLL-CH and CTLL-DP and rat spleen conditioned medium used as a positive control, were kindly provided by Dr. Dolores Byrne.

RESULTS

RNA Synthesis in Macrophages

Adherent macrophages incorporate [3H]uridine into TCA precipitable material as a linear function of time (Fig. 1) and as a function of cell number (Fig. 2). That the incorporation represents RNA synthesis is evidenced by the fact that actinomycin D (5 μg/ml) completely inhibits the incorporation (o in Fig. 1).

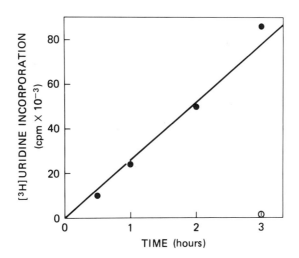

Figure 1. Time course of RNA synthesis in macrophages and the effect of actinomycin D. Macrophages purified by active adherence from 2×10^6 PE cells were allowed to incorporate [3H]uridine (5 μCi) for varying periods of time. Incorporation into TCA-precipitable material was determined as described under Methods. A similar incubation was carried out for 3 hr in the presence of 5 μg/ml actinomycin D (o).

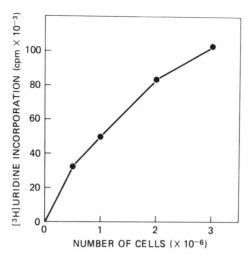

Figure 2. RNA synthesis in macrophages as a function of cell number. Macrophages, purified by active adherence from varying numbers of peritoneal exudate cells, were allowed to incorporate [^{3}H]uridine (5 µCi) for 2 hr.

Effect of Benzene on RNA Synthesis in Macrophages

When macrophages purified by adherence are exposed to benzene in complete medium, a dose-dependent inhibition of RNA synthesis is observed (Table 1). Because of benzene's extreme volatility, there are inherent technical difficulties in exposing adherent cells to benzene in a petri dish. Even with adequate sealing, the large headspace above the incubation mixture greatly reduces the effective concentration of benzene in the incubation. Therefore, in order to perform the experiments described above, macrophages were purified by adherence to a glass petri dish and then removed with a rubber policeman and transferred to a 2 ml glass-stoppered tube for exposure to benzene. Since removal of the cells with the rubber policeman caused a 25% decrease in viability, the results were confirmed using peritoneal exudate which consists predominantly of macrophages and which was exposed to benzene directly in a glass-stoppered tube. The results of a typical experiment presented in Fig. 3 are similar to those obtained with the adherent macrophages; exposure to 5 mM benzene inhibited RNA synthesis 45% in purified macrophages and 40% in peritoneal exudate cells. This concentration of benzene had no effect on the viability of the cells.

TABLE 1

Effect of Benzene on RNA Synthesis by Adherent Macrophages

Conditions	[^{3}H]Uridine incorporation	
	cpm	% of control
Control incubation[*]	23,700 ± 5200	–
+ Benzene (2 mM)	17,300 ± 4000	73
+ Benzene (5 mM)	12,900 ± 1200	55

[*] PE cells (2×10^6) were allowed to adhere to a glass dish for 2 hr at 37°C. The non-adherent cells were removed and the macrophages washed with RPMI 1640 medium and then removed with a rubber policeman and transferred to a 2-ml glass-stoppered tube. The macrophages were incubated with benzene for 2 hr in medium in a final volume of 2 ml. [^{3}H]Uridine (5 μCi) was then added and RNA synthesis was allowed to proceed for 1 hr.

Exposure of Macrophages to Benzene Metabolites

Exposure of adherent macrophages in complete medium plus 10% FCS to varying concentrations of phenol, hydroquinone or p-benzoquinone resulted in a dose-dependent inhibition of RNA synthesis (Fig. 4). The IC_{50}s were computed from these data and are presented in Table 2 (right hand column). These metabolites are reactive and can interact with the protein and sulfhydryl components of the complete medium containing FCS, thus giving higher IC_{50}-values than would otherwise be observed. To determine the magnitude of this difference, macrophages were exposed to metabolites in PBS containing glucose (2 mg/ml) which was essential for RNA synthesis. Under these conditions, inhibition of RNA synthesis occurred at a significantly lower metabolite concentration (Fig. 5). This can be readily be seen in a comparison of IC_{50} – values presented in Table 2.

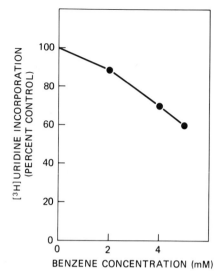

Figure 3. Effect of benzene on RNA synthesis in peritoneal exudate cells. Cells were preincubated in the presence or absence of varying concentrations of benzene for 1 hr. at 37°C in an incubation volume of 1 ml in a 2-ml glass stoppered flask. [^{3}H]Uridine was then added and the cells were allowed to synthesize RNA for 3 hr. 10^6 cells/incubation; 10 μCi [^{3}H]uridine; control incorporation 25,900 cpm. The results shown are typical of those obtained in 4 experiments.

TABLE 2

Effect of Medium on the Ability of Benzene Metabolites
to Inhibit RNA Synthesis in Macrophages

Compound	$IC_{50}[M]$*	
	PBS	RPMI 1640 + 10% FCS
Benzene	–	5×10^{-3}
Phenol	1.6×10^{-3}	2.5×10^{-3}
Hydroquinone	1×10^{-5}	2.5×10^{-5}
p-Benzoquinone	2×10^{-6}	6×10^{-6}

* $IC_{50}[M]$, molar concentration giving 50 percent inhibition.

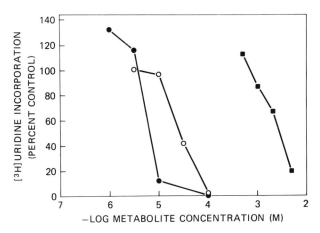

Figure 4. The inhibition of RNA synthesis in macrophages exposed to benzene metabolites in RPMI 1640 media + 10% FCS. Adherent macrophages were preincubated for 1 hr. at 37°C with varying concentrations of the appropriate metabolite and then incubated with [^{3}H]uridine (5μCi) for 1.5 hr. ■, phenol. Control value – 29,700 cpm [^{3}H]uridine incorporated. o , hydroquinone. Control value – 32,600 cpm [^{3}H]uridine incorporated ● , p-benzoquinone. Control value – 20,719 cpm [^{3}H]uridine incorporated.

RNA Synthesis in Lymphocytes

Lymphocytes derived from spleen incorporate [^{3}H]uridine into TCA-precipitable material as a linear function of time for at least 4 hr. (data not shown). That the incorporation represents RNA synthesis is evidenced by the fact that actinomycin D completely inhibits the reaction

When lymphocytes are exposed to benzene in complete medium, a dose-dependent inhibition of RNA synthesis is observed (Fig. 6). Furthermore, exposure of lymphocytes to the benzene metabolites, hydroquinone or p-benzoquinone, in PBS also resulted in a dose-dependent inhibition. As can be seen, p-benzoquinone, the oxidation product of hydroquinone and putative toxic metabolite of benzene [14,15], was the most potent inhibitor. At the highest concentrations of benzene or metabolites tested, there was less than 10% loss in cell viability as measured by trypan blue dye exclusion.

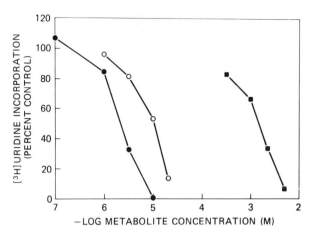

Figure 5.   The inhibition of RNA synthesis in macrophages exposed to
benzene metabolites in phosphate buffered saline + glucose (2
mg/ml).   Adherent macrophages were preincubated for 1 hr. at 37°C
with varying concentrations of the appropriate metabolite and then
incubated with [3H]uridine (5 µCi) for 1.5 hr.

■, phenol.   Control value - 26,900 cpm [3H]uridine incorporated.

o , hydroquinone.   Control value - 4500 cpm [3H]uridine incorporated.

● , p-benzoquinone.   Control value - 15,039cpm [3H]uridine incorporated.

Effect of p-Benzoquinone on Interleukin-2 Production

The effect of p-benzoquinone on IL-2 production by Con A-stimulated
T-lymphocytes was determined by measuring the ability of conditioned
medium from such an incubation to support the proliferation of two
different IL-2 dependent T-lymphocyte clones.   Lymphocytes were incubated with or without 5 µM p-benzoquinone for 30 min.   After washing the
cells to remove the metabolite, they were cultured for 24 hr. to obtain
conditioned medium containing IL-2.   The presence of IL-2 was determined
by monitoring the incorporation of [3H]thymidine into two IL-2 dependent
cell lines incubated with conditioned medium.

Incubation of lymphocytes with Con A results in production of IL-2
and the subsequent stimulation of [3H]thymidine incorporation in the two
IL-2 dependent T-lymphocyte clones.   Exposure of lymphocytes to 5 µM
p-benzoquinone prior to the addition of Con A completely suppresses the
mitogen stimulated production of IL-2 (Table 3).

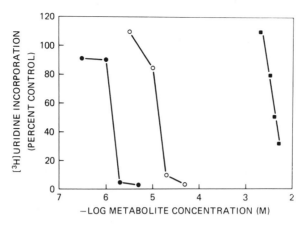

Figure 6. The effects of benzene, hydroquinone and p-benzoquinone on RNA synthesis in lymphocytes _in vitro_. Spleen lymphocytes (2x10^5 cells) were preincubated with varying concentrations of benzene or its quinone metabolites and then incubated with [^{3}H]uridine to monitor their effects on RNA synthesis as described under Methods. ■ benzene; o hydroquinone; ● p-benzoquinone. Each concentration of compound was tested in triplicate and the experiment was performed three times.

DISCUSSION

Mouse peritoneal macrophages and spleen lymphocytes are capable of RNA synthesis _in vitro_ which is dependent on cell number and is linear with time. Elicited peritoneal macrophages were chosen for study because they are a potent source of hemopoietic growth factors and a relatively pure cell population can be more easily obtained than from bone marrow. By inhibiting RNA synthesis in these cell types benzene, through its metabolites, could prevent the production of growth factors required for the survival, proliferation and differentiation of the hemopoietic progenitor cells (Nicola and Vadas, 1984). A deficiency of growth factors could explain the observed preferential toxicity of benzene towards differentiating blood cells as opposed to pluripotent

Table 3

Effect of p-benzoquinone (PBQ) on IL-2 production by

Con A-stimulated mouse T-lymphocytes

Source of IL-2 (Conditioned Medium)*	T-cell line	
	CTLL-CH	CTLL-DP
	[^{3}H]thymidine incorporation, cpm	
Medium (RPMI 1640)	1,747	59
Rat spleen	42,668	13,629
Lymphocyte	4,286	122
+ Con A (5 µg/ml)	54,935	23,256
+ Con A (5 µg/ml)+PBQ (5 µM)	1,850	36
+ PBQ (5 µM)	2,420	86

* The data presented were obtained using a 1:1 dilution of conditioned medium. Further serial dilutions to 1:64 showed a decreasing stimulation of [^{3}H]thymidine uptake.

stem cells (Lee et al., 1974; Irons et al., 1979; Tunek, et al., 1981; Baarson et al., 1984), and would also be an expected consequence of damage to the bone marrow's microenvironment.

Benzene must be activated to cause toxicity (Andrews et al., 1977) and metabolites concentrate in bone marrow after exposure (Andrews et al., 1977; Rickert et al., 1979). Phenol, the major metabolite, is hydroxylated to hydroquinone (Sawahata and Neal, 1983) which oxidizes to p-benzoquinone, a highly electrophilic molecule which covalently binds to cellular macromolecules (Tunek et al., 1978; Lunte and Kissinger, 1983). We found that the putative toxic metabolite, p-benzoquinone, was the most potent inhibitor of RNA synthesis in both macrophages and lymphocytes; the order of potency being p-benzoquinone > hydroquinone > phenol > benzene.

Benzene and its metabolites, hydroquinone and p-benzoquinone, completely inhibit RNA synthesis in macrophages and lymphocytes at concentrations which have essentially no effect on viability. The concentrations of hydroquinone $(1-2 \times 10^{-5} M)$ and p-benzoquinone $(1-2 \times 10^{-6} M)$ which inhibit RNA synthesis in lymphocytes by 50 percent are similar to those which inhibited RNA synthesis 50 percent in mouse macrophages, and which also decreased the ability of mouse bone marrow stromal cells to support the development of CFU-G/M by 50 percent (Gaido and Wierda, 1984).

Because of its high reactivity, the level of p-benzoquinone in bone marrow after benzene exposure has not been measured. However, a report by Rickert et al. (1979) indicated that $6 \times 10^{-4} M$ (70 µg/g) hydroquinone was present in bone marrow of rats exposed to 500 ppm benzene for 6 hours. This concentration is well above our reported IC_{50} for hydroquinone in macrophages of $2.5 \times 10^{-5} M$.

Toxicity to bone marrow and lymphoid organs correlates with the concentrations of hydroquinone accumulated in those tissues (Greenlee et al., 1981a, Greenlee et al, 1981b. Hydroquinone inhibits lectin-stimulated lymphocyte activation in culture (Pfeifer and Irons, 1981) and interferes with microtubule assembly (Irons and Neptune, 1980). Suppression of lymphocyte activation by hydroquinone has been postulated (Irons et al., 1981) to be mediated through interference with the cytoskeleton specifically by interaction of its oxidation product, p-benzoquinone, with intracellular sulfhydryl groups essential for lymphocyte agglutination and blastogenesis and with similar groups responsible for maintaining a functional microtubule apparatus. While this represents an intriguing explanation for benzene toxicity, a alternative explanation might be the inhibition of synthesis of growth factors by T cells. We have reported here that the exposure of T-lymphocytes to 5 µM p-benzoquinone completely inhibited Con A - stimulated T-cell proliferation and the production of the T-cell growth factor interleukin-2.

The mechanism whereby quinone metabolites inhibit RNA synthesis in macrophages and lymphocytes is unknown. Metabolites of benzene have been shown to covalently bind to nuclear (Gill and Ahmed, 1981) and mitochondrial (Rushmore et al., 1984) DNA of bone marrow cells. p-Benzoquinone and hydroquinone have been shown to specifically inhibit the activity of wheat germ RNA polymerase II, the enzyme responsible for the synthesis of mRNA (Nagaraja and Shaw, 1982). At the present time, it has not been ascertained whether the inhibition of transcription by these metabolites results from their covalent binding to the genome and the subsequent impedance of RNA polymerase along the template or from the inactivation of the enzyme _per se_ as a result of covalent binding to a sensitive site on the enzyme or both.

Since both macrophages and T-lymphocytes represent major sources of factors that regulate hemopoiesis, an inhibition of mRNA synthesis by hydroquinone/p-benzoquinone and the subsequent loss of factors required for both survival and development of hemopoietic precursor cells could contribute to the lymphocytopenia and aplastic anemia following benzene exposure.

ACKNOWLEDGEMENTS

We wish to thank Dr. Dolores Byrne for her considerable assistance with the IL-2 experiments. This work was supported by grant ES-02391 from the NIEHS.

REFERENCES

Andrews, L., Lee, E. Witmer, C., Kocsis, J., and Snyder, R., 1977, Effects of toluene on the metabolism, disposition, and hematoopoietic toxicity of [^{3}H]-benzene, Biochem. Pharmacol., 26:293.

Baarson, K., Snyder, C. A., and Albert, R. E., 1984, Repeated exposures of C57Bl mice to 10 ppm inhaled benzene markedly depressed erythropoietic colony formation, Tox.Lett., 20:337.

Cronkite, E., Bullis, J., Inoue, T., and Drew, R., 1984, Benzene inhalation produces leukemia in mice, Toxicol. Appl. Pharmacol., 75:358.

D'Agostino, M. A., Lowry, K. M., and Kalf, G. F., 1975, DNA biosynthesis in rat liver mitochondria: inhibition by sulfhydryl compounds and stimulation by cytoplasmic proteins, Arch. Biochem. Biophys., 166:400.

Frash, V. N., Yushkov, B. G., Karaulov, A. V., and Skuratov, V. L., 1976, Mechanism of action of benzene on hematopoiesis (investigation of hematopoietic stem cells), Bull. Exper. Biol., 82:985.

Gaido, K. and Wierda, P., 1984, In vitro effects of benzene metabolites on bone marrow stromal cells, Toxicol. Appl. Pharmacol., 76:45.

Gill, D. P. and Ahmed, A., 1981, Covalent binding of [^{14}C]benzene to cellular organelles and bone marrow nucleic acids, Biochem. Pharmacol., 30:1127.

Goldstein, B. D., 1984, Clinical hematotoxicity of benzene, in: "Carcinogenicity and Toxicity of Benzene," M.A. Mehlman ed., Princeton Scientific Publishers, Princeton.

Green, J. D., Snyder, C. A., LoBue, J., Goldstein, B. D., and Albert, R. E., 1981, Acute and chronic dose/response effect of benzene inhalation on the peripheral blood, bone marrow, and spleen cells of CD-1 male mice, Toxicol. Appl. Pharmacol., 59:204.

Greenlee, W. F., Gross, E., and Irons, R., 1981a, Relationship between benzene toxicity and the disposition of ^{14}C-labelled benzene metabolites in the rat, Chem.-Biol. Interact., 33:285.

Greenlee, W. F., Sun, D., and Bus, J. 1981b, A proposed mechanism of benzene toxicity: formation of reactive intermediates from polyphenol metabolites, Toxicol. App. Pharmacol., 59:187.

Harigaya, K., Miller, M., Cronkite, E., and Drew, R., 1981, The detection of in vitro liquid bone marrow cultures, Toxicol. App. Pharmacol., 60:346.

Hesketh, P.J., Sullivan, R., Valeri, C.R., and McCarroll, L.A., 1984, The production of granulocyte/monocyte colony-stimulating activity by isolated human T lymphocyte subpopulations, Blood, 63:1141.

Irons, R., Heck, H., Moore, B., and Muirhead, K., 1979, Effects of short term benzene administration on bone marrow cell cycle kinetics in the rat, Toxicol. Appl. Pharmacol., 51:399.

Irons, R.D., and Moore, B.J., 1980, Effect of short term benzene administration on circulating lymphocyte subpopulations in the rabbit: evidence of a selective B-lymphocyte sensitivity, Res. Commun. Chem. Path. Pharmacol., 27:147.

Irons, R., Dent, J., Baker, T., and Rickert, D., 1980, Benzene is metabolized and covalently bound in bone marrow in situ, Chem.-Biol. Interact., 30:241.

Irons, R. D. and Neptun, D. A., 1980, Effects of the principal hydroxy-metabolites of benzene on microtubule polymerization, Arch. Toxicol., 45:397.

Irons, R. D., Neptun, D. A., and Pfeifer, R. W., 1981, Inhibition of lymphocyte transformation and microtubule assembly by quinone metabolites of benzene: evidence of a common mechanism, J. Reticuloendothelial Soc., 30:359.

Kurland, J. I., 1984, Granulocyte-monocyte progenitor cells, in: "Hematopoiesis," D.W. Golde, ed., Churchill Livingstone, New York.

Lee, E., Kocsis, J., and Snyder, R., 1974, Acute effects of benzene on ^{59}Fe incorporation into circulating erythrocytes. Toxicol. Appl. Pharmacol., 27:431.

Lunte, S. M. and Kissinger, P. T., 1983, Detection and identification of sulfhydryl conjugates of p-benzoquinone in microsomal incubations of benzene and phenol, Chem.-Biol. Interact., 47:195.

Moore, M. A. S., 1978, Regulatory role of the macrophage in haemopoiesis, in "Stem Cells and Tissue Homeostasis," B. I. Lord, C. S. Potten, and R. J. Cole, eds., Cambridge Univ. Press, Cambridge.

Nagaraja, K. V. and Shaw, P. D., 1982, Inhibition of wheat germ RNA polymerase II by 2,6-dibromobenzoquinone and related compounds from Aplysina fistularis. Arch. Biochem. Biophys., 215:544.

Nicola, N.A. and Vadas, M., 1984, Hemopoietic colony-stimulating factors, Immunol. Today, 5:76.

Pfeifer, R.W. and Irons, R.D., 1981, Inhibition of lectin-stimulated lymphocyte agglutination and mitogenesis by hydroquinone: reactivity with intracellular sulfhydryl groups, Exper. Mol. Pathol., 35:189.

Post, G. B., Snyder, R., and Kalf, G. F., 1984. Inhibition of mRNA synthesis in rabbit bone marrow nuclei in vitro by quinone metabolites of benzene, Chem.-Biol. Interactions, 50:203.

Prystowsky, M.B., Otten, G., Naujokas, M.F., Fardiman, J., Ihle, J.W., Goldwasser, E., and Fitch, F.W., 1984, Multiple hemopoietic lineages are found after stimulation of mouse bone marrow precursor cells with interleukin 3, Am. J. Pathol., 117:171.

Rickert, D., Baker, T., Bus, J., Barrow, C. and Irons, R., 1979, Benzene disposition in the rat after exposure by inhalation. Toxicol. Appl. Pharmacol. 49:417.

Rushmore, T., Snyder, R., and Kalf, G., 1984, Covalent binding of benzene and its metabolites to DNA in rabbit bone marrow mitochondria in vitro, Chem-Biol. Interact., 49:133.

Sawahata, T. and Neal, R., 1983, Biotransformation of phenol to hydroquinone and catechol by rat liver microsomes, Molec. Pharmacol., 23:453.

Snyder, R., Lee, E., Kocsis, J., and Witmer, C., 1977, Bone marrow depressant and leukemogenic actions of benzene, Life Sciences, 21:1709.

Snyder, R., Longacre, S.L., Witmer, C.M., and Kocsis, J.J., 1982, Metabolic correlates of benzene toxicity, in "Biological Reactive Intermediates," R. Snyder, D. V. Parke, J. J. Kocsis, D. J. Jollow, D. G. Gibson, and C. M. Witmer, eds., Plenum Press, New York.

Tavassoli, M. and Friedenstein, A., 1983, Hemopoietic stromal microenvironment, Am. J. Hematol., 15:195.

Tunek, A., Platt, K.L., Bently, P., and Oesch, F., 1978, Microsomal metabolism of benzene to species irreversibly binding to microsomal protein and effects of modifications of this metabolism, Mol. Pharmacol., 14:920.

Tunek, A., Olofsson, T., and Berlin, M., 1981, Toxic effects of benzene and benzene metabolites on granulopoietic stem cells and bone marrow cellularity in mice. Toxicol. Appl. Pharmacol., 59:149.

Verma, D.S., Johnston, D.A., McCredie, K.B., 1984, Identification of T lymphocyte subpopulations that regulate elaboration of granulo-cyte-macrophage colony stimulating factor. Br. J. Haematol., 57:505.

Wierda, D., and Irons, R., 1982, Hydroquinone and catechol reduce the frequency of progenitor B lymphocytes in mouse spleen and bone marrow. Immunopharm. 4:41.

p-NITROSOPHENETOLE: A REACTIVE INTERMEDIATE OF PHENACETIN THAT BINDS TO

PROTEIN

Jack A. Hinson and Joann B. Mays

National Center for Toxicological Research
Jefferson, AR   72079

INTRODUCTION

Although phenacetin is an effective analgesic, in a small number of chronic abusers it has been proven to be toxic causing interstitial nephritis and progressive reduction in kidney size and function (1). Moreover, a signficant number of tumors of the urinary tract have been reported in the same population (2). Phenacetin also produces methemoglobinemia and hemolysis in genetically susceptible individuals (3). Because of its toxicities, and in particular its known tumorigenicity in animals and suspected carcinogenicity in human, it has been removed from the U.S. market.

We have been interested in mechanisms whereby phenacetin may be converted to reactive metabolites (4). In a recent work, it was shown that the phenacetin metabolite, N-hydroxy-p-phenetidine, covalently bound to DNA but apparently did not react with glutathione (GSH). In contrast, its oxidation product, p-nitrosophenetole, did not bind to DNA; however, it reacted with GSH to form N-(glutathion-S-yl)-p-phenetidine (5).

In this work, we have examined the possibility that p-nitrosophenetole may covalently bind to protein. This possibility seemed reasonable since other aryl nitroso derivatives have been reported to bind to protein, and p-nitrosophenetole reacted with GSH (5-9).

MATERIALS AND METHODS

Chemicals. [ring-^{3}H]p-Phenetidine was synthesized by Dr. R. Roth, Midwest Research Institute (Kansas City, MO). [ring-^{3}H]p-Nitrosophenetole was synthesized by oxidation of the corresponding amine with metachloroperoxybenzoic acid^3. All radiolabeled chemicals were greater than 98% pure. All other chemicals were of the purest grade commercially available.

Assays. Protein covalent binding assays contained either 10 mg bovine serum albumin or 4.5 mg liver microsomes, isolated from 250 g male SD rats, and varying amounts of substrate in 3 ml of 50 mM NaPO$_4$, pH 7.4. Unless otherwise stated, a NADPH generating system was not present. Protein covalent binding assays were performed after extensive methanol or acetone washing of protein (10). When the effect of HCl or GSH on

Table 1. Binding of [^{3}H]p-Nitrosophenetole to Bovine Serum Albumin and
Effect of Acid and GSH on Protein-Bound Residue

Treatment of Protein	Binding (nmoles)	Percent of Control
Buffer	44.7	–
	47.9	–
HCl	13.4	29%
	13.5	29%
GSH	43.6	94%
	47.1	102%

[^{3}H]p-Nitrosophenetole (50 nmoles) was added in 10 μl methanol to
incubations containing 10 mg bovine serum albumin, and 0.05 M NaPO$_4$, pH
7.4, in a total volume of 3.0 ml. After 1 hr incubation at 37°, the
protein was precipitated and washed with acetone. After unbound
radioactivity was removed, the protein was treated with either 0.05 M
Bis-Tris-HCl, pH 7.4, the same buffer plus 1 N HCl or the same buffer
plus 10 mM GSH (1 ml) at 37° for 1 hour. Protein binding was
subsequently determined by standard methods (10).

protein binding was examined, the protein was first washed with solvent
to remove unbound radioactivity and subsequently treated with HCl or GSH.
HPLC assays were performed as previously described (5).

RESULTS

The previous finding that p-nitrosophenetole reacted rapidly with GSH
to form a conjugate (5) suggested that it may also covalently bind to
protein sulfhydryl groups. To test this possibility, 50 nmoles [^{3}H]p-
nitrosophenetole was incubated with bovine serum albumin. Subsequently
the protein was precipitated and washed with solvent to remove any
unbound radiolabel. As shown in Table 1, 89-96% of the radiolabel was
bound to protein. In a similar experiment, when [^{3}H]p-nitrosophenetole
was incubated with microsomal protein (without NADPH), the radiolabel
quantitatively bound to protein.

To determine if p-nitrosophenetole bound to sulfhydryl groups of
protein, microsomal protein was preincubated with N-methylmaleimide which
binds to sulfhydryl groups. Subsequent addition of [^{3}H]p-nitrosophene-
tole to the protein incubation mixtures resulted in a 90% decrease in the
covalent binding of radiolabel to protein compared to control incubation
mixtures which had not been preincubated with N-methylmaleimide.

Experiments were subsequently performed to determine the effect of
GSH and acid on the bound residue. Various aryl nitroso compounds have
been previously shown to react with cysteine derivatives such as GSH to

form two different types of conjugates (Fig. 1): one being a sulfinanilide, the other a sulfinanilide S-oxide. Whereas only the sulfinanilide
can be reduced by GSH to the parent amine, both conjugates can be acid
hydrolyzed to the parent amine (5-9). As shown in Table 1 GSH did not
alter the amount of [^{3}H]p-nitrosophenetole bound to the solvent-washed
protein. However, the amount of [^{3}H]p-nitrosophenetole bound to the
solvent-washed protein was decreased by 71% when treated with HCl (Table
1). HPLC analysis of this HCl extract revealed that approximately 75% of
the radioactivity eluted with a retention time identical to the parent
amine, p-phenetidine. In similar experiments in which microsomes were
used as a source of protein, GSH decreased [^{3}H]p-nitrosophenetole protein
bound residues by only a minor amount; however, HCl decreased binding by
72% and caused p-phenetidine to be released. These data suggest that
p-nitrosophenetole binds to sulfhydryl groups on protein via
sulfinanilide S-oxide linkages.

In other experiments, we investigated the possibility that p-phenetidine, the deacetylated metabolite of phenacetin, may be converted by
microsomes to p-nitrosophenetole which binds to protein. p-Phenetidine
is known to be N-hydroxylated and aryl hydroxylamines are easily oxidized
to the corresponding nitroso derivatives (11). Table 2 shows that microsomes catalyzed the NADPH-dependent binding of [^{3}H]p-phenetidine to protein. As shown, treatment of the [^{3}H]p-phenetidine-modified protein with
HCl, decreased binding by 45%. Treatment of the [^{3}H]p-phenetidine-
modified protein with GSH did not alter the amount of binding (Table 2);
however, when GSH (1 mM) was included in the microsomal incubation
mixtures, binding of [^{3}H]p-phenetidine was inhibited by 70% (data not
shown). These data implicate p-nitrosophenetole in the majority of the
NADPH-dependent protein binding of [^{3}H]p-phenetidine.

DISCUSSION

The involvement of reactive metabolites in the toxicities of various
chemicals has been well documented. Whereas DNA binding is believed to
be causative in the initiation of tumors (12), protein binding may be an
important parameter in various cell specific toxicities (13). For

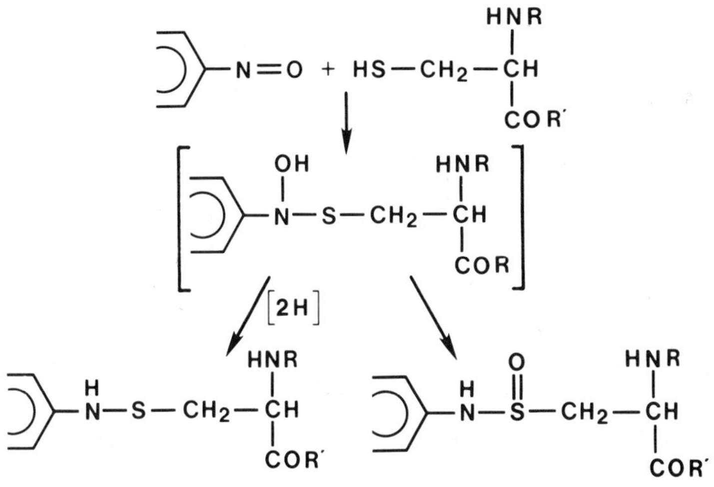

Figure 1. Postulated mechanism of reaction of aryl nitroso compounds
with cysteine derivatives to form conjugates. The reductive
pathway yields the sulfanilide and the isomerization pathway
yields the sulfanilide S-oxide.

Table 2. NADPH-Dependent Binding of [³H]p-Phenetidine to Microsomal
         Protein and Effect of Acid and GSH Treatment on the Protein-
         Bound Residue

Treatment of Protein	Binding (nmoles)	Percent of Control
Buffer	10.2	–
HCl	5.6	55%
GSH	10.0	98%

[³H]p-Phenetidine (1 mM) was incubated with 4.5 mg rat liver microsomes,
NaPO₄, pH 7.4, and a NADPH generating system for 10 min at 37°C in a
total volume of 3.0 ml. The protein was precipitated by addition of 3 ml
methanol and extensively washed with methanol to remove unbound
radioactivity. Subsequently, the protein was treated with either 1 ml of
50 mM Bis-Tris-HCl, pH 7.4 (labeled H₂O), 1 ml of 1 N HCl, or 1 ml of 10
mM GSH in 50 mM Bis-Tris, pH 7.4, for 1 hour. The data are the average
of duplicate incubation mixtures. Variation between duplicates did not
exceed ± 5%.

example, reactive intermediates that may bind to protein have been
implicated as toxic metabolites in the hepatic necrosis caused by large
doses of acetaminophen, chloroform and bromobenzene (13). The mechanism
of toxicity induced by other chemicals, such as the nephrotoxicity of
phenacetin, has remained elusive. This is in part a result of the
absence of an animal model for the toxicity. Nevertheless, we have been
investigating mechanisms by which phenacetin may be converted to reactive
metabolites (4). In recent work it was shown that p-nitrosophenetole
rapidly reacted with GSH to form the sulfinanilide conjugate, N-(glutathion-S-yl)-p-phenetidine (5). In an extension of this work, we have
examined the possibility that p-nitrosophenetole may covalently bind to
protein.

As shown in Table 1, 89–96% of the [³H]p-nitrosophenetole was bound
to bovine serum albumin. Similar results were obtained when [³H]p-nitro-
sophenetole was incubated with microsomal protein. Moreover, derivatization of microsomal protein sulfhydryls with N-methylmaleimide prior to
addition of [³H]p-nitrosophenetole resulted in a 90% decrease in binding
of radiolabel to protein. These results indicate that p-nitrosophentole
is a reactive metabolite that binds to protein sulfhydryl groups.

The nature of the bound adduct was investigated by treating the
[³H]p-nitrosophenetole-modified protein with HCl and GSH. Glutathione
reduces sulfanilides, whereas acid treatment hydrolyzes both sulfanilides
and sulfanilide S-oxides to the corresponding amines (5-9). The finding
that treatment of the solvent-washed protein with GSH did not alter the
amount of protein bound residue, whereas HCl decreased it by 71%,
suggests that p-nitrosophenetole reacts with protein sulfhydryl groups to
form a sulfanilide S-oxide bond. Similar data were obtained with

694

microsomal protein. This mechanism seem plausible since sulfanilide formation requires reduction of an intermediate (Fig. 1), and reducing equivalents were not present in the protein incubations. Moreover, this finding suggests the reason that only the sulfanilide was found in incubations of p-nitrosophenetole and GSH may have been the rate of reduction of the intermediate N-hydroxy-N-glutathione conjugate greatly exceeded isomerization of the intermediate to the sulfanilide S-oxide (Fig. 1).

The possibility was investigated that in a microsomal incubation mixture, p-phenetidine may be converted to p-nitrosophenetole and subsequently bind to protein. It was envisaged that p-phenetidine, the deacetylated metabolite of phenacetin, would be N-hydroxylated by cytochrome P-450. Once formed, the N-hydroxy-p-phenetidine could be easily oxidized to p-nitrosophenetole and this metabolite could either bind to protein or be reduced to N-hydroxy-p-phenetidine by NADPH. As shown in Table 2, microsomes catalyzed the NADPH-dependent conversion of [^{3}H]p-phenetidine to a covalently-bound metabolite(s). The finding that HCl treatment of the protein residue decreased binding by 45% suggested that much of the binding may be via p-nitrosophenetole. Figure 2 summarizes mechanisms whereby p-nitrosophenetole may be formed from phenacetin.

The role that p-nitrosophenetole may have in the toxicities of phenacetin is unknown. The fact that reactive metabolites that bind to protein are important in other toxicities and the finding herein that p-nitrosophenetole binds to protein suggests that it may be important. However, this conclusion requires more extensive investigation.

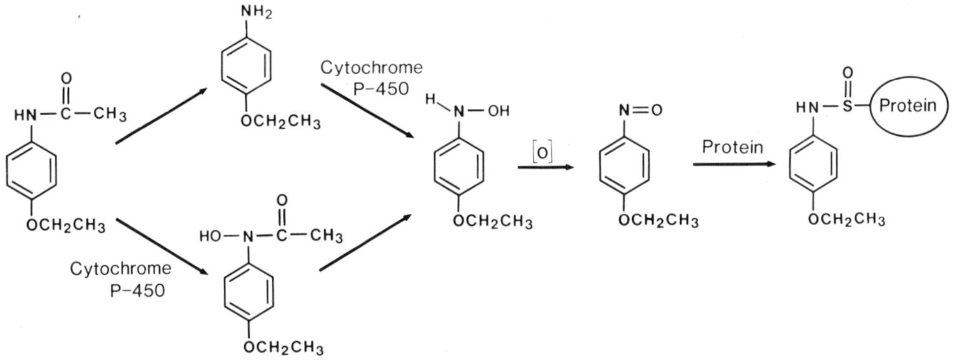

Figure 2. Mechanism of metabolism of phenacetin to p-nitrosophenetole and binding of p-nitrosophenetole to protein.

REFERENCES

1. T. Murray and M. Goldberg, Analgesic abuse and renal disease, Ann. Rev. Med. 26:537-550 (1975).
2. U. Bengtsson, S. Johansson, and L. Angerwall, Malignancies of the urinary tract and their relation to analgesic abuse, Kidney Int. 13:107-113 (1978).
3. N. T. Shahidi and A. Hemaidan, Acetophenetidin-induced methemoglobinemia and its relation to the excretion of diazotizable amines, J. Lab. Clin. Med. 74:581-585 (1969).
4. J. A. Hinson, Reactive metabolites of phenacetin and acetaminophen: A review, Environ. Health Persp. 49:71-79 (1983).
5. G. J. Mulder, F. F. Kadlubar, J. B. Mays, and J. A. Hinson, Reaction of mutagenic phenacetin metabolites with glutathione and DNA. Possible implication for toxicity, Mol. Pharmacol. 26:342-347 (1984).
6. B. Dolle, W. Topner, and H.-G. Neumann. Reaction of arylnitroso compounds with mercaptans, Xenobiotica 10:527-536 (1980).
7. G. J. Mulder, L. E. Unruh, F. E. Evans, B. Ketterer, and F. F. Kadlubar, Formation and identification of glutathione conjugates from 2-nitrosofluorene and N-hydroxy-2-aminofluorene. Chem.-Biol. Interact. 39:111-127 (1982).
8. P. Eyer and M. Schneller, Reactions of the nitroso analogue of chloramphenicol with reduced glutathione, Biochem. Pharmacol. 32:1029-1036 (1983).
9. J. P. Uetrecht, Reactivity and possible significance of hydroxylamine and nitroso metabolites of procainamide, J. Pharmacol. Exp. Ther. 232:420-425 (1985).
10. J. A. Hinson, S. D. Nelson, and J. R. Mitchell, Studies on the microsomal formation of arylating metabolites of acetaminophen and phenacetin, Mol. Pharmacol. 13:625-633 (1977).
11. T. Nohmi, K. Yoshikawa, M. Nakadate, and M. Ishidate, Species difference in the metabolic activation of phenacetin by rat and hamster liver microsomes, Biochem. Biophys. Res. Comm. 110:746-752 (1983).
12. E. C. Miller and J. A. Miller, Reactive metabolites as key intermediates in pharmacologic and toxicologic responses. Examples from chemical carcinogenesis, Adv. Exp. Med. Biol. 126:1-21 (1982).
13. J. R. Mitchell, W. Z. Potter, J. A. Hinson, W. R. Snodgrass, J. A. Timbrell, and J. R. Gillette, Toxic drug reactions, in: "Handbook of Experimental Pharmacology," O. Eichiles, A. Farah, H. Herken, and A. D. Welch, eds., Vol. XXVIII/3, Springer Verlag, Berlin, pp. 383-419 (1975).

MECHANISM OF FASTING-INDUCED SUPPRESSION OF ACETAMINOPHEN GLUCURONIDATION

IN THE RAT

Veronica F. Price, Jennifer M. Schulte, Stephen M. Spaethe
and David J. Jollow

Department of Pharmacology
Medical University of South Carolina
Charleston, SC 29425

RELATIONSHIPS BETWEEN ACETAMINOPHEN GLUCURONIDATION AND HEPATOTOXICITY

Acetaminophen overdosage, taken either as a suicidal gesture or as an accidental poisoning, is a significant clinical entity and one that appears to be growing in incidence. Results from a recent study have shown that the percentage of total drug-related deaths associated with acetaminophen in England and Wales rose dramatically from 3.8% in 1973 to 21% in 1980; further, the number of acetaminophen-associated deaths was the highest of all drug-related deaths for the years 1978 to 1980 (1). Thus there is a need to define more clearly the reasons for differences in patient responsiveness and to develop new treatment regimens to decrease the severity of liver failure.

Presuicide individuals commonly show depression, ethanol abuse and/or marked weight loss (2,3). The weight loss presumably results from poor eating habits or prolonged fasting. Of particular interest, in rats (4) and mice (5-7), an acute fast (18 to 42 hr) is well known to potentiate acetaminophen hepatotoxicity. Thus, it is reasonable to suspect that changes in nutritional status may contribute significantly to the development of acetaminophen liver injury.

In man, the susceptibility and severity of acetaminophen hepatic necrosis cannot be predicted from either the dose or the plasma concentration of the drug; the most useful predictor of severity of liver injury appears to be the combination of plasma levels and extent of lengthening of the plasma half-life of the drug (8). In both man (9) and experimental animals (10-12), glucuronidation is the major pathway of clearance of high doses of acetaminophen. It follows that after high dose levels the half-life in man and animals largely reflects the glucuronidation capacity and hence that a marked prolongation of the half-life reflects a relative deficiency in glucuronidation. Previous studies in animals have shown that a deficiency in their glucuronidation capacity, as for example occurs when galactosamine is coadministered, is associated with marked lengthening of the half-life of acetaminophen, increase in the amount of acetaminophen converted to its toxic metabolite, and increased incidence and severity of liver injury (13,14). Thus, individuals with lesser capacity for acetaminophen glucuronidation may be considered at greater risk of liver injury after acetaminophen overdosage. Therefore, it is of importance to understand the factors which determine the glucuronidation capacity. Such an understanding may provide insight into a new possible metabolic basis for protective therapy against this reactive metabolite-mediated toxicity.

RELATIONSHIPS BETWEEN ACETAMINOPHEN GLUCURONIDATION AND HEPATIC INTER-MEDIARY METABOLISM OF GLUCOSE

Previous studies have shown that streptozotocin-induced diabetic rats, in which the hepatic metabolism of glucose via insulin-sensitive pathways (glucose phosphorylation and glycogen synthesis) is markedly depressed, are surprisingly more resistant to acetaminophen liver injury; this relative resistance is largely the result of enhanced clearance of the drug by glucuronidation (12). Previous studies in normal fed rats have indicated that at high dose levels of acetaminophen the glucuronidation capacity is determined by production of the cosubstrate, UDP-glucuronic acid (UDPGA) rather than saturation of glucuronyl transferase by acetaminophen (15). In turn, UDPGA, and precursors and cofactors required for its synthesis (glucose, NAD$^+$ and UTP) are produced and utilized by the normal pathways of intermediary metabolism of the liver cell (gluconeogenesis, glycolysis, glycogen metabolism) (Fig. 1). It follows that acetaminophen glucuronidation and susceptibility to liver injury may be determined by the rate-limiting interactions in inter-mediary metabolism that determine the rate of supply of these precursors and cofactors. Further, agents or conditions which alter intermediary metabolism and hence the metabolic poise of the liver cell may be expected to influence profoundly the glucuronidation and hepatotoxicity of acetaminophen.

An acute fast, like diabetes, is known to cause profound alterations in the hepatic metabolism of glucose. Of importance, fasting is well known to deplete liver levels of glycogen (16), the postulated source of glucose for the production of UDPGA (17). The regulation of the synthesis of UDPGA and its relationship to glycogen levels, to gluconeogen-

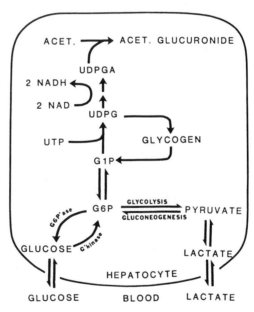

Fig. 1. Interrelationships Between Glucose Metabolism and Acetaminophen Glucuronidation.

esis and/or to other factors in hepatic carbohydrate metabolic homeostasis (Fig. 1) are hence major determinants of the susceptibility of the
fasted liver to acetaminophen hepatotoxicity.

GLUCURONATE-XYLULOSE PATHWAY

UDPGA arises as an intermediate in the glucuronate-xylulose pathway
(Fig. 2). This overall metabolic sequence is thought to play a minor
role in carbohydrate homeostasis although, as pointed out by Sochor et al.
(18), when combined with the pentose phosphate pathway, it can be formulated to yield ATP from G6P; viz,

$$G6P + 30 \ ADP \rightarrow 6CO_2 + 30 \ ATP$$

It is more commonly considered that the pathway normally functions to
provide intermediates for special purposes: UDP-glucose (UDPG) for glycogen; UDPG, UDPGA, UDP-galactose and related hexosamines for glycoprotein and protein polysaccharide formation; gulonate for formation of
ascorbate; various pentoses for nucleic acid and other syntheses; and
UDPGA, the essential cosubstrate for drug glucuronidation.

The regulation of the glucuronate-xylulose pathway is not well understood. Glycogen levels are controlled at the levels of glycogen synthesis
(19) and glycogenolysis (20) and apparently are not dependent directly on
UDPG formation.   In in vitro studies, UDPGA production from UDPG was
observed to be inhibitable by NADH (21,22) which led to the suggestion
that this enzyme system (and by implication the overall pathway) might be
regulated by the redox state of the cell.

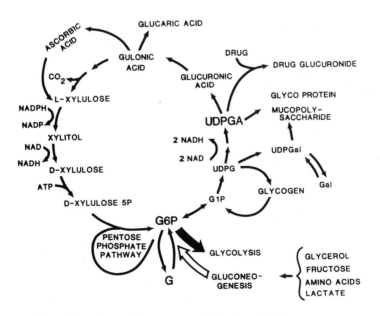

Fig. 2.   Glucuronate-Xylulose Pathway.

The interdependence of the glucuronate-xylulose pathway with other major pathways of carbohydrate metabolism is complex (Fig. 2). In the diabetic state the activity of the pathway is believed to be enhanced four-fold whereas most other carbohydrate pathways (glycolysis, pentose phosphate shunt and glycogen synthesis) are depressed (23-28). These relationships emphasize that the activity of a minor pathway may be altered dramatically by small changes in flux of major pathways, and that the activity of a minor pathway may be significantly enhanced even if the overall carbohydrate metabolism is significantly depressed.

ACETAMINOPHEN GLUCURONIDATION IN THE FASTED RAT

As noted above, an acute fast is known to cause profound alterations in hepatic metabolism of glucose. Low blood glucose levels trigger glucagon release and consequent enhanced glycogenolysis as well as enhanced gluconeogenesis from amino acids, glycerol and lactate. The glucose-6-phosphate (G6P) produced is rapidly dephosphorylated due to enhanced glucose-6-phosphatase (G6P'ase) activity and released from the liver (Figs. 1 & 2).

An acute fast is also well known to potentiate acetaminophen liver injury in rats; this relative potentiation has been associated with a depression in their hepatic glutathione protective capacity (4). However since fasting depletes glycogen levels, it is reasonable to suspect that fasting would also depress UDPGA synthesis and hence glucuronidation of acetaminophen. Depression of clearance of the drug by glucuronidation would, as noted above, result in a lengthening of the half-life of the drug, an increase in the proportion of the dose converted to the reactive metabolite hence potentiation of liver injury.

Experimentally, an overnight fast decreased the overall elimination rate (β) of high doses of acetaminophen. This was largely the result of a decrease (ca. 40%) in the glucuronidaton capacity (estimated by the apparent rate constant for glucuronide formation) (29). The rate of formation of the reactive metabolite (estimated by the apparent rate constant for formation of acetaminophen mercapturate, K'_{MA}) was not affected by fasting. The proportion of the dose converted to the reactive metabolite (K'_{MA}/β) was significantly increased in fasted rats due to the reduction in β. That is, due to the decreased capacity of the liver to clear acetaminophen by nontoxic glucuronidation, a greater amount of the reactive metabolite was formed in cells which also had a decreased glutathione protective capacity. Of interest, although fasted rats had a significantly decreased glucuronidation capacity after high subtoxic and toxic dose levels of acetaminophen, the activity remained at approximately 60% that of fed rats (29).

In additional studies we have examined the mechanism underlying the fasting-induced decreased glucuronidation capacity by assessing the rate-determining roles of glucuronyl transferase activity, hepatic levels of UDPGA, and various factors involved in production of UDPGA (Fig. 1) during metabolism of a hepatotoxic dose of acetaminophen (30). These studies revealed that fasting did not decrease the activity/levels of acetaminophen glucuronyl transferase as measured in isolated microsomal fractions, suggesting that it is unlikely that the rate-limiting factor for glucuronidation in vivo is the amount of glucuronyl transferase enzyme present in the endoplasmic recticulum.

Neither basal levels of hepatic UDPGA nor the extent of its depletion by acetaminophen were different in fasted rats (Table 1). In marked contrast, the recovery of UDPGA levels back to predrug levels took significantly longer in fasted rats (30), suggesting a decreased capacity to

TABLE 1.   EFFECT OF ACETAMINOPHEN ON VARIOUS PARAMETERS OF CARBOHYDRATE
METABOLISM IN FED AND FASTED RATS

Parameter	Animals	Vehicle[a] (basal)	+Acet[b] (nadir)	+Acet[c] (recovery)
UDPGA (nmol/g liver)	Fed	307 ± 13	83 ± 5	543 ± 15
	Fasted	341 ± 71	70 ± 12	252 ± 30*
UDPG (nmol/g liver)	Fed	558 ± 39	230 ± 35	249 ± 28
	Fasted	346 ± 22*	141 ± 32*	126 ± 12*
Glycogen (mg/g liver)	Fed	28.9 ± 1.0	3.6 ± 1.2	12.4 ± 2.9
	Fasted	0.37 ± 0.10*	0.23 ± 0.06*	0.24 ± 0.03*
Blood Glucose (mg/dl)	Fed	108 ± 8	210 ± 8	159 ± 7
	Fasted	76 ± 2*	207 ± 26	208 ± 9*

[a] Vehicle: 20% Tween 80 in normal saline

[b] Animals received acetaminophen (700 mg/kg, ip); values are those at
nadir of depletion (between 0.5 and 2 hr after acetaminophen).

[c] Animals received acetaminophen (700 mg/kg, ip); values are those during
recovery phase (4 to 6 hr after acetaminophen).

* Significantly different from fed rats, $p < 0.05$; values are means ± S.E.,
n=4, representative of two or three separate experiments.

synthesize UDPGA under metabolic demand.   Calculation of the average rate
of UDPGA synthesis during the first three half-lives of acetaminophen
(equivalent to the metabolism of 87.5% of the dose) revealed that the
UDPGA synthetic rate was approximately 35% slower in fasted rats than that
in fed rats (378 ± 12 vs 581 ± 9 μmol/kg rat/hr) (30).

Recognition of the rate-determining role of UDPGA synthesis for glucuronidation has focused our attention on factors which may regulate
UDPGA synthesis, including: activity of UDPG dehydrogenase, cellular
levels of NAD and/or NADH, and supply of the precursor, UDPG (Fig. 1).

Fed and fasted rats were found to have similar activity/levels of
UDPG dehydrogenase, as measured in isolated hepatic cytosolic fractions
(30).   It is therefore unlikely that the lower rate of UDPGA synthesis
seen in fasted rats is the result of decreased levels of the enzyme
responsible for its formation.

Since in vitro studies (21,22) have suggested that UDPG dehydrogenase activity in vivo may be allosterically enhanced by cellular levels
of NAD^+, or depressed by NADH, or controlled by the $NADH/NAD^+$ ratio, the
rate of UDPGA synthesis may be regulated by the redox state of the cell.
In agreement with previous workers (31), fasted rats were found to have
modestly but significantly increased total basal levels of NADH; total
basal levels of NAD were not affected by fasting (30).   During the metabolism of a hepatotoxic dose of acetaminophen fasting had no significant
effect on NAD or NADH levels.   Overall, these data suggest that the
higher predrug levels of NADH would probably be only a minor contributor
to the slower rate of UDPGA synthesis seen at two through six hr after
administration of acetaminophen.

Since the in vivo activity of UDPG dehydrogenase may be determined by the supply of its substrate, UDPG levels were examined (Table 1). Basal levels of UDPG were significantly lower in fasted rats (ca. 60% of levels in fed rats). After administration of a hepatotoxic dose of acetaminophen, UDPG levels rapidly fell to approximately 40% of predrug levels in both groups of rats, and remained low at six hr after drug administration. At all times after acetaminophen, UDPG levels were significantly lower in fasted rats indicating that the lower rate of UDPGA synthesis was the result of lower supply of the precursor, UDPG.

Of interest, the ratio of hepatic levels of UDPGA/UDPG after acetaminophen administration were calculated for fed and fasted rats. This ratio could be considered as an estimate of the in vivo activity of UDPG dehydrogenase during metabolic demand for UDPGA. As shown in Fig. 3, the UDPGA/UDPG ratio was generally similar in fed and fasted rats. This observation suggests that conversion of UDPG to UDPGA during acetaminophen metabolism was similar in the two groups of rats, although differences in either rate of supply of UDPG or rate of utilization of UDPGA are also possible.

UDPG is a common precursor for the formation of both UDPGA and glycogen. Since hepatotoxic doses of acetaminophen are known to deplete hepatic glycogen levels in fed animals (32-34), and supply of glucose from glycogen for UDPGA synthesis has been suggested to be the major

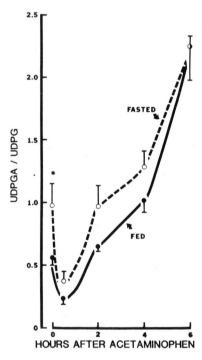

Fig. 3.   Effect of acetaminophen (700 mg/kg, ip) on the ratio of hepatic levels of UDPGA/UDPG in fed (●) and fasted (0) rats. Values are means ± S.E., n=4, representative of two separate experiments. (*Significantly different from fed rats, p < 0.05).

determinant of the rate of glucuronidation (17), it was of interest to examine whether acetaminophen had any detectable effect on the essentially depleted glycogen levels in fasted rats. As shown in Table 1, acetaminophen had no significant effect on the very low glycogen levels in fasted rats. Thus a good correlation does not exist between levels of UDPG (ca. 60% of levels in fed rats) and glycogen (ca. 2% of levels in fed rats). Collectively, the data suggest that the glucose required for UDPG hence UDPGA syntheses in fasted rats must be derived from sources other than glycogen.

Hepatotoxic doses of acetaminophen also induce marked increases in blood glucose levels in fed animals (34,35). This observation suggests that the glucose-1-phosphate (G1P) which is released from glycogen during the metabolism of acetaminophen not only is available for UDPG and UDPGA production, but also a portion of that G1P is converted to G6P, dephosphorylated and leaves the liver (Fig. 1) resulting in the increase in blood glucose levels. Basal blood glucose levels are significantly lower in fasted rats. Surprisingly, these low blood glucose levels increased markedly after acetaminophen administration in fasted rats to the same levels seen in fed rats and remained elevated longer (Table 1). Since glycogen levels are depleted, the source of the increased blood glucose levels presumably is gluconeogenesis which is known to operate at a higher rate in fasted rats. These data suggest that the glucose produced in the fasted liver via gluconeogenesis is directed toward liberation into the blood stream and · is less available for utilization in the liver. It follows that during fasting the activity of G6P'ase may be higher relative to that of phosphoglucomutase and/or UDPG pyrophosphorylase (Fig. 1), and that the amount of G6P derived from enhanced gluconeogenesis in fasted rats not only supports acetaminophen glucuronidation at approximately 60% of the rate in fed rats, but also that a significant amount of G6P is dephosphorylated and released into the blood.

Additional support for this postulate is suggested from studies in which we attempted to reverse the effects of fasting on acetaminophen glucuronidation and hepatotoxicity by administering glucose (36). In these studies glucose was given in a dosing regimen to cause significant increases in blood glucose levels (ca. 200 mg/dl) and glycogen deposition (ca. 16 mg/g liver) in non-acetaminophen treated fasted rats. However, glucose treatment failed: to reverse fasting-induced potentiation of acetaminophen liver injury, to enhance acetaminophen glucuronidation, or to increase hepatic levels of UDPG or UDPGA. Acetaminophen significantly reduced glucose-induced glycogen deposition to approximately 5 mg/g liver, and appeared to block removal of glucose from the blood and/or the entry of glucose into tissues as evidenced by a marked increase (ca. 500 mg/dl) in blood glucose levels in the presence of the drug. Of interest, blood glucose levels were extremely high during the time period in which acetaminophen glucuronidation is limited by the supply of UDPG for UDPGA synthesis. Thus fasting-induced suppression of acetaminophen glucuronidation appears to be due to altered activities of enzymes of carbohydrate metabolism rather than lack of glucose per se (Fig. 1).

During an acute fast lower blood glucose levels result in lower insulin levels favoring lower glucokinase activity, and higher glucagon levels favoring higher G6P'ase activity (Fig. 1). The data are consistent with acetaminophen inhibiting glucokinase activity or stimulating G6P'ase activity. To bypass the glucokinase reaction we have administered the gluconeogenic precursors, lactate, alanine and fructose to fasted rats. It was postulated that the G6P formed from these compounds may be available for UDPG and UDPGA syntheses, thereby to enhance acetaminophen glucuronidation, and to reverse fasting-induced potentiation of hepatotoxicity. Experimentally (36), we observed that although these compounds did increase blood glucose levels, they did not: cause significant glycogen deposition, enhance acetaminophen glucuronidation, or reverse fasting-induced potentiation of liver injury. These studies indicate that the G6P formed via gluconeogenesis was dephosphorylated, and released into the blood, and hence was not available for UDPG/UDPGA production for glucuronidation. These results provide further support for the hypothesis that fasting-induced decrease in UDPG formation may be due to enhanced activity of G6P'ase relative to that of phosphoglucomutase and/or UDPG pyrophosphorylase (Fig. 1).

The insulin to glucagon ratio may largely determine the alterations in flux of glucose through the various pathways of hepatic intermediary metabolism and alterations in the complex interrelationships between acetaminophen and intermediary metabolism described above (Figs. 1 and 2). These in turn may ultimately determine the rate of UDPGA synthesis, hence the glucuronidation capacity of the liver and its susceptibility to hepatotoxicity after acetaminophen overdoses.

SUMMARY

These studies have revealed the occurrence of important relationships among nutritional status, hepatic intermediary metabolism, acetaminophen glucuronidation and susceptibility to hepatotoxicity. During an acute fast hepatic metabolism of glucose is altered profoundly. The altered metabolic poise of the fasted liver appears to favor higher G6P'-ase activity relative to UDPG pyrophosphorylase activity, resulting in decreased production of UDPG secondary to depleted glycogen levels. Although the rate of gluconeogenesis is enhanced and maintains UDPG levels at approximately 60% of those in fed animals, the decreased production of UDPG limits the rate of UDPGA synthesis for glucuronidation of high doses of acetaminophen. Since glucuronidaton is the major pathway of clearance of these high doses of the drug, UDPG synthesis is rate-limiting for acetaminophen elimination; the resulting prolongation of the drug half-life is associated with increased amount of reactive metabolite formed and potentiation of liver injury. Glucuronidation is also the major pathway of clearance in the human overdose situation and if UDPG production occupies a similar rate-determining role, then enhancement of UDPG production might be of significant value in the therapy of acetaminophen overdosage. Thus, determination of factors which control UDPG production in the liver under different physiological (nutritional/hormonal) conditions has both fundamental and practical value.

ACKNOWLEDGEMENTS

This work was supported by a grant from the United States Public Health Service (GM 30546) and a Medical University of South Carolina Biomedical Research Grant. The authors thank Mischelle Johnston and John Peters for excellent technical assistance and Marie Meadowcroft for typing this manuscript.

REFERENCES

1. M.D. Osselton, R.C. Blackmore, L.A. King and A.C. Moffat, Poisoning-associated deaths for England and Wales between 1973 and 1980, Human Toxicol. 3: 201 (1984).
2. R.A. Flood and C.P. Seager, A retrospective examination of psychiatric case records of patients who subsequently committed suicide, Brit. J. Psychiat. 114: 443 (1968).

3. B. Barraclough, J. Bunch, B. Nelson and P. Sainsbury, A hundred cases of suicide: Clinical aspects, Brit. J. Psychiat. 125: 355 (1974).
4. D. Pessayre, A. Dolder, J.-Y. Artigou, J.-C. Wandscheer, V. Descatoire, C. Degott and J.-P. Benhamou, Effect of fasting on metabolite-mediated hepatotoxicity in the rat, Gastroenterology, 77: 264 (1979).
5. A. Wendel, S. Feuerstein and K.-H. Konz, Acute paracetamol intoxication of starved mice leads to lipid peroxidation in vivo, Biochem. Pharmacol. 28: 2051 (1979).
6. O. Strubett, E. Dost-Kempf, C.-P. Seigers, M. Younes, M. Volpel, U. Preuss and J.G. Dreckmann, The inlfuence of fasting on the susceptibility of mice to hepatotoxic injury, Tox. Appl. Pharmacol. 60: 66 (1981).
7. R.M. Walker, T.E. Massey, T.F. McElligott and W.J. Racz, Acetaminophen toxicity in fed and fasted mice, Can. J. Physiol. Pharmacol. 60: 399 (1982).
8. L.F. Prescott, N. Wright, P. Roscoe and S.S. Brown, Plasma-paracetamol half-life and hepatic necrosis in patients with paracetamol overdosage. Lancet, 1: 519 (1971).
9. L.F. Prescott, Kinetics and metabolism of paracetamol and phenacetin, Brit. J. Pharmacol. 10: 291S (1980).
10. D.J. Jollow, S.S. Thorgeirsson, W.Z. Potter, M. Hashimoto and J.R. Mitchell, Acetaminophen-induced hepatic necrosis. VI. Metabolic disposition of toxic and nontoxic doses of acetaminophen, Pharmacology, 12: 251 (1974).
11. D.J. Jollow, S. Roberts, V. Price and C. Smith, Biochemical basis for dose response relationships in reactive metabolite toxicity, in "Biological Reactive Intermediates II: Chemical Mechanisms and Biological Effects", R. Snyder, D.V. Parke, J.J. Kocsis, G. Gibson amd D.J. Jollow, ed., Plenum Press, New York (1982).
12. V.F. Price and D.J. Jollow, Increased resistance of diabetic rats to acetaminophen-induced hepatotoxicity, J. Pharmacol. Exp. Ther. 220: 504 (1982).
13. C. Smith, and D.J. Jollow, Potentiation of acetaminophen induced liver necrosis in hamsters by galactosamine, Pharmacologist 18: 156 (1976).
14. V.F. Price and D.J. Jollow, Acetaminophen hepatotoxicity: Reduced susceptibility of diabetic rats, Pharmacologist 22: 245 (1980).
15. V.F. Price and D.J. Jollow, Role of UDPGA flux in acetaminophen clearance and hepatotoxicity, Xenobiotica 7: 553 (1984).
16. N.V. Carroll, R.W. Longley and J.H. Roe, The determination of glycogen in liver and muscle by the use of anthrone reagents, J. Biol. Chem. 220: 583 (1956).
17. R.G. Thurman and F.C. Kauffman, Factors regulating drug metabolism in intact hepatocytes, Pharmacol. Rev. 31: 229 (1980).
18. M. Sochor, N.Z. Baquer, and P. McLean, Glucose overutilization in diabetes: Evidence from studies on the changes in hexokinase, the pentose phosphate pathway and glucuronate-xylulose pathway in rat kidney cortex in diabetes, Arch. Biochem. Biophys. 198: 632 (1979).
19. S. Hizukuri and J. Larner, Studies on UDPG: α-1,4-glucan α-4-glucosyltransferase. VII. Conversion of the enzyme from glucose-6-phosphate-dependent to independent form in liver, Biochemistry 3: 1783 (1964).
20. H.G. Hers, The control of glycogen metabolism in the liver, Ann. Rev. Biochem. 45: 167 (1976).
21. J. Zalitis and D.S. Feingold, The mechanism of action of UDPG dehydrogenase, Biochem. Biophys. Res. Commun. 31: 693 (1968)
22. P. Moldeus, B. Anderson and A. Norling, Interaction of ethanol oxidation with glucuronidation in isolated hepatocytes, Biochem. Pharmacol. 27: 2583 (1978).

23. A.I. Winegrad and C.L. Burden, Hyperactivity of the glucuronic acid pathway in diabetes mellitus. Trans. Assoc. Am. Physicians 75: 158 (1965).

24. A.I. Winegrad and C.L. Burden, L-Xylulose metabolism in diabetes mellitus, New England J. Med. 274: 298 (1966).

25. J.W. Anderson, Alterations in the metabolic fate of glucose in the liver of diabetic animals, Am. J. Clin. Nutr. 27: 746 (1974).

26. J.W. Anderson, Metabolic abnormalities contributing to diabetic complications. I. Glucose metabolism in insulin-insensitive pathways, Am. J. Clin. Nutr. 28: 273 (1975).

27. Y. Hinolara, S. Takanashi, R. Nagashima and A. Shioya, Glucuronic acid pathway in alloxan diabetic rabbits. I. Urinary excretion of metabolite related to the glucuronic acid pathway. Jpn. J. Pharmacol. 24: 869 (1974).

28. D.R.P. Tulsiani and O. Touster, Studies on dehydrogenases of the glucuronate-xylulose cycle in the livers of diabetic mice and rats, Diabetes 28: 793 (1979).

29. V.F. Price, M.G. Miller and D.J. Jollow, Mechanism of fasting-induced potentiation of acetaminophen hepatotoxicity in the rat, Biochem. Pharmacol. Submitted.

30. V.F. Price and D.J. Jollow, Mechanism of fasting-induced suppression of acetaminophen glucuronidation in the rat, Biochem. Pharmacol. Submitted.

31. H.A. Krebs, The redox state of nicatinamide adenine dinucleotides in the cytoplasma and mitochondria of rat liver, Adv. Enzyme Res. 5: 409 (1967).

32. J.R. Mitchell, D.J. Jollow, W.Z. Potter, D.C. Davis, J.R. Gillette and B.B. Brodie, Acetaminophen-induced hepatic necrosis. I. Role of drug metabolism, J. Pharmacol Exp. Ther. 187: 185 (1973).

33. M.F. Dixson, B. Dixson, S.R. Aparicio and D.P. Loney, Experimental paracetamol-induced hepatic necrosis: A light- and electron-microscope, and histochemical study, J. Path. 116: 17 (1975).

34. J.A. Hinson, J.B. Mays and A.M. Cameron, Acetaminophen-induced hepatic glycogen depletion and hyperglycemia in mice, Biochem. Pharmacol. 32: 1979 (1983).

35. J.A. Hinson, H. Han-Hsue, J.B. Mays, S.J. Holt, P. McLean and B. Ketterer, Acetaminophen-induced alterations in blood glucose and blood insulin levels in mice, Res. Commun. Chem. Path. Pharmacol. 43: 381 (1984).

36. V.F. Price and D.J. Jollow, Acute glucose does not reverse fasting-induced suppression of acetaminophen glucuronidation in the rat. Biochem. Pharmacol. Submitted.

EFFECT OF DIETHYL ETHER ON THE BIOACTIVATION, DETOXIFICATION, AND HEPATOTOXICITY OF ACETAMINOPHEN IN VITRO AND IN VIVO

Peter G. Wells and Esther C.A. To

Faculty of Pharmacy, University of Toronto, 19 Russell Street Toronto, Ontario, Canada, M5S 1A1

INTRODUCTION

Diethyl ether (ether) is used widely as a general anesthetic in animal research, and it remains a convenient human anesthetic in some third world countries. Acetaminophen (N-acetyl-p-aminophenol, APAP, Tylenol®) is a widely used analgesic/antipyretic drug which in high doses can cause centrilobular hepatic necrosis in animals (Boyd and Bereczky, 1966) and humans (Proudfoot and Wright, 1970). Hepatotoxicity is thought to result from cytochromes P-450-catalysed bioactivation of a small amount of acetaminophen to a toxic, reactive intermediary metabolite (Mitchell et al., 1973a), while up to 60% of acetaminophen is eliminated by a competing glucuronidation pathway (fig. 1). Under normal conditions, the reactive intermediate is evanescent, being immediately detoxified by enzymatic conjugation with hepatic reduced glutathione (GSH) and subsequently excreted as cysteine and N-acetylcysteine conjugates (Mitchell et al., 1973b). However, if the amount of acetaminophen bioactivated is increased, or if detoxification of the reactive intermediate is reduced, then the available reactive intermediate arylates hepatic macromolecules, which is thought to initiate processes leading to hepatocellular necrosis (Mitchell et al., 1973a; Jollow et al., 1973).

We have shown that pretreatment with ether or halothane anesthesia greatly potentiates acetaminophen hepatotoxicity if acetaminophen is given several hours after the anesthetic (Wells and Ramji, 1983; Wells et al., 1985). Our hypothesis was that while ether initially could inhibit both the nontoxifying glucuronidation pathway (Eriksson and Strath, 1981; Watkins and Klaassen, 1982) and the toxifying or bioactivating cytochromes P-450 pathway (Umeda and Inaba, 1978; Johannessen et al., 1981), the toxifying P-450 pathway might recover first (fig. 2). In that case, acetaminophen hepatotoxicity would be potentiated by anesthesia only if acetaminophen was given during the time window when P-450 activity had recovered while glucuronidation remained inhibited. Since conjugation with sulfate is both a minor and saturable pathway of elimination, it might be expected that even a small reduction by ether in the major glucuronidating pathway could result in a major percentage increase in the quantitiatively minor but toxicologically critical P-450 bioactivating pathway.

The present study was conducted to confirm this delayed toxicologic synergism and to examine the possible biochemical mechanisms in light of the proposed hypothesis.

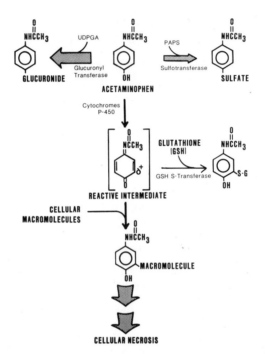

Fig. 1.   Presumed relation of acetaminophen biotransformation to hepatotoxicity. As much as 60% of acetaminophen is eliminated via enzymatic conjugation with glucuronic acid, producing a nontoxic, water soluble metabolite. A small amount (5-10%) is bioactivated by the hepatic cytochromes P-450 to a reactive intermediary metabolite which, if not detoxified by conjugation with glutathione, can bind covalently to essential hepatocellular macromolecules, thereby initiating a process leading to cellular necrosis (see text).

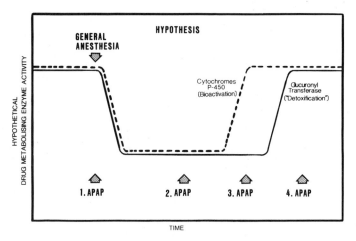

Fig. 2.   Biochemical hypothesis for the potentiation of acetaminophen (APAP) hepatotoxicity by general anesthetics.   General anesthesia inhibits the enzymatic pathways for both the bioactivation (P-450 toxification) and "detoxification" (elimination via the alternative, nontoxifying glucuronyl transferase pathway) of acetaminophen.   If bioactivation recovers first, then administration of acetaminophen at this time (case 3) would cause enhanced hepatotoxicity.   Acetaminophen given at other times after anesthesia, either when both bioactivation and detoxification were unaffected (case 1), inhibited (case 2) or recovered (case 4), would cause the same hepatotoxicity as acetaminophen given alone.

METHODS

Male CD-1 mice (Charles River Canada Inc., St. Constant, Quebec) weighing 25-30 g were housed in plastic cages containing ground corn cob bedding.   A 12 hr light cycle was maintained automatically, and animals were provided ad libitum with food (Purina Rodent Chow®) and tap water.   All animals were maintained for one week before receiving any treatment.

Ether (diethyl ether, BDH Laboratories, Toronto, Ontario) was administered for 5 min to animals housed in an inhalation chamber, with anesthetic efficacy monitored by abolition of the righting reflex.   For the confirmatory toxicological study, acetaminophen (N-acetyl-p-aminophenol, Sigma Chemical Co., St. Louis, Missouri), 300 mg/kg i.p., was administered at various times after ether. Hepatotoxicity was monitored continuously over 48 hr by repetitive measurement of plasma glutamic-pyruvic transaminase (GPT) concentrations in each mouse (Wells and To, 1985; Wells and To, submitted).

The biochemical and pharmacological effects of ether pretreatment potentially involved in the enhancement of acetaminophen hepatotoxicity were assessed by the variables illustrated in fig. 1 as described below:

a)   the in vitro activities of enzymes responsible for the elimination, bioactivation and detoxification of acetaminophen (fig. 1). A pilot study was conducted using 4 mice per group to evaluate the temporal biochemical effects of ether. This was followed by a more comprehensive biochemical study using 8 mice per group at times after ether administration when there was minimal (2 hr) and maximal (8 hr) potentiation of acetaminophen hepatotoxicity.

b)   the in vivo elimination of parent acetaminophen and production of plasma and urinary metabolites of acetaminophen reflecting toxicologically critical pathways of acetaminophen metabolism (fig. 1). Acetaminophen and metabolites were analysed by high-performance liquid chromatography (To and Wells, 1985).

c)   the covalent binding of radiolabeled acetaminophen to hepatocellular protein in vivo (fig. 1) (Wells et al., 1980).

RESULTS

Ether significantly potentiated the hepatotoxicity of acetaminophen as determined by changes in the peak plasma GPT concentrations (table 1). A maximal, 20-fold toxicologic enhancement was observed if acetaminophen was given 6 hr after ether. This enhancement was evident whether the timing of either ether or acetaminophen was held constant. Ether alone did not alter the plasma GPT concentration.

In the pilot study, ether alone produced maximal biochemical effects between 6 and 10 hr after administration. Based upon this study and the toxicological study, the full study was carried out with acetaminophen being administered 6 hr after ether, and all determinations were made either 2 hr after acetaminophen administration (8 hr after ether), or 2 hr after ether administration. Ether pretreatment was followed 8 hr later by decreases both the in vitro activity of glucuronyl transferase (20%) and GSH S-transferase (24%), and the hepatic GSH content (15%) (table 2), with a concomitant 3-fold increase in the covalent binding of acetaminophen to hepatocellular protein (table 2). At 2 hr after ether, there was a small (16%) but significant decrease in the hepatic P-450 content which had returned to control values by 8 hr. This decreased content at 2 hr was not reflected in a change in activity, at least as measured by aniline hydroxylation.

Ether caused small decreases in the in vivo production of plasma glucuronide and sulfate conjugates of acetaminophen, with a correspondingly small increase in acetaminophen concentrations, and significant, up to 2-fold increases in the production of GSH and cysteine conjugates reflecting detoxification of the reactive intermediate of acetaminophen (table 3). The plasma half-life of acetaminophen was increased 29% by ether pretreatment, from 0.31 ± 0.10 to 0.40 ± 0.14 hr (mean $\pm$ S.D.) (p = 0.31). For plasma area-under-the-curve data among individual mice administered acetaminophen with or without ether pretreatment, decreasing glucuronide production correlated significantly with increasing acetaminophen concentrations (r = 0.58, n = 16, $p < 0.01$), which correlated significantly with decreasing hepatic GSH content (r = 0.66, $p < 0.01$), which correlated significantly with increasing covalent binding of acetaminophen at 2 hr (r = 0.69, $p < 0.01$).

Etner decreased by 22% the cumulative 6 hr urinary recovery of the sulfate conjugate, while increasing the recovery of unmetabolised acetaminophen and the cysteine conjugate by 100% and 36% respectively (table 4).

Table 1. Effect of Ether Pretreatment on Acetaminophen (APAP) Hepatotoxicity

Treatment[1]	Time Between Treatments (hr)	Peak Plasma GPT[2] (I.U./L)
Ether control		18 ± 6
APAP control		206 ± 370
Ether + APAP	2	1710 ± 1310*
	6	4150 ± 2836*
	10	3800 ± 1920*

1. Ether and APAP were administered to groups of 6 male CD-1 mice as described in the Methods.
2. The peak plasma concentration of glutamic-pyruvic transaminase (GPT) (mean ± S.D.) regardless of its time of occurence was determined in each animal as described in the Methods.
3. Asterisks denote values which are significantly different from the APAP controls ($p < 0.05$).

Table 2. Biochemical Effects of Ether

	Time After Ether[1]			
	2 hr		8 hr	
	Control	Ether	Control	Ether
GT[2]	11.71 ± 1.55	12.10 ± 1.00	10.24 ± 2.13	8.19 ± 1.79*
P-450 content	1.34 ± 0.14	1.12 ± 0.18*	1.00 ± 0.13	1.09 ± 0.14
GST (CDNB)	2.01 ± 0.38	2.49 ± 0.48*	2.45 ± 0.39	2.49 ± 0.42
(DCNB)	0.18 ± 0.03	0.17 ± 0.02	0.14 ± 0.03	0.11 ± 0.02*
GSH content	2.35 ± 0.14	2.28 ± 0.19	1.95 ± 0.38	1.65 ± 0.18*
APAP covalent binding[3]	—	—	17.11 ± 3.53	58.40 ± 19.14*

1. Groups of 8 male CD-1 mice were sacrificed at 2 hr or 8 hr after in vivo exposure to ether. These are the times after which APAP was given which produced minimal and maximal hepatotoxic synergism respectively.
2. Mean ± S.D. from in vitro studies for GT, glucuronyl transferase activity, nmole/min/mg protein using 5.0 mM APAP as the substrate (To and Wells, 1984); P-450 content, nmole/mg protein (Omura and Sato, 1964); GST, glutathione (GSH) S-transferase) activity, μmoles/min/mg protein, using 3,4-dichloronitrobenzene (Booth et al., 1961) and 1-chloro-2,4-dinitrobenzene (Habig et al., 1974) as substrates; GSH, mg/g liver (Sedlak and Lindsay, 1968). No differences between ether-treated and control groups were observed for the activities of glucuronyl transferase using 1-naphthol as the substrate (To and Wells, 1984), sulfotransferase using 2-naphthol as the substrate (To and Wells, 1984), or cytochrome P-450 using aniline as the substrate (Imai et al., 1966).
3. In vivo study in a separate group of animals measuring the covalent binding of APAP to hepatocellular protein 2 hr after i.p. administration of radiolabeled APAP, pg equivalents/mg protein. Values represent the mean ± S.D. for 8 animals.
4. Asterisks denote values which are significantly different from the respective control group ($p < 0.05$).

Table 3. Effect of Ether on the Plasma Concentration of Acetaminophen (APAP) and Metabolites In Vivo

Metabolite[1]	Treatment[2]	Plasma Concentration[3] (µg/ml)		
		0.5 hr	1.0 hr	2.0 hr
APAP	APAP	223.6 ± 53.9	91.5 ± 41.9	7.0 ± 1.9
	Ether + APAP	213.4 ± 37.4	107.3 ± 37.8	17.9 ± 9.9
Glucuronide	APAP	117.1 ± 17.7	99.7 ± 17.4	23.5 ± 13.9
	Ether + APAP	109.7 ± 16.7	91.3 ± 19.0	8.8 ± 17.5
Sulfate	APAP	18.3 ± 4.7	15.5 ± 3.0	13.4 ± 5.0
	Ether + APAP	12.3 ± 2.4*	14.2 ± 4.4	8.3 ± 4.3
Glutathione	APAP	3.4 ± 1.2	4.4 ± 0.8	4.2 ± 1.2
	Ether + APAP	6.0 ± 1.8*	6.6 ± 1.5*	4.3 ± 1.7
Cysteine	APAP	4.3 ± 0.6	3.3 ± 1.0	2.6 ± 1.2
	Ether + APAP	4.5 ± 1.5	5.0 ± 1.0*	2.7 ± 1.5

1. APAP and its conjugated metabolites.
2. APAP was administered with or without ether pretreatment to 8 male CD-1 mice as described in the Methods.
3. Mean ± S.D. Each animal was sampled repetitively over 2 hr.
4. Asterisks denote values which are significantly different from the respective control group (p < 0.05).

Table 4. Effect of Ether on the % Urinary Recovery of Acetaminophen (APAP) and Metabolites[1]

	Control	Ether Treated
APAP[2]	4.2	8.6
Glucuronide	51.0	53.1
Sulfate	12.2	9.6
Glutathione	3.4	3.7
Cysteine	17.4	19.7

1. Percentage recovery from cumulative urinary samples obtained 0 to 6 hr following APAP administration to groups of 3 male CD-1 mice housed together in a metabolic cage.
2. APAP and its conjugated metabolites.

DISCUSSION

The 20-fold potentiation of acetaminophen hepatotoxicity by earlier pretreatment with ether confirms the results of similar studies using ether or halothane (Wells and Ramji, 1983; Wells et al., 1985).

The significant decrease in glucuronyl transferase activity measured in vitro 8 hr after ether, together with the small in vivo reductions by ether in the plasma concentrations of acetaminophen glucuronide and sulfate conjugates, were consistent with the small albeit nonsignificant increase in the plasma concentrations of acetaminophen, the increase in the half-life of acetaminophen, and the increased urinary recovery of unmetabolised acetaminophen. The up to 2-fold increase in the in vivo plasma concentrations of GSH-derived metabolites, together with the 3-fold increase in the amount of acetaminophen covalently bound to hepatocellular protein, support the hypothesis that minor reductions in the major elimination pathways of glucuronidation and to a lesser extent sulfation can result in a major percentage increase in the production of the acetaminophen reactive intermediate. The plasma observations also were consistent with the decreased and increased urinary recoveries of the sulfate and cysteine conjugates respectively.

The above trends from mean data were supported by the correlations observed among the above parameters, all of which were studied in each mouse. The correlations of decreasing plasma acetaminophen glucuronide production with increasing unmetabolised acetaminophen, decreasing hepatic GSH content and increasing acetaminophen covalent binding were compatible with a causal inter-relationship. Furthermore, in two additional in vivo studies, the individual peak plasma GPT concentrations, which accurately reflect the maximal hepatotoxicity in each mouse, correlated with the amount of acetaminophen covalently bound to hepatocellular protein in the same animals 36 hr after acetaminophen administration ($r = 0.83$, $n = 20$, $p < 0.05$) (Wells and To, 1985; Wells and To, submitted).

The reductions 8 hr after ether in the in vitro activity of GSH S-transferase using DCNB, and in the hepatic content of GSH, suggest that additional mechanisms could be contributing to the potentiation of acetaminophen hepatotoxity by ether. the possibilities would include additive depletion of intracellular GSH pools leading to derangement in calcium homeostasis (Orrenius et al., 1983), and/or an impairment in the detoxification of the acetaminophen reactive intermediate. However, one cannot be certain that DCNB is reflective of the substrate activity for the reactive intermediate of acetaminophen, given the isoenzymatic constituency of the GSH S-transferases (Habig et al., 1974). The increased transferase activity observed for CDNB at 2 hr after ether could have contributed to the lack of potentiation of acetaminophen hepatotoxicity at that time.

While the initial reduction in the hepatic content of the cytochromes P-450 at 2 hr after ether followed by recovery to control levels by 8 hr would fit with our hypothesis (fig. 2), this was not accompanied by a congruent change in enzymatic activity, at least not as measured by aniline hydroxylation. Unless aniline hydroxylase does not reflect the isoenzymatic form of the cytochromes P-450 which bioactivates acetaminophen, it would appear that the effects of ether on the cytochromes P-450 are not toxicologically limiting in this delayed, synergistic interaction.

SUMMARY

Our working hypothesis (fig. 2) for designing this study involved early inhibition by ether of P-450-dependent bioactivation and glucuronyl transferase-dependent "detoxification", with an earlier recovery of bioactivation. The combined in vivo and in vitro results from the same animals indicate that the increased susceptibility to acetaminophen hepatotoxicity may have been due to a combination of delayed decreases induced by ether in the activities of glucuronyl transferase, sulfotransferase and GSH S-transferase, along with a depletion of hepatic GSH. The small decrease in hepatic content of cytochromes P-450 at 2 hr when toxicologic enhancement was minimal, together with repletion at 8 hr when enhancement was maximal, while the above detoxification pathways were inhibited, is compatible with our hypothesis (fig. 2). However, the lack of an accompanying change in the activity of P-450 suggests either that a different P-450 isoenzyme is involved, or that P-450 activity was not toxicologically limiting. The toxicological imbalance in the bioactivation and detoxification of acetaminophen observed after ether pretreatment was evidenced by significant increases both in the plasma concentrations of GSH and cysteine conjugates, and in the covalent binding of acetaminophen to hepatocellular protein.

ACKNOWLEDGEMENTS

This work was supported by grants from the Atkinson Foundation, the Banting Research Foundation, the Hospital for Sick Children Foundation, the Connaught Fund and the Medical Research Council of Canada. E.C.A. To was the recipient of an Ontario Graduate Scholarship.

REFERENCES

Booth, J., Boyland, E. and Sims, P. (1961) An enzyme from rat liver catalysing conjugations with glutathione. Biochem. J. 79: 516-524.
Boyd, E.M. and Bereczky, G.M. (1966) Liver necrosis from paracetamol. Br. J. Pharmacol. 26: 606-614.
Eriksson, G. and Strath, D. (1981) Decreased UDP-glucuronic acid in rat liver after ether narcosis. FEBS Lett. 124: 39-42.
Habig, W.H., Pabst, M.J. and Jakoby, W.B. (1974) Glutathione S-transferases. J. Biol. Chem. 249: 7130-7139.
Imai, Y., Ito, A., and Sato, N. (1966) Evidence for biochemically different types of vesicles in the hepatic microsomal fraction. J. Biochem. 60: 417-428.
Johannessen, W., Gadeholt, G. and Aarbakke, J. (1981) Effects of diethyl ether anaesthesia on the pharmacokinetics of antipyrine and paracetamol in the rat. J. Pharm. Pharmacol. 33: 365-368.
Jollow, D.J., Mitchell, J.R., Potter, W.Z., Davis, D.C., Gillette, J.R. and Brodie, B.B. (1973) Acetaminophen-induced hepatic necrosis. II. Role of covalent binding in vivo. J. Pharmacol. Exp. Ther. 187: 195-201.
Mitchell, J.R., Jollow, D.J., Potter, W.Z., Davis, D.C., Gillette, J.R. and Brodie, B.B. (1973a) Acetaminophen-induced hepatic necrosis. I. Role of drug metabolism. J. Pharmacol. Exp. Ther. 187: 185-194.
Mitchell, J.R., Jollow, D.J., Potter, W.Z., Davis, D.C., Gillette, J.R. and Brodie, B.B. (1973b) Acetaminophen-induced hepatic necrosis. IV. Protective role of glutathione. J. Pharmacol. Exp. Ther. 187: 211-217.
Omura, T. and Sato, R. (1964) The carbon monoxide pigment of liver microsomes. I. Evidence for its haemoprotein nature. J. Biol. Chem. 239: 2370-2385.
Orrenius, S., Jewell, S.A., Bellomo, G., Thor, H., Jones, D.P. and Smith, M.T. (1983) Regulation of calcium compartmentation in the hepatocyte - a critical role of glutathione. In: Functions of Glutathione: Biochemical, Physiological, Toxicological and Clinical Aspects, A. Larsson, S. Orrenius, A. Holmgren and B. Mannervik (eds.), pp. 261-271, Raven Press, New York.
Proudfoot, A.T. and Wright, N. (1970) Acute paracetamol poisoning. Br. Med. J. 3: 557-558.

Sedlak, J. and Lindsay, R.H. (1968) Estimation of total, protein-bound and nonprotein sulphydryl groups in tissues with Ellman's reagent. Analyt. Biochem. 25: 192-205.

To, E.C.A. and Wells, P.G. (1984) Rapid and sensitive assays using high-performance liquid chromatography to measure the activities of phase II drug metabolising enzymes: glucuronyl transferase and sulfotransferase. J. Chromatogr. 301: 282-287.

To, E.C.A. and Wells, P.G. (1985, accepted). Repetitive microvolumetric sampling and analysis of acetaminophen and its toxicologically relevant metabolites in murine plasma and urine using high-performance liquid chromatography. J. Analyt. Toxicol.

Umeda, T. and Inaba, T. (1978) Effects of anesthetics on diphenylhydantoin metabolism in the rat: possible inhibition by diethyl ether. Can. J. Physiol. Pharmacol. 56: 241-244.

Watkins, J.B. and Klaassen, C.D. (1982) Determination of hepatic uridine 5'-diphosphoglucuronic acid concentration with diethylstilbestrol. J. Pharmacol. Methods 7: 145-151.

Wells, P.G., Boerth, R.C., Oates, J.A. and Harbison, R.D. (1980) Toxicologic enhancement by a combination of drugs which deplete hepatic glutathione: acetaminophen and doxorubicin (Adriamycin®). Toxicol. Appl. Pharmacol. 54: 197-209.

Wells, P.G. and Ramji, P. (1983) Modulation of acetaminophen biotransformation and hepatotoxicity by diethyl ether. Proc. Can. Fed. Biol. Soc. 26: 228.

Wells, P.G., Ramji, P. and Ku, M.S.W. (1985, accepted) Delayed enhancement of acetaminophen hepatotoxicity by general anesthesia using diethyl ether or halothane. Fundam. Appl. Toxicol.

Wells, P.G. and To, E.C.A. (1985) Murine hepatotoxicity: dependence of individual peak plasma glutamic-pyruvic transaminase (GPT) concentrations on in vivo covalent binding of acetaminophen. Proc. Can. Fed. Biol. Soc. 28: 228.

PIPERONYL BUTOXIDE REDUCES SALICYLATE-INDUCED NEPHROTOXICITY AND COVALENT

BINDING IN MALE RATS

Marlene E. Kyle and James J. Kocsis

Department of Pharmacology
Thomas Jefferson University
Philadelphia, PA 19107

INTRODUCTION

The salicylates are among the most widely used therapeutic agents in
the world. Their effectiveness in treating pain, fever and inflammation
all contribute to the extensive use of these drugs. Although generally
regarded as safe, a number of side effects have been associated with the
administration of salicylates. In particular, the development of kidney
toxicity is of concern as this damage may lead to renal failure and death
(NIH Consensus Development Conference, 1984).

Two distinct forms of salicylate-induced renal disease have been
identified both in humans and experimental animals: renal papillary
necrosis and proximal tubular necrosis of cortical nephrons (PTN). Renal
papillary necrosis appears to develop after long term use of relatively
high doses of salicylates in combination with other nonsteroidal antiinflammatory drugs, suggesting that synergistic interactions between
salicylates and other non-steroidal anti-inflammatory drugs may markedly
increase toxicity (NIH Consensus Development Conference, 1984). In
animal models, RPN can be produced by the chronic administration of a
single agent as well as by combinations of drugs (Nanra, 1974; Molland,
1978).

Acute exposure to high doses of either salicylic acid (SAL) or
aspirin, in contrast, results in PTN (Shelly, 1978). Damage to the
medullary region of the kidney is generally not observed under these conditions (Robinson et al., 1967). Clinically, SAL-induced PTN is characterized by elevations in blood urea nitrogen (BUN) as well as the excretion of protein, sugar, blood and cells into the urine (Stygles and
Iuliucci, 1981). These effects are usually transient; however, in
severe cases, renal failure can occur (Mitchell et al., 1977). Administration of SAL to laboratory animals produces a similar profile of biochemical and histological alterations (Calder et al., 1971; Arnold et al.,
1973). The mechanism by which SAL causes PTN is not clear, although it
has been suggested that SAL is biotransformed into products which mediate
the toxicity (Mitchell et al., 1977).

Recent evidence from this laboratory has shown that following the
administration of 500 mg/kg (^{14}C) SAL to male rats, significant covalent
binding to renal cortical protein occurs before evidence of proximal

tubular damage (Kyle and Kocsis, in press). These results suggest that
SAL may be converted to a reactive metabolite, perhaps by the kidney
itself. The purpose of the present study was to examine the effects of
mixed function oxidase inhibition on SAL-induced renal toxicity and
covalent binding in order to clarify further the relationship between
metabolism and PTN.

METHODS

Animals and Treatment Protocol

Male Sprague-Dawley rats (310-350 g; Ace Breeders, Douglassville, PA)
were allowed food and water ad libitum. After a 48 hr accommodation
period, rats were treated ip with 1000 mg/kg piperonyl butoxide (ICN Phar-
maceuticals, Plainville, NY) or corn oil 1 hr before SAL administration.
(^{14}C)Salicylate was purchased from New England Nuclear, Boston, MA.
(Specific activity 250 mCi/mmol). The radioactive material was diluted
with cold salicylate (Merck) and administered to rats as a single ip dose
of 500 mg/kg. Control rats received 0.9% saline in place of SAL. The
animals were then individually housed in metabolism cages for urine collection. Six hr following SAL treatment, rats were decapitated, kidneys
and livers quickly removed and placed in ice cold saline.

Pathology

Sections of the left kidney were quickly removed and fixed in 10%
formalin phosphate. The tissues were cut into 5 um sections and stained
with hematoxylin and eosin. Histological assessment was carried out in a
blind fashion by one observer.

BUN and SPGT Analysis

Blood samples were collected from each animal in small tubes containing 100 units heparin. Plasma was obtained by centrifuging the
samples at 2100 rpm for 20 min at 5°C. BUN and SPGT were then determined
on duplicate aliquots using standard diagnostic kits (Sigma, St. Louis,
MO).

Covalent Binding Determination

Kidney tissues (after excision of the medulla) were homogenized in
10 volumes of a 0.27 M sucrose, 1 mM EDTA, 5 mM Tris-HCL buffer, pH 7.4.
Renal mitochondria were isolated by the method of Johnson and Lardy
(1967) as modified by Weinberg (1981). Livers were homogenized in an
0.34 M sucrose, 2 mM Tris HCL buffer, pH 7.4 at 4°C and mitochondria prepared by the method of Kalf and O'Brien (1976). Kidney and liver microsomes were also isolated by centrifuging the post-mitochondrial supernatants at 100,000 g for 70 min (Beckman 970L Ultracentrifuge). Covalent
binding in the whole homogenate as well as its mitochondrial, microsomal
and cytosolic subfractions was determined by exhaustive solvent extraction
using 80% methanol (Jollow et al., 1973). The washed protein pellets were
solubilized in 1 N NaOH and an aliquot counted in a Packard Tri-Carb scintillation counter after the addition of scintillation fluid. Protein concentrations were determined by the method of Lowry et al., (1951) with
bovine serum albumin serving as a protein standard.

Determination of Urinary Metabolites of Salicylate

Urine volumes were measured and the metabolism cages washed with
water to collect any residual urine. Salicylate metabolites were deter-

mined as described previously (Kyle and Kocsis, in press) using a modification of the HPLC method of Cham et al., (1980). Urinary radioactivity was also determined by scintillation counting.

Renal ATP Levels

In another set of experiments, rats were treated as described above and renal ATP levels determined 3 hr after SAL treatment using firefly luciferase (Lundin et al., 1976). Animals were anesthetized with sodium pentobarbital (Nembutal), left kidneys gently exposed and then freeze-clamped in situ using Wollenberger clamps. Frozen tissues were pulverized and the powder added to 9 volumes of ice cold trichloroacetic acid (TCA). The homogenate was centrifuged at 2800 rpm for 20 min at $5^{\circ}C$. The supernatants were extracted 3x with water-saturated ether to remove the TCA and the residual ether evaporated from the aqueous phase with water-saturated nitrogen. Small aliquots of extracted supernatant were diluted in 100 mM Tris, 2 mM EDTA buffer, pH 7.0 and the ATP concentration measured in a Lumi-Aggregometer (Chrono-Log, Havertown, PA) after the addition of Chronolume-Luciferase/Luciferin (Chrono-Log). The disodium salt of ATP was used as the standard (Sigma, St. Louis, MO).

Statistics

One-way Analysis of Variance with a Neuman‑Keul ad hoc test (Steel and Torrie, 1960) was used for statistical analysis. Only p values less than 0.05 were considered significant.

RESULTS

Renal Toxicity

SAL administration produced focal areas of coagulative proximal tubule necrosis in 4 out of 5 animals. Cast formation in the tubular lumen was also noted. This histological damage was accompanied by a significant increase in BUN as compared to controls (Table 1). Pretreating rats with piperonyl butoxide, however, significantly decreased the SAL-induced elevation in BUN and also completely abolished PTN in 2 of the 3 animals examined. In contrast, no significant changes in SGPT levels were noted after any of the SAL treatments.

Renal ATP levels were also used as an index of renal toxicity in this study since previous evidence suggests that the mitochondria may play a pivotal role in the development of SAL-induced PTN (Kyle and Kocsis, in press). ATP levels found in the kidneys from control rats ranged between 1-1.2 μmol/g wet wt tissue and agree with renal ATP levels previously reported in the literature (Nishiitsutsuji-Uwo et al., 1967; Bowman, 1970). As seen in Fig. 1, renal ATP levels were significantly depressed 3 hr after SAL treatment. Pretreatment with piperonyl butoxide, however, restored renal ATP almost to control values, indicating that piperonyl butoxide protects against the SAL-induced decreases in renal ATP levels.

Covalent Binding Analysis

Covalently bound radioactivity was measured in both kidney and liver homogenates following (^{14}C)SAL administration. Binding to kidney, however, was more than 3-fold greater than binding to liver. When various subcellular fractions were isolated from kidney homogenates, 57% of the total bound radioactivity was associated with the mitochondrial fraction (Table 2). Moreover, 65% of bound radiolabel in the mitochondria

Table 1. Effect of Piperonyl Butoxide on BUN and SGPT
Levels After Salicylate Treatment in Rats

	BUN (mg%)	SGPT (units/ml)
Saline control	13.6±2	31.1±4
Salicylate Alone	37.2±5*	55.0±18
Salicylate + Piperonyl Butoxide	23.2±2**,+	40.6±16

Rats were treated as described under Methods. Plasma
BUN and SGPT levels were determined 6 hr following
SAL administration using standard diagnostic kits.
Data represents X ±SD for 5 rats.
 *Significantly different from saline control (p <0.01)
**Significantly different from saline control (p <0.05)
 +Significantly different from SAL alone (p <0.01)

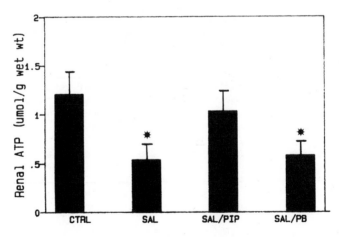

Fig. 1. Effect of piperonyl Butoxide (PIP) on renal
ATP levels following (SAL) administration. Rats were
treated as described under Methods. Renal ATP levels
were then determined in triplicate 3 hr after salicylate using a firefly luciferase method. CTRL
designates saline treated rats.
*Significantly different from control ATP levels
 (p <0.01), n=3

remained after removal of the outer membrane with digitonin (Schnaitman and Greenwalt, 1968), indicating that SAL equivalents bound more extensively to the matrix and/or inner membrane protein of mitochondria. Renal microsomes and cytosol contained 12 and 22% of the total bound radioactivity, respectively. In contrast to the kidney, most of the total binding to liver was found within the microsomes while only 10% of the binding was located in hepatic mitochondria. Covalent binding to liver cytosolic proteins represented 20% of the total, a level similar to that observed in the kidney.

Table 2. Intracellular Distribution of Covalent Binding in Kidney and Liver After (^{14}C)Salicylate Administration in Rats

Fraction	Kidney	Liver
	% of total bound radioactivity[a]	
Mitochondria	57±10	10±3
Microsomes	12±4	52±7
Cytosol	22±6	19±5
Recovery	91±7	81±6

[a]Rats were treated with 500 mg/kg (^{14}C)salicylate and then sacrificed 6 hr later. Covalent binding was determined in liver and kidney homogenates as well as in the mitochondria, microsomes and cytosol by exhaustive solvent extraction.
Data represents X ± SD for 5 rats.

Piperonyl butoxide significantly reduced covalent binding to kidney homogenates (Table 3); the most pronounced effect occurred in the mitochondrial fraction where binding was reduced by 71%. Piperonyl butoxide also decreased covalent binding to liver macromolecules, however, only the microsomal fraction was significantly affected (Table 4).

Table 3. Effect of Piperonyl Butoxide on Renal Covalent Binding After (^{14}C)Salicylate Administration

	Homogenate	Mitochondria	Microsomes	Cytosol
	pmol SAL equivalents bound/mg protein[a]			
Salicylate Alone	95±34	216±68	51±18	49±18
Salicylate + Piperonyl Butoxide	41±8**	64±29*	38±12	15±10**

[a]Rats were treated as described under Methods. Covalent binding was determined in kidney homogenates and in mitochondrial, microsomal and cytosolic fractions. Data represent X ± SD for 5 rats.
* Significantly different from SAL alone (p <0.01)
**Significantly different from SAL alone (p <0.05)

Table 4. Effect of Piperonyl Butoxide on Hepatic Covalent
Binding After (^{14}C)Salicylate Administration

	Homogenate	Mitochondria	Microsomes	Cytosol
	pmol SAL equivalents bound/mg protein[a]			
Salicylate Alone	28±11	77±68	149±43	19±8
Salicylate/ Piperonyl Butoxide	15±7	61±31	41±28*	N.S.

[a]Rats were treated as described under Methods. Covalent
binding was determined in liver homogenates and in the
mitochondrial, microsomal and cytosolic fractions. N.S.
indicates no significant binding was found. Data represents X ±SD for 5 rats.
*Significantly different from SAL alone (p <0.01)

Urinary Metabolites

To determine the effect of pretreatments on SAL metabolism, urinary
levels of SAL and SAL metabolites were measured using HPLC with fluorescence detection. SAL metabolites identified in rat urine included:
1) salicyluric and gentisuric acids; 2) phenolic and acyl glucuronides
and 3) 2,5-dihydroxybenzoic acid (gentisate) and 2,3-dihydroxybenzoic
acid. Piperonyl butoxide pretreatment significantly decreased the
excretion of gentisate and 2,3-dihydroxybenzoate as compared to SAL
alone (Table 5). Piperonyl butoxide also generally decreased glucuronide
excretion although this reduction was not statistically significant. The
reduced excretion of SAL metabolites, however, was accompanied by an
equivalent increase in the excretion of unchanged SAL, indicating that
piperonyl butoxide inhibited SAL metabolism.

DISCUSSION

It has been previously been suggested that salicylate-induced proximal
tubular necrosis may result from the formation of nephrotoxic metabolites
(Mitchell et al., 1977). In the present report, significantly more
covalently bound radiolabel was found in the kidney cortex than in the
liver after treating rats with SAL, suggesting that the kidney may have
a greater capacity to form reactive intermediates from SAL. Mitochondria
in the kidney were found to contain approximately 60% of the total bound
radioactivity, even though these organelles constitute only 15–18% of
the total kidney protein. In contrast, most of the covalent binding in
liver was localized in the microsomal fraction while relatively little
binding was observed in the mitochondria. The disparity between liver
and kidney covalent binding suggests that the nature of the reactive
intermediate(s) in kidney may differ from that in liver and/or that its
site of synthesis may be different. The susceptibility of the kidney to
the acute cytotoxic action of SAL may, therefore, be the result of inherent differences in SAL metabolism between liver and kidney tissue.

Table 5. Effect of Piperonyl Butoxide on Urinary Metabolite Levels After ([14C])Salicylate Administration

	Treatment	
	Salicylate Alone	Salicylate + Piperonyl Butoxide
	μmol excreted[a]	
Gentisate	8.85±3.0	1.94±0.76*
2,3-Dihydroxy-benzoate	5.58±1.8	1.71±0.50*
Salicylurate	0.69±0.03	1.31±0.68
Gentisurate	0.31±0.13	0.45±0.26
Salicylate	141±26	185±85
Salicyl-glucuronides[b]	130±30	98±30
Total Metabolites[c]	290±76	275±84

[a]Individual metabolites quantified using HPLC with fluorescence detection. Data represents X ±SD for 5 rats.

[b]Glucuronide estimation based on increase in SAL equivalents released after treatment of urine samples with β-glucuronidase (Sigma).

[c]Total metabolites determined by dividing the amount of radio-activity excreted in the urine by the specific activity of the administered drug.

*Significantly different from SAL alone (p <0.01)

Pretreatment of rats with piperonyl butoxide, which has been demon-strated to inhibit certain cytochrome P-450 activities (Anders, 1968; Friedman and Couch, 1974), significantly reduced the degree of morpho-logical and biochemical damage produced by SAL administration. These results are in agreement with those of Mitchell et al., (1977) who re-ported that pretreating rats with piperonyl butoxide and cobaltous chloride protected against the nephrotoxic effects of SAL. Indeed, ad-ministration of piperonyl butoxide decreased the production of the ortho and para hydroxylation products of SAL, indicating an inhibition of P-450 mediated oxidation. Piperonyl butoxide also significantly decreased covalent binding that was particularly evident in the renal mitochondria. Thus, inhibition of SAL metabolism decreased renal toxicity accompanied by a decreased binding to renal mitochondria.

A relationship between renal mitochondrial covalent binding and renal toxicity was also demonstrated by the significant correlation (r=0.87, p <0.01 where n=10) between mitochondrial binding in the kidney and in-creased BUN levels (results not shown). Although extensive liver damage from SAL treatment was generally not observed under these conditions, a significant correlation (r=0.74, p <0.05 where n=14) again existed between mitochondrial binding in the liver and elevation of plasma SGPT levels. In other experiments where rats were pretreated with the mixed function oxidase inducing agents, Aroclor and phenobarbital, no effects on SAL-induced nephrotoxicity nor on the levels of mitochondrial covalent binding in the kidney were noted (results not shown). Taken together, these results indicate that irreversible binding of a reactive metabolite(s) of SAL to renal mitochondria, generated perhaps via SAL oxidation, may play an important role in SAL-induced nephrotoxicity.

High concentrations of salicylate have been shown to disrupt ATP synthesis which may be due to uncoupling of oxidative-phosphorylation (Smith, 1968; Tokumitsu et al., 1977). The present study shows that SAL significantly decreases ATP levels in the rat kidney 3 hr after treatment, a time when no morphological changes were evident (Kyle and Kocsis, in press). Renal ATP levels when piperonyl butoxide was given prior to SAL, however, were no different from saline control levels. These results indicate that a metabolite may mediate SAL's effect on mitochondrial ATP synthesis. Furthermore, the ATP "sparing" effect of piperonyl butoxide correlates with its protective action against SAL-induced PTN.

Mitochondrial injury has been shown to precede damage to renal proximal tubular cells caused by a number of agents, including mercuric chloride (Weinberg et al., 1982), gentamicin (Simmons et al., 1980) and cephaloridine (Tune and Fravert, 1980). In the present study, the observation that SAL administration interfered with ATP synthetic reactions early in the development of PTN further supports the hypothesis that mitochondrial dysfunction is involved in the pathogenesis of SAL-induced nephrotoxicity. The molecular mechanism by which SAL causes mitochondrial injury is unknown; however, the present results indicate that covalent binding of a reactive SAL metabolite(s) to mitochondrial protein is closely associated with toxicity. It is generally assumed that binding of reactive intermediates occurs in proximity to the site at which such intermediates are produced. If this holds true for SAL derived reactive intermediates, then the binding of radiolabel from (^{14}C)SAL to renal mitochondria suggests that mitochondria themselves possess the capacity to convert SAL to reactive products. Niranjan and Avadhani (1984) have recently shown that hepatic mitochondria contain a P-450-like system which metabolically activates benzopyrene, dimethylnitrosoamine and aflatoxin B. Kalf et al. have likewise shown that benzene is converted to reactive intermediates by both isolated bone marrow and liver mitochondria (1982). Studies are now in progress in this laboratory to determine whether renal mitochondria contain a similar system for SAL bioactivation.

ACKNOWLEDGMENTS

We would like to thank Dr. William Goldschmidt, Department of Pathology, School of Veterinary Medicine, University of Pennsylvania for his help with slide preparation and renal morphology interpretation. We would also wish to thank Dr. George Kalf for helpful discussions and Drs. Carol Wojenski and Diane Reibel for their assistance with the ATP determinations. Elaine Collins and Denise Wojciechowski are also thanked for secretarial assistance.

REFERENCES

Anders, M.W., 1968, Inhibition of microsomal drug metabolism by methylene dioxybenzenes, Biochem. Pharmacol., 17:2367.
Arnold, L., Collins, C. and Starmer, G., 1973, The short term effects of analgesics on the kidney with special reference to acetylsalicylic acid, Pathol., 5:123.
Bowman, R.H., 1970, Gluconeogenesis in the isolated perfused rat kidney, J. Biol. Chem., 245:160&.
Calder, J.C., Fender, C.C., Green, G.R., Ham, K.N. and Tange, T.D., 1971, Comparative nephrotoxicity of aspirin and phenacetin derivatives, Br. J. Med., 4:518

Cham, B.E., Bochner, F., Imhoff, D.M., Johns, D. and Rowland, M., 1980, Simultaneous liquid-chromatographic quantitation of salicylic acid, salicyluric acid and gentisic acid in urine, Clin Chem., 26:111

Friedman, M.A. and Couch, D.B., 1974, Inhibition by piperonyl butoxide of phenobarbital mediated induction of mouse liver microsomal enzyme activity, Res. Comm. Chem. Pathol. Pharmacol., 8:515

Haas, R., Parker, Jr, W.D., Stumpf, D. and Eguren, L.A., 1985, Salicylate-induced loose coupling: protonmotive force measurements, Biochem. Pharmacol., 34:900

Johnson, D. and Lardy, H., 1967, Isolation of liver and kidney mitochondria, Methods Enzymol., 10:94

Jollow, D.J., Mitchell, J.R., Potter, W.Z., Davis, D.C., Gillette, J.R. and Brodie, B.B., 1973, Acetaminophen-induced hepatic necrosis. Role of covalent binding in vivo, J. Pharmacol. Exp. Ther., 187:195

Kalf, G.F., Rushmore, T. and Snyder, R., 1982, Benzene inhibits RNA and protein synthesis in mitochondria from regenerating liver and bone marrow, Chem. Biol. Interact., 42:353

Kyle, M.E. and Kocsis, J.J., The effect of age on salicylate induced nephrotoxicity in male rats, Toxicol. Appl. Pharmacol., in press.

Lundin, A. and Thore, A., 1975, Analytical information obtainable by evaluation of the time course of firefly bioluminescence in the assay of ATP, Anal. Biochem., 66:47

Lowry, O., Rosebrough, N.J., Farr, A.L. and Randall, R.J., 1951, Protein measurement with the Folin phenol reagent, J. Biol. Chem., 193:265

Mitchell, J.R., McMurty, R.J., Statham, C.N. and Nelson, D.S., 1977, Molecular basis for drug-induced nephropathies, Am. J. Med., 62:518

Molland, E.A., 1978, Experimental renal papillary necrosis, Kidney Int., 13:5

Nanra, R.S., 1974, Pathology, aetiology and pathogenesis of analgesic nephropathy, Roy. Aust. Coll. Physicians, 4:602.

National Institutes of Health Consensus Development Conference, 1984, Analgesic-associated kidney disease, February 27-29.

Niranjan, B.G., Wilson, N.M., Jefcoate, C.R. and Avadhani, N.G., 1984, Hepatic mitochondrial cytochrome P-450 system. Distinctive features of cytochrome P-450 involved in the activation of Aflatoxin B1 and benzo(a)pyrene, J. Biol. Chem., 259;12495

Nishiitsutsuji-Uwo, J.M., Ross, B.D. and Krebs, H.A., 1967, Metabolic activities of the isolated perfused rat kidney, Biochem. J., 103:852.

O'Brien, T.W. and Kalf, G.F., 1967, Ribosomes from rat liver mitochondria. Isolation procedure and contamination studies, J. Biol. Chem., 242:2172

Robinson, M.J., Nichols, E.A. and Taitz, L., 1967, Nephrotoxic effect of acute sodium salicylate intoxication in the rat, Arch. Pathol., 84:224

Schnaitman, C.A. and Greenawalt, J.W., 1968, Enzymatic properties of the inner and outer membranes of rat liver mitochondria, J. Cell Biol., 38:158

Shelly, J.H., 1978, Pharmacological mechanisms of analgesic nephropathies, Kidney Int., 13:15

Simmons, F., Bogusky, R.T. and Humes, H.D., 1980, Inhibitory effects of gentamicin on renal mitochondrial oxidative phosphorylaon, J. Pharmacol. Exp. Ther., 214:709

Steel, R.G.D. and Torrie, J.H., 1960, "Principles and Procedures of Statistics," McGraw-Hill, New York.

Stygles, V.G. and Iuliucci, J.D., 1981, Structural and functional alterations in the kidney following intake of nonsteroidal antiinflammatory analgesics, in: "Toxicology of the Kidney," J.B. Hook, ed., Raven Press, New York.

Tokumitsu, Y., Lee, S. and Ui, M., 1978, In vitro effects of nonsteroidal anti-inflammatory drugs on oxidative phosphorylation in rats, Biochem. Pharmacol., 26:2101.

Tune, B.M. and Fravert, D., 1980, Mechanisms of cephalosporin nephro-
    toxicity: A comparison of cephaloridine and cephaloglycin, <u>Kidney
    Int.</u>, 18:591
Weinberg, J.M., Harding, P.G. and Humes, H.D., 1982, Mitochondrial bio-
    energetics during the initiation of mercuric chloride-induced renal
    injury, <u>J. Biol. Chem.</u>, 257:60

CROSS-LINKING OF PROTEIN MOLECULES BY THE REACTIVE METABOLITE OF ACETAMINOPHEN, N-ACETYL-p-BENZOQUINONE IMINE, AND RELATED QUINOID COMPOUNDS

Anthony J. Streeter[*], Peter J. Harvison, Sidney D. Nelson
and Thomas A. Baillie

Department of Medicinal Chemistry, BG-20
University of Washington, Seattle, WA 98195, U.S.A.

INTRODUCTION

Acetaminophen (4'-hydroxyacetanilide; 4HAA; Fig. 1) is a widely used analgesic and antipyretic agent which, while considered to be safe at therapeutic dose levels, has been found to cause acute hepatic centrilobular necrosis in both humans and experimental animals when consumed in large doses (Prescott et al., 1971; Boyd and Bereczky, 1966). Evidence from a variety of animal studies (Mitchell et al., 1975; Dahlin et al., 1984) has indicated that cytochrome P-450 plays an important role in the oxidation of acetaminophen to a chemically-reactive and potentially toxic electrophilic metabolite, N-acetyl-p-benzoquinone imine (NAPQI; Fig. 1), which binds covalently to hepatic protein. Recently, we have shown that the major covalent adduct formed between 4HAA and proteins is a 3'-cystein-S-yl conjugate of the drug (Streeter et al., 1984b; Hoffmann et al., 1985a,b). This finding supports the contention that 4HAA is first metabolized to NAPQI, which then arylates proteins by selective reaction with cysteinyl thiol residues.

Early studies on 4HAA-induced liver injury demonstrated that the localization of covalently-bound drug within liver tissue, and the extent of irreversible binding, correlated well with the site and severity, respectively, of hepatic necrosis (Potter et al., 1974). Based in part upon this observation, it was proposed that covalent binding of the reactive metabolite of 4HAA to liver proteins is the critical event which initiates the processes leading to cell death. However, despite this correlation, the underlying mechanisms of these processes remain poorly understood, and in recent years a number of apparent anomalies have arisen. For example, it has been found that in the presence of certain thiols (Labadarios et al., 1977) and flavones (Devalia et al., 1982), covalent binding of 4HAA to cellular proteins was decreased only minimally, while hepatotoxicity was prevented. Moreover, N-acetylcysteine has also been found to protect against 4HAA-induced liver necrosis without significantly altering covalent binding, even if administered after the binding was maximal (Gerber et al., 1977). Other studies (Fernando et al., 1980) tested the toxicity of 3',5'-dimethyl-4'-hydroxyacetanilide (3,5-dimethyl-4HAA; Fig. 1) and 2',6'-dimethyl-

---

[*]Present address: Department of Pharmacy, School of Pharmacy
University of California San Francisco
San Francisco, CA 94143, U.S.A.

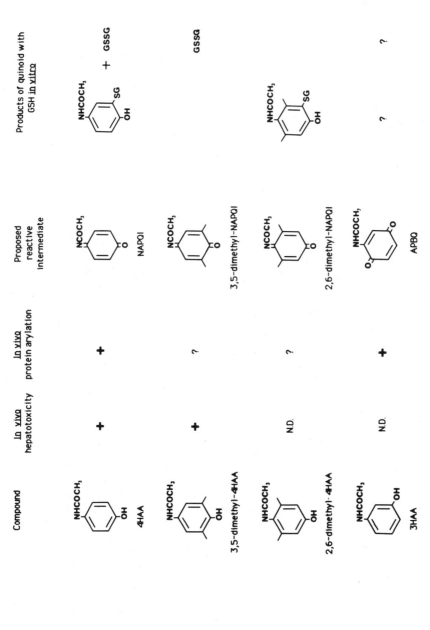

Fig. 1. Structures of the various hydroxyacetanilides cited in the text, together with the structures of their quinoid metabolites and the products of reaction of these quinoids with glutathione. The ability of the respective parent compounds to cause liver damage in vivo and to arylate proteins is also given and is taken from literature sources cited in the text. N.D. = not detected; ? = unknown.

4'-hydroxyacetanilide (2,6-dimethyl-4HAA; Fig. 1) and found that while the former agent displayed a similar level of hepatotoxicity to 4HAA in mice when dosed intragastrically, no liver damage could be demonstrated with the latter compound. On the assumption that 4HAA and its methylated analogs would all be oxidized to their respective quinone imines in vivo (Fig. 1), a subsequent study (Rosen et al., 1984) investigated the interactions between the quinone imines and reduced glutathione (GSH) in aqueous solution. It was found that NAPQI produced both oxidized glutathione (GSSG) and a gluta-thione adduct of 4HAA. 2,6-Dimethyl-NAPQI, on the other hand, produced an adduct but no GSSG, while the putative hepatotoxic quinoid 3,5-dimethyl-NAPQI produced GSSG but no glutathione adduct. Finally, it has been re-ported that an isomer of 4HAA, 3'-hydroxyacetanilide (3HAA; Fig. 1), is not hepatotoxic to hamsters (Roberts and Jollow, 1978) or to mice (Nelson, 1980), even though it binds extensively to liver protein both in vivo (Roberts and Jollow, 1979) and in vitro (Streeter et al., 1984). We have demonstrated recently that 3HAA is metabolized effectively to 2',5'-dihy-droxyacetanilide by mouse liver microsomes, that this hydroquinone is readily oxidized to 2-acetamido-p-benzoquinone (APBQ; Fig. 1) and that APBQ goes on to bind covalently to protein to a very large extent in vitro (Streeter and Baillie, 1985).

In view of the apparent dissociation of protein arylation from the hepatotoxicity of substituted acetanilide derivatives, the studies reported here were undertaken in an attempt to find differences in the interactions between proteins and the four above-mentioned quinoid compounds, viz. NAPQI, 3,5-dimethyl-NAPQI, 2,6-dimethyl-NAPQI and APBQ. Bovine serum albumin (BSA) was chosen as the model protein for these investigations since it is sol-uble (and thus easily manipulated), consists of a single polypeptide chain (65,500 daltons) of known sequence and has been shown to contain only one cysteine residue (cys-34) which is not involved in intramolecular disulfide bonds (Brown, 1976). As such, BSA may be viewed as a macromolecular thiol analog of GSH.

MATERIALS AND METHODS

Materials

BSA (crystalline; stated to be > 99% pure by electrophoresis) was pur-chased from Calbiochem (La Jolla, CA) and used without further purification. Dimethyl sulfoxide (DMSO) was supplied by Mallinckrodt (St. Louis, MO), and was dried by stirring overnight over calcium hydride, distilled under reduced pressure at 90–100°C and stored over 4 Å molecular sieve beads until used. APBQ (Wunderer, 1972), NAPQI (Dahlin and Nelson, 1982) and 3,5- and 2,6-dimethyl-NAPQI (Fernando et al., 1980) were synthesized as described previously. [Ring-U-^{14}C]NAPQI (0.17 mCi mmol^{-1}) was a gift from Dr. D. C. Dahlin of our laboratories.

Methods

Incubations were carried out in sealed 20 ml glass vials for 20 min at 37°C in a shaking water bath. Each vial contained 15 mg of BSA dissolved in 1.0 ml of 0.1 M potassium phosphate buffer (pH 7.4), to which was added 10 µl of DMSO containing either no quinoid (control), or quinoid (freshly dissolved immediately before addition) at a final concentration of 3.0 mM. At the end of the incubation, the vials were frozen at −20°C, at which tem-perature they were held until shortly before their contents were analyzed by chromatography.

The amount of cross-linking of BSA was determined by separating the monomer from the various cross-linked forms (dimer and higher oligomers)

according to molecular weight, using Sephadex G-150 gel filtration chromatography (Andersson, 1966; Pedersen, 1962). In each case, the whole of each
individual incubation mixture was applied, by gravity flow, to the top of a
95 x 0.8 cm column. Chromatography was carried out at room temperature
using a Pharmacia P1 peristaltic pump to elute the column with degassed 0.1
M potassium phosphate buffer (pH 7.4) as mobile phase. The effluent was
monitored at 280 nm using a Pharmacia UV-1 single path detector, and 25 min
fractions (approximately 3.0 ml) were collected using a Pharmacia FRAC-100
fraction collector. Aliquots of 250 µl of each fraction were assayed for
protein content by the method of Lowry et al. (1951), and the percentages of
the total protein which were recovered from each sample as the monomeric and
cross-linked species were determined.

In some experiments, the reaction of NAPQI with BSA was followed by the
addition of dithiothreitol (to a final concentration of 6.5 mM), and
reaction was allowed to continue at 4°C for 30 min.

RESULTS

Sephadex G-150 gel filtration chromatography of the incubation mixtures to which no quinoid had been added (control; Fig. 2B) revealed that
the majority of the protein remained in the monomeric form (peak with retention time of approximately 13 hr, Fig. 2). The remaining protein eluted
earlier (at around 10 hr) and represented higher molecular weight material
taken to be the dimer of BSA (Andersson, 1966; Pedersen, 1962). The inclusion of 3 mM APBQ in the incubation medium did not alter the observed molecular weight distribution of the protein products (Fig. 2C), whereas the
inclusion of 3 mM NAPQI resulted in a large increase in the proportion of
higher molecular weight material formed (Fig. 2A). Furthermore, it was
noted that this polymeric, or cross-linked, material consisted of protein
components of several different molecular weights, indicating that more than
two BSA monomers had become linked in some cases. This finding was not
totally unexpected since a similar phenomenon has been observed during the
oxidation of BSA with ferricyanide, when it was suggested that some of the
17 intramolecular disulfide bonds of the BSA molecule opened and subsequently reacted to form intermolecular S-S linkages (Andersson, 1966). When
the results of the present experiment were quantified (Table 1), it was
found that the putative quinoid metabolites of both of the hepatotoxic compounds (NAPQI and 3,5-dimethyl-NAPQI) brought about significant increases in
the proportion of higher molecular weight or cross-linked protein molecules,
while the corresponding quinoid metabolites of the non-hepatotoxic compounds
(2,6-dimethyl-NAPQI and APBQ) had no effect on the molecular weight distribution of the protein. No differences were seen in the total recoveries of
protein after the chromatographic step (Table 1).

In each case, chromatography of incubation mixtures which had contained
quinoids revealed a late peak (retention time approximately 25 hr) of brown-
colored material which elicited almost or no response in the assay for protein content. This material was assumed to represent the product of polymerization of the quinoid molecules themselves by analogy with the known behavior of similar compounds under these incubation conditions (Rotman et
al., 1976).

Two experiments were performed in order to gain some information on the
nature of the cross-linked protein material formed in the incubations containing NAPQI. In the first, incubations of NAPQI with BSA were followed
immediately by treatment with dithiothreitol. This procedure would be expected to cleave BSA oligomers if such products had been generated as a
result of intermolecular disulfide bond formation (Andersson, 1966; Rotman
et al., 1976). In the event, however, dithiothreitol treatment did not
change the proportions of crosslinked BSA obtained (Table 2).

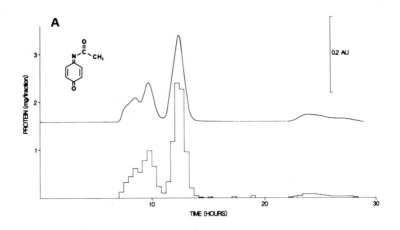

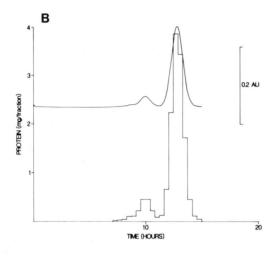

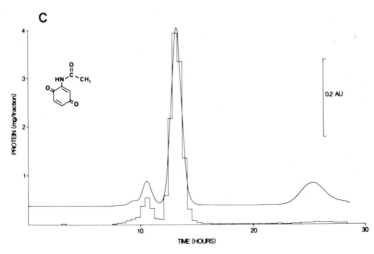

Fig. 2.   Chromatography on Sephadex G-150 of the products of incubation (20 min) of BSA (15 mg ml^{-1}) with (A) NAPQI (3 mM), (B) no quinoid (control) and (C) APBQ (3 mM).   The upper trace in each chromatogram represents UV absorbence (280 nm) of the column effluent while the lower trace indicates the protein content of successive fractions.

731

Table 1. Cross-linking of BSA (15 mg ml^{-1}) by Various Quinoid Compounds (3 mM).

Quinoid	N	Percentage of cross-linked BSA present after 20 min incubation	Percentage of incubated protein recovered from the Sephadex G-150 column
None	4	10.5 ± 0.5	89.2 ± 2.7
NAPQI	3	36.4 ± 4.8[a]	88.4 ± 2.4
3,5-dimethyl-NAPQI	4	17.2 ± 1.7[a]	97.8 ± 4.3
2,6-dimethyl-NAPQI	4	11.1 ± 0.2	93.3 ± 1.9
APBQ	3	11.8 ± 0.9	91.0 ± 3.9

Results are expressed as mean values $\pm$ S.E.M.

[a] Significantly different from control incubation ($p < 0.01$).

Table 2. Effect of Dithiothreitol (6.5 mM) on the Cross-linked Protein Formed During Incubation of BSA with NAPQI (3 mM).

Experiment	N	Dithiothreitol	Percentage of cross-linked BSA
1	1	-	53.2
2	3	+	55.3 ± 6.2

In the second experiment, 3 mM [^{14}C]NAPQI was incubated, as before, with 15 mg of BSA to investigate whether NAPQI molecules were binding covalently to the cross-linked species and, if so, to what extent. It was found that at the concentrations of reactants employed, approximately 8 equivalents of NAPQI were bound per equivalent of BSA monomer in the protein material (monomer and oligomers) recovered from the incubation (Fig. 3).

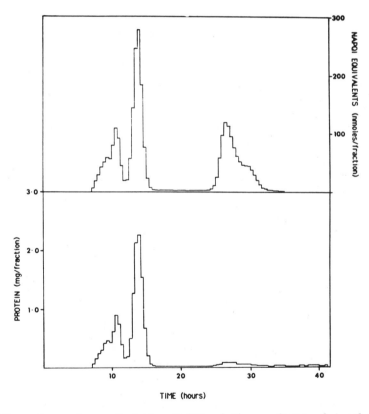

TIME (hours)

Fig. 3. Chromatography on Sephadex G-150 of the products of incubation (20 min) of BSA (15 mg ml^{-1}) with [^{14}C]NAPQI (0.17 mCi mmol^{-1}; 3 mM). The upper trace denotes radioactivity eluting from the column (expressed in terms of equivalents of NAPQI per fraction) while the lower trace indicates the protein content of each fraction.

DISCUSSION

In a previous study (Rotman et al., 1976), it was demonstrated that when a number of hydroquinones and catechols, including 6-hydroxydopamine and 5,6-dihydroxytryptamine, were incubated under aerobic conditions with BSA, extensive cross-linking of BSA molecules occurred to yield higher molecular weight products. The authors suggested that the cross-linking processes were mediated by quinones formed by oxidation of the substrates; moreover, they proposed that the cross-linking properties of the quinones, together with the specific neuronal uptake mechanisms operative for 6-hydroxydopamine and 5,6-dihydroxytryptamine, may be responsible for the cytotoxicity of the parent compounds in vivo. Rotman et al. (1976) went on to demonstrate that cysteine residues were the most likely sites of cross-linking and that, due to the lack of reversal of cross-linking by reducing agents such as dithiothreitol, the mechanism of the process probably was not merely disulfide bond formation but rather involved the attachment of two protein molecules through a single quinone bridge. The latter mechanism also has been suggested to account for the cross-linking of the two chains of human hemoglobin by N,N-dimethyl-p-aminophenol. This molecule has been proposed to undergo oxidation to a quinoid structure which reacts with a cysteine residue on the hemoglobin in a Michael-type addition. The resulting protein-bound adduct is then postulated to undergo an oxidation reaction, possibly effected by a second free quinoid molecule, thus generating a protein-bound quinone which finally reacts with another sulfhydryl group on an adjacent polypeptide chain (Eyer et al., 1983). In addition, similar bis(amino acid)-drug adducts have been suggested to be formed in the reaction of NAPQI with N-acetylcysteine (Huggett and Blair, 1983).

In the present studies, we have demonstrated that the reactive quinoid metabolites NAPQI and 3,5-dimethyl-NAPQI, derived from the hepatotoxic acetanilide derivatives 4HAA and 3,5-dimethyl-4HAA, cross-link BSA molecules in vitro, whereas the quinoids 2,6-dimethyl-NAPQI and APBQ derived from the non-hepatotoxic compounds 2,6-dimethyl-4HAA and 3HAA fail to do so. The ability of certain quinoids to cross-link protein molecules to a greater extent than others probably depends on a number of factors, the most important of which is likely to be their respective redox potentials. Once formed, however, covalent protein-drug-protein bridges are likely to be chemically very stable and to result in irreversible linking of two or more protein monomers. It would be anticipated that such cross-linking of polypeptide chains might have much more profound effects on the structure and function of proteins than simple arylation events, and that such alterations in protein structure may have important toxicological consequences.

The exact nature of the cross-linked products formed in these model studies with BSA remains to be established, but the lack of reversal by dithiothreitol would argue against simple disulfide bond formation, at least in the case of NAPQI (Fig. 4A). A more likely sequence of events involves initial formation of a 3'-cysteinyl adduct with BSA (Fig. 4B; Hoffmann et al., 1985a), followed by oxidation of the bound residue with a second molecule of NAPQI to form a protein-bound quinoid; the latter species, in turn, could react with another cysteinyl residue on a separate polypeptide chain to form the cross-link (Fig. 4C). Such a sequence of events may, however, be an oversimplification of the reactions actually taking place since when [^{14}C]NAPQI was added to solutions of BSA, approximately 8 equivalents of NAPQI became bound per BSA monomer throughout all of the protein fractions. (The scheme proposed above would require only 0.5 equivalents of NAPQI bound per BSA monomer). This large amount of binding to BSA is probably related to the accompanying polymerization of NAPQI molecules, as evidenced by the presence of the slowly-chromatographing brown material noted during Sephadex gel filtration. Polymerization reactions of this type have been suggested to reflect the generation of semiquinone imine free radicals by the

734

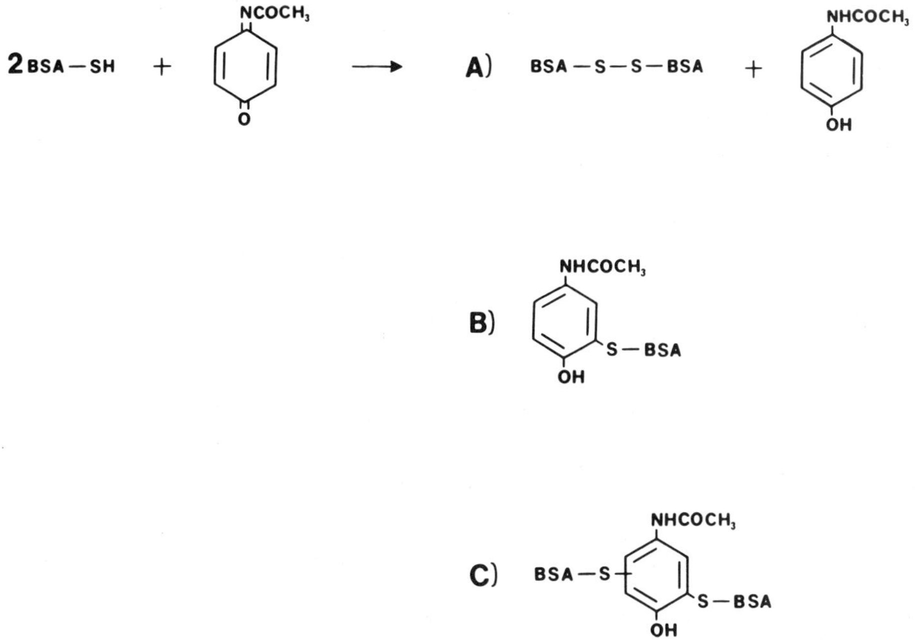

Fig. 4. Some possible interactions between BSA and NAPQI.

comproportionation of NAPQI and 4HAA. Any reactive radical species thus
formed would be expected to be less discriminating in their choice of binding sites on protein molecules than would be the corresponding quinoids.
Further studies are necessary in order to determine the nature of the quin-
oid-modified BSA derivatives and to establish the toxicological significance
of these observations with respect to the mechanism of acetaminophen-induced
liver injury.

ACKNOWLEDGEMENTS

We are grateful to Dr. D. C. Dahlin for a gift of [^{14}C]NAPQI. These
studies were supported by research grants AM 30699 and GM 25418 from the
National Institutes of Health.

REFERENCES

Andersson, L.-O., 1966, The heterogeneity of bovine serum albumin,
    Biochim. Biophys. Acta, 117:115.
Boyd, E. M., and Bereczky, G. M., 1966, Liver necrosis from paracetamol,
    Br. J. Pharmacol. Chemother., 26:606.
Brown, J. R., 1976, Structural origins of mammalian albumin, Fed. Proc.,
    35:2141.
Dahlin, D. C., Miwa, G. T., Lu, A. Y. H., and Nelson, S. D., 1984,
    N-Acetyl-p-benzoquinone imine: a cytochrome P-450 mediated oxida-
    tion product of acetaminophen, Proc. Natl. Acad. Sci. U.S.A.,
    81:1327.
Dahlin, D. C., and Nelson, S. D., 1982, Synthesis, decomposition kinetics,
    and preliminary toxicological studies of pure N-acetyl-p-benzoquin-
    one imine, a proposed toxic metabolite of acetaminophen, J. Med.
    Chem., 25:885.

Devalia, J. L., Ogilvie, R. C., and McLean, A. E. M., 1982, Dissociation
of cell death from covalent binding of paracetamol by flavones in a
hepatocyte system, Biochem. Pharmacol., 31:3745.

Eyer, P., Lierheimer, E., and Strosar, M., 1983, Site and mechanism of
covalent binding of 4-dimethylaminophenol to human hemoglobin, and
its implications to the functional properties, Mol. Pharmacol.,
23:282.

Fernando, C. R., Calder, I. C., and Ham, K. N., 1980, Studies on the
mechanism of toxicity of acetaminophen: synthesis and reactions of
N-acetyl-2,6-dimethyl- and N-acetyl-3,5-dimethyl-p-benzoquinone
imines, J. Med. Chem., 23:1153.

Gerber, J. G., MacDonald, J. S., Harbison, R. D., Villeneuve, J.-P.,
Wood, A. J. J., and Nies, A. S., 1977, Effect of N-acetylcysteine
on hepatic covalent binding of paracetamol (acetaminophen), Lancet,
1:657.

Hoffmann, K.-J., Streeter, A. J., Axworthy, D. B., and Baillie, T. A.,
1985a, Structural characterization of the major covalent adduct
formed in vitro between acetaminophen and bovine serum albumin,
Chem.-Biol. Interact., 53:155.

Hoffmann, K.-J., Streeter, A. J., Axworthy, D. B., and Baillie, T. A.,
1985b, Identification of the major covalent adduct formed in vitro
and in vivo between acetaminophen and mouse liver proteins, Mol.
Pharmacol., 27:566.

Huggett, A., and Blair, I. A., 1983, The mechanism of paracetamol-induced
hepatotoxicity: implications for therapy, Human Toxicol., 2:399.

Labadarios, D., Davis, M., Portmann, B., and Williams, R., 1977, Paracetamol-induced hepatic necrosis in the mouse: relationship between
covalent binding, hepatic glutathione depletion, and the protective
effect of α-mercaptopropionylglycine, Biochem. Pharmacol., 26:31.

Lowry, O. H., Rosebrough, N. J., Farr, A. L., and Randall, R. J., 1951,
Protein measurement with the Folin phenol reagent, J. Biol. Chem.,
193:265.

Mitchell, J. R., Potter, W. Z., Hinson, J. A., Snodgrass, W. R., Timbrell, J. A., and Gillette, J. R., 1975, Toxic drug reactions,
in: Concepts in Biochemical Pharmacology, Vol. XXVIII/3, J. R.
Gillette and J. R. Mitchell, eds., Springer, New York, p.383.

Nelson, E. B., 1980, The Pharmacology and toxicology of meta-substituted
acetanilide. I. Acute toxicity of 3-hydroxyacetanilide in mice,
Res. Commun. Chem. Path. Pharmacol., 28:447.

Pedersen, K. O., 1962, Exclusion chromatography, Arch. Biochem. Biophys.,
suppl. 1:157.

Potter, W. Z., Thorgeirsson, S. S., Jollow, D. J., and Mitchell, J. R.,
1974, Acetaminophen-induced hepatic necrosis. V. Correlation of
hepatic necrosis, covalent binding and glutathione depletion in hamsters, Pharmacology, 12:129.

Prescott, L. F., Wright, N., Roscoe, P., and Brown, S. S., 1971, Plasma
paracetamol half-life and hepatic necrosis in patients with paracetamol overdosage, Lancet, 1:519.

Roberts, S. A., and Jollow, D. J., 1978, Acetaminophen structure-toxicity
relationships: why is 3-hydroxyacetanilide not hepatotoxic?, Pharmacologist, 20:259.

Roberts, S. A., and Jollow, D. J., 1979, Acetaminophen structure-toxicity
studies: in vivo covalent binding of a non-hepatotoxic analog,
3-hydroxyacetanilide, Fed. Proc., 38:426.

Rosen, G. M., Rauckman, E. J., Ellington, S. P., Dahlin, D. C., Christie,
J. L., and Nelson, S. D., 1984, Reduction and glutathione conjugation
reactions of N-acetyl-p-benzoquinone imine and two dimethylated
analogues, Mol. Pharmacol., 25:151.

Rotman, A., Daly, J. W., and Creveling, C. R., 1976, Oxygen-dependent reaction of 6-hydroxydopamine, 5,6-dihydroxytryptamine, and related compounds with proteins in vitro: a model for cytotoxicity, Mol. Pharmacol., 12:887.

Streeter, A. J., and Baillie, T. A., 1985, 2-Acetamido-p-benzoquinone: a reactive arylating metabolite of 3'-hydroxyacetanilide, Biochem. Pharmacol., in press.

Streeter, A. J., Bjorge, S. M., Axworthy, D. B., Nelson, S. D., and Baillie, T. A., 1984a, The microsomal metabolism and site of covalent binding to protein of 3'-hydroxyacetanilide, a nonhepatotoxic positional isomer of acetaminophen, Drug Metab. Dispos., 12:565.

Streeter, A. J., Dahlin, D. C., Nelson, S. D., and Baillie, T. A., 1984b, The covalent binding of acetaminophen to protein. Evidence for cysteine residues as major sites of arylation in vitro, Chem.-Biol. Interact., 48:349.

Wunderer, H., 1972, Darstellung und Eigenschaften von Jodamino-p-benzochinonen, Chem. Ber., 105:3479.

ACETAMINOPHEN AS A COSUBSTRATE AND INHIBITOR OF PROSTAGLANDIN H SYNTHASE

Peter J. Harvison,[1] R. W. Egan,[2] P. H. Gale,[2] and Sidney D. Nelson[1]

[1]Department of Medicinal Chemistry, BG-20
University of Washington, Seattle, WA 98195 and
[2]Merck Institute for Therapeutic Research
Rahway, NJ 07065

INTRODUCTION

Recently, several reports (Marnett et al., 1983; Nordenskjöld et al., 1984) have implicated prostaglandin H synthase (PHS) in the bioactivation of xenobiotics to potentially toxic metabolites. Benzidine (Zenser et al., 1983), p-aminophenol (Josephy et al., 1983), and phenacetin (Andersson et al., 1982) are among the compounds known to undergo metabolic activation by PHS. Of particular interest to us is the fact that this enzyme can metabolize acetaminophen (APAP) to a reactive species that can bind to proteins or form a glutathione conjugate (Moldeus and Rahimtula, 1980; Boyd and Eling, 1981; Mohandas et al., 1981; Moldeus et al., 1982). In fact, it has been suggested (Boyd and Eling, 1981; Mohandas et al., 1981) that the nephrotoxicity sometimes associated with APAP overdosage may be due in part to its metabolism by PHS which is present in high levels in the renal inner medulla.

PHS has two related, inseparable activities: a cyclooxygenase, which catalyzes the bis-dioxygenation of arachidonic acid (AA) to prostaglandin G_2 (PGG_2); and a hydroperoxidase, which subsequently reduces the perhydroxyl group at C-15 to a hydroxyl group, yielding prostaglandin H_2 (PGH_2, Fig. 1, Marnett et al., 1980; Egan et al., 1981; Moldeus et al., 1982). It has recently been shown (Mason et al., 1980; Hemler and Lands, 1980) that free radicals are involved in the oxidation of AA by PHS. APAP and other easily oxidizable compounds can apparently serve as cosubstrates for the hydroperoxidase, and are consequently metabolized, via a one-electron oxidation, to reactive free radicals (Egan et al., 1981; Moldeus et al, 1982; Zenser and Davis, 1984).

APAP itself has been shown to inhibit renal prostaglandin production in a manner that is considerably different than aspirin (Zenser et al., 1978; Mattammal et al., 1979). In fact, it has been reported (Robak et al., 1978) that APAP can either stimulate PHS at low concentration or inhibit it at high concentration. It has been suggested (Hemler and Lands, 1980) that the stimulatory activity may be due to a reduction of unproductive, high oxidation states of PHS by APAP. The inhibition at high concentration may be due to a reduction of hydroperoxide levels which are required for enzyme activation (Egan et al., 1980).

The studies described here were undertaken for two reasons: 1) to investigate the metabolic activation of APAP to radical derived products by PHS; and 2) to further study the effect of APAP and several analogues on arachidonic acid metabolism. The formation of radical products from APAP by horseradish peroxidase (HRP) has recently been described (Potter et al., 1985).

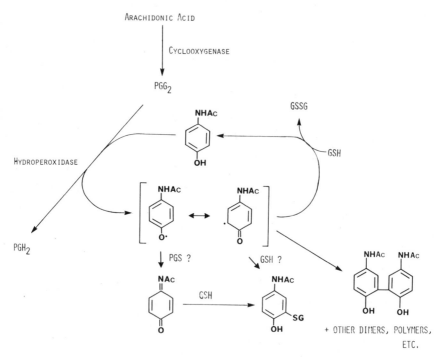

Fig. 1. Metabolic scheme for bioactivation of acetaminophen by prostaglandin H synthase.

MATERIALS AND METHODS

Materials

[Ring-U-^{14}C]acetaminophen was obtained from Pathfinder Laboratories, St. Louis, MO and [1-^{14}C]arachidonic acid was purchased from New England Nuclear, Boston, MA. NAPQI (Dahlin and Nelson, 1982) and 2,6-dimethyl- and 3,5-dimethylacetaminophen (Fernando et al., 1980) were synthesized as previously described. All other chemicals were of the highest grade commercially available.

Methods

Assays to determine the effect of APAP and analogues on ^{14}C-arachidonic acid metabolism were performed as previously described (Egan et al., 1976).

APAP metabolism was measured by incubating 0.2 mM ^{14}C-APAP (sp. act. 0.41 mCi/mmol), the appropriate inhibitor, ram seminal vesicle microsomal protein (RSVM, 0.85 mg/ml), and arachidonic acid (0.1 mM) or H_2O_2 (0.1 mM) at 30°C. After work-up the reaction mixtures were analyzed by reversed phase HPLC (μBondapak C$_{18}$ column) using a non-linear phosphate buffer-methanol gradient. Radioactivity was measured by liquid scintillation counting. Under these conditions the following retention times were obtained: APAP,

9.6 min; GS–APAP, 14.5 min; metabolite A (di–APAP), 18.0 min; metabolite B, 21.8 min; metabolite C, 24.5 min; and metabolite D, > 35 min (eluted with 100% methanol flush).

Experiments to detect NAPQI were performed similarly except that a cold carrier solution of NAPQI was added directly to each incubation immediately prior to HPLC analysis. Using isocratic elution (Microsorb C_{18} column) the following retention times were obtained: APAP, 4.3 min; metabolite X, 5.6 min; and NAPQI, 8.2 min.

RESULTS

In the presence of 0.1 mM arachidonic acid (AA) PHS catalyzed the formation of 4 distinct APAP metabolites that were separated and quantified by HPLC. A total of ca. 30 nmol/mg RSVM/5 min of radical derived products were formed in complete incubations (Fig. 2). The major metabolite (metabolite A, 12.2% of recovered radioactivity) co-chromatographed with an authentic sample of 4,4'-dihydroxy-3,3'-diacetanilide, an APAP dimer (see Fig. 1). Similar results were obtained in incubations supported by H_2O_2, although the overall level of metabolism was about 50% higher (Fig. 3). Essentially no activity was present in control incubations (i.e. no microsomes or initiator present) in both cases (see Figs. 2 and 3).

A variety of antioxidants and PHS inhibitors were included in the incubations to further probe the system. The cyclooxygenase inhibitors indomethacin and aspirin were both effective in incubations supported by AA (Fig. 2). However, indomethacin had no significant effect on metabolism in H_2O_2-supported incubations (Fig. 3). Methimazole (2-mercapto-1-methylimidazole) which has previously been reported to inhibit PHS and other peroxidases (Zenser et al., 1983; and loc. cit.), was relatively effective in either case. The antioxidants ascorbic acid and butylated hydroxyanisole (BHA) were also investigated and the latter completely abolished the reaction. Metyrapone, a cytochrome P-450 inhibitor, had no effect on the reaction whether supported by AA or H_2O_2 (Figs. 2 and 3).

An acetaminophen glutathione conjugate (GS–APAP) was produced in incubations that contained GSH (Fig. 4). This metabolite co-chromatographed with a synthetically prepared standard. Interestingly, ascorbic acid had a much less pronounced effect on the formation of this metabolite than the radical derived products (Fig. 4).

The two electron oxidation product of APAP, N-acetyl-p-benzoquinone imine (NAPQI), was detected in incubations under different chromatographic conditions (Fig. 5). A second, as yet unidentified metabolite (metabolite X), was also detected in these experiments. The formative of NAPQI was enhanced significantly in the presence of AA. There was no significant difference in the amount of metabolite X formed either in the presence or absence of AA. As above, neither metabolite was formed in the absence of enzyme (Fig. 5, -RSVM).

A parallel series of experiments was performed to investigate the effect of APAP on PHS activity. As anticipated, APAP stimulated PHS at low concentrations, but inhibited the enzyme at higher concentrations (Table 1). The regioisomer, 2'-hydroxyacetanilide, and the analogue, 3,5-dimethylacetaminophen (3,5-diMeAPAP), were approximately as equipotent as APAP in their effects on AA metabolism (Table 1). In contrast, the meta isomer, 3'-hydroxyacetanilide, and 2,6-dimethylacetaminophen (2,6-diMeAPAP) were relatively poor in stimulating or inhibiting the system. The ability of these compounds to stimulate or inhibit PHS approximately correlated with their half-wave oxidation potentials (Table 1).

741

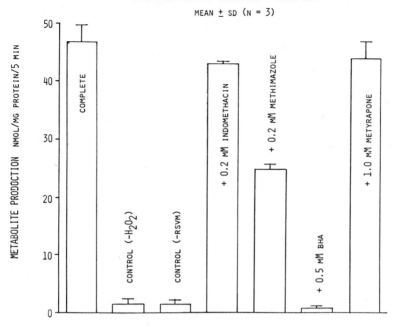

Fig. 2. Production of radical derived APAP metabolites (total
A-D) by PHS in incubations supported by arachidonic acid.

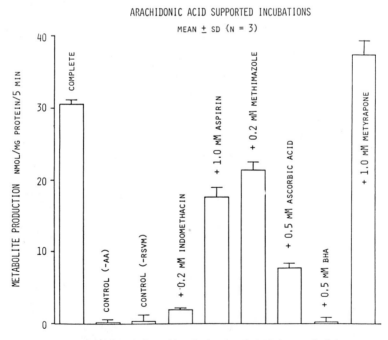

Fig. 3. Production of radical derived APAP metabolites
(total A-D) by PHS in incubations supported by H_2O_2.

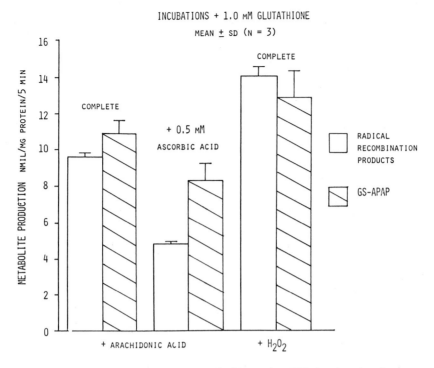

Fig. 4. Production of APAP metabolites by PHS in incubations
containing glutathione.

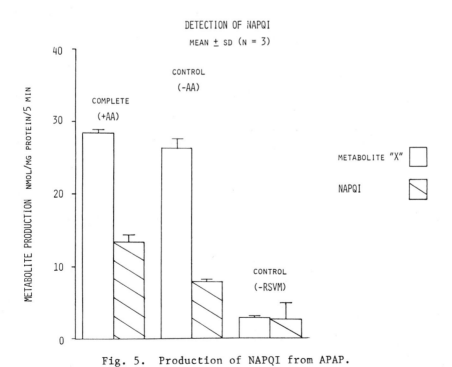

Fig. 5. Production of NAPQI from APAP.

Table 1. Effect of Acetaminophen on Arachidonic Acid Metabolism.

Compound	Concentration (uM) [a] PGH$_2$ Max.	Maximum A.A. [b] Stimulation	IC$_{50}$ [c] (mM)	Redox [d] Potential (V)
Acetaminophen	100	75%	25	0.39
2,6-Dimethylacetaminophen	2000	98%	>40	0.59
3,5-Dimethylacetaminophen	100	129%	20	0.26
2'-Hydroxyacetanilide	100	104%	7	0.45
3'-Hydroxyacetanilide	10000	36%	>40	0.78

[a] The Lowest Concentration of the Compound at Which PGH$_2$/(PGG$_2$ + PGH$_2$) Reached 0.95.

[b] The Maximum Increase in Arachidonic Acid Metabolism.

[c] The Concentration That Decreased Arachidonic Acid Metabolism to 50% of its value in the Absence of Compound.

[d] Half-Wave Redox Potentials Determined by Cyclic Voltammetry (Oxidation Wave) in Phosphate Buffer of Ionic Strength 0.2 m, pH 7.4. Scans Were Initiated in the Positive Direction at 150 mV/S. Working Electrode: Glassy Carbon, Reference Electrode: Ag/AgCl/3 M NaCl.

DISCUSSION

Previous workers have demonstrated that PHS can metabolically activate APAP to a reactive and potentially toxic species (Moldeus and Rahimtula, 1980; Boyd and Eling, 1981; Mohandas et al., 1981; Moldeus et al., 1982). It has been suggested that this activation proceeds via a one-electron oxidation to yield a semiquinone free radical (Moldeus and Rahimtula, 1980; Moldeus et al., 1982). The potential toxicological significance of these reactive intermediates has been reviewed (Mason, 1979).

The studies described in this report provide further, albeit indirect, evidence that PHS can metabolize APAP to a free radical. Indeed, the detection of an APAP dimer, which could form as a result of a radical recombination reaction (Fig. 1), lends credence to this hypothesis. The recent report that PHS and horseradish peroxidase (HRP) can oxidize p-phenetidine to polymeric products is especially significant in this regard (Ross et al., 1985). Electron spin resonance (ESR) data has demonstrated the formation of identical semiquinone free radicals from p-aminophenol catalyzed by either PHS or HRP (Josephy et al., 1983). Furthermore, an unstable APAP free radical produced by HRP has been detected by fast flow ESR (West et al., 1984). The signal from this radical rapidly changed to that of a polymeric form. Finally, HRP has recently been shown (Potter et al., 1985) to oxidize APAP to a variety of polymers, including the dimer detected here. The similarity in the reactions catalyzed by these two enzymes provides good evidence that the metabolites detected herein, do indeed, arise via a one-electron oxidation.

It is apparent that the activity seen was enzymatic in nature since no metabolism was evident in the absence of microsomes (-RSVM, Figs. 2 and 3). Furthermore, the lack of inhibition in the presence of metyrapone indicates that no cytochrome P-450 activity was present. Since either AA or H$_2$O$_2$ can support the reaction, it can be concluded that PHS is responsible for the metabolism.

The potent cyclooxygenase inhibitor indomethacin abolished the reaction in AA supported incubations, but had no effect when H_2O_2 was used as the initiator. These results confirm that the hydroperoxidase activity of PHS is responsible for the metabolic activation of APAP, a finding reported by others (Moldeus and Rahimtula, 1980; Boyd and Eling, 1981; Mohandas et al., 1981; Moldeus et al., 1982). As expected the peroxidase inhibitor methimazole was effective in either case. The antioxidant BHA had a potent inhibitory activity (Figs. 2 and 3) indicating the oxidative nature of the process.

The detection of GS-APAP in incubations containing GSH (Fig. 4) is interesting in that it could conceivably arise via two pathways (Fig. 1). However, it was found that ascorbic acid had a much greater effect on the formation of the radical derived products than on GS-APAP. This suggests that this metabolite does not arise by a direct conjugation with the semiquinone free radical, but rather with NAPQI (see Fig. 1). Indeed, NAPQI, itself a two-electron oxidation product of APAP, was also detected in PHS incubations (Fig. 5). Whether this is a product of two successive one-electron oxidations cannot be said with certainty.

APAP, 3,5-diMeAPAP, and 2'-hydroxyacetanilide were found to be effective in stimulating or inhibiting PHS depending on concentration (Table 1). This is in accord with previous observations (Robak et al., 1978). On the other hand, 2,6-diMeAPAP and 3'-hydroxyacetanilide were essentially inactive. Interestingly, the relative potencies of these compounds was roughly correlated with their oxidation potentials as determined by cyclic voltammetry.

In conclusion, this work has provided evidence that APAP can function as a cosubstrate for PHS and, in turn, is oxidized to a free radical.

ACKNOWLEDGEMENTS

These studies were supported by grant GM 25418 from the National Institutes of Health.

REFERENCES

Andersson, B., Nordenskjöld, M., Rahimtula, A., and Moldeus, P., 1982, Prostaglandin synthetase-catalyzed activation of phenacetin metabolites to genotoxic products, Mol. Pharmacol., 22:479.

Boyd, J. A., and Eling, T. E., 1981, Prostaglandin endoperoxide synthetase-dependent cooxidation of acetaminophen to intermediates which covalently bind in vitro to rabbit renal medullary microsomes, J. Pharmacol. Exp. Ther., 219:659.

Dahlin, D. C., and Nelson, S. D., 1982, Synthesis, decomposition kinetics, and preliminary toxicological studies of pure N-acetyl-p-benzoquinone imine, a proposed toxic metabolite of acetaminophen, J. Med. Chem., 25:885.

Egan, R. W., Paxton, J., and Kuehl, F. A., Jr., 1976, Mechanism for irreversible self-deactivation of prostaglandin synthetase, J. Biol. Chem., 251:7329.

Egan, R. W., Gale, P. H., Beveridge, G. C., Marnett, L. J., and Kuehl, F. A., Jr., 1980, Direct and indirect involvement of radical scavengers during prostaglandin biosynthesis, in: "Advances in Prostaglandin and Thromboxane Research, Vol. 6," B. Samuelson, P. W. Ramwell, and R. Paoletti, eds., Raven Press, New York, p. 153.

745

Egan, R. W., Gale, P. H., Baptista, E. M., Kennicott, K. L., VandenHeuvel, W. J. A., Walker, R. W., Fagerness, P. E., and Kuehl, F. A., Jr., 1981, Oxidation reactions by prostaglandin cyclooxygenase-hydroperoxidase, J. Biol. Chem., 256:7352.

Fernando, C. R., Calder, I. C., and Ham, K. N., 1980, Studies on the mechanism of toxicity of acetaminophen: synthesis and reactions of N-acetyl-2,6-di-methyl- and N-acetyl-3,5-dimethylbenzoquinone imines, J. Med. Chem., 23:1153.

Hemler, M. E., and Lands, W. E. M., 1980, Evidence for a peroxide-initiated free radical mechanism of prostaglandin biosynthesis, J. Biol. Chem., 255:6253.

Josephy, P. D., Eling, T. E., and Mason, R. P., 1983, Oxidation of p-aminophenol catalyzed by horseradish peroxidase and prostaglandin synthase, Mol. Pharmacol., 23:461.

Marnett, L. J., Bienkowski, M. J., Pagels, W. R., and Reed, G. A., 1980, Mechanism of xenobiotic cooxygenation coupled to prostaglandin H$_2$ biosynthesis, in: "Advances in Prostaglandin and Thromboxane Research, Vol. 6," B. Samuelson, P. W. Ramwell, and R. Paoletti, eds., Raven Press, New York, p. 149.

Marnett, L. J., Dix, T. A., Sacks, R. J., and Sieldlik, P. H., 1983, Oxidation by fatty acid hydroperoxides and prostaglandin synthase, in "Advances in Prostaglandin, Thromboxane, and Leukotriene Research, Vol. 11," B. Samuelson, R. Paoletti, and P. Ramwell, eds., Raven Press, New York, p. 79.

Mason, R. P., 1979, Free radical metabolites of foreign compounds and their toxicological significance, in: "Reviews in Biochemical Toxicology, Vol. 1," E. Hodgson, J. R. Bend, and R. M. Philpot, eds., Elsevier North Holland, New York, p. 151.

Mason, R. P., Kalyanaraman, B., Tainer, B. E., and Eling, T. E., 1980, A Carbon-centered free radical intermediate in the prostaglandin synthetase oxidation of arachidonic acid, J. Biol. Chem., 255:5019.

Mattammal, M. B., Zenser, T. V., Brown, W. W., Herman, C. A., and Davis, B. B., 1979, Mechanism of inhibition of renal prostaglandin production by acetaminophen, J. Pharmacol. Exp. Ther., 210:405.

Mohandas, J., Duggin, G. G., Harvath, J. S., and Tiller, D. J., 1981, Metabolic activation of acetaminophen (paracetamol) mediated by cytochrome P-450 mixed-function oxidase and prostaglandin endoperoxide synthetase in rabbit kidney, Toxicol. Appl. Pharmacol., 61:252.

Moldeus, P., and Rahimtula, A., 1980, Metabolism of paracetamol to a glutathione conjugate catalyzed by prostaglandin synthetase, Biochem. Biophys. Res. Commun., 96:469.

Moldeus, P., Andersson, B., Rahimtula, A., and Berggren, M., 1982, Prostaglandin synthetase catalyzed activation of paracetamol, Biochem. Pharmacol., 31:1363.

Nordenskjöld, N, Andersson, B., Rahimtula, A., and Moldeus, P., 1984, Prostalandin synthase-catalyzed metabolic activation of some aromatic amines to genotoxic products, Mutat. Res., 127:107.

Potter, D. W., Miller, D. W., and Hinson, J. A., 1985, J. Biol. Chem., in press.

Robak, J., Wieckowski, A., and Gryglewski, R., 1978, The effect of 4-acetamidophenol on prostaglandin synthetase activity in bovine and ram seminal vesicle microsomes, Biochem. Pharmacol., 27:393.

Ross, D., Larsson, R., Andersson, B., Nilson, U., Lindquist, T., Lindeke, B., and Moldeus, P., 1985, The oxidation of p-phenetidine by horseradish peroxidase and prostaglandin synthase and the fate of glutathione during such oxidations, Biochem. Pharmacol., 34:343.

West, P. R., Harman, L. S., Josephy, P. D., and Mason, R. P., 1984, Acetaminophen: enzymatic formation of a transient phenoxyl free radical, Biochem. Pharmacol., 33:2933.

Zenser, T. V., Mattammal, M. B., Herman, C. A., Joshi, S., and Davis, B. B., 1978, Effect of acetaminophen on prostaglandin E_2 and prostaglandin $F_{2\alpha}$ synthesis in the renal inner medulla of rat, <u>Biochim. Biophys. Acta</u>, 542:486.

Zenser, T. V., Mattammal, M. B., Wise, R. B., Rice, J. R., and Davis, B. B., 1983, Prostaglandin H synthase-catalyzed activation of benzidine: a model to access pharmacologic intervention of the initiation of chemical carcinogenesis, <u>J. Pharmacol. Exp. Ther.</u>, 227:545.

Zenser, T. V., and Davis, B. B., 1984, Enzyme systems involved in the formation of reactive metabolites in the renal medulla: cooxidation via prostaglandin H synthase, <u>Fund. Appl. Toxicol.</u>, 4:922.

THE FATE OF 4-CYANOACETANILIDE IN RATS AND MICE:

THE MECHANISM OF FORMATION OF A NOVEL ELECTROPHILIC METABOLITE

C. J. Logan, D. Hesk and D. H. Hutson*

Shell Development Company, Westhollow Research Center
P. O. Box 1380, Houston, Texas 77001, U.S.A.
*Shell Research Limited, Sittingbourne Research Centre
Sittingbourne, Kent, ME9 8AG, U.K.

INTRODUCTION

4-Cyanoacetanilide (PCAA, b in Scheme 1) is a metabolite of 4-cyano-N,N-dimethyl-aniline (CDA, Scheme 1). Whilst the metabolism of CDA, a designed analogue of butter yellow (Ashby, Styles, Paton, 1980) was being investigated, a novel involvement of glutathione was detected (Hutson, Lakeman and Logan, 1984). From the structure of the mercapturic acid conjugate (C), which was isolated from the urine of the rats that had been dosed with CDA, it was postulated that the CDA had been de-methylated (A), acetylated (b), and that the methyl group of the acetyl group had been activated to form an electrophilic intermediate, the intermediate then being trapped, at least in part, by glutathione

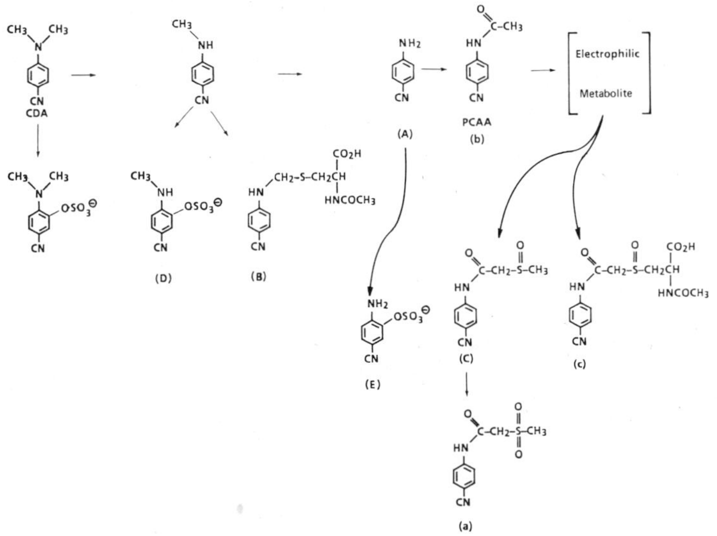

Scheme 1. The Metabolic Fate of CDA

and eventually excreted as the mercapturic acid (C). Interestingly, mice seemed to be unable to form this electrophilic intermediate.

As DNA from the livers of rats dosed with [^{14}C]CDA contained appreciably higher levels of radioactivity than did the DNA from the livers of mice that had been similarly dosed (Edwards et al., 1985) it was felt that further investigation of this intermediate was warranted. This report details the mechanism of formation of the novel electrophilic metabolite in the rat, and presents reasons why the mouse fails to produce it.

RESULTS

The postulated route of formation of the novel electrophilic metabolite is shown in Scheme 2.

In the first experiment rats were pretreated with pentachlorophenol, an inhibitor of sulphation (Mulder and Scholtens, 1977) and then with [^{14}C]CDA. The amounts of metabolites C, a, and c, (the products of the reactive intermediate) that were present in the urine, were all reduced, in comparison to non-pretreated animals. This suggested that sulphation may have been an important step in the production of the intermediate.

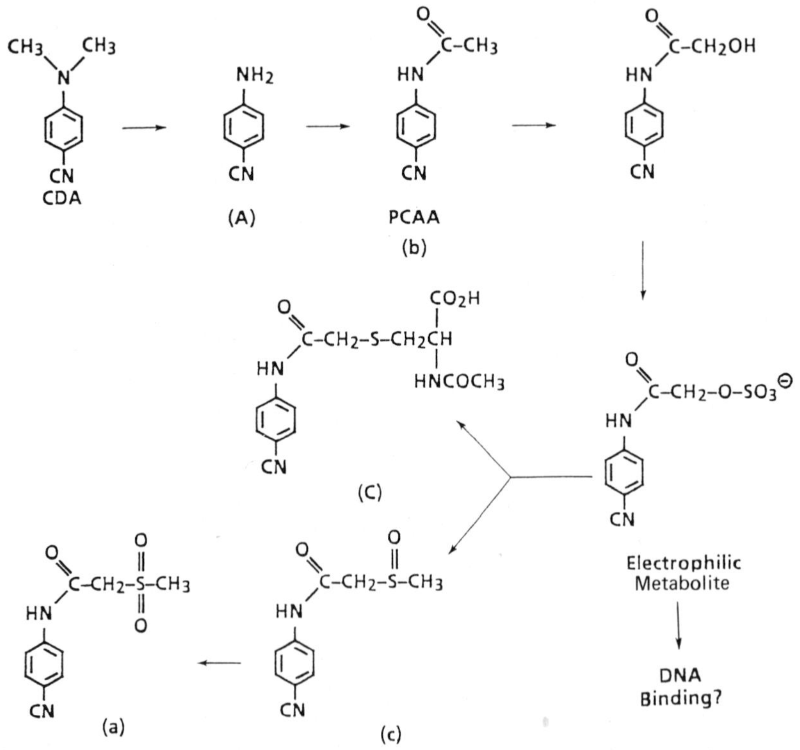

Scheme 2. Route of Formation of Electrophilic Metabolite

In order to test if acetylation was required to produce the intermediate rats and mice were dosed with 4-cyano-[1-^{14}C,1,2-^{13}C] acetanilide (PCAA, b). Comparison of the TLC profiles of the urines from rats dosed with CDA or PCAA showed that several of the metabolites were similar (Fig. 1). In fact, all of the metabolites produced from PCAA were also produced from CDA, indicating that indeed acetylation is a required step in the production of many of the metabolites of CDA, including the thioether metabolites such as C.

The presence of two bands of low Rf on the TLC of urine from PCAA dosed animals was unexpected. No acetyl containing metabolites had been isolated from this area of the TLC in the CDA metabolism study. Isolation of the compounds followed by fast atom bombardment mass spectrometry (FAB-MS) indicated that they were N-acetyl-O-sulphates, but did not give any information as to the position of the O-sulphate moiety in the metabolites (i.e. whether it was on the aromatic ring or in the amide side chain). Hence a further sample of the N-acetyl-O-sulphates was isolated and treated with sulphatase. The two resultant alcohols were purified by reverse phase HPLC and analysed by MS. The spectra indicated that the structures were as shown in Figs. 2 and 3. The presence of the ^{13}C isotope pattern at m/z 176, 177, 178 in both spectra indicated that both metabolites had retained the carbon backbones of their acetate groups. The isotope pattern at m/z 145, 146 in figure 2 indicated that this metabolite had been metabolised in the side chain, whilst the lack of any isotope pattern at m/z 134 in figure 3 clearly showed that the second sulphate metabolite had been produced via oxidation of the aromatic ring. The precise position of the O-sulphate on the aromatic ring was confirmed by nuclear magnetic resonance spectroscopy.

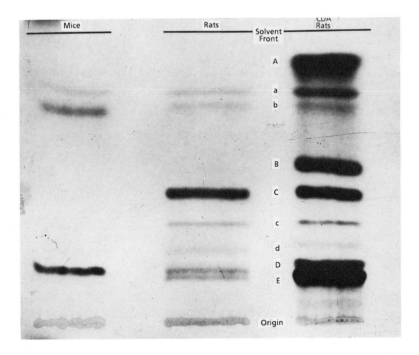

Figure 1. Autoradiograph of Day 1 Urine from Rats and Mice Orally Dosed with (^{14}C) CDA

The mice dosed with [^{14}C]PCAA essentially excreted only unchanged PCAA and the aromatic ring O-sulphate (fig. 3). There was no indication of any biotransformation of the acetyl group.

CONCLUSION

CDA is bio-activated by the following series of transformations: N-demethylation, N-acetylation, oxidation of the CH$_3$ of the acetyl group and conjugation of the resultant alcohol with sulphate. The sulphate is electrophilic enough to react with glutathione and may be reactive enough to react with cellular macro-molecules.

The mouse does not produce the reactive O-sulphate because it neither carries out the N-acetylation nor the oxidation of the acetyl group.

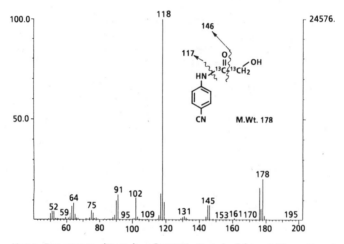

Figure 2. Mass Spectrum (E.I.) of PCAA Metabolite After Treatment with Sulphatase

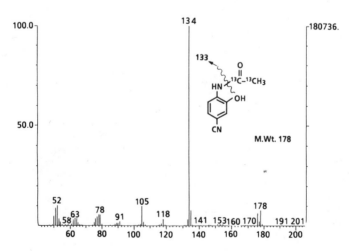

Figure 3. Mass Spectrum (E.I.) of PCAA Metabolite After Treatment with Sulphatase

752

ACKNOWLEDGEMENTS

The authors would like to thank the U.K.E.M.S. for the provision of the [^{14}C]CDA, Dr. P. D. Regan for determining the n.m.r. spectra and Dr. J. A. Page and M. Selby for the MS service.

REFERENCES

J. Ashby, J. A. Styles, and D. Paton, 1980, Carcinogenesis 1, 1.

S. Edwards, D. Hesk, C. J. Logan, and R. Waters, 1985, in: 2nd U. K. Environmental Mutagen Society Ring-Test, eds. C. Arlot, J. Ashby, and J. Parry. Pub. McMillan, London.

D. H. Hutson, S. K. Lakeman, and C. J. Logan, 1984, Xenobiotica, 14, 925.

G. J. Mulder, and E. Scholtens, 1977, Biochem. J., 165, 553.

MECHANISTIC STUDIES OF ACETAMIDE HEPATOCARCINOGENICITY

Erik Dybing[1], W. Perry Gordon[2], Erik J. Søderlund[1],
Jørn A. Holme[1], Edgar Rivedal[3], and Snorri S. Thorgeirsson[4]

National Institute of Public Health, Oslo, Norway[1]
University of Washington, Seattle, USA[2], The Norwegian Radium Hospital, Oslo, Norway[3], and National Cancer Institute, Bethesda, USA[4]

INTRODUCTION

Feeding male Leeds rats a diet containing 5 % acetamide (AA) for 17 months resulted in the development of hepatic cell neoplasms in all animals (Flaks et al., 1983). Many studies have reported that AA does not show genotoxic effects. However, Pienta et al. (1977) found that AA transformed Syrian hamster embryo cells in culture. N-Hydroxyacetamide (N-OH-AA), a possible metabolite of AA, has been approved by the US Food and Drug Administration under the orphan drug products program for treatment of infection-induced renal stones (in: Putcha et al., 1984). This study was undertaken attempting to shed light on the mechanism of AA hepatocarcinogenicity.

MATERIALS AND METHODS

Chemicals

$(1-^{14}C)AA$ (7.45 mCi/mmol) was purchased from California Bionuclear Co., Sun Valley, CA. $(1-^{14}C)N$-OH-AA was synthesized from $(1-^{14}C)$ethyl acetate generated from $(1-^{14}C)$acetyl chloride (57 mCi/mmol, New England Nuclear, Boston, MA) by the method of Fishbein et al. (1969). $L-(U-^{14})$Leucine (348 mCi/mmol) was obtained from The Radiochemical Centre, Amersham, England and $(1-^{14}C)$acetic acid, sodium salt (56 mCi/mmol) from New England Nuclear.

Animals and treatments

Male Wistar rats (180-250 g) were obtained from Møllegaard Breeding Centre, Ejby, Denmark. Some animals were pretreated with PCB (Aroclor 1254) 500 mg/kg bw i.p. 5 days before death.

In vitro assays

Mutagenic activity was assayed with S. typhimurium TA 98 and TA 100 (Ames et al., 1975). Covalent binding to liver microsomal and cytosolic as well as hepatocyte proteins was measu-

Table 1. Mutagenicity Testing of AA and N-OH-AA
in Salmonella Typhimurium

TEST COMPOUND	µG PER PLATE	S 9 MIX	TA 98 REV./PLATE	TA 100 REV./PLATE
AA	50	-	0	3
	100	-	0	0
	500	-	0	0
	1000	-	5	0
	50	+	7	0
	100	+	1	0
	500	+	15	7
	1000	+	1	0
N-OH-AA	50	-	0	0
	100	-	13	0
	500	-	21	25
	1000	-	70	69
	50	+	7	25
	100	+	5	15
	500	+	39	48
	1000	+	74	68

red according to Søderlund et al. (1982). Determination of DNA damage in cultured Reuber H4-II-E rat hepatoma cells was achieved by alkaline elution (Kohn et al., 1981). Cytotoxicity and DNA repair of isolated rat hepatocytes in monolayer culture was performed according to Holme and Søderlund (1984). In vitro transformation of primary Syrian hamster embryo cells was studied as described by Rivedal and Sanner (1982).

In vivo assays

Rats were treated with various doses of AA and N-OH-AA i.p. for 24 hr, whereafter heparinized blood was collected for SGOT analysis and liver sections were fixed, cut and stained for light microscopic examination. Some animals were pretreated with 600 mg/kg bw i.p. diethyl maleate (DEM) 30 min before death.

RESULTS

AA did not show any mutagenic effect in S. typhimurium TA 98 or TA 100, without or in the presence of a metabolizing system (Table 1). On the other hand, N-OH-AA showed a weak, direct mutagenic activity.

When N-OH-AA was studied for DNA damage in cultured Reuber rat hepatoma cells, a slight, concentration dependent increase in alkaline elution of DNA was observed (Fig. 1). N-

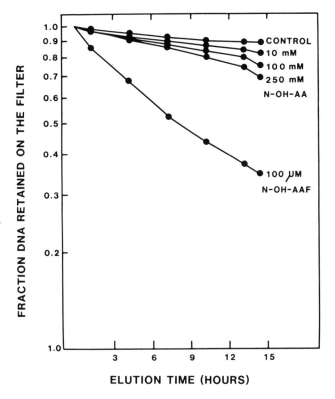

Fig. 1. DNA damage in Reuber rat hepatoma
cells after exposure to N-OH-AA

Hydroxy-2-acetylaminofluorene was used as a positive control.
N-OH-AA was also found to induce DNA repair synthesis (UDS) in
monolayer cultures of isolated hepatocytes. At the same concentrations, N-OH-AA was cytotoxic to the cells (LDH leakage).
These effects were not altered in cells from PCB-pretreated animals. AA did not cause cytotoxicity or DNA repair at concentrations up to 250 mM.

DEM pretreatment of the hepatocytes markedly enhanced cytotoxicity (Fig. 3), whereas UDS was reduced in DEM-pretreated
cells. On the other hand, N-OH-AA did not deplete cellular glutathione.

[14]C-AA did not bind covalently to rat liver microsomes fortified
with NADPH or to cytosol (Table 2). A very low or absent rate
of binding was seen with ^{14}C-N-OH-AA and microsomes plus NADPH,
or microsomes plus xanthine/xanthine oxidase, cytosol, cytosol
plus acetyl CoA or cytosol plus proline and ATP. Neither AA nor
N-OH-AA showed alkylating activity in the nitrobenzylpyridine
test (data not shown).

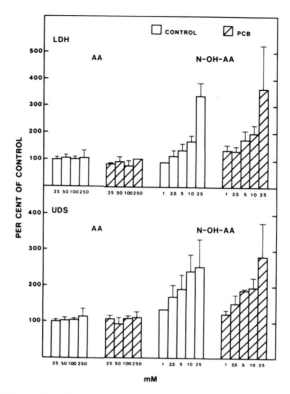

Fig. 2. Cytotoxicity and DNA repair synthesis in isolated hepatocytes
from control or PCB-pretreated
rats exposed to AA or N-OH-AA

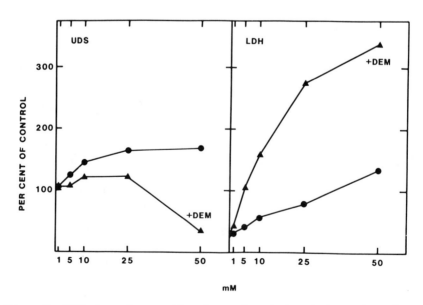

Fig. 3. DNA repair synthesis and cytotoxicity caused
by N-OH-AA in control or DEM-treated hepatocytes

Table 2.   Covlaent Binding of ^{14}C-AA-and ^{14}C-N-OH-AA to Subcellular Rat liver fractions.

CONDITIONS	COVALENT BINDING pmol/mg protein/min
AA + MICROSOMES + NADPH	0.0
AA + CYTOSOL	0.0
N-OH-AA + MICROSOMES + NADPH	2.8
N-OH-AA + MICROSOMES - NADPH	0.1
N-OH-AA + CYTOSOL	0.2
N-OH-AA + CYTOSOL + ACETYL CoA	0.2
N-OH-AA + CYTOSOL + PROLINE + ATP	0.0
N-OH-AA + MICROSOMES + XANTHINE + XANTHINE OXIDASE	1.9

Both ^{14}C-AA and ^{14}C-N-OH-AA were incorporated into macromolecules of isolated hepatocytes in monolayer culture (Table 3). However, this binding was blocked by the addition of cycloheximide to the same extent as was incorporation of ^{14}C-acetate.

When AA and N-OH-AA were tested up to cytotoxic concentrations in primary Syrian hamster embryo cells in culture, it was not possible to demonstrate any transforming activity (Table 4).

Administration of AA (100-400 mg/kg bw) or N-OH-AA (100-1000 mg/kg bw) to rats i.p. did not cause any elevation of SGOT-values (Table 5). Microscopically no signs of hepatocellular necrosis were evident. Pretreatment of animals with DEM did not result in any hepatotoxic effects of N-OH-AA.

DISCUSSION

N-OH-AA, a possible metabolite of AA, showed slight genotoxic effects in 3 different test systems. These effects may be the result of further metabolic conversion of N-OH-AA, since it did not show any direct alkylating activity. However, the genotoxicity of N-OH-AA could not be ascribed to the formation of an electrophile, since any significant covalent binding of ^{14}C-N-OH-AA under different conditions could not be found.

The cytotoxic effect of N-OH-AA in isolated hepatocytes was presumably also not due to the formation of an electrophile, even if DEM pretreatment enhanced cytotoxicity. This latter effect may simply be related to the increased lipid peroxidation known to occur at low glutathione levels in hepatocytes (Anundi et al., 1979).

In contrast to in vitro, N-OH-AA did not cause liver cell toxicity in vivo. This may be due to the rapid elimination of single doses of N-OH-AA in vivo (Puthca et al., 1984). AA itself did not appear to be necrogenic in long term feeding studies (Jackson and Desau, 1961; Flaks et al., 1983).

Table 3. Incorporation of Radioactivity in Monolayers of
Rat Hepatocytes

CONDITIONS	INCORPORATION nmol/mg protein/hr	PER CENT INHIBITION
^{14}C-AA	0.114	–
^{14}C-AA + CYCLOHEXIMIDE	0.018	84.3
^{14}C-N-OH-AA	0.654	–
^{14}C-N-OH-AA + CYCLOHEXIMIDE	0.114	82.6
^{14}C-ACETATE + CYCLOHEXIMIDE	–	85.3
^{14}C-LEUCINE + CYCLOHEXIMIDE	–	94.5

Table 4. Transformation Testing of AA and N-OH-AA in
Syrian Hamster Embryo Cells

TEST COMPOUND	UG PER ML	CLONING EFFICIENCY PER CENT	NUMBER OF COLONIES	NUMBER OF TRANSFORMED COLONIES	PER CENT TRANSFORMED COLONIES
AA	200	26	468	0	0
	1000	26	468	0	0
	5000	25	450	1	0.2
N-OH-AA	2	27	486	0	0
	10	28	504	0	0
	50	20	360	0	0
DMSO	–	24	432	1	0.2

Table 5. SGOT Levels in Rats after AA and
N-OH-AA Treatment

TREATMENT	DOSE mg/kg b.w.	SGOT units/ml
CONTROL	-	144 ± 20
AA	100	148 ± 6
AA	400	84 ± 11
N-OH-AA	100	98 ± 10
N-OH-AA	400	191 ± 48
CONTROL	-	258 ± 54
N-OH-AA	500	273 ± 32
N-OH-AA + DEM	500	221 ± 41
N-OH-AA	1000	232 ± 15
N-OH-AA + DEM	1000	218 ± 81

Values are means ± S.D. of 5 animals

Whether N-OH-AA plays a role in AA hepatocarcinogenesis remains to be established. Presently, initiation/promotion of rat liver by AA is being studied.

CONCLUSION

AA was not mutagenic in S. typhimurium or caused DNA damage in rat hepatoma cells or in rat hepatocytes. N-OH-AA showed a slight, direct mutagenic effect in S. typhimurium and elicited DNA damage in rat hepatoma cells and in isolated rat hepatocytes. Radiolabelled N-OH-AA or AA did not bind covalently to any significant degree to macromolecules in the presence of various activating systems. N-OH-AA caused a concentration-dependent cytotoxicity in isolated hepatocytes, this effect was enhanced in cells pretreated with DEM. However, N-OH-AA did not deplete cellular glutathione. Neither N-OH-AA nor AA caused liver cell damage in vivo. Neither AA nor N-OH-AA transformed primary Syrian hamster embryo cells in culture.

ACKNOWLEDGEMENT
Supported by Nato Grant RG. 112/84.

REFERENCES

Ames, B.N., McCann, J., and Yamasaki, E., 1975, Methods for detecting carcinogens and mutagens with the Salmonella/mammalian-microsome mutagenicity test, Mutation Res., 31:347.
Anundi, I., Högberg, J., and Stead, A.H., 1979, Glutathione depletion in isolated hepatocytes: its relation.to lipid peroxidation and cell damage, Acta Pharmacol. Toxicol., 45:45.
Fishbein, W.N., Daly, J., and Streeter, C.L., 1969, Preparation and some properties of stable and carbon-14 and tritium labelled short-chain aliphatic hydroxamic acids, Anal. Biochem., 28:13.

Flaks, B., Trevan, M.T., and A. Flaks, 1983, An electron microscope study of hepatocellular changes in the rat during chronic treatment with acetamide. Parenchyma, foci and neoplasms, <u>Carcinogenesis</u>, 4:1117.

Holme, J.A. and Søderlund, E.J., 1984, Unscheduled DNA synthesis of rat hepatocytes in monolayer culture, <u>Muta tion Res.</u>, 126:205.

Jackson, B. and Desau, F.I., 1961, Liver tumors in rats fed acetamide, <u>Lab. Invest.</u>, 10:909.

Kohn, K.W., Erickson, L.C., Ewig, R.A.G., and Qwelling, L.A., 1981, Measurement of strand breaks and cross-links by alkaline elution, <u>in</u>: "DNA Repair: A Laboratory Manual of Recent Procedures", E.C. Friedberg and P.C. Hanawalt, eds., Marcel Dekker, New York.

Pienta, R.J., Poiley, J.A., and Lebherz, W.B., 1977, Morphological transformation of early passage of golden Syrian hamster embryo cells dervied from cryopreserved primary cultures as a reliable in vitro bioassay for identifying diverse carcinogens, <u>Int. J. Cancer</u>, 19:642.

Putcha, L., Griffith, D.P., and Feldman, S., 1984, Disposition of ^{14}C-acetohydroxamic acid and ^{14}C-acetamide in the rat, <u>Drug Metab. Disp.</u>, 12:438.

Søderlund, E.J., Nelson, S.D., von Bahr, C., and Dybing, E., 1982, Species differences in kidney toxicity and metabolic activation of tris(2,3-dibromopropyl)phosphate, <u>Fund. Appl. Toxicol</u>, 2:187.

REACTIONS OF GLUTATHIONE WITH OXIDATIVE INTERMEDIATES OF ACETAMINOPHEN

David W. Potter and Jack A. Hinson

National Center for Toxicological Research
Jefferson, AR

INTRODUCTION

Acetaminophen (4'-hydroxyacetanilide) is an analgesic and antipyretic drug, which is reportedly safe at low doses, but when administered in large doses causes hepatic necrosis and renal damage in both humans and laboratory animals (1-8). Acetaminophen is thought to be metabolized to reactive intermediate(s) that cause toxic reactions and current thinking suggests that acetaminophen may be activated via a one-electron oxidation mechanisms to give N-acetyl-p-benzosemiquinone imine (9) or a two-electron oxidation to give N-acetyl-p-benzoquinone imine (10).

We have been interested in examining the in vitro reactions of both the one- and two-electron oxidation products of acetaminophen. In examining these reactions, it has been our hope to come to an understanding as to the reaction products formed via the different oxidative intermediates. In previous studies we found that horseradish peroxidase catalyzed acetaminophen to polymers via N-acetyl-p-benzosemiquinone imine intermediates (11). Additional studies indicated that when GSH was added to the horseradish peroxidase system, GSH-acetaminophen conjugates were formed, suggesting the possible involvement of N-acetyl-p-benzosemiquinone imine in the production of the GSH-acetaminophen conjugates (12).

In this study, we have investigated how GSH reacts with N-acetyl-p-benzosemiquinone imine. When GSH was added to reaction mixtures, polymerization decreased, GSSG formed, and low levels of GSH conjugation occurred. Without acetaminophen, horseradish peroxidase slowly oxidized GSH to GSSG; with acetaminophen, GSH oxidation was more rapid and GSSG formation was accompanied by decreased polymerization. HPLC analyses of horseradish peroxidase reaction mixtures without GSH demonstrated that N-acetyl-p-benzoquinone imine was a minor product. These results suggest that GSH reacts with N-acetyl-p-benzosemiquinone imine to form GSSG and acetaminophen and were consistent with N-acetyl-p-benzosemiquinone imine disproportionation to N-acetyl-p-benzoquinone imine which in turn conjugates with GSH.

MATERIALS AND METHODS

[phenyl-UL-^{14}C]Acetaminophen (1.65 mCi/mmol) was obtained from Dr.

Robert W. Roth of Midwest Research Institute, Kansas City, MO and was purified by thin-layer chromatography and HPLC to >99% (11). Horseradish peroxidase, EC number 1.11.1.7 (type VI), acetaminophen, 30% H_2O_2, GSH, and ascorbic acid were from Sigma Chemical Co., St. Louis, MO. Silver oxide (I) was purchased from Aldrich Chemical Co., Milwaukee, WI.

Synthesis of N-acetyl-p-benzoquinone imine. N-acetyl-p-benzoquinone imine was synthesized by a modification of the method of Dahlin and Nelson (13). Acetaminophen (100 mg) was oxidized with 100 mg silver oxide (I) in 5 ml freshly-redistilled chloroform containing 10 mg anhydrous sodium sulfate for 2 hr at room temperature. Activated charcoal (10 mg) was added to chloroform and the suspension was filtered. The chloroform layer was applied to a Waters silica Sep-Pak and the N-acetyl-p-benzoquinone imine was eluted with anhydrous diethyl ether. Dry dimethylsulfoxide (Me_2SO) (200 µl) was added to the ether, and the ether was evaporated under reduced pressure to give the N-acetyl-p-benzoquinone imine in Me_2SO. N-acetyl-p-benzoquinone imine was stored in liquid nitrogen and was stable for at least one month.

Reaction Procedure. Reaction mixtures of 1 ml contained 100 mM potassium phosphate, pH 7.4, and various amounts of GSH and acetaminophen. Specific details of the reaction mixtures contents are given in the legends to the figures and tables. Concentrations of GSSG were determined spectrophotometrically by the method of Sies and Summer (14). Reaction mixtures contained 0 or 1 mM acetaminophen, 1 mM GSH, 80 nM horseradish peroxidase, 100 µM H_2O_2 and 100 mM potassium phosphate, pH 7.4. NADPH followed by glutathione reductase were added to incubation mixtures 1 min after reactions were initiated with H_2O_2.

Liquid Chromatography. Acetaminophen polymers and GSH-acetaminophen conjugates were analyzed by reversed-phase HPLC (11,12). The analytical HPLC system consisted of two Model 6000 HPLC pumps, a Model 440 UV detector (254 nm) and a Model 660 microprocessor from Waters Associates, Inc., a Micromeritics Model 725 automatic injector, a Model 3390A Reporting Integration from Hewlett Packard, and 5μ C_{18} Ultrasphere ODS reversed-phase column (4.6 X 250 mm) from Altex. A binary solvent system with a flow rate of 1.0 ml/min was used for acetaminophen metabolite separation. Solvent A, which consisted of 87.9% water, 10% methanol, 2% acetic acid, and 0.1% ethyl acetate, was maintained at 100% for 10 min, followed by a 15 min linear gradient to give 81% A and 19% methanol.

N-acetyl-p-benzoquinone imine formation was determined by HPLC (10) using an isocratic solvent system which consisted of 20% methanol and 10 mM potassium phosphate, pH 7.4.

H_2O_2 and Horseradish Peroxidase Quantitation. Deionized glass-distilled water was used to prepare stock solutions of H_2O_2 and horseradish peroxidase. Concentrations were estimated by optical absorbance spectroscopy using the extinction coefficient of 43.6 M^{-1} cm^{-1} at 240 nm for H_2O_2 (15) and of 89.5 x 10^3 M^{-1} cm^{-1} at 403 nm for HRP (16).

RESULTS

Horseradish peroxidase and H_2O_2 have been previously shown to catalyze the polymerization of acetaminophen (11). These products, shown in Fig. 1, were isolated by HPLC and identified by a combination of 500-MHz ^{1}H NMR spectroscopy and mass spectrometry (11). As shown in Table 1 the acetaminophen dimer (A_2) was the major polymerization product. The amount of the other polymers formed was dependent on the initial acetaminophen concentration. In reaction mixtures containing 1 mM acetaminophen, the acetaminophen trimer (A_3) and the acetaminophen tetramer were

more readily formed than the N-acetaminophen dimer (N-A$_2$), N"-acetamino-
phen trimer (N"-A$_3$), or N-acetaminophen trimer (N-A$_3$). As the concentration of acetaminophen was increased the concentration of the two dimers
increased while the concentration of all other polymers decreased.

GSH decreased the horseradish peroxidase-mediated polymerization of
acetaminophen and minor amounts of GSH-acetaminophen conjugates were
formed (Fig. 2). The GSH-acetaminophen conjugates have been previously
identified as 3-(glutathion-S-yl)acetaminophen (GS-A) and 3-(glutathion-

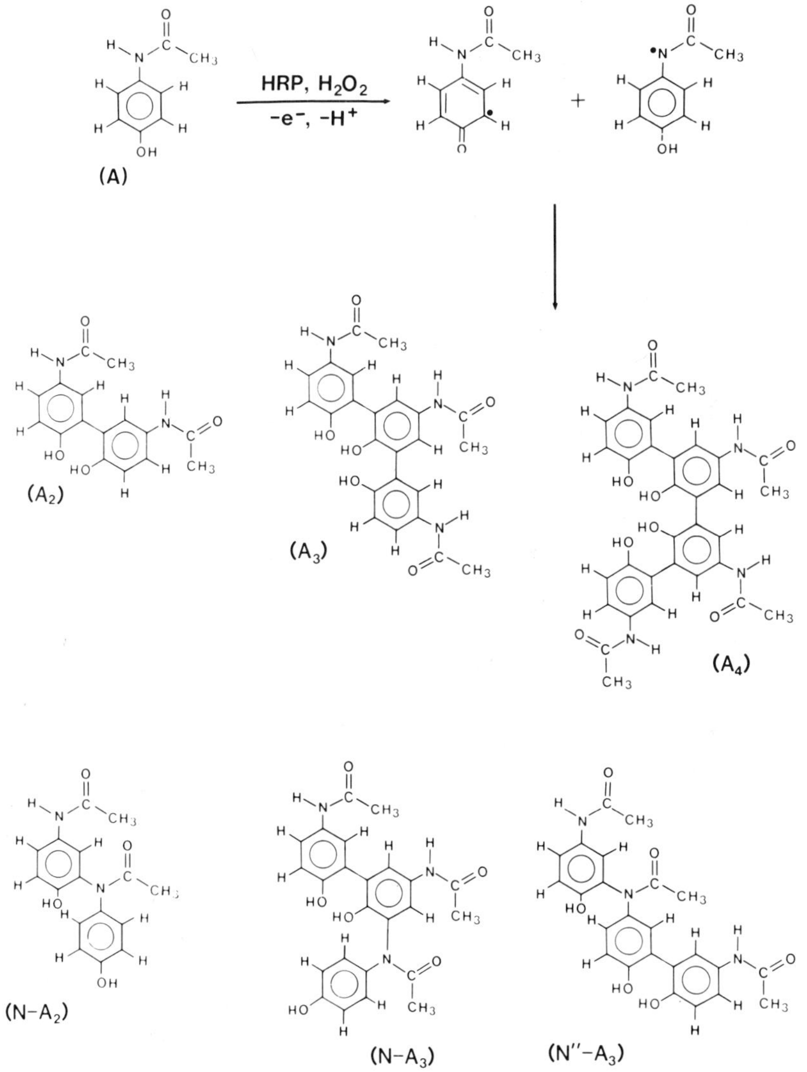

Fig. 1.   Schematic representation of acetaminophen polymerization
catalyzed by horseradish peroxidase (HRP) and H$_2$O$_2$. Polymers
were previously identified by 500-MHz NMR spectroscopy and by
mass spectrometry (11). The products were identified as
acetaminophen (A), acetaminophen dimer (A$_2$), acetaminophen
trimer (A$_3$), acetaminophen tetramer (A$_4$), N-acetaminophen dimer
(N-A$_2$), N"-acetaminophen trimer (N"-A$_3$), and N-acetaminophen
trimer (N-A$_3$).

Table 1. Effect of Acetaminophen Concentration on Horseradish
Peroxidase-Catalyzed Formation of Acetaminophen Polymers

Acetaminophen (mM)	Reaction Products[a]					
	A_2	A_3	$N-A_2$	$N''-A_2$	A_4	$N-A_3$
	μM					
1.0	76.2	37.6	4.7	1.8	6.0	3.2
2.0	76.7	24.4	6.7	1.1	3.4	5.2
4.0	87.2	17.6	9.1	ND[b]	1.5	6.3
8.0	108.2	9.9	13.1	ND	0.4	5.2
10.0	136.4	9.3	14.8	ND	0.4	5.1

[a]Reaction mixtures contained 1-10 mM acetaminophen, 76 μM horseradish
peroxidase, 200 μM H_2O_2, and 100 mM potassium phosphate, pH 7.4.
Samples were incubated for 1 min and reactions were terminated with 1 ml
of ice-cold methanol:water (90:10, V/V) containing 2 μmol ascorbate (to
quench reaction). Assay conditions are described in "Methods" section.

[b]Not detected. The limit of dectection was judged to be 0.1 μM.

S-yl)diacetaminophen (GS-A_2) (12). Formation of the two GSH-acetaminophen conjugates was optimal at approximately 40 μM GSH. Higher concentrations of GSH decreased GSH-acetaminophen conjugate formation by the
same proportion that acetaminophen polymer formation decreased. Thus, as
the concentration of GSH increased, both conjugation and polymerization
were decreased to approximately the same extent with the ratio of the two
types of products remaining constant.

The effect of acetaminophen concentration on horseradish peroxidase-
mediated catalysis of acetaminophen in the presence of 100 μM GSH was
examined in incubation mixtures similar to those described above. For
acetaminophen polymerization, the results were the same as those shown in
Table 1 except that the total amount of polymerization was less. The
amount of 3-(glutathion-S-yl)acetaminophen formation increased with an
increase in acetaminophen concentration (Table 2). In contrast, 3-
(glutathion-S-yl)diacetaminophen was maximal at 1 mM acetaminophen and
above 4 mM acetaminophen, a decrease in 3-(glutathion-S-yl)diacetamino-
phen formation was observed.

To determine if GSH decreased polymerization by competing with
acetaminophen to enzymatically reduce H_2O_2 to H_2O or if GSH reduced the
free radical of acetaminophen, we examined the oxidation of GSH to GSSG
in reaction mixtures containing horseradish peroxidase and H_2O_2 with or
without acetaminophen. Without acetaminophen, 18 nmol of GSSG was formed
in 1 ml incubation mixtures containing 1 μmol GSH, 80 pmol horseradish
peroxidase and 100 nmol H_2O_2. With 1 μmol acetaminophen, 60 nmol GSSG
was formed in similar reaction mixtures. This difference in GSH

oxidation suggests that GSH decreased at least 70% of the acetaminophen polymerization by reducing the acetaminophen free radical after it was formed by horseradish peroxidase.

As shown in Fig. 2, GSH had another effect which was to form conjugates with acetaminophen. Two mechanisms for GSH-acetaminophen conjugation were considered: 1) GSH reacted with the two-electron oxidation product of acetaminophen, N-acetyl-p-benzoquinone imine; or, 2) that the GSH radical reacted with the one-electron oxidation product of acetaminophen, N-acetyl-p-benzosemiquinone imine. The data presented in Table 3 strongly suggest that the first mechanism in which GSH reacted with N-acetyl-p-benzoquinone imine, was the principle pathway in the formation of 3-(glutathion-S-yl)acetaminophen. Table 3 shows the formation of acetaminophen polymers and GSH-acetaminophen conjugates at various incubation times after GSH was added to incubation mixtures. Acetaminophen polymerization increased most rapidly during the first min after incubation was initiated. At 1 min, polymerization was nearly maximal. Formation of 3-(glutathion-S-yl)acetaminophen was maximal when GSH was added 15 sec after initiation of incubation. As the incubation time was increased before addition of GSH, the formation of 3-(glutathion-S-yl)acetaminophen decreased which suggested that the intermediate reacting with GSH to form the GSH-acetaminophen conjugate was decreasing with time.

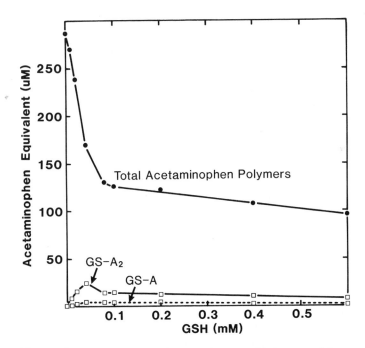

Fig. 2. Effect of GSH concentration on horseradish peroxidase-mediated catalysis of acetaminophen. Reaction mixtures contained 1 mM acetaminophen, 0-800 μM GSH, 80 nM horseradish peroxidase, 100 μM H_2O_2 and 100 mM potassium phosphate pH 7.4. Assay conditions are described in "Methods" section. The acetaminophen equivalent was determined for each polymer by multiplying polymer concentration by number of acetaminophen molecules per polymer.

Table 2. Effect of Acetaminophen Concentration on Horseradish Peroxidase-Catalyzed Formation of Acetaminophen Polymers in the
Presence of GSH

Acetaminophen (mM)	Reaction Products[a]							
	GS-A	GS-A_2	A_2	A_3	N-A_2	N''-A_3	A_4	N-A_3
				μM				
1.0	4.6	7.5	46.5	8.8	1.0	1.6	1.8	1.2
2.0	4.4	7.6	67.5	7.5	2.5	0.8	0.6	1.5
4.0	10.5	7.5	69.3	6.1	4.9	ND[b]	ND	2.1
8.0	10.8	5.7	74.8	3.2	7.9	ND	ND	2.0
10.0	13.3	5.1	78.4	3.1	9.4	ND	ND	1.7

[a]Reaction condition were the same as those described in Table 1 except
that incubation mixtures contained 100 μM GSH.

[b]Not detected. The limit of detection was judged to be 0.1 μM.

Since these data indicated that the intermediate in polymer formation
might be different than the intermediate in 3-(glutathion-S-yl)acetamino-
phen formation, the formation of N-acetyl-p-benzoquinone imine was
examined in similar incubation mixtures without addition of GSH (Table
3). Reaction products were analyzed by the HPLC system described for the
separation of N-acetyl-p-benzoquinone imine. In these incubation
mixtures, a product was formed that coeluted with N-acetyl-p-benzoquinone
imine but not with any of the other products. This product was collected
in vials containing either ascrobate or GSH and reanalyzed by the HPLC
system described for separation of acetaminophen polymers and conjugates.
The isolated product reacted with ascorbate to give acetaminophen and
reacted with GSH to give acetaminophen and 3-(glutathion-S-yl)acetamino-
phen. These results are consistent with the isolated product being N-
acetyl-p-benzoquinone imine.

The amount of N-acetyl-p-benzoquinone imine formed at various times
was nearly the same as the amount of 3-(glutathion-S-yl)acetaminophen
formed under similar conditions (Table 3). As with 3-(glutathion-S-
yl)acetaminophen formation, the concentration of N-acetyl-p-benzoquinone
imine was maximal at 15 sec and slowly decreased with longer incubation
times. Neither N-acetyl-p-benzoquinone imine nor 3-(glutathion-S-yl)-
acetaminophen were detected after a 5 min incubation.

DISCUSSION

Horseradish peroxidase catalyzes the oxidation of a number of
electron-donating hydroxyl- and amino-substituted aromatic compounds
while reducing H_2O_2 to water. Generally, these oxidations are thought to
proceed via a one-electron transfer mechanism (17-20); however, as has
been discussed by Yamazaki (20), it is difficult to determine whether a
substrate is oxidized by one- or two-electron mechanisms.

Table 3. Horseradish Peroxidase-Catalyzed Conversion of Acetaminophen to Products with or without Addition of GSH at Different Time Points

				Reaction Products					
Time (min)	NAPQI[a]	GS-A[b]	GS-A$_2$	A$_2$	A$_3$	N-A$_2$	N''-A$_3$	A$_4$	N-A$_3$
					μM				
0	–	6.2	3.8	43.2	7.9	3.5	ND[c]	ND	ND
0.25	31.0	25.8	9.5	54.7	20.2	6.5	2.5	4.2	3.7
0.5	20.2	19.8	8.7	57.3	23.5	12.6	3.0	12.1	5.0
1	13.6	14.2	4.0	58.0	24.4	10.0	3.8	12.0	9.5
2	6.0	6.0	ND	59.1	27.2	10.2	4.0	12.2	9.7
3	2.7	4.0	ND	59.0	27.4	9.5	4.1	13.0	9.9
5	ND	ND	ND	60.0	27.4	8.2	4.8	14.2	10.6

[a]NAPQI (N-acetyl-p-benzoquinone imine). Reactions were incubated without GSH for various lengths of time and N-acetyl-p-benzoquinone imine formation was determined by HPLC (10).

[b]Acetaminophen polymers and GSH-acetaminophen conjugates were formed in reaction mixtures contained 1 mM acetaminophen, 80 μM horseradish peroxidase, 220 μM H$_2$O$_2$ and 100 mM potassium phosphate, pH 7.4. These conditions were the same as those for determination of N-acetyl-p-benzoquinone imine except GSH (2 mM) was added at different time points (0-5 min) after incubation was initiated and a methanol:water solution (90:10) containing 2 μmol ascorbate was added 15 sec after GSH addition. Assay conditions are described in "Methods" section.

[c]Not detected. The limit of detection was judged to be 0.1 μM.

We have been interested in understanding the chemistry of the one- and two-electron oxidation products of acetaminophen. Recently, we reported that acetaminophen was polymerized by horseradish peroxidase to yield six acetaminophen polymers which is indicative of a free radical termination mechanism (11). This is consistent with ESR data of West et al. (21) who showed that N-acetyl-p-benzosemiquinone imine is formed by peroxidation of acetaminophen.

In this work, we have examined the mechanism by which GSH reacts with the one- and two-electron intermediates of acetaminophen. GSH rapidly reacted with N-acetyl-p-benzoquinone imine to give primarily 3-(gluta-thion-S-yl)acetaminophen and to give secondarily acetaminophen. In the horseradish peroxidase system, GSH decreased acetaminophen polymerization and the formation of minor amounts of GSH-acetaminophen conjugates was observed. GSH apparently decreased polymerization by reacting with N-acetyl-p-benzosemiquinone imine to form GSSG and acetaminophen since acetaminophen increased the rate of GSH oxidation to GSSG in the horse-

radish peroxidase systems. Other experiments indicated that the formation of 3-(glutathion-S-yl)acetaminophen was via conjugation of GSH with
the N-acetyl-p-benzoquinone imine rather than the N-acetyl-p-benzosemiquinone imine intermediate since similar formation of 3-(glutathion-S-
yl)acetaminophen and N-acetyl-p-benzoquinone imine were formed in
reaction mixtures with and without GSH, respectively (Table 3). The data
further suggest that N-acetyl-p-benzoquinone imine was not formed via a
direct horseradish peroxidase catalyzed two-electron oxidation of
acetaminophen but rather that acetaminophen was oxidized by a one-
electron transfer and that the resulting N-acetyl-p-benzosemiquinone
imine disproportionated to give acetaminophen and N-acetyl-p-benzoquinone
imine. If N-acetyl-p-benzoquinone imine was formed via a direct two-
electron oxidation, the formation of 3-(glutathion-S-yl)acetaminophen
should either remain the same or increase as the concentration of GSH was
increased; however, GSH–acetaminophen conjugation decreased in a manner
similar to the decrease in acetaminophen polymerization (Fig. 2).

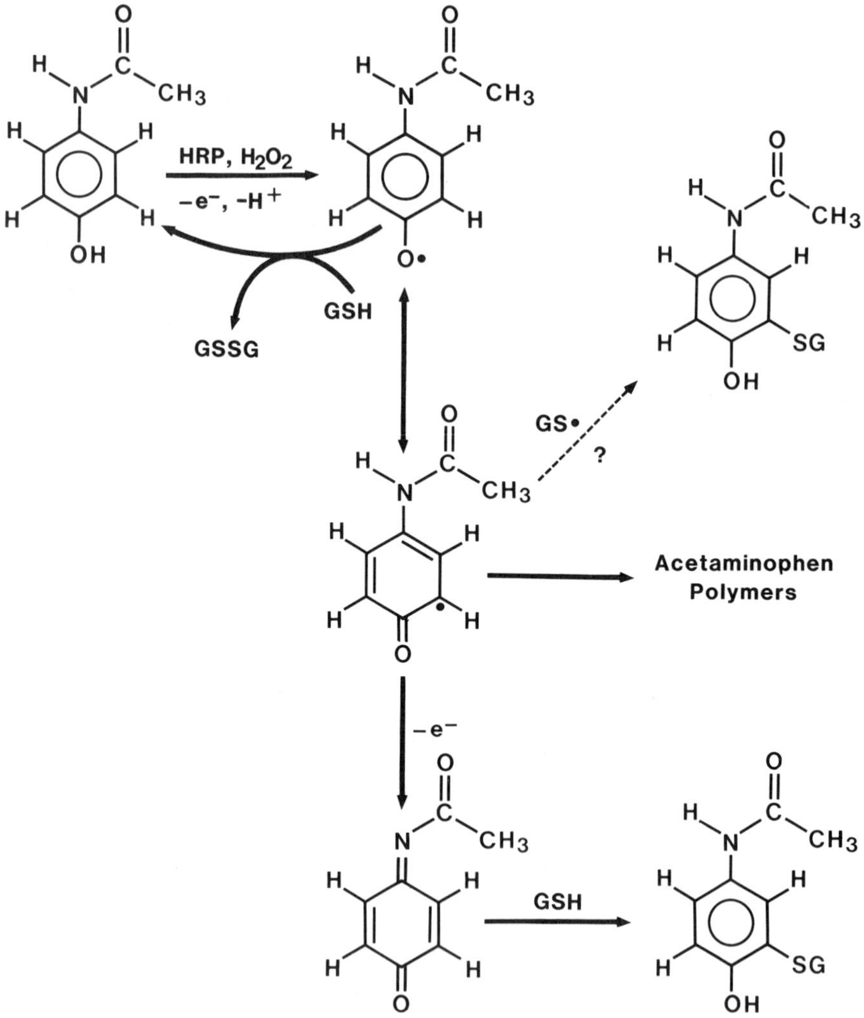

Fig. 3.   Schematic representation by horseradish peroxidase mediated
          oxidation of acetaminophen.

In current work, we are investigating the mechanism of 3-(glutathion-S-yl)diacetaminophen formation. At present two mechanisms are being considered. Either 3-(glutathion-S-yl)acetaminophen could be further oxidized to the corresponding semiquinone imine followed by reaction with N-acetyl-p-benzosemiquinone imine or the quinone imine of the acetaminophen dimer was formed and reacted with GSH.

A summary of the horseradish peroxidase reactions of acetaminophen is shown in Fig. 3. Horseradish peroxidase catalyzed the one-electron oxidation of acetaminophen. GSH can then reduce the N-acetyl-p-benzo-semiquinone intermediate back to acetaminophen with concomitant formation of GSSG. The phenyl- or the nitrogen-centered radicals are the likely intermediates that form polymers primarily through a covalent bond between carbons ortho to the hydroxyl group and, to a lesser extent, between the carbon ortho to the hydroxyl group and the amino group of another acetaminophen molecule. Alternatively, two molecules of N-acetyl-p-benzosemiquinone imine may disproportionate to N-acetyl-p-benzoquinone imine and acetaminophen. The evidence suggests that the disproportionation is the major pathway leading to GSH-acetaminophen conjugate formation. If GSH-acetaminophen conjugates are formed via a radical mechanism, the data suggest that it is a minor reaction.

REFERENCES

1. E. M. Boyd and G.M. Bereczky, Liver necrosis from paracetamol, Brit. J. Biochem. 26:606-607 (1966).
2. D. G. D. Davidson and W. N. Eastham, Acute liver necrosis following overdose of paracetamol, Brit. Med. J. 2:497-499 (1966).
3. L. F. Prescott, N. Wright, P. Roscoe, and S.S. Brown, Plasma paracetamol half-life and hepatic necrosis in patients with paracetamol overdose, Lancet 1:519-522 (1971).
4. J. R. Mitchell, D.J. Jollow, W. Z. Potter, J. R. Gillette, and B. B. Brodie, Acetaminophen-induced hepatic necrosis. IV. Protective role of glutathione, J. Pharmacol. Exp. Ther. 187:211-217 (1973).
5. R. Clark, R. P. H. Thompson, V. Borirakchanyavat, B. Widdop, A. R. Davidson, R. Goulding, and R. Williams, Hepatic damage and death from overdose of paracetamol, Lancet 1:66-70 (1973).
6. D. C. Davis, W. Z. Potter, D. J. Jollow, and J. R. Mitchell, Species differences in hepatic glutathione depletion, covalent binding and hepatic necrosis after acetaminophen, Life Sci., 14:2099-2109 (1974).
7. D. J. Jollow, J. R. Mitchell, W. Z. Potter, D. C. Davis, J. R. Gillette, and B.B. Brodie, Acetaminophen-induced hepatic necrosis. II. Role of covalent binding in vivo, J. Pharmacol. Exp. Ther., 187:195-202 (1973).
8. W. Z. Potter, S. S. Thorgeirsson, D. J. Jollow, and J. R. Mitchell, Acetaminophen induced hepatic necrosis. V. Correlation of hepatic necrosis, covalent binding and glutathione depletion in hamsters, Pharmacology 12:129-143 (1974).
9. J. De Vries, Hepatotoxic metabolic activation of paracetamol and its derivatives phenacetin and benorilate: oxygenation or electron transfer?, Biochem. Pharmacol. 30:399-402 (1984).
10. D. C. Dahlin, G. T. Miwa, A. Y. H. Lu, and S. D. Nelson, N-acetyl-p-benzoquinone imine: A cytochrome P-450 mediated oxidation product of acetaminophen, Proc. Natl. Acad. Sci. USA 81:1327-1331 (1984).
11. D. W. Potter, D. W. Miller, and J. A. Hinson, Identification of acetaminophen polymerization products catalyzed by horseradish peroxidase, J. Biol. Chem. in press (1985).

12. D. W. Potter, D. W. Miller, and J. A. Hinson, Horseradish peroxidase-catalyzed oxidation of acetaminophen to intermediates that form polymers or conjugate with glutathione, Submitted (1985).

13. D. C. Dahlin and S. D. Nelson, Synthesis, decomposition kinetics, and preliminary toxicological studies of pure N-acetyl-p-benzoquinone imine, a proposed toxic metabolite of acetaminophen, J. Med. Chem. 24:988-993 (1982).

14. H. Sies and K. H. Summer, Hydroperoxide-metabolizing systems in rat liver, Eur. J. Biochem. 57:503-512 (1975).

15. A. G. Hildebrandt and I. Roots, Reduced nicotinamide dinucleotide phosphate (NADPH)-dependent formation and breakdown of hydrogen peroxide during mixed function oxidation reactions in liver microsomes, Arch. Biochem. Biophys. 171:385-397 (1975).

16. A. C. Maehly, Plant peroxidase, Methods Enzymol. 2:801-813 (1955).

17. P. George, Chemical nature of the secondary hydrogen peroxide compound formed by cytochrome-C peroxidase and horseradish peroxidase, Nature (London) 169:612-613 (1952).

18. B. Chance, The kinetics and stoichiometry of the transition from the primary to the secondary peroxidase complex, Arch. Biochem. Biophys. 41:416-424 (1952).

19. I. Yamazaki, H. S. Mason, and L. Piette, Identification, by electron paramagnetic resonance spectroscopy, of free radicals generated from substrates of peroxidase, J. Biol. Chem. 235:2444-2449 (1960).

20. I. Yamazaki, One-electron and two-electron transfer mechanisms in enzymatic oxidation-reduction reactions, Advan. in Biophys. 2:33-76 (1971).

21. P. R. West, L. S. Harmon, P.D. Josephy, and R. P. Mason, Acetaminophen: enzymatic formation of a transient phenoxyl free radical, Biochem. Pharmacol. 33:2933-2936 (1984).

FLAVIN-CONTAINING MONOOXYGENASE ACTIVITY IN HUMAN LIVER MICROSOMES

M.E. McManus, D.J. Birkett, W.M. Burgess, I. Stupans,
J.A. Koenig, A.R. Boobis[1], D.S. Davies[1], P.J. Wirth[2],
P.H. Grantham[2], and S.S. Thorgeirsson[2]

Department of Clinical Pharmacology, Flinders University
Bedford Park 5042 South Australia; Royal Postgraduate
Medical School, London W12 OHS U.K.[1], and National Cancer
Institute, Bethesda, Maryland, 20205, U.S.A.[2]

INTRODUCTION

In laboratory animals both the flavin-containing and cytochrome P450 monooxygenase enzymes have been shown to be involved in the metabolism of nitrogen and sulfur containing compounds (1). Multiple forms of both these enzymes have now been reported and the composition of these monooxygenases is known to differ greatly between animal species, among individuals of the same species, and between organs in the same species (2,3). The differential distribution of these monooxygenases has been used to explain species, individual and organ selective toxicity of drugs and chemical carcinogens (4). Clearly, characterization of the monooxygenases involved in the metabolism of nitrogen and sulfur containing compounds, and definition of the regulatory control of these processes, is essential for understanding the biological disposition of these chemicals.

In man while the cytochrome P450 system has been extensively investigated, the flavin-containing monooxygenase (FCMO) has been relatively neglected (5,6). Attempts to study the FCMO have in certain respects been hampered by the fact that many of its substrates are also metabolised by the cytochrome P450 system. Recently we have shown the N-oxidation of guanethidine to be a useful probe for studying FCMO activity in rat hepatocytes and microsomes (7). In the present study we have therefore used guanethidine N-oxidation to study FCMO activity in human liver microsomes and as a measure of cytochrome P450 mediated N-oxygenation we have determined the capacity of the same microsomes to N-hydroxylate 2-acetylaminofluorene (9,10). In addition, since the FCMO has been implicated in the metabolic activation of aromatic amines to mutagens in laboratory animals (8), we have also investigated the involvement of the human FCMO enzyme in this process.

METHODS

Human Tissue Samples

Microsomal fractions of human liver were obtained either from wedge biopsy samples taken at laparotomy or from samples of liver of renal transplant donors maintained on life-support systems until the kidneys could be removed. The use of each tissue in these studies had local Research Ethics Committee permission and, where appropriate, Coroner's permission. Samples were stored at -80° until required, during which time there was no loss of activity.

Enzyme Assays

Guanethidine N-oxidation and 2-acetylaminofluorene metabolism were measured as previously described (7,11). Dimethylaniline N-oxidation and thiobenazmide S-oxidation were determined according to the methods of Ziegler & Pettit (12) and Cashman and Hanzlik (13), respectively. Cytochrome P450 content was measured as previously described (14).

Mutagenesis Assay

The mutagenesis assay was performed essentially according to the method of Ames et al. (15) except a preincubation step was included to prevent inactivation of the FCMO by the high temperature of the molten-top agar (16). A 0.5 ml incubation volume contained 0.25 mg microsomal protein, 1-2 X 10^{8} cells of bacteria tester strain Salmonella typhimurium 1538, 1.2 μmol NADPH and varying amounts of 2-aminofluorene and/or modulators dissolved in dimethyl sulfoxide (DMSO). The final concentration of DMSO in the incubation mixture was 2%. All solutions were filter sterilized through 0.2 μm Swimnex filter units (Millipore) prior to use except subcellular fractions which were filtered through a 0.45 μm filter. The concentrations of protein in the filtrates were determined after filtration to estimate losses during this process. After a preincubation of 20 min at 37°C 2.0 ml of molten-top agar at 45° was added and the mixture was poured onto petri dishes containing 30 ml of minimal-glucose agar containing a limited amount of L-histidine. The colonies on each plate (histidine-independent revertants) were scored after a 48 hr incubation at 37°.

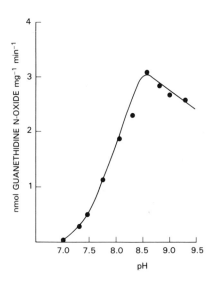

Figure 1. The pH dependency of guanethidine N-oxidation in human liver microsomes pooled from three subjects.

Results and Discussion

The rate of guanethidine N-oxidation by human liver microsomes was linear at both pH 7.4 and 8.5 with time to 30 min and with protein concentration up to 1 mg/ml of incubation mixture. As observed in other species (17) N-oxidase activity is optimal near pH 8.5 and at pH 7.4 guanethidine N-oxidation proceeds at only 16% of the maximal rate (Figure 1). Figure 2 shows the rate of guanethidine N-oxidation in human liver microsomes from 25 subjects determined at both pH 7.4 and 8.4. All samples exhibited a marked increase in the rate of guanethidine N-oxide formation at the higher pH and enzyme activity varied 17- and 11-fold at pH 7.4 and 8.4, respectively. However, the pH dependent increase in activity varied markedly between subjects. In the same microsomes the N-hydroxylation of 2-acetylaminofluorene, a cytochrome P450 dependent pathway, varied 69-fold and cytochrome P450 content varied 3.6-fold, with a mean of 0.37 ± 0.11 nmol per mg liver microsomal protein (mean $\pm$ SD). No correlation existed between guanethidine N-oxidation at pH 7.4 and 8.4 ($r = 0.41$) suggesting that different enzyme may be involved at the two pH values. Further, as expected no correlation existed between guanethidine N-oxidation at pH 7.4 and 2-acetylaminofluorene N-hydroxylation ($r = 0.08$), which strongly suggests that different enzymes are involved in these pathways.

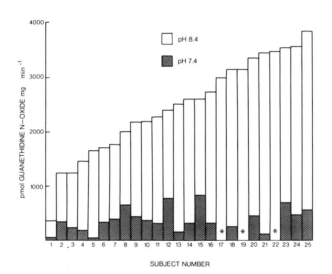

Figure 2. Distribution of guanethidine N-oxidase activity in human liver microsomes at pH 7.4 and 8.4. The asterisk indicates that these values were not determined.

Both dimethylaniline N-oxidation (12,17) and thiobenzamide S-oxidation (13,18) have been classically used to determine flavin-containing monooxygenase activity in liver microsomal preparations. Table 1 shows that except for subject 1 guanethidine and dimethylaniline N-oxidations and thiobenazmide S-oxidation are all increased in human liver microsomes as a function of pH. Interestingly, microsomes from subject 1 exhibited the lowest FCMO activity, but the highest cytochrome P450 content, and the lack of increased thiobenzamide S-oxidation at pH 8.4 may reflect the predominant involvement of cytochrome P450 in this reaction. Cytochrome P450 has been shown to contribute significantly to thiobenzamide S-oxidation in rat and mouse liver and lung microsomes (18).

Table 1. Guanethidine and dimethylaniline N-oxidations and thiobenzamide S-oxidation by human liver microsomes

Subject	Cytochrome P450 (nmole mg^{-1} protein)	Guanethidine N-oxidation (nmole mg^{-1} min^{-1})		Dimethylaniline N-oxide (nmole mg^{-1} min^{-1})		Thiobenzamide S-Oxide (nmole mg^{-1} min^{-1})	
		pH 7.4	pH 8.4	pH 7.4	pH 8.4	pH 7.4	pH 8.4
1	0.57	0.05	0.25	0.50	1.0	0.68	0.50
2	0.21	0.34	1.24	2.02	4.64	1.95	3.15
3	0.35	0.23	1.24	1.59	3.47	1.82	2.87
4	0.21	0.18	1.46	1.30	3.35	1.96	2.40
6	0.34	0.33	1.70	1.82	4.08	1.58	3.34

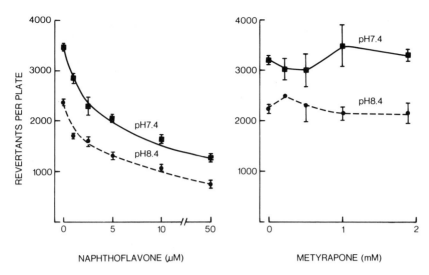

Figure 3. The effect of cytochrome P450 inhibitors -naphthoflavone and metyrapone on 2-aminofluorene mutagenesis mediated by human liver microsomes. Values represent the mean ± SD n = 3.

Methods used to differentiate between FCMO and cytochrome P450 monooxygenase activities have generally employed inhibitors of cytochrome P450 (19). Alpha-naphthoflavone significantly inhibited 2-aminofluorene mutagenesis mediated by human liver microsomes at both pH 7.4 and 8.4, whereas metyrapone had no effect (Fig. 3). These data suggest the involvement of selective forms of cytochrome P450 in the activation of 2-aminofluorene in human liver microsomes. The cytochrome P450 inhibitor SKF-525A (2-diethylaminoethyl-2,2-diphenyl valerate) also inhibited 2-aminofluorene mutagenesis in human liver microsomes (Table 2). Another indication that a reaction is FCMO mediated is stimulation by the primary amines n-octylamine and DPEA (2,4-dichloro-6-phenylphenoxyethylamine) (17). Both n-octylamine and DPEA in the present study significantly reduced 2-aminofluorene mutagenesis with human liver microsomes further suggesting cytochrome P450 is the predominant enzyme mediating this reaction (Table 2). However, the lack of complete inhibition of 2-acetylaminofluorene mutagenesis by α-naphthoflavone is consistent with the observations of Frederick et al (8) that the FCMO plays a minor role in the metabolic activation of this aromatic amide.

The relative contribution of the flavin containing and cytochrome P450 monooxygenases in the metabolic processing of nitrogen and sulfur containing chemicals is as yet to be clearly defined. From the present study the N-oxidation of the tertiary amine guanethidine appears to be a good probe for FCMO activity in human liver microsomes. However, clear delineation of the contribution of the FCMO v cytochrome P450 in nitrogen and sulfur metabolism will require development of specific inhibitors for each enzyme system. To this effect we have purified the human liver NADPH-cytochrome P450 reductase (mwt 72,000, specific activity 41 μmol cytochrome c reduced mg^{-1} min^{-1}) and anticipate to be able to inhibit cytochrome P450 activity with an antibody to this enzyme. Studies are also in progress in our laboratory to purify the human flavin containing monooxygenase.

Table 2. The effect of 2-diethylaminoethyl-2,2-diphenyl valerate
(SKF-525A), n-octylamine and 2,4-dichloro-6-phenyl-
phenoxyethylamine (DPEA) on 2-aminofluorene mutagenesis
mediated by human liver microsomes.

Treatment	Concentration	pH	Number of revertants per plate
Control		7.4	4401
SKF-525A	30 μM	"	3115
	60 μM	"	2136
n-octylamine	1 mM	"	1639
	2 mM	"	774
DPEA	30 μM	"	2452
	60 μM	"	2763
Control		8.4	2754
SKF-525A	30 μM	"	1360
	60 μM	"	748
n-octylamine	1 mM	"	1447
	2 mM	"	894
DPEA	30 μM	"	1858
	60 μM	"	1700

References

1.   D.M. Ziegler, Metabolic oxygenation of organic nitrogen and sulfur
      compounds in; "Drug Metabolism and Drug Toxicity", J.R. Mitchell
      and M.G. Horning ed. Raven Press, New York, (1984).

2.   D.E. Williams, D.M. Ziegler, D.J. Nordin, S.E. Hale and B.S.S.
      Masters. Biochem. Biophys. Res. Commun, 125, 116 (1984).

3.   E.F. Johnson, Rev. Biochem. Pharmacol., 1, 1 (1979).

4.   S.S. Thorgeirsson and D.W. Nebert. Adv. Cancer Res. 25, 149
      (1977).

5.   A.R. Boobis and D.S. Davies. Xenobiotica 14, 151 (1984).

6.   M.S. Gold and D.M. Ziegler. Xenobiotica 3, 179 (1973).

7.   M.E. McManus, P.H. Grantham, J.L. Cone, P.P. Roller, P.J. Wirth
      and S.S. Thorgeirsson. Biochem. Biophys. Res. Commun. 112, 437
      (1983).

8.   C.B. Frederick, J.B. Mays, D.M. Ziegler, F.P. Guengerich and F.F.
      Kadlubar. Cancer Res., 42, 2671 (1982).

9.   M.E. McManus, R.F. Minchin, N. Sanderson, D. Schwartz, E.F.
     Johnson and S.S. Thorgeirsson, Carcinogenesis 5, 1717 (1984).

10. S.S. Thorgeirsson, I.B. Glowinski and M.E. McManus. Rev. Biochem.
     Toxicol. 5, 349 (1983).

11. M.E. McManus, A.R. Boobis, R.F. Minchin, D.M. Schwartz, S. Murray,
     D.S. Davies and S.S. Thorgeirsson. Cancer Res. 44, 5692 (1984).

12. D.M. Ziegler and F.H. Pettit. Biochem. Biophys. Res. Commun., 15,
     188 (1964).

13. J.R. Cashman and R.P. Hanzlik. Biochem. Biophys. Res. Commun.,
     98, 147 (1981).

14. M.E. McManus, A.R. Boobis, G.M. Pacifici, R.Y. Frempong, M.J.
     Brodie, G.C. Kahn, C. Whyte and D.S. Davies. Life Sci. 26, 481
     (1980).

15. B.N. Ames, J. McCann and E. Yamasaki. Mutation Res. 31, 347
     (1975).

16. L.L. Poulsen, R.M. Hyslop and D.M. Ziegler. Arch. Biochem.
     Biophys., 198, 78 (1979).

17. D.M. Ziegler. Microsomal flavin-containing monooxygenase:
     Oxygenation of nucleophilic nitrogen and sulfur compounds, in:
     "Enzymatic basis of detoxication", W.B. Jakoby ed, Academic Press,
     New York (1980).

18. R.E. Tynes and E. Hodgson. Biochem. Pharmacol., 32, 3419 (1983).

19. H. Ulhleke. Xenobiotica 1, 327 (1971).

S9-DEPENDENT ACTIVATION OF 1-NITROPYRENE AND 3-NITROFLUORANTHENE IN

BACTERIAL MUTAGENICITY ASSAYS

L.M. Ball, K. Williams[*], M.J. Kohan[*] and J. Lewtas[*]

Department of Environmental Sciences and Engineering
University of North Carolina, Chapel Hill, NC 27514
[*]Genetic Bioassay Branch, Health Effects Research
Laboratory, U.S. Environmental Protection Agency
Research Triangle Park, NC 27711, U.S.A.

INTRODUCTION

Nitro-substituted polycyclic aromatic hydrocarbons (NO_2PAH) such as 1-nitropyrene (NP) and 3-nitrofluoranthene (NFA; see Fig. 1) have been detected in diesel and other combustion emissions (Nishioka et al., 1982; Gibson, 1982; Schuetzle et al., 1982). Many of these compounds are potent mutagens, and some are animal carcinogens (Hirose et al., 1984; Ohgahki et al., 1982). The mutagenicity of NO_2PAH in the Ames Salmonella typhimurium plate incorporation assay is generally dependent on nitroreductase activity in the bacterial tester strains utilised (Mermelstein et al., 1981). This was shown to be the case for NP (Mermelstein et al., 1981; Ball et al., 1984a). When the bacterial tester strain was deficient in the required nitroreductase enzyme, metabolism by mammalian S9 or microsomal fractions was also able to activate this compound, although to a lesser extent (Kohan and Claxton, 1983; Ball et al, 1984a). Oxidative metabolism of NP both in vivo (Ball et al., 1984b) and in vitro (El-Bayoumy and Hecht, 1983; Ball et al., 1984a) was shown to form phenols and dihydrodiols some of which were themselves mutagenic to Salmonella. Binding to DNA, also an indication of potential for genotoxic damage, was catalysed both by reductive (Messier et al., 1981; Howard et al., 1983) and by NADPH-dependent oxidative metabolism of NP (Ball and Lewtas, 1985). Therefore both oxidative and reduc - tive pathways apppear to be capable of producing genotoxic intermediates from NP; further investigations will show whether this holds true for other

1-NITROPYRENE   3-NITROFLUORANTHENE

Fig. 1.   Structures of 1-nitropyrene and 3-nitrofluoranthene

NO_2PAH, and what factors determine the relative importance of each route of metabolism. Thus we describe here a comparison of the activation pathways of NP and NFA (a NO_2PAH which has not yet been as well characterised as NP) in the Ames assay for reversion to histidine independence in <u>Salmonella</u> strain TA98 and in a forward mutation to 8-azaguanine resistance in <u>Salmonella</u> strain TM677 (Skopek et al., 1978, modified by Goto et al., 1985).

MATERIALS AND METHODS

<u>Chemicals</u>

1-Nitro[4,5,9,10-^{14}C]pyrene (^{14}C-NP; 10 mCi/mmole), unlabelled NP and NFA, and 3-nitro[3-^{14}C]fluoranthene (^{14}C-NFA; 13.8 mCi/mmole) were synthesised by Midwest Research Institute, Kansas City, MO. 3-Aminofluoranthene (AFA) was purchased from Aldrich Chemical Co., Milwaukee, WI, purified by recrystallisation, and acetylated to yield 3-acetylaminofluoranthene (AAFA). Other standards and reference compounds were obtained as described previously (Ball et al., 1984a, 1984b). Routine chemicals and solvents (reagent grade or better) were purchased from commercial suppliers, principally Sigma Chemical Co., St Louis, MO, and Burdick and Jackson Inc., Muskegon, MI.

<u>Mutagenicity Assays</u>

The <u>Salmonella typhimurium</u> plate-incorporation assay was carried out in strains TA98 (obtained from Dr. B. Ames, University of California, Berkeley, CA, USA), TA98NR and TA98/1,8-DNP6 (from Dr. H.S. Rosenkranz, Case Western Reserve University, Cleveland, OH, USA) according to the procedures described by Ames et al. (1975), with the following modifications: minimal histidine was added to the base agar rather than to the soft agar overlay, and plates were counted at 72 rather than 48 hr. The forward mutation to 8-azaguanine resistance in <u>Salmonella typhimurium</u> strain TM677 (grown from a clone selected for maximum activity towards benzo(a)pyrene from a culture supplied by Dr. W.G. Thilly, Massachusetts Institute of Technology, Cambridge, MA) was performed scaled down ten-fold from the method of Skopek et al., 1978, as described by Goto et al., 1985. Chemicals were dissolved in DMSO for assay.

<u>S9 Metabolic Activation System</u>

The S9 fraction (9000 x g supernatant) used to provide exogenous metabolic activation was obtained from the livers of Aroclor-1254-treated male CD-1 rats (Charles River, Wilmington, MA, USA) as described by Ames et al., 1975, and stored frozen at -70°C until used. The NADPH-generating co-factor mix added to the S9 was made up as described by Ames et al., 1975.

<u>Measurement of 1-Aminopyrene Formation</u>

NP (20 μg/ml) was incubated with <u>Salmonella typhimurium</u> strain TA98, with S9 alone (final vol. 1 ml), and with TM677 in the presence and absence of S9 (final vol. 0.1 ml), under the conditions and concentrations utilised for the respective forward and reverse mutation assays (conducted under air, with no attempt at generating an anaerobic atmosphere). After 1 hr at 37°, an equal vol. of MeOH was added and mixed by vortexing. The incubation mixture was centrifuged (2000 rpm, 10 min) to precipitate protein, then the supernatant was sampled (20 μl) for high pressure liquid chromatography (HPLC) analysis. (Zorbax-ODS 4.6 x 250 mm column, Dupont Instrument Co., Wilmington, DE, eluted at 0.8 ml/min with a gradient of 80 to 100% MeOH in H_2O over 15 min). The 1-aminopyrene (AP) content of the supernatant was determined by fluorescence spectroscopy at 360 nm excitation, 430 nm emission (MPF-2A fluorometer, Perkin-Elmer Corp., Norwalk, CT.) The area of the peak

eluting with the retention time of authentic AP was measured electronically
(Vista 4000 Data System, Varian Instrument Co., Palo Alto, CA) and calibrated
by comparison with a standard curve constructed by injection of authentic AP.
NP was quantitated analogously, by measurement of its UV absorbance at 280 nm.

Incubation of 3-Nitro[3-^{14}C]fluoranthene with S9

^{14}C-NFA (50 μM) dissolved in DMSO was added to Aroclor-treated rat liver
S9 fraction (1 mg/ml in 0.1M HEPES buffer, pH 7.4) prepared as described
(Ames et al., 1975). The reaction was started by addition of NADPH (1mM)
and terminated (after 1 hr at 37°C) by extraction as described below.

Analysis and Quantitation of 3-Nitrofluoranthene Metabolites

The incubation mixture was extracted with ethyl acetate/acetone (2:1
v/v; 3 x 1 vol). The organic extract was concentrated by rotary evaporation,
taken to dryness under a stream of N_2 gas, redissolved in MeOH (0.5 ml), then
injected into an HPLC system consisting of a Varian Model 5050 HPLC (Varian
Instrument Co., Palo Alto, CA) equipped with a Zorbax-ODS 4.6 x 250 mm column
(Dupont Instrument Co., Wilmington, DE). The column was eluted at 1 ml/min
with a linear gradient of 70% MeOH in H_2O rising to 100% MeOH over 20 min.
The eluate was monitored for UV absorbance at 254 nm (UV-100 detector,
Varian Instrument Co., Palo Alto, CA) and for fluorescence at 360 nm excita-
tion, 430 nm emission (MPF-2A fluorometer, Perkin-Elmer Corp., Norwalk, CT).
The eluate was collected in 30-sec fractions (Aliquogel B-200, Gilson Medical
Electronics, Middleton, WI); the ^{14}C content of the fractions was determined
by liquid scintillation counting (Model 2660 liquid scintillation counter,
Packard Instrument Co., Downers Grove, IL).

RESULTS

The mutagenicity of NFA was characterized in the Ames histidine
reversion assay and in the forward mutation assay, and compared to that of
NP in the same two systems. Results are shown in Table 1. (Data on the
activity of NP in the Ames assay are from Ball et al., 1984a). Since NP
and NFA are isomers (M. Wt. 247), comparisons can equally well be made on
the basis of direct weight or of molecular weight. Both compounds exhibited
the same general pattern of responses; both were more active in the absence
of S9 in the reversion assay, but required the presence of S9 to induce significant mutagenicity in the forward assay. Within each assay, NFA gave
maximal responses at doses approximately one-tenth of those required for NP.
Since mutagenicity of NFA in the Ames assay thus appeared to follow the
nitroreductase-requiring pattern previously observed for NP, the activity
of its reduced derivatives and potential metabolites was also examined in
this system (Table 2). Direct-acting mutagenicity was lower with the amino
derivative of NFA than with the parent compound, and was quite abolished by
N-acetylation; a similar pattern had previously been shown for NP (Ball et
al., 1984a). The S9-dependent activity of NP was also decreased by reduction, but further acetylation then increased activity beyond that of the
parent compound; 6-hydroxy-N-acetylaminopyrene (a metabolite formed in vivo)
was found to have ten times the S9-dependent activity of NP (Ball et al.,
1984b) and may therefore be implicated in activation of NP in the whole
animal. In contrast, both 3-aminofluoranthene and N-acetyl-3-aminofluoran-
thene were less active with S9 than was the parent compound; the activity
of any further ring-hydroxylated derivatives remains to be ascertained.

The results obtained in the forward mutation assay with Salmonella typhimurium strain TM677 (activity of NP and NFA almost entirely S9-dependent)
were in contrast to the general pattern seen in other Salmonella strains

Table 1. Comparison of the Mutagenicity of 1-Nitropyrene and
3-Nitrofluoranthene in Forward and Reverse Bacterial
Mutagenicity Assays

Compound	Histidine[+] Reversion (Ames et al., 1975) in Salmonella Typhimurium strain TA98[a]			8-Azaguanine Resistance (Skopek et al., 1978; Goto et al., 1985) in Salmonella TM677[b]		
	Dose	− S9	+ S9	Dose	− S9	+ S9
Vehicle (DMSO)		20–30	40–50		1–2	2–4
1-Nitropyrene	0.1	152	71[c]	12.5	0.9	32
	0.5	771	144	50	1.2	54
	1.0	1475	246	100	2.0	40
3-Nitrofluoranthene	0.1	1670	118	1.25	0.8	6
	0.5	3265	460	2.5	1.6	41
	1.0	3055	1067	5	1.6	88
				10	3.8	94

[a] Dose is given as μg/plate, and results are expressed as His[+] revertants per plate.
[b] Dose is given as μg/ml, and results are expressed as 8-Azaguanine resistant mutants per 10^5 survivors.
[c] Data on NP in the Ames assay from Ball et al., 1984a.

in the Ames assay, where the presence of S9 reduces the activity of these NO_2PAH. Examination of the formation of AP from NP in strain TM677 under the exact conditions of the 1 hr incubation involved in the forward mutation assay indicated that this strain did not appear to metabolise NP at all, and only when S9 was included in the incubation mixture was AP formed in detectable amounts (Table 3). In contrast, both Salmonella TA98 and S9 alone (incubated under air, in concentrations corresponding to those used for the Ames assay) consumed at least 20 and 50% of the NP substrate respectively and produced measurable quantities of AP in 1 hr.

Table 2. Comparison of the Mutagenicity of Reduced Metabolites
and Derivatives of 1-Nitropyrene and 3-Nitrofluoranthene in the Ames Histidine[+] Reversion Assay

Substituent (R)	1-R-Pyrene[a]		3-R-Fluoranthene	
	− S9	+ S9	− S9	+ S9
Nitro-	339[b]	60	2420	222
Amino-	93	33	43	113
N-Acetylamino-	≈ 0	414	≈ 0	61
6-Hydroxy-N-acetylamino	≈ 0	600	?	?

[a] Data from Ball et al., 1984a, 1984b.
[b] Mutagenicity values are given as revertants per nmole, calculated from the linear portion of the dose-response curve.

Table 3. Metabolism of 1-Nitropyrene to 1-Aminopyrene by <u>Salmonella</u>
Strains TA98 and TM677 and by Aroclor-treated rat liver S9

Strain/Tissue	Percent Disappearance of substrate (1-Nitropyrene)	Formation of 1-Aminopyrene (pmoles/ml)
Salmonella TM677	< 5[b]	< 9[b]
Salmonella TM677 + S9	≈ 50	200
Salmonella TA98	20-30	51
Aroclor-treated Rat Liver S9	50-80	59

[a] Incubations were conducted under air for 1 hr at 37°C. Concentrations of organisms and/or tissue were as used in the corresponding mutagenicity assay. The substrate was 20 µg/ml of
1-nitropyrene (80 µM) throughout, corresponding to dose levels
which produced a definite positive response in the forward mutation assay, but little signs of toxicity. The reaction mixture
was quenched and extracted with an equal volume of MeOH, centrifuged
to precipitate protein, then analysed by HPLC. 1-Aminopyrene was
detected and quantitated by fluorescence spectrometry (Ex. 360 nm,
Em. 430 nm), 1-nitropyrene by its UV absorbance at 280 nm.
[b] These values represent the limits of detection of the assay.

Rat liver S9 has been shown to metabolise NP to phenols, a dihydrodiol,
AP and N-acetyl-1-aminopyrene (El-Bayoumy and Hecht, 1983; Ball et al.,
1984a). ^{14}C-NFA was incubated with Aroclor-treated rat liver S9 in order
to investigate the nature and extent of the metabolic transformations undergone by this NO_2PAH. After 1 hr at 37°C, 70-80% of the ^{14}C was extractable
into ethyl acetate/acetone. Figure 2 shows the profile obtained by HPLC
examination of the organic-extractable metabolites formed. Over 75% of the
chromatographed ^{14}C eluted as unchanged NFA. Less than 1% of the ^{14}C eluted
with the retention times of reference 3-aminofluoranthene and 3-acetylamino-
fluoranthene. This contrasts with previous findings for NP, which indicated
that reduction and acetylation represented 17 % of the metabolism (over 5% of
total ^{14}C) under air (Ball et al., 1984a). The predominant metabolite fraction (R_T 17-18 min) represented ≈ 15% of the organic-extractable ^{14}C. Other
minor peaks amounted to 0.5-2% of the extractable radioactivity. Thus,
including non-extractable material, a total of 40-45% of the NFA underwent
transformation, which would correspond to ≈ 20 nmoles of substrate consumed
per mg of S9 protein in 1 hr.

The peak eluting between 17 and 18 min was collected for further
analysis. On rechromatographing this material resolved into one major and
several minor peaks. The UV spectrum of the principal component was shifted
to longer wavelengths in alkaline solution, indicative of a phenol; the
250 MHz proton n.m.r. spectrum showed the presence of a single component
with 8 aromatic protons, with resonances shifted upfield from the corresponding peaks in the n.m.r. spectrum of NFA itself (results not shown). Assignment of the individual resonance peaks (confirmed by examination of the
coupling constants and comparison with the spectrum of NFA) indicated the
presence of a hydroxyl group, and excluded all positions of substitution
other than carbons 8 or 9. Consideration of the reactivity of fluoranthene
substituted on carbon 3 with a <u>meta</u>-directing group such as NO_2 towards

electrophilic substitution reactions (Campbell and Keir, 1955) would suggest that the latter position (carbon 9) would be more favoured. We therefore conclude that the principal metabolite fraction formed from NFA by Aroclor-treated rat liver S9 is 9-hydroxy-NFA.

The other minor metabolites eluting between 5 and 10 min may represent dihydrodiols, since on treatment with 1M HCl their retention times shift to 15-20 min, but this identification is not definitive.

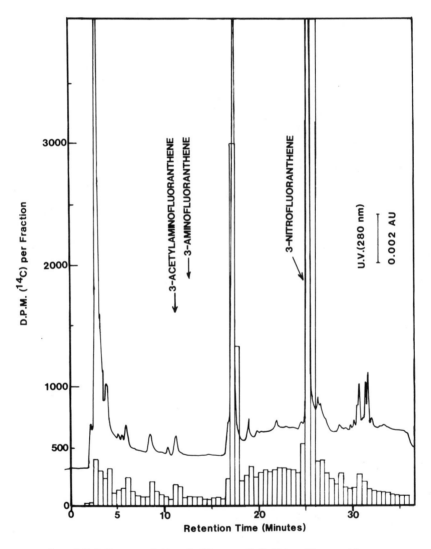

Fig. 2. HPLC Trace of Metabolites of 3-Nitrofluoranthene formed by Aroclor-treated rat liver S9 fraction. 3-Nitro[14C]fluoranthene was incubated with 1 ml/ml protein and 1 mM NADPH for 1 hr at 37°C. Metabolites were extracted with ethyl acetate/acetone, concentrated then analysed by reverse phase HPLC and liquid scintillation counting.

DISCUSSION

In the Ames Assay NP and NFA both exhibit their maximal response in the absence of S9. The direct product of reduction (i.e. the amino derivative) is less mutagenic than the parent compound, confirming earlier conclusions that the active genotoxic intermediates involved are formed predominantly during the process of reduction of the nitro group. Unlike NP, acetylation of the aminofluoranthene does not lead to enhanced S9-dependent mutagenic activity. This suggests that activation in mammalian tissues through formation of the acetamido derivative, and generation of an electrophilic arylnitrenium ion by further metabolism along the pathways defined for carcinogenic aromatic amines such as 2-aminofluorene (Miller and Miller, 1981), may not be a mechanism applicable to all NO_2PAH.

In the forward mutation assay the presence of S9 is required for both NP and NFA to increase the frequency of mutation above control levels. This is in sharp contrast to the situation in the standard Ames assay, where the presence of S9 generally results in a decrease in the level of activity of most NO_2PAH. If nitroreductase activity (as measured by AP formation) is indeed required for generation of active intermediates from NO_2PAH, low reductase activity in Salmonella TM677 without S9, and enhanced activity in this strain in the presence of S9, might explain the requirement for S9 for expression of the mutagenicity of NP and NFA in the forward mutation assay.

The absence of detecteable levels of AP formation in TM677 suggests that this strain may lack the "classical" bacterial oxygen-insensitive nitroreductase which is the primary enzyme involved in activation of NP (Mermelstein et al., 1981). The Salmonella strain from which TM677 is derived is TA1535, which is fully competent with respect to nitroreduction (Messier et al., 1981). A substantial loss of enzymic activity would therefore have to have taken place in the process of formulating the new strain. If indeed TM677 is deficient in oxygen-insensitive nitroreductase, then the presence of active S9 fraction in the incubation mix might in itself act as an oxygen scavenger, to allow expression of a bacterial oxygen-sensitive nitroreductase. This possibility is supported by the fact that AP production is higher in TM677 + S9 than in S9 alone. Alternatively, oxidative metabolism by rat liver S9 fraction can form products which themselves exhibit mutagenic activity. This was shown to be the case for NP (El-Bayoumy and Hecht, 1983; Ball et al., 1984a); the genotoxic characteristics of the principal metabolite of NFA formed by rat liver S9, here identified as 9-hydroxy-NFA, remain to be evaluated in this respect.

Therefore both enhanced nitroreduction in the presence of S9 and S9-specific oxidative metabolism may contribute to the activation of the NO_2PAHs NP and NFA in the forward mutation assay with Salmonella typhimurium strain TM677.

ACKNOWLEDGMENTS

This work was partly supported by EPA #CR-811817-01-0. This paper has been reviewed by the Health Effects Research Laboratory, U.S. Environmental Protection Agency, and approved for publication. Approval does not signify that the contents reflect Agency views and policies, nor does mention of trade names constitute endorsement or recommendation for use.

REFERENCES

Ames, B. N., McCann, J., and Yamasaki, E., 1975, Methods for detecting

carcinogens and mutagens with the Salmonella/mammalian microsome
mutagenicity test, Mutat. Res., 31:347-363.

Ball, L. M. and Lewtas, J, 1985, Rat liver subcellular fractions catalyse
aerobic binding of 1-nitro[^{14}C]pyrene to DNA, Environ. Health Perspect., 62, in press.

Ball, L. M., Kohan, M. J., Claxton, L. D. and Lewtas, J., 1984a, Mutagenicity of Derivatives and Metabolites of 1-Nitropyrene: Activation
by Rat Liver S9 and Bacterial Enzymes, Mutat. Res., 138:113-125.

Ball, L. M., Kohan, M. J., Inmon, J., Claxton, L. D. and Lewtas, J., 1984b,
Metabolism of 1-Nitro[^{14}C]pyrene in vivo in the rat and mutagenicity
of urinary metabolites, Carcinogenesis, 5:1557-1564.

Campbell, N. and Keir, N. H., 1955, The orientation of disubstituted fluoranthene derivatives, J. Chem. Soc., 1233-1237.

El-Bayoumy, K. and Hecht, S. S., 1983, Identification and mutagenicity of
metabolites of 1-nitropyrene formed by rat liver, Cancer Res., 43:
3132-3137.

Gibson, T. L., 1982, Nitroderivatives of polynuclear aromatic hydrocarbons in
airborne and source particulate matter, Atmos. Environ. 16:2037-2040.

Goto, S., Williams, K. and Lewtas, J., (1985), Further development and application of a micro-forward mutation assay in Salmonella Typhimurium
TM677, manuscript in preparation.

Hirose, M., Lee, M. S., Wang, C. Y., and King, C. M., 1984, Induction of
rat mammary gland tumors by 1-nitropyrene, a recently-recognised
environmental mutagen, Cancer Res., 44:1158-1162.

Howard P. C., Heflich R. H., Evans F. E. and Beland F. A., 1983, Formation
of DNA adducts in vitro and in Salmonella typhimurium upon metabolic
reduction of the environmental mutagen 1-nitropyrene, Cancer Res.,
43:2052-2058.

Kohan, M. J. and Claxton, L., 1983, Mutagenicity of diesel exhaust particle
extract, 1-nitropyrene, and 2,7-dinitrofluorene in Salmonella
typhimurium under various metabolic activation conditions, Mutat.
Res., 124:191-200.

Mermelstein, R., Kiriazides, D. K., Butler, M., McCoy, E. and Rosenkranz,
H. S., 1981, The extraordinary mutagenicity of nitropyrenes in
bacteria, Mutat. Res., 89:187-196.

Messier, F., Lu, C., Andrews, P., McCarry, B. E., Quilliam, M. A., and
McCalla, D.R., 1981, Metabolism of 1-nitropyrene and formation of
DNA adducts in Salmonella typhimurium, Carcinogenesis, 2:1007-1011.

Miller, E. C. and Miller, J. A., 1981, Searches for ultimate chemical carcinogens and their reactions with cellular macromolecules, Cancer,
47:2327-2345.

Nishioka, M. G., Peterson, B. A. and Lewtas, J., 1982, Comparison of nitro-
aromatic content and direct-acting mutagenicity of diesel emissions.
In: M. Cooke, A.J. Dennis, and G.L. Fisher (eds.), Polynuclear
Aromatic Hydrocarbons: Phsical and Biological Chemistry, Battelle
Columbus Press, Columbus, OH, pp. 603-613.

Ohgaki, H., Matsukara, N., Morino, K., Kawachi, T., Sugimura, T., Morita,
K., Tokiwa, H. and Hirota, T., 1982, Carcinogenicity in rats of
the mutagenic compounds 1-nitropyrene and 3-nitrofluoranthene.
Cancer Lett., 15:1-7.

Schuetzle, D., Riley, T. L., Prater, T. J., Harvey, T. M., and Hunt, D. F.,
1982, Analysis of nitrated polycyclic hydrocarbons in diesel parti-
culates, Analyt. Chem., 54:265-271.

Skopek, T. R., Liber, H. L., Krolewski, J. J. and Thilly, W. G., 1978,
Quantitative forward mutation assay in Salmonella typhimurium
using 8-azaguanine resistance as a genetic marker, Proc. Natl.
Acad. Sci. (U.S.A.), 75:410-414.

FORMATION OF A RING-OPENED PRODUCT FROM BENZENE IN A HYDROXYL RADICAL

GENERATING SYSTEM

L. Latriano, B. D. Goldstein, and G. Witz

Dept. of Environmental and Community Medicine
UMDNJ-Rutgers Medical School
Busch Campus, Piscataway, NJ 08854

INTRODUCTION

Benzene, a ubiquitous compound in our petrochemical society, has been shown to cause hematoxicity and leukemia in both animals and humans. Current evidence strongly suggests that a metabolite(s) of benzene mediates its toxicity (Snyder et al., 1981). The closed-ring metabolites on the main path of benzene metabolism, e.g. phenol, catechol and hydroquinone, have been studied by many investigators, and though there have been many promising leads, the origin and identification of the toxic metabolite(s) remain uncertain.

Our laboratory has proposed that a ring-opened form of benzene, such as trans,trans-muconaldehyde, may be responsible in part for benzene hematoxicity (Goldstein et al., 1982). This hypothesis is based on several lines of experimental evidence. Pulse radiolysis of aqueous solutions of benzene results in the opening of the benzene ring and the formation of muconaldehyde presumably via hydroxyl radicals ($\cdot$OH) (Daniels et al., 1956). Metabolism of benzene in vivo to a ring-opened product has been demonstrated by the isolation of trans,trans-muconic acid, the corresponding diacid of muconaldehyde, from the urine of rabbits to which ^{14}C-benzene was administered (Parkes et al., 1953). More recently, trans,trans-muconic acid was also identified in the urine of CD-1 mice to which benzene had been administered (Gad-el Karim et al., 1985). Studies in our laboratory have demonstrated that administration of 2 mg/kg of trans,trans-muconaldehyde to CD-1 mice daily for 10 and 16 days resulted in statistically significant decreases in red blood cells, hematocrit, bone marrow cellularity, hepatic total and free sulfhydryl groups (Witz 1985b et al). Figure 1 depicts a likely pathway for the formation of trans,trans-muconic acid via muconaldehyde.

The present paper contains preliminary evidence that benzene ring opening occurs in vitro in a system where hydroxyl radicals are chemically generated, and that trans,trans-muconaldehyde is a product of benzene ring fission. The chemical hydroxyl radical generating system used in these studies is the iron-catalyzed oxidation of ascorbic acid. This system was developed by Brodie et al. (1954) for studies on the hydroxylation of aromatic compounds. These investigators found that substrates such as aniline, tyramine and quinoline were hydroxylated to products which are identical to those formed in vivo. In this model system, hydrogen peroxide (H_2O_2) is produced during the autooxidation of

789

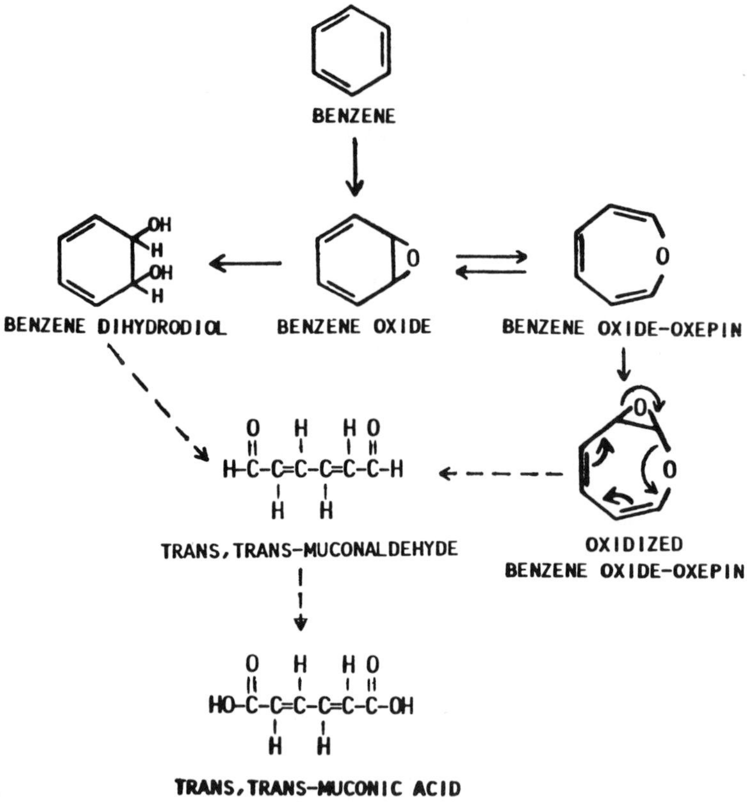

Fig. 1. Potential pathways for the formation of muconaldehyde from benzene

ascorbic acid to dehydroascorbic acid in the presence of ferrous iron
salts and oxygen. This combination of H_2O_2 and Fe^{++} is known as
"Fenton's reagent." It has subsequently been shown that the actual
oxidant in such a system is the hydroxyl radical (Walling, 1975). More
recently, studies by Cohen et al. (1983) suggest that molecular oxygen is
activated and incorporated into phenol during a Fenton-type attack on
benzene.

METHODS

The chemical hydroxyl radical generating system used in this study
consists of a reaction in which ascorbic acid is oxidized in an iron-
catalyzed process. In this system, 19mM benzene is incubated with 14mM
ascorbic acid, 6.5mM disodium EDTA and 1.3mM $FeSO_4$ in 0.1M sodium phosphate
buffer, pH 6.7 at 37° C for two hours. The mixture is subsequently reacted
with 2-thiobarbituric acid (TBA), i.e. one ml of the reaction mixture is
incubated with 0.5 ml of TBA reagent (0.67% TBA in 1N acetic acid) in a
boiling water bath for one hour. Trans, trans-muconaldehyde,
synthesized and characterized in our laboratory (Goldstein et al. 1982),
was similarly reacted with the TBA reagent. The resulting
muconaldehyde/TBA adduct was identified in the mixture and served as the
standard for the identification of muconaldehyde formed from benzene in
the ·OH generating system.

Previous studies in this laboratory have shown that trans,trans-muconaldehyde reacts with TBA to form an adduct with a 495 nm absorbance maximum (Witz et al. 1983). Separation of the muconaldehyde/TBA adduct from other components in the mixture is accomplished by high performance liquid chromatography (HPLC) using a 25 cm C-18 reverse phase column and a mobile phase of 40% methanol and 60% 0.1M phosphate buffer, pH 6.5, with detection by absorbance at 495 nm (Latriano et al. 1985). The muconaldehyde/TBA peak was identified by its absorption spectrum recorded on line. This spectrum matched the spectrum of the muconaldehyde/TBA adduct purified by a solid phase extraction procedure as well as that of a sample purified by thin layer chromatography (Witz et al. 1985a).

An investigation into the effect of hydroxyl radical scavengers mannitol and ethanol on the production of muconaldehyde from benzene in the ·OH generating system was carried out. In these studies benzene was reacted in the ·OH generating system in the presence of hydroxyl radical scavenger for 2.5 hours. Optical density values are obtained from difference spectra recorded over the region of 600-400 nm. The reference cuvette contained a reaction mixture consisting of the model ·OH generating system and scavenger (but no benzene) which had been reacted with TBA.

RESULTS

Reaction of benzene in the chemical hydroxyl radical generating system and subsequent reaction of the mixture with TBA resulted in the formation of a chromogen with an absorbance maximum at 495 nm which was not present when benzene was omitted (Fig. 2). This maximum is similar to the maximum seen after the reaction of TBA and trans,trans-muconaldehyde in 0.1M phosphate buffer, pH 6.7 (Fig. 2). In addition to the 495 nm absorption maximum, two other maxima were observed at approximately 450 and 530 nm. These maxima were present in both the ·OH generating system incubated with benzene and the ·OH generating system without benzene. A faint shoulder at approximately 460 nm is present in the spectrum of the muconaldehyde standard reacted with TBA.

In order to determine the role of the various components of the model ascorbic acid system in producing the chromogen which gives a 495 nm maximum characteristic of the muconaldehyde/TBA adduct, incubations of the complete system minus one of the components were carried out (Table 1). When either ascorbate, ferrous sulphate or EDTA was omitted, i.e. components necessary for the generation of hydroxyl radicals in this system, the 495 nm absorption maximum was reduced to a level slightly higher than background.

Table 1. Effect of Various Components of the Hydroxyl Radical
Generating System on the Production of TBA Chromogens

| Components | Optical Density[a] | | |
	450 nm	495 nm	530 nm
Complete[b]	0.260	0.510	0.420
Complete (-) EDTA	0.070	0.100	-
Complete (-) FeSO$_4$	0.065	0.075	-
Complete (-) Ascorbate	0.110	0.080	0.125

[a] Observed values after reaction with 2-thiobarituric acid

[b] Complete system consists of 19 mM benzene, 14 mM ascorbate, 6.5 mM EDTA and 1.3 mM FeSO$_4$ in 0.1M phosphate buffer, pH 6.7 in a total volume of 3 ml.

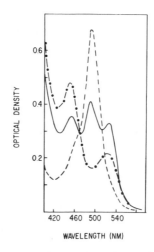

Figure 2. Absorption spectra of samples from the hydroxyl radical generating system and the trans,trans-muconaldehyde standard reacted with 2-thiobarbituric acid. With 19 mM benzene ————; without benzene —·—·—; 0.01 mM muconaldehyde in 0.1M phosphate buffer, pH 6.7 -----.

Separation of the muconaldehyde/TBA adduct from unreacted TBA and from reaction byproducts was accomplished by high performance liquid chromatography. In the chemical hydroxyl radical generating system containing benzene, a peak with a retention time of approximately 8 minutes was observed (Fig. 3C). This peak cochromatographs with a peak which was identified as the muconaldehyde/TBA adduct obtained from the reaction of trans,trans-muconaldehyde with thiobarbituric acid (Fig. 2B). HPLC studies of the muconaldehyde/TBA reaction mixture using stop-flow, on-line scanning techniques have confirmed the identity of this peak as a muconaldehyde/TBA adduct (Latriano et al. 1985). When benzene is omitted from the ·OH generating system, the peak with a retention of 8 minutes is present in very small amounts (Fig. 3A). Under the HPLC conditions employed in these studies, the trans,trans-muconaldehyde /TBA adduct is well separated from other benzene metabolites such as phenol, catechol and hydroquinone.

To determine whether hydroxyl radical scavengers are involved in the oxidation of benzene to trans,trans-muconaldehyde in the model ascorbic acid system, the ·OH scavengers mannitol and ethanol were added to the reaction mixture. The addition of either 10 mM mannitol or ethanol resulted in an 24% and 64% inhibition, respectively, of the formation of muconaldehyde.

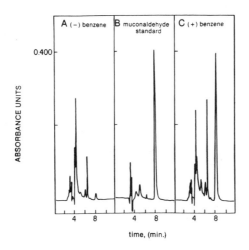

Figure 3. HPLC analysis of hydroxyl radical generating system after the addition of TBA. Separation was achieved on C-18 reverse phase column using 0.01M phosphate buffer, pH 6.5, and methanol (60:40). Detection is by absorbance at 495 nm. A. Hydroxyl radical generating system without benzene. B. Chromatogram of 10^{-5} M trans,trans-muconaldehyde as the TBA adduct. C. Hydroxyl radical generating system with benzene.

DISCUSSION

Incubation of benzene in a system containing ascorbic acid and a source of ferrous ions results in a product which, after reaction with TBA, has a 495 nm absorption maximum which is characteristic of a trans,trans-muconaldehyde/TBA adduct. Previous studies in this laboratory on the spectral characteristics of TBA adducts show that the 495 nm absorption maximum is characteristic of the adduct formed between thiobarbituric acid and certain alpha, beta-unsaturated aldehydes such as muconaldehyde, acrolein, and crotonaldehyde (Witz et al., 1985a). Trans,trans-muconaldehyde is unique among the reactive aldehydes studied in that its adduct with TBA lacks a 530 nm absorption and exhibits a slight shoulder at approximately 450 nm. Acrolein and crotonaldehyde, on the other hand, exhibit pronounced absorption maxima at these two wavelengths, which however are less intense than that at 495 nm. The 450 and 530 nm absorbance in the complete ·OH generating system without benzene after reaction with TBA, might be derived from the reaction of oxidized ascorbate with TBA (Gutteridge and Wilkins, 1982). The present studies therefore suggest that in a model hydroxyl radical generating system, benzene undergoes a ring-opening reaction to form a product with an activated aldehydic functional group, similar to trans,trans-muconaldehyde. This finding is supported by HPLC studies. The benzene-derived intermediate which reacts with TBA to form a 495 nm chromogen cochromatographs with a product derived from the reaction between authentic muconaldehyde and thiobarbituric acid. Additional studies to confirm the identity of the ring-opened product are currently underway.

There are a number of possible pathways by which oxidative ring opening might occur, leading to the formation of muconaldehyde. In a system where ·OH are generated, it is not unlikely that benzene is converted to a dihydrodiol, which could undergo ring opening by a free radical reaction involving beta scission of the carbon-carbon single bond. Alternatively, oxidation of an arene oxide species followed by rearrangement could also lead to muconaldehyde. In vivo, muconaldehyude could be further oxidized to muconic acid. Using the chemical hydroxyl radical generating system described for the current studies, the route for the formation of muconaldehyde from benzene is being investigated.

Involvement of hydroxyl radicals in benzene ring-opening and subsequent production of muconaldehyde is suggested by the inhibition of muconaldehyde formation when one of the components necessary for ·OH production is omitted or when ·OH scavengers are added to the reaction mixture. Recent studies on benzene metabolism in rabbit liver microsomes have suggested that hydroxyl radicals mediate benzene metabolism by the enzyme cytochrome P-450 (Johansson and Ingelman-Sundberg, 1983). In these studies the formation of aromatic hydroxylated metabolites as well as several unidentified metabolites was inhibited by ·OH scavengers. Our laboratory is currently investigating whether benzene ring fission and muconaldehyde formation occur in microsomes and, if so, whether the ring-opening process is hydroxyl radical mediated.

That trans,trans-muconaldehyde is hematoxic in a manner similar to benzene has been recently demonstrated by this laboratory (Witz et al., 1985b). Toxicity studies with trans,trans muconaldehyde show that intraperitoneal administration of 2 mg/kg of muconaldehyde daily for 10 and 16 days to male CD-1 mice results in statistically significant decreases in red cell count, hematocrit, bone marrow cellularity and hepatic total and free sulfhydryl content. (Witz et al., 1985). There was also an increase in white blood cell count and spleen weight after 16 days of administration of 2 mg/kg muconaldehyde. Though the mechanism(s) of the toxic effects is unknown, work by other investigators on the activated aldehydes has shown that this class of compounds reacts readily with nucleophiles. Schauenstein et al., (1972) has shown that the faster an alpha, beta-unsaturated aldehyde reacts with sulfhydryl groups, the more toxic it is. The toxic effects reported include inhibition of cellular metabolism, as well as inhibition of DNA, RNA and protein synthesis; effects which they believe are caused by the deactivation of essential cell sulfhydryl groups. The reactivity of the alpha, beta-unsaturated carbonyl system toward glutathione decreases in the order of acrolein > 4-hydroxynonenal > crotonaldehyde (Schauenstein et al., 1972). Studies in this laboratory show that muconaldehyde is 1.5 times more reactive than 4-hydroxynonenal and 3 times less reactive than acrolein (unpublished data).

The studies reported here indicate that in a chemical hydroxyl radical generating system, reactive species are generated with enough energy to cause benzene ring opening. The nature of the chromogen formed upon TBA addition suggests that the ring-opened product has an activated aldehyde group. That the product is trans,trans-muconaldehyde is suggested by spectral and chromatographic analysis. Further studies are underway to confirm that the ring-opened product is trans, trans-muconaldehyde.

ACKNOWLEDGEMENTS

We would like to thank Ms. Nancy Lawrie for her assistance in the preparation of the art work. Supported by NIH grant No. ES02558 and N.J. DEP Institute of Hazardous Waste Mgmt. Spill Fund.

REFERENCES

Brodie, B.B., Axelrod, J., Shore, P.A., Udenfriend, S., 1954, Ascorbic acid in aromatic hydroxylation. II. Products formed by reaction of substrates with ascorbic acid, ferrous ion, and oxygen. J. Biol. Chem. 208:731.

Cohen, G., Ofodile S., 1983, Activation of molecular oxygen during a Fenton reaction: A study with $^{18}O_2$ in "Oxy Radicals and Their Scavenger System, Vol. 1.", G. Cohen and R.A. Greenwald ed., Elsevier Biomedical, N.Y.

Daniels, M., Scholes, G., Weiss, J., 1956, Chemical action of ionizing radiation in solution, part XV. Effect of molecular oxygen in the irradiation of aqueous benzene solutions with X-rays. J. Chem. Soc., 832.

Gad-El Karim, M.M., Ramanujum, S., Legator, M.S., 1985, trans,trans-Muconic acid, an open-chain urinary metabolite of benzene in mice. Quantification by high pressure liquid chromatography. Xenobiotica, 15: 211.

Goldstein, B.D., Witz, G., Javid, J., Amoruso, M., Rossman, T., Wolder, R., 1982, Muconaldehyde, a potential toxic intermediate of benzene metabolism. in "Biologically Reactive Intermediates II, Part A" Snyder, Parke, Kocsis, Gibson, Jollow and Witmer, eds., Plenum Publ. Corp., N.Y.

Gutteridge, J.M.C., Wilkins, S., 1982, Copper-dependent hydroxyl radical damage to ascorbic acid. FEBS Lett. 137:327.

Johansson, L., Ingelman-Sundberg, M., 1983, Hydroxyl radical-mediated, cytochrome P-450-dependent metabolic activation of benzene in microsomes and reconstituted enzyme systems from rabbit liver. J. Biol. Chem. 258:7311.

Latriano, L. Duncan, S.B.T. Hartwick, R., Witz, G., 1985, Determination of muconaldehyde, a potential metabolite of benzene, by high pressure liquid chromatography. Toxicologist 5:359.

Parkes, D.V., Williams, R.T., 1953, Studies in detoxification 49. The metabolism of benzene $^{14}C_1$-benzene. Biochem. J. 54:231.

Schauenstein, E., Esterbauer, H., Zollner, H., (1972) in: "Aldehydes in Biological Systems. Their Natural Occurrence and Biological Activities." Methuen, Inc. New York, pg. 25 -102.

Snyder, R., Kocsis, J. J., Witmer, C.M., 1981, The biochemical toxicology of benzene. in: "Reviews in Biochemical Toxicology Volume 3." Hodgson, Bend, Philpot, eds. Elsevier/North Holland, New York.

Walling, C., 1975, Fenton's reagent revisted. Accts. Chem. Res. 8:125.

Witz, G., Rao, G., Latriano, L., Goldstein, B.D., 1983, The reaction of 2-thiobarbituric acid with trans,trans-muconaldehyde, a potential benzene metabolite. Toxicol. Letters 18:Suppl.1, 93.

Witz, G., Zaccaria, A., Lawrie, N.J., Ferran, H.J. Jr., Goldstein, B.D., 1985a, Adduct formation of toxic alpha, beta-unsaturated aldehydes with 2-thiobarbituric acid Toxicologist 5:91 2.

Witz, G., Rao, G., Goldstein, B.D., 1985b, Short term toxicity of trans,trans-muconaldehyde. Toxicol. Appl. Pharmacol.:80 (in press).

H-2 MODULATION OF ARYL HYDROCARBON HYDROXYLASE INDUCTION AND THE MUTA-

GENICITY OF BENZO(a)PYRENE AFTER 3-METHYLCHOLANTHRENE TREATMENT

B.A. Brooks, D.L. Wassom, and J.G. Babish

Department of Preventive Medicine
NYS College of Veterinary Medicine
Cornell University
Ithaca, NY 14853

INTRODUCTION

Pretreatment of animals with polycyclic aromatic compounds such as 3-methylcholanthrene (MC) and 2,3,7,8-tetrachlorodibenzo-p-dioxin can alter the subsequent biotransformation of foreign compounds both quantitatively and qualitatively (Conney, 1982; Nebert et al., 1981; Lu and West, 1980). These changes in metabolism are a result of an increase (termed induction) in several drug metabolizing enzymes including cytochrome P_1-450 (Negishi and Nebert, 1979), and P_3-450 (Negishi and Nebert, 1979; Negishi et al., 1981). Mouse P_1-450 and P_3-450 are defined (Gonzalez et al., 1984) as those forms of MC-induced cytochrome P-450 having the highest turnover number for induced aryl hydrocarbon hydroxylase (AHH) and acetanilide 4-hydroxylase activity, respectively. Additionally, P_1-450 is associated with the production of mutagenic and carcinogenic metabolites of benzo(a)pyrene (BP) as well as other polycyclic hydrocarbons (reviewed in Pelkonen and Nebert, 1982).

The induction of these enzymes by polycyclic aromatics is controlled by the Ah gene system (reviewed in Nebert and Gonzalez, 1985). The Ah locus encodes for a cytosolic receptor that specifically binds these chemicals, translocates them into the nucleus, and activates numerous structural genes; P_1-450 and P_3-450 are products of two of these structural genes (Nebert et al., 1982).

In a study of the genetic regulation of AHH activity, Legraverend et al. (1984) reported that the presence or absence of benzo(a)anthracene induced AHH activity in mouse x hamster somatic cell hybrid clones correlated with the presence or absence of chromosome 17. They concluded that a major gene regulating AHH inducibility is located on the distal portion of mouse chromosome 17.

Mouse chromosome 17 is of particular interest because the major histocompatibility complex is located near the center of this chromosome. The major histocompatibility complex, or H-2 complex in the mouse, is responsible in part for the regulation of the magnitude and type of immune response generated (Klein et al., 1981).

797

A relationship between immune function and AHH inducibility is suggested by the observations that many AHH inducers are also immunosuppressive (Nebert et al., 1982b), and that many of the nonresponsive strains of mice have immunologic deficiencies. Additionally, Lubet et al. (1984) found that AHH inducible mice were more sensitive to the immunosuppressant effect of MC after ip administration than were AHH non-inducible mice.

Using H-2 congenic mice with an AHH responder background (C57BL/10), it is possible to study the influence of the H-2 genes on P_1-450 induction following treatment with MC. If differences in AHH inducibility are seen among H-2 congenic strains, they must be referable to H-2 genes, since these strains differ only at the H-2 gene loci.

The purpose of this work was to investigate the influence of the H-2 on changes in total hepatic cytochrome P-450, AHH activity, and the biotransformation of BP to mutagenic metabolites following MC treatment.

MATERIALS AND METHODS

Animals

B10.WB(H-2^j), B10.M(H-2^f), B10.RIII(H-2^r), B10.S(H-2^s), B10.D2(H-2^d) and C57BL/10 (B10) female mice 6-8 wk of age were bred in the colony of DLW at the N.Y.S. College of Veterinary Medicine, Cornell University. DBA/2J and C57BL/6J (B6) were obtained from the Jackson Laboratory, Bar Harbor, ME. Mice were fed Prolab RMH 1000 rat, mouse and hamster food (Agway) and given water *ad libitum*. Bedding consisted of hardwood chips. All mice were housed 3-5 per cage and maintained on a photoperiod of 12 hr light and 12 hr darkness.

Chemicals

All chemicals were purchased from commercial sources and were of the highest purity available.

Treatment of animals and assays

Mice were killed 24 hr after an intraperitoneal (ip) injection of MC (40 mg/kg) or corn oil alone; the volume of the injections ranged between 0.2 and 0.3 ml per mouse. All animals were fasted overnight prior to being killed.

An hepatic 9,000 x g supernatant fraction (S9) was prepared according to the procedure of Garner (1972). Livers were aseptically removed, washed in ice-cold, sterile 0.15 M KCl and homogenized in a Potter-Elvehjem tissue homogenizer with 4 volumes of ice-cold KCl. The liver homogenate was centrifuged at 9,000 x g for 20 min at 0-4°C in a Beckman J2-21 model centrifuge. Aliquots of the S9 were transfered to sterile cryotubes, frozen in liquid nitrogen and stored at -80°C. Samples were assayed within 3 wk of preparation. Protein concentration was determined by the method of Lowry et al. (1951).

Cytochrome P-450 was determined in the S9 using the dithionite-difference technique as described by Schoene et al. (1972). This method minimizes the interference of hemoglobin present in the S9. An extinction coefficient of 91 mM^{-1} cm^{-1} for the difference in absorbance between the Soret maximum and the baseline at 490 was used for quantitating P-450 (Omura and Sato, 1964). Work in this laboratory has shown excellent agreement between the dithionite-difference technique and the CO-difference procedure.

AHH activity was determined according to a modification of the procedure of Yang and Kicha (1978). These modifications consisted of lowering both the substrate (BP) and protein concentrations. Approximately 40-200 ug of S9 protein were added to 1.95 ml of buffer (100 mM potassium phosphate, 5 mM $MgCl_2$, and 0.1 mM EDTA, pH 7.4) in a 1 cm quartz cuvette. After equilibration at room temperature, baseline fluorescence was determined. One-hundred pmol BP in 5 ul acetone were added and the increase in fluorescence was recorded. After a 1 min equilibration period, NADPH (0.05 umol in 50 ul of assay buffer) was added and the rate of decrease in fluorescence was recorded over a period of 5 min when reaction rates were linear. Measurements of fluorescence were made with a Turner Model 430 spectrofluorometer equipped with 15 nm excitation and emission bandwidth slits. Excitation and emission wavelengths for BP were 367 and 407, respectively.

Salmonella typhimurium strain TA100 was obtained from Dr. B.N. Ames (Biochemistry Dept., University of California, Berkeley) and stored at -80°C. Assays were based on plate incorporation techniques described by Ames et al. (1975) and modified by Batzinger et al. (1978). These modifications included the addition of histidine (6.6 ug/ml) and biotin (1.1 ug/ml) to the bottom agar (15 ml) rather than to the top soft agar layer. Additionally, Vogel-Bonner medium (1956) replaced the 0.5% sodium chloride solution in the top agar. These modifications have been reported to reduce the day-to-day variability of the assay (Batzinger et al., 1978).

Optimum protein (0.9 mg/plate) and promutagen dose (5 ug/plate) were determined initially and used for studies reported here. Five ug BP in 50 ul DMSO were pipetted into sterile disposable culture tubes followed by S9 cofactors (0.45 ml), 2 ml agar, 0.1 ml of an overnight culture of bacterial tester strain and 0.05 ml of S9 from individual mice. The tubes were mixed and poured onto Vogel-Bonner minimal glucose plates. A solvent control was included for each animal to determine spontaneous reversion rate in the presence of S9 and to check S9 sterility. Mean spontaneous reversion rates did not differ between control and MC-treated animals; all spontaneous rates were within the quality control limits of this laboratory (120-210) for TA100. Each mutagen assay was done 2 to 3 times per experiment. All plates were incubated for 72 hr at 37°C. Revertant colonies were counted using the NBS Biotran II colony counter (New Brunswick Scientific Co., Edison, NJ).

Two separate experiments were performed utilizing the 5 congenic and 2 reference strains. Six to 10 animals/strain/treatment were used (e.g. corn oil or MC) in each experiment. All response variables (P-450, AHH activity and mutagenicity of BP) were determined based on the individual animal as the unit of observation.

Cytochrome P-450 values were expressed as pmol/mg S9 protein. Relative increases in P-450 were determined by dividing the individual P-450 value of an MC-treated animal by the mean constitutive value for that strain within each experiment. One unit of AHH activity was defined as 1 pmol BP metabolized/min/mg S9 protein. Relative increases were calculated as described for P-450; absolute increases were calculated by subtracting the strain mean constitutive value from the individual value of an MC-treated animal. Relative changes in the activation of BP to a mutagenic metabolite(s) were calculated by dividing the excess revertants due to BP/mg S9 protein (excess revertants due to BP = BP plate response – spontaneous response) for the individual MC-treated animal by the mean strain constitutive value for BP induced excess revertants/mg S9 protein. Absolute changes were calculated by subtracting the strain mean constitutive value from the individual value of an MC-treated animal.

Data were analyzed by analysis of variance procedures; the log transformation was used to normalize the distributions of absolute and relative changes of all variables; differences among treatments were determined using Tukey's HSD test. Treatment differences were deemed significant when the probability of a type I error was less than 5%. All statistical procedures were as described by Snedecor and Cochran (1972). Results presented in this paper are from the second experiment; conclusions of both experiments are the same.

RESULTS

Treatment of the congenic and inbred strains with MC resulted in a 1.6-fold average increase in hepatic cytochrome P-450 in all strains except DBA/2J, B10.S and B10.RIII, where no increase was observed (Figure 1). B10.D2 showed a 2.2-fold increase, which was higher than the other strains with the exception of the B10.M. Constitutive P-450 values did not differ among the congenic strains and averaged 69 pmol/mg S9 protein.

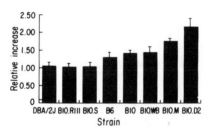

Figure 1. RELATIVE HEPATIC CYTOCHROME P-450 INCREASES IN CONGENIC AND REFERENCE MICE FOLLOWING TREATMENT WITH 3-METHYLCHOLANTHRENE. Values are means +/- SEM of 5 to 9 observations per treatment calculated as described in MATERIALS AND METHODS.

As seen in Figure 2, pretreatment with MC enhanced BP metabolism in all congenic strains relative to their corn oil controls. This relative increase in BP metabolism ranged from a low of 12-fold in the B10.S strain to a high of 66-fold in the B10.D2 strain. The B6 strain, with a 37-fold increase in AHH activity, exhibited the median response. Both the low relative response of the B10.S and the high relative response of the B10.D2 were different from the relative increase of the B6. No other strains exhibited relative increases in AHH activity which differed from the B6.

It was found that constitutive AHH values varied more between experiments than induced AHH activity. Constitutive AHH activity varied more than 2-fold among the congenic strains within and between experiments. These control animals may have been marginally induced by some factor in the environment such as diet (Babish and Stoewsand, 1975, Levragerend et al., 1980). Since a small change in constitutive activity can markedly change relative induction values, absolute increases in AHH activity were computed by subtracting the strain mean control AHH activity from individual MC-induced values.

Absolute increases in AHH activity for the five congenic strains and two reference strains are presented in Figure 3. Strains B10.M, B10.D2, and B10.WB demonstrated absolute increases in specific AHH activity greater than either the B6 or B10 strains, while the absolute response of the B10.S and B10.RIII strains did not differ from the B6 or B10. The enhanced absolute increase in AHH activity of the B10.M and the B10.D2 relative to the B10.S and B10.RIII parallels the greater P-450 response in these strains. DBA/2J did not exhibit either a relative or absolute increase in specific AHH activity, and therefore is not listed in Figures 2 or 3.

As with AHH activity, the ability of the congenic strains to metabolically activate BP to mutagenic metabolites was examined in both relative and absolute terms. Figure 4 shows the relative increase in TA100 revertants following MC treatment for the five congenic and three reference strains. The congenic strains B10.M and B10.D2 segregated with the B6 strain with an average relative increase of 15-fold. With increases in ability to activate BP to mutagenic metabolites of 4.1 and 6.6-fold respectively, the B10.S and B10.RIII strains showed the lowest change following MC administration. The congenic strain B10.WB exhibited an intermediate change in capacity to activate BP of 9.6-fold. No differences were seen in the ability of constitutive or MC-treated DBA/2J mice to metabolically activate BP.

As can be seen in Figure 5, which shows the absolute excess revertants produced by MC-treated mice, the B10.M, B10.WB and B10.D2 were more effective in producing mutagenic metabolites of BP than were the B10.RIII and B10.S. The high responding congenic strains produced an average of 871 excess revertants per mg of S9 protein, while the low responding congenics averaged only 393 excess revertants per mg S9 protein. It is interesting to note that although MC-treated B10.RIII mice had 2-fold higher AHH activity than B10.S animals, P-450 values and the ability to activate BP to mutagenic metabolites did not differ.

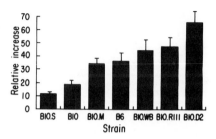

Figure 2. RELATIVE HEPATIC AHH ACTIVITY INCREASES IN CONGENIC
AND REFERENCE MICE FOLLOWING TREATMENT WITH 3-METHYLCHOLANTHRENE.
Values are means +/- SEM of 6 to 10 observations per treatment
calculated as described in MATERIALS AND METHODS.

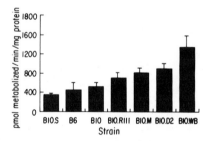

Figure 3. ABSOLUTE HEPATIC AHH ACTIVITY INCREASES IN CON-
GENIC AND REFERENCE MICE FOLLOWING TREATMENT WITH 3-METHYL-
CHOLANTHRENE. Values are means +/- SEM of 6 to 10 observa-
tions calculated as described in MATERIALS AND METHODS.

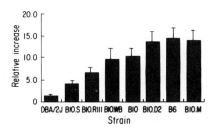

Figure 4. RELATIVE INCREASES IN BENZO(a)PYRENE INDUCED REVER-
TANTS OF <u>SALMONELLA</u> <u>TYPHIMURIUM</u> TA100 BY HEPATIC S9 OF CON-
GENIC AND <u>REFERENCE</u> MICE FOLLOWING TREATMENT WITH 3-METHYL-
CHOLANTHRENE. Values are means +/- SEM of 6 to 10 observa-
tions calculated as described in MATERIALS AND METHODS.

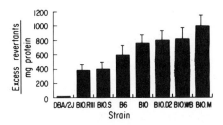

Figure 5. ABSOLUTE INCREASES IN BENZO(a)PYRENE-INDUCED
REVERTANTS OF <u>SALMONELLA</u> <u>TYPHIMURIUM</u> TA100 BY HEPATIC S9
OF CONGENIC AND REFERENCE MICE FOLLOWING TREATMENT WITH 3-
METHYLCHOLANTHRENE. Values are means +/- SEM of 6 to 10
observations calculated as described in MATERIALS AND METHODS.

DISCUSSION

Like the non-responsive DBA/2J, the B10.S (s haplotype) and B10.RIII (r haplotype) showed no increase in hepatic cytochrome P-450 levels and a diminished capacity to form mutagenic metabolites with BP following MC treatment. Unlike the DBA/2J however, AHH induction with MC was similar to the background reference strains, B10 and B6, for these two congenic strains.

Of note is the observation that the high AHH activity in the B10.RIII following induction with MC did not correlate with the low degree of metabolic activation of BP in the bacterial mutagen assay. It is possible that the spectrum of metabolites produced by the B10.RIII strain following induction are qualitatively different than the other strains, perhaps due to the production of a different isozyme or combination of isozymes of P-450. Alternatively, the results in the mutagenicity assay could be due to differences in enzymes other than P_1-450. The Ah locus is known to regulate more than 2 dozen microsomal and non-microsomal enzymes (Nebert et al., 1982b).

The immunological significance of a regulatory gene of the Ah loci being linked to the H-2 complex is not known. It is possible that this association may in part explain the immunosuppression observed with exposure to certain polycyclic aromatics (Lubet et al., 1984) or the reported inverse relationship between interferon levels and inducibility of hepatic cytochrome P-450 by polynuclear aromatics (Mannering et al., 1980).

Alternatively, the linkage may not be related to the function of the immune system. Non-immunological traits controlled by the H-2 include mating preference (Yamazaki et al., 1975), fertility (Ivanyi, 1978 and Kunz et al., 1980) and DNA repair (Hall et al., 1981). In humans, congenital adrenal hyperplasia due to 21-hydroxylase deficiency has been linked to the human major histocompatibility complex, the HLA (Dupont et al., 1977). The 21-hydroxylase deficiency results from a deletion of the cytochrome P-450 gene specific for steroid 21-hydroxylation termed P-450$_{c21}$ (White et al., 1984).

In this study, 2 reference, AHH-inducible strains, the B6 and B10, were compared to 5 congenic strains on the basis of P-450 and AHH inducibility and the metabolic activation of BP to mutagenic metabolites following treatment with MC. The B6 genotype is the same as the B10 except for a difference at an H locus on chromosome 4, an H-9 difference on chromosome 17 and an Ea difference (Heiniger and Dorsey, 1980). No differences were found between these 2 strains in any of the variables examined. The congenic strains, differing only within the H-2 complex (other differences which may be linked to the H-2 are possible), exhibited qualitatively and quantitatively different responses to MC treatment in terms of P-450 expression, AHH inducibility and activation of BP to a mutagenic metabolite(s).

In addition to repeating these experiments, further work includes the study of these 5 haplotypes on another Ah-responsive background and the quantitation of hepatic Ah receptor levels in congenic strains. Preliminary evidence suggests that Ah receptor levels differ among the congenic strains with the B10 background (unpublished observations).

SUMMARY

All strains except the DBA/2J, B10.S and B10.RIII exhibited an increase in hepatic cytochrome P-450 levels following MC treatment; futhermore, the P-450 response of the B10.D2 strains was greater than that seen in the B10 and B6 strains.

MC treatment increased BP metabolism in all congenic strains relative to their corn oil controls. Both relative and absolute changes in AHH activity were higher in the B10.D2 than the B6 or B10 reference strains.

All congenic strains exhibited an enhanced capacity to produce mutagenic metabolites of BP following treatment with MC; the B10.M, B10.WB and B10.D2 segregated with the B6 and B10 in this regard, while the B10.S and B10.RIII changed least following MC administration.

CONCLUSION

From these data it appears that a gene or genes located within or linked to the H-2 complex modulates the effect of inducers on the metabolism of polynuclear aromatic hydrocarbons and the expression of hepatic cytochrome P-450.

REFERENCES

Ames, B.N., McCann, J. and Yamasaki. 1975. Methods for detecting carcinogens and mutagens with the Salmonella mammalian microsome mutagenicity test. Mutat. Res. 31:347.

Babish, J.G. and Stoewsand, G.S. 1975. Hepatic microsomal enzyme induction in rats fed varietal cauliflower leaves. J. Nutr. 105:1593.

Batzinger, R.P., Ou, S.-Y.L., and Bueding, E. 1978. Anti-mutagenic effect of 2(3)-tert-butyl-4-hydroxyanisole and of antimicrobial agents. Cancer Res. 38:4478.

Conney, A.H. 1982. Induction of microsomal enzymes by foreign chemicals and carcinogenesis by polycyclic aromatic hydrocarbons. Cancer Res. 42:4875.

Dupont, B., Oberfield, S.E., Smithwick, E.M., Lee, T.D. and Levine, L.S. 1977. Close genetic linkage between HLA and congenital adrenal hyperplasia (21-hydroxylase deficiency) Lancet ii, 1309.

Garner, R.C., Miller, E.C. and Miller, A. 1972. Liver microsomal metabolism of aflatoxin B_1 to a reactive derivative toxic to Salmonella typhimurium TA 1530. Cancer Res. 32:2058.

Gonzalez, F.J., Tukey, R.H. and Nebert, D.W. 1984. Structural gene products of the Ah locus. Transcriptional regulation of cytochrome P_1-450 and P_3-450 mRNA Levels by 3-methylcholanthrene. Mol. Pharm. 26:117.

Hall, K.Y., Bergmann, L., Walford, R.L. 1981. DNA repair, H-2 and aging in N2B and CBA mice. Tissue Antigens 16:104.

Heiniger, H.J. and Dorey, J.L., eds. 1980. Handbook on genetically standardized JAX mice. 3rd ed. The Jackson Laboratory, Bar Harbor Maine.

Ivanyi, P. 1978. Some aspects of the H-2 system, the major histocompatibility system in the mouse. Proc. R. Soc. Lond. [Biol.] 201:117.

Klein, J., Juretic, A., Baxevanis, C.N. and Nagy, Z. 1981. The traditional and new version of the mouse H-2 complex. Nature. 291:455.

Kunz, H.W., Gill, T.J., Dixon, B.D., Taylor, F.H. and Greiner, D.L. 1980. Growth and reproduction complex in the rat. Genes linked to the major histocompatibility complex that affect development. J. Exp. Med. 152:1506.

Legraverend, C., Karenlampi, S.O., Bigelow, S.W., Lalley, P.A., Kozak, C.A., Womack, J.E. and Nebert, D.W. 1984. Aryl hydrocarbon hydroxylase induction by benzo(a)anthracene: Regulatory gene localized to the distal portion of mouse chromosome 17. Genetics 107:447.

Legraverend, C., Nebert, D.W., Boobis, A.R. and Pelkonen, O. 1980. DNA binding of benzo(a)pyrene metabolites. Effects of substrate and microsomal protein concentration in vitro, dietary contaminants and tissue differences. Pharmacology 20:137.

Lowry, O.H., Rosenbrough, N.J., Farr, A.I. and Randall, R.J. 1951. Protein measurement with the Folin phenol reagent. J. Biol. Chem. 193:265.

Lu, A.Y.H. and West, S.B. 1980. Multiplicity of mammalian microsomal cytochromes P-450. Pharmacol. Rev. 31:277.

Lubet, R.A., Brunda, M.J., Taramelli, D., Dansie, D., Nebert, D.W. and Kouri, R.E. 1984. Induction of immunotoxicity by polycyclic hydrocarbons: role of the Ah locus. Arch. Toxicol. 56:18.

Mannering, G.J., Renton, K.W., El Azhary, R. and Deloria, L.B. 1980. Effects of interferon-inducing agents on hepatic cytochrome P-450 drug metabolizing systems pp. 314-331. In: "Regulatory Functions of Interferons", Vol. 350. Edited by J. Vilcek, I. Gresser and T. Merrigan. New York Academy of Science, New York.

Nebert, D.W., Eisen, H.J., Negishi, M., Lang, M.A., Hjelmeland, L.M. and Okey, A.B. 1981. Genetic mechanisms controlling the induction of polysubstrate monooxygenase (P-450) activities. Ann. Rev. Pharmacol. Toxicol. 21:431.

Nebert, D.W., Negishi, M., Lang, M., Hjelmeland, L.M. and Eisen, H.J. 1982. The Ah locus, a multigene family necessary for survival in a chemically adverse environment: comparison with the immune system. Adv. Genet. 21:1.

Nebert, D.W., Jensen, N.M., Shinozuka, H., Kunz, H.W. and Gill, T.J. 1982b. The Ah phenotype survey of forty-eight rat strains and twenty inbred mouse strains. Genetics 100:79.

Nebert, D.W. and Gonzalez, F.J. 1985. Cytochrome P-450 gene expression and regulation. Trends Pharmacol. Sci. 6:160.

Negishi, M. and Nebert, D.W. 1979. Structural gene products of the Ah locus: genetic and immunochemical evidence for two forms of mouse liver cytochrome P-450 induced by 3-methylcholanthrene. J. Biol. Chem. 254:11015.

Negishi, M., Jensen, N.M., Garvia, G.S. and Nebert, D.W. 1981. Structural gene products of the murine Ah locus: differences in ontogenesis, membrane location and glucosamine incorporation between liver microsomal cytochromes P_1-450 and P-448 induced by polycyclic aromatic compounds. Eur. J. Biochem. 115:585.

Omura, T. and Sato, R. 1964. The carbon monoxide-binding pigment of liver microsomes. II. Solubilization, purification, and properties. J. Biol. Chem. 239:2379.

Pelkonen, O. and Nebert, D.W. 1982. Metabolism of polycyclic aromatic hydrocarbons: etiologic role in carcinogenesis. Pharmacol. Rev. 34:189.

Schoene, B.R., Fleischmann, R., Remmer, R., and Olderhauser, H.G. 1972. Determination of drug metabolizing enzymes in needle biopsies of human liver. Eur. J. Clin. Pharmacol. 4:61.

Snedecor, G.W. and Cochran, W.G. 1972. Statistical Methods, 6th ed. Ames: Iowa State Univ. Press.

Vogel, H.J. and Bonner, D.M. 1956. Acetylorthinase of Escherichia coli: partial purification and some properties. J. Biol. Chem. 218:97.

White, P.C., New, M.I. and Dupont, B. 1984. HLA-linked congenital adrenal hyperplasia results from a defective gene encoding a cytochrome P-450 specific for steroid 21-hydroxylation. Proc. Natl. Acad. Sci. 81:7505.

Yamazaki, K., Yamaguchi, M., Baranoski, L., Bard, J., Boyse, E.A., and
    Thomas, L. 1975. Recognition among mice. Evidence from the use
    of a Y-maze differentially scented by congenic mice of different
    major histocompatibility types. J. Exp. Med. 150:755.
Yang, C.S. and Kicha, L.P. 1978. A direct fluorometric assay of
    benzo(a)pyrene hydroxylase. Anal. Biochem. 84:154.

STEREOSELECTIVITY OF RAT LIVER CYTOCHROME P-450 ISOZYMES: DIRECT DETER-
MINATION OF ENANTIOMERIC COMPOSITION OF K-REGION EPOXIDES FORMED IN THE
METABOLISM OF BENZ[A]ANTHRACENE AND 7,12-DIMETHYLBENZ[A]ANTHRACENE

S. K. Yang, M. Mushtaq, P.-L. Chiu, and H. B. Weems

Department of Pharmacology, F. Edward Hébert School of
Medicine, Uniformed Services University of the Health
Sciences, Bethesda, Maryland 20814-4799 U.S.A.

ABSTRACT

The K-region 5,6-epoxides of benz[a]anthracene (BA) and 7,12-dime-
thylbenz[a]anthracene (DMBA) were isolated by normal-phase HPLC from
metabolites formed by incubation of the respective parent compound with
liver microsomes from untreated (control), phenobarbital (PB)-treated,
and 3-methylcholanthrene (MC)-treated rats in the presence of an epoxide
hydrolase inhibitor, 3,3,3-trichloropropylene 1,2-oxide. The enantiomeric contents of the metabolically formed K-region 5,6-epoxides of BA
and DMBA were directly determined by chiral stationary phase HPLC. The
K-region 5,6-epoxides formed in the metabolism of BA have
$(5R,6S):(5S,6R)$ enantiomer ratios of 25:75 (control), 21:79 (PB), and
4:96 (MC), respectively. In contrast, the $(5R,6S):(5S,6R)$ enantiomeric
ratios of the K-region 5,6-epoxides formed in the metabolism of DMBA are
76:24 (control), 80:20 (PB), and 97:3 (MC), respectively. These and
earlier results on the stereoselective K-region metabolism studies of 7-
methylbenz[a]anthracene and 12-methylbenz[a]anthracene indicate that (i)
cytochrome P-450 isozymes exhibit different stereoselectivities in the
K-region epoxidations of BA and DMBA and (ii) a methyl substituent at
the C_{12} position of BA alters the stereoheterotopic interactions between
cytochrome P-450 isozymes and the BA molecule.

INTRODUCTION

Epoxides (arene oxides) are the initial products formed in the
metabolism of polycyclic aromatic hydrocarbons by cytochrome P-450 isozymes of drug-metabolizing enzyme systems (1,2). Epoxides are converted
to *trans*-dihydrodiols by microsomal epoxide hydrolase (1,2). Some non-K-
region *trans*-dihydrodiols are further metabolized by cytochrome P-450
isozymes to vicinal dihydrodiol-epoxides, some of which are the ultimate
carcinogenic metabolites of the parent hydrocarbons (3,4).

Due to stereoheterotopic interactions between the substrate and the
cytochrome P-450 isozymes, some epoxides formed in the metabolism of
polycyclic aromatic hydrocarbons are optically active (1-10). Although
the K-region epoxide metabolite of some polycyclic aromatic hydrocarbons
can be isolated (1), the enantiomeric compositions of metabolically
formed K-region epoxides of only a few polycyclic aromatic hydrocarbons

have been determined (7-11). The enantiomeric compositions of K-region epoxide, formed in the metabolism of benz[a]anthracene (BA) (8) and of benzo[a]pyrene (BP) (7) in a reconstituted enzyme system containing highly purified cytochrome P-450c (P-448), were determined as diastereomeric glutathione conjugates. This method established that a BA 5*S*,6*R*-epoxide and a BP 4*S*,5*R*-epoxide are each the predominant stereoisomer formed in the cytochrome P-450c (P-448)-catalyzed metabolism at the K-region of BA and of BP, respectively (7,8). Recently a simple and direct chiral stationary phase HPLC (CSP-HPLC) method was developed to resolve the enantiomers of K-region epoxides of chrysene, BA, 1-methyl-BA, 7-methyl-BA, 12-methyl-BA, DMBA, dibenz[a,h]anthracene, and BP, and a non-K-region 7,8-epoxide of BP (11). This CSP-HPLC method has been applied to the determination of the enantiomeric compositions of K-region epoxides of BA (9), 1-methyl-BA (11), 7-methyl-BA (11), 12-methyl-BA (11), and BP (9) formed by metabolism of the respective parent hydrocarbon with rat liver microsomes. This report describes the results on the stereoselective formation of K-region epoxides in rat liver microsomal metabolisms of BA and DMBA and the effects of pretreatment of rats with phenobarbital and 3-methylcholanthrene.

MATERIALS AND METHODS

Materials

Racemic K-region 5,6-epoxides of BA and DMBA, and BA *trans*-5,6-dihydrodiol were obtained from the Chemical Repository of National Cancer Institute. 3,3,3-Trichloropropylene 1,2-oxide (TCPO) was purchased from Aldrich Chemical Co. (Milwaukee, WI). HPLC grade solvents were purchased from Mallinckrodt, Inc. (Paris, KY). DMBA *trans*-5,6-dihydrodiol was obtained by incubation of racemic DMBA 5,6-epoxide with liver microsomes from PB-treated rats in a pH 8.9 buffer and the absence of NADPH (12).

Male Sprague-Dawley rats weighing 80-100 g were treated i.p. with phenobarbital (PB) (75 mg/kg body weight, injected in 0.5 mL of water) once daily on each of three consecutive days, or with 3-methylcholanthrene (MC) (25 mg/kg body weight, injected in 0.5 mL corn oil) once daily for each of four consecutive days. The rats were sacrificed the next day after the last injection of the drug and liver microsomes were prepared as described (13). Microsomal protein was determined by the method of Lowry *et al.* (14), with bovine serum albumin as the standard. Optically pure BA *trans*-5,6-dihydrodiol enantiomers were prepared by resolution of their diastereomeric di(-)menthoxyacetates (15). Optically pure DMBA *trans*-5,6-dihydrodiol enantiomers were obtained by CSP-HPLC (12,16).

CSP-HPLC

The enantiomers of BA 5,6-epoxide were separated with a semi-preparative HPLC column (10 mm i.d. x 25 cm, Regis Chemical Co., Morton Grove, IL) packed with spherical particles of 5 micrometer diameter of γ-aminopropylsilanized silica to which (*R*)-*N*-(3,5-dinitrobenzoyl)-phenylglycine ((*R*)-DNBPG) was bonded ionically (solvent: 1% (v/v) of ethanol/acetonitrile (2:1, v/v) in hexane; flow rate: 3 mL/min) (9). The enantiomers of DMBA 5,6-epoxide were separated with an analytical HPLC column (4.6 mm i.d. x 25 cm, Regis Chemical Co., Morton Grove, IL) packed with spherical particles of 5 micrometer diameter of γ-amino-propylsilanized silica to which (*S*)-*N*-(3,5-dinitrobenzoyl)leucine ((*S*)-DNBL) was bonded covalently (solvent: 0.5% (v/v) of ethanol/acetonitrile (2:1, v/v) in hexane; flow rate: 2 mL/min) (10,11). HPLC was performed using a Waters Associates (Milford, MA) liquid chromatograph consisting of a Model 6000A solvent delivery system, a Model M45 solvent delivery system, a Model 660 solvent programmer, and a Model 440 absorbance (254 nm) detector. Samples were injected via a Valco model N60 loop injector (Valco Instruments, Houston, TX). Retention times and ratios of enantiomers, determined by areas under the peaks, were recorded with a Hewlett-Packard Model 3390A integrator. Large amounts of optically pure epoxides were obtained by repetitive chromatography. Solvent was removed from the resolved enantiomers by evaporation under nitrogen.

Preparation of Biosynthetic Epoxides

Enzymatically formed K-region epoxides of BA and DMBA were each isolated from a mixture of products formed by incubation of the respective parent hydrocarbon with liver microsomes prepared from untreated, PB-treated, and MC-treated rats, respectively. A 100-mL reaction mixture contained 100 mg protein equivalent of rat liver microsomes, 5 mmol Tris-HCl (pH 7.5), 0.3 mmol of $MgCl_2$, 10 units of glucose-6-phosphate dehydrogenase (type XII, Sigma Chemical Co., St. Louis, MO), 10 mg of $NADP^+$, 48 mg of glucose-6-phosphate, and 0.06 mmol of the microsomal epoxide hydrolase inhibitor, TCPO. The reaction mixture was pre-incubated at 37°C for 2 min in a water shaker bath. BA or DMBA (8 µmol in 4 mL of acetone) was then added and incubated for 30 min. BA and its metabolites were extracted four times with 200 mL of hexane containing 2.5% (v/v) ethyl acetate. DMBA and its metabolites were extracted by sequential additions of acetone (100 mL) and ethyl acetate (200 mL) containing 2% (v/v) triethylamine. Organic solvent extracts were dehydrated with anhydrous $MgSO_4$, filtered, and evaporated to dryness under reduced pressure. BA 5,6-epoxide (retention time 6.0 min) was isolated on a DuPont (DuPont Co., Wilmington, DE) Zorbax SIL column (6.2 mm i.d. x 25 cm) eluted with 20% (v/v) of tetrahydrofuran in hexane at 2 mL/min. DMBA 5,6-epoxide (retention time 12.0 min) was also isolated on a Zorbax SIL column eluted with 10% (v/v) of ethyl acetate in hexane containing 0.3% (v/v) triethylamine at 2 mL/min. The epoxides isolated were dried under a stream of nitrogen and analyzed by CSP-HPLC as described above. Identities of resolved epoxide enantiomers were confirmed by ultra-violet-visible, CD, and mass spectral analyses, and were further validated by their ability to be enzymatically converted to K-region *trans*-dihydrodiols by rat liver microsomal epoxide hydrolase.

In order to establish if dihydrodiol(s) was formed in the liver microsomal metabolism of BA (or DMBA) in the presence of TCPO, the polar products were re-analyzed by reversed-phase HPLC after the 5,6-epoxide was removed by normal-phase HPLC as described above. BA *trans*-5,6-dihydrodiol was not detected when the products of microsomal incubation were extracted by either of the two extraction procedures described above. However, a minor amount of DMBA *trans*-5,6-dihydrodiol was found to be formed when DMBA was incubated with rat liver microsomes in the presence of TCPO under the conditions described above. The possibility that a fraction of the enzymatically formed DMBA 5,6-epoxide is converted non-enzymatically to DMBA *trans*-5,6-dihydrodiol is currently under investigation.

Enzymatic Hydration of Epoxide Enantiomers

Optically pure epoxide enantiomers (dissolved in acetone) were converted to *trans*-dihydrodiols by incubation with rat liver microsomes in the absence of NADPH. The reaction mixture (25 or 50 mL) which contained 0.05 M Tris-HCl (pH 7.5) and 2 mg protein equivalent of rat liver microsomes per mL of incubation mixture, was incubated at 37°C for 1 hr in a water shaker bath. Hydration products were extracted with 1 volume of acetone and 2 volumes of ethyl acetate. The resulting organic phase was dehydrated with anhydrous $MgSO_4$, filtered, and evaporated to dryness under reduced pressure. The dihydrodiol product was isolated by reversed-phase HPLC using a Vydac C_{18} column (4.6 mm i.d. x 25 cm; The Sep/a/ra/tions Group, Hesperia, CA) eluted with methanol/water (3:1, v/v) at a flow rate of 1.2 mL/min. Purified dihydrodiols were dried and redissolved in methanol for CD spectral measurements. The optical purity of BA *trans*-5,6-dihydrodiol was determined by comparison with CD spectral data of a BA *trans*-5,6-dihydrodiol enantiomer of known optical purity and absolute configuration. The optical purity of DMBA *trans*-5,6-dihydrodiol was determined by CSP-HPLC (12,16).

Spectral Analysis

Mass spectral analysis was performed on a Finnigan Model 4000 gas chromatograph-mass spectrometer-data system by electron impact with a solid probe at 70 eV and 250°C ionizer temperature. Ultraviolet-visible absorption spectra of samples (epoxides in hexane and dihydrodiols in methanol) were determined using a 1-cm path length quartz cuvette with a Varian Model Cary 118C spectrophotometer. CD spectra of samples (epoxides in hexane and dihydrodiols in methanol) were measured in a quartz cell of 1-cm path length at room temperature using a Jasco Model 500A spectropolarimeter equipped with a Model DP-500 data processor. The concentration of the sample is indicated by $A_{\lambda 2}$/mL (absorbance units at wavelength $\lambda 2$ per mL of solvent). CD spectra are expressed as ellipticity ($\Phi_{\lambda 1}/A_{\lambda 2}$, in millidegrees) for solutions that have an absorbance of 1.0 unit per mL of solvent at wavelength $\lambda 2$ (usually the wavelength of maximal absorption). Under conditions of measurements indicated above, the molecular ellipticity ($[\theta]_{\lambda 1}$, in deg.cm^2.dmole^{-1}) and ellipticity ($\Phi_{\lambda 1}/A_{\lambda 2}$, in millidegrees) are related to the extinction coefficient ($\epsilon_{\lambda 2}$, in cm^{-1}M^{-1}) as follows:

$$[\theta]_{\lambda 1} = 0.1 \ \epsilon_{\lambda 2} \ (\Phi_{\lambda 1}/A_{\lambda 2})$$

The CD spectral properties of optically pure enantiomers are: BA 5*R*,6*R*-dihydrodiol, Φ_{264}/A_{266} = −6.5 mdeg (methanol); DMBA 5*S*,6*S*-dihydrodiol, Φ_{237}/A_{269} = +26.1 mdeg (methanol); BA 5*S*,6*R*-epoxide, Φ_{316}/A_{269} = +3.0 mdeg (hexane); DMBA 5*R*,6*S*-epoxide, Φ_{275}/A_{275} = −14.1 mdeg (hexane). CD spectra of enantiomeric epoxides of BA (9) and DMBA (10) are reported earlier.

RESULTS AND DISCUSSION

With the exception of naphthalene 1*S*,2*R*-epoxide (17), all non-K-region epoxides of polycyclic aromatic hydrocarbons studied to date are enzymatically hydrated to dihydrodiols with *trans*-water attack at the nonbenzylic carbons of the epoxides (2,5-7,17,18). However, the position of *trans*-water attack of enantiomeric K-region epoxides in epoxide hydrolase-catalyzed hydrations cannot be reliably predicted (9,10,18,19). For example, both enantiomeric K-region epoxides of BP are hydrated by *trans*-water attack preferentially at the *S*-center (9,19). As the result, racemic BP 4,5-epoxide is hydrated to a *trans*-4,5-dihydrodiol highly enriched in the (−)4*R*,5*R* enantiomer (9,18,20).

Racemic BA 5,6-epoxide was hydrated to BA *trans*-5,6-dihydrodiol containing 60, 66, and 62% of the 5*R*,6*R* enantiomer by liver microsomes from untreated, PB-treated, and MC-treated rats, respectively (Table 1). The *trans*-5,6-dihydrodiols formed in the metabolism of BA by three rat liver microsomal preparations contain 77-84% of the 5*R*,6*R* enantiomer (Table 1). These results suggest that the metabolically formed 5,6-

Table 1. Enantiomeric Compositions of *trans*-5,6-Dihydrodiols Formed in the Hydration of Racemic 5,6-Epoxides of BA and DMBA by Rat Liver Microsomal Epoxide Hydrolase and in the Metabolisms of BA and DMBA by Rat Liver Microsomes.

Substrate	Control[a]	PB[a]	MC[a]
	$5R,6R$-Dihydrodiol (%)		
(±)BA 5,6-epoxide[b]	60	66 (71)[d]	62
(±)DMBA 5,6-epoxide[c]	8	9	8
BA[e]	77	81 (68)[f]	84 (81)[f]
DMBA[e]	11	5	6

[a] Liver microsomes prepared from untreated, PB-treated, and MC-treated male Sprague-Dawley rats, respectively.
[b] BA 5,6-epoxide (80 nmol per mL of incubation mixture) was incubated at 37° for 30 min with rat liver microsomes (2 mg protein per mL of incubation mixture) in the absence of NADPH.
[c] DMBA 5,6-epoxide (10 nmol per mL of incubation mixture) was incubated at 37° for 60 min with rat liver microsomes (3 mg protein per mL of incubation mixture) in the absence of NADPH.
[d] The hydration reaction was catalyzed by purified epoxide hydrolase in detergent solution (19).
[e] The enantiomeric compositions of the K-region dihydrodiol formed in the metabolism of BA (9) and of DMBA (12) was determined as described.
[f] Data from Thakker *et al.* (15).

epoxides are enriched in an enantiomer which is subsequently stereoselectively hydrated to $5R,6R$-dihydrodiol by microsomal epoxide hydrolase. The $(5S,6S):(5R,6R)$ enantiomer ratio of the *trans*-5,6-dihydrodiol formed by microsomal epoxide hydrolase-catalyzed hydration of BA $5S,6R$-epoxide is 1:9 (Fig. 1). This enantiomer ratio is 13:7 from the hydration of BA $5R,6S$-epoxide (Fig. 1). On the basis of these results, the enantiomeric compositions of BA 5,6-epoxides formed in the metabolism of BA by liver microsomes from untreated, PB-treated, and MC-treated rats can be calculated to have $(5S,6R):(5R,6S)$ enantiomer ratios of 26:74, 16:84, and 11:89, respectively. As far as predicting the major epoxide enantiomer is concerned, these ratios are very close to the enantiomeric compositions of the K-region epoxides formed in the metabolism of BA by rat liver microsomes (Fig. 1).

BA (or DMBA) 5,6-epoxide, formed in the metabolism of BA (or DMBA) by rat liver microsomes in the presence of the epoxide hydrolase inhibitor TCPO, was isolated from a mixture of metabolites by normal-phase HPLC using a Zorbax SIL column. The identification of metabolically formed BA (or DMBA) 5,6-epoxide is based on the following criteria: The metabolically formed epoxide was identical to the authentic standard with respect to its retention times on both reversed-phase and normal-phase HPLC, ultraviolet-visible absorption spectrum, and mass spectrum. Furthermore, when the metabolically formed BA 5,6-epoxide was incubated with liver microsomes from PB-treated rats in the absence of NADPH, a BA (or DMBA) *trans*-5,6-dihydrodiol was obtained.

Stereoselective Metabolism at the K-region 5,6-Double Bond of
DMBA (or BA) by Rat Liver Microsomes

CH$_3$

CH$_3$
12
5
7 6
CH$_3$

RLM
P-450

CH$_3$
CH$_3$
O
5R,6S

EH → (65%) 95% (35%) 5%

CH$_3$
H$_3$C OH OH
5S,6S

+

CH$_3$
H$_3$C OH OH
5R,6R

CH$_3$
CH$_3$
O
5S,6R

EH → 40% (10%) 60% (90%)

		5R,6S	5S,6R		5S,6S	5R,6R
DMBA (BA)	RLM (control)	76% (25%)	24% (75%)	EH	89% (23%)	11% (77%)
DMBA (BA)	RLM (PB)	80% (21%)	20% (79%)	EH	95% (19%)	5% (81%)
DMBA (BA)	RLM (MC)	97% (4%)	3% (96%)	EH	94% (16%)	6% (84%)

Fig. 1. The enantiomeric compositions of (i) the 5,6-epoxide formed in
the metabolism of BA (data in parentheses) and of DMBA by rat
liver microsomes in the presence of TCPO, (ii) the *trans*-5,6-
dihydrodiol formed in rat liver microsomal epoxide hydrolase-
catalyzed hydration of optically pure epoxide enantiomers of BA
(data in parentheses) and of DMBA, and (iii) *trans*-5,6-dihydro-
diol formed in the metabolism of BA (data in parentheses) and of
DMBA by rat liver microsomes. RLM (control), RLM (PB), and RLM
(MC) abbreviate for liver microsomes from untreated, PB-treated,
and MC-treated rats, respectively. Hydration of epoxide enantiomers was carried out with 2 mg protein equivalent of liver
microsomes from PB-treated rats per mL of incubation mixture.
The enantiomeric composition of the *trans*-5,6-dihydrodiol formed
in enzymic hydration of DMBA 5S,6R-epoxide is dependent on the
concentration of microsomal epoxide hydrolase (10).

Racemic DMBA 5,6-epoxide was hydrated to DMBA *trans*-5,6-dihydrodiol
containing 92, 91, and 92% in the 5S,6S enantiomer by liver microsomes
from untreated, PB-treated, and MC-treated rats, respectively (Table 1).
The *trans*-5,6-dihydrodiols formed in the metabolism of DMBA by rat liver
microsomes contain 89-95% in the 5S,6S enantiomer (Table 1). These
results suggest that the metabolically formed 5,6-epoxides are enriched
in the 5R,6S enantiomer which is subsequently stereoselectively hydrated
to the 5S,6S-dihydrodiol by microsomal epoxide hydrolase. The
(5S,6S):(5R,6R) enantiomer ratio of the *trans*-5,6-dihydrodiol formed by
microsomal epoxide hydrolase-catalyzed hydration of DMBA 5R,6S-epoxide
is 19:1 (Fig. 1) and is independent of the concentration of liver microsomes (10). Although the (5S,6S):(5R,6R) enantiomer ratio of the *trans*-
5,6-dihydrodiol formed in the hydration of DMBA 5S,6R-epoxide is 2:3
under the experimental conditions described (Fig. 1), it is highly
dependent on the concentration of microsomal protein present in the

incubation mixture (10 and Fig. 1). Furthermore, DMBA 5,6-epoxide may undergo non-enzymatic hydrolysis to DMBA *trans*-5,6-dihydrodiol and DMBA 5*S*,6*R*-epoxide may react with microsomal proteins (unpublished observations). Thus it is not possible to reliably calculate the enantiomeric compositions of DMBA 5,6-epoxide formed in the metabolism of DMBA by rat liver microsomes. However, the results do suggest that the 5*R*,6*S*-epoxide is the major enantiomer formed in the metabolism of DMBA by various rat liver microsomal preparations. These are confirmed by CSP-HPLC analyses of the 5,6-epoxides formed in the metabolism of DMBA by rat liver microsomes in the presence of TCPO (Fig. 1).

The enantiomeric compositions of the K-region epoxides formed in the metabolism of BA by liver microsomes from MC-treated rats (Fig. 1) are similar to those found by using purified cytochrome P-450c (P-448) in a reconstituted rat liver enzyme system (7). Cytochrome P-450c is the major (>70%) P-450 isozyme present in liver microsomes from MC-treated rats and is the most active form of P-450 isozyme in catalyzing the oxidative metabolism of BP (21,22). Thus the high degree of stereoselectivity in catalyzing the formation of K-region epoxide is due to the high content of P-450c (P-448) isozyme in liver microsomes from MC-treated rats. Since DMBA 5*R*,6*S*-epoxide is the predominant enantiomer formed in the metabolism of DMBA by liver microsomes from MC-treated rats, the stereoheterotopic interactions between the 5,6-double bond of BA and of DMBA with the catalytic site of cytochrome P-450c must be opposite (Fig. 1).

More than half of the cytochrome P-450 isozymes contained in liver microsomes from PB-treated rats is cytochrome P-450b and less than 2% is cytochrome P-450c (P-448) (22). The K-region epoxide formed in the metabolism of BA and of DMBA by liver microsomes from PB-treated rats has a (5*R*,6*S*):(5*S*,6*R*) enantiomer ratio of 21:79 and 80:20, respectively (Fig. 1). Results of immunochemical studies indicate that the oxidative metabolism of BP by liver microsomes from PB-treated rats is primarily catalyzed by cytochrome P-450b isozyme (22). The results shown in Fig. 1 thus indicate that cytochromes P-450b and P-450c have similar degrees of stereoselectivity but opposite absolute stereoselectivity in catalyzing the epoxidation reactions at the K-region 5,6-double bonds of BA and DMBA.

More than 85% of cytochrome P-450 isozymes contained in liver microsomes from untreated rats have not been characterized (22). Cytochromes P-450b and P-450c each constitutes 2-4% of the total cytochrome P-450 isozymes (22). Since the catalytic activities of the majority of cytochrome P-450 isozymes contained in the livers of untreated rats toward the metabolisms of BA and DMBA are not known, the enantiomeric epoxidation products formed by liver microsomes from untreated rats cannot be ascribed to either a specific cytochrome P-450 isozyme or a group of cytochrome P-450 isozymes. As a whole, the cytochrome P-450 isozymes contained in the livers of untreated rats have a lower degree of stereoselectivity than those of PB- or MC-treated rats in catalyzing the metabolism at the K-regions of BA and DMBA. The cytochrome P-450 isozymes contained in the livers of untreated rats also prefer opposite stereoheterotopic faces of the K-region double bonds of BA and DMBA for epoxidation reactions.

In summary, the enantiomeric compositions of K-region epoxides formed in the metabolism of BA and of DMBA by rat liver microsomes have been determined by CSP-HPLC method and by CD spectral analyses. Enantiomeric compositions of the metabolically formed epoxides were found to be dependent on the cytochrome P-450 isozyme content present in liver microsomes. The major cytochrome P-450 isozymes contained in liver

microsomes from untreated, PB-treated, and MC-treated rats, respectively, differ in their stereoselective preferences toward the stereoheterotopic K-region face of BA and of DMBA.

On the Model of the Substrate Binding Site of Cytochrome P-450c

In the livers of rats pretreated with 3-methylcholanthrene, greater
than 70% of the total cytochromes P-450 isozymes is cytochrome P-450c
(P-448) (22). The finding that BP is highly stereoselectively metabolized to dihydrodiols and bay-region 7,8-dihydrodiol-9,10-epoxides
(5,6,23) by liver microsomes from 3-methylcholanthrene-treated rats as
well as results from metabolism studies of other unsubstituted
polycyclic aromatic hydrocarbons led Jerina *et al.* (24) to propose a
model of the substrate binding site for cytochrome P-450c which
correctly predicted the stereochemistries of some non-K-region dihydrodiols formed in the metabolism of phenanthrene, chrysene, BA, and BP.
The model also correctly predicted the major K-region epoxide enantiomers formed from the metabolisms of BA and BP by cytochrome P-450c in a
reconstituted rat liver enzyme system (7-9,19). In view of the findings
that the 5R,6S-epoxide is the predominant stereoisomer formed in the K-
region metabolisms of 12-methyl-BA (11,25) and DMBA (10 and Fig. 1), the
proposed model (24) therefore cannot be generalized to include hydrocarbons such as 12-methyl-BA and DMBA. A 5S,6R-epoxide is the major enantiomer formed in the K-region metabolism of 7-methyl-BA (11,26). Thus
the formation of 5R,6S-epoxide enantiomer in the K-region metabolisms of
12-methyl-BA and DMBA are due to the presence of C_{12}-methyl group which,
due to steric crowding, forces the 1,2,3,4 benzo ring (A-ring) to pucker
up (or down) in an angle of 15.5° relative to the anthracene nucleus
(27). Apparently this ring puckering favors a particular three-dimen-
sional interaction between the cytochrome P-450 isozyme and DMBA (or 12-
methyl-BA) resulting in the enzymatic formation of 5R,6S-epoxide as the
major enantiomer.

The 1,2- and 3,4- positions of BA, monomethyl-BA, and DMBA do not
fit the catalytic binding site of the proposed model. Unless the
proposed boundary of the substrate binding site is expanded, the
formations of 1,2- and 3,4-dihydrodiols in the metabolisms of BA and
methyl-substituted BAs cannot be explained. Furthermore, the 1,2-double
bond of BA *trans*-3R,4R-dihydrodiol, the 3,4-double bond of BA *trans*-1,2-
dihydrodiol, the 7,8- and 9,10- double bonds of cholanthrene, the 1,2-
and 2,3- double bonds of BP all do not fit the catalytic binding site in
the proposed model, although metabolism is known to occur at these
double bonds (3,4,28-32). BA *trans*-1,2-dihydrodiol enantiomers can be
stereoselectively metabolized to the 1,2-dihydrodiol-3,4-epoxide (29).
BA 3R,4R-dihydrodiol is metabolized to the bay-region *anti*-3,4-dihydro-
diol-1,2-epoxide (33). BP 2,3-epoxide is a major intermediate in the
metabolism of BP (31). The formation of 1-hydroxy-BP as a metabolite of
BP suggests that BP 1,2-epoxide is also a metabolic intermediate (32).
Trans-1,2- and *trans*-3,4- dihydrodiols are among the major metabolites
formed from the metabolism of 8-methyl-BA (34,35) and of 8-hydroxy-
methyl-BA (36). Thus these and other compounds mentioned above must fit
into the catalytic binding site of cytochrome P-450c in some way, but
these were not considered by the model originally proposed (24). In
order to accommodate substrates such as those mentioned above, the
minimal size of the catalytic binding site of cytochrome P-450c must be
enlarged considerably. If enlarged, the proposed model would have lost
its predictive value. A model for the substrate binding site for cytochrome P-450c (or any one of the cytochrome P-450 isozymes) must be able
to account for the formations of known metabolites, regardless whether
the metabolites formed are chiral or achiral.

ACKNOWLEDGMENTS

This work was supported by National Cancer Institute grant no. CA29133 and Uniformed Services University of the Health Sciences Protocol no. RO7502. The opinions or assertions contained herein are the private ones of the authors and are not to be construed as official or reflecting the views of the Department of Defense or the Uniformed Services University of the Health Sciences. The experiments reported herein were conducted according to the principles set forth in the "Guide for the Care and Use of Laboratory Animals", Institute of Animal Resources, National Research Council, DHEW Pub. No. (NIH) 78-23.

REFERENCES

1   Sims, P., and Grover, P. L. (1974) *Adv. Cancer Res.* 20, 165-274
2. Jerina, D. M., and Daly, J. W. (1974) *Science* 185, 573-581
3. Gelboin, H. V. (1980) *Physiol Rev.* 60, 1107-1166
4. Conney, A. H. (1982) *Cancer Res.* 42, 4875-4917
5. Yang, S. K., McCourt, D. W., Leutz, J. C., and Gelboin, H. V. (1977) *Science* 196, 1199-1201
6. Yang, S. K., Roller, P. P., and Gelboin, H. V. (1977) *Biochemistry* 16, 3680-3686
7. Van Bladeren, P. J., Armstrong, R. N., Cobb, D., Thakker, D. R., Ryan, D. E., Thomas, P. E., Sharma, N. D., Boyd, D. R., Levin, W., and Jerina, D. M. (1982) *Biochem. Biophys. Res. Commun.* 106, 602-609
8. Armstrong, R. N., Levin, W., Ryan, D., Thomas, P. E., Mah, H. D., and Jerina, D. M. (1981) *Biochem. Biophys. Res. Commun.* 100, 1077-1084
9. Yang, S. K., and Chiu, P.-L. (1985) *Arch. Biochem. Biophys.*, 240, 546-552
10. Mushtaq, M., Weems, H. B., and Yang, S. K. (1984) *Biochem. Biophys. Res. Commun.* 125, 539-545
11. Weems, H. B., Mushtaq, M., and Yang, S. K. (1985) *Anal. Biochem.*, 148, 328-338
12. Yang, S. K., and Fu, P. P. (1984) *Biochem. J.* 223, 775-782
13. Alvares, A. P., Schilling, G., Garbut, A., and Kuntzman, R. (1970) *Biochem. Pharmacol.* 19, 1449-1455
14. Lowry, O. H., Rosebrough, N. J., Farr, A. L., and Randall, R. J. (1951) *J. Biol. Chem.* 193, 265-275
15. Thakker, D. R., Levin, W., Yagi, H., Turujman, S., Kapadia, D., Conney, A. H., and Jerina, D. M. (1979) *Chem.-Biol. Interac.* 27, 145-161
16. Yang, S. K., and Weems, H. B. (1984) *Anal. Chem.* 56, 2658-2662
17. Van Bladeren, P. J., Vyas, K. P., Sayer, J. M., Ryan, D. E., Thomas, P. E., Levin, W., and Jerina, D. M. (1984) *J. Biol. Chem.* 259, 8966-8973
18. Thakker, D. R., Yagi, H., Levin, W., Lu, A. Y. H., Conney, A. H., and Jerina, D. M. (1977) *J. Biol. Chem.* 252, 6328-6334
19. Armstrong, R. N., Kedzierski, B., Levin, W., and Jerina, D. M. (1981) *J. Biol. Chem.* 256, 4726-4733
20. Yang, S. K., Roller, P. P., and Gelboin, H. V. (1977) *Biochemistry* 16, 3680-3686
21. Lu, A. Y. H., and West, S. B. (1980) *Pharmacol. Rev.* 31, 277-295
22. Thomas, P. E., Reik, L. M., Ryan, D. E., and Levin, W. (1981) *J. Biol. Chem.* 256, 1044-1052
23. Thakker, D. R., Yagi, H., Akagi, H., Koreeda, M., Lu, A. Y. H., Levin, W., Wood, A. W., Conney, A. H., and Jerina, D. M. (1977) *Chem.-Biol. Interac.* 16, 281-300
24. Jerina, D. M., Michaud, D. P., Feldman, R. J., Armstrong, R. N., Vyas, K. P., Thakker, D. R., Yagi, H., Thomas, P. E., Ryan,

D. E., Levin, W. (1982) in: "Microsomes, Drug Oxidation, and Drug Toxicity", Sato, R. and Kato, R., eds., Japan Scientific Societies Press: Tokyo, pp. 195-201

25. Yang, S. K., Weems, H. B., and Fu, P. P. Ninth International Pharmacology Congress, 1984, abstract no. 510, London

26. Mushtaq, M., and Yang, S. K. (1985) An abstract presented at the 13th International Congress of Biochemistry, Amsterdam, The Netherlands, August 25-30, 1985

27. Jones, D. W., and Sowden, J. M. (1976) *Cancer Biochem. Biophys.* 1, 281-287

28. Yang, S. K., Roller, P. P., and Gelboin, H. V. (1978) in: "Carcinogenesis - A Comprehensive Survey. Polynuclear Aromatic Hydrocarbons", Freudenthal, R. and Jones, P. W., eds., Raven Press, New York, pp. 285-301

29. Chou, M. W., Chiu, P.-L., Fu, P. P., and Yang, S. K. (1983) *Carcinogenesis* 4, 629-638

30. Li, X. C., Fu, P. P., Chou, M. W., and Yang, S. K. (1983) in: "Polynuclear Aromatic Hydrocarbons: Formation, Metabolism, and Measurement", Cooke, M. and Anthony, J. D., eds., Battelle Press: Columbus, Ohio, pp. 583-598

31. Yang, S. K., Roller, P. P., Fu, P. P., Harvey, R. G., and Gelboin, H. V. (1977) *Biochem. Biophys. Res. Commun.* 77, 1176-1182

32. Selkirk, J. K., Croy, R. G., and Gelboin, H. V. (1976) *Cancer Res.* 36, 922-926

33. Thakker, D. R., Levin, W., Yagi, H., Tada, M., Ryan, D. E., Thomas, P. E., Conney, A. H., and Jerina, D. M. (1982) *J. Biol. Chem.* 257, 5103-5110.

34. Yang, S. K., Chou, M. W., Weems, H. B., and Fu, P. P. (1979) *Biochem. Biophys. Res. Commun.* 90, 1136-1141

35. Yang, S. K., Chou, M. W., Fu, P. P., Wislocki, P. G., and Lu, A. Y. H. (1982) *Proc. Natl. Acad. Sci., USA* 79, 6802-6806

36. Yang, S. K., Chou, M. W., Evans, F. E., and Fu, P. P. (1984) *Drug Metab. Disp.* 12, 403-413

INDUCTION OF UNSCHEDULED DNA SYNTHESIS IN HUMAN MONONUCLEAR
LEUKOCYTES BY OXIDATIVE STRESS

Ralf Morgenstern, Ronald W. Pero,
and Daniel G. Miller

Preventive Medicine Institute-Strang Clinic
55 East 34th Street, New York, N.Y. 10016

SUMMARY

The effect on unscheduled DNA synthesis (UDS) of different
agents that induce oxidative stress was investigated using human
mononuclear leukocytes (HML). It was found that Xanthine plus
Xanthine oxidase increaseed UDS by 47%. Hydrogen peroxide had no
significant effect. Cumene hydroperoxide increases UDS by 50%
with a large interindividual variation (7-100%). Bleomycin and
Mitomycin C increase UDS 189% and 295%, respectively. The validity
of UDS induced by these agents for screening interindividual differences in susceptibility to oxidative stress is discussed.

INTRODUCTION

Current theories hold that reactive oxygen species play a
role in tumor promotion, cancer and aging (1,2). Reactive species of oxygen are being formed in vivo as a consequence of normal
metabolism, radiation and exposure to redox-active compounds (2).
For example it has been shown that liver produces 5 nmol hydrogen
peroxide/min/g liver and that peroxisome proliferators which
greatly increase this production gives greater DNA damage (3).
The reactive forms of oxygen; i.e. singlet oxygen, hydrongen peroxide, superoxide anion radical and the hydroxyl radical, have all
been shown to be formed in vivo (2). It has also been shown that
they can cause chromosomal abberations (4), cell transformation
(5) and tumor promotion (1) either directly or by inducing secondary reactive compounds via initiating lipid peroxidation (see
2 for refs.). Cellular defense mechanisms against oxidative
stress include catalase, glutathione peroxidase, superoxide dismutase as well as cellular antioxidants including Vitamins C and
E and B-carotene (see 1,2 for refs.). Uric acid which is found
in high concentrations in human plasma has also been suggested to
have a protective role as an antioxidant (6).

Experiments with certain human hereditary diseases with
indications of abnormal oxygen metabolism (ataxia telangiectasia,
Fanconi's anemia and Bloom's syndrome) show that some cell types
from these patients are hypersensitive to DNA damage resulting
from agents that induce reactive oxygen (x-rays, UV, bleo-

mycin,mitomycin C, increased oxygen tension) (7).

General non-invasive procedures need to be developed to probe individual sensitivity to the DNA damaging effects of reactive oxygen species in humans. Such procedures would allow the monitoring of populations in order to determine if increased DNA damage induced by oxidative stress is a risk factor for cancer. Moreover the relation of reactive oxygen sensitivity to aging could also be determined. We have used human mononuclear leukocytes (HML) in our studies subjecting the isolated cells to different types of oxidative stress and measuring unscheduled DNA synthesis (UDS). As models of oxidative stress we have chosen hydrogen peroxide, xanthine-xanthine oxidase, cumene hydroperoxide (CuOOH) and bleomycin. Bleomycin is particularly interesting since it has been shown to give higher UDS in hepatocytes from older rats as compared to younger (8) as well as inducing more DNA damage in fibroblasts from patients with ataxia telangiectasia as compared to controls (7). It has also been shown that bleomycin induces more chromosomal damage in lymphocytes from patients with cancer as compared to controls (9).

MATERIALS AND METHODS

Heparinized vacutainers containing 143 U.S.P. of heparin per 10 ml tube were used to obtain 30-40 ml samples of peripheral venous blood from healthy donors. Mononuclear leukocytes were prepared as described (10). Incubations consisted of 5 x 10^6 cells that were incubated with the various agents inducing oxidative stress in 5 ml of Eagle's minimal essential media fortified with Hank's salt solution and 1% autologous plasma for 30 minutes at 37°C. Optimal concentrations giving highest UDS without cytotoxicity are shown. When chemical scavengers were used these were added immediately before the incubation whereas catalase and superoxide dismutase were preincubated with the cells 30 minutes 37°C before the oxidant was added. After exposure the cells were incubated in fresh culture medium for an additional 17 hr in the presence of 10 mM hydroxyurea and ^{3}H-dThd (10 μCi/ml) (25 Ci/mmol, Amersham Corporation, Arlington Heights, IL). All chemicals and enzymes used were from Sigma Chemical Co. St. Louis, MO. except bulk chemicals which were from Fisher Scientific N.J.

RESULTS AND DISCUSSION

Table 1 shows the effects on UDS by the different agents used to induce oxidative stress. As can be seen the xanthine-xanthine oxidase system increases UDS by 47%. Hydrogen peroxide alone at 1-200 μM (only 50μM shown in Table 1) does not significantly increase UDS. This indicates that the increase in UDS elicited by xanthine and xanthine oxidase requires superoxide anion radical, perhaps in combination with hydrogen peroxide. It has been shown that this reactive oxygen generating system can degrade DNA in vitro (11) and give rise to malignant transformation of cells in culture (5).

Organic hydroperoxides are known to cause oxidative stress (2,11) and as shown in Table 1 CuOOH causes a 50% increase in UDS as compared to controls. When different agents were used to block the UDS increase elicited by CuOOH the following were found to be without effect: catalase (0.1 mg/ml), superoxide dismutase (0.1 mg/ml), vitamin C (50 μM), vitamin E (10 μM) or

TABLE 1. Unscheduled DNA Synthesis In Human Mononuclear Leukocytes Induced By Oxidative Stress.

	n	UDS (cpm ^{3}H dThd/μg DNA)	% of Control	% Range
Hydrogen peroxide (50 μM)	3	180 ± 28[d]	109 ± 12	
Control	3	164 ± 8		
Xanthine (1 mM) + Xanthine- Oxidase (30 mU)	4	153 ± 35[c]	147 ± 18	(132-170)
Control	4	105 ± 28		
Cumene Hydroperoxide (50 μM)	15	190 ± 47[a]	150 ± 25	(107-200)
Control	15	127 ± 31		
Bleomycin (50 μg/ml)	4	210 ± 61[b]	189 ± 31	(168-236)
Control	4	110 ± 21		
Mitomycin C (100 μg/ml)	5	350 ± 81[a]	295 ± 146	(170-463)
Control	5	132 ± 43		
2-Aminofluorene (20 μM) + Hydrogen peroxide (50 μM)	2	191 ± 11[d]	128 ± 20	
Control	2	152 ± 32		

Probability Levels:
a= $p < 0.001$, b= $p < 0.05$, c= $p < 0.1$, d= not significant
Values are means± standard deviation.

the combination of the proteins on the one hand and the vitamins on the other. This indicates that secondary effects involving the production of hydrogen peroxide and superoxide are not responsible for the increase and that the capacity of exogenously added vitamins to scavenge any reactive intermediate is limited. Mannitol (50 mM) had some protective effect (21%, 52%) but further experiments are required to demonstrate if it is significant. CuOOH could be a useful agent to monitor genotoxicity from oxidative agents in individuals because of the broad range of response (7-100%). t-Butyl hydroperoxide also gives a similar increase in UDS (not shown). Studies are planned to monitor lung cancer patients as compared to matched controls.

Bleomycin and mitomycin C yield high numbers of mutants in the salmonella TA104 tester strain which is sensitive to a wide variety of oxidative mutagens (12). These agents also induce higher levels of UDS (Table 1). Since mitomycin C is also an alkylating agent, bleomycin has been chosen for further studies. Bleomycin also has other advantages already discussed in the introduction.

2-Aminofluorene has been shown to be metabolized by peroxidases, in the presence of hydrogen peroxide, to reactive intermediates that bind to macromolecules (13). This approach was used to quantitate if peroxidase metabolism in HML could give rise to increased UDS but no significant increase was obtained (Table 1). HML are very low in cytochrome P-450, and therefore, could be of potential value in probing interindividual differences in one-electron oxidation pathways of foreign compounds. As of yet, however, we have found no system sensitive enough.

N-acetoxy-N-2-fluorenylacetamide (N-AcO-2-FAA) is an example of a compound that gives a very high UDS in HML (10). This compound is deacetylated to N-OH-N-2-fluorenylacetamide that binds to DNA and gives a high UDS because of high DNA binding. The UDS value is also large because the bulky residues introduced into the DNA give a long-patch type of repair (14). In contrast the UDS increases by oxidative stress are probably low for 2 reasons: (i) the repair is probably of the short-patch type (10-30 fold shorter) as has been shown for instance with gamma radiation (13). If this were true then a low UDS value would reflect very significant DNA damage. (ii) at higher doses cytotoxicity becomes a problem. There are of course alternate explanations for increases in UDS other than direct DNA damage. For instance inhibition of ADP-ribosyltransferase activity can bring about increases in UDS (15).

In conclusion we have shown that various conditions of oxidative stress also give rise to increases in UDS in human mononuclear leukocytes. These observations form the basis for further studies into the human population as well as screening certain cancer populations. It is of great interest whether increased susceptibility to oxygen toxicity is a marker for cancer.

ACKNOWLEDGEMENTS

We are deeply indebted to John W. Kluge and the Cancer Prevention Fund at the Preventive Medicine Institute-Strang Clinic for financial support of this research project.

822

REFERENCES

1. P.A. Cerutti, 1985, Science, 227: 375.
2. B.N. Ames, 1983, Science, 221: 1256.
3. W.E. Fahl, N.D. Lalwani, T. Watanabe, S.K. Goll and J.K. Reddy, 1984, Proc. Natl. Acad. Sci.USA, 81: 7827.
4. See Carcinogenesis, A Comprehensive Survey, E. Hecker, N. Fusening, W. Marks, H. Thielman, Eds. Raven Press, New York (1982) Vol. 7.
5. S.A. Weitzman, A.B. Weilberg, E.P. Clark and T.P. Stossel, Science, 227: 1231.
6. B.N. Ames, R. Catheart, E. Schwiers and P. Hochstein, 1981, Proc. Natl. Acad. Sci. USA, 78: 6858.
7. P. Cerutti, in: Progress in Mutation Research, A. Natarajan, J. Altman, eds. Elsevier, Amsterdam, 4: 103 (1982).
8. H.E. Kennah, M.L. Coetzee and P. Ove, (1985) Mechanisms of Ageing and Development, 29: 283.
9. T.C. Hsu, L.M. Cherry and N.A. Samoan (1985) Cancer Genet. Cytogenetics. In Press .
10. R.W. Pero, C. Bryngelsson, F. Mitelman, T. Thulin and A. Norden, 1976, Proc. Natl. Acad. Sci. USA 73: 2496.
11. B. Halliwell and J.M.C. Gutteridge, (1984) Biochem. J. 219: 1.
12. D.E. Levin, M. Hollstein, D.F. Christman, E.A. Schwiers and B.N. Ames, 1982 Proc. Natl. Acad. Sci. USA 79: 7445.
13. J.A. Boyd, D.J. Harvan and T.E. Eling, (1983) J. Biol.Chem. 258: 8246.
14. A.A. Francis, R.D. Snyder, W.C. Dunn and J.D. Regan, (1981) Mutation Res. 83: 159.
15. J.E. Cleaver (1985) Cancer Res. 45: 1163

DEOXYGUANOSINE ADDUCTS FORMED FROM BENZOQUINONE AND HYDROQUINONE

L. Jowa, S. Winkle[1], G. Kalf[2], G. Witz and R. Snyder

Joint Program in Toxicology, [1] Department of Chemistry Rutgers Univ. and UMDNJ/Rutgers Medical/School, Piscataway N.J. 08854 and [2] Department of Biochemistry, Thomas Jefferson University, Philadelphia, PA. 19107

INTRODUCTION

The mechanism by which benzene produces bone marrow damage is not known with certainty, but it is generally accepted that the toxicity is mediated by one or more benzene metabolites. Radiolabeled metabolites of benzene administered in vivo have been shown to covalently bind to DNA in rat liver (Lutz and Schlatter, 1977) and mouse bone marrow (Gill and Ahmed, 1981). Rabbit bone marrow mitoplasts have been found to metabolize benzene to products which bind to mitochondrial DNA (Rushmore, et al., 1984). Hydrolysis of the DNA to nucleosides yielded 6-7 possible adducts to guanosine. It is hypothesized that the covalent binding of benzene metabolites to DNA may be an etiological factor in benzene-induced bone marrow depression.

Rushmore et al. (1984) suggested that adducts of deoxyguanosine (dG) may arise from reactions between DNA and phenol, hydroquinone (HQ), catechol, benzoquinone (BQ) and trihydroxy benzene. However the structures of the adducts were not identified. The aim of the work reported here was to characterize the adducts of dG formed during reactions with HQ and BQ. This study will show that HQ and BQ can react with deoxyguanosine (dG) to produce several adducts. One of these adducts has now been characterized.

METHODS

Deoxyguanosine Adduct Formation

Adducts of HQ with dG were formed by incubating 0.02 mg of $[^{14}C]$HQ (300 uCi/mm) with 1 mg of $[^{3}H](1',2')$dG (0.032 uCi/mm) for 18 hours at 37°C in 0.05M potassium phosphate buffer, pH 7.2, including 1.5 mM ferrous chloride. Upon termination of the reaction, the mixture was analyzed by HPLC using a C-18 column, a 5-100% gradient and a flow of 1 ml/min. Identical mixtures but omitting iron were used as controls. Samples were collected and counted in a liquid scintillation spectrometer.

Adducts of BQ with dG were formed by incubating BQ (0.5 mg/ml) with dG (1 mg/ml) in 0.05M potassium phosphate buffer, pH 7.2, for 18 hours at

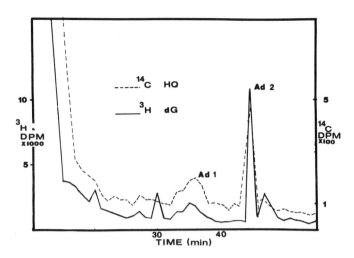

Fig. 1.   Adducts formed from HQ and dG.   One mg of
(1',2') [³H] dG, sp. act. 0.32 uCi/mM was
incubated with 1.5mM FeCl₃ and 5 uCi OF [¹⁴C]
hydroquinone in 0.05M potassium phosphate
buffer pH 7.2 at 37° for 18 hrs.   A 20 uL
sample was applied to a C18 HPLC column and
run in a water-methanol gradient.   After 10
min.   One ml samples were collected and
counted.

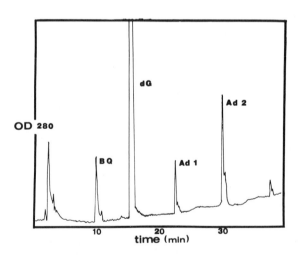

Fig. 2.   Adducts formed from BQ and dG. Deoxyguanosine
(1 mg/ml) was incubated with benzoquinone
(0.5 mg/ml) in 0.05M potassium phosphate
buffer pH7.2 for 18 hours at 37°.   A 0.02 ml
sample was applied to a C18 column and run in
a water-methanol gradient. The HPLC profile
was obtained by monitering at 280nm (scale is
0.100).

37°. HPLC was performed as described above except that reactants and products were detected by UV absorption at 280 nm. In preparative studies BQ and dG were reacted (1:1) in 100ml Kphosphate buffer pH 7.2. Ethanol was added 1:1 and the precipitate which was formed was discarded. The mixture was concentrated by evaporation, purified by preparative HPLC and lyophilized.

Characterization of Adduct 2

A series of UV spectra were obtained of the adduct at several different pHs. Fluorescence spectra were obtained for HQ, dG and the adduct. The adduct was suspended in D_2O and scanned repeatedly in a Varian XL400 NMR. A mass spectrum was obtained of adduct 2 after derivatization with trimethylsilane (H/TMS) and deuterated trimethylsilane (D/TMS). The sample was then inserted on a direct probe into a Finnegan MAT212 mass spectrometer in electron impact mode at 90ev. The mass of the molecular ion was determined by the the peak matching method with perfluorokerosine.

Formation of Adducts with DNA

One mg of purified calf thymus or __Micrococcus lysodeikticus__ DNA was incubated with $[^{14}C]HQ$ for 24 hours in the presence of iron as described above. The DNA was denatured and subjected to complete enzymatic hydrolysis according to Yamazaki, et al(1977). The hydrolysate was then chromatographed as described above.

RESULTS

Two dG adducts were formed with HQ and BQ (Figures 1,2) with the retention time of 22 min, (adduct 1) and 30 min, adduct 2. The presence of iron was required for HQ to form adducts with dG probably because iron oxidized HQ to the more electrophilic BQ or a semiquinone species. Adduct 2 was chosen to be isolated and characterized further.

Adduct 2 has a unique UV spectrum which differs from the spectra obtained from dG or BQ/HQ alone at several pHs (Figure 3a). The adduct is also fluorescent whereas dG and HQ/BQ are not (Figure 3b).

Adduct 2 was also studied using proton NMR. Figure 4 shows an NMR spectrum of Adduct 2 run in deuterium oxide. Reading from left to right the signals correspond with the C8 proton, three protons on the quinone (labeled BQ), and the remainder are assigned to the ribose (R1', R2', etc.). These data suggest that it is unlikely that N-7 on the purine was the point of atack by the reactive metabolite since that would probably have caused the loss of the ribose.

Adduct 2 was derivatized with trimethylsilane and a mass spectrum was obtained using the electron ionization mode (Figure 5). A peak probably cooresponding to the molecular ion was found with 573 m/z. The adduct was also derivatized with D/TMS (not shown) in which all the TMS hydrogens were substituted with deuterium atoms. By comparing the m/z of the molecular ion from the H/TMS and that from D/TMS it was possible to ascertain the number of TMS bound groups. The difference between 600 (D/TMS) and 573 (H/TMS) was 27. There are 9 additional mass units for each TMS when deuterium is substituted for hydrogen; so the molecular ion had 3 TMS groups. The most prominent ion in the H/TMS spectrum was 313 m/z, and had one TMS group.

The formula derived for the molecular ion was was $C_{16}H_{15}N_5O_5+$ 3 TMS for which a calculated molecular weight would be 573.2259; the actual molecular weight was 573.2246, which agreed well with the calculated

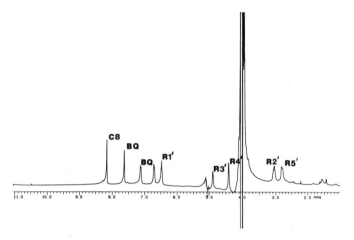

Fig. 4. NMR spectrum of deoxyguanosine adduct 2. The deoxyguanosine adduct 2 was suspended in D_2O with dioxane as the internal reference. The spectrum was obtained on a Varian XL 400 spectrometer; number of transmissions = 500 and T=22°C.

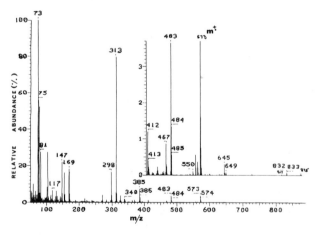

Fig. 5. Mass spectrum of adduct 2. Adduct 2 was derivatized with TMS before and inserted by direct probe into a Finnigan MAT mass spectrometer.

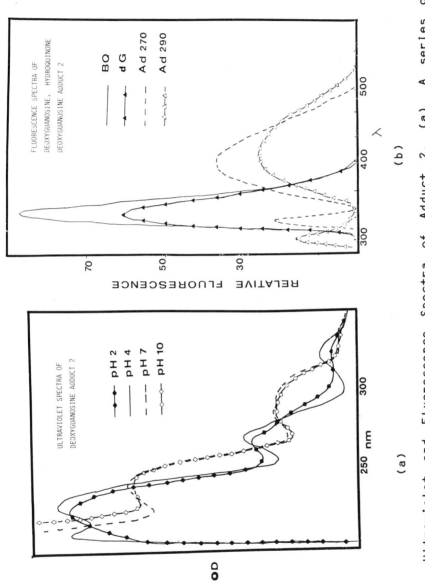

Fig. 3. Ultraviolet and Fluorescence Spectra of Adduct 2. (a) A series of UV spectra was obtained on 1 ug of deoxyguanosine adduct #2 suspended in water adjusted to pH 2,4, 7 and 10. Absorbance was measured from 200 to 350 nm. (b) Benzoquinone (1ug) with deoxyguanosine (1ug) were suspended in water, pH4.0 and excited at 290 and 270, respectively. Approximately 1 ug of deoxyguanosine Adduct 2 was excited at 290 (Ad 290) and at 270 (Ad 270). Flourescence emission was scanned from 290.to 530 nm.

829

weight. Then the molecular weight of the adduct without TMS would be 357. If the molecular weights of deoxyguanosine, 267 and BQ, 108 where added together the total would be 375. The difference between 375 and 357 is 18, the molecular weight of water. The molecular weight of the largest peak cooresponds to the molecular weight of guanine and a C_6H_4O moiety. Since TMS reacts primarily with OH groups, four TMS derivatives would have been expected if there were two OHs on the ribose and two on the reduced quinone. Since only three were observed, it was possible that an OH group was lost during the reaction. A conclusion that can be drawn from these results is that there is an alkylation reaction occurs between dG and BQ which results in the loss of one molecule of water.

On the basis of both proton NMR and mass spectral results, a possible reaction mechanism and structure was postulated (Figure 6). The first step in the reaction might be the nucleophilic attack by the exocyclic amine nitrogen (N-2) of dG on the electrophilic alpha carbon of the quinone followed by stabilization through enolization. The ring is formed by nucleophilic attack by N-1 of dG on the remaining keto group of the quinoid intermediate generated by the previous step. The resulting planar, nearly aromatic compound is depicted in Figure 6. Additional support for this structure comes from an observation made during the proton NMR studies where a drop in temperature from 40° to 5° leads to an upfield shift of 0.2 ppm, a feature that is consistent with the stacking of aromatic rings. Peak broadening at low temperatures also indicates aggregation which occurrs with planar molecules.

Although it was clear that Adduct 2 could be formed using the nucleoside as the active metabolite trap, it was important to determine whether it was formed with DNA (Figure 7). Adduct 2 could be produced from purified DNAs under the same conditions which were required when dG was the receptor. More Adduct 2 was produced from calf thymus DNA than from M. lysodeikticus DNA which is probably due to the different configuration of the DNA.

DISCUSSION

In this study we produced a (3'OH)benzetheno(1,N-2)deoxyguanosine adduct from the reaction of HQ/BQ with deoxyguanosine. Adducts to the N-2 position have been reported with benzo(a)pyrene, methylaminobenzene, acetoaminofluorene among others (Singer and Grunberger, 1983). Adducts to the N-2 and the O-6 position are known to be extremely stable, being less susceptible to depurination than C-8 or N-7 adducts. Adducts to the N-2 maybe resistent to repair processes. Benzo(a)pyrenediol epoxide N-2 adduct was reported to be stable for at least half a year, whereas the cooresponding N-7 adduct had a half life of 3 hours (Osborne, 1984).

Shapiro and Hachman identified a cyclic N-1, N-2 adduct of guanine formed by reacting with glyoxal, a dicarbonyl compound. Chloroacetaldehyde, a purported mutagenic metabolite of vinyl chloride, also formed a cyclic(1,N-2 ethenoguanosine) adduct with guanine (Singer and Grunberger, 1983). Both of these adducts were formed slowly in mild conditions, similar to the benzoquinone adduct.

The reaction to both the N-1 and N-2 sites of guanosine is predictably difficult. Both sites are buried within the helix; the N-2 is only exposed in the minor groove. Binding to these sites requires the disruption of hydrogen bonding and probably would result in the distortion of the helix. We beleive that such an adduct may be formed when the DNA is at least partially unwound, as during the processes of transription and replication. The benzoquinone would attack the N-2 positions in the minor grove or areas where the N-2 would be exposed (as in Z-DNA). There it may

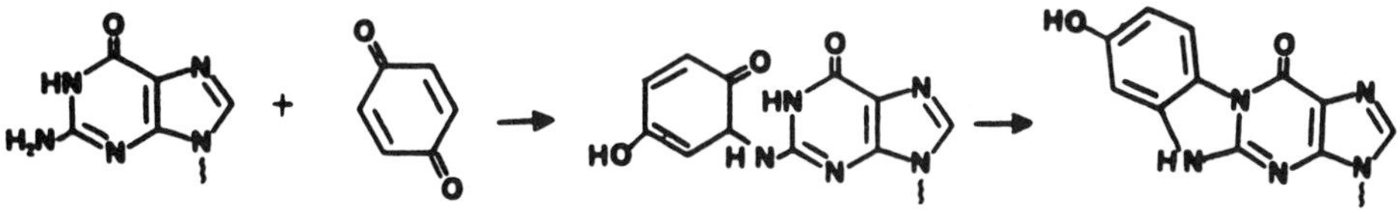

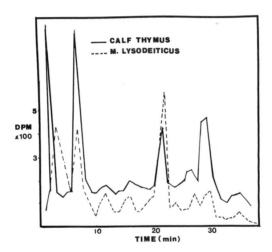

Fig. 6. Proposed mechanism of adduct 2 formation. Deoxyguanosine reacts with benzoquinone to form a N-2 adduct intermediate. A reaction with the N-1 position forms adduct 2.

Fig. 7. Hydroquinone adducts derived from DNA. Calf thymus and __Micrococcus lysodeikticus__ DNA (1mg/ml) was suspended in potassium phosphate buffer 0.05M, pH 7.2 with 1.5mM $FeCl_2$, and 0.005mCi of $[C^{14}]$ Hydroquinone. Incubation was carried out at 37° for 24 hrs. The DNAs were then subjected to a sequential enzymatic digestion to produce deoxynucleosides. Samples were then chromatographed in a water-methanol gradient, 1 ml samples were collected and counted.

remain dormant until N-1 position looses its H bonding, and becomes exposed long enough to form the cyclic adduct. The persistence of the cyclic adduct would then depend on the efficiency of DNA repair. In the rapidly replicating cells of the bone marrow, the repair process may be not effective in removing the adduct and miscoding or misspairing may result. Further work is needed to determine whether the formation of the BQ-adduct 2 is related to benzene induced leukemia or aplastic anemia.

REFERENCES

Gill, D.P. and Ahmed, A (1981). Covalent binding of [^{14}C] benzene to cellular organelles and bone marrow nucleic acids. Biochem. Pharmacol., 30 :1127.

Lutz, W.K. and Schlatter, C. H. (1977) Mechanism of carcinogenic action of benzene; irreversible binding to rat liver DNA. Chem. Biol. Interact., 18:241.

Osborne, M.R. (1984) DNA interactions of reactive intermediates derived from carcinogens, In:"Chemical Carcinogens" 2nd ed. Searle, C ed. ACS monograph 182, American Chemical Society, Washington, D.C. p508.

Rushmore. T., Snyder, R and Kalf, G. (1984). Covalent binding of benzene and its metabolites to DNA in rabbit bone marrow mitochondria In vitro. Chem. Biol. Interact., 49:133.

Shapiro, R. and Hachman, S. (1969). The reaction of guanine derivatives with 1,2 dicarbonyl compounds. Biochemistry, 8:238.

Singer, B. and Grunberger, D. (1983) "Molecular Biology of Mutagens and Carcinogens" Plenum Press. N.Y.

Yamazaki, H., Pulkrolock, K.,Grunberger, D. and Weinstein, J.B. Differential excission from DNA of C8 and N2 guanosine adducts of N-acetyl-2-amino fluorene by single strand specific endonucleases. Cancer Res., 37:3756.

3-HYDROXY-TRANS-7,8-DIHYDRO-7,8-DIHYDROXY-BENZO(a)PYRENE, A METABOLITE OF 3-HYDROXYBENZO(a)PYRENE

O. Ribeiro, C.A. Kirkby, P.C. Hirom, and P. Millburn

Dept. of Biochemistry, St. Mary's Hospital
Medical School, London, U.K.

INTRODUCTION

3-Hydroxybenzo(a)pyrene(3-OH-BP) (I) is a major
metabolite of the environmental pro-carcinogen BP. It is
produced in a wide variety of biological systems (1-4) and
is excreted as a glucuronide in bile (5, 6). Under
appropriate conditions it binds to DNA (7) but it is not
carcinogenic (8) and is a poor tumour initiator (9).

I

Glatt and Oesch (10) found, however, that 3-OH-BP is
weakly mutagenic to Salmonella typhimurium TA98 without
metabolic activation. Under identical conditions BP is not
mutagenic. These results were interpreted as being due to
the generation of mutagenic radicals during spontaneous
oxidation of 3-OH-BP to BP-3,6-quinone. When the Ames Test
is performed in the presence of microsomes derived from
C57BL/6N mice induced with 3-methylcholanthrene, Owens et
al. (11) found that 1-OH- and 3-OH-BP are highly mutagenic,
being comparable in mutagenicity to BP-7,8-diol. They
postulated that 1-OH- and 3-OH-BP are activated via a 7,8-
diol-9,10-epoxide. Very little is known about the
metabolites derived from 3-OH-BP other than that it forms a
highly lipophilic sulphate (12,13) and a glucuronide,
which undergoes enterohepatic circulation (14). Treatment
of the latter conjugate with β-glucuronidase yields
intermediates which are capable of binding to DNA (15).
BP-3,6-quinone has been implicated as an oxidative

metabolite of 3-OH-BP (16, 17) but this cannot account for its mutagenicity since the quinone, though weakly cytotoxic, is not mutagenic (10). We have therefore investigated the metabolism of 3-OH-BP in the rat and present our findings on the metabolites found in the bile.

MATERIALS AND METHODS

Chemicals and biochemicals

BP-4,5-diol, BP-7,8-diol, BP-3,6-quinone, and [^{3}H]3-OH-BP were obtained from the National Cancer Institute, Carcinogenesis Research Program, Bethesda, Md. USA. Unlabelled 3-OH-BP was prepared by the method of Cook et al. (18) as modified by Yagi et al. (19), and was identical to an authentic sample supplied by the National Cancer Institute. 5-OH-BP was prepared by acidic dehydration of BP-4,5-diol (20). The rest of the chemicals were of the highest purity available. Helix pomatia type H-2 β-glucuronidase (containing aryl sulphatase) and Tween 80 were purchased from Sigma Chemical Co. Ltd. (Poole, Dorset, UK).

Spectra

U.v. spectra were recorded in methanol using a Beckman DU-6 spectrophotometer. 250 MHz proton magnetic resonance (p.m.r.) spectra were recorded on a Bruker WH250 Fourier transform spectrometer equipped with an Oxford Instruments superconducting magnet. Mass spectra were measured on a ZAB 1F analytical mass spectrometer. It was operated in the electron impact (EI) mode with electron energy of 70ev, and source temperature of 200°C.

Chromatographic methods

H.p.l.c. was carried out using a Waters apparatus: two Model 6000A solvent delivery systems, a Model 660 solvent programmer, a Model 440 u.v. absorbance detector and a Model U6K injector (Waters Chromatography Division, Millipore (UK) Ltd., Harrow, Middlesex, UK). The u bondapack C_{18} column (0.39 x 30 cm, from Waters) was eluted with a linear gradient of 60-90% methanol in water over 40 min at a flow rate of 1 ml/min. Eluants were monitored for u.v. absorbing material at 254 nm.

P.l.c. was performed on 20 x 20cm (2 mm thickness) plates of silica gel 60 F_{254} (Merck, Darmstadt, F.R.G.), which fluoresce green when viewed under u.v. light: CC-20G chromato-view cabinet operating in the short wave mode (Ultra-violet Products Inc., San Gabriel, Ca., USA). The plates were developed with toluene:ethanol (9:1 v/v), dried under nitrogen and re-run in the same solvent system.

Animals and treatment

Female Lewis or male Wistar rats (250-300g) were anaesthetised with sodium pentobarbitone (Sagatal, May and Baker Ltd., Dagenham, Essex, UK; 60 mg/kg) given by i.p. injection and the bile ducts cannulated with vinyl tubing

(0.4 mm i.d., 0.8 mm o.d.; Portex Ltd., Hythe, Kent, UK).
^{3}H-Labelled 3-OH-BP (50 mg/kg; 32 uCi/kg) dissolved in 0.2
ml of Tween 80:ethanol (1:2 v/v) was injected i.p. Bile was
collected at hourly intervals for 5 hours. The rats were
kept anaesthetised by further i.p. injections of sodium
pentobarbitone when necessary.

Isolation of metabolites.

For each rat, the five hourly bile samples were pooled
for metabolite analysis. Samples (15 ml) of the bile were
added to an equal volume of Sigma H-2 β-glucuronidase/
arylsulphatase and incubated at 37°C overnight in the dark.
Incubates were then diluted with an equal volume of 0.1M
sodium acetate buffer (pH 5) and extracted with an equal
volume of ethyl acetate (x4). Phase separation could be
achieved only by centrifugation at 10,000 rev/min for 30
min. The combined organic fractions were dried (Na_2SO_4) and
evaporated to dryness. The residue was dissolved in MeOH
(0.5ml). A small sample was analysed by h.p.l.c. The bulk
of the crude aglycones were applied to two p.l.c. plates and
developed as described above. The silica gel bands with Rf
values of 0.2 and 0.4, which had a blue fluorescence, were
scraped off and eluted with methanol (25 ml). The eluates
were examined by h.p.l.c. to check the purity of the
isolated metabolites.

RESULTS

Excretion of [^{3}H]3-OH-BP metabolites in bile

When [^{3}H]3-OH-BP (1 mg/kg i.v.) was given to bile-duct
cannulated male Wistar rats, 68.0±5.4% of the dose was
excreted in bile in 4 hours, 50.3±5.5% in 1h. On analysis
by h.p.l.c. unhydrolysed 0-1h bile samples gave three major
radiolabelled, u.v. peaks representing (i) 18.8±1.2%, (ii)
38.5±0.7% and (iii) 16.0±1.9% of biliary ^{3}H respectively.
The h.p.l.c. retention time of (i) was extended to 20 or 21
mins respectively by hydrolysis with sulphatase or
glucuronidase. Hydrolysis with both enzymes simultaneously
gave a retention time of 34 mins. Peaks (ii) and (iii) of
the unhydrolysed bile were BP-3-0-glucuronide and BP-3-0-
sulphate respectively since on enzymic hydrolysis 3-OH-BP
was released.

Following i.p. administration of [^{3}H]3-OH-BP to female
Lewis rats at the higher dose of 50 mg/kg about 13% of the
dose was excreted in bile in 5h. The rate of excretion of
^{3}H was found to be between 1 and 4.5% of the dose per h. The
biliary radioactivity was predominantly in conjugated form
since hydrolysis with β-glucuronidase and sulphatase was
necessary before approximately 70% of the ^{3}H-labelled
biliary metabolites could be extracted into ethyl acetate.
Figure 1 is the h.p.l.c. elution profile of the aglycones.
Peaks I and II are identical with bands of Rf 0.2 and 0.4,
respectively, found on p.l.c. of the aglycones. Peak III
contains a product that is indistinguishable in its
chromatographic and u.v. spectral properties from BP-3,6-
quinone. Peak IV is 3-OH-BP.

Identification of Peak I as 3-OH-BP-7,8-diol

The u.v. spectrum of this metabolite was found to shift to longer wavelengths on the addition of a drop of alkali and this is, of course, a characteristic of aromatic phenols. The similarity between the spectrum of Peak I and that of BP-7,8-diol was quite striking and suggests that Peak I is probably a derivative of BP-7,8-diol. The shift of the absorption maxima of Peak I to slightly longer wavelengths suggests substitution on the pyrene moiety of BP-7,8-diol by an electron donating substituent. Hence Peak I is probably 3-OH-BP-7,8-diol (II). It was predicted that the shift to the longer wavelengths caused by the hydroxy-function would be nullified by acetate formation. This was confirmed by the spectrum of Peak 1-acetate which is virtually identical to that of BP-7,8-diol.

The p.m.r. spectrum (Figure 2) is consistent with Peak I being 3-OH-BP-7,8-diol. The detailed assignments were made by comparison with the spectra of known dihydrodiols (21, 22) and by decoupling studies. The alkyl and vinyl protons were readily assigned and the large coupling constant between the carbinol protons ($J_{7,8}$ = 10.7Hz) suggests that this is a <u>trans</u>-isomer and that the hydroxyl functions are in a predominantly <u>quasi</u>-diequatorial conformation. Normally, due to edge deshielding effects, it would be expected that the aromatic hydrogen in the bay region (H_{11}) would be the proton lowest downfield (12, 19 and references therein). In this case we believe we have an exception and that the proton with the lowest downfield signal is, in fact, the H_4 proton. The logic used to assign the H_4 and H_{11} protons is as follows: Introduction of a phenolic function into the 3-position of BP causes an upfield displacement in all the protons (H_{11} = 0.26 p.p.m. upfield) with the exception of the H_4 proton which is peri to the 3-position and has a downfield shift of 0.34 p.p.m (19). The introduction of a phenolic function into the

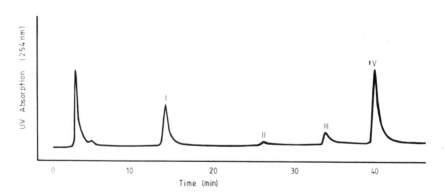

Figure 1   H.p.l.c. of 3-OH-BP metabolites after treatment of bile with β-glucuronidase and aryl sulphatase. The conditions for this separation are described in the text.

3-position of BP-7,8-diol should have a similar effect. For
BP-7,8-diol, Harvey (personal communication) has assigned
the chemical shift of 8.46 p.p.m. to the H_{11} proton and
that of 8.15 (or 8.12) p.p.m. to the H_4 proton, so the
introduction of a 3-hydroxyl-function will result in the
chemical shift for the H_{11} proton being less than 8.46
p.p.m. and that for the H_4 proton being greater than 8.15
(or 8.12) p.p.m.   Thus, the doublet at 8.29 p.p.m. is the H_4
proton and that at 8.12 p.p.m. is the H_{11} proton;
introduction of the 3-phenolic function into BP-7,8-diol
causes a 0.14 p.p.m. shift downfield for the H_4 proton and
an upfield shift of 0.34 p.p.m. for the H_{11} proton.

II

      Accurate mass determination of Peak I and low
resolution EI mass spectra (Table I) are consistent with
Peak I being a dihydrodiol with the mass ion at m/z 302.
Fragments at m/z 285 (M-17), 284 (M-18), 271 (M-31), 268
(M-34), 256 (M-46), 255 (M-47) and 253 (M-49) follow the
pattern of fragmentation observed for other dihydrodiols by
McCaustland et al. (23).   The m/z 273 (M-29) fragment is a
strong indication of the presence of a phenolic moiety in
the molecule.   The fragment m/z 226 (M-76) is probably due
to loss of 2 (COH) from the base peak m/z 284 (relative
abundance 100%) and would be expected from phenolic-
dihydrodiols derived from BP.

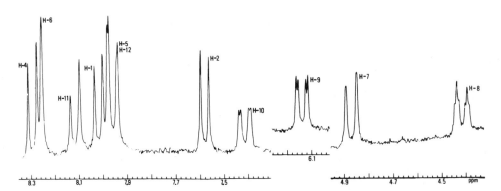

Figure 2   250 MHz p.m.r. spectra of the 3-OH-BP metabolite,
            Peak I in DMSO-d_6.   The coupling constants (Hz)
            are: $J_{1,2}$ = 8.4; $J_{4,5}$ = 9.1; $J_{7,8}$ = 10.7;
            $J_{8,9}$ = 2.1; $J_{8,10}$ = 2.0; $J_{9,10}$ = 10.3;
            $J_{11,12}$ = 9.5.

Table I. Relative abundance of peaks (>4%) with
a m/z >200 in the mass spectrum of 3-OH-BP-7,8-diol

M^+ found 302.0949; $C_{20}H_{14}O_3$ requires 302.0943

m/z	%		m/z	%
303	8.1		241	5.6
302	28.3	(M^+)	240	5.6
287	6.0		239	17.2
286	6.0		237	4.7
285	25.7		231	9.9
284	100	$([M-H_2O]^+)$	227	15.0
283	13.7		226	32.2
273	6.9		225	7.7
272	4.3		224	8.2
271	10.7		219	6.4
269	6.4		218	6.7
268	15.9		217	6.4
267	6.4		215	6.0
266	10.3		213	12.8
257	9.0		211	4.3
256	36.5		210	8.6
255	48.9		206	8.6
253	5.6		205	45.1
252	6.0		202	6.4
251	41.2		201	7.3
243	5.6			

From its u.v., p.m.r. and mass spectral properties, it
is quite conclusive that Peak I is 3-OH-BP-7,8-diol (II).

Tentative identification of Peak II as 3,5-di-OH-BP

This highly labile metabolite is much less polar than
3-OH-BP-7,8-diol, therefore the other possible benzo-ring
dihydrodiol metabolite 3-hydroxbenzo(a)pyrene-9,10-diol can
be ruled out on the basis of polarity. Its u.v. spectrum is
very similar to the spectra of 4-OH-BP and 5-OH-BP, but
shifted to longer wavelengths, hence Peak II could be either
3,4-di-OH-BP or 3,5-di-OH-BP. The u.v. spectra of these
phenols are highly characteristic and virtually identical,
the only difference being the molar extinction coefficients
(20). Thus, it is not possible to distinguish between 3,4-
di-OH-BP and 3,5-di-OH-BP by u.v. spectral properties alone.
However, it has been shown that rearrangement of
benzo(a)pyrene-4,5-epoxide yields exclusively 5-OH-BP, also
acidic dehydration of BP-4,5-diol yields 5-OH-BP as the sole
product (20). Therefore, on balance, we believe that Peak
II is probably 3,5-di-OH-BP.

DISCUSSION

Oxidative metabolites of BP are excreted as conjugates
in the bile and persist in vivo via enterohepatic recycling
(6, 14, 24). These conjugates include glucuronides,

sulphates (14) and glutathione derivatives (25, 26). Bile
from rats and rabbits dosed with BP is mutagenic to
Salmonella typhimurium in the presence of β-glucuronidase
(27, 28) and one source of the biliary mutagen(s) is BP-3-0-
glucuronide (29).

The present investigation shows that when 3-OH-BP
itself is administered to rats it is excreted in bile partly
as a conjugate of 3-OH-BP-7,8-diol. This conjugate appears
to be a double conjugate with sulphate on position 3 and
glucuronic acid on either positions 7 or 8. In the
intestinal tract this biliary conjugate could be hydrolysed
to the free 3-OH-BP-7,8-diol which may then undergo further
metabolism within the intestinal cells to a triol-epoxide
(III) capable of binding to macromolecules, or the
triol may pass via the portal blood and liver to other
tissues of the body. Jernstrom et al. have postulated that
this triol-epoxide may be responsible for the covalent
binding of BP-7,8-diol and BP-7,8-diol-9,10-epoxide to
protein in microsomes (30) and isolated hepatocytes (31)
from 3-MC treated rats.

The microsome-mediated binding of 3-OH-BP to DNA (7)
yields adducts which, on chromatography on Sephadex LH20,
have elution volumes similar to those of DNA-BP-diol-
epoxide adducts. However, these adducts could be derived
from the proposed triol-epoxide of BP since both the diol-
and triol- epoxide DNA adducts derived from chrysene are
known to have very similar elution volumes from Sephadex
LH20 (32).

Further, it may be significant that the triol-epoxide
thought responsible for the binding of chrysene to DNA in
mouse skin (32) has the same structural arrangement of the
phenol, dihydrodiol and epoxide substituents, as the triol-
epoxide described in this paper (III). Hulbert and
Grover have recently proposed that highly reactive quinone-
methides may be involved in the metabolic activation of
chrysene and other chemial carcinogens (33). If this is the
case, then the triol-epoxide of BP should also be capable of
rearrangment to a quinone-methide (IV).

III   IV

The in vivo production of 3,5-di-OH-BP suggests that
the presence of the 3-OH group in the BP ring structure
renders an epoxide on the 4,5-position relatively unstable
in that it rearranges to a phenol. In contrast, the 'K-
region' epoxide of BP itself (BP-4,5-epoxide) is relatively
stable and requires either epoxide hydrase or a glutathione
transferase for its further metabolism (34).

ACKNOWLEDGEMENTS

We would like to thank K. Welham and D. Carter (School of Pharmacy, University of London) for performing the mass spectral studies, Jane Elliot (Kings College, University of London) for the p.m.r. spectral analyses, R.G. Harvey (The University of Chicago, The Ben May Laboratory for Cancer Research) for making the p.m.r. spectrum of BP-7,8-diol available to us, and P.L. Grover (Chester Beatty Research Institute, London) for helpful discussions. The help of Joy Dexter and Pauline McAree in the preparation of this manuscript is gratefully acknowledged. This work was supported by a grant from the Cancer Research Campaign.

REFERENCES

1. Jernstrom,B., Vadi,H. and Orrenius,S. (1976) Formation in isolated rat liver microsomes and nuclei of benzo(a)pyrene metabolites that bind to DNA, Cancer Res., 36, 4107-4113.
2. Burke,D.M., Vadi,H., Jernstrom,B. and Orrenius,S. (1977), Metabolism of benzo(a)pyrene with isolated hepatocytes and the formation and degradation of DNA-binding derivatives, J. Biol. Chem., 252, 6424-6431.
3. Okano,P., Miller,H.N., Robinson,R.C. and Gelboin,H.V. (1979), Comparison of benzo(a)pyrene and (-)-trans-7,8-dihydroxy-7,8-dihydrobenzo(a)pyrene metabolism in human monocytes and lymphocytes, Cancer Res., 39, 3184-3193.
4. Camus,A.M., Aitio,A., Sabadie,N., Wahrendorf,J. and Bartsch,H. (1984), Metabolism and urinary excretion of mutagenic metabolites of benzo(a)pyrene in C57 and DBA mice strains, Carcinogenesis, 5, 35-39.
5. Chipman,J.K., Frost,G.S., Hirom,P.C. and Millburn,P. (1981), Biliary excretion, systemic availability and reactivity of metabolites following intraportal infusion of [^{3}H]benzo(a)pyrene in the rat, Carcinogenesis, 2, 741-745.
6. Chipman,J.K., Bhave,N.A., Hirom,P.C. and Millburn,P. (1982), Metabolism and excretion of benzo[a]pyrene in the rabbit, Xenobiotica, 12, 397-404.
7. Owens,I.S., Legraverend,C. and Pelkonen,O. (1979), Deoxyribonucleic acid binding of 3-hydroxy- and 9-hydroxybenzo(a)pyrene following further metabolism by mouse liver microsomal cytochrome P$_1$-450, Biochem. Pharmacol., 28, 1623-1629.
8. Wislocki,P.G., Chang,R.L., Wood,A.W., Levin,W., Yagi,H., Hernandes,O., Mah,H.D., Dansett,P.M., Jerina,D.M. and Conney,A.H. (1977), High carcinogenicity of 2-hydroxybenzo(a)pyrene on mouse skin, Cancer Res., 37, 2608-2611.
9. Slaga,T.J., Bracken,M.W., Dresner,S., Levin,W., Yagi,H., Jerina,D.M. and Conney,A.H. (1978), Skin tumor-initiating activities of the twelve isomeric phenols of benzo(a)pyrene, Cancer Res., 38, 678-681.
10. Glatt,H.R. and Oesch,F. (1976), Phenolic benzo(a)pyrene metabolites are mutagens, Mutation Res., 36, 379-384.
11. Owens,I.S., Koteen,G.M. and Legraverend,C. (1979), Mutagenesis of certain benzo(a)pyrene phenols in vitro following further metabolism by mouse liver, Biochem.

Pharmacol., 28, 1615-1622.
12. Cohen,G.M., Haws,S.M., Moore,B.P. and Bridges,J.W. (1976), Benzo(a)pyren-3-yl hydrogen sulphate, a major ethyl acetate-extractable metabolite of benzo(a)pyrene in human, hamster and rat lung cultures, Biochem. Pharmacol., 25, 2561-2570.
13. Cohen,G.M., Moore,B.P. and Bridges,J.W. (1977), Organic solvent soluble sulphate ester conjugates of monohydroxybenzo(a)pyrenes, Biochem. Pharmacol., 26, 551-553.
14. Chipman,J.K., Hirom,P.C., Frost,G.S. and Millburn, P. (1981), The biliary excretion and enterohepatic circulation of benzo(a)pyrene and its metabolites in the rat, Biochem. Pharmacol., 30, 937-944.
15. Kinoshita,N. and Gelboin,H.V. (1978), -Glucuronidase catalyzed hydrolysis of benzo[a]pyrene-3-glucuronide and binding to DNA, Science, 199, 307-309.
16. Wiebel,F.J. (1975), Metabolism of monohydroxybenzo(a) pyrenes by rat liver microsomes and mammalian cells in culture. Archs. Biochem. Biophys., 168, 609-621.
17. Jernstrom,B., Vadi,H. and Orrenius, S. (1978), Formation of DNA-binding products from isolated benzo(a)pyrene metabolites in rat liver nuclei, Chem. Biol. Interactions, 20, 311-321.
18. Cook,J.W., Ludwiczak,R.S. and Schoental,R. (1950), Poly- cyclic aromatic hydrocarbons. Part XXXVI. Synthesis of the metabolic oxidation products of 3:4-benzpyrene, J. Chem. , 1112-1121.
19. Yagi,H., Holder,G.M., Dansette, P.M., Hernandez,O., Yeh,H.J.C., LeMahieu,R.A. and Jerina,D.M. (1976), Synthesis and spectral properties of the isomeric hydroxybenzo(a)pyrenes, J. Org. Chem., 41, 977-985.
20. Yang,S.K., Roller,P.P. and Gelboin,H.V. (1977), Enzymatic mechanism of benzo(a)pyrene conversion to phenols and diols and an improved high-pressure chromatographic separation of benzo(a)pyrene derivatives, Biochemistry, 16, 3680-3687.
21. Jerina,D.M., Selaner,H., Yagi,H., Wells,M.C., Davey,J.F., Mahadevan,V. and Gibson,D.T. (1976), Dihydrodiols from anthracene and phenanthrene, J. Am. Chem. Soc., 98, 5988-5996.
22. Hadfield,S.T., Abbott,P.J., Coombs,M.M. and Drake,A.F. (1984), The effect of methyl substituents on the in vitro metabolism of cyclopenta[a]phenanthren-17-ones: Implication for biological activity, Carcinogenesis, 5, 1395-1399.
23. McCaustland,D.J., Fisher,D.L., Kolwyck,K.C., Duncan,W.P., Wiley,J., Menon,C.S., Engel,J.F., Selkirk,J.K. and Roller,P.P., (1976), Polycyclic aromatic hydrocarbon derivatives: synthesis and physiochemical characterisation, in Freudenthal,R. and Jones,P.W. (eds.) Carcinogenesis - A comprehensive survey, Vol. 1, Raven, New York, pp. 349- 411.
24. Boroujerdi,M., Kung,H., Wilson,A.G.E. and Anderson,M.W. (1981), Metabolism and DNA binding of benzo(a)pyrene in vivo in the rat, Cancer Research, 41, 951-957.
25. Elmhirst,T.R.D., Chipman,J.K., Hirom,P.C. and Millburn.P. (1984), Enterohepatic circulation of benzo(a)pyrene-4,5-oxide in the rat, Biochem. Soc. Trans. 12, 677.

26. Elmhirst,T.R.D., Chipman,J.K., Ribeiro,O., Hirom,P.C. and Millburn,P. (1985), Metabolism and enterohepatic circulation of benzo(a)pyrene-4,5-oxide in the rat, Xenobiotica, in press.

27. Connor,T.H., Forti,G.C., Sitra,P. and Legator,M.S. (1979), Bile as source of mutagenic metabolites produced in vivo and detected by Salmonella typhimurium, Environ. Mutagen., 1, 269-276.

28. Chipman,J.K., Millburn,P. and Brooks,T.M. (1983), Mutagenicity and in vivo disposition of biliary metabolites of benzo(a)pyrene, Toxicol. Lett., 17, 233-240.

29. Chipman,J.K., Millburn,P. and Brooks,T.M. (1983), Mutagenicity of benzo(a)pyrene-3-0-glucuronide, a biliary metabolite of benzo(a)pyrene, Toxicol. Lett., 17, 361-362.

30. Jernstrom,B., Dock,L. and Martinez,M. (1984), Metabolic activation of benzo(a)pyrene-7,8-dihydrodiol and benzo(a) pyrene-7,8-dihydrodiol-9,10-epoxide to protein-binding products and the inhibitory effect of glutathione and cysteine, Carcinogenesis, 5, 199-204.

31. Jernstrom,B., Martinez,M., Svensson,S-A. and Dock,L. (1984), Metabolism of benzo(a)pyrene-7,8-dihydrodiol and benzo(a)pyrene-7,8-dihyrodiol-9,10-epoxide to protein-binding products and glutathione conjugates in isolated rat hepatocytes, Carcinogenesis, 5, 1079-1085.

32. Hodgson,R.M., Weston,A. and Grover,P.L. (1983), Metabolic activation of chrysene in mouse skin: evidence for the involvement of a triol-epoxide, Carcinogenesis, 4, 1639-1643.

33. Hulbert,P.B. and Grover,P.L. (1983), Chemical rearrangement of phenol-epoxide metabolites of polycylic aromatic hydrocarbons to quinone-methides, Biochem. Biophys. Res. Commun., 117, 129-134.

34. Grover,P.L., Hewer,A. and Sims,P. (1972), Formation of K-region epoxides as microsomal metabolites of pyrene and benzo(a)pyrene, Biochem. Pharmacol., 21, 2713-2726.

TIME COURSE OF THE EFFECT OF 4-HYDROPEROXYCYCLOPHOSPHAMIDE

ON LIMB DIFFERENTIATION IN VITRO

Barbara F. Hales and Pierre Brissette

Department of Pharmacology and Therapeutics and
Centre for The Study of Reproduction
McGill University
Montreal, Quebec

INTRODUCTION

Cyclophosphamide is a commonly used anti-tumor and immunosuppressive agent. It is also mutagenic and teratogenic in a variety of species. Studies of the embryo have demonstrated that exposure to cyclophosphamide during organogenesis results in a spectrum of malformations that include exencephaly or hydrocephaly, open eyes, cleft palate, phocomelia, adactyly, syndactyly, polydactyly and kinky tail as well as disturbances in skeletal ossification (Gibson & Becker, 1968; Mirkes, 1985). Cyclophosphamide is usually teratogenic in a narrow dose range - lower doses have no apparent effects while higher doses are highly embryolethal.

It is now well documented that cyclophosphamide itself has little alkylating, antineoplastic or teratogenic activity. Cyclophosphamide must be metabolically activated to biologically reactive intermediates to be effective as an antineoplastic (Brock, 1976; Foley et al, 1961) or to . be teratogenic in vitro (Fantel et al, 1979; Klein et al, 1980). A scheme of the major pathway for metabolic activation is shown in Figure 1 The first step in this activation is 4-hydroxylation of cyclophosphamide to 4-hydroxycyclophosphamide. The activation occurs in the presence of liver microsomal fractions, O_2 and NADPH; it is induced by phenobarbital pretreatment and inhibited by carbon monoxide, SKF 525A and metyrapone (Hales & Jain, 1980; Greenaway et al, 1982). 4-Hydroxycyclophosphamide is unstable and equilibrates with its open ring tautomer, aldophosphamide; this compound, in turn, undergoes a β-elimination reaction to yield two cytotoxic metabolites, phosphoramide mustard and acrolein. An analog of cyclophosphamide, 4-hydroperoxycyclophosphamide, has been synthesized; in solution, this compound breaks down to 4-hydroxycyclophosphamide and thus does not require metabolic activation (Takazimawa et al, 1975; Voelker et al, 1974). 4-Hydroperoxycyclophosphamide is a relatively stable, non-alkylating compound with a half-life in tissue culture medium at 37°C of less than 2 hours (Hilton, 1984).

In any in vivo system it is difficult to determine the role of reactive metabolites in mediating the teratogenic effects of drugs and to characterize the enzymes catalyzing such metabolic activations. Two in vitro systems, the rat whole embryo culture system and the mouse limb bud culture system, have been used to study effects of metabolism on drug-induced teratogenicity.

In the rat whole embryo culture system cyclophosphamide is not teratogenic but cyclophosphamide, in the presence of a metabolic activating system, is a potent teratogen (Fantel et al, 1979; Kitchin et al, 1981). Mirkes et al (1980) demonstrated that phosphoramide mustard, a metabolite of cyclophosphamide with alkylating activity, produces growth retardation and malformations such as hypoplasia of the prosencephalon, mandibular arches, limb buds and tail in the rat embryo culture system. Acrolein, a second toxic metabolite of cyclophosphamide, is also teratogenic in cultured embryos but only in concentrations approximately 10-fold higher than those for phosphoramide mustard (unless there is a pre-incubation period in the absence of serum) (Mirkes et al 1984; Slott and Hales, 1985). An analysis of the time of exposure required for the rat embryo response to cyclophosphamide in the presence of an activating system revealed that exposure periods of up to 2.5 hours had no significant effects on growth, incidence of malformed embryos or mitotic indices; a 5-hour exposure period was sufficient to produce embryos indistinguishable from those exposed to activated cyclophosphamide continuously for the entire culture period (24-26 hours) (Mirkes, 1983).

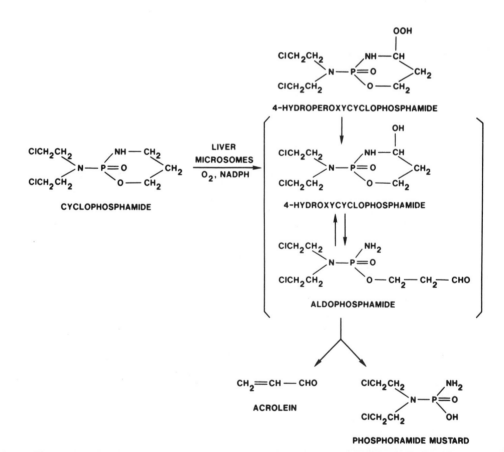

Figure 1. Metabolic Activation of Cyclophosphamide.

Cyclophosphamide itself is also not teratogenic in the limb bud culture system (Manson & Simons, 1979; Manson & Smith, 1977). Manson and her co-workers (1977) found some disruption of limb morphogenesis by 4-ketocyclophosphamide but not by cyclophosphamide. Barrach and his co-workers (Barrach et al, 1978; Barrach & Neubert, 1980) found that the addition of "pre-activated" cyclophosphamide, i.e. 4-hydroperoxy-cyclosphosphamide, to the medium dramatically affected the differentiation of the cartilaginous bone anlagens of cultured mouse limbs. They investigated the importance of exposure time in 24 hour periods or windows, i.e. exposure during the first, second or third day of culture. Abnormal limb development resulted from exposure to this compound only during the first 24 hours of the culture period.

The mechanisms of cyclophosphamide teratogenicity in terms of what occurs between exposure during gestation and delivery are not well understood. This period has been referred to as a "black box" (Neubert, 1982; Mirkes, 1985). To study the effects of activated cyclophosphamide on a developing organ, the limb, we determined the time course of the effects of exposure to 4-hydroperoxycyclophosphamide on cultured mouse limb morphology. We determined the effects of this treatment on growth by monitoring DNA, RNA and protein content. To test for differential effects of this drug on myogenesis versus chondrogenesis we measured the activities of two enzymes, creatine phosphokinase and alkaline phosphatase. Creatine phosphokinase activity increases rapidly when myoblasts fuse to form myotubes and has previously been used to gauge the extent to which myogenesis has proceeded in developing control and treated limbs (Kwasigroch & Neubert, 1978). Alkaline phosphatase activity is essential for osteogenesis to proceed and this activity can be used as a distinctive marker of osteogenic expression (Osdoby & Caplan, 1981).

METHODS

Timed gestation pregnant ICR mice were purchased from Charles River Canada Inc., St. Constant, Quebec. Females were exposed to males for two hours between 12 p.m. and 2 p.m. and then examined for vaginal sperm plugs. Those with observable plugs were designated as day 0 pregnant; they were housed in the McIntyre Animal Center, Montreal, Quebec. Twelve days later, the females were killed, the uteri removed and the embryos dissected out in sterile Tyrode's saline solution; the embryos at this time were 5 to 6 mm in length and possessed about 40 somite pairs. Forelimbs and hindlimbs were excised just lateral to the somites and pooled separately. The limbs were cultured in 60 ml Wheaton serum bottles (approximately 10-12 limbs per bottle) in 6 ml of culture medium consisting of 75% Bigger's medium, (BGJ$_b$ Gibco, Burlington, Ont.) and 25% fetal calf serum supplemented with ascorbic acid (150 µg/ml), streptomycin (12.5 µg/ml) and penicillin (7.5 µg/ml) as described by Kochhar (1983). 4-Hydroperoxycyclophosphamide (a gift from Dr. N. Brock, Asta-Werke, Germany) was added to the designated cultures at a concentration of 10µg/ml. Each bottle was sealed with a rubber stopper and aluminum clasp. The cultures were rotated at approximately 25 r.p.m. at 37.5° C. The medium was changed at the end of the desired exposure time and/or after three days and the cultures were terminated after six days. Three separate identical experiments were done.

At the end of the culture period some of the limbs were fixed in Bouin's fixative overnight. Fixed limbs were stained in 0.1% toluidine blue in 70% ethanol for one day and then washed and serially dehydrated in 70%, 90% and 100% ethanol. Stained limbs were cleared and stored in 100% cedarwood oil. The other limbs were homogenized with a Polytron

(Brinkmann Instruments Canada, Ltd.) at a setting of 5 for two 30-second periods in 2 ml of Tyrode's saline solution. Each sample was divided for determination of DNA, RNA and protein (1 ml) and creatine phosphokinase and alkaline phosphatase activities (1 ml).

The method described by Chung and Coffey (1971) was used for the extraction of DNA, RNA and protein. Each sample (1 ml) was mixed with 2 ml of cold 20% trichloroacetic acid, allowed to stand 30 minutes and centrifuged at 15,000 r.p.m. for 15 minutes. The precipitate was washed twice with 2 ml of cold 10% trichloroacetic acid prior to washing twice in 3 ml of 1% potassium acetate in ethanol: ether (3:1). This pellet was dissolved in 0.5 ml of 0.3N potassium hydroxide and incubated at 37° C for 1 hour. Cold 0.8N perchloric acid (1.5ml) was added and the sample was centrifuged at 15,000 r.p.m. for 5 minutes. The supernatant was removed for determination of RNA by the orcinol reaction (Schneider, 1957). The pellet was resuspended in 2 ml of 0.8N perchloric acid; the sample was incubated at 70° C for 20 minutes and then centrifuged. The resultant supernatant was removed for determination of DNA by the diphenylamine method (Burton, 1968). The remaining pellet was dissolved in 2 ml of 0.1N sodium hydroxide and heated in a shaking water bath overnight at 37° C. After centrifugation at 20,000 r.p.m. for 10 minutes, the supernatant was removed and used to assess protein content as described by Lowry et al (1951) with bovine serum albumin as standard.

Creatine phosphokinase activity was measured spectrophotometrically using the procedure based on the hexokinase, glucose-6-phosphate dehydrogenase enzyme coupled system (Kit No. 45-UV) from Sigma Chemical Co. (St. Louis, MO). The reduction of NADP is monitored at 340 nm. (Oliver, 1955). Alkaline phosphatase activity was measured by the method of Garen and Levinthal (1960) in which the rate of release of p-nitrophenol from p-nitrophenyl phosphate is determined by following absorbancy changes at 410 nm.

RESULTS

Limb Morphology

Control and 4-hydroperoxycyclophosphamide-exposed forelimbs are shown in Figure 2. These limbs are representative of those stained in each of the three experiments. The effects of exposure of hindlimbs to 4-hydroperoxycyclophosphamide were almost identical to those with forelimbs and are not presented. A scapula, humerus, radius, ulna, carpals, metacarpals and phalanges are all usually present in the control forelimb (upper left hand panel) after six days of culture. With as little as one hour of exposure to 4-hydroperoxycyclophosphamide at the beginning of the culture period, there is some decrease in the chondrification of the long bone and paw skeleton. The scapula is not affected. In addition, the soft tissue of the limb in the paw area is pointed in appearance. The paw bone anlagens are also pointed with three hours of exposure to 4-hydroperoxycyclophosphamide. The radius and ulna are reduced in size and only two phalanges are present. With six hours of exposure to 4-hydroperoxycyclophosphamide the overall forelimb area is decreased. Bone formation is inhibited such that long bones and paw structures are not present. With 20 hours of drug exposure, many of the limbs appear swollen or edematous. A scapula and a few paw bone anlagens are present but none of the long bones can be visualized. With 72 hours of exposure to 4-hydroperoxycyclophosphamide the total forelimb area is dramatically reduced and no toluidine-blue stainable cartilaginous bones are apparent.

846

Limb DNA, RNA and Protein Content

The effects on forelimb DNA, RNA and protein contents of exposure in culture to 4-hydroperoxycyclophosphamide for various time periods are shown in Table 1. With an exposure period of 1 hour the DNA and protein contents of the forelimbs are decreased to 68 and 65% of control, respectively, while the RNA content is 85% of control. With three hours of exposure to 4-hydroperoxycyclophosphamide the DNA content is decreased to less than 50% of control but limb contents of both RNA and protein are about 70% of control. Doubling the exposure time to six hours does further decrease the limb RNA and protein contents to approximately 50% of control values, but the most dramatic effect is again on DNA; limb DNA content decreases to 17% of control. A further decrease in all three parameters is observed with longer periods of drug exposure.

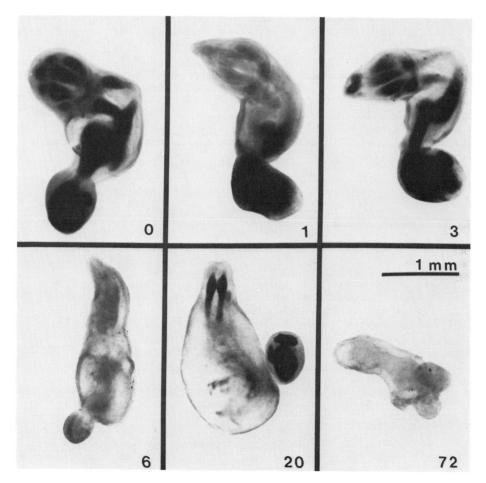

Figure 2: Mouse forelimbs cultured for six days were exposed to 4-hydroperoxycyclophosphamide (10µg/ml) for the designated time periods (hours) at the beginning of the culture period.

Table 1. Effects on cultured mouse forelimb DNA, RNA and protein
contents of exposure to 4-hydroperoxycyclophosphamide
(10µg/ml) for various time periods.

Length of Exposure (hrs)	DNA (µg/limb)	RNA (µg/limb)	Protein (µg/limb)
0	5.9 ± 1.4[a]	3.9 ± 1.1	38.7 ± 5.9
1	4.0 ± 1.5	3.3 ± 0.6	25.3 ± 1.3
3	2.5 ± 0.7	2.7 ± 0.3	28.0 ± 6.4
6	1.8 ± 0.6	2.1 ± 0.2	19.2 ± 6.1
20	<1.5[b]	1.5 ± 0.6	9.6 ± 4.2
72	<1.5[b]	<1.3[b]	8.9 ± 3.1

[a] Values represent means ± standard errors of the mean (n=3).
[b] Less than the limit of detection of the assay.

Table 2. Effects on forelimb alkaline phosphatase and creatine phosphokinase activities of exposure of cultured mouse forelimbs to
4-hydroperoxycyclophosphamide (10µg/ml) for various time
periods.

Length of Exposure (hrs.)	Alkaline Phosphatase[a]	Creatine Phosphokinase[b]
0	9.4 ± 3.6[c]	7.1 ± 1.0
1	4.2 ± 1.3	4.9 ± 0.7
3	4.3 ± 0.8	4.9 ± 0.3
6	1.6 ± 0.3	4.4 ± 0.5
20	1.0 ± 0.3	4.4 ± 0.5
72	<1.0[d]	4.2 ± 0.5

[a] Alkaline phosphatase activity is expressed as µmoles of
p-nitrophenylphosphate hydrolyzed per minute per limb x 10^{-4}.
[b] Creatine phosphokinase activity is expressed as µmoles of NADP
reduced per minute per limb x 10^{-3}.
[c] Values represent means ± standard errors of the mean (n=3).
[d] Less than the limit of detection of the assay.

Limb alkaline phosphatase and creatine phosphokinase activity

The effects of exposure of cultured forelimbs to 4-hydroperoxy-
cyclophosphamide on the limb alkaline phosphatase and creatine phosphokinase activities are shown in Table 2. Alkaline phosphatase activity
per limb decreases to less than 25% of control with 6 hours of exposure
to 4-hydroperoxycyclophosphamide. With 72 hours of exposure to 4-hydro-
peroxycyclophosphamide, alkaline phosphatase activity in treated forelimbs further decreases by nearly 50%. In contrast, the decrease in
creatine phosphokinase activity in drug-treated limbs was less than 40%
of control with 6 hours of exposure. Even with 72 hours of exposure to
4-hydroperoxycyclophosphamide, creatine phosphokinase activity per limb
is maintained at about 60% of the control activity.

DISCUSSION

Treatment of cultured mouse limbs with 4-hydroperoxycyclophosphamide dramatically affects their development. These effects are not "non-specific". Differential effects are observed both with respect to limb morphology and biochemistry. 4-Hydroperoxycyclophosphamide produces limb reduction malformations – total limb bone area is reduced. Forelimbs have a claw-like paw with no humerus, no radius, ulna or carpals. As a consequence, the percentage of total limb bone area contributed by paw structures in the drug-treated limbs increases relative to control limbs. DNA, RNA and protein contents of the limbs exposed to 4-hydroperoxycyclo-phosphamide are all reduced relative to control. Alkaline phosphatase activity is greatly reduced after even short periods of exposure to 4-hydroperoxycyclophosphamide. Alkaline phosphatase is a marker for presumptive periosteum (Osdoby & Caplan, 1981). That 4-hydroperoxycyclo-phosphamide treatment decreases the amount of alkaline phosphatase activity per limb almost 10-fold while only a 1.7 fold decrease is observed in creatine phosphokinase activity suggests that this drug preferentially inhibits the differentiation of embryonic bone and cartilage with much less effect on muscle development. It is possible that cyclo-phosphamide produces malformations by selectively influencing gene expression.

Increasing the time period of exposure of cultured mouse limbs to 4-hydroperoxycyclophosphamide produces progressively greater effects on all the parameters of limb development measured. However, the half-life of 4-hydroperoxycyclophosphamide in solution is usually about 1 hour (Low et al, 1982). Previous reports have provided evidence that toxic products of 4-hydroperoxycyclophosphamide do persist for long periods of time. Using yeast, Fleer and Brendel (1982) demonstrated that 4-hydro-peroxycyclophosphamide retains a considerable amount of its cytotoxic and DNA cross-linking activity even after a 12 hour preincubation period at pH 7 and 36° C.

The relative roles of DNA, RNA and protein as the "critical" targets of cyclophosphamide and/or its metabolites have been investigated in a variety of test systems. In some studies, the data are consistent with embryonic DNA as the "critical" target for cyclophosphamide-induced teratogenicity (Mirkes et al, 1984; Murthy et al, 1973; Short et al, 1972). However, other studies have provided evidence that supports a role for RNA as the critical target (Kohler & Merker, 1973).

It is attractive to hypothesize that the altered enzyme levels observed in limbs treated with 4-hydroperoxycyclophosphamide result from drug-induced changes in genes and/or their expression and lead to the production of the limb reduction malformations observed here. Treatment of mice with cyclophosphamide on gestational days 9 through 14 does not result in the absence of fetal long bones but does produce shortened fetal long bones in addition to a variety of digital anomalies (Gibson & Becker, 1968). It would be of interest to determine if exposure to cyclophosphamide in vivo alters the developmental profile of limb alkaline phosphatase or creatine phosphokinase activity.

ACKNOWLEDGEMENTS

This work was supported by the Medical Research Council of Canada (Grant MA 7078). B.F.H. is a Scholar of the Medical Research Council of Canada and P.B. is the recipient of a summer studentship from the FCAC

Centre Grant for Research in Reproductive Biology. We thank Professor Norbert Brock for providing the 4-hydroperoxycyclophosphamide used in these experiments, Ranjana Jain for expert technical assistance and Roy Raymond and Robert Lamarche for the photography.

REFERENCES

Barrach, H. J., Baumann, I., and Neubert, D., 1978, The applicability of in vitro systems for the evaluation of the significance of pharmacokinetic parameters for the induction of an embryotoxic effect, in: "Role of Pharmacokinetics in Prenatal and Perinatal Toxicology", D. Neubert, H. J. Merker, H. Nau & J. Langman, eds., Georg Thieme Publishers, Stuttgart, pp 323-349.

Barrach, H. J. and Neubert, D., 1980, Significance of organ culture techniques for evaluation of prenatal toxicity, Arch. Toxicol., 45:161-187.

Brock, N., 1976, Comparative pharmacologic study in vitro and in vivo with cyclophosphamide (NSC-26271), cyclophosphamide metabolites, and plain nitrogen mustard compounds, Cancer Treat. Rep., 60:301-307.

Burton, K., 1968, Determination of DNA concentration with diphenylamine, Methods Enzymology, 12:163-166.

Chung, L. W. K. and Coffey, D. S., 1971, Biochemical characteristics of prostatic nuclei. II Relationship between DNA synthesis and protein synthesis, Biochim. Biophys. Acta., 247:584-596.

Fantel, A. G., Greenaway, J. C., Juchau, M. R., and Shepard, T. H., 1979, Teratogenic bioactivation of cyclophosphamide in vitro, Life Sci., 25:67-72.

Fleer, R. and Brendel M., 1982, Toxicity, interstrand cross-links and DNA fragmentation by activated cyclophosphamide in yeast - comparative studies on 4-hydroperoxy-cyclophosphamide, its monofunctional analog on, acrolein, phosphoramide mustard, and nor-nitrogen mustard, Chem. Biol. Interact., 39: 1-15.

Foley, G. E., Friedman, O. M., and Drolet, B. P., 1961, Studies on the mechanism of action of cytoxan - evidence of activation in vivo and in vitro, Cancer Res., 21:57-63.

Garen, A. and Levinthal, C., 1960, A fine structure genetic and chemical study of the enzyme alkaline phosphatase of E. coli. I. Purification and characterization of alkaline phosphatase, Biochim. Biophys. Acta., 38:470-483.

Gibson, J. E. and Becker, B. A., 1968, The teratogenicity of cyclophosphamide in mice, Cancer Res., 28:475-480.

Greenaway, J. C., Fantel, A. G., Shepard, T. H.,Juchau, M. R.,1982, The in vitro teratogenicity of cyclophosphamide in rat embryos, Teratology, 25:335-343.

Hales, B. F., and Jain, R., 1980, Characteristics of the activation of cyclophosphamide to a mutagen by rat liver, Biochem. Pharmacol.,29:256-259.

Hilton, J., 1984, Deoxyribonucleic-acid crosslinking by 4-hydroperoxycyclophosphamide in cyclophosphamide-sensitive and cyclophosphamide-resistant L1210 cells, Biochem. Pharmacol., 33: 1867-1872.

Kitchin, K. T., Schmid, B. P., and Sanyal, M. K., 1981, Teratogenicity of cyclophosphamide on a coupled microsomal activating/embryo culture system, Biochem. Pharmacol., 30:59-64.

Klein, N. W., Vogler, M. A., Chatot, C. L., and Pierro, L. J.,1980, The use of cultured rat embryos to evaluate the teratogenic activity of serum:cadmium and cyclophosphamide, Teratology, 21:199-208.

Kochhar, D. M., 1983, Embryonic organs in culture, in: "Handbook of Experimental Pharmacology,",Johnson, E. M., and Kochhar, D. M., eds. Springer-Verlag, Heidelberger Platz, pp. 301-314.

Kohler, E. and Merker, H. J., 1973, The effect of cyclophosphamide
pretreatment of pregnant animals on the activity of nuclear
DNA-dependent RNA-polymerases in different parts of rat embryos,
Naunyn-Schmeid. Arch. Pharmacol., 277:71-88.

Kwasigroch, T. E.,and Neubert, D., 1978, A simple method to test
chondrogenic and myogenic tissues for differential effects of drugs,
in: "Role of Pharmacokinetics in Prenatal and Perinatal
Toxicology", D. Neubert, H.-J. Merker, H. Nau, and J. Langman,
eds.,Georg Thieme Publishers, Stuttgart, pp. 621-630.

Low, J. E., Borch, R. F., and Sladek, N. E., 1982, Conversion of
4-hydroperoxycyclophosphamide and 4-hydroxycyclophosphamide to
phosphoramide mustard and acrolein mediated by bifunctional
catalysts, Cancer Res., 42:830-837.

Lowry, O. H., Rosebrough, N. J., Farr, A. L. and Randall, R. J., 1951,
Protein measurement with the folin phenol reagent, J. Biol. Chem.,
193:265-275.

Manson, J. M., and Simons, R., 1979, In vitro metabolism of
cyclophosphamide in limb bud culture, Teratology, 19:149-158.

Manson, J. M., and Smith, C. C., 1977, Influence of cyclophosphamide and
4-ketocyclophosphamide on mouse limb development, Teratology,
15:291-300.

Mirkes, P. E., 1985, Cyclophosphamide teratogenesis - a review,
Teratogen. Carcinogen. Nutagen., 5:75-88.

Mirkes, P. E., and Greenaway, J. C., 1985, Uptake and binding of tritium
from [chloroethyl³H] cyclophosphamide by rat embryos in vitro,
Teratology, 31:373-380.

Mirkes, P. E., Greenaway, J. C., Rogers, J. G., and Brundrett, R. B.,
1984, Role of acrolein in cyclophosphamide teratogencity in rat
embryos in vitro, Toxicol. Appl. Pharmacol., 72:281-291.

Mirkes, P. E., Fantel, A. G., Greenaway, J. C., and Shepard, T. H., 1981,
Teratogenicity of cyclophosphamide metabolites: phosphoramide
mustard, acrolein and 4-ketocyclophosphamide in rat embryos cultured
in vitro, Toxicol, Appl, Pharmacol., 58:322-330.

Mirkes, P. E., Greenaway, J. C., and Shepard, T. H.,1983, A kinetic
analysis of rat embryo response to cyclophosphamide exposure in
vitro, Teratology, 28:249-256.

Murthy, V. V., Becker, B. A., and Steele, W. S., 1973, Effects of dosage,
phenobarbital and 2 diethylaminoethyl 2, 2-diphenylvalerate on the
binding of cyclophosphamide and/or its metabolites to the DNA, RNA
and protein of the embryo and liver of pregnant mice, Cancer Res.,
33:664-670.

Neubert, D., 1982, The use of culture techniques in studies on prenatal
toxicity, Pharmac. Ther., 18:397-434.

Oliver, I. T., 1955, A spectrophotometric method for the determination of
creatine phosphokinase and myokinase, Biochem. J., 61:116-122.

Osdoby, P., and Caplan, A. I., 1981, First bone formation in the
developing chick limb, Develop. Biol., 86:147-156.

Schneider, W. C., 1957, Determination of nucleic acids in tissues by
pentose analysis, Methods Enzymology,3:680-691.

Short, R. D., Rao, K. S. and Gibson, J. E., 1972, The in vivo
biosynthesis of DNA, RNA and proteins by mouse embryos after a
teratogenic dose of cyclophosphamide, Teratology, 6:129-138.

Slott, V. and Hales, B. F., 1985, Effect of glutathione (GSH) depletion
by buthionine sulfoximine (BSO) on the in vitro teratogenicity and
embryolethality of acrolein (AC), Teratology, 31:33A.

Takazimawa, A., Matsumoto, S., Iwata, T., Tochino, Y., Katagiri, K.,
Yamaguchi, K., and Shiratori, O., 1975, Studies on cyclophosphamide
metabolites and their related compounds.2. Preparation of an active
species of cyclophosphamide and related compounds, J. Med. Chem.,
18:376-383.

Voelker, G., Draeger, U., Peter, G. and Hohorst, H.-J., 1974, Studien zum

spontanzerfall von 4-hydroxycyclophosphamid and
4-hydroperoxycyclophosphamid mit hilfe der
dunnschichtchromatographie, Arzneim. Forsch./Drug Res.,
24:1172-1176.

HEPATIC PATHOLOGICAL CHANGES DUE TO HYDROXYALKENALS

D.W. Wilson, H.J. Segall* and M.W. Lame*
Departments of Veterinary Pathology, Pharmacology and
    Toxicology*
School of Veterinary Medicine
University of California - Davis

INTRODUCTION

The nonenzymatic autoxidation of polyunsaturated fatty acids is known to cause increased formation of alkenals, alk-2-enals, malonaldehyde and 4-hydroxyalkenals (1-3). A number of alkenals are formed during microsomal lipid peroxidation and some of these alkenals, such as the hydroxyalkenals, are effective nonradical products which could be responsible for part of the effects associated with lipid peroxidation (4-10). Esterbauer et al and Benedetti et al have demonstrated that alkenals are produced during stimulated (ADP-Fe^{+2} or NADPH-Fe^{+2}) microsomal lipid peroxidation and that these alkenals (in particular 4-hydroxyalkenals) could be responsible for part of the destructive effects caused by lipid peroxidation on cells and cell constituents (5,6). Cytotoxic alkenals originating due to lipid peroxidation were shown to damage cellular membranes of red blood cells, inhibit protein synthesis, and decreased the activity of membrane bound enzymes such as microsomal glucose-6-phosphatase, cytochrome P-450, and aminopyrine N-demethylase (7-13). Further investigations have determined that alkenal compounds may play an important role in liver injury caused by CCl_4 and $BrCCl_3$ (14,15). Alkenal groups are formed in phospholipids of liver microsomes as a consequence of the peroxidative cleavage of phospholipid-bound unsaturated fatty acids after in vivo intoxication with CCl_4 and $BrCCl_3$ (14).

Studies characterizing alkenal compounds originating from the peroxidation of liver microsomal lipids have proven that the most abundant alkenal was trans-4-hydroxy-2-nonenal with trans-4-hydroxy-2-octenal, trans-4-hydroxy-2-decenal, trans-4-hydroxyundecenal and trans-4, 5-dihydroxy-2-decenal present in small amounts (5,6). Recently we showed that trans-4-hydroxy-2-hexenal (t-4HH) was a metabolite derived from microsomal metabolism of the macrocyclic pyrrolizidine alkaloid (PA) senecionine and that this hydroxyalkenal caused hepatic necrosis when injected into the portal vein of rats (16). Many PAs are potent hepatotoxins and may cause an irreversible hemorrhagic necrosis, hepatic fibrosis and megalocytosis (17-24).

Zonal distribution of hepatocellular necrosis as a consequence of chemical injury is postulated to occur as a result of metabolic heterogeneity of the hepatic acinus (25) combined with the relative hypoxia of the zone 3 or central vein region and in some instances, microcirculatory changes in bloodflow induced by the administered chemical (26). In the case of toxins activated to reactive intermediates by the mixed function oxidase system (MFO), necrosis of hepatocytes near terminal hepatic (or central) veins caused by these agents is attributed to the high concentrations of these enzymes in this region. Agents such as allyl alcohol (25) or N-hydroxy-2-acetylaminofluorene (27) not requiring MFO activation to exert their toxic effects cause necrosis of hepatocytes in zone 1 (periportal necrosis).

Using the hypothesis that chemical structure and/or requirement for metabolic activation of hepatotoxins determines the distribution and extent of hepatic lesions caused by these compounds, we undertook to compare hepatic lesions caused by the parent compound senecionine with those caused by it's putative metabolite t-4HH and to determine whether chain length or the presence of the hydroxyl group on hydroxalkenals affect their hepatotoxicity.

MATERIALS AND METHODS

Chemicals

Trans-2-nonenal (t-2N) was obtained from Aldrich Chemical Co. and redistilled prior to use. Trans-4-hydroxy-2-hexenal (t-4HH) was synthesized by the method of Erickson (28). Trans-4-hydroxy-2-nonenal (t-4HN) was similarly synthesized except that propionaldehyde used in the synthesis of t-4HH was replaced with hexanal. Senecionine was extracted and purified from Senecio vulgaris plants as previously described (29).

Animal experiments

The lower half of a 15 ml tube was coated with a thin layer of phosphatidylcholine and a solution of alkenal plus buffer was added along with two No. 6 glass beads, chilled to 0°C, vortexed for 1 min and maintained at 0°C for 10 min. This procedure was repeated five times, the solution sonicated (under N_2) for short bursts and maintained at 0°C during this period. The solution was filtered through a 0.2μm polycarbonate filter (Nucleopore) and warmed to 37°C. A 10 - 12 mm incision was made on the ventral surface of the midline of 120-140 g male rats and injection (0.1-0.2 ml) made directly into the hepatic portal vein. Two rats at each dose were injected with one of three doses in the range given in table 1. Rats were killed 48 hr following surgery and the livers fixed in formalin. All surgical techniques were performed according to the University of California, Davis·and NIH animal handling procedures.

Pathologic evaluation

Animals were killed 48 hours after dose administration by intraperitoneal injection of pentobarbital. After gross examination of the abdominal and thoracic viscera for lesions associated with injection, the liver was removed in toto and the nature and distribution of gross lesions recorded. The liver was sliced into 2 mm sections and fixed by immersion in 10% neutral buffered formalin. Sections for histologic evaluation were made from each of the major lobes such that the section extended from the entrance of the portal veins at the hilus through the midportion of the lobe to the periphery. Hematoxylin and eosin stained sections were evaluated without prior knowledge of treatment group assignment. The location of lesions within acini, the relative numbers and distribution of acini affected and the severity or extent of degenerative versus irreversible changes in hepatocytes in affected regions were recorded.

RESULTS

Gross Pathology

While the median lobes of the liver were most consistently affected in all treated animals, frequently the right and left lateral lobes and occasionally the caudate lobe were involved. Both t-4HH and t-4HN consistently produced radiating yellow tan streaks of hepatic necrosis that appeared to follow the course of the portal vascular tree in the center of affected lobes and frequently extended from the middle of a lobe to one capsular surface. These lesions appeared more widely distributed at the periphery of affected lobes when compared with regions near the hilum where necrosis appeared more limited to the central portion of the lobe near the larger vessels. At higher doses (24-32 mg/kg) necrotic tracts at the periphery of lobes coalesced forming large well demarcated regions of necrosis.

The surface of the livers of animals injected with senecionine had a diffusely mottled appearance due to irregular regions of dark red collapsed parenchyma which alternated with regions of normal parenchyma. Animals infected with t-2N had both focal pale tan regions of necrosis similar to rats injected with the hydroxylated compounds and diffuse mottling due to parenchymal collapse as seen in senecionine treated animals. Diffuse parenchymal collapse was seen in all doses while focal necrosis was evident primarily at higher doses.

Histopathologic evaluation

Sections of liver from animals administered senecionine had coagulative necrosis of hepatocytes in zone 3 surrounding nearly all terminal hepatic (central) veins from affected lobes (Fig 2). Necrosis was most extensive in the right medial and lateral lobes reflecting the gross distribution previously described.

Animals given t-4HH and t-4HN had similar microscopic lesions which were characterized by focal aggregates of several acini which had undergone coagulative necrosis. These necrotic foci were well demarcated from adjacent more normal parenchyma by a thin band of mixed inflammatory cells and frequently extended from portal triads to only a portion of parenchyma surrounding an individual terminal hepatic (central) vein. (Fig 3 and 4) This pattern is generally referred to as paracentral necrosis. Often in these regions of necrosis, a thin band of surviving hepatocytes surrounded portal triads.

Rats injected with t-2N had generalized alteration in periportal (zone 1) hepatocytes and portal triads. Nearly all portal regions examined were prominent due to reactive hyperplasia and hypertrophy of bile duct epithelium, periportal edema, and degeneration of periportal hepatocytes with loss of the usual intracytoplasmic glycogen accumulations seen in control animals. (Fig 5) In many portal regions, hepatocytes in zone 1 had undergone coagulative necrosis (Fig 6).

Livers from animals injected with the lipid suspension alone were unaffected (Fig 1). A summary of dose, gross distribution and microscopic pattern of necrosis is given in table 1.

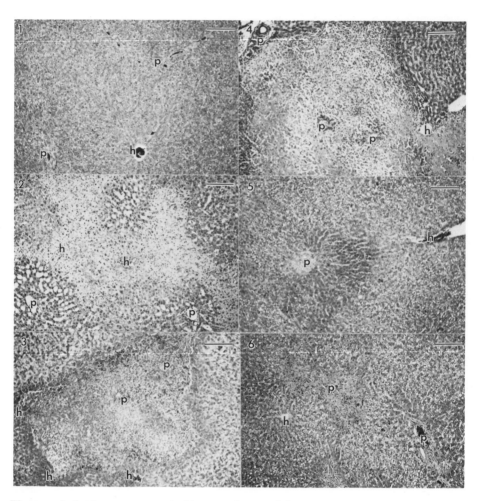

Figures 1-6 show representative regions of hepatic parenchyma from rats given intraportal injections of 1) lipid carrier, 2) senecionine, 3) trans-4-hydroxy-2-hexenal, 4) trans-4-hydroxy-2-nonenal, 5) and 6) trans-2-nonenal. Portal triads are indicated as p and terminal hepatic (central) veins by h. Bar = 100 m. See text for detailed description.

TABLE 1

Compound	Dose range (mg/kg)	Gross Distribution	Microscopic pattern of necrosis
Trans-2-nonenal	15-29	Diffuse	Periportal
Senecionine	8-25	Diffuse	Periacinar (centrilobular)
Trans-4-hydroxy-2-hexenal	12-24	Focal	Paracentral (panacinar)
Trans-4-hydroxy-2-nonenal	16-32	Focal	Paracentral (panacinar)

DISCUSSION

These results demonstrate that a different pattern of hepatic necrosis results from administration of a compound (senecionine) which requires metabolic activation by the MFO system to exert its toxic effect than that which results from administration of it's putative metabolite (t-4HH). A similar compound of longer chain length (t-4HN) produced similar lesions while a related compound not containing a hydroxyl group (t-2N) caused milder but more generalized changes. Among the short chain alkenals, the hydroxylated compounds t-4HH and t-4HN appear to cause striking hepatocellular necrosis involving most of the affected acinus but in limited distribution while the non-hydroxylated nonenal causes necrosis limited to periportal hepatocytes but widely distributed in the liver. These results suggest that the alkenals are not necessarily dependent on the high concentrations of MFO activity in pericentral regions to exert their effect but instead may not require further metabolism for toxicity to occur. The more diffuse distribution and lessened severity of t-2N induced lesions suggests that this compound is less reactive then the hydroxylated alkenals and may as well have a wider distribution in the liver following portal injection due to its relatively lower aqueous solubility.

The paracentral necrosis seen in animals treated with the hydroxylated alkenals represents necrosis of one acinus with sparing of adjacent acini (30). Vascular casting techniques show that different acini within an individual classic hepatic lobule can be supplied by different portal vein radicals (31). Selective necrosis of some acini within lobules could then be explained by differing concentrations of compound entering branches of the portal vein due to differential blood flow. Indeed the frequent, well demarcated, pyramidal shaped necrotic foci shown in figures 2 and 3 often have terminal hepatic veins at their periphery and probably represent necrosis of entire complex acini supplied by a single portal vein radical.

The role of vascular compromise in the pathogenesis of these lesions is uncertain. Thrombosis from endothelial cell damage or reflex vasoconstriction could cause ischemic necrosis of hepatocytes and may well be producing the large well-demarcated lesions seen at higher doses. The relative importance of compound distribution vs vascular compromise in the genesis of these lesions is a subject of continuing investigation.

ACKNOWLEDGEMENTS

The technical assistance of C.H. Brown, E. Avery and D. Morin is gratefully acknowledged. Supported by the Livestock Disease Research Laboratories and NIH Grant ES-03343.

REFERENCES

1. E. Schauenstein, H. Esterbauer and H. Zolner. "Aldehydes in Biological Systems", Pion Limited, London (1977).

2. E. Schauenstein. Autoxidation of polyunsaturated esters in water: chemical structure and biological activity of the products, J. Lipid Res. 8:417 (1967).

3. M. Loury. Possible mechanisms of autoxidative rancidity, Lipids 7:671 (1973).

4. H. Esterbauer. Aldehydic products of lipid peroxidation. in: "Free Radicals, Lipid Peroxidation and Cancer," D.C.H. McBrien and T.F. Slater, Eds. pp. 101, Academic Press, Lond. (1982)

5. H. Esterbauer, K.H. Cheeseman, M.U. Dianzani, G. Poli, and T.F. Slater. Separation and characterization of the aldehydic products of lipid peroxidation stimulated by ADP-Fe^{+2} in rat liver microsomes, Biochem. J 208:129 (1982).

6. A. Benedetti, M. Comporti and H. Esterbauer. Identification of 4-hydroxynonenal as a cytotoxic product originating from the peroxidatio of liver microsomal lipids, Biochim. Biophys. Acta 620:281 (1980)

7. A. Benedetti, A.F. Casini, M. Ferrali, R. Fulceri and M. Comporti. Cytotoxic effects of carbonyl compounds (4-hydroxyalkenals) originating from the peroxidation of microsomal lipids. in: Recent Advances in Lipi Peroxidation and Tissue Injury. T.F. Slater and A. Garner, Eds., pp. 56, Brunel Printing Services, Uxbridge (1981).

8. M. Ferrali, R. Fulceri, A. Benedetti and M. Comporti. Effects of carbonyl compounds (4-hydroxyalkenals) originating from the peroxidation of liver microsomal lipids on various microsomal enzyme activities of th liver, Res. Commun. Chem. Pathol. Pharmacol. 30:99 (1980).

9. A. Benedetti, L. Barbieri, M. Ferrali, A.F. Casini, R. Fulceri and M. Comporti. Inhibition of protein synthesis by carbonyl compounds (4-hydroxyalkenals) originating from the peroxidation of liver microsoma lipids, Chem. Biol. Inter. 35:331 (1981).

10. A.F. Casini, A. Benedetti, M. Ferrali, R. Fulceri and M. Comporti. Estrazione di prodotti tossici dializzabili originatisi dalla perossidazione dei lipidi microsomali epatici, Bull. Soc. Ital. Biol. Sper. 54:893 (1978).

11. M.K. Roders, E.A. Glende Jr., and R.O. Recknagel. Prelytic damage of re cells in filtrates from peroxidizing microsomes, Science 196:1221 (1977).

12. A. Benedetti, A.F. Casini, and M. Ferrali. Red cell lysis coupled to the peroxidation of liver microsomal lipids. Compartmentalization of the hemolytic system, Res. Commun. Chem. Pathol. Pharmacol. 17:519 (1977).

13. A. Benedetti, A.F. Casini, M. Ferrali, and M. Comporti. Extraction and partial characterization of dialysable products originating from the peroxidation of liver microsomal lipids and inhibiting microsomal glucose 6-phosphatase activity, Biochem. Pharmacol. 28:2909 (1979).

14. A. Benedetti, R. Fulceri, M. Ferrali, L. Ciccoli, H. Esterbauer and M.

Comporti. Detection of carbonyl functions in phospholipids of liver microsomes in CCl_4 and $BrCCl_3$ poisoned rats, Biochim. Biophys. Acta. 712:628 (1982).

15. A. Benedetti, H. Esterbauer, M. Ferrali, R. Fulceri and M. Comporti. Evidence for aldehydes bound to liver microsomal protein following CCl_4 or $BrCCL_3$ poisoning, Biochim. Biophys. Acta. 711:345 (1982).

16. H.J. Segall, D.W. Wilson, J.L. Dallas and W.F. Haddon. Trans-4 hydroxy-2-hexenal: A reactive metabolite from the macrocyclic pyrrolizidine alkaloid senecionine, Science in press.

17. L.B. Bull, and A.T. Dick. The chronic pathological effects on the liver of the rat of the pyrrolizidine alkaloids heliotrine, lasiocarpine and their N-oxide, J. Path. Bact. 78:483 (1959).

18. C.E. Green, H.J. Segall, and J.L. Byard. Metabolism, cytotoxicity and genotoxicity of senecionine in primary cultures of rat hepatocytes, Toxicol. Appl. Pharmacol. 60:176 (1981).

19. D.J. Svoboda and J.K. Reddy. Malignant tumors in rats given lasiocarpine, Cancer Res. 32:908 (1972).

20. J.R. Allen, I. Hsu, and L.A. Carstens. Dehydroretronecine-induced rhabdomyosarcomas in rats, Cancer Res. 35:997 (1975).

21. M.N. Pearson, J.J. Karchesy, A. Denney, M.L. Deinzer and G.S. Beaudreau. Induction of endogenous avian tumor virus gene expression by pyrrolizidine alkaloids, Chem. Biol. Interact. 49:341 (1984).

22. U. Candrian, J. Lüthy, U. Graf, and Ch. Schlatter. Mutagenic activity of the pyrrolizidine alkaloids seneciphylline and senkirkine in drosophilia and their transfer into rat milk, Fd. Chem. Toxic. 22:223 (1984).

23. T.W. Petry, G.T. Bowden, R.J. Huxtable and I. Sipes. Characterization of hepatic DNA damage induced in rats by the pyrrolizidine alkaloids monocrotoline, Cancer Res. 44:1505 (1984).

24. D.F. Eastman, G.P. Dimenna and H.J. Segall. Covalent binding of senecionine and seneciphylline to hepatic macromolecules and their distribution, excretion and transfer into milk of lactating mice, Drug. Metab. Disposit 10:236 (1982).

25. R. James, A. Desmond, A. Kupfer, S. Schenker and R.A. Branch. The differential localization of various drug metabolizing systems within the rat liver lobule as determined by the hepatotoxins allylalcohol, carbon tetrachloride and bromobenzene, J. Pharmacol. Exp. Ther. 217:127 (1981).

26. A.M. Rappaport. Physioanatomical basis of toxic liver injury. in: "Toxic Injury of the Liver," E. Tarker and M.M. Fischer Eds., Marcell Dekker Inc, New York (1978).

27. M.M. Groothuis, D.K.F. Meyer, and M.J. Hardonk. Morphological studies on selective acinar liver damage by N-hydroxy-2-acetylaminofluorene and carbon tetrachloride, Naunyn-Schmiedeberg's Arch. Pharmacol. 322:298 (1983).

28. B.W. Erickson. α-hydroxy-α, β-unsaturated aldehydes via 1,3-bis (methylthio)allyllithium: Trans-4-hydroxy-2-hexenal, Organic Synthesis 54:19 (1974).

29. H.J. Segall. Preparative isolation of pyrrolizidine alkaloids derived from Senecio vulgaris, J. Liq. Chromatog. 2:1319 (1979).

30. W.R. Kelly. The liver and bilary system in: "Pathology of Domestic Animals". K.V.F. Jubb, P.C. Kennedy and N. Palmer, eds., Academic Press New York (1985).

31. A.M. Rappaport. The microcirculatory acinar concept of normal and pathological hepatic structure, Beitr. Path. Bd. 159:215 (1976).

CLOFIBRATE SELECTIVELY INDUCES AZOREDUCTION OF

DIMETHYLAMINOAZOBENZENE (DAB) BY RAT LIVER MICROSOMES

W. G. Levine and H. Raza

Department of Molecular Pharmacology
Albert Einstein College of Medicine
Bronx N. Y. 10461

INTRODUCTION

Reduction of azo linkages in xenobiotics has long been known in biological systems. Fifty years ago, the antibacterial dye, prontosil, was shown to be reduced to the active product, sulfanilamide, by colonic bacteria[1]. Hernandez et al.[2] found that reduction of prontosil and neotetrazolium by rat liver microsomes is partially inhibited by CO, implying the involvement of cytochrome P-450. Purified NADPH-cytochrome c reductase also catalyzes reduction and accounts for CO-insensitive activity[3]. Fujita and Peisach[4] reported that azoreduction of amaranth is almost completely inhibited by CO and is induced by treatment with phenobarbital (PB) or 3-methylcholanthrene (MC). Under all conditions, the rate of azoreduction of amaranth is proportional to total microsomal cytochrome P-450. Antibodies against PB- and MC-induced cytochrome P-450 almost completely inhibit azoreduction of amaranth in PB- or MC-induced microsomes, respectively[6]. Recently, it was shown that purified cytochrome P-450 from PB-induced rat livers readily reduces amaranth[7]. Thus, there are several distinct azoreductases in microsomes which can be distinguished by their substrate specificities. FMN and FAD markedly stimulate azoreduction of amaranth[5] and neoprontosil[8]. The stimulated activity is unaffected by CO, implying, although not necessarily proving, that cytochrome P-450 is not involved. However, it was later demonstrated that these flavines could transfer electrons from both NADPH-cytochrome c reductase and cytochrome P-450 to dye[9].

Azoreductase activity of a different nature is found in rat liver cytosol[10]. It is induced by MC but exhibits narrow substrate specificity and is inhibited by dicumarol. Structural requirements for the substrates is highly selective in that dimethylaminoazobenzene (DAB) is not reduced whereas methyl red (3'-COOH-DAB) is rapidly reduced. Similarly, an azoreductase isolated from Pseudomonas KF46[11] reduces Orange II but not similar structures missing the 2-OH substituent.

The azodye carcinogen, DAB, is metabolized by N-demethylation, ring hydroxylation, N-hydroxylation, and azoreduction. Azoreduction is NADPH-dependent[12] and induced by PB and MC[13]. The study of azoreduction often is complicated by oxygen inhibition. This may be due to reoxidation of the hydrazo[3] or free radical[14] intermediate. Reduction of DAB has been studied aerobically[12,15-18] and anaerobically[15,18,19] and lack of inhibition by oxygen has been reported. The reaction probably does not involve a free radical intermediate, or one that exists long enough to be reoxidized, and proceeds directly to the fully reduced product. CO may not inhibit DAB reduction[15,17], giving rise to the question of whether the reaction is truly catalyzed by cytochrome P-450.

Conventional spectrophotometric methods for measuring DAB azoreduction by loss of color may yield erroneous results if performed aerobically[20]. Demethylated and hydoxylated products exhibit spectral absorbance somewhat less than that for DAB, giving a false high rate of azoreduction. This problem can be avoided by measuring substrate disappearance anaerobically, by chromatographically separating all products, or by measuring fluorometrically only the reduction products With these precautions, it was shown that DAB azoreduction is much less sensitive to oxygen than are a number of other dyes. Compared to the rate in a nitrogen atmosphere, azoreduction of DAB is depressed only 15% in air and 50% in 100% oxygen[20].

Regulation of the metabolism of DAB by glutathione (GSH) has been suggested by several studies. GSH stimulates oxidative DAB metabolism[21] but depresses azoreduction[20]. This implies that mechanisms for oxidative and reductive pathways may be independently regulated by GSH. The sulfhydryl reagents, N-ethylmaleamide and p-hydroxymercuribenzoate, suppress reduction of both MAB and DAB, strongly implying a role for thiol groups in catalytic activity. Further study is needed to define such a role more thoroughly.

Clofibrate and other hypolipidemic drugs induce hepatic peroxisomal proliferation and long term treatment leads to increased incidence of liver tumors in rodents[22]. These drugs also induce hepatic cytochrome P-450[23-26]. At least one isozyme of cytochrome P-450 isolated from livers of clofibrate-treated rats exhibits characteristics different from those associated with classical inducers such as PB or BNF. Its reduced CO spectrum has a peak at 452 nm and it does not metabolize ethoxyresorufin, benzphetamine or testosterone[24].

In view of the rather different nature of oxidative versus reductive pathways for DAB, the effects of various inducers of cytochrome P-450 activity on the azoreduction of DAB were determined. The results indicate a highly selective response, and suggest that DAB is reduced by few, perhaps only one isozyme of cytochrome P-450.

METHODS

Animal Treatment

Male Wistar rats (200-225g) were treated with one of the following compounds: clofibrate, I.P. 300 mg/kg/day, 7 days; nafenopin in Emulphor EL-620, P.O. 300 mg/kg/day, 4 days; PB in 0.9% NaCl, I.P. 75 mg/kg/day, 4 days; BNF in corn oil, I.P. 40 mg/kg/day, 3 days; isosafrol in corn oil, I.P. 150 mg/kg/day, 4 days; PCN suspended in 0.9% NaCl plus a few drops of Tween 80, P.O. 50 mg/kg/day, 5 days; control rats received vehicle alone. Liver microsomes were prepared 48 hr after nafenopin and 24 hr after the last dose of other drugs. Protein concentration was measured according to the method of Lowry et al.[28].

Anaerobic Metabolism of DAB and MAB

Microsomes were incubated with dye in a nitrogen atmosphere using a

NADPH generating system[21]. Primary amine products were measured fluoro-metrically according to a method described by Huang et al.[10]. Neither azo dye interfered in the assay. Both reduction products of DAB, aniline and N,N-dimethyl-p-phenylenediamine (DMPPD), were extracted and measured in the assay.

Aerobic Metabolism of DAB and MAB

A similar microsomal system was incubated in air using ^{14}C-DAB or ^{3}H-MAB and all metabolites were separated by thin-layer chromatography and quantified by scintillation counting[21]. Azoreduction of DAB and MAB is only slightly inhibited by oxygen[20], permitting the same assay system to be used for determination of reduced as well as oxidized metabolites.

Lauric Acid Hydroxylation

This was measured by a modification of the method of Parker and Orton[25] in the same microsomal system used for DAB metabolism. ^{14}C-Laurate and metabolites were extracted and separated on ITLC plates (Gellman) with hexane:diethylether:acetic acid (60:19:1). After spraying with 2'7'-dichlorofluorescein, the laurate and hydroxylaurate spots were cut from the chromatogram, and counted. This measured total hydroxy-lation of lauric acid because 12- and 11-hydroxylaurate have the same R_f values.

NADPH-cytochrome c Reductase

Activity was measured in microsomes according to the method of Dignam and Strobel[29].

RESULTS

Azoreduction of DAB was performed anaerobically and primary amine reduction products were determined fluorometrically. Microsomes were prepared from rats treated with clofibrate, nafenopin, PB, isosafrol or PCN. Clofibrate induced activity several fold while nafenopin, a struc-turally related hypolipidemic compound, depressed activity 90% (fig. 1). Treatment with other compounds had relatively little effect. Nafenopin or its metabolites did not directly inhibit DAB azoreduction since addition of 1 mM nafenopin to a microsomal reaction mixture did not affect oxidative or reductive metabolism, even after 10 minutes of preincubation. With MAB as substrate, a 2-fold induction of azoreductase by clofibrate and 90% suppression by nafenopin was again observed (data not shown).

Oxidative and reductive metabolism of DAB was measured aerobically and products of both pathways were separated by thin-layer chromato-graphy. With this method, only DMPPD and MPPD, azoreduction products of DAB and MAB, respectively, could be quantified. Other reduction products, PPD and aniline, were not measured. In agreement with the anaerobic experiments (fig. 1), azoreduction (DMPPD and MPPD formation) was induced by clofibrate treatment (Table 1), confirming the relative lack of oxygen sensitivity for this reaction. However, both N-demethylation (MAB and AB formation) and ring-hydroxylation (4'-OH-DAB formation) of DAB were suppressed. Nafenopin treatment again led to greatly inhibited azore-duction while N-demethylation and ring-hydroxylation were only slightly affected (Table 1). With MAB as substrate, azoreduction was again induced. However, clofibrate had little effect on N-demethylation (Table 2), suggesting that only N-demethylation of the tertiary amine (DAB) is altered by clofibrate treatment.

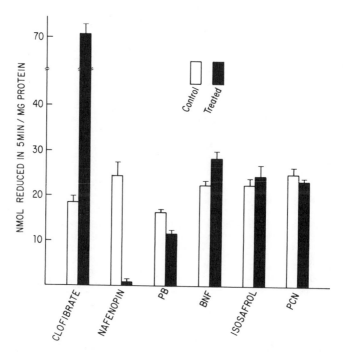

Fig. 1. <u>Effect of treatment with various agents on the anaerobic azoreduction of DAB by rat liver microsomes</u>. Liver microsomes were prepared from rats treated with the compounds indicated. They were incubated anaerobically with DAB, as described in METHODS and the reduced products assayed fluorometrically[10].

TABLE 1

AEROBIC METABOLISM OF DAB. EFFECT OF CLOFIBRATE AND NAFENOPIN.

nmol metabolites present in 10 min/mg protein[a]

	MAB[b]	AB[b]	4'-OH-DAB[c]	DMPPD[d]	MPPD[d]
Control	14.0 + 0.9	11.8 + 0.3	9.5 + 0.8	7.5 + 0.7	4.4 + 0.7
Clofibrate	5.9 + 0.1[e]	4.7 + 0.4[e]	5.3 + 0.6[e]	24.0 + 1.1[e]	11.5 + 1.1[e]
Control	12.4 + 0.7	6.7 + 0.8	12.1 + 0.8	5.1 + 0.7	6.7 + 1.0
Nafenopin	10.7 + 1.3	5.2 + 0.4	8.1 + 1.4[e]	1.1 + 0.1[e]	0.7 + 0.1[e]

[a]Values are the mean + SEM for three trials.
[b]N-Demethylation products.
[c]Ring-hydroxylation product.
[d]Azoreduction products.
    DMPPD = N,N-dimethyl-p-phenylenediamine.
    MPPD = N-methyl-p-phenylenediamine.
[e]Values differ significantly from those of the corresponding controls (P< 0.05).

TABLE 2

AEROBIC METABOLISM OF MAB. EFFECT OF CLOFIBRATE.

nmol metabolites present in 10 min/mg protein[a]

	AB[b]	4'-OH-MAB[c]	MPPD[d]
Control	55.6 + 1.3	3.8 + 0.2	1.3 + 0.4
Clofibrate	51.8 ± 3.9	6.4 ± 0.2[e]	4.0 ± 0.6[e]

[a]Values are the mean ± SEM for three trials.
[b]N-demethylation product.
[c]Ring-hydroxylation product.
[d]Azoreduction product.
     MPPD=N-methyl-p-phenylenediamine
[e]Values differ significantly from those of the corresponding controls (P< 0.05).

NADPH-cytochrome c reductase activity was unaffected by pretreatment with clofibrate or nafenopin (data not shown).

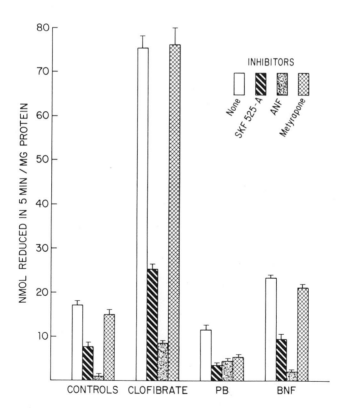

Fig. 2. Effects of various inhibitors of cytochrome P-450 activity on the anaerobic azoreduction of DAB by microsomes prepared from rats treated with various inducing agents. Animals were administered one of the inducing agents while inhibitors were added in vitro.

Evidence has thus far favored cytochrome P-450 as the catalyst for DAB azoreduction. However, the report of insensitivity to CO[15,17] has raised some question as to this identity. Therefore, the effect of known inhibitors of cytochrome P-450 activity was determined (fig. 2). SKF 525-A and alpha-naphthoflavone were inhibitory in control and induced microsomes. Alpha-naphthoflavone is generally more effective against activity induced by MC[30] although there are notable exceptions[27]. Metyrapone, on the other hand, inhibited only PB-induced microsomes, which is in keeping with known activity[31].

Laurate hydroxylation was markedly induced by treatment with clofibrate (fig. 3), confirming other reports[23]. Nafenopin exhibited the same effect (fig. 3) despite its opposite response with azoreduction activity (fig. 1). Induction by PB also confirmed the work of others[32], however, no increase in laurate hydroxylation was seen after treatment with isosafrol, PCN or BNF (fig. 3).

DISCUSSION

Reductive biotransformation of xenobiotics by cytochromes P-450 has received relatively little attention compared to the abundant information on oxidative metabolism. Reduction of carcinogenic azodyes may partially determine their carcinogenic potential and for many compounds is considered a detoxication pathway[33]. On the other hand, some products of azo reduction are mutagenic and therefore potentially carcinogenic[34]. Azoreduction of DAB is mediated by NADPH-dependent enzymes in liver microsomes and is probably catalyzed mainly by cytochrome P-450[15,17,20]. The role of selective forms of cytochrome P-450 in oxidative metabolism of DAB has been reported earlier[27]. The present study suggests that clofibrate

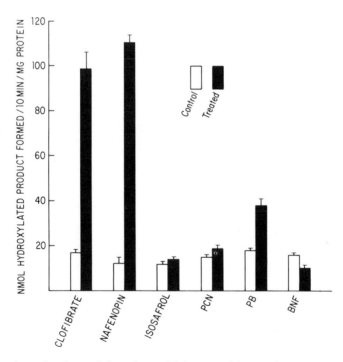

Fig. 3. Hydroxylation of lauric acid by rat liver microsomes prepared from animals treated with various agents.

induces one or more forms of cytochrome P-450 which catalyze the reductive metabolism of DAB. After clofibrate treatment, azo reduction of DAB and MAB was markedly induced, whereas oxidative pathways, N-demethylation and ring-hydroxylation, were depressed. PB, BNF, isosafrol and PCN, selective inducers of cytochrome P-450[35-37], did not appreciably induce azoreduction. Depression of the oxidative pathways of DAB by clofibrate is unusual in view of its induction of spectrally determined cytochrome P-450 and several pathways of xenobiotic metabolism. Catalytic properties of clofibrate-induced cytochrome P-450 differ from those of other types of cytochrome P-450. Omega-hydroxylation of lauric acid in rats is readily induced by clofibrate but not by PB or BNF, whereas the metabolism of benzphetamine and ethoxyresorufin is not induced by clofibrate[24]. Selective hydroxylation of progesterone and estradiol is also induced by clofibrate although cholesterol hydroxylation is unaffected[38,39].

Individual steps in the oxidative metabolism of DAB and MAB are selectively induced by PB and BNF[27,40], implying the involvement of specific forms of cytochrome P-450. Reductive metabolism of DAB also appears to involve a cytochrome P-450 with unique properties, including a relatively narrow spectrum of substrate reactivity. Previously, the reduction of neoprontosil was attributed to microsomal cytochrome P-450 and to NADPH-cytochrome c reductase[3], while reduction of amaranth seemed to involve total microsomal cytochrome P-450[4]. In this study, azoreduction of a specific substrate, DAB, has been associated with a very narrow range or perhaps even a specific isoform of cytochrome P-450. It is apparent that azoreduction cannot be considered a general microsomal activity. There are specific enzymes with selective substrate specificity.

In contrast to the oxygen sensitivity exhibited by sulfonazo III reduction[14], DAB reduction is somewhat less sensitive to oxygen[18,20]. This implies that formation of a free radical intermediate, which is detected in the reduction of sulfonazo III, may not occur during reduction of DAB. If metabolites are measured individually, DAB azoreduction can be assayed aerobically in conjunction with oxidative pathways. The validity of the use of either aerobic or anaerobic method is born out by the observation that clofibrate-induced azoreduction of DAB is seen in both cases.

Treatment with nafenopin does not increase aniline hydroxylation, aminopyrine demethylation or total hepatic cytochrome P-450[41]. Unlike clofibrate, nafenopin treatment inhibited DAB and MAB azoreduction almost completely, while in vitro addition of drug did not appreciably inhibit activity. This suggests that nafenopin, or one of its metabolites, may depress synthesis of the specific form of cytochrome P-450 which catalyzes DAB azoreduction. Nafenopin treatment also moderately (40%) depressed ring-hydroxylation whereas N-demethylation was not significantly affected.

Although azoreduction of certain dyes is catalyzed in part by NADPH-cytochrome c reductase[2,3]. We found no significant stimulation of activity by either clofibrate or nafenopin, suggesting that DAB azoreductase activity was only associated with cytochrome P-450. We observed that SKF 525-A and ANF non-specifically inhibited azoreductase activity in control and induced microsomes. Metyrapone, which has a preferred affinity for PB type cytochrome P-450[31], as expected, inhibited azoreductase activity only in PB-treated microsomes. PB induces omega-1 hydroxylation of lauric acid but not omega hydroxylation[32] while clofibrate induces principally omega hydroxylation[23]. Omega-1, but not omega hydroxylation, is inhibited by ANF, SKF 525-A and metyrapone. DAB azoreduction is blocked by these inhibitors, suggesting enzymic identity with omega-1 activity. However, DAB azoreduction is not induced by PB and it is unclear at the present time how these activities relate, if at all. Another complication is that nafenopin treatment induced laurate hydroxylation but inhibited azoreduction. The induction of laurate hydroxylation by nafenopin has also been reported recently[42].

In conclusion, we have clearly demonstrated a selective induction by clofibrate of azoreduction of the hepatocarcinogen, DAB. This may be associated with a specific isoform of cytochrome P-450 since PB, BNF, isosafrol and PCN do not affect activity. Another hypolipidemic drug, nafenopin, profoundly depresses azoreduction but induces laurate hydroxylation. Thus, the isoform of cytochrome P-450 which catalyzes the azoreduction may not be identical with those which catalyze oxidative metabolism.

ACKNOWLEDGEMENTS: The excellent technical contribution of Craig Schwartz is acknowledged. Many helpful suggestions were contributed to this study by Dr. Jack Peisach. This work was supported in part by the following grants from the National Institutes of Health: CA-14231, CA-13330, AM-17702.

REFERENCES

1. J. Trefouel, J. Trefouel, F. Netti, and D. Bovet, Activite de p-aminophenylsulfamide sur les infections streptococciques experimentales de la souris et du lapin, C.R. Seanc. Soc. Biol., 120:756 (1935).
2. P.H. Hernandez, P. Mazel and J.R. Gillette, Studies on the mechanism of action of mammalian hepatic azoreductase-II: The effects of phenobarbital and 3-methylcholanthrene on carbon monoxide sensitive and insensitive azoreductase activities, Biochem. Pharmacol., 16:1877 (1967).
3. P.H. Hernandez, J.R. Gillette and P. Mazel, Studies on the mechanism of action of mammalian hepatic azoreductase-I: Azoreductase activity of reduced nicotinamide adenine dinucleotidephosphate-cytochrome c reductase, Biochem. Pharmacol., 16:1859 (1967).
4. S. Fujita and J. Peisach, Liver microsomal cytochromes P-450 and azoreductase activity, J. Biol. Chem., 252:4512 (1978).
5. S. Fujita and J. Peisach, The stimulation of microsomal azoreduction by flavins, Biochim. Biophys. Acta. 719:178 (1982).
6. S. Fujita, Y. Okada, and J. Peisach, Inhibition of azoreductase activity by antibodies against cytochrome P-450 and P-448, Biochem Biophys. Res. Commun., 102:492 (1981).
7. A.K. Mallett, L.J. King, and R. Walker, Solubilization purification and reconstitution of hepatic microsomal azoreductase activity, Biochem. Pharmacol., 34:337 (1985).
8. L. Shargel and P. Mazel, Influence of 2,4 dichloro-6-phenoxymethyl-amine (DPEA) and β-diethylaminoethyl diphenylpropylacetate (SKF 525-A) on hepatic microsomal azoreductase activity from phenobarbital or 3-methylcholanthrene induced rats, Biochem. Pharmacol., 21:69 (1972).
9. A.K. Mallett, L.J. King, and R. Walker, A continuous spectrophotometric determination of hepatic microsomal azoreductase activity and its dependence on cytochrome P-450, Biochem. J., 201:589 (1982).
10. M-T. Huang, G.T. Miwa, N. Cronheim, and A.Y.H. Lu, Rat liver cytosolic azoreductase: Electron transport properties and the mechanism of dicumarol inhibition of the purified enzyme, J. Biol Chem., 254:11223 (1979).

11. T. Zimmerman, H.G. Kulla, and T. Leisinger, Properties of purified Orange II azoreductase, the enzyme initiating azo dye degradation by Pseudomonas KF46, Eur. J. Biochem., 129:197 (1982).
12. G.C. Mueller and J.A. Miller, The reductive cleavage of 4-dimethylaminoazobenzene by rat liver: The intracellular distribution of the enzyme system and its requirement for triphosphopyridine nucleotide, J. Biol. Chem., 180:1125 (1949).
13. A.H. Conney, E.C. Miller, and J.A. Miller, The metabolism of methylated aminoazo dyes. V. Evidence for induction of enzyme synthesis in the rat by 3-methylcholanthrene, Cancer Res., 16:450 (1956).
14. R.P. Mason, F.J. Peterson, and J.L. Holtzman, Inhibition of azoreductase by oxygen: the role of the azo anion free radical metabolite in the reduction of oxygen to superoxide, Mol. Pharmacol., 14:665 (1978).
15. H. Autrup and G.P. Warwick, Some characteristics of two azoreductase systems in rat liver. Relevance to the activity of 2-[4'-di(2'-bromopropyl)-aminophenylazo] benzoic acid (CB10-252), a compound possessing latent cytotoxic activity, Chem. Biol. Interact., 11:329 (1975).
16. F. De Cloitre, J. Chauveau, and M. Martin, Influence of age and 3-methylcholanthrene on azodye carcinogenesis and metabolism of p-dimethylaminoazobenzene in rat liver, Int. J. Cancer, 11:676 (1973).
17. P.S. De-Araujo, E. De-Andrade-Silva, and I. Raw, Effect of drugs and hormones on rat liver dimethylaminoazobenzene reductase activity, Braz. J. Med. Biol. Res., 15:17 (1982).
18. B.M. Elliot, Azoreductase activity of Sprague Dawley and Wistar-derived rats towards both carcinogenic and non-carcinogenic analogues of 4-dimethylaminophenylazobenzene (DAB). Carcinogenesis, 5:1051 (1984).
19. B. Ketterer, P. Ross-Mansell, and H. Davidson, The effect of the administration of N,N-dimethyl-4-amino-azobenzene (DAB) on the activity of DAB-reductase and NADPH-cytochrome c reductase, Chem. Biol. Interact., 2:183 (1970).
20. W.G. Levine, Studies on microsomal azoreduction of N,N-dimethyl-4-aminoazobenzene (DAB) and its derivatives, Biochem. Pharmacol., in press (1985).
21. W.G. Levine and S.B. Lee, Effect of glutathione on the metabolism of N,N-dimethyl-4-aminoazobenzene by rat liver microsomes, Drug Metab. Dispos., 11:239 (1983).
22. J.K. Reddy and N.D. Lalwani, Carcinogenesis by hepatic peroxisome proliferators: Evaluation of the risk of hypolipidemic drugs and industrial plasticizers to humans, CRC Crit. Rev. Toxicol., 12:1 (1983).
23. T.C. Orton and G.L. Parker, The effect of hypolipidemic agents on the hepatic microsomal drug metabolizing enzyme system of the rat. Induction of cytochrome (s) P-450 with specificity toward terminal hydroxylation of lauric acid, Drug Metab. Dispos., 10:110 (1982).
24. P.P. Tamburini, H.A. Masson, S.K. Bains, R.J. Makowski, B. Morris and G.G. Gibson, Multiple forms of hepatic cytochrome P-450. Purification, characterization and comparison of a novel clofibrate-induced isozyme with other major forms of cytochrome P-450, Eur. J. Biochem. 139:235 (1984).
25. G.L. Parker, and T.C.Orton, Induction by oxyisobutyrates of hepatic and kidney microsomal cytochrome P-450 with specificity towards hydroxylation of fatty acids, in "Biochemistry, Biophysics and Regulation of Cytochrome P-450," J.A. Gustafsson, J.C. Duke, A. Mode and J. Rafter, eds. Elsevier/North Holland Press. Amsterdam (1980).

26. R.A. Salvador, S. Haber, C. Atkins, B.W. Gommi and R.M. Welch, Effect of clofibrate and 1-methyl-4-piperidylbis (p-chlorophenoxy) acetate (Sandoz 42-348) on steroid and drug metabolism by rat liver microsomes, Life Sci. 9 part II:397 (1970).

27. W.G. Levine and A.Y.H. Lu, Role of isozymes of cytochrome P-450 in the metabolism of N,N-dimethyl-4-aminoazobenzene in the rat, Drug Metab. Dispos. 10:102 (1982).

28. O.H. Lowry, N.J. Rosebrough, A.L. Farr, and R.J. Randall, Protein measurement with the Folin phenol reagent, J. Biol. Chem. 193:265 (1951).

29. J.D. Dignam and H.W. Strobel, NADPH-cytochrome P-450 reductase from rat liver: Purification by affinity chromatography and characterization, Biochemistry 16:1116 (1977).

30. F.J. Weibel, J.C. Lentz, L. Diamond, and H.V. Gelboin, Aryl hydrocarbon (benzo[a]pyrene) hydroxylase in microsomes from rat tissues: differential inhibition and stimulation by benzoflavones and organic solvents, Arch. Biochem. Biophys., 144:78 (1971).

31. F. Mitani, E.A. Shepard, I.R. Phillips, and B.R. Rabin, Complexes of cytochrome P-450 with metyrapone. A convenient method for the quantitative analysis of phenobarbital-inducible cytochrome P-450 in rat liver microsomes, FEBS Lett 148:302 (1982).

32. R.T. Okita and B.S.S. Masters, Effect of phenobarbital treated and cytochrome P-450 inhibitors on the laurate omega and (omega -1)-hydroxylase activities of rat liver microsomes, Drug Metab. Dispos. 8:147 (1980).

33. R. Walker, The metabolism of azo compounds: A review of the literature, Fd. Cosmet. Toxicol., 8:659 (1970).

34. K-T. Chung, The significance of azo-reduction in the mutagenesis and carcinogenesis of azo dyes, Mut. Res. 14:269 (1983).

35. A.Y.H. Lu and S.B. West, Multiplicity of mammalian microsomal cytochromes P-450, Pharmacol. Rev., 31:277 (1980).

36. E.G. Schuetz, S.A. Wrighton, J.L. Barwick, and P.S. Guzelian, Induction of cytochrome P-450 by glucocorticoids in rat liver. I. Evidence that glucocorticoids and pregnenalone 16α-carbonitrile regulate de novo synthesis of a common form of cytochrome P-450 in cultures of adult rat hepatocytes and in the liver in vivo, J. Biol. Chem., 259:1999 (1984).

37. T.R. Fennell, M. Dickins, and J.W. Bridges, Interaction of isosafrole in vivo with rat hepatic microsomal cytochrome P-450 following treatment with phenobarbitone or 20-methylcholanthrene, Biochem. Pharmacol., 28:1427 (1979).

38. K. Einarsson, J.K. Gustafsson, and K. Hellstrom, Effect of clofibrate on the metabolism of progesterone and oestradiole in rat liver microsomal fraction, Biochem. J., 136:623 (1973).

39. B. Angelin, I. Bjorkhem, and K. Einarsson, Effect of clofibrate on some microsomal hydroxylations involved in the formation and metabolism of bile acids in rat liver, Biochem. J., 156:445 (1976).

40. W.G. Levine, Induction and inhibition of the metabolism and biliary excretion of the azo dye carcinogen, N,N-dimethyl-4-aminoazobenzene (DAB) in the rat, Drug Metab. Dispos., 8:212 (1980).

41. W.G. Levine, Effects of hypolipidemic drug nafenopin (2-methyl-2-[p-(1,2,3,4-tetrahydro-1-naphthyl)phenoxy]propionic acid, TPIA, Su-13,4337) on the hepatic disposition of foreign compounds in the rat, Drug Metab. Dispos., 2:178 (1974).

42. B.G. Lake, T.J.B. Gray, et al., The effect of hypolipidemic agents on peroxisomal β-oxidation and mixed-function oxidase activities in primary cultures of rat hepatocytes. Relationship between induction of palmitoyl-CoA oxidation and lauric acid hydroxylation, Xenobiotica, 14:269 (1984).

INTRACELLULAR DISSIPATIVE STRUCTURES (IDS) AS ULTIMATE TARGETS OF

CHEMICAL CYTOTOXICITY

S. Ji, S. Ray and R. Esterline

Department of Pharmacology and Toxicology
Rutgers University, Piscataway, NJ 08854

The primary objective of this communication is to present a general
theory of chemically induced cell injury and cell death that is based
partly on experimental observations on acetaminophen hepatotoxicity (Ji
et al., 1985a) and partly on a recently developed theoretical model of the
living cell called the Bhopalator (Ji, 1985a, b). First, we will
summarize experimental data on acetaminophen hepatotoxicity. Second, the
Bhopalator model of the living cell will be described. Finally, the idea
that the universal mechanistic features of all chemical cytotoxicity can
be recognized only at the level of intracellular dissipative structures
(IDS) will be elaborated.

I. A general model of acetaminophen hepatotoxicity

Acetaminophen is one of the most thoroughly investigated
hepatotoxins known (Mitchell et al., 1973; Jollow et al., 1973; Dahlin et
al., 1984; West et al., 1984; Black, 1984). The initial fate of
acetaminophen administered to living animals is its conversion to the
gulucuronide and sulfate conjugates in the liver for excretion into the
urine and the bile (see steps 1 and 2 in Figure 1). In parallel with or
following the saturation of the conjugating pathways, some of the drug is
metabolized to form the reactive intermediate, N-acetyl-p-benzoquinone
imine via the cytochrome P-450 enzyme system (Dahlin et al, 1984) (step 3
in Figure 1). As long as cells contain adequate amounts of reduced
glutathione (GSH), this reactive intermediate is harmless since it can be
rapidly removed as GSH conjugates (step 4) before it can cause cell
injury. However, when the intracellular GSH level is insufficient for
this task, the biological reactive intermediate can bind covalently to
essential enzymes and proteins inside the cell, thereby adversely
altering metabolic fluxes leading to cell injury (steps 5 and 10 in
Figure 1). In addition, N-acetyl-p-benzoquinone imine can undergo one-
electron reduction to the semiquinone radical derivative which can
subsequently donate the electron to molecular oxygen to generate the
superoxide anion radical and other reactive oxygen products including
H_2O_2 and HO· (Kappus, 1985).   These reactive oxygen metabolites are
regarded as members of "reactive intermediates" shown in Figure 1. Such
reactive oxygen metabolites are injurious to hepatocytes since they can
cause lipid peroxidation (step 6) and oxidize critical protein-bound
sulfhydryl groups to disulfide states (step 7) leading to an influx of
calcium into the cell (Jewell et al., 1982).

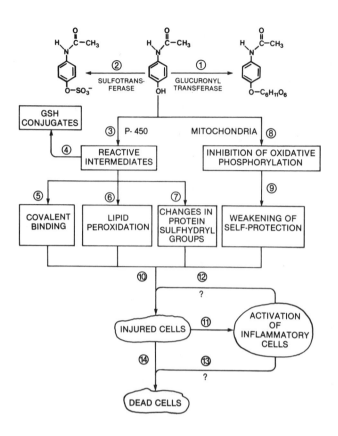

Figure 1. A general model of the mechanism of acetaminophen
hepatotoxicity. See text for detailed explanations.

During the past three years, two new aspects of the acetaminophen
hepatotoxicity problem have emerged. First, acetaminophen has been found
to inhibit mitochondrial respiration at coupling site 1 (Esterline and
Ji, 1985), thereby inhibiting oxidative phosphorylation in perfused liver
(Ji et al., 1984, 1985a) and in isolated mitochondria (Cheng et al.,
1984; Esterline and Ji, 1985) (step 8). Although we do not yet have
direct experimental proof, we conjecture, on theoretical grounds, that
this acetaminophen-induced inhibition of oxidative phosphorylation will
contribute significantly to the weakening of the endogenous self-
protection and self-repair mechanisms of hepatocytes, thus potentiating
the cytotoxicity of the acetaminophen-derived reactive intermediates
(steps 9 and 10). Second, we have found that treating fed rats with
acetaminophen (1.2g/kg) overnight did not lead to any overt histological
signs of cellular necrosis but caused cell injury nevertheless, since the
histological sections of the liver so treated revealed the accumulation
of polymorphonuclear leukocytes into the pericentral regions of the liver
lobule (Ji et al., 1984; Ji et al, 1985a) (Figure 2E and F). This
initial observation led to a series of investigations by Laskin et al.
(1984) and Pilaro et al. (1985) who have demonstrated that the treatment
of cultured hepatocytes with low doses of acetaminophen (50 to 150 μM)
released chemoattractants and activating factors which induced the
chemotaxis of isolated Kupffer cells and stimulated the rate of
superoxide anion production and the phagocytic activity of these cells.
These observations have led us to postulate that acetaminophen-induced
activation of inflammatory cells will make important contributions to

Table 3. An Analogy between the Living Cell and the Digital Computer

Parameter	Cell	Computer
Current Carrier	Electrons	Electrons
Current Conductor	Chemicals	Wires
Current controller	Enzymes	Transistors
Mechanism of control of electron flow	Conformons	Photons
Electron source	Chemicals	External voltage source
Electron sink	O_2 or equivalent	Ground
Memory device	Nucleic acids Polypeptides, IDS's	Flip-flops capacitors
Structural rigidity	Thermally fluctuating	Thermally rigid
Size	Microscopic (10^{-3} cm)	Macroscopic (10^2 cm)
Circuit dimensionality	n-Dimensional in chemical space*	3-Dimensional in Euclidean space
Physical structure	Cellular morphology	Computer architecture
Self-reproducibility	Yes	Not yet
Mathematical construct	None yet	Turing machine

*n= the number of metabolic pathways. See Ji (1985b).

endoplasmic reticulum, sacroplasmic reticulum and lysosomes. This contrasts with the space-dependent dissipative structures that are seen in the Belousov-Zhabotinsky reaction without any spatial compartmentation (Winfree, 1974).

In addition to IDS's composed of inorganic ions, there can exist IDS's built from organic molecules such as various substrates, intermediate metabolites and products of intracellular metabolism. Any subcellularly compartmented organic molecules such as ATP, NAD^+, NADH, $NADP^+$, NADPH, etc. can serve as examples of organic molecular IDS's.

Although the power of resolution is too low to measure subcellular concentration profiles of reduced pyridine nucleotides (NADH, NADPH), the tissue fluorometric technique developed by B. Chance and his coworkers (1965; Ji, 1979b) can be employed to measure continuously the intracellular concentrations of reduced pyridine nucleotides averaged over a large number ($10^2 \sim 10^3$) of cells in biological tissues. We will refer to the averages of as many IDS's as can be measured by Chance's tissue photometric technique as the "multicellular average dissipative structures (MADS)". We have recently carried out a series of experiments aimed at characterizing MADS by continuously monitoring the reduced pyridine nucleotide fluorescence from the surface of the perfused rat liver employing a bifurcated lightguide described elsewhere (Ji, 1979b) (Figure 5). The tip of the the lightguide in contact with the liver surface was 2 mm in diameter and the estimated number of hepatocytes being observed fall in the range of 10^2 to 10^3 cells.

Representative recordings of tissue fluorescence are given in Figure 6. As can be seen, when a toxic dose (25 mM) of acetaminophen is infused into the liver, the tissue fluorescence rises rapidly, the maximal increase being reached in less than 100 seconds. The fluorescence signal then declined more or less linearly with the time during the following 10 minutes. The rate of oxygen uptake (the third panel) decreased immediately after adding acetaminophen. When the same dose of acetaminophen was added to the liver following the infusion of 40 μM menadione, quite a different result was observed - the tissue fluorescence signal revealed a triphasic change (see the second panel) and the rate of hepatic oxygen uptake increased by about 20% rather than

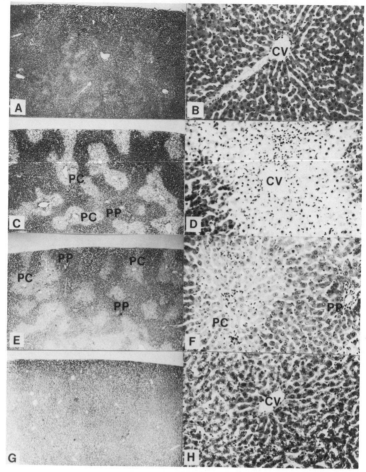

Figure 2. The sublobular zonation of the acetaminophen—induced liver
injury. Rats (Sprague-Dawley, female, 200g) were treated with
acetaminophen (1.29/Kg) via gastic intubation. Livers were removed under
pentobarbital anestheisa (50 mg/kg) and perfused with oxygenated Krebs-
Henseleit buffer before fixing in buffered 10% formaldehyde. The panels
on the right (B, D, F, H) were magnified by 400 and those on the left (A,
C, E, G) by 25 fold. A and B are controls; C and D are hematoxylin- and
eosin-stained to visualize necrotic cells; E and F are stained with
periodic acid Schiff base to visulaize glycogen. G and H are control H &
E stained sections from rats treated with sesame seed oil (in
collaboration with Dr. R. L. Trelstad).

cell injury and cell death (see steps 11, 12 and 13).

As evident in Figure 1, the mechanisms of acetaminophen
hepatotoxicity are complex and implicate multiple contributory pathways.
The model in Figure 1 is consistent with the emerging concept that the
covalent binding of biological reactive intermediates of hepatotoxins to
essential cellular macromolecules is often necessary but not always
sufficient for causing cell death. In fact, the model predicts that the
necessary and sufficient conditions for cell injury consist of the
"activation" of not only one of the cell-injuring steps 5, 6, and 7 but
also step 9. That is, without a simultaneous weakening of the endogenous
self-protection mechanisms of hepatocytes, the activation of cell-
damaging mechanisms alone will be insufficient to cause cell injury.

Another important feature of the acetaminophen hepatotoxicity model proposed in Figure 1 is the explicit recognition of at least three distinct states of the cell – the normal (N), injured (I), and dead (D) states. Step 10 represents the transition of hepatocytes from the normal to the injured state, and step 14 indicates the irreversible transition of hepatocytes from the injured state to the dead state. The experimental evidence for the existence of the injured state is provided by the afore-mentioned observation that acetaminophen given to fed rats led to the accumulation of inflammatory cells to glycogen-depleted pericentral regions without any histological signs of cell necrosis (Ji et al., 1984, 1985a) (Figure 2E and F).

In order to better characterize the molecular processes underlying the N to I and the I to D cell-state transitions of hepatocytes induced by acetaminophen, it is necessary to have a working model of the living cell itself. This is described next.

II. The Bhopalator – a molecular model of the living cell

The Bhopalator, named after the city of Bhopal, India where it was first presented in 1983, is a molecular model of the cell designed to link molecular genetics, enzymology, and biochemistry within a coherent theoretical framework. It is clear that these three disciplines are essential in understanding the internal workings of the cell in molecular terms, and yet no published record appears to exist wherein these three major branches of biology have been integrated into a model of the living cell. It is our opinion that there are three basic physical principles necessary and sufficient for describing the living cell on the molecular level. These are listed in Table 1. Before discussing the Bhopalator in detail, it is desirable to first explain two of the three basic concepts listed in Table 1, namely conformons and intracellular dissipative structures (IDS).

A. Conformons

Crucial to the understanding of the molecular biology of the living cell is the fundamental role played by enzyme-catalyzed intracellular chemical reactions. We postulate that the enzyme-catalyzed intracellular chemical fluxes that are organized in space and in time under the influences of both the genetic information of DNA and the information

Table 1. The postulated set of physical principles neccessary and sufficient for explaining the phenomenon of life on the cellular level in molecular terms.

Biological discipline	Physical Principle
1. Molecular Genetics	Watson-Crick principle of structural complementarity
2. Enzymology	Conformons as the source of energy and genetic information to drive biomacromolecular machines (enzymes, DNA, RNA)
3. Biochemistry	IDS (intracellular dissipative structures) as the source of energy and genetic information necessary and sufficient for the cell to communicate with its environment

received by the cell from its environment are synonymous with cell life on the molecular level. We will call this idea the "organized chemical flux - life correspondence postulate." All net fluxes entail dissipation of free energy (Prigogine, 1977, 1978, 1980), and any organization requires information in the sense of Shannon (1948; Raisebeck, 1964). Depending upon how energy dissipation is organized in space and time, either the living state or the non-living state can result. This analysis reveals the fundamental and complementary roles played by energy (Gibbs free energy) and genetic information in the phenomenon of life. The basic units of energy and genetic information that are necessary and sufficient for realizing spatio-temporally organized intracellular chemical fluxes are called "conformons", a term formed from two roots "conform-" (conformation of biopolymers) and "-on" meaning a mobile structural entity (Green and Ji, 1972; Ji, 1985a, b). Conformons are visualized as genetically determined conformational strains of biopolymers resulting from the recruitment of critical monomeric units into a spatio-temporally organized structure within the active centers of enzymes and nucleic acids (Ji, 1985a, b). Conformational strains associated with such recruitment processes are the source of potential energy (capable of doing work) and the selection of the critical monomeric units into the active center in correct spatial and temporal order reflects the informational aspect of conformons. A preliminary estimation indicates that a typical conformon possesses a free energy content of 5 to 10 Kcal/mole and an information content of 50 to 100 bits (Ji, 1985a, b).

B. Intracellular Dissipative Structures (IDS)

When we think of the term "structure" in biology, we immediately imagine the pictures of the ATP molecule, polypeptides, DNA double helix, biomembranes, etc. By and large, we associate "structures" with those arrangements of atoms and molecules in the 3-dimensional space that maintain their shapes by virtue of strong interactions among their constituent parts, either through covalent binding or non-covalent associations involving sizable interaction energies ($>>0.5$ kcal/mole); that is, the component parts of traditional structures are held together by forces strong enough to resist the randomizing influences of thermal fluctuations.

The realization that these so-called "equilibrium structures" are not the only kind of structures in nature and that there exists another class of structures called "dissipative structures" represents a major conceptual breakthrough in physical sciences, for which the originator of the concept, I. Prigogine, received a Nobel prize in chemistry in 1977 (Prigogine, 1977, 1978). A dissipative structure can be defined as a dynamic, non-random distribution of matter in space and time that is kept far from equilibrium by a continuous dissipation of free energy. Examples of dissipative structures include the flame of a candle, self-organizing chemical and biochemical reactions such as the Belousov-Zhabotinsky reaction (Winfree, 1974) and oscillating glycolytic reactions (Hess, 1983; Hess and Markus, 1985; Goldbeter, 1973), membrane potentials, biological rhythms, the living cell itself, organisms, and human societies, etc. The universal feature of all dissipative structures is the absolute requirement of the dissipation of free energy (Prigogine, 1977, 1978). When the input of free energy is discontinued, all dissipative structures disappear. Another common feature of dissipative structures is that the thermodynamic system under consideration (e.g., the flame, the reaction mixture of the Belousov-Zhabotinsky reaction in a test tube, the living cell) exist far from equilibrium·so that the differential equations describing the dynamics of the system exhibit nonlinear behaviors (Glansdorff and Prigogine, 1971;

Prigogine, 1977, 1978). In this so-called "far-from-equilibrium" regime, the equations of motion of the system no longer possess a unique solution but lead to multiple solutions. The value of the parameter of the environment at which the thermodynamic system undergoes a transition from a linear to a nonlinear regime is called a bifurcation point (Prigogine, 1980; Prigogine and Stengers, 1984). Beyond a bifurcation point, the system can exhibit far-from-equilibrium properties such as (1) spatial ordering of chemicals, (2) temporal oscillations of chemical concentrations, (3) amplification of microscopic (molecular) fluctuations into macroscopically observable changes, leading to an exquisite sensitivity of the system towards weak signal inputs from its environment, and (4) the ability of the system to remember its history (e.g., the order of perturbations experienced by the system).

The dissipative structures extensively investigated so far by theoreticians and experimentalists are of macroscopic dimensions - e.g., the Belousov-Zhabotinsky reactions carried out in test tubes or petri dishes, and glycolytic oscillations of yeast extracts performed in reaction chambers, etc. However, there appears to be no theoretical reason why similar dissipative structures cannot exist inside the living cell. Therefore we propose to define "intracellular dissipative structures (IDS)" as those far-from equilibrium distributions of chemicals (i.e., ions, free radicals, small molecules, macromolecules) inside the cell that are maintained through dissipation of free energy. So defined, the concept of IDS is rooted in the theory of dissipative structures developed by Prigogine and his school. Although the dissipative structures inherent in the Belousov-Zhabotinsky reaction and glycolytic oscillations provide concrete examples upon which the concept of IDS has been suggested, it is likely that the properties of IDS will differ significantly from those of the Belousov-Zhabotinsky-type dissipative structures. For example, IDS could result either from the non-linearity of associated reaction-diffusion equations or from the mere presence of enzymes organized in space (e.g., enzymes anchored on cytoskeletons) and time (e.g., protein synthesis driven by oscillating hormonal levels) within the living cell.

It is not unrealistic to imagine that the living cell can regulate the spatio-temporal organizations of chemical substances within it by controlling the primary structure of enzymes and their intracellular locations. The primary structure of an enzyme is known to determine the 3-dimensional folding pattern under a given micro-environmental condition, and the 3-dimensional shape of an enzyme is thought to influence the nature and the frequency of the formation of conformons that is postulated to be the ultimate molecular entity responsible for controlling enzymic rates (Ji, 1974, 1979a, 1985a, b). Thus it is possible to visualize how the primary structure of an enzyme, in interaction with its prevailing micro-environmental conditions, could specify the formation of conformons, hence regulate the velocity of each elementary step in enzymic catalysis, and thus regulate the spatio-temporal organization of matter inside the cell. This line of reasoning has led to the formulation of a molecular model of the living cell called the Bhopalator.

C. The Bhopalator

The Bhopalator model of the living cell has been formulated on the basis of the following key postulates:

(1) The living cell results from interactions among three major classes of substances - nucleic acids, polypeptides, and other biochemicals; and these interactions are mediated through the exchange of energy and

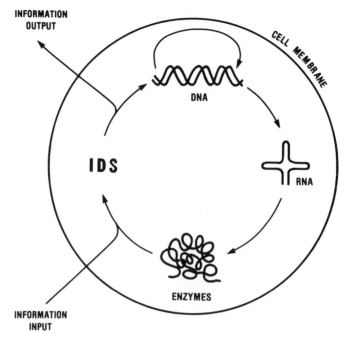

INFORMATION
OUTPUT

CELL MEMBRANE

DNA

RNA

IDS

ENZYMES

INFORMATION
INPUT

Figure 3. A simplified representation of the Bhopalator, a molecular
model of the living cell. The arrows indicate the flow of information
intracellularly and between the cell and its environment. This figure
emphasizes the information processing capacity of the living cell. Not
shown explicitly is the role of energy without which no information will
flow. The energy "released" during binding interactions among chemicals,
proteins and nucleic acids as well as that released during enzyme-
catalyzed chemical reactions powers information flow and transduction in
the living cell.

genetic information (Figure 3).

(2) The cell is the smallest autonomous unit of all living systems that
has the capacity to receive information from its environment, to
transduce (or process) it according to the genetic programs stored in
DNA, and finally to output appropriately processed signals into its
environment in order to realize teleonomically designed functions (Figure
3). The information is transferred between the living cell and its
environment through the exchange of biomolecules and inorganic ions
driven by intracellular dissipative structures (IDS).

(3) IDS's and not polypeptides are the ultimate form of the expression
of genetic information stored in DNA (Figure 3). The genetic information
written in the form of a linear sequence of nucleotides in DNA is
successively transduced into the information of m-RNA, linear sequence of
polypeptides, 3-dimensional foldings of enzymes, conformons, and
intracellular dissipative structures (Figure 4). IDS's can be viewed as
space-and time-dependent concentration gradients of biochemicals inside
the cell whose structures are controlled by both genetic information
transmitted via conformons and environmental information carried by
diffusible chemicals impinging on the cellular structures (e.g.,
membrane, cytosolic, or nuclear receptors, and enzymes). Furthermore,
IDS's have the capacity to perform work on its environment, i.e., they

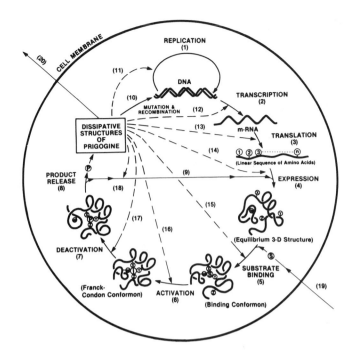

Figure 4. The Bhopalator in a more detailed representation. The solid arrows indicate information flow and the dotted arrows symbolize feedback controls. The dissipative structures of Prigogine residing inside the living cell are called intracellular dissipative structures or IDS. The conformational strains (conformons) responsible for determining the energy level of the transition state of the enzyme-substrate complex is called the Franck-Condon conformon (see residues W, X, Y, and Z interacting with I, a chemical intermediate). For more details, see text.

can transmit intracellularly processed information to its environment. It is postulated that without IDS's, the living cell cannot communicate with its environment and hence cannot live, adapt, nor evolve.

(4) The reception, processing, and outputting of information by the living cell are powered (or driven) by the free energies made available from binding interactions between ligands and their appropriate receptors and enzymes and from covalent rearrangements accompanying enzymic catalysis. Without being driven by exergonic chemical and physical processes, no information processing nor exchange are possible by the cell. To do otherwise would be tantamount to violating the Second Law of Thermodynamics. Thus, information transduction and transfer and energy transduction and transfer are synonymous on the molecular level and represent two aspects of the same molecular reality. Conformons are defined to be the smallest units of such coupled entity of energy and genetic information (Ji, 1985a, b).

(5) The present-day living cell, being a dissipative structure of Prigogine, has the capacity to memorize its evolutionary (phylogenic) and developmental (ontogenic) histories in the form of its metabolic states. In addition, cells interact through their IDS's to construct telonomically selected multicellular systems, including organs and whole organisms. The information production and transduction that links the

Table 2.   The hierarchy of the mechanisms of encoding biological
           information

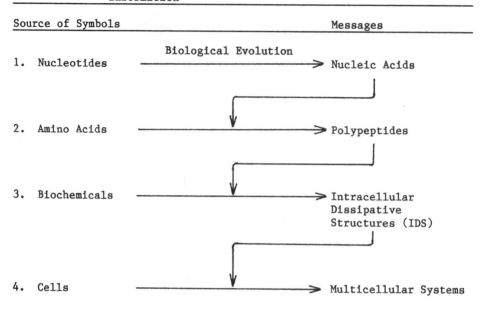

Source of Symbols   Messages

1. Nucleotides   Biological Evolution   Nucleic Acids

2. Amino Acids   ─────────────────────────►   Polypeptides

3. Biochemicals   ─────────────────────────►   Intracellular
                                                      Dissipative
                                                      Structures (IDS)

4. Cells   ─────────────────────────►   Multicellular Systems

present-day multicellular systems to the first form of life and
subsequent biological evolution are schematized in Table 2.

The living cell as represented by the Bhopalator has certain similarities
with the modern-day digital computer (Table 3). Both are machines driven
by the free energy made available by the flow of electrons from higher to
lower energy levels. The equilibrium structures (e.g., the lipid
bilayers, certain aspects of cytoskeletons, metabolic equilibria, etc.)
may be regarded as the analogue of the hardware of the digital computer.
The electron flow through the computer hardware is controlled externally
through the action of phontons, whereas the electron flow within the
living cell is regulated by the action of conformons at the active sites
on enzymes (Table 3). Thus, conformons are again analogous to photons as
already alluded to above. Just as short-circuiting electron flow in a
computer leads to its malfunction, chemical-induced derangements of
normal electron flow through the intracellular metabolic network results
in cell injury and death.

III. Experimental Evidence for IDS

       The best known IDS is the membrane potential of the living cell; to
maintain high intracellular concentration of the K^+ ion and low
intracellular concentrations of Na^+ and Ca^{++} ions relative to the
corresponding concentrations in the extracellular space, the cell must
continuously utilize free energy of ATP and other substrates to actively
transport these ions across the plasma membrane in appropriate
directions. When free-energy dissipation is discontinued or blocked, the
transmembrane gradients rapidly disappear by passive diffusion of ions
through appropriate transmembrane channels.

       Intracellular ions, H^+, K^+, Na^+, and Ca^{++}, can be distributed
inhomogenously within the intracellular space, again through the
dissipation of metabolic energy. Some space-dependent ionic IDS's
require the presence of subcellular organelles such as mitochondria,

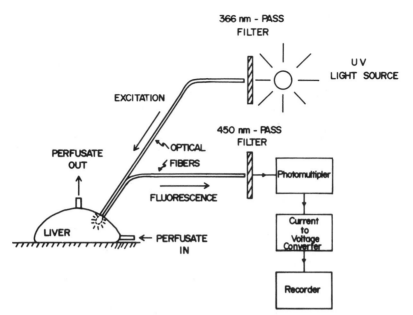

Figure 5. A schematic diagram showing the experimental setup for tissue
photometry applied to isolated perfused liver. The Y-shaped light guide
can be constructed from two strands of glass or quartz fibers (70 to 100
microns in diameter) for sublobular measurements or from bundles of
optical fibers with a measuring tip diameter of 1 to 2 mm. The rat
liver was isolated under pentobarbital anesthesia and perfused at $37°C$
with the Krebs-Henseleit bicarbonate buffer equilibrated with 95% O_2 and
5% CO_2. The perfusate was pumped through the liver at a constant flow
rate (3 to 4 mls/min/g liver) via a cannula inserted into the portal
vein. The liver tissue was illuminated with the near-UV light peaking
around 360 nm and the resulting fluorescence was measured at 450 nm. The
signal was detected by an EMI photomultiplier and further amplified using
a Johnson Foundation (Univ. of Pennsylvania) fluorometer. The 450 nm
fluorescence primarily reflects the tissue contents of reduced pyridine
nucleotides.

decrease for as yet unknown reasons (see bottom panel). The tissue
fluorescence recordings shown in Figure 6 represent MADS reflecting
changes in intracellular reduced pyridine nucleotides. In collaboration
with Drs. A. Ghosh and F. Matschinsky at the University of Pennsylvania,
we have measured tissue metabolites as a function of time following
acetaminophen infusion. We found that acetaminophen infusion into the
liver caused rapid and profound metabolic alterations involving adenine
and pyridine nucleotides, and glycolytic intermediates and products (Ji
et al., 1985). For example, the tissue contents of ATP decreased to 40%
of the control and those of ADP and AMP increased by 2 to 3 fold within
100 seconds following the acetaminophen infusion. There was a reasonably
good correlation between the tissue pyridine nucleotide fluorescence
intensity and tissue contents of reduced pyridine nucleotides after the
first 10 seconds of the drug addition. For some unknown reasons, the
tissue fluorescence increased rapidly during the first 10 seconds
following the acetaminophen addition without any accompanying increase in
reduced pyridine nucleotides.

Clearly, the intracellular concentration of reduced pyridine
nucleotides as measured by tissue fluorometry can be regarded as an

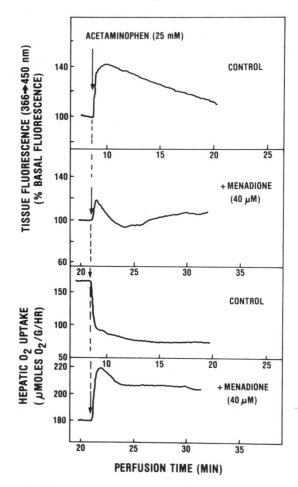

Figure 6. Typical fluorescence responses of the perfused rate liver upon
addition of acetaminophen (25mM). The liver from a fed female Sprague-
Dawley rat was isolated and perfused with the Krebs-Henseleit buffer as
described in Figure 5. Acetaminophen was dissolved in the Krebs-
Henseleit buffer and was infused into the liver by switching the
perfusate source from the control to the acetaminophen containing buffer
via a three-way stopcock. The addition of menadione was achieved in a
similar manner. The rate of oxygen uptake was measured with a Clark-type
oxygen electrode inserted into the venous effluent.

example of intracellular dissipative structure. This is because these
pyridine nucleotide levels are maintained by the hepatocytes within a
fairly narrow concentration ranges (Chance et al., 1965). This "pyridine
nucleotide homeostasis" is possible if and only if hepatocytes can
reoxidize reduced pyridine nucleotides back to the oxidized forms by
transferring reducing equivalents to molecular oxygen through the
mitochondrial electron transport chain. Without this irreversible flow
of electrons from reduced pyridine nucleotides to oxygen, no free energy
dissipation occurs and hence no homeostasis of the intracellular pyridine
nucleotides is possible.

To test whether or not the metabolic processes responsible for a
steady-state tissue fluorescence intensity represent a structure far from
equilibrium (i.e., dissipative structure of Prigogine) or merely a near-

equilibrium phenomenon, we perturbed the tissue fluorescence intensity by
a successive addition of two chemicals, ethanol and 7-hydroxycoumarin.
Ethanol causes an increase in the intracellular levels of NADH through
alcohol and aldehyde dehyrogenases, whereas the fluorescence intensity of
7-hydroxycoumarin, which cannot be distinguished from the fluorescence of
NADH under our experimental conditions, primarily reflects the capacity
of hepatocytes to conjugate it to form the nonfluorescent glucuronide and
sulfate derivatives. Since glucuronidation and sulfation require a
continuous supply of ATP, the tissue fluorescence increase observed after
adding 7-hydroxycoumarin correlates negatively with the intracellular ATP
levels.

   Independent of the detailed mechanism of the metabolic processes
underlying the tissue fluorescence changes, it is possible to check
whether the fluorescence intensity reflects a dissipative structure or
near-equilibrium processes. This is possible because, under appropriate
experimental conditions, dissipative structures exhibit characteristic
properties not observed in equilibrium or near-equilibrium systems. One
such property is the possibility of successive metabolic bifucations, the
general principle of which was recently discussed by Prigogine and
Stengers (1984). If the tissue fluorescence intensity can change in two
different ways by the addition of a chemical perturbant (a metabolic
bifurcation); and if we chose two chemicals that lead to two different
metabolic states of cells, then it is possible that the tissue

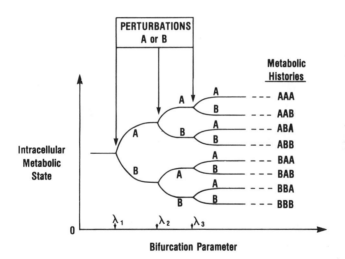

Figure 7. A bifurcation tree for cell metabolism. The y-axis represents
the intracellular concentration of metabolites such as NADH or 7-
hydroxycoumarin. The x-axis indicates the extent of the metabolic
perturbations induced by substrate additions and measured by the
increased free energy of the system relative to equilibrium. The points
where metabolism bifurcate are called bifurcation points (λ_1, λ_2, λ_3).
Depending upon the nature of perturbants (A or B) the hepatocellular
metabolism undergoes successive bifurcations to reach new metabolic
states represented by symbols AAA, BAA, etc. These symbols reflect the
order of perturbations, namely the metabolic history of the hepatocyte.

fluorescence intensity can respond differently depending on the order of addition of these two chemicals to the perfused liver. For example, if one plots the tissue fluorescence level on the y-axis and the degree of perturbation by chemicals A and B on the x-axis, one can envision a bifurcation tree depicted in Figure 7.

We found that, depending on whether 7-hydroxycoumarin is added first followed by ethanol or vice versa the resulting tissue fluorescence was either stable or unstable, respectively (Figure 8). Although the difference between the fluorescence behavior resulting from the two ways of adding substrates was not always as clear as in Figure 8, the average rates of fluorescence change after adding these two chemicals to the liver in different orders showed a statistically significant difference in the direction indicated (Ji et al., 1985b). We interpret these observations as concrete experimental evidence that the intracellular concentrations of some metabolites including reduced pyridine nucleotides are kept far from equilibrium; that is, there exist dissipative structures inside the cell. These so-called IDS's will play important roles in the application of the Bhopalator to toxicology as will become evident in the following sections.

IV. An Algebraic representation of the Bhopalator

It is our assertion that the Bhopalator is the smallest material entity that has the following three properties - (1) self-moving, (2) self-thinking, and (3) self-reproducing. Structural entities possessing the first two properties are called "automata"; but there exist no single term that conveys all of the above three qualities. We will call such an

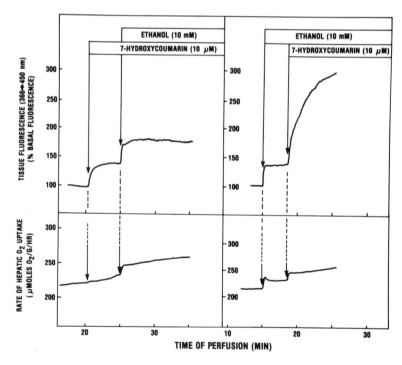

Figure 8. The non-commutativity of the order of substrate addition. The liver was perfused as in Figure 5. Ethanol was given through the portal vein by an infusion pump, and 7-hyrdroxycoumarin dissolved in the Krebs-Henseleit buffer was added as described in Figure 6. The rate of oxygen uptake was measured with a Clark-type electrode.

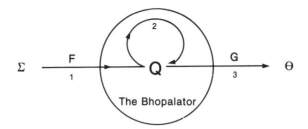

Figure 9. An algebraic representation of the Bhopalator. Σ = set of input signals; Q = set of cell states; Θ = set of output signals; F = state transition function; G = output function. Please see text for details.

entity "self-reproducing" automaton. Therefore, we can identify the Bhopalator as the <u>molecular self-reproducing automaton</u>, the adjective "molecular" reflecting the idea that the Bhopalator cannot be reduced to simpler structures.

Beginning with the pioneering work of von Neumann in the 1950's, mathematicians have developed algebraic techniques to describe the properties of machines called automata theory (Hopcroft and Ullman, 1979). M. Holcombe recently applied such an algebraic technique to analyze the Krebs cycle (1982). We concur with Holcombe's idea that the algebraic approach will be invaluable and even indispensable in the description and analysis of complex molecular machines such as enzymes, biochemical networks and the living cell.

To represent the Bhopalator in algebraic terms, it is necessary for us to postulate the following;

(1) The Bhopalator can exist in any one of a set of n internal states;
$$Q = \{ q_1, q_2, q_3, \cdots, q_n \} \qquad \cdots \cdots \cdots \quad (1)$$

(2) The Bhopalator can recognize (or interact with) a set of m input signals originating from its environment;
$$\Sigma = \{ \sigma_1, \sigma_2, \sigma_3, \cdots, \sigma_m \} \qquad \cdots \cdots \cdots \quad (2)$$

(3) The Bhopalator can influence its environment by producing one or more of a set of p output signals;
$$\Theta = \{ \theta_1, \theta_2, \theta_3, \cdots, \theta_p \} \qquad \cdots \cdots \cdots \quad (3)$$

(4) The Bhopalator undergoes a transition from one internal state to another upon interacting with an input signal according to a set of well-defined rules;
$$F : Q \times \Sigma \longrightarrow Q \qquad \cdots \cdots \cdots \quad (4)$$

where "x" indicates interaction, and F is a function or a set of rules that pairs each member of set $Q \times \Sigma$ with one or more members of set Q. $F(q, \sigma)$ signifies a new internal state that the Bhopalator assumes upon receiving input signal σ while at internal state q.

(5) The Bhopalator sends out output signals whose character is determined by its internal state according to the rule;

$$G : Q \times \Sigma \longrightarrow \Theta \quad \cdots \quad \cdots \quad (3)$$

where G is the output function that pairs the internal states plus received signals of the Bhopalator with an output signal. $G(q,\sigma)$ signifies the output signal selected by function G that is compatible with internal state q and signal σ. These five statements are schematically shown in Figure 9.

Figure 9 clearly reveals the machine-like characteristics of the Bhopalator; it is a "machine" that is designed by evolution to receive signals from its environment (i.e., step 1 in Fig. 9), to transduce input signals according to the genetic programs stored in the cell (step 2), and to output the processed signal (step 3). All these steps are powered and driven by IDS (intracellular dissipative structures) created by the action of conformons, the mediators of the interaction between enzymes and nucleic acids (Ji, 1985a, b). The Bhopalator machine is not exactly analogous to man-made machines. The discussion of this difference is beyond the scope of the present article.

As evident above, the internal states of the Bhopalator Q emerges as a parameter of paramount importance in determining the properties of the Bhopalator. To obtain some guidance in defining the molecular characteristics of the cell state, it may be useful to draw an analogy between the cell and the molecule. Just as the molecules can exist in multiple dynamic states characterized by the energy levels of intramolecular electronic, vibrational and rotational motions, the cell can exist in numerous "cell states". Some of these cell states constitute the N (normal), I (injured), or D (dead) states. We postulate that the cell state structures consist of IDS's which are in turn composed of a set of biochemical and biophysical fluxes, and these fluxes are determined by the action of an appropriate set of conformons (section III).

Unlike molecular states that are determined by dynamic motions of constituent parts, cell states are determined by rates of constituent chemical reactions (i.e., chemical fluxes). Careful analysis of this interesting difference may lead to a deeper understanding of the properties of the living cell.

V. The Bhopalator and Molecular Toxicology

Molecular toxicology can be defined as a subdiscipline of toxicology whose ultimate aim is to account for all toxic effects of chemicals on living systems in molecular terms. Since the cell is the simplest of all living systems, the basic principles of molecular toxicology may be most readily discerned through the study of toxicant-cell interactions.

Based on the Bhopalator model of the living cell presented above, we propose the following postulates concerning the mechanism of cell-toxicant interactions;

Postulate 1: No chemicals can injure the living cell without causing cell-state transitions.

Postulate 2: There exist a finite number of cell states with associated output signals recognized experimentally as signs of cell injury and cell death.

These postulates again highlight the primary importance of the concept of cell states in toxicology. An alternative expression of postulate 1 is that the ultimate target of all cytotoxicants is the cell state; and, since the cell state is determined by IDS's, postulate 1 implies that the ultimate targets of all cytotoxicants are those IDS's that lead to adverse cell state transitions (i.e., N to I or N to D).

A corollary to postulate 1 is that not all covalent binders (i.e., chemicals whose biological reactive intermediates bind covalently to cellular components) will be cytotoxic since it is conceivable that not all covalent binding will perturb metabolic fluxes to the extent of causing cell-state transitions, either because of the weakness of perturbations or because of a simultaneous activation of endogenous self-protecting mechanisms that ameliorate the metabolic effects of covalent binding. Another corollary is that it is not necessary to have covalent binding to observe cell injury, because it is again conceivable that the IDS's can be altered purely by physical means such as by ionophores that alter the membrane permeability to metal ions leading to the dissipation of membrane potentials and consequently to cell injury and death (Schanne et al, 1979). These corollaries are compatible with the recent experimental observations indicating the independence of covalent binding of metabolic intermediates and cell injury (DeValia et al., 1982; Labadarios et al., 1977; Gerson et al., 1985).

Postulate 1 raises the question whether or not the covalent binding hypothesis can be regarded as a universal principle in toxicology. The lack of 1:1 correspondence between cell injury and covalent binding of reactive metabolites reported in numerous publications cited above justifies our attempt to formulate a more general hypothesis than the covalent binding theory.

Conclusions

Molecular toxicology is in its very early infancy. To unravel the basic physical principles underlying chemically induced cell injury and cell death on the molecular level, new concepts and theories as well as new languages may have to be developed, taking into account the relevant new developments in physics, chemistry, and mathematics. In this article, we have described a molecular model of the living cell (the Bhopalator) and its algebraic representation, and we have applied these to the mechanism of acetaminophen hepatotoxicity. Although the level of sophistication is limited at this stage of development, it is hoped that the approaches adopted herein will lead to fruitful applications in molecular toxicology.

Acknowledgements

We would like to acknowledge stimulating discussion with and helpful suggestions from Drs. R. Snyder, M. Iba, D. Laskin, G. Welch, and M. Holcombe. Supported in part by grants from the National Institute of Alcohol Abuse and Alcoholism, AA-05848.

References

Black, M., 1984, Acetaminophen Hepatotoxicity. Ann. Rev. Med. 35:577

Chance, B. Williams, J.R., Jamieson, D. And Schoener, B., 1965. Properties and kinetics of reduced pyridine nucleotide fluorescence of isolated and in vivo rat heart, Biochem. Zeit. 341:357

Cheng, L. and Ji, S., 1984, Inhibition of Mitochondrial respiration by acetaminophen: A new mechanism of acetaminophen hepatotoxicty. The Toxicologist. 4:78.

Dahlin, D.C., Miwa, G.T., Lu, A.Y., and Nelson, S.D., 1984, N-acetyl-p-benzoquinone imine: A cytochrome P-450-mediated oxidation product of acetaminophen. Proc. Nat. Acad. Sci. (U.S.) 81:1327.

DeValia, J.L. Ogilvie, R.C. and McLean, A.E.M., 1982. Dissociation of cell death from covalent binding of paracetamol by flavones in a hepatocyte system, Biochem. Pharmacol. 31:3745.

DiMonte, D., Bellomo, G., Thor, H., Nicotera, P. and Orrenius, S., 1984, Menadione-Induced Cytotoxicity Is Associated with Protein Thiol Oxidation and Alteration in Intracellular Ca Homeostasis. Arch. Biochem. Biophys. 235:343.

Esterline, R. and Ji, S., 1985, The Site of Inhibition of the Mitochondrial Respiration by Acetaminophen (in preparation).

Gerson, R.J., Casini, A., Gilfor, D., Serroni, A. and Farber, J., 1985. Oxygen-Mediated Cell Injury in the Killing of Cultured Hepatocytes by Acetaminophen, Biochem. Biophys. Res. Commun. 126:1129.

Glansdorff, P. and Prigogine, I., 1971, "Thermodynamic Theory of Structure, Stability and Fluctuation." Wiley-Interscience, London.

Goldbeter, 1973. Patterns of spatiotemporal organization in an allosteric enzyme model, Proc. Nat. Acad. Sci. (U.S.) 70:3255.

Green, D.E. and Ji, S., 1972, Electromechanochemical model of mitochondrial structure and function in: "The Molecular Basis of Electron Transport", J. Schultz and B.F. Cameron, eds. Academic Press, New York, p. 1.

Hess, B., 1983. Non-equilibrium dynamics of biochemical processes, Hoppe-Seyler's Z. Physiol. Chem. 364:S.1.

Hess, B. and Markus, M. 1985, Ber. Bunsen Phys. Chem. 89:642.

Holcombe, W.M.L., 1982. "Algebraic Automata Theory." Cambridge University Press, Cambridge.

Hopcroft, J.E. and Ullman, J.D., 1979, "Introduction to Automata Theory, Languages and Computation," Addison-Wesley Publishing Co., Reading, Mass.

Jewell, S.A., Bellomo, G., Thor, H. and Orrenius, S., 1982. Bleb Formation in Hepatocytes During Drug Metabolism Is Caused by Disturbances in Thiol and Calcium Ion Homeostasis, Science 217:1257.

Ji, S., 1974. Energy and negentropy in enzymic catalysis, Ann. N.Y. Acad. Sci. 227:419.

Ji, S., 1979a, The principles of ligand-protein interactions and their applications to the mecahnism of oxidative phosphorylation, in: "Structure and Function of Biomembranes," K. Yagi, ed., Japan Scientific Societies Press, Tokyo, p. 25.

Ji, S., 1979b, Some quantitative aspects of micro-light guide photometry of biological tissues, in: "Advance Technobiology," B. Rybak, ed., Sijthoff and Noordhoff, Alphen aanden Rign, The Netherlands, p. 237.

Ji, S., 1985a. The Bhopalator - a molecular model of the living cell, Asian J. Exp. Sci. 1:1.

Ji, S., 1985b. The Bhopalator - a molecular model of the living cell based on the concepts of conformons and dissipative structures, J. theor. Biol. 116:399.

Ji, S., Cheng, L., Ghosh, A., Matschinsky, F., Maliniak, C. and Trelstad, R.L., 1984, Histological and Metabolic Injuries of Rat Liver Following Acetaminophen Treatment in vivo and in vitrto. The Toxicologist 4:44.

Ji, S., Desvouges, I., Cheng, L., Ghosh, A., Matschinsky, F., Maliniak, L., and Trelstad, R., 1985a. Inhibition of hepatic respiration by acetaminophen: A potential contributory pathway to acetaminophen hepatotoxicity, (submitted).

Ji, S., Ray, S. and Esterline, R., 1985b, Noncommutativity of the Order of Substrate Addition to Perfused Rat Liver: Evidence for Successive Bifurcations in Hepatocellular Meabolism (in preparation).

Jollow, D.J., Mitchell, J.R., Potter, W.Z., Davis, D.C., Gillette, J.R., and Brodie, B.B., 1973. Acetaminophen induced hepatic necrosis II. Role of covalent binding in vivo., J. Pharmacol. Exp. Ther. 187:195.

Kappus, H., 1985, Lipid Peroxidation: Mechanisms, Analysis, Enzymology and Biological Relevance, in: "Oxidative Stress," H. Sies, ed., Academic Press, London, p. 273.

Labadarios, D., Davis, M., Portmann, B., and Williams, R., 1977, Paracetamol-induced hepatic necrosis in the mouse - Relationship between covalent binding, hepatic glutathione depletion and the protective effect of d-mercaptopropionylglycine, Biochem. Pharmacol. 26:31.

Laskin, D., Pilaro, A.M., Cheng, L., Trelstad, R.L. and Ji, S., 1984. Accumulation of phagocytes in rat liver following acetaminophen treatment in the absence of cellular necrosis. Fed. Proc. 43(3):544.

Mitchell, J.R., Jollow, D.J., Potter, W.Z., Davis, D.C., Gillette, J.R., and Brodie, B.B., 1973. Acetaminophen-induced hepatic necrosis. I. Role of drug metabolism. J. Pharmacol. Exp. Ther. 187:185.

Pilaro, A.M., Ji S. and Laskin, D., 1985, Accummulation of Activated Macrophages in the Liver Following Acetaminophen Treatment of Rats, The Toxicologist 5:75.

Prigogine, I., 1977, Dissipative structures and biological order, Adv. Biol. Med. Phys. 16:99

Prigogine, I., 1978, Time, structure and fluctuations, Science 201:777.

Prigogine, I., 1980, "From being to becoming," W.H. Freeman and Co., San Francisco.

Prigogine, I. and Stengers, I., 1984, "Order Out of Chaos," Bantam Books, Toronto.

Raisebeck, G., 1964, "Information Theory," The M.I.T. Press, Mass. Institute of Technol., Cambridge.

Schanne, F.A.X., Kane, A.B., Young, E.E. and Farber, J.L., 1979, Calcium Dependence of Toxic Cell Death: A Final Common Pathway, Science 206:700.

Shannon, C., 1948. Mathematical theory of communication, Bell System Technical J. 27:379 and 623.

Vinogradov, A., Scarpa, T., and Chance, B., 1972, Pyridine nucleotide-membrane interactions, Arch Biochem. Biophys. 115:450.

West, P.R., Harman, L.S., Josephy, P.D. and Mason, R.P., 1984, "Acetaminiphen" Enzymatic formation of a transcient phenoxyl free radical." Biochem. Pharmacol. 33:2933.

Winfree, A.T., 1974, Rotating chemical reactions, Sci Am., 230:82.

HELENALIN: MECHANISM OF TOXIC ACTION

J. Merrill, H.L. Kim and S. Safe

Veterinary Physiology and Pharmacology
Texas A&M University
College Station, TX 77843

INTRODUCTION

Several range plants, including <u>Hymenoxys odorata</u> (bitterweed) and <u>Helenium microencephalum</u> DC (smallhead sneezeweed) contain toxic principles believed to be the etiologic agents responsible for severe losses of grazing livestock[1]. Hymenoxon, a sesquiterpene lactone, has been isolated from <u>Hymenoxys odorata</u> and studies with laboratory animals have demonstrated that this compound elicits toxic effects which resemble those reported for field case studies. Helenalin, a structurally related sesquiterpene lactone isolated from <u>Helenium microencephalum</u> DC, exhibits toxic properties which resemble those reported for hymenoxon and suggests that these compounds act via a common mechanism.

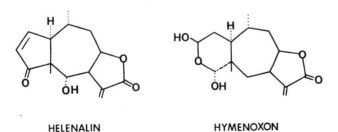

HELENALIN   HYMENOXON

Figure 1. Structures of toxic sesquiterpene lactones.

Both compounds also contain a common structural moiety, namely an α,β, -unsaturated-γ-lactone. (Figure 1). Previous in vitro studies[2] have demonstrated that these sesquiterpene lactones can act as electrophiles via the rapid 1,2-Michael addition of nucleophiles to the exocyclic double bond. Cysteine provides some protection for sheep and mice from the toxicity of hymenoxon and this suggests that cellular thiol levels may play a role in the mechanism of action of hymenoxon and helenalin. Several other studies have reported the following data; a) drug-metabolizing enzyme inducers such as phenobarbital and Aroclor 1254 did not alter the toxicity of hymenoxon in laboratory animals whereas pretreatment of mice with carbon tetrachloride, an agent which inactivates microsomal monooxygenases, protected the animals from the acute effects of hymenoxon[3]; b) the antioxidants, ethoxyquin[4] and cysteine[5], protect sheep and mice from hymenoxon toxicity but butylated hydroxyanisole increased the LD_{50} value of hymenoxon in mice but not sheep[4]; (c) recent studies in our laboratory[2] have demonstrated that hymenoxon does not alter hepatic cytochrome P-450 levels, hepatic microsomal monooxygenases, glutathione reductase, glutathione peroxidase or glutathione S-transferase activities.

This paper probes the mechanism of action of sesquiterpene lactone toxins by determining the effects of hymenoxon on hepatic lipoperoxidation in mice. In addition the toxicity of helenalin is investigated in animals which have been pretreated with L-2-oxothiazolidine-4-carboxylate (OTC), a chemical which increases hepatic levels of glutathione[7] and protects against the toxic effects of acetaminophen[8] via the increased formation of relatively non-toxic conjugates.

MATERIALS AND METHODS

Helenalin (molecular weight 262.3) was extracted from dried ground sneezeweed by the procedure of Kim[9]. The test compound was dissolved in minimal DMSO and diluted with isotonic saline. In this vehicle the LD_{50} in mice was found to be 25 mg/kg when administered intraperitoneally. OTC (molecular weight 146.1, m.p. 171-173°C) was synthesized in our laboratory by the method of Boettcher and Meister[10]. OTC (6.5 mmol/5 ml) was dissolved in isotonic saline and neutralized with equimolar NaOH. The mice were maintained on food and water ad libitum, housed in polycarbonate cages on a diurnal 12 h light/12 h dark cycle and fasted 12 h before sacrifice. Mice were killed by cervical dislocation and the livers were immediately perfused with ice cold isotonic saline (pH 7.3). Biochemical determinations were performed on the fresh preparations.

OTC (6.5 mmol/kg) was administered as a single dose 24 h before, 6 h before or 6 h after an LD_{50} dose of helenalin and the influence of OTC on lethality was measured at 6 days. The effect of coadministering OTC/helenalin on mouse liver GSH was measured using reverse phase high-performance liquid chromatography (HPLC)[11]. An 0.25 g sample of tissue homogenate was protein denatured by the addition of 1.5 ml of 0.2 N HCl, vortexed and held at room temperature for 10 minutes. The sample was deproteinated by the addition of 0.25 ml 0.3 M sodium tungstate containing 0.1 M EDTA, vortexed and incubated for 10 minutes. Samples were centrifuged (1000 x g) for 10 minutes and the supernatant was added (1:1) to 2.5 M sodium acetate buffer (pH 5.0). Fifty microliters of 50 mM 1-fluoro-2,4-dinitrobenzene in methanol was added, mixed and the sample incubated in the dark for 8 h. The mobile phase was 74% 0.1% phosphoric acid and 26% acetonitrile at a flow rate of 1.0 ml/min. The instrument was a Beckman 334 HPLC with an LDC/Milton Roy variable wavelength detector set at 330 nm. The HPLC column was C_{18} N-capped column, 4.6mm x 25cm.

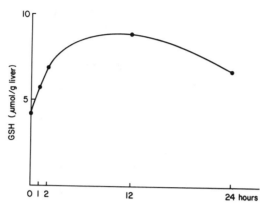

Figure 2.   Time course elevation of GSH after treatment of mice with OTC.

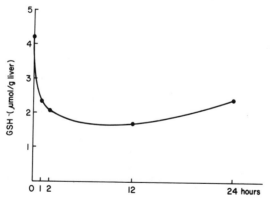

Figure 3.   Time course decrease of GSH after treatment of mice with helenalin.

The extent of microsomal lipoperoxidation was determined by measuring malondialdehyde production as thiobarbituric acid-reactive substances at 535 nm as described[12]. The malondialdehyde formation in mouse liver tissue was determined 4, 8, 12 and 72 h after administration of a toxic dose of hymenoxon (20 mg/kg in corn oil/DMSO).

RESULTS AND DISCUSSION

Hymenoxon did not alter hepatic malondialdehyde levels 4, 8, 12 and 72 h after initial exposure (Table 1) and this suggests that lipoperoxidation does not play a significant role in the toxic mechanism

Table 1 The effect of hymenoxon on mouse hepatic lipoperoxidation

Lipoperoxidation	Hours Post Treatment			
	4	8	12	72
control	2.62 + 0.28	2.59 + 0.20	2.38 + 0.17	2.79 + 0.40
hymenoxon	2.78 + 0.38	2.47 + 0.18	2.62 + 0.21	2.57 + 0.25

[a]nmoles TBA-reactive products/mg protein/10 min.

of action. It is known that some anitoxidants such as butylated hydroxyanisole not only inhibit lipoperoxidation but modulate enzymes associated with the maintenance of cellular glutathione levels. Administration of OTC to male ICR mice resulted in a time-dependent increase in hepatic glutathione levels as determined by HPLC analysis (Figure 2). In contrast, administration of a toxic dose of the sesquiterpene lactone, helenalin. (25 mg/kg), resulted in a rapid two-fold decrease in hepatic glutathione two hours after initial exposure (Figure 3). Moreover the glutathione levels remained depressed for 12 h. The effects of helenalin, OTC and helenalin following OTC on hepatic GSH levels are summarized in Table 2. Control levels of glutathione (4.21 umol/g wet liver) were depressed to 1.70 umol/g wet liver 12 h after treatment with helenalin. Administration of helenalin 6 or 12 h after OTC pretreatment and sacrificed 12 h post helenalin treatment resulted in "normal" levels of hepatic glutathione (4.36 and 4.03 umol/g wet liver respectively).

OTC modified the toxicity of helenalin in mice but resulted in no deaths when given on its own (Table 3). The efficacy of OTC administration was dependent on the timing and sequence of administration with respect to helenalin exposure. OTC administered 6 h prior to helenalin dosing resulted in a decrease in lethality (16.7% compared to 66.7%). However, this protective effect is not seen when helenalin exposure was delayed until 24 h after initial OTC administration. The sequence of exposure is critical to the protective effect of OTC. Mice which had been pretreated with helenalin and subsequently exposed to OTC were not protected from helenalin-mediated toxicity. This finding suggests the toxic effects of helenalin develop rapidly after initial exposure and are irreversible at the dose level used in this study. This is consistent with a mechanism which involves the alkylation of critical tissue thiols which lead to the development of the toxic and lethal responses in mice.

Table 2 Effect of helenalin, OTC and coadministration OTC/helenalin
on hepatic GSH in mice.

Treatment Schedule	GSH $\dfrac{\text{umoles}}{\text{gram wet liver}}$
control	4.21 ± 0.54 [a]
Helenalin; sacrificed at 12 h	1.70 ± 0.59 [b]
OTC; sacrificed at 12 h	8.88 ± 0.66 [c]
OTC; sacrificed at 24 h	6.75 ± 0.66 [d]
OTC; followed by helenalin at 12 h and sacrificed at 24 h	4.03 ± 0.59 [a]
OTC; followed by helenalin at 6 h and sacrificed at 18 h	4.36 ± 0.59 [a]

Values with different superscripts are significantly different from each
other at $P < 0.05$.

Table 3 Effect of OTC on helenalin lethality.

Treatment Schedule	% died at 6 days	
OTC	(0/5)	0
Helenalin	(4/6)	66.7
OTC; followed by helenalin at 6 h	(1/6)	16.7
OTC; Oh followed by helenalin 24 h	(5/6)	83.3
Helenalin; followed by OTC at 6 h	(4/5)	80.0

The effects of OTC on the toxicity of helenalin suggests that the
induction of glutathione and other non-protein sulfhydryls play a key
role in detoxifying this compound. Presumably OTC-mediated induction
results in the availability of excess cellular thiols which subsequently
react with the exocyclic methylene and cyclopentanone groups of helenalin
to form a non-toxic adduct. Current research in our laboratory is
investigating other aspects of the mechanism of action of toxic range
plant sesquiterpene lactones and the role of tissue sulfhydryls in
modulating their toxic effects.

ACKNOWLEDGEMENTS

The financial assistance of the Texas Agricultural Experiment
Station is gratefully acknowledged.

REFERENCES

1. J.M. Kingsbury, "Poisonous Plants of the United States and Canada" Prentice-Hall, Englewood Cliffs (1964).

2. J.C. Merrill, H.L. Kim and S.H. Safe, Hymenoxon: biologic and toxic effects. Biochem. Pharmacol. (in press).

3. D.H. Jones and H.L. Kim, Toxicity of hymenoxon in Swiss white mice following pretreatment with microsomal enzyme inducers, inhibitors and carbon tetrachloride. Res. Comm. Chem. Pathol. Pharmacol. 33:361 (1981).

4. H.L. Kim, A.C. Anderson, B.W. Herrig, L.P. Jones and M.C. Calhoun, Protective effects of antioxidants on bitterweed (Hymenoxys odorata DC) toxicity in sheep. Am. J. Vet. Res. 43:1945 (1982).

5. L.D. Rowe, H.L. Kim and B.J. Camp, The antagonistic effect of L-cysteine in experimental hymenoxon intoxication in sheep. Am. J. Vet. Res. 41:484(1980).

6. S.M. Kupchan, D.C. Fessler, M.A. Eakin, and T.J. Giacobbe, Reactions of alpha methylene lactone tumor inhibitors with model biological nucleophiles. Science 168:376 (1970).

7. J.M. Williamson and A. Meister, Stimulation of hepatic glutathione formation by administration of L-2-oxothiazolidine-4-carboxylate, a 5-oxo-L-prolinase substrate. Proc. Natl. Acad. Sci. 78:936 (1981).

8. J.M. Williamson, B. Boettcher and A. Meister, Intracellular cysteine delivery system that protects against toxicity by promoting glutathione synthesis, Proc. Natl. Acad. Sci. 79:6246 (1982).

9. H.L. Kim, Toxicity of sesquiterpene lactones, Res. Comm. Chem. Path. Pharm. 28:189 (1980).

10. B. Boettcher and A. Meister, Synthesis of L-2-oxothiazolidine-4-carboxylic acid. Analyt. Biochem. 138:449 (1984).

11. E.G. Schanus, R.S. Pobiel, and R.E. Lovrien, Quantitation of biological sulfhydryls and disulfides via Sahger's reagent and reverse phase high-performance liquid chromatography, Analyt. Biochem. (submitted).

12. J.A. Buege and S.D. Aust, Microsomal lipid peroxidation, Methods in Enzymology 52:302 (1978).

EVIDENCE FOR STEREOSELECTIVE PRODUCTION OF PHENYTOIN

(5,5-DIPHENYLHYDANTOIN) ARENE OXIDES IN MAN

James H. Maguire and Judy S. McClanahan

Division of Medicinal Chemistry and Natural Products
School of Pharmacy
University of North Carolina
Chapel Hill, NC   27514   U.S.A.

INTRODUCTION

Phenytoin (5,5-Diphenylhydantoin, PHT), a widely-used antiepileptic
drug, is metabolized via arene oxides (AO), which have been implicated in
teratogenesis and idiosyncratic reactions associated with the drug (Martz
et al., 1977; Speilberg et al., 1981; Gerson et al., 1983). PHT is a
prochiral molecule and can be stereoselectively metabolized on either the
pro-S- or pro-R- phenyl substituents to arene oxides, designated as (S)-
or (R)-AO (Figure 1). As PHT arene oxides have not yet proved amenable
to isolation and characterization, determination of stereochemistry of
PHT AO's has to be by indirect determination of the stereochemistry of the
corresponding diastereomeric trans-dihydrodiols (5-(3,4-dihydroxy-1,5-
cyclohexadien-1-yl)-5-phenylhydantoin, DHD) or enantiomeric p-phenols (5-
(4-hydroxyphenyl)-5-phenylhydantoin, p-HPPH). Enantiomeric content will
reflect arene oxide isomeric content only if all p-HPPH is formed via
arene oxides and if the non-enzymatic rates of phenol formation (K_R and
K_S) from AO's are the same. Previous studies had indicated that the
majority of human stereoselective metabolism of PHT is via the pro-S-
phenyl substituent (Butler, 1957; Butler et al., 1976; Maguire et al.,
1980; Chang and Glazko, 1982). In order to investigate variations in
stereoselective metabolism, HPLC methods were employed to determine
diastereomeric content of DHD and enantiomeric content of p-HPPH. These
methods were used to estimate stereoselective metabolite production in
volunteers and in a patient population.

MATERIALS AND METHODS

Chemicals

Authentic samples of (S)- and (R)-DHD were available from previous
studies (Maguire et al., 1980) as were samples of the enantiomers of
p-HPPH.

HPLC System

Two Altex Model 110A pumps were coupled to an Axxiom Model 710
gradient controller, a Rheodyne Model 7125 injector, and an Isco Model
V4 variable wavelength detector with a Hewlett-Packard Model 3390A
recording integrator. C18 columns for the initial separation of DHD and

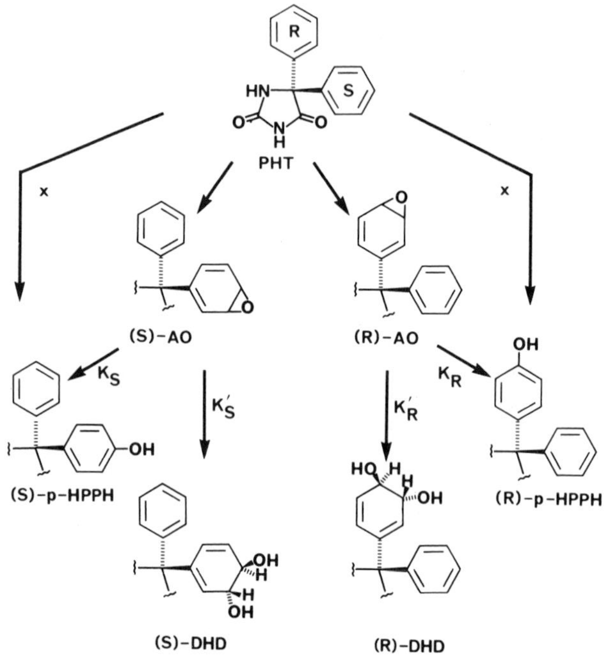

Fig. 1. Scheme depicting stereoselective metabolic pathways
involved in the production of phenolic (p-HPPH) and
dihydrodiol (DHD) metabolites of PHT. The hydantoin
ring remains intact in all of the metabolites.
Possible stereoselective direct hydroxylation pathways
are indicated with an "X".

p-HPPH and for the subsequent separation of DHD diastereomers were a
4.6 x 250 mm 10 μm Lichrosorb C18 (Merck) and a 4.6 x 250 mm 5 μm
spherical (Alltech), respectively. The β-cyclodextrin column, 4.6 x 250
mm, (Cyclobond I) was purchased from Astec (Whippany, NJ).

DHD and p-HPPH assays

The HPLC assay of DHD diastereomeric content in human urinary metabolite samples has been previously described (Maguire and Wilson, 1985). The
method employs separate collection of DHD and p-HPPH metabolite fractions,
and further assay of the DHD fraction. For assay of p-HPPH enantiomeric
content, the p-HPPH fraction collected in the above procedure was injected
onto a Cyclobond I column. Detection was at 210 nm, the column temperature
was 22°, and the eluent was 20% acetonitrile/80% water at a flow rate of
0.6 ml/min. Injection of authentic p-HPPH enantiomers allowed assignment
of retention times for (R)-(+)-p-HPPH (14.6 min.) and (S)-(-)-p-HPPH
(16.1 min.). The enantiomeric content was determined by integration of
the peak areas. A complete description of this assay methodology is forth-
coming (McClanahan and Maguire, manuscript in preparation, 1985). All
urine samples assayed by these HPLC methods had been previously assayed
for total DHD and p-HPPH content by a GLC method (Maguire et al., 1979).

Human Studies

1. Human urine samples were obtained from pediatric patients on chronic phenytoin therapy through the pediatric neurology clinics at North
Carolina Memorial Hospital (University of North Carolina, Chapel Hill, NC)
and at Duke University Medical Center (Durham, NC). Parental consent was
obtained and all studies were approved by the local human studies review
boards.
2. Volunteers given a single iv dose (5 mg/kg) of PHT participated
in a study at the Department of Clinical Pharmacology of the Karolinska
Institute (Huddinge, Sweden), and 12-hour urine samples corresponding to
0 through 72 hours post-dose were made available by Dr. Gunnar Alván.
3. Urine samples from a 12-year old female experiencing a hypersensitivity reaction to PHT were obtained from Dr. Anders Rane of the
Department of Clinical Pharmacology of the Karolinska Institute (Huddinge,
Sweden).

RESULTS

Stereoselectivity in Volunteers

Adult volunteers given a single iv dose of PHT had sequential 12-
hour samples collected and assayed for DHD and p-HPPH isomeric content.
In six of seven volunteers, no changes were observed in isomeric content
of either DHD or p-HPPH as a function of time after dose. The average
values for samples observed are indicated in Table 1, together with the
range of values observed for p-HPPH enantiomeric content as a function of
time for one volunteer (*). The p-HPPH samples from this volunteer grad-
ually increased in %S content as a function of time after dose. No such
changes were observed in DHD isomeric content for this volunteer. This
behavior was apparently not correlated with plasma half-life of PHT, as
other volunteers with similar half-lives did not show such variation.

Table 1 indicates the variations observed in DHD and p-HPPH stereochemical composition as a function of PHT half-life and debrisoquine
metabolizer status (extensive, EM; poor, PM). Little variation in
stereoselectivity is observed in the volunteers, with the exception of
the one previously mentioned (*). There is a stereoselectivity in
favor of high percentages of (S)-p-HPPH (average of 94% S-), as is
consistent with previous reports of human PHT metabolism. The isomeric
content of the DHD samples (average of 78% S-) indicates a lower stereoselectivity, and indicates the presence of at least two isomeric arene
oxides from which the DHD isomers are formed.

Table 1. Stereoselective Metabolism of Phenytoin To
Urinary Dihydrodiol (DHD) and Phenolic (p-
HPPH) Metabolites in Volunteers.

Volunteer	Debrisoquine Status[a]	PHT $t_{1/2}$ (hours)	DHD[b] (% S-)	p-HPPH[b] (% S-)
A.B.	PM	9.9	77.3 (2.5)	94.5 (1.5)
G.T.	EM	13.1	77.2 (1.5)	94.6 (0.7)
P.R.	PM	13.9	79.3 (1.3)	92.4 (1.1)
E.P.	PM	12.4	76.7 (3.1)	94.8 (0.7)
S.B. (*)	PM	14.3	77.3 (2.1)	86-93
E.S.	EM	10.0	78.2 (2.0)	95.4 (0.6)
N.K.	EM	14.5	79.0 (3.2)	93.7 (0.8)
			77.0	94.2

[a]PM= poor metabolizer, EM= extensive metabolizer.
[b]Mean value (standard deviation).

Table 2. Stereoselective Metabolism of Phenytoin to
Urinary Dihydrodiol (DHD) and Phenolic (p-
HPPH) Metabolites in Pediatric Patients.

Patient Number	Sex	Age (Years)	DHD (% S-)	p-HPPH (% S-)
3-1	M	14	75.2	88.3
3-2	M	19	68.8	92.3
3-3	M	6	74.1	90.2
3-4	M	19	69.6	82.7
3-5	M	16	76.7	84.8
4-1	M	12	76.1	89.0
4-23	F	14	72.9	92.9
4-2	M	17	77.2	88.0
4-29	M	15	73.9	95.3
			73.8[a] (3.0)	89.3[a] (4.0)

[a]Mean value (standard deviation).

Stereoselectivity in Patients

As the volunteer study and other studies indicated that the stereochemical composition of DHD and p-HPPH metabolites were largely unaffected by time after dose, random urine samples obtained from the
patients on chronic PHT therapy should reflect the stereoselective
metabolism that would be observed in a 24-hour urine collection.

As is shown in Table 2, the average stereoselectivity observed for
DHD (74% S-) and p-HPPH (89% S-) production in the patients were similar
to those observed for the volunteers (Table 1). Also consistent with the
volunteer study is the decreased stereoselectivity observed in DHD formation as compared to p-HPPH formation.

Stereoselectivity in a Patient with a Toxic Reaction to PHT

Urine samples from a patient on PHT therapy who experienced a hypersensitivity reaction to the drug were assayed for DHD and p-HPPH isomeric
content. Samples were obtained following the last dose of PHT. p-HPPH
stereoselectivity was similar to that observed in volunteers, with an
average value of 96% S- obtained from three separate 12-hour urine samples from this patient. Similarly, the average DHD stereochemistry was
found to be 73% S-, again a value similar to that observed in both the
volunteers and patients (Tables 1 and 2).

DISCUSSION

In both the volunteer study and the patient population, two isomeric
DHD's can be detected, and by inference, two isomeric AO's of PHT must have
been formed. The stereoselective production of these DHD's is similar in
the two groups (Tables 1 and 2), despite the chronic treatment of patients
with PHT, which can induce its own metabolism. Based solely upon these
data, we must conclude that, if PHT induces its own metabolism, the induction apparently does not alter the stereoselective metabolism of the drug,
either to DHD or p-HPPH metabolites. Similarly, no correlations have been
observed between debrisoquine metabolizer status, phenytoin half-lives,
and stereoselective metabolism as reflected in Table 1.

The patients in the study have been treated with PHT for at least
6 months, and idiosyncratic reactions that could be associated with PHT
AO's were not apparent in these patients. We can conclude that the isomeric AO's being produced are being detoxified in these individuals. In
the case of one patient who experienced a hypersensitivity reaction to PHT
(see results), the stereoselective metabolism of PHT was not substantially
different from that observed in the volunteers or in the patient population.
If hypersensitivity reactions are associated with covalent binding of
phenytoin arene oxides to cellular constituents, in this one case there is
not an obvious difference in urinary metabolite stereoselectivity associated with the toxicity. We must assume that other factors, such as the
susceptibility of target organs/tissues to the AO's, are responsible for
the toxicity, as has been suggested by others (Spielberg et al., 1981;
Gerson et al., 1983).

Observed differences in stereoselectivity for the DHD and p-HPPH
metabolites may be explained if one assumes that all p-HPPH is formed via
AO's, and that the rates of rearrangement of (R)- and (S)-AO's, K_R and K_S
(Figure 1), are similar. Under these conditions, p-HPPH enantiomeric
content should reflect AO isomeric content. If this is so, then the production of DHD isomers by epoxide hydrolase (rate constants K_R and K_S) from
(R)- and (S)- AO's must be different, and favor the conversion of (R)-AO
to (R)-DHD (Figure 1). This situation would give lower %S composition of
DHD as compared to p-HPPH.

An alternative explanation would be that there is a mixture of arene
oxide and direct hydroxylation pathways for PHT hydroxylation, and that
the observed differences in DHD and p-HPPH stereoselectivity are combinations of stereoselective direct hydroxylation and the pathways

mentioned above (Figure 1). At the present time, evidence for direct hydroxylation pathways for PHT is not available, whereas evidence for arene oxide-dependent pathways is numerous, and "NIH-shifts" have been observed for human metabolism of PHT (Claesen et al., 1982).

The methods used in this study have provided the opportunity to study stereoselective arene oxide pathways in human PHT metabolism. We are continuing studies of the relationship of human stereoselective metabolism to toxic effects observed with PHT therapy in order to evaluate theories of the role of specific reactive intermediates in PHT toxicities.

ACKNOWLEDGEMENTS

We wish to thank Drs. Gunnar Alván and Anders Rane of the Department of Clinical Pharmacology, Karolinska Institute (Huddinge, Sweden) for supplying some of the urine samples used in this study. Support for this study was from NIDR/NIH grant DE 06541, an R.J. Reynolds Junior Faculty Development award (to J.H.M.), and a NIEHS/NIH postdoctoral traineeship (National Research Service Award 5T32 ES 07126) to J.S.M.

REFERENCES

Butler, T.C., 1957, The Metabolic Conversion of 5,5-Diphenylhydantoin to 5-(p-hydroxyphenyl)-5-phenylhydantoin, J. Pharmacol. Exp. Ther., 119:1.

Butler, T.C., Dudley, K.H., Johnson, D., and Roberts, S.B., 1976, Studies of the Metabolism of 5,5-Diphenylhydantoin Relating Principally to the Stereoselectivity of the Hydroxylation Products in Man and the Dog, J. Pharmacol. Exp. Ther., 199:82.

Chang, T., and Glazko, A.J., 1982, Phenytoin: Biotransformation, In"Antiepileptic Drugs"(D.M. Woodbury, J.K. Penry and C.E. Pippenger, eds.), Raven Press, N.Y., p. 209.

Claesen, M., Moustafa, M.A.A., Adline, J., Vandervorst, D., and Poupaert, J.H., 1982, Evidence for an Arene Oxide-NIH Shift Pathway in the Metabolic Conversion of Phenytoin to 5-(4-hydroxyphenyl)-5-phenylhydantoin in the Rat and in Man, Drug Metab. Dispos., 10:667.

Gerson, W.T., Fine, D.G., Spielberg, S.P., and Sensenbrenner, L.L., 1983, Anticonvulsant-Induced Aplastic Anemia: Increased Susceptibility to Toxic Drug Metabolites In Vitro, Blood, 61:889.

Maguire, J.H., Kraus, B.L., Butler, T.C., and Dudley, K.H., 1979, Determination of 5-(3,4-dihydroxy-1,5-cyclohexadien-1-yl)-5-phenylhydantoin (Dihydrodiol), and Studies of Phenytoin Metabolism in Man, Ther. Drug Mon., 1:359.

Maguire, J.H., Butler, T.C., and Dudley, K.H., 1980, Absolute Configuration of Phenytoin Dihydrodiol Metabolites from Rat, Dog, and Human Urines, Drug Metab. Dispos., 8:325.

Maguire, J.H., and Wilson, D.C., 1985, Urinary Dihydrodiol Metabolites of Phenytoin: HPLC Assay of Diastereomeric Composition, J. Chromatogr., 342:323.

Martz, F., Failinger, C., and Blake, D.A., 1977, Phenytoin Teratogenesis: Correlation between Embryopathic Effect and Covalent Binding of Putative Arene Oxide Metabolite in Gestational Tissue, J. Pharmacol. Exp. Ther., 203:231.

Spielberg, S.P., Gordon, G.B., Blake, D.A., Goldstein, D.A., and Herlong, H.F., 1981, Predisposition of Phenytoin Hepatotoxicity Assessed In Vitro, New Eng. J. Med., 305:722.

TISSUE DISTRIBUTION AND COVALENT BINDING OF [14C]1,1-DICHLOROETHYLENE IN

MICE: IN VIVO AND IN VITRO STUDIES

Laud K.N. Okine and Theodore E. Gram

National Cancer Institute
National Institutes of Health
Bethesda Maryland 20205
U.S.A

INTRODUCTION

1,1-Dicholorethylene (DCE), a widely used compound in the manufacture of plastics, has been shown to be toxic to kidney, liver and lung in laboratory animals[1-5], and to exhibit some mutagenic and carcinogenic potential[6-8]. Its toxicity has been suggested to result from covalent binding of reactive metabolites to cellular macromolecules[3,9,10]; these metabolites are normally detoxified by conjugation with reduced glutathione (GSH)[11-13]. However the exact mechanisms of its expressed toxicity and its application to man are not understood. Species differences among animals in toxicity of DCE are thought to be due to the relative amounts of bioactivated and covalently bound DCE produced in each species[9,14]. We present in this paper the in vivo tissue distribution and covalent binding as well as the in vitro metabolism and covalent binding of DCE in the mouse, the species thought to be most susceptible to its toxicity[9,10], with particular reference to the kidney, liver and lung.

METHODS

Male C57Bl/6N mice (20-25g) were treated with [1,2-14C] DCE (125 mg/kg i.p., 4 µCi) and killed at various time intervals (0-96 hr). Some animals were pretreated with 3-methylcholanthrene (3-MC), phenobarbital (PB), piperonyl butoxide or SKF-525A prior to [14C] DCE administration, and killed 24 hr later. Tissues were removed and homogenized in 150 mM KCl - 50 mM Tris - HCl buffer pH 7.4 to determine DCE covalently bound to tissue constituents. Tissue GSH levels were measured 0-8 hr after DCE treatment. Studies on the in vitro metabolism and covalent binding of DCE were conducted with kidney, liver and lung microsomes or other subcellular fractions from control mice or animals pretreated with PB, 3-MC, β-naphthoflavone (β-NF) or pregnenolone 16 α-carbonitrile (PCN). The optimum conditions for covalent binding of DCE in vitro were studied. Protein concentration[15] and cytochrome P-450 content[16] were also determined.

RESULTS AND DISCUSSION

Administration of [14C] DCE to mice revealed that covalent binding occured predominantly in kidney, liver and lung. Maximum binding was observed 6-12 hr after administration (Fig.1), and this preceeded any demonstrable toxic effects reported with DCE[4],[5]. This suggests possible correlation between covalent binding and toxicity, particularly in the lung where extensive necrosis of the Clara cell of the bronchiolar epithelium coupled with reductions in pulmonary monooxygenases were observed at 24 hr after DCE treatment[5]. The lack of toxicity in kidney or liver even though these organs showed higher levels of covalent binding than lung (2 to 3-fold) suggests differences in cytotoxicity among organs in response to covalently bound DCE species and may also reflect differences in tissue GSH levels (Fig.2). Subcellular distribution studies in organs from [14C] DCE-treated mice indicated that covalently bound material was distributed uniformly and non-specifically across all subcellular fractions in kidney, liver and lung (Fig.3). In contrast, in vitro studies with mouse liver subcellular fraction indicated that metabolism of DCE to covalently bound species takes place predominantly in the microsomes (Table 1).

The effects of pretreatment of mice with inducers (PB and 3-MC) and inhibitors (SKF-525A and piperonyl butoxide) of the mixed function oxidases as well as diethylmateate on the covalent binding of DCE in vivo (Fig.4) suggests that activation of DCE involves cytochrome P-450, and that GSH is involved in its detoxification as has been reported in earlier studies in rats[11-13]. Diethylmaleate pretreatment also increased the lethality of DCE (results not shown). The anomalous increase in covalent binding of DCE in kidney of SKF-525A pretreated mice may be due to

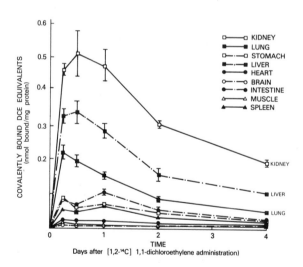

Fig. 1: Distribution of covalently bound [14C] DCE in mouse tissues. Values shown are the means ± SEM of four determinations. Tissues pooled from 3-6 animals were used for each determinations.

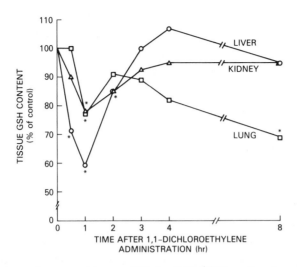

Fig.2: Depletion of mouse tissue GSH by DCE (125 mg/kg i.p.).
Values shown are the means of four determinations. Tissues from
2 animals were used for each determination.

possible inhibitory effects on the detoxification pathway, and may
explain the observed toxicity of DCE in SKF-525A pretreated animals[1,17].

Renal microsomes incubated with [^{14}C] DCE and NADPH were unable to
catalyze covalent binding of detectable levels of radiolabel while liver
microsomes actively bound DCE in amounts about 4 times those bound by
lung microsomes (Table 2). However, that covalent binding of DCE in the
kidney in vivo was higher than in the liver and lung (Fig.1), suggests
the liver as the primary site of DCE metabolism, and that covalent
binding in kidney in vivo may be due to reactive species produced
elsewhere, possibly in the liver.

The differential increase in total cytochrome P-450 (2-fold) and
covalent binding (3-fold) by liver microsomes from mice pretreated with
PB, 3-MC, PCN or β-NF (Table 2) indicated that different cytochrome
P-450 isozymes can metabolize DCE to covalenty bound species, which
support earlier findings using purified cytochrome P-450 isozymes[18].
Furthermore, the exhibition of detectable levels of covalent binding by
kidney microsomes from mice pretreated with 3-MC or PCN, to levels
similar to control lung microsomes (Table 2), suggests that these
pretreatments induced mouse renal cytochrome P-450 isozymes capable of
metabolizing DCE to covalently bound materials and may explain the
observed mutagenicity of DCE species in the Ames' Salmonella test using
renal microsomes from 3-MC pretreated animals[6]. Since DCE has been
shown to induce renal cytochrome P-450 and its dependent monooxygenases
(5 to 10-fold) within 24 hr[5], it is possible that DCE-induced renal
microsomal cytochrome P-450 isozymes may metabolize and bind DCE
in vivo.

Table 1: NADPH-dependent covalent binding of ^{14}C-DCE by subcellular fractions of mouse liver *in vitro*

Subcellular Fraction (n=4)	Cytochrome P-450 (nmol/mg protein)	Covalent Binding (nmol equiv/mg protein/10 min)
Nuclei	0.17	0.18 (1.0)
Mitochondria	0.19	0.15 (0.8)
Microsomes	0.65	0.49 (0.8)
Cytosol	<0.02	0.05

Complete incubation system comprised an NADPH-generating system (1.64 mM NADP[+], 17.2 mM glucose 6-phosphate, 1 unit/ml glucose 6-phosphate dehydrogenase), protein (1 mg/ml), [1,2-^{14}C] DCE (1.1 mM, 0.6 μCi) in KCl - Tris buffer pH 7.4 in a final volume of 2.005 ml and in the presence of O_2. All incubations were done for 10 min at 37°. Figures in parentheses (nmol equiv./nmol P-450/10 min)

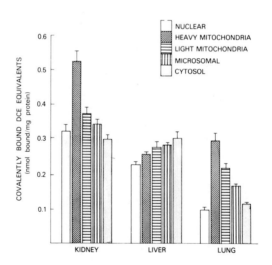

Fig.3: Distribution of covalently bound [^{14}C] DCE in subcellular fractions of mouse tissue.
Values shown are the means ± SEM of three determinations. Tissues pooled from 4-8 animals were used for each determination.

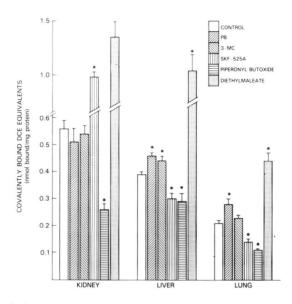

Fig.4: Effect of drug pretrements on the covalent binding of [^{14}C] DCE
in mouse tissue.
Values shown are the means ± SEM of 4-8 determinations Tissues
pooled from 2-3 animals were used for each determinations.
*Significantly different from control; p<0.01 (Students t-test)

Studies of the optimum conditions for covalent binding of DCE
(Table 3) indicated that it is an NADPH-dependent enzymatic phenomenon
which may not require oxygen. There may thus be an alternative cytochrome
P-450dependent pathway of DCE activation under anerobic conditions. The
inhibition of the in vitro covalent binding by SKF-525A and piperonyl
butoxide as well as GSH, support our earlier suggestions of the
involvement of cytochrome P-450 and GSH in the activation and detoxification of DCE respectively. The latter also confirms the intermediacy
of activated electrophilic species.

In summary, DCE is metabolized by mouse liver and lung microsomes
by an NADPH-dependent enzymatic reaction to covalently bound species
which may be detoxified by conjugation with GSH. The metabolism involves
different isozymes of cytochrome P-450 and may not require O_2. Although
in vivo, kidney exhibited the highest levels of covalent binding, in
vitro studies revealed that renal microsomes were unable to catalyze the
conversion of [^{14}C] DCE to covalently bound species. These findings
suggest that liver and lung can metabolize and covalently bind DCE per se
whereas covalent binding in kidney is likely to be due to reactive
intermediates transported from the liver via the blood.

Table 2: NADPH-dependent covalent binding of ^{14}C-DCE in vitro by microsomes from mice pretreated with inducing agents.

Treatment	Liver (n=4)		Kidney (n=2)		Lung (n=2)	
	Cytochrome P-450	Covalent Binding	Cytochrome P-450	Covalent Binding	Cytochrome P-450	Covalent Binding
Control	0.55	0.8 (1.5)	0.23	<0.001	0.13	0.23 (1.7)
PB	1.2*	2.4*(2.0)	0.22	<0.001	0.13	0.23 (1.7)
βNF	1.1*	2.3*(2.1)	0.14	<0.001	0.12	0.30 (2.5)
PCN	1.1*	1.1*(1.0)	0.14	0.20 (1.5)	0.10	0.14 (1.4)
3-MC	1.1*	2.7*(2.1)	0.16	0.25 (1.6)	0.12	0.15 (1.3)

Incubation conditions, and units for cytochrome P-450 and covalent binding are as in Table 1.
Figures is parentheses (nmol equiv./nmol P-450/10 min)
*Statistically significant from control; $p < 0.02$ (Student's t-test)

Table 3: Optimum conditions for covalent binding of ^{14}C-DCE to mouse liver and lung microsomes in vitro.

System	Covalent Binding (nmol equiv./mg protein/10 min)	
	Liver (n=4)	Lung (n=2)
Complete	0.77	0.28
microsomes boiled	0.18* (77)	0.13 (54)
- NADPH	0.06* (92)	0.05 (81)
+ N$_2$ (-O$_2$)	0.60* (22)	0.35 (-)
+ CO:O$_2$ (4:1 v/v)	0.25* (68)	0.07 (77)
+ SKF-525A (2.0 mM)	0.39* (50)	0.17 (41)
+ Piperonyl butoxide (2.0 mM)	0.34* (56)	0.13 (54)
+ GSH (5.0 mM)	0.26* (67)	0.17 (42)

Complete system and incubation conditions are as in Table 1.
Figures in parentheses (% Inhibition).
*Statistically significant from complete system; $p < 0.01$ (Student's t-test)

REFERENCES

1. R.D. Short, J.M. Winston, J.L. Minor, J. Seifter, and C.C. Lee, Effect of various treatments on toxicity of inhaled vinylidene chloride, Environ. Health Perspect. 21: 125 (1977).
2. L.J. Jenkins, and M.E. Andersen, 1,1-Dichloroethylene nephrotoxicity in the rat, Toxicol Appl. Pharmacol. 46: 131 (1978).
3. M.E. Anderson, J.E. French, M.L. Gargas, R.A. Jones, and L.J. Jenkins, Saturable metabolism and the acute toxicity of 1, 1-dichloroethylene, Toxicol. Appl. Pharmacol. 47: 385 (1979).
4. Y. Masuda, and N. Nakayama, Protective action of diethyldithiocarbamate and carbon disulfide against acute toxicities induced by 1, 1-dichloroethylene in mice, Toxicol. Appl. Pharmacol. 71: 42 (1983).
5. K.R. Krijgsheld, M.C. Lowe, E.G. Mimnaugh, M.A. Trush, E. Ginsburg, and T.E. Gram, Selective damage to non-ciliated bronchiolar epithelial cells in relation to impairment of pulmonary monooxygenase activities by 1, 1-dichloroethylene in mice, Toxicol. Appl. Pharmacol. 74: 201 (1984).
6. H. Bartsch, C. Malaveille, R. Montesano, and L. Tomatis, Tissue-mediated mutagenicity of vinylidene chloride and 2-chlorobutadiene in Salmonella typhimurium, Nature (London) 255: 641 (1975).
7. C. Malton, Recent findings on the carcinogenicity of cholorinated olefins, Environ. Health Perspect. 21: 1 (1977).
8. C.C. Lee, J.C. Bhandri. J.M. Winston, W.B. House, R.L. Dixon, and J.S. Wood, Carcinogenicity of vinyl chloride and vinylidene chloride, J. Toxicol. Environ. Health 4: 15 (1978).

9. M.J. McKenna, P.G. Watanabe, and P.J. Gehring, Pharmacokinetics of vinylidene chloride in the rat, Environ. Health Perspect. 21: 99 (1977).

10. R.J. Jaeger, L.G. Shoner, and L. Coffman, 1,1-Dichlorethylene hepatotoxicity: proposed mechanism of action and distribution and binding of ^{14}C radioactivity following inhalation exposure in rats, Environ. Health Perspect. 21: 113 (1977).

11. R.J. Jaeger, L.G. Conolly, and S.D. Murphy, Effect of 18 hr fast and glutathione depletion on 1,1-dichloroethylene-induced hepatotoxicity and lethality in rats, Exp. Mol. Pathol. 20: 187 (1974).

12. M.E. Anderson, O.E. Thomas, M.L. Gargas, R.A. Jones, and L.J. Jenkins Jr., The significance of multiple detoxification pathways for reactive metabolites in the toxicity of 1,1-dichloroethylene, Toxicol. Appl. Pharmacol. 52: 422 (1980).

13. D.C. Liebler, M.J. Meredith, and F.P. Guengerich, Formation of glutathione conjugates by reactive metabolites of vinylidene chloride in microsomes and isolated hepatocytes, Cancer Res. 45: 186 (1985).

14. B.K. Jones, and D.E. Hathway, Differences in metabolism of vinylidene chloride between mice and rats, Brit. J. Cancer 37: 411 (1978).

15. O.H. Lowry, N.J. Rosebrough. A.L. Farr, and R.J. Randall, Protein measurement with oflin phenol reagent, J. Biol. Chem. 193: 265 (1951).

16. T. Omura, and R. Sato, The carbon monoxide-binding segment of liver microsomes II. Solubilization, purification and properties, J. Biol. Chem. 239: 2379 (1964).

17. M.E. Andersen, and L.J. Jenkins, Jr., The oral toxicity of 1, 1-dichloroethylene in the rat: effects of sex, age and fasting, Environ. Health Perspect. 21: 157 (1977).

18. D.C. Lielber, and F.P. Guengerich, Olefin oxidation by cytochrome P-450: evidence for group migration in catalytic intermediates formed with vinylidene cholride and trans-1-phenyl-1-butene, Biochemistry 22: 5482 (1983).

INHIBITION OF RESPIRATION IN RABBIT PROXIMAL TUBULES BY

BROMOPHENOLS AND 2-BROMOHYDROQUINONE

Rick G. Schnellmann and Lazaro J. Mandel

Department of Physiology
Duke University Medical Center
Durham, N.C. 27710

INTRODUCTION

The kidney cortex is a highly aerobic and metabolically active tissue rich in mitochondria and Na- and K-dependent adenosine triphosphatase (Na,K-ATPase) activity. The basal respiration rate of a suspension of rabbit proximal tubules utilizes 50-60 percent of the mitochondrial respiratory capacity (Harris et al., 1981). Since Na,K-ATPase-mediated ion transport consumes more metabolic energy than any other single enzymatic process within the mammalian organism, it is not surprising that rabbit proximal tubules utilize half of their basal respiration for this process (Cohen and Kamm, 1976; Harris et al., 1981). The remaining half of their basal respiration is divided between ATP generation for other processes and nonphosphorylating respiration. Using isolated mitochondria, non-phosphorylating respiration has been shown to be 5-8 percent of the basal respiration (Davis et al., 1974).

Using various chemical agents, it is possible to examine mitochondrial respiration under a variety of conditions. The addition of nystatin, a monovalent ionophore, to a suspension of rabbit proximal tubules causes sodium to enter the tubules and thereby stimulates Na,K-ATPase activity with a consequent increase in ATP turnover and respiration (Fig. 1). Under these conditions, nystatin stimulates respiration to the maximum rate at which the tubular mitochondria are capable of producing ATP (Harris, et al., 1981). The oxygen consumption (QO_2) achieved with nystatin is normally identical to state 3 mitochondrial respiration (Chance and Williams, 1956). By adding ouabain, an inhibitor of Na,K-ATPase, it is possible to determine that portion of respiration responsible for Na,K-ATPase activity.

LUMEN   BLOOD

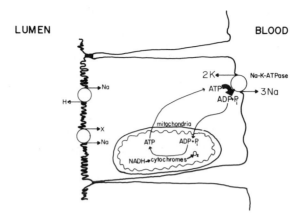

Figure 1. A schematic of a renal proximal tubular cell
           illustrating the coupling of mitochondrial
           respiration and Na,K-ATPase activity.

Since the kidney is a highly metabolic tissue and spends a large
portion of its energy on transport processes, it is very sensitive to any
toxicant that may disrupt oxidative metabolism. The aim of this work was
to examine whether 2-, 3-, 4-bromophenol or 2-bromohydroquinone (BHQ) cause
nephrotoxicity by interfering with oxidative metabolism.

MATERIALS AND METHODS

Renal proximal tubules from female New Zealand White rabbits (3-4 kg)
were isolated according to Soltoff and Mandel (1984) with dextran being
omitted from all buffers following the in situ perfusion. Tubules were
resuspended (4-5 mg cellular protein/ml) in a buffer containing (in mM):
115 NaCl, 5 KCl, 25 NaHCO$_3$, 2 NaH$_2$PO$_4$, 1 CaCl$_2$, 1 MgSO$_4$, 5 glucose, 4 sodium
lactate, 1 alanine, 5 sodium malate, and 2 sodium butyrate (pH 7.4).

Tubules resuspended in the above buffer were incubated at 37°C and
continuously gassed with 95% O$_2$/5% CO$_2$ in a gyrating shaker bath for 15
min. Various concentrations of 2-, 3-, 4-bromophenol or 2-bromohydro-
quinone were added (0 min) and the tubules allowed to incubate for an
additional 60 min. Additional sodium butyrate was added at 30 min for a
final concentration of 2 mM. 2-, 3-, 4-Bromophenol and BHQ were dissolved
in DMSO. Controls received DMSO alone. Following the 60 min incubation,
aliquots of the tubule suspension were taken for O$_2$ consumption and LDH
release.

The rate of oxygen consumption (QO$_2$) was monitored using a Clark-type
oxygen electrode, an oxymeter and a 1.6 ml magnetically-stirred chamber

thermostatically set at 37°C. After obtaining a basal QO_2, the monovalent ionophore, nystatin (1400 units), was added. To obtain the QO_2 coupled to the Na ,K -ATPase, ouabain (10^{-4} M final concentration) was added to the chamber.

LDH release is a common marker for plasma membrane damage and cellular toxicity and was determined as previously described (Schnellmann and Mandel, 1985a). LDH activity was determined by the method of Bergmeyer (1963). Protein content was determined by the biuret method (Gornall et al., 1949) with bovine serum albumin serving as the protein standard.

The data are presented as the mean ± SEM. Data were analyzed by the paired Students t test. A P value of less than 0.05 was considered significant.

The sources of all chemicals have been reported previously (Soltoff and Mandel, 1984; Schnellmann and Mandel, 1985a; Schnellmann and Mandel, 1985b).

RESULTS

Incubation of 1 mM 3-bromophenol with rabbit proximal tubules for 60 min resulted in a decrease in basal QO_2 (18%) and nystatin-stimulated QO_2 (33%), (Figure 2). In contrast, ouabain-insensitive QO_2 increased 56 percent at the same concentration. The addition of 2 mM 3-bromophenol almost abolished QO_2 to 2.5 nmol O_2/mg protein/min and greatly elevated LDH release (52%). The effects of 4-bromophenol on tubular function were similar to those of 3-bromophenol (Figure 2). The addition of 1 mM 4-bromophenol

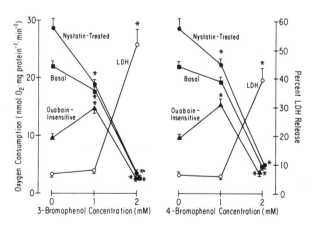

Figure 2. The concentration-dependent effects of 3- and 4-bromophenol on oxygen consumption and lactate dehydrogenase (LDH) release in rabbit proximal tubules. Data are the mean ± SEM, N = 4. *, Significantly different from controls (P≤0.05).

decreased nystatin-stimulated QO_2 (19%), stimulated ouabain-insensitive QO_2 (59%), but did not change basal QO_2. Incubation of 2 mM 4-bromophenol greatly inhibited QO_2 to 4.5 nmol O_2/mg protein/min and elevated LDH release (39%). These results show that 3- and 4-bromphenol are similar in toxicity and that decreases in QO_2 precede plasma membrane damage, as measured by LDH release.

2-Bromophenol inhibited QO_2 and elevated LDH release in a dose-dependent manner (Figure 3). Nystatin-stimulated QO_2 was inhibited 23, 48, and 93 percent with doses of 1, 2, and 5 mM 2-bromophenol, respectively. Basal QO_2 was inhibited in a qualitatively similar manner. Ouabain-insensitive QO_2 increased slightly (7%) when 2 mM 2-bromophenol was added and decreased 88 percent when 5 mM was added. LDH release increased over controls at doses of 2 and 5 mM. These results show that 2-bromophenol is approximately 1/2 as potent a cytotoxicant as 3- and 4-bromophenol, and like 3- and 4-bromophenol, decreases in QO_2 are seen prior to increases in LDH release.

Incubation of 2-bromohydroquinone (0.1 mM) with proximal tubules inhibited nystatin-stimulated QO_2 (24%) without affecting basal and ouabain-insensitive QO_2 and LDH release (Figure 4). Higher doses of 2-bromohydroquinone resulted in a further decrease in nystatin-stimulated QO_2, decreases in basal and ouabain-insensitive QO_2 and an increase in LDH release. These results show that 2-bromohydroquinone is approximately 10 times more potent than 3- and 4-bromophenol, and 20 times more potent than 2-bromophenol in inhibiting QO_2 and elevating LDH release.

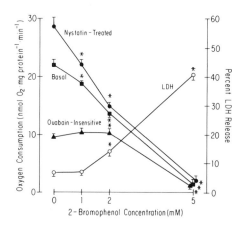

Figure 3. The concentration-dependent effects of 2-bromophenol on oxygen consumption and lactate dehydrogenase (LDH) release in rabbit proximal tubules. Data are the mean $\pm$ SEM, N = 3-7. *, Significantly different from controls (P$\leq$0.05).

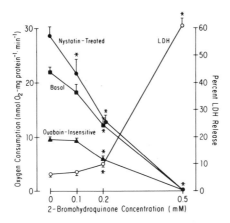

Figure 4.   The concentration-dependent effects of 2-
bromohydroquinone on oxygen consumption
and lactate dehydrogenase (LDH) release
in rabbit proximal tubules. Data are mean
$\pm$ SEM, N = 7. *, Significantly different
from controls ($P \leq 0.05$).

DISCUSSION

Examination of proximal tubular QO_2 in the presence of cytotoxicants
proved to be a very sensitive measure of cellular viability. All four
compounds inhibited basal QO_2 in a dose-dependent manner with a rank order
of potency of: 2-bromophenol < 3- and 4-bromophenol < 2-bromohydroquinone.
Using the monovalent ionophore, nystatin, it is possible to stimulate the
cell to state 3 mitochondrial respiration. Under this stressed condition
it is possible to observe toxicity prior to changes in basal QO_2. Nystatin-
stimulated QO_2 was inhibited by all four compounds prior to or to a greater
extent than basal QO_2. Furthermore, in all cases, changes in QO_2 were seen
prior to increases in LDH release. These results show that nystatin-
stimulated QO_2 is a very sensitive measure of cellular viability. In
addition, nystatin-treatment can uncover hidden impaired processes in the
tubules.

Using ouabain, it is possible to divide QO_2 into two components. The
QO_2 that is ouabain-sensitive is that QO_2 associated with Na,K-ATPase
activity. The QO_2 that is insensitive to ouabain is the QO_2 associated
with other oxygen consuming reactions and has not been well characterized.
Following cytotoxicant exposure, when the tubules are stress by increasing
intracellular sodium, via nystatin, the treated tubules may exhibit a
decreased nystatin-stimulated QO_2. By adding ouabain, it is possible to
determine whether the reduced nystatin-stimulated rate is a result of

decreased Na,K-ATPase activity or an increase in oxygen consumption by non-Na,K-ATPase-mediated cellular processes. A decreased Na,K-ATPase activity can result from: 1) inhibited mitochondrial function, 2) lack of communication between the mitochondria and the Na,K-ATPase or 3) direct inhibition of the Na,K-ATPase.

At concentrations (1 mM) that did not increase LDH release, 3-, and 4-bromophenol inhibited nystatin-stimulated ouabain-sensitive QO_2 and stimulated ouabain-insensitive QO_2. These results suggest that 3- and 4-bromophenol are probably acting at two points in causing toxicity, non-Na,K-ATPase cellular processes and Na,K-ATPase processes. The large increase (56-60%) in the ouabain-insensitive QO_2 is suggestive that these compounds may be acting as uncouplers of oxidative phosphorylation.

2-Bromohydroquinone inhibited nystatin-stimulated QO_2 at 1/10 the dose of the bromophenols and unlike the bromophenols did not stimulate ouabain-insensitive QO_2 at any dose. Preliminary results from experiments which examined mitochondrial function in rabbit proximal tubules in situ show that 2-bromohydroquinone inhibits coupled and uncoupled respiration.

The current hypothesis of bromobenzene-induced nephrotoxicity is that bromobenzene is metabolized to 2-bromophenol and then to 2-bromohydroquinone in the liver. 2-Bromohydroquinone or a conjugate of BHQ is transported to the kidney where it elicits toxicity (Lau et al., 1984). This study supports this hypothesis and suggests that the mechanism of toxicity of 2-bromohydroquinone is inhibition of mitochondrial function.

In summary, this study shows that nystatin-stimulated QO_2 as well as ouabain-sensitive and ouabain-insensitive QO_2 are very sensitive measures of cellular viability. In addition, this study suggests that 2-bromohydroquinone or a conjugate thereof may be responsible for bromobenzene-induced nephrotoxicity and that the mechanism of toxicity is impaired mitochondrial function.

ACKNOWLEDGEMENTS

This study was supported by NIH grant ES-05329 and AM-28616 and a grant from the American Heart Association, North Carolina Affiliate. The authors would like to thank Fontaine Ewell and Susan Murdaugh for their excellent technical assistance and Gay Blackwell for processing the manuscript.

REFERENCES

Bergmeyer, H.-U., Bernt, E., and Hess, B., 1973, Lactic dehydrogenase, in: "Methods of Enzymatic Analysis," H.-U. Bergmeyer, ed., Academic Press, London.

Chance, B. and Williams, G.R., 1956, The respiratory chain and oxidative phosphorylation. Adv. Enzymol., 17:65.

Cohen, J.J. and Kamm, D.E., 1976, Renal metabolism: relation to renal function, in: "The Kidney," B.M. Brenner and F.C. Rector, eds., Saunders, Philadelphia.

Davis, J.E., Lumeng, L., and Bottoms, D., 1974, On the relationships between the stoichiometry of oxidative phosphorylation and the phosphorylation potential of rat liver mitochondria as functions of respiratory rate, FEBS Lett., 39:9.

Harris, S.I., Balaban, R.S., Barrett, L., and Mandel, L.J., 1981, Mitochondrial respiratory capacity and Na^+- and K^+-dependent adenonsine triphosphatase-mediated ion transport in the intact renal cell, J. Biol. Chem., 256:10319.

Gornall, A.G., Bardawill, C.J., and David, M.M., 1949, Determination of serum proteins by means of a biuret reaction, J. Biol. Chem., 177:751.

Lau, S.S., Monks, T.J., and Gillette, J.R., 1984, Identification of 2-bromohydroquinone as a metabolite of bromobenzene and o-bromophenol: Implications for bromobenzene-induced nephrotoxicity, J. Pharmacol. Exp. Ther., 230:360.

Schnellmann, R.G. and Mandel, L.J., 1985a, Multiple effects of presumed glutathione depletors on rabbit proximal tubules, Kidney Int. (in press).

Schnellmann, R.G. and Mandel, L.J., 1985b, Cellular toxicity of bromobenzene and bromobenzene metabolites to rabbit proximal tubules: The role and mechanism of 2-bromohydroquinone, J. Pharmacol. Exp. Ther. (submitted).

Soltoff, S.P. and Mandel, L.J., 1984, Active ion transport in the renal proximal tubule, J. Gen. Physiol., 84:601.

IN VITRO COVALENT BINDING OF ^{14}C-MIBOLERONE

TO RAT LIVER MICROSOMES

P. S. Jaglan

Biochemistry and Residue Analysis
The Upjohn Company
Kalamazoo, Michigan

Mibolerone (17-Hydroxy-7,17-dimethylestr-4-en-3-one; 7α-17αdimethyl-19-nortestosterone) is being marketed by The Upjohn Company for the inhibition of estrus in bitches. The aim of this study was to determine the extent of covalent binding of mibolerone to rat liver microsomes. Liver microsomes were obtained from Control and phenobarbitol-treated female Fisher rats, and were incubated with ^{14}C-mibolerone at 37°C for 10 minutes. No covalent binding to macromolecules was observed when ^{14}C-mibolerone was incubated with rat liver microsomes. Under identical conditions, ^{14}C-estradiol was covalently bound to macromolecules. Slightly higher covalent binding of estradiol was observed with microsomes from phenobarbitol-treated rats. Ascorbic acid and glutathione inhibited covalent binding of estradiol to macromolecules in the in vitro microsomal system.

INTRODUCTION

Many chemical compounds are metabolized by the liver and various other tissues to potent alkylating and arylating intermediates.[1-11] Such studies demonstrated how chemically stable compounds can result in serious tissue lesions such as hepatic, renal, pulmonary necrosis, bone marrow aplasia, neoplasm and other injuries in man and experimental animals.[12] Most drugs and foreign compounds that enter the body are converted to chemically stable compounds that are readily excreted into urine and bile or are expired. However, in some cases chemically reactive intermediates may bind to the macromolecules.

Based on studies where an animal model was developed for a particular chemical-induced tissue lesion, a relation could often be made between the severity of tissue lesion and the amount of metabolite that was covalently bound to the damaged tissue.[12] In conjunction with these animal studies, experiments were performed in vitro with microsomal enzymes isolated from the target organ tissues. Covalent binding of the reactive intermediates was found to be a useful index of product formation when various additions and deletions were made to the system or when animals are pretreated with various enzyme inducers or inhibitors.

Mibolerone (17-Hydroxy-7,17-dimethylestr-4-en-3-one; 7α-17αdimethyl-19-nortestosterone) is being marketed by The Upjohn Company for the inhibition of estrus in bitches. Therefore, the aim of this study was to determine the extent of covalent binding of mibolerone to rat liver microsomes. Estradiol, which is known to bind covalently with liver microsomes, was used as a positive control.[13] Perturbed systems, such as liver microsomes from phenobarbitol-treated rats and adjuncts such as ascorbic acid and glutathione were also used to define their effects.

EXPERIMENTAL

Liver Microsome

Control Rats. Female Fisher rats about 110 days old were killed by cervical dislocation and livers were immediately removed. They were rinsed in ice cold 1.15% KCl solution, blotted dry, weighed, minced with scissors and ground in 20 ml of ice cold 1.15% KCl in a Potter homogenizer. The homogenates were centrifuged at 10,000 g for 25 minutes at 4°C. The supernatants were centrifuged at 105,000 g for 1 hour at 2°C. The supernatant was rejected and the pellet resuspended in 20 ml 1.15% cold KCl and recentrifuged at 105,000 g for 1 hour. The pellet was ground in 0.1M phosphate buffer (40 ml), using an all-glass Potter homogenizer.

Phenobarbitol Treated Rats. Rats as described above were given water ad libitum containing 0.1% phenobarbitol for 5 days. Preparation of liver microsomes was similar to that described above.

Cytochrome P_{450}

About 7 ml liver microsomes were stirred with 50 mg sodium dithionate. The mixture was transferred to two 1 cm cuvettes and then transferred to a Cary 15 spectrophotometer. The beam was balanced between 500-400 nm. Then carbon monoxide was bubbled slowly into the mixture in the sample cuvette for 2 minutes and spectra recorded between 400-500 nm. The absorbance at 450 nm was then determined. Concentration of cytochrome P_{450} was calculated based on molar extinction coefficient $E^{450}_{max} = 91000$.

Determination of Protein

The concentration of protein was determined by the method of Lowry et at.[14] Standard curve was prepared with bovine albumin.

Mibolerone ^{14}C (17α-methyl)

Mibolerone (17-Hydroxy-7,17-dimethylestr-4-en-3-one; 7α-17αdimethyl-19-nortestosterone) specific activity 53.5 mCi/mmole, New England Nuclear, 25 µl containing 227 µg and 546,505 DPM (2408 DPM/µg) in dimethylsulfoxide was used in each incubation (0.25 mM).

Estradiol

(4-C^{14}) (specific activity 50 mCi/mmole New England Nuclear). Twenty-five μl dimethylsulfoxide solution of ^{14}C-estradiol containing 227 μg and 499,999 DPM (2203 DPM/μg) was used per incubation (0.28 mM).

Cofactor and Inhibitor Solutions

Cofactor solutions were prepared as shown in Table 1. Nicotanimade dinucleotide phosphate (NADP) and glucose-6-phosphate dehydrogenase (Sigma-Chemical Co.), glucose-6-phosphate (Nutritional Biochemical Corporation), magnesium chloride (A. R. Mallinckrodt), and glutathione (Aldrich) were purchased. Ascorbic acid was procured from The Upjohn Company.

Procedure for Incubation

(a) Control microsomes. Two ml of microsomes used per incubation contained six mg protein and specific activity of cytochrome P_{450} was 146 picomole/mg protein.

(b) Microsomes from phenobarbitol-treated rats. Two ml of microsomes contained 6 mg protein and specific activity of cytochrome P_{450} was 575 picomole/mg protein.

Different incubation systems are described below (A-E) and given in Table 1.

(A) Two ml of liver microsomes as described above were incubated with one ml of cofactor solution and designated as complete system.

(B) Two ml of liver microsomes as described above were heated in boiling water to destroy enzyme activity and incubated with one ml of cofactor solution and designated as heated system. This system was supposed to give blank values.

(C) Two ml of microsomes as described above were incubated with one ml of 0.1 ml phosphate buffer without any cofactors and designated as no cofactors.

(D) Two ml of microsomes as described above were incubated with one ml solution of cofactor containing ascorbic acid and designated as complete system + ascorbic acid.

(E) Two ml of microsomes as described above were incubated with one ml of cofactor solution containing glutathione and designated as complete system + glutathione.

Incubation with ^{14}C-Mibolerone/^{14}C-Estradiol

All incubations were started by adding 25 μl of dimethyl sulfoxide solution of mibolerone (0.25 mM). All incubations were done in quadruplicate at 37°C for 10 minutes in a constant temperature shaking bath. Reaction was stopped by immediately cooling to 0°C and by adding 10 ml of cold methanol. The solutions were centrifuged for 10 minutes at 2000 rpm. Supernatant methanol was decanted off and the pellet was resuspended in 5 ml of cold methanol and recentrifuged as above. The pellet was further washed with 5 ml cold methanol and centrifuged. The pellet was further treated as above with 5 ml of cold 3:1 ethanol-ether three more times.

The pellet remaining after the above washings was dissolved in 5 ml of 1N sodium hydroxide. An aliquot of 0.5 ml in triplicate was counted in 15 ml diotol for 10 minutes three times. DPM were computed after adding ^{14}C-toluene internal standard.

The protein concentration was determined in the above pellet as described under protein determination.

Picomoles of compound bound/mg protein/min were then calculated.

RESULTS AND DISCUSSION

Table 1 describes the various systems used for the incubation of rat liver microsomes with ^{14}C-mibolerone and ^{14}C-estradiol. Concentration of protein, compounds and cofactors was similar to that used by Nelson et al.[13] Ascorbic acid (2 mM) was used as inhibitor of covalent binding to macromolecules since it is believed to reduce formation of superoxide anions.[15]

Table 1. Systems Used for Incubation of Mibolerone and Estradiol with Rat Liver Microsomes[a]

SYSTEM	DESCRIPTION	COFACTORS				INHIBITORS	
		NADP	Glucose 6 Phosphate	Glucose 6 Phosphate Dehydrogenase	MgCl$_2$	Ascorbic Acid	Glutathione
		(mM)	(mM)	(Units)	(mM)	(mM)	(mM)
A	Complete System	.67	10	6	10	--	--
B	Complete System Heated	.67	10	6	10	--	--
C	No Cofactors	--	--	--	--	--	--
D	Complete System + Ascorbic Acid	.67	10	6	10	2	--
E	Complete System + Glutathione	.67	10	6	10	--	2

[a]Each incubation contained two ml microsome solution containing 6 mg protein and concentration of Mibolerone or Estradiol was 0.25, 0.28 mM, respectively

Glutathione, a naturally occurring sulfhydryl containing tri-peptide was also used as an inhibitor since it is believed to form adducts with reactive intermediates of xenobiotics.[16]

Table 2 lists nanomoles of mibolerone and estradiol bound per mg protein/min under different conditions. These data show that although estradiol was covalently bound to macromolecules as observed before[13], mibolerone was not bound. System B where liver microsomes were destroyed by heating was treated as blank value for these data. Slightly higher covalent binding of estradiol was observed with microsomes from phenobarbitol-treated rats although the mg of protein per incubation was similar (6 mg) to control microsomes. This is probably due to the high specific activity of cytochrome P_{450} in microsomes from the phenobarbitol-treated rat livers (575 picomole/mg protein as compared to 146 in control).

Table 2.   In Vitro Covalent Binding of Mibolerone/Estradiol to Rat Liver Microsomes

| | NANOMOLES BOUND/MG PROTEIN/MIN | | | |
| | Control | | Phenobarbitol Treated | |
System	Mibolerone	Estradiol	Mibolerone	Estradiol
A	0.032	0.144	0.022	0.179
B	0.033	0.023	0.034	0.016
C	0.030	0.050	0.016	0.149
D	0.018	0.031	0.011	0.057
E	0.022	0.038	0.013	0.066

Both ascorbic acid and glutathione inhibited covalent binding of estradiol to macromolecules.

Serum L-alanine aminotransferase (ALT) levels were unchanged in female rats fed medicated diets containing 10, 40, and 400 µg/Kg/day of mibolerone for 97 days indicating the drug is nonhepatotoxic at the level used which supported the above in vitro work.

REFERENCES

1. J. A. Miller, Cancer Res., 30, 559-576 (1970).
2. E. C. Miller and J. A. Miller, Pharmacol. Rev., 18, 805-838 (1966).
3. P. N. Magee and J. M. Barnes, Adv. Cancer Res., 10, 163-246 (1967).
4. R. O. Recknagel, Pharmacol. Rev., 19, 145-208 (1967).
5. G. J. Traiger and G. L. J. Plaa, Pharmacol. Exp. Therap., 183, 481 (1972).
6. J. R. Mitchell, D. J. Jollow, J. R. Gillette, and B. B. Brodie, Drug Metab. Dispos., 1, 418-423 (1973).
7. J. R. Gillette, J. R. Mitchell, and B. B. Brodie, Ann. Rev. Pharmacol., 14, 271-288 (1974).
8. J. R. Mitchell, D. J. Jollow, and J. R. Gillette, Israel J. Med. Sci., 10, 339-345 (1974).
9. J. R. Mitchell and D. J. Jollow, Israel J. Med. Sci., 10, 312-318 (1974).
10. J. R. Mitchell and D. J. Jollow, Gastroenterology, 68, 392-410 (1975).
11. D. M. Jerina and J. W. Daly, Science, 185, 573-582 (1974).
12. S. D. Nelson, M. R. Boyd, and J. R. Mitchell, Drug Metabolism Concepts: ACS Symposium Series, 44, 155-185 (1977).

13. S. D. Nelson, J. R. Mitchell, E. Dybing, and H. A. Sasame, Biochem. Biophys. Res. Comm., 70, 1157-1165 (1976).
14. O. H. Lowry, N. J. Rosenbrough, A. Farr, and R. Randall, J. Biol. Chem., 193, 265-271 (1951).
15. N. Nishikimi, Biochem. Biophys. Res. Comm., 63, 463-468 (1975).
16. T. R. Tephly, C. Webb, P. Trussler, F. Kniffer, E. Hasegawa, and W. Pipor, Drug Metab. Dispos., 1, 259-266 (1973).

CHEMISTRY OF METABOLITES OF THIOUREAS

Audrey E. Miller and Judith J. Bischoff

Department of Chemistry, University of Connecticut
Storrs, Ct. 06268

Thiourea and some monosubstituted thioureas are toxic and tumorogenic
while 1,3-diarylthioureas are not. There are strong indications that the
biological activity of these compounds is due to oxidation at sulfur followed by reactions in which the sulfur moiety is separated from the rest
of the molecule. That is, the metabolism of toxic thioureas, involves
this loss of sulfur while nontoxic compounds are either excreted unchanged
or undergo alternate metabolism.[1-3] The oxidation at sulfur is probably
catalyzed by FAD-containing monooxygenase (FADMO). Model studies show
that this enzyme oxidizes thioureas to the S-monoxide and S,S-dioxide.
The slower oxidation in solution to the corresponding trioxides does not
appear to be catalyzed by FADMO.[4,5]

One possible alternative metabolic pathway for nontoxic thioureas is
oxidation and subsequent reduction in vivo. For example, in vitro oxidations of thiourea, phenylthiourea and 1,3-diphenylthiourea all occur readily[6] and there are indications that the oxidized products can be reduced
by glutathione back to the parent compound in vivo.[7] The rapid reduction
of the related thiobenzamide-S-oxides to thiobenzamides by thiols also
gives support to this possibility.[8]

As a working hypothesis we suggest that the S-oxides of toxic thioureas react readily with biological nucleophiles to give intermediates which
lead to toxicity while the S-oxides of nontoxic thioureas either are reduced by glutathione to the thiourea more readily than they undergo nucleophilic substitution or undergo elimination to nontoxic metabolites.
Since most of the metabolic work has been done with thiourea, N-aryl-
thioureas and N,N'-diarylthioureas, representatives of these compounds,
thiourea, phenylthiourea and N,N'-diphenylthiourea were chosen as models.
Unfortunately the monoxides of these compounds are unstable except in solution,[4,9] and the dioxide of N,N'-diphenylthiourea is very unstable.[10] On
the other hand, the trioxides have proved to be quite stable compounds and
our initial studies have involved them. For these compounds the hypothesis predicts that the relative rates of substitution versus reduction and
elimination for aminoiminomethanesulfonic acid (AIMSO) and
N-phenylaminoiminomethanesulfonic acid (PAIMSO) will be considerably higher than for N,N'-diphenylaminoiminomethanesulfonic acid (DPAIMSO).

AIMSO and PAIMSO have been reported in the literature.[11,12] DPAIMSO
was synthesized in a similar manner and characterized by infrared spec-

troscopy and elemental analysis. The results of the reactions of these compounds with amino acids as nucleophiles are given in the Table. The reactions with AIMSO are finished within a few minutes, which can be seen not only by the precipitation of the product in most cases but also by tlc analysis of the reaction mixture at time intervals. The reactions with PAIMSO take a few hours and with DPAIMSO, a few days. Since the electronic effect of the phenyls should be to strengthen the electrophilicity of

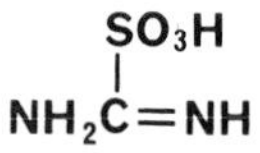

$$SO_3H$$
$$NH_2C{=}NH$$

AIMSO

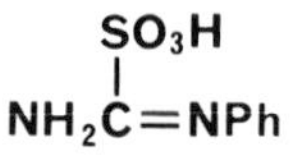

$$SO_3H$$
$$NH_2C{=}NPh$$

PAIMSO

$$SO_3H$$
$$PhNHC{=}NPh$$

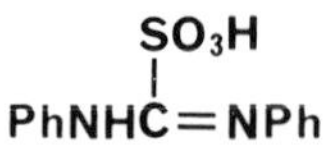

DPAIMSO

the carbon undergoing nucleophilic attack, it appears that the relative reactivities must be due primarily to a steric effect. While the relative reactivities of the three AIMSO's support the working hypothesis, an added bonus is that the reaction represents a convenient synthesis of many guanidino acids under exceptionally mild conditions.[13]

Table 1. Reactions with Amino Acids

Amino Acid	% Yield of Guanidino Compounds from		
	AIMSO	PAIMSO	DPAIMSO
β-alanine	75	70	25
DL-alanine	15		
p-aminobenzoic acid	80	65	25
γ-aminobutyric acid	50	none	50
δ-aminovaleric acid	55		
glycine	75	85	35
L-isoleucine	5		
L-leucine	none	none	
DL-methionine	60		
L-phenylalanine	45		
L-proline	none		
DL-serine	50		
DL-valine	55		

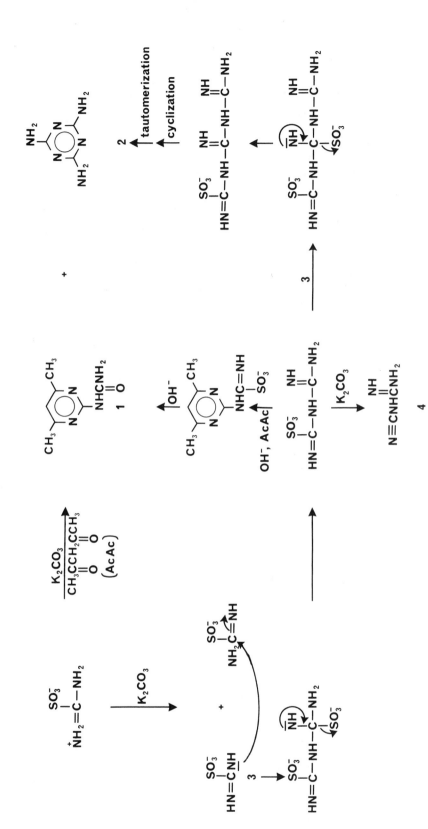

Fig. 1. Possible Reaction Pathways for AMSO Plus Potassium Carbonate and Trapping with 2,4-Pentanedione

Although more complex than the reactions with amino acids, the reactions of AIMSO, PAIMSO and DPAIMSO with base also support the working hypothesis. While AIMSO cannot be put on a reactivity scale in terms of nucleophilic substitution, its reactions in base are very fast. Also PAIMSO reacts considerably faster than DPAIMSO.

Initial experiments were run with an equivalent of 1 M potassium carbonate. PAIMSO reacted to give an 85% yield of phenylurea while DPAIMSO gave 33% of the corresponding potassium salt which precipitated rapidly from the reaction mixture as well as a mixture of diphenylthiourea and what appeared to be diphenylcarbodiimide. Even in 10^{-2} M aqueous solution, AIMSO and potassium carbonate gave a 50% yield of melamine, 2, and a 7% yield of cyanoguanidine, 4, the dimer of cyanamide. Tlc showed that there were traces of cyanamide in solution but a control showed that cyanamide was not the source of melamine. Nonetheless, an intermediate could be trapped; that is, from a solution 1 M in AIMSO, base and 2,4-pentanedione a 50% yield of 2-ureido-4,6-dimethylpyrimidine, 1, as well as 20% melamine were obtained. The figure shows a possible reaction pathway for each of these reactions of AIMSO.

Work is in progress to study the reduction of the AIMSO's with glutathione models as well as to synthesize the corresponding aminoiminomethanesulfinic acids and study their reactivities with nucleophiles and reducing agents.

REFERENCES

1. R. R. Scheline, R. L. Smith, and R. T. Williams, The Metabolism of ^{14}C- and ^{35}S-Labelled 1-Phenyl-2-thiourea and its Derivatives, J. Med. Pharm. Chem. 4: 109-135 (1961).

2. R. L. Smith and R. T. Williams, The Metabolism of Arylthioureas IV. p-Chlorophenyl- and p-Tolylthiourea, ibid. 147-161.

3. R. L. Smith and R. T. Williams, The Metabolism of 1,3-Diphenyl-2-thiourea (Thiocarbanilide) and its Derivatives, ibid. 97-107.

4. L. L. Poulsen, R. M. Hyslop, and D. M. Ziegler, S-Oxygenation of N-substituted Thioureas Catalyzed by the Pig Liver Microsomal FAD-Containing Monooxygenase, Arch. Biochem. Biophys., 198: 78-88 (1979).

5. L. L. Poulsen, R. M. Hyslop, and D. M. Ziegler, S-Oxidation of Thioureylenes Catalyzed by a Microsomal Flavoprotein Mixed-Function Oxidase, Biochem. Pharmacol. 23: 3431-3440 (1974).

6. L. L. Poulsen, Organic Sulfur Substrates for the Microsomal Flavin-Containing Monooxygenase, in: "Reviews in Biochemical Toxicology", Vol. 3, E. Hodgson, J. R. Bend, R. M. Philpott, Eds., Elsevier/North Holland, New York (1981), pp 33-49.

7. P. A. Krieter, D. M. Ziegler, K. E. Hill, R. F. Burk, Increased Biliary GSSG Efflux from Rat Livers Perfused with Thiocarbamide Substrates for the Flavin-Containing Monooxygenase, Mol. Pharm. 26: 122-127 (1984).

8. J. R. Cashman, R. P. Hanzlik, Oxidation and Other Reactions of Thiobenzamide Derivatives of Relevance to Their Hepatotoxicity, J. Org. Chem. 47: 4645-4650 (1982).

9. W. Walter and G. Randau, Oxidation Products of Thioamides XIX Thiourea-S-monoxides, *Ann*. 722: 52-79 (1969).

10. A. E. Miller and J. J. Bischoff, unpublished observations.

11. J. Boeseken, Studies on the Oxides of Thioureas II. The Trioxide of Thiourea, *Rec. Trav. Chim.* 55:1044-1045 (1936).

12. W. Walter and G. Randau, Oxidation Products of Thioamides XXI Thiourea-S-Trioxides (Guanylsulfonic Acid Betaines), *Ann*. 722: 98-109 (1969).

13. We appreciate helpful discussions with scientists at McNeil Pharmaceutical who have[14] also been actively involved in using AIMSO's as guanylating agents.

14. C. A. Maryanoff, R. C. Stanzione, J. N. Plampin and J. E. Mills, A Convenient Synthesis of Guanidines from Thioureas, in: "Abstracts of Papers", 190th National Meeting of the American Chemical Society, Chicago, Illinois, ORGN 112 (1985).

CONVERSION OF PHENYTOIN TO REACTIVE METABOLITES IN RAT LIVER MICROSOMES

Ruth E. Billings and Shirlette G. Milton

Department of Pharmacology
University of Texas Health Sciences Center
Houston, TX 77225

INTRODUCTION

The widely used anticonvulsant drug phenytoin (PHT) is primarily metabolized by aromatic hydroxylation. It is presumed that liver microsomal cytochrome(s) P450 catalyze the formation of an epoxide intermediate which gives rise to the major metabolite, 5-(4-hydroxyphenyl)-5- phenylhydantoin (4-HPPH)[1,2] and a 3,4-dihydrodiol metabolite (DHD).[3] In addition, PHT is converted to a catechol metabolite [5-(3,4-dihydroxyphenyl)-5-phenylhydan-toin,CAT].[4-7] CAT is formed from both 4-HPPH and DHD, although the predominate route is via 4-HPPH.[8]

Reactive metabolites have been repeatedly postulated to contribute to PHT toxicity, including its teratogenicity,[9] and covalent protein binding in vitro to liver microsomal proteins has also been observed.[10] The objective of the present investigation was to study further the formation of reactive PHT metabolites by measuring covalent binding to liver microsomal proteins. In particular, the aim of the studies was to investigate whether CAT is oxidized to reactive free radicals (semiquinone/quinone) and to determine, by the addition of modulators, whether such metabolites might contribute to the covalent binding of PHT. Other potential reactive metabolites include a 3,4-epoxide of PHT and a 2,3-epoxide of 4-HPPH. Covalent binding and CAT formation from the initial metabolites, 4-HPPH and DHD, were also determined. The studies described in this paper were conducted with rat liver microsomes as a model system, although preliminary studies[11] have also employed mouse liver microsomes and intact cell systems.

METHODS

PHT and 4-HPPH were purchased from Aldrich Chemical Co., Inc.,
(Milwaukee, WI). [4-^{14}C]PHT (48mCi/mmol) was obtained from New England
Nuclear (Boston, MA). [^{14}C]4-HPPH and [^{14}C]DHD were isolated from incubations of [^{14}C]PHT with rat liver cells as previously described.[7] [^{14}C]CAT
was obtained from incubations of [^{14}C]4-HPPH with rat liver microsomes in
the presence of ascorbate as described later. All isolations were done by
reverse-phase high pressure liquid chromatography (HPLC)[7] and the purity of
the products was determined by reinjection on the HPLC to be greater than
98%. Initial experiments showed that significant amounts of NADPH-
independent binding occurred with some lots of [^{14}C]PHT; this chemical was
therefore also purified by HPLC.

Male Sprague-Dawley rats (250-350g) were used for all experiments. The
rats were killed by decapitation and livers were immediately removed and
placed in three volumes of ice-cold 1.15% KCl containing 10mM phosphate
buffer, pH 7.4. Following homogenization, microsomes and cytosol were prepared by differential centrifugation by standard procedures. For incubations
the microsomal pellet was resuspended in 0.1M phosphate buffer, pH 7.4, and
used on the day of preparation. The standard incubation mixture contained
[^{14}C]-labeled substrate (approximately 2,000 dpm/nmol), microsomes (1-4mg
protein), 5mM $MgCl_2$, 50mM phosphate buffer (pH 7.4), 0.5mM $NADP^+$, 2.5mM
glucose-6-phosphate and 13ug glucose-6-phosphate dehydrogenase. The total
volume was 1-2ml, and the protein concentration was 1-2mg/ml. Protein
concentrations were determined by the Lowry method.[12]

The incubations were conducted at 37°C in a shaking water bath (120
osc/min). They were terminated by the addition of 5ml of ethyl acetate; 1ml
of 3M Tris buffer (pH 7.5) was added and the samples were extracted as
previously described.[7] The ethyl acetate extract was used for HPLC analysis
of DHD, 4-HPPH, and CAT.[7] The ethyl acetate extract was used for HPLC
analysis of DHD, 4-HPPH and CAT.[7] The metabolites were quantified by
collecting the HPLC effluent and determining the radioactivity in the peak
by liquid scintillation spectrometry.

After metabolite extraction, protein was precipitated by addition of
trichlorocetic acid (TCA) to give a TCA concentration of 10%. The TCA
precipitate was collected and washed twice with additional 10% TCA.

The precipitate was then extracted with methanol until the extracts contained only background levels of radioactivity; this was usually six extractions. The precipitate was then heated at 60° for 1 hour with 1N NaOH to digest the protein; this solution was subsequently analyzed for protein by the Lowry method[12] and for radioactivity by liquid scintillation spectrometry.

RESULTS AND DISCUSSION

Incubation of $[^{14}C]$CAT with rat liver microsomes results in covalent binding of radioactivity to protein (Table 1). While little binding was observed in the absence of cofactors, it was increased 5-fold by inclusion of NADPH in the incubation medium and 7-fold by the addition of a super-oxide-generating system, namely xanthine oxidase. Both the NADPH-supported and xanthine oxidase-supported binding were inhibited by superoxide dismutase; this indicates that reactive metabolite formation requires superoxide in both cases. The NADPH-supported reaction was also inhibited by the cytochrome P450 inhibitor, DPEA. These data suggest that cytochrome P450 participates in the formation of reactive metabolites of CAT by providing superoxide anion. The results are entirely consistent with previous observations with other catechols[13] and suggest that CAT is oxidized by superoxide anion to the semiquinone free radical which may then enter the semiquinone/quinone pathway; this is a self-perpetuating redox cycle in which the semiquinone ion is oxidized to the quinone in reactions which generate superoxide anion for use in oxidation of additional catechol to the semiquinone. Within this pathway, both the quinone and semiquinone ion are metabolites which may bind to proteins.[14]

That the superoxide-mediated binding of CAT is due to a semiquinone and/or quinone is also supported by the data shown in Table 2. Thus, nucleophilic trapping agents, reduced glutathione (GSH) and N-acetylcysteine, decreased by about 90% the covalent binding of CAT metabolites, but neither had an effect on overall CAT metabolism (as indicated by CAT remaining at the end of the incubation). These results suggest that the protective effect of these nucleophiles is due to the formation of conjugates with the reactive CAT metabolite(s) - a conclusion which was further supported by the observation that the TCA-soluble radioactivity (the fraction expected to contain such conjugates) was increased proportionate to the decrease in covalent binding (data not shown).

The addition of ascorbate nearly eliminated covalent binding and also increased CAT remaining at the end of the incubation. This result indicates further that the covalent binding is due to a semiquinone/quinone since ascorbate is expected to reduce the quinone back to CAT. Addition of liver cytosol was very effective in preventing the covalent binding of CAT. This effect could be due to the presence of reducing agents and/or nucleophiles in the cytosol or to enzymes such as quinone reductase which have been shown to ameliorate the covalent binding of phenol metabolites.[15]

Table 1. Conversion of CAT to Reactive Metabolites

Co-Factor Added	Addition	Covalent Binding pmol/min/mg protein
None	None	31
NADPH	None	140
NADPH	DPEA	86
NADPH	SOD	69
Xanthine Oxidase	None	230
Xanthine Oxidase	SOD	12

CAT (5 μM) was incubated with liver microsomes (1-2 mg protein) for 5-15 min. Additions were: NADPH, NADP$^+$ plus glucose-6-phosphate plus glucose-6-phosphate dehydrogenase as indicated in Methods; dichlorophenoxyethylamine (DPEA), 0.5 mM; superoxide dismutase (SOD), 100 ug; xanthine osidase, 0.1 units, plus xanthine, 0.25 mM.

Table 2.   Inactivation of Reactive Metabolites Formed with CAT

Addition	CAT Recovered	Covalent Binding
	-----nmol/incubation------	
None	0.82	2.20
Ascorbate	3.90	0.02
GSH	0.45	0.25
NAC	0.58	0.18
Cytosol	2.98	0.08

CAT (5 μM) was incubated with liver microsomes (1 mg protein) for 15 min in a volume of 1 ml. Additions were: Ascorbate, 2 mM; reduced glutathione (GSH), 1 mM; N-acetylcysteine (NAC), 1 mM; cytosol, 1 mg protein.

Table 3.   Comparison of Covalent Binding and CAT Recovery
with PHT and its Metabolites

Substrate (μM)	Covalent Binding pmol/min/mg protein	CAT Recovered (μM)
PHT (200)	30	0.2
4-HPPH (50)	75	3.9
DHD (50)	19	0.7
CAT (1)	41	0.3
CAT (5)	140	2.1

Substrates were incubated for 15 min or 5 min (CAT) with microsomes (4 mg protein) in a volume of 2 ml.

PHT and its initial metabolites, 4-HPPH and DHD, were all found to be converted to metabolites which covalently bind to protein, and this binding was dependent upon NADPH. Table 3 shows a comparison among the substrates of the extent of covalent binding and the amount of CAT recovered at the end of the incubation (as an indication of the amount of CAT formed with the various substrates). It is apparent from these data that the NADPH-dependent binding of CAT is sufficient to account for the binding observed with PHT and its initial metabolites. These data do not rule out, however, that other reactive metabolites such as epoxides may contribute to the binding. It is noteworthy, nevertheless, that more covalent binding was found with 4-HPPH than with PHT; additional data with different concentrations of the substrates support this conclusion (data not shown). This suggests that subsequent metabolites of the phenol may be the predominate binding species formed with PHT.

To further define the covalent binding of PHT metabolites, various modulators were added and the results are shown in Table 4. SKF 525A, a prototype cytochrome P450 inhibitor, reduced covalent binding as well as formation of all stable metabolites. Epoxide hydrolase inhibitors, trichloropropene oxide and cyclohexene oxide, decreased DHD recovery, but they had little effect on covalent binding. GSH, ascorbate, and cytosol all nearly eliminated covalent binding as previously observed with CAT (Table 2). Dialysis of the cytosol removed about 50% of its protective effect. This is similar to results with covalent binding of phenol metabolites,[15] and it was also observed with 4-HPPH (data not shown). Catechol O-methyltransferase was unexpectedly ineffective in reducing either CAT recovery or covalent binding even though it was present at saturating concentrations; additional experiments are required to substantiate that this enzyme is not effectively protective against catechol-mediated covalent binding. The free radical scavengers, alpha-tocopherol and mannitol, were much less effective than ascorbate in reducing the covalent binding; this suggests that the effectiveness of ascorbate is due to its ability to reduce reactive quinones/semiquinones to CAT rather than inactivating other types of free radicals, such as hydroxyl radical. These results, coupled with the lack of effectiveness of epoxide hydrolase inhibitors, indicate that the covalent binding observed with PHT may be due to oxidation products of CAT rather than to an epoxide as previously suggested.[9,16] These earlier studies employed TCPO in vivo or with intact cells, and it is possible that the increased covalent binding and/or toxicity observed in the presence of TCPO was due to its known effects as a glutathione-depleting agent.

936

Conclusive evidence, however, as to the nature of the binding species requires isolation of the conjugate; this work is in progress.

Table 4. The Metabolism of PHT to Stable and Reactive Metabolites: Effect of Modifiers

Addition	Metabolite (Precent Control)			
	DHD	4 HPPH	CAT	Covalent Binding
SKF525A	55	21	35	22
Trichloropropene Oxide	75	98	120	111
Cyclohexene Oxide	58	104	70	102
GSH	131	125	205	15
Ascorbate	117	97	665	12
Mannitol	94	106	100	49
Alpha-Tocopherol	100	101	117	87
COMT	76	101	47	73
Cytosol	87	136	409	4
Dialyzed Cytosol	63	105	197	62

Incubations (2 ml) were conducted for 15-30 min with 2 mg microsomal protein and 200 µM PHT. Additions were: SKF525A, 1 mM; trichloro-propene oxide, 0.5 mM; cyclohexeneoxide, 0.25 mM; COMT (catechol 0-methyltransferase + S-adenosylmethionine, 100 M); cytosol, 2 mg protein; GSH, 1 mM: ascorbate, 2 mM; Mannitol, 100 mM; alpha-toco-pherol, 200 µM. Cytosol was dialyzed for 24 hours against 100 mM phosphate buffer, pH 7.4. Catechol 0-methyltransferase was added at a concentration which gave maximal formation of the 0-methoxy deriv-ative.[11]

Whether the covalent binding of PHT contributes to its toxic effects, such as its teratogenicity, remains to be established. The observation that liver cytosol readily inactivates the reactive metabolite(s) may explain why PHT is not hepatotoxic in spite of the extensive binding observed with liver microsomes.

Covalent binding in isolated rat hepatocytes is also very low.[11] However, it has been found that, in pregnant mice, the covalent binding is greater in embryonic tissue than in maternal liver.[11]

In conclusion, the data reported in this paper show that CAT is readily oxidized by superoxide anion to metabolites, probably a semiquinone and/or quinone, which covalently bind to proteins. The extensive binding observed with low concentrations of CAT as well as the effects of modulators suggest that CAT contributes to the covalent binding from PHT. The role of the covalent binding in PHT toxicity remains to be determined.

REFERENCES

1. T. C. Butler, The metabolic conversion of 5,5-diphenylhydantoin to 5-(p-hydroxyphenyl)-5-phenylhydantoin, J. Pharmacol. Exp. Ther., 119:1 (1957).

2. E. W. Maynert, The metabolic fate of diphenylhydantoin in the dog, rat and man, J. Pharmacol. Exp. Ther., 130:275 (1960).

3. T. Chang, A. Savoy and A. J. Glazko, A new metabolite of 5,5-diphenylhydantoin (Dilantin), Biochem. Biophys. Res. Commun., 38:444 (1970).

4. T. Chang, R. A. Okerholm and A. J. Glazko, Identification of 5-(3,4-dihydroxphenyl)-5-phenylhydantoin: A metabolite of 5,5-diphenylhydantoin in rat urine. Anal. Lett., 5:195 (1972).

5. K. K. Midha, K. W. Hindmarsh, I. J. McGilveray and J. K. Cooper, Identification of urinary catechol and methylated catechol metabolites of phenytoin in humans, monkeys and dogs by GLC and GLC-mass spectrometry, J. Pharm. Sci., 66:1596 (1977).

6. S. A. Chow and L. J. Fischer, Phenytoin metabolism in mice, Drug. Metab. Dispos., 10:156 (1982).

7. R. E. Billings, Sex differences in rats in the metabolism of phenytoin to 5-(3,4-dihydroxyphenyl)-5-phenylhydantoin, J. Pharmacol. Exp. Ther., 225:630 (1983).

8. R. E. Billings and L. J. Fischer, Oxygen-18 incorporation studies of the metabolism of phenytoin to the catechol, Drug. Metab. Dispos., 13:312 (1985).

9. F. Martz, C. Failinger, III, and D. A. Blake, Phenytoin teratogenesis: Correlation between embryopathic effect and covalent binding of putative arene oxide metabolite in

gestational tissue, J. Pharmacol. Exp. Ther., 203:231 (1977).

10. C. Pantarotto, M. Arboix, P. Sezzano and R. Abbruzzi, Studies of
    5,5-diphenylhydantoin irreversible binding to rat liver
    microsomal proteins, Biochem. Pharmacol., 31:1501 (1982).

11. S. G. Milton, D. K. Hansen and R. E. Billings, Covalent binding of
    phenytoin metabolites in liver and embryonic tissue, Fed. Proc.,
    44:4116 (1985).

12. O. N. Lowry, N. J. Rosebrough, A. L. Farr and R. J. Randall, Protein
    determination with the Folin phenol reagent, J. Biol. Chem.,
    193:265 (1951).

13. E. Dybing, S. D. Nelson, J. R. Mitchell, H. A. Sasame and J. R.
    Gillette, Oxidation of alpha-methyldopa and other catechols by
    cytochrome P450-generated superoxide anion: Possible mechanism
    of methyldopa hepatitis, Mol. Pharmacol., 12:911 (1976).

14. E. C. Horning, J. P. Thenot and E. D. Helton, Toxic agents resulting
    from the oxidative metabolism of steroid hormones and drugs. J.
    Toxicol. Environ. Hlth., 4:341 (1978).

15. R. C. Smart and V. G. Zannoni, DT-Diaphorase and peroxidase
    influence the covalent binding of the metabolites of phenol, the
    major matabolite of benzene. Mol. Pharmacol., 26:105 (1984).

16. S. P. Spielberg, G. B. Gordon, D. A. Blake, E. D. Mellits and Dean
    S. Bross, Anticonvulsant toxicity in vitro: Possible role of
    arene oxides. L. Pharmacol. Exp. Ther., 217:386 (1981).

COMPARISON OF N-METHYLFORMAMIDE-INDUCED HEPATOTOXICITY AND METABOLISM IN

RATS AND MICE

K. Tulip, J.K. Nicholson and J.A. Timbrell

Toxicology Unit, Department of Pharmacology
The School of Pharmacy, University of London
29/39 Brunswick Square, London, WC1N 1AX

INTRODUCTION

N-Methylformamide (NMF) is a solvent and a potential anticancer drug
which is hepatotoxic in man (1); mice (2) and rats (3) causing centrilobular hepatic necrosis. However, the mechanism underlying this hepatotoxicity is currently unknown. NMF is extensively metabolised in mice in
vivo (4)(Fig. 1). However, no metabolism has been detected in vitro in
liver fractions or isolated hepatocytes (2). N-hydroxymethyl formamide
(NMFOH) has been indirectly identified as a urinary metabolite of NMF (4).
NMF causes depletion of non-protein thiols in vivo in mice (5) and in mouse
hepatocytes (2) and also causes lipid peroxidation in hepatocytes (2).
Although NMF has been reported to be hepatotoxic in rats (3) as well as
mice, in our hands, rats were very much less sensitive. Consequently we
have investigated the metabolism and toxicity of NMF in both rats and mice,
to determination the possible correlation between metabolism and toxicity.

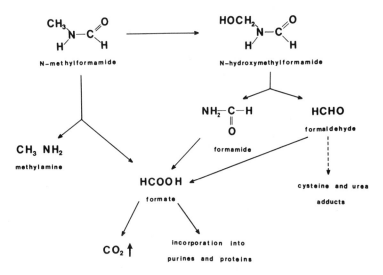

Fig. 1. Metabolism of N-methylformamide.

METHODS

Male Sprague Dawley rats (200-250 g) and male Balb/c mice (20-22 g) were given NMF and $[^{14}C$ methyl]-NMF dissolved in saline by i.p. injection. Animals dosed with $[^{14}C$-methyl]-NMF were housed in metabolism cages for the separate collection of urine, faeces and expired air. Urine was collected over ice and expired $^{14}CO_2$ was trapped in ethanolamine/methoxyethanol (1:2; v/v) as previously described (6). Radioactivity was determined by scintillation counting.

$[^{14}C$ methyl]-NMF (Sp. act 57.8 µCi mMol^{-1} radiochemical purity 97.4%) was synthesised as described (7). Radioactive metabolites in urine were determined by thin layer chromatography and quantitated by scintillation counting of chromatogram sections (radio t.l.c.) in two solvent systems.

Non labelled metabolites were quantitated by gas chromatography (g.c.)(NMF and formamide), high performance liquid chromatography (methylamine) and high resolution proton NMR (formate). Proton NMR spectra of urine were run on a Bruker 400 MHz instrument. Hepatic damage was assessed histologically by light microscopy and by determination of the serum transaminases, alanine aminotransferase (ALT) and aspartate aminotransferase (AST).

Hepatic soluble non protein sulphydryl (NPS) levels were determined by the Ellman method (8). In vitro covalent binding to microsomal protein was determined by exhaustive washing of the protein pellet after termination of the reaction and precipitation of the protein with methanol.

RESULTS

Rats treated with doses of 2000 mg/kg of NMF showed only slight hepatic damage both as determined histologically (Fig. 2) and enzymatically (Table 1) with a slight elevation of plasma AST. In mice however, doses of 300 mg/kg caused significant centrilobular hepatic damage and 400 mg/kg caused extensive necrosis (Fig. 3). Both plasma transaminases were significantly elevated and in a dose related fashion. (Table 2).

Table 1: Aspartate transaminase and alanine transaminase plasma levels 24 hr after administration of N-methylformamide (i.p.) to Sprague-Dawley rats (mean + SEM).

dose mg kg^{-1}	AST units l^{-1}	ALT units l^{-1}
control	3.1±1.2	19.3±4.3
1000	29.6±6.3	18.4±2.1
2000	9.8±1.0	22.8±7.4

Table 2 : Aspartate transaminase and alanine transaminase plasma levels 24 hr after administration of N-methylformamide (i.p.) to Balb/c mice (mean + SEM).

dose mg kg^{-1}	AST units l^{-1}	ALT units l^{-1}
control	2.87±0.59	17.87±1.82
100	1.76±1.40	14.9±5.30
200	14.30±5.19	74.28±24.06
300	24.38±4.38	126.7±41.07
400	49.5±14.8	145.0±110.3
800	174.0±31.0	203.3±18.1

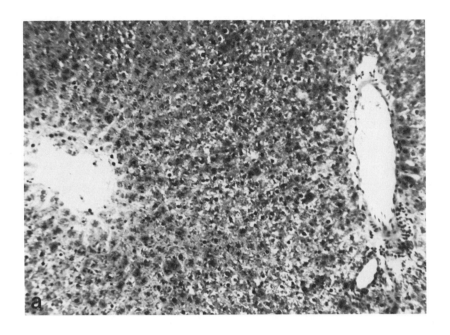

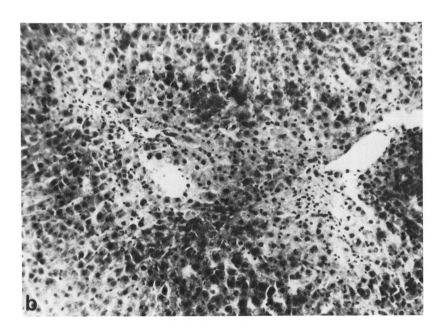

Fig. 2 Sections of liver from Sprague Dawley rats treated with
a) saline or b) NMF (1000 mg/kg).

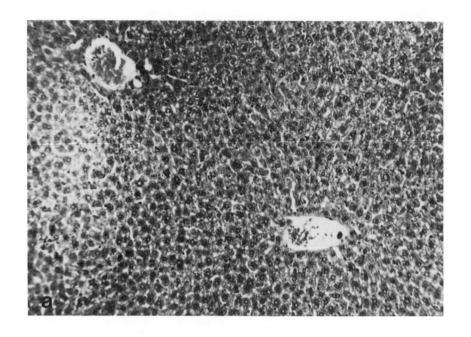

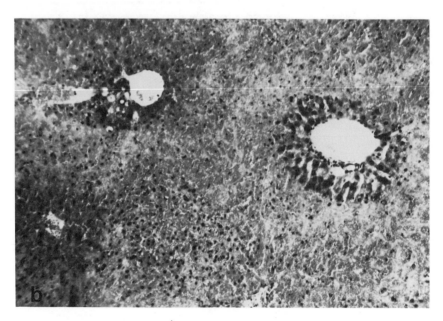

Fig. 3. Sections of liver from Balb/c mice treated i.p. with a) saline or b) NMF (400 mg/kg).

Hepatotoxic doses of NMF (400 mg/kg) also produced a depletion of hepatic non-protein sulphydryl levels (NPS) in mice but not in rats (Fig 4). In mice the depletion of NPS was dose dependent (Fig. 5)

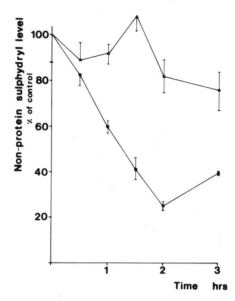

Fig. 4. Effect of NMF on hepatic non-protein sulphydryls after doses of 400 mg/kg to mice (●) or 1000 mg/kg to rats (▲). Mean ± SE.

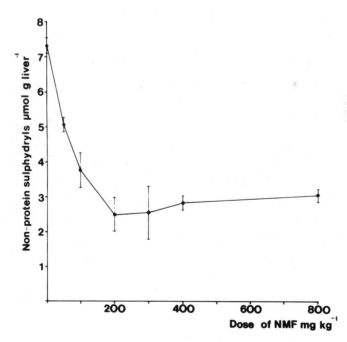

Fig. 5. Effect of dose of NMF on depletion of hepatic non-protein sulphydryls in mice. Mean ± SE.

Disposition of radioactivity in rats and mice after a dose of ^{14}C
NMF is shown in Fig. 6. Mice excreted considerably more radioactivity in
urine, expired air and faeces in 24 hrs post dosing than rats. However,
by 48 hrs post-dosing the differences in excretion between rats and mice
had narrowed and rats excreted 53% of the dose compared to 56% in mice.

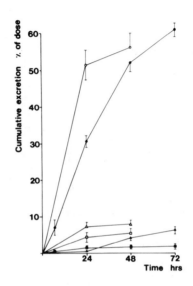

Fig. 6. Cumulative urinary excretion of
radioactivity in urine (O), faeces (■) and
expired air (Δ) after administration of
^{14}C-NMF to rats (1000 mg/kg; closed symbols)
or mice (300 mg/kg; open symbols). Mean + SE.

Direct examination of the urine of rats and mice dosed with NMF using
high resolution proton NMR revealed a number of metabolites identifiable
by their chemical shift values and by comparison with authentic standards.
Compounds detected in both species were NMF, methylamine and formamide and
in rats formate was also detected (Fig. 7). A resonance characteristic of
the formyl proton of NMF OH was also detected in urine from both species,
but the methylene proton signal would be obscured by the water resonance
therefore complete characterisation is difficult. Other resonances were
detected in urine from treated animals which are characteristic of
N-acetyl groups and an ABX resonance system characteristic of the
cysteinyl residue was also detected. The demonstration of the previously
undetected metabolites, methylamine and (formamide by) NMR illustrates the
potential of this technique.

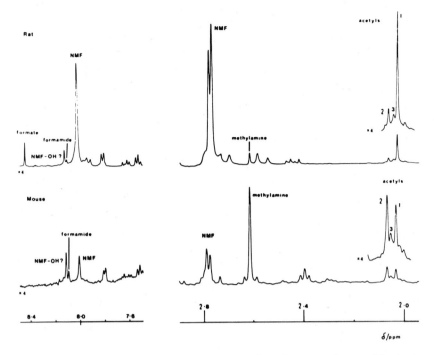

Fig. 7. 400 MHz proton NMR spectra of urine samples collected 24 hrs after administration of NMF to rats (1000 mg/kg) or mice (300 mg/kg).

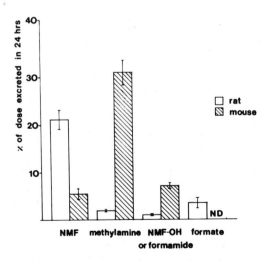

Fig. 8. Urinary metabolites of NMF excreted 24 hrs afterdosing to rats (1000 mg/kg) or mice (300 mg/kg) Mean ± SE.

Analysis of these metabolites by g.c. hplc and radio t.l.c. indicated that in the rat 21% of the dose is excreted as unchanged NMF (Fig. 8), quantitated by g.c. (and radio t.l.c). In mice however, only 5.5% of the dose is excreted as unchanged NMF in the urine (Fig. 8). Rats excreted a small amount of methylamine (1.9% dose) quantitated by hplc (and radio t.l.c.) and formate detected and quantitated by high resolution proton NMR (3.8% dose)(Fig. 7). In the mouse the major urinary metabolite was methylamine accounting for 31% of the dose. No formate was detected in mouse urine. A radioactive metabolite detected by t.l.c. and corresponding to NMFOH was found in rat and mouse urine. A quantitatively similar amount of formamide, 1% of the dose in rats, 7% of the dose in mice was detected by g.c. probably arising from thermal breakdown of NMFOH which is known to take place in the gas chromatograph.

Incubation of [^{14}C methyl]-NMF with mouse liver microsomes resulted in covalent binding of radiolabel to microsomal protein (Table 3). This binding was greater at 5 mM compared to 1 mM and was dependent upon an NADPH generating system. The binding was also inhibited by glutathione. Further studies revealed that enzyme activity was necessary as heat inactivated microsomes were unable to catalyse covalent binding (Table 4). SKF 525A did not significantly inhibit covalent binding but methimazole caused a significant reduction in covalent binding to microsomal protein (Table 4).

Table 3: Covalent binding of ^{14}C methyl NMF to mouse liver microsomes.

Treatment	Covalent binding nmoles mg protein^{-1}	
	1 mM NMF	5 mM NMF
+ NADPH	0.67±0.05	1.37±0.12
− NADPH	0.02±0.01	0.15±0.02
GSH (2 mM) + NADPH	0.03 (0.02, 0.04)	0.15±0.02

^{14}C methyl NMF was incubated with freshly prepared mouse liver microsomes (1.0 mg protein/ml for 60 mins in the presence or absence of an NADPH generating system.

Table 4: Covalent binding of ^{14}C methyl NMF to mouse liver microsomes.

Treatment	Covalent binding nmoles mg protein^{-1} 2.5 mM NMF
+ NADPH	1.17±0.09
− NADPH	0.05±0.004
Heat inactivated microsomes + NADPH	0.06±0.01
SKF 525A (0.25 mM) + NADPH	1.06±0.03
Methimazole (0.25 mM) + NAPH	0.40±0.04

^{14}C methyl NMF was incubated with freshly isolated mouse liver microsomes (2 mg protein/ml) for 60 mins in the presence or absence of an NADPH generating system and a possible inhibitor.

DISCUSSION

The results clearly show that NMF is considerably more hepatotoxic to mice than to rats. Even doses of 2000 mg/kg to rats cause little hepatic damage yet doses of 400 mg/kg given to mice cause very extensive centrilobular necrosis, with in some cases only a narrow band around the portal tract spared (Fig. 2 & 3). The plasma transaminase values also reflect this, showing less elevation in the rat after 2000 mg/kg compared with a 20 X increase in AST and 10 X increase in ALT in the mouse after 400 mg/kg (Tables 1 & 2). This difference in susceptibility is also shown in the depletion of hepatic non protein sulphydryls (NPS) after NMF is administered. After 1000 mg/kg NMF is administered to rats there is no significant depletion of hepatic NPS over 3 hours (Fig. 4). In the mouse however, a dose of 400 mg/kg NMF rapidly depleted NPS to about 20% of control levels in 2 hours. This depletion was dose dependent, and the dose for maximal depletion is between 100 mg/kg and 200 mg/kg as previously shown by Gescher et al (5). This is also the threshold dose range for hepatotoxicity (Table 2). These results suggest that depletion of glutathione is an essential prerequisite for the toxicity of NMF. The species difference is further manifested in the metabolism; with the mouse metabolising ^{14}C NMF more rapidly to $^{14}CO_2$ and urinary metabolites and so excreting less unchanged compound in the urine in 24 hrs. The major metabolites in the mouse, methylamine and NMFOH are both produced in greater quantities than in the rat. Methylamine presumably results from hydrolysis of NMF and should yield equimolar amounts of formate (Fig. 1). This formate was detected in the urine of rats dosed with NMF, but not mice, presumably reflecting species differences in the further metabolism of formate. The importance of the hydrolytic pathway in the hepato-toxicity is currently unknown but methylamine itself is not hepatotoxic in mice (unpublished observations). The other metabolic pathway important in the mouse is hydroxylation of the N-methyl group (Fig. 1). This leads to the carbinolamine intermediate, which although relatively stable, will rearrange to yield formamide on g.c. Formamide was detected and quantitated by g.c. and the amount was similar to that of a radiolabelled metabolite having the same R_f value as the synthetic NMFOH. This is consistent with NMFOH being the radiolabelled metabolite and the precursor of the formamide detected on g.c. Again mice produced more of this metabolite than rats but currently the role of this metabolic pathway in the hepatotoxicity of NMF is unknown. The NMFOH neither causes lipid peroxidation nor depletes glutathione in hepatocytes (2). The in vitro studies with mouse liver microsomes revealed that micrososmal enzyme mediated metabolism yields a metabolite which binds covalently to protein and this is decreased by glutathione and methimazole. The decrease in binding caused by glutathione may be the result of reaction of the reactive intermediate with glutathione, the latter acting as an alternative nucleophile.

Methimazole is an inhibitor of microsomal amine oxidases (9) and clearly decreases the binding substantially, whereas the cytochrome P-450 inhibitor SKF 525A does not significantly decrease the binding. The covalent binding to protein therefore may be the result of nitrogen oxidation to yield a reactive intermediate. Whether this is a metabolic pathway in vivo remains to be determined.

In conclusion, the results are consistent with the contention that the greater hepatotoxicity of NMF in mice than rats is due to greater metabolism in the susceptible species, possibly giving rise to a reactive intermediate which reacts covalently with and depletes, non protein sulphydryl groups. The identity of the metabolite is currently unknown.

REFERENCES

1. W.P.L. Myers, D. Karnofsky and J.H. Burchenal. The hepatotoxic action of N-methylformamide in man. Cancer, 9: (1956).

2. H. Whitby, A Gescher and L. Levy. An investigation of the mechanism of hepatotoxicity of the antitumour agent N-methylformamide in mice. Biochem. Pharmacol. 33: 295 (1984).

3. I. Lundberg, S. Lundberg and T. Kronevi. Some observations on dimethylformamide hepatotoxicity. Toxicology 22: 1 (1981).

4. C. Brindley, A. Gescher, E.S. Harpur, D. Ross, J.A. Slack, M.D. Threadgill and H. Whitby. Studies on the pharmacology of N-methylformamide in mice. Cancer Treatments Reports 66: 1957 (1982).

5. A. Gescher, N.W. Gibson, J.A. Hickman, S.T. Langdon, D. Ross and G. Atassi. N-Methylformamide: Antitumour activity and metabolism in mice. Brit. J. Cancer 45: 843 (1982).

6. H. Jeffay and J.Alvarez. Liquid scintillation counting of carbon carbon-14. Anal. Chem. 33: 612 (1961).

7. M.D. Threadgill and E.N. Gate. Labelled compounds of interest as antitumour agents I: N-Methylformamide and N,N-dimethylformamide. J. Lab. IId Cmpds. Radiopharm. XX: 447 (1983).

8. G.L. Ellman. Tissue sulphydryl groups. Arch. Biochem. Biophys. 82: 70 (1955).

9. R.A. Prough and D.M. Zeigler. The relative participation of liver microsomal amine oxidase and cytochrome P450 in N-demethylation reactions. Arch. Biochem. Biophys. 180: 363 (1977).

EPOXIDATION OF 1,3-BUTADIENE IN LIVER AND LUNG TISSUE OF MOUSE, RAT,

MONKEY AND MAN

U. Schmidt and E. Loeser

Bayer AG
Institute of Toxicology
D-5600 Wuppertal 1

ABSTRACT

When 1,3-butadiene is incubated with <u>liver</u> postmitochondrial fractions from mouse, rat, monkey or man and a NADPH-regenerating system the formation rate of butadiene monoxide is different in the four species. With the exception of rhesus monkey the amount of epoxide is proportional to the monooxygenase activity. The sequence of epoxide formation is $B6C3F_1$-mouse, Sprague Dawley rat, man, rhesus monkey. The relation between mouse and monkey was about 7:1. When 1,3-butadiene is incubated with homogenates from <u>lung</u> tissue, only tissues from mouse and rat produces measurable butadiene monoxideconcentrations. The monooxygenase activity in lung tissue of the mouse was only 1/30 that in mouse liver. By contrast, lung tissue formed the epoxide concentrations comparable to those formed by liver tissue, whereas monkey and human lung tissue did not produce any measurable levels of butadiene monoxide. The data might suggest that the results of recent rodent inhalation studies with 1,3-butadiene could not automatically be extrapolated to man.

INTRODUCTION

1,3-Butadiene, is used as a chemical intermediate or as monomer in the production of various polymers, e.g. in the synthetic rubber industry.

Butadiene is mutagenic in the Ames-test in the presence of S-9 mix only (de Meester et al. 1980).

Malvoisin et al. (1979) have demonstrated butadiene monoxide as the main metabolite of butadiene by rat liver microsomes in vitro. Malvoisin and Roberfroid (1982) have studied the roles of microsomal mixed function oxidases and epoxide hydrolases in the metabolic fate of butadiene monoxide. The monoxide has also been detected in exhaled air of rats exposed to butadiene (Bolt et al. 1983, Filser et al. 1984).

Recent long-term inhalation studies in rats (Hazleton, 1981, Loeser, 1982) and mice (NTP, 1984) have demonstrated that rats survive higher butadiene exposure levels than mice. The inhalation exposure of mice to 625 or 1250 ppm butadiene resulted in excessive high mortality in early stages of the long-term study. Fatal tumors occurred at high incidence in all exposed groups. Rats, however, survived levels up to 8000 ppm for approx. 2 years showing minimal toxic or tumor response.

AIM OF THE STUDY

Since mutagenicity studies as well as in vitro and in vivo metabolic studies have demonstrated the formation of the butadiene monoxide as a critical reactive intermediate, we initiated studies to demonstrate possible qualitative and quantitative differences in butadiene epoxidation and inactivation of the epoxide by epoxide hydrolases and/or glutathione trans-ferases, using postmitochondrial prepartions from liver and lung tissues of mice, rats, monkey and man. The data obtained may serve to develop a suitable extrapolation from animal data to human.

MATERIALS

Chemicals

1,3-Butadiene, technical grade of 99,6 % purity, was obtained from Erdoelchemie GmbH, Dormagen; butadiene monoxide and butadiene diepoxide, both of 97 % purity, was from EGA Chemie. All other chemicals, analytical grade, were obtained commercially.

Tissues of lung and liver

Tissue material was obtained from Sprague Dawley rats (Charles River, Germany), Wistar rats and NMRI-mice (Winkelmann, Germany), B6C3F$_1$-mice (Charles River, Germany), rhesus monkeys (own breed). Human lung and liver tissue has been obtained upon surgery.

Tissue homogenates

Animals were sacrificed and liver and lung tissues (1 - 1.5 g wet weight) were quickly excised. They were given into ice cold 0.15 M KCL solution and were homogenized while being chilled. Lung tissues of 5 mice were pooled. The human material was processed after surgery as quickly as possible. The homogenates were centrifuged at 4°C and 10,000 g for 20 min. 1 Milliliter of the obtained supernatant was equivalent to approx. 200 mg tissue wet weight.

METHODS

Monooxygenase activity

Monooxygenase activity was determined in the 10,000 g supernatants from liver ($\hat{=}$ 1 mg tissue) and lung ($\hat{=}$ 10 mg tissue) homogenates, using 7-ethoxycoumarin as substrate according to Ullrich and Weber (1972), with modifications introduced by Greenle and Poland (1978), Aitio (1978). O-Deethylase activities were calculated as nmol 7-hydroxycoumarin/g wet tissue per min.

Epoxide hydrolase

The activity of the epoxide hydrolase was determined according to the method of Guiliano et al. (1980), which uses 3-(p-nitrophenoxy)-1,2-propeneoxide as substrate. The hydrolysis product, 3-(p-nitrophenoxy)-1,2-propanediol, was been determined by HPLC. The epoxide hydrolase activity was calculated as the mean of two determinations as nmol 3-(p-nitrophenoxy)-1,2 propanediol/g tissue per min.

Epoxidation of 1,3-butadiene in liver and lung homogenates

Incubation mixtures, in head-space vials (about 30 ml), contained 500 µl tissue homogenate (from 100 mg tissue) and a NADPH-regenerating system.

After 10 min, preincubation at 37°C the headspace viales were capped and 1 ml 1,3-butadiene gas (ca. 2,41 mg) was added with a Hamilton gas syringe. The final concentration in the vial was about 30,000 ppm butadiene. Maximally 3 gas samples of 0.5 ml each were withdrawn at different intervals and injected into a gas chromatograph to analyse for butadiene monoxide. The separation was done on a 4 m glass column (diameter 1/4", with 10% P 2000) at 70°C column temperature. Under these conditions (using an FID) the following retention times were recorded: 1,3-butadiene, r_t =1.1 min; butadiene monoxide, r_t = 5.9 min. The limit of detection was about 0.1 µg butadiene monoxide/30 ml air, corresponding to 2.5 ppm. The butadiene monoxide was calculated as nmol butadiene monoxide produced/g tissue.

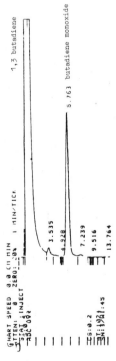

Fig. 1: Typical gaschromatogram of a incubationsample of 1,3 butadiene with liver postmitochondrial preparation ($B_6C_3F_1$-mouse). Conditions: glass column 4m, diameter 1/4" with 10 % P 2000 at 70°C column temperature, injection 0,5 ml head-space, (FID).

TABLE 1

ACTIVITIES OF BUTADIENE OXIDASE, ETHOXYCOUMARIN DEETHYLASE AND EPOXIDE HYDROLASE
IN LIVER POSTMITOCHONDRIAL PREPARATIONS

| | LIVER | | | | | | | | |
| | BMO[1] $\left[\dfrac{nmol}{g \cdot 45\ min.}\right]$ | | | EOD[2] $\left[\dfrac{nmol}{g \cdot min.}\right]$ | | | EH[3] $\left[\dfrac{nmol}{g \cdot min.}\right]$ | | |
	N	$\bar{x}$	$\pm$S.D.	N	$\bar{x}$	$\pm$S.D.	N	$\bar{x}$	$\pm$S.D.
MICE, MALE	5	125	49	5	17	7	5	430	74
MICE, FEMALE	5	217	33	5	37	6	5	340	26
RAT, MALE	5	82	45	5	15	4	5	320	133
RAT, FEMALE	5	67	11	5	7	2	5	220	102
MONKEY, MALE	2	32	3	2	75	27	2	5150	700
MONKEY, FEMALE	2	34	6	2	79	20	2	6200	600
HUMAN	1	79	-	1	17	-	1	6800	-

1) BMO = Butadiene monoxidase
2) EOD = Ethoxycoumarin deethylase
3) EH = Epoxide hydrolase

TABLE 2

ACTIVITIES OF BUTADIENE OXIDASE, ETHOXYCOUMARIN DEETHYLASE AND EPOXIDE HYDROLASE
IN LUNG POSTMITOCHONDRIAL PREPARATIONS

| | LUNG | | | | | | | | |
| | BMO[1] $\left[\dfrac{nmol}{g \cdot 45\ min.}\right]$ | | | EOD[2] $\left[\dfrac{nmol}{g \cdot min.}\right]$ | | | EH[3] $\left[\dfrac{nmol}{g \cdot min.}\right]$ | | |
	N	$\bar{x}$	$\pm$S.D.	N	$\bar{x}$	$\pm$S.D.	N	$\bar{x}$	$\pm$S.D.
MICE, MALE	2*	251	28	2	2.3	0.1	2	40	6
MICE, FEMALE	4*	198	14	2	1.3	0.4	-	-	-
RAT, MALE	4	41	11	4	1.4	0.2	2	20	4
RAT, FEMALE	3	27	12	3	1.2	0	1	30	-
MONKEY, MALE	2	N.D.	-	2	0.02	-	2	80	33
MONKEY, FEMALE	3	N.D.	-	3	0.02	-	3	110	38
HUMAN	1	N.D.	-	1	0.01	-	1	100	-

*   5 lungs each were pooled.
    n.d. = not detectable
1) BMO = Butadiene monoxidase
2) EOD = Ethoxycoumarin deethylase
3) EH = Epoxide hydrolase

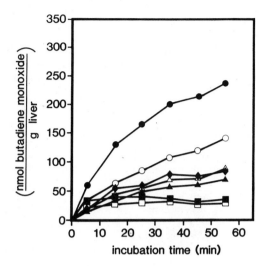

Fig. 2: Formation fo butadiene monoxide from 1,3-butadiene in liver
tissue (10 000 g fraction) of rat (△ male, ▲ female), mouse
(○ male, ◉ female), monkey (□ male, ■ female), man (♦).

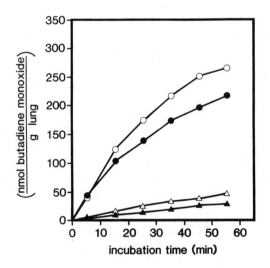

Fig. 3: Formation of butadiene monoxide from 1,3-butadiene in lung
tissue (10 000 g fraction) of rat (△ male, ▲ female) and
mouse (○ male, ● female).

RESULTS

As a monooxygenase marker the ethoxycoumarin-o-deethylase (EOD) has
been determined in the liver and lung homogenates of the different species
examined. Similarly, epoxide hydrolase (EH) activities have been measured.
The results are compiled in table 1 (liver) and table 2 (lung).

The formation of butadiene monoxide by tissue homogenates from the
different species has been followed up to 60 min. The data are shown
in fig. 2 (liver) and fig. 3 (lung) for homogenates from the mouse, rat,
monkey and man.

Interspecies comparison

From an interspecies comparison (fig. 2) it appears that the female
mouse liver produced the highest amounts of butadiene monoxide, next
followed by the male mouse liver. The formation of the monoxide in rat and
human liver was less, but was similar in these both species. Rhesus monkey
liver produced the lowest amounts of epoxide among the species tested.

With respect to EH activities monkey and human liver, some 20 times
higher activities were measured than in the liver of rodents.

With the exception of the monkey liver there was an agreement between
the relative EOD activities and monoxide formation in vitro. However, monkey
liver was exceptionally inactive in producing butadiene monoxide, despite
very high EOD activities in this tissue. This might suggest dissimilar
patterns of isoenzymes of monooxygenases and/or epoxide hydrolases in
monkey and man.

Although the EOD activities were much similar in the lung tissues of
rats and mice, butadiene monoxide formation in mouse lung homogenates was
some 5 - 6 times higher than in rat lung preparations. In agreement with
the very low EOD activities in monkey and human lung tissue no butadiene
monoxide could be detected on incubation of butadiene with these tissues.

CONCLUSION

The present data demonstrate a high capability of liver and especially
lung postmitrochondrial preparations of mice to produce butadiene monoxide
after incubation with butadiene. With rat liver and lung clearly less of
butadiene monoxide was produced, which may account for the different
response in toxicity and carcinogenicity of both species in outstanding
long-term bioassays. Monkey and human postmitochondrial liver preparations
catalysed only a slow formation of the epoxide; with lung preparations no
epoxide was detected. These results could to be of some importance in
suggesting that the rodent studies with 1,3-butadiene, especially that with
mice, may not truly reflect the human situation. Further studies are therefore necessary to quantitate the possible formation of butadiene epoxide in
various species in vivo and to evaluate the contribution of different
tissues, e.g. lung and liver, to the bioactivation of butadiene.

REFERENCES

Aitio, A., 1978, A simple and sensitive assay of 7-ethoxycoumarin
    deethylation. Anal Biochem 85:488.

Bolt, H,M., Schmiedel, G., Filser, J.G., Rolzhäuser, H.P., Lieser, K., Wistuba, D., and Schurig, V., 1983, Biological activation of 1,3-butadiene to vinyl oxirane by rat liver microsomes and expiration of the reactive metabolite by exposed rats. J Cancer Res Clin Oncol , 106:112.

Filser, J.G. and Bolt, H.M., 1984, Inhalation pharmacokinetics based on gas uptake studies: VI. Comparative evaluation of ethylene oxide and butadiene monoxide as exhaled reactive metabolites of ethylene and 1,3-butadiene in rats. Arch Toxicol 55:219.

Giuliano, K.A., Lau, E.P. and Fall, R.R., 1980, Simplified liquid chromatographie assay for epoxidhydrolase. J Chromatogr 202:447.

Greenle, W.F. and Poland, A., 1978, An improved assay of 7-ethoxycoumarin-o-deethylase activity: induction of hepatic enzyme activity in C57BL/6J and DBA/2 J mice by phenobarbital, 2-methylcholanthrene and 2,3,7,8-tetrachlorodibenzo-p-dioxin. J Pharmacol Exp Ther 205: 596.

Hazleton Laboratories Europe LTD, 1981, The toxicity and carcinogenicity of butadiene gas administered to rats by inhalation for approximately 24 months.

Loeser, E., 1982, IISRP sponsored toxicity studies on 1,3-butadiene. Paper presented at the Twenty-Third Annual Meeting of International Institute of Synthetic Rubber Producers.

Malvoisin, E., Lhoest, G., Poncelet, F., Roberfroid, M. and Mercier, M., 1979, Identification and quantitation of 1,2-epoxybutene-3 as the primary metabolite of 1,3-butadiene. J Chromatogr 178:419.

Malvoisin, E., Mercier, M. and Roberfroid, M., 1982, Enzymatic hydration of butadiene monoxide and importance in the metabolism of butadiene. Adv Exp Med Biol 138A:437.

De Meester, C., Poncelet, F., Roberfroid, M. and Mercier, M., 1980, The mutagenicity of butadiene towards Salmonella Typhimurium. Toxicol Lett 6:125.

National Toxicology Program, Technical Report, 1984, Series No. 288. Toxicology and carcinogenesis: Studies of 1,3-butadiene in B6C3F$_1$ mice (inhalation studies).

Ullrich, V. and Weber, P., 1972, The o-dealkylation of 7-ethoxycoumarin by liver microsomes. A direct fluorimetric test. Hoppe-Seyler's Z Physiol Chem 353:1171.

ENZYME CATALYSED CLEAVAGE OF THE N-N BOND OF N-NITROSAMINES

K.E. Appel, M. Bauszus, I. Roots*, C.S. Rühl, and
A.G. Hildebrandt

Max von Pettenkofer Institute, German Health Office, D-1000
Berlin 33 and* Klinikum Steglitz, D-1000 Berlin 41 (F.R.G.)

The bioactivation pathway of N-nitrosamines is very well understood and has
been investigated. The rate limiting step is the cytochrome P-450 dependent
hydroxylation on the α-carbon (1, 2). Very little is known however about possible
inactivation pathways.

We have looked especially for a possible denitrosation pathway of N-nitrosamines.
We could show that after incubation of various N-nitrosamines with liver microsomes from mice or rats and an NADPH-regenerating system nitrite was formed.
On the basis of induction and inhibition experiments it was concluded that cytochrome P-450 might be involved in this metabolic pathway (3, 4, 5). Some inhibition experiments are shown in table 1.

Table 1. Generation of nitrite from nitrosodimethylamine in mouse liver
microsomes. The concentration of subtrate was 35 mM, that of protein
was 2 mg/ml.

	Denitrosation (nmole nitrite/ mg protein x 50 min)	Per cent inhibition
Basic system	8.3 ± 1.3	---
Boiled microsomes	0.02 ± 0.01	99
NADPH omitted	0.05 ± 0.03	99
Under CO (30 sec)	5.4 ± 0.5	34
Under nitrogen	0.07 ± 0.03	99
Piperonyl butoxide (0.2 mM) added	3.8 ± 0.3	54
Metyrapone (0.1 mM) added	4.7 ± 0.5	43

In the following experiments it should be confirmed, that cytochrome P-450 can catalyse this novel pathway in nitrosamine metabolism. We therefore used a reconstituted system comprising purified cytochrome P-450 and NADPH cytochrome P-450 reductase. Both components were isolated from pig liver, which had been induced by phenobarbital. Particular emphasis was given to the possibility that NADPH cytochrome P-450 reductase alone could also catalyse this metabolic step. From the kinetics of nitrite formation from nitrosomorpholin and di-n-propylnitrosamine it can be demonstrated, that NADPH cytochrome P-450 reductase alone was not able to catalyse nitrite generation. The complete reconstituted system only was able to catalyse nitrite formation, which demonstrates that the hemoprotein is necessarry for this metabolic step (6).

Figure 1 a. demonstrates optical difference spectra of nitrosomorpholine obtained with the reconstituted monooxygenase system after reduction with NADPH. The

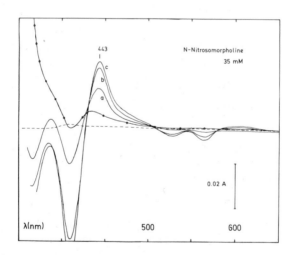

Figure 1a. Difference spectra obtained after addition of NNM to the reconstituted monooxygenase system. Both cuvettes contained 3.1 nmol P-450 PLM IV and 0.56 nmol NADPH P-450 reductase per ml Tris/KCI buffer (pH 7.4). After establishing a baseline (----), nitrosomorpholine solution was added to the sample cell and an equal amount of solvent to the reference cell. When the oxidized spectrum had been recorded (o-o) NADPH was added to both cuvettes (5 mM). The reduced spectra were repeatedly recorded (a, b, c; cycle time 2 min).

peak at around 444 nm represents the ferrous cytochrome P-450 NO-complex. It is known that the ferrous cytochrome P-450 NO-complex is highly unstable at room temperature. Under anaerobic conditions it is kinetically converted to a ferrous cytochrome P-420 NO-complex.

The same spectrum, the cytochrome P-450 NO-complex (peak at 444 nm) can be obtained with nitrite which is reduced to nitric oxide under these conditions (Figure 1 b.).

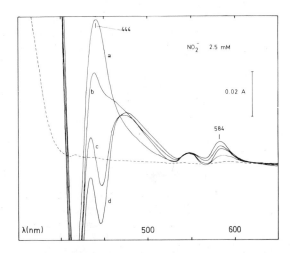

Figure 1b. Optical difference spectra after addition of nitrite to the reconstituted monooxygenase system. Both cuvettes contained 3.1 nmol P-450 PLM IV and 0.56 nmol NADPH P-450 reductase per ml Tris/KCl buffer (pH 7.4). 10 μl of NaNO$_2$ solution was added to the sample cell and an equal amount of water to the reference cell before the oxidized spectrum was recorded (---). After addition of NADPH in both cuvettes (5 mM) the reduced spectra were repeatedly recorded (a, b, c, d; cycle time 2 min).

Electroparamagnetic resonance (EPR) is sensitive to changes in the coordination sphere of the iron of cytochrome P-450 and can reveal ligand binding reactions. The interaction of a nitrosamine with ferric cytochrome P-450 produced characteristic spectral changes (Figure 2.).

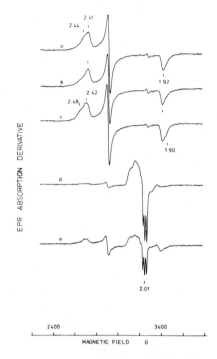

Figure 2. Effect of di-n-propylnitrosamine (DPNA) and diphenylnitrosamine (NDphA) on the EPR spectrum of microsomal cytochrome P-450. Curve a represents the spectrum of liver microsomes from PB-pretreated mice (86 mg protein/ml, 1.8 nmol cytochrome P-450/mg protein; at 100 K, microwave power 30 mW, modulation amplitude 10 G, Gain 1000). On addition of 1 mg NDphA or 3 µl DPNA to 0.3 ml of a micosomal suspension spectra b and c respectively were obtained. After thawing the sample b containing NDphA a few grains of sodium dithionite were added and the sample incubated for 5 min at 20 C. It was subsequently frozen to - 173°C and spectrum d recorded. Curve e represents sample c containing DPNA after an analogous reduction with dithionite.

Curve a in Figure 2. represents the spectrum of liver microsomes from phenobarbital pretreated mice. Cytochrome P-450 is in the oxidized form and is present mainly in the low spin form. On addition of diphenylnitrosamine or di-n-propylnitrosamine to the microsomes spectra b and c respectively were obtained, which represent spectra of a changed low spin or high spin status of the cytochrome. After addition of NADPH or dithionite to EPR samples containing the nitrosamines the EPR spectra showed sharp 3-line signals around $g = 2{,}01$ of increasing intensity and a disappearing low-spin signal (4). In model studies with nitric oxide and cytochrome P-450 it was shown, especially by Ebel et al. (7) and O'Keefe et al. (8) that a proportion of the microsomal enzyme decomposes to a cytochrome P-420 NO-species, with EPR lines identical to those produced by nitrosamines. Mainly on the basis of these spectroscopic measurements we suppose, that the nitrosamine molecule is reductively cleaved to nitric oxide and its secundary amine. The heme moiety is essential for the transfer of one electron to the molecule. Under aerobic conditions NO is displaced by O_2 from the 6th coordination site. Nitric oxide is then probably in parts converted to nitrite (6, 9). The secundary amine may remain bound to the active site of the cytochrome and may be hydroxylated by the following reaction cycle either to the corresponding hydroxylamine or to the α-C-hydroxy-compound leading to N-dealkylation (Figure 3.).

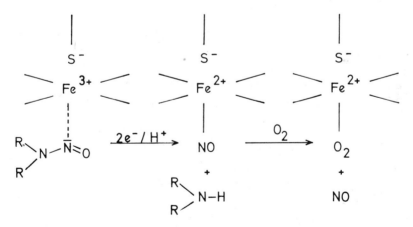

Figure 3. Proposed scheme of denitrosation of N-nitrosamines by cytochrome P-450.

Under anaerobic conditions however, nitric oxide cannot be displaced by O_2 and the cytochrome P-450-NO-complex is converted to a catalytically inactive cytochrome P-420-NO-complex.

In summary we propose that cytochrome P-450 catalyses both oxidative dealkylation (activation) and reductive denitrosation (inactivation) of carcinogenic N-nitrosamines.

To further substantiate this assumption we have done various experiments especially in order to proof that oxidative dealkylation and reductive denitrosation are different metabolic pathways.

We used acetoxymethylmethylnitrosamine and acetoxymethylbutylnitrosamine in order to demonstrate that the intact molecule and not an metabolite from the oxidative pathway, for instance the α-C-hydroxylated compound or the alkyl-diazohydroxide is the substrate leading to nitrite formation.

As is shown in Figure 4. after incubation of these compounds with liver micro-somes, neither with nor without NADPH we were able to detect nitrite (10). But if we inhibit the microsomal esterases by diisopropylfluorphosphate or pa-raoxon, as slight nitrite formation could be seen (Appel et al. in preparation).

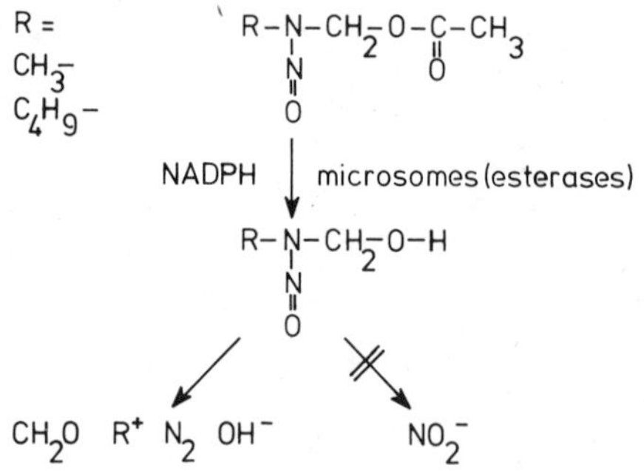

R =
CH_3-
C_4H_9-

$R-N-CH_2-O-C-CH_3$
       | ‖
       N   O
       ‖
       O

NADPH | microsomes (esterases)

$R-N-CH_2-O-H$
   |
   N
   ‖
   O

CH_2O R^+ N_2 OH^- NO_2^-

Figure 4. Metabolic scheme of acetoxymethylmethylnitrosamine and acetoxy-methylbutylnitrosamine in microsomal incubation systems.

We have tried to measure oxidative demethylation and reductive denitrosation in the same incubation mixture to investigate if there is any correlation between these two pathways or not. We incubated nitrosomethylaniline with microsomes from female rats or mice and an NADPH-regenerating system.

The time dependent kinetics of nitrite and formaldehyde formation were inducible by pretreatment of mice with phenobarbital and also by pretreatment of rats with butylhydroxytoluene.

However the extent of formaldehyde and nitrite generation was different after pretreatment with phenobarbital and butylhydroxytoluene (table 2.).

	NMRI - mice ♀			Wistar - rats ♀		
	Co	PB	%	Co	BHT	%
nmol CH_2O / mg prot. min	2.04	3.06	+ 50	0.81	2.42	+ 199
nmol NO_2^- / mg prot. min	0.19	0.66	+ 247	0.28	0.43	+ 53

Table 2. Effects of induction by phenobarbital (PB) or butylhydroxytoluene (BHT) on the generation of nitrite and formaldehyde from nitrosomethylaniline in a microsomal incubation system. The concentration of substrate was 3 mM, that of protein was 2 mg/ml.

Pretreatment of mice with phenobarbital caused a great enhancement of nitrite formation (+ 250 %) while butylhydroxytoluene induction was less active in rats (+ 53 %). In contrast CH_2O formation was stimulated only by 50 % after phenobarbital induction. It was however enhanced by 200 % in microsomes from butylhydroxytoluene induced rats (11).

When NADH in varying concentrations was added to the NADPH-regenerating system, both formaldehyde and nitrite generation could be enhanced when phenobarbital or butylhydroxytoluene induced microsomes were used. However, at higher concentrations of NADH (in the range of 2-5 mM), nitrite formation was not further enhanced but even depressed in contrast to the oxidative demethylation activity.

The same phenomen could be observed when incubations were performed with varying concentrations of NADH alone. Formaldehyde was increased with increasing concentrations of NADH (Figute 5.). For nitrite formation a maximum was obtained at 1 mM NADH. With higher concentrations (2-5 mM) a decrease was observed in both phenobarbital (PB) and butylhydroxytoluene (BHT) induced microsomes (11).

The synergistic effect of NADH on the dealkylation activity is known from other substrates of the monooxygenase system. This increase in product formation is postulated to occur because NADH is able to maintain a higher steady state level of reduced cytochrome P-450. In contrast to oxidative dealkylation however denitrosation needs one electron only as was suggested by us. So, under certain circumstances -e.g. when cytochrome P-450 is very rapidly reduced by the second electron via cytochrome b5 and cytochrome b5 reductase resulting in the fact that oxidative demethylation is the preferential pathway rather than reductive denitrosation.

Furthermore we have used diphenylnitrosamine and other N-nitrosamines to detect denitrosation in vivo. Diphenylnitrosamine possesses no oxidizable hydrogens on the carbon atoms in alpha position to the N-nitroso function. The molecule is therefore unsusceptible to the generally accepted oxidative bioactivation pathway of N-nitrosamines. After administration of diphenylnitrosamine to rats both nitrite and nitrate could be detected in the urine. In comparison to nitrite con-

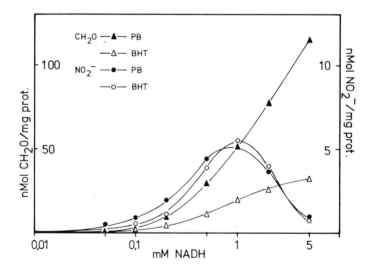

Figure 5. Influence of various NADH concentrations on the formaldehyde and nitrite formation from nitrosomethylaniline in a microsomal system. Substrate concentration was 3 mM, incubation period 30 min. The values represent means from 2-4 experiments.

centration the concentration of nitrate was relatively high. Diphenylamine was also detected, generally the amounts were relatively low. Nearly the same amounts of diphenylamine was found when diphenylamine itself was adminsitered to rats in an equimolar dose. In a somewhat higher concentration, a monohydroxylated diphenylamine was detected (9, 12).

Figure 6. demonstrates the time-dependent nitrite and nitrate concentrations in the urine of rats after a single dose of diphenylnitrosamine (NDphA) given orally or by i.p. injection. The columns in the back represent the results after oral administration (1 mmol NDphA/rat), in the front after i.p. injection (0,5 mmol NDphA/rat). I.p. administration obviously leads to an increased rate of denitrosation probably as a consequence of an altered availability for the liver. About 50 % of the administered dose was detected as nitrite/nitrate after 96 h (12).

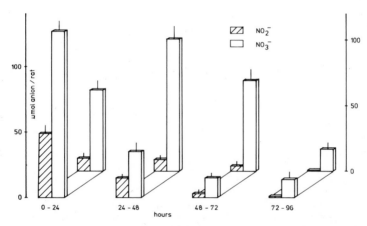

Figure 6. Urinary excretion of nitrite and nitrate by rats treated with a single dose of NDphA. The columns in the back represent the results after oral application (1 mmol/rat), in the front after i.p. injection (0.5 mmol/rat). Nitrite/nitrate was determined as described by Appel et al. (12).

Figure 7. shows the dose dependency of nitrite/nitrate generation in vivo for various N-nitrosamines, diphenylnitrosamine (NDphA), nitrosomethylaniline (NMA) and di-n-butylnitrosamine (DBNA). The lowest denitrosation rate was observed with DBNA. In relation to the administered dose about 10 % of nitrite/nitrate was recovered in the urine.

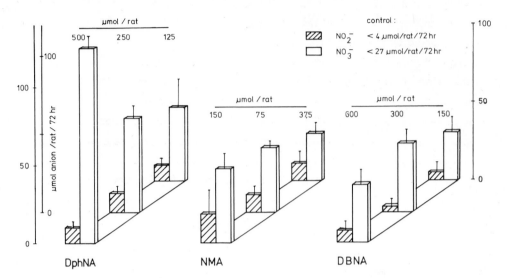

Figure 7. Urinary excretion of nitrite and nitrate by rats treated with various doses of DphNA, NMA and DBNA. Nitrite/nitrate was determined as described by Appel et al. (12).

Table 3. Enzymatic activities of oxidative dealkylation and reductive denitrosation of nitrosodimethylamine (NDMA) and nitrosomethylaniline (NMA) in liver microsomes from various species including man. The concentrations of substrates were 3 mM, that of protein 2 mg/ml.

		NDMA				NMA	
		CH_2O $\frac{NMOL}{MG\ MIN}$	NO_2^- $\frac{NMOL}{MG\ MIN}$	$\frac{CH_2O}{NO_2^-}$	CH_2O $\frac{NMOL}{MG\ MIN}$	NO_2^- $\frac{NMOL}{MG\ MIN}$	$\frac{CH_2O}{NO_2^-}$
HAMSTER SYR.	♀	2,30	0,10	23	5,30	1,30	4
MOUSE NMRI	♀	2,27	0,09	25	2,04	0,15	14
RAT WISTAR	♀	1,20	0,09	13	0,81	0,24	3
HUMAN SUBJECT		0,42	0,08	5	0,81	0,26	3

In table 3. some experiments on activation and inactivation of nitrosodimethyl-amine (NDMA) and nitrosomethylaniline (NMA) by liver microsomes of various species including man are summarized. The main results are: a) there is a great species variability for activation and inactivation; the hamster shows the highest activity for both reactions; b) man is also able to denitrosate both nitrosamines; c) the rat seems to be the species which mostly resembles man in respect of both enzymatic activities.

References

1. Magee, P.N. and Farber, E. (1982) Toxic Liver Injury and Carcinogenesis. Methylation of Rat Liver Nucleic Acids by Dimethylnitrosamine in vivo. Biochem. J., 83, 114-124.

2. Magee, P.N., Montesano, R. and Preussmann, R. (1976) N-Nitroso Compounds and Related Carcinogens. In: Searle, C.E., ed., Chemical Carcinogenesis (Am. Chem. Soc. Monogr. Ser., No. 173), Washington D.C., American Chemical Society, pp. 491-625.

3. Appel, K.E., Rickart, R., Schwarz, M. and Kunz H.W. (1979) Influence of Drugs on Activation and Inactivation of Hepatocarcinogenic Nitrosamines. Arch. Toxicol. Suppl., 2, 471-477.

4. Appel, K.E., Ruf, H.H., Mahr, B., Schwarz, M., Rickart, R. and Kunz W. (1979) Binding of Nitrosamines to Cytochrome P-450 of Liver Microsomes. Chem. Biol. Interactions, 28, 17-33.

5. Appel, K.E., Schrenk, D., Schwarz, M., Mahr, B., and Kunz W. (1980) Denitrosation of N-Nitrosomorpholine by Liver Microsomes; Possible Role of Cytochrome P-450. Cancer Letters, 9, 13-20.

6. Appel, K.E. and Graf, H. (1982) Metabolic Nitrite Formation from N-Nitrosamines: Evidence for a Cytochrome P-450 dependent Reaction. Carcinogenesis, 3, 293-296.

7. Ebel, R.E., O'Keefe, D.H. and Peterson, J.A. (1975) Nitric Oxide Complexes of Cytochrome P-450. FEBS Lett., 55, 198-200.

8. O'Keefe, D.H., Ebel, R.E. and Peterson, J.A. (1978) Studies on the Oxygen Binding Site of Cytochrome P-450. J. Biol. Chem., 253, 3509-3515.

9. Appel, K.E., Rühl, C.S. and Hildebrandt, A.G. (1984) Metabolic Inactivation of N-Nitrosamines by Cytochrome P-450 in vitro and in vivo. In N-Nitroso Compounds: Occurence, Biological Effects and Relevance to Human Cancer; ed. I.K. O'Neill, R.C. von Borstel, C.T. Miller, J. Long and H. Bartsch; IARC Scientific Publication No. 57, 443-451.

10. Appel, K.E., Frank, N. and Wiessler, M. (1981) Metabolism of Nitrosoacetoxy-methylmethylamine in Liver Microsomes. Biochem. Pharmacol., 30, 2767-2772.

11. Appel, K.E., Rühl, C.S., Hildebrandt, A.G. (1985) Oxidative Dealkylation and reductive Denitrosation of Nitrosomethylaniline in vitro. Chem.Biol. Interactions, 53, 69-76

12. Appel, K.E., Rühl, C.S., Spiegelhalder, B. and Hildebrandt, A.G. (1984) Denitrosation of Diphenylnitrosamine in vivo. Toxicol. Letters, 23, 353-358.

ACKNOWLEDGEMENT

This research was in part supported by a grant from the Deutsche Forschungs-gemeinschaft.

USE OF THE CHEMILUMINIGENIC PROBES LUMINOL AND LUCIGENIN FOR THE DETECTION

OF ACTIVE OXYGEN SPECIES IN HEPATIC MICROSOMES AND IN INTACT HEPATOCYTES

Alfred G.Hildebrandt, Andreas Weimann and Regine Kahl

Max von Pettenkofer Institute of the Federal Health Office
Postfach 33 00 13, D-1000 Berlin 33, F.R.G.

INTRODUCTION

During microsomal NADPH oxidation, reduced oxygen species are produced which must be considered a normal by-product of the NADPH-cytochrome P-450 reductase / cytochrome P-450 system (Gillette, Brodie and LaDu, 1957; Hildebrandt, Speck and Roots, 1973; Estabrook, Kawano, Werringloer et al., 1979; Cederbaum and Dicker, 1983). Formation of such reactive oxygen molecules is enhanced in the presence of many monooxygenase substrates (Hildebrandt and Roots, 1975; Hildebrandt, Heinemeyer and Roots, 1982) and may contribute to the toxicity of foreign compounds undergoing microsomal biotransformation (Farber and Gerson, 1984). It has been suggested that superoxide anion radicals are released during autoxidation of the oxycytochrome P-450 complex and are subsequently dismutated to yield hydrogen peroxide (Ullrich and Kuthan, 1980); hydroxyl radical formation may then result from hydrogen peroxide cleavage in the presence of superoxide anion by a Haber-Weiss reaction (Haber and Weiss, 1934).

Microsomal oxygen activation has been studied with a wide variety of methods all making use of the oxidizing resp. reducing properties of the reactive oxygen metabolites. However, little attention has been paid in microsome studies to the ability of reduced oxygen species to transfer via oxygenation common chemiluminigenic probes (luminol and lucigenin) to an excited state and thus to elicit photon emission from these probes, a test procedure widely used in inflammation studies to follow the production of reactive oxygen species during the "respiratory burst" of phagocytizing leukocytes (Allen, 1982). This communication deals with the effect of synthetic antioxidants on the availability of reactive oxygen species for luminol and lucigenin oxygenation resulting in light emission from suspensions of liver microsomes or intact hepatocytes.

MATERIALS AND METHODS

Conventionally prepared microsomes and intact hepatocytes from the livers of male Wistar rats were analyzed in a "Biolumat LB 9505" (Berthold, Wildbad, F.R.G.). This photon counter is equipped with six photomultipliers to allow for simultaneous sample measurements. In microsome experiments, 2 mg/ml of microsomal protein (obtained from livers of phenobarbital pretreated animals) was incubated in a final volume of

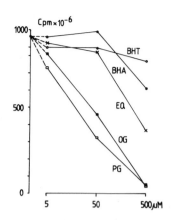

Fig.1   Antioxidant inhibition of lucigenin-dependent chemiluminescence
        induced by NADPH addition to rat liver microsomes. Photon yield
        between 0 and 20 min is shown. Data are means of three experi-
        ments,S.E.M. (omitted from the figure for clarity) was < 15 % of
        most means, but up to 40% for very low photon yields. p < 0.05 for
        5 µM octyl gallate (OG), p < 0.01 for 50 µM propyl gallate (PG)
        and for 500 µM ethoxyquin (EQ), p < 0.001 for 500 µM PG and OG.
        Statistically significant inhibition was not obtained with
        butylated hydroxyanisole (BHA) and with butylated hydroxytoluene
        (BHT).

200 µl consisting of 66 mM Tris buffer pH 7.5, an NADPH generating system
and 10 µl luminol (1 mM in NaOH, pH 12) or lucigenin (8 mM in aq.bid.)
at 37°. The reaction was started by NADPH addition (final concentration
500 µM). In hepatocyte experiments, livers from untreated animals were
used. Cells harvested after collagenase perfusion of the organ were extensively washed to remove cofactors released from defect hepatocytes. Preparations were used if Trypan Blue exclusion was > 80%. A total of 400.000
cells were incubated in a final volume of 200 ul Hepes buffer pH 7.6. The
reaction was started by the addition of 10 µl lucigenin (8 mM in
aq.bid.).

    Microsomal lipid peroxidation and formaldehyde formation from dime-
thyl sulfoxide were measured as described previously (Bernheim, Bernheim
and Wilbur, 1948; Klein, Cohen and Cederbaum, 1981)

RESULTS AND DISCUSSION

When NADPH is added to a microsomal preparation containing 0.4 mM lucigenin or 0.05 mM luminol a chemiluminescence signal is generated with peak
values between 3 and 6 min which slowly fades during the following 30 min.
This signal could partially be suppressed by the addition of a) 300 U/ml
superoxide dismutase (60%),b) 1mg/ml catalase (30%) and c) 150 mM dimethyl
sulfoxide (30-40%). A specificity of both the lucigenin assay and the
luminol assay as to an indivicual active oxygen species is not obvious at
present.

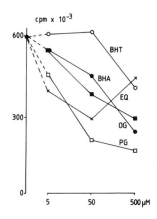

Fig.2 Antioxidant inhibition of luminol–dependent chemiluminescence induced by NADPH addition to rat liver microsomes. Photon yield
between 0 and 20 min is shown. Data are means of three experiments; S.E.M. (omitted from the figure for clarity) was < 15% of
the means; $p < 0.05$ for 5 µM propyl gallate (PG) and for 50µM
butylated hydroxyanisole (BHA), $p < 0.01$ for 5 and 50 µM
ethoxyquin (EQ) and for 500 µM butylated hydroxytoluene (BHT),
$p < 0.001$ for 50 and 500 µM PG and octyl gallate (OG) and for
500 µM BHA.

    Figs.1 and 2 show that some synthetic antioxidants which are com-
mercially used for inhibition of lipid peroxidation in food by means of
their radical scavenging properties are also able to inhibit the chemiluminescence due to lucigenin resp. luminol oxygenation. In the lucigenin
assay (Fig.1) the gallic acid ester antioxidants, propyl gallate (PG)
and octyl gallate (OG), are the most potent inhibitors, suppressing half
of the photon yield below 50 µM and leading to complete inhibition at
500 µM; ethoxyquin (EQ) is intermediate, and the two phenolic antioxidants
butylated hydroxytoluene (BHT) and butylated hydroxyanisole (BHA) do not
lead to statistically significant inhibition in concentrations up to
500 µM. In the luminol test (Fig.2) ranging of the antioxidants was less
distinct but still the gallic acid ester antioxidant PG was the most
effective inhibitor while BHT was the least effective inhibitor. The
failure of high concentrations of EQ to suppress the luminol–dependent
chemiluminescence cannot be explained at present but was consistent in all
experiments performed.

    It should be noted that light emission from lucigenin was 1.000 times
higher than from luminol (compare ordinates of Figs. 1 and 2). This is
different from data from phagocyte studies where both probes lead to
comparable chemiluminescence yields (Allen, 1982). The reasons for this
discrepancy have not yet been elucidated. However, it is obvious from

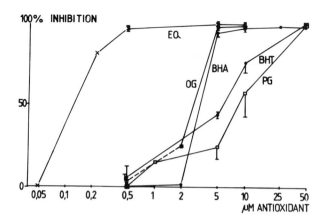

Fig.3   Dose response curve of the inhibition of lipid peroxidation in rat
        liver microsomes by antioxidants. Suppression of malondialdehyde
        formation as measured by means of the thiobarbituric acid reaction
        is plotted. Values are means + S.E.M. (n = 3-5). EQ: ethoxyquin,
        OG: octyl gallate, BHA: butylated hydroxyanisole, BHT: butylated
        hydroxytoluene, PG: propyl gallate.

these data that lucigenin is a very sensitive probe when used in microsomal preparations.

        From our data no conclusion can be drawn to the agents which are
removed from the microsomal membrane by the antioxidants to result in
suppression of photon emission from the two chemiluminigenic probes.
Obviously, these agents must be able to oxygenate lucigenin and luminol.
In toxicological studies, inhibition of a process by an antioxidant is
often interpreted to suggest a critical role of lipid peroxidation because
termination of the chain reactions involved in lipid peroxidation is the
mechanism of action most commonly associated to this class of agents.
Microsomal lipid peroxidation is well known to produce excited species
which can be measured by low level chemiluminescence, i.e. light emission
in the absence of chemiluminescence "enhancers" like lucigenin and luminol
(Wright, Rumbaugh, Colby et al.,1979; Cadenas and Sies, 1982), and it
has been demonstrated that this event is markedly accelerated in microsomes from vitamin E-deficient animals (Cadenas, Ginsberg, Rabe et al.,
1984) indicating that the naturally occurring antioxidant, vitamin E,
decreases the availability of such excited species within the microsomal
membranes. A synthetic antioxidant, di-tert-butylquinol, also suppresses
organic hydroperoxide-induced low level chemiluminescence in microsomes
(Cadenas and Sies, 1982).

        In our experimental protocol which was not designed to measure low
level chemiluminescence but to make use of so-called chemiluminescence
enhancers, i.e. to detect oxygenating agents produced during NADPH oxidation, antioxidant inhibition of chemiluminescence is most likely not due
to inhibition of lipid peroxidation because both actions take place at
widely differing concentrations. Fig. 3 summarizes data on antioxidant
inhibition of iron-ADP-stimulated lipid peroxidation which reveal that
this process is in general much more sensitive to the antioxidants than

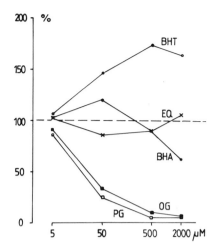

Fig. 4 Influence of antioxidants an the formation of formaldehyde from
dimethyl sulfoxide in rat liver microsomes. Data are means from
3-4 experiments, S.E.M (omitted from the figure for clarity) was
< 20%. $p < 0.05$ for 500 µM and 2 mM butylated hydroxytoluene
(BHT), $p < 0.01$ for 50 µM propyl gallate (PG) and octyl gallate
(OG), $p < 0.001$ for 500 µM and 2 mM PG and OG. No statistically
significant modulation of activity was obtained with butylated
hydroxyanisole (BHA) and ethoxyquin (EQ)

lucigenin- or luminol-dependent chemiluminescence. Complete inhibition is
obtained at 0.5 µM EQ, at 5 µM OG and BHA and at 50 µM PG and BHT. The
chemiluminescence experiments of Figs. 1 and 2 were not performed in the
presence of iron-ADP; however, if iron-ADP was added to the chemiluminescence assay to provide peroxidizing conditions, similar inhibition data

Tab.1 Correlation coefficients for antioxidant effects on chemiluminescence, dimethyl sulfoxide oxidation and extra production of
hydrogen peroxide

Luminol-induced chemiluminescence	
vs. lucigenin-induced chemiluminescence	+ 0.48
vs. DMSO oxidation by OH· radicals	+ 0.65
vs. extra production of hydrogen peroxide	+ 0.02
Lucigenin-induced chemiluminescence	
vs. DMSO oxidation by OH· radicals	+ 0.95 ($p < 0.05$)
vs. extra production of hydrogen peroxide	− 0.14
DMSO oxidation by OH· radicals	
vs. extra production of hydrogen peroxide	− 0.30

Effects of 500 µM propyl gallate, octyl gallate, ethoxyquin,
butylated hydroxyanisole, and butylated hydroxytoluene on the four
reactions listed were correlated. DMSO: dimethyl sulfoxide.

were obtained. Therefore, except for PG, the only agent suppressing both processes in the same concentration range, the antioxidant-inhibitable step involved in chemiluminescence cannot be intimately linked with lipid peroxidation, since lipid peroxidation does no longer take place at the antioxidant concentrations required to inhibit lucigenin- or luminol oxygenation.

Microsomal dimethyl sulfoxide oxidation yielding formaldehyde is assumed to be elicited by hydroxyl radicals or an agent possessing the oxidizing properties of hydroxyl radicals (Klein, Cohen and Cederbaum, 1981). This reaction can also be inhibited by antioxidants (Fig.4), and inhibition occurs at the same concentration range as does chemiluminescence inhibition. Hydroxyl radicals probably arise from hydrogen peroxide via a Haber Weiss reaction (Haber and Weiss, 1934). Since most of the antioxidants used in this study lead to extra production of hydrogen peroxide in rat liver microsomes (Rössing, Kahl and Hildebrandt, 1985; Cummings and Prough, 1983), it is conceivable that increased supply of hydroxyl radicals ensuing increased formation of the precursor will compromise the capacity of the antioxidants to cope with the radical. In this case, the ability of an individual antioxidant to induce extra production of hydrogen peroxide should be inversely related to its ability to inhibit hydroxyl radical-mediated oxidation reactions such as formaldehyde production from dimethyl sulfoxide. On the other hand, the radical scavenging properties of the antioxidants are likely to counteract the potentially harmful consequences of increased hydrogen peroxide formation which may result in lipid peroxidation and cell damage. In Tab.1 antioxidant effects on chemiluminescence, dimethyl sulfoxide oxidation and extra production of hydrogen peroxide are related by means of correlation coefficients. The table shows that the inverse relation between the two latter reactions is only marginal, at least when tested by their inhibition by the five antioxidants used in this study. This is due to the fact that BHT on one hand is inactive in stimulating hydrogen peroxide formation and on the other hand will not remove the dimethyl sulfoxide oxidizing agent but rather enhances·its availability (Fig.4); correlation is better if BHT is omitted from the calculation. Inhibition of dimethyl sulfoxide oxidation was highly correlated with inhibition of lucigenin-dependent chemiluminescence; however, evidence from the literature (Allen, 1982; Müller-Peddinghaus, 1984) as well as observations in our laboratory do not inequivocally support the conclusion that hydroxyl radicals are the oxygen species responsible for lucigenin oxygenation. Inhibition of luminol-induced chemiluminescence is less clearly correlated with inhibition of hydroxyl radical action on dimethyl sulfoxide, and correlation between antioxidant influence in the luminol assay on one hand and in the lucigenin assay on the other hand is still less obvious. The actual oxygenating agents involved in the chemiluminescence events described here remain to be elucidated.

The steady state concentration of reactive oxygen species in the intact hepatocyte should be much lower than in microsomal preparations because of the operation of highly efficient endogenous antioxidative enzymes able to remove superoxide and hydrogen peroxide. However, hydrogen peroxide formation can, by indirect means, be followed within the intact cell, and its intracellular stimulation by drug substrates has been demonstrated (Orrenius, Thor, Eklöw et al., 1982). Superoxide anion is not normally released into the medium by the intact hepatocyte but secretion can be induced by the uptake of quinones undergoing redox cycling in the cell and thereby producing appreciable amounts of superoxide (Powis, Svingen and Appel, 1982). We have attempted to apply "enhanced" chemiluminescence for the detection of oxygen activation in suspensions of hepatocytes. Indeed, a chemiluminescence signal can be obtained in such experiments with lucigenin (though not with luminol) as the chemiluminigenic

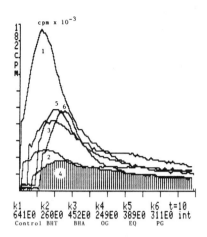

Fig.5   Antioxidant inhibition of lucigenin-dependent chemiluminescence in
suspensions of intact rat hepatocytes. An original recording is
shown. k1 - k6 give the integrals of photon yield between 0 and 10
min; data must be multiplied with 1.000. Antioxidant concentration
was 50 µM, Trypan Blue exclusion was 84%. 1: no antioxidant present, 2: butylated hydroxytoluene (BHT), 3: butylated hydroxyanisole (BHA),4: octyl gallate (OG), 5: ethoxyquin (EQ), 6: propyl
gallate (PG).

probe (Fig.5). The interpretation of this finding is rendered difficult by
the fact that a hepatocyte preparation containing no defect cells cannot
be obtained at least by the method we use. This means that a "homogenate-
like" effect resembling the processes obtained in microsome studies will
be contributed to the overall chemiluminescence recordings by the defect
cells. This part of total light emisssion can be large though percentage
of defect cells may be small, because photon yield in the cell free system
is about 1.000 times higher with lucigenin than in the intact cell system
(compare Figs.1 and 5). Availability of NADPH released from defect cells
will enhance this homogenate-like effect, and extensive washing of the
preparation to remove the cofactor does indeed decrease photon emission
from defect cells (data not shown). Nevertheless, a positive correlation
between the percentage of intact cells (i.e. cells that exclude Trypan

Tab.2 Antioxidant inhibition of lucigenin-dependent chemiluminescence in suspensions of intact hepatocytes

Light emission (cpm x 10^{-3},   Integral between 0 and 20 min)

Control	Antioxidants (50 µM)				
	BHT	BHA	OG	EQ	PG
590 ± 66	425 ± 86	467 ± 76	$315 \pm 87^{*}$	488 ± 73	$336 \pm 50^{**}$

Means $\pm$ S.E.M. (n = 6-7), * $p < 0.05$, ** $p < 0.01$

977

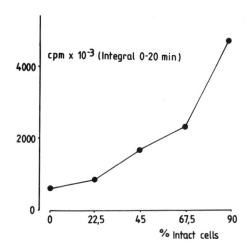

Fig.6 Dependence of lucigenin-induced chemiluminescence in hepatocyte
suspensions on the percentage of intact cells. A single experiment
is plotted. Statistical evaluation leads to the following data
(light emission at 80-90% of intact cells considered as 100%):

% Intact cells	% Light emission
80 – 90	100
60 – 70	86 ± 7
40 – 60	59 ± 6
20 – 30	43 ± 4
0	39 ± 5

(Means ± S.E.M., n = 8-9)

Blue) and lucigenin-dependent chemiluminescence is obtained (Fig.6). The
statistics given in the legend of Fig.6 reveal that about 40% of the
chemiluminescence signal elicited by a preparation containing 80-90% of
intact cells can be obtained in a preparation in which all cells have been
purposely destroyed. It is conceivable that intracellular processes can be
measured by this technique provided the chemiluminigenic probe is capable
to penetrate the cell membrane. Our attempts to detect uptake of lucigenin
into the hepatocyte have so far failed, and our data do not allow for
discrimination of photon emission within the cell and photon emission from
an oxygen species released into the medium.

Antioxidants lead to inhibition of chemiluminescence in intact hepatocytes (Fig.5, Tab.2); however, statistically significant suppression of
lucigenin-dependent chemiluminescence at the 50 µM antioxidant level was
only obtained with the gallic acid ester antioxidants PG and OG which had
also been the most efficient agents in the microsomal lucigenin assay
(Fig.1). It should, however, be considered that part of the chemiluminescence signal recorded is most likely due to an extracellular event (Fig.6)
so that evaluation of antioxidant potency towards intact hepatocytes is
only approximative at present.

A role of reactive oxygen species in cytotoxicity (Fridovich, 1976) and recently also in tumor promotion (Copeland, 1983) has been postulated. Antioxidants provide protection against a wide variety of toxic and carcinogenic chemicals (for a review see Kahl, 1984). Part of this protective action has been ascribed to the ability of the antioxidants to modulate the metabolism of the toxic or carcinogenic compound (Kahl, 1984); however, part of the beneficial biological effects cannot be attributed to this mechanism of action and may actually rather be related to the ability of the antioxidants to interfere with oxygen activation. Chemiluminescence assays may provide a convenient tool for testing the influence of these agents on production and availability of reactive oxygen species in a variety of biological systems.

Acknowledgement: This study was performed with the financial support of the Deutsche Forschungsgemeinschaft, Bonn, F.R.G.
The able technical assistance of Mrs. H.Jauer is gratefully acknowledged.

REFERENCES

Allen, R.C., 1982, Biochemiexcitation: Chemiluminescence and the study of biological oxygenation reactions, in: " Chemical and Biological Generation of Excited States", Academic Press, New York, p.309.

Bernheim, F., Bernheim, M.L., and Wilbur, K.M., 1948, The reaction between thiobarbituric acid and the oxidation products of certain lipids, J.biol.Chem., 174:257.

Cadenas, E., and Sies, H., 1982, Low level chemiluminescence of liver microsomal fractions initiated by tert-butyl hydroperoxide. Relation to microsomal hemoproteins, oxygen dependence, and lipid peroxidation, Eur.J.Biochem., 124:349.

Cadenas, E., Ginsberg, M., Rabe, U., and Sies, H., 1984, Evaluation of - tocopherol antioxidant activity in microsomal lipid peroxidation as detected by low-level chemiluminescence, Biochem.J., 223:755.

Cederbaum, A.I., and Dicker, E., 1983, Inhibition of microsomal oxidation of alcohols and of hydroxyl radical-scavenging agents by the iron-chelating agent desferrioxamin, Biochem.J., 210:10.

Copeland, E.S., 1983, Meeting report: Free radicals in promotion - a chemical pathology study section workshop, Cancer Res., 43:5631.

Cummings, S.W., and Prough, R.A., 1983, Butylated hydroxyanisole-stimulated NADPH oxidase activity in rat liver microsomal fractions, J.biol.Chem., 258:12315.

Estabrook, R.W., Kawano, S., Werringloer, J., Kuthan, H., Tsyi, H., Graf, H., and Ullrich, V., 1979, Oxycytochrome P-450: its breakdown to superoxide for the formation of hydrogen peroxide, Acta biol. med. Germ., 38:423.

Farber, J.L., and Gerson, R.J., 1984, Mechanisms of cell injury with hepatotoxic chemicals, Pharmacol.Rev. 36:71S.

Fridovich, I, 1976, Oxygen Radicals, hydrogen peroxide, and oxygen toxicity, in: "Free Radicals in Biology", W.A.Pryor, ed., Academic Press, New York, p.239.

Gillette, J.R., Brodie, B.B., and LaDu, B.N., 1957, The oxidation of drugs by liver microsomes: on the role of TPNH and oxygen, J.Pharmacol.exp. Ther., 119:532.

Haber, F., and Weiss, J., 1934, The catalytic decomposition of hydrogen peroxide by iron salts, Proc.Roy.Soc.Lond.A, 147:332.

Hildebrandt, A.G., Speck, M., and Roots, I., 1973, Possible control of hydrogen peroxide production and degradation in microsomes during mixed function oxidation reactions, Biochem.biophys.Res.Commun., 54:968.

Hildebrandt, A.G., and Roots, I., 1975, Reduced nicotinamide dinucleotide phosphate (NADPH)-dependent formation and breakdown of hydrogen peroxide during mixed-function oxidation reactions in liver microsomes, Arch.Biochem.Biophys., 171:385.

Kahl, R., 1984, Synthetic antioxidants: Biochemical actions and interference with radiation, toxic compounds, chemical mutagens and chemical carcinogens, Toxicology, 33:185.

Klein, S.M., Cohen, G., and Cederbaum, A.I., 1981, Production of formaldehyde during metabolism of dimethyl sulfoxide by hydroxyl radical generating systems, Biochemistry, 20:6006.

Müller-Peddinghaus, R., 1984, Invitro determination of phagocyte activity by luminol- and lucigenin-amplified chemiluminescence, Int.J.Immunopharmacol., 6:455.

Orrenius, S., Thor, H., Eklöw, L., Moldeus, P., and Jones, D.P., 1982, Drug-stimulated H_2O_2 formation in hepatocytes. Possible toxicological implications, in: "Biological Reactive Intermediates", R.Snyder, D.V.Parke, J.J.Kocsis, D.Jollow, C.G.Gibson, C.M.Witmer, eds., Plenum Press, New York and London, p.395.

Powis, G., Svingen, B.A., Appel, P., 1982, Factors affecting the intracellular generation of free radicals from quinones, in: "Biological Reactive Intermediates", R.Snyder, D.V.Parke, J.J.Kocsis, D.J.Jollow, C.G.Gibson, C.M.Witmer, eds., Plenum Press, New York and London, P.349.

Rössing, D., Kahl, R., and Hildebrandt, A.G., 1985, Effect of synthetic antioxidants on hydrogen peroxide formation, oxyferro cytochrome P-450 concentration and oxygen consumption in liver microsomes, Toxicology, 34:67.

Ullrich, V., and Kuthan, H., 1980, Autoxidation and uncoupling in microsomal monooxygenations, in: "Biochemistry, Biophysics and Regulation of Cytochrome P-450, J.Gustafsson, J. Carlstedt-Duke, A.Mode, J.Rafter, eds., Elsevier, Amsterdam, p.267.

Wright, J.R., Rumbaugh, R.C., Colby, H.D., and Miles, P.R., 1979, The relationship between chemiluminescence and lipid peroxidation in rat hepatic microsomes, Arch. Biochem.Biophys., 192:344.

ALTERED EXPRESSION OF ONCOGENES IN MOUSE EPIDERMIS FOLLOWING EXPOSURE TO

BENZO(A)PYRENE DIOL EPOXIDES

Jill C. Pelling,[1] Sharon M. Ernst,[1] George Patskan,[2] Rodney
S. Nairn,[2] Douglas C. Hixson,[3] Solon L. Rhode,[1] and Thomas J.
Slaga[2]

[1]Eppley Institute for Research in Cancer and Allied Diseases
The University of Nebraska Medical Center, Omaha, Nebraska
[2]The University of Texas System Cancer Center, Science Park
Smithville, Texas
[3]Department of Medical Oncology, Rhode Island Hospital
Providence, Rhode Island

INTRODUCTION

There now exists a great deal of evidence which indicates that all
normal eukaryotic cells contain endogenous, highly-conserved DNA sequences
known as proto-oncogenes (1). The normal functions of these cellular oncogenes are not yet known, although it has been hypothesized that they may
be important in cellular differentiation, fetal development and control of
cell proliferation (2-4). Since a significant proportion of human cancers
are presumed to be the result of exposure to environmental chemicals,
extensive research efforts have focused on determining the effects of
chemical carcinogens and tumor promoting agents on the expression of these
c-onc sequences. Studies by Barbacid and coworkers using the rat mammary
carcinoma model have shown that the Ha-ras oncogene is activated in rat
mammary carcinomas induced by N-methylnitrosourea (5). Sequencing of the
activated rat c-Ha-ras oncogene in individual mammary adenocarcinomas
indicated that the rat proto-oncogene had undergone a point mutation in
the 12th codon, resulting in a glycine-for-valine substitution in the ras
P21 protein product. In similar studies, Balmain and Pragnell employed the
two stage model of initiation and promotion in mouse skin to demonstrate
that a percentage of papillomas and carcinomas induced by 7,12-dimethyl-
benz(a)anthracene (DMBA) and promotion with 12-0-tetradecanoyl phorbol-13-
acetate (TPA) contained elevated levels of Ha-ras transcripts compared
with normal mouse epidermis. Furthermore, DNA from papillomas and squamous
cell carcinomas caused morphological transformation of NIH/3T3 cells in
vitro (6,7). Southern blot hybridization studies demonstrated that the
transforming properties of the DNA were due to transfection of an activated cellular Ha-ras oncogene.

We have expanded these earlier studies by investigating the various
stages of mouse skin carcinogenesis for alterations in patterns of proto-
oncogene expression, in order to establish at what stage in tumor development elevated Ha-ras expression first occurs. For this purpose, optically
pure enantiomers of benzo(a)pyrene-7,8-diol-9,10-epoxide-anti (BPDE-anti)
were used to initiate SENCAR mice, after which the animals were treated

981

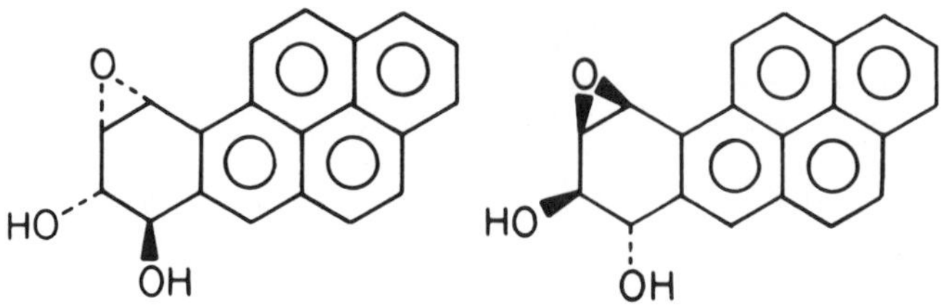

(+)anti BPDE   (−)anti BPDE

Fig. 1. Structures and absolute stereochemistry of the pure (+) and (−) enantiomers of BPDE-anti.

with twice-weekly doses of TPA. The (+) enantiomer of BPDE (Figure 1) has previously been shown by others to be the ultimate carcinogenic metabolite of benzo(a)pyrene (8,9) as well as the major source of DNA adducts in tissues or cells exposed to benzo(a)pyrene or the proximate metabolites benzo(a)pyrene 7,8-diol and BPDE (10-12). We have compared levels of Ha-ras expression in "early" BPDE-induced epidermal papillomas and pre-tumor epidermis with those in "older" papillomas and squamous cell carcinomas. We have also compared the expression of myc-related sequences with that of Ha-ras at various times following initiation and tumor promotion of mouse epidermis, since studies in other cell systems and tissues have predicted that the myc oncogene may play an important role in regulation of cell proliferation. In addition to ras and myc, we have evaluated tumors for the expression of the env gene of Murine leukemia virus (MuLV), since numerous studies have shown that these endogenous proviral genes can be induced to replicate or to undergo alterations in expression in the presence of chemical carcinogens (13).

METHODOLOGY

Adult female SENCAR mice (a tumor sensitive stock) were treated topically with an initiating dose of 200 nmol of the (+) enantiomer of BPDE-anti. As a control for the toxic effects of polycyclic aromatic hydrocarbon on the epidermis, other mice were "sham"-initiated with 200 nmol of the (−) enantiomer, which is inactive as a tumor initiator but reacts chemically in a manner identical to the (+) enantiomer (except against optically active substrates). Beginning one week following initiation, some animals received twice-weekly applications with the tumor promoter TPA. Groups of animals (30 mice each) were sacrificed at the following stages: 1) seven days after initiation; 2) following 1 week of treatment with TPA; 3) following 6 weeks of treatment with TPA; 4) following development of "early" (1-4 mm diameter) papillomas (12-14 weeks); 5) following development of "late" papillomas (6-8 months); and 6) following development of squamous cell carcinomas (8-12 months). Tumors were excised, frozen in liquid nitrogen, and ground to a fine powder with a mortar and pestle. Epidermis from mice which had not yet developed tumors was removed, spread on foil, frozen in liquid nitrogen, and scraped with a razor blade.

RNA was purified in guanidine isothiocyanate-cesium chloride gradients as described by Ullrich et al. (14). DNA was isolated from the same density gradients and digested with proteinase K and RNAse A, followed by phenol-chloroform extraction.

Samples of total cellular RNA were electrophoresed through 1% aga-

rose-formaldehyde gels as described by Goldberg (15). Gels were blotted
onto nitrocellulose and hybridized with [32-P]-nick translated DNA probes
(15,16). A 460 bp DNA probe (5-9) specific for the Ha-ras oncogene (17)
and a v-myc probe (18) were graciously provided by Dr. Allen Balmain of
the Beatson Institute. The c-myc probe used in these studies represented
the 5.6 kb Bam HI fragment of the murine c-myc proto-oncogene encompassing
the second and third exons of the gene (19). The MuLV env-probe was 1.7 kb
in length and represented the 3' end of the env-gene (20).

Southern blot hybridization analysis of purified DNA was carried out
in 0.8% agarose gels blotted onto nitrocellulose (21). Nick-translated
probes specific for the Ha-ras and c-myc oncogenes were prepared as described above. Autoradiography was carried out for 3-15 days using Kodak
XAR-5 film.

RESULTS

Expression of Ha-ras RNA During Epidermal Carcinogenesis

In order to determine at what stage in epidermal carcinogenesis the
elevation in Ha-ras expression first occurs, we have purified epidermal
RNA from groups of mice at specific points in the initiation/promotion
regimen, as described above. Samples of total cellular RNA (20 μg) were
electrophoresed on formaldehyde-agarose gels and blotted onto nitrocellulose paper prior to hybridization with a nick-translated Ha-ras DNA probe.
Results of Northern blot hybridization are presented in Figure 2. Lane 1
represents the level of Ha-ras-specific RNA sequences present in untreated
mouse epidermis. This level does not differ significantly from that present in epidermis following initiation with the (+) enantiomer (lane 2) or
initiated epidermis following 1 or 6 weeks of promotion (lanes 3 and 4,
respectively). Significantly higher levels of Ha-ras expression occurred
in "early" papillomas which developed to a size of 3-5 mm after 14 weeks
of promotion with TPA (two of which are shown in lanes 5 and 6). Elevated

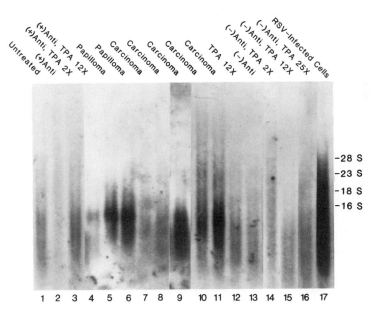

Fig. 2. Ha-ras Northern blot hybridization of RNA
purified from SENCAR mouse epidermis and
tumors at various stages of tumorigenesis.

expression of Ha-ras was also observed in 6 of 13 late stage papillomas (data not shown) and in 3 of 5 carcinomas (lanes 7-11). Epidermal RNA isolated from mice treated with TPA alone for 6 weeks (lane 12), the (-) enantiomer alone (lane 13), or the (-) enantiomer followed by TPA for 1 week, 6 weeks, or 12 weeks (lanes 14-16), showed no enhanced expression of Ha-ras RNA. Lane 17 contains RNA from rat sarcoma virus infected cells (positive control). These results are summarized in Table 1.

Our results indicated that elevated expression of the Ha-ras oncogene can occur relatively early in the two-stage model of skin carcinogenesis, and is detectable in very small papillomas which have developed after 14 weeks of TPA treatment. Elevated levels of Ha-ras transcripts were also observed in some, but not all, older papillomas and squamous cell carcinomas.

Southern Blot Analysis of Tumor DNA for Evidence of Ha-ras Gene Amplification

In order to determine if there was a genetic change in the tumors and if this change was associated with enhanced Ha-ras oncogene expression, we used Southern blot analysis to analyze the Ha-ras gene locus in several tumor samples with high levels of Ha-ras transcripts. Aliquots of DNA were digested with Bam HI restriction endonuclease, electrophoresed on 0.8% agarose gels, and blotted onto nitrocellulose according to the method of Southern (21) and hybridized with the same Ha-ras nick translated probe used in the above Northern blot studies. The resulting autoradiogram is shown in Figure 3. Panel 3B presents the Southern blot of DNA samples from epidermis and tumors, and panel 3A represents the corresponding lanes of a Northern blot illustrating the relative levels of Ha-ras transcripts expressed by the same tissue samples. Densitometric analysis of the autoradiogram in Figure 3B indicated that no significant difference existed between the intensity of the 3.6 kb Ha-ras Bam HI DNA band in tumors expressing enhanced levels of Ha-ras RNA or the Bam HI DNA band from initiated/promoted skin and untreated epidermis which produced low levels of Ha-ras RNA. This result indicates that the increased expression of Ha-ras observed in early papillomas and in some late-stage papillomas and carcinomas is not due to amplification of the Ha-ras gene.

Table 1. Ha-ras Expression at Various Stages of Epidermal Two-Stage Carcinogenesis

Stage of Tumorigenesis	Elevated Expression of Ha-ras RNA
Untreated epidermis	negative
Epidermis treated with (+)BPDE-anti	negative
Epidermis treated with (+)BPDE-anti and 1 week TPA	negative
Epidermis treated with (+)BPDE-anti and 6 weeks TPA	negative
Early papillomas [induced with (+)BPDE-anti and 14 weeks TPA]	positive (2/2)
Older papillomas [(induced with (+)BPDE-anti and 26 weeks TPA (>6 months old)]	50% positive (6/13)
Squamous cell carcinomas [(induced with (+)BPDE-anti, 26 weeks TPA, and no treatment for an additional 5 months]	60% positive (3/5)

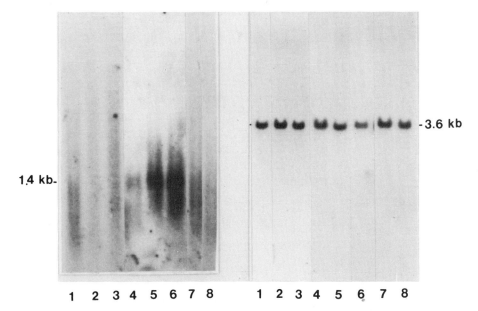

1 2 3 4 5 6 7 8   1 2 3 4 5 6 7 8

Fig. 3. Comparison of Northern and Southern blot hybridization at different stages of tumorigenesis in
SENCAR mouse epidermis. Panel A represents selected lanes from Northern hybridization studies
illustrating low and high expression of Ha-ras
RNA. Panel B represents the corresponding samples
of DNA from the same epidermis and tumors represented in lanes 1 through 8 of Panel A. Lane 1,
untreated epidermis; lane 2, initiated epidermis;
lane 3 and 4, epidermis treated with initiation
plus 1 or 6 weeks of TPA, respectively; lane 5,
"early" papilloma; lane 6, squamous cell carcinoma; lane 7 and 8, epidermis treated with
"sham" initiator (-) BPDE-anti, followed by 1 or
6 weeks of TPA, respectively.

Expression of myc Sequences During Initiation and Promotion

In view of the fact that a number of laboratories have presented
evidence for the involvement of more than one oncogene during tumorigenesis and cell transformation (22,23,24), we have screened epidermis and
tumors previously shown to possess elevated expression of Ha-ras RNA for
concurrent expression of myc proto-oncogene sequences. Figure 4 illustrates the results of Northern blot hybridization studies using a radiolabeled probe for the v-myc oncogene (18). This probe identifies a 2.4 kb
species of RNA present at high levels in two early papillomas but absent
in 6 carcinomas tested. The papillomas represented in lanes 2 and 3 of
Figure 4 correspond to lanes 5 and 6 of Figure 2 (Ha-ras Northern blot)
above. Significant levels of myc sequences were also expressed in normal
(untreated) SENCAR mouse epidermis (Fig 4, lane 1). These results demonstrate that more than one proto-oncogene can be expressed concurrently in
the same tumors at very early stages of murine epidermal tumor development. Additional experiments are currently underway in our laboratory to
establish what role(s) if any, these sequences might play in tumorigenesis.

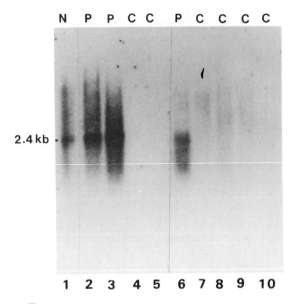

Fig. 4. Myc-Northern blot hybridization of RNA from epidermis at various stages of tumorigenesis. N =
untreated epidermis; P = papillomas; C = squamous
cell carcinoma.

Southern blot hybridization analysis was also carried out on samples
of DNA from individual tumors to determine whether the c-myc gene had
undergone gene amplification. Hybridization with the radiolabeled c-myc
DNA probe detected a single 5.6 kb fragment in mouse DNA as expected (19).
Densitometry studies indicated no evidence of amplification of the murine
c-myc sequence, as presented in Table 2.

Concurrent Expression of Endogenous Murine Retrovirus Genes and Oncogenes
During Skin Carcinogenesis

Recent evidence suggests that the process of carcinogenesis may be
influenced by interactions between viruses, environmental chemical carcinogens, and noncarcinogenic tumor promoters (13). The RNA tumor viruses
(retroviruses) are of particular interest in this regard, since the cells
of most vertebrates contain endogenous information for producing C-type
RNA retroviruses (25). These proviral genes can be induced to replicate or
to undergo alterations in expression in the presence of various chemical
carcinogens (13). With these points in mind, we were interested in determining whether the two-stage protocol of initiation with BPDE-anti followed by TPA promotion was inducing alterations in the expression of endogenous murine leukemia virus (MuLV) genes concurrently with or prior to
oncogene expression. To address this question, we used the procedure of

Table 2. Summary of c-myc Blot Hybridization

Stage of Epidermal Tumorigenesis	Relative Intensity of myc Fragment in Southern Blot (relative to untreated epidermis)
Untreated epidermis	1.0
Epidermis treated with (+)BPDE-anti, plus 6 weeks TPA	0.9
Papilloma induced by (+)BPDE-anti and 14 weeks of TPA (Showing elevated myc expression)	1.1
Papilloma induced by (+)BPDE-anti and 14 weeks of TPA (showing elevated myc expression)	1.0
Carcinoma induced by (+)BPDE-anti and 26 weeks of TPA (showing no detectable myc expression)	1.0

Thomas (26) to wash Northern blots previously hybridized with oncogene probes and re-hybridized them with a radiolabeled probe (20) for the env gene of MuLV which codes for a 70,000 MW viral envelope glycoprotein. Figure 5 presents a comparison of hybridization patterns of the Ha-ras probe, the v-myc probe, and the MuLV env probe on the same Northern blot containing RNA from normal mouse epidermis (lane 1), 14-week old papillomas (lanes 2,3,6,7), and carcinomas (lanes 4,5,8-12). The lower panel in Figure 5 indicates that significant levels of MuLV env RNA were expressed in an early papilloma (lane 3) and in a squamous cell carcinoma (lane 5). This early papilloma was concurrently expressing elevated levels of the Ha-ras and myc-specific RNA sequences as well. In view of the fact that MuLV genomes process long terminal repeat (LTR) regions which contain genetic promoters capable of turning on the expression of downstream sequences (27,28), we are currently carrying out further studies to determine whether induction of MuLV genes may subsequently influence the expression of neighboring sequences during skin tumorigenesis.

DISCUSSION

Activated ras oncogenes have been demonstrated in a wide variety of human tumors (29,31), as well as in carcinogen-induced mammary carcinomas in rats (5) and in papillomas and squamous carcinomas induced by the two-stage initiation/promotion protocol in mice (6,7). Earlier studies by Balmain and coworkers have shown the presence of an activated Ha-ras oncogene in a portion of papillomas and squamous cell carcinomas induced by exposure to DMBA and TPA. However, these tumors were screened after long periods of tumor promotion, and the 'ages' of the individual tumors varied or were undeterminable. Our present studies were undertaken to investigate the time course of expression of the Ha-ras oncogene in SENCAR mouse epidermis undergoing multi-stage tumorigenesis, in order to establish at what stage in epidermal carcinogenesis the enhanced expression of Ha-ras occurs.

In this article we present data indicating that elevated levels of Ha-ras RNA are expressed in papillomas induced by the two-stage protocol of initiation and promotion after only 14 weeks of TPA treatment. Our results are an extension of those of Balmain et al., who have presented

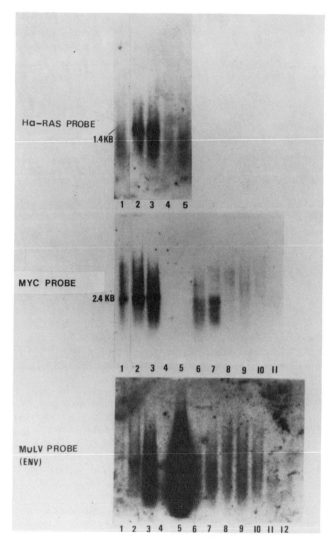

Fig. 5. Comparison of Northern blot hybridization studies using probes for Ha-ras, v-myc and MuLV env. Lane 1, normal epidermis; lanes 2,3,6 and 7, papillomas; lanes 4,5,8-12, squamous cell carcinomas.

evidence that Ha-ras activation can occur in pre-malignant tumors (i.e., papillomas) (6,7). Furthermore, we have shown that this enhancement of expression apparently is not the result of amplification of the murine Ha-ras proto-oncogene, since Southern blot analysis of DNA indicates that the same gene copy number is present in tumors exhibiting enhanced expression of Ha-ras as in untreated epidermis and epidermis immediately following initiation and a brief period of promotion. We conclude from these studies that enhanced expression of Ha-ras sequences can be a very early event in tumor development, occurring in papillomas which become grossly visible on the dorsal epidermis of the mouse after only 7-9 weeks of promotion with TPA. The increased expression of Ha-ras is not due merely to the increased proliferative activity of papillomas, since treatment of the mouse epidermis with TPA alone for 6 weeks (which induces epidermal hyperplasia) did not produce elevation of Ha-ras RNA. Furthermore, when epidermis surrounding individual papillomas and carcinomas was screened for Ha-ras sequences by the dot blot hybridization method, we observed no elevated levels of Ha-ras RNA in areas adjacent to tumors exhibiting enhanced Ha-ras expression (data not shown).

In addition to our studies on Ha-ras expression at different stages of papilloma development, we examined epidermis and tumors for evidence of concurrent expression of another oncogene, the myc proto-oncogene. Northern blots were washed free of labeled Ha-ras probe and re-hybridized with a v-myc probe (18), which detected a 2.4 kb species present at significant levels in normal epidermis as well as in the two early papillomas which had previously been shown to exhibit enhanced expression of Ha-ras. Interestingly, none of the carcinomas exhibited substantial levels of myc expression. Although the function of the c-myc proto-oncogene has not yet been established, it has been speculated that myc may belong to a class of oncogenes whose role is essential for development of tissues, not only to stimulate proliferation of immature stem cells, but also to act as a regulator of differentiation (4). Our data suggest that myc expression occurs normally in untreated epidermis and at early stages of tumor development, but may diminish later in tumorigenesis. In view of the fact that myc has been postulated to have a role in cell proliferation/differentiation control, further experiments are warranted to examine the level of differentiation of individual tumors in an attempt to correlate myc expression with the relative degree of differentiaion in tumors at various stages of development.

As a third criterion for studying altered gene expression during epidermal carcinogenesis, we have presented evidence for the induction of endogenous MuLV transcripts in tumors produced by the two-stage protocol of initiation and promotion. These experiments were of interest to us in light of the recent evidence suggesting that carcinogenesis may be influenced by interactions between viruses, environmental chemical carcinogens, and noncarcinogenic tumor promoters (13). In this paper we have presented a comparison of MuLV env gene expression and expression of both the Ha-ras and myc oncogenes. Our results indicate that an early papilloma shown previously to be expressing elevated Ha-ras and myc sequences was also producing enhanced levels of env transcripts, compared to levels in normal epidermis. Although a role for MuLV gene expression in chemical carcinogenesis has not been established, these viral sequences are extremely well characterized in the mouse genome, and therefore may represent a molecular marker(s) for studying the effects induced by chemical carcinogens and tumor promoters at the molecular level. However, additional studies are required to establish whether activation or altered patterns of MuLV gene expression following carcinogen exposure coincide with particular stages of carcinogenesis and tumor development in mouse epidermis in vivo.

In summary, we have employed the two-stage skin carcinogenesis model in SENCAR mice to investigate the respective roles played by carcinogenic initiators and noncarcinogenic tumor promoters in inducing alterations in gene expression in epidermis at various stages of carcinogenesis in vivo. For this purpose we have examined the patterns of expression of the Ha-ras and myc oncogenes, and the MuLV env gene, all of which represent endogenous murine sequences which have been highly conserved in the mouse genome. We have presented evidence that enhanced expression of Ha-ras may be an early event in tumor development, occurring in papillomas induced by 14 weeks of TPA treatment. Evidence has also been presented that myc RNA is expressed concurrently with Ha-ras sequences in some papillomas, although the molecular effects of this simultaneous expression is unknown at present. Furthermore, these elevated levels of expression are not apparently due to amplification of either the Ha-ras or c-myc proto-oncogenes in the mouse. Studies are continuing in our laboratory in hopes of establishing whether altered patterns of oncogene expression are required for particular stages of tumor development in vivo.

ACKNOWLEDGMENTS

We wish to thank Dr. Allen Balmain, of the Beatson Institute, for providing plasmid clones of the Ha-ras and v-myc oncogenes. The helpful discussion and criticisms offered by Dr. Samuel Cohen and Dr. Edward Bresnick are gratefully acknowledged. This research was supported by NCI grants CA40847, CA34962, by a Basic Research Support Grant from the University of Texas System Cancer Center, M.D. Anderson Hospital and Tumor Institute, and by an LB506 grant from the Nebraska Board of Health.

REFERENCES

1. J. M. Bishop, Cellular oncogenes and retroviruses, Ann. Rev. Biochem. 52:301 (1983).
2. R. W. Craig, and A. Bloch, Early decline in c-myb oncogene expression in the differentiation of human myeloblastic leukemia (ML-1) cells induced with 12-0-tetradecanoylphorbol-13-acetate, Cancer Res. 44:442 (1984).
3. R. Muller, D. J. Slamon, J. M. Tremblay, M. J. Cline, and I. M. Verma, Differential expression of cellular oncogenes during pre- and post-natal development of the mouse, Nature 299:640 (1982).
4. C.-H. Heldin, and B. Westermark, Growth factors: Mechanism of action and relation to oncogenes, Cell 37:9 (1984).
5. S. Sukumar, V. Notario, D. Martin-Zanca, and M. Barbacid, Induction of mammary carcinomas in rats by nitrosomethylurea involves malignant activation of H-ras-1 locus by single point mutations, Nature 306:658 (1983).
6. A. Balmain, and I. Pragnell, Mouse skin carcinomas induced in vivo by chemical carcinogens have a transforming Harvey-ras oncogene, Nature 303:72 (1983).
7. A. Balmain, M. Ramsden, G. T. Bowden, and J. Smith, Activation of the mouse cellular Harvey-ras gene in chemically induced benign skin papillomas, Nature 307:658 (1984).
8. T. J. Slaga, W. J. Bracken, G. Gleason, W. Levin, H. Yagi, D. M. Jerina, and A. H. Conney, Marked differences in the skin tumor initiating activities of the optical enantiomers of the diastereomeric benzo(a)pyrene 7,8-diol-9,10-epoxides, Cancer Res. 39:67 (1979).
9. M. K. Buening, P. G. Wislocki, W. Levin, H. Yagi, D. Thakker, H. Akagi, N. Koreeda, D. M. Jerina, and A. H. Conney, Tumorigenicity of the optical enantiomers of the diastereomeric benzo(a)pyrene-7,8-diol-9,10-epoxides in newborn mice: Exceptional activity of (+)7β,8α-dihydroxy-9α,10α-epoxy-7,8,9,10-tetrahydrobenzo(a)pyrene, Proc. Natl. Acad. Sci. U.S.A. 75:5358 (1978).
10. S. W. Ashurst, G. M. Cohen, J. DiGiovanni, and T. J. Slaga, Formation of benzo(a)pyrene-DNA adducts and their relationship to tumor initiation in mouse epidermis, Cancer Res. 43:1024 (1983).
11. M. Koreeda, P. D. Moore, P. G. Wislocki, W. Levin, A. H. Conney, H. Yagi, and D. M. Jerina, Binding of benzo(a)pyrene 7,8-diol-9,10-epoxides to DNA, RNA and protein of mouse skin occurs with high stereoselectivity, Science 199:778 (1978).

12. J. C. Pelling, T. J. Slaga, and J. DiGiovanni, Formation and persistence of DNA, RNA, and protein adducts in mouse skin exposed to pure optical enantiomers of 7β,8α-dihyroxy-9α,10α-epoxy-7,8,9,10-tetrahydrobenzo(a)pyrene in vivo, Cancer Res. 44:1081 (1984).

13. R. W. Tennant, and R. J. Rascati, Mechanisms of cocarcinogenesis involving endogenous retroviruses, in: "Carcinogenesis", Vol. 5: Modifiers of Chemical Carcinogenesis, T.J. Slaga ed., Raven Press, New York (1980).

14. A. Ullrich, J. Shine, J. Chirgwin, R. Pictet, E. Tischer, W. Rutter, and H. M. Goodman, Rat insulin genes: Construction of plasmids containing the coding sequences, Science 196:1313 (1977).

15. D. A. Goldberg, Isolation and partial characterization of the Drosphila alcohol dehydrogenase gene, Proc. Natl. Acad. Sci. U.S.A. 77:5794 (1980).

16. P. W. Rigby, M. Dieckman, C. Rhodes, P. Berg, Labeling deoxyribonucleic acid to high specific activity in vitro by nick translation with DNA polymerase I, J. Mol. Biol. 113:237 (1977).

17. R. W. Ellis, D. DeFeo, J. M. Maryak, H. A. Young, T. Y. Shih, E. Chang, D. R. Lowy, and E. M. Scolnick, Dual evolutionary origin for the rat genetic sequences of Harvey murine sarcoma virus, J. Virol. 36:408 (1980).

18. B. Vennstrom, C. Moscovici, H. M. Goodman, and J. M. Bishop, Molecular cloning of the avian myelocytomatosis virus genome and recovery of infectious virus by transfection of chicken cells, J. Virol. 39:625 (1981).

19. G. L. Shen-Ong, E. J. Keath, S. P. Piccoli, and M. D. Cole, Novel myc oncogene RNA from abortive immunoglobulin gene recombination in mouse plasmacytomas, Cell 31:443 (1982).

20. S. K. Chattopadhyay, M. W. Cloyd, D. L. Linemeyer, M. R. Lander, E. Rands, and D. R. Lowy, Cellular origin and role of mink cell focus-forming viruses in murine thymic lymphomas, Nature 295:25 (1982).

21. E. M. Southern, Detection of specific sequences among DNA fragments separated by gel electrophoresis, J. Mol. Biol. 98:503 (1975).

22. H. Land, L. F. Parada, and R. A. Weinberg, Tumorigenic conversion of primary embryo fibroblasts requires at least two cooperating oncogenes, Nature 304:596 (1983).

23. P. Yaswen, M. Goyette, P. R. Shank, and N. Fausto, Expression of c-Ki-ras, C-Ha-ras, and c-myc in specific cell types during hepatocarcinogenesis, Molec. Cell. Biol. 5:780 (1985).

24. D. G. Thomassen, T. M. Gilmer, L. A. Annab, and J. C. Barrett, Evidence for multiple steps in neoplastic transformation of normal and preneoplastic Syrian hamster embryo cells following transfection with Harvey murine sarcoma virus oncogene (v-Ha-ras), Cancer Res. 45:726 (1985)

25. G. Todaro, in: "Viruses in Naturally Occurring Cancer," Book B, M. Essex, G. Todaro, H. Zur Hausen, eds., Cold Spring Harbor Congress on Cell Proliferation, Vol. 7 (1980).

26. P. Thomas, Hybridization of denatured RNA and small DNA fragments transferred to nitrocellulose, Proc. Natl. Acad. Sci. U.S.A. 77:5201 (1980).

27. H. M. Temin, Function of the retrovirus long terminal repeat, Cell 28:3 (1982).

28. T. G. Wood, M. L. McGeady, D. G. Blair, and F. G. Vande Woude, Long terminal repeat enhancement of v-mos transforming activity: Identification of essential regions, J. Virol. 46:726 (1983).

29. C. J. Der, T. G. Krontiris, and G. M. Cooper, Transforming genes of human bladder and lung carcinoma cell lines and homologous to the ras genes of Harvey and Kirsten sarcoma viruses, Proc. Natl. Acad. Sci. U.S.A. 79:3637 (1982).

30. L. F. Parada, C. J. Tabin, C. Shih, and R. A. Weinberg, Human EJ bladder carcinoma oncogene is homologue of Harvey sarcoma virus ras gene. <u>Nature</u> 297:474 (1982).

31. E. Santos, S. R. Tronick, S. A. Aaronson, S. Pulciani, and M. Barbacid, T24 human bladder carcinoma oncogene is an activated form of normal human homologue of Balb- and Harvey-MSV transforming genes, <u>Nature</u> 298:343 (1982).

ACETAMINOPHEN TOXICITY IN ISOLATED HEPATOCYTES

D.S. Davies, L.B.G. Tee, C. Hampden, and A.R. Boobis

Department of Clinical Pharmacology
Royal Postgraduate Medical School
Ducane Road, London W12.

INTRODUCTION

Acetaminophen is an analgesic and antipyretic agent which is readily available and extensively used throughout the world. Therapeutic doses of up to 4g daily produce few side-effects but doses in excess of 10-15g can cause hepatic necrosis whilst larger doses are often fatal [1].

The metabolic processes leading to hepatic damage following exposure to high doses of acetaminophen were discovered by Gillette, Mitchell and colleagues in the 1970's [2-6]. Acetaminophen is largely eliminated by conjugation (sulphate and glucuronide), a minor pathway (6-8% of dose) being oxidation by cytochrome P-450 to yield an electrophilic intermediate which forms an adduct with reduced glutathione (GSH) and is excreted as the mercapturic acid or cysteine conjugate. Following overdosage, sufficient of the electrophilic intermediate is formed to deplete hepatic GSH. In the absence of GSH the reactive intermediate is free to damage the cell.

Over the past 10 years considerable evidence has accummulated that the reactive metabolite of acetaminophen is N-acetyl-p-benzoquinonei-mine (NABQI) [7-10] but the mechanism of hepatotoxicity remains to be established. In the absence of GSH, NABQI covalently binds to tissue macromolecules, particularly proteins. This led to suggestions that this is the cause of cell death. However, a number of groups have shown a dissociation between covalent binding and toxicity.

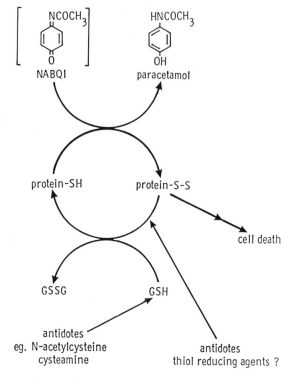

Figure 1. A mechanism for the toxicity
of acetaminophen (paracetamol) due to
oxidative damage by NABQI

When a convenient synthesis of NABQI was devised [8] and its properties
investigated it was found that not only is it a potent electrophile which
forms adducts with GSH but that it also oxidizes GSH to GSSG with
regeneration of acetaminophen. The oxidizing properties of NABQI provide
an alternative mechanism to covalent binding for its toxicity, [11] possibly
by oxidation of thiol groups to disulphide bonds in key enzymes, thereby
rendering them inactive (Figure 1). Studies conducted by A.C. Huggett [12]
in this laboratory demonstrated that NABQI inactivated model SH-containing
enzymes and that this inactivation was reversible with disulphide
reductants such as dithiothreitol (DTT). The target for NABQI in the
hepatocyte might be membrane bound ATP-dependent Ca^{2+}-translocases as
described for t-butyl hydroperoxide [13]. Inhibition of these enzymes
causes a rise in cytosolic Ca^{2+} leading to disruption of the cytoskeleton,
plasma membrane blebbing and cell death [14].

    If acetaminophen, through NABQI, causes toxicity by direct oxidation then it should be possible to reverse the progression to cell death by the introduction of a disulphide reducing agent, such as DTT. The studies described here were set up to examine this hypothesis and are presented in detail elsewhere [15].

REVERSAL OF ACETAMINOPHEN TOXICITY WITH DTT

In preliminary studies, it was established that incubation of freshly isolated hamster hepatocytes for 90 minutes with 2.5mM acetaminophen did not significantly decrease cell viability (Figure 2) but did cause a significant depletion of GSH. Following incubation for 90 minutes, the hepatocytes were washed three times and then resuspended for further incubation in buffer alone, in buffer containing 2.5mM acetaminophen or 1.5mM DTT. Using carbon-14 labelled acetaminophen it was shown that the washing procedure removed all but the covalently bound radioactivity.

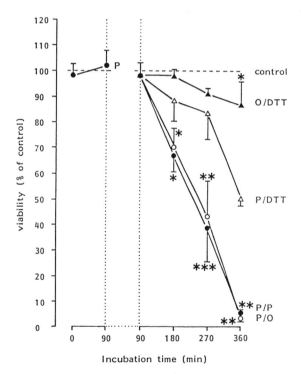

Figure 2. The effects of acetaminophen (2.5mM) on the viability of hamster hepatocytes. Cells were incubated with acetaminophen for 90 minutes washed and resuspended for further incubation in buffer (P/O), 2.5mM acetaminophen (P/P), 1.5mM DTT (P/DTT). Cells not previously exposed to acetaminophen were also incubated with DTT (0/DTT) after washing. * p<0.05 ** p<0.01 *** p<0. 001 Compared to p/DTT

During the second phase of incubation cell viability fell dramatically in the presencee or absence of acetaminophen (Figure 2). This demonstrated that the events leading to cell death in this phase of the study were independent of the continuing presence of acetaminophen or its non-bound metabolites. Addition of DTT during the second phase of incubation largely prevented the loss in cell viability. The mechanism whereby DTT prevents cell death in the Phase 2 incubation is not established. However, it was shown that addition of DTT did not replenish the depleted GSH nor did it cause a reversal of covalent binding of acetaminophen metabolites. Preliminary data suggest that it repairs the oxidative damage induced by NABQI on the Ca^{2+}-sequestering enzymes which control the concentration of intracellular calcium.

STUDIES WITH ANTIDOTES TO ACETAMINOPHEN TOXICITY

The studies of Gillette and colleagues [2-6] established that GSH depletion was a pre-requisite for acetaminophen-induced hepatotoxicity. They suggested that precursors of GSH such as N-acetylcysteine or methionine, by replenishing depleted stores of hepatic GSH which would then inactive the toxic metabolite, would be useful antidotes. This was found to be true in animals and in over-dosed patients [16]. N-Acetylcysteine and, to a lesser extent, methionine are widely used in the treatment of poisoning with acetaminophen. However, N-acetylcysteine has proved of value when given 12-15 hours after ingestion of acetaminophen, when metabolism is often complete, a fact incompatible with the provision of GSH to inactivate the reactive metabolite.

The protocol described above for studies with DTT was used to examine the effects, if any, of N-acetylcysteine and methionine on cell viability, GSH levels and covalent binding during Phase 2 incubation. Freshly isolated hamster hepatocytes were incubated for 90 minutes with 2.5mM acetaminophen, the cells were washed to remove the drug and metabolites and resuspended in buffer alone or buffer containing 1.25mM N-acetylcysteine. During the Phase 2 incubation, cell viability fell dramatically whether in the absence or presence of acetaminophen (figure 3). However, addition of N-acetylcysteine to the buffer largely prevented the loss in viability. Since there was no acetaminophen present during the Phase 2 incubation the beneficial effects of N-acetylcysteine in this model are clearly unrelated to its role as a pre-cursor of GSH for the inactivation of NABQI. These experiments were repeated with the

substitution of 1.25mM methionine for N-acetylcysteine. Methionine, in
this model failed to prevent the loss of cell viability during the Phase 2
incubation (Figure 4). Neither antidote had an effect on the level of
covalent binding of radioactivity which remained constant during Phase 2
incubation. However measurement of the effects on glutathione levels
produced interesting results.

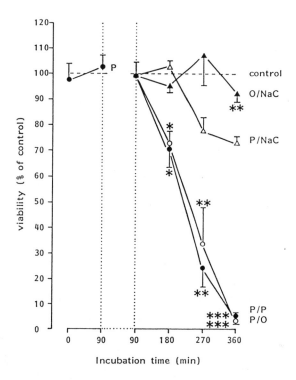

Figure 3. The effects of acetaminophen on cell viability.
Details as for Fig. 2 with substitution of 1.25mM
N-acetylcysteine (NaC) for DTT

Methionine produced a 2-fold increase in GSH in cells not previously
exposed to acetaminophen but failed to change the depleted GSH levels in
cells incubated with the drug during Phase 1. In control cells,
N-acetylcysteine caused a 3-fold increase in GSH but, unlike methionine,
it also completely restored GSH in acetaminophen exposed cells. Thus, it
is possible that the beneficial effect of N-acetylcysteine in this model
is mediated through the generation of GSH, the cell's endogenous
disulphide reducing agent.

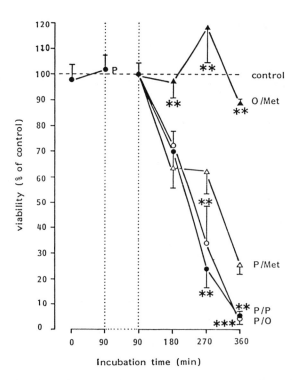

Figure 4. The effects of acetaminophen on cell viability.
Details as Fig. 2 with substitution of 1.25mM methionine
(Met) for DTT.

The failure of methionine to generate GSH in this model or to protect
cells requires further investigation. However, it is of interest to note
that the generation of L-cysteine from methionine via homocysteine and
cystathionine involves two enzymes, cystathionine synthase and
cystathioninase, which are readily inactivated by S-H oxidizing agents
[17,18]. This inactivation is reversed by compounds such as DTT. These
enzymes may be inhibited in hepatocytes which are oxidatively damaged by
NABQI.

The success with N-acetylcysteine also needs further study. The rate
limiting step in the synthesis of glutathione from L-cysteine is catalyzed
by α-glutamyl-cysteine synthetase, an enzyme which requires free
sulphhydryl groups for activity [19] and which might be expected to be
inactivated by NABQI. If α-glutamyl-cysteine synthetase is inactivated by
NABQI this might be reversed by N-acetylcysteine directly or by the
cysteine it generates. Either way this would lead to a replenishment of
glutathione and the repair of the NABQI-induced oxidative damage in the cell.

SPECIES AND INTER-INDIVIDUAL DIFFERENCES IN SENSITIVITY TO ACETAMINOPHEN
HEPATOTOXICITY

It is well established that there are large species differences in
sensitivity to the hepatotoxic effects of acetaminophen. Mice and
hamsters are particularly sensitive, exhibiting marked liver changes with
doses of 150 to 200mg/kg, whilst rats are relatively resistant even at
doses of 1500mg/kg. Man's sensitivity is difficult to determine because
of the poor quality of the information on dose size and time of ingestion
in poisoned patients. However, it is generally considered that doses of
15-20g (200 to 300mg/kg) will produce liver damage.

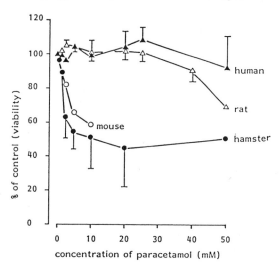

Figure 5. Effects of acetaminophen on the viability
of hepatocytes during incubation for 180 minutes.

Freshly isolated hepatocytes from mouse, rat, hamster and man [20] were
obtained by collagenase perfusion. Incubation of hepatocytes with
increasing concentrations of acetaminophen for 180 minutes caused
considerable loss of viability in hepatocytes from hamster and mouse but
cells from rats and humans were resistant (Figure 5). However,
hepatocytes from all four species were equally sensitive to the toxic
effects of NABQI, suggesting that rate of formation of NABQI dictates
species sensitivity. This is confirmed on examination of rates of
oxidation of acetaminophen measured in vitro (Table 1) or in vivo (Table
2). Man, like rat, is poor at generating the reactive metabolite of
acetaminophen.

These data on rates of acetaminophen oxidation and sensitivity of hepatocytes do not conform to the clinical impression that man, like hamster and mouse, is sensitive to 200mg/kg of acetaminophen. However, further examination of acetaminophen oxidation by man in vitro and in vivo reveals large inter-individual differences. Man's reported susceptibility may be biased by the over-representation in the sample of those subjects with high activity who develop profound liver damage.

Table 1. Species differences in the covalent binding of acetaminophen to microsomal proteins in vitro

Species	N	Covalent binding (nmol/mg/min)
Human	28	0.0258 ± 0.0023
Mouse	3	0.1880 ± 0.0030
Rat	8	0.0203 ± 0.0040
Hamster		0.180

Values are mean ± S.E.M.

Table 2. Species differences in the clearance of acetaminophen to its mercapturic acid in vivo

Species	Clearance (ml/min/kg)	% Dose as mercapturate	Clearance to mercapturate (ml/min/kg)
Human	4.1	5.0	0.205
Mouse	25.7	13.3	3.418
Rat	10.5	3.6	0.378

Data are from references 2 and 5.

CONCLUSIONS

Acetaminophen is hepatoxic because of two distinct, but complementary, effects of its reactive metabolite, NABQI.

1.   NABQI depletes GSH in hepatocytes rendering them susceptable to oxidative attack.

2.   NABQI is a potent oxidizing agent in biological systems and can inactivate key SH-dependent enzymes such as those controlling intra-cellular calcium concentrations. This oxidative damage is lethal in glutathione depleted cells.

Some antidotes to acetaminophen toxicity may have a dual mechanism of action. Firstly, they act as precursors for the generation of GSH which inactivates NABQI as it is formed. Thus, N-acetylcysteine and methionine can prevent hepatoxicity when given with, or soon after, acetaminophen. However, the present studies demonstrate that N-acetylcysteine has an action beyond the metabolic stage which probably involves the repair of oxidative damage induced by NABQI, either directly or through the generation of cysteine and/or glutathione. Methionine does not show this latter property and probably ought not to be used to treat acetaminophen poisoning except in the first few hours immediately after ingestion.

Finally, man is probably more resistant to the hepatoxic effects of acetaminophen than is apparent from clinical data. The rate of generation of NABQI is probably the major factor determining susceptibility to acetaminophen toxicity and this exhibits large inter-individual differences in man. Subjects with high activity will be as sensitive as mice or hamsters while those with low activity will be as resistant as rats. It remains to be established whether there is a genetic polymorphism in the oxidation of acetaminophen in man.

REFERENCES

1.   J. L. Davidson, and W. N. Eastham, Acute liver necrosis following overdose of paracetamol, Brit. Med. J. 2:497-499 (1966).
2.   J. R. Mitchell, D. J. Jollow, W. Z. Potter, D. C. Davis, J. R. Gillette, and B. B. Brodie, Acetaminophen-induced hepatic necrosis I. Role of drug metabolism, J. Pharmacol. Exp. Ther. 187:185-194 (1973).

3. J. R. Mitchell, D. J. Jollow, W. Z. Potter, J. R. Gillette, and B. B. Brodie, Acetaminophen-induced hepatic necrosis. IV. Protective role of glutathione, J. Pharmacol. Exp. Ther. 187:211-217 (1973).

4. D. J. Jollow, J. R. Mithcell, W. Z. Potter, D. C. Davis, J. R. Gillette, and B. B. Brodie, Acetaminophen-induced hepatic necrosis. II. Role of covalent binding in vivo, J. Pharmacol. Exp. Ther. 187:195-202 (1973).

5. D. J. Jollow, S. S. Thorgeirsson, W. Z. Potter, M. Hashimoto, and J. R. Mitchell, Acetaminophen-induced hepatic necrosis. VI. Metabolism disposition of toxic and nontoxic doses of acetaminophen, Pharmacology 12:251-271 (1974).

6. W. Z. Potter, S. S. Thorgeirsson, D. J. Jollow, and J. R. Mitchell, Acetaminophen-induced hepatic necrosis. V. Correlation of hepatic necrosis, covalent binding and glutathione depletion in hamsters. Pharmacology 12:129-143 (1974).

7. D. J. Miner, and P. T. Kissinger, Evidence for the involvement of N-acetyl-p-quinoneimine in acetaminophen metabolism, Biochem. Pharmcol. 28:3285-3291 (1979).

8. I. A. Blair, A. R. Boobis, D. S. Davies, and T. M. Cresp, Paracetamol oxidation: synthesis and reactivity of N-acetyl-p-benzoquinoneimine, Tetrahedron Lett. 21:4947-4950 (1980).

9. J. A. Hinson, L. R. Pohl, and J. R. Gillette, N-Hydroxyacetaminophen: a microsomal metabolite of N-hydroxyphenacetin but apparently not of acetaminophen, Life Sci. 24:2133-2138 (1979).

10. D. C. Dahlin, G. T. Miwa, A. Y. H. Lu, and S. D. Nelson, N-Acetyl-p-benzoquinoneimine: A cytochrome P-450-mediated oxidation product of acetaminophen. Proc. Natl. Acad. Sci. USA 81:1327-1331 (1984).

11. D. S. Davies, Drug hepatotoxicity: formation and importance of reactive metabolites, in: "Drug Reactions and the liver", M. Davis, J. M. Tredger, and R. Williams, eds, Pitman Medical, London, (1981).

12. A. C. Huggett, The Mechanism of Paracetamol Toxicity, PhD Thesis, University of London.

13. G. Bellomo, H. Thor, and S. Orrenius, Increase in cytosolic Ca^{2+} concentration during t-butyl hydroperoxide metabolism by isolated hepatocytes involves NADPH oxidation and mobilization of intracellular Ca^{2+} stores, FEBS Letts 168:38-42 (1984).

14. S. A. Jewell, G. Bellomo, H. Thor, S. Orrenius, and M. T. Smith, Bleb formation in hepatocytes during drug metabolism is caused by disturbances in thiol and calcium ion homeostasis, Science 217:1257-1259 (1982).

15. L. B. G. Tee, A. R. Boobis, A. C. Huggett, and D. S. Davies, Reversal of paracetamol toxicity in isolated hepatocytes by dithiothreitol, (Submitted for publication 1985).

16. L. F. Prescott, Glutathione: a protective mechanism against hepatotoxicity, Biochem. Soc. Trans. 10:84-85 (1982).

17. S. Kashiwamata, and D. M. Greenberg, Studies on cystathionine synthase of rat liver properties of the highly purified enzyme, Biochim. et Biophys. acta 212:488-500 (1979).

18. D. Deme, O. Durieu-Trautmann, and F. Chatagner, The thiol groups of rat liver cystathionas: Influence of pyridoxal phosphate,1-homoserine and 1-alanine on the effect of p-chloromercuribenzoate and 5,5'-Dithiobis-(2-nitrobenzoate) on the enzyme, Eur. J. Biochem. 20:269-274 (1971).

19. G. F. Seeling, and A. Meister, α-Glutamylcysteine synthetase. Interactions of an essential sulfhydryl group. J. Biol. Chem. 259:3534-3538 (1984).

20. L. B. G. Tee, T. Seddon, A. R. Boobis, and D. S. Davies, Drug metabolising activity of freshly isolated human hepatocytes, Brit. J. Clin. Pharmac. 19:279-294 (1985).

MECHANISM OF ORGAN SPECIFICITY IN NITROSAMINE CARCINOGENESIS

Michael C. Archer, K. Charles Silinskas and Peter F. Zucker

Department of Medical Biophysics, University of Toronto
Ontario Cancer Institute
500 Sherbourne Street
Toronto, Canada, M4X 1K9

The nitrosamines exhibit remarkable species and tissue specificity in their carcinogenic action (Shank and Magee, 1981). The biological basis of this specificity, however, is poorly understood. A primary target for many nitrosamines, particularly unsymmetrical dialkylnitrosamines, is the rat esophagus. These nitrosamines induce high incidences of basal and squamous cell tumours, independent of the route of administration.

In order to study the mechanism of the tissue specificity of nitrosamines, we have been investigating the mode of action of nitrosomethylbenzylamine (NMBzA) and nitrosodimethylamine (NDMA) in both target and non-target tissues. In the rat, NMBzA is a potent esophageal carcinogen that produces no liver tumours, while NDMA is a potent hepatocarcinogen that produces no esophageal tumours (Shank and Magee, 1981). The tissue specificity of these compounds is not caused by preferential uptake of the nitrosamines into their respective target organs, since whole body autoradiography (Johansson and Tjalve, 1978; Kraft and Tannenbaum, 1980) and nitrosamine analysis of isolated tissues (Magee, 1956; Hodgson et al., 1980) have demonstrated that NDMA and NMBzA are widely distributed in the body. The liver is a major tissue for uptake of both compounds.

We began our studies by investigating whether there is tissue-specific metabolic activation of the nitrosamines. The activation pathway for these compounds is considered to be cytochrome P450-dependent hydroxylation at the a-carbon atom, a step which ultimately yields an alkylating agent and an aldehyde in stoichiometric amounts (Lai and Arcos, 1980). The alkylating agent can react with tissue nucleophiles, or with water to form the corresponding alcohol. We found that rat esophageal mucosa contains an enzyme with the properties of a cytochrome P450 which metabolizes NMBzA at a high rate (Labuc and Archer, 1982). NDMA is a poor substrate for this enzyme. Hepatic microsomes have long been known to metabolize NDMA, but we found that NMBzA is also a good substrate for this enzyme, the rates of metabolism being somewhat higher than for NDMA. Mucosal microsomes oxidize NMBzA at the benzylic carbon to yield benzaldehyde and a methylating agent 100 times faster than at the methyl carbon. This differential is 10-fold in the case of hepatic metabolism. These results support the studies of Hodgson et al. (1980) who showed that esophageal DNA was methylated following a single i.v. dose of NMBzA, whereas benzylation of DNA was undetectable.

The difference in carcinogenic activities of NMBzA and NDMA in rat esophagus may be related to the capacity of this tissue to activate NMBzA at a high rate, and its inability to activate NDMA. However, the difference in the carcinogenic activities of NMBzA and NDMA in the liver is clearly unrelated to differences in the metabolic activation of the two nitrosamines in this tissue. This conclusion is supported by studies that have shown that both nitrosamines methylate hepatic DNA to similar extents in the whole animal (Hodgson et al., 1980; Pegg, 1977) and in liver slices (Fong et al., 1979). Since the liver is a major tissue for uptake of NMBzA, and since the nitrosamine is rapidly metabolized to a reactive methylating agent which methylates DNA in the liver, it seemed likely that NMBzA would possess the ability to initiate neoplasia in the liver, but failed to produce tumours for some other reason. In order to test this hypothesis, we decided to utilize the established assays of Tsuda et al. (1980) and Pitot et al. (1978) for agents which initiate neoplasia in rat liver.

In our first experiments (Labuc and Archer, 1982), we decided to administer NMBzA at the same molar dose that would produce a positive response in the initiation assays by the hepatocarcinogen NDMA. A problem in these experiments, however, is the toxicity of NMBzA. A dose of 5 mg (33.5 μmol)/kg administered 18 hours post-partial hepatectomy killed about half of the rats within 3-4 days from esophageal necrosis. A lower dose would have been more desirable, but it was precluded by the uncertainty of obtaining a positive result with NDMA even at 33.5 μmol/kg. Two minor modifications of the original protocol of Tsuda et al. (1980) were made. First, in order to decrease the mortality of the NMBzA-treated rats, the time between nitrosamine administration and commencement of 2-acetylaminofluorene (AAF) feeding was increased from 2 to 3 weeks. Second, because NMBzA-treated rats consume less food than control or NDMA-treated animals, a pair feeding regimen was used to ensure that all rats received the same amount of dietary AAF.

Using this protocol, a dose of 33.5 μmol/kg NDMA produced 14.6 ± 4.1 γ-glutamyltranspeptidase (GGT)-positive foci/cm^2 liver, which was significantly ($p < 0.05$) greater than the background (0.7 ± 0.3 foci/cm^2). The same dose of NMBzA, however, produced 1.0 ± 0.4 foci/cm^2 which was not above background.

For the phenobarbital selection assay of Pitot et al. (1978), the concentration of phenobarbital in the drinking water was continuously adjusted so that all rats ingested comparable amounts of phenobarbital throughout the entire period. With this procedure, NDMA at 33.5 μmol/kg produced 4.0 ± 0.8 foci/cm^2, significantly ($p < 0.05$) above the background of 1.1 ± 0.3 foci/cm^2. Again NMBzA at the same dose produced no effect (0.6 ± 0.1 foci/cm^2).

We showed that the activity of microsomes prepared from rats 18 h following partial hepatectomy was somewhat lower than microsomes from intact liver, but the decrease in activity was similar for both compounds. Furthermore, neither NDMA nor NMBzA inhibited the first wave of DNA synthesis in regenerating liver. Thus our results indicated the somewhat surprising result that in contrast to NDMA, NMBzA lacks initiating activity in rat liver, and we concluded that while carcinogen activation and DNA modification per se may be necessary, they are not sufficient for initiation of neoplasia.

In order to substantiate this conclusion, we felt that it was important to determine whether the failure of NMBzA to initiate foci was caused by its inability to produce O^6-methylguanine in regenerating rat liver. Pereira et al. (1983) administered seven different methylating agents to partially hepatectomized rats and showed that although six compounds produced 7-methylguanine, only those four that produced O^6-methylguanine initiated GGT-positive foci. Pereira concluded that formation of O^6-methylguanine is required for initiation. We therefore administered NDMA and NMBzA to rats 18 h after partial hepatectomy at the same molar doses (33.5 μmol/kg) we had used in the initiation assays (Silinskas et al., 1984).

Figure 1A shows that the 7-methylguanine profiles in hepatic DNA were similar for both nitrosamines with maximum accumulations of the base occurring at about 8 h following administration. An almost linear loss of 7-methylguanine occurred over the next 64 h. Figure 1B shows that both NDMA and NMBzA produce O^6-methylguanine. Maximum accumulations of this base also occurred at 8 h. It is noteworthy that at all time points the levels of O^6-methylguanine after NMBzA administration were about half those following NDMA.

Since equimolar doses of NDMA and NMBzA produced dissimilar amounts of O^6-methylguanine, we considered the possibility that there may be a threshold level of O^6-methylguanine that is required to initiate foci, and that this level was not reached in our experiments with NMBzA. The data of Pereira et al (1983) for dimethylhydrazine, however, suggests that this is not the case. Dimethylhydrazine produced foci at a dose of 200 μmol/kg. At this dose, however, it produced only 3.7 μmol O^6-methylguanine/mol guanine in regenerating rat liver at 2.5 h. From Figure 1B it is clear that NMBzA gave about 12 μmol O^6-methylguanine/mol guanine after 2.5 h, which is more than three times that produced by dimethylhydrazine.

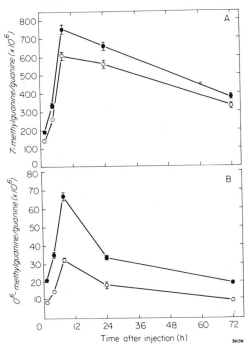

Figure 1.   Formation and loss of 7-methylguanine (A) and O^6-methylguanine (B) from rat liver DNA after administration of 33.5 μmol/kg NDMA (●) or NMBzA (O). Mean ±S.E.M.; N=3. Reprinted with permission from Silinskas et al., 1984.

In order to clarify this problem, it was clearly necessary to determine whether NDMA and NMBzA produced foci at doses which gave equal amounts of 0^6-methylguanine in hepatic DNA. Since increasing the dose of NMBzA resulted in unacceptable animal death due to esophageal necrosis, we were forced to approach the problem by lowering the dose of NDMA (Silinskas et al, 1985). We first showed that a dose of 26 μmol/kg NDMA gave approximately 28 μmol 0^6-methylguanine/mol guanine 8 h following administration of the nitrosamine to rats that had received a 2/3 partial hepatectomy 18 h previously (Fig. 2). This level of 0^6-methylguanine is not different from that produced by 33.5 μmol/kg NMBzA (Fig. 2). This non-linear production of 0^6-methylguanine with dose of NDMA is in agreement with the results of Pegg (1977) for the livers of intact rats at similar nitrosamine levels. Figure 2 illustrates the production of 7-methylguanine from NDMA is approximately linear with dose of the nitrosamine, again in agreement with the data of Pegg (1977). Figure 2 also illustrates that the effects of NDMA and NMBzA on the production of methylated bases are additive.

Figure 3 illustrates that when the dose of NDMA was lowered from 33.5 to 26 μmol/kg, the number of GGT-foci was reduced from about 14 to 7.5/cm^2 per liver. The latter value, however, was still well above the background level. It is clear, therefore, that at doses which produce comparable levels of 7-methylguanine and 0^6-methylguanine in hepatic DNA, NDMA initiates formation of GGT-positive foci while NMBzA does not. Furthermore, co-administration of NDMA and NMBzA produced no more foci that NDMA alone, even though the combined effect of the two nitrosamines on DNA methylation was additive. The inability of NMBzA to produce foci then, could not be explained by the existence of a threshold level of 0^6-methylguanine below which no foci would be produced.

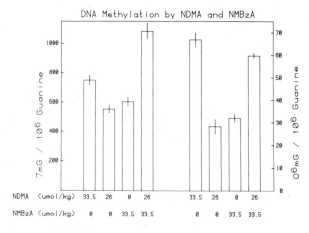

Figure 2. Effect of NDMA and NMBzA, alone or in combination, on methylated guanine levels (left panel 7-methylguanine, right panel 0^6-methylpanel guanine) in rat liver DNA 8 h after treatment.

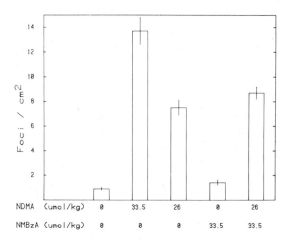

Figure 3. Production of GGT-positive foci in rat liver by NDMA and NMBzA
alone or in combination using the procedure of Tsuda et al.,
1980.

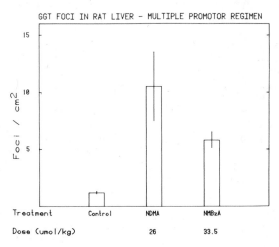

Figure 4. Production of GGT-positive foci in rat liver by NDMA or NMBzA
using as promotor a choline-methionine diet containing 1% orotic
acid and 0.6% phenobarbital.

The difference in initiating activity of NDMA and NMBzA in rat liver could be caused by the two nitrosamines attacking different cell populations (Labuc and Archer, 1982). The similar abilities of the two compounds to methylate hepatic DNA suggested that they should initiate hepatocytes to similar extents and therefore their differences in activity might be due to the differences in the behaviour of different populations of initiated cells towards promotors. To test this hypothesis, we decided to use a new selection procedure for initiated hepatocytes that combines several different promotor types (D.S.R. Sarma, personal communication).

In this assay, 26 μmol/kg NDMA or 33.5 μmol/kg NMBzA were administered to rats 18 h following a partial hepatectomy. Two weeks later, the animals were placed on a choline and methionine deficient diet containing 1% orotic acid and 0.06% phenobarbital. After 8 weeks on this regimen, the animals were sacrificed, and GGT-foci in liver sections were enumerated. The results are shown in Figure 4. As before, NDMA increased the number of foci significantly ($p < 0.05$) over the background. For the first time, however, NMBzA also produced GGT-foci. The number of foci produced by NMBzA was not significantly different from that produced by NDMA.

The results of this assay therefore, suggest that NMBzA is an initiating agent in rat liver, but clearly the ability of NMBzA to produce preneoplastic foci in the liver depends on the nature of the promotor used. Neither AAF/CCl$_4$ nor phenobarbital stimulate cells initiated by NMBzA to grow into foci, whereas some component(s) of the multiple promotor regimen is active in this regard. There are at least two models which could explain the results. In the first, NDMA and NMBzA initiate discrete populations of hepatocytes which differ in their sensitivities towards various promotors. In the second, NDMA and NMBzA initiate the same population of hepatocytes, but the nature of the initiating lesion(s) is different for the two nitrosamines so that the initiated cells behave differently towards the various promotors. Differences in the chemistry of interaction of reactive intermediates from the two nitrosamines with DNA could lead to such differences in the initiating lesion. Our experiments do not distinguish between these two models.

The results of the experiments described here suggest that the differences in carcinogenic activity of NDMA and NMBzA in rat liver is caused, not by differences in the ability of the two nitrosamines to initiate neoplasia in the liver, but rather by differences in their abilities to promote the cells they have initiated. NDMA is known to be a potent hepatotoxin, whereas NMBzA produces no toxicity in rat liver (Mehta et al., 1984). This difference in toxic response may account for the different promoting activities of the two nitrosamines. The mechanism of production of liver injury by nitrosamines is not understood.

ACKNOWLEDGEMENTS

These investigations were supported by grant MT 7025 from the Medical Research Council of Canada, and by the Ontario Cancer Treatment and Research Foundation.

REFERENCES

Fong, L.Y., Lin, H.J. and Lee, C.L., 1979, Methylation of DNA in target and non-target organs of the rat with methylbenzylnitrosamine and dimethylnitrosamine, Int. J. Cancer, 23:679-682.
Hodgson, R.M., Wiessler, M. and Kleihues, P., 1980, Preferential methylation of target organ DNA by the oesophageal carcinogen N-nitroso-methylbenzylamine, Carcinogenesis, 1:861-866.

Johansson, E.B. and Tjalve, H., 1978, The distribution of (^{14}C) dimethylnitrosamine in mice. Autoradiographic studies in mice with inhibited and non-inhibited dimethylnitrosamine metabolism and a comparison with the distribution of (^{14}C) formaldehyde, Tox. Appl. Pharmacol., 45:565-575.

Kraft, P.L. and Tannenbaum S.R., 1980, Distribution of N-nitrosomethylbenzylamine evaluated by whole-body radioautography and densitometry, Cancer Res., 40:1921-1927.

Labuc, G.E. and Archer, M.C., 1982a, Esophageal and hepatic microsomal metabolism of N-nitrosomethylbenzylamine and N-nitrosodimethylamine in the rat, Cancer Res., 42:3181-3186.

Labuc, G.E. and Archer, M.C., 1982b, Comparative tumor initiating activities of N-nitrosomethylbenzylamine and N-nitrosodimethylamine in rat liver, Carcinogenesis, 3:519-523.

Lai, D.Y. and Arcos, J.C., 1980, Dialkylnitrosamine bioactivation and carcinogenesis, Life Sci., 27:2149-2165.

Magee, P.N., 1956, Toxic liver injury. The metabolism of dimethyl-nitrosamine, Biochem. J., 64:676-682.

Mehta, R., Labuc, G.E., Urbanski, S.J. and Archer, M.C., 1984, Organ specificity in the microsomal activation and toxicity of N-nitrosomethylbenzylamine in various species, Cancer Res., 44:4017.

Pegg, A.E., 1977, Formation and metabolism of alkylated nucleosides: possible role in carcinogenesis by nitroso compounds and alkylating agents, Adv. Cancer Res.:25, 195-269.

Pereira, M.A., Lin, L.-H.C. and Herren, S.L., 1983, Role of O^6-methylation in the initiation of GGTase-positive foci, Chem. Biol. Interactions, 20:313-322.

Pitot, H.C., Barsness, L., Goldsworthy, T. and Kitagawa, T., 1978, Biochemical characterisation of stages of hepatocarcinogenesis after a single dose of diethylnitrosamine, Nature, 271:456-458.

Shank, R.C. and Magee, P.N., 1981, Toxicity and carcinogenicity of N-nitroso compounds, in Shank, R.C. (ed), Mycotoxins and N-nitroso compounds: Environmental Risks, 1, CRC Press, Florida, USA, pp. 185-217.

Silinskas, K.C., Zucker, P.F., Labuc, G.E. and Archer, M.C., 1984, Formation of O^6-methylguanine in regenerating rat liver by N-nitrosomethylbenzylamine is not sufficient for initiation of preneoplastic foci, Carcinogenesis, 5:541-542.

Silinskas, K.C., Zucker, P.F. and Archer, M.C., 1985, Formation of O^6-methylguanine in rat liver DNA by nitrosamines does not predict initiation of preneoplastic foci, Carcinogenesis, 6:773-775.

Tsuda, H., Lee, G. and Farber, E., 1980, Induction of resistant hepatocytes as a new principle for a possible short-term in vivo test for carcinogens, Cancer Res., 40:1157-1164.

BIOLOGICAL REACTIVE METABOLITES IN HUMAN TOXICITY

Hermann M. Bolt, Hans Peter, and Hans-Jürgen Wiegand

Institut für Arbeitsphysiologie
an der Universität Dortmund
Ardeystrasse 67, D-4600 Dortmund 1, F.R.G.

INTRODUCTION

Accidental exposures of employees to some acutely and chronically toxic industrial chemicals have brought up the question of effectiveness of antidotes which may facilitate detoxication. Because of recent severe (Vogel and Kirkendall, 1984) or even fatal (BG Chemie, 1980) accidents initial emphasis was focussed on acrylonitrile. Later, the investigations also included methacrylonitrile, methyl bromide, and chromates.

Much clinical experience has been accumulated concerning the use of N-acetylcysteine in cases of acute paracetamol overdoses (Prescott et al., 1977, 1979; Oh and Shenfield, 1980; Prescott, 1981). N-Acetylcysteine, as a biologically active sulfhydryl compound, can react directly with reactive intermediates, and is utilized for biosynthesis of glutathione. Clinically, an intravenous initial (within 15 min) dose of 150 mg/kg, followed by further infusions of 50 mg/kg (within 4 h), and 100 mg/kg (within 16 h), has been successfully applied.

According to current regulations (BG Chemie, 1980) suitable antidotes must be available where certain toxic chemicals are handled in industrial settings. This was the reason to study the antidotal effectiveness of N-acetylcysteine in animals (rats) poisoned by toxicants which are reactive or are transformed into reactive metabolites.

ACRYLONITRILE

Acrylonitrile is metabolized via two pathways. The major one is direct conjugation with glutathione; a minor portion is oxidatively biotransformed via the epoxide (glycidonitrile) to cyanide which is finally excreted as thiocyanate (Langvaard et al., 1980; Geiger et al., 1983). DNA adducts are thought to be caused by glycidonitrile (Geiger et al., 1983) which is strongly mutagenic (Peter et al., 1983). However, binding to proteins (Peter and Bolt, 1981) is also possible with the parent compound itself (Geiger et al., 1983).

In most species, the main toxicity of acrylonitrile is not due to cyanide liberation, but to the reactivity of the entire molecule. There are indications, based on metabolic studies with liver tissue from different species including man, that this is also valid for humans (Appel et

al., 1981). The importance of detoxication of acrylonitrile by glutathione is visualized by the effectiveness of N-acetylcysteine as an antidote: application of an intravenous dose of 300 mg/kg Fluimucil[R] to rats,
10 min after an otherwise lethal acrylonitrile exposure (30 min at
4,800 ppm), saved all treated animals. By contrast, cyanide antidotes
(4-dimethylaminophenol/sodium thiosulfate) were not effective (Buchter et
al., 1984).

METHACRYLONITRILE

The methyl analog of acrylonitrile is also produced industrially in
large quantities. Recent biochemical data (Tanii and Hashimoto, 1984)
demonstrate that hepatic microsomes (from mice) catalyze much more cyanide
liberation from methacrylonitrile than from acrylonitrile. This points to
a greater importance of the oxidative pathway (via an hypothetical epoxide).
This is consistent with the clinical symptoms observed in animals. Rats
acutely intoxicated with acrylonitrile initially show marked salivation,
lachrymation, diarrhoea and convulsions, consistent with increased levels
of acetylcholine (Peter and Bolt, 1984); the lethal outcome (30 min exposure
to 4,800 ppm) was mostly between 3 and 4 h after exposure. By contrast,
methacrylonitrile (same dose range) caused rapid unconsciousness with
convulsions, and lethality occured between 30 and 60 min after exposure
(Peter and Bolt, 1985); this clinical picture is identical with that of a
cyanide intoxication.

In accordance with these observations, the acute toxicity of
methacrylonitrile can be antagonized by cyanide antidotes (which trap the
ultimately toxic metabolite cyanide) as well as by N-acetylcysteine
(which prevents the compound from being oxidatively metabolized; Peter
and Bolt, 1985).

In view of the carcinogenicity of acrylonitrile which is thought to
be due to epoxide formation (Geiger et al., 1983) it ought to be predicted
that methacrylonitrile were even more carcinogenic. However, carcinogenicity (or even mutagenicity) studies with this industrial compound have not
been published so far.

METHYL BROMIDE

Methyl bromide is industrially used as fumigant, desinfectant, etc.
Toxicity symptoms in man are nonspecific; alterations of CNS functions,
liver, lung and kidneys have been reported. Part of the clinical symptoms
appear with a latency time of up to 48 h (Weller, 1982). The analogous
methyl chloride is metabolized, via pathways involving glutathione, to
formaldehyde and formate (Kornbrust and Bus, 1982). A possible neurotoxic
intermediate could be methanethiol (Kornbrust and Bus, 1982).

Methyl bromide is reactive (Djalali-Behzad et al., 1981) and causes
local malignancies in the rat forestomach (Danse et al., 1984).

N-Acetylcysteine exerts a significant antidotal effect when
administered to rats after an otherwise lethal exposure (30 min at
3,910 ppm) to methyl bromide. The effective dose of the antidote was:
200 mg/kg Fluimucil[R] i.v., immediately after exposure; the same dose
repeated (i.p.) after 2 h (Peter et al., 1985).

CHROMATES

Hexavalent chromate exposure of humans occurs in various industries. Acute poisoning by soluble chromates leads to a severe and often fatal tubular renal failure (Kaufmann et al., 1970).

Recent investigations have confirmed that only chromium (VI) compounds pass through cell membranes (Norseth, 1981; Wiegand et al., 1984a, 1985a). However, intracellularly a reduction to chromium (III) occurs which most probably represents the ultimately toxic chromium species (Norseth, 1981; Wiegand et al., 1984b). This explains the differential toxicities and distribution patterns of chromium (III) and chromium (VI) compounds (Tsapakos et al., 1983; Bryson and Goodall, 1983).

On this basis, it has been proposed to use high doses of ascorbic acid in cases of acute human poisoning (Korallus et al., 1984) to prevent the hexavalent chromium from intracellular bioactivation.

As reduction of chromium (VI) is also possible by sulfhydryl compounds (Wiegand et al., 1984b), N-acetylcysteine was also examined for antidotal effects. It was found that a standard dose of 300 mg/kg N-acetylcysteine, administered i.p. after an i.p. injection of 20 mg/kg potassium chromate, reduced significantly the number of fatal outcomes (Wiegand et al., 1985b).

CONCLUSIONS

The examples presented here serve to demonstrate that knowledge of the nature of the ultimately toxic principles of chemicals is helpful in designing therapeutic regimens to prevent human toxicity. This refers to chemicals which are reactive by themselves or are converted into biologically reactive metabolites; it also refers to organic as well as to (selected) inorganic compounds.

REFERENCES

Appel, K.E., Peter, H., and Bolt, H.M., 1981, Interaction of acrylonitrile with hepatic microsomes of rats and men, Toxicol. Lett., 7:335.
Berufsgenossenschaft der Chemischen Industrie, 1980, Merkblatt Acrylnitril, Ausgabe 10, Verlag Chemie, Weinheim.
Bryson, W.G., and Goodall, C.M., 1983, Differential toxicity and clearance kinetics of chromium (III) or (VI) in mice, Carcinogenesis, 4:1535.
Buchter, A., Peter, H., and Bolt, H.M., 1984, N-Acetyl-Cystein als Antidot bei akzidenteller Acrylnitril-Intoxikation, Int. Arch. Occup. Environ. Hlth., 53:311.
Danse, L.H.J.C., van Velsen, F.L., and van der Heijden, C.A., 1984, Methyl bromide: carcinogenic effects in the rat forestomach, Toxicol. Appl. Pharmacol., 72:262.
Djalali-Behzad, G., Hussain, S., Osterman-Golkar, S., and Segerbäck, D., 1981, Estimation of genetic risk of alkylating agents. VI. Exposure of mice and bacteria to methyl bromide, Mut. Res., 84:1.
Geiger, L.E., Hogy, L.L., and Guengerich, F.P., 1983, Metabolism of acrylonitrile by isolated rat hepatocytes. Cancer Res., 43:3080.
Kaufmann, D.B., Dinicola, W., and McIntosh, R., 1970, Acute potassium dichromate poisoning, Am. J. Dis. Child, 119:374.
Korallus, U., Harzdorf, C., and Lewalter, J., 1984, Experimental bases for ascorbic acid therapy of poisoning by hexavalent chromium compounds, Int. Arch. Occup. Environ. Hlth., 53:247.

Kornbrust, D.J., and Bus, J.S., 1982, Metabolism of methyl chloride to formate in rats. Toxicol. Appl. Pharmacol., 65:122.

Langvaard, P.W., Putzig, C.L., Braun, W.H., and Young, J.D., 1980, Identification of the major urinary metabolites of acrylonitrile in the rat, J. Toxicol. Environ. Hlth., 6:273.

Norseth, T., 1981, The carcinogenicity of chromium, Env. Hlth. Perspect., 40:121.

Oh, T.E., and Shenfield, G.M., 1980, Intravenous N-acetylcysteine for paracetamol poisoning, Med. J. Aust., 1:664.

Peter, H., and Bolt, H.M., 1981, Irreversible protein binding of acrylonitrile, Xenobiotica, 11:51.

Peter, H., Schwarz, M., Mathiasch, B., Appel, K.E., and Bolt, H.M., 1983, A note on synthesis and reactivity towards DNA of glycidonitrile, the epoxide of acrylonitrile, Carcinogenesis, 4:235.

Peter, H., and Bolt, H.M., 1984, Experimental pharmacokinetics and toxicology of acrylonitrile, G. Ital. Med. Lavoro, 6:77.

Peter, H., and Bolt, H.M., 1985, Effect of antidotes on the acute toxicity of methacrylonitrile, Int. Arch. Occup. Environ. Hlth., 55:175.

Peter, H., Hopp, D., Huhndorf, U., and Bolt H.M., 1985, Untersuchungen zur Methylbromid-Vergiftung und ihrer Behandlung mit N-Acetylcystein, in: "Aktuelle arbeitsmedzinische Probleme in der Schwerindustrie. - Theorie und Praxis biologischer Toleranzwerte für Arbeitsstoffe (BAT-Werte). - Bedeutung neuer Technologien für die arbeitsmedizinische Praxis." H.M. Bolt, J. Rutenfranz, C. Piekarski, eds., Verh. Dtsch. Ges. Arbeitsmed., Vol. 25, Genter Verlag, Stuttgart (in press).

Prescott, L.F., Park, J., Ballantyne, A., and Adriaenssens, P., 1977, Treatment of paracetamol (acetaminophen) poisoning with N-acetyl-cysteine, Lancet, 2:432.

Prescott, L.F., Illingworth, R.N., Critchley, J.A., Stewart, M.J., Adam, R.J., and Proudfood, A.T., 1979, Intravenous N-acetylcysteine: the treatment of choice for paracetamol poisoning, Brit. Med. J., 2:1097.

Prescott, L.F., 1981, Treatment of severe acetaminophen poisoning with intravenous acetylcysteine, Arch. Int. Med., 141:386.

Tanii, H., and Hashimoto, K., 1984, Studies on the mechanism of acute toxicity of nitriles, Arch. Toxicol., 55:47.

Tsapakos, M.J., Hampton, T.H., and Wetterhahn, K.E., 1983, Chromium (VI)-induced DNA lesions and chromium distribution in rat kidney, liver and lung, Cancer Res., 43:5662.

Vogel, R.A., and Kirkendall, W.M., 1984, Acrylonitrile (vinyl cyanide) poisoning: a case report, Texas Med., 80:48.

Weller, D., 1982, "Methylbromid; Toxikologie und Therapie," Degesch GmbH, Frankfurt/Main.

Wiegand, H.J., Ottenwälder, H., and Bolt, H.M., 1984a, Disposition of intratracheally administered chromium (III) and chromium (VI) in rabbits, Toxicol. Lett., 22:273.

Wiegand, H.J., Ottenwälder, H., and Bolt, H.M., 1984b, The reduction of chromium (VI) to chromium (III) by glutathione: an intracellular redox pathway in the metabolism of the carcinogen chromate, Toxicology, 33:341.

Wiegand, H.J., Ottenwälder, H., and Bolt, H.M., 1985a, Fast uptake kinetics in vitro of 51-Cr(VI) by red blood cells of man and rat, Arch. Toxicol., 57:31.

Wiegand, H.J., Ottenwälder, H., and Bolt, H.M., 1985b, Die Verwendbarkeit von N-Acetyl-Cystein bei Cr(VI)-Vergiftung, in: "Aktuelle arbeitsmedizinische Probleme in der Schwerindustrie. - Theorie und Praxis biologischer Toleranzwerte für Arbeitsstoffe (BAT-Werte). - Bedeutung neuer Technologien für die arbeitsmedizinische Praxis." H.M. Bolt, J. Rutenfranz, C. Piekarski, eds., Verh. Dtsch. Ges. Arbeitsmed., Vol. 25, Gentner Verlag, Stuttgart (in press).

METABOLIC HOST FACTORS AS MODIFIERS OF REACTIVE INTERMEDIATES

POSSIBLY INVOLVED IN HUMAN CANCER

E. Hietanen*, H. Bartsch and H. Vainio

International Agency for Research on Cancer
Division of environmental carcinogenesis
F-69372 Lyon Cedex 08, France

INTRODUCTION

The current theory of the mechanism of chemical carcinogenesis is that precarcinogens are metabolized by mostly mixed-function oxidases into reactive intermediates that act as ultimate carcinogens (Miller and Miller, 1966) in the liver and in extrahepatic organs (Vainio and Hietanen, 1980). The wide variation among human individuals in the enzymes involved in metabolic activation of xenobiotics may be due in part to host factors and to interactions with environmental exposures (Table 1; Farrell et al., 1979; Sotaniemi et al., 1980). These factors include different life styles, genetic differences and the various clinical and subclinical diseases that may alter liver metabolism in man during his relatively long life span (Vesell, 1982). Exposure to cancer-causing foreign compounds appears to be one of the most important environmental risk factors whereas the host-response to xenobiotics may modify individual risk for cancer arising from such exposures. Some current evidence for such an hypothesis is rerieved herein.

METHODS FOR ASSESSING CARCINOGEN METABOLISM IN MAN

In evaluating the role of host factors in human cancer, both in vitro studies of samples from humans and in vivo studies have been used; the latter include the identification in body fluids of carcinogen adducts formed after exposure to occupational, therapeutic or other agents and the use of harmless probe drugs as indirect markers of carcinogen metabolism. Numerous studies have been published in which the rates of elimination of drugs in vivo have been correlated with metabolism of the drug in vitro in liver biopsies. Correlations have been made, for instance, between glymidine clearance and aryl hydrocarbon hydroxylase (AHH) activity (r=0.59) (Held, 1980) antipyrine (AP) half-life and AHH activity (r=-0.51) (Farrell et al., 1979), AP half-life and cytochrome P-450 concentration (r=-0.47; -0.43) (Farrell et al., 1979; Sotaniemi et al., 1980), AP half-life and ethylmorphine demethylation (r=0.51) (Farrell et al., 1979), and AP half-life and glucaric acid excretion (r=0.31) (Sotaniemi et al., 1980). Measurement of in vivo drug metabolism has also been used to study the effects of occupational exposure to chemical mixtures (Dossing, 1982) and of smoking (Jusko, 1978; Vestal and Wood, 1979). However,

*Present address: Department of Physiology, University of Turku, SF-20520 Turku, Finland.

Table 1. Some Host and Environmental Factors that Modify Risk for Cancer in Man

Host Factors	Environmental Factors
Age	Exposure to Xenobiotics
Sex	Occupational
Genetic Pattern (controlling carcinogen	Smoking
metabolism and DNA repair)	Environmental Pollution
Hormonal Status	Diet
Diseases	
Nutritional Status	
Immunological Factors	
Microbial Flora	

the correspondence between drug elimination in vivo and metabolism in vitro has not been fully evaluated, and sometimes very weak associations have been found.

Various non-invasive techniques have been evaluated for determining hepatic drug and carcinogen metabolism. The aminopyrine breath test has been used to measure hepatic microsomal drug oxidation rates (Schneider et al., 1980). Kellermann and Luyten-Kellermann (1978) found that salivary AP excretion correlated well with urinary excretion of AP-4-OH, AP-CH$_2$OH and AP-NH$_2$metabolites.

Assessment of the pattern of in vivo drug metabolism is of practical importance to predict the response of individuals to therapeutic drugs (Idle et al., 1978; Sotaniemi et al.,1980; Kalow, 1982; Penno and Vesell, 1983). In order to study carcinogen metabolism in man in vivo, marker drugs have been used to measure the capacity for drug metabolism, mainly in the liver. Marker drugs can be chosen that simulate the metabolism of certain carcinogens. Debrisoquine (DB) 4-hydroxylation in vivo has been proposed as a model to assess individual oxidative drug metabolizing capacity, and to classify people into poor and extensive metabolizers (Mahgoub et al., 1977; Price-Evans et al., 1980). Oxidation of DB appears to be under monogenic control, and humans display genetic polymorphism in respect of this drug (Mahgoub et al., 1977). It has been suggested that poor metabolizers (1-30% of the population) (Kalow et al., 1980) are at a lower risk for lung, colon and liver cancers when exposed to environmental carcinogens than are the extensive metabolizer phenotypes (Ayesh et al., 1984; Hetzel et al., 1980; Idle et al., 1981; Ritchie and Idle, 1982).

Isolated blood cells, mainly lyphocytes, have frequently been used as models to represent the metabolic properties of organs. Kellermann and Luyten-Kellermann (1977) found that phenobarbital and AP half-lives and AHH inducibility in mitogen-stimulated lymphocytes were closely associated, giving correlation coefficients of -0.88 and -0.94, respectively. In other studies, however, no such association was found (cf. Pelkonen and Kärki, 1982). In numerous (but not all) studies, the inducibility of cultured lymphoblasts by polycyclic aromatic hydrocarbons (PAH) has been related to risk for lung cancer (cf. Bartsch et al., 1982). No association between lymphocyte AHH inducibility and leukaemia has been established (Levine et al., 1984). One limitation of such an approach is that lymphocytes may not reflect faithfully genetic regulation of AHH inducibility, because man undergoes many complex exposures during his life; however, in vitro cultures of lymphocytes may be free of certain exogenous factors and thus metabolic reactions are more under genetic control.

Cytochrome P-450 patterns in human tissues are shown to vary as shown by the use of monoclonal antibodies (Mab) toward P-450 isozymes (Fujino et al., 1982). The dominating cytochrome P-450 isozyme in extrahepatic organs may be quite different from that in the liver; but whether this reflects differences in the cancer susceptibility of these organs has not been determined. The isozyme patterns of tissues can be characterized by inhibiting the catalytic properties of cytochrome P-450 isozymes. For example, using a 3-methylcholanthrene (MC)-type (clone 1-7-1) antibody, Fujino et al. (1982) were able to show that in human subject cytochrome P-450 isozyme in lymphocytes (both control and induced) was sensitive to inhibition by MC-type antibody, but that this sensitivity was not present in monocytes, indicating a wide variation between blood cells in isozyme pattern. The availability of Mab and of polyclonal antibodies towards cytochrome P-450 isozymes opened new venues in cancer research, by making it possible to relate eventually an individual metabolic phenotype with a characteristic sensitivity to a drug and to carcinogens.

In order to predict hepatic carcinogen metabolism on the basis of parameters measured in urine or blood, it must be established (1) how well the metabolites of the probe drug represent in vitro drug and carcinogen metabolism in liver; and (2) whether there is a correlation between extrahepatic drug/carcinogen metabolism, especially in lymphocytes, and hepatic drug/carcinogen metabolism.

ANIMAL MODELS FOR STUDIES OF HOST FACTORS

In order to evaluate the role of metabolic host factors in human cancer, we have compared certain parameters in animal models and in human tissues. The B6 (C57 BL/6) and D2 (DBA/2) mouse strains are widely used models for the study of hepatic monooxygenase inducibility and its genetic regulation. In D2 mouse liver, the monooxygenase enzymes are not inducible by PAH (Poland et al., 1974). In our study, we used a Mab (Mab-clone 1-7-1, provided by H.V. Gelboin, NCI, Bethesda ,USA) produced against the rat liver cytochrome P-450 isozyme that catalyses the PAH-inducible monooxygenases. A clear-cut strain different in enzyme inhibition was observed: no inhibition in AHH activity was produced in MC-induced D2 mouse livers, while strong inhibition was seen in those of MC-induced B6 mice (Hietanen et al., 1985 d) (Table 2); however, some inhibition of ethoxycoumarin O-deethylase and ethoxyresorufin O-deethylase acitivities was observed in MC-induced D2 strain mice (Table 2).

Such inhibition studies using Mab to characterize further the enzymes that yield reactive intermediates may provide information on the relationship between antigenic sites in enzyme proteins and catalytic activity. Because of the potential use of Mab to characterize the isozyme pattern in human tissues responsible for activation of carcinogens, it might thus become possible to associate risks for cancer due to exposure to certain chemicals, with an (individual) isozyme pattern. One present limitation is that screening has been carried out mainly on hepatic tissue, for which a liver biopsy is necessary. In an attempt to obviate this procedure, the metabolic patterns of probe drugs excreted in urine have been analysed in order to estimate indirectly hepatic isozyme composition (Kaminsky et al., 1984). Since some cytochrome P-450 isozymes are more likely than others to activate carcinogens, this kind of approach might become promising when enough data have been accumulated on the relationship between the formation of ultimate carcinogens and hepatic isozyme patterns.

Table 2. Inhibition of Monooxygenase Activities in the Livers of Male C57
BL/6 (B6) and DBA/2 (D2) Mice by Monoclonal Antibody (Mab) Clone
1-7-1[a]

Mouse strain	Treatment[c]	Inhibition of Monooxygenase Activities[b]		
		AHH	ECDE	ERDE
B[6]	Controls	+	+	++
	MC	++	++	++
D[2]	Controls	0	+	++
	MC	0	+	++

[a]This clone inhibits AHH and ECDE activities induced by 3-methylcholanthrene
(MC) in rat and human tissues (Fujino et al., 1982; Park et al., 1982).
[b]AHH, aryl hydrocarbon hydroxylase; ECDE, ethoxycoumarin O-deethylase; ERDE,
ethoxyresorufin O-deethylase. Intibition was graded at a maximal Mab concentration of 5 mg Ig/nmol cytochrome P-450 as follows: 0, inhibition <25%;
+, 25-49%; ++, >50%.
[c]Mice were treated with a single i.p. injection of 3-methylcholanthrene (MC)
(40 mg/kg) 48 h earlier.

One specific approach that has been developed is the use of DB as a
probe drug (Idle et al., 1978). This drug is hydroxylated, mainly into
4-hydroxy DB, by a distinct cytochrome P-450 isozyme that is not inducible
by known enzyme inducers (Larrey et al., 1984); it has similar antigenic
characteristics in man and rats (Distlerath and Guengerich, 1984). It has
been suggested that this drug could be used to separate populations into poor
and fast metabolizers and that fast metabolizers (90% of the population) are
at elevated risk for bronchial cancer (Ayesh et al., 1984). DB was also used
to screen people for an increased susceptibility to the kidney disease, Balkan
nephropathy leading to kidney and urinary-tract tumours (Ritchie et al., 1983),
which may be related to the exposure to a mycotoxin, ochratoxin A (OA ; Fig. 1)
present in mouldy grain (Röschenthaler et al., 1984). The nature of the
reactive intermediate(s) of OA possibly involved in the disease is not known,
although in rats OA is selectively hydroxylated by microsomal monooxygenases
at C-4 position.

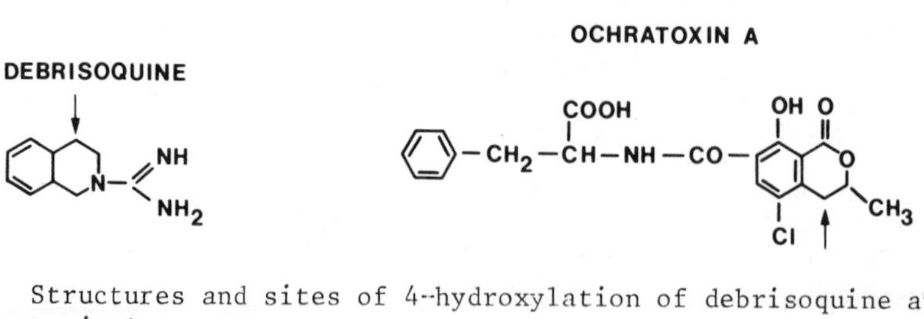

Fig. 1. Structures and sites of 4-hydroxylation of debrisoquine and ochratoxin A.

We have evaluated a possible link between DB and OA metabolism both in rat strains with genetically low DB metabolism and in human liver samples (donated by M. Roberfroid, Louvain, Belgium). In female homozygous DA rats there was significantly less OA 4-hydroxylase activity than in female Lewis rats (Table 3), which corresponded to differences in DB 4-hydroxylase activity. However, OA but not DB 4-hydroxylase activity was inducible by MC and pheno-barbital, suggesting that the oxidation of these two compouds is catalyses by different isozymes (Hietanen et al., 1985b,c).The OA 4-hydroxylase activity in human liver samples was much lower than that in either rat strain (Table 3). The immunochemical inhibition of OA 4-hydroxylation by DB-type polyclonal type Ab (generously provided by F.P. Guengerich, Nashville, USA) in human and rat livers was clearly different (Table 3): no inhibition was observed in untreated rat liver microsomes, but there was some inhibition in MC-treated DA rat livers, and marked inhibition was seen in human liver microsomes. Such strain and species differences in substrate specificity indicate further the importance of characterizing cytochrome P-450 isozymes directly in man, since they might limit extrapolation of animal data to man.

Table 3. Ochratoxin 4-Hydroxylase Activity [a] and its Inhibition by Anti-bodies (Ab) against Rat Liver Debrisoquine (DB) 4-Hydroxylase

Pretreatment[b]	DB-Ab[c]	Rat strain Lewis	DA	Man
None (control)	−	100 (2.5)	100 (1.1)	100 0.5)
	+	85	106	60
3-MC	−	100 (9.8)	100 (20.3)	
	+	111	67	

[a]Activities shown as % of activity with preimmune protein; in parentheses, mean activity (pmol/min/mg protein)
[b]Rats were treated with a single i.p. injection of 3-methylcholanthrene (MC) (40 mg/kg) 48 h earlier.
[c]Assays were carried out in the presence of preimmune (−) or immune (+) protein (6 mg/kg microsomal protein).

ORGAN SPECIFICITIES

Important host factors in the formation of reactive intermediates are tissue- and organ-specific differences in the capacity to metabolize carcino-gens (Vainio and Hietanen, 1980). The isozyme patterns in extrahepatic human organs vary considerably, as observed by enzyme inhibition in the presence of Mab againgst cytochrome P-450 isozymes (Fujino et al., 1982), which might explain some of the organ-specific differences in cancer susceptibility in man. In rats, a typical tissue-specific difference in isozyme composition is seen in intestine and liver: even in control rats, intestinal cytochrome P-450-catalyzed AHH and ethoxycoumarin O-deethylase activities are inhibited mainly by Mab clone 1-7-1, which in the liver inhibits mainly MC-inducible activities (Hietanen et al., 1985a). The creation of a library of antibodies may permit to screen human samples for the isozyme patterns of various tissues to estimate possible organ-specific risks for a chemical-induced cancer and the formation of reactive intermediates.

In humans, there exists a number of chronic, yet manageable diseases that increase interindividual variation in the activities of enzymes that catalyse the formation of reactive intermediates. It is well known that disease

Table 4. Biotransformation Enzyme Activities in Human Small Intestinal Biopsies (with values observed in rat tissues, for comparison)

Parameter[a]	Histological Diagnosis			Rat (Range)
	Normal	Partial Villus Atrophy	Total Villus Atrophy	
Age (years)	6.72±0.48	5.62±0.95	8.94±1.33	
N(M/F)	141 (77/64)	33 (15/18)	20 (14/6)	
AHH (pmol/min/mg protein)	64.2±9.3 (50)	63.5±14.3 (18)	55.6±18.0 (10)	5-10
ECDE (pmol/min/mg protein)	0.02±0.01 (9)	0.02±0.01 (5)	0.04±0.02 (5)	2-14
Epoxide hydrolase (BP 4,5-oxide) (pmol/min/mg protein)	43.6±2.0 (90)	39.2±4.0 (15)	26.3±4.5*** (10)	
Glutathione peroxidase (pmol/min/mg protein)	27.3±2.1 (90)	21.4±4.0 (15)	4.8±1.6*** (10)	
UDP glucuronosyltransferase (4-methylumbelliferone) (pmol/min/mg protein)	8.7±2.7 (9)	3.9±2.5 (4)	no data	250-1500
Alkaline phosphatase (nmol/min/mg protein)	161±9 (139)	118±17* (31)	61±10*** (20)	200-2000

Statistical comparisons with normal histology: *, 2P<0.05; ***, 2P<0.001. Mean±SEM. Actual number of samples in each group shown in parentheses.

[a]AHH, aryl hydrocarbon hydroxylase; ECDE, ethoxycoumarin O-deethylase; BP, benzo(a)pyrene.

related liver changes effect the pharmacokinetics of therapeutic drugs; however, there have been fewer studies on whether cancer risk is affected by differences in the level or type of reactive intermediates formed in the liver, due to changes in oxidative enzyme activities.

Some diseases involve extrahepatic tissues that metabolize xenobiotics into reactive intermediates. Pathological changes in the small intestinal mucosa may alter its morphology. In a study in progress of intestinal biopsies from children with atrophy of the duodenal villi, no decrease in AHH activity was found, but the activities of many of the enzymes involved in the conjugation and detoxification of reactive intermediates were decreased, in parallel with a decrease in the activity of a marker enzyme, alkaline phosphatase (Table 4). This decreased detoxification rate in intestinal biopsies suggests reduced inactivation of reactive intermediates and, therefore, increased susceptibility to toxic insults.

DIET-HOST FACTOR INTERACTIONS

The host-mediated formation of reactive intermediates is generally recognized to be effected by the general nutritional status and habits. The metabolism of drugs and hormones in humans is strongly dependent on individual dietary habits, as shown in studies on human volunteers (Kappas et al., 1983) and the effects may be more pronounced in severly ill people suffering from nutritional imbalances. High-lipid diets increase monooxygenase activities (Hietanen et al., 1982), and diets containing PAH as contaminants affect strongly the mucosal metabolism of probe drugs in the rat intestine (Pantuck et al., 1976). Changes in the incidence of cancer in both liver and extrahepatic organs of experimental animals after administration of high-lipid diets and chemical carcinogens are well established; however, the mechanisms of how diet modifies cancer incidence are less well studied. In a study in progress, changes in organ specificity to carcinogenesis were related to diet (Table 5 unpublished data); it appeared that high-fat diets may result in more cancer in extrahepatic tissues. The lower activity of hepatic monooxygenases in animals given the low-fat diet would change the rate of formation of reactive intermediates and result in higher levels of carcinogens in extrahepatic organs; which of these two mechanisms is responsible remains to be studied.

CONCLUSIONS

The role of host factors either inherited or environmentally ecquired in human cancer and their effect on the formation of reactive intermediates has yet to be explored. However, it appears that such host factors are among the determinants causing variations in cancer incidence among individuals exposed to the same level of environmental carcinogens, e.g. tobacco smokers. It is often difficult, however, to separate host factors from environmental factors because of the multitude of interactions. This may explain the conflicting results that have been obtained in efforts to correlate lung cancer incidence with individual levels of activation of PAH in blood cells (cf. Pelkonen and Kärki, 1982). More consistent data are becoming available on the significance of host factors in the regulation of the metabolism of drugs; and these findings have led to the development of methods to use some probe drugs in assessing individual metabolic capacity to produce reactive intermediates that are possibly related to cancer causation. Additional approaches include the development and application of non-invasive in vivo methody for the detection of carcinogens DNA adducts, e.g. for those chemicals known to produce cancer in man (International Agency for Research on Cancer, 1982), the use to safe nitrosatable substrates for assessing endogenous formation of N-nitroso carcinogens from their precursors (Bartsch et. al., 1983), and studies on the significance of indices of reactive oxygen

Table 5. Distribution in Rats fed High- and Low-fat Diets of Tumours Induced by N-Nitrosodiethylamine[a]

| | Diet[b] | |
	High fat	Low fat
Number of rats	22	23
Number of rats with tumours	13	16
Number of rats with tumours in:		
Liver	13	8
Nasal cavities	0	7
Kidneys	0	7
Colon	0	1
Oesophagus	1	0

[a] N-Nitrosodiethylamine was given intragastrically at 3 mg/kg on 5 days/week for 10 weeks from the age of 5 weeks.
[b] High-fat diet contained 30 % fat as cotton-seed oil; the low-fat group was first fed a fat-free diet, then from the age of 14 weeks onwards sunflower oil was added to bring the fat content to 2 % (w/w) fat . Animals were fed these diets on an equicaloric basis for 40 weeks before analysis.

species and break-down products formed in vivo in the general population, the latter seems important (Kauppila et. al., 1984) as the role of genetic factors in controlling the formation of reactive oxygen species and the DNA-damage derived theory has been demonstrated in some cancer-prone genetic disorders (cf. Cerutti, 1985).

ACKNOWLEDGEMENTS

We thank Mrs Elisabeth Heseltine for editorial help and Ms Maarit Kallio for secretarial assistance.

REFERENCES

Ayesh, R., Idle, J. R., Ritchie, J. C., Crothers, M. J., and Hetzel, M. R. 1984, Metabolic oxidation phenotypes as markers for susceptibility to lung cancer, Nature, 312:169.
Bartsch, H., Aitio, A., Camus, A. M., Malaveille, C., Ohshima, H., Pignatelli, B., and Sabadie, N., 1982 Carcinogen-metabolizing enzymes and susceptibility to chemical carcinogens, in, "Host Factors in Human Carcinogenesis (IARC Scientific Publications No. 39), International Agency for Reseach on Cancer, H. Bartsch, and B. Armstrong, eds., Lyon, p. 337.
Bartsch, H., Ohshima, H., Munoz, N., Pignatelli, B., Friesen, M., O'Neill, J., Crespi, M., and Lu, S. H., 1983, Assessment of endogenous nitrosation in humans in relation to the risk of cancer of the digestive tract, in: "Developments in the Science and Practice of Toxicology," A. W. Hayes, R. C. Schnell, and T. S. Miya, eds., Elsevier, Amsterdam, p. 299.
Cerutti, P. A., 1985, Prooxidant state and tumor promotion, Science, 227: 375.
Distlerath, L. M., and Guengerich, F. P., 1984, Characterization of a human

liver cytochrome P-450 involved in the oxidation of debrisoquine and other drugs using antibodies raised to the analogus rat enzyme, Proc. Natl. Acad. Sci. USA, 81:7348.

Dossing, M., 1982, Changes in hepatic microsomal enzyme function in workers exposed to mixtures of chemicals, Clin. Pharmacol. Ther., 32:340.

Farrell, G. C., Cooksley, W. G. E., and Powell, L. W., 1979, Drug metabolism in liver disease: Activity of hepatic microsomal metabolizing enzymes, Clin. Pharmacol. Ther., 26:488.

Fujino, T., Park, S. S., West, D., and Gelboin, H. V., 1982, Phenotyping of cytochrome P-450 in human tissues with monoclonal antibodies, Proc. Natl. Acad. Sci. USA, 79:3682.

Held, H., 1980, Correlation between the activity of hepatic benzo(a)pyrene hydroxylase in human needle biopsies and the clearance of glymidine, Hepato-Gastroenterol., 27:266.

Hetzel, M. R., Law, M., Keal, E. E., Sloan, T. P., Idle, J. R., and Smith, R. L., 1980, Is there a genetic component in bronchial carcinoma in smokers?, Thorax, 35:709.

Hietanen, E., Ahotupa, M., Heikelä, A., and Laitinen, M., 1982, Dietary lipids as modifiers of monooxygenase induction, in: "Cytochrome P-450, Biochemistry, Biophysics and Environmental Inplications," E. Hietanen, M. Laitinen, and O. Hänninen, eds., Elsevier, Amsterdam, p. 705.

Hietanen, E., Bartsch, H., Ahotupa, M., Park, S. S., and Gelboin, H. V., 1985 a, Tissue specificity of extrahepatic monooxygenases in the metabolism of xenobiotics, in: "Proc. 9th European Workship on Drug Metabolism," G. Siest, ed., Pergamon Press, London, p. 141.

Hietanen, E., Bartsh, H., Gastegnaro, M. Malaveille, C., Michelon, J., and Broussole, L., 1985 b, Use of antibodies against cytochrome P-450 isozymes to study genetic polymorphism in drug oxidations. Submitted.

Hietanen, E., Malaveille, C., Camus, M.-C., Béréziat, J. C., Brun, G., Gastegnaro, M., Michelon, J., Idle, J. R., and Bartsch, H., 1985 c, Interstrain comparison of hepatic and renal microsomal carcinogen metabolism and liver S9 mediated mutagenicity in DA and Lewis rats phenotyped as poor and extensive metabolizers of debrisoquine. Submitted.

Hietanen, E., Malaveille, C., Friedman, F. K., Park, S. S., Béréziat, J. C., Brun, G., Bartsch, H., and Gelboin, H. V., 1985 d, Monoclonal antibody directed analysis of cytochrome P-450 dependent monooxygenases and mutagen activation in the livers of D2 and B6 mice. Submitted.

Idle, J. R., Mahgoub, A., Lancaster, R., and Smith, R. L., 1978, Hypotensive response to debrisoquine and hydroxylation phenotype, Life Sci., 22:979.

Idle, J. R., Mahgoub, A., Lancaster, R., Smith, R. L., Mbanefo, C. O., and Bababunmi, E. A., 1981, Some observations on the oxidation phenotype status of Nigerian patients presenting with cancer, Cancer Lett., 11:331.

International Agency for Research on Cancer, 1982, IARC Monographs on the Evaluation of the Carcinogenic Risk of Chemicals to Humans, Suppl. 4, Lyon.

Jusko, W., 1982, Role of tobacco smoking in pharmacokinetics, J. Pharmacokin. Biopharmaceut., 6:7.

Kalow, W., 1982, The metabolism of xenobiotics in different populations, Can. J. Physiol. Pharmacol., 60:1.

Kalow, W., Otton, S. V., Kadard, D., Endrenyi, L., and Inaba, T., 1980, Ethnic differences in drug metabolism: debrisoquine 4-hydroxylation in Caucasians and Orientals, Can. J. Physiol. Pharmacol., 58:1142.

Kaminsky, L. S., Dunbar, D. A., Wang, P. P., Beaune, P., Larrey, D., Guengerich, F. P., Schnellman, R. G., and Sipes, I. G., 1984, Human hepatic cytochrome P-450 composition as probed by in vitro microsomal metabolism of warfarin, Drug Metab. Dispos., 12:470.

Kappas, A., Anderson, K. E., Ionney, A. H., Pantuck, E. J., Fishman, J., and Braddow, H. L., 1983, Nutrition-endocrine interacitons: Induction of reciprocal changes in the Δ^4-5α-reducton of testosterone and the cyto-chrome P-450-dependent oxidation of estradiol by dietary macronutrients in man, Proc. Natl. Acad. Sci. USA, 80:7646.

Kellermann, G., and Luyten-Kellermann, M., 1977, Phenobarbital-induced drug metabolism in man, Toxicol. Appl. Pharmacol., 39:97.

Kellermann, G. H., and Luyten-Kellermann, M., 1978, Antipyrine metabolism in man, Life Sci., 23:2485.

Kauppila, A., Sundström, H., Korpela, H., Viinikka, L., and Yrjönheikki, E., 1984, Serum selenium and gynecological cancer, in: "Icosanoids and Cancer," H. Thaler-Dao, A. Crasted de Paulet, and R. Paoletti, eds., Raven Press, New York, p. 263.

Larrey, D., Distlerath, L. M., Dannan, G. A., Wilkinson, G. R., and Guen-gerich, P. F., 1984, Purification and characterization of the rat liver microsomal cytochrome P-450 involved in the 4-hydroxylation of debriso-quine, a prototype for genetic variation in oxidative drug metabolism, Biochemistry, 23:2784.

Levine, A. S., McKinney, C. E., Echelberger, C. K., Kouri, R. E., Edwards, B. K., and Nebert, D. W., 1984, Aryl hydrocarbon hydroxylase inducibili-ty among primary relatives of children with leukemia or solid tumours, Cancer Res., 44:358.

Mahgoub, A., Idle, J. R., Dring, L. G., Lancaster, R., and Smith, R. L., 1977, Polymorphic hydroxylation of debrisoquine in man, Lancet, 2:584.

Miller, E. C., and Miller, J. A., 1966, Mechanisms of chemical carcinogene-sis: Nature of proximate carcinogen and interactions with macromole-cules, Pharmacol. Rev., 18:805.

Pantuck, E. J., Hsiao, K. C., Conney, A. H., Garland, W. A., Kappas, A., Anderson, K. E., and Alvares, A. P., 1976, Effect of charcoal-broiled beef on phenacetin metabolism in man, Science, 194:1055.

Park, S. S., Fujino, T., West, D., Guengerich, F. P., and Gelboin, H. V., 1982, Monoclonal antibodies that inhibit enxyme activity of 3-methyl-cholanthrene-induced cytochrome P-450, Cancer Res., 42:1798.

Pelkonen, O., and Kärki, N. T., 1982, Is aryl hydrocarbon hydroxylase induc-tion "systemically" regulated in man, in: "Host Factors in Human Carcinogenesis" (IARC Scientific Publications No. 39) H. Bartsch and B. Armstrong, eds., International Agency for Research on Cancer, Lyon, p. 421.

Penno, M. B., and Vesell, E., S., 1983, Monogenic control of variations in antipyrine metabolite formation. New polymorphism of hepatic drug oxidation, J. Clin. Invest., 71:1698.

Poland, A., Glover, E., Robinson, J. R., and Nebert, D. W., 1974, Genetic expression of aryl hydrocarbon hydroxylase activity: induction of mono-oxygenase activities and cytochrome P-450 formation by 2,3,7,8,-tetra-chlorodibenzo-p-dioxin in mice genetically "nonresponsive" to other aromatic hydrocarbons, J. Biol. Chem., 249:5599.

Prince-Evans, D. A., Mahgoub, A., Sloan, T. P., Idle, J. R., and Smith, R. L., 1980, A family and population study of the genetic polymorphism of debrisoquine oxidation in a white British population, J. Med. Genet., 17:102.

Ritchie, J. C., Crothers, M. J., Idle, J. R., Freig., J. B., Connors, T. A., Nikolov, J. G., and Chernozemsky, J. N, 1983, Evidence for an inherited metabolic susceptibility to endemic (Balkan) nephropathy, in: "Current Research in Endemic (Balkan) Nephropathy," S. Strahijic and V. Stefano-vic, eds., NIS,Bulgaria, pp. 23-27.

Ritchie, J. C., and Idle, J. R., 1982, Population studies of polymorphism in drug oxidation and its relevance to carcinogenesis, in: "Host Factors in Human Carcinogenesis" (IARC Scientific Publications No. 39). Bartsch, and B. Armstrong, eds., International Agency for Research on Cancer, Lyon, p. 338.

Röschenthaler, R., Creppy, E. E., and Dirheimer, G. G., 1984, Ochratoxin: mode of action of a ubiquitous mytoxin, J. Toxicol. Toxin Rev., 3:53.

Schneider, J. F., Baker, A. L., Haines, N. W., Hatfield, G., and Boyer, J. L., 1980, Aminopyrine N-demethylation: A prognostic test of liver function in patients with alcoholic liver disease, Gastroenterology, 79:1145.

Sotaniemi, E. A., Pelkonen, R. O., and Puukka, M., 1980, Measurement of hepatic drug-metabolizing enzyme activity in man, Eur. J. Clin. Pharmacol., 17:267.

Vainio, H., and Hietanen, E., 1980, Role of extrahepatic metabolism, in: "Concepts in Drug Metabolism," B. Testa and P. Jenner, eds., vol. 1, Marcel Dekker, New York, p. 281.

Vesell, E. S., 1982, Complex, dynamically interactig host factors that affect the disposition of drug and carcinogens, in: "Host Factors in Human Carcinogenesis" (IARC Scientific Publications No. 39), H. Bartsch and B. Armstrong, eds., International Agency for Research on Cancer, Lyon, p. 427.

Vestal, R. E., and Wood, A. J. J., 1979, Influence of age and smoking in drug kinetics in man. Studies using model compounds, Clin. Pharmacokin., 5: 309.

CLOSING REMARKS

Elizabeth C. Miller

McArdle Laboratory for Cancer Research
University of Wisconsin Medical School
Madison, WI 53706

In closing this stimulating and most interesting conference I first
wish to express the thanks of all of the participants to the organizers,
especially Dr. Snyder and Dr. Nelson, for their foresight and skill in
planning and developing this Third International Symposium on Biological
Reactive Intermediates. The Symposium has fulfilled the most important
goal of such a conference -- i.e., to bring together in a strongly interactive manner persons carrying out research at the forefront of a field.
Based on the quality of the talks and the discussions, both formal and
informal, I am confident that this meeting, like its predecessors in 1975
and 1980, will have a major role in stimulating further important advances
in our knowledge of biological reactive intermediates and the use of this
knowledge for solving problems relevant to human health.

The preceding speakers at this meeting have already pointed out many
of the approaches that are critical to the continuing expansion of knowledge of biological reactive intermediates and to an understanding of these
compounds in relation to human exposures to xenobiotic chemicals. Their
talks emphasized that the variety of chemicals known to give rise to
strongly reactive electrophilic intermediates in biological systems is
now quite broad and that a relatively large number of functional groups
may be involved in the generation of these reactive intermediates. Further, the talks emphasized that the biological consequences of interactions
of cells with reactive chemicals, either contacted directly in the environment or formed from other chemicals, include, as a minimum, mutation,
carcinogenesis, teratogenesis, immunological responses, and cell death.
Of these carcinogenesis and mutagenesis have received the most detailed
study, and, especially for mutagenesis, there is now some understanding
at the molecular level. Some insight into chemical carcinogenesis at the
molecular level also appears to be emerging, although as emphasized by
Weinstein (this volume), a complete understanding of this process still
seems very distant. Nevertheless, the progress in the past five years on
the chemistry and biochemistry of electrophilic reactants, the development
of molecular biological technics, and the understanding of relevant biological systems provides a very strong foundation for a rapid escalation
of knowledge and understanding of the toxicities caused by many chemicals.

METABOLIC OXIDATIONS AND CONJUGATIONS

The progress and the promise for future progress are exemplified by
the massive expansion of the knowledge of the mono-oxygenases and the
families of cytochrome P-450's. As demonstrated by the detailed and very
exciting reports presented here, these enzymes, first recognized almost
30 years ago, are now known in detail which could not even be imagined at
that time (Adesnik and Atchison, 1985; Black and Coon, 1985; and papers
in this volume). At least five families of cytochrome P-450 proteins are
clearly distinguished by the chemicals that induce their synthesis, by
the spectra of their ferrous-carbonyl complexes, and by their substrate
specificities. Cloning of the c-DNA's for the genes coding for these
enzymes and structural analyses at the gene, mRNA, and protein levels
have demonstrated the structural relatednesses and diversities of the
hemoproteins within each family and between families. These data have
further demonstrated the degrees of divergence and conservation of the
genes and proteins during evolution. The specific inhibitions of particular isozymes by individual antibodies have helped to delineate the isozymes
responsible for particular activities in tissues and the differential
induction of the isozymes under various conditions (Friedman, this volume;
Thomas, this volume). The induction of increased levels of these enzymes
by administration of certain chemicals, first recognized by Allan Conney
in graduate work in our laboratory in the 1950's (Conney et al., 1956,
1957), is now understood in terms of a receptor-mediated transfer of an
inducer to the nucleus and a resultant increased synthesis of particular
mRNA's (Poland and Knutson, 1982). However, the molecular mechanisms by
which the receptor-inducer complexes cause the enhanced expression of the
genes is not yet clear.

As also discussed at this meeting, other oxidative systems also extensively metabolize xenobiotic chemicals. These include a variety of
peroxidases (Marnett, 1983; Mason, this volume) and the microsomal flavin-
monooxygenase non-haem system, first described by Ziegler and his colleagues (Ziegler, 1980), which oxidizes nitrogen and sulfur in some compounds. The peroxidases and the redox systems in which they participate
generate a variety of reactive oxygen species, especially O_2^-, H_2O_2, $^1O_2$
and OH· (Fridovich, 1983; DiGuiseppi and Fridovich, 1984; Orrenius et
al., 1985; Aust, this volume; Sies, this volume). Like the cytochrome P-
450 systems these latter oxidation systems are widely distributed and
most probably are of major importance for the metabolism of some xenobiotic
chemicals in some tissues. Examining each of these and other enzyme activities in relation to their tissue localizations and chemical specificities is a research endeavor of considerable importance for understanding the species- and tissue-specific toxicities of xenobiotic chemicals.

Primary oxidative metabolism of xenobiotic chemicals is frequently
followed by enzymatically-catalyzed hydrolytic or conjugation reactions
(Jakoby, 1980; Bock, this volume; Oesch, this volume; Pickett, this volume). Although these enzyme systems appear to function primarily as detoxification systems and/or to facilitate excretion, observations made in
the past few years have greatly revised our understanding of the toxicological importance of these systems. Thus, in many situations their
activities protect the host, but in other cases their activities make the
host more susceptible to damage from a xenobiotic chemical. For instance,
epoxide hydrolyases detoxify epoxides, but at the same time they may, as
in the case of the hydrolysis of benzo(a)pyrene 7,8-oxide to benzo(a)-
pyrene 7,8-dihydrodiol, set the stage for a second epoxidative reaction
(Sims et al., 1974). Likewise, some aromatic hydroxamic acids, aromatic
hydroxylamines, and benzylic allylic alcohols are substrates for the
enzymatic formation of strongly electrophilic sulfuric acid or acetic
acid esters. Sulfuric acid esters derived from some of these carcinogens

have been strongly implicated as the ultimate carcinogens responsible for
the initiation of hepatic tumors in infant male B6C3F$_1$ mice (Miller and
Miller, this volume). A role for acyl transferases and acetylcoenzyme A-
dependent acetyl transferase has been established for the mutagenicity of
some aryl hydroxylamines, and it seems likely that these enzymes may also
be important under some conditions in carcinogenesis by these compounds
(King and Allaben, 1980; Kato, this volume; Kadlubar, this volume). Similarly, although glucuronide conjugates of xenobiotic chemicals are usually
not electrophilic reactants, the glucuronosyl transferases can also convert
some xenobiotic chemicals to electrophilic reactants. Thus, the O-glucuronide of N-hydroxy-2-acetylaminofluorene is weakly reactive (Miller et
al., 1968), while that of N-hydroxy-2-aminofluorene is a potent electrophilic derivative which is also strongly mutagenic for transforming DNA
(Irving and Russell, 1970; Maher and Reuter, 1973). Further, electrophilic reactants can also result from cysteine or glutathione conjugation
of certain difunctional agents, such as 1,2-dihaloalkanes (Sundheimer et
al., 1982; Ozawa and Guengerich, 1983; Anders, this volume; Reed, this
volume; Sipes, this volume).

In addition to oxidation and conjugation reactions, enzymatic reductions, dehalogenations and other activities have roles in the metabolic
activation of xenobiotic chemicals. Features that are common to most, if
not all, of the enzyme systems involved in the metabolism of xenobiotic
chemicals are : (a) their importance in some aspects of normal intermediary
metabolism as well as in the metabolism of xenobiotic chemicals, (b) the
occurrence of enzymes (sometimes structurally related) with different
catalytic activities and/or cellular locations, and (c) differences in
enzyme activities as functions of factors such as age, tissue, species,
sex, or diet. Thus, extrapolation of data for a given enzyme activity
from one biological situation to another, although necessary, must be
made with great care and with as much knowledge of the variables as can
be obtained.

REACTION OF ELECTROPHILIC REACTANTS WITH CELLULAR CONSTITUENTS

Protein adducts

The toxic effects of strong electrophilic reactants usually, if not
always, result from their covalent reactions with cellular constituents.
In those cases in which the toxicity is dependent on the depletion of a
readily renewable molecule, such glutathione, toxic manifestations generally require very significant reductions in the cellular level of the target
molecule. The toxic effects of reactions with proteins depend to a large
extent on the specific proteins modified and a number of, as yet, poorly
defined properties that probably include the amounts of the proteins and
their accessibility to modification. The specifities of protein modifications in relation to the resultant toxicities are exemplified by the
reports of Jollow and his colleagues (this volume) on the effects of aromatic hydroxylamines on red blood cells and of Pohl and his associates
(this volume) on the mechanisms involved in the development of acute hepatitis on exposure to halothane. In the latter case the antigenic nature
of the modified protein amplified the biological response, so that a low
level of reaction elicited a major toxic response.

DNA Adducts

The inferences from early studies that modifications of the bases in
DNA could give rise to mutations resulted in a special interest in the
adducts formed in DNA by reactions with exogenous chemicals (Miller and
Miller, 1981; Singer and Grunberger, 1983). This interest was heightened

by the deduction that, at least theoretically, a single adduct should be sufficient for the formation of a mutagenic lesion. Furthermore, as first noted by Loveless (1969) for methylating and ethylating agents, the mutagenic effectiveness of various adducts in DNA differ remarkably. The evidence is now very strong that methylation or ethylation of the oxygen atoms on the DNA bases is much more likely to induce a mutation than is alkylation at the N-7 position of guanine, the major site of reaction (Singer and Kusmierek, 1982; Pegg, 1984; Singer, 1984). Much less is known about the relative mutagenic efficiencies of the several adducts formed from most other mutagens. In those relatively few cases in which only one type of DNA adduct appears to be formed from a mutagen in vivo that adduct is probably the critical adduct, although the possible importance of a quantitatively very minor adduct can not be ruled out. However, for those cases in which two or more structurally different adducts are found in the DNA, more discriminating studies, especially those that use polydeoxynucleotides with specific known modifications, are needed. Molecular biological approaches are already beginning to facilitate these studies. Thus, the use of cloned DNA fragments for mutagenesis, the use of specific DNA probes for recovery of the desired DNA sequences, and the ability to sequence accurately very small amounts of long polynucleotides are all of critical value for these studies (Singer and Grunberger, 1983; Fuchs, 1984; Loechler et al., 1984).

The same requirements and uncertainties just discussed for analysis of mutagenic lesions also apply to analysis of the structural changes in the DNA that are involved in carcinogenesis. From the available data ethylation or methylation of the oxygen atoms of the bases in DNA appear to be much more critical for carcinogenesis than is N-7-alkylation of guanine residues (Pegg, 1984; Singer, 1984). For other carcinogens deductions have been made on the basis of the apparent mutagenic activities of specific adducts and their resistance to removal from the DNA of the target cells in vivo. More studies are needed in which the structures of the adducts, as well as their rates of formation and removal, are examined under the conditions used for the induction of tumors. For this purpose carcinogenesis systems that yield appreciable tumor incidences after a single dose of carcinogen are particularly desirable, so that there is little probability that the yields of tumors are modified by effects of the carcinogens other than the formation of DNA adducts (e.g., promoting effects).

For the past decade the analysis of DNA adduct formation in vivo has been largely dependent on the use of tritiated carcinogens of high specific activity. However, the development in the past few years of new technologies involving the ^{32}P-post-labeling of nucleotides derived from DNA (Randerath et al., 1981, 1984) and the utilization of antibodies specific for individual DNA adducts (Poirier, 1981; Müller and Rajewsky, 1981) make possible very sensitive assays for the study of carcinogens not readily available at high specific radioactivity. These latter systems have the potential of very high sensitivity, and the antibody system has the further advantage that, with adequately characterized antibodies, the adduct is, at least presumptively, identified by the antibody which recognizes it. These systems also have the important advantage, as contrasted to radioactively-labeled carcinogens, that they can be readily applied to analysis of the DNA from human tissues.

There has been concern for some time on the inability to detect very short-lived adducts formed from electrophilic reactants in DNA. As demonstrated recently by Moreno et al. (1984) for the metabolism of certain aromatic nitro derivatives in the outer membrane of hepatic mitochondria, attention must also be directed to the possible formation during metabolism of xenobiotic chemicals of reactive oxygen species. These reactive oxygen species may also be formed during "normal" intermediary metabolism and as a response to ionizing radiation. Since these oxygen species damage DNA, their roles in "spontaneous", chemically induced, and radiation-induced mutagenesis and carcinogenesis and in the "aging" process are important areas of investigation (DiGuiseppi and Fridovich, 1984). Furthermore, reactive oxygen species may have roles in the promotion of carcinogenesis as well as in its initiation (Troll and Wiesner, 1985).

NEW CONCEPTS IN THE BIOLOGY OF CARCINOGENESIS

One of the most dramatic new developments in biology is the finding in the DNA of normal mammalian cells of proto-oncogenes with homologies to oncogenes of transforming viruses (Bishop, 1982; Weinberg, 1982). The further finding of specific modifications of certain of these proto-oncogenes (especially those of the ras family) in a variety of tumors has emphasized the likelihood that in some cases these modifications play a key role in the induction of carcinogenesis (Zarbl, 1985). The characterization of the protein product encoded by the proto-oncogene c-sis as platelet-derived growth factor (Aaronson et al., 1985) and the structural homology of the v-erb-B oncogene product to that of the epidermal growth factor receptor (Downward et al., 1984) have underscored this possibility, since the key known biological difference between tumors and their tissues of origin is in the relative lack of responsiveness of the tumors to factors that control normal growth. On the other hand, no evidence for a mutated or over-expressed proto-oncogene has been found in some other tumors induced under similar or identical conditions to those with demonstrable alterations (Zarbl, 1985). In these latter cases it is not evident whether the lack of evidence for the occurrence of an altered proto-oncogene in the malignant tissue is the result of methodological short-comings, is indicative of as yet undiscovered proto-oncogenes, or suggests another etiology of these tumors. Clearly much further research is required for a complete understanding of the roles of proto-oncogenes and their modifications in carcinogenesis, whether the tumors develop spontaneously or are induced by known carcinogens (Balmain, 1985; Klein and Klein, 1985). In any case, the occurrence of altered proto-oncogenes homologous to the oncogenes of infectious transforming viruses in some tumors suggests an important intersection of studies on chemical and viral carcinogenesis. As noted by Pelling (this volume), examination of changes in proto-oncogenes in tissues exposed to carcinogens, but in which tumors are not yet evident, may facilitate analysis of the role(s) of the alteration of the proto-oncogenes in the induction of both benign and malignant tumors.

A different genetic mechanism appears to be involved in the etiologies of retinoblastoma, Wilm's tumor of the kidney, and some other human cancers (Cavenee et al., 1983; Koufos et al., 1984; Murphree and Benedict, 1984; Klein and Klein, 1985). In these cases malignancy has been associated with the absence of a specific dominant gene, which has been designated an anti-oncogene. The anti-oncogene, when present on one of a pair of chromosomes, seems to repress the expression of a gene required for the development of a specific type of tumor. Loss of the anti-oncogene has apparently resulted, at least in some cases, from specific chromosomal translocations or cross-over events. Since chromosomal translocations have long been associated with chemical or radiation damage to cells, determination of the natures of the damage to DNA or chromatin that predispose to these specific modifications become critical areas for research. The interactions of successive genetic changes in the induction and progression of tumors have been discussed recently by Klein and Klein (1985).

Finally, it is important to consider briefly how we can apply some of our knowledge and ideas on biological reactive intermediates in the analysis of human problems, since a principal objective in studies on the toxicities of chemicals is to determine and understand those toxic effects of particular importance to human populations. Much of the essential basic knowledge is best obtained through studies on experimental animals, microorganisms, cell culture, and broken cell preparations. However, for monitoring the actual levels of exposures of human populations and for the most meaningful extrapolations of the data on animals to human populations, studies on humans are obviously essential. Ethical and legal constraints rightly limit severely the toxicological studies that can be carried out in the human. Nevertheless, the development of very sensitive chemical, physical, immunological and microbiological assays for specific chemicals and for specific biological responses are providing windows through which to approach these problems. Imaginative coupling of these methods with epidemiologic studies on specific human populations or their application to human samples in tissue and serum banks and excreta can facilitate the development of toxicological data on humans at the levels at which exposures occur. One hoped-for outcome is much better data on the range of responses of human individuals to various chemicals as a function of their genetic and environmental backgrounds, including their diets, social habits, age, fitness, and other factors (Conney, 1982; Omenn and Gelboin, 1984). Such data will provide a much more meaningful basis than is currently available for consideration of the range of toxic manifestations that might occur from various exposures. An important goal is to learn how to integrate lifetime exposures to chemicals and biological agents for an understanding of their roles in the aging process and in the development of syndromes which can not be understood from more acute exposures to one or a few agents.

A VIEWPOINT

From our view of research over the past 40-plus years, we are convinced that solutions to problems currently under investigation and new technologies to be developed will alert investigators to new problems and new avenues of research that will lead to a still better understanding and control of biological reactive intermediates. Ultimately research is limited only by the initiative and ingenuity of the investigators in defining problems, searching for and developing adequate methods, applying them appropriately to analyzable segments, and finally bringing the parts together to provide a comprehensive solution. Rarely is a problem of any magnitude solved by a single investigator or group of investigators, but the interactions between investigators in different areas of science and in different places have a marvelous interactive and synergistic effect in the solution of major problems.

IN APPRECIATION

Let me end on a personal note. To be the honorees of this meeting of our peers has been a most exciting and heart-warming experience for Jim and me. We accept this honor with much humility, and we are most appreciative of the friendship and the confidence in our scientific work that this recognition implies. We were fortunate to be carrying out studies that put us in a receptive position when the first data on biological reactive intermediates was generated, and we have had the unusual advantage of working together in a stimulating research environment at the McArdle Laboratory. Lastly, we have had the privilege of working over many years with a group of bright and stimulating graduate students and post-doctoral fellows who were key contributors to the research from our laboratory.

REFERENCES

Aaronson, S. A., Robbins, K. C., and Tronick, S. R., 1985, Human proto-oncogenes, growth factors, and cancer, in: "Mediators in Cell Growth and Differentiation", p. 241, R. J. Ford and A. L. Maizel, eds., Raven Press, New York.

Adesnik, M., and Atchison, M., 1985, Genes for cytochrome P-450 and their regulation, CRC Reviews in Biochemistry, in press.

Balmain, A., 1985, Transforming ras oncogenes and multistage carcinogenesis, Brit. J. Cancer, 51:1.

Bishop, J. M., 1982, Retroviruses and cancer genes, Adv. Cancer Res., 37:1.

Black, S. D., and Coon, M. J., 1985, Comparative structures of P-450 cytochromes, in: "Cytochrome P-450: Structure, Mechanism, and Biochemistry," in press, P. R. Ortiz de Montellano, ed., Plenum Press, New York.

Cavenee, W. K., Dryja, T. P., Phillips, R. A., Benedict, W. F., Godbout, R., Gallie, B. L., Murphree, A. L., Strong, L. C., and White, R. L., 1983, Expression of recessive alleles by chromosomal mechanisms in retinoblastoma, Nature (London), 305:779.

Conney, A. H., 1982, Induction of microsomal enzymes by foreign chemicals and carcinogenesis by polycyclic aromatic hydrocarbons: G. H. A. Clowes memorial lecture, Cancer Res., 42:4875.

Conney, A. H., Miller, E. C., and Miller, J. A., 1956, The metabolism of methylated aminoazo dyes. V. Evidence for induction of enzyme synthesis in the rat by 3-methylcholanthrene, Cancer Res., 16:450.

Conney, A. H., Miller, E. C., and Miller, J. A., 1957, Substrate-induced synthesis and other properties of benzpyrene hydroxylase in rat liver, J. Biol. Chem., 228:753.

DiGuiseppi, J., and Fridovich, I., 1984, The toxicology of molecular oxygen, CRC Crit. Rev. Toxicol., 12:315.

Downward, J., Yarden, Y., Mayes, E., Scrace, G., Totty, N., Stockwell, P., Ullrich, A., Schlessinger, J., and Waterfield, M. D., 1984, Close similarity of epidermal growth factor receptor and v-erb-B oncogene protein sequences, Nature (London), 307:521.

Fridovich, I., 1983, Superoxide radical: an endogenous toxicant, Ann. Rev. Pharmacol. Toxicol., 23:239.

Fuchs, R. P. P., 1984, DNA binding spectrum of the carcinogen N-acetoxy-N-2-acetylaminofluorene significantly differs from the mutation spectrum, J. Mol. Biol., 177:173.

Irving, C. C., and Russell, L. T., 1970, Synthesis of the O-glucuronide of N-2-fluorenylhydroxylamine. Reaction with nucleic acids and with guanosine-5'-monophosphate, Biochemistry, 9:2471.

Jakoby, 1980, "Enzymatic Basis of Detoxication", Academic Press, New York/London.

King, C. M., and Allaben, W. T., 1980, Arylhydroxamic acid acyltransferase, in: "Enzymatic Basis of Detoxication," vol. 2, p. 187, W. B. Jakoby, ed., Academic Press, New York.

Klein, G., and Klein, E., 1985, Evolution of tumours and the impact of molecular oncology, Nature (London), 315:190.

Koufos, A., Hansen, M. F., Lampkin, B. C., Workman, M. L., Copeland, N. G., Jenkins, N. A., and Cavenee, W. K., 1984, Loss of alleles at loci on human chromosome 11 during genesis of Wilms' tumor, Nature (London), 309:170.

Loechler, E. L., Green, C. L., and Essigmann, J. M., 1984, In vivo mutagenesis by O^6-methylguanine built into a unique site in a viral genome, Proc. Natl. Acad. Sci., U.S.A., 81:6271.

Loveless, A., 1969, Possible relevance of O-6 alkylation of deoxyguanosine to the mutagenicity and carcinogenicity of nitrosamines and nitrosamides, Nature (London), 223:206.

Maher, V. M., and Reuter, M. A., 1973, Mutations and loss of transforming activity caused by the O-glucuronide of the carcinogen, N-hydroxy-2-aminofluorene, Mutation Res., 21:63.

Marnett, L. J., 1983, Cooxidation during prostaglandin biosynthesis: a pathway for the metabolic activation of xenobiotics, in: "Reviews in Biochemical Toxicology," vol. 5, p. 135, E. Hodgson, J. R. Bend, and R. M. Philpot, eds., Elsevier Biomedical, New York.

Miller, E. C., Lotlikar, P. D., Miller, J. A., Butler, B. W., Irving, C. C., and Hill, J. T., 1968, Reactions in vitro of some tissue nucleophiles with the glucuronide of the carcinogen N-hydroxy-2-acetylaminofluorene, Mol. Pharmacol., 4:147.

Miller, E. C., and Miller, J. A., 1981, Searches for ultimate chemical carcinogens and their reactions with cellular macromolecules, Cancer, 47:2327.

Moreno, S. N. J., Mason, R. P., and Docamp, R., 1984, Reduction of Nifurtimox and nitrofurantoin to free radical metabolites by rat liver mitochondria. Evidence of an outer membrane-located nitroreductase, J. Biol. Chem., 259:6298.

Müller, R., and Rajewsky, M. F., 1981, Antibodies specific for DNA components structurally modified by chemical carcinogens, J. Cancer Res. Clin. Oncol., 102:99.

Murphree, A. L., and Benedict, W. F., 1984, Retinoblastoma: clues to human oncogenesis, Science, 223:1028.

Omenn, G. S., and Gelboin, H. V., 1984, eds. "Genetic Variability in Responses to Chemical Exposure." Banbury Report 16," Cold Spring Harbor Laboratory, Cold Spring Harbor.

Orrenius, S., Thor, H., Dimonte, D., Bellomo, G., Nicotera, P., Ross, D., and Smith, M. T., 1985, Mechanisms of oxidative cell injury studied in intact cells, in: "Microsomes and Drug Oxidations," p. 238, A. R. Boobis, J. Caldwell, F. De Matteis, and C. R. Elcombe, eds., Taylor & Francis, London.

Ozawa, N., and Guengerich, F. P., 1983, Evidence for formation of an S-[2-(N[7]-guanyl)ethyl]glutathione adduct in glutathione-mediated binding of the carcinogen 1,2-dibromoethane to DNA, Proc. Natl. Acad. Sci., U.S.A., 80:5266.

Pegg, A. E., 1984, Methylation of the O^6 position of guanine in DNA is the most likely initiating event in carcinogenesis by methylating agents, Cancer Investigation, 2:223.

Poirier, M. C., 1981, Antibodies to carcinogen-DNA adducts, J. Natl. Cancer Inst., 67:515.

Poland, A., and Knutson, J. C., 1982, 2,3,7,8-Tetrachlorodibenzo-p-dioxin and related halogenated aromatic hydrocarbons: Examination of the mechanism of toxicity, Ann. Rev. Pharmacol. Toxicol., 2:517.

Randerath, K., Haglund, R. E., Phillips, D. H., and Reddy, M. V., 1984, ^{32}P-post-labelling analysis of DNA adducts formed in the livers of animals treated with safrole, estragole, and other naturally-occurring alkenylbenzenes. I. Adult female CD-1 mice, Carcinogenesis (London), 5:1613.

Randerath, K., Reddy, M. V., and Gupta, R. C., 1981, ^{32}P-labeling test for DNA damage, Proc. Natl. Acad. Sci., U.S.A., 78:6126.

Sims, P., Grover, P. L., Swaisland, A., Pal, K., and Hewer, A., 1974, Metabolic activation of benzo(a)pyrene proceeds by a diol-epoxide, Nature (London), 252:326.

Singer, B., 1984, Alkylation of the O^6 of guanine is only one of many chemical events that may initiate carcinogenesis, Cancer Investigation, 2:233.

Singer, B., and Grunberger, D., 1983, "Molecular Biology of Mutagens and Carcinogens," Plenum Press, New York/London.

Singer, B., and Kusmierek, J. T., 1982, Chemical mutagenesis, Ann. Rev. Biochem., 51:655.

Sundheimer, D. W., White, R. D., Brendel, K., and Sipes, I. G., 1982, The bioactivation of 1,2-dibromoethane in rat hepatocytes: covalent binding to nucleic acids, Carcinogenesis, 3:1129.

Troll, W., and Wiesner, R., 1985, The role of oxygen radicals as a possible mechanism of tumor promotion, Ann. Rev. Pharmacol. Toxicol., 25:509.

Weinberg, R. A., 1982, Oncogenes of spontaneous and chemically induced tumors, Adv. Cancer Res., 36:149.

Zarbl, H., Sukumar, S., Arthur, A. V., Martin-Zanca, D., and Barbacid, M., 1985, Direct mutagenesis of Ha-ras-1 oncogenes by N-nitroso-N-methylurea during initiation of mammary carcinogenesis in rats, Nature (London), 315:382.

Ziegler, D. M., 1980, Microsomal flavin-containing monooxygenase: Oxygenation of nucleophilic nitrogen and sulfur compounds, in: "Enzymatic Basis of Detoxication," vol. 1, p. 201, W. B. Jakoby, ed., Academic Press, New York.

Benzo(a)pyrene (continued)
  carcinogenicity, 13
  7,8-diol-8,10-epoxide, 13
  3-hydroxylation, 208
  metabolism, 12
Benzo(a)pyrene-7,8-dihydrodiol,
      311-318, 1030
Benzo(a)pyrene-7,8-diol, 982
Benzo(a)pyrene diol epoxide, 13,
      981-992
Benzo(a)pyrene hydoxylase, 124-128,
      130-139, 203
Benzo(a)pyrene-4,5-oxide, 484
Benzo(a)pyrene-7,8-oxide, 1030
Benzo(a)pyrene phenol, 123
p-Benzoquinone, 232, 646, 675, 678,
      682-687
S-2-Benzothiazolyl-L-cysteine, 384
Benzphetamine, 233, 265, 266
  demethylation, 664
Benzylic alcohol, 337
Benzylmethylsulfide, 266
Bhopalator, 871,877-880, 884-887
  as automation, molecular, self-
      reproducing, 885
  as molecular model of the living
      cell, 875, 878-879
  representation, algebraic, 884-886
  and toxicology, molecular, 886-887
BHT, see Hydroxytoluene, butylated
BHT-quinone methide, see 2,6,-Di-
      tert- butyl-4-methylene-
      2,5-cyclohexadienone
Bile, 564-567
Bile acid, 403
Bilirubin, 172, 185, 431, 566, 574
Binding, covalent, 31-38, 63-82,
      354, 393-396, 436-439, 583,
      759, 872, 903-910, 919-924,
      935, 936, 942, 948, 1000
  and bromobenzene, 31-36
  carcinogenicity, 63-82
  chemistry, 31-40
  determination, 718-722
  NADPH-dependent, 906-909
  and lipid, 63-82
  and protein, 63-82
  and thiobenzamide, 36-38
Biphenyl, chlorinated, 103
Bisulfite ion, 32, 33
Biuret method, 913
Bladder carcinoma, 537-543
Blastogenesis, 645-649
Bleomycin, 273-280, 291-300, 311,
      313, 314, 819-822
  -iron complex, 273-280
  toxicity, pulmonary, 291
Bleomycin hydrolase, 291, 294-296
Blood, arterial, 80
Bloom's syndrome, 819
BOH formation, 340, 432

$B(OH)_2$ formation, 341-342
Bone marrow
  aplasia, 919
  degeneration, 675
  depression, 223, 825
  toxicity, 687
BOOH formation, 339-341
  destruction, 339-345
Bovine serum albumin, 729
Bradford protein method, 244
Broiling and mutagen formation, 621
Bromide, 461, 463
Bromoacetaldehyde, 459, 460
Bromobenzene, 31-36, 46, 74, 457-458
  binding, covalent, 31-36
  detoxification, 458
  early work, 31
  hepatotoxicity, 34-36
  labeled, 34-36
  tritiated, 34, 35
Bromobenzene-3,4-oxide, 31, 73
S-2-Bromoethylglutathione, 460
2-Bromohydroquinone, 911-917
  nephrotoxicity, 911-917
Bromophenol, 31, 911-917
  nephrotoxicity, 911-917
Bromoquinone, 31
Butadiene diepoxide, 952
Butadiene monoxide, 951, 953, 956
1,3-Butadiene, 951-957
  Ames test, 951
  epoxidation, 951-957
Tert-Butylhydroperoxide, 46, 341, 506,
      822
Butylhydroxytoluene, 965, 966
N-(tert-Butyl)-thiobenzamide, 37
BZC, see S-2-Benzothiazolyl-L-cysteine

Calcium, 41-42, 46-49, 634, 713, 871
Calmodulin, 46, 47
Cancer, human, 1017-1027
  and diet, 1023
  and environmental factors, listed,
      1018
  and host factors, listed, 1018
    in animal model, 1019-1021
  and metabolism, carcinogenic,
      1017-1019
  and organ specificity, 1021-1023
  and oxygen stress, 819
  and smoking, 1017
  and xenobiotics, 1018
  see Carcinogen, Carcinogenesis,
      Carcinogenicity
Carbamate, 528
Carbinolamine, 949
Carbon dioxide, 237
Carbon disulfide, 237-241
  desulfuration in liver, 237
  metabolism, 237-241
  toxicity, 237-241